GROUND-WATER RESOURCES

OF THE

UNITED STATES

Premier Press Books

Gibson and Singer – *Water Well Manual*
Miller – *Waste Disposal Effects on Ground Water*
National Water Well Association – *Water Well Specifications*
Warner and Lehr – *Subsurface Waste Water Injection*
Todd – *Ground-Water Resources of the United States*

Ground-Water Resources of the United States

An Illustrated Compendium of the Nation's Underground Water Resources Compiled from Publications of the United States Geological Survey

Compiled by
David Keith Todd

PREMIER PRESS
Berkeley, California

P.O. Box 4428, Berkeley, CA 94704

This book is a photographic reproduction of substantial portions of a series of reports published under the general title *Summary Appraisals of the Nation's Ground-Water Resources* and covering twenty regions of the United States. These reports were prepared and issued by the United States Geological Survey as Professional Papers 813-A to 813-T during the period from 1974 to 1982.

Publication of these reports in book form has been undertaken to make these valuable materials available in a single convenient format for general and reference use.

Library of Congress Catalog Card Number: 80-80553
ISBN: 0-912722-02-9

Printed in the United States of America

PREFACE

In the decade ending in 1982 the United States Geological Survey undertook to appraise the ground-water resources of the United States. The country was divided into twenty water resource regions with a report being prepared for each region. The resulting series of reports was published under the general title *Summary Appraisals of the Nation's Ground-Water Resources* (U.S. Geological Survey Professional Papers 813-A to 813-T). These reports have been reproduced as chapters of this book. On the following page a map shows the regional locations.

Collectively, the regional reports provide a comprehensive description of the nation's underground water resources. The reports vary in scope from region to region, expressing not only differences in water resources but also interpretations of the individual authors. The reports were intended to assist planners and managers concerned with utilization of water resources; therefore, they are written in a readable style and accompanied by numerous maps, diagrams, and illustrations. The information should be of value to anyone seeking a greater awareness of ground water in the United States.

A typical regional report begins with a description of the area, including hydrology and geology. This is followed by a quantitative evaluation of the availability of ground water and a discussion of its chemical quality. The utilization of ground water and management techniques are then presented. Finally, regional problems and specialized applications of ground-water resources are described. Each report ends with a list of references for readers wishing more detailed information in a particular locality.

Only a large staff of skilled professionals could have accomplished the prodigious effort culminating in this volume. The thirty men and women of the U.S. Geological Survey who authored these chapters form a group knowledgeable in ground water that exists nowhere else in the United States. The hundreds of illustrations and tables and the thousands of references contained in this book attest to the careful documentation that went into its preparation. Furthermore, much of the material is of timeless value as it focuses on the fundamental occurrence and movement of ground water in each region.

Each chapter of this book consists of a photographic reproduction of one of the U. S. Geological Survey reports. The chapters are arranged geographically, beginning with the New England Region and ending with the Hawaii Region. Preceding each chapter is a map showing the location of the region. Where plates were included in a report, these have been added at the end of the particular chapter. Modifications were necessary in a few plates in order to fit them into a book format.

Publication of the report series in book form was undertaken to make this wealth of useful material available in a convenient single volume. Hopefully, this publication will promote better understanding and increased management of some of our nation's most important water resources.

David Keith Todd

Berkeley, California
July 1983

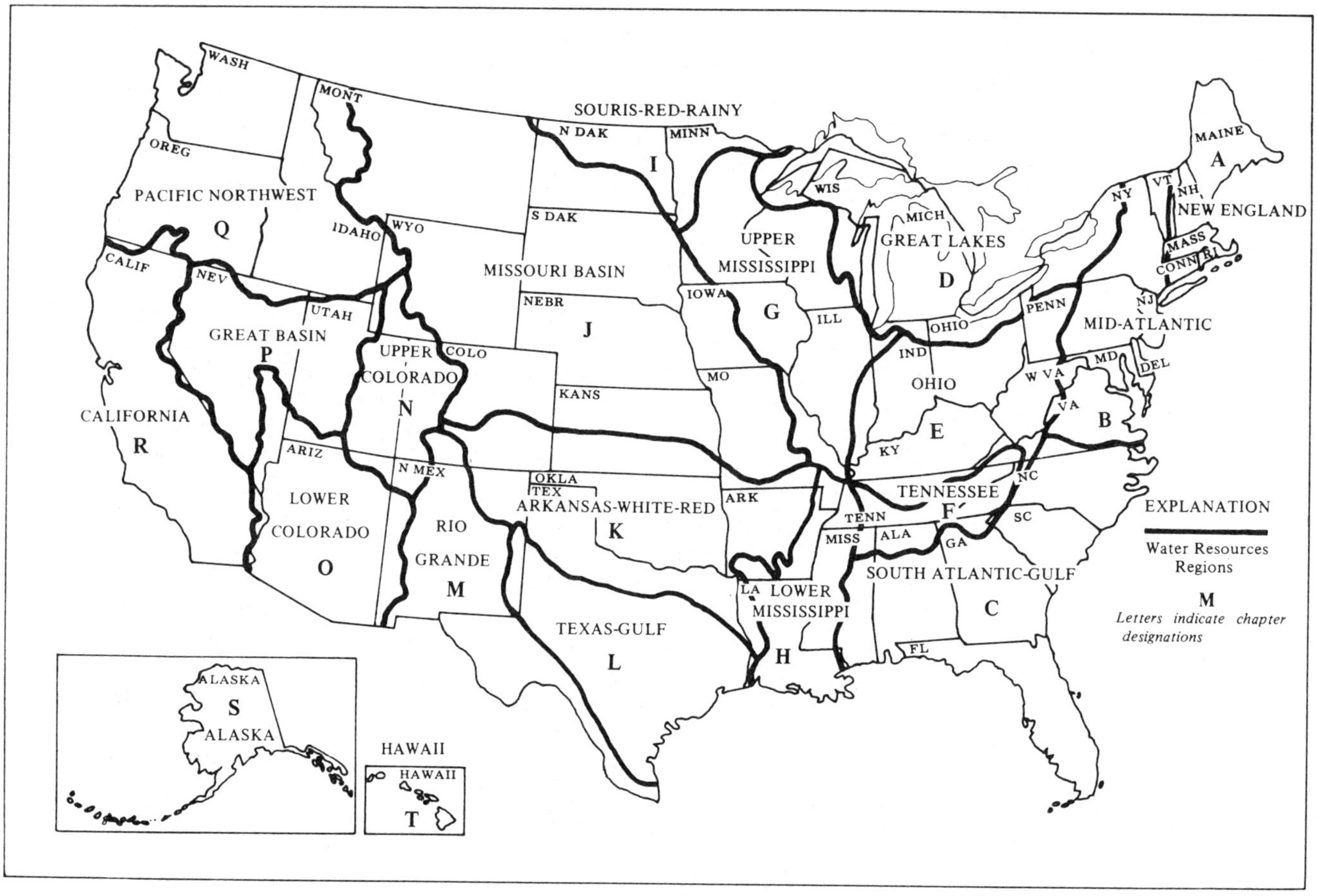

Geographic index to the water-resources regions in the United States.

GROUND-WATER RESOURCES
OF THE
UNITED STATES

CONTENTS

PAGE

A. NEW ENGLAND REGION 1
– Allen Sinnott

B. MID-ATLANTIC REGION 25
– Allen Sinnott and Elliott M. Cushing

C. SOUTH ATLANTIC-GULF REGION 59
– D. J. Cederstrom, E.H. Boswell, and G. R. Tarver

D. GREAT LAKES REGION 95
– William G. Weist, Jr.

E. OHIO REGION 127
– Richard M. Bloyd, Jr.

F. TENNESSEE REGION 169
– Ann Zurawski

G. UPPER MISSISSIPPI REGION 205
– Richard M. Bloyd, Jr.

H. LOWER MISSISSIPPI REGION 229
– J. E. Terry, R. L. Hosman, and C. T. Bryant

I. SOURIS-RED-RAINY REGION 271
– Harold O. Reeder

J. MISSOURI BASIN REGION 297
– O. James Taylor

K. ARKANSAS-WHITE-RED REGION 345
– M. S. Bedinger and R. T. Sniegocki

L. TEXAS-GULF REGION 377
– E. T. Baker, Jr. and James R. Wall

PAGE

M. RIO GRANDE REGION 415
– S. W. West and W. L. Broadhurst

N. UPPER COLORADO REGION 455
– Don Price and Ted Arnow

O. LOWER COLORADO REGION 507
– E. S. Davidson

P. GREAT BASIN REGION 543
– Thomas E. Eakin, Don Price, and J. R. Harrill

Q. PACIFIC NORTHWEST REGION 591
– Bruce L. Foxworthy

R. CALIFORNIA REGION 631
– H. E. Thomas and D. A. Phoenix

S. ALASKA REGION 689
– Chester Zenone and Gary S. Anderson

T. HAWAII REGION 719
– K. J. Takasaki

APPENDIX: Metric-English Equivalents 749

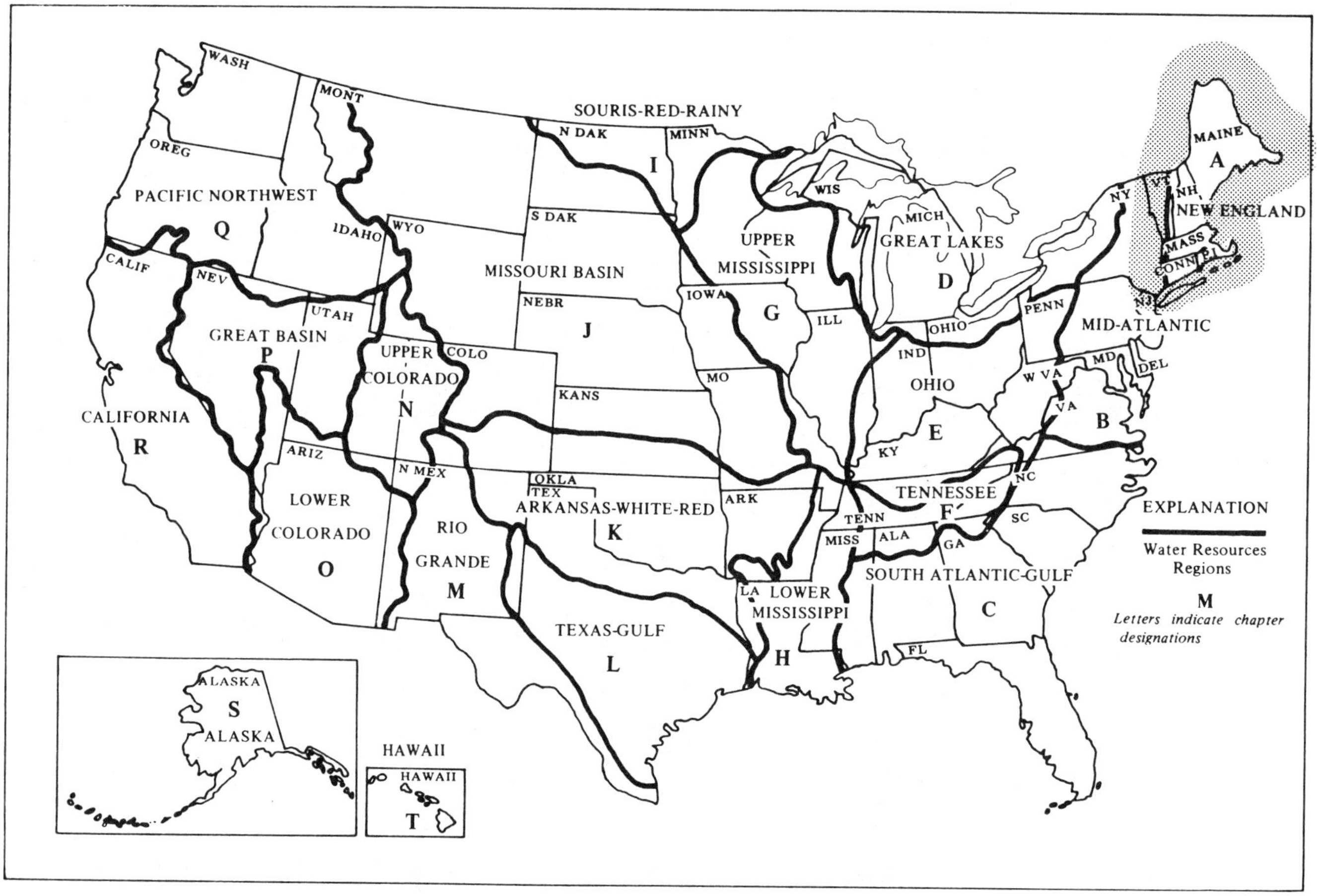

A — Location Map for the New England Region

A

NEW ENGLAND REGION

By ALLEN SINNOTT

ABSTRACT

The New England Region has a total area of about 62,400 square miles (160,000 km^2) and includes the States of Maine and New Hampshire, eastern Vermont, most of Massachusetts and Connecticut, all of Rhode Island, and a small part of southeastern New York. The longest stream is the Connecticut River, which extends from northern Vermont and New Hampshire, through western Massachusetts and central Connecticut, and drains into Long Island Sound. Other major streams are the Penobscot and Kennebec Rivers in Maine, the Androscoggin in Maine and New Hampshire, the Merrimack in New Hampshire and Massachusetts, and the Housatonic in western Massachusetts and Connecticut. Of the smaller streams, some, like the Charles River in the Boston area, are widely known because of their proximity to large population centers.

Ground water occurs in two types of geologic materials: consolidated rocks and unconsolidated sedimentary rocks. The consolidated rocks underlie the entire region. They include crystalline igneous and metamorphic rocks and consolidated sedimentary rocks—shale, sandstone, and limestone and other carbonate rocks. The most productive unconsolidated rocks are sand and gravel of glacial origin. These deposits occur all over Cape Cod and nearby islands in southeastern Massachusetts and in many valleys throughout the region.

Ground water is derived from precipitation. It can be intercepted for use by pumping from wells (1) before it discharges to the streams as base flow and (2) before it drains directly into coastal wetlands, bays, Long Island Sound, or the ocean.

Withdrawals of fresh ground water in 1975 aggregated about 220 billion gallons (830 hm^3), or about 12 percent of the total freshwater withdrawals (from all sources) of 1,800 billion gallons (6,800 hm^3). In view of the available ground-water reserves, considerable additional water, for the anticipated continuing increase in population and economic activity, could be developed.

INTRODUCTION

Early development of ground water in the United States was brought about largely by individuals seeking small domestic supplies. Later, use of ground water extended to small farms, towns, and small commercial and industrial establishments, all of which required only moderate quantities of water. Such developments rarely interfered with each other, and problems were few. However, with increasing population and with gradual urbanization and industrial growth, local overdevelopment of ground water became more common. Local concentration of withdrawals resulted in excessive pumping lifts, decline of water levels below pump intakes or even below the bottoms of wells that had been drilled when water levels stood higher. Reduction in yields resulted in increasing costs for deepening wells and increasing pump capacities with new equipment. In addition, local deterioration in water quality became troublesome. These and related problems gradually encouraged the acquisition of knowledge about ground water that has today become a major scientific discipline.

Because water is a vital element in any long-range plan for development, it should be accorded its proper place in planning. Major projects utilizing surface water have the benefit of long experience and much information on structural design and concepts to enable optimum development. On the other hand, comparatively few large-scale systems have been developed for ground-water supplies, and much less design information is available for such projects. The comparative lack of experience in such developments also has hampered consideration of conjunctive use of ground water with surface water in order to meet specific water requirements. In certain areas, ground water may be the only source available or may provide water for several years until more distant supplies can be developed. In other places, ground water may be used conjunctively with surface water in meeting peak demands, such as augmenting seasonal low streamflow, or it may serve for day-to-day supply.

This report is concerned with one of the 21 Water Resources Regions designated by the U.S. Water Resources Council (1970)—the New England Region in northeastern United States. The chief purposes of this study are (1) to present the current overall situation with respect to ground-water resources, (2) to discuss important problems, (3) to point out areas that have good potential for additional development, and (4) to identify those areas where additional information is needed on ground water and its occurrence and availability, so as to provide an improved basis for present and future management of all water resources.

PHYSICAL AND ECONOMIC SETTING

GEOGRAPHY AND CLIMATE

Almost the entire New England Region (fig.1), as defined by the U.S. Water Resources Council (1970),

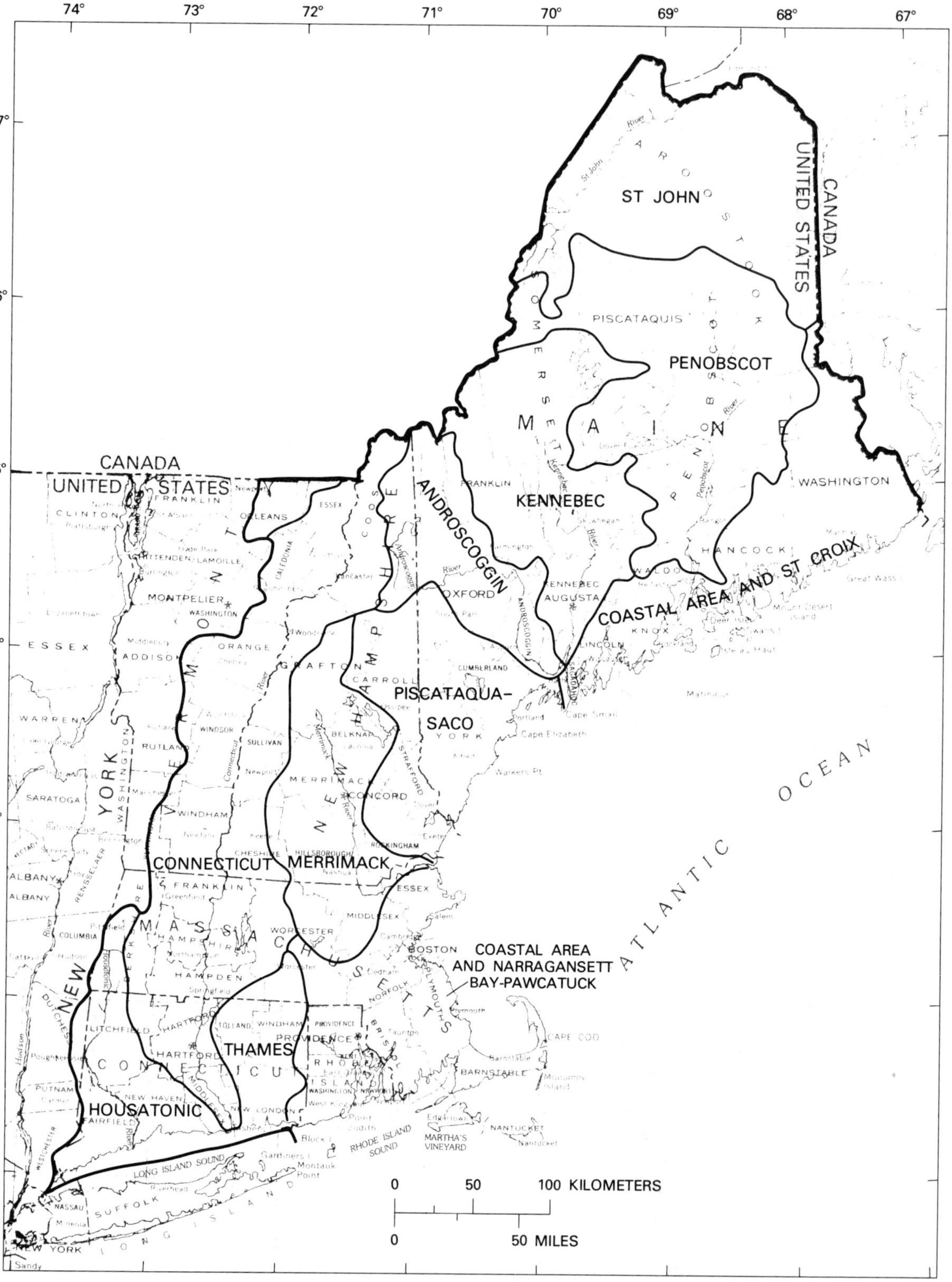

FIGURE 1.—The New England Region and its subregions constitute all or parts of the six New England States and a small part of southeastern New York. Its major stream is the Connecticut River.

lies within the Glaciated Appalachians ground-water region (Thomas, 1952, p. 63). Cape Cod and the nearby islands, Martha's Vineyard and Nantucket, lie in the Coastal Plain ground-water region. Within the New England Region, these two ground-water regions correspond to the New England and Coastal Plain physiographic provinces, respectively, as defined by Fenneman and others (1946).

The New England Region includes all of Maine and New Hampshire, eastern Vermont, most of Massachusetts, virtually all of Connecticut, and all of Rhode Island. In addition, it includes a strip as wide as 10 mi (16 kilometers) in southeastern New York.

The longest stream is the Connecticut River, which extends from the Canadian border through eastern Vermont and western New Hampshire to the coast of Connecticut. Other major streams include the Housatonic River, extending from northwestern Massachusetts to coastal Connecticut; the Merrimack, from New Hampshire to northeastern Massachusetts; the Androscoggin, from northern New Hampshire to south-central Maine; and the Kennebec and Penobscot Rivers, in central and eastern Maine, respectively.

The New England Region extends 125 mi (200 km) inland from the coastline for most of its southern two-thirds; in Maine, it reaches about 200 mi (320 km) in width. The distance from northernmost Maine to the southwest corner of Connecticut is 528 mi (850 km); the total area of the region is about 62,400 mi^2 (160,000 km^2).

As elsewhere in the northeastern United States, the New England Region is generally humid and has four distinct seasons, with frequent changes of weather. The climate along the coast is generally less severe than it is inland, because of the moderating influence of the ocean and Long Island Sound.

Average annual temperature ranges from 38°F (3.3°C) in northern Maine to about 50°F (10°C) in Connecticut, Rhode Island, and southeastern Massachusetts. Average annual precipitation ranges from about 40 in. (1,000 mm) in Maine to 45 in. (1,140 mm) in the three southernmost States.

POPULATION AND ECONOMIC DEVELOPMENT

The New England Region covers only 2 percent of the land area of the United States but contains about 6 percent of its population—11,803,000 in 1975. Compared with the national average of 54 people/mi^2 (21/km^2) for the United States, the region averages about 190 people/mi^2 (75/km^2). The largest population center is Boston, but the region includes 24 other Standard Metropolitan Statistical Areas. (See fig. 2.)

The New England Region generates about 6 percent of the national personal income, and its per capita personal income is 7 percent above the national average. Thus, economic development is generally high—among the six highest of the 21 Water Resources Regions. (See U.S. Water Resources Council, 1972, summary tables.)

GROUND WATER

Nationally, ground water has developed from a minor but important source of domestic supply during the early years to a source of 23 percent of the nation's total freshwater withdrawals in 1975 (Murray and Reeves, 1977, table 17). Withdrawals of ground water in the New England Region totaled about 600 million gal/d (2.3 million m^3/d; fresh ground-water withdrawals amounted to 12 percent of the total freshwater withdrawals.

During the protracted northeast drought of 1962–1965, many municipal water supplies were greatly reduced as levels in major surface reservoirs became dangerously low (Barksdale and others, 1966). This was the largest, longest, and most severe drought in the history of the Northeast United States. The effect was cumulative from year to year as reserves were depleted and streamflow and ground-water levels dropped to record lows (Schneider and Spieker, 1969, p. 4, 5). Many local water shortages and problems developed by 1965. For example, in Maine 21 supply systems restricted water use, and more than 50 municipalities in Massachusetts imposed similar restrictions.

CONSOLIDATED ROCKS

Crystalline igneous and metamorphic rocks and sedimentary rocks, including carbonates, occur in extensive areas.

Among the rocks of igneous origin are granite, rhyolite, diabase, pegmatite and basalt; those of metamorphic origin include gneiss, schist, phyllite, slate, marble, quartzite, and argillite. These mostly hard rocks are the most widespread and are commonly tapped for small domestic or livestock supplies. They lie at the surface in belts throughout Maine, in much of New Hampshire and eastern Vermont, in the broad uplands flanking the Connecticut Valley lowland in Massachusetts and Connecticut, and in most of Rhode Island except the Narragansett Bay area.

Among the sedimentary rocks, carbonate rocks occur notably in the Aroostook Valley in northern Maine and in the Housatonic River valley in western Massachusetts and western Connecticut. They include limestone, dolomite, and calcareous shale. Wells drilled in these rocks or in marble, their metamorphic equivalent, can obtain substantial supplies if large solution openings in the saturated zone are reached. Dry

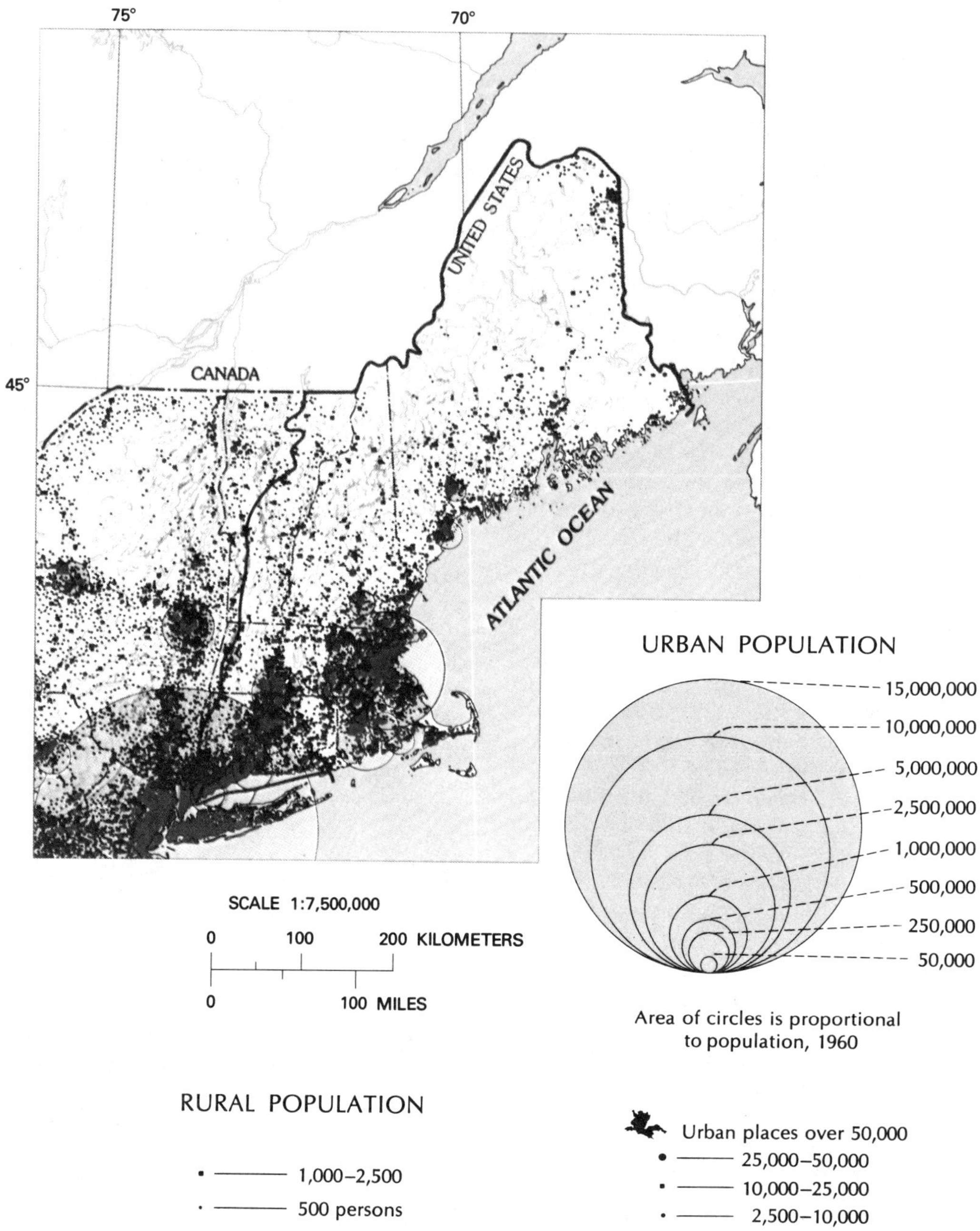

FIGURE 2.—Boston, which has an important place in the Nation's history, is by far the largest population center in the New England Region.

holes, however, are common, and so test drilling is usually required.

The Connecticut Valley lowland in west-central Massachusetts and central Connecticut (fig. 3) is underlain by sandstone, conglomerate, shale, and interbedded trap rock. The last is a poor aquifer, but the sedimentary rocks yield small to moderate supplies—more than 300 gal/min (about 19 L/s) to individual wells.

Elsewhere, the partly metamorphosed sedimentary rocks include quartzitic conglomerate, slate, phyllite, argillite, and marble. These rocks occur widely in Maine and in several areas in New Hampshire, in western Massachusetts and western Connecticut, northwest of Boston, and in the Boston and Narragansett basins in eastern and southeastern Massachusetts and southeastern Rhode Island. They yield small to moderate supplies of water. In some places large

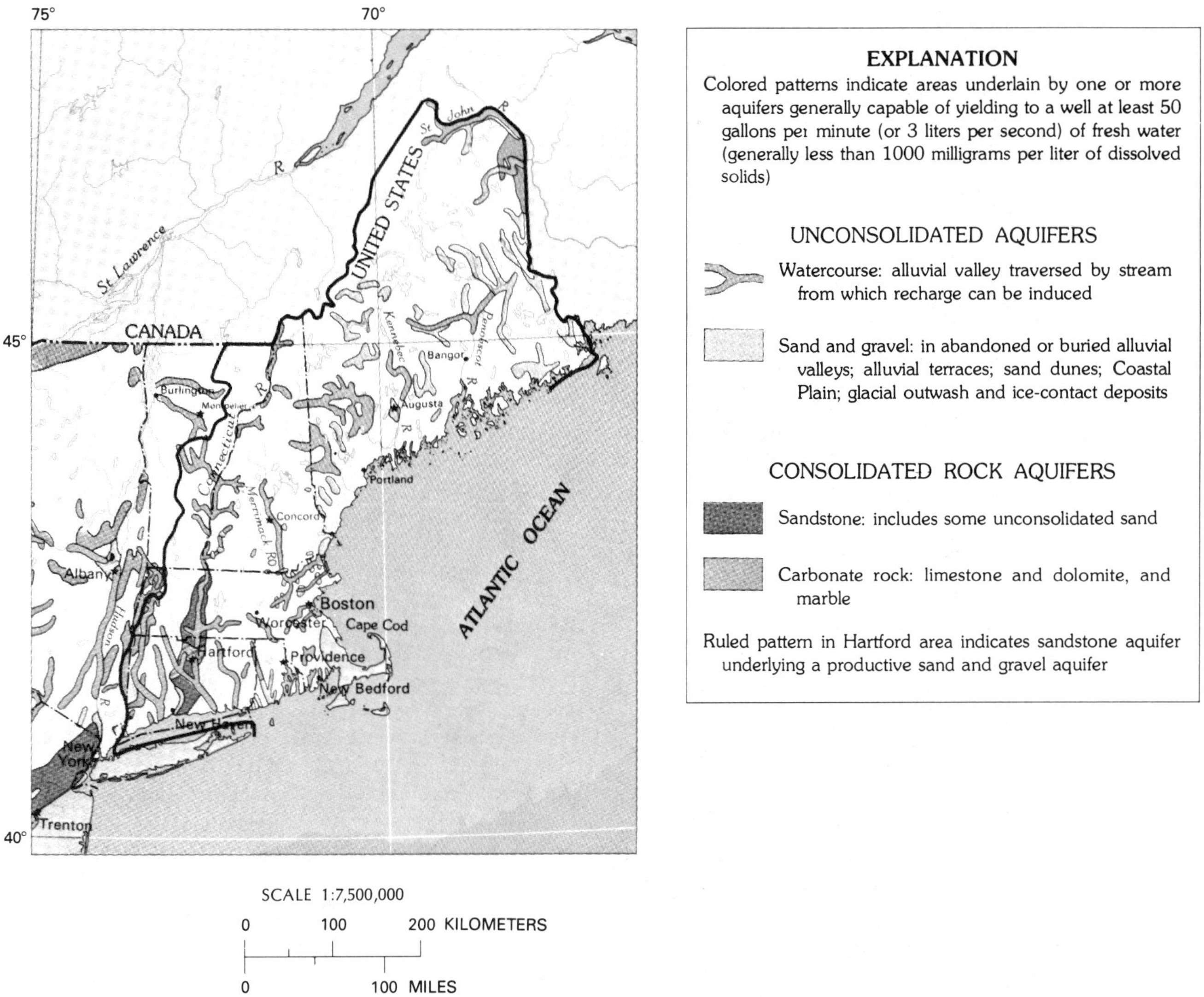

FIGURE 3.—Cape Cod, the lower Connecticut Valley, and many smaller areas in the New England Region are underlain by at least one aquifer capable of yielding at least 50 gal/min (3L/s) of water to individual wells.

supplies, up to perhaps 500 gal/min (32 L/s), can be obtained where these rocks are strongly fractured, as in the shear zones of faults.

Preglacial sedimentary rocks have been reached at depth in wells drilled near Provincetown on Cape Cod and near Duxbury, in southeastern Massachusetts; they also crop out on Martha's Vineyard, south of the Cape. These sediments evidently are related to the extensive Coastal Plain deposits outside the region in New Jersey and farther south; they have not been tapped for water because ample supplies are available from the more accessible overlying drift.

UNCONSOLIDATED ROCKS

The geologically younger rocks in the New England Region are commonly the more productive sources of ground water. These include the unconsolidated sedimentary rocks that were deposited during the glacial epoch, or Ice Age. As the continental glaciers, which covered all of New England, retreated, their meltwaters deposited bodies of so-called stratified drift. Except for lake deposits, the stratified drift contains little clay, and the sand and gravel beds are moderately to highly permeable.

Where meltwater streams flowed in contact with glacier ice, the deposits of stratified drift thus formed are called ice-contact deposits. Bradley (1964) found that in southeastern New Hampshire they are generally coarser and more permeable than outwash deposits formed beyond the glacier terminus.

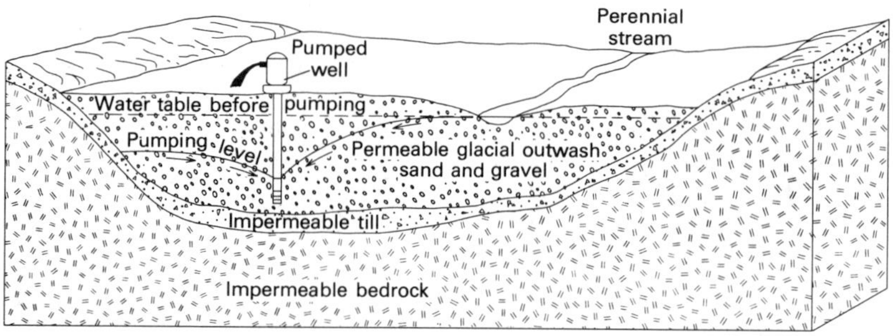

FIGURE 4.—Diagram of a watercourse aquifer showing a pumped well obtaining part of its water from a perennial stream. Situations like this are common in many valleys in the New England Region.

The unstratified drift, or till, consists of poorly sorted clay, sand, gravel, and boulders, deposited directly by the ice, and yields only very small supplies of water to dug wells, chiefly for domestic use.

WATERCOURSE AQUIFERS

Gravel and sand beds in river valleys are among the best sources of large quantities of ground water. These so-called watercourses are of particular importance to ground-water development in the entire region (fig. 3).

Thomas (1952, p. 10) defines watercourses as hydrologic units that include both surface water and ground water; that is, the water in a stream channel plus the ground water in the alluvium that underlies the channel and forms the bordering flood plains. The pervious beds in the stream valley are commonly in hydraulic continuity with the stream, and pumping water from them induces recharge from the stream. (See fig. 4.) In many places, this results in exceptionally large yields, provided that the permeability of the aquifer between the river and the well field is high. Where the alluvial deposits (or, in the New England Region, commonly glacial outwash deposits) are of fine texture, yields are proportionately lower; indeed, they may be very low.

In recent usage, the water-bearing sediments in the watercourses have come to be referred to commonly as watercourse aquifers; this term serves to suggest their elongated shape as well as the potential or recharge from streams. Well yields adequate for municipal water supplies are obtained from sand and gravel in stratified drift and alluvium in watercourses. Individual wells may yield from a few hundred gallons per minute to as much as 2,000 gallons per minute (125 L/s).

Genetically related to watercourse aquifers are buried valleys, which are valleys no longer occupied by the streams that formed them. As noted by Thomas (1952, p. 10–11), some alluvial channel deposits have an extent, thickness, and permeability far greater than the present streams could possibly produce, and some valley deposits have been buried so that they no longer form any part of the present drainage system. Thus, although the materials may be of equal or greater permeability than a comparable watercourse aquifer, the opportunity for recharge may be limited to direct infiltration from precipitation, and if so, the perennial yield is likely to be much less.

As of mid-century, many of the principal watercourses of New England had not been explored for their ground-water potential (Thomas, 1952, p. 64). However, substantial progress has since been made to determine the watercourse aquifer distribution in many parts of the New England Region. Among the larger river systems where permeable sands and gravels have been identified and described are the St. John, Penobscot, Kennebec, Androscoggin, Connecticut, and Housatonic systems. (See Prescott, 1966, 1968, 1969, 1973a; Cederstrom and Hodges, 1967; Norvitch, 1966; and Norvitch and others, 1968.)

QUANTITY OF GROUND WATER

The component of streamflow that is base flow, or ground-water discharge, has been determined for various geologic terranes by many investigators, generally by inspection of stream hydrographs. The peaks in such hydrographs represent increased runoff caused by precipitation; the lower parts of the curves of the hydrographs represent the component of discharge from the aquifers that are in hydraulic continuity with the stream.

Bue (1970) shows that the mean streamflow from the New England Region during 1931–60 was 91,240 ft^3/s (2,580 m^3/s). Adding to this rate the estimated streamflow from the region into Canada, 14,030 ft^3/s (397 m^3/s) (Lois A. Swallow, U.S. Geol. Survey, unpub. data), the mean streamflow from the region is about 105,300 ft^3/s (3,000 m^3/s), or about 68 billion gal/d (258 hm^3/d).

Assuming, conservatively, that base flow is only 40 percent of the mean streamflow in the region, the average total yield of water from the rocks in the entire region, under present hydrologic conditions, is about

27 billion gal/d (102 hm^3/d). This estimate of ground-water yield compares favorably with the result from an alternative method wherein an average recharge rate of 0.2 million gal/d/mi^2 (290 $m^3/d/km^2$) was assumed for areas of till and bedrock and 1 million gal/d/mi^2 (1,460 $m^3/d/km^2$) for the areas of stratified drift and alluvial aquifers.

It is of course impractical, nor would it be desirable, to intercept all of this ground water before it is discharged into streams, for they would be dry or nearly so during times of minimal precipitation. In addition, a part of this water is flowing through low-permeability earth materials that would yield water only slowly to wells. However, a substantial part of this estimated total is nevertheless a potentially valuable resource.

WITHDRAWALS AND USE OF GROUND WATER

The rate of withdrawal of freshwater from all sources during 1975, in the New England Region, was 5,000 million gal/d (18.9 million m^3/d). Of this, 12 percent, or 600 million gal/d (2.3 million m^3/d) was from ground-water sources (Murray and Reeves, 1977, table 17). This is equivalent to a total volume of freshwater from all sources for 1975 of 1,800 billion gal (6,800 hm^3) and from ground-water sources 220 billion gal (830 hm^3). For the regional population of 11,803,000 in 1975, the rate of use of freshwater was about 420 gal/d (1.6 m^3/d) per person from all sources (Murray and Reeves, 1977, table 17). The rate of use of ground water by industry in 1975 (Murray and Reeves, 1977, table 15) was 200 million gal/d (760 thousand m^3/d)—that is, only 4 percent of the total rate of withdrawal of freshwater from all sources. Although the larger demands for water are met from surface-water sources, there is a growing awareness of the development potential of the region's ground-water resources. Considerable additional use of ground water may be expected in the future as economic activity continues to increase.

CONJUNCTIVE USE OF GROUND WATER AND SURFACE WATER

Withdrawal of water from the streams under present conditions has not noticeably affected the amount of water available in the aquifers. However, substantial steady withdrawals of water from a shallow water-table aquifer near a stream will in time reduce base flow in the stream, especially during dry periods when flow is sustained almost wholly by ground water.

Ground water and surface water might be developed conjunctively in many places. Permeable glacial outwash deposits are widespread in valleys. Where these deposits are in watercourses (Thomas, 1952, p. 10), thus forming what have been termed watercourse aquifers, water can be pumped from them so as to induce recharge from the adjacent perennial stream. This is a simple form of conjunctive use.

Large-scale conjunctive management of both ground water and surface water is not yet being practiced in the New England Region so far as is known. Such programs would involve balancing the use of ground-water and surface-water reservoirs according to the availability of water and to the demand for it. Surface water might, in times of surplus, be used to replenish the ground-water reservoir, and ground water might be used to augment surface reservoirs during periods of low flow or drought.

WATER-RELATED PROBLEMS

Despite abundant precipitation, the region has its share of water-related problems. These arise in part from natural hydrogeologic conditions and partly from growing urbanization, increasing population, and industrial activity.

Several problems relate primarily to the natural occurrence of ground water. For example, the development potential for some aquifers is limited because they are narrow, thin, and discontinuous. Other problems involve inferior quality and the susceptibility of ground water to saltwater encroachment in coastal reaches, if substantial amounts are pumped.

In addition, many problems arise from man's activities. Ground-water contamination from landfills is a major problem in some areas; others include contamination from applying deicing salts on highways and contamination from suburban septic tanks and cesspools.

RESTRICTED SIZE AND SHAPE OF AQUIFERS

The most productive aquifers in the New England Region are largely confined to stream valleys. Thus, they are long, thin, and narrow compared with the thick blanket-type deposits of the Coastal Plain in States farther south. This distribution imposes constraints on the location of larger municipal or industrial well fields. Moreover, apparently desirable sites in these valleys require test drilling to make sure that the aquifers are suitable for development. Some aquifers that occupy stream valleys contain only marginally permeable materials. Alternatively, they may not be in hydraulic continuity with the streams and so are restricted as to recharge.

The interstream upland areas are occupied by generally dense and not very permeable crystalline igneous and metamorphic rocks, largely overlain by relatively impermeable till. Such interstream areas yield only small supplies of ground water to wells.

GROUND WATER OF INFERIOR QUALITY

The chemical character of the ground water varies widely, but its overall quality is generally good to excellent. Most ground water in the region is suitable for most purposes with little or no treatment (Miller and others, 1974, p. 78). However, in some localities the water is undesirable for various uses.

Because the chemical quality of ground water is influenced by the mineral character of the rocks through which it flows, troublesome concentrations of certain dissolved constituents are found locally in some parts of the region.

Most of the common dissolved solids in ground water are undesirable in excessive concentrations. Thus, for example, fluoride, desirable in small concentrations as a factor in reducing dental caries, occurs locally in greater amounts in east-central New Hampshire north and south of Lake Winnepesaukee. Water from at least one well reportedly had a concentration of more than 10 mg/L of fluoride (Miller and others, 1974, p. 97).

Iron, usually also associated with manganese, is widespread, and has been found in ground water from several different kinds of rock. Above concentrations of about 0.3 mg/L, iron is troublesome because it stains plumbing fixtures, glassware, and laundered clothing, and impairs the taste of beverages. Manganese causes similar effects at even lower concentrations.

Radioactivity higher than usual is found in ground water in Carroll County in eastern New Hampshire and from the Rangely Lakes area in the uplands of western Maine southward to Cumberland County in the southern part of the State. It is attributed to radon gas from natural decay of radioactive minerals in granite.

In a few places, certain wells tap strongly mineralized natural water. For example, samples of water from two wells near the crest of Talcott Mountain, a trap ridge in north-central Connecticut, contained more than 3,100 mg/L of dissolved solids (Randall, 1964, p. 97). However, as pointed out by Feth and others (1965, p. 2 and sheet 1), in the entire New England Region no mineralized ground water—water containing more than 1,000 mg/L of dissolved solids—is known to occur naturally in producible quantities. It is likely that such water would be found wherever permeable rocks extend to depths of a few thousand feet, but in view of the generally dense nature of bedrock in the region, yields from deep sources are likely to be small. The general abundance—and scant development to date—of fresh ground water would seem to preclude deep drilling for supplementary water supplies.

Where mineralized water, converted to fresh water, may be required to augment other supplies, the most readily available sources are the ocean, Long Island Sound, and brackish water bodies, such as Narragansett Bay.

SALINE GROUND WATER

Saline ground water has long been a problem in many parts of the country, and the New England Region is no exception. It may be found in place during exploratory drilling, or it may be drawn into the ground-water reservoir by continued pumping of wells or well fields in places subject to salt-water encroachment.

The region has a long embayed coastline, with many estuaries that contain brackish water. The lowered water level caused by pumping from wells may reverse the normal seaward gradient, and draw saline water from brackish-water bodies or from the ocean into the shallow freshwater aquifers. Encroachment of saline and brackish water along coastal reaches and estuaries is a growing problem. Some typical instances of encroachment are listed in table 1.

TABLE 1.—*Locations of saltwater encroachment in coastal areas of the New England Region*

[Modified from Miller and others, 1974, table 44]

Location	Nature of problem	Action taken
Maine:		
Town of Bowdoinham, Sagadahoc County.	Salty water from bedrock of Kennebec River estuary contaminated a well 300 ft deep.	Well abandoned.
New Hampshire:		
Portsmouth.	Minor lateral intrusion from tidal water in Piscataqua River.	Unknown.
Massachusetts:		
Provincetown, Scituate, and Somerset.	Minor lateral intrusion from ocean and saltwater marshes affected water from wells tapping shallow aquifers.	New wells drilled farther inland; pumping from old wells reduced.
Connecticut:		
Long Island Sound coastal area including the cities of New Haven and Bridgeport.	Lateral intrusion of salty water from harbors and tidal river estuaries contaminated water from several dozen industrial and municipal wells tapping glacial sand and gravel, older sandstone and shale, and crystalline rock aquifers in areas of heavy pumping.	Wells relocated inland or pumping reduced; some wells abandoned, and at least one scavenger well installed to intercept salty water.
Rhode Island:		
City of Providence; Town of Barrington.	Salty water from estuaries and bays contaminated water from some municipal wells tapping the glacial outwash aquifer because of heavy pumping.	Pumping reduced; replacement of ground-water supply by surface water considered.

Saltwater encroachment in coastal areas near brackish-water bodies is, thus, both a current and a potential problem. However, where encroachment has already occurred, it may still be possible to utilize the ground water in one of two ways: (1) without treatment, in applications where salinity can be tolerated—for cooling or washing or flushing in certain industries or (2) by converting it to freshwater, rendering it suitable for a wide range of uses, including drinking.

GROUND-WATER CONTAMINATION

Contamination is used here to mean any appreciable deterioration of ground-water quality by man's activities. The southernmost three States of the New England Region have a rapidly growing population, advancing urbanization, and heavy industrial growth, and it is inevitable that the environment will be affected to some degree.

INDIVIDUAL SEWAGE-DISPOSAL SYSTEMS

Nearly all homes beyond the reach of central sewage facilities have individual sewage-disposal systems—septic tanks and cesspools. The spread of housing developments in expanding suburban communities has resulted in many local concentrations of individual sewage-disposal systems in the New England Region.

A study in Massachusetts (Morrill and Toler, 1973) is an example of the approach being made to assess the effects of individual onsite sewage-disposal systems. The Ipswich and Shawsheen River basins north of Boston were studied with tracers to develop a correlation between unsewered-housing density and increase in dissolved solids in the base flow of streams. Making a correction for the effect of application of highway deicing salts, the authors found that the discharge of degraded ground water from individual septic tanks was the cause of the increase in the dissolved-solids concentration in the rivers. The increase is roughly proportional to the unsewered-housing density. For example, a density of about 400 houses per square mile (150 houses per square kilometer) caused an increase in dissolved-solids concentration of about 60 mg/L in the base flow of streams; for a density of 900 houses per square mile (350 houses per square kilometer), the increase was found to be about 130 mg/L.

INDUSTRIAL AND MUNICIPAL LANDFILLS

Among the major contributors to localized deterioration of ground-water quality in the New England Region are industrial and municipal landfills. Of these, the industrial landfills are the more abundant. As noted by Miller and others (1974, p. 213),

> the location and even the existence of industrial landfills rarely are recorded with any public agency. Thus, they are not inspected on a routine basis, and problems do not become evident unless ground-water contamination is obviously taking place. In the case of the municipal landfills, although their locations are generally known to regulatory agencies, few are monitored and, again, contamination of ground water normally takes place unobserved.

Ground water contaminated by leachates (solutions that drain by gravity through landfills) ultimately discharges to streams. Although many streams are routinely monitored for water quality, deterioration of stream quality by discharge of contaminated ground water is not detected until it is too late. The aquifer beneath the landfill or other contaminating source will itself have become contaminated before detection and therefore before remedial measures can be put into effect. Installation of water-quality monitoring wells around a landfill together with a frequent or continuous sampling program would allow time for an early judgment to be made in each situation, so that appropriate management decisions could be rendered with sufficient facts and with some degree of confidence. Moreover, careful selection of landfill sites according to criteria that include local hydrogeology might result in anticipation of the adverse effects of contamination, allowing it to take place in such a manner that it could be successfully managed.

WINTER HIGHWAY DEICING PROGRAMS

Another significant source of contamination in the New England Region is the general use of highway deicing salts in the winter. The compounds used are sodium chloride and calcium chloride. Inevitably, this use has resulted in increasing chloride concentrations in ground-water supplies near the highways—some of which have received more than 20 tons of chemicals per lane-mile (11,300 kilograms per lane-kilometer) in a typical winter. The magnitude of the problem is well shown by table 2, which gives the quantities of deicing chemicals applied to highways by State highway departments during a single winter in four of the New England States. In addition, towns and cities add to the chloride burden with their own deicing programs.

The immediate advantage of the use of ice-removing salts is offset to a considerable degree, however, by the delayed cost to other public agencies forced to replace unusable water sources. As an example, New Hampshire spent about $140,000 annually to replace wells contaminated from this source. It is conceivable that desalination plants might ultimately be required to manage this problem (Terry, 1974, p. 65).

TABLE 2.—*Use of ice-removal chemicals in four New England States during the winter of 1965–1966*

[From Hanes and others, 1970]

State	Calcium chloride (tons)	Sodium chloride (tons)	Rate[1] (in tons per single-lane mile)
Connecticut	8,000	74,600	8.9
Massachusetts	5,855	120,304	20.7
New Hampshire	540	82,737	11.9
Vermont	500	83,122	18.2

[1] Rounded.
Conversions:
Tons to kilograms, multiply by 907.
Tons per mile to kilograms per kilometer, multiply by 563.

OTHER SOURCES OF CONTAMINATION

In their report on ground-water contamination in the northeastern States, Miller and others (1974) discuss several other significant sources of contaminated ground water. Among these may be mentioned (1) surface impoundments, (2) buried pipelines and storage tanks, (3) accidental spills, and (4) surface discharge of toxic materials.

Controls designed to reduce or prevent contamination from these various sources are being formulated by regulatory agencies. In reference to surface impoundments of industrial discharges, Miller and others (1974, p. 229) point out that most of the existing sites are difficult to inventory because, like industrial landfills, they are on private property. In much of New England, particularly in industrialized Massachusetts, Connecticut, and Rhode Island, these sources of contamination are common. Ongoing investigations and planned additional studies should locate most of these sources, so that they can be avoided when planning municipal water developments.

LEGAL CONSTRAINTS ON GROUND-WATER DEVELOPMENT

All the States represented in the New England Region except New Hampshire and New York follow the English rule of the riparian doctrine, which provides for unrestricted use of ground water by each landowner. Ordinarily, this does not pose serious legal problems because of the abundance of ground water in this humid region. However, in areas of increasing population and heavy and competing demands, the effects of pumping wells will probably be felt more and more, and some modification is perhaps inevitable.

As among the States in the adjacent Mid-Atlantic Region (Sinnott and Cushing, 1978, p. 2), wherever specific legal tests have been made, the present trend in the New England Region seems to be toward the American modification of the English rule—the so-called doctrine of reasonable use. This doctrine restricts the water-use rights of the landowner in relation to the needs of other landowners. By carefully planning new ground-water developments so that they are realistically scaled for actual need, much litigation may be avoided. Moreover, if use is challenged, realistic defense may be made.

As pointed out by McGuinness (1951, p. 3, 24), the American doctrine of reasonable use was first formulated in New Hampshire in 1862.

GROUND-WATER STUDIES AND NEEDS FOR ADDITIONAL INFORMATION

The general trend in growth in the New England Region is likely to continue, both in population and in economic activity. Increasing urbanization in the less populated northern States is expected. In order to meet the inevitable increased demand for additional water, more information about potential ground-water supplies is necessary. In addition, it will be necessary to control the contamination of water supplies that results from various land-use practices and municipal and industrial waste disposal.

Present programs of ground-water investigations in the region are being carried on largely by the U.S. Geological Survey, in cooperation with State and local agencies. Many of the cooperating agencies are themselves making ground-water studies.

Studies thus far in the New England Region have been based primarily on political boundaries, such as counties or designated areas around key cities. In recent years, however, many studies have been oriented toward hydrologic units—drainage basins, individual aquifers, or aquifer systems. A comparison of figure 5 (work done or nearing completion) and figure 6 (areas where additional work is indicated) suggests a trend toward hydrologic units, rather than political boundaries.

However, funding and program authorization have remained, as would be expected, with the sources of cooperative funds—namely, the State and smaller political entities. Thus, programs are primarily State-oriented, and the discussion that follows considers present coverage and anticipated needs in each of the States involved.

The map (fig. 5) shows the areas covered by recent ground-water investigations. The nature of the coverage is shown by three categories:

1. Detailed quantitative studies in which ground water has been treated as a dynamic fluid resource. These studies include analyses of aquifer tests and provide guidance in predicting response of aquifers to changes in pumping regimen.
2. Detailed reconnaissances in which most of the available ground-water information has been collected, including quantitative data and interpretations that provide a basis for determining, at least in a general way, effects of possible future developments.
3. General reconnaissances in which a representative sampling of readily available information has been collected and interpreted and which can provide guidance for planning either small ground-water developments or more detailed investigations before major developments.

Some areas in Massachusetts are not colored in figure 5. For these areas, formal reports have not been published, nor are studies currently underway. However, basic data are on file for most localities in these

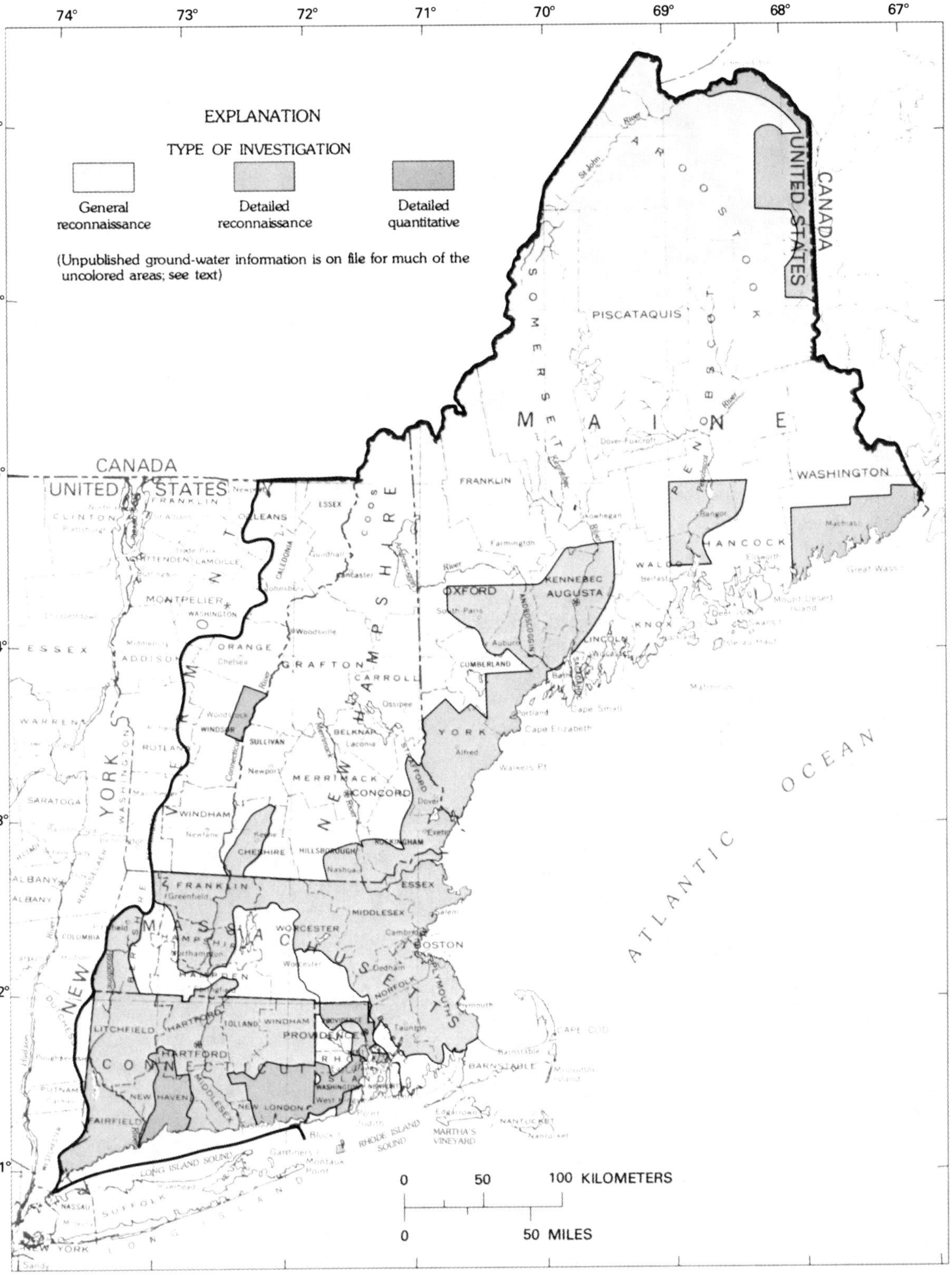

FIGURE 5.—Status of ground-water investigations in the New England Region.

areas and are available for inspection in the New England District Office of the U.S. Geological Survey in Boston. Inquiries may also be made to the Survey offices in other States for basic data and other unpublished information in the open files.[1]

Two categories of necessary ground-water information are indicated in figure 6:

1. Detailed quantitative information, as defined above. This level of ground-water information is desirable for areas where the need is immediate or foreseeable. The detailed information provided by investigations at this level allows immediate use for planning and developing major projects, such as municipal or industrial well fields, with only minimal additional site investigation.

2. Detailed reconnaissance information, also as previously defined. This level of information is indicated for areas where only moderate future developments are likely and where substantial detailed information would suffice for small developments. Such information would provide an adequate base for additional work prior to large-scale developments.

MAINE

Maine, the largest State in the region, has an area of about 33,200 mi^2 (86,000 km^2)—more than half the area of the region. Its entire area was glaciated during the Ice Age, and it is largely mantled by drift. The bedrock consists mainly of crystalline igneous and metamorphic rocks. Of the sedimentary rocks, most are consolidated. Ground water occurs in fractures and, in the calcareous sedimentary rocks, in solution cavities—many of which are fractures enlarged by solution.

Early ground-water studies in Maine were mostly reconnaissances, with information on a few selected drilled wells and lists of data on the principal springs. (See for example, Bayley, 1905.) The most comprehensive early report is that of Clapp (1909) describing ground-water conditions in southern Maine.

The report by Prescott (1963a), a reconnaissance of ground water in Maine, is the first that covers the entire State. After that reconnaissance, Prescott prepared maps of ground-water favorability where needs were most urgent. These maps are chiefly of the more populated and industrialized coastal areas and also include eastern Aroostook County in northern Maine, where ground water is used for irrigation. (See Prescott, 1963b, 1966, 1968, 1969, 1972, 1974a, b, 1977a, b, 1980.) The maps and their accompanying texts are all at the detailed reconnaissance level. (See fig. 5.) In general, the detailed reconnaissance has been directed toward those areas where population and industry are greatest.

As development expands, adjoining areas may expect an increase in economic activity and will need ground-water information for anticipated demands. Areas close to large population centers may eventually share in future industrial developments. These include the coastal reaches southeast and east of Augusta and south of Bangor—in Lincoln, Knox, Waldo and Hancock Counties in eastern coastal Maine. Also included are the upper reaches of the Saco, Presumpscot, and Royal River basins in southwestern Maine and the upper reaches of the Penobscot, Kennebec, and Androscoggin River basins in central and western Maine. These areas are gradually growing in population amid other manifestations of increasing economic activity. (See fig. 6.)

At the present time, detailed reconnaissance information is the priority need throughout the State. When, however, the need for detailed quantitative information becomes urgent as a result of pressing water-supply demands, the Portland area—encompassing the lower reaches of the Royal River draining into Casco Bay at Yarmouth—would have high priority. Integrated ground-water and surface-water information would be most useful in this area.

General baseline information on ground-water quality in Maine is necessary for defining water-quality problems. Salt used in highway deicing is probably the most widespread contaminant. In places, the salt is stored uncovered allowing contamination even when the salt is not being used. Of a more local nature are other problems such as contamination by effluent from septic tanks in rural areas.

Among the local problems in natural ground water are high radioactivity, excessive iron or manganese, and the presence of saline water in coastal areas.

Ground water with a high degree of radioactivity has been found locally in the granite of western Maine, extending from the Rangely Lakes to the coastal area of Cumberland and Sagadahoc Counties (Chandler and others, 1959, p. 21). The highest radioactivity was found to be more than 290 times the allowable maximum concentration defined by the U.S. National Bureau of Standards (1953). Detailed study of this problem is warranted soon. So far as known, however, no public supplies from ground-water sources contain excessive concentrations of radioactivity.

Ground water in Maine is generally of good chemical quality, and the problems are localized.

[1]Inquiries may be addressed to the following officials:

District Chief, New England District, U.S. Geological Survey, 150 Causeway Street, Suite 1001, Boston, Mass. 02114 (for Massachusetts, Maine, New Hampshire, Vermont, and Rhode Island).

District Chief, U.S. Geological Survey, 135 High Street, Room 235, Hartford, Conn. 06103 (for Connecticut).

District Chief, U.S. Geological Survey, Post Office Box 1350, 236 U.S. Post Office & Court House, Albany, N.Y., 12201 (for New York).

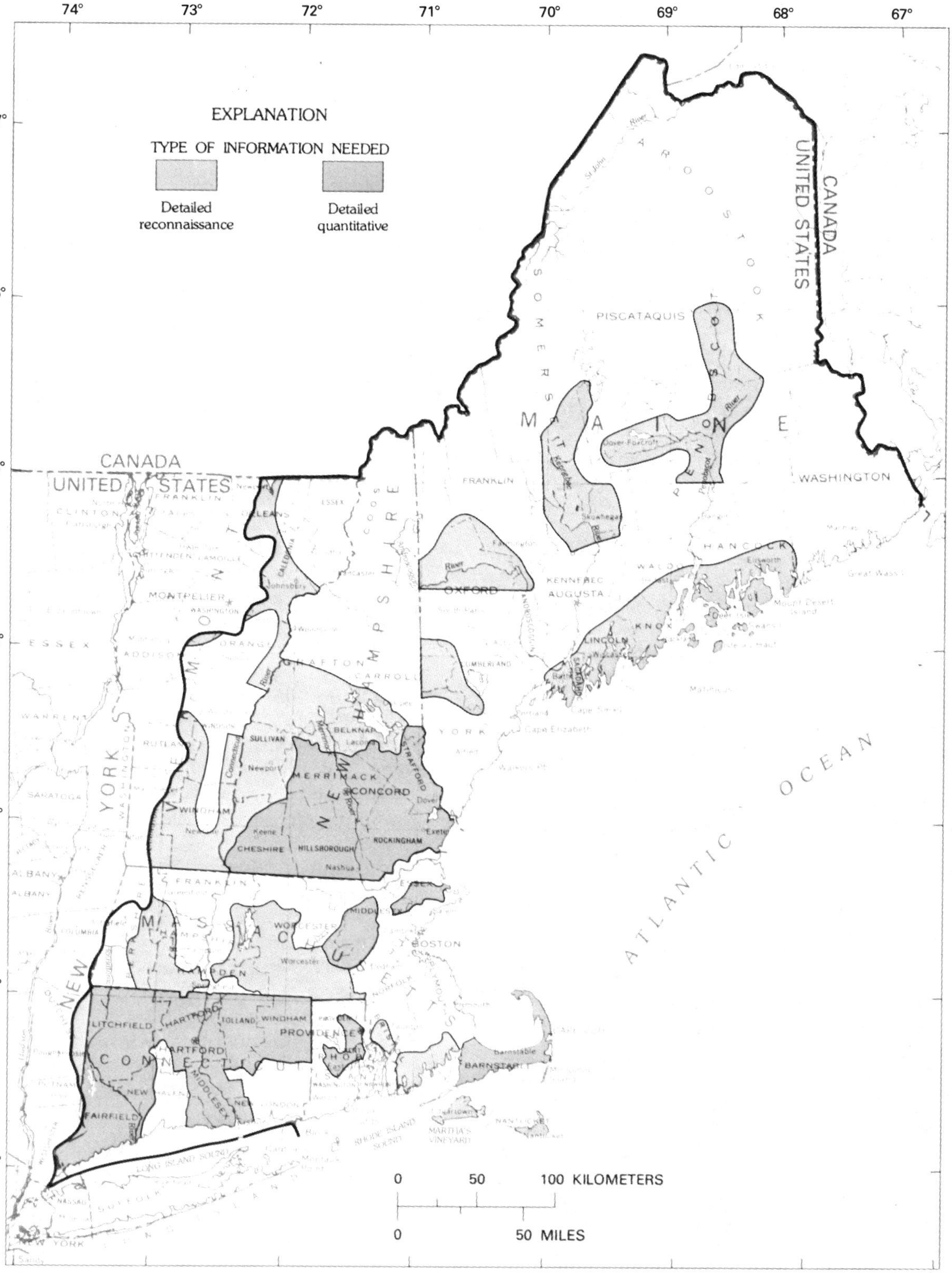

FIGURE 6.—In parts of the New England Region, additional ground-water information is necessary to assist water-resources development and management.

NEW HAMPSHIRE

New Hampshire has an area of about 9,300 mi^2 (24,000 km^2) and has a predominantly rugged topography, culminating in the White Mountains in the north-central part of the State. Mount Washington has an altitude of 6,288 ft (1,917 m) and is the highest mountain in New England.

The precipitation is highest on these mountains, more than 70 in. (1,780 mm) compared with the general average in the State of 43 in. (1,090 mm).

Much of the bedrock consists of ancient sedimentary rocks, which are partially metamorphosed; about a third of the State is underlain by intrusive igneous rocks. Yields from wells in the bedrock are generally small because of its low permeability.

Mantling the bedrock are till and stratified drift deposited during the Ice Age. Till is generally thin in the uplands, and in places the bedrock is exposed. The unsorted till is of low permeability and yields only small supplies of water to many domestic dug wells. These are gradually being replaced by wells drilled and cased into the bedrock, affording greater sanitary protection.

Stratified drift consists of water-sorted materials deposited by the meltwaters of the waning ice sheet and commonly contains permeable sand and gravel with some clay and silt. Sand and gravel were deposited by streams, clay and silt accumulated in glacial lakes, and coastal deposits were of marine origin. Stratified drift is widespread in the lowlands; it is particularly important where it occurs in watercourse aquifers that are in hydraulic continuity with perennial streams. These aquifers are the best in the State.

Reconnaissance ground-water field work has been completed for the entire State, and the results were published in nine reports (Cotton, 1975 a–d; 1975 a, b; 1977 a–c).

An interstate general reconnaissance of the Connecticut River basin along the western boundary of the State by Cederstrom and Hodges (1967), published as a hydrologic atlas, provides additional coverage in the western part of New Hampshire.

Three detailed reconnaissances have been completed. Bradley (1964) described the ground-water resources of southeastern New Hampshire, and Weigle (1968) completed a hydrologic atlas for the area immediately adjacent to the west—the lower Merrimack River basin. Farther west, Whitcomb (1973) studied the ground-water resources of the Ashuelot River basin, also published as a hydrologic atlas.

On the basis of estimates of population and public water-supply demands through the year 2020 (Anderson-Nichols & Co., Inc., 1969), detailed quantitative studies are necessary for the southern part of New Hampshire, as indicated in figure 6. The population there is expected to quadruple, from half a million in 1966 to over 2 million in 2020. In this area, public water-supply use is expected to increase from about 50 million gal/d (190 thousand m^3/d) to about 400 million gal/d (1½ million m^3/d).

A detailed reconnaissance of central and west-central New Hampshire (fig. 6) may be somewhat less urgent. Here the population is expected to increase from about 106,000 in 1966 to about 250,000 by the year 2020. The use of water for public supplies is expected to increase from 7.6 million gal/d (29,000 m^3/d) to 43 million gal/d (163,000 m^3/d); a substantial part of the total would be ground water. The capacities of present systems would be exceeded by 26 million gal/d (98,000 m^3/d) (Anderson-Nichols & Co., Inc., 1969). More information on ground-water quality in New Hampshire would be helpful, especially analyses of representative samples from all parts of the State. Of particular interest are hardness, iron and manganese, fluoride, pH, and radioactivity. Among practices likely to have a major effect on chemical quality of ground water, two are widespread: (1) the construction and operation of landfills and (2) application of deicing salts to roads. Both result in contamination problems, at least locally. Extensive sampling of ground water for chemical analyses would pinpoint areas where problems are likely to need attention and would provide a measure of the nature and scope of the water-quality information required.

In recent years, ground water has provided at least a part of the supply in 60 percent of the municipal water systems, and a major role for ground water in meeting future demands is nearly assured.

The foregoing estimates of future ground-water investigational needs are based on anticipated public-supply demands. However, such investigations may serve other needs, in areas not designated in figure 6. Land-use planning will undoubtedly proceed at an accelerated rate throughout New Hampshire during the next 40 years. Adequate water-resources information, based on systematic hydrogeologic studies, is necessary for long-range planning. Thus, additional ground-water investigations will be needed.

VERMONT

Vermont is primarily rural, with few industries and a relatively small population. A little less than half of its total area of 9,600 mi^2 (25,000 km^2) lies within the New England Region.

Withdrawals of water from all sources are small (Murray and Reeves, 1977, table 10), but its use of ground water is proportionately high—largely because of its rural economy.

All of eastern Vermont that lies within the New

England Region is covered by general reconnaissances (fig. 5) published as ground-water favorability maps. (See Hodges, 1967, 1968a, b, c, and d.) As McGuinness (1963, p. 891) points out, the paucity of information on ground water reflects the general abundance of water and the small population and water demand.

Besides the maps, a quantitative study of the developing White River Junction area was made by Hodges and Butterfield (1972b).

Ground-water information at the detailed reconnaissance level will be necessary for the Interstate Highway corridor in the northern part of Vermont and along the Connecticut River from about the latitude of St. Johnsbury southward to the Massachusetts boundary. Much of southern Vermont would also require study because of increasing economic activity. (See fig. 6.)

The growing communities clustered around St. Johnsbury in the north and Brattleboro in the south might eventually require detailed quantitative studies.

Ground water in Vermont is ample for current needs. There is no heavy industry; tourism, skiing, and other outdoor recreation are the major attractions. Consequently, the principal need is for drinking water and water for other domestic purposes.

The quality of the water is typically good to excellent and environmental concerns are not now serious.

MASSACHUSETTS

About 8,090 mi² (21,000 km²) of Massachusetts lies within the New England Region. This area constitutes the entire State except the small area (164 mi²) (425 km²) in the extreme northwest corner along the Hoosic River, which lies in the Hudson River Basin in the Mid-Atlantic Region.

Most of Massachusetts is in the New England physiographic province. Cape Cod, and the islands of Martha's Vineyard and Nantucket, all in the extreme eastern part of the State, are in the Coastal Plain province. The entire area was glaciated during the Ice Age.

The Connecticut Valley lowland extends through the west-central part of the State, from New Hampshire and Vermont southward to central Connecticut. It is underlain by readily eroded sandstone and shale, interbedded with volcanic rocks and mantled by drift. The lowland is flanked on the west by uplands, rising westward to the Berkshire Hills—ridges of schist, slate, and other rocks that have been strongly folded and faulted. East of the Connecticut Valley are uplands of folded sedimentary and metamorphic rocks, and still farther eastward is a broad belt of igneous rocks and hard sedimentary rocks, extending to the glacial deposits of Cape Cod.

The western half of Massachusetts is covered by several ground-water studies (fig. 5), published as hydrologic atlases or as ground-water-favorability maps. (See Cederstrom and Hodges, 1967; Norvitch and others, 1968; Collings and others, 1969; and Gay and others, 1974.) Several areas in the central and eastern parts are covered by detailed reconnaissances.

The Assabet River basin, a tributary of the Merrimack River in central Massachusetts, has been studied (Pollock and others, 1969).

In the eastern part of the State, parts of the coastal drainage basins studied include the area between Hingham, southeast of Boston, and Wareham, at the head of Buzzard's Bay (Williams and Tasker, 1974a, b). The Taunton and Tenmile River basins in the southeastern part, west of the coastal area, were studied by Williams (1968) and Williams and others (1973). The Neponset and Weymouth River basins south of Boston have also been studied (Brackley and others, 1973), as well as the Charles River basin west of Boston (Walker and others, 1975).

In northeastern Massachusetts, Sammel and others (1964) made a detailed reconnaissance of the Ipswich River basin, and immediately to the north, Sammel (1967) made a similar study of the Parker and Rowley River basins. Baker and others (1964) wrote a water-supply paper on the Wilmington-Reading area.

In western Massachusetts, flanking the west side of the Connecticut Valley, a detailed reconnaissance may be necessary in the Westfield River basin and in the upstream Massachusetts part of the Farmington River basin that extends into Connecticut. East of the Connecticut Valley, similar information may be necessary in the Chicopee River basin and in the French and Quinebaug basins adjacent on the south; also, in the Blackstone River Basin adjacent to the latter basins on the east. No systematic work has been done in the Blackstone River basin, and studies only at the general reconnaissance level have been made for the others. Economic activity is increasing in all these basins, which may require more ground-water information for anticipated new developments.

Other areas in Massachusetts, where detailed reconnaissances may be required, include the Shawsheen and Mystic River basins north of Boston and the North Shore coastal region between Boston and Gloucester. In addition, comprehensive ground-water information is lacking in southeastern Massachusetts, including the coastal area south of the Taunton River basin, and on Martha's Vineyard and Nantucket south of Cape Cod. Coverage in these areas, together with those in the western and central parts of the State, would help to implement the systematic long-range basin-by-basin program to provide an adequate base of ground-water information.

Detailed hydrologic information in the upper Charles River basin is a prerequisite for planned treatment and disposal of sewage from several towns and for discharge of treated sewage into and out of the basin.

In the Ipswich River basin, growing needs for public water supplies require detailed quantitative information for guidance in dealing with problems of expansion and water management. A proposal to build a 10 billion gal (38 million m^3) reservoir in the basin also requires detailed ground-water facts for construction and management. Conjunctive use of ground water and surface water also is a possible management alternative.

Cape Cod is one of the most scenic parts of Massachusetts, and has some of the best beaches in New England. In 1961, the Atlantic shore of the northern part of the Cape was designated a National Seashore Park, and the influx of tourists into this vacationland has since increased considerably. Detailed quantitative ground-water information has been collected, and a model analysis for the entire Cape has been completed.

NEW YORK

The boundary of the New England Region extends only about 10 mi (16 km) into southeastern New York, mostly in Dutchess County west of Connecticut but also much smaller areas in Putnam, Westchester, and Columbia Counties. These areas aggregate about 230 mi^2 (nearly 600 km^2). Ground-water coverage is at the general reconnaissance level. A detailed reconnaissance will probably be necessary, as development increases.

CONNECTICUT

Connecticut has an areas of 5,009 mi^2 (12,973 km^2), of which about 20 mi^2 (50 km^2), along the western boundary, lies in the adjacent Mid-Atlantic Region.

Most of the State is underlain by crystalline rocks, chiefly metamorphic. Along the western boundary, in the upper Housatonic River basin, there are areas of limestone, dolomite, and some marble. Extending from north to south, in the central part, lies the broad Connecticut Valley lowland, which is underlain by sandstone, shale, and interbedded basalt. The entire State was glaciated during the Ice Age, and glacial deposits overlie the bedrock nearly everywhere.

The sand and gravel deposits laid down during the waning phases of glaciation are the most productive aquifers. They consist of valley-train and other outwash deposits and ice-contact deposits, commonly in lowlands and valleys. Where they are near perennial streams, from which water can be induced by pumping wells, they are excellent aquifers.

Quantitative work in the State—and indeed in the United States as a whole—was pioneered by the classic study by Meinzer and Stearns (1929) in the Pomperaug basin, a tributary of the Housatonic River in western Connecticut. Another major early paper is that by Brown (1925) on coastal ground water, with special reference to Connecticut. Most of the other early reports covered the Connecticut Valley lowland and adjacent areas; they were based to a large extent on records of dug wells and springs.

As shown in figure 5, the entire State has been covered by detailed reconnaissance ground-water investigations; in addition, two areas, southeastern Connecticut and the Quinnipiac River basin (New Haven-Meriden area), are covered by detailed quantitative studies. These studies are products of the Federal-State cooperative program. Detailed quantitative studies of the entire State is the present goal.

More than two-thirds of Connecticut has not been studied quantitatively (fig. 6). The need for studying the northwestern and northeastern parts of the State in the immediate future is somewhat less urgent than it is elsewhere.

The Farmington River basin, tributary to the Connecticut River basin in northwest-central Connecticut, may warrant detailed quantitative ground-water study because its rapid urbanization has resulted in conflicting land-use practices. Also, anticipating continuing population growth and economic activity, quantitative information will probably be needed on the productive glacial sand and gravel deposits in the Connecticut River Valley. The valley of the Naugatuck River, a major tributary of the Housatonic River in west-central Connecticut, is becoming more and more industrialized and urbanized, and like the Farmington River basin, problems resulting from land use are becoming widespread. Quantitative ground-water information can help to assign realistic and potentially beneficial priorities toward specific developments.

Information is necessary also on topical problems not necessarily closely related to particular areas. Among these are (1) general statewide coverage of the chemical quality of ground water, (2) urban area pilot studies where ground water would be likely to play a major role, and (3) studies relating to land-use planning.

Comprehensive information may also be needed on the sedimentary bedrock in the Connecticut Valley. Although the bedrock is not as productive as the overlying glacial aquifers, it is a potentially valuable additional source of ground water about which not much is yet known. Detailed studies would determine its suitability for storage of gas or wastes—a use heretofore unexplored. Model studies of the relation between ground water and surface water in promising areas would be a valuable prerequisite to local conjunctive management of water.

RHODE ISLAND

Rhode Island covers an area of 1,214 mi^2 (3,144 km^2), including Narragansett Bay. It is one of the most densely populated and highly industrialized States.

Most of Rhode Island is underlain by crystalline rocks of granitic composition. However, Narragansett Bay and surrounding land areas are underlain by consolidated sedimentary rocks, which grade from essentially unmetamorphosed in the north to highly metamorphosed in the south.

The bedrock is covered nearly everywhere by glacial deposits; these may be subdivided into till and stratified drift. Till covers the uplands and averages about 20 ft (6 m) thick; it is generally of low permeability. Stratified drift occupies the lowlands; it is commonly 50 to 100 ft (15 to 30 m) thick and is moderately to highly permeable.

Wells tapping till and crystalline bedrock typically yield 10 gal/min (⅔ L/s) or less and are generally adequate to supply individual homes. Yields of wells tapping unmetamorphosed sedimentary rocks in northern Rhode Island, in areas where the glacial deposits are thick, are somewhat greater; a small percentage of these yield 100 gal/min (6 L/s) or more, or enough to supply small commercial and industrial needs. The major aquifers consist of coarse-grained stratified drift, which is locally present in most of the major stream valleys; they are the source of all large industrial and municipal ground-water supplies. Yields of large-diameter wells in these deposits are commonly 500 gal/min (30 L/s) or more, and many will yield more than 1,000 gal/min (60 L/s).

The chemical quality of the ground water is generally good (Allen, 1953, p. 44–50). The dissolved-solids content is commonly less than 150 mg/L, and with the exception of iron and manganese, concentrations of individual mineral constituents are invariably well within the limits considered acceptable for drinking water by the U.S. Public Health Service. Undesirably high concentrations of manganese and, in a few instances, iron occur in water from a significant percentage of heavily pumped wells in stratified drift. The initial concentration of manganese in water from new wells is generally very low. In some wells concentration has increased gradually—generally over a period of years. It is suspected that local manganese enrichment of ground water is related to physiochemical changes in sand and gravel aquifers as a result of infiltration induced from nearby streams, especially where streamflow is heavily laden with organic waste matter (H. E. Johnston, oral commun., 1976).

Few other widespread problems of ground-water quality have been reported. Locally, however, the quality of ground water is being degraded by landfills, deicing salts applied to highways, agricultural fertilizers, and land deposition of sewage sludge.

Ground-water studies at the reconnaissance level were completed for the entire State between 1944 and 1964, and detailed quantitative investigations, begun in 1958, have been completed for most of the principal ground-water reservoirs. Results of these studies are described in more than 50 State and Federal publications. A list of publications through 1973 is available in pamphlet form from the U.S. Geological Survey offices in Boston, Mass., and Providence, R.I. Since 1973, results of a quantitative study of the Branch River basin in northern Rhode Island (Johnston and Dickerman, 1974b) have been published. In addition, a quantitative study and a basic-data report for the Blackstone River area, also in the northern part of the State, have been published (Johnston and Dickerman, 1974a and 1974c, respectively). (See fig. 7.)

A general reconnaissance of the entire State was made by Allen (1953), and 21 quadrangle studies have been published in a State Ground-Water Map series. The ground-water maps show, on 7½-minute quadrangle sheets, the distribution of outwash deposits (stratified drift) and till, geologic sections, contours on the water-table and bedrock surfaces, and locations of wells for which data are available. They include a brief description of the availability and quality of ground water.

Detailed reconnaissances have been completed for the Providence, Woonsocket, and Wallum Lake areas, in the northern part of the State; for Block Island, off the southern shore; and for 6 of 27 quadrangle study areas. Results of these studies are published in a State Geological Bulletin series. These reports include detailed description of ground-water conditions, as well as basic data and maps; the latter provide information similar to that given in the State Ground-Water Map series.

The need for quantitative information was appraised by Lang (1961), who subdivided the State into 17 proposed study areas. Intensive quantitative studies of several of these areas have since been made, especially where economic development was extensive or potentially so. These studies assess the ground water available for development in areas containing thick stratified drift and show by a model the magnitude of yields theoretically obtainable from selected arrays of wells. Maps of the water table, aquifer thickness, and

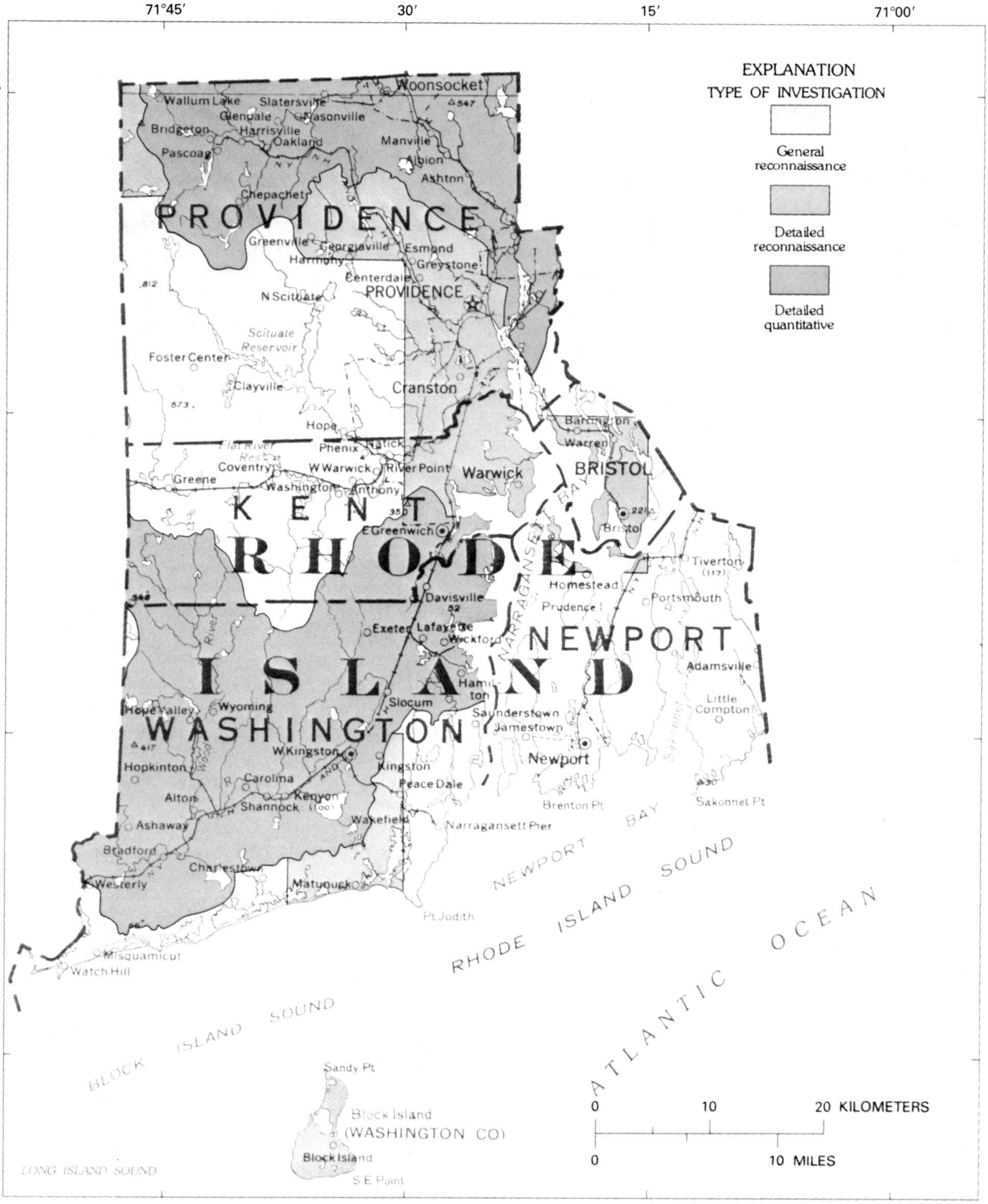

FIGURE 7.—Map showing areas of ground-water investigations in Rhode Island.

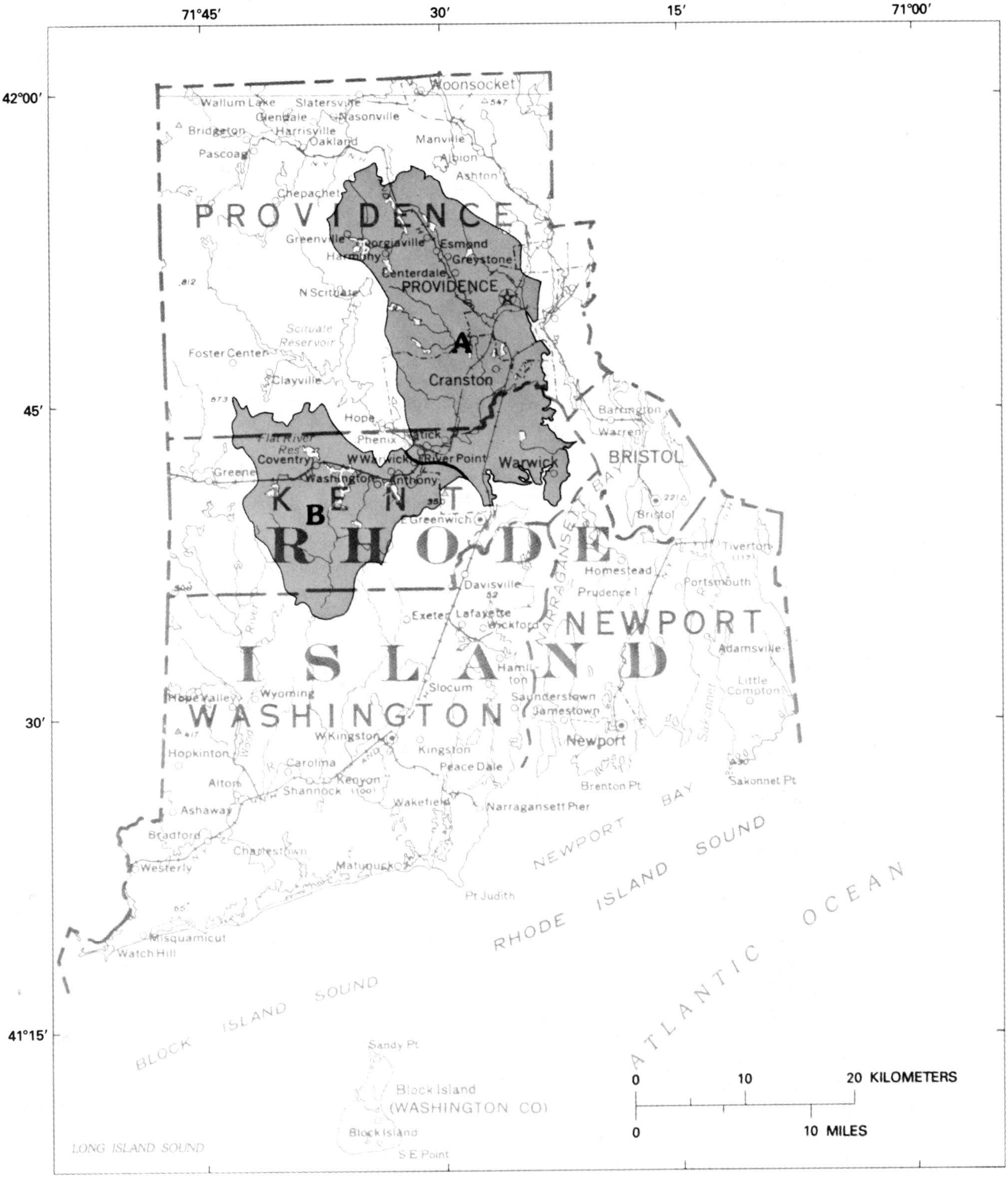

FIGURE 8.—Map showing areas in Rhode Island where quantitative ground-water information may be needed. A, Providence-Warwick area; B, South branch Pawtuxet River basin area.

aquifer permeability (or transmissivity) obtained during the studies provide information necessary for constructing digital or analog models of the ground-water system.

The Rhode Island Water Resources Board is currently (1981) engaged in an extensive exploratory drilling and aquifer-test program in the Pawcatuck River basin in southern Rhode Island. Its purpose is to locate sites favorable for developing large supplies of ground water. The most favorable will be developed as future demands dictate. The program resulted from a long-range plan to provide for orderly development and management of the State's water resources. The U.S. Geological Survey is assisting in the collection and analysis of data and will provide a management model of the stream-aquifer system in the Pawcatuck River basin. Digital and analog models of the stream-aquifer system in several subbasins are being constructed by the University of Rhode Island.

Quantitative ground-water information may be necessary in two additional areas in northern and central Rhode Island. (See fig. 8.) The Providence-Warwick area (A, fig. 8), which includes the most densely populated and highly industrialized parts of the State, is underlain by more than 50 mi^2 (130 km^2) of stratified drift. This drift, which is more than 50 ft (15 m) thick in more than half the area and locally more than 200 ft (60 m) thick, constitutes the largest ground-water reservoir in the State. The drift has yielded as much as 8 million gal/d (30,000 m^3/d) (Lang, 1961, p. 13) and may be capable of yielding more than 20 million gal/d (75,000 m^3/d). Parts of the drift deposits are in hydraulic continuity with Narragansett Bay and its brackish estuaries. Consequently, optimum development will require judicious location of pumping centers and careful management of withdrawal to prevent invasion by salt water. Development schemes and subsequent management of the reservoir can be effectively appraised by analog or digital models.

Quantitative ground-water information in the South Branch Pawtuxet River basin area would also be helpful (B, fig. 8). The northeastern part of the area is underlain by stratified drift as thick as 100 ft (30 m). Wells tapping this drift have an aggregate pumping capacity of about 3 million gal/d (11,000 m^3/d), and an additional 3 million gal/d (11,000 m^3/d) can possibly be developed (New England River Basin Commission, 1975). Excessive ground-water withdrawal may deplete low streamflows and lower pond levels. These adverse effects may be aggravated by regulation after the Big River reservoir in the headwaters of the southern part of the area is constructed. Conjunctive operation of the ground-water and surface-water reservoirs, so as to minimize stream depletion and lowering of pond levels, will require quantitative studies of the ground-water reservoir and its interaction with overlying water bodies.

CONCLUSIONS

New England is favored by an abundance of water—both surface water and ground water. It is also a region of increasing economic activity and population growth. Inevitably, additional supplies of water will be needed, not only for domestic use but also for public supplies and to meet industrial needs. Management of the water resources will require more and more ground water.

Advantages of using ground water are several: (1) In favorable areas it is available near the point of need, obviating the need for pipelines and storage reservoirs. (2) The temperature of ground water is nearly constant—roughly the same as the average air temperature. (3) Although it has higher concentrations of dissolved solids than surface water, these concentrations are relatively stable, facilitating planning and maintenance for many industrial and public supplies. (4) It is less likely to be contaminated than surface water, provided that well equipment has been properly installed; thus, it is especially desirable for personnel in industry and for public supplies. (5) Its storage volume is great—vastly more than the volume of fresh surface water and thus less likely to be affected by prolonged droughts.

These and other factors make ground water particularly attractive for various management options.

Maine, New Hampshire, and Vermont have relatively small populations. The trend, however, is growth. New Hampshire, where the population is expected to quadruple by early in the 21st century, is likely to grow most. Ground water will doubtless be important in the future development of that State.

Coastal Maine, particularly in the south, is expected also to grow considerably. Vermont, on the other hand, will probably remain more rural and much less industrialized and will continue to depend largely on tourism and recreation. Much smaller quantities of water will, therefore, be needed. The use of ground water in Vermont is high in relation to total water use; although the volume used is small, this relation is expected to continue.

Massachusetts, Connecticut, and Rhode Island are much more densely populated, urbanized, and industrialized than the northern three States. Here, the importance of ground water is already manifest. General investigations have been made to date, and more detailed ones are planned by various agencies.

The humid climate of the New England Region assures abundant supplies of water. The most pervasive problem relating to ground water, however, is the threat of local deterioration of its chemical quality. Curtailment and control of the many actual and potential sources of contamination—highway salting, landfills, and other land-use practices—are major goals for water-management programs.

Increasing awareness of environmental concerns in recent years encourages the view that generally satisfactory management of ground-water quality and contamination control will eventually be achieved. Ground water will very likely continue to play an increasingly important part in the economy of the New England Region and warrants continuing study geared to anticipated needs.

SELECTED REFERENCES

Allen, W. B., 1953, The ground-water resources of Rhode Island—A reconnaissance, with a section on surface-water resources, by H. B. Kinnison: Rhode Island Devel. Council Geol. Bull. 6, 170 p.

Allen, W. B., Hahn, G. W., and Brackley, R. A., 1966, Availability of ground water, upper Pawcatuck River basin, Rhode Island: U.S. Geol. Survey Water-Supply Paper 1821, 66 p.

Anderson-Nichols and Co., Inc., 1969, Public water supply study, Phase One Report: New Hampshire Dept. Resources and Economic Devel., 550 p.

Asselstine, E. S., and Grossman, I. G., 1955, The ground water resources of Westchester County, New York; Pt. 1, Records of wells and test holes: New York Water Power and Control Comm. Bull. GW–35, 79 p.

Baker, J. A., Healy, H. G., and Hackett, O. M., 1964, Geology and ground-water conditions in the Wilmington-Reading area, Massachusetts: U.S. Geol. Survey Water-Supply Paper 1694, 80 p.

Barksdale, H. C., O'Bryan, Deric, and Schneider, W. J., 1966, Effect of drought on water resources in the Northeast: U.S. Geol. Survey Hydrol. Inv. Atlas HA–243.

Bayley, W. S., 1905, Maine, *in* Fuller, M. L. (geologist in charge), Underground waters of eastern United States: U.S. Geol. Survey Water-Supply Paper 114, p. 41–56.

Black, R. F., 1977. Regional stagnation of ice in northeastern Connecticut: An alternative model of deglaciation for part of New England: Geol. Soc. America Bull., v. 88, no 9, p. 1331–1336 (Sept.)

Brackley, R. A., Fleck, W. B., and Meyer, W. R., 1973, Hydrology and water resources of the Neponset and Weymouth river basins, Massachusetts: U.S. Geol. Survey Hydrol. Inv. Atlas HA–484, 3 sheets.

Bradley, Edward, 1964, Geology and ground-water resources of southeastern New Hampshire: U.S. Geol. Survey Water-Supply Paper 1695, 80 p.

Brown, J. S., 1925, A study of coastal ground water, with special reference to Connecticut: U.S. Geol. Survey Water-Supply Paper 537, 101 p.

———1928, Ground water in the New Haven area, Connecticut: U.S. Geol. Survey Water-Supply Paper 540, 206 p.

Bue, C. D., 1970, Streamflow from the United States into the Atlantic Ocean during 1931–60: U.S. Geol. Survey Water-Supply Paper 1899–I, 36 p.

Burns, A. W., Frimpter, M. H., and Willey, R. E., 1975, Evaluation of data availability and examples of modeling for ground-water management on Cape Cod, Massachusetts: U.S. Geol. Survey Water-Resources Inv. WRI 16–75, 22 p.

Cederstrom, D. J., 1972, Evaluation of yields of wells in consolidated rocks, Virginia to Maine: U.S. Geol. Survey Water-Supply Paper 2021, 38 p.

———1973, Cost analysis of ground-water supplies in the North Atlantic Region, 1970: U.S. Geol. Survey Water-Supply Paper 2034, 48 p.

Cederstrom, D. J., Baker, J. A., and Tarver, G. R., 1971, Ground water in the North Atlantic Region; Appendix D, North Atlantic Regional Water Resources Study: U.S. Geol. Survey open-file report, 240 p.

Cederstrom, D. J., and Hodges, A. L., Jr., 1967, Ground-water favorability of the Connecticut River basin, New England States: U.S. Geol. Survey Hydrol. Inv. Atlas HA–249, 2 sheets.

Chandler, R. C., Fuller, J. W., and Campbell, E. W., 1959, Radioactivity in water: Maine Water Utilities Assoc. Jour., v. 35, no. 2, p. 19–25.

Clapp, F. G., 1909, Underground waters of southern Maine: U.S. Geol. Survey Water-Supply Paper 223, 268 p.

Collings, M. R., Wiesnet, D. R., and Fleck, W. B., 1969, Water resources of the Millers River basin, north-central Massachusetts and southwestern New Hampshire: U.S. Geol. Survey Hydrol. Inv. Atlas HA–293, 4 sheets.

Cotton, J. E., 1975a, Availability of ground water in the Saco River basin, east-central New Hampshire; U.S. Geol. Survey Water-Resources Inv. WRI 39–74.

———1975b, Availability of ground water in the Androscoggin River basin, northern New Hampshire: U.S. Geol. Survey Water-Resources Inv. WRI 22–75.

———1975c, Availability of ground water in the Pemigewasset and Winnipesaukee River basins, central New Hampshire: U.S. Geol. Survey Water-Resources Inv. WRI 47–75.

———1975d, Availability of ground water in the upper Connecticut River basin, northern New Hampshire: U.S. Geol. Survey Water-Resources Inv. WRI 53–75.

———1976a, Availability of ground water in the middle Connecticut River basin, west-central New Hampshire: U.S. Geol. Survey Water-Resources Inv. WRI 76–18.

———1976b, Availability of ground water in the middle Merrimack River Basin, central and southern New Hampshire: U.S. Geol. Survey Water-Resources Inv. WRI 76–39.

———1977a, Availability of ground water in the lower Merrimack River basin, southern New Hampshire: U.S. Geol. Survey Water-Resources Inv. WRI 77–69.

———1977b, Availability of ground water in the Piscataqua and other Coastal River basins, southeastern New Hampshire: U.S. Geol. Survey Water-Resources Inv. WRI 77–70.

———1977c, Availability of ground water in the lower Connecticut River basin, southwestern New Hampshire: U.S. Geol. Survey Water-Resources Inv. WRI 77–79.

Cushman, R. V., 1964, Ground-water resources of north-central Connecticut: U.S. Geol. Survey Water-Supply Paper 1752, 96 p.

Fenneman, N. M., and others, 1946, Physical divisions of the United States: U.S. Geol. Survey map prepared in cooperation with Physiographic Committee, scale 1:7,000,000 (reprinted, 1957).

Feth, J. H., and others, 1965, Preliminary map of the conterminous United States showing depth to and quality of shallowest ground water containing more than 1,000 parts per million dissolved solids: U.S. Geol. Survey Hydrol. Inv. Atlas HA–199.

Fuller, M. L. (geologist in charge), 1905, Underground waters of Eastern United States: U.S. Geol. Survey Water-Supply Paper 114, 285 p.

Gay, F. B., Toler, L. G., and Hansen, B. P., 1974, Hydrology and water resources of the Deerfield River basin, Massachusetts: U.S. Geol. Survey Hydrol. Inv. Atlas HA–506, 2 sheets.

Geraghty, J. J., Miller, D. W., Van der Leeden, Frits, and Troise, F.

L., 1973, Water Atlas of the United States: Water Information Center, Inc., Port Washington, N.Y.

Gonthier, J. B., Johnston, H. E., and Malmberg, G. T., 1974, Availability of ground water in the Lower Pawcatuck River basin, Rhode Island: U.S. Geol. Survey Water-Supply Paper 2033, 40 p.

Gregory, H. E., 1909, Underground water resources of Connecticut, with a study of the occurrence of water in crystalline rocks, by E. E. Ellis: U.S. Geol. Survey Water-Supply Paper 232, 200 p.

Grossman, I. G., 1957, The ground water resources of Putnam County, New York: New York Water Power and Control Comm. Bull. GW–37, 78 p.

Hanes, R. E., Zelazny, L. W., and Blaser, R. E., 1970, Effects of deicing salts on water quality and biota: National Acad. Sci., Highway Research Board, Nat'l. Coop. Highway Research Program Rept. 91.

Hodges, A. L., 1967, Ground water favorability map of the Nulhegan-Passumpsic River basin, Vermont: Vermont Dept. Water Resources.

———1968a, Ground water favorability map of the Wells-Ompompanoosuc River basin, Vermont: Vermont Dept. Water Resources.

———1968b, Ground water favorability map of the White River basin, Vermont: Vermont Dept. Water Resources.

———1968c, Ground water favorability map of the Ottauquechee-Saxton River basin, Vermont: Vermont Dept. Water Resources.

———1968d, Ground water favorability map of the West-Deerfield River basin, Vermont: Vermont Dept. Water Resources.

Hodges, A. L., Jr., and Butterfield, D., 1972a, Ground-water availability in the Barre-Montpelier area, Vermont: Vermont Dept. Water Resouces duplicate rept., 27 p.

———1972b, Ground-water availability in the White River Junction area, Vermont: Vermont Dept. Water Resources duplicate rept., 27 p.

Johnston, H. E., and Dickerman, D. C., 1974a, Availability of ground water in the Blackstone River area, Rhode Island and Massachusetts: U.S. Geol. Survey Water-Resources Inv. WRI 4-74, 2 sheets (map, text, tables).

———1974b, Availability of ground water in the Branch River basin, Providence County, Rhode Island: U.S. Geol. Survey Water-Resources Inv. WRI 18–74, 39 p.

———1974c, Geologic and hydrologic data for the Blackstone River area, Rhode Island: Rhode Island Water Resouces Board Hydrol. Bull. 7, 42 p.

Kammerer, J. C., and Baldwin, Helene L., 1962, Water problems in the Springfield-Holyoke area, Massachusetts: U.S. Geol. Survey Water-Supply Paper 1670, 68 p. [revised 1963].

Kohout, F. A., ed., 1970, Saline water; a valuable resource: Water Resources Research, v. 6, no. 5, p. 1441–1531.

Kohout, F. A., Walker, E. H., Bothner, M. H., and Hathaway, J. C., 1976, Fresh ground water found deep beneath Nantucket Island, Massachusetts: U.S. Geol. Survey Jour. Research, 4, no. 5, Sept.–Oct., p. 511–515.

Lang, S. M., 1961, Appraisal of the ground-water reservoir areas in Rhode Island: Rhode Island Water Resources Coordinating Board, Geol. Bull. 11, 38 p.

Lang, S. M., Bierschenk, W. H., and Allen, W. B., 1960, Hydraulic characteristics of glacial outwash in Rhode Island: Rhode Island Water Resources Coordinating Board Hydrol. Bull. 3, 38 p.

La Sala, A. M., 1964, Geology and ground-water resources of the Bristol-Plainville-Southington area, Connecticut: U.S. Geol. Survey Water-Supply Paper 1578, 70 p.

McGuinness, C. L., 1951, Water law, with special reference to ground water: U.S. Geol. Survey Circ. 117, 30 p.

———1963, The role of ground water in the national water situation: U.S. Geol. Survey Water-Supply Paper 1800, 1121 p.

Meinzer, O.E., and Stearns, Norah D., 1929, A study of ground water in the Pomperaug basin, Connecticut, with special reference to intake and discharge: U.S. Geol. Survey Water Supply Paper 597–B, p. 73–146.

Miller, D. W., DeLuca, F. A., and Tessier, T. L., 1974, Ground water contamination in the Northeast States: U.S. Environmental Protection Agency, Environmental Protection Technology Ser. EPA–660/2–74–056, June, 325 p.

Morrill, G. B., III, and Toler, L. G., 1973, Effect of septic-tank wastes on quality of water, Ipswich and Shawsheen River basins, Massachusetts: U.S. Geol. Survey Jour. Research, v. 1, no. 1, [Jan.-Feb.] p. 117–120.

Moulder, E. A., 1970, Freshwater bubbles: A possibility for using saline aquifers to store water: Water Resources Research, v. 6, no. 5, p. 1528–1531.

Murray, C. R., and Reeves, E. B., 1972, Estimated use of water in the United States in 1970: U.S. Geol. Survey Circ. 676, 37 p.

———1977, Estimated use of water in the United States in 1975: U.S. Geol. Survey Circ. 765, 39 p.

National Academy of Sciences, National Academy of Engineering, 1972 [1974], Water Quality Criteria 1972: U.S. Govt. Printing Office, 594 p.

New England River Basin Commission, 1975, Report of the Southeastern New England Study—8. Pawtuxet planning area report, New England River Basin Commission: Boston, Mass., no. 8, chapt. 4, p. 4–1 to 4–3.

Norvitch, R. F., 1966, Ground-water favorability map of the Housatonic River basin, Massachusetts: Massachusetts Water Resources Comm. Hydrol. Inv. Atlas, 1 map.

Norvitch, R. F., Farrell, D. F., Pauszek, F. H., and Petersen, R. G., 1968, Hydrology and water resources of the Housatonic River basin, Massachusetts: U.S. Geol. Survey Hydrol. Inv. Atlas HA–281, 4 sheets.

Pollock, S. J., Farrell, D. F., and Caswell, W. W., 1969, Water resources of the Assabet River basin, central Massachusetts: U.S. Geol. Survey Hydrol. Inv. Atlas HA–312, 1 sheet.

Pollock S. J., and Toler, L. G., 1972, Effects of highway deicing salts on ground water and water supplies in Massachusetts: Massachusetts Dept. Public Works and U.S. Geol. Survey open-file rept.

———1973, Effects of highway deicing salts on ground water and water supplies in Massachusetts: Highway Research Record, no. 425, 6 p. (in Environmental degradation by deicing chemicals and effective countermeasures: Highway Research Board, Natl. Research Council, Natl. Acad. Sci.-Natl. Acad. Eng.)

Prescott, G. C., Jr., 1963a, Reconnaissance of ground-water conditions in Maine: U.S. Geol. Survey Water-Supply Paper 1669–T, 52 p.

———1963b, Geologic map of the surficial deposits of part of southwestern Maine and their water-bearing characteristics: U.S. Geol. Survey Hydrol. Inv. Atlas HA–76.

———1966, Surficial geology and availability of ground water in part of the lower Penobscot River basin, Maine: U.S. Geol. Survey Hydrol. Inv. Atlas HA–225.

———1968, Ground-water favorability areas and surficial geology of the lower Androscoggin River basin, Maine: U.S. Geol. Survey Hydrol. Inv. Atlas HA–285.

———1969, Ground-water favorability areas and surficial geology of the lower Kennebec River basin, Maine: U.S. Geol. Survey Hydrol. Inv. Atlas HA–337.

———1972, Ground-water favorability and surficial geology of the lower Aroostook River basin, Maine: U.S. Geol. Survey Hydrol. Inv. Atlas HA–443.

———1973a, Ground-water favorability and surficial geology of the lower St. John River valley, Maine: U.S. Geol. Survey Hydrol. Inv. Atlas HA–485.

———1973b, Ground-water favorability and surficial geology of parts of the Meduxnekeag River and Prestile Stream basins, Maine: U.S. Geol. Survey Hydrol. Inv. Atlas HA–486.

———1974a, Ground-water favorability and surficial geology of the Cherryfield-Jonesboro area, Maine: U.S. Geol. Survey Hydrol. Inv. Atlas HA–529.

———1974b, Ground-water favorability and surficial geology of the Machias-Lubec area, Maine:U.S. Geol. Survey Hydrol. Inv. Atlas HA–535.

———1977a, Ground-water favorability and surficial geology of the Portland area, Maine: U.S. Geol. Survey Hydrol. Inv. Atlas HA–561.

———1977b, Ground-water favorability and surficial geology of the Windham-Freeport area, Maine: U.S. Geol. Survey Hydrol. Inv. Atlas HA–564.

———1980, Ground-water favorability and surficial geology of the Royal, upper Presumpscot, and upper Saco River basins, Maine: U.S. Geol. Survey Water-Resources Inv. WRI-79-1287, 3 sheets (text, maps, tables).

Randall, A. D., 1964, Geology and ground water in the Farmington-Granby area, Connecticut: U.S. Geol. Survey Water-Supply Paper 1661, 130 p.

Rosenshein, J. S., Gonthier, J. B., and Allen, W. B., 1968, Hydrologic characteristics and sustained yield of principal ground-water units, Potowomut-Wickford Area, Rhode Island: U.S. Geol. Survey Water-Supply Paper 1775, 38 p.

Ryan, Barbara J., 1980, Cape Cod aquifer, Cape Cod, Massachusetts: U.S. Geol. Survey Water-Resources Investigations Open-File Rept. 80–571, 23 p.

Sammel, E. A., 1967, Water resources of the Parker and Rowley River basins, Massachusetts: U.S. Geological Survey Hydrologic Investigation Atlas HA-247, 1 sheet; text, 9 p.

Sammel, E. A., Brackley, R. A., and Palmquist, W. N., Jr., 1964, Synopsis of water resources of the Ipswich River basin, Massachusetts: U.S. Geol. Survey Hydrol. Inv. Atlas HA–196, 2 sheets.

Schneider, W. J., and Spieker, A. M., 1969, Water for the cities—the outlook: U.S. Geol. Survey Circ. 601–A, 6 p.

Silvey, W. D., and Johnston, H. E., 1977, Preliminary study of sources and processes of enrichment of manganese in water from University of Rhode Island supply wells: U.S. Geol. Survey Open-File Rept. 77–561, 33 p.

Simmons, E. T., Grossman, I. G., and Heath, R. C., 1961, Ground water resources of Dutchess County, New York: New York Water Resources Comm. Bull. GW–43, 82 p.

Sinnott, Allen, and Cushing, E. M., 1978, Summary appraisals of the nation's ground-water resources—Mid-Atlantic Region: U.S. Geol. Survey Prof. Paper 813–I, 32 p.

Terry, R. C., Jr., 1974, Road salt, drinking water, and safety—improving public policy and practice: Cambridge, Mass., Arthur D. Little, Inc., Ballinger Publishing Co., 161 p.

Thomas, H. E., 1951, The conservation of ground-water: New York, McGraw-Hill Book Co., Inc., 321 p.

———1952, Ground-water regions of the United States—their storage facilities: U.S. 83d Cong., House Interior and Insular Affairs Committee, The physical and economic foundation of natural resources, v. 3, 78 p.

U.S. Department of Commerce, National Weather Service, 1968, Climatic atlas of the United States: 80 p.

U.S. Geological Survey, 1970, National Atlas of the United States of America, Arch O. Gerlach, ed.: Washington, D.C., 417 p.

U.S. National Bureau of Standards, 1953, Maximum permissible amounts of radioisotopes in the human body and maximum permissible concentrations in the air and water: Handb. 52, 45 p.

U.S. Public Health Service, 1962, Drinking water standards: U.S. Public Health Service Pub. 956, 61 p.

U.S. Water Resources Council, 1968, The Nation's water resources—the first national assessment of the Water Resources Council: 403 p.

———1970, Water resources regions and subregions for the national assessment of water and related land resources: 188 p.

———1972, OBERS projections—regional economic activity in the United States; V.1, Concepts, methodology, and summary data: 109 p. and appendices.

Walker, E. H., and Caswell, W. W., 1977, Map showing availability of ground water in the Connecticut River lowlands, Massachusetts: U.S. Geol. Survey Hydrologic Inv. Atlas HA–563, 2 sheets with text and tables.

Walker, E. H., Wandle, S. W., Jr., and Caswell, W. W., 1975, Hydrology and water resources of the Charles River basin, Massachusetts: U.S. Geol. Survey Hydrologic Inv. Atlas HA–554.

Weigle, J. W., 1968, Ground-water resources of the lower Merrimack River valley, south-central New Hampshire: U.S. Geol. Survey Hydrologic Inv. Atlas HA–277.

Whitcomb, H. A., 1973, Ground-water resources of the Ashuelot River basin, southwestern New Hampshire: U.S. Geol. Survey Hydrologic Inv. Atlas HA-441.

Williams, J. R., 1968, Availability of ground water in the northern part Tenmile and Taunton River basins, southeastern Massachusetts: U.S. Geol. Survey Hydrol. Inv. Atlas HA–300, 1 sheet.

Williams J. R., Farrell, D. F., and Willey, R. E., 1973, Water resources of the Taunton River basin, southeastern Massachusetts: U.S. Geol. Survey Hydrol. Inv. Atlas HA–460, 3 sheets.

Williams, J. R., and Tasker, G. D., 1974a, Water resources of the coastal drainage basins of southeastern Massachusetts, Weir River, Hingham, to Jones River, Kingston: U.S. Geol. Survey Hydrol. Inv. Atlas HA–504, 2 sheets.

———1974b, Water resources of the coastal drainage basins of southeastern Massachusetts, Plymouth to Weweantic River, Wareham: U.S. Geol. Survey Hydrol. Inv. Atlas HA–507, 2 sheets.

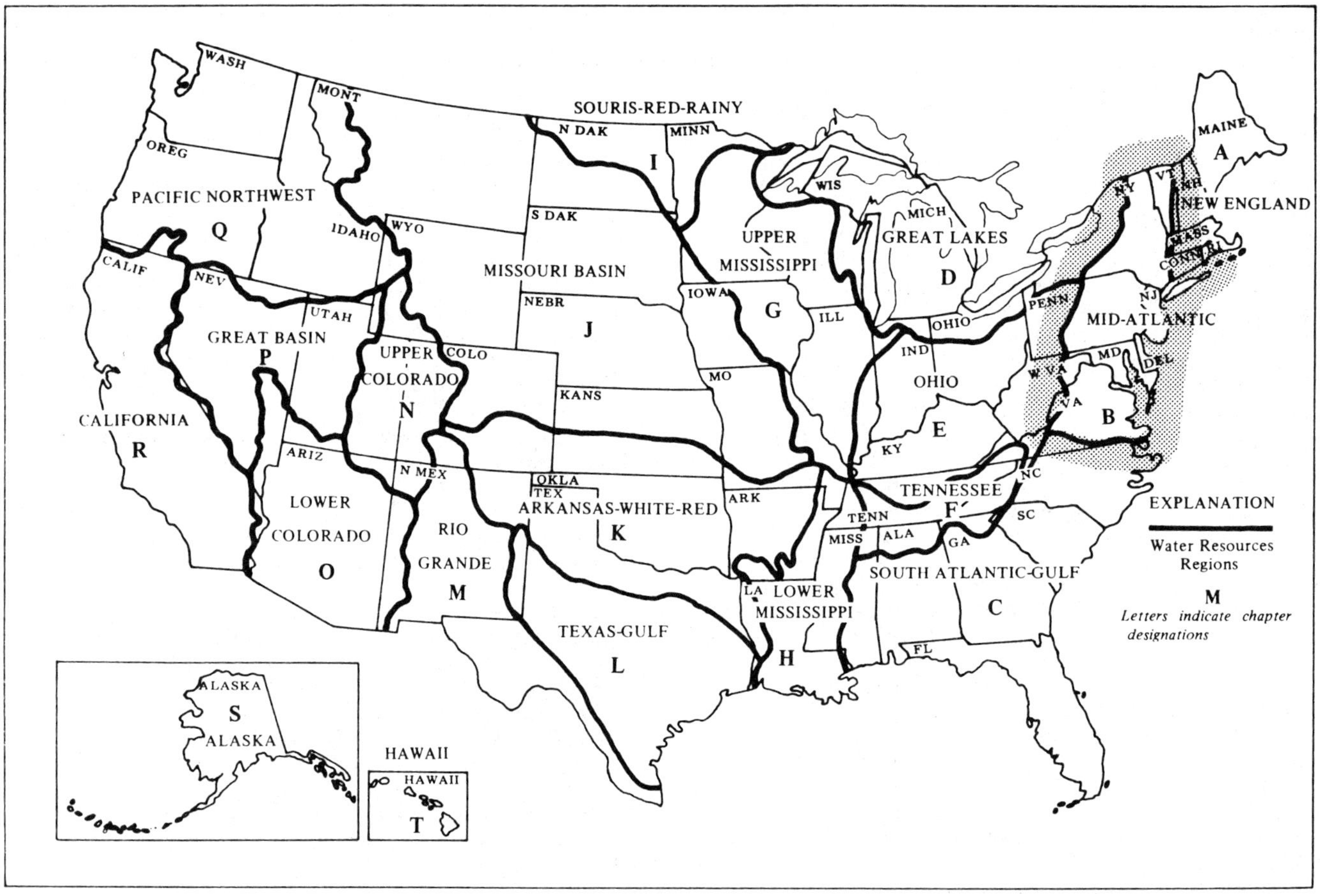

B — Location Map for the Mid-Atlantic Region

B

MID-ATLANTIC REGION

By ALLEN SINNOTT and ELLIOTT M. CUSHING

ABSTRACT

The Mid-Atlantic Region covers a total area of about 108,000 square miles. It includes parts of Vermont, Massachusetts, New York, Pennsylvania, Maryland, West Virginia, and Virginia, the entire States of New Jersey and Delaware, and the District of Columbia. It encompasses the entire drainage basins (within the United States) of the Hudson, Delaware, Susquehanna, Potomac, and the James River and includes Long Island and the coastal drainage of New Jersey, Delaware, Maryland, and Virginia.

Ground water in the region occurs in three broad types of geologic terrane: (1) unconsolidated deposits in the Coastal Plain seaward of the Fall Line; (2) hard, consolidated sedimentary rocks and crystalline igneous and metamorphic rocks in the remainder of the region; and (3) unconsolidated sand, gravel, and other deposits of glacial origin that overlie the older rocks extensively in the glaciated northern part of the region.

Ground water is derived primarily from precipitation; it can be intercepted for use by pumping from wells before it reaches the streams or before it drains into coastal wetlands or the ocean.

The natural discharge from the aquifers in the region is estimated to be at about 38.6 billion gallons per day; in addition, at least 140–350 trillion gallons is stored in the rocks.

About 949 billion gallons of fresh ground water was withdrawn in 1970. This quantity represents about 9 percent of the total freshwater use of 10,220 billion gallons. Available ground-water reserves indicate that a considerable part of the additional supplies needed for the anticipated increase in economic activity in the region could be developed from ground water.

INTRODUCTION

This report describes the ground-water conditions in one of the 21 Water Resources Regions designated by the U.S. Water Resources Council (1970)—the Mid-Atlantic Region in northeastern United States (fig. 1). The purposes of the report are (1) to summarize the broad facts about the overall ground-water resources of the Mid-Atlantic Region, (2) to describe current development and management problems of important ground-water reservoirs in the region, (3) to discuss options for future ground-water development, and (4) to identify the additional ground-water information needed to provide a sound basis for better management of the region's water resources.

The report is one of a series being published by the Geological Survey to document the Nation's ground-water resources, region by region, and to analyze the role of the resource in National water supply.

PHYSICAL, GEOGRAPHICAL, AND ECONOMIC SETTING

The Mid-Atlantic Region includes the entire drainage basins of five major streams in the heavily populated and industrial northeastern United States (fig. 2). These are the Hudson River, in New York, Vermont, and Massachusetts; the Delaware River, chiefly in New York, Pennsylvania, and New Jersey; the Susquehanna River, chiefly in Pennsylvania and New York; the Potomac River, in West Virginia, Pennsylvania, Maryland, and Virginia; and the James River, chiefly in Virginia. Two smaller tidewater basins, those of the Rappahannock and York Rivers, lie between the Potomac and James Rivers in Virginia. The region as a whole extends from the international boundary with Canada southward nearly to Virginia's southern boundary.

In addition to these five basins, the region includes also a small part of the St. Lawrence River drainage area in northeastern New York and northwestern Vermont, and lesser stream basins and coastal reaches from southeastern Virginia to New York, including all of Long Island.

The Mid-Atlantic Region extends inland from the coast an average of about 200 mi, and it is nearly 700 mi long. It covers an area of about 108,000 mi ^{2}, and encompasses all or parts of nine States and the District of Columbia. One of the largest—and certainly the most economically important—of the

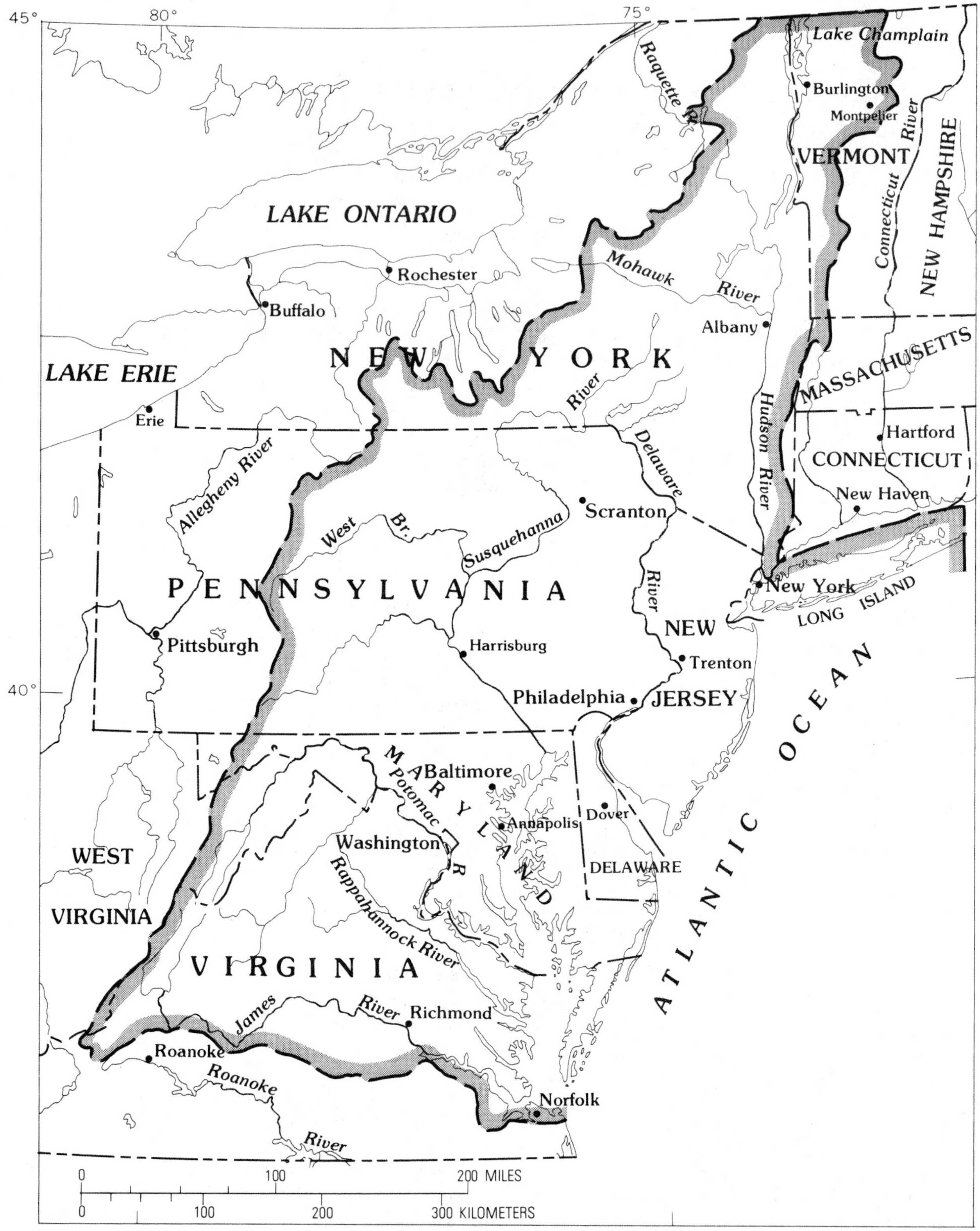

FIGURE 1.—The Mid-Atlantic Region includes much of the heavily populated and industrially developed northeastern United States.

world's estuaries, Chesapeake Bay, lies within the southeastern coastal part of the region.

The Mid-Atlantic Region is generally humid, with four distinct seasons, characterized by frequent weather changes. The climate along the coast is generally less severe than inland, because the ocean is warmed near the southern part of the region by the northeastward-flowing currents of the Gulf Stream.

Average annual temperatures range from about 44° F (6.6° C) in northern Vermont to about 60° F (15.6° C) in southern Virginia.

Precipitation is the ultimate source of virtually all the freshwater available for use in the Mid-Atlantic Region, although water stored in deep-lying rocks may have been emplaced many thousands of years ago. The annual precipitation on the region averages about 40 inches but varies somewhat, depending on location. It is about 40 inches or more along the coast and over 50 inches in upstate New York, southern Vermont, and West Virginia. It is a little less than 36 inches in northern Vermont, south-central New York, two small areas in south-central Pennsylvania, and near the Virginia-West Virginia boundary. (See fig. 3.)

Although the Mid-Atlantic Region covers only about 3 percent of the land area of the United States, it contains about 19 percent of the national population—more than 38,500,000 people. The regional average population density is about 350 people/mi ^{2}, compared to about 54 people/mi ^{2} for the entire United States. Several large population centers are within its boundaries, including three of the largest metropolitan areas of the country, New York City, Philadelphia, and Washington, D.C. (See fig. 4.)

The region generates nearly 22 percent of the national personal income, and its per-capita personal income is 15 percent above the national average. Thus, the Mid-Atlantic Region has a high level of economic development—the highest among the Water Resources Regions in the country. (See U.S. Water Resources Council, 1972, summary tables.) Despite this fact, however, there are economically depressed areas in the region with declining population, particularly the abandoned coal-mining districts in Pennsylvania and West Virginia. The present increasing demand for coal as a source of energy may reactivate many former coal-producing districts.

PHYSIOGRAPHIC DIVISIONS

Several physiographic provinces, as defined by Fenneman and others (1946), lie within the Mid-Atlantic Region. (See fig. 5.) The most important of these, from the standpoint of ground-water supplies, is the Coastal Plain province, which flanks the entire eastern part of the region along the Atlantic Ocean. The other major physiographic divisions include the Piedmont province, immediately west of the Coastal Plain; the narrow Blue Ridge province and its northeastern correlative, the Reading Prong of the New England province; the Valley and Ridge province; and, in the west and northwest, the Appalachian Plateaus province. In addition, in the northernmost part of the region, there are the eastern part of the Adirondack province, the Champlain Valley, and the extreme western flank of the New England province.

OCCURRENCE AND AVAILABILITY OF GROUND WATER

The Mid-Atlantic Region is underlain by a wide variety of geologic formations that together control the storage, movement, and chemical quality of the ground water. The rocks range from the unconsolidated clays, sands, and gravels of the Coastal Plain and watercourses to much older rocks west and northwest of the Fall Line (fig. 6). These older rocks include consolidated sedimentary rocks and some ancient intrusive and volcanic rocks in the Piedmont Lowland; limestones, dolomites, sandstones, and shales of the Valley and Ridge province; crystalline igneous and metamorphic rocks of the Piedmont Upland; and those of the Blue Ridge and the Adirondack provinces. Most of these older rocks are relatively impermeable, but yield small to moderate supplies of ground water from fractures or joints to depths of about 200 to 300 feet below the surface.

The rocks of the region can be grouped into three general classes. These are (1) the unconsolidated deposits that underlie the Coastal Plain, (2) the crystalline igneous and metamorphic rocks and consolidated sedimentary rocks found in the remainder of the region, and (3), surficial deposits of glacial origin, occurring widely in the northern part of the region. The glacial deposit of greatest importance with respect to ground water is stratified drift, much of which consists of sand and gravel that commonly occur in stream valleys forming what are called watercourse aquifers.

The general hydrogeologic characteristics of the rocks in each of the principal physiographic prov-

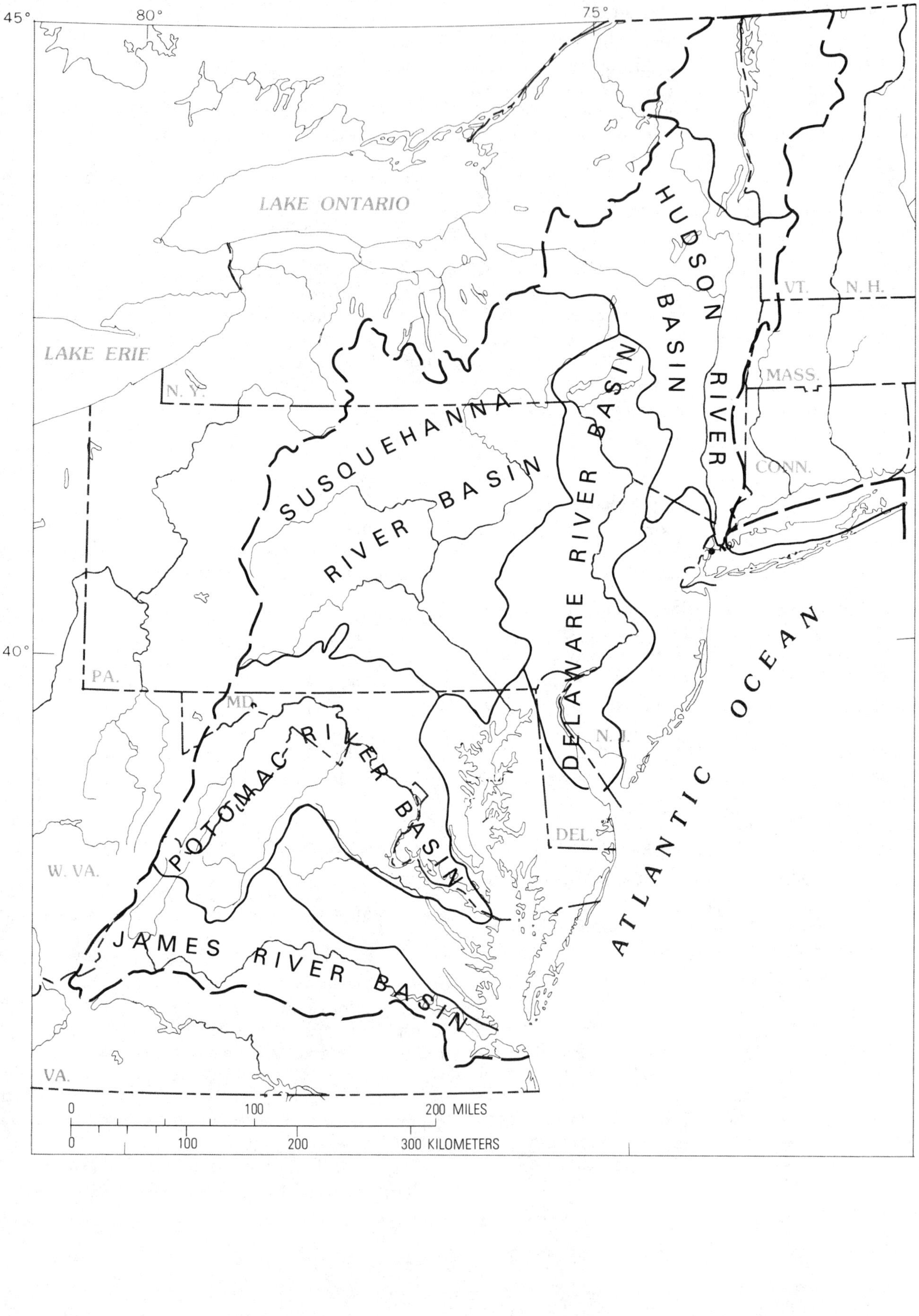
45°
80°
75°
40°
LAKE ONTARIO
LAKE ERIE
N. Y.
PA.
MD.
W. VA.
VA.
DEL.
N. J.
VT.
N. H.
MASS.
CONN.
HUDSON RIVER BASIN
SUSQUEHANNA RIVER BASIN
DELAWARE RIVER BASIN
POTOMAC RIVER BASIN
JAMES RIVER BASIN
ATLANTIC OCEAN
0
100
200 MILES
0
100
200
300 KILOMETERS

inces of the Mid-Atlantic Region are described in the following pages.

COASTAL PLAIN

The Coastal Plain lies east of the Fall Line, and forms the southeastern part of the Mid-Atlantic Region. It extends from southern Virginia northeastward through Maryland, Delaware, Pennsylvania, New Jersey, and Long Island. (See fig. 6.) Topographically its surface is nearly flat or gently undulating, and altitudes range from sea level along the Atlantic coast to roughly 250 ft at the Fall Line.

The Coastal Plain is underlain by unconsolidated deposits. The formations consist mostly of sand and clay with minor amounts of gravel, and locally some beds of "marl" or shell rock. These deposits begin as a featheredge along the Fall Line and dip gently southeastward in a thickening wedge that rests directly on the underlying crystalline igneous and metamorphic rocks and other hard rocks that make up the basement complex—the seaward extension of the hard rocks of the adjacent Piedmont province. Configuration of the buried surface of the basement complex is shown in figure 7, from which it can be seen that the overlying deposits reach a maximum total thickness of more than 8,000 ft.

Overall, the general geology of the Coastal Plain deposits appears to be simple. There is, however, considerable complexity in many of its details. (See Brown and others, 1972, for a description of the structural and stratigraphic framework of the Atlantic Coastal Plain, and the relation of the tectonic and geologic history to present distribution of permeability.) The formations are areally extensive. In general, the shallow formations dip more gently than the deeper ones, and the seaward (southeastward) dip is on the order of 10 to 20 ft/mi.

Surficial deposits of the Coastal Plain are largely permeable unconsolidated sand and gravel that is thickest on Long Island and in New Jersey, Delaware, and eastern Maryland (fig. 8). Large quantities of water can be obtained from these deposits by wells generally less than 300 ft deep. Yields of individual wells may be as much as 2,000 gal/min where the greatest thicknesses of coarse material are present. Elsewhere in eastern Maryland, and in eastern Virginia, the surficial deposits of the Coastal Plain are thin and the yield of individual wells is small.

Deeper artesian aquifers of the Coastal Plain are among the most productive in the Mid-Atlantic Region. The depth of wells may be as much as 1,300 ft, and yields of more than 2 million gal/d may be available from some wells.

Beneath the ocean shoreline, and seaward at various distances, the downdip reaches of the artesian aquifers contain brackish water. (See Back, 1966.) Large quantities of ground water withdrawn from wells close to an interface between freshwater and brackish water may cause encroachment of brackish water into the freshwater part of the aquifer and eventually into wells.

Near the Fall Line, where the Coastal Plain sediments thin out against the basement rocks that are exposed in the Piedmont Province, only small to moderate supplies generally are available. Most wells there will not yield more than a few hundred gallons per minute. Locally, as in the Philadelphia-Camden area, somewhat larger yields have been obtained.

A number of distinct stratigraphic units have been delineated in the Coastal Plain, ranging from about 10 in Virginia to more than 20 in New Jersey. Many of these formations are permeable and have been defined as aquifers. Most are of wide areal extent and those that are of particular economic importance as major sources of ground water have been investigated and described. Large rates of withdrawal from one aquifer commonly are reflected in others proximate to it because of vertical leakage through intervening less-permeable beds. Hydraulic continuity through the complex interbedding is characteristic of Coastal Plain sediments. Most of the aquifers are not separate and independent hydrologic units; rather, each is part of a complex hydrologic system, and this system may even include streams and lakes where the aquifers are at or near the land surface.

Because the leaky nature of this complex system allows water to flow from one formation to another, it is generally not relevant to estimate quantities of ground water that are available from the individual aquifers. Rather, it is more meaningful to estimate the quantity available from the entire system. As will be shown later, for the Coastal Plain part of the region the minimum quantity of ground water that is being discharged under present hydrologic conditions is estimated to be a little more than 9 billion gal/d.

FIGURE 2.—The Mid-Atlantic Region comprises the entire drainage basins of five major rivers in the northeastern United States, including also smaller coastal river basins, and the United States part of the upper St. Lawrence River basin. It covers all of New Jersey and Delaware, parts of several other States, and the District of Columbia.

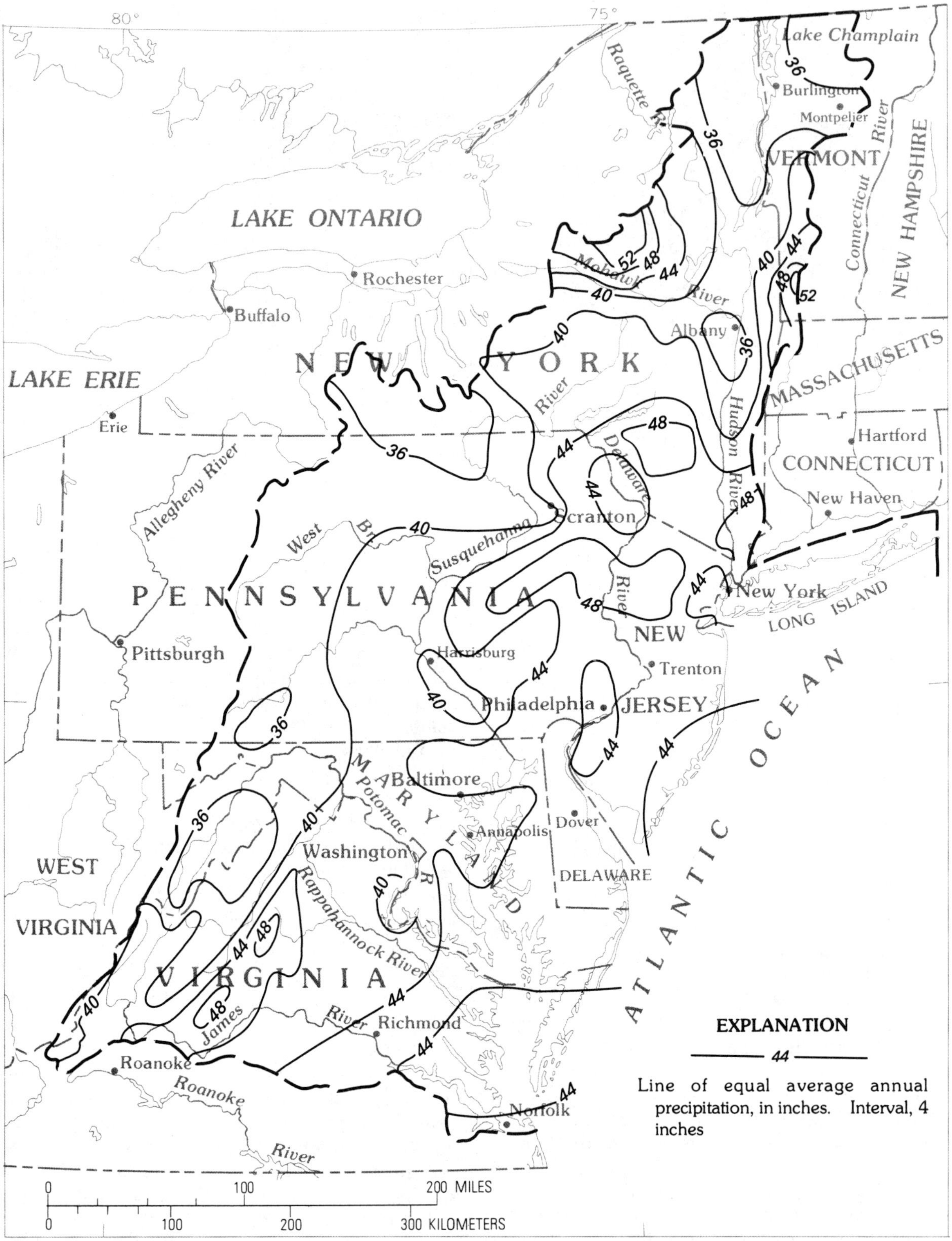

FIGURE 3.—Precipitation in the Mid-Atlantic Region averages about 40 inches and is the ultimate source of virtually all recharge of the aquifers in the region. (Based on period 1931–60. Source: National Weather Service.)

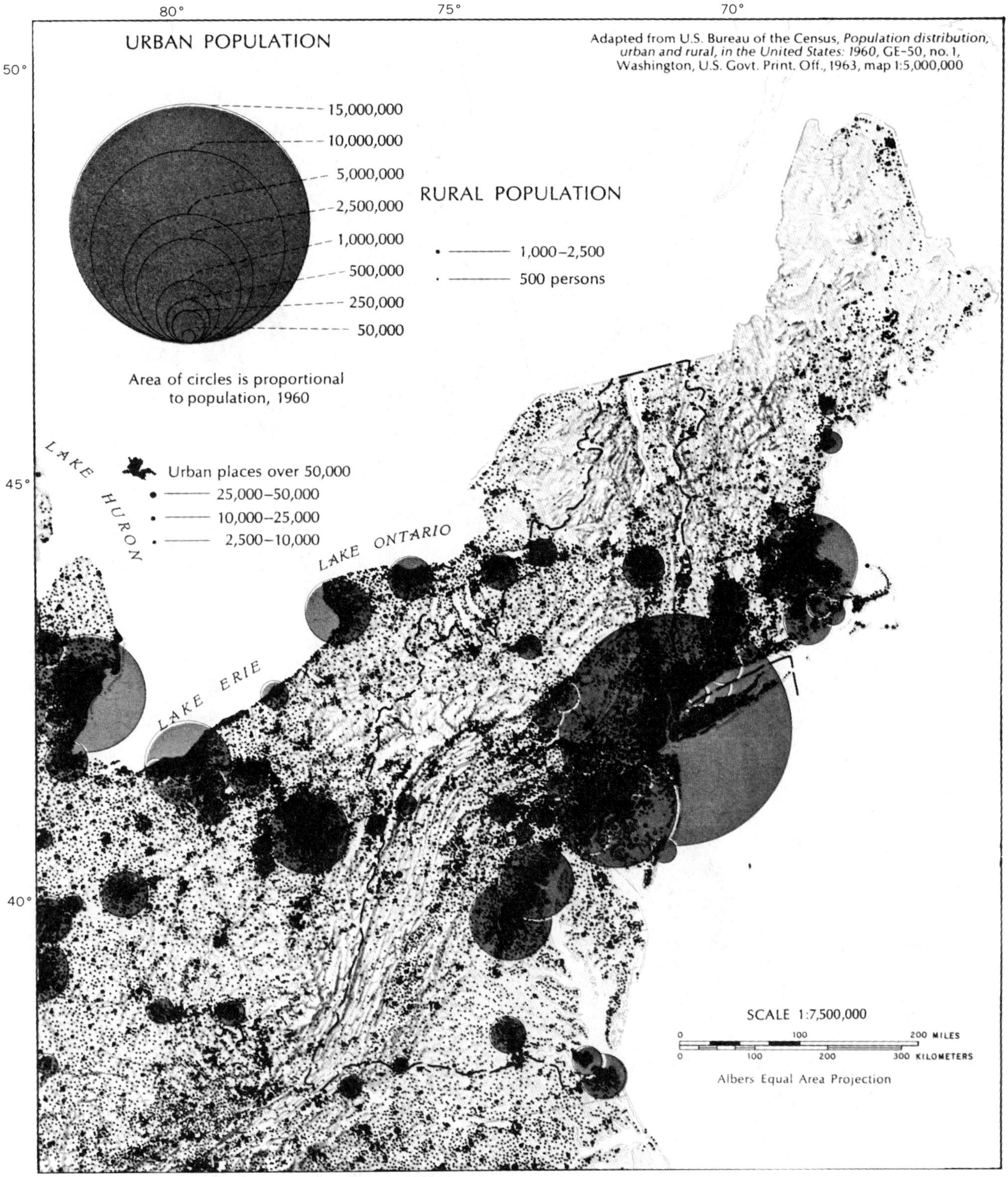

FIGURE 4.—The Mid-Atlantic Region includes large urban centers of population. (Map, from U.S. Geological Survey, 1970, National Atlas of the United States: p. 243.)

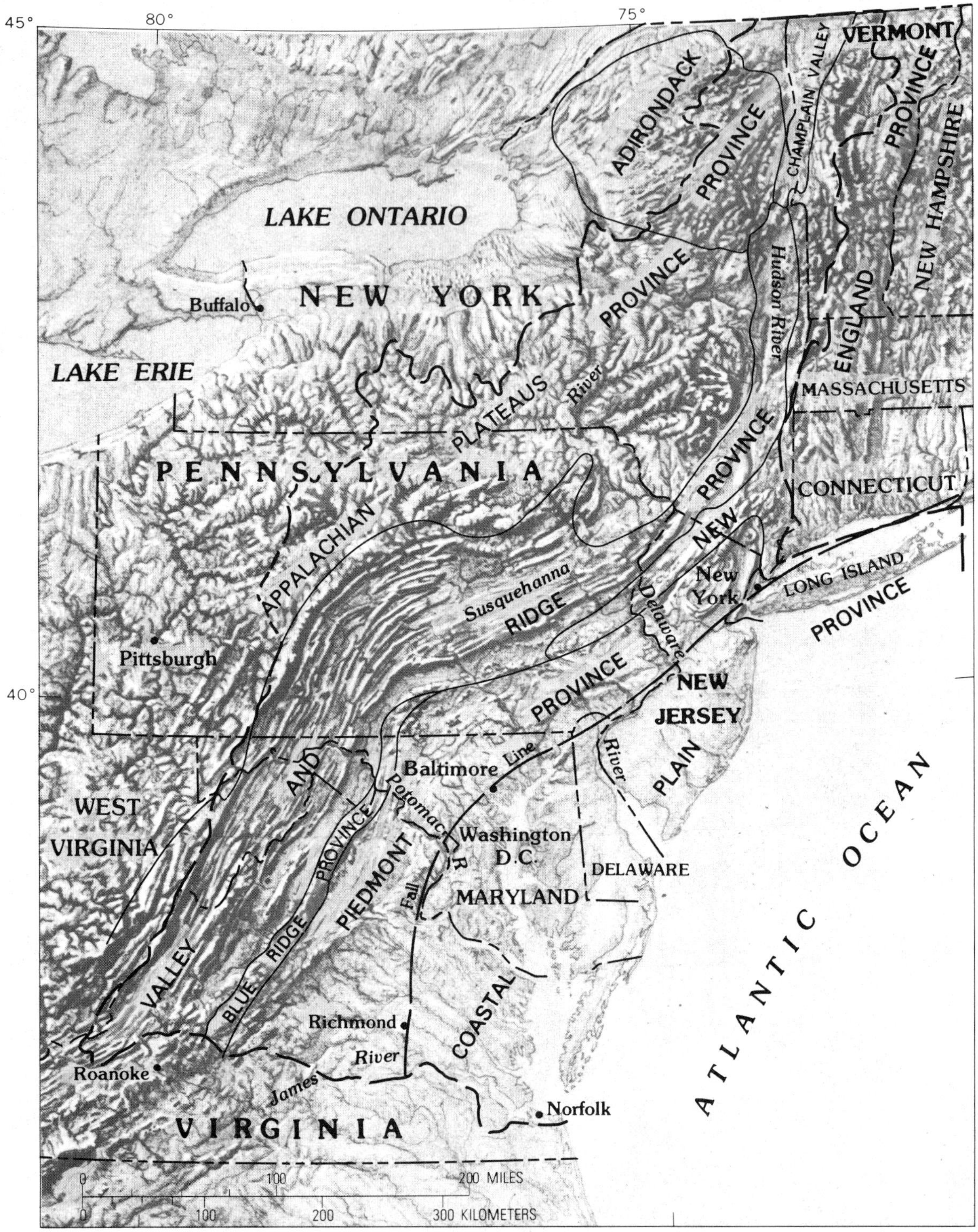

FIGURE 5.—The major physiographic divisions of the Mid-Atlantic Region also define, in a general way, the principal geologic terranes. (After Fenneman and others, 1946.)

The surficial deposits of sand and gravel that underlie the southeastern part of the Region probably constitute the most potentially productive aquifer in the Coastal Plain (fig. 6). These surficial deposits, together with the updip parts of deeper aquifers, constitute a water-table aquifer that blankets about 20,000 mi^2 on Long Island, N.Y., New Jersey, Delaware, eastern Maryland, and eastern Virginia. Large supplies of more than about 5 million gal/d are obtainable from this water-table aquifer. Withdrawals of this magnitude from the deeper artesian aquifers would entail large drawdowns and pumping lifts, and possible dewatering of these aquifers.

Large ground-water withdrawals from the surficial deposits should be located as far as possible from the ocean, consistent with other requirements, in order to minimize the threat of saline encroachment. The water-table aquifer in the surficial deposits also is vulnerable to surface pollutants.

PIEDMONT PROVINCE

The basement complex that lies beneath the wedge of Coastal Plain deposits crops out at the surface along the Fall Line, an imaginary line marking the western limit of the Coastal Plain. (See fig. 6.) The Piedmont province is a plateau of gently undulating hills and locally extensive lowlands lying between the Fall Line and the foot of the Blue Ridge in Virginia and Maryland and the Reading Prong of the New England province farther to the northeast. It averages about 50 mi in width, being narrowest in the northeast and widest southward in southern Virginia.

The province is underlain by weathered crystalline metamorphic and igneous rocks, by consolidated sedimentary rocks and some intrusive and volcanic rocks, and by carbonate rocks. The Piedmont contains two subprovinces, the Piedmont Upland and the Triassic Lowland. The upland consists mostly of crystalline rocks, and the lowland consists of relatively soft shale and sandstone and hills or ridges of harder rocks, such as diabase, basalt, and argillite.

Adequate quantities of ground water are available in the Piedmont province for domestic supply, light industry, and small municipalities. In the crystalline rock areas, the most dependable supplies of water are obtained by wells of moderate yield drilled into the lower part of the weathered zone and the upper part of the underlying fractured rock. Wells commonly yield up to about 15 gal/min per well, and some wells produce as much as 400 gal/min. Yields from wells in the consolidated sedimentary rocks commonly are somewhat greater than those in the crystalline rocks.

The quality of the ground water in the Piedmont is generally excellent, although in places it is high in iron or is hard where withdrawn from limestone aquifers or other sedimentary rocks containing calcareous materials.

BLUE RIDGE PROVINCE

Along the western flank of the Piedmont province in Virginia, Maryland, and southern Pennsylvania lies the narrow northeast-trending Blue Ridge province (fig. 5). Within the Mid-Atlantic Region the average width of this province is less than 10 mi.

The rocks that underlie the Blue Ridge are metamorphosed sedimentary and igneous rocks very much like the crystalline rocks of the Piedmont. The Blue Ridge rocks, however, are more resistant to erosion and stand higher than the carbonate rocks to the northwest and the Piedmont rocks to the southeast. Because of their elevated position, they tend to be stripped of surficial weathered material and fractured rock, and drained of ground water (McGuinness, 1963). Expected well yields are less than 50 gal/min. Very little is known about the maximum water-yielding potential of the Blue Ridge rocks. Four wells drilled at Catoctin Mountain Park, Md., have yields that range up to 74 gal/min and average 36 gal/min. Possibly 150 to 200 gal/min could be pumped from the highest yielding well. These wells were drilled in metarhyolite and Catoctin Metabasalt and their depths range from 120 to 247 feet. They were located in topographic depressions believed to represent fracture zones in the underlying bedrock (Nutter, 1973). The average yield of these wells is about four times better than the average yield of the domestic wells in the vicinity. Water from the Blue Ridge rocks generally is of good quality, containing low concentrations of dissolved solids.

VALLEY AND RIDGE PROVINCE

The Valley and Ridge province lies just west of the Blue Ridge and its northeastern counterparts from Virginia to southern New York. In addition, it extends northward as the Hudson-Champlain section to the Canadian boundary; in this northward extension it lies west of the New England province (fig. 5). The province is bounded on the west by the Appalachian Plateaus and Adirondack provinces. In its southern half it is 50 to 100 mi wide; in eastern Pennsylvania and New York it narrows to about 25 mi.

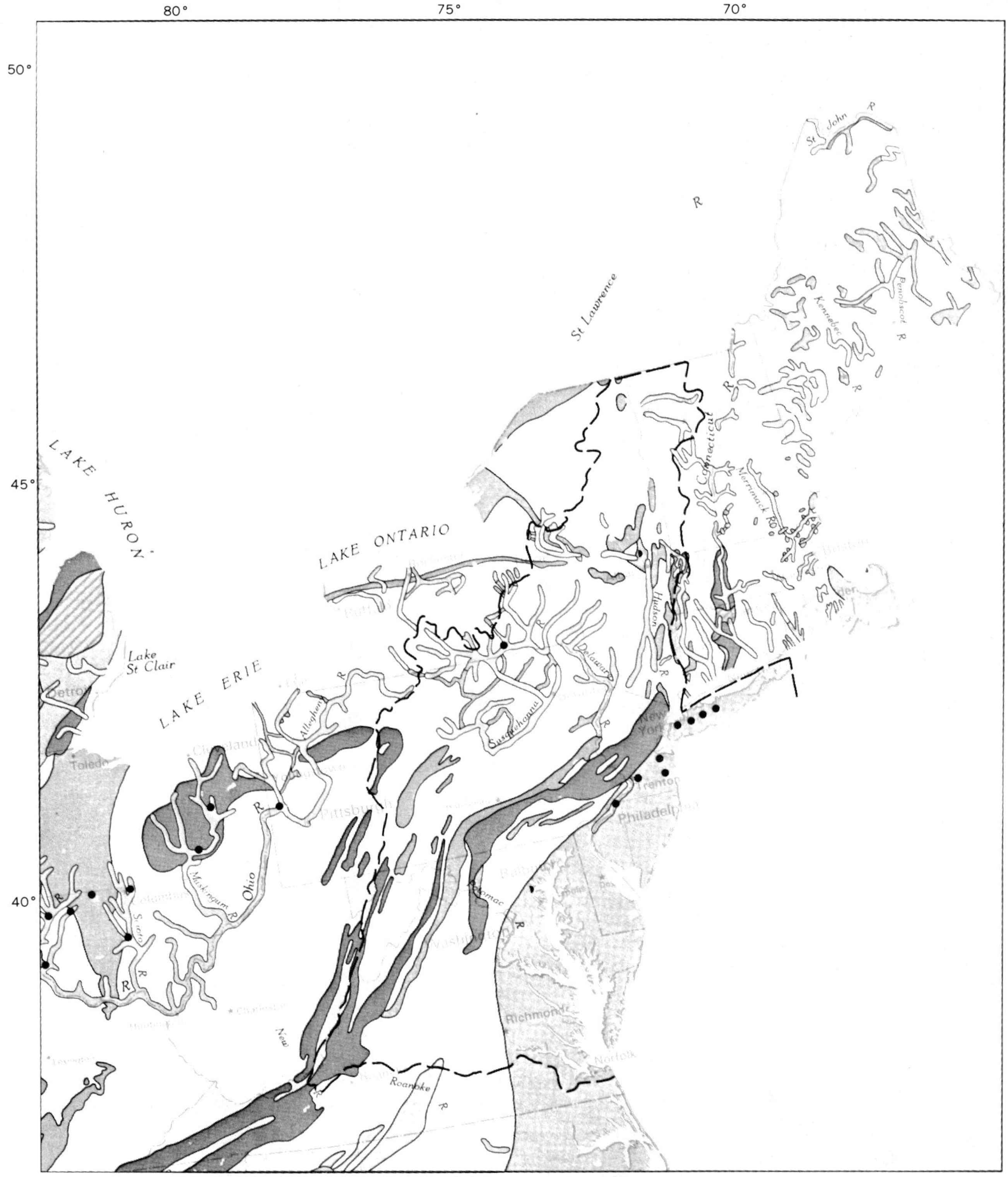

SCALE 1:7,500,000

0 100 200 300 MILES

0 100 200 300 400 KILOMETERS

Map from U.S. Geological Survey United States National Atlas, 1970

EXPLANATION

Colored patterns indicate areas underlain by one or more aquifers generally capable of yielding to a well at least 50 gallons per minute (or 3 liters per second) of fresh water (generally less than 1000 milligrams per liter of dissolved solids)

UNCONSOLIDATED AQUIFERS

Watercourse: alluvial valley traversed by stream from which recharge can be induced

Sand and gravel: in intermontane valleys; abandoned or buried alluvial valleys; alluvial terraces; sand dunes; Coastal Plain; glacial outwash and ice-contact deposits of glaciated regions

CONSOLIDATED ROCK AQUIFERS

Sandstone: includes some unconsolidated sand

Carbonate rock: limestone and dolomite

Sandstone and carbonate rocks

Crystalline rocks, igneous and metamorphic

Ruled pattern of any of these colors indicates that the corresponding aquifer underlies a productive sand and gravel aquifer as explained above

WITHDRAWALS FROM WELLS

• Represents withdrawal of 81,000 acre-feet or 100 million cubic meters annually, equivalent to 72 million gallons a day

• Represents half of the quantity described for preceding symbol

FIGURE 6.—Many areas in the Mid-Atlantic Region are underlain by at least one aquifer capable of yielding supplies of at least 50 gallons per minute of ground water to most wells. (Map, from U.S. Geological Survey, 1970, National Atlas of the United States: p. 123.)

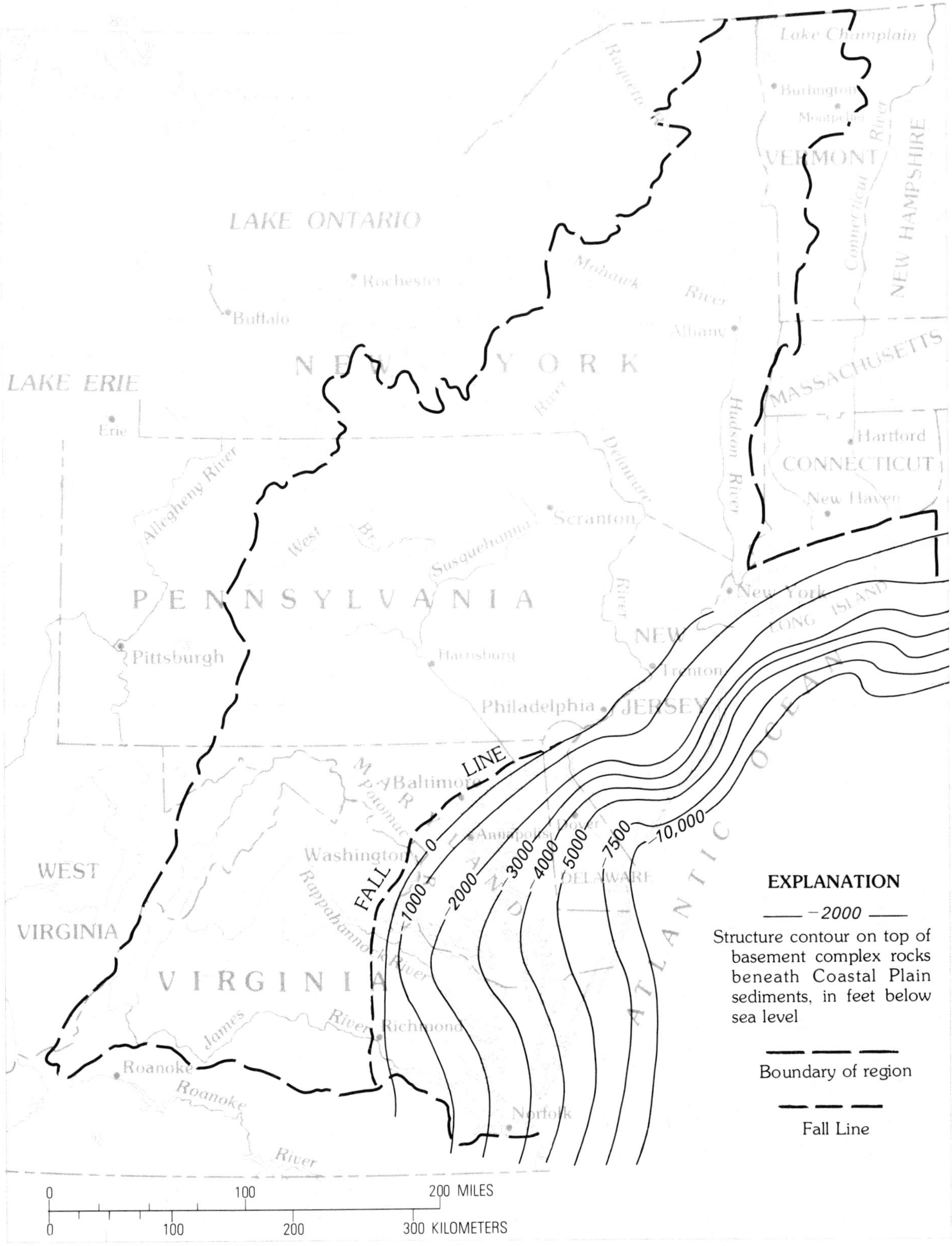

FIGURE 7.—The buried surface of the basement complex slopes seaward from the Fall Line to depths greater than 8,000 ft below sea level. (Adapted from Maher, 1971, pl. 4.)

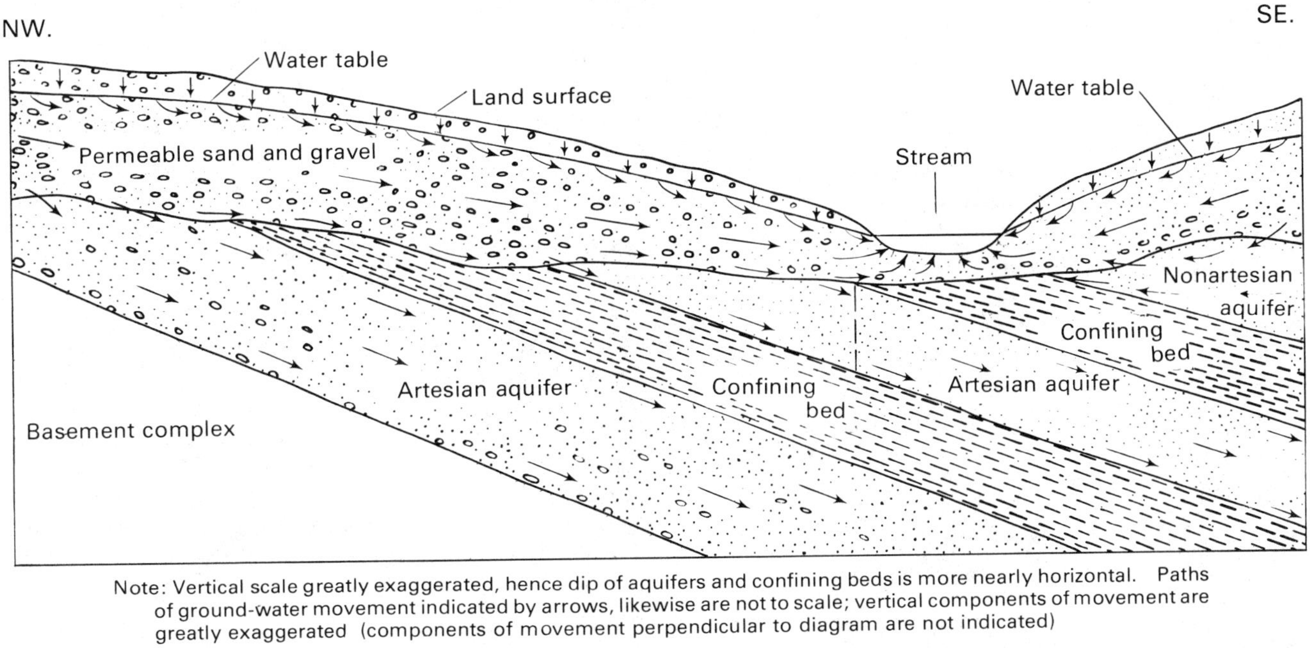

FIGURE 8.—Seaward-dipping artesian aquifers of the Coastal Plain are covered by surficial deposits of water-bearing sand and gravel on Long Island, N.Y., and in New Jersey, Delaware, eastern Maryland, and eastern Virginia. (Adapted from Parker and others, 1964, fig. 13.)

The province is underlain by folded and faulted sandstones, limestones, dolomites, shales, and quartzites. The structural trend in its southern part is generally northeastward, changing to northward up the Hudson Valley. Differential erosion has resulted in resistant beds standing as ridges, with intervening alluvium-filled valleys.

The chief sources of ground water in the Valley and Ridge province are permeable fractured sandstones and cavernous limestones. These are readily recharged from precipitation and stream infiltration. Much smaller quantities of ground water are available from the less permeable shales and siltstones. In all these rocks, secondary openings and fractures tighten with depth; hence, the shallow rocks, to depths of several hundred feet, are the most productive to wells.

Dry wells are common in the areas underlain by limestone. Some of these may be attributed to the fact that the water table is extremely deep because of underground drainage through interconnecting solution cavities. Water can usually be obtained by deepening the well to reach the water table. Many dry wells in limestone are caused simply by failure of the wells to penetrate any water-bearing solution cavities or fractures. A copious supply might be found in another well—perhaps only a few feet away— that taps such openings. Thus, exploratory drilling is commonly required.

In the Susquehanna River basin, well yields in the more mountainous part of the basin area average about 125 gal/min and may reach 700 to 1,000 gal/min (Seaber and Hollyday, 1965; Seaber, 1968). In the Great Valley section of the same basin, the predominant rocks, limestone and dolomite, are even more productive, and yields of wells may exceed 1,000 gal/min.

In the Hudson-Champlain section, the valley-fill deposit in the Glens Falls, N.Y., area is a productive aquifer (Giese and Hobba, 1970). Well yields average 67 gal/min but reach 400 gal/min. The bedrock, near Plattsburgh, N.Y., consists of sandstone and carbonate rocks. It is overlain by relatively impermeable glacial deposits and is in an artesian system. Well yields average about 15 gal/min from sandstone and 35 gal/min from carbonate bedrock. Some wells in the carbonate rocks produce as much as 200 gal/min.

The chemical quality of the ground water in the Valley and Ridge province is closely related to the lithology. Water from sandstone is usually soft and generally low in dissolved solids. Water from carbonate rocks generally is hard and is contaminated locally by nitrate. Shales in the eastern panhandle of West Virginia (Hobba and others, 1973) are reported to yield hard water containing excessive iron. In the broad, major downfolded (synclinal) basins, high chloride or sulfate content is common. Where coal beds occur, many wells yield acid water that may also contain high iron and sulfate.

Many wells constructed in permeable weathered zones in carbonate-rock aquifers yield turbid water from the residual clay in solution channels and caverns. Hobba and others (1973) note that in the West Virginia panhandle, "lowering the water table removes support of clay plugs in the roofs of water-filled caverns in the bedrock, and the plugs may collapse to cause sink holes at the land surface."

The chemical quality of ground water in the Hudson-Champlain section is generally excellent or good, although hardness and alkalinity are problems locally.

APPALACHIAN PLATEAUS PROVINCE

In the western part of the Mid-Atlantic Region, the Appalachian Plateaus province extends from north-central Pennsylvania into southern New York (fig. 5). A small area in the extreme northwestern part of the Potomac River basin in West Virginia and Maryland also lies in the province.

The underlying bedrock consists of nearly horizontal beds of alternating sandstone, shale, some limestone, and seams of coal.

In the Pennsylvania part of the province, the average yield per well may be about 200 gal/min. In New York, yields are significantly lower, averaging about 60 gal/min. The quality of water from the bedrock at depths of less than about 250 feet is generally good except near coal mines, where iron is locally troublesome. Saline water commonly occurs at depths of only 300 to 500 feet (Becher, 1970).

The parts of the Appalachian Plateaus province that lie in the Mid-Atlantic Region in northern Pennsylvania and New York are glaciated. Outwash deposits from the meltwaters of Pleistocene continental glaciers are found in the lowlands. In addition, postglacial alluvial sands and gravels in the stream valleys form watercourse aquifers (Thomas, 1952, p. 10). The glacial outwash deposits and the watercourse aquifers are the chief sources of dependable ground-water supplies in the glaciated part of the province. (See fig. 6.) Many wells in sand and gravel deposits yield moderate to large supplies of water—some more than 1,000 gal/min. Recharge from the streams can be induced by pumping wells drilled in the watercourse aquifers, but induced infiltration of stream water reduces streamflow.

The quality of the ground water from the glacial deposits is generally good. However, it may be hard where the drift contains limestone pebbles, and iron is troublesome locally, particularly in New York.

ADIRONDACK PROVINCE

The eastern part of the Adirondack province lies within the Mid-Atlantic Region in northern upstate New York (See fig. 5.) It is just north of the Appalachian Plateaus province and west of the Hudson-Champlain section of the Valley and Ridge province. The province is a nearly circular uplift of ancient granite, gneiss, and other crystalline rocks surrounded by shale, sandstone, limestone, and dolomite. Valleys contain glacial outwash and alluvium. Many peaks exceed 4,000 ft in altitude; Mount Marcy, the highest mountain in New York, has an altitude of 5,344 ft.

The population of the province is sparse, and the demand for water, accordingly, is small. Yields of wells are generally moderate, 15 to 35 gal/min; the water occurs mostly in fractures in both the crystalline rocks and the consolidated sedimentary rocks. Larger yields are obtained locally from solution openings developed along fractures in the carbonate rocks. Locally, substantial yields are available also in glacial outwash deposits in valleys.

The quality of ground water is generally good, and that from the crystalline rocks is excellent—less than 100 mg/L of dissolved solids.

NEW ENGLAND PROVINCE

In the Mid-Atlantic Region, the New England province is represented by the western part of Vermont, the extreme northwestern part of Massachusetts, and a narrow area extending southwestward across southern New York and northern New Jersey into eastern Pennsylvania as far as Reading. (See fig. 5.) This extension is known as the Reading Prong; in New Jersey it is called the Highlands.

The bedrock consists of ancient crystalline, igneous, and metamorphic rocks, some of which are carbonate rocks. Locally the carbonate rocks are faulted and contain solution openings, forming aquifers that may yield a few hundred gallons per minute. Similar yields may be obtained from deposits of stratified glacial drift in stream valleys (watercourse aquifers).

Ground water of generally good quality can be developed from the various kinds of bedrock and from the stratified drift. The carbonate bedrock, however, yields water that is moderately hard.

AVAILABILITY OF GROUND WATER TO MEET FUTURE NEEDS

C. L. McGuinness, in his comprehensive treatise on the "Role of ground water in the National water situation" (1963, p. 110–111) recounted the circumstances leading to the gradual increase in demand for ground water as a source of water supply during the early development of the country. Ground water in the Mid-Atlantic Region has assumed increasing importance through the years, following much the same pattern as that described by McGuinness for the country as a whole.

Nationally, ground water has developed from a minor but important source of domestic supply to one that provides about 21 percent of the Nation's total water withdrawals from all sources in 1970 (Murray and Reeves, 1972, table 17). In the Mid-Atlantic Region, however, ground-water withdrawals in 1970 amounted to only 9.3 percent of the total water withdrawals, attesting to the relatively minor role still played by ground water in this important region in the humid East where surface water is the major source of supplies.

Many water-resources investigations and reports deal separately with ground water and surface water. Even though they are physically interrelated, ground water and surface water require different techniques of study, observation, and development. Ground water and the base flow, or dry-weather flow, of streams are two phases of a single resource. Base flow is ground water that is discharged continually from the ground-water reservoir. Overland runoff from precipitation greatly increases streamflow intermittently but flow then subsides during an ensuing period of dry weather.

WATER DISCHARGED FROM AQUIFERS

Many investigators of the water resources in areas within the Mid-Atlantic Region have made determinations of the percentage of total streamflow that is base flow, or ground-water discharge. (See, for example, Cushing and others, 1973; Nutter, 1973; Franke and McClymonds, 1972; Trainer and Watkins, 1974; Johnston, 1971; Olmsted and Hely, 1962; Rasmussen and Andreasen, 1959; Dingman and Ferguson, 1956; Dingman and others, 1954.) In the Coastal Plain province, base flow ranges from about 40 to 95 percent of streamflow, and in the part of the Mid-Atlantic Region underlain by consolidated rocks, from about 25 to 90 percent.

Bue (1970) showed that the mean streamflow from the Mid-Atlantic Region to the Atlantic Ocean during 1931–60 was 128,900 ft^3/s. Adding to this quantity the amount estimated to flow from the region into Canada, the mean streamflow from the region (area, 108,000 mi^2) is about 139,000 ft^3/s, or about 90 billion gal/d.

Assuming, conservatively, that only 55 percent of the streamflow in the Coastal Plain part of the region is base flow, then the natural discharge from unconsolidated aquifers is computed to be about 9.2 billion gal/d. Again, assuming that only 40 percent of the streamflow in the remainder of the region, underlain chiefly by hard consolidated rocks, is base flow, the discharge from the aquifers there is about 29.4 billion gal/d. Thus, the average potential yield of ground water from the entire Mid-Atlantic Region under present hydrologic conditions is at least 38.6 billion gal/d. These estimates are summarized in table 1.

TABLE 1.—*Estimates of natural discharge of ground water from aquifers in the Mid-Atlantic Region*

	Area (mi ^{2})	Streamflow (billion gal/d)	Portion of streamflow estimated to be base flow (percent)	Estimated discharge of ground water from aquifers (billion gal/d)
Coastal Plain part of region (mostly unconsolidated, or "soft," rocks)	20,000	16.7	55	9.2
Remaining part of region (mostly consolidated, or "hard," rocks)	88,000	73.5	40	29.4
Totals for Mid-Atlantic Region	108,000	90.2	--	38.6

It would be, of course, impractical to intercept all this water before it is discharged into streams. This estimated total, nevertheless, indicates the magnitude of the resource. For example, assuming an average daily consumption of 150 gallons of water per person, the estimated total available supply of 38.6 billion gal/d of ground water, if developed for domestic use, would provide for about 257 million people—more than six times the population of the region in 1970. For the region at large, the potential supply is sufficient to provide for anticipated population growth for the foreseeable future.

During extended periods of greater than average precipitation, more ground water would be available than the estimated 38.6 billion gal/d. Conversely, during drought conditions, substantially less would be available.

WATER IN STORAGE

In addition to the quantity of ground water estimated above, at least another 140 to 350 trillion gal-

TABLE 2.—*Estimates of ground water in storage in hydrogeologic units of the Mid-Atlantic Region*

Hydrogeologic unit	Approximate area (mi²)	Average productive thickness (ft)	Average available drawdown (ft)	Estimated average storage coefficient (dimensionless)	Estimated storage available (trillions of gallons)
Coastal Plain province (sand, clay, minor gravel)	20,000	250	150	0.1–0.2	63–125
Piedmont, New England, and Adirondack provinces (crystalline igneous and metamorphic rocks)	29,800	250	200	.001–.05	1.2–62
Blue Ridge province (crystalline rocks)	2,100	200	150	.001–.02	.07–1.3
Valley and Ridge province (sandstone and limestone)	28,700	250	200	.05–.1	60–120
Champlain Valley (sandstone and limestone)	2,700	150	100	.05–.1	2.8–5.6
Appalachian Plateaus province (sandstone and shale, coal seams)	24,700	200	150	.02–.05	15–39
Totals	108,000	---	---	-------	140–350 (rounded)

lons of water is estimated to be in storage in the major unconfined shallow aquifers of the region. As shown in table 2, this total is derived from estimates of the generally available drawdown (the difference between the static water level and maximum pumping levels) in the rocks in each of the major physiographic divisions of the region. In the manner developed by Rasmussen (1955) for Delaware, the areas of each division were multiplied by the estimated available drawdown, and by the storage coefficient (a decimal fraction representing the quantity of water per unit volume of the aquifer that can be withdrawn or replenished), and the results converted to trillions of gallons.

In addition to the estimated quantity of water in storage in the major unconfined aquifers, substantial quantities also are stored in other units in the Mid-Atlantic Region. Among these are (1) artesian aquifers, chiefly in the Coastal Plain province; (2) certain glacial deposits, roughly in the northern half of the region that was covered by continental glaciers; (3) aquifers in watercourses—stream valleys containing glacial outwash sand and gravel, or alluvial deposits, in hydraulic continuity with streams, (4) sand and gravel deposits in former river valleys that were buried beneath by later glacial deposits.

Except for the artesian aquifers beneath the Coastal Plain, it is difficult to make meaningful estimates of the quantities of ground water in these other deposits. It is particularly difficult for watercourse aquifers, which are quickly replenished by streams as ground water is withdrawn from the aquifer. However, a very conservative estimate of ground water stored in these various kinds of deposits would be at least another 20 trillion gallons, in addition to the 140 to 350 trillion gallons that are estimated to be in storage in the major unconfined aquifers.

A substantial part of the quantity that is in storage—perhaps at least 50 percent—would be potentially available as a reserve supply during emergency periods, such as the northeast drought of the middle 1960's. However, only a small part of this potential reserve supply could be developed during a drought because of the lead time required for exploratory drilling, aquifer testing, drilling of production wells, and development of suitable distribution systems by which to transport the water to municipal or industrial centers where it is needed. Thus, to cope with possible emergencies, these reserve sources of ground water ideally could be explored and water-supply systems installed well in advance of critical demand for emergency supplies.

WITHDRAWALS AND USE

In 1970, the rate of withdrawals of freshwater from all sources in the Mid-Atlantic Region was 28,000 million gal/d, of which 9.3 percent, or 2,600 million gal/d, was ground water (Murray and Reeves, 1972, table 17). This is equivalent to a total volume of 10,220 billion gal from all sources during 1970, of which 949 billion gal was from ground water. In terms of the regional population of 38,401,000 in 1970, the rate of use of freshwater was about 730 gal/d per person in the region from all sources, and the rate of use for fresh ground water was about 68 gal/d per person (Murray and Reeves, 1972, table 17). Thus, the present demand for water in the region is met primarily from surface-water sources. The quantity of ground water that is estimated to be available from the aquifers of the region indicates that substantial additional utilization of ground water is a valid management option.

Of the total freshwater withdrawn in the region from all sources in 1970, only 5 percent was consumed (lost to the atmosphere by evaporation or incorporated with manufactured products), and 95 percent was returned to the streams or aquifers (Murray and Reeves, 1972, table 17.)

Industrial use of ground water was only 1,100 million gal/d, or about 4 percent of the total freshwater withdrawals. Thus, industry, which is the principal economic activity in the region, depends largely on surface-water supplies. These supplies are vulnerable to drought and pollution and their extensive land-use requirements may not be entirely compatible with the interests of environmental protection and conservation. Accordingly, future industrial growth may depend more and more on ground-water resources, particularly small or light industries whose requirements for process water may be small and which require water of good quality.

CHEMICAL QUALITY

DESCRIPTION OF CHEMICAL QUALITY

The chemical quality of ground water in the Mid-Atlantic Region differs from place to place in accordance with the hydrologic history of the water and the nature of rock materials through which it flows. However, in most parts of the Mid-Atlantic Region freshwater of satisfactory chemical quality for virtually all uses occurs to depths of several hundred feet, and in some areas to a depth of a thousand feet or more. In overview, the Coastal Plain province contains a large, complex blanket of fresh ground water overlying a much greater volume of deep-lying brackish and saline ground water extending downdip (downslope) seaward. The boundary between the updip freshwater and the downdip saline water is mappable (Cushing and others, 1973). Freshwater, in the great mass of nonmarine formations constituting the oldest sedimentary rocks in the province, extends from several to more than 60 miles downdip from outcrop areas along the western side of the province. The quality of water in these deposits is generally soft and slightly acidic in the outcrop area, with local excessive amounts of iron, and total dissolved solids content on the order of 50 to 150 mg/L. Downdip and at greater depths below the land surface, in parts of the formation confined under younger deposits, the water is more mineralized, is commonly of the calcium bicarbonate type and slightly hard, with dissolved solids on the order of 300 to 500 mg/L. As the boundary between fresh and saline ground water is approached, dissolved solids exceed 1,000 mg/L.

A large mass of marine deposits rests on and crops out east of the nonmarine strata. These marine deposits contain fresh ground water with dissolved-solids content on the order of 100 to 250 mg/L. Water from these deposits, too, contains excessive concentrations of iron locally, and owing to the abundance of calcium carbonate shells and other calcareous material, moderate to excessive hardness is encountered in places. The relationship with downdip saline ground water in these sediments is less well defined than for the nonmarine deposits.

Water in the glacial, watercourse, and alluvial deposits is generally soft, containing less than 50 mg/L up to 150 mg/L of dissolved solids, although in places concentrations may be as much as 400 mg/L. Excessive hardness and high-iron concentrations occur in places. Saline water from tidal rivers, bays, and the Atlantic Ocean, where these waters border or contact low-lying deposits, may infiltrate the fresh ground water, either naturally or by induced landward movement owing to well pumpage. Peripheral to these saltwater bodies aquifers may contain water with dissolved-solids concentrations of 1,000 mg/L and much more. The quality of ground water in aquifers subject to freshwater recharge from rivers, lakes, or fresh reaches of estuaries may vary with the quality of the water infiltrating from these surface sources.

Owing to the great variety of rock materials in the remaining provinces, west and north of the Coastal Plain, ground-water quality varies widely. Freshwater in the silicate-type rocks (most of the metamorphic crystalline rocks and quartzose sedimentary rocks) that form highland areas, mountains, and ridge areas is different from that in the calcareous carbonate rocks (primarily limestone and dolomite) that commonly underlie major regional lowland valleys. Although there are exceptions, ground water in the silicate-type rocks is generally soft with low dissolved solids, usually less than 200 mg/L. Metamorphic crystalline rocks of the Blue Ridge province, for example, are not readily soluble to ground water, and water derived from wells or springs characteristically is very low in dissolved solids. In contrast, water in the calcareous rocks, such as in the Shenandoah Valley of Virginia, the Valley and Ridge Province, and the Taconic Province of New York and New England, is typically hard, with dissolved-solids content of 200 to 600 mg/L.

Throughout much of the area of the Appalachian provinces underlain by sedimentary belts of rocks, the fresh ground water is underlain by salty ground water at depths of several hundred to several thousand feet.

PROBLEMS ASSOCIATED WITH GROUND-WATER QUALITY

Salty ground water presents some problems and some opportunities. Many aquifers throughout the Mid-Atlantic Region are partly filled with salty water. Although unsatisfactory as a source of water supply for most uses without desalination, under favorable hydrogeologic conditions these saline reservoirs may serve as storage media for liquid wastes, petroleum, or even artificially recharged freshwater where excess surface water is available for storage.

Hard water is a problem and inconvenience, particularly in the regional valley areas of the Appalachian provinces where hardness of the water is highest. Treatment (softening) units are available for municipal, domestic, and industrial use.

Objectionable concentrations of iron are prevalent throughout the region. The problem is most pronounced in the shallow terrace and glacial deposits, and in Appalachian areas where mining of coal has affected ground-water quality.

Other natural quality problems of lesser importance include hydrogen-sulfide gas, excessive concentrations of fluoride, and contamination by oil and by organically derived natural gases such as methane.

Many quality-of-water problems are associated with man's activities. One prominent cause of ground-water quality deterioration is pumping of wells near sources of contamination and inducing water of poor quality into the aquifer. For example, salty water moved landward in one of the principal aquifers on Long Island, and in a similar situation in Atlantic City, N.J., because large quantities of ground water were pumped near the freshwater-saltwater boundary.

Pollution resulting from man's actions is a recognized source of degradation of ground-water quality. Nitrate concentrations exceeding 45 mg/L have been identified in water from wells affected by cesspools, septic tanks, and fertilized agricultural land. High concentrations of chloride have been identified in ground water affected by winter-road salting and near storage sites for road salt. Leachates from landfills and dumps have affected ground-water quality. Accidental or intentional disposal of industrial wastes may also influence ground-water quality detrimentally.

The ground-water quality problems of the Mid-Atlantic Region are many, and these are only the major ones. Miller, DeLuca, and Tessier (1974) compiled records of hundreds of cases of ground-water contamination in the northeastern States, reporting that they represent only a very small percentage of those that actually exist. Ground-water-quality management is especially important in this region of intense industry and agriculture and heavy population density.

RELATION OF GROUND-WATER TO SURFACE-WATER RESOURCES

Ground water and the base flow of streams are two different phases of a single fluid resource. It follows, then, that pumping water from wells in some areas could effectively diminish the amount of water available from streams, especially during dry periods when the base flow of streams is sustained almost wholly by ground water. Normally, the withdrawal of water from the streams under present conditions does not noticeably affect the amount of water available in the aquifers of the Mid-Atlantic Region. In the Coastal Plain, pumping from these shallow aquifers reduces the quantity of ground water that is discharged directly into bays and the ocean in addition to reducing the base flow of streams.

Development of the aquifers to their full potential would eventually reduce the base flow of streams to zero, and it would probably allow incursion of brackish water or seawater into the lower reaches of streams and into aquifers at places where they are in direct hydraulic continuity with tidal streams, bays, or the ocean.

There are many places in the Mid-Atlantic Region where ground water and surface water could be developed conjunctively but there are as yet no large-scale conjunctive developments. Such development projects would involve balancing the use of ground-water and surface-water reservoirs according to availability and demand: surface water might, in times of surplus, be used to recharge the ground-water reservoir; conversely, ground water would be used to augment surface-reservoir supplies during low flows or during droughts.

Permeable glacial outwash deposits occur in many places in the northern part of the region, particularly in upstate New York and northern Pennsylvania and New Jersey (fig. 5). Where these deposits are watercourses (Thomas, 1952, p. 10), water can be pumped from them so as to induce recharge from the adjacent perennial stream. This might be regarded as a simple form of conjunctive use, in which water withdrawn from the ground-water body is derived from streamflow.

Artificial recharge of aquifers by pumping surplus surface water into them in the winter has been practiced by certain industries in Newark, N.J., and this practice has helped to control saltwater encroachment (Thomas, 1951, p. 127).

In southeastern Virginia, ground water from four wells drilled in Lower Cretaceous sands in the Coastal Plain is pumped directly into the Lake Prince reservoir of the City of Norfolk to augment the surface-water supplies during periods of low flow.

WATER-RELATED PROBLEMS IN THE MID-ATLANTIC REGION

IMPACT OF ECONOMIC DEVELOPMENT ON MANAGEMENT OF GROUND-WATER RESOURCES

One of the pervasive characteristics of the Mid-Atlantic Region is growth in economic activity (U.S. Water Resources Council, 1972, summary tables). Although it has attained a high average population density—more than 340/mi^2—its population is still growing. This growth is probably in response to the continuing general increase in economic development. Industrial growth expands employment opportunities, increases per capita income, and intensifies commercial agriculture, which historically has tended to concentrate in regions and areas of comparative economic advantage (U.S. Water Resources Council, 1972, p. 19). Other factors, including availability of marketing and transportation facilities and excellent harbors, contribute to industrial growth and agricultural production.

An increasing need for adequate water supplies attends the growth in economic activity. Water supplies have generally kept pace with the increasing demands through the years, but the severe northeast drought of the mid-1960's brought the realization that the margin of reserve surface-water supplies was not as comfortable as had been thought. Meetings were held in major cities, many articles and alarmist statements appeared in magazines and in the press, and the public soon became water conscious. Ground water quickly became a topic of interest, and there was much speculation as to its potential for increased development, not only as a dependable reserve supply in similar water emergencies but also as a full-fledged water source that should be accorded higher status in developmental planning.

There are ample quantities of water, particularly ground water, in the Mid-Atlantic Region. Problems relate primarily to its development, management, and distribution. Industries that require large quantities of water would be best served in the Coastal Plain part of the region. In the hard-rock areas, the best locations are along major streams, in watercourse aquifers where ground water is effectively recharged because of the hydraulic continuity between a perennial stream and the sand and gravel of the valley-fill deposits. Of course, an industry could locate some distance from the source of supply and provide water through a system of pipelines and booster pumps. The cost of such an arrangement might be advantageous when compared to the cost of locating near the source of supply.

Light industry, on the other hand, whose need for water is primarily for personnel rather than for process water, could locate almost anywhere in the Mid-Atlantic Region, consistent with other requirements.

If economic activity continues to increase, as predicted by recent studies (U.S. Water Resources Council 1972, summary tables), more consideration will undoubtedly be given to water-supply alternatives. Among these, the increased use of ground water and the conjunctive use of ground water and surface water are of greatest potential. Conjunctive use of freshwater and saline water may ultimately prove to be entirely feasible, as desalination becomes less expensive through improved technology. Indeed, certain industries, such as the dye industry, may consider locating in places where saline water is available because of the need for it in process water. Use of brackish water, with minimal cost for mixing with locally available freshwater, may result in substantial savings in certain situations.

Ground-water developments commonly entail little modification in existing land-use patterns, because the resource is obtained from beneath the surface by wells with surface pumping plants which, after construction, can be made inconspicuous by landscaping and other means. Therefore, most ground-water developments can be made with a minimum of adverse environmental impact. The coming decades may witness a greater preference for ground water than for surface water, simply because it is possible to develop ground water with less environmental damage.

Large-scale ground-water developments, even in areas of proven abundance like the Coastal Plain, may not always be possible without a considerable pipeline distribution system and booster pumping stations. During the northeast drought of the mid-1960's, when surface-water resources were strained to the limit, such large-scale ground-water transmission systems were considered. One such plan involved development of ground water from the aquifers of the New Jersey Coastal Plain, where about 20 trillion gallons of water are estimated to be in storage, and conveying it by pipeline to industrial and population centers in the metropolitan complex in Newark and vicinity. This scheme and several others were discussed in the public press, in magazine articles, and at various meetings—not, of course, with the idea of immediate solution to the drought-related water shortage then prevalent, but rather to generate positive thinking toward provision for adequate water supplies and their satisfactory distribution in preparation for the next prolonged drought, whenever it might occur. Most of the enthusiasm and discussion of these proposals quickly subsided with the end of the drought in 1967. Since then, there has been little, if any, public discussion or concern about large conjunctive or standby ground-water development projects.

Continued increase in ground-water development may be expected in the region as the modest level of development to date can be increased many times.

ACID MINE DRAINAGE

One of the difficult problems in regions where coal has been extensively mined is pollution of streams and ground water by acid mine drainage. In the Mid-Atlantic Region, there are many coal-bearing areas in Virginia, West Virginia, Maryland, and Pennsylvania. Water draining from the coal mines in these areas into streams has resulted in serious contamination. The general problem of stream and ground-water contamination by drainage of mines is well summed up by McGuinness (1963, p. 727):

> The large population and heavy industralization have combined to produce large quantities of pollutants. Mines, chiefly coal mines, are one of the main sources. Coal and certain ores contain sulfur compounds which, when exposed to the air in mines and waste piles, become oxidized to form sulfuric acid. The acid contaminates both streams and ground water, and is generated in such large quantities that the problem is far from solved in spite of the intensity with which it has been attacked.

Streams are affected more readily than ground water by drainage of water from mines, but a few good aquifers have been contaminated in the Susquehanna and Delaware River basins in Pennsylvania.

In a recent study of acid waters draining from the underground coal mines of western Maryland, Hollyday and McKenzie (1973) summarized some of the effects of acid mine drainage—the degradation of the water supply for recreational, industrial, and municipal uses; the increase in costs of treatment; more frequent replacement of treatment facilities; the staining of porcelain fixtures and of laundered clothing; endangerment of aquatic life; and detrimental effects on engineering structures and navigational equipment. They suggest pursuing the hypothesis that neutralization of acid mine waters can take place by "using the natural underground environment in place of or in conjunction with mine-mouth water treatment plants." Thus, in appropriate hydrogeologic settings, convenient natural laboratories are available in which to test the hypothesis of natural neutralization of acid mine waters.

SALINE GROUND WATER

Until recent years, saline ground water has almost everywhere been regarded as a problem. Its encroachment was to be prevented so far as possible. Its elimination, once encroachment occurred, was almost impossible to achieve. Its usefulness was limited to a very few applications, such as cooling and certain cleaning or flushing operations in some industries. More recently, however, especially in the past 10 years, saline water and particularly saline ground water and saline aquifers have come to be looked upon as a potential resource that should be included for consideration by those concerned with large-scale water management.

As mentioned previously, saline (sodium chloride) ground water occurs in the downdip reaches of all the aquifers of the Coastal Plain. (See Back, 1966, p. A2, fig. 2 and pl. 1; Cushing and others, 1973, p. 9.)

In their summary report on a comprehensive study of Long Island, Franke and McClymonds (1972, p. F 55) point out that within the next 25 to 50 years the technology for economically converting saline water to freshwater may be greatly improved. Thus, it is important to consider saline ground water in the Mid-Atlantic Region as a potentially usable component of the total ground-water resources.

Saline ground water may be considered from three points of view: (1) as a source of water, usable without treatment, by itself or in a blend with freshwater, (2) as a source of additional freshwater by employment of one of several methods of desalting, and (3) as an occupant of aquifer space usable for storage of freshwater or wastes.

Although the utility of untreated saline ground water is low, it is nevertheless an important resource in certain applications. One such application is the augmentation of freshwater supplies by mixing saline water with existing sources of supply during periods of peak demand. For example, Cederstrom (1957, p. 42) notes that during World War II a surface reservoir for the city of Newport News, on the lower York-James peninsula in Virginia, was successfully augmented with ground water containing 400 to more than 1,000 mg/L of chloride.

Saline ground water is useful also for cooling and for many industrial washing purposes. Moreover, in certain industries (as for example dye manufacturing), the chloride concentration in process water is required to be higher than that desirable for general use, and the needed concentration can be adjusted by controlled dilution.

Various methods for the conversion of saline water to suitable freshwater have been under study by the Federal Government for more than 20 years and, since 1958, on a relatively substantial scale. Desalting costs for seawater have been reduced gradually, and in 1973 were less than $1 per 1,000 U.S. gallons under favorable conditions. Further conversion cost reduction would stimulate use of saline water more widely. Many aspects of the development and feasible utilization of saline ground water have been reviewed in detail. (See Kohout, 1970.)

CONSTRAINTS ON GROUND-WATER DEVELOPMENT AND MANAGEMENT

Perhaps the most obvious constraint on the optimum development and management of ground-water resources is the lack of sufficient information upon which to base wise decisions. Needs for additional information in each of the States of the Mid-Atlantic Region are discussed in a later section of this report.

LEGAL CONSTRAINTS

Constraints imposed by law are complex and are among the various factors tending to inhibit larger scale ground-water developments. Early doctrines relating to the withdrawal and use of water are based primarily on English common law, which was gradually developed at a time when the population was widely scattered and water was far more abundant than the demand for it. Inevitably, with increasing population growth, industrialization, and urbanization, conflicts developed and some modifications were made. The need for modification is indeed genuine, as pointed out by Thomas (1951, p. 243), who stated, "many established concepts are now known to be scientifically unsound and should be revised in the light of present knowledge."

Of the States represented in the Mid-Atlantic Region, Vermont, Massachusetts, New Jersey, and Pennsylvania generally follow the "English common-law version of the riparian doctrine, which gives absolute ownership of ground water to the land owner" (Geraghty and others, 1973, pl. 34). The remaining States, New York, Maryland, Delaware, West Virginia, and Virginia, modify this doctrine, applying the American rule of reasonable use, which "restricts the land owner's rights in relation to others" (Geraghty and others, 1973, pl. 34). The present trend, however, appears to be in the direction of application of the American modification—the doctrine of reasonable use—for all the States in the Mid-Atlantic Region.

In addition to these general doctrines, some States in the Mid-Atlantic Region have established more specific controls on ground-water use. For example, Virginia and Maryland have enacted laws governing the drilling of wells. Since the late 1950's, following a long period of gradually declining artesian pressure, Virginia has required owners of flowing wells to cap them, and records for commercial and industrial wells must be furnished to the State. New Jersey and Vermont require approval of anticipated withdrawals above specified limits. On Long Island, N.Y., ground water used for cooling purposes must be returned to the ground through recharge wells, provided it has not been contaminated.

GROUND-WATER STUDIES AND NEEDS FOR ADDITIONAL INFORMATION IN THE MID-ATLANTIC REGION

There has been a general upward growth trend in the Mid-Atlantic Region, not only in population but in economic activity, manifested by industrial development, increasing urbanization, and increasing water requirements. More information will be needed regarding the occurrence and availability of additional ground-water supplies in the region, and the hydrologic effects of their developments, to facilitate efficient management of the total water resources.

Systematic programs of ground-water investigations are being carried on largely by the Geological

Survey in cooperation with State or local agencies. The discussion that follows considers the existing coverage and future needs in each of the States in the Mid-Atlantic Region.

Figure 9 shows the areas that have been studied by ground-water investigations.

Three categories of coverage are shown:

1. *Detailed quantitative* studies, in which ground water has been treated as a dynamic fluid resource. These studies include analysis of the results of aquifer-performance tests or other analyses that provide hydraulic information that serves as a guide for making predictions about the response of aquifers to changes in the pumping regimen that would result from possible additional developments.
2. *Detailed reconnaissance* studies, in which most of the available ground-water information has been collected, including quantitative data and interpretations that provide a basis for general understanding of the effects of possible future developments.
3. *General reconnaissance* studies, in which a representative sampling of readily available information has been collected and interpreted and which can provide general guidance for water developments.

Some areas in New York, New Jersey, and Virginia are not colored on figure 9. For these areas, formal reports have not been published, and no studies nearing completion are currently (1976) underway. However, Federal and State agencies have basic data on file for most of these areas. Specific inquiry may be made directly to the district office of the Geological Survey or to State agencies responsible for water studies for the appropriate State.

The following two categories of ground-water information are needed as of 1976, in the areas indicated on figure 10:

1. *Detailed quantitative* information, as previously defined. This level of ground-water information is indicated for areas where there is either an immediate need or a foreseeable need. The detailed information provided by investigations at this level allows immediate use for planning and development of major projects, such as municipal or industrial well fields, with only minimal additional study, if any.
2. *Detailed reconnaissance* information, as previously defined. This level of information is indicated for areas where only moderate future developments are likely and where a moderate amount of detailed information would suffice for small developments. Such information would provide an adequate base for additional work prior to any large-scale developments that may materialize.

VERMONT

Largely because Vermont is a rural State with no large cities and only minimal industrial development, the total use of water from all sources is the smallest among the northeastern States, amounting to only 110 million gal/d in 1970 (Murray and Reeves, 1972, table 10). However, use of ground water (42 million gal/d) is proportionately the highest among these same States—38 percent of the total water used (Geraghty and others, 1973). Despite this fact, the need for additional ground-water information in Vermont is only moderately urgent, the chief demand being related primarily to tourism and recreation.

In the part of western Vermont that lies within the Mid-Atlantic Region (fig. 9), ground-water favorability maps have been published for the several basins draining into Lake Champlain and the Hudson River valley (Hodges, 1966; 1967a, b, c, and d). These maps and accompanying texts are classified as general reconnaissance studies.

In addition, Hodges and Butterfield (1972) have published a detailed study that defines the quantity and quality of the ground water in the sand and gravel aquifers of the Barre-Montpelier area, in the western upstream reaches of the Winooski River basin, which drains into Lake Champlain. This study appraises the potential ground-water sources to meet expanded needs of many of the towns in Washington County. A detailed reconnaissance study is nearing completion immediately north and northeast of the Barre-Montpelier area, also in the Winooski basin.

Detailed reconnaissance ground-water information will be needed soon in the eastern parts of three basins—the Missisquoi, Lamoille, and Winooski—as a sequel to that which was provided by the work in the Barre-Montpelier area. The burgeoning recreation industry in those areas in recent years, associated mainly with the excellent skiing country, justifies such information (A. L. Hodges, Jr., written commun., 1975). Likewise, ground-water information is generally inadequate for anticipated development in southwestern Vermont, from Rutland to the Massachusetts boundary.

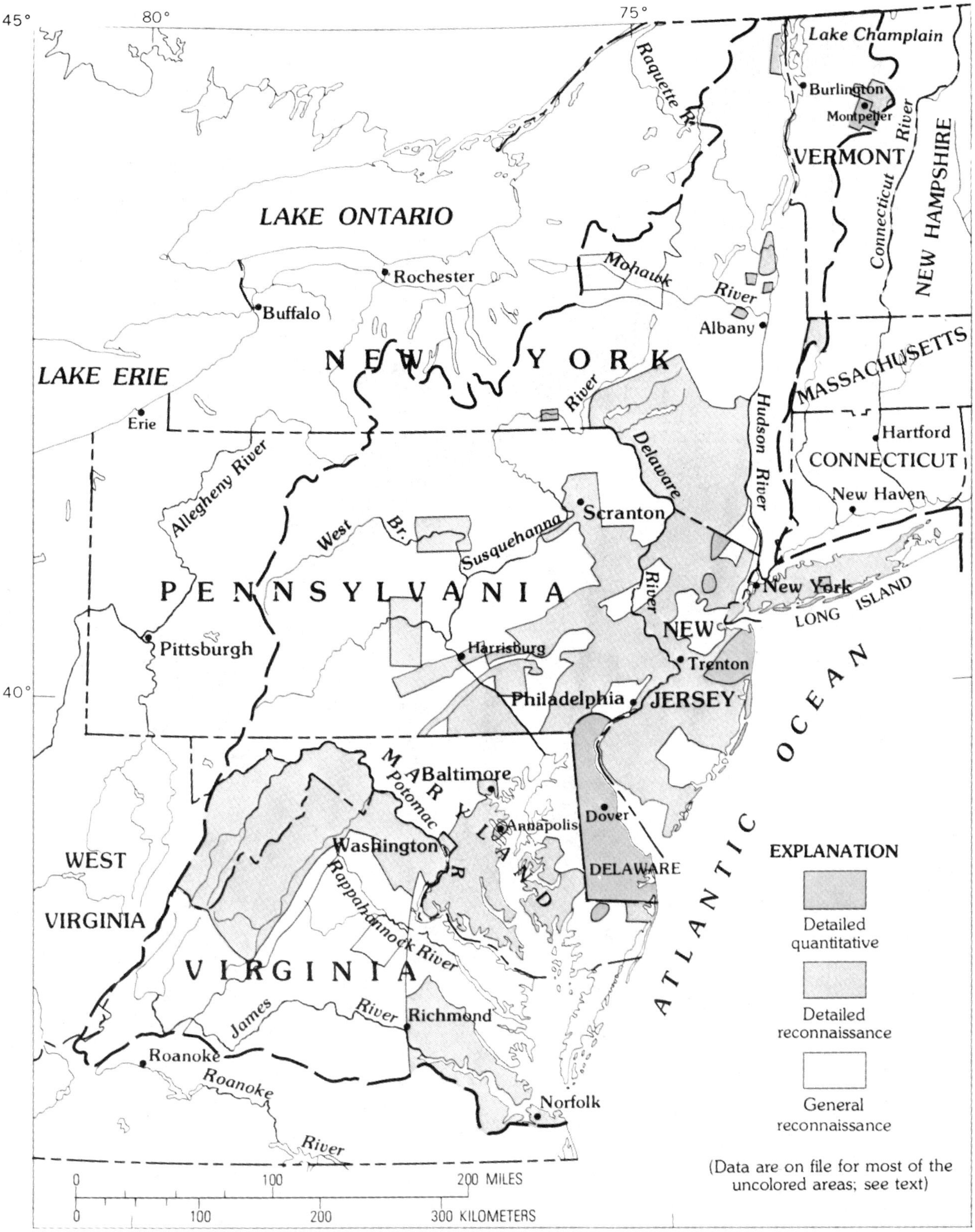

FIGURE 9.—Status of ground-water investigations in the Mid-Atlantic Region.

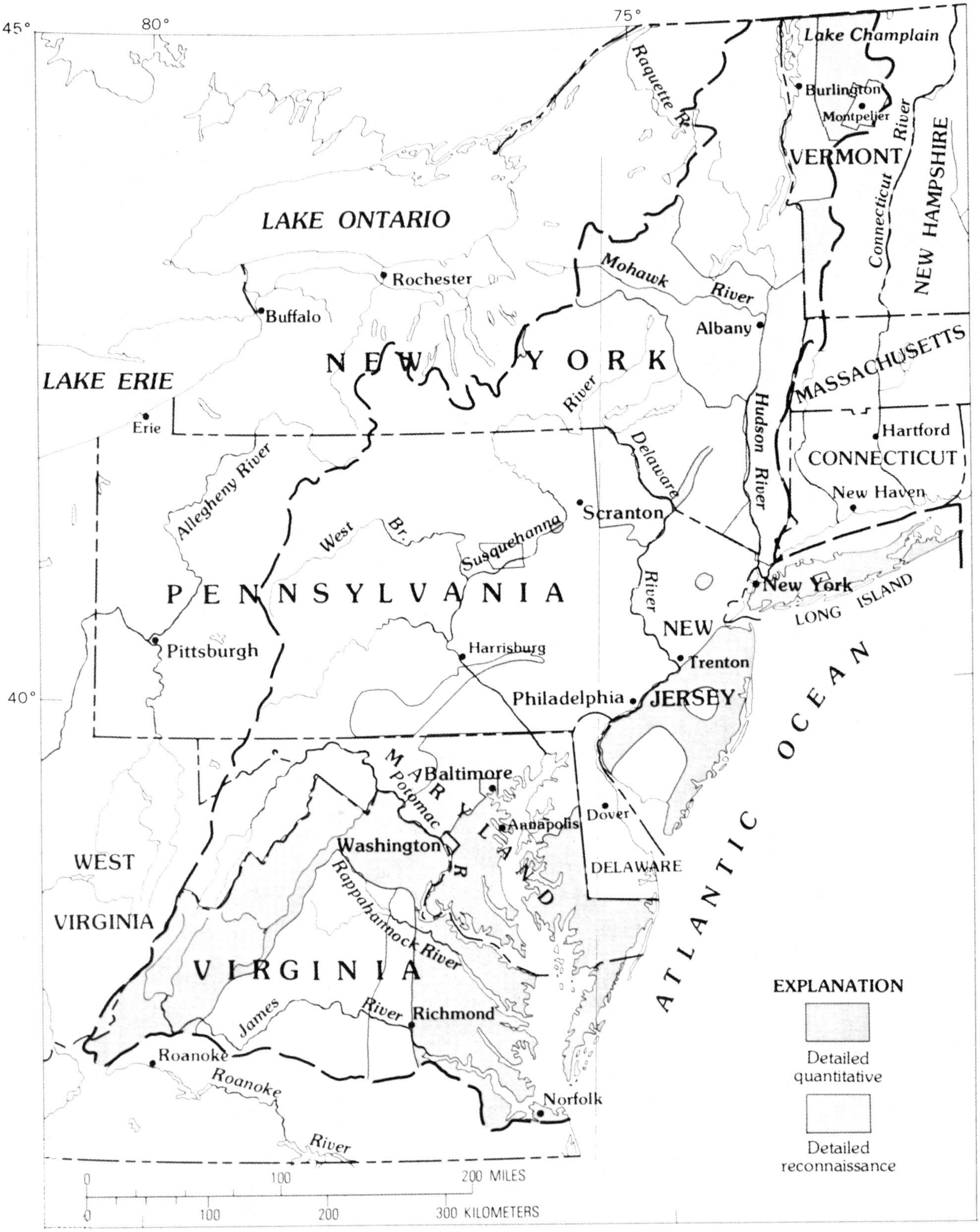

FIGURE 10.—In certain areas in the Mid-Atlantic Region, additional ground-water information is needed in order to assist water-resources development and management.

In the intervening region between Montpelier and Rutland, the need for water is believed now to be relatively stable. Adequate surface-water supplies have been developed from Lake Champlain, and additional supplies from this source should be adequate for the foreseeable future without recourse to ground-water sources, assuming that trends in economic activity continue at present rates.

MASSACHUSETTS

Only a small part of northwestern Massachusetts lies within the Mid-Atlantic Region. The Hoosic River, whose basin contains 205 mi^2 upstream from the Massachusetts-Vermont State boundary, drains 164 mi^2 in Massachusetts, 39 mi^2 in Vermont, and 2 mi^2 in New York. A tributary of the Hudson River, it flows through southwestern Vermont into New York.

A detailed reconnaissance study of the Hoosic River basin was published recently as a hydrologic atlas (Hansen, Toler, and Gay, 1973). It is sufficiently detailed to provide adequate information on the hydrogeology of the principal ground-water reservoirs in the basin. These are glacial outwash deposits forming watercourse aquifers in deep bedrock valleys along the main stem of the river. In places these deposits are 300 ft thick. Test drilling is needed to determine the best locations for additional development along the favorable river reaches described in the atlas.

The Hoosic River basin in northwestern Massachusetts is mostly rural in character and is likely to remain so for the near future. There are three principal communities—Adams, North Adams, and Williamstown. The study by Hansen, Toler, and Gay (1973) is believed adequate for needs of the near-term future, as it provides sufficient information to furnish a guide for exploratory drilling for additional ground-water developments.

NEW YORK

More than 17,000 mi^2 of New York lies within the Mid-Atlantic Region: part of the St. Lawrence Valley, nearly the entire Hudson River basin, and the northern parts of the Delaware and Susquehanna River basins. This area was glaciated during the ice age, and extensive outwash deposits occur in many of the larger valleys. The sand and gravel deposits in these valleys are the most widely used and promising sources of ground water in upstate New York. They are generally highly permeable and their hydraulic continuity with the perennial streams in the valleys allows ample replenishment for ground-water withdrawals from these deposits.

Except for parts of the upper Hudson River basin, the entire area of New York within the region has been adequately reviewed in reports prepared under the Federal-State cooperative program—largely reconnaissance studies (fig. 9). Also, a few quantitative studies of small areas have been completed. The most recent studies were made near Binghampton (A. R. Randall, oral commun., 1975), and on Long Island (Kimmel and Braids, 1974). In addition, an important series of reports covering many aspects of the hydrology of Long Island has been published; these reports contain much quantitative information. (See especially McClymonds and Franke, 1972; Franke and McClymonds, 1972.)

Many short reports have been prepared on small areas in New York by private consultants. For the most part, however, these are unpublished. Consultants also have prepared water-resources reports for all of the counties under a program with the New York Department of Health. Some of these reports describe reconnaissance ground-water investigations.

Although much information on the ground-water resources of New York has been developed in several decades of cooperative study by the U.S. Geological Survey and State and County agencies, additional quantitative information on the hydrologic systems is needed as the economy and population of the State grow. The need exists for continued detailed ground-water studies on Long Island, because most of Long Island depends almost entirely on ground water to support a large and expanding population. Additional studies now under way and others proposed for the future include (1) a continuation and refinement of the present hydrologic data-collection activities, in particular more intensive water-quality study and monitoring in parts of Long Island; (2) more complete definition of the shape and areal extent of the several aquifers, particularly in eastern Suffolk County, which is rapidly being developed and where deep public-supply wells are being drilled; and (3) predictive modeling studies to aid management decisions. These and other needs are discussed in the summary report by Franke and McClymonds (1972, p. F55–56).

In a ground-water investigation of Orange and Ulster Counties in lower upstate New York, Frimpter (1972, p. 74) suggests additional information may soon be needed for two important segments of a watercourse aquifer. One segment, northeast of Port Jervis, is in the valleys of the lower Neversink River and Basher Kill; the other segment is in the

Sandburg Creek valley near Ellenville. These segments are believed to be parts of a continuous sand and gravel body that lies beneath the present divide between the two drainages in intervening Sullivan County (M. H. Frimpter, oral commun., 1975). (See fig. 10.) More industry and housing developments are anticipated in these areas, and therefore increasing development of ground-water supplies is foreseen. Frimpter suggests that artificial recharge of ground water with excess surface-water supplies might aid optimum utilization of the water resources.

Other small aquifers probably will need study within the next few years if present trends in economic activity continue. For example, aquifers in Westchester and Cortland Counties (fig. 10) are being considered for quantitative study. Cortland County has an extensive shallow watercouse aquifer system that is highly susceptible to pollution from suburban and agricultural wastes.

Additional general ground-water investigations have been proposed in recent years for the lower Hudson River basin. Reconnaissance, in sufficient detail to provide adequate preliminary information for possible later detailed studies in local critical areas, is under consideration.

NEW JERSEY

New Jersey lies entirely within the Mid-Atlantic Region; its area is 7,836 mi^2. About three-fifths of the State is part of the Atlantic Coastal Plain, where excellent aquifers store vast quantities of ground water. The hard rocks in the northern part of the State provide sources for small to moderate ground-water supplies; where stratified glacial drift is present, larger supplies might be developed locally.

As shown in figure 9, ground-water studies have been made nearly everywhere in New Jersey, including both general and detailed reconnaissance studies. In the northern part of the State, where ground water is important but much less abundant than in the Coastal Plain, Somerset County and parts of Warren and Bergen Counties lack formal investigation. General reconnaissance coverage has been completed in Middlesex County, southwest of New York City, and in Atlantic County farther south in the Coastal Plain. Detailed reconnaissance studies have been made in the remainder of the State.

Recently, quantitative studies were completed for two areas in northern New Jersey (Vecchioli and Nichols, 1966; Vecchioli and others, 1967). Another quantitative study, of the Ramapo River, also in northern New Jersey, was made by Vecchioli and Miller (1973). A digital model is nearing completion for the Englishtown aquifer in Monmouth and Ocean Counties in the Coastal Plain (W. D. Nichols, oral commun., 1975). A comprehensive investigation of the chemical character of ground water in the Englishtown Formation has been published (Seaber, 1965).

Much of the information collected during a comprehensive study of the Wharton Tract, in southern Camden and Burlington Counties and northwestern Atlantic Coutny, has been published (Rhodehamel, 1973). In addition, detailed quantitative data from an extensive pumping test made in 1961 in the Wharton Tract are in the open files of the Geological Survey.

New Jersey ranks high among the States in volume of ground-water withdrawals, and it has probably developed its ground-water reservoirs more completely than any other State in the humid eastern part of the country (Thomas, 1951, p. 127). A number of problems include the threat of saltwater encroachment, which results because of local concentration of pumping adjacent to saltwater bodies or near the freshwater-saltwater interface in aquifers. Management of these problems would involve changing the locations of pumping centers and obtaining detailed quantitative knowledge about the ground-water reservoirs.

New Jersey lies along the axis of the Boston-Washington transportation corridor. Increasing urbanization and industrial development is taking place, following roughly along Interstate Routes 95 (the New Jersey Turnpike) and 295. There is, therefore, need for quantitative ground-water information along this corridor. For similar reasons, quantitative information is warranted in Monmouth and Ocean Counties, and in the coastal parts of Burlington, Atlantic, and Cape May Counties. Such information would augment and refine the basic preliminary work already accomplished in these areas and would assist in developing ways to minimize saltwater encroachment and other potential developmental problems.

That part of the Pine Barrens in central New Jersey, between Camden and Atlantic City, also is an area of increasing economic activity and population growth resulting from the completion, in 1964, of the expressway between the two cities. Detailed quantitative knowledge of the ground water, including model analysis to predict the impacts of ground-water development, is warranted in anticipation of the planned development of the Wharton Tract as a major source of water supply for southern New Jersey. Such a study would facilitate the conjunctive development of ground water and surface water in the Tract.

PENNSYLVANIA

Pennsylvania has the largest area of the States within the Mid-Atlantic Region—over 20,000 mi^2. About two-thirds of this area lies within the Susquehanna River basin, most of the remainder is in the Delaware River basin, and only a small part is in the Potomac River basin. (See fig. 2.)

In the glaciated northern part of the State, alluvial and outwash deposits occur in most large valleys. These deposits also extend some distance south into the unglaciated area. As watercourses, they provide an abundant source of ground water because of their hydraulic continuity with the streams that occupy the valleys. Recharge from the stream channels can be induced when these deposits are developed for ground-water supplies.

Elsewhere in the Pennsylvania part of the region, the older hard rocks yield small to moderate supplies of water. In the relatively small area of the Coastal Plain in the extreme southeastern part of the State, the unconsolidated sedimentary rocks include several productive aquifers (Greenman and others, 1961).

Pennsylvania was covered by general reconaissance ground-water studies during the 1930's (Piper, 1933; Hall, 1934; Leggette, 1936; Lohman, 1937, 1938, and 1939). Since then, the Federal-State cooperative program has carried out more detailed investigations in those parts of the State that required refined coverage for various reasons. including industrial expansion and associated population growth. This cooperative program of ground-water investigations continues to be active, and the present status of ground-water coverage in terms of published reports and active studies nearing completion is indicated on figure 9.

The general direction of effort in the continuing cooperative program is toward providing ground-water information at least at the level of detailed reconnaissance in those areas of the State where such information is required because of anticipated or potential expansion in economic activity. (See fig. 10). For the immediate future, the detailed reconnaissance level of ground-water information probably would fulfill the anticipated requirements adequately.

Ground-water investigations in Pike and Wayne Counties, which lie within the upper Delaware River basin in Pennsylvania, are scheduled in the Federal-State cooperative program. The information obtained from the proposed studies will be important in view of the anticipated acceleration of development in these areas for recreation. Recreational, tourist, industrial, and commercial activity has been enhanced by the Northeast Extension of the Pennsylvania Turnpike, completed in 1957, and by the recently completed Interstate Highway 80 across Pennsylvania to New Jersey and New York City. The planned Delaware Water Gap National Recreation Area will require water information presently only partially available.

Saline ground water occurs naturally at depths of 300 to 500 ft in the Appalachian Plateaus province in Pennsylvania (fig. 6). (Becher, 1970). More comprehensive information is needed on the occurrence of saline water in this area, and elsewhere within the region in Pennsylvania. Such information would help define the depths to which potable freshwater occurs and also provide knowledge of the chemical nature and physical extent of the underlying saline water, which, despite its quality, has potential as a possible future resource.

MARYLAND, DELAWARE, AND DISTRICT OF COLUMBIA

In view of their close proximity and current joint Federal-State cooperative programs, the District of Columbia and the States of Maryland and Delaware may conveniently be considered together. Much of their combined area lies within the Chesapeake Bay region and the eastern part of the Potomac River basin; the balance consists of coastal reaches in the Maryland Eastern Shore and southern Delaware, and minor parts of the Susquehanna and Delaware River basins.

Beginning in 1942, with a study of the Baltimore industrial area, reconnaissance studies of counties or groups of counties were made of the entire State of Maryland as part of the Federal-State cooperative ground-water program. These studies were completed in the early 1960's. Since then, detailed reconnaissance studies have been made in the five Maryland counties east of Washington, between Baltimore and Virginia on the western shore and in Talbot and Dorchester Counties on the Maryland Eastern Shore. (See fig. 9.) Detailed quantitative ground-water studies have been published for the entire State of Delaware. (See Sundstrom and Pickett, 1968, 1969, 1970, and 1971.) A comprehensive reconnaissance investigation of the water resources of the Delmarva Peninsula, which includes the Eastern Shore parts of Virginia, Maryland, and

Delaware, was completed in 1973 (Cushing and others, 1973).

Because of continued urbanization and industrial development, ground-water information is needed at the quantitative level in the area in north-central Maryland between Baltimore and the Pennsylvania boundary, which includes Howard, Carroll, Baltimore, and Harford Counties, and the city of Baltimore. It may be desirable to test the hypothesis put forth by Hollyday and McKenzie (1973) that natural neutralization of acid mine drainage waters take place in Maryland coal fields under suitable hydrogeologic conditions.

Quantitative information on ground water is needed in the remainder of Sussex County in Delaware and the Eastern Shore counties of Maryland, as a followup of the detailed reconnaissance recently completed for the Delmarva Peninsula. Economic activity in both of these areas is increasing and is expected to continue, with the usual attendant increase in demand for water.

WEST VIRGINIA

The eastern panhandle of West Virginia lies within the Mid-Atlantic Region, in the western reaches of the Potomac River basin. A hydrologic map, sections, and text have been published (Hobba and others, 1973), and a comprehensive report is in preparation. This coverage is believed adequate for this area for the near future, except for local areas where more detailed information may be needed as economic activity develops along the Interstate Highway system.

VIRGINIA

More than 17,000 mi^2 of Virginia lies within the Mid-Atlantic Region. The Coastal Plain occupies somewhat less than one-third of that area. It is underlain by unconsolidated sediments containing the State's most productive aquifers. The balance of the State is underlain by crystalline igneous and metamorphic rocks, limestones, and other carbonate rocks, sandstones, and shales.

Reconnaissance ground-water studies have been completed in the Virginia part of the region, except for most of the Piedmont province which has been investigated only in a few areas. (See fig. 9.) Virginia's paper industry is centered in the Franklin area, just south of the Mid-Atlantic Region, along the James River, in the southeastern corner of the State. It has been a major user of ground water since the early 1940's. This area has been the subject of quantitative studies (Sinnott, 1968; Brown and Cosner, 1974). The effect of the present (1975) rate of withdrawal, about 40 million gal/d, extends northward more than 25 mi to the York-James Peninsula in the southern part of the Mid-Atlantic Region.

A systematic test-drilling program in the Coastal Plain would establish a basic network of factual data for regional subsurface correlations of the water-bearing formations. Samples of water for chemical analysis, from individual aquifers, collected while drilling test wells, and at selected wells drilled for municipal or industrial supplies, would yield a better understanding of the State's ground-water quality.

Expansion of the present water-level observation-well network is desirable for determining the effects of pumping in individual aquifers near centers of major withdrawals. Also, water-quality monitoring at selected sites in the Coastal Plain should be expanded to detect changes in the position of the freshwater-saltwater interface in major aquifers.

Reconnaissance information is needed for most of the Virginia Piedmont that lies within the region, in order to determine places most favorable for light industry. Although little or no process water is required for this kind of activity, small dependable supplies of water for personnel or other limited demands are generally available in the Piedmont. Moreover, the physical environment of the Piedmont counties will continue to attract light industry, especially in established and growing municipal centers, such as the Charlottesville area.

Exploration for possible productive watercourses in the Virginia Piedmont might also be undertaken, as was done in 1950 in North Carolina (Mundorff, 1950).

CONCLUSIONS

The rocks that constitute the ground-water reservoirs in the Mid-Atlantic Region can be grouped broadly into three general classes. These are (1) the unconsolidated sedimentary rocks that underlie the Coastal Plain; (2) the igneous, metamorphic, and consolidated sedimentary rocks that underlie the Piedmont, Blue Ridge, Valley and Ridge, Appalachian Plateaus, Adirondack, and New England provinces; and (3) the unconsolidated glacial and alluvial deposits that overlie the consolidated rocks in the northern part of the region. Among these rocks, the most permeable have become the important aquifers in the region.

Large volumes of water move through and are stored in the aquifers in the Mid-Atlantic Region, and much of this water is available for use. Determinations of the base flow of streams in the region indicate that as much as 38.6 billion gal/d of ground water is being discharged into the ocean. In addition, another 140 to 350 trillion gallons of good-quality water is estimated to be in storage in the aquifers.

Current problems of ground-water development and management in the region are primarily caused by pumping large amounts of ground water from fairly small areas and by poor ground-water quality. These problems are directly or indirectly related to the large and growing population and industrialization of the region, and to natural ground-water-quality problems that occur from place to place. Ground water with naturally occurring excessive amounts of troublesome constituents, such as chloride, iron, hydrogen sulfide, calcium, and magnesium, must be dealt with throughout the region. Local contamination of aquifers in the region is being caused by substances placed on the land or put into the ground accidentally or intentionally. Activities such as deep mining of coal and coal stripping have resulted in serious contamination. Many ground-water supplies in the region have been ruined by pumping wells too near natural or manmade sources of contamination and drawing water of poor quality into the wells. This is particularly true along the coastal part of the region where salt-water encroachment is a constant threat.

If the economic activity of the Mid-Atlantic Region continues to grow as predicted, the need for water will increase also. There are several ways to meet this anticipated need in the region. Among these, greater use of ground water and the conjunctive use of ground water and surface water offer the highest potential. Ground water development can be increased many times over the modest level of development to date. Large additional quantities of ground water could best be developed in the Coastal Plain part of the region. Elsewhere the largest amounts could be obtained from areally-extensive, permeable carbonate rocks and coarse-grained glacial and alluvial deposits, particularly watercourse aquifers along major streams.

Future development should be based on the knowledge of past and present ground-water management problems in order to avoid the same pitfalls. Large-scale ground-water developments should not be restricted to small areas and the major ground-water reservoirs should be protected as much as possible from contamination. The severe water shortages experienced during the Northeast drought of the middle 1960's suggest that substantial reserve supplies should be developed to meet the needs of such emergency periods. Perhaps 50 percent of the usable ground water stored in the aquifers of the region could be developed for reserve supplies. However, only a small part of the potential reserve could be utilized during a drought because of the lead time required for exploratory drilling, aquifer testing, drilling of production wells, and construction of suitable distribution systems to transport the water to municipal or industrial centers. Thus, to cope with possible emergencies, these reserve sources of ground water ideally should be explored and water-supply systems installed well in advance of critical demand for alternate sources of supply.

To ensure the wise development and management of the water resources of the Mid-Atlantic Region more information will be required about the occurrence, availability, and quality of ground water. Additional detailed reconnaissance data are essential in north-central and southwestern Vermont; east-central and southeastern New York; northeastern, central, and south-central Pennsylvania; central and northeastern Maryland; and east-central Virginia. Moreover, supplementary detailed quantitative information is desirable for the aquifers underlying the Coastal Plain in Long Island, New Jersey, Maryland, and Virginia; for some unconsolidated glacial and alluvial aquifers in Pennsylvania, New York, and New Jersey; and for some Piedmont and Valley and Ridge aquifers—particularly the carbonate-rock aquifers—in Virginia.

Collection of the necessary knowledge about ground water, as discussed in this report, and the implementation of lines of development indicated by this knowledge, will contribute in a major way toward the full use and optimal management of ground water as an indispensable component of the total water resources of the Mid-Atlantic Region.

SELECTED REFERENCES

Back, William, 1966, Hydrochemical facies and ground-water flow patterns in northern part of Atlantic Coastal Plain: U.S. Geol. Survey Prof. Paper 498–A, 42 p.

Barksdale, H. C., 1970, A program for the investigation and management of ground water in the Delaware River basin: Delaware River Basin Comm., Trenton, N.J., 120 p.

Barksdale, H. C., Greenman, D. W., Lang, S. M., Hilton, G. S., and Outlaw, D. E., 1958, Ground-water resources in the tri-State region adjacent to the lower Delaware River: New Jersey Div. Water Policy and Supply Spec. Rept. 13, 174 p.

Barksdale, H. C., O'Bryan, Deric, and Schneider, W. J., 1966, Effect of drought on water resources in the Northeast: U.S. Geol. Survey Hydrol. Inv. Atlas HA–243.

Becher, A. E., 1970, Ground water in Pennsylvania: Pennsylvania Geol. Survey, 4th ser., ES-3.

Biesecker, J. E., Lescinsky, J. B., and Wood, C. R., 1968, Water resources of the Schuylkill River basin: Pennsylvania Dept. Forests and Waters, Water Resources Bull. 3, 198 p.

Brown, D. L., and Silvey, W. D., 1973, Underground storage and retrieval of fresh water from a brackish-water aquifer, *in* Second International Symposium on Underground Waste Management and Artificial Recharge, New Orleans, La., Reprints, v. 1, p. 379–419: American Assoc. of Petroleum Geologists.

Brown, G. A., and Cosner, O. J., 1974, Ground-water conditions in the Franklin area, southeastern Virginia: U.S. Geol. Survey Hydrol. Inv. Atlas HA-538, 3 sheets.

Brown, P. M., Miller, J. A., and Swain, F. M., 1972, Structural and stratigraphic framework, and spatial distribution of permeability of the Atlantic Coastal Plain, North Carolina to New York: U.S. Geol. Survey Prof. Paper 796, 79 p.

Bue, C. D., 1970, Streamflow from the United States into the Atlantic Ocean during 1931–60; U.S. Geol. Survey Water-Supply Paper 1899–I, 36 p.

Cederstrom, D. J., 1943, Chloride in ground water in the Coastal Plain of Virginia: Virginia Geol. Survey Bull. 58.

——— 1945, Geology and ground-water resources of the Coastal Plain in southeastern Virginia: Virginia Geol. Survey Bull. 63, 384 p.

——— 1946, Genesis of ground waters in the Coastal Plain of Virginia: Econ. Geology, v. 41, no. 3, p. 239–244.

——— 1947, Artificial recharge of a brackish water well: The Commonwealth (Virginia), v. 14, no. 12, Dec. 1947.

——— 1957, Geology and ground-water resources of the York-James peninsula, Virginia: U.S. Geol. Survey Water-Supply Paper 1361, 237 p.

——— 1972, Evaluation of yields of wells in consolidated rocks, Virginia to Maine: U.S. Geol. Survey Water-Supply Paper 2021, 38 p.

——— 1973, Cost analysis of ground-water supplies in the North Atlantic Region, 1970: U.S. Geol. Survey Water-Supply Paper 2034, 48 p.

Cederstrom, D. J., Baker, J. A., and Tarver, G. R., 1971, Ground water in the North Atlantic Region, Appendix D, North Atlantic Regional Water Resources Study: U.S. Geol. Survey open-file report, 240 p.

Cushing, E. M., Kantrowitz, I. H., and Taylor, K. R., 1973, Water resources of the Delmarva Peninsula: U.S. Geol. Survey Prof. Paper 822, 58 p.

Darton, N. H., and Fuller, M. L., 1905, Virginia, *in* Fuller, M. L. (geologist in charge), Underground waters of eastern United States: U.S. Geol. Survey Water-Supply Paper 114, p. 127–135.

DeBuchananne, G. D., 1968, Ground-water resources of the James, York, and Rappahannock River basins of Virginia west of the Fall Line: U.S. Geol. Survey Hydrol. Atlas HA-283.

——— 1938, Ground-water resources of the Eastern Shore of Virginia and the James, York, and Rappahannock River basins of Virginia east of the Fall Line: U.S. Geol. Survey Hydrol. Atlas HA-284.

DeKay, R. H., 1972, Development of ground-water supplies in Shenandoah National Park, Virginia: Virginia Div. Mineral Resources, Mineral Resources Rept. 10, 158 p.

Dingman, R. J., and Ferguson, H. F., 1956, The ground-water resources of the Piedmont part, *in* The water resources of Baltimore and Harford Counties (Maryland): Maryland Dept. Geology, Mines, and Water Resources Bull. 17, 128 p.

Dingman, R. J., Meyer, Gerald, and Martin, R. O. R., 1954, The water resources of Howard and Montgomery Counties (Maryland): Maryland Dept. Geology, Mines, and Water Resources Bull. 14, 260 p.

Fenneman, N. M., and others, 1946, Physical divisions of the United States: U.S. Geol. Survey map prepared in coop. with Physiog. Comm., U.S. Geol. Survey, scale 1:7,000,-000. (Reprinted 1957.)

Feth, J. H., and others, 1965, Preliminary map of the conterminous United States showing depth to and quality of shallowest ground water containing more than 1,000 parts per million dissolved solids: U.S. Geol. Survey Hydrol. Inv. Atlas HA-199.

Franke, O. L., and McClymonds, N. E., 1972, Summary of the hydrologic situation on Long Island, New York, as a guide to water-management alternatives: U.S. Geol. Survey Prof. Paper 627–F, 9 p.

Frimpter, M. H., 1972, Ground-water resources of Orange and Ulster Counties, New York: U.S. Geol. Survey Water-Supply Paper 1985, 80 p.

Geraghty, J. J., Miller, D. W., VanderLeeden, Frits, and Troise, F. L., 1973, Water atlas of the United States: Water Information Center, Inc., Port Washington, N.Y.

Giese, G. L., and Hobba, W. A., Jr., 1970, Water resources of the Champlain-Upper Hudson basins in New York State: New York Conserv. Dept. Water Resources Comm. Bull. 153 p.

Greenman, D. W., Rima, D. R., Lockwood, W. N., and Meisler Harold, 1961, Ground-water resources of the Coastal Plain area of southeastern Pennsylvania, with special reference to the effects of human activities on the quality of water in the Coastal Plain sediments: Pennsylvania Geol. Survey, 4th ser., Bull. W-13, 375 p.

Hall, G. M., 1934, Ground water in southeastern Pennsylvania: Pennsylvania Geol. Survey, 4th ser., Bull. W-2, 255 p.

Hansen, B. P., Toler, L. G., and Gay, F. B., 1973, Hydrology and water resources of the Hoosic River basin, Massachusetts: U.S. Geol. Survey Hydrol. Inv. Atlas HA-481.

Heath, R. C., 1964, Ground water in New York: New York Conserv. Dept. Water Resources Comm. Bull. GW-51.

Hobba, W. A., Jr., Friel, E. A., and Chisholm, J. L., 1973, Ground-water hydrology of the Potomac River basin, West Virginia: U.S. Geol. Survey (map and summary text).

Hodges, A. L., Jr., 1966, Ground-water favorability map of the Batten Kill, Walloomsac River, and Hoosic River basins, Vermont: Vermont Dept. Water Resources.

——— 1967a, Ground water favorability map of the Missisquoi River basin, Vermont: Vermont Dept. Water Resources.

——— 1967b, Ground-water favorability map of the Lamoille River basin, Vermont: Vermont Dept. Water Resources.

——— 1967c, Ground-water favorability map of the Winooski River basin, Vermont: Vermont Dept. Water Resources.

——— 1967d, Ground-water favorability map of the Otter Creek basin, Vermont: Vermont Dept. Water Resources.

Hodges, A. L., Jr., and Butterfield, David, 1972, Ground-water availability in the Barre-Montpelier area, Vermont: Vermont Dept. Water Resources dupl. rept., 27 p.

Hollyday, E. F., 1969, An appraisal of the ground-water resources of the Susquehanna River basin in New York State: U.S. Geol. Survey open-file report, 52 p.

Hollyday, E. F., and McKenzie, S. W., 1973, Hydrogeology of the formation and neutralization of acid waters draining from underground coal mines of western Maryland: Maryland Geol. Survey Rept. Inv. 20, 50 p.

Johnston, R. H., 1971, Base flow as an indicator of aquifer characteristics in the coastal plain of Delaware, *in* Geological Survey research 1971: U.S. Geol. Survey Prof. Paper 750–D, p. 212–215.

Kimmel, G. E., and Braids, O. C., 1974, Leachate plumes in a highly permeable aquifer: Ground Water, v. 12, no. 6, Nov.-Dec. (Special issue, v. 1, Proc., 2d National Ground Water Quality Symposium, Sept., 1974, Denver, Colo.)

Kohout, F. A., ed., 1970, Saline water, a valuable resource: Water Resources Research, v. 6, no. 5, p. 1441–1531. (Series of 16 papers reprinted as a special symposium volume.)

League of Women Voters of Pennsylvania, 1962, The Susquehanna: Harrisburg, 26 p.

Leggette, R. M., 1966, Ground water in northwestern Pennsylvania: Pennsylvania Geol. Survey, 4th ser., Bull. W–3, 215 p.

Lichtler, W. F., and Wait, R. L., 1974, Summary of the ground-water resources of the James River basin, Virginia: U.S. Geol. Survey open-file report 74–139, Norfolk, Va., 54 p.

Lohman, S. W., 1937, Ground water in northeastern Pennsylvania: Pennsylvania Geol. Survey, 4th ser., Bull. W–4, 300 p.

——— 1938, Ground water in south-central Pennsylvania: Pennsylvania Geol. Survey, 4th ser., Bull. W–5, 315 p.

——— 1939, Ground water in north-central Pennsylvania: Pennsylvania Geol. Survey, 4th ser., Bull. W–6, 220 p.

Lohman, S. W., and others, 1972, Definitions of selected ground-water terms—revisions and conceptual refinements: U.S. Geol. Survey Water-Supply Paper 1988, 21 p.

McClymonds, N. E., and Franke, D. L., 1972, Water-transmitting properties of aquifers on Long Island, New York: U.S. Geol. Survey Prof. Paper 627–E, 24 p.

McGuinness, C. L., 1963, The role of ground water in the national water situation: U.S. Geol. Survey Water-Supply Paper 1800, 1121 p.

MacNish, R. D., Randall, A. D., and Ku, H. F. H., 1969, Water availability in urban areas of the Susquehanna River basin (in New York)—A preliminary appraisal: New York Conser. Dept., Water Resources Comm. Rept. Inv. RI–7, 24 p.

Maher, J. C., 1971, Geologic framework and petroleum potential of the Atlantic Coastal Plain and Continental Shelf: U.S. Geol. Survey Prof. Paper 659, 92 p.

Miller, D. W., DeLuca, F. A., and Tessier, T. L., 1974, Ground water contamination in the Northeast States: U.S. Environmental Protection Agency, 660/2–74–056.

Moulder, E. A., 1970, Freshwater bubbles: a possibility for using saline aquifers to store water: Water Resources Research, v. 6, no. 5, p. 1528.

Mundorff, M. J., 1950, Flood-plain deposits of North Carolina Piedmont and mountain streams as a possible source of ground-water supply—preliminary report: North Carolina Dept. Conserv. Devel., Div. Mineral Resources Bull. 59, 20 p.

Murray, C. R., and Reeves, E. B., 1972, Estimated use of water in the United States in 1970: U.S. Geol. Survey Circ. 676, 37 p.

Nutter, L. J., 1973, Hydrogeology of the carbonate rocks, Frederick and Hagerstown Valleys, Maryland: Maryland Geol. Survey Rept. Inv. 19.

Olmsted, F. H., and Hely, A. G., 1962, Relation between ground water and surface water in Brandywine Creek basin, Pennsylvania: U.S. Geol. Survey Prof. Paper 417–A, 21 p.

Page, L. V., and Seaber, P. R., 1970, Water resources investigations in the Susquehanna River Basin: Pennsylvania Dept. Forests and Waters, Tech. Bull. 2.

Papadopulos, S. S., Bennett, R. R., Mack, F. K., and Trescott, P. C., 1974, Water from the Coastal Plain aquifers in the Washington, D.C., metropolitan area: U.S. Geol. Survey Circ. 697, 11 p.

Parker, G. G., and others, 1964, Water Resources of the Delaware River basin: U.S. Geol. Survey Prof. Paper 381, 200 p.

Piper, A. M., 1933, Ground water in southwestern Pennsylvania: Pennsylvania Geol. Survey, 4th ser., Bull. W–1, 410 p.

Rasmussen, W. C., 1955, Magnitude of the ground waters of Delaware: Proc. 28th Ann. Conf. Maryland-Delaware Water and Sewage Assoc. (May 26, 27, 1955), p. 58–66.

Rasmussen, W. C., and Andreasen, G. E., 1959, Hydrologic budget of the Beaverdam Creek basin, Maryland: U.S. Geol. Survey Water-Supply Paper 1472, 106 p.

Rhodehamel, E. C., 1973, Geology and water resources of the Wharton Tract and Mullica River basin in southeastern New Jersey: New Jersey Dept. Environmental Protection, Div. Water Resources Spec. Rept. 36, 58 p.

Richards, H. G., 1967, Stratigraphy of Atlantic Coastal Plain between Long Island and Georgia—Review: Am. Assoc. Petroleum Geologists Bull., v. 51, no. 12, p. 2400–2429.

Sanford, Samuel, 1913, The underground water resources of the Coastal Plain province of Virginia: Virginia Geol. Survey Bull. 5, 361 p.

Seaber, P. R., 1965, Variations in chemical character of water in the Englishtown Formation, New Jersey: U.S. Geol. Survey Prof. Paper 498–B, 35 p.

——— 1968, An appraisal of the ground-water resources of the upper Susquehanna River basin in Pennsylvania (an interim report): U.S. Geol. Survey open-file report, 75 p.

Seaber, P. R., and Hollyday, E. F., 1965, An appraisal of the ground-water resources of the lower Susquehanna open-file report, 75 p.
River basin (an interim report): U.S. Geol. Survey

——— 1966, An appraisal of the ground-water resources of the Juniata River basin (an interim report): U.S. Geol. Survey open-file report, 58 p.

Simmons, E. T., Grossman, I. G., and Heath, R. C., 1961, Ground water resources of Dutchess County, New York: New York Water Resources Comm. Bull. GW–43, 82 p.

Sinnott, Allen, 1968, Results of aquifer tests in sands of the Potomac Group in the Franklin area, southeastern Virginia (1949–50): U.S. Geol. Survey open-file report, 80 p.

Sinnott, Allen, and Tibbitts, G. C., Jr., 1968, Ground-water resources of Accomack and Northampton Counties, Virginia: Virginia Div. Mineral Resources, Mineral Resources Rept. 9, 113 p.

Stuart, W. T., Schneider, W. J., and Crooks, J. W., 1967, Swatara Creek basin of southeastern Pennsylvania—an evaluation of its hydrologic system: U.S. Geol. Survey Water-Supply Paper 1829, 79 p.

Subitzky, Seymour, Bettendorf, J. A., and Anderson, P. W., 1964, Water resources investigations in the Delaware River Basin: U.S. Geol. Survey Index Map.

Sundstrom, R. W., and others, 1967, The availability of ground water from the Potomac Formation in the Chesapeake and Delaware Canal area, Delaware: Newark, Del., Univ. Delaware, Water Resources Center, 95 p.

Sundstrom, R. W., and Pickett, T. E., 1968, The availability of ground water in Kent County, Delaware: Newark, Del., Univ. Delaware, Water Resources Center, 123 p.

——— 1972, OBERS projections—regional economic activity in the United States, v. 1 *of* Concepts, methodology, and summary data: 109 p. and appendices.

Thomas, H. E., 1951, The conservation of ground water: New York, McGraw-Hill Book Co., 327 p.

——— 1952, Ground-water regions of the United States—their storage facilities: U.S. 83d Cong., House Interior and Insular Affairs Comm., The physical and economic foundation of natural resources, v. 3, 78 p.

Trainer, F. W., and Watkins, F. A., Jr., 1974, Use of base-runoff recession curves to determine areal transmissivities in the upper Potomac River basin: U.S. Geol. Survey Jour. Res., v. 2, no. 1, Jan.-Feb., p. 125–131.

——— 1975, Geohydrologic reconnaissance of the upper Potomac River basin: U.S. Geol. Survey Water-Supply Paper 2035, 68 p.

Upson, J. E., 1966, Relationships of fresh and salty ground water in the northern Atlantic Coastal Plain of the United States, *in* Geological Survey research 1966: U.S. Geol. Prof. Paper 550–C, p. C235–C243.

URS/Madigan-Praeger, Inc. (New York, N.Y.) and Conklin and Rossant (New York, N.Y.), 1975, A comprehensive study of the Tocks Island Lake Project and alternatives, Draft Summary, 48 p.

U.S. Department of Commerce, National Weather Service, 1968, Climatic atlas of the United States, 80 p.

U.S. Geological Survey, 1970, National atlas of the United States of America, A. O. Gerlach, ed.: Washington, D.C., 417 p.

——— 1975, Geological Survey Research 1975: Prof. Paper 975, 373 p.

U.S. Public Health Service, 1962, Drinking water standards, 1962: U.S. Public Health Service Pub. 956, 61 p.

U.S. Water Resources Council, 1968, The Nation's water resources—the first national assessment of the Water Resources Council: 403 p.

——— 1970, Water resources regions and subregions for the national assessment of water and related land resources: 188 p.

Vecchioli, John, and Miller, E. G., 1973, Water resources of the New Jersey part of the Ramapo River basin: U.S. Geol. Survey Water-Supply Paper 1974, 77 p.

Vecchioli, John, and Nichols, W. D., 1966, Results of the drought-disaster test-drilling program near Morristown, New Jersey: New Jersey Dept. Conserv. and Econ. Devel. Div. Water Policy and Supply, Water Resources Circ. 16, 48 p.

Vecchioli, John, Nichols, W. D., and Nemickas, Bronius, 1967, Results of the second phase of the drought-disaster test drilling program near Morristown, New Jersey: New Jersey Dept. Conserv. and Econ. Devel., Div. Water Policy and Supply, Water Resources Circ. 17, 23 p.

Virginia State Water Control Board, 1974, Groundwater of southeastern Virginia: Virginia State Water Control Board, Bureau of Water Control Management, Planning Bull. 261–A, 38 p.

Voigt, William, Jr., 1972, The Susquehanna Compact: Rutgers University Press, New Brunswick, N.J., 336 p. (Chap. 2 on Water Quality, p. 11–29).

Wyrick, G. G., 1968, Ground-water resources of the Appalachian Region: U.S. Geol. Survey Hydrol. Inv. Atlas HA–295 (4 map sheets with text).

——— 1969, The availability of ground water in eastern Sussex County, Delaware: Newark, Del., Univ. Delaware, Water Resources Center, 136 p.

——— 1970, The availability of ground water in western Sussex County, Delaware: Newark, Del., Univ. Delaware, Water Resources Center, 118 p.

——— 1971, The availability of ground water in New Castle County, Delaware: Newark, Del., Univ. Delaware, Water Resources Center, 156 p.

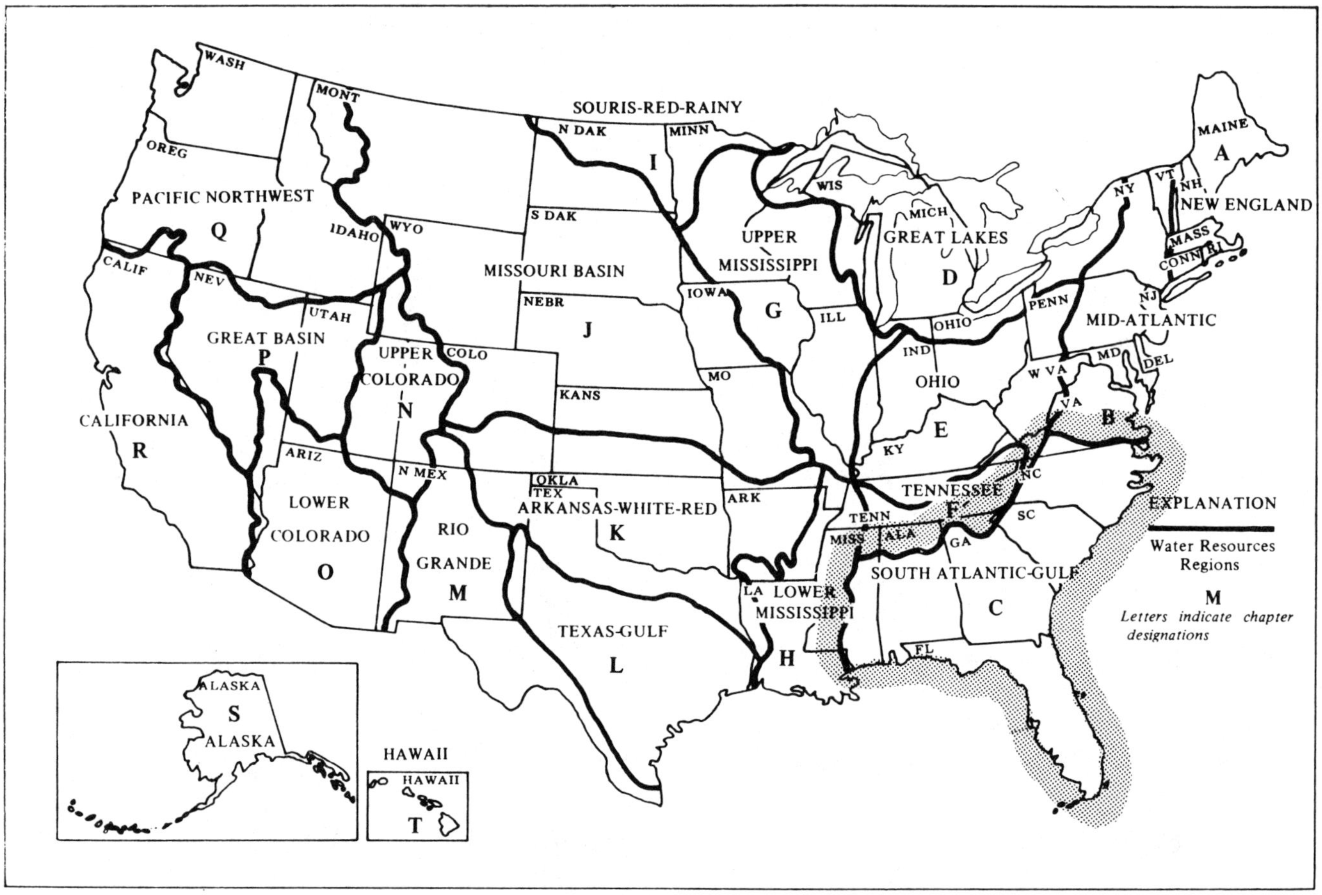

C – Location Map for the South Atlantic-Gulf Region

C

SOUTH ATLANTIC-GULF REGION

By D. J. CEDERSTROM, E. H. BOSWELL, and G. R. TARVER

ABSTRACT

Precipitation in the 270,000-square-mile South Atlantic-Gulf Region ranges from 44 to 80 inches, and the average runoff is about 15 inches. The ground-water discharge that forms the base flow of streams is conservatively estimated to be about 78,000 million gallons per day—the equivalent of about 6 inches of precipitation. On this basis, the regional sustained ground-water supply is about 286,000 gallons per day per square mile. Projected water use through 2020 indicates that about 10 percent of the supply will meet the region's requirement for ground water.

Enormous quantities of ground water are available in the South Atlantic-Gulf States in the extensive aquifers that underlie the Coastal Plain province. Large supplies of ground water can be developed also in places in the Cumberland Plateau and the Valley and Ridge provinces. In the Piedmont and Blue Ridge provinces ground water is available only in modest quantities.

The principal Coastal Plain aquifers consist largely of deltaic sand and gravel deposits of Cretaceous to Quaternary age; however, a notable exception is the highly permeable Tertiary limestone aquifer, which underlies parts of four States. Ground water in the Blue Ridge and the Valley and Ridge provinces is available from openings along fissures and joints in consolidated rocks and from the unconsolidated sand and gravel in the residuum and alluvium that overlie the bedrock. In parts of these areas faulting and folding has created favorable conditions for the development of moderate supplies of ground water. In the Piedmont province modest quantities of water are available from joints and fractures. Solution cavities in massive limestone units in some areas underlying the Cumberland Plateau section yield very large quantities of ground water.

Most of the major Coastal Plain aquifers are recharged where they are exposed and water moves downdip to the south. The general direction of ground-water movement in the Coastal Plain aquifers is seaward. There is, however, movement of water vertically and laterally that affects pressure and quality in every aquifer.

Ground water of good chemical and physical quality is available generally except in a small area in western Alabama and eastern Mississippi and in some small coastal areas. Freshwater extends to depths of slightly more than 3,000 feet in some areas. In many areas, water can be produced that does not require any treatment for general use. The more common objectionable constituents or properties are excessive iron, fluoride, hardness, color, and acidity. Contamination or pollution of ground water is not a major regional problem, but it is significant in local areas. Many of these local problems of contamination were the result of earlier improper disposal of oil-field and industrial waste. Saltwater intrusion either directly from the sea or from estuaries has been reported in every State in the region, and the potential for the proliferation of this problem is very large.

The relationship of ground water and surface water is significant at many places in the region where large ground-water withdrawals affect swamps, lakes, and the flow of streams and springs. There is a potential in some areas for using ground water during dry periods for low-flow augmentation of stream supplies. In other areas it is feasible to recharge ground-water reservoirs artificially by means of injection wells, pits, and water spreading.

Although large water-level declines caused by withdrawals have occurred locally, water-level changes have been small or modest when considered on a regional basis owing to the abundant rainfall and favorable recharge conditions.

The saline-water resources of the region are known to be extremely large. In some areas where the availability of freshwater is limited, saline ground water is available for desalinization to supplement freshwater supplies. Also, the saline aquifers are potential storage sites for injected freshwater.

Ground-water problems generally are not severe. Critical situations are restricted to areas where large quantities of ground water are being withdrawn or where aquifers are contaminated by oil-field or industrial waste. Large withdrawals in coastal areas have caused some saltwater intrusion. In other localities, highly mineralized water may have migrated along fault zones to freshwater aquifers. Many of the present problems can be resolved or ameliorated by redistributing withdrawals or developing alternative water sources.

INTRODUCTION

This report describing the ground-water resources of the South Atlantic-Gulf Region is one of a series of regional ground-water appraisals that summarize data in many published reports and unpublished data in the files of Federal, State, and other agencies. The report discusses the enormous quantities of ground water that can be obtained by conventional development methods in many areas in the region and suggests that, by conjunctive use of surface water and ground water, larger volumes of water can be made available.

About three-fourths of the region is in the Atlantic and Eastern Gulf Coastal Plains where the geologic units that make up the aquifers are limestone or unconsolidated sand and gravel. The remainder of the region is underlain by the older consolidated rocks of the Appalachian Highlands which contain ground water in fractures and solution cavities.

Freshwater at some localities is known to occur at depths greater than 3,000 feet. There are only a few areas of very limited extent where freshwater is not available in usable quantities. Water-quality problems are those usually amenable to simple treatment.

The largely undeveloped ground-water resources in the region will be one of the controlling factors in future economic development. Industrial expansion and intensive agricultural development will place increasing demands on water. It is, therefore, essential that sound appraisals of the water resources be available to planners and managers to meet the need for additional demands for water.

GEOGRAPHIC SETTING

The South Atlantic-Gulf Region, which comprises an area of about 270,000 square miles in the southeastern part of the United States, is drained by streams flowing to the Atlantic Ocean and the Gulf of Mexico between the James River drainage in Virginia and the lower Mississippi River drainage. It includes all of Alabama, Florida, Georgia, North Carolina, South Carolina, and parts of Mississippi and Virginia (fig. 1). The region includes parts of the Appalachian Highlands and the Coastal Plain—two of the eight major physiographic divisions of the United States (Fenneman, 1938).

The Coastal Plain province occupies about 70 percent of the region, the Piedmont province about 25 percent, and the Blue Ridge province, Valley and Ridge province, and Cumberland Plateau section together constitute about 5 percent. The Coastal Plain is separated from the rest of the region by the Fall Line, the irregular topographic boundary between the coastal deposits and the older rocks of the Appalachian Highlands.

Precipitation extremes in the South Atlantic-Gulf Region range from about 44 inches in parts of the Coastal Plain to 80 inches in small areas in the mountains of North Carolina and South Carolina (fig. 2). Thunderstorms are particularly common in Florida, occurring 70 days or more a year. As much as 5 inches of rain in 24 hours along the Gulf Coast and as much as 3½ inches in 24 hours on the Coastal Plain along the Atlantic and inland from the Gulf is not uncommon. Such storms frequently result in floods of short duration.

Average runoff in the region is 15 inches per year (Murray and Reeves, 1977, p. 19). Runoff ranges from about 10 inches per year along the Atlantic Coast up to 30 inches along the panhandle of Florida and the Alabama-Mississippi coast, where thick permeable sands in the coastal areas absorb much of the heavy rainfall. Runoff is about 15 inches in the more inland areas where the annual rainfall is 45 to 50 inches. In the mountainous areas, runoff is about half the total precipitation. The difference of about 35 inches between average runoff and the average rainfall over the Coastal Plain and Piedmont provinces is lost by evaporation and transpiration.

The flow of streams varies with precipitation and is generally highest in February and March and lowest in October, except in central and southern Florida where flows are highest in October and low in late fall and in May.

Average annual air temperature ranges from about 60°F (16°C) in southern Virginia to 70°F (21°C) in the northern Gulf Coast, and to 77°F (25°C) in southern Florida.

GEOLOGIC FRAMEWORK

STRUCTURE

The Piedmont province is made up of ancient sedimentary rocks that have been intruded by igneous rocks and subjected to repeated stress. Throughout the area rocks have been fractured, faulted, and folded. The Triassic basins within the Piedmont province are infaulted blocks of younger sedimentary rocks.

The Coastal Plain has been subjected to repeated somewhat gentle differential movements. There are now a series of highs and intervening sags in the basement rock surface and overlying sediments along the entire coastline (fig. 3). Irregularities in the basement surface may be due to movement in pre-Cretaceous times; however, persistent movement in places has affected the younger sediments.

The northernmost major structural high in the South Atlantic-Gulf Region is the Cape Fear Arch (fig. 3). To the south a definite sag in the basement is known as the Southeast Georgia Embayment. Crossing the neck of Florida the basement rises again across the Ocala Uplift and then falls away in the Southwest Georgia Embayment. The Chattahoochee Anticline, more or less parallel to the river of the same name, passes to the west of the Southwest Georgia Embayment.

The major structures have created irregularities in the outcrop pattern, the most marked being the pattern over the Cape Fear Arch.

GEOLOGY

The South Atlantic-Gulf Region may be divided into two broad components: (1) the Coastal Plain province, about 196,000 square miles, made up of soft unconsolidated sand, gravel, and clay and consolidated or semiconsolidated limestone and (2) the remainder (Piedmont province, Blue Ridge province, Valley and Ridge province, and Cumberland Plateau section), about 74,000 square miles, consisting for the most part of hard consolidated rocks—indurated and metamorphosed sedimentary rocks and crystalline igneous rocks.

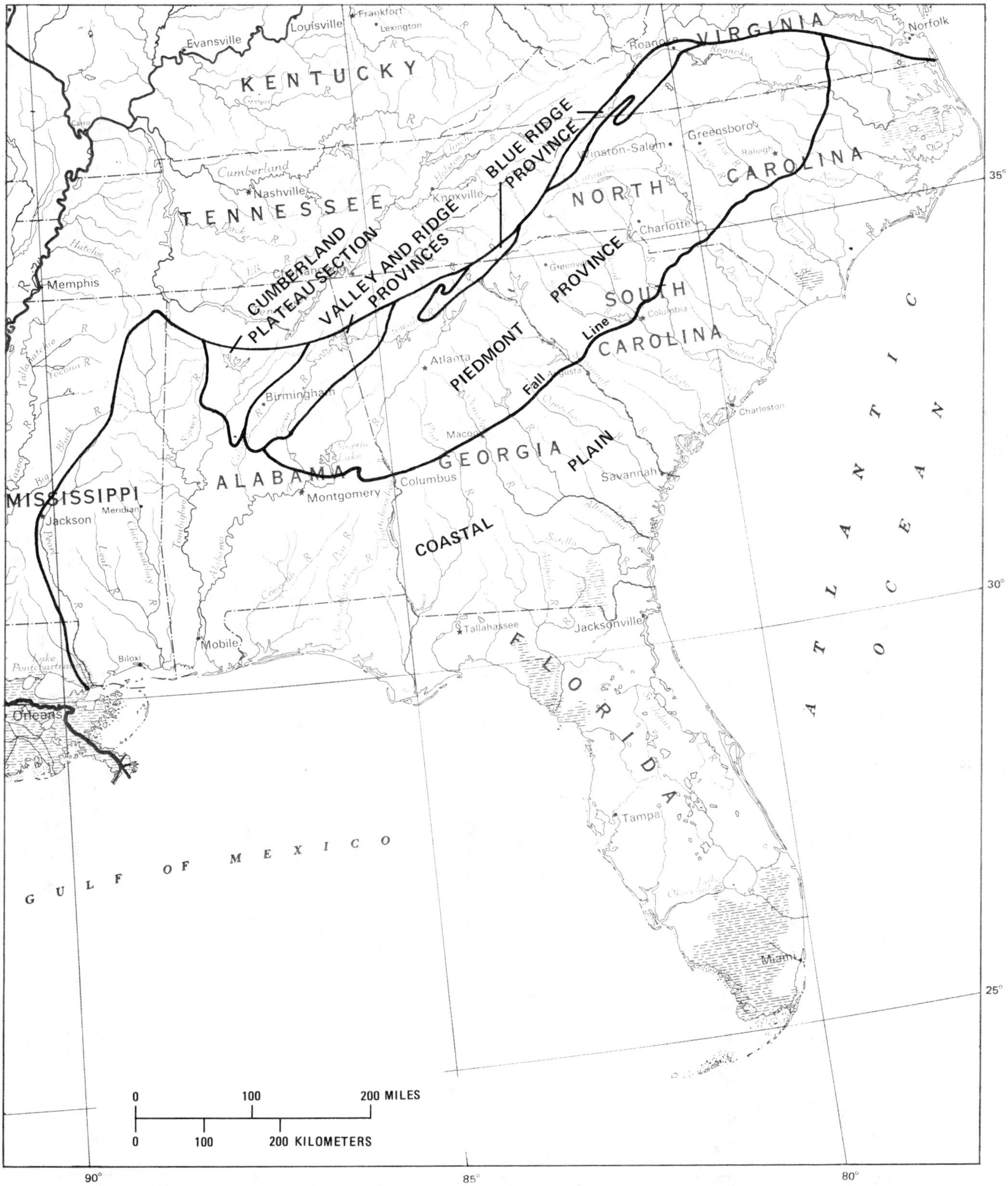

FIGURE 1.—Map showing location, major drainage, and principal physiographic subdivisions of the South Atlantic-Gulf Region.

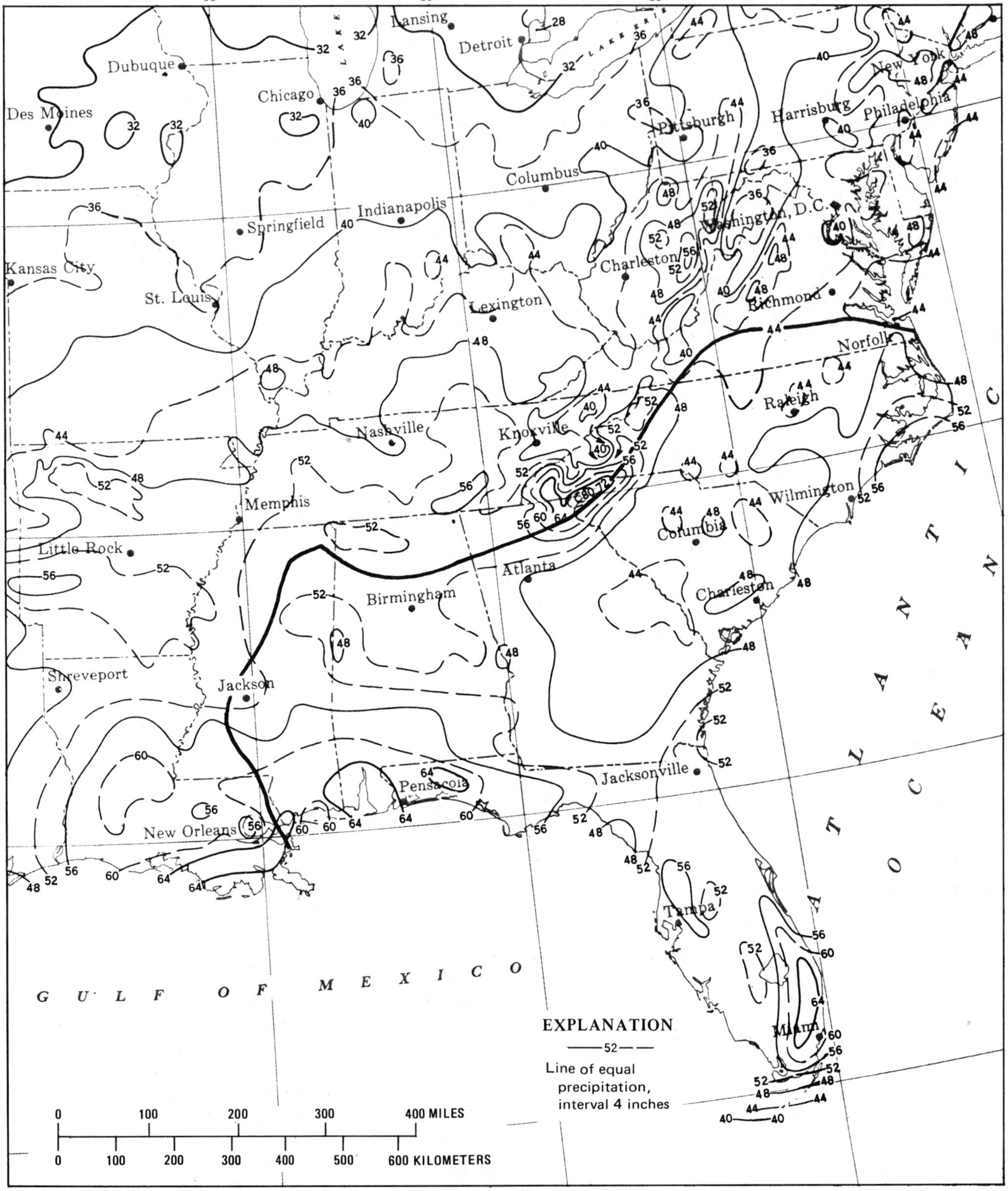

FIGURE 2.—Average annual precipitation in the South Atlantic-Gulf Region.

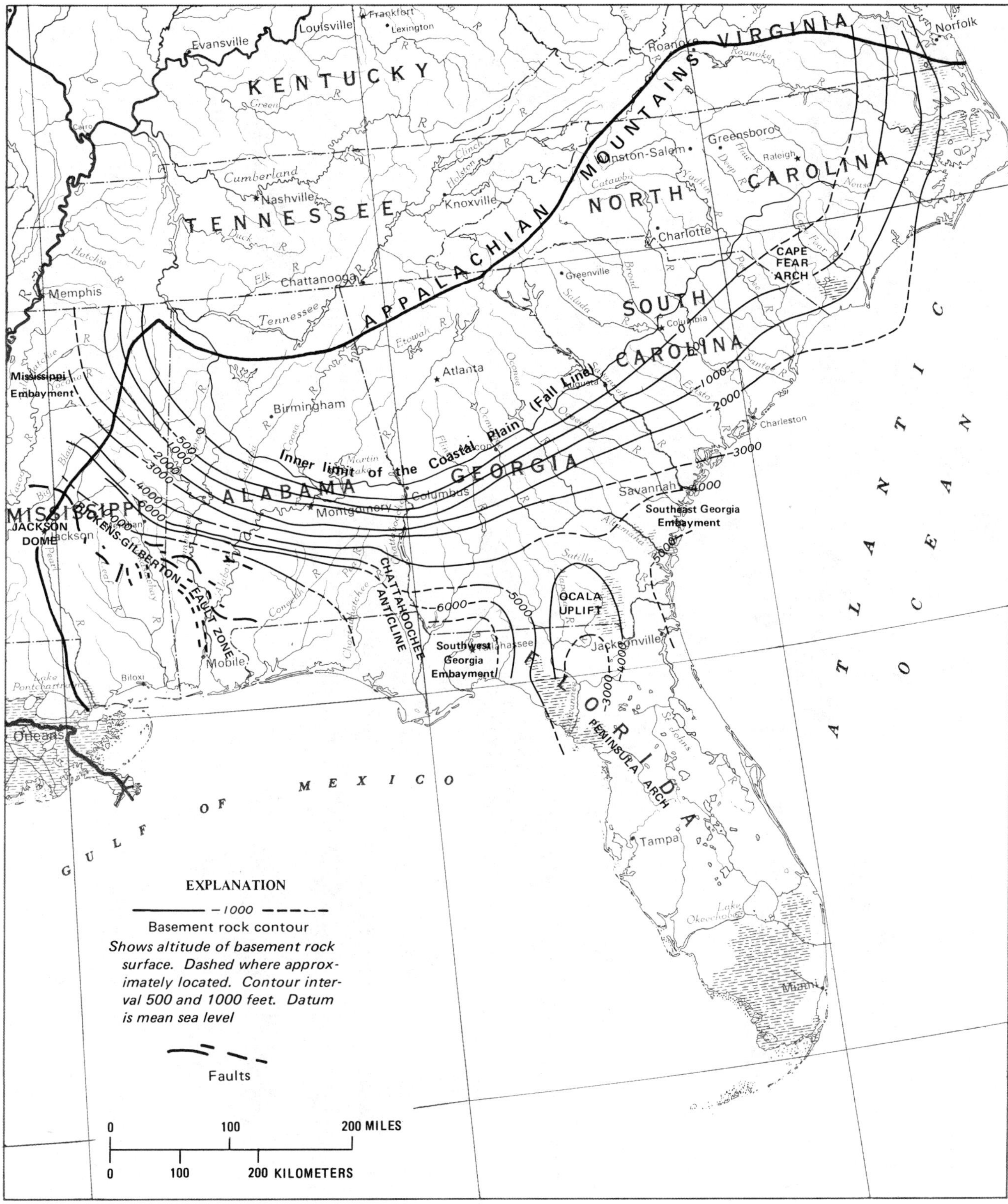

FIGURE 3.—Major structural features and depth below sea level of basement in the Coastal Plain province.

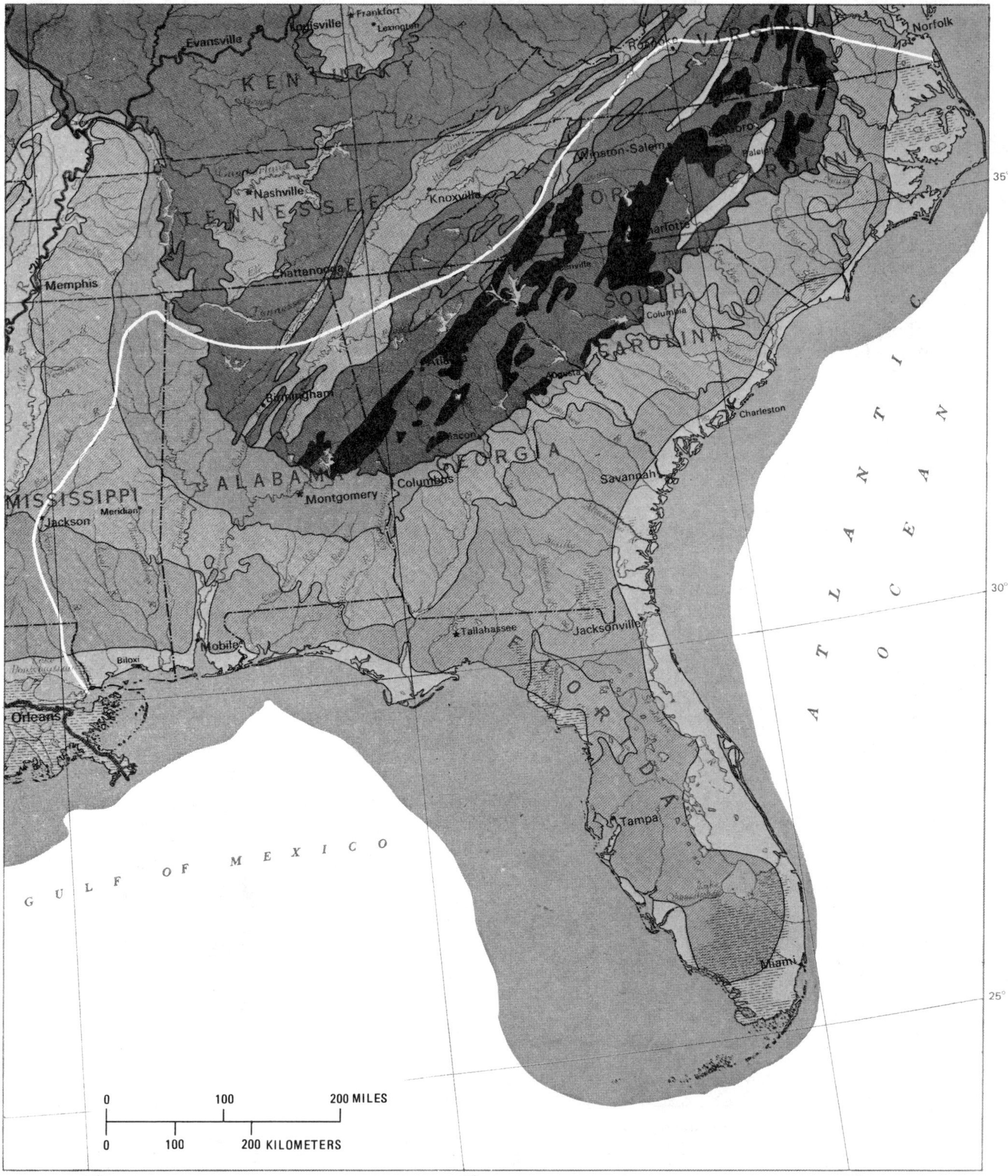

FIGURE 4.—Geologic map of the South Atlantic-Gulf Region.

The Coastal Plain province is underlain by unconsolidated clay, sand, and gravel and consolidated or semiconsolidated limestone. The deposits, which range in age from Cretaceous to Holocene, form a hugh arc extending from southern Virginia through the Carolinas, Georgia, Alabama, and into eastern Mississippi (fig. 4). From a thin edge at the Fall line, most of the deposits thicken seaward; a southward projection forms the peninsula of Florida.

In most places, the Coastal Plain beds have a gentle dip seaward. Each formation was generally overlapped by the next younger formation and their eroded edges are now exposed in a succession of older (inland) to younger (seaward) arcuate belts. The surface exposures of some formations are irregular. For instance, in northern North Carolina the inner edges of older strata were beveled by the advancing seas, and younger formations extend to the Fall Line.

The Coastal Plain formations are rarely uniform laterally (along the strike) or downdip. Nearshore sandy deltaic continental sediments thicken downdip and grade into deeper water silty or limy marine deposits. Laterally, sediments may also change in proportions of sand and clay, or may become limy.

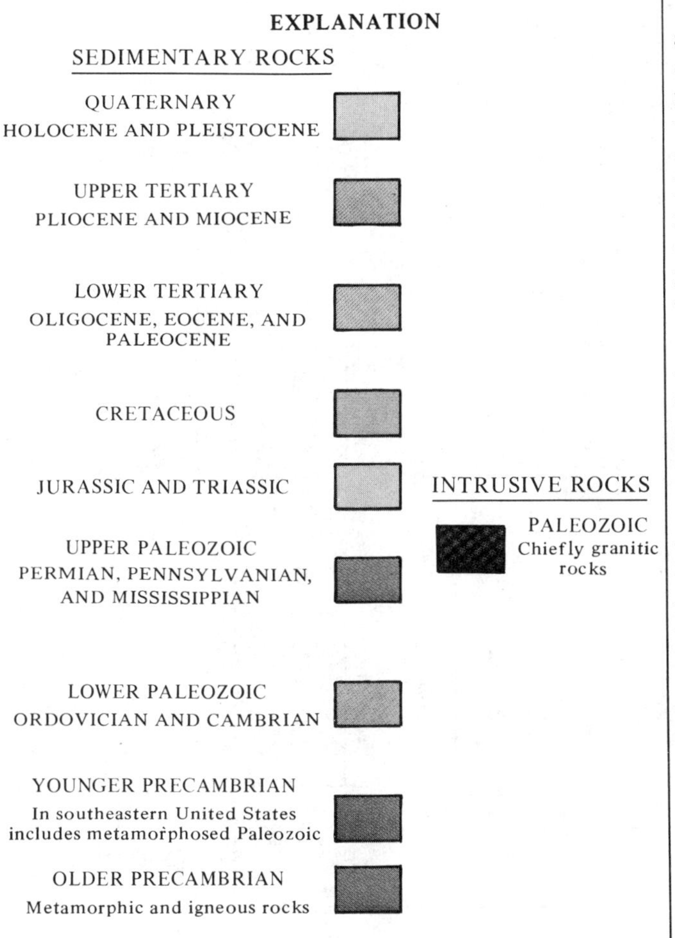

FIGURE 4.—Continued.

Sandy terrace deposits were superimposed upon the older formations throughout the Atlantic and Gulf Coastal regions during the Pleistocene Epoch.

Table 1 shows the stratigraphic units and aquifers of Cretaceous to Quaternary age that underlie the Coastal Plain province.

The Blue Ridge province is underlain by igneous and metamorphic rocks of older age which have been subjected to repeated stress.

The provinces outside the Coastal Plain are underlain by consolidated sedimentary rocks. In the Valley and Ridge province, these rocks have been greatly folded and faulted. In the adjacent Cumberland Plateau section the rocks have been subjected to less folding and faulting.

GROUND-WATER SUPPLIES

Most of the South Atlantic-Gulf Region is underlain by aquifers that are generally capable of yielding to wells 50 gal/min or more of water containing less than 1,000 milligrams per liter (mg/L) of dissolved solids (fig. 5 and table 1). Lower yields are characteristic of aquifers in the consolidated rocks that underlie the Blue Ridge, Valley and Ridge, and Piedmont provinces. In the Piedmont, Blue Ridge, Valley and Ridge provinces, and the Cumberland Plateau section, wells may intersect openings that result from structural deformation and solution, and yields range from substantial to meager.

The most prolific sources of ground water in the South Atlantic-Gulf Region are the highly permeable clastic and limestone aquifers in the Coastal Plain province. In small areas in western Alabama, eastern Mississippi, southeastern North Carolina, and northeastern South Carolina, these aquifers contain only saline water.

The aquifers occurring within each physiographic province in the South Atlantic-Gulf Region generally are not separate units operating independently; rather, each is a part of a complex system. Because of this interrelation, it is generally irrelevant to estimate quantities of water that are available from the individual aquifers. It is more meaningful to estimate the quantity available from the entire system.

Because the base flow of streams is supported almost entirely by ground-water outflow, the quantity of ground water available can be estimated by separating the total amount of streamflow from a region into its component parts—overland flow and base flow. The base-flow component is the approximate yield of the ground-water reservoirs under the existing hydrologic conditions.

TABLE 1.—*Generalized correlation of Coastal Plain stratigraphic units and aquifers*

System	Series	Mississippi	Alabama: Western	Alabama: Southeastern	Florida: Panhandle	Florida: Peninsula	Georgia	South Carolina	North Carolina	Southern Virginia
Quaternary	Holocene and Pleistocene	Alluvium and terrace deposits	Alluvium and terrace deposits		Terrace deposits	Alluvium and terrace deposits upper part of Biscayne aquifer	Alluvium and terrace deposits	Alluvium and terrace deposits Waccamaw Formation	Alluvium and terrace deposits	Alluvium and terrace deposits
Tertiary	Pliocene	Citronelle Formation Graham Ferry Formation[1]	Citronelle Formation	Absent	Citronelle Formation	Caloosahatchee Marl Alachua Formation \ Tamiami Formation	Charlton Formation \ Duplin Marl	Duplin Marl	Duplin Marl / Yorktown Formation	Yorktown Formation
	Miocene	Pascagoula Formation Hattiesburg Formation Catahoula Sandstone ? — ? — ?	Undifferentiated		Unnamed coarse clastic Shoal River Formation Pensacola clay Chipola Formation Tampa Limestone and equivalents	Miccosukee Formation Hawthorn Formation	Hawthorn Formation Tampa Formation	Hawthorn Formation Tampa (?) Limestone[2]	Pungo River Formation	St. Marys Formation Choptank Formation Calvert Formation
	Oligocene	Paynes Hammock Sand Chickasawhay Limestone Vicksburg Group: Byram Formation Marianna Limestone Forest Hill Sand Red Bluff Clay	Paynes Hammock Sand Chickasawhay Limestone Vicksburg Group: Byram Formation Marianna Limestone	Absent Absent Red Bluff Clay	Chickasawhay Limestone Suwannee Limestone Byram Formation Marianna Limestone		Flint River Formation Suwannee Limestone	Flint River Formation / Cooper Marl (upper part)	Undifferentiated	Undifferentiated
	Upper Eocene	Jackson Group: Yazoo Clay Moodys Branch Formation	Jackson Group: Yazoo Clay Moodys Branch Formation	Ocala Limestone	Ocala Limestone		Cooper Marl (lower part) Ocala Limestone (Lower part grades into Barnwell Formation up dip)	Barnwell Formation (up dip) Cooper Marl (lower part)	Absent	Chickahominy Formation[2]
	Middle Eocene	Claiborne Group: Cockfield Formation Cook Mountain Formation Sparta Sand Zilpha Clay Winona Sand Tallahatta Formation	Claiborne Group: Gosport Sand Lisbon Formation Tallahatta Formation			Avon Park Limestone Tallahassee Limestone[2] Lake City Limestone[2]	Claiborne Group[3]: Gosport (?) Sand Lisbon Formation (McBean Formation up dip) Tallahatta Formation	McBean Formation (up dip) Santee Limestone Warley Hill Marl (up dip) Congaree Formation (up dip)	Castle Hayne Limestone Absent	Absent Nanjemoy Formation
	Lower Eocene	Wilcox Group: Hatchetigbee Formation Tuscahoma Sand Nanafalia Formation	Wilcox Group: Hatchetigbee Formation Tuscahoma Sand Nanafalia Formation		Undifferentiated	Oldsmar Limestone[2]	Wilcox[3] Group: Hatchetigbee Formation Tuscahoma Sand Nanafalia Formation	Black Mingo Formation	Absent	Aquia Formation
	Paleocene	Midway Group: Naheola Formation Porters Creek Clay[4] Clayton Formation	Midway Group: Naheola Formation Porters Creek Clay Clayton Formation	Absent	Undifferentiated	Cedar Keys Limestone[2]	Clayton Formation	Undifferentiated	Beaufort Formation	
Cretaceous	Upper Cretaceous	Selma Group: Prairie Bluff Chalk Ripley Formation Demopolis Chalk Coffee Sand[6] Mooreville Chalk Eutaw Formation McShan Formation Tuscaloosa Group: Gordo Formation Coker Formation[2]	Selma Group: Prairie Bluff Chalk Ripley Formation Demopolis Chalk Mooreville Chalk Eutaw Formation McShan Formation Tuscaloosa Group: Gordo Formation Coker Formation	Providence Sand Ripley Formation Demopolis (?) chalk Mooreville (?) Chalk and Blufftown Formation Eutaw Formation Undifferentiated[2]	Undifferentiated[2]		(Much as in southeastern Alabama, northeastward the post-Tuscaloosa Cretaceous Formations do not crop out and are difficult to differentiate in the subsurface.) Tuscaloosa Formation	Peedee Formation Black Creek Formation (Middendorf Formation up dip) Ellenton Formation[2] Cape Fear Formation	Peedee Formation Black Creek Formation (Middendorf Formation up dip) Cape Fear Formation	Mattaponi Formation[2]
	Lower Cretaceous	Undifferentiated[2]	Undifferentiated[2]				Undifferentiated[2]	Undifferentiated[2]	Undifferentiated[2]	Potomac Group

[1] In southeastern Mississippi only.
[2] Does not crop out.
[3] In southeast Georgia only.
[4] The Meridian Sand Member of the Tallahatta Formation and the uppermost sand beds of the Wilcox Group form the Meridian-upper Wilcox aquifer.
[5] The basal sand beds of the Wilcox Group and sand beds in the Naheola Formation form the lower Wilcox aquifer.
[6] In northeastern Mississippi only.

(Shaded pattern) Fresh-water aquifers

(Hatched outline) Tertiary limestone aquifer. Referred to in Florida as the Floridan aquifer; also has been referred to in Alabama, Florida, Georgia, and South Carolina as the principal artesian aquifer (Stringfield, 1966).

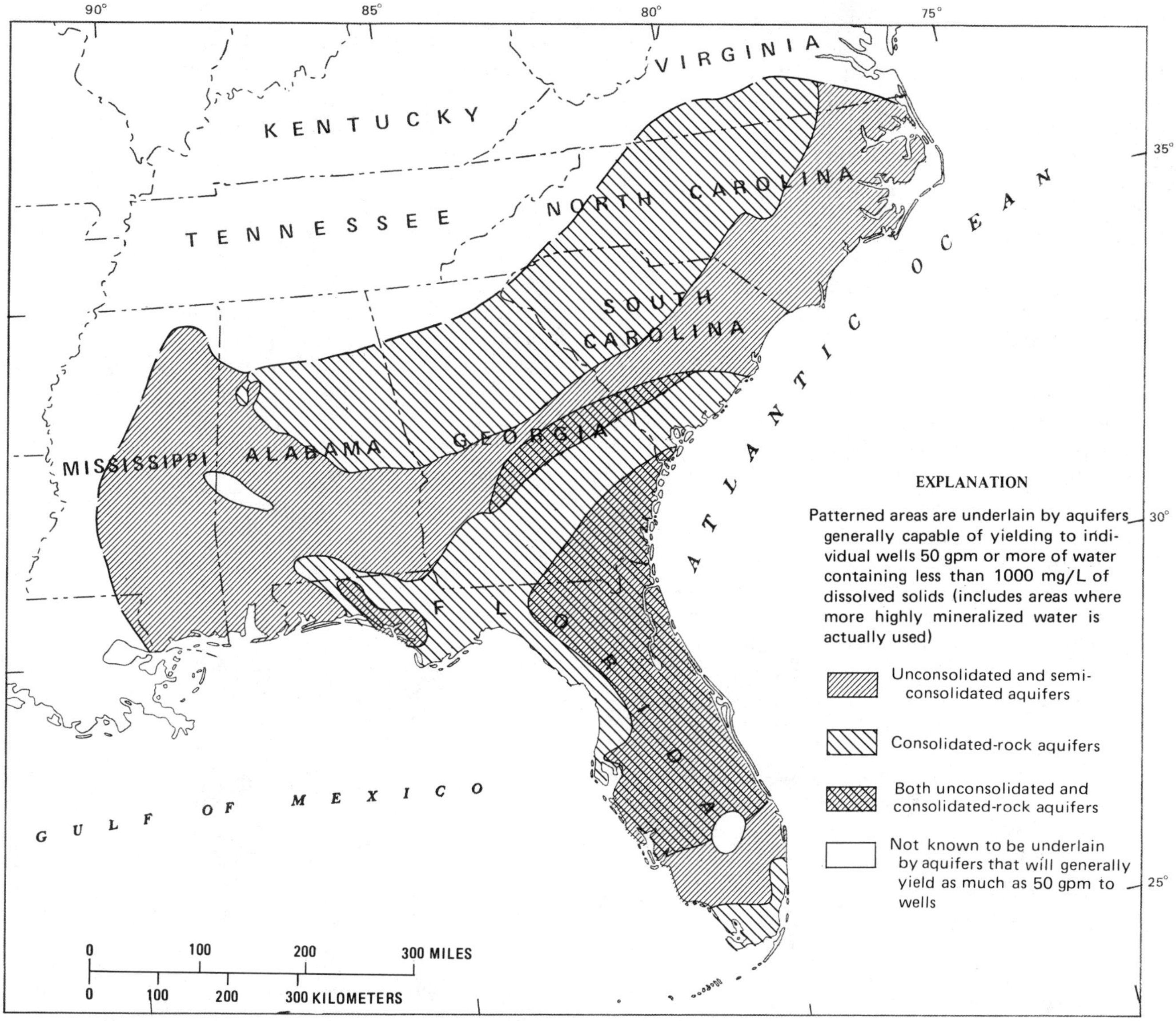

FIGURE 5.—Types of aquifers in the South Atlantic-Gulf Region.

Many investigators of the water resources in areas within the Mid-Atlantic Region have estimated the percentage of streamflow that is base flow, or ground-water discharge. In the Coastal Plain, base flow ranges from about 40 to 95 percent of the total streamflow; a conservative average would be about 55 percent. In that part of the region underlain by consolidated rocks, base flow ranges from about 25 to 90 percent of the streamflow, a conservative average being about 40 percent.

Wilson and Iseri (1967) show that the total annual streamflow from the South Atlantic-Gulf Region averages about 303,000 ft^3/s or 196,000 Mgal/d, 60 percent of which flows into the Gulf of Mexico. However, the proportion of this streamflow that originates in the Coastal Plain as opposed to the remainder of the region was not determined. Using only 40 percent (the conservation average percent of streamflow that is baseflow in the consolidated rocks part of the region) and applying this figure to the entire streamflow from the region, the average total yield of ground water from the South Atlantic-Gulf Region under present hydrologic conditions is at least 78,000 Mgal/d or about 286,000 (gal/d)/mi^2—the equivalent of about 6 inches of annual recharge. It is impractical and, in fact, environmentally undesirable to intercept all of this water before it is discharged into streams; however, the amount indicates the order of magnitude of ground water available as a potential resource.

There is also in storage in the ground-water reservoirs of the region about 600,000 billion cubic feet of water having a dissolved-solids concentration of less than 3,000 mg/L (M. I. Kaufman, oral commun., 1977). This volume of water is sufficient to cover the entire region to a depth of about 80 feet; however, much of this water cannot be withdrawn economically using present techniques. Moreover, the environmental effects of removing excessive amounts of ground water from storage may not be acceptable.

The hydrologic properties of an aquifer provide a means of measuring or predicting the aquifer's water-supply potential. In this regional appraisal of ground-water resources, data are not available covering the regional variation in physical properties and hydrologic characteristics, and instead ranges or averages for hydraulic properties are used. Terms used to describe hydraulic characteristics in this study include the following:

Transmissivity.—The rate at which water is transmitted through a unit width of an aquifer under a unit hydraulic gradient. Depends on the hydraulic conductivity and thickness of the aquifer. Expressed in units of square feet per day (ft^2/d).

Hydraulic conductivity.—The volume of water that will move in unit time under a unit hydraulic gradient through a unit area measured at right angles to the direction of flow. Expressed in units of feet per day (ft/d).

Storage coefficient.—The volume of water that an aquifer releases from or takes into storage per unit surface area of the aquifer per unit change in head—dimensionless.

The values of transmissivity and hydraulic conductivity as defined above expressed in square feet per day and feet per day may be converted to units of transmissibility (gallons per day per foot) and permeability (gallons per day per square foot) by multiplying by 7.48. Explanations of the derivation and application of values for hydrologic properties are given by Lohman (1972).

PRESENT AND PROJECTED GROUND-WATER USE

The withdrawal of fresh and saline water for all uses from all sources in the South Atlantic-Gulf Region in 1975 was reported to be 43,000 Mgal/d. Surface water accounted for 38,000 Mgal/d or about 88 percent of the total, and ground water accounted for 12 percent. Consumptive use was 3,700 Mgal/d (Murray and Reeves, 1977, table 17).

Ground-water withdrawals in the region for all purposes in 1975 totaled about 5,500 Mgal/d (Murray and Reeves, 1977). The largest withdrawal, 1,953 Mgal/d, was for self-supplied industrial use. Public water supplies accounted for 1,500 Mgal/d, and rural and livestock, 660 Mgal/d. About 53 Mgal/d or 1 percent of the total was saline ground water used by self-supplied industries. Included in the total is 1,500,000 acre-ft of water withdrawn for irrigation (about 1,340 Mgal/d on an annual basis).

Ground water supplies about 48 percent of all water used for public supplies and 42 percent of all irrigation water in the region. The total of ground-water withdrawal is only about 7 percent of the estimated 78,000 Mgal/d available from the region.

Projections for future water needs made by the Water Resources Council (1968, table 1–2) are shown in the following table:

Projected water requirements, South Atlantic-Gulf Region

[Million gallons daily. From U.S. Water Resources Council, 1968]

Withdrawals, surface and ground water			*Consumptive use*		
1980	*2000*	*2020*	*1980*	*2000*	*2020*
53,180	87,440	130,190	3,395	5,655	8,265

Assuming that ground water continues to supply 14 percent of all water withdrawn in the region, projected ground-water use would be as shown in the following table:

Projected ground-water requirements based on the 1970 rate

[Million gallons daily]

Projected withdrawals			*Projected consumptive use*		
1980	*2000*	*2020*	*1980*	*2000*	*2020*
7,445	12,242	18,227	475	792	1,157

If one-fourth, or about 19,000 Mgal/d, of the estimated available ground-water supply can be withdrawn without serious environmental effects, the projected regional ground-water needs through 2020 can be met.

Large-capacity wells can be constructed in perhaps 40 percent of the region (fig. 5); hence, most of the ground-water development in the region will probably be in the most favorable area.

GROUND-WATER QUALITY

In the South Atlantic-Gulf Region good-quality ground water is generally taken for granted. In most of the region enough water of good quality is available for current requirements; however, simple treatment for excessive iron, corrosiveness, and hardness is common.

In this report the following classification is used for water hardness in terms of the amount of calcium carbonate (or its equivalent) that it contains:

[Values in milligrams per liter]

0–60	Soft.
61–120	Moderately hard.
121–180	Hard.
More than 180	Very hard.

In the sandy coastal plain formations there is normally a progressive increase in the mineral content of ground water downdip. Along the outcrop of the aquifers the water is low in mineral content but may be somewhat corrosive, and water from many wells is high in iron.

The high chloride concentrations in water in some Upper Cretaceous aquifers appear to be related to the permeability of the sediments. In places, saltwater has been replaced by freshwater that has moved farther downdip in high permeable sediments than in the shallower but less permeable sediments. The movement of water is slower in less permeable sediments.

Freshwater contains less than 1,000 mg/L of dissolved solids; however, some water supplies use water that has a dissolved-solids concentration higher than 1,000 mg/L, and some water containing up to 2,000 mg/L is suitable for irrigating some crops. In this report saline water is classified as follows according to the dissolved-solids concentration:

Description	*Dissolved solids mg/L)*
Fresh	Less than 1,000.
Slightly saline	1,000 to 3,000.
Moderately saline	3,000 to 10,000.
Very saline	10,000 to 35,000.
Brine	More than 35,000.

The younger Tertiary aquifers contain freshwater that extends to or even beyond the coast in most areas (Hathaway and others, 1976). In much of the coastal area the Tertiary limestone aquifer is protected from seawater intrusion by a cover of Miocene clays and marls of low permeability. However, deep saline water has migrated into the freshwater zone in a few places in response to sharp lowering of pressure.

The Pleistocene and Holocene unconsolidated sediments commonly yield slightly acidic water that is low in dissolved solids.

Colored ground water occurs in a few places in Florida where turbid stream waters can move into the aquifers. In Mississippi, some Eocene aquifers yield colored water in their downdip extensions.

Outside the Coastal Plain, water is generally limited in quantity but of good quality. As the depth of ground-water occurrence commonly does not exceed a few hundred feet and the aquifers have long been free of seawater, the high salinity that is common in the deep zones of the Coastal Plain aquifers is not a problem. The aquifer materials with the exception of the carbonate rocks are relatively insoluble, and therefore the dissolved-solids concentration in the ground water is low.

In moving downdip, the slightly acidic water becomes neutralized, iron content decreases, and the hardness of water increases. Because of ion exchange at greater depth, the water becomes a soft sodium bicarbonate type (Cederstrom, 1957, p. 36). Still farther downdip the water becomes saline owing to the presence of seawater with which the formations were once saturated.

A similar progression is seen in water in some limestone aquifers. In those aquifers water may be initially harder than that in sandy formations, but hardness does not everywhere become excessive downdip owing to base-exchange action. Stringfield (1966) notes that although raw water for public supplies in Florida ranges up to 1,000 mg/L of hardness, only 9 percent of those supplies has a hardness greater than above 250 mg/L. Bicarbonate, however, may increase above 300 mg/L.

Low bicarbonate concentrations may be related to rapid recharge through sinkholes and solution openings by water having a low carbon dioxide concentration, whereas high bicarbonate concentrations may be expected where recharge water is high in carbon dioxide as a result of percolation through soil that is rich in organic matter (Trainer and Heath, 1976).

Although the quality of ground water in the Coastal Plain is generally good, contamination problems exist or may develop in certain areas. In most places ground water does not receive the amount of contamination that streams do. Surface contaminants tend to be degraded, diluted, and filtered before reaching the aquifer. Water moving miles downdip through sandy aquifers is almost everywhere free of organic contamination.

Ground water can be degraded, however. Heavy application of fertilizers, a concentration of septic tanks, sanitary landfills, or feed lots are sources of organic contaminants that may pollute both ground water and streams.

In some industrial areas, heavy metals or nondegradable contaminants may be present in waste water entering the ground-water reservoir. These are not filtered out by movement through sandy beds.

Limestone terranes that are not protected by a thick sandy or silty soil cover are particularly susceptible to contamination from the surface. Although such water may be diluted, no filtering action occurs, and contaminants may persist for miles.

GROUND WATER IN THE COASTAL PLAIN PROVINCE

Except in the Florida Peninsula, the Coastal Plain strata are generally present as a series of overlapping beds that dip gently seaward. The exposed inland outcrop areas absorb rainfall that moves downgradient, ordinarily seaward. Water accumulates in interfluve areas and moves slowly to the streams in response to gravity. In such areas, water occurs under unconfined (water-table) conditions.

Downdip the water is confined within the aquifer by permeable overlying material, such as clay, and is under pressure from the water filling the aquifer back to the outcrop area. Water will rise in wells drilled in such an aquifer, and the water occurs under confined (artesian) conditions.

This simple explanation suits the confined systems as a whole and is applicable without notable qualifications throughout the Coastal Plain. In much of the Florida Peninsula and in some smaller areas, recharge is local rather than at "inland" exposures of the formations.

A part of the potential recharge water in the outcrop area of the formations is rejected and becomes part of the sreamflows. Part of the water entering these strata, however, moves downdip into the confined parts of the aquifers. Some of the recharge that is not discharged as base flow to streams in the outcrop area leaks upward through confining beds downgradient from the outcrop area (Siple, 1967b).

Numerous aquifers are present in the Coastal Plain province of the South Atlantic-Gulf Region. They vary in areal distribution and extent, and some differ in name from one State to another even though they are lateral equivalents (table 1). Perhaps the most prolific aquifer is the Tertiary limestone aquifer which crops out or is near the surface in Florida, southern Alabama, central Georgia, and southeastern South Carolina and is made up of Paleocene to Miocene limy formations that function as one aquifer. The sand aquifers in the Tuscaloosa Group and correlative units of Cretaceous age crop out generally along the Fall Line and extend seaward beneath the younger formations from North Carolina to Mississippi. In the Western part of the region the thick permeable sand beds of the Miocene Series form an extensive aquifer system.

CRETACEOUS AQUIFERS

The Cretaceous deposits of the region have an outcrop area of about 40,000 mi^2 from Mississippi to Virginia (fig. 4) and extend seaward under a cover of younger sediments. Although the downdip limit of freshwater (water containing less than 1,000 mg/L dissolved solids) in the Cretaceous aquifers is not known in parts of the region, the areal extent of freshwater in these aquifers probably exceeds 70,000 square miles.

Except in North Carolina and Virginia, the Lower Cretaceous sedimentary rocks (including the aquifers) are overlapped in the subsurface and do not crop out. These rocks contain salty water in the region except in North Carolina and Virginia and in a limited area in Mississippi (where the quality of water is inferred from geophysical logs). At Franklin, Isle of Wright County, Va., industrial and municipal pumpage from Lower Cretaceous deposits has caused water-level declines over a 1,300 square-mile area that extends into North Carolina (Brown and Cosner, 1974). Yields of individual wells are as high as 2,500 gal/min in Virginia, and the aquifer is a source of large ground-water supplies in North Carolina.

Upper Cretaceous formations crop out along the inner edge of the Coastal Plain from Mississippi to North Carolina. The Tuscaloosa aquifers, which extend over 500 miles from Mississippi to North Carolina, form one of the more extensive ground-water reservoirs in the region (fig. 6). In Mississippi and southwestern Alabama the Tuscaloosa Group consists of the Coker and Gordo Formations. In Georgia, units within the Tuscaloosa are not differentiated. The correlative units of the Tuscaloosa in North and South Carolina are the Cape Fear Formation and, in the updip areas, the overlying Middendorf Formation. Yields of 1,000 gal/min from the Tuscaloosa aquifers are common, and, where the full thickness is developed, yields of 3,000 gal/min may be possible.

In Mississippi and northwestern Alabama, the McShan and Eutaw Formations overlie the Tuscaloosa Group. Sand beds in these formations are fair aquifers in some areas.

Above the Eutaw Formation, except in east-central Mississippi, are sand beds in the Selma Group. In northern Mississippi, these beds are in the Coffee Sand and the Ripley Formation. In eastern Alabama and western Georgia, the Selma includes in ascending order the Blufftown Formation, Ripley Formation, including the Cusseta Sand Member, and Providence Sand. These units are only fair aquifers near their outcrop, but downdip in Alabama high yields are available from combinations of these strata. In eastern Georgia, South Carolina, and North Carolina, the aquifers correlative with Selma sequences are the Black Creek and Peedee Formations.

The yields from wells in the Cretaceous aquifers will vary from place to place in each aquifer and from aquifer to aquifer. Table 2 shows the maximum reported yield from each aquifer in several counties in each State. These values do not imply that such yields can be obtained everywhere in the aquifer or that such yields are the maximum possible yields.

The highest average transmissivities for Cretaceous aquifers are those for the Tuscaloosa Group (Formation). The Coker Formation is an extremely productive aquifer in eastern Mississippi where the highest transmissivity, 80,200 ft^2/d, was reported for a well 1,857 feet deep in Noxubee County, Miss. (Newcome, 1971, p. 36). Results

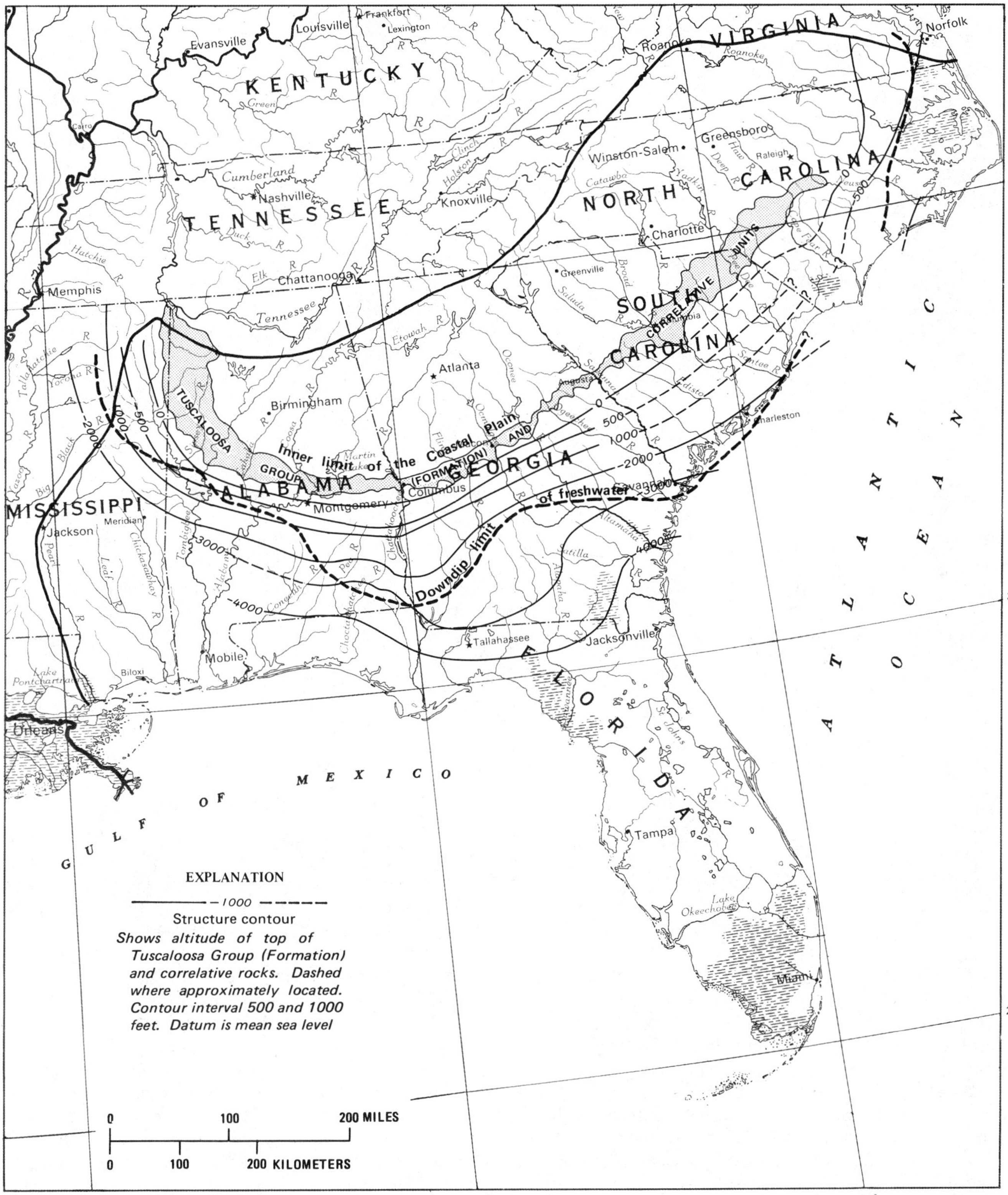

FIGURE 6.—Outcrop area, depth below sea level to the top, and downdip limit of freshwater in the Tuscaloosa Group (Formation) and correlative rocks.

TABLE 2.—*Maximum reported yields of wells in Upper Cretaceous aquifers, Mississippi to North Carolina*

State and county	Yield (gal/min)	Drawdown (ft)	Aquifer
Alabama:			
Autauqa	500	—	Coker.
Barbour	350	—	Blufftown-Providence Sand.
Bullock	515	212	Gordo.
Dale	[1]1,400	—	Ripley.
Greene	11,000	—	Coker.
	11,000	—	Gordo.
	460	32	McShan.
Houston	700	—	Providence Sand.
Montgomery	1,000	—	Coker.
	200	—	Gordo.
Pickens	640	55	Gordo.
	600	38	McShan.
Tuscaloosa	500	—	Coker.
Georgia:			
Bibb	500	40	Blufftown-Providence Sand.
Houston	1,850	—	Do.
Macon	1,230	70	Do.
Peach	656	23	Do.
Screven	870	flow	Providence Sand-Clayton.
Sumpter	800	—	Blufftown-Providence Sand.
Terrell	1,000	—	Providence Sand-Clayton.
Mississippi:			
Chickasaw	600	—	Eutaw-McShan.
Clay	900	75	Gordo.
Lee	770	43	Eutaw-McShan.
Lowndes	600	—	Gordo.
	2,300	flow	Coker.
Monroe	1,600	—	Gordo.
Noxubee	1,200	57	Coker.
Oktibbeha	1,000	42	Gordo.
North Carolina:			
Bertie	550	—	Tuscaloosa (?).
Bladen	650	—	Black Creek.
	700	—	Peedee.
Colombus	500	—	Do.
Craven	2,000	57	Tuscaloosa(?)-Black Creek.
Duplin	700	71	Black Creek-Peedee.
	750	90	Black Creek.
Greene	550	122	Tuscaloosa(?).
Hertford	1,000	67	Do.
Hoke	900	—	Do.
Johnston	300	—	Do.
Lenoir	1,300	—	Black Creek.
	1,000	52(?)	Tuscaloosa(?)-Black Creek.
Martin	700	—	Black Creek-Peedee.
Onslow	500	80	Black Creek-Peedee(?).
Pender	600	40	Black Creek.
Pitt	380	—	Peedee.
	1,100	—	Tuscaloosa(?).
			Black Creek(?).
	800	—	Lower Cretaceous.
			Tuscaloosa(?)-Black Creek.
Robeson	775	—	Tuscaloosa(?).
	800–1,000		Do.
Sampson	650	—	Black Creek.
Wayne	720		Do.
South Carolina:			
Aiken	3,300	122	Upper Cretaceous (Tuscaloosa?).
Beaufort	100	flow	Lower Black Creek-Tuscaloosa.
Colleton	1,200	flow	Black Creek(?)-Tuscaloosa.
Hampton	500	flow	Upper Cretaceous-Peedee-Black Creek(?).
Horry	1,200	120	Black Creek.
Sumter	1,690	—	Upper Cretaceous (Tuscaloosa?).
Williamsburg	1,100	—	Peedee-Black Creek.

[1]Estimated.

of about 30 tests in Cretaceous aquifers in Mississippi indicate an average transmissivity of 1,200 ft²/d for the Eutaw and McShan Formations and 6,500 ft²/d for the Gordo Formation (Newcome, 1971). Similar conditions can be expected in adjoining parts of Alabama. Siple (1960, p. 165) reported a transmissivity of 26,700 ft²/d for a well developed in his "Tuscaloosa Formation" (Cape Fear Formation) in South Carolina, a value accepted by Callahan (1964, p. 10) for use in evaluating the Tuscaloosa aquifer in Georgia and adjoining states.

In northeastern North Carolina transmissivities in Cretaceous aquifers range from 190 to 27,000 ft²/d (Lindskov, 1972, table 4). A reasonable value for the average transmissivity is 10,000 ft²/d. In the opinion of R. C. Heath (written commun., 1977), most of the water withdrawn from the Cretaceous aquifers in North Carolina is derived from compression of the aquifer, expansion of the water, and drainage of the confining beds. In northeastern South Carolina, the hydraulic conductivity averages about 30 ft/d (A. L. Zack, oral commun., 1977).

Water in and near the outcrops and in the updip parts of the Cretaceous aquifers is generally low in dissolved solids, high in iron, and slightly acid (corrosive). Although much higher concentrations of iron have been reported in places in Mississippi and Alabama, 2 or 3 mg/L is common in most areas. Downdip the water becomes neutral or slightly alkaline, and the iron content decreases sharply.

Water from some wells in North Carolina is hard, presumably because recharge is by seepage down through overlying limy Miocene sediments. Samples high in bicarbonate and low in hardness from deeper wells in the same area have presumably traveled for some distance downdip from their point of origin. In South Carolina, water from the Black Creek aquifer is characterized by low iron and high fluoride.

At varying distances downdip and at depth, water in the Cretaceous aquifers is highly mineralized owing to the fact that the sediments were once saturated with saline waters.

LOWER TERTIARY AQUIFERS

TERTIARY LIMESTONE AQUIFER

The Tertiary limestone aquifer, the most extensive and widely used aquifer in the South Atlantic-Gulf Region, underlies an area of about 90,000 square miles which includes Florida (where it is referred to as the Floridan aquifer), southern and coastal Georgia, and adjacent parts of Alabama and South Carolina (fig. 7). Local names for this aquifer are given in table 1. The aquifer is the source of some of the largest ground-water supplies in the United States (Stringfield, 1966, p. 1). It has been commonly referred to as the "principal artesian aquifer," although in many areas it is an unconfined aquifer. The aquifer includes parts of several geologic units (table 1) that are principally Tertiary limestone but contain intercalated beds and lenses of dolomite, sand, silt, marl, and clay in which porosity and permeability differ greatly from place to place. The aquifer thickness increases in a seaward direction from about 50 feet in its

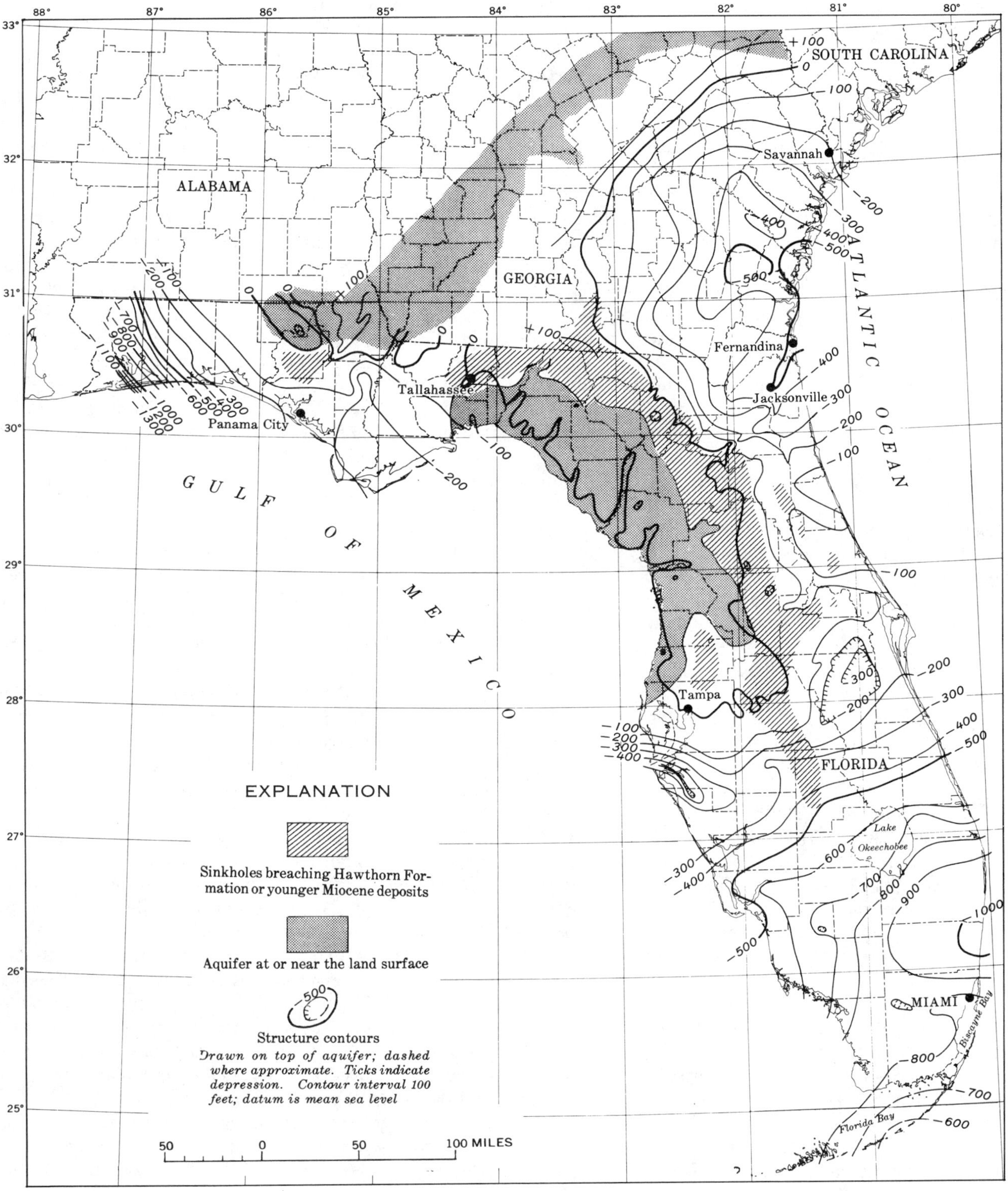

FIGURE 7.—Distribution and structure contour map of the top of the Tertiary limestone aquifer. (From Stringfield, 1966, fig. 23).

updip areas in Georgia to more than 2,500 feet in parts of Florida (Klein, 1971).

Most of the massive limestone of Tertiary age has been removed by erosion from the Cap Fear Arch. Nevertheless, the Tertiary limestone of North Carolina is considered a hydrologic equivalent of this aquifer.

The several limestone units constituting the aquifer (table 1) are principally of marine origin and generally grade into or interfinger with clastic deposits toward the continental landmass to the north from which the clastics were derived. These peripheral clastic facies are not considered part of the aquifer. The aquifer is near the surface and in hydraulic continuity with the unconfined shallow aquifers over an extensive area where overlying rocks of lesser permeability are perforated by innumerable sinkholes, particularly around the Ocala Uplift (fig. 3). Elsewhere the impermeable clays of the Hawthorn Formation overlie the aquifer; however, this formation is breached in a few places, as near Parris Island and Brickyard Point, Beaufort County, South Carolina.

Ground-water development is extensive, including withdrawal of hundreds of millions of gallons per day to meet the water-supply needs of numerous major cities, such as Savannah, Jacksonville, and St. Petersburg; the phosphate, pulp, paper, and allied chemical and other industries; and for citrus and other crop irrigation. Throughout the area, the aquifer yields water to tens of thousands of domestic, industrial, and irrigation wells, but most of the large industrial supplies are developed in coastal areas.

The wide range in hydraulic characteristics in the Tertiary limestone aquifer reported by Stringfield (1966, p. 101–109) is representative of an aquifer composed of sediments that range from interbedded sand and clay to cavernous limestone. The reported transmissivities range from 3,700 to 134,000 ft^2/d, hydraulic conductivities from 35 to 334 ft/d, and storage coefficients from 0.00019 to 0.021. These values do not indicate the true potential; for example, Callahan (1964, table 4) estimated transmissivity values in Georgia as high as 268,000 ft^2/d for his determinations of ground-water flow through the aquifer.

Large springs having yields ranging up to hundreds of millions of gallons per day from the Tertiary limestone aquifer are numerous in Florida and Georgia. Silver Springs and Rainbow Springs in central Florida are the largest, their combined flow averaging about 1,000 Mgal/d (Rosenau and Faulkner, 1974). Well yields of 2,000 gal/min are common. Locally, yields up to 11,000 gal/min are known, and a natural flow of 20,000 gal/min is reported for a well in central Florida (Pascale, 1975). The base flow of streams is well sustained as a result of ground-water discharge through springs.

Under natural conditions, the aquifer contains potable water as defined by Klein (1971) to depths of more than 2,000 feet in parts of the area (fig. 8). The extent and thickness of the freshwater body is dependent upon the rainfall and the geologic and hydrologic properties of the rocks.

Since the last emergence of the land from the sea, the aquifer has been flushed of seawater in most areas. In southern Florida and in places along the eastern seaboard where the potentiometric head is low, the aquifer has not been completely flushed, and the water is saline. Increases in salinity with depth are commonly abrupt immediately below an impermeable stratum underlying a freshwater zone.

In the Georgia and South Carolina coastal plains, water at depth inland is fresh through the full thickness of the aquifer. At Savannah, Ga., the section is about 600 feet thick, but 70 percent of the water produced in that area comes from two upper zones that have a combined thickness of about 100 feet (McCollum and Counts, 1964). Saline water occurs in the lower part of the aquifer at Hilton Head Island, S. C., and the entire aquifer is saline in most of Beaufort County, S. C. (L. R. Hayes, oral commun., 1977). At Brunswick, Ga., the 500- to 1,000-foot zone is fresh; saline water is present from 1,000 to 1,400 feet and low chloride water occurs from 1,400 to 1,700 feet (R. L. Wait, 1965, p. 67).

The freshwater in the aquifer is generally hard. In the recharge area, the water may contain less than 20 mg/L hardness as calcium and magnesium bicarbonate, but almost everywhere downdip the hardness is generally in excess of 100 mg/L.

At centers of heavy pumping, lowered head tends to induce upward flow of deeper, possibly saline, water. Many wells have been drilled through several zones of the aquifer, thereby providing direct flow paths for deeper more highly mineralized water to migrate upward (or laterally) and contaminate the freshwater reservoir. Upward movement may occur in a short time and is, therefore, the greater hazard.

In southwestern Florida, the mineral content of ground water along the Peace River is high. Kaufman and Dion (1967) suggested that here deep ground water rises "along a linear zone (fault?) of greater vertical permeability." In eastern Florida, "upward leakage of highly mineralized water from deep artesian zones" is in part "along fault zones" (Klein, 1971).

Saline water is moving laterally from points of vertical leakage into well fields at Brunswick, Ga. (Wait and Gregg, 1967). At Jacksonville, Fla., saline water has migrated into the lowermost zone of the Tertiary limestone aquifer, thus eliminating it as a source of potable water.

Contamination by seawater is possible in coastal areas where the overlying confining bed is breached. Seawater is presently migrating toward the Savannah wells from the Hilton Head Island, S.C., area (Counts and Donsky, 1963, p. 72); however, under existing hydraulic conditions, salt water will require many years to reach the center of heaving pumping.

LOWER TERTIARY AQUIFERS OTHER THAN THE TERTIARY LIMESTONE AQUIFER

The Paleocene, Eocene, and Oligocene sediments that make up the lower Tertiary aquifers (exclusive of the Tertiary limestone aquifer) in the region have an outcrop area of 45,000 square miles from Mississippi to Virginia (fig. 4). They dip seaward under a cover of younger sed-

FIGURE 8.—Depth to the base of potable water in the Tertiary limestone aquifer (Floridan aquifer) in Florida.

iments. Although the individual lower Tertiary aquifers are not as extensive areally as the Cretaceous aquifers, they are major sources of ground water where they occur in the region (table 1).

The lower Tertiary aquifers include limestone beds that grade laterally into the Tertiary limestone aquifer (table 1). These strata have poor hydraulic connection with one another, and development of maximum supplies in any one place would require either several wells ending in different aquifers or a well screened at various levels.

Without exception, water in the lower Tertiary aquifers is fresh in the area of outcrop and for a few tens of miles downdip, beyond which the water is saline. Maximum development at present is ordinarily confined to whatever productive stratum lies nearest the surface. Additional supplies are available at greater depths in many places, but in other places the deeper beds yield only saline water.

The lower Tertiary aquifers in Mississippi, Alabama, and western Georgia (table 1) are highly productive. Yields ranging up to 2,000 gal/min to individual wells are common. A well field in Tertiary limestone yielded as much as 15 Mgal/d at Ft. Gaines, Ga. (Callahan, 1964, p. 19). Well yields of 2,000 gal/min or more could be reasonably expected from efficiently constructed wells at many places in the Pascagoula River basin in Mississippi (Newcome, 1967b, p. 1). Deeper parts of some of the Tertiary aquifers in Mississippi, in areas where freshwater extends to depths of 3,000 feet (fig. 9), remain virtually untapped and are potential sources of future water supplies.

The Castle Hayne Limestone, separated from the Tertiary limestone aquifer by erosion across the Cape Fear Arch, is the principal source of water supplies in the coastal plain area of North Carolina. Withdrawals of water from the Castle Hayne in Beaufort County, N. C., for mining exceeded 60 Mgal/d in 1974 (North Carolina Department of Natural Resources, 1974).

The lower Tertiary aquifers exhibit a very wide range in hydraulic properties. The estimated average transmissivity for the Castle Hayne Limestone in North Carolina is about 25,000 ft^2/d (R. C. Heath, written commun., 1977), and the range for the lower Tertiary aquifers in South Carolina is from 350 to 13,400 ft^2/d (C. A. Spiers, oral commun., 1977). Callahan (1964, p. 19) reported that transmissivities ranged from 940 to 20,000 ft^2/d in Georgia.

A compilation of about 110 aquifer tests made in lower Tertiary aquifers in Mississippi (Newcome, 1971) shows a range in hydraulic conductivity from near 1 to about 240 ft/d. Corresponding values for transmissivity in an aquifer 100 feet thick would be as much as 24,000 ft^2/d. The average transmissivity for a 100-foot aquifer would be about 6,400 ft^2/d. Values for lower Tertiary aquifers in Alabama are estimated to be about the same.

UPPER TERTIARY AQUIFERS

Miocene and Pliocene sediments make up the upper Tertiary aquifers in the South Atlantic-Gulf Region. These sediments crop out over an area of about 70,000 square miles from Mississippi to Virginia (fig. 4).

Miocene beds are widely distributed in the region and extend from Mississippi to Virginia, although absent over the Cape Fear Arch. From Georgia to central South Carolina, Miocene limestones are significant parts of the upper Tertiary limestone aquifer.

In the Florida panhandle, Alabama, and Mississippi, the Miocene beds dip toward the south and southwest and thicken rapidly. The total Miocene section is as much as 5,000 feet thick in southwestern Mississippi, but only the higher beds contain freshwater. The depth to base of freshwater in the area (fig. 9) ranges from 500 feet near the northern limit of those formations to as much as 3,000 feet near the coast and near the mouth of Pearl River in Mississippi.

Units of Pliocene age overlie the Miocene deposits and are part of the freshwater hydrologic system of the Gulf Coastal area. The Citronelle Formation of Pliocene age is the source of water for many domestic and public-supply wells, but most large yield wells have been drilled into the underlying Miocene aquifers where water is under higher artesian pressure.

Practially all domestic and municipal water supplies and most industrial supplies along the Gulf Coast in western Florida, Alabama, and Mississippi are obtained from Miocene to Pliocene aquifers. Well yields commonly exceed 1,000 gal/min. At Pensacola, in western Florida, municipal wells yield 1,500 to 2,000 gal/min, and pumpage exceeded 17 Mgal/d in 1970. In the Pascagoula River basin in Mississippi, with careful programs of exploration and development, individual well yields of 2,000 gal/min could become commonplace, and yields of 5,000 gal/min are not unreasonable to expect in some places (Newcome, 1967b, p. 35).

The upper Tertiary aquifers are characterized by high hydraulic conductivity and transmissivity values. Lindskov (1972, table 4) reported transmissivity values ranging from 44 to 2,710 ft^2/d for aquifers in the Yorktown Formation in North Carolina. Callahan (1964, p. 47) refers to transmissivities ranging from 4,540 to 66,800 ft^2/d in northwestern Florida and southern Alabama and indicates that conditions are similar in the Atlantic Coastal area of Georgia in the Pliocene to Holocene aquifers (Callahan, 1964, p. 50). In southern Mississippi the upper Tertiary aquifers, mostly of Miocene age, are among the most productive in the State. Newcome (1971) summarized about 130 aquifer tests that indicated hydraulic conductivities ranging from 5 to 334 ft/d. The average hydraulic conductivity for all tests was 94 ft/d. Assuming that the average aquifer is 100 feet thick, the average transmissivity would be 9,400 ft^2/d. The largest transmissivity value for the Miocene aquifer system in Mississippi was 84,000 ft^2/d at a site in Harrison County (Newcome, 1971, p. 25).

There appears to be little contamination of deep aqui-

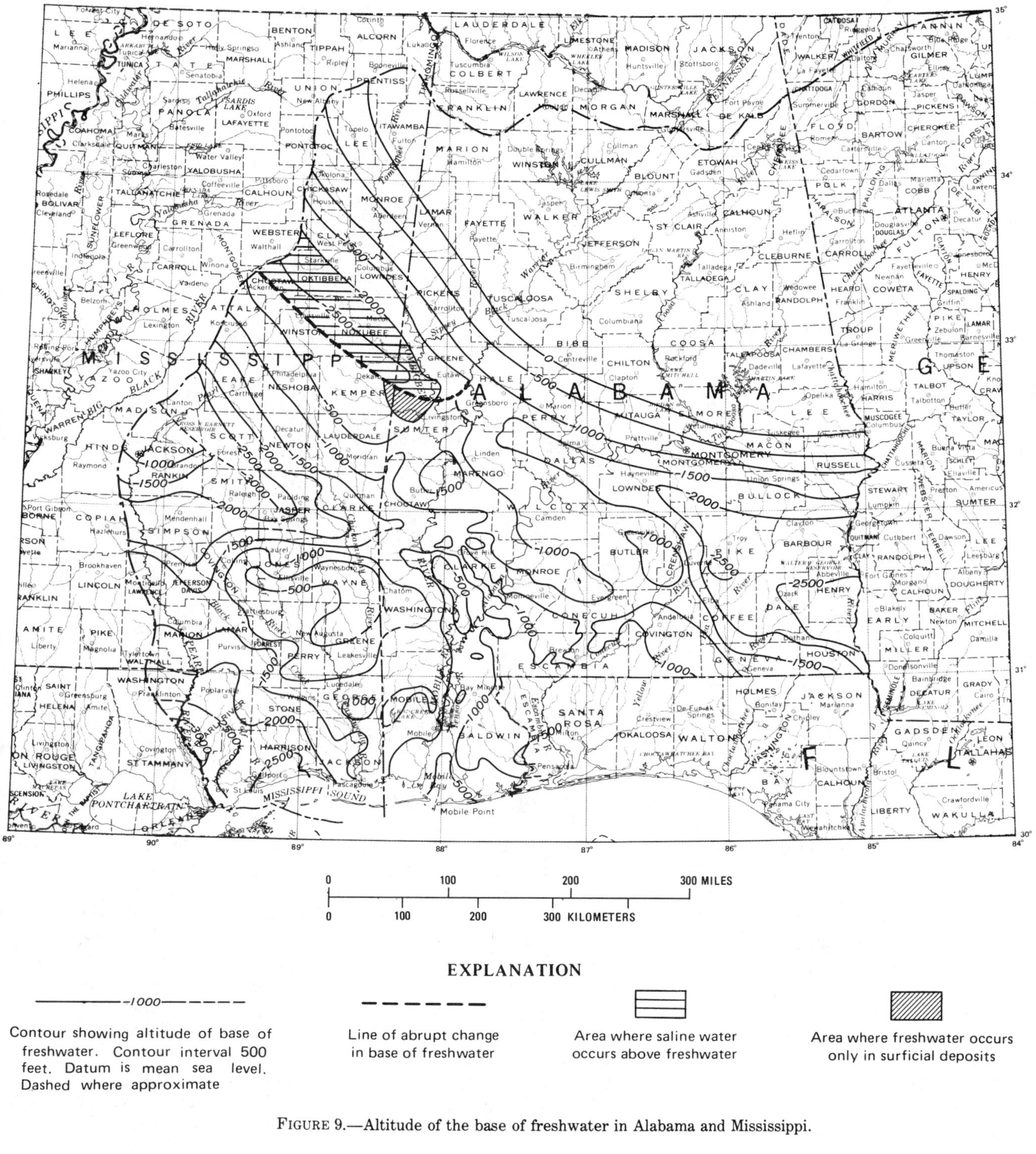

FIGURE 9.—Altitude of the base of freshwater in Alabama and Mississippi.

fers by seawater, but contamination of the shallower aquifers that are hydraulically connected to estuarine streams probably occurs in a few places (Newcome, 1967b, p. 29).

QUATERNARY AQUIFERS

Pleistocene and Holocene sediments make up the Quaternary aquifers in the South Atlantic-Gulf Region. The sediments crop out over an area of about 40,000 square miles, primarily in the coastal area from Mississippi to Virginia (fig. 4).

In the western Florida panhandle, the upper Tertiary and Quaternary units (Miocene to Holocene) constitute a large unconfined (water table) aquifer. The sand and gravel deposits are thickest to the south and west and thin to the north and east. Wells 200 to 400 feet deep, capable of producing 1,000 to 2,000 gal/min, are common. The quality of the water is excellent except for the erratic occurrence of excessive iron. The aquifer is susceptible to saltwater encroachment near estuaries.

During the drought of 1954 to 1956, the minimum discharge or ground-water outflow at 10 of 15 gaging stations in the Florida panhandle was 0.308 (Mgal/d)/mi^2 (Pride and Crooks, 1962, table 2, p. 34–35), equal to about 6.5 inches of precipitation. Thus, in the 6,500 square mile panhandle area, a minimum of about 2,000 Mgal/d would be available for development if the environmental effects of the withdrawals were acceptable.

Deposits of Miocene to Holocene age also make up an important freshwater aquifer in southwestern Florida. The aquifer is about 130 ft thick in western Collier County, but thins eastward and wedges out toward Dade and Broward Counties. Most of the wells near the Gulf are generally less than 100 feet deep, and the municipal water-supply wells are properly spaced and discharged at low rates to prevent saltwater encroachment (McCoy, 1962). Wells inland yield up to 1,300 gal/min. The quality of the water is good except for a high iron content.

The highly productive Biscayne aquifer, predominantly highly permeable limestone of Quaternary age, is the most important source of freshwater for populous southeastern Florida, including the Miami area (Parker and others, 1955). The aquifer underlies about 3,000 square miles of Dade, Broward, and parts of Palm Beach Counties. It is as much as 400 feet thick along the shoreline, but is more commonly less than 200 feet thick and thins to a featheredge 40 miles to the west. It is estimated that about two-thirds of the 60 inches annual rainfall infiltrates the zone of saturation in Dade County (Parker and others, 1955, p. 221).

Well yields commonly exceed 2,000 gal/min (Pascale, 1975). An estimated 120 billion gallons of water was pumped from a 600 mi^2 area in the coastal ridge in 1965, of which 22 billion gallons was consumptive use. Although copious supplies are available, one of the factors limiting withdrawals is the susceptibility of the aquifer to saltwater encroachment from the ocean and from tidal canals.

Prior to the 1930's, little attention was paid to the detrimental effects of the tidal canals; however, water management practices initiated in 1945 have effectively stabilized the saltwater instrusion. These are (1) the importation of freshwater through canals to maintain higher water levels, (2) the construction of salinity-barrier dams in tidal canals to prevent the intrusion of salty water and furnish additional water to infiltrate into the aquifer, and (3) the shifting of pumping to new well fields to the west.

North of Palm Beach County, a shallow sand aquifer extends along most of the coastal area. In Duval County, more than 40,000 wells produce 10 to 25 Mgal/d (Fairchild, 1972) from a shallow aquifer system consisting of Miocene to Holocene deposits.

The coastal aquifers in the Quaternary deposits (fig. 4) generally yield only small supplies to wells. They are relatively low in permeability, and the storage capacity at any one place usually is small; however, these aquifers absorb rainfall readily, and in many areas they are a perennial source of small water supplies and recharge to streams and underlying aquifers.

Where Quaternary aquifers are extensive in humid areas, both storage and recharge potential are favorable. At Elizabeth City, N. C., for instance, 14 shallow wells yield a total of 450 gal/min—more than half the city supply (R. C. Heath, oral commun. Nov. 30, 1972). The greater cost in such developments may be the cost of land and pipelines rather than that of the wells. Along some of the larger rivers in the region, alluvium as much as 75 feet thick may yield up to 1,000 gal/min. Sustained yields of this magnitude should not be expected except where wells are properly spaced and recharge conditions are favorable.

Data on hydraulic characteristics of the Quaternary deposits are limited because few aquifer tests have been made. Large wells in Quaternary aquifers are not common because of limited available drawdown, and in many localities, the water requires treatment for most uses. A noteworthy exception is the Biscayne aquifer, the major source of water supplies in southeastern Florida. Parker and others (1955, p. 2) reported that the Biscayne is among the most permeable aquifers investigated by the Geological Survey. Transmissivities for the aquifer range from 400,000 to 2,670,000 ft^2/d, and the average is 670,000 ft^2/d. The average storage coefficient is 0.20. A reasonable value for transmissivity of Quaternary deposits in North Carolina is about 750 ft^2/d (R. C. Heath, written commun., 1977). Transmissivity values for Quaternary aquifers in South Carolina range from

850 to 2,000 ft^2/d (C. A. Spiers, oral commun., 1977), and Callahan (1964, p. 50) estimated that values could be as large as 66,800 ft^2/d in the Pliocene to Holocene section in Georgia.

The chief value of the Quaternary deposits in the hydrologic system is their function of absorbing and storing large volumes of precipitation for gradual release to underlying aquifers and to streams.

GROUND WATER IN THE PIEDMONT PROVINCE

In describing the occurrence of ground water in the Piedmont and Blue Ridge provinces, LeGrand (1967, p. 8) states

> The water table, representing the top of the reservoir, generally lies in the clay, or disintegrated rock materials. In the lower part of the reservoir, water occurs in interconnecting fractures in bedrock; the fractures diminish in number and size with increasing depth. Water enters the fractures by seeping through the overlying clay, and drilled wells draw water from these fractures. The source of this water is precipitation in the general area of a well and not in some remote place.

Minor valleys generally mark zones of fracture and, in places, fault zones, and yields are significantly higher there than elsewhere (Baker, 1957). The majority of wells in the Piedmont, however, are located for convenience on high ground, commonly under the community water tank. These wells nearly always are low yielding; consequently, the Piedmont is considered an area where well yields are low (fig. 5).

LeGrand (1967, p. 9) stated that 90 percent of all ground water in the area occurs in the first 100 feet below the water table and that generally two wells 200 feet deep each will yield more water than one well 400 feet deep. Yields of up to 500 gal/min have been reported from wells in crystalline rocks (Johnson and others, 1968), but many deep wells yield less than 75 gal/min.

The source of recharge in the Piedmont is the precipitation that saturates the saprolite (weathered rock) that overlies unweathered rock. Where the saprolite cover is thick or sandy, wells will have the better yields. A well in crystalline rocks ordinarily draws water from a limited area in the general vicinity of the well, and fissure zones supplying water to the well are commonly limited in extent. Wells must be widely spaced in order to avoid mutual interference, and successive wells located along the same rock lineament will create more mutual interference than if successive wells are located at right angles to the "grain" of a rock mass. Wells near a stream may have significantly higher yields than wells distant from such a source of recharge.

The 7-day, 2-year low flow in Pickens County, S.C., ranges from 0.3 to 0.7 $(Mgal/d)/mi^2$ (Johnson and others, 1968). This measurement of ground-water discharge to the stream, at the time of low hydraulic gradients, indicates that the average annual recharge is at least 6.3 to 17.5 inches.

An example of ground-water development in the Piedmont province is at Tega Cay, S. C., where 22 wells averaging 580 feet in depth were drilled to provide water for a resort (McCall, 1972). The well sites were located on the basis of a careful evaluation of geology and hydrology. Yields ranged from less than 10 gal/min to 370 gal/min and averaged 109 gal/min. From this group, 14 wells yielding an average of 160 gal/min were selected as supply wells. Most of the water is obtained from below a depth of 450 feet. This well field is on a peninsula jutting into Lake Wylie, a possible source of recharge that may have been a factor in the high yields obtained even though the path to the wells was circuitous and the source of recharge somewhat distant. Tests for coliform bacteria in the well waters were negative, suggesting that if the lake is a source of much of the recharge, slow filtration through the fine-grained saprolite has been highly effective in removing the bacteria. However, in areas where the saprolite cover is thin, filtration of surface-water seepage would not be very effective.

GROUND WATER IN THE VALLEY AND RIDGE PROVINCE AND CUMBERLAND PLATEAU SECTION

A small part of the study area lies within the folded Appalachians, largely that area in Alabama and Georgia drained by the headwaters of the Coosa River. The underlying rocks are Paleozoic limestone, dolomite, sandstone, and shale, mostly overlain by a mantle of residual material. In the Valley and Ridge province (fig. 1), the rocks are not only crumpled by compressive stresses but are broken into several huge segments, some of which have slid over the more stable masses. Lesser displacements occur along high-angle faults. The Paleozoic rocks of the Cumberland Plateau section, adjacent to the Valley and Ridge province on the northwest, are gently warped.

Synclines (downwarps of rock strata) may be thought of as favorable locations for ground-water development in the sense that water funnels into troughs, but circulation is poor due to the compression of fractures in deep synclinal troughs and the water may be of inferior quality. Anticlines (upwarps), on the other hand, tend to crack the rock into open fractures along their crests during folding and are favorable recharge areas.

In this area of large-scale deformation, the more brittle rocks were shattered and sheared; subsequently, dissolution by percolating water enlarged the fractures in limestone and dolomite (the carbonate rocks). Fractures in sandstone, unless calcareous, are not enlarged by later solution. True (noncalcareous) shales, initially poor aquifers, scarcely improve in water-bearing character as a result of deformation and later dissolution.

The hydrology of shattered consolidated rocks is greatly different from that of the unconsolidated sandy formations. In shattered limestone, water can move nearly as fast as through a pipe, and in both fractured limestone and sandstone filtering action is lacking. The cone of depression around a well in a highly fractured rock mass enlarges quickly, and water may move rapidly to a well from long distances. Stored water in the carbonate rocks may be that filling an open cavern, an interconnected series of smaller openings, or in the unconsolidated materials overlying the limestone that is hydraulically connected to water in the caverns. Although the rate of recharge in a carbonate rock terrane overlain by a clay mantle may be low, water can flow rapidly to a well from a large surrounding area and high yields with small drawdowns are possible.

In Talladega County, Ala., the Cambrian dolomites are highly productive (Causey, 1965). One well yields 1,100 gal/min with 27 feet of drawdown and has been pumped at 1,600 gal/min. A well in Precambrian(?) marble yields 900 gal/min with 72 feet of drawdown.

Cressler (1970) notes that in northwestern Georgia, 1,000 gal/min should be available in many places from properly located wells. Near Huntsville, Ala., just north of this report area in the Cumberland Plateau section, a recently completed well 160 feet deep yielded 2,900 gal/min with 20 feet of drawdown (W. J. Powell, oral commun., March 1, 1972).

In Calhoun County, Ala., a 95-foot well in Cambrian dolomite yielded 257 gal/min with only 1.6 ft of drawdown; whereas, a 550-foot well yielded about as much with 15 feet of drawdown (Warman and Causey, 1962). It is entirely possible that drilling to 400 or 500 feet in strongly deformed areas may intersect a low-or high-angle fault from which large quantities of water may be obtained; therefore, it seems likely that deep drilling is desirable where maximum production is sought. Drilling in the most convenient locations relative to points of use can easily result in failure; whereas, sites a mile or a few miles distant that are selected on the basis of structural control may provide ample water at very low cost.

It is probable that the ground-water potential of the Valley and Ridge province is generally underrated, although many examples are known of failures to develop productive wells.

GROUND WATER IN THE BLUE RIDGE PROVINCE

In the rugged Blue Ridge province weathered rock cover (residuum) is commonly thin or absent except in valleys. Ground water generally occurs in residuum or alluvium and in fractures in the underlying igneous and metamorphic rocks. Small water supplies are available from the springs that drain the fracture systems and residuum. Well yields depend on the number and degree of interconnection of fractures penetrated by the well but commonly are only a few gallons per minute. Where all the available water is stored in fractures, the yield diminishes as stored water is pumped and the water level declines. LeGrand (1967) determined that wells are more productive and have a more stable yield where there is a thick mantle of residuum and that most interconnected fractures occur no deeper than 150 feet below the surface. Few wells produce significant quantities of water from depths greater than 300 feet.

Ground water in the province is generally of good quality. The dissolved-solids concentration is very low, and the principal quality problems are excessive iron and acidity (low pH). Aquifers in rock fractures are subject to contamination, however, in areas where the residuum is too thin to be an effective filter.

SALINE-WATER RESOURCES

Vast quantities of saline water are available in the South Atlantic-Gulf Region. This saline-water resource is mostly in aquifers in the unconsolidated rocks that underlie the Coastal Plain above the Paloezoic "basement" rocks to depths that approach 20,000 feet in southern Florida and southwestern Mississippi. All permeable deposits below the deepest freshwater (fig. 9) are saturated with saline water, some of which has a much higher dissolved-solids concentration than seawater.

Some of the dissolved minerals are of economic value; however, the most likely reason for immediate development will be utilization of slightly saline and moderately saline water. Some water containing up to about 2,000 mg/L of dissolved solids can be used for irrigation. Slightly saline water can be blended with freshwater to augment ground-water supplies for public and industrial uses. Desalting operations for public water supplies and other uses are more economical where the raw water has a low concentration of dissolved solids.

Saline aquifers offer reservoir space for injection and temporary storage of freshwater. The Norfolk, Va., injection project demonstrated that only 15 percent of the injected water was not potable upon recovery (Brown and Silvey, 1977, p. 51). Injection and storage could be used to provide a supplemental source of water to meet peak demands or as an emergency source.

DEVELOPMENT AND MANAGEMENT OF GROUND-WATER RESOURCES

OBJECTIVES

The manager has the responsibility of examining the alternatives for developing sources of supply; that is, a ground-water supply based on conventional development and management procedures as against supplies based on unconventional technologies. The planner is necessarily responsive to public expression regarding acceptable types of development and to the economic and social aims and interests of the particular States concerned. One of the major considerations confronting both manager and planner is the effect of ground-water development on the environment, particularly if a drastic lowering of water levels or large reduction in streamflow is involved. Where these constraints take precedence, dollar costs will be of minor significance in reaching final decisions.

In dealing with total available supply, ground water is not an entity separate from surface water. Much of the ground water that is not lost by evapotranspiration becomes the base flow of streams. Capture by wells of ground water moving toward a stream or induced infiltration of streamflow is highly desirable in many places but does not increase the total water supply. The increase in the ground-water supply is balanced by decrease in streamflow.

Where there is salvage of evapotranspiration loss, that salvage is an addition to the total water supply available for man's use; however, the effects on the environment must be considered. Reuse of water, the use of treated sewage effluent, or the use of poor-quality water for cooling or other purposes will conserve the total available supply. Use of ground water that would otherwise discharge into saline estuaries is an addition to the total supply in most situations.

The above mentioned management techniques are intended to show that the total available supply can be increased by utilizing aquifer storage during the low streamflow months and making up the resultant underground reservoir deficiency by capture of streamflow during the high-flow period.

Although the value of conventional approaches to many problems is granted (wider spacing of wells, for instance), emphasis here is placed on procedures that are not generally in use in the region but which have been demonstrated to be effective in other areas. Objections to any of the suggested management procedures may be based on costs; however, alternate sources would also be expensive.

NATURAL RECHARGE

The overland and ground-water runoff in the southeastern Coastal Plain is about one-third the total rainfall, roughly 15 inches, and the remainder is assigned to evapotranspiration loss and ground-water recharge (Pierce, 1966). Estimates of ground-water recharge have ranged from 1 to 11 inches (Callahan, 1964). Recharge of 6 inches would amount to about 0.29 (Mgal/d)/mi^2.

Some extreme exceptions to this generalization are highly permeable localities in areas underlain by the sandy Cretaceous formations. Callahan (1964, p. 12) states that in places the Cretaceous sand aquifers may discharge to streams the equivalent of 40 inches per year. Thomson and others (1956) showed that in one place the 7-day, 2-year flow (1937–55) of streams fed by Cretaceous aquifers was about 1.04 (Mgal/d)/mi^2—equivalent to about 22 inches of recharge.

DEVELOPMENT OF UNCONFINED SANDY AQUIFERS

Unconfined aquifers are similar to surface reservoirs in that water pumped from storage can be replenished by natural or artifical recharge. Recharge may be seasonal precipitation or it may be from another natural source (streams, lakes, or leakage from deeper aquifers). Overdraft, within limits, during dry periods leaves storage space for the surplus water available for recharge during wet periods—a condition that can be planned. During the period of high recharge, the intercepted ground-water outflow to the stream, although much greater in volume, is insignificant in relation to the total flow of moderate sized or large streams. The amount of intercepted ground-water flow to streams during low-flow periods is small because the normal ground-water contribution to the stream at that time is small. Planning the development of unconfined aquifers can involve many combinations of numbers and locations of wells and well fields relative to streams or other sources of recharge. Additional flexibility is provided by various methods of artifical recharge.

In widely separated well fields the recharge that falls upon a well field in an unconfined aquifer is not the only water available to those wells. Water flows to the pumping wells from all directions; thus the wells draw in water replenished by recharge over many square miles.

Removing water from storage in unconfined aquifers lowers the water level, thereby reducing the losses to evapotranspiration.

DEVELOPMENT OF CONFINED SANDY AQUIFERS

Only part of the recharge water in the outcrop area is removed by evapotranspiration, discharged as streamflow, or intercepted by withdrawals. Some moves downdip into the confined part of the aquifer.

Applying Darcy's Law and assuming a transmissivity of 27,000 ft^2/d and an average hydraulic gradient of 2.5 ft/mi, 0.5 Mgal/d will move downdip in a typical aquifer across each mile of the section. Increasing the hydraulic gradient in the confined aquifer (by establishing a cone of depression or by raising the head in the recharge area) would increase the downdip flow proportionately. Further application of Darcy's Law indicates that a 5-mile width of aquifer would transmit 10 Mgal/d if the hydraulic gradient were increased to 10 ft/mi.

The transmissivity of the aquifer establishes the first constraint on ground-water development. Acceptable pumping lifts in deepening cones of depression establishes the second constraint. Larger volumes of water would be made available by increasing the drawdown and the eventual extent of the cone of depression. Leakage from aquifers above and below might occur, providing additional water to that which moves laterally and diminishing the supply in the other aquifers. A temporary bonus is water released from storage as a result of lowering the pressure during the water-level decline. Artificial recharge in either the unconfined or confined zones of the aquifer will increase the available water supply. Where enough data are available, and through the use of analog and digital modeling, the optimum combination of recharge, hydraulic gradient, and pumping lift can be planned and problems can be minimized.

Certain effects on the environment, such as the potential for land subsidence, should be considered when large withdrawals from confined aquifers are planned.

DEVELOPMENT OF THE TERTIARY LIMESTONE AQUIFER

Callahan (1964, p. 26) points out that in the Valdosta, Ga., area the base flow of streams was relatively low in the fall of 1954, 0.014 to 0.054 ft^3/s/mi^2 or 0.009 to 0.035 (Mgal/d)/mi^2, but that much of the infiltrating rainfall of 1.25 ft^3/s/mi^2 (0.81 (Mgal/d)/mi^2) may have moved downdip in the principal artesian aquifer rather than emerging as base flows of streams. According to H. B. Counts (oral commun., Feb. 18, 1977), in November 1972 the total flow of the Withlacoochee River, an estimated 12 Mgal/d, drained into two sinkholes and much more could have been accepted.

Estimates of rates of recharge to the Tertiary limestone aquifer have been made by determining outflow through the aquifer (table 3) according to various published values or estimates of transmissivity and hydraulic gradients. These are given below for the several large areas.

In the Alachua-Bradford-Clay-Union County, Fla., area the recharge computation of 0.08 (Mgal/d)/mi^2 (table 3) is based on a transmissivity value derived from development of the upper 200 feet of the aquifer (Clark and others, 1964) in an area where the aquifer is about 1,000 feet thick.

In the Green Swamp (Florida) recharge area the outflow is limited by the cross-sectional area through which water must flow (Pride and others, 1966). Hence, although the transmissivity is higher in the western basin, the cross-sectional area through which discharge takes place is smaller, and the apparent recharge is smaller.

TABLE 3.—*Recharge rates to the Tertiary limestone aquifer based on estimates of flow through the aquifer*

	Area (mi^2)	Aquifer flow (Mgal/d)	Recharge rate per square mile		Equivalent precipitation (in/yr)
			(ft^3/s)	(Mgal/d)	
Florida:					
Alachua-Bradford Clay-Union Counties above the 75-ft contour.	525	45	0.132	0.085	1.8
Orange County	950	210	.34	.22	4.62
Polk County:					
Alfia and Peace Rivers	640	163	.395	.255	5.36
Kissimmee River	415	151	.536	.364	7.65
Green Swamp area	870	114.8	.204	.132	2.7 min.
		121.8	.217	.140	2.9 avg.
		166.2	.28	.191	4.0 max.
Western basin	160	22.1	.214	.138	2.9
Eastern basin	208	67.4	.501	.324	6.8
Hillsborough County:					
Local	1,040	450	.67	.432	9.1
Inflow	566	100	.274	.177	3.72

It was estimated that in Hillsborough County, Fla., in the late 1950's about 100 Mgal/d entered the county from the north and east through the aquifer, but there was additional local recharge of about 0.43 (Mgal/d/mi^2, equivalent to 9.1 inches of precipitation per year (Menke and others, 1961, p. 15). Results of other determinations are given in the table.

Although only 0.2 to 0.5 Mgal/d of water per square mile of recharge area moves downdip at present, potential production from the artesian beds in the aquifer is much greater. When gradients are increased, the flow is increased proportionally, and additional ground water that is now discharged to streams or springs would be captured. For instance, there are large springs northwest of where the Miocene deposits overlap older strata (fig. 10). These springs may represent potential recharge to the Tertiary limestone aquifer that cannot pass southeastward under the Miocene cover and through fault zones under existing hydraulic gradients.

Some or much recharge through the top of the aquifer will also be available if gradients are increased, and in some areas water from deeper sources may rise along faults. In some instances additional water would be available through capture of evapotranspiration loss and from inflow from streams originating outside the area.

Florida's two largest springs, Silver Springs and Rainbow Springs, each have a ground-water drainage area of about 625 square miles underlain by a thin or broken clay cover. The average combined flow for the two springs was almost 1,000 Mgal/d in 1968 (Faulkner, 1970). Discharge ranged from 0.5 to 1.3 (Mgal/d)/mi^2 (table 4). An annual recharge rate of 10.6 to 28 inches of rainfall is thus indicated. Additional water was probably lost by evapotranspiration from the water table.

Freshwater is discharged through springs from the Tertiary limestone aquifer in more than one-half (30,000 square miles) of the Florida land area. Estimated flow of springs cataloged by Callahan (1964) and

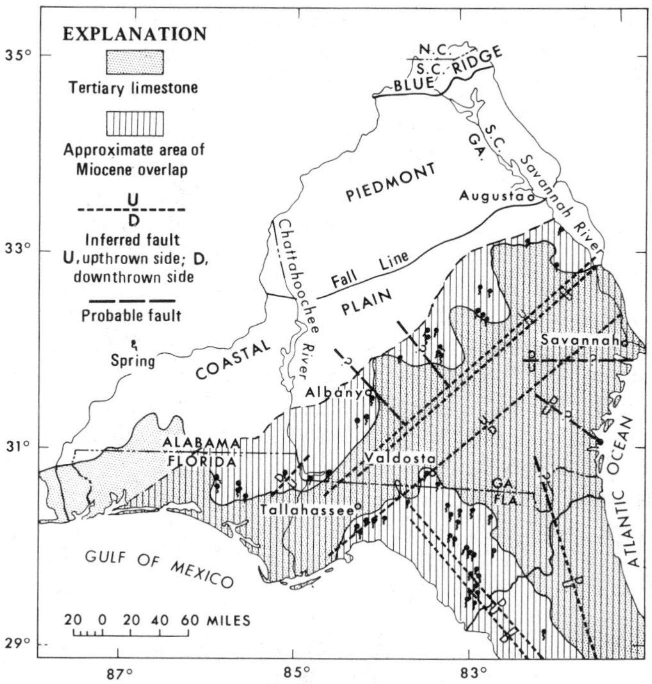

FIGURE 10.—Location of large springs in Georgia and northern Florida. (Modified from Callahan, 1964, figs. 5 and 7.)

Stringfield (1966) probably represents less than half the total discharge of all springs. With all cataloged springs considered at maximum flow, recharge is estimated to be about 0.32 (Mgal/d)/mi^2 or about 7 inches of rainfall.

The minimum flow of cataloged springs (half the total) is equal to 82 percent of the 3,000 Mgal/d of ground water pumped in all of Florida during 1970 (Murray and Reeves, 1977).

Spring discharge, like base-flow measurements, does not take into account water "lost" by underground flow downdip or by evapotranspiration. Hence, the value of 0.32 (Mgal/d)/mi^2 noted above is only an indication of the volumes available.

Recharge to the Tertiary limestone aquifer may occur through the Miocene cover at a very low rate. The rate is dependent on the thickness, the hydraulic conductivity of the overlying stratum, and the gradients involved. For instance, a highly conservative estimate of recharge through the thick Miocene cover in the Okefenokee Swamp area is about 0.035 (Mgal/d)/mi^2 under maximum development (Callahan, 1964, p. 27). An extremely large volume of water is in storage in the Tertiary limestone aquifer in Florida and Georgia. Seventy thousand square miles of Florida and Georgia are underlain by water-bearing strata with thicknesses ranging from several hundred feet to a maximum of about 2,000 feet (fig. 8). In northwestern Florida a part of the aquifer thickness shown includes aquifers below the Tertiary limestone

TABLE 4.—*Recharge rates to the Tertiary limestone aquifer based on spring discharge*

	Area (mi^2)	Total flow (Mgal/d)	Flow per square mile (ft^3/s)	Flow per square mile (Mgal/d)	Precipitation (in./yr)
Florida Springs area	30,000	2,456	0.127	0.082	1.72 min.
		9,824	.5	.328	6.88 max.
Silver Springs	625	348	.862	.557	11.7 min.
		531	1.314	.849	17.8 avg.
		833	2.06	1.333	28 max.
Rainbow Springs	625	315	.780	.504	10.6 min.
		468	1.16	.748	15.7 avg.
		684	1.67	1.094	22.9 max.
Northwestern Peninsular Florida above the 30–40-ft contours	1,950	95	.0754	.0487	1.0
North Central-Peninsular Florida above the 50–60-ft contours	3,770	181	.0743	.048	1.0
South Central Lake District above the 70-ft contour	6,000	544	.139	.09	1.9

aquifer. In the other States the volume of water stored in the Tertiary limestone aquifer is much less but is significant. Withdrawing much of this water in storage cannot be considered today in view of undesirable environmental effects that would occur.

Recharge to the Tertiary limestone aquifer would increase if water levels were lowered. About 70 percent of the 48 to more than 55 inches of rainfall in the area is lost by evapotranspiration. Where the area is riddled with sinkholes (fig. 7), it must be assumed that a high percentage of the rainfall will enter the ground and that much of it will be available to wells, either in the area of recharge or downdip in the confined part of the aquifer.

The volume of water that would move downdip through the upper 500 feet of a mile-wide section of the aquifer assuming a transmissivity of 60,000 ft^2/d and a hydraulic gradient of 2 ft/mi is estimated to be 0.9 Mgal/d.

If gradients were increased to 4 ft/mi and if 1,000 feet of the aquifer were penetrated, 3.6 Mgal/d would flow through each linear mile of the aquifer (taken along the strike). In any one well field of limited areal extent, however, water will be drawn into that center of discharge laterally and from downdip, and "leakage" will take place through the roof and the floor. On the basis of calculations of maximum recharge as derived from the flow of springs (table 4), maintaining a discharge of 3.6 Mgal/d might require as little as 3 square miles of recharge area. With lowering of the water table and capture of some evapotranspiration loss, the recharge area required would be even smaller.

Maximum production of water supplies from wells will necessarily take into account the tendency to induce a lateral flow of saline water into the well field from the sea. Also, in certain areas there is a tendency for sinkholes to occur as a result of lowering water levels. In any event, monitor wells and mathematical models of the aquifer would be beneficial in understanding the behavior of the hydrologic system where maximum stress is approached.

The Tertiary limestone aquifer is exceedingly complex and differs greatly in its physical makeup from place to place. The generalizations given here can only bring out several of the pertinent factors that must be considered.

DEVELOPMENT OF AQUIFERS IN THE APPALACHIAN MOUNTAINS AND THE PIEDMONT

Success in developing ground-water supplies in the part of the region outside the Coastal Plain can be enhanced by applying scientific principles and modern techniques. Although many unsuccessful or low-yielding water wells are known, the ground-water potential of the upland area has probably been underrated. Numerous reports of high-yielding wells in this part of the region indicate that well-site selection based on various geologic and hydrologic criteria and exploratory techniques can result in very productive wells.

It is now know that remote-sensing techniques can identify the lines of structural deformation with which many highly productive wells are associated. Remote sensing may be useful in identifying soil or vegetation types that may be useful in locating ground water. Geophysical techniques may help in identifying and delineating solution openings in carbonate rocks.

Ground reconnaissance by trained personnel will improve the chances for locating satisfactory well sites. In a study of the occurrence of ground water in the Piedmont and Blue Ridge provinces, LeGrand (1967) found that topography and soil thickness are good indices for siting high-yielding wells. The most favorable locations are in valleys or draws that have a thick cover of alluvium, residuum, or weathered rock.

Except in cavernous carbonate-rock aquifers, the volume of ground water that is available is directly related to (1) the number, size, and depth of fracture openings and (2) the thickness and permeability of the soil cover. The soil cover, where present, is the source of recharge to the underlying fractures and also serves as a storage reservoir when saturated. Experience has shown that in many areas most interconnecting fractures occur within 150 feet of the land surface, and drilling deeper than about 300 feet is generally not justified (LeGrand, 1967, p. 5).

OTHER MANAGEMENT PROCEDURES

Management procedures other than those already discussed may be employed where the supply of potable ground water appears to be limited. Briefly they fall under these major headings: (1) use of multiple sources of water, (2) use of saline water for special purposes, (3) artificial recharge, and (4) conjunctive use of ground water and surface water.

MULTIPLE SOURCES OF FRESH GROUND WATER

Where fairly thick sand terraces border large losing streams, an inexpensive source of ground water can be developed (Paulson and others, 1962). Assuming that the river terrace is rather wide, maximum advantage might be taken of such a shallow aquifer by pumping naturally filtered river water from wells near the river during periods of high flow and by pumping from wells as far back from the river as practicable in the low-flow period.

In a situation where a supply from a confined aquifer is declining, wells in the terrace deposits might be relied upon to a maximum extent in the higher rainfall months during which time the deep aquifer would be drawn upon less. In the following rather brief low-flow river period, the confined aquifer might then supply much or all of the water needed.

Where wells in confined aquifers yield a reliable supply of highly mineralized water, river terrace water might be used for blending, and furthermore at times of maximum availability, excess freshwater from the shallow aquifer might be used to recharge the confined aquifer.

In most places where higher or inland terrace deposits are present, it is questionable whether those deposits would yield supplies large enough to function economically in a combination system; however, there is a possibility that higher inland terrace deposits may be as thick as 150 feet in some areas. The use of shallow well water as a source of municipal supply at Elizabeth City, N. C., has been noted above. In Norfolk, Va., as much as 60 gal/min has been developed from wells in terrace sands (Cederstrom, 1945).

USE AND MANAGEMENT OF SALINE WATER

In most places freshwater occurs in near-surface formations, whereas at greater depths the water is saline. Use of saline water after desalinization may relieve the draft on a freshwater aquifer.

In most aquifers, particularly those that are confined, pressure head increases with depth; therefore, where the shallow freshwater stratum is drawn upon heavily, the poor-quality water at greater depth may migrate upward. Where the beds underlying the freshwater aquifer are silts, lenticular clays, or massive rocks riddled with solution channels or broken by faults, contamination may become a real concern. Utilization of the deeper water for special purposes, therefore, would tend to reduce the head in the lower aquifer and minimize that concern as well as add to the total supply.

One problem of using deep poor-quality water is that hydraulic connection may be established in the borehole between the deep and shallow aquifers. Saline-water wells should, therefore, be grouted through the freshwater strata.

In special cases where contamination of the upper stratum has taken place, pressure-relief wells may be drilled to diminish head in the deeper stratum. This technique was attempted in Brunswick, Ga., to minimize the upward movement of saline water in the Tertiary limestone aquifer. However, sufficient reduction of pressure head to completely halt the upward flow of saltwater could not be effected owing to the high transmissivity of the saltwater aquifer (Harlan Counts, oral commun., Nov. 30, 1972).

In many downdip localities in Alabama, Mississippi, and the Carolinas, the deeper Cretaceous aquifers yield water that is fresh, whereas the overlying formations are saline. In such places the water developer—farmer, municipality, or industrialist—must bear the cost of a well 1,000 feet or more in depth or use shallower water of inferior quality. Additionally, an obvious disadvantage is the danger of contamination of freshwater aquifer by movement of saline water from above as heavy pumping from the deeper aquifer continues. To guard against degradation of water in the deeper aquifer, water users should be encouraged to use the more cheaply developed inferior shallow water to the fullest extent possible in order to maintain a low-pressure head in the formation.

In some areas where relatively shallow water is of inferior quality and freshwater is available only at considerable depth, nearly all users initially will develop the shallower source of inferior water. In these areas, taking advantage of normal pressure differentials (where deeper aquifers have higher heads), connector wells which would allow freshwater to move upward to dilute the saline water aquifer might be desirable.

In an ideal situation a connector well would be centrally located with reference to several heavily pumped shallower wells or to a local supply consisting of many private wells. With maintenance of lower head in the upper formation through heavy withdrawals for air conditioning or other purposes not requiring the highest quality water, normal upward flow of freshwater through the connector well would be accelerated, and in time all wells in the general area might benefit from an improvement in the quality of water. Owing to normal head differences, even during periods of no pumping, flow would continue upward and outward into the shallower stratum, although the flow would be less than when pumps are operating.

Where the deep aquifers contain saline water and there is surplus water under higher pressure in shallow aquifers, the system could be reversed, thereby storing freshwater in a saline-water reservoir.

In most confined aquifers there is a downdip transition from freshwater to saline water. Planners may consider measures to forestall lateral or updip movement of saline water into centers of freshwater withdrawal. An example is the situation at Montgomery, Ala., where water from both the Eutaw and Gordo Formations is of good quality (Knowles and others, 1963). West of Montgomery, the chloride content of water in the Eutaw Formation increases rapidly. Flowing wells tapping the Eutaw Formation west of Montgomery along the Alabama River lower the head of water in the Eutaw Formation and bleed off highly saline water, thereby affording a measure of long-term protection to the Montgomery supply. The flowing wells discharge highly saline water (1,150, 1,330, and 1,600 mg/L chloride).

Where a well field is threatened by lateral intrusion of poor-quality water, including the intrusion of seawater, artificial recharge by injection wells, connector wells, or a combination of water spreading and connector wells, might be employed to create freshwater barriers between the well field and the source of contamination.

ARTIFICIAL RECHARGE OF GROUND WATER

Ground-water discharge accounts for a large part of streamflow; hence, development of ground-water supplies may not add greatly to the total available supply of water. Additional water can be made available to the system, however, by reducing losses due to evapotranspiration, by recycling of water, and by interception of high-flow components of streamflow. Interception of high streamflow in the past has been almost entirely accomplished by the construction of surface reservoirs. However, in recent years in California the recharging of underground reservoirs with water pumped from distant streams is widely practiced. The practice is less common in the mid-western States and unusual in the eastern States.

Methods range from collecting storm runoff in pits to constructing canals and pipelines that bring water to the recharge basins or injection wells. Difficulties attend these efforts in places, depending on type of water used, hydraulic characteristics, climatological conditions, and the methodology employed. Nevertheless, the practicality of artificial recharge is well established, and we may expect that, generally, technology will provide satisfactory answers to most problems.

In the arid West, ground-water levels generally occur well below the level of ephemeral streams, and there is storage space for whatever recharge water can be made available. In the humid Southeast, however, ground-water levels generally are fairly close to the surface. Addition of water to the unconfined aquifers will be limited by space requirements and may only promote more rapid lateral migration of that water to the streams.

Storage space can be provided by pumping during the dry periods, which are times of maximum demand and low streamflow. Water levels will decline, and space will then be provided for receiving and storing the excess water that becomes available during wet periods.

The most likely sites in the region for artificial recharge are those places where large cones of depression or significant water-level declines have been caused by ground-water withdrawals.

The techniques of water spreading by flooding, or by ditches, furrows, and pits, are outlined generally by Muckel and Schiff (1955). The effectiveness of recharge by spreading water on the surface, as opposed to pits or injection wells, depends upon the soil type, the type of cover material, and of course, the permeability of the formation beneath the soil cover. Bare soils tend to need treatment, but soil covered with undisturbed vegetation is highly receptive to infiltration. Intermittent flooding and drying and chemical treatment of the soil are beneficial in some areas.

Losing streams are uncommon in the East except in limestone terrain. Where streams lose water naturally or as a result of pumping from connected aquifers, small low-head dams would facilitate infiltration.

From Mississippi to southern North Carolina the sandy sediments store a tremendous volume of water. The amount of water that will drain from a sandy column upon lowering of the water table (specific yield) ranges from 15 to 20 percent in medium- to coarse-grained sediments. Assuming a value of 20 percent, a square mile of unconfined sandy reservoir in which the water level could be lowered 10 feet would have a capacity of 418 million gallons—equal to 24 inches of water or a pumping rate of 1.14 Mgal/d for 12 months. In this situation, assuming an annual recharge rate of 6 inches, there is an annual recharge deficit of 18 inches of water or 0.86 Mgal/d that can be provided by artificial recharge during a high rainfall season.

The thick weathered rock mantle (saprolite) that lies above consolidated rocks in the southern Piedmont has a high storage capacity (Stewart, 1962), but depending on the clay content, vertical drainage may be slow. Repeated flooding of the saprolite surface may be highly effective and preferable to trying to increase storage by injection wells that penetrate fresh rock at some depth. Flooding might be accomplished by constructing distributaries from large streams and rivers that would divert part of the high flows. In some instances, abandoned quarries in the Piedmont may lose water through fissures in the walls and floor and thus function as large recharge pits.

Where confined aquifers are intensively developed, the potentiometric surface may be greatly depressed. An eventual natural limitation on ground-water development in some areas may be forestalled by employing artificial recharge. Recharge by injection of water through wells built for the purpose is an alternate method to spreading water. Injection through wells is the only method for immediately recharging confined aquifers. The method is especially attractive if accomplished by gravity rather than under pressure, thereby conserving energy. The injection-well method of storing water or recharging aquifers is feasible where surplus water that is chemically and physically suitable for the operation is available. The characteristics of the available water must be such that clogging of the well screen and the aquifer are minimal.

Where rivers are bordered by sandy terrace deposits or other thin shallow aquifers, gravity recharge connector wells might be constructed to conduct naturally filtered water to deep confined aquifers in which the head is depressed. The operation might be restricted to periods when river stage rises above some predetermined point. Such recharge wells would be particularly suited to areas underlain by the highly permeable limestone aquifers.

This method might be considered for parts of Florida and along the Atlantic coast into South Carolina. Heavy draft is imposed upon the limestone aquifer in those areas, and additional draft may be expected in the future. Hence, maintaining a relatively high head upgradient from major well fields might do much to forestall potential saltwater intrusion and in any event would help minimize pumping costs.

Cool water is necessary in the development of water supplies for some industrial uses. In Hopewell, Va., cold surface water has been recharged during the winter months through wells that are ordinarily pumped to supply a chemical plant. Cold water stored temporarily in the ground and pumped out during the hot summer season is particularly advantageous in processing operations (Cederstrom, 1945).

Injection wells ending in sandy sediments might need to introduce water under pressure in order to achieve a significant (economically practical) rate of acceptance. High efficiencies might be achieved if the injection and formation waters are compatible and if the water injected is free of gases and particulate matter. However, chemical or biochemical actions that form slimes may occur, which would reduce the efficiency. Where hard waters are injected into aquifers, carbonate precipitates would be likely to form unless the waters were pretreated. Turbidity would also have to be reduced to insignificant amounts. Where the aquifer to be injected is a cavernous or highly shattered limestone or dolomite, as in the Appalachian Highlands and in some Coastal Plain formations, there may be little or no difficulty with clogging. In all instances, treatment to remove pollutants would be desirable.

RECHARGE USING EFFLUENT FROM TREATED SEWAGE

About 1,300 sewage-disposal systems in the United States in 1967 disposed of 19.8 billion gallons of treated sewage effluent by land-disposal methods (Wilson and Beckett, 1968). About two-thirds of these systems were used specifically for waste-water treatment as such, and one-third for crop benefits gained.

This practice may be considered as a method of continuous artificial recharge. The controlled spreading or spraying of treated sewage effluent on a soil surface allows oxidation and plant and biochemical action to take place, which upgrades the character of that water. After percolation down to the unconfined aquifer, the purified effluent may move into the confined aquifers or laterally to springs and streams. The aquifers are recharged or streamflow is augmented, presumably without deleterious ecological and health effects.

Disposal of effluent proceeds more slowly than other methods of recharge, allowing ample time for purification by plants and soils. Ground-water recharge should take place, however, even when only light applications of effluent are available, because the soil is kept near the saturation point and summer rains will tend to percolate through to the water table instead of being retained to satisfy soil-moisture requirements.

The sandy beds in the outcrop areas of the post-Paleozoic aquifers seem ideal for land disposal of effluent from treated sewage. The climate is favorable for functioning of digestion lagoons all year, the permeability of the surficial materials is high, the sands are thick, and in many places the water level is low enough to favor the process. Tremendous areas covered by the coastal sandy terrace formations likewise offer ideal sites for land disposal of effluent.

In some instances, as a result of the land disposal of treated sewage effluent, the quality of stream water might be upgraded to the extent that less treatment would be required by some of the downstream consumers. More water would be available for downstream consumers or, conversely, less water from upstream would have to be provided to maintain flow at a standard required for dilution of contaminants.

Land disposal of treated sewage effluent in relatively impermeable terrain may be feasible even though the geologic setting is quite different from that in the sandy areas. Here the objective would be to provide water for dry-weather irrigation and augmentation of flow of small streams.

Briefly, on the presumption that there are no adverse environmental effects, the process of land disposal of treated sewage offers a constant source of water for ground-water recharge or low-flow augmentation of streams, provides a solution to the problem of finding space for disposal of solid sewage waste, eliminates gross pollution of surface streams, and may upgrade the quality of soil in disposal areas.

FRESHWATER STORAGE IN SALINE AQUIFERS

Artificial recharge offers some significant advantages in those parts of the Coastal Plain where sandy confined strata are now saturated with saline water, as in northeastern North Carolina. Freshwater can be injected through wells and stored in the saline-water confined beds and pumped back for later use. The feasibility of storage under such conditions has been demonstrated (Cederstrom, 1957, p. 45–52), and further work to determine quantitative and other factors involved is now underway (Brown and Silvey, 1977; Kimbler and others, 1973).

Some coastal areas where there is a short but fairly high summer demand may find this method of increasing their supply advantageous as an alternative to piping in freshwater from inland wells or surface reservoirs. The large volume of water available from shallow wells in a high-rainfall winter season may suffice for recharge operations to the confined saline-water beds.

LOW-FLOW AUGMENTATION OF STREAMS

Low-flow augmentation is a technique for increasing the volume of streamflow by releasing stored water from reservoirs or by discharging water from wells into streams or reservoirs. This method is commonly practiced to increase streamflow during dry periods, thereby increasing the water supply for downstream users. It is also a technique for improving the capacity of the stream to accept and dilute wastes or to provide a better recreational environment.

Within the belt of Upper Cretaceous formations in Mississippi and Alabama, the chalk beds of the Selma Group are an extreme example of low-permeability formations. Streams originating within the chalk belt or impermeable clay belts are intermittent, flowing during and shortly after heavy rains. In dry weather the stream bed becomes dry, and in wet weather surface runoff is almost the equivalent of rainfall (Pierce, 1966). Chalk and clay offer a minimum of storage and thus increase the effect of floods. Streams originating in these areas cannot be sources of water supply during the dry seasons without storage.

Where justified by economic, environmental, or general esthetic considerations, this situation could be ameliorated by development of the aquifers underlying the streams. Widely spaced wells might be located along the streams and discharge directly into the streams or into minor tributaries. Examples of aquifers that might be used in this way are the Eutaw Formation and the Tuscaloosa Group.

Ground water from some aquifers would make a particularly desirable dilutant for highly contaminated streams; also, because ground water is somewhat oxygenated it would act positively in reducing biochemical-oxygen demands.

In downdip locations deep wells yielding mineralized water might be constructed where such water is considered acceptable for augmenting the flow of streams with a low discharge. If the discharge of the receiving stream is low, addition of high-chloride water may not be tolerable, but in a stream with a higher discharge the increase in chloride concentration may be acceptable. Although esthetics would be served, the extent to which the ecosystem (if aquatic life is present) would be affected by dilution of a highly contaminated stream with saline ground water would have to be considered.

Areas that have desirable locations for users requiring moderate volumes of water for cooling purposes might benefit from low-flow augmentation. A rise of stream-water temperature might be acceptable in areas where the streams would otherwise be dry during low-rainfall seasons and where aquatic life is absent. Another effect of augmenting low flow with ground water that might be desirable is the dilution of small polluted flows.

Ground water may also be used in conjunction with surface-water reservoirs intended for low-flow augmentation. Depending on the geological setting and environmental considerations, ground water may be available to augment impounded surface water.

One advantage of the use of supplemental wells is that their yield is not dependent upon rainfall in any particular year; thus even a relatively small amount of water pumped over a period of many months may furnish a very desirable increment to a surface-water supply. Furthermore, where wells are used for brief periods of time, ground-water withdrawals may be at rates greater than the aquifer can support continuously. Thus a well field might be pumped at two or three times the sustained yield of the aquifer if there is ample recharge during intervening periods of no pumping.

GROUND WATER FOR PUMPED STORAGE

In some areas there are extensive aquifers that are readily recharged. Transfer of water from these aquifers to pumped-storage reservoir sites might help alleviate peak power shortages if the combination of aquifers and storage sites can be found. In localities where power is generated, the power excess generated in off-peak hours could be used for pumped storage of water. In water-short areas diversion of streamflow for storage may not be desirable where that flow is needed for dilution or where the flow is allocated to downstream users.

In the Piedmont, pumped storage of ground water is generally not practicable. In the Valley and Ridge Province, pumped storage might be practicable if high-yield wells can be constructed tapping major fractures of fault zones.

CONJUNCTIVE USE OF GROUND WATER AND SURFACE WATER

Conjunctive use is the technique for integrating withdrawals of surface water and ground water in order to provide maximum utilization of these resources. In many areas an economical and satisfactory water supply might be developed by taking advantage of both ground- and surface-water sources where neither source is sufficient to supply the total demand at all times.

With both sources developed, surface reservoirs may be drawn upon heavily in the high rainfall season. In the dry season, a ground-water reservoir not subject to short-term vagaries of precipitation might provide all or nearly all the supply and preclude possible expensive additions to the surface-supply system. Most aquifers may be overpumped for a period of weeks or a few months, discharging much more water than if they were in continuous use and at the same time providing more storage space for recharge.

Recharging aquifers by injecting excess water available from surface sources may be an economical operation where a water- treatment plant is already in operation for treatment of the surface water and if the cost of the treatment and well maintenance are not prohibitive.

Where surface supplies carry a heavy waste load, the cost of treatment of that water may be excessively high at times of low flow. At such times maximum reliance may be placed upon a well supply, granting the presence of suitable aquifers. Conversely, surface water of good quality may be mixed with poor-quality ground water to increase the total supply.

GROUND-WATER PROBLEMS

One of the most critical ground-water problems in the South Atlantic-Gulf Region is the regional lowering of potentiometric surfaces and the development of extensive cones of depression in areas where large ground-water withdrawals have been made. In some areas of northeastern Mississippi withdrawals of water from the Eutaw Formation are exceeding the volume of water that the aquifer can transmit. At Mobile, Ala., and in some areas in Florida, concentrated withdrawals have caused intrusion of seawater. Contamination of the Tertiary limestone aquifer by saline water at Brunswick, Ga., is attributed to incomplete flushing of connate saline water (Gregg and Zimmerman, 1974, p. 36). In other areas, the lowered heads in cones of depression have induced the upward migration of saline water from underlying aquifers—a potential threat where confining beds are leaky or where there is faulting.

Notable ground-water declines have occurred outside the region in the Franklin, Va., area, where water levels have declined about 185 feet since 1937–39 (Brown and Cosner, 1974) and the cone of depression extends into northern North Carolina. At Brunswick, Ga., pumpage of over 100 Mgal/d has induced upward migration of saline water (Krause and Gregg, 1972). In central Florida large industrial and agricultural withdrawals have lowered water levels in confined aquifers more than 50 feet since 1934 (Kaufman, 1967). In North Carolina dewatering for open-pit mining has lowered the potentiometric surface below sea level in an area of about 800 square miles (Peek, 1969, p. 12).

There is potential for saltwater encroachment along much of the Atlantic Coast and in some areas on the Gulf Coast where shallow aquifers are hydraulically connected to the sea or to estuaries. Withdrawals can cause lateral movement of saline water, and natural factors such as inundation by hurricane flooding can change the position of the freshwater-saltwater interface in estuaries.

Head differentials permit the movement of water through boreholes or damaged casing from one aquifer to another. Generally, the direction of movement is upward, and where the deeper confined aquifers contain water of undesirable quality, the shallower aquifers become contaminated. Deep water wells and oil-test wells are the common potential sources of this type of contamination.

Lowered water levels result in correspondingly lower pumping levels. Obviously, there is a direct cost relation between total pumping head and energy consumption. Excessive pumping lift can be attributed either to inefficient well construction or to water-level declines. Many water wells in the region have excessive pumping lifts due to poor well design or improper completion and development methods.

Land subsidence is a potential threat where large volumes of ground water are pumped from aquifers in uncompacted sediments. The areas in the South Atlantic-Gulf Region that have the highest subsidence potential are Late Eocene and Quaternary deposits near the coastline where the effects would be most noticeable. Subsidence of as much as 4 inches was reported at Savannah, Ga. (Davis and others, 1963). More spectacular are sinkhole collapses related to ground-water withdrawals. Examples of this activity, some of which were catastrophic, have been reported by Newton and Hyde (1971, p. 1) and Kaufman (1967, p. 30). Sinkholes are restricted to those areas underlain by carbonate rocks.

Mining at some places causes problems of ground-water quantity and quality. Most public attention today is directed toward strip and pit mining. These operations may result in actual destruction of parts of shallow aquifers, and water-soluble mining products and byproducts can contaminate ground-water supplies. The need for low-grade coal and lignite to help meet energy requirements will intensify the problem caused by mining.

Landfills, considered an economical method for disposing of solid wastes for many years, commonly have been operated without consideration of the character and movement of leachate and without regard for the potential effect on water supplies. Improperly located or managed landfills can be expected to be a major source of ground-water contamination.

The technology of deep-well injection is now quite advanced but not without problems. Most of the injection operations are for waste disposal, and some wastes are highly toxic. The U.S. Geological Survey and other agencies and industries are conducting research to evaluate techniques and potential effects of injection. Current studies in the region, involving the Geological Survey, include ascertaining the feasibility of storing materials, liquids, and gases in artificial cavities in salt domes and in the extensive impermeable clay and chalk strata. The protection of ground-water resources is a primary consideration in planning injection for disposal or storage.

Problems with natural ground-water quality in the region are generally the result of excessive concentrations of iron, chloride, hydrogen sulfide, fluoride, or color. Acidic (low pH) and hard waters are common. Some of these objectional constituents and properties are amenable to simple water treatment; however, the cost of treatment for excessive color, chloride, or fluoride may be prohibitive.

The unequal distribution of the ground-water resources in the South Atlantic-Gulf Region is a significant factor in future economic development. Generally, the largest undeveloped ground-water supplies are in the Coastal Plain. Generally, the available supply increases as aquifers thicken and become more numerous seaward. Because water-resources development engenders water problems, most of the major problems in the future can be expected in coastal areas.

DATA NEEDS

This description of the round-water resources of the South Atlantic-Gulf Region is based on data and interpretations resulting from geologic and water-resources studies. The volume of data available and the complexity of evaluating, planning, and managing water-resources development dictate the use of computer analysis and modeling for optimum utilization of these resources. A basic need is for additional data in areas where coverage is sparse and for studies that will point out areas where water problems may be anticipated.

An immediate need is for management of the ground-water system in areas that have problems related to concentrated withdrawals. Digital and electric-analog models are useful tools for approximating solutions to hydrologic and related problems. Almost as urgent is the need for modeling entire aquifers or aquifer systems. Models are capable of simulating effects of past, present, and projected ground-water withdrawals—basic needs in planning and managing water-resources development; however, model projections should be used with caution.

More data are needed about the magnitude of and the potential for saltwater encroachment in the Coastal Plain aquifers. Modeling can be an effective tool in studies related to this problem.

Artificial recharge, subsurface storage of liquids and gases, and subsurface disposal of wastes employ similar methodologies. Studies are needed to develop more satisfactory methods of operation and management in these fields. Many more data are needed on the properties of potential reservoir materials below the freshwater zone and on the alternation, diffusion, and dispersion of injected substances.

Research is underway to ascertain the offshore extent of freshwater aquifers along the Atlantic coast. There is a need to intensify these studies and to expand the work to the Gulf Coast. The importance of these potential sources of water in the development of the coastal areas cannot be overemphasized.

Saline water as an economic resource has received little attention. Highly mineralized water and brines may contain minerals of economic value.

Further studies of land subsidence and sinkhole collapse, particularly with reference to prediction and prevention of catastrophic occurrences, are needed.

Use of reclaimed water will become more significant in total development of the resource. Studies are needed to find methods of reclaiming economic products from waste water. Also, the increasing cost of energy suggests the need for methods of recovering heat from cooling water.

Surface water and ground water are closely related in the highly permeable formations in the Coastal Plain. Special studies are needed to evaluate and predict the effects of changes in either regimen.

SUMMARY AND CONCLUSIONS

The rapid population increase and industrial growth now underway in the South Atlantic-Gulf Region will accelerate as the energy situation becomes more critical. The reason is that economic development will favor areas in the Nation that have moderate climate, access to ample energy, abundant labor, and adequate water supplies.

With planning and management, the ground-water supply in the South Atlantic-Gulf Region is adequate to meet needs through 2020 if one-forth of the estimated ground-water yield of 78,000 Mgal/d in the region can be withdrawn. In 1970, 4,500 Mgal/d, about 14 percent of all water pumped in the region, was obtained from ground-water sources. If the same use ratio for ground water and surface water continues, projections indicate that about 12,200 Mgal/d would be needed in 2000 and about 18,200 Mgal/d in 2020. The projections indicate that consumptive use will account for only about 6 percent of the total, so most of the water will be available for reuse.

Although virtually the entire region is underlain by aquifers that will yield more than 50 gal/min to wells, the most prolific sources of ground water are (1) the Tertiary limestone aquifer, (2) the Miocene, Pliocene, and Quaternary sand and gravel aquifers that are more or less interconnected, and (3) the Tuscaloosa sand and gravel aquifers. The Tertiary limestone aquifer and the Miocene-through-Quaternary aquifers are the most prolific sources of water owing to extremely high transmissivities and favorable conditions for direct recharge from precipitation and streams. The Tuscaloosa aquifers, although capable of large yields, have a more limited potential owing to the relatively small recharge areas and the generally restricted interconnection with other aquifers.

Rapid advances in computer science and in mathematical and electronic modeling will provide useful tools for planners and managers; however, these tools will be only as accurate as the data on which their output is based. The collection, assimilation, and analysis of dependable ground-water data are prerequisites for successful planning and management.

Management of the major and minor aquifers will require establishing the optimum sustained yield of the aquifers. The optimum sustained yield of the aquifer is the condition whereby the maximum long-term withdrawal is accomplished with a minimum of undesirable effects. Undesirable effects include excessive diversion of streamflow either by capture of ground-water discharge to streams or by increasing ground-water recharge from streams; water-level declines that result in uneconomical pumping lifts or land subsidence; and intrusion of poor-quality water from the sea, estuaries, surface sources, other aquifers, or other parts of the same aquifer.

Conjunctive use of surface and ground water and the optimum sustained development of aquifers, or any controlled development of water resources, are administered by water-management agencies. The most efficient utilization of the resource may require transferring water from one management area (for example, a basin) to another, suggesting the need for coordination among agencies.

Among the first problems confronting water managers will be the redistribution of withdrawals to minimize or terminate saltwater encroachment, reduce the extent of extremely deep cones of depression, and control pollution. Extraordinary coordination of water-management agencies may be required in the instances where cones of depression extend over hundreds of square miles.

Efficient use of ground water by the primary users is a means of conserving water resources by reducing the draft on the aquifer. One of the major considerations in water-deficient areas is the reuse of the waste water. Some waste water requires minimal treatment for many uses, and there are some uses for waste water containing a large load of suspended and dissolved materials. Beneficial use for water discharged from heat-exchange applications is highly probably.

Widespread artificial recharge of aquifers with surplus surface water and treated waste water (or in special instances, untreated waste water) is a inevitable, and high priority should be given to developing and applying this technique. Artificial recharge by wells is especially attrative if accomplished by gravity rather than under pressure, thereby conserving energy. Artificial recharge by spreading can be environmentally acceptable.

Less than 10 percent of the region's mean annual runoff (197,000 Mgal/d), if used for artificial recharge, could meet the entire projected requirement for ground water without significantly affecting streamflow. Systematic recharge of major aquifers during wet seasons would offset the effects of the removal of large quantities of ground water from storage, limiting the spread of cones of depression and avoiding undesirable effects such as large pumping lifts and saltwater intrusion.

One of the basic considerations in developing ground-water sources is the effect on the low flow of streams and on vegetation. These environmental considerations are among the constraints that should be considered in the optimum development of aquifers.

SELECTED REFERENCES

Baker, Jack, 1957, Geology and ground-water resources of the Piedmont area of Alabama: Geol. Survey Alabama, Spec. Rept. 23, 99 p.

Boswell, E. H., 1963, Cretaceous aquifers of northeastern Mississippi: Mississippi Board Water Comm. Bull. 63–10, 202 p.

Boswell, E. H., Moore, G. K., MacCary, L. M., and others, 1965, Cretaceous aquifers in the Mississippi embayment, *with discussions of* quality of the water by H. G. Jeffery: U.S. Geol. Survey Prof. Paper 448–C, 37 p.

Brown, D. L., and Silvey, W. D., 1977, Artificial recharge to a freshwater-sensitive brackish-water sand aquifer, Norfolk, Virginia: U.S. Geol. Survey Prof. Paper 939, 53 p.

Brown, F. G., Foster, V. M., Adams, R. W., Reed, E. W., and Padgett, H. D., 1944, Geology and ground-water resources of the coastal area in Mississippi: Mississippi State Geol. Survey Bull. 60, 229 p.

Brown, G. A., and Cosner, O. J., 1974, Ground-water conditions in the Franklin area, southeastern Virginia: U.S. Geol. Survey Hydrol. Inv. Atlas HA-538, 3 sheets.

Brown, P. M., 1959, Geology and round-water resources in the Greenville area, North Carolina: North Carolina Dept. Conserv. Devel. Bull. 73.

Brown, P. M., Miller, J. A., and Swain, F. M., 1972, Structural and stratigraphic framework, spatial distribution of permeability of the Atlantic Coastal Plain, North Carolina to New York: U.S. Geol. Survey Prof. Paper 796, 79 p.

Callahan, J. T., 1964, The yield of sedimentary aquifers of the coastal plain Southeast River Basins: U.S. Geol. Survey Water-Supply Paper 1669-W, 56 p.

Causey, L. V., 1965, Availability of ground water in Tallageda County, Alabama, a reconnaissance: Geol. Survey of Alabama Bull. 81, 63 p.

Cederstrom, D. J., 1945, Geology and ground-water resources of the Coastal Plain in southeastern Virginia: Virginia Geol. Survey Bull. 63.

———1957, Geology and ground-water resources of the York-James Peninsula, Virginia: U.S. Geol. Survey Water-Supply Paper 1361, 237 p.

Clark, W. E., Musgrove, R. H., Menke, C. G., and Cagle, J. W., Jr., 1964, Water resources of Alachua, Bradford, Clay, and Union Counties, Florida: Florida Geol. Survey Rept. Inv. 35.

Counts, H. B., and Donsky, Ellis, 1963, Salt-water encroachment, geology, and ground-water resources of the Savannah area, Georgia and South Carolina: U.S. Geol. Survey Water-Supply Paper 1611, 100 p.

Cressler, C. W., 1970, Large supplies of ground-water of good quality, northwestern Georgia: U.S. Geol. Survey Prof. Paper 700–A, p. 63.

Cushing, E. M., 1966, Map showing altitude of the base of fresh water in coastal plain aquifers of the Mississippi embayment: U.S. Geol. Survey Hydrol. Inv. Atlas HA-221, 1 sheet.

Cushing, E. M., Boswell, E. H., and Hosman, R. L., 1964, General geology of the Mississippi embayment: U.S. Geol. Survey Prof. Paper 448-B, 28 p.

Cushing, E. M., Boswell, E. H., Speer, P. R., Hossman, R. L., and others, 1970, Availability of water in the Mississippi embayment: U.S. Geol. Survey Prof. Paper 448-A, 13 p.

Dalsin, G. J., 1972, Saline ground-water resources of Mississippi: Mississippi Board Water Comm. Water Resources Map 72-1, 1 sheet.

Davis, G. H., Small, J. B., and Counts, H. B., 1963, Land subsidence related to decline of artesian pressure in the Ocala Limestone at Savannah, Georgia: Geol. Soc. America Eng. Geology Case Histories, no. 4, p. 1-8.

Ellard, J. S., 1977, Map of fresh and slightly saline ground-water resources in the coastal plain of Alabama: Geol. Survey Alabama Map 179.

Fairchild, R. W., 1972, The shallow aquifer system in Duval County, Florida: Florida Geol. Survey Rept. Inv. 59, 50 p.

Faulkner, G. L., 1970, Geohydrology of the Cross-Florida barge canal area with special reference to the Ocala vicinity: U.S. Geol. Survey open-file rept., 117 p.

Fenneman, N. M., 1938, Physiography of the eastern United States: New York, McGraw Hill Book Co., Inc., 714 p.

Flawn, P. T., chm., 1967, Basement map of North America, between latitudes 24° and 60° N., by the American Association of Petroleum Geologists and United States Geological Survey: U.S. Geol. Survey, scale 1:5,000,000, 1 sheet.

Gregg, D. O., and Zimmerman, E. A., 1974, Geologic and hydrologic control of chloride contamination in aquifers at Brunswick, Glynn County, Georgia: U.S. Geol. Survey Water-Supply Paper 2029-D, 44 p.

Hathaway, J. C., and others, 1976, Preliminary summary of the 1976 Atlantic Margin Coring Porject: U.S. Geol. Survey Open-File Report 76-844, 217 p.

Herrick, S. M., and Vorhis, R. C., 1963, Subsurface geology of the Georgia Coastal Plain: Georgia Geol. Survey Inf. Circ. 25, 80 p.

Hosman, R. L., Long, A. T., Lambert, T. W., and others, 1968, Tertiary aquifers in the Mississippi embayment, *with discussions of* quality of the water by H. G. Jeffery: U.S. Geol. Survey Prof. Paper 448-D, 29 p.

Johnson, P. A., Siple, G. E., and Cummings, T. R., 1968, A reconnaissance of the water resources of Pickens County, South Carolina: South Carolina Water Resources Comm. Rept. 1, 69 p.

Kaufman, M. I., 1967, Hydrologic effects of ground-water pumpage in the Peace and Alafia River basins, Florida, 1934-1965: Florida Div. Geology Rept. Inv. 49, 32 p.

Kaufman, M. I., and Dion, N. F., 1967, Chemical character of water in the Floridian aquifer in southern Peace River basin, Florida: U.S. Geol. Survey Map Series 27, 1 sheet.

Kimbler, O. K., Kazmann, R. G., and Whitehead, W. R., 1973, Saline aquifers—future storage reservoirs for fresh water: Underground Waste Management and Artificial Recharge, 2d Internat. Symposium, v. 1, AAPG, New Orleans, La., p. 192–206.

Klein, Howard, 1971, Depth to base of potable water in the Floridan aquifer: Florida Dept. Nat. Resources, Bur. Geology Map Series 42, 1 sheet.

Knowles, D. B., Reade, H. L., Jr., and Scott, J. C., 1963, Geology and ground-water resources of Montgomery County, Alabama: U.S. Geol. Survey Water-Supply Paper 1606, 76 p.

Krause, R. E., and Gregg, D. O., 1972, Water from the principal artesian aquifer in coastal Georgia: Georgia Dept. Nat. Resources Hydrol. Atlas 1.

LaMoreaux, P. E., 1949, Ground water in southeast Alabama, *in* Carter, R. W., Williams, M. R., LaMoreaux, P. E., and Hastings, W. W., Water resources and hydrology of southeastern Alabama: Alabama Geol. Survey Spec. Rept. 20, 165 p.

LeGrand, H. E., 1967, Ground water of the Piedmont and Blue Ridge provinces in the southeastern states: U.S. Geol. Survey Circ. 538, 11 p.

Leve, G. W., 1966, Ground water in Duval and Nassau Counties, Florida: Florida Geol. Survey Rept. Inv. 43, 1 sheet.

Lindskov, K. L., 1972, Water resources of northeast North Carolina above Cape Lookout: U.S. Geol. Survey open-file rept., 71 p.

Lohman, S. W., 1972, Ground-water hydraulics: U.S. Geol. Survey Prof. Paper 708, 70 p.

McCall, G. C., 1972, Planning and developing ground-water supplies, *in* Proc. Water Well Seminar for Professional Engineers: South Carolina Soc. Prof. Engineers and South Carolina Water Resources Comm.

McCarthy, G. R., 1936, Magnetic anomalies and geologic structures of the Carolina Coastal Plain: Jour. Geology, v. 44, no. 3, p. 396–406.

McCollum, M. J., and Counts, H. B., 1964, Relation of salt-water encroachment to the major aquifer zones, Savannah area, Georgia and South Carolina: U.S. Geol. Survey Water-Supply Paper 1613-D, 26 p.

McCoy, H. J., 1962, Ground-water resources, Collier County, Florida: Florida Geol. Survey Rept. Inv. 31, 82 p.

McMaster, W. M., and Hubbard, E. F., 1970, Water resources of the Great Smoky Mountains National Park, Tennessee and North Carolina: U.S. Geol. Survey Hydrol. Inv. Atlas HA-420, 2 sheets.

Menke, C. G., Meredith, E. W., and Wetterhall, W. S., 1961, Water resources of Hillsborough County, Florida: Florida Geol. Survey Rept. Inv. 25, 101 p.

Moore, D. B., and Joiner, T. J., 1969, A subsurface study of southeast Alabama: Alabama Geol. Survey Bull. 88, 33 p.

Muckel, D. C., and Schiff, Leonard, 1955, Replenishing ground water by spreading: U.S. Dept. Agriculture Yearbook "Water," p. 302.

Murray, C. R., and Reeves, E. B., 1977, Estimated use of water in the United States in 1975: U.S. Geol. Survey Circ. 765, 39 p.

Murray, G. E., 1961, Geology of the Atlantic and Gulf Coastal Province of North America: New York, Harper and Brothers, 692 p.

Newcome, Roy, Jr., 1965, Configuration of the base of the fresh-ground-water section in Mississippi: Mississippi Board Water Comm. Water Resources Map 65–1, 1 sheet.

———1967a, Development of ground-water supplies at Mississippi Test Facility, Hancock County, Mississippi: U.S. Geol. Survey Water-Supply Paper 1839–H, 28 p.

———1967b, Ground-water resources of the Pascagoula River basin, Mississippi and Alabama: U.S. Geol. Survey Water-Supply Paper 1839-K, 36 p.

———1971, Results of aquifer tests in Mississippi: Mississippi Board Water Comm. Bull. 71-2, 44 p.

Newton, J. G., and Hyde, L. W., 1971, Sinkhole problems in and near Roberts Industrial Sabdivision, Birmingham, Alabama: Geol. Survey Alabama, Circ. 68, 42 p.

North Carolina Department of Natural and Economic Resources, 1974, Status report on ground water conditions in capacity use area no. 1, central coastal plain, North Carolina: Ground Water Bull. 21, 146 p.

Parker, G. G., Ferguson, G. E., Love, S. K., and others, 1955, Water resources of southeastern Florida: U.S. Geol. Survey Water-Supply Paper 1255, 969 p.

Pascale, C. A., 1975, Estimated yield of fresh-water wells in Florida: Florida Bur. Geol. Map Series 70, 1 sheet.

Paulson, Q. F., Miller, J. D., Jr., and Drennen, C. W., 1962, Ground-water resources and geology of Tuscaloosa County, Alabama: Alabama Geol. Survey County Rept. 6, 97 p.

Peek, H. M., 1951, Cessation of flow of Kissengen Spring in Polk County, Florida: Florida Bur. Geology Rept. Inv. 7, pt. 3, p. 73-82.

———1969, Effects of large-scale mining withdrawals of ground water: Ground Water, v. 7, no. 4, p. 12-20.

Pierce, L. B., 1966, Surface water in southwestern Alabama, *with a section on* chemical quality of surface water by Stanley M. Rogers: Alabama Geol. Survey Bull. 84, 182 p.

Pride, R. W., and Crooks, J. W., 1962, The drought of 1954–56, its effect on Florida surface water resources: Florida Geol. Survey, Rept. Inv. 26, 65 p.

Pride, R. W., Meyer, F. W., Cherry, R. N., 1966, Hydrology of Green Swamp area in central Florida: Florida Geol. Survey Rept. Inv. 42, 137 p.

Rosenau, J. C., and Faulkner, G. L., 1974 (revised 1975), An index to

springs of Florida: Florida Bur. Geol. Map Series 63, 1 sheet.
Siple, G. E., 1960, Piezometric levels in the Cretaceous sand aquifer of the Savannah River basin: Georgia Mineral Newsletter, v. 13, no. 4, p. 163-166.
———1967a, Geology and ground water in the Savannah River plant and vicinity, South Carolina: U.S. Geol. Survey Water-Supply Paper 1841, 113 p.
———1967b, Salt-water encroachment of Tertiary limestone along coastal South Carolina: Internat. Assoc. Sci. Hydrology Pub. 74, p. 439–453.
Stewart, J. W., 1962, Water yielding potential of weathered crystalline rocks at the Georgia Nuclear Laboratory: U.S. Geol. Survey Prof. Paper 450-B.
Stringfield, V. T., 1966, Artesian water in Tertiary limestone in the southeastern states: U.S. Geol. Survey Prof. Paper 517, 226 p.
Tarver, G. R. 1968, Hydrology of the Biscayne Aquifer in the Pompano Beach area, Broward County, Florida: Florida Geol. Survey Rept. Inv. 36.
Thomas, H. E., 1951, The conservation of ground water: New York, McGraw-Hill Book Co., 327 p.
Thomson, M. T., Herrick, S. M., Brown, Eugene, and others, 1956, The availability and use of water in Georgia: Georgia Dept. Mining and Geology Bull. 65, 329 p.
Trainer, F. W., and Heath, R. C., 1976, Bicarbonate content of groundwater in carbonate rock in eastern North America: Jour. Hydrology, v. 31, p. 37–55.
U.S. Army Corps of Engineers, 1972, North Atlantic regional water resources study, Appendix D, Geology and Ground Water: N.A.R. Water Resources Study Coordinating Committee.
U.S. Department of Agriculture, 1955, Water: Washington, U.S. Govt. Printing Office, 751 p.
U.S. Department of Commerce, Environmental Science Services Administration, 1968, Climatic atlas of the United States: Washington, U.S. Govt. Printing Office, 80 p.
U.S. Department of the Interior, Geological Survey, 1970, The National Atlas of the United States of America, p. 417.
U.S. Water Resources Council, 1968, The Nation's water resources, summary report (part 1): Washington, U.S. Govt. Printing Office, 32 p.
Wait, R. L., 1965, Geology and occurrence of fresh and brackish ground water in Glynn County, Georgia: U.S. geol. Survey Water-Supply Paper 1613-E, 94 p.
Wait, R. L., and Gregg, D. O., 1967, Hydrology and chloride contamination of the principal artesian aquifer in Glynn County, Georgia: U.S. Geol. Survey open-file rept., 214 p.
Warman, J. C., and Causey, L. V., 1962, Geology and ground-water resources of Calhoun County, Alabama: Alabama Geol. Survey County Rept. 7, 77 p.
Wilson, Alfonso, and Iseri, K. T., 1967 (revised 1969), River discharge to the sea from the shores of the conterminous United States, Alaska, and Puerto Rico: U.S. Geol. Survey Hydrol. Inv. Atlas HA–282, 2 sheets.
Wilson, C. W., and Beckett, F. E., (symposium eds.), 1968, Municipal sewage effluent for irrigation: Louisiana Polytechnic Inst., Ruston, La., 169 p.
Wyrick, G. G., 1966, Ground-water resources of Martin County, North Carolina: North Carolina Dept. Water Resources Ground-Water Bull. 9, 85 p.

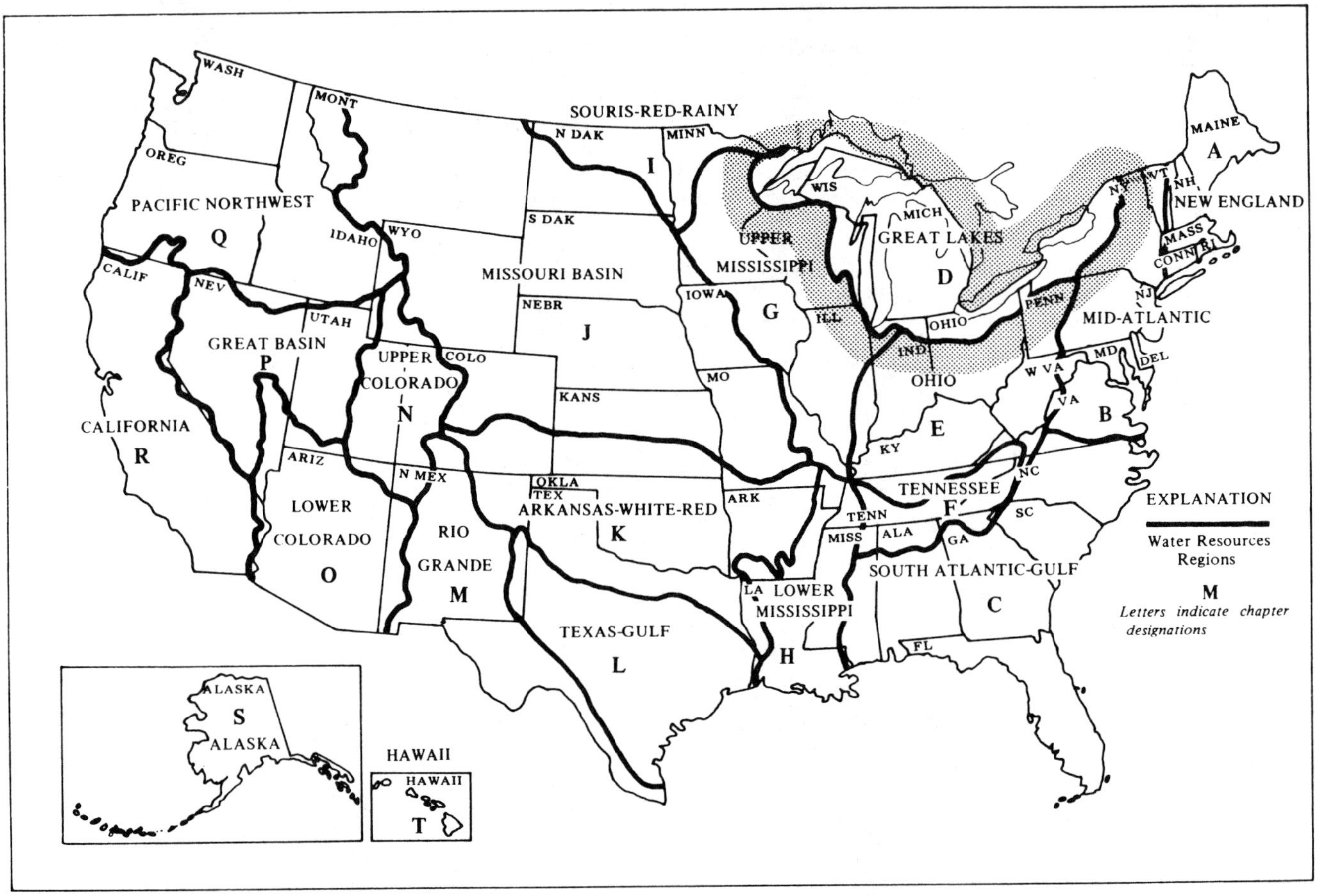

D – Location Map for the Great Lakes Region

D

GREAT LAKES REGION

By WILLIAM G. WEIST, JR.

ABSTRACT

The Great Lakes Regions, as a whole, has abundant supplies of water. Nearly 805,000 billion cubic feet of water is contained in the Great Lakes. An additional 35,000 billion cubic feet of potable ground water is available from storage in the region. Estimated ground-water discharge to the streams and lakes of the region is 26 billion gallons per day.

Despite this abundance of water, the United States part of the Great Lakes basin is faced with many water-related problems, most of which involve water quality and water supply. Other problems concern periods of low flow in streams, preservation of wetlands, detrimental effects of erosion, and flooding. The significance of ground water in these problems is often overlooked.

Ground water can be an alternative to surface water as a source of supply, or it can be used conjunctively with surface water to provide flexibility in water-supply management. Ground water supplied approximately 1,800 million gallons per day of the 39,900 million gallons per day used in the Great Lakes Region in 1970. The ground-water contribution was only 4.5 percent of the water used. Thus, ground water represents a potential source of supply for much of the region. It also can be used, where conditions permit, to maintain lake levels and flow in streams, to dilute poor quality surface water, and to maintain or create wetlands and ponds.

In managing water resources, ground water and surface water should be considered parts of a single system. Management includes not only planning and controlling the development but also monitoring the effects of this development. Recent advances in ground-water hydrology have provided methods to resolve some of the development and management questions that formerly slowed the development of ground water.

All of the States in the Great Lakes Region have some regulations to control the development or protect the quality of the ground water. These regulations, however, are not as comprehensive as those governing surface water. Future legislation could be designed to encourage the development of ground water and, at the same time, to protect the resource.

Efficient development and management of ground-water resources requires a through knowledge of the system. Reports on ground water are available for about 80 percent of the Great Lakes Region. Most of these reports, however, are not sufficiently detailed to be useful in comprehensive planning. As ground-water development continues, quantitative ground-water studies, utilizing models as predictive tools, will enable this development to proceed in an efficient manner.

INTRODUCTION

In 1971, the U.S. Geological Survey began a broad-perspective analysis of the Nation's ground-water resources. This analysis was designed to emphasize the extensive occurrence and availability of ground water and the significant role it can play in regional development. The results are being published in the Geological Survey professional paper series, "Summary Appraisals of the Nation's Ground-Water Resources." This comprehensive summary of ground water in the United States should aid in orderly planning and development of the Nation's water resources.

This report discusses ground-water resources for the part of the Great Lakes basin that is in the United States. The report is intended neither as a source of data nor a comprehensive discussion of water availability in the different aquifers in the region. Rather, it builds on previous reports by focusing on water-related problems and the role ground water can play in meeting these problems. A comprehensive summary of ground-water availability in the Great Lakes basin, including an extensive bibliography, was recently published by the Great Lakes Basin Commission (Waller and Allen, 1975).

Most of the material in this report is based on published reports and on information gathered through meetings with State and Federal personnel and members of regional planning commissions throughout the Great Lakes Region—the people who are gathering and using the water-resources data. The author gratefully acknowledges their contributions.

DESCRIPTION OF THE REGION

The Great Lakes basin (fig. 1) is an international basin containing 295,800 mi^2 in Canada and the United States. The Great Lakes Region discussed in this report includes only the 173,470 mi^2 in the United States, of which 60,602 mi^2 is water surface.

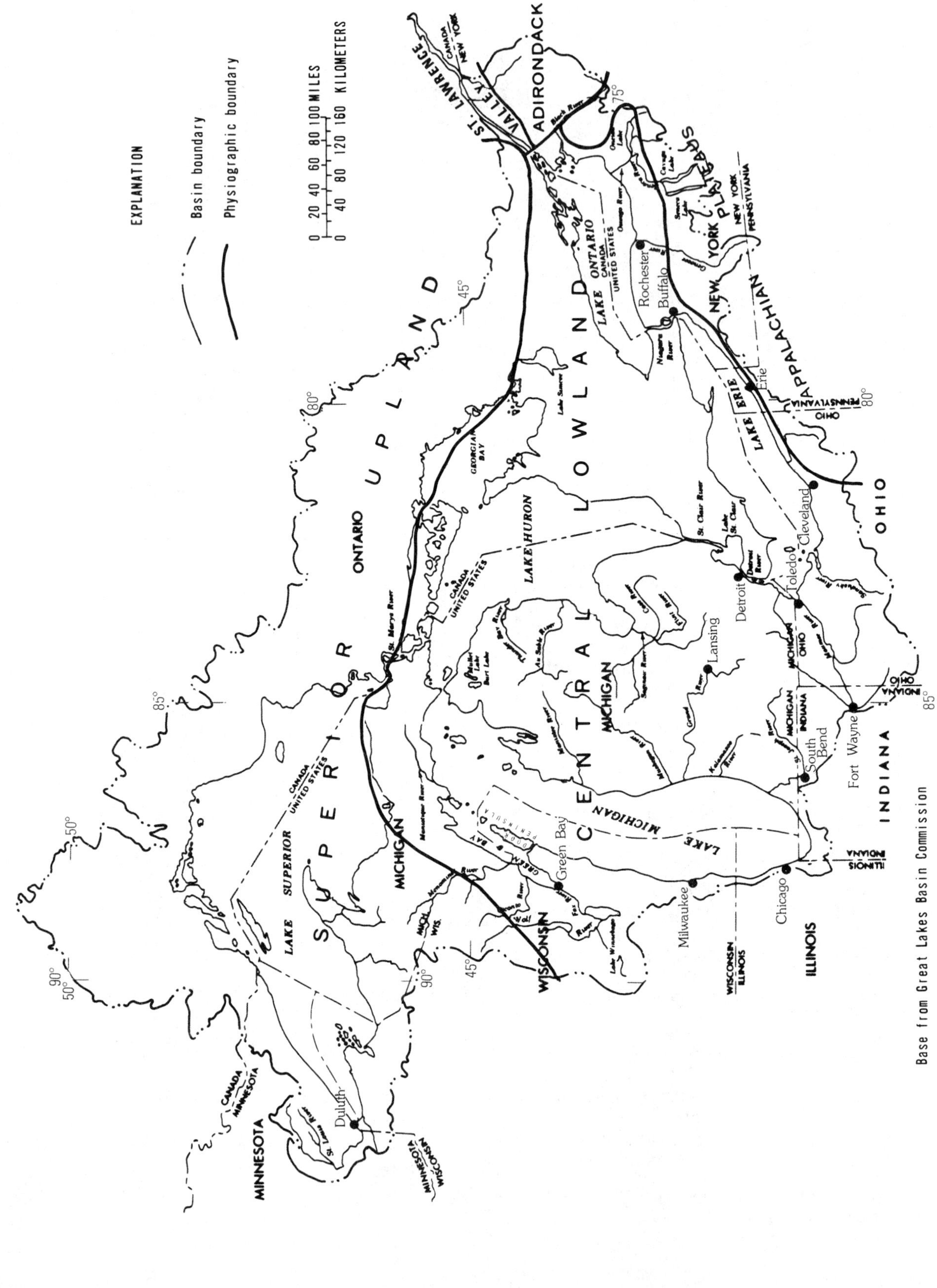

FIGURE 1.—The Great Lakes basin.

PHYSIOGRAPHIC SETTING

Most of the Great Lakes Region lies in the Central Lowland physiographic province (fig. 1). Parts of the region in Wisconsin, Minnesota, and the Upper Peninsula of Michigan are in the Superior Upland province, and parts of Ohio, Pennsylvania, and New York are in the Appalachian Plateau, Adirondack, and St. Lawrence Valley provinces. Most of the region is characterized by generally flat or gently rolling lowlands and lake plains. Areas along the western, eastern, and southeastern borders are dissected plateaus with varied relief and prominent escarpments. Most of the region is less than 1,000 ft above mean sea level. The highest point (4,621 ft) is in the Adirondack Mountains, and the lowest (150 ft) is along the St. Lawrence River. The highest elevation in the headwaters is 2,301 ft, in Minnesota.

The region underwent four major periods of glaciation. As much as 1,100 ft of glacial deposits overlie the bedrock surface. Streams have partly reworked these deposits, leaving alluvium along their channels. Local relief and much of the drainage was modified by the glaciation. The irregularity of the glacial deposits has resulted in an imperfect drainage pattern, with numerous lakes, ponds, marshes, and bogs. Major drainage is toward the Great Lakes and the St. Lawrence River. Twelve tributary river basins have drainage areas of about 6,000 mi² each; the rest have drainage areas ranging from a few to several hundred square miles (Waller and Allen, 1975, p. 1).

Bedrock consists of a series of sedimentary formations overlying crystalline rocks, except where the crystalline rocks are exposed in parts of Minnesota, Wisconsin, Michigan, and New York. The major structural bedrock feature is a deep sedimentary basin centered in Michigan.

HYDROLOGIC SETTING

The Great Lakes Region has an abundant supply of water (table 1). Because so much water is available in the Great Lakes and the streams that drain to them, the large amount of available ground water is frequently overlooked.

All the water resources of the region are part of a single hydrologic system, in which a change in one phase will cause changes in the other phases. Because this system is very complex, precipitation, surface water, and ground water are generally investigated and described separately. The next three sections give a brief picture of these three phases in the Great Lakes Region.

CLIMATE

Because of its large water areas, the region has a more moderate climate than regions of similar latitude to the west. In January, normal daily average temperatures range from 10° F (−12° C) in Minnesota to 28° F (−2° C) in Ohio. In August they range from 65° F (18° C) in Minnesota to 73° F (23° C) in Ohio (Environmental Data Service, 1968).

Precipitation is affected by the Great Lakes and by the highlands along the eastern border of the region. Normal annual precipitation ranges from 28 in in the Duluth area, along part of western Lake Michigan and Lake Huron, and in the Detroit area, to more than 52 in in the Adirondacks (fig. 2).

TABLE 1.—*Water resources of the Great Lakes Region*

Lake basin	Average annual precipitation on basin,[1] 1900–69 (inches)	Average runoff into lake,[2] 1935–64 (ft³/s)	Volume of water in lake[3] (billion ft³)	Average outflow of lake through natural channel,[4] 1860–1970 (ft³/s)	Estimated volume of ground water containing less than 3,000 mg/L dissolved solids available from storage[5] (billion ft³)
Superior	29.56	50,300	431,400	75,000	3,500
Michigan	31.16	37,400	173,500	52,000	14,900
Huron	31.26	51,300	124,800	187,300	4,800
Erie	33.79	24,500	17,200	201,900	8,200
Ontario	34.18	28,100	57,800	239,200	3,600
Region	31.46	141,600	804,700	---	35,000

[1] Data from Leonard and Raoul, 1975, p. 36.
[2] Data from Leonard and Raoul, 1975, p. 37.
[3] Data from Leonard and Raoul, 1975, p. 7.
[4] Data from Leonard and Raoul, 1975, p. 13.
[5] Data from Norris and Fidler, unpub. rept., 1975.

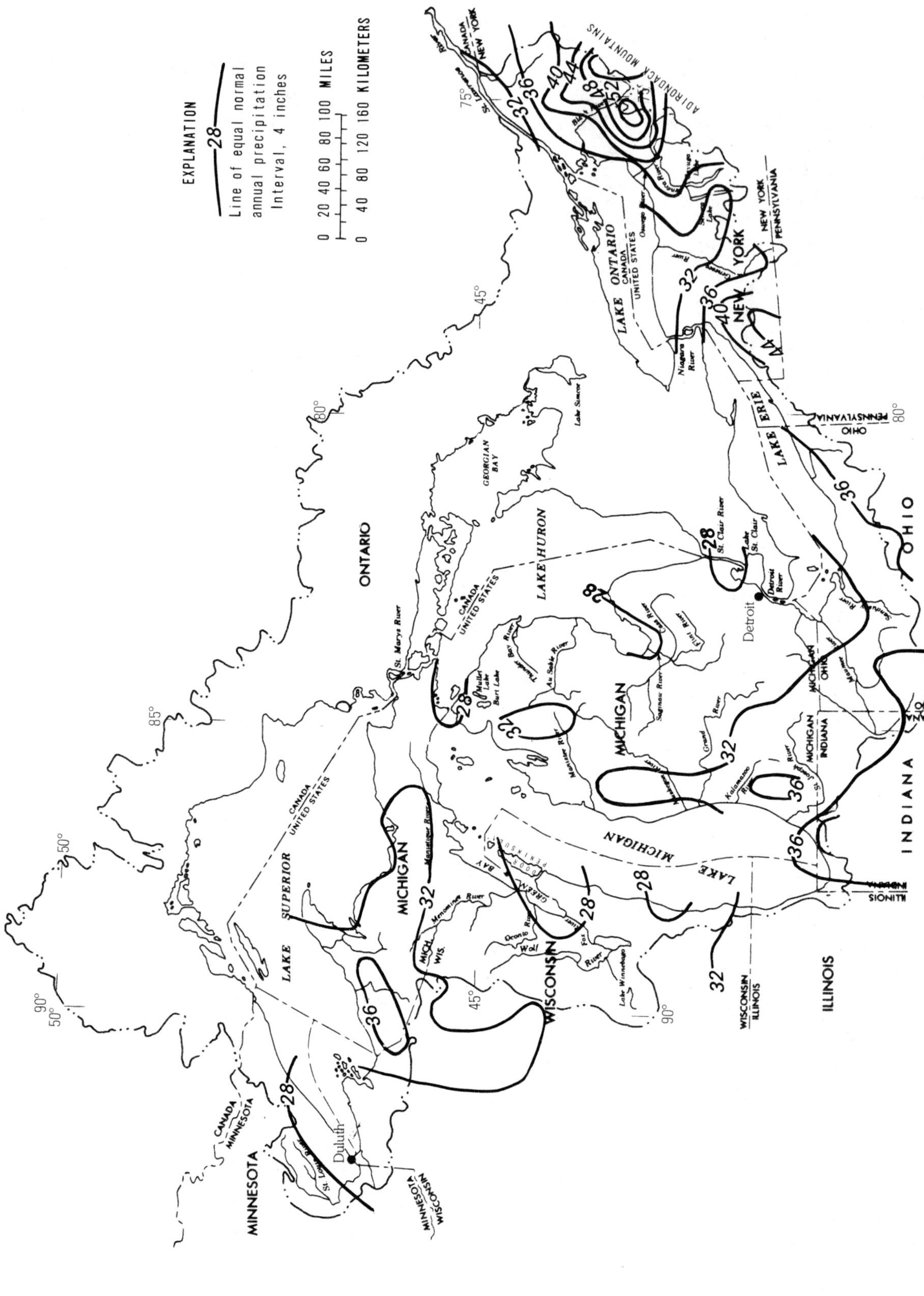

Base from Great Lakes Basin Commission

From Environmental Data Service (1968, p. 43)

FIGURE 2.—Normal annual precipitation, 1931–60.

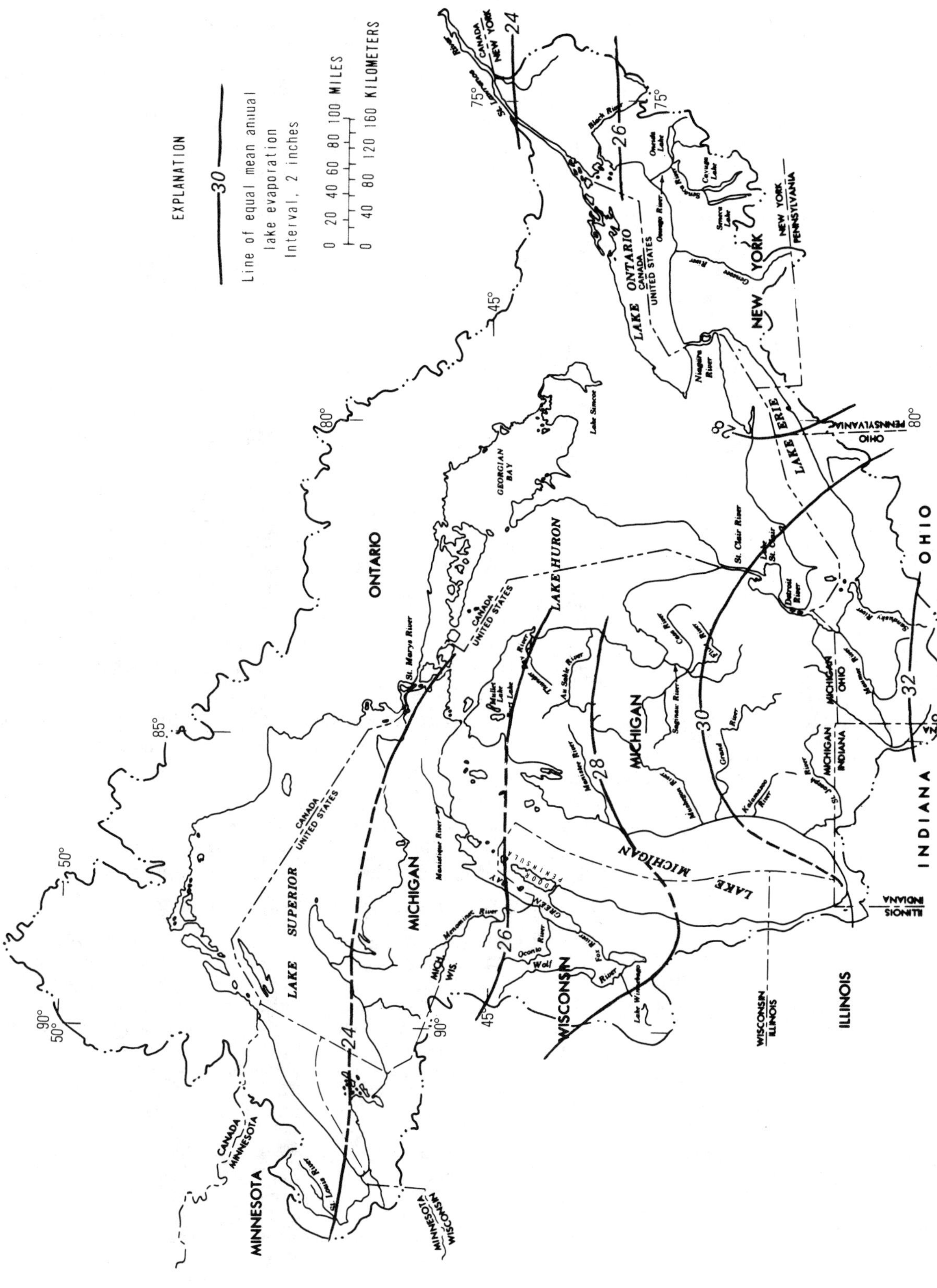

Base from Great Lakes Basin Commission

From Environmental Data Service (1968, p. 63)

FIGURE 3.—Mean annual lake evaporation, 1946–55.

Annual precipitation as a whole averages 31 in. Areas downwind (east) of the lakes generally receive more precipitation than areas upwind. The Adirondacks rise from the lake plains of the eastern border of the region and deflect moisture-laden air upward, so that the Lake Ontario basin receives the most precipitation. Precipitation is fairly evenly distributed throughout the year. February generally has the least (1.78 in) and September the most (3.24 in) (Leonard and Raoul, 1975).

Nearly two-thirds of the precipitation returns to the atmosphere through evaporation and transpiration. As shown in figure 3, mean annual lake evaporation ranges from 24 to 32 in, being greatest along the southernmost edge of the region. About 80 percent of the evaporation occurs between May and October (Environmental Data Service, 1968).

SURFACE WATER

Average annual runoff in the Great Lakes Region is 11.6 in, ranging from 9 to 38 in (Great Lakes Basin Commission, 1975a, p. 25). This wide range is due to differences in geology, surficial features, and land use as well as to differences in precipitation. The Great Lakes form the major drainage system; average flow out of Lake Ontario is more than 239,000 ft^3/s (table 1). Developed storage on streams and lakes provides a sustained water-supply yield of more than 18 billion gallons per day (Great Lakes Basin Commission, 1975b).

Although flood flows get more publicity than low flows because of the damage and suffering they can cause, low flows of streams are more important than floods in terms of water supply, waste assimilation, and fish and wildlife. Floods are a function of precipitation, soil-moisture conditions, snow cover, topographic relief, and other physical features, whereas low flows are a function primarily of ground-water contribution to streamflow. When surface runoff ends, flow in streams is maintained by ground-water discharge. Where streams are fed by extensive aquifers, flow will be maintained even during prolonged periods of no rain; but where streams are adjacent to small aquifers or deposits having low permeability, such as till and lake sediments, flow will decrease fairly rapidly, and may even cease.

The ground-water contribution to streamflow can be estimated from flow-duration curves, which are cumulative frequency curves showing the percentage of time specified discharges are equaled or exceeded. Waller and Allen (1975, p. 8) determined that flow at 70-percent duration represents a conservative value for average ground-water discharge in the Great Lakes basin.

Runoff at 70-percent duration ranges from 0.01 $(ft^3/s)/mi^2$ in streams in parts of the eastern Lake Ontario basin to 0.90 $(ft^3/s)/mi^2$ in the Mannistee River in Michigan and the Black River in New York. Runoff of many of the streams is in the range from 0.20 to 0.40 $(ft^3/s)/mi^2$ (Waller and Allen, 1975).

GROUND WATER

The occurrence of ground water in the region is described in detail in the ground-water appendix to the Great Lakes Basin Framework Study (Waller and Allen, 1975). Therefore, only a brief summary is presented in this report.

Ground water is unevenly distributed across the Great Lakes Region. Land-surface features, precipitation, surface and subsurface drainage, and the hydrologic properties of the underlying rocks combine to determine the availability of ground water in any area. The hydrologic property of the rocks is probably the most important factor in the Great Lakes Region. Unconsolidated outwash sand and gravel deposits are highly permeable and represent the best aquifers in the region. The most productive aquifers are in the valleys of major streams. These aquifers can be recharged by induced infiltration of streamflow, and the sand and gravel act as natural filters to remove many impurities.

In areas of the Great Lakes Region where the unconsolidated deposits are poorly permeable, thin, or absent, ground-water sources are limited to the consolidated (bedrock) aquifers. Supplies from these aquifers are generally adequate and dependable, except locally where shale or crystalline rocks crop out. Fortunately, these areas are few and scattered.

Waller and Allen (1975, p. v) state:

> Carbonate (limestone and dolomite) aquifers constitute the most common bedrock aquifers in the Basin. They occur along the northern and western shore of Lake Michigan, from Illinois to Cleveland, and along the southern shore of Lake Ontario. The carbonates are most productive, with well yields as much as 1,000 gpm, where they extrude or are overlain by unconsolidated deposits. Solution processes have developed good permeability in these areas. Sandstone aquifers are the next most common bedrock aquifers. A thick sequence of productive sandstone units (well yields as much as 1,300 gpm) is present along the western and northern part of the Lake Michigan basin. Such productive units with well yields as much as 500 gpm are also present in parts of Michigan and in Ohio, Pennsylvania, and New York. As aquifers, shale beds are the least productive sedimentary unit. Shales are abundant in the southern part of the Great Lakes Basin from Indiana to the Adirondack Mountains.

Crystalline rocks constitute the bedrock west of Lake Superior and in the Adirondacks region. Wells in these rocks provide adequate yields for domestic and rural supplies and, in exceptional areas, for small commercial and industrial supplies. Water is obtained from fractures and from the upper, weathered part of the rocks.

Yields of as much as 5,000 gal/min are obtainable from wells in some thick deposits of highly permeable sand and gravel, and yields in excess of 500 gal/min have been reported from bedrock in some areas. On the other hand, yields of 5 gal/min may be difficult to obtain in some areas underlain by till or crystalline bedrock. Table 2 summarizes the occurrence of ground water in the Great Lakes Region. Figure 4 shows the distribution of the bedrock units in the region and figures 5–9 show the availability of ground water in unconsolidated deposits in each Great Lake basin.

Well yields and water quality vary across the region, even within the same aquifer system, and a detailed study of the water resources is needed before heavy development can be planned in any area. An important aquifer in one area may yield only small amounts of water elsewhere or water that is too saline for most uses. For example, the Cambrian-Ordovician aquifer system is an important source of water for the Chicago area. However, in northern Indiana the water is so saline that the system is relegated to deep-well waste disposal. In the Lake Superior basin, the Silurian aquifer system yields 50 to 100 gal/min to wells, whereas in parts of the Lake Michigan basin the system yields as much as 1,000 gal/min. In general, the highest yields are obtained from fractured rock aquifers or those that contain solution channels.

The long-term water-yielding potential of an aquifer is dependent on the availability of recharge. In the Great Lakes Region most recharge from snowmelt and rainfall occurs as water infiltrates the soil and percolates to the water table. However, highest well yields are generally obtained from aquifers

TABLE 2.—*Occurrence of ground water in the Great Lakes Region*

[Data from Waller and Allen, 1975]

Aquifer system	Yields of highest-capacity wells (gal/min)	Principal water-bearing units	Areas of greatest potential
		Unconsolidated deposits	
Glacial lake deposits	10–20	Sandy zones within the silt and clay.	
Outwash sand and gravel	50–5,000	Highest yields are obtained in areas with good sources of recharge.	Indiana—South Bend area. Michigan—Jackson, Kalamazoo, area; Au Sable, Manitee, and Muskegon River valley. New York—Genesee River basin. Ohio—northeastern. Pennsylvania.
Till	10–20	Sand and gravel lenses	
		Bedrock	
Pennsylvanian	50–700	Sandstone and shale. Brine and sulfates frequently encountered in lower part.	Southern Michigan.
Mississippian	50–1,800	Sandstone and shale. May contain oil, gas, and brine.	Southern Michigan, northeastern Ohio, Pennsylvania.
Devonian	50–700	Mostly carbonate rocks with some sandstone. May contain oil, gas, and brine.	Ohio, southern Michigan, northeastern Indiana.
Devonian-Silurian	50–500	Carbonate rocks	North Michigan.
Silurian	50–1,000	Mostly carbonate rocks. Water is saline in places.	Eastern Wisconsin, Illinois, northeastern Indiana, northwestern Ohio.
Ordovician	50–500	Carbonate rocks and sandstone. Saline water at depth.	Western New York.
Cambrian-Ordovician	50–1,300	Carbonate rocks and sandstone. Saline water at depth.	Illinois, eastern Wisconsin.
Cambrian	50–500	Sandstone. Saline water at depth.	Illinois.
Cambrian-Precambrian	50–500	Sandstone	North Michigan.
Precambrian	100–250	Sandstone and crystalline rocks	Minnesota—Mesabi Range.

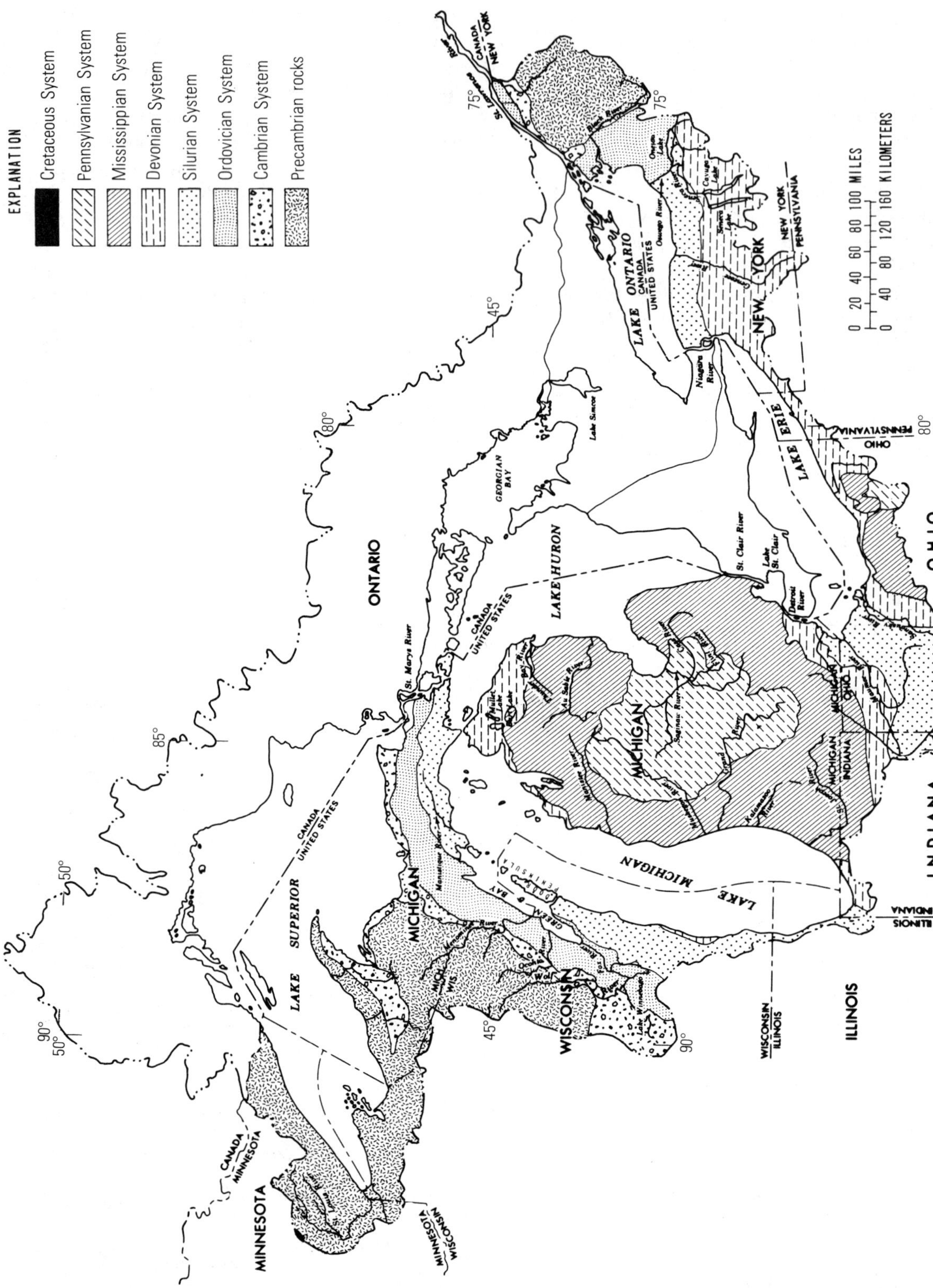

Base from Great Lakes Basin Commission

Adapted from Waller and Allen (1975)

FIGURE 4.—Bedrock geology of the Great Lakes Region.

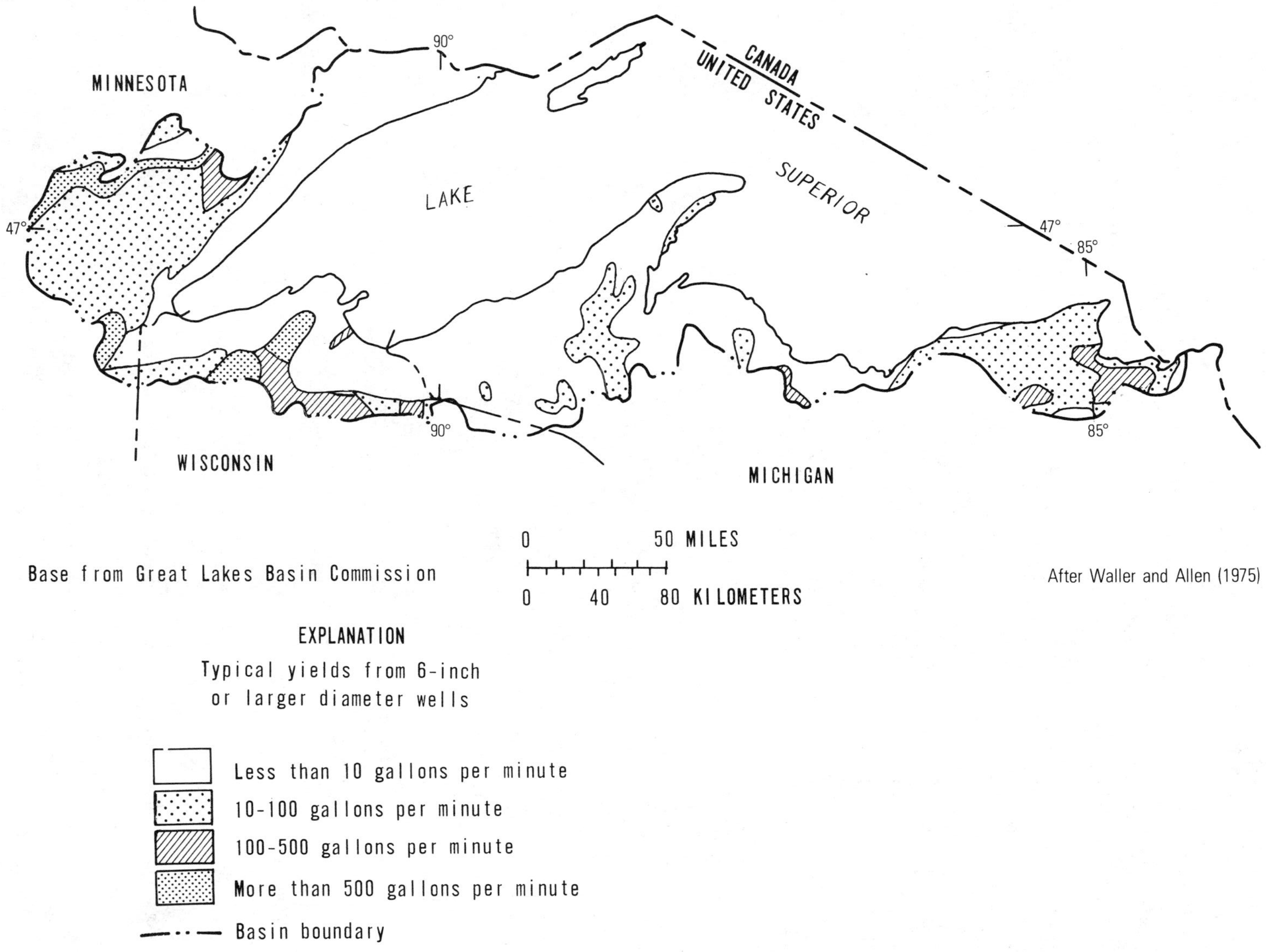

Figure 5.—Availability of ground water in unconsolidated deposits in the United States part of the Lake Superior basin.

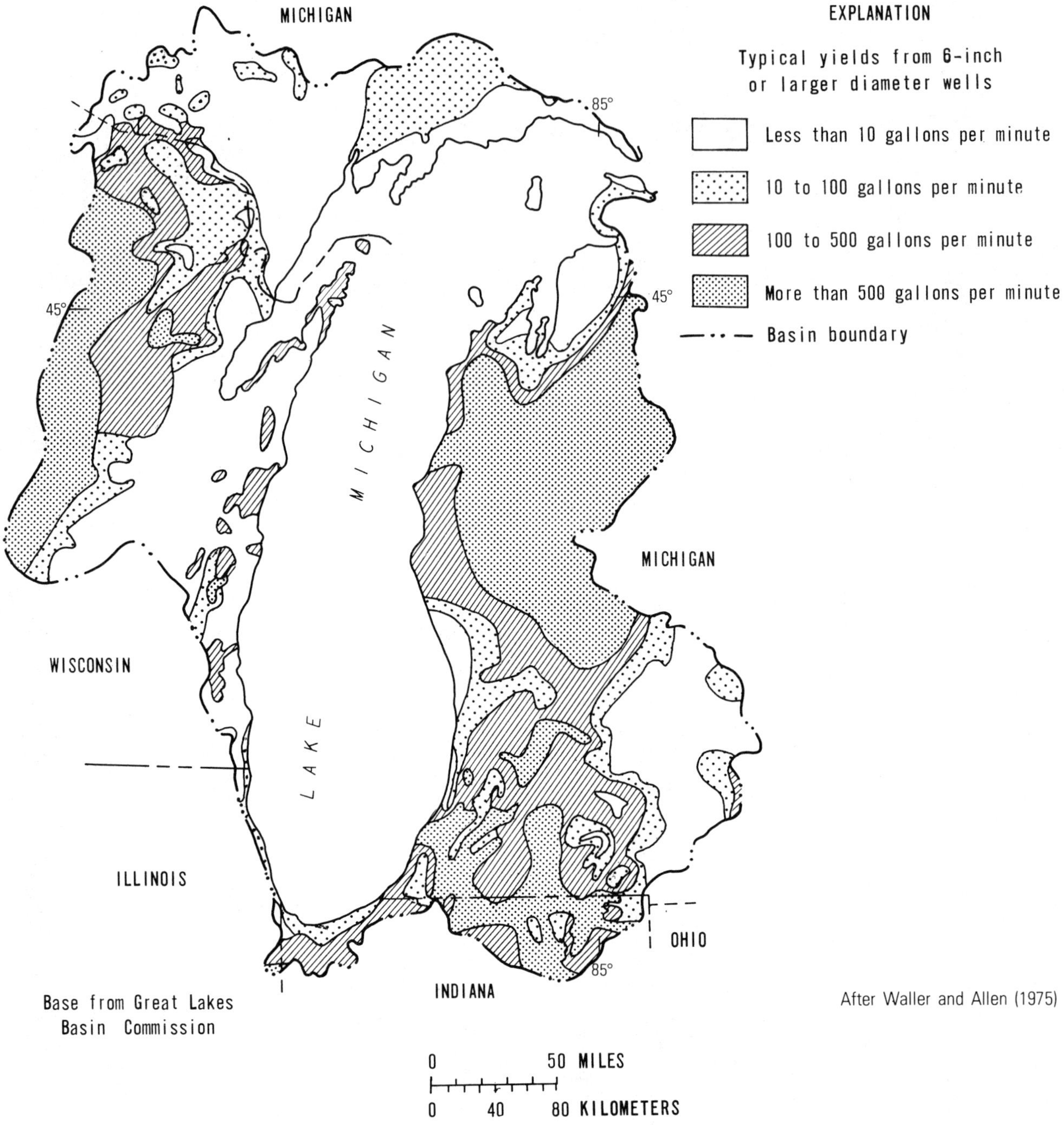

FIGURE 6.—Availability of ground water in unconsolidated deposits in the United States part of the Lake Michigan basin.

that are in direct hydraulic continuity with lakes or streams. Under such conditions, heavy pumping can induce recharge from these water bodies. This will, of course, result in reduced streamflow or lower lake levels.

Natural ground-water discharge maintains streamflow during dry periods and helps to maintain lake levels and wetlands. This discharge varies seasonally. At least 30 percent of the time, during low flow, nearly all water in streams comes from ground-water discharge. This amounts to nearly 26 billion gallons per day for the region (Great Lakes Basin Commission, 1975b). At higher streamflows

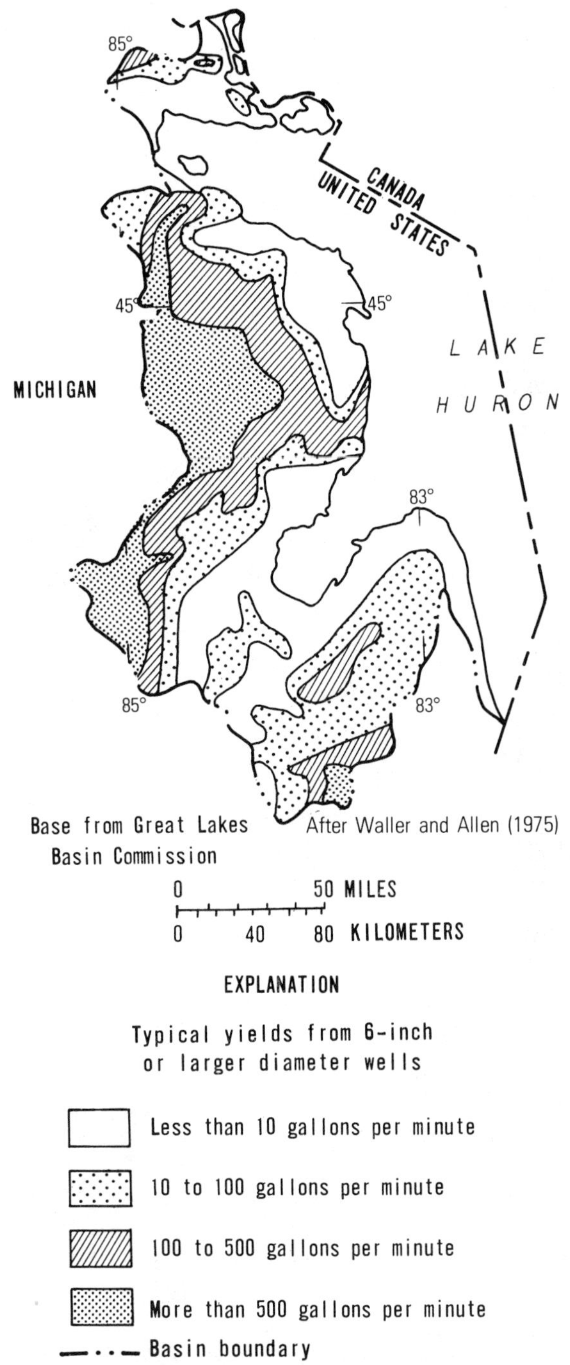

FIGURE 7.—Availability of ground water in unconsolidated deposits in the United States part of the Lake Huron basin.

the ground-water discharge may be greater, but data are not readily available to determine amounts.

Ground-water divides are generally assumed to coincide with surface-water divides. However, this is not always so in the Great Lakes Region. Near Chicago the Great Lakes drainage area is only a mile or so wide, but the bedrock aquifers extend several tens of miles beyond. Heavy pumping from these aquifers in the Chicago area has lowered the pressure in the aquifers, causing water to flow into the area from outside the Great Lakes Region.

ECONOMIC SETTING

DEVELOPMENT

The Great Lakes Region, which constitutes only 4 percent of the Nation's land area, contained slightly more than 14 percent of the population, or 29,322,300 people, in 1970. Of these, 85 percent were concentrated in the Lake Michigan and Lake Erie basins and only 2 percent in the Lake Superior basin.

Major industries include steel production (47 percent of the Nation's total in 1970), petroleum refining, and manufacturing of chemicals, paper, and food products. Urban and industrial areas occupy 8.4 percent of the land.

More than 38 percent of the land is devoted to agriculture, which ranks behind the metal, food, and chemical industries in number of employees. Major crops include corn, wheat, oats, soybeans, and hay. Locally important crops include vegetables, grapes, fruit, and dairy products.

Nearly 50 percent of the land is forested. Recreational use of the land is increasing rapidly, particularly in northern Wisconsin, northern Michigan, and parts of New York. Mining, especially production of sand and gravel, crushed stone, iron ore, and portland cement, is important in some areas. Significant quantities of petroleum are pumped in scattered fields in the region.

WATER USE

The magnitude of water use in the Great Lakes Region is a function of economic development. Heaviest use is in the urban and industrial areas; lightest use is in the forested areas. Steel production, petroleum, refining, and manufacturing of chemicals, paper, and food products account for about 80 percent of the region's industrial water requirements, as well as most water-quality problems (Great Lakes Basin Commission, 1975d, p. 4–5). Locally, mining and ore processing require large amounts of water. With the large quantities of water readily available and most of the major urban areas located on their shores, the Great Lakes are the major source of water. Water use in the region in 1970 is summarized in table 3.

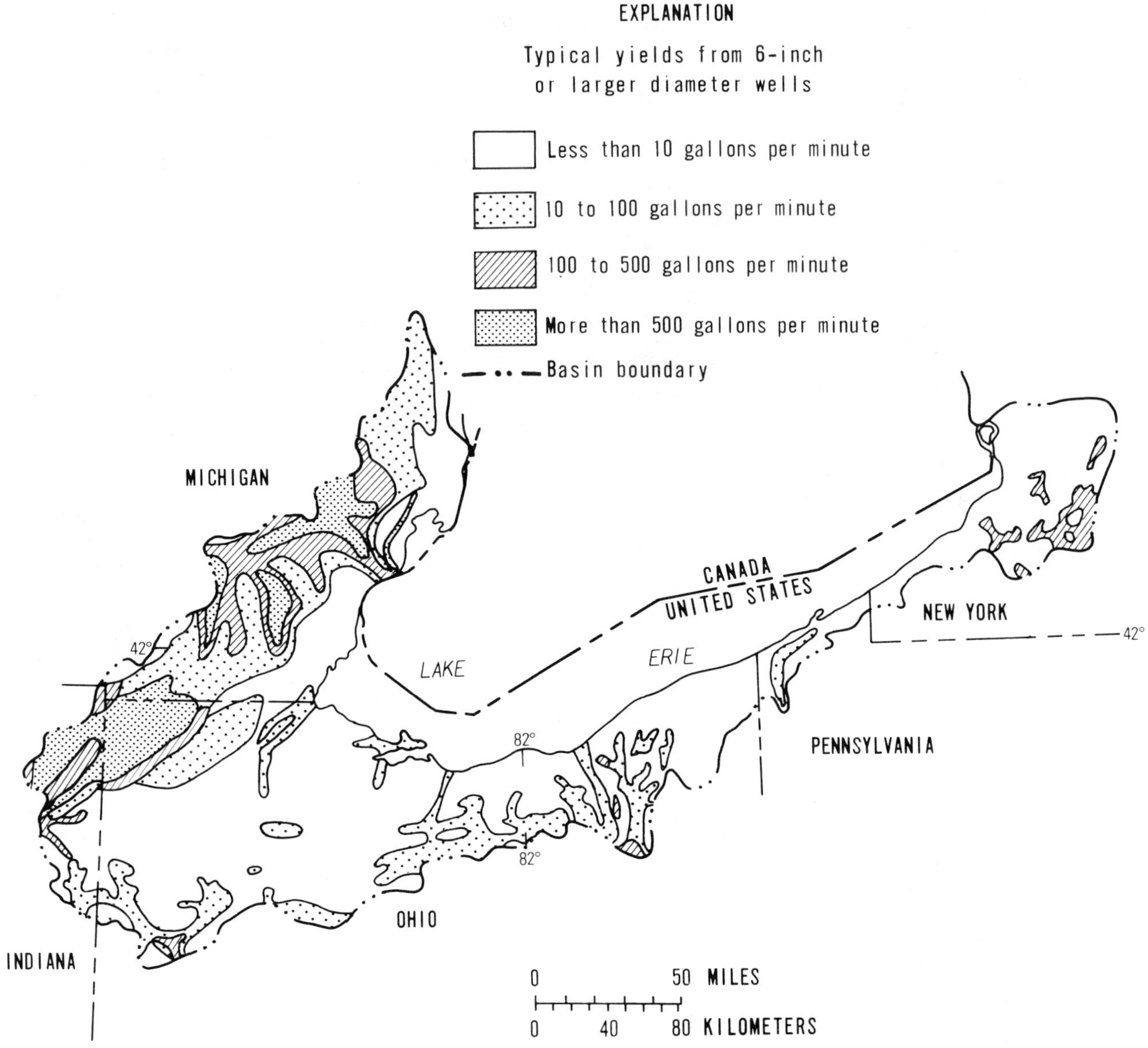

Base from Great Lakes Basin Commission

After Waller and Allen (1975)

FIGURE 8.—Availability of ground water in unconsolidated deposits in the United States part of the Lake Erie basin.

TABLE 3.—*Water use in the Great Lakes Region, 1970, in millions of gallons per day*

[Partial figures may not add to totals because of independent rounding]

Use	Total	Source[1] Surface water	Ground water
Public supply	4,400	3,700	700
Self-supplied industrial	[2]35,000	34,300	340 fresh 400 saline
Rural domestic	280	7	270
Livestock	90	24	62
Irrigation	90	53	37
	39,900	38,100	1,800

[1] Data from Murray and Reeves, 1972.
[2] Includes thermoelectric power use.

ROLE OF GROUND WATER IN THE WATER-RELATED PROBLEMS OF THE REGION

The main objective of this report is to discuss water-related problems and concerns of the Great Lakes Region, how ground water relates to them, and the role ground water can play in meeting or helping to alleviate them. In discussions with water officials and investigators of the region, the author asked what they felt were their major water-related problems. Table 4 summarizes their replies and also

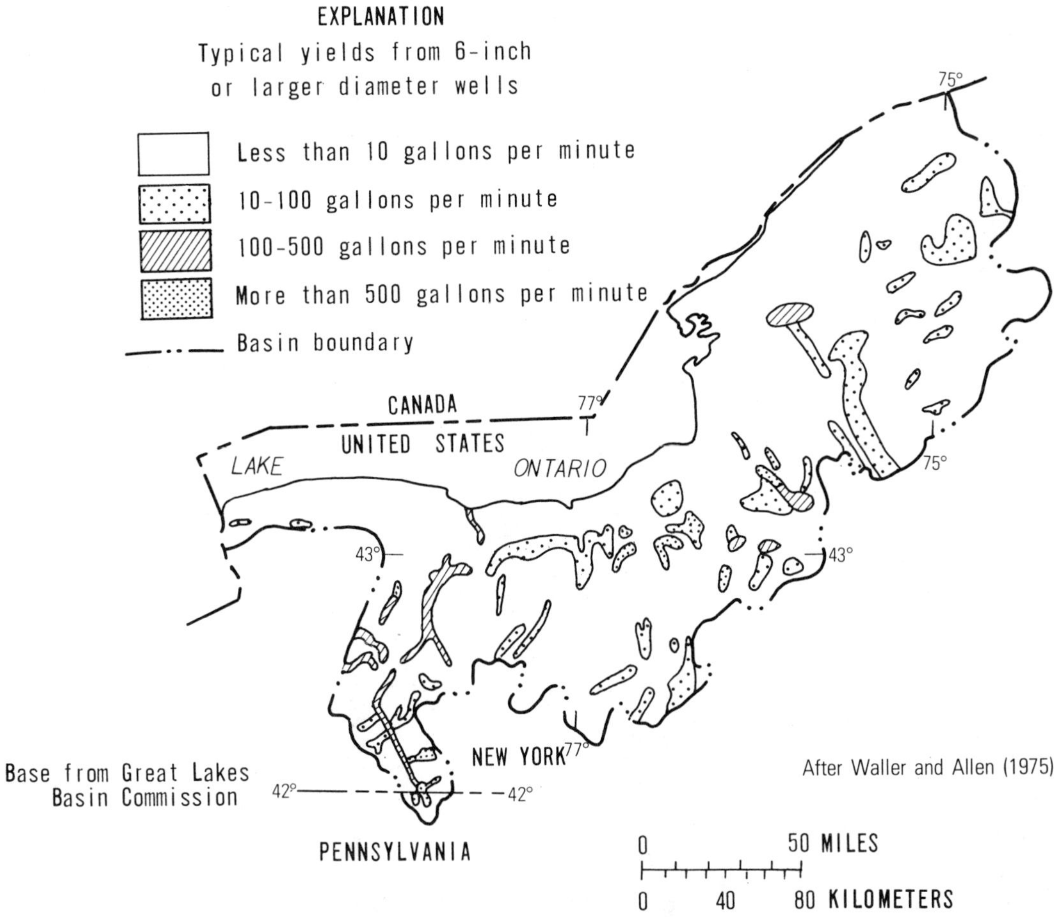

FIGURE 9.—Availability of ground water in unconsolidated deposits in the United States part of the Lake Ontario basin.

summarizes problems identified in reports prepared by various water-resource agencies. The concerns most frequently mentioned relate to management, lack of data, water quality, waste disposal, and water supply. Other, in general lesser concerns include flooding, erosion, wetlands, and low streamflow. Management and lack of data are covered under separate major headings, "Managing ground-water resources" and "Future Considerations," respectively. All the other concerns are discussed in the sections that immediately follow.

WATER QUALITY

NATURAL QUALITY

Natural quality of ground water in the Great Lakes Region is variable, though generally acceptable for most uses. Ground water is hard (more than 120 mg/L) (milligrams per liter) in most of the region, and iron concentrations greater than 0.3 mg/L are common, particularly in water from the sandstone aquifer in Wisconsin and Illinois and from the glacial deposits. High sulfate concentrations (more than 100 mg/L) are found in water from some of the unconsolidated aquifers, particularly in New York, Ohio, Indiana, and Michigan, and in water from Silurian and older bedrock aquifers throughout the region. In general, the water becomes more highly mineralized with depth.

Because ground-water discharge maintains dry-weather flow in the streams and helps to maintain the levels of most lakes and ponds, the chemical quality of these surface bodies of water during dry weather reflects the quality of water in adjacent shallow aquifers. During periods of high flow and in areas affected by pumping, water may move from the streams into the adjacent aquifers.

Ground water usually has a more constant quality than does surface water. During periods of low streamflow, when the quality of surface water may

TABLE 4.—*Water-related concerns*

Agency or investigator	Erosion	Flooding	Lack of data	Low stream-flow	Manage-ment	Recrea-tion	Waste disposal	
							Deep well	Landfills
Illinois								
Illinois Water Survey	--	X	--	--	X	--	--	--
Northeastern Illinois Planning Commission	--	X	--	--	--	X	--	--
Metropolitan Sanitary District, Chicago	--	X	--	--	--	--	--	--
Indiana								
U.S. Geological Survey	X	--	X	--	X	--	--	X
Northwestern Indiana Regional Planning Commission	--	X	--	--	X	--	--	--
Michiana Area Council of Governments	X	--	X	--	X	--	--	--
Michigan								
U.S. Geological Survey	--	--	--	X	X	X	--	--
Michigan Geological Survey	--	--	X	--	X	--	--	X
Soil Conservation Service	X	X	--	--	--	X	--	--
Michigan Department of Natural Resources	--	X	X	--	X	--	--	X
Michigan Bureau of Water Management	--	--	--	--	--	--	--	--
Dr. C. R. Humphreys (Mich. State Univ.)	--	X	X	--	X	X	X	X
Minnesota								
U.S. Geological Survey	X	X	X	X	X	X	--	--
Minnesota Department of Natural Resources	X	--	X	--	X	--	--	X
Arrowhead Regional Development Commission	--	--	X	X	--	--	--	--
New York								
U.S. Geological Survey	--	--	--	--	X	--	--	--
New York State Geological Survey	--	--	--	--	X	--	X	X
New York State Department of Public Service	--	--	--	X	--	--	X	--
New York State Department of Environmental Conservation	--	--	--	X	--	--	X	--
Central New York Regional Planning and Development Board	--	--	X	--	X	--	--	X
Genesee/Finger Lakes Regional Planning Board	--	X	X	X	X	--	X	X
Erie and Niagara Counties Regional Planning Board	--	--	--	--	X	--	--	X
Corps of Engineers, Buffalo District	--	--	X	X	--	--	--	X
Black River Basin Regional Water Resources Planning Board	--	X	--	X	--	X	--	--
Ohio								
U.S. Geological Survey	--	--	--	--	X	--	--	--
Ohio Environmental Protection Agency	--	--	--	--	--	--	X	X
Ohio Department of Natural Resources	X	X	X	X	X	X	X	X
Pennsylvania								
U.S. Geological Survey	X	--	X	--	--	--	X	X
Erie County Department of Health	--	--	--	X	--	--	--	--
Wisconsin								
U.S. Geological Survey	X	X	X	X	--	--	--	--
Wisconsin Geological Survey	X	--	X	--	X	--	--	X
Wisconsin Department of Natural Resources	X	X	X	X	X	X	--	X
Dr. David Stevenson (Univ. of Wisconsin)	X	--	X	X	X	--	X	X
Southeast Wisconsin Regional Planning Commission	X	X	--	X	X	--	--	X
Bay-Lake Regional Planning Commission	X	X	--	--	X	X	--	--
East-Central Wisconsin Regional Planning Commission	X	X	--	--	X	X	--	X
North-Central Wisconsin Regional Planning Commission	--	--	--	--	--	X	--	X
Entire Basin								
Great Lakes Basin Commission	X	X	X	X	X	X	X	X
National Oceanic and Atmospheric Administration	X	X	X	--	--	--	--	--
W. J. Drescher, U.S. Geological Survey	X	--	--	--	X	--	X	X

not meet water-quality standards, it may be feasible to utilize ground water to improve the quality of the streams. (See section on "Low Flow.") Ground water also can be used to improve water quality in small ponds. (See section on "Wetlands and Lakes.")

MINERALIZED GROUND WATER

The term "mineralized water" means different things to different people. This report follows the classification of Krieger, Hatchett, and Poole (1957, p. 5):

Terminology	Dissolved solids (mg/L)
Slightly saline	1,000–3,000
Moderately saline	3,000–10,000
Very saline	10,000–35,000
Brine (sea water)	More than 35,000

The U.S. Public Health Service (1962) recommends that water containing more than 1,000 mg/L of dissolved solids not be used for interstate commerce.

of the Great Lakes Region

Waste disposal—Continued			Water quality						Water supply				
Spreading	Effluents	Septic tanks	Hardness	Iron	Salinity	Sulfate	Industrial operations	Road deicing	Domestic	Industrial	Irrigation	Municipal	Wetlands
Illinois—Continued													
--	--	--	--	--	--	--	--	--	--	--	--	X	X
--	X	X	--	--	--	--	--	--	--	--	--	X	--
--	X	--	--	--	--	--	--	--	--	--	--	X	--
Indiana—Continued													
--	X	X	X	X	X	X	--	X	--	--	--	X	X
--	--	--	--	--	--	--	--	--	--	--	--	--	--
--	X	--	--	--	--	--	--	--	--	--	--	X	--
Michigan—Continued													
X	X	X	--	--	X	--	--	X	--	(1)	X	X	--
--	X	--	--	--	--	--	X	--	--	--	X	--	--
--	--	--	--	--	--	--	--	X	--	--	X	--	--
X	--	X	--	--	--	--	X	X	--	--	--	X	--
--	--	--	--	--	--	--	--	--		--	--	X	--
--	--	--	--	--	X	--	--	--	--	--	--	X	--
Minnesota—Continued													
X	X	--	--	--	X	--	(1)	--	X	(1)	X	X	--
--	--	--	--	--	--	--	--	--	--	X	--	X	--
--	--	--	--	--	--	--	--	--	--	(1)	--	X	--
New York—Continued													
--	--	--	--	--	--	--	--	--	--	--	--	--	--
--	--	--	--	--	--	--	--	--	--	(2)	--	--	--
--	--	--	--	--	--	--	X	--	--	(2)	--	--	--
--	--	--	--	--	X	--	--	--	--	--	X	--	--
--	--	--	--	--	X	--	--	X	--	--	--	--	X
--	--	X	X	--	--	X	--	X	--	X	X	X	X
--	--	--	X	--	--	X	--	--	--	--	--	--	--
--	--	--	--	--	--	--	--	--	--	--	--	--	--
--	--	--	--	--	--	X	--	--	--	X	X	X	X
Ohio—Continued													
--	--	--	--	--	--	--	--	--	--	--	--	--	--
--	--	X	--	--	--	--	X	X	--	--	--	--	--
--	X	--	X	--	--	--	--	--	--	--	X	X	--
Pennsylvania—Continued													
--	--	X	--	--	X	--	X	--	--	--	--	X	--
--	X	--	--	X	--	--	--	--	X	--	X	X	--
Wisconsin—Continued													
--	X	X	X	X	X	X	--	X	X	X	X	X	X
X	X	X	X	X	--	X	X	X	X	X	X	X	X
--	X	X	X	X	--	X	X	--	X	X	X	X	X
--	X	X	--	--	--	--	--	--	--	--	--	--	--
--	X	X	--	--	X	--	X	--	--	--	--	X	--
--	X	X	--	--	--	X	X	--	--	X	--	X	X
--	--	X	--	--	--	--	X	--	--	--	--	--	--
--	--	X	--	--	--	--	X	--	--	--	--	--	--
Entire Basin—Continued													
--	X	X	X	X	X	X	X	X	--	X	X	X	X
--	--	--	--	--	--	--	--	--	--	--	--	--	--
--	--	--	--	--	--	--	--	--	--	X	--	X	X

[1] Mining.
[2] Power.

However, people can become accustomed to water with a higher dissolved-solids concentration, and water containing as much as 3,000 mg/L can be used for a domestic supply, where less mineralized water is not available.

Slightly saline water occurs at depths less than 500 ft below land surface throughout much of the region (fig. 10). In parts of eastern Michigan and Pennsylvania, saline water is at depths less than 100 ft below land surface. Most mineralized water at shallow depths is due to natural migration upward and outward from deeper bedrock formations and to leakage through uncased or improperly constructed wells.

Mineralized ground water is generally considered a nuisance in most of the Great Lakes Region. However, it does have considerable economic potential. In areas where supplies of potable ground water are scant, such as in parts of eastern Michigan and central New York, mineralized ground water is a

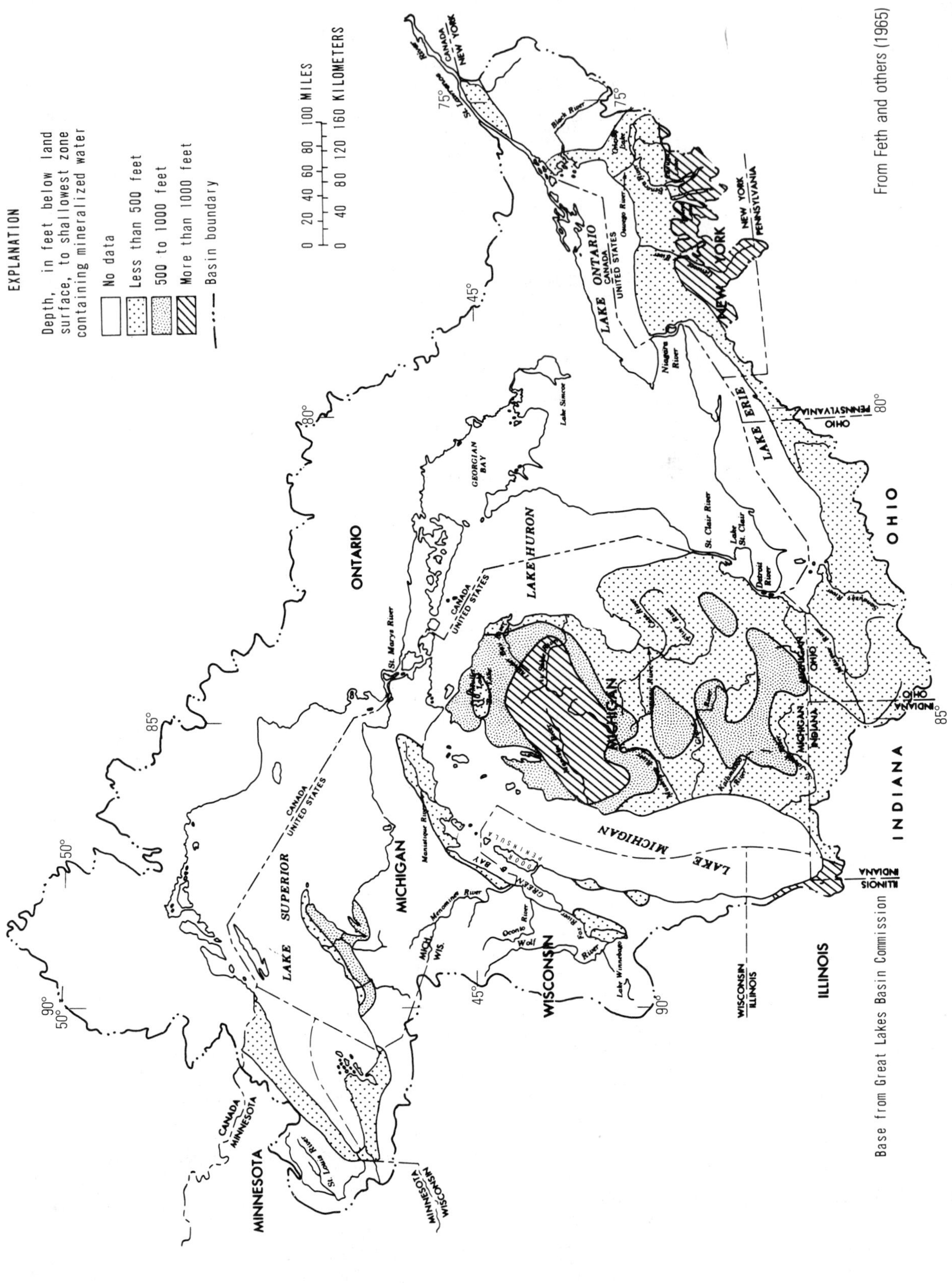

FIGURE 10.—Areas where dissolved-solids concentration of ground water exceeds 1,000 mg/L.

potential source of water supply. Advances in desalination processes are making saline water economically competitive with other alternatives, such as reservoirs and pipelines. Industry can use mineralized water for cooling and for processing, where the water does not become part of the finished product.

A concept that is attracting increasing interest is the use of saline aquifers to store freshwater or natural gas. Kimbler, Kazmann, and Whitehead (1975), working with hydrologic models (physical and mathematical) of unconsolidated aquifers, found that a zone of stagnation can be created within a saline aquifer to reduce mixing of injected fresh and host saline waters. They report that the percentage of the freshwater recovered uncontaminated with saline water improves with each cycle of injection and recovery. Storing freshwater in a saline aquifer would be particularly attractive in heavily populated areas, where all feasible surface reservoir sites have been developed or where land is too expensive for surface reservoirs.

Because most investigations so far have been confined to aquifers having intergranular permeability and porosity, storage of freshwater in a saline aquifer may have only slight application in the Great Lakes Region, where most saline aquifers are in fractured bedrock. Additional studies, however, may show that the principles described by Kimbler, Kazmann, and Whitehead (1975, p. 73–74) can be modified to apply to these aquifers.

Saline aquifers are also used for disposal of liquid wastes. Caution is necessary, however, for much still needs to be learned about the processes and risks involved. Disposal of liquid wastes is discussed in more detail in the next section.

POLLUTION

Ground water can become polluted in many ways. Some pollution is related to development of the resource, as when heavy pumping induces flow into an aquifer from an adjacent saline water aquifer; some is related to waste disposal, as when leachate enters an aquifer from a landfill; and some is related to other activities of man, such as oil and gas production. The major sources of ground-water pollution in the Great Lakes Region are listed in table 5.

TABLE 5.—*Major sources of ground-water pollution in the Great Lakes Region*

Source of pollution	Primary pollutants
Waste disposal	
Septic tanks and cesspools	Bacteria, nitrate, detergent, chloride.
Landfills and dumps	Heavy metals, PCB's (polychlorinated biphenyls), chloride, phenols, nutrients.
Spreading of water-treatment and waste-treatment effluents.	Chloride, nutrients, heavy metals.
Holding ponds	Variety of industrial wastes such as phenols, PCB's, heavy metals.
Disposal wells	Radioactive wastes, industrial chemicals.
Industrial operations	
Oil and gas recovery	Chloride, petroleum products.
Salt mining	Chloride.
Accidental spills, leakage	Petroleum products, industrial chemicals.
Road deicing	Chloride.
Agricultural activities	
Feedlots	Bacteria, nutrients.
Use of fertilizers	Nutrients.
Use of pesticides	Various hydrocarbon compounds.
Management problems	
Overdevelopment of aquifer	Chloride.
Abandoned wells, leaky casings, improperly cased wells.	Chloride, nitrate.

Waste disposal and the effects it may have on water quality is a major concern in the region. Under the Federal Water Pollution Control Act Amendments of 1972, the States are mandated to eliminate stream pollution. Implementation of this law could place additional stress on the ground-water system through an increase in use of landfills and deep disposal wells and through the surface spreading of liquid-waste effluents.

SEPTIC TANKS

The most widespread means of waste disposal in the region is through septic tanks. Unfortunately, soils in much of the region are not suitable for this type of disposal. Septic tanks do not function well where the water table is near land surface, where the soil is clayey, or where the soil cover is thin. These conditions, singly or in combination, are common throughout much of the Great Lakes Region. Numerous septic tanks clustered together under these conditions, as in summer-home developments, can cause pollution.

Vogt (1972) and Pettyjohn (1972) describe two areas where ground water has become polluted from septic-tank wastes. In both areas only a few feet of soil overlies a highly permeable limestone, which is the source of household water for the areas. At Posen, Mich. (Vogt, 1972), septic-tank effluent carrying hepatitis virus was discharged into the soil zone. Percolating water from melting snow and spring rains carried the effluent and virus into the ground-water reservoir, spreading the virus to nearby wells and causing an epidemic of infectious hepatitis.

Residents of Bellevue, in north-central Ohio, used septic tanks that discharged into a thin soil zone overlying a very permeable limestone aquifer. They also drilled waste-disposal wells into the limestone (Pettyjohn, 1972). The result was a plume of polluted ground water extending 15 mi to Lake Erie through a highly productive aquifer (well yields greater than 500 gal/min).

Wisconsin's Door Peninsula is another area where pollution of ground water is widespread owing to the disposal of septic-tank effluent and agricultural, industrial, and municipal wastes into a thin soil zone (Sherrill, 1975). The wastes percolate through the soil, enter a cavernous limestone aquifer, and then spread rapidly through the aquifer. Thus, the shallow ground water is polluted, and the State now requires at least 100 ft of casing in new wells. This pollution could increase and spread laterally and vertically unless other means of waste disposal are used.

Increased eutrophication of many lakes in the region is, in part, a result of ground-water pollution from septic wastes. Nutrients in septic-tank effluents are transported by ground water and discharged into lakes. Conditions conductive to eutrophication are particularly prevalent in lakes of northern Wisconsin and northern Michigan, where lake shores are occupied by year-round communities and subdivisions.

LANDFILLS

Although not as numerous or widespread as septic tanks, landfills are another potential source of ground-water pollution. For example, three landfills in northern Ohio have polluted a shallow aquifer, which is the source of water for nearby residents (David Johe, Ohio Environmental Protection Agency, oral commun., 1975). Johe reports that two of the landfills have been closed but that polluted water will continue to enter the aquifer for years, and new sources of household water will have to be found. Because ground water moves so slowly and diffusion rates are so low, polluted ground water may not be noticed for many years. Other landfills in the Great Lakes Region are probably polluting ground-water bodies.

A waste-disposal site is usually considered suitable if the surface will remain unsaturated. In humid areas, such as the Great Lakes Region, sites where excavation will not intersect the zone of saturation are difficult to find. For this reason, many landfills are located in materials of low permeability, such as till. Generally, test drilling is necessary to identify the materials at the potential landfill site, which can be expensive. For preliminary evaluation of a site, data from nearby water wells can be used to estimate the characteristics of the subsurface materials and the depth to water.

WASTEWATER EFFLUENTS

Disposal of effluents from water-treatment and sewage-treatment plants is of increasing concern as more wastes are generated. Statutes restrict the discharge of effluent to streams. Early results from an effluent-disposal project near Muskegon, Mich., indicate that spray-irrigation of liquid wastes may be practical. Effluent from Muskegon and from a nearby paper company are being used to spray irrigate about 6,000 acres of principally very sandy soil. Irrigation water that percolates through the upper few feet of soil is collected in perforated tile drains and discharged into nearby streams. Thus far the return water has been of acceptable quality. If return water of good quality can be maintained, spray irrigation may not only provide a means of waste disposal but also may become a source of recharge to the local aquifer.

DEEP-WELL DISPOSAL

Deep-well disposal of liquid wastes, although not new, has received increasing interest in recent years (Cook, 1972; Braunstein, 1973). This approach has an out-of-sight, out-of-mind appeal and is sometimes the easiest method of disposal, but it may be impractical or unwise in many areas. The effects of injecting wastes into deep aquifers are complex and varied. Injection is not a means of permanent disposal but, rather, a means of storage (Piper, 1969, p. 6), and the wastes could become a problem in the future. Most of the wastes are injected into aquifers containing highly saline water. Although the saline water has no use at present, it may be of beneficial use in the future.

The Great Lakes Basin Commission reports 86 waste-disposal wells in the region (Eugene A. Jarecki, oral commun., 1975). The U.S. Environmental Protection Agency (1974) published records of 33 operating wells and 19 inactive wells in the region in 1973. Of those in use, 21 are in Michigan, 8 in northwestern Indiana, 3 in northwestern Ohio, and 1 in New York. One well in Michigan injects wastes into Mississippian sandstone at a depth of 1,030 ft. The Cambrian Mount Simon Sandstone receives wastes from 3 wells in Michigan, 3 in Ohio, and 6 in Indiana. Well depths range from 2,160 ft in Indiana to 4,608 ft in Michigan. The rest of the wells inject waste into Devonian limestone, dolomite, and sandstone. Well depths range from 255 ft in Indiana to 4,986 ft in Michigan. Most of the wastes are from steel mills (spent pickling liquor), chemical and drug manufacturing, and brine processing but also include wastes from oil refining and from a laundromat. Rates of injection range from about 1 gal/min to 556 gal/min.

Because of the wide variation in geologic and hydrologic settings and in the volume and nature of wastes, adoption of uniform nationwide standards and criteria for deep-well disposal is not fully practical. At present (1976), Minnesota and Wisconsin do not allow deep-well disposal of liquid wastes. Where such disposal is considered in the other States in the Great Lakes Region, careful studies help to determine the physical and chemical nature of the wastes, compatibility of the wastes with fluids in the injection zone, and possible reactions between the wastes and the rock materials constituting the aquifers. Stringent well-construction standards and careful monitoring of the injection process would help to protect potable ground-water.

OTHER SOURCES OF POLLUTION

In addition to waste disposal, other activities cause ground-water pollution in the region. Improperly cased wells, improperly plugged abandoned wells, or unplugged abandoned wells with deteriorated casings have allowed water of poor quality to move into aquifers that contained good water. Accidental spills, leaky storage tanks, and broken pipelines have allowed petroleum products to contaminate ground water. The effects of these and similar activities are usually local, but they are very long lasting.

One widespread source of ground-water pollution that is increasingly recognized is the salt used in deicing roads. The salt not only imparts a bitter taste to the water but also may be hazardous to persons on a low-sodium diet.

WATER SUPPLY

Considering the amount of water available in the Great Lakes Region, the idea of water-supply problems may seem anomalous. Nevertheless, the problems are real. Although many supply problems relate to water quality, as previously discussed, some areas face a potential shortage of water. In other areas, water-supply problems are the result of local overdevelopment.

The water-supply situation in the Great Lakes Region is summarized in the following table (Great Lakes Basin Commission, 1975b, p. 30):

	1970 (Mgal/d)	2020 (Mgal/d)
Public supply:		
Use	4,400	9,200
1970 capacity/2020 needs	7,400	1,800
Industrial supply:		
Use	[1] 10,600	12,800
1970 capacity/2020 needs	10,600	2,200
Rural supply:		
Use	470	740
1970 capacity/2020 needs	470	270

[1] Does not include thermoelectric power use.

The entry "1970 capacity" is the developed capacity to supply water as of 1970. The entry "2020 needs" refers to the additional capacity that must be developed to meet the demands anticipated by the year 2020. This capacity could involve developing additional water supplies, increasing the capacity of water-treatment plants, installing new distribution lines, or other similar measures.

Many people are aware of some of the advantages and disadvantages of ground water as a source of supply because many of them obtain water from their own wells or from municipally owned wells. Nevertheless, a brief discussion seems indicated. Ground water generally costs less to develop than an equal amount of surface water. Howard (1974) reported that costs for developing a ground-water supply range from $100 to $300 per gallon per minute, whereas the costs for developing a surface-water supply ranges from $1,100 to $1,500 per gallon per minute. A ground-water supply can be developed gradually as demand warrants, whereas an intake structure or reservoir for a surface-water supply must be designed and constructed to meet not only present demands but also anticipated future demands. Properly managed, a ground-water supply does not lose its capacity or deteriorate, nor does it require a large land area. A surface-water reservoir, on the other hand, is subject to loss of capacity from siltation, eutrophication, pollution from water-based recreation, and deterioration of the physical facilities.

Ground water usually has a narrow temperature range, little or no sediment or turbidity, is relatively free from nuclear fallout, and usually is not polluted by bacteria or viruses. (A study of public water supplies in New York has shown that ground water there is free from organic pollution (U.S. Geol.

Survey, 1974). There were no traces of PCB's, pesticides, or organic mercury in any samples of ground-water supplies.) In addition, ground-water loss through evaporation is minimal, and ground water is widely available throughout the region, at least in moderate quantities.

On the other hand, ground water is generally more mineralized than surface water and, once polluted, is very difficult, if not economically impossible, to clean up. Where a ground-water source is heavily developed, water levels are lowered in nearby wells, pumping costs are increased, and streamflow may be decreased as a result of induced flow into the ground-water system. If water of poor quality is present in an adjacent aquifer, heavy pumping may induce this water to flow into the developed aquifer. In parts of the Great Lakes Region, available quantities of ground water are insufficient to support large-scale development.

PUBLIC SUPPLIES

Most large metropolitan areas in the Great Lakes region obtain water from surface reservoirs, streams, or one of the Great Lakes. Smaller communities generally obtain their supply from ground water. In general, in spite of local problems, ample water is available to meet current municipal demands.

Water for public supplies in 1970 was withdrawn at a rate of about 4,400 Mgal/d (Great Lakes Basin Commission, 1975b, p. 30), of which 3,400 Mgal/d, or 77 percent, came from the Great Lakes; 390 Mgal/d, or 9 percent, from inland lakes and streams; and 560 Mgal/d, or 13 percent, from ground-water bodies. Developed water-supply capacity was 7,400 Mgal/d. The Great Lakes Basin Commission estimates that by the year 2020 nearly 9,200 Mgal/d will be needed to meet public-supply demands—an increase of 1,800 Mgal/d over present developed capacity.

The city of Chicago obtains most of its water from Lake Michigan. Many of the surrounding communities and industries, however, obtain water from deep bedrock aquifers. This deep aquifer system is fairly uniform throughout the area, and well yields and water quality are predictable. Pumping has lowered water levels several hundred feet at major pumping centers. Pumping has been heavy also in the Milwaukee area, and the drawdown cones of the Chicago and Milwaukee areas have begun to overlap. Further development of the deep aquifer system would probably accelerate the decline in water levels and may induce water of poor quality to move updip into the area from the east.

Less is known about the hydrology of the shallow aquifers in the Chicago area because they have not been developed as much as the deep aquifers. Increased development of the shallow aquifers could relieve some of the stress on the deep aquifer system. Digital models of the aquifer systems have been made by the Illinois Water Survey for the Chicago area and by the U.S. Geological Survey for the Milwaukee area. These models will enable planners to manage the development and use of the aquifers more effectively.

Prior to 1957, the city of Green Bay, along with neighboring communities and industries, obtained its water supply from a sandstone aquifer. By 1957 the water level in wells tapping the aquifer had declined more than 200 ft in downtown Green Bay (Knowles, 1964), and the city had to seek another source of supply. Other sources in the immediate area—Green Bay, Fox River, and Lake Winnebago—are polluted. Thus, the city turned to Lake Michigan for its supply and since 1957 has obtained water from the Lake by pipeline across the Door Peninsula. When Green Bay discontinued pumping from the sandstone aquifer, water levels recovered rapidly. However, subsequent development of the aquifer has caused the water levels to decline severely again.

Although the sandstone aquifer is heavily used, it has potential for additional development, but careful planning would be necessary to prevent interference between wells and consequent local depletion of supplies. Local officials are concerned that concentrated heavy pumping will cause water of poor quality from the south to move into the aquifer.

As glaciers moved across the Great Lakes Region, many valleys were filled with debris. These buried valleys that contain extensive deposits of saturated sand and gravel are a potential source of public-water supply.

Fort Wayne, Ind., obtains its water supply from the St. Joseph River. The U.S. Geological Survey, in cooperation with the Indiana Department of Natural Resources and the city of Fort Wayne, is making a study of the availability of ground water for future supply. Preliminary results indicate that a buried valley to the northwest may be able to supply sufficient water to the city for many years.

Buried valleys are also potential sources of water for communities in the Erie, Pa., area and parts of northeastern Ohio. These valleys have not been systematically studied, so little is known about their hydrologic character. The U.S. Geological Survey, in cooperation with the Pennsylvania Geological Survey, is starting a study of Erie County that will include investigation of the ground-water potential of some of these valleys.

For cities near one of the Great Lakes, the conjunctive use of ground water and surface water may have advantages. Ground water could be used during summer and fall, when it is cooler than lake water. The lake water could be used during winter and spring while the aquifers are recharged. Mixing water from both sources could reduce treatment costs. Alternatively, areas near the lake could be supplied with lake water while inland areas are supplied with ground water. This could reduce pumping and transmission costs.

INDUSTRIAL SUPPLIES

The Great Lakes Region is heavily industrialized, and industrialization is expected to increase during the next few decades. In addition to manufacturing, other large industrial uses of water include power generation, sand and gravel production, and ore processing. Self-supplied industrial use of water in 1970 was 10,600 Mgal/d. This is expected to increase to nearly 12,800 Mgal/d by the year 2020 (Great Lakes Basin Commission, 1975b, p. 30).

Many industries in the region purchase water from municipalities; many others, however, provide their own supply. Industries tend to cluster together, usually in or near large cities. Their closeness can lead to overdevelopment of the water sources and to deterioration of water quality. Nearly everywhere that municipalities in the region are faced with a water-supply problem, industries also have a problem.

Although outside the Great Lakes Region, the Mill Creek valley in southwestern Ohio is a good example of what industries can do by cooperating in developing their water supplies. The Mill Creek valley is part of the greater Cincinnati industrial area. Industrial development started in the late 1800's and increased steadily, with both World Wars providing impetus for rapid growth. Ground-water withdrawals rose comparably, and, by the late 1940's, water levels were extremely low and well yields sharply reduced. In an effort to improve the situation, 11 industries formed the Southwestern Ohio Water Company. They developed a ground-water system in an adjacent valley and piped the water into their industrial complex. As a result, withdrawals from the aquifer in Mill Creek valley declined, and water levels and yields recovered. However, increased industrial development brought the threat of further problems, and the water users requested a study of the feasibility of artificial recharge to the aquifer. This study (Fidler, 1970), made in cooperation with the Ohio Department of Natural Resources, Division of Water, concluded that artificial recharge through injection wells would reverse the declining water levels.

Formation of a cooperative to import water may not be feasible everywhere in the Great Lakes Region. Nevertheless, it shows what cooperation and ingenuity can do to meet a water-supply problem.

IRRIGATION

Irrigation of crops does not yet play a major role in the Great Lakes Region. Out of a total of 11,972,000 potentially irrigable acres, 196,800 acres were irrigated in 1968 (Great Lakes Basin Commission, 1975c, p. 8–9). More than half the irrigation is in Michigan. Irrigated acreage is estimated to increase to more than 520,000 acres by 2020. Agricultural irrigation in the region required 106,700 acre-ft in 1968 and probably will require more than 480,000 acre-ft in 2020. Similar amounts were used and will be used to irrigate golf courses.

In recent years, agriculture has depended on improved hybrid crops and fertilizers to increase production. A plateau may have been reached in their effectiveness, however, and farmers may be turning more to irrigation to ensure high productivity. If so, demand for water will increase.

Even in the humid climate of the Great Lakes Region, soil moisture can limit crop growth. Studies of the relation of evapotranspiration to climate show that yield-depressing droughts are common, even though average precipitation during the growing season may exceed the amount needed by the crop (Richards and Richards, 1957, p. 49).

A steady, dependable source of good-quality water is needed for irrigation, which means that most of the increased demand will be supplied by ground water. This water may come directly from wells, from ponds excavated into the water table, or by pumping from wells into a holding pond and then distributing the water as needed. This last method will permit irrigation in areas where high yielding wells are not feasible.

LOW FLOW

Low flow (dry-weather flow or base flow) in streams consists of ground-water discharge into the streams. In the Great Lakes Region, low flows are greatest per unit area in streams that drain highly permeable aquifers. In streams that drain poorly permeable aquifers, flow will decrease fairly rapidly after rains and may even cease altogether between rains. Flow-duration curves for two streams in Ohio are shown in figure 11. The data were adjusted to the 1931–60 period (Cross, 1968). Curve A is for the Ottawa River at Allentown, Ohio. The river drains an area of fairly coarse-grained glacial materials that readily yield ground water to maintain streamflow. The stream also receives discharge from the underlying carbonate-rock aquifer. Curve B is for the Ashtabula River near Ashtabula, Ohio. The river drains fine-grained glacial and lake deposits, which do not yield much ground water to maintain streamflow. Using figure 11, one can calculate that, at the 95 percentile, the Ottawa River will have a flow of 11 ft^3/s, whereas the Ashtabula will have a flow of 0.1 ft^3/s.

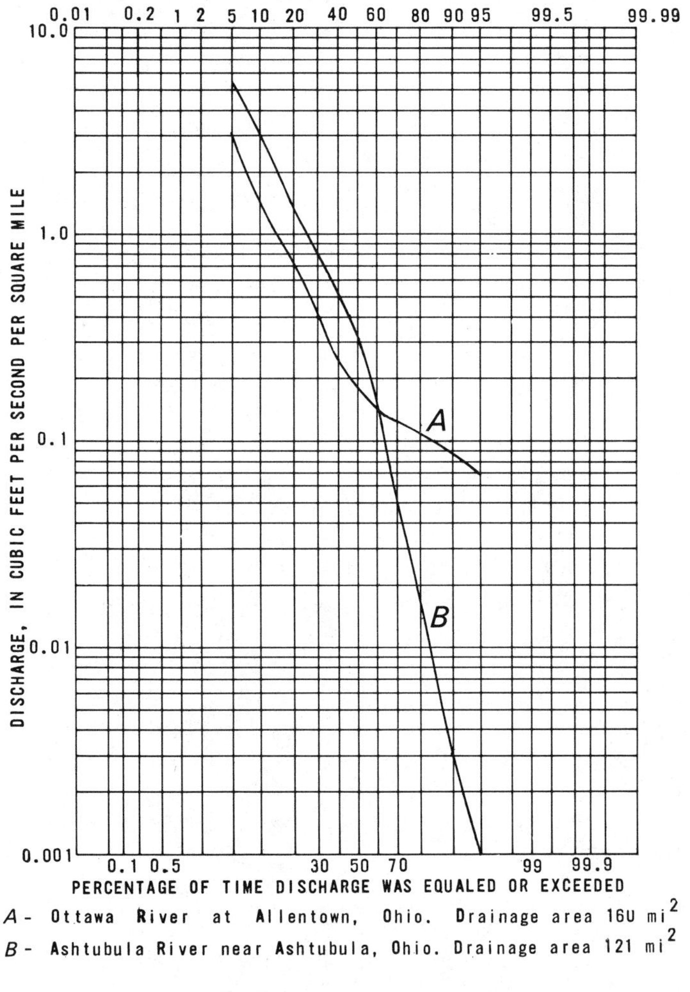

FIGURE 11.—Flow-duration curves for two streams in Ohio, 1931–60.

Augmentation of low flow has been considered for some streams, usually to maintain the quality of the water. Implementation of the Federal Water Pollution Control Act Amendments of 1972 should lessen the need for augmentation to control water quality. However, nonpoint sources of pollution will continue to affect the quality of many streams. Other reasons for low-flow augmentation include maintaining streamflow as a source of water for industries and municipalities, maintaining fish and wildlife habitats, and recreation.

Many of the small communities along the Great Lakes obtain their water supply from small streams that drain directly into the Lakes. During dry periods, flow in many of these streams is insufficient to meet needs of the communities. Even on larger streams, a combination of low flow and heavy development can cause water shortages. In addition to maintaining sufficient flow to avoid damage to fish and wildlife habitats, low-flow augmentation can be used to improve water-quality (temperature, dissolved oxygen) conditions in trout streams (Novitski, 1973).

Most low-flow augmentation is done with water from reservoirs. In some areas, ground water is an alternative source. Where the aquifer is separated from a stream by a less permeable zone, pumping ground water into a stream for flow augmentation may be especially practical. Even where the aquifer is in good hydraulic connection with the stream, proper well location can prevent adverse effects of pumping on streamflow during the critical low-flow period. Most periods of critical low flow, when augmentation would be needed, last only a few days. By locating the well(s) several hundred to a few thousand feet (depending on hydrologic character of the aquifer) from the stream, the effects of diverting ground-water flow will not reach the stream until the critical period is over. By then, pumping will have stopped, and the water table can recover.

WETLANDS AND LAKES

The relationship between ground water and wetlands is frequently misunderstood. Many people think of wetlands as areas of ground-water recharge, whereas most wetlands are maintained because they are areas of ground-water discharge. Similarly, ground-water discharge maintains the water level in most of the smaller lakes and ponds in the region. Permanently draining a wetland, pond, or lake would decrease the availability of ground water in the area because it would lower the water table.

Wetlands, lakes, and ponds do recharge aquifers during floods, when the water level in the wetlands and lakes is higher than in adjacent aquifers. However, the recharged water generally quickly dissipates once flood levels recede, as water flows from the aquifers back into the wetlands and lakes. This same process along streams is termed "bank storage."

In some places, ground water can be used to maintain the water level artificially in a wetlands or a small lake or to create a new wetlands or lake. If the aquifer is not hydraulically connected with the lake or wetlands, water could be pumped from the aquifer to maintain levels (fig. 12*A*). However, if the aquifer is hydraulically connected with the lake (fig. 12*B*), pumping from a well near the lake could cycle the water from the lake to the aquifer and back to the lake.

During recent years, the Detroit municipal water system, which obtains water from Lake Huron, has enlarged its service area to include about 2,000 mi^2 in southeastern Michigan, including Pontiac and Flint. Much of this area formerly used ground-water supplies but switched to the Detroit system as water levels declined and pumping costs increased. Ground-water levels have now recovered, and, in some areas, wells are flowing. This ground water could be diverted or pumped to create recreational ponds or lakes.

Water quality is becoming a problem in many lakes in the region, particularly because of high nutrient concentration. Ground water, particularly that from deep aquifers, generally contains a much smaller nutrient load than surface water. It may be feasible to remove much of the nutrients in small, highly eutrophic lakes by draining and refilling the lakes, either by natural ground-water discharge or by pumping from a deep aquifer. Born and others (1973) described an attempt to rehabilitate Snake Lake, in northern Wisconsin. The lake, which covers 12.3 acres, had received direct discharge of municipal waste for more than 20 years, destroying the lake's recreational and esthetic value. The lake was drained and then allowed to refill from natural ground-water discharge and from precipitation. Although dissolved-oxygen-depletion problems continued, there was a substantial initial reduction in nutrient concentrations, as well as a reduction in chloride concentration, conductance, and water color. The subsequent return of these constituents to near predraining levels is attributed to their release from bottom sediments. If the nutrient-rich bottom sediments had been leached or removed before the lake was refilled, the reduction in nutrients, chloride, and other mineral concentrations, would probably have been permanent. Other beneficial effects were the elimination of a heavy growth of duckweed and a deepening of part of the littoral zone through sediment compaction.

EROSION

The role of ground water in shoreline erosion is not generally recognized. The action of ground water may not be as spectacular as that of a stream cutting into its banks or waves beating on a cliff. Nevertheless, ground water can erode. Where the water table in a bluff is above the level of the lake or stream, perhaps a perched water table held up by a clay layer in unconsolidated deposits, the water loosens the deposits, allowing them to slump down to the shore and be carried away by wave action. Even a small trickle of water seeping out of a hillside can gradually wash away the deposits over which it flows (fig. 13).

In some areas this erosion may be reduced or stopped by dewatering the bluff along the shore. Depending on local conditions, this might be done by several large capacity wells or by a line of drive-point wells. The pumped water could be used for local public-supply, industrial, or irrigation needs.

FLOODING

The ground-water system can be used to alleviate flooding. Where conditions permit, flood peaks could be reduced by diverting some or all the flow into recharge areas, such as abandoned sand and gravel pits or specially prepared areas. Waller and Ayer (1975, p. 163–164) have suggested that floodflows in the Black River basin in western New York be diverted into ditches running across the top of sand deltas at the base of the Adirondacks. The water would either recharge the aquifers or go into temporary storage that would later supply baseflow to the numerous streams that rise in the deltas.

MANAGING GROUND-WATER RESOURCES

Effective and efficient management of ground-water resources requires a knowledge of ground-water hydrology and the local hydrologic system and cooperation of all involved in water use. It includes planning the development of the resource, managing withdrawals, monitoring the results of development, and modifying pumping schedules and

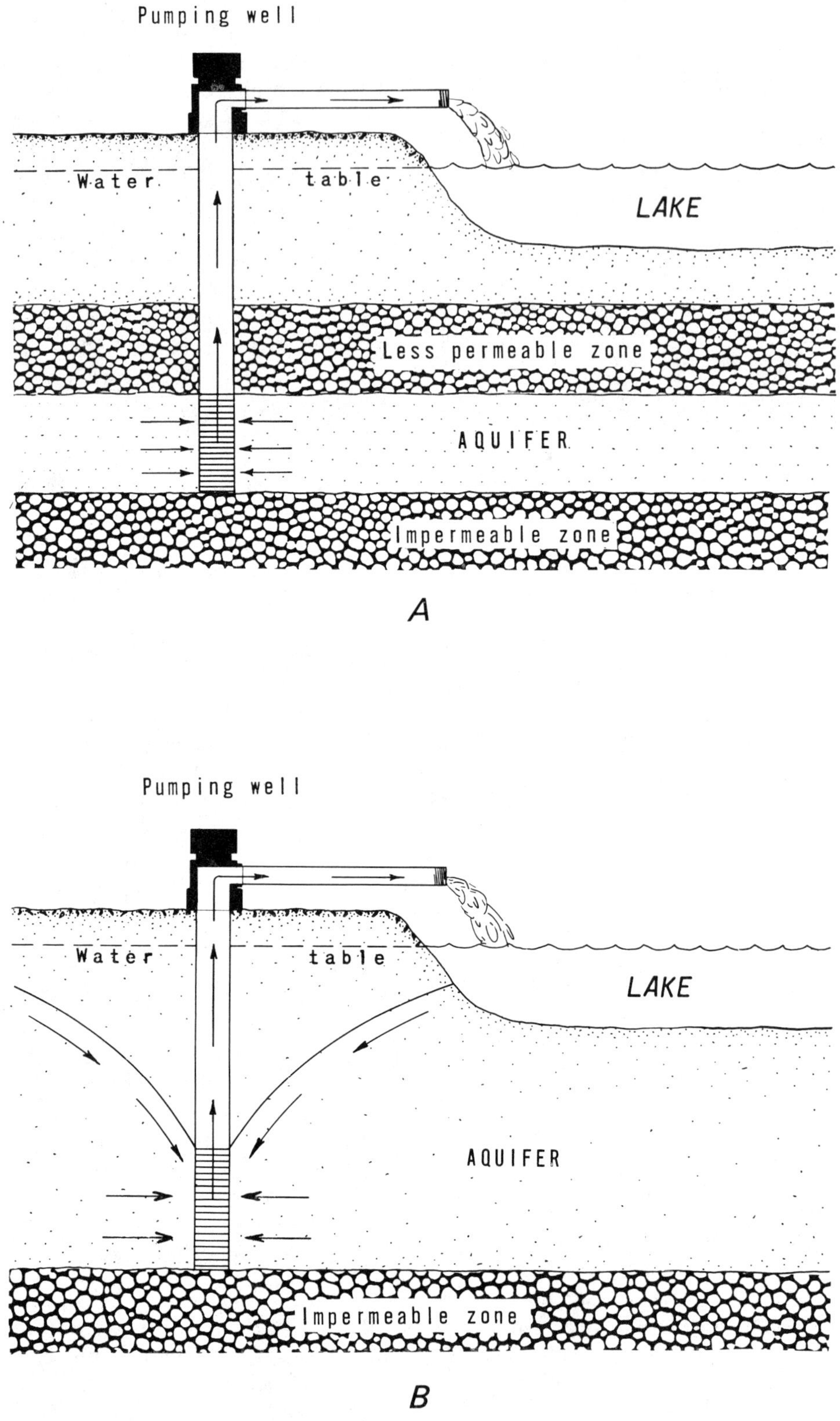

FIGURE 12.—Feasible (*A*) and infeasible (*B*) conditions for pumping ground water to maintain lake levels.

FIGURE 13.—Erosion from ground-water discharge along Lake Michigan in Indiana.

additional development on the basis of past experience and analytical studies. Ground-water management combined with surface-water management enables the two sources to be treated as a single resource.

The discussions in the preceding sections of this paper indicate the probability of water-related problems even in an area of abundant water such as the Great Lakes Region. Recognizing the need for a reference on management principles, the American Society of Civil Engineers has published a manual on ground-water management (Coe and others, 1972). The manual discusses in detail the concepts that are only briefly mentioned here and contains an extensive bibliography of pertinent reports.

COORDINATING DEVELOPMENT

The first step in managing ground-water resources is to coordinate the development of supplies by using knowledge of the hydrologic system. This requires a good data base, including information on recharge rates, boundary conditions, water quality, and geology. A good development plan would have wells spaced to minimize boundary and well-interference effects without overly increasing the costs of transmission lines. It could also include conjunctive use of surface water, where feasible. For example, coordinated development in the area along the Mesabi Iron Range, particularly around Hibbing and Virginia, Minn., could be of great benefit to both municipal and industrial water users. The surface-water potential in the area is only fair, and ground water, which is used for most industrial and municipal supplies, may not be sufficient for future demands.

Mining companies in the area have been using 1,200 to 6,000 gal of water/ton of ore processed, mostly from surface sources. At the same time, much of the large quantities of ground water being pumped to dewater deep mines is discharged to nearby streams (Cotter and others, 1965; Winter, 1973). Through coordination, the pumped ground water could be made available to nearby municipalities or used for ore processing, leaving the surface water for municipal use. The suggestion has been made that it might be possible to pipe in water, either from Lake Superior or from a distant well field, to meet future needs. If all those who would benefit from such a plan would finance it jointly, the cost would be minimal for any single entity.

Perhaps the most important factor in ground-water management is mutual planning and development by municipalities and industries. Several areas in the Great Lakes Region are faced with water-supply problems because cooperation and coordination has been inadequate. The Lansing, Mich., area is an example.

The city of Lansing installed its first well in 1885 in unconsolidated deposits along the Grand River. In 1974, the city pumped 8,054 million gal from about 125 wells tapping both unconsolidated glacial deposits and the Saginaw Formation. Four other water-supply systems in the area pumped 4,438 million gal from about 50 wells (fig. 14). Additional quantities were pumped by several large industrial wells in the area. This heavy development of ground water has produced a cone of depression under at least 100 mi^2 of the Lansing area. Near the center of the cone, water levels have declined as much as 160 ft (F. R. Twenter, written commun., 1976). As a result, many privately owned wells have gone dry.

The principal aquifer underlying the Lansing area is the Pennsylvanian Saginaw Formation, but large supplies of water also can be obtained from the overlying Pennsylvanian Grand River Formation and

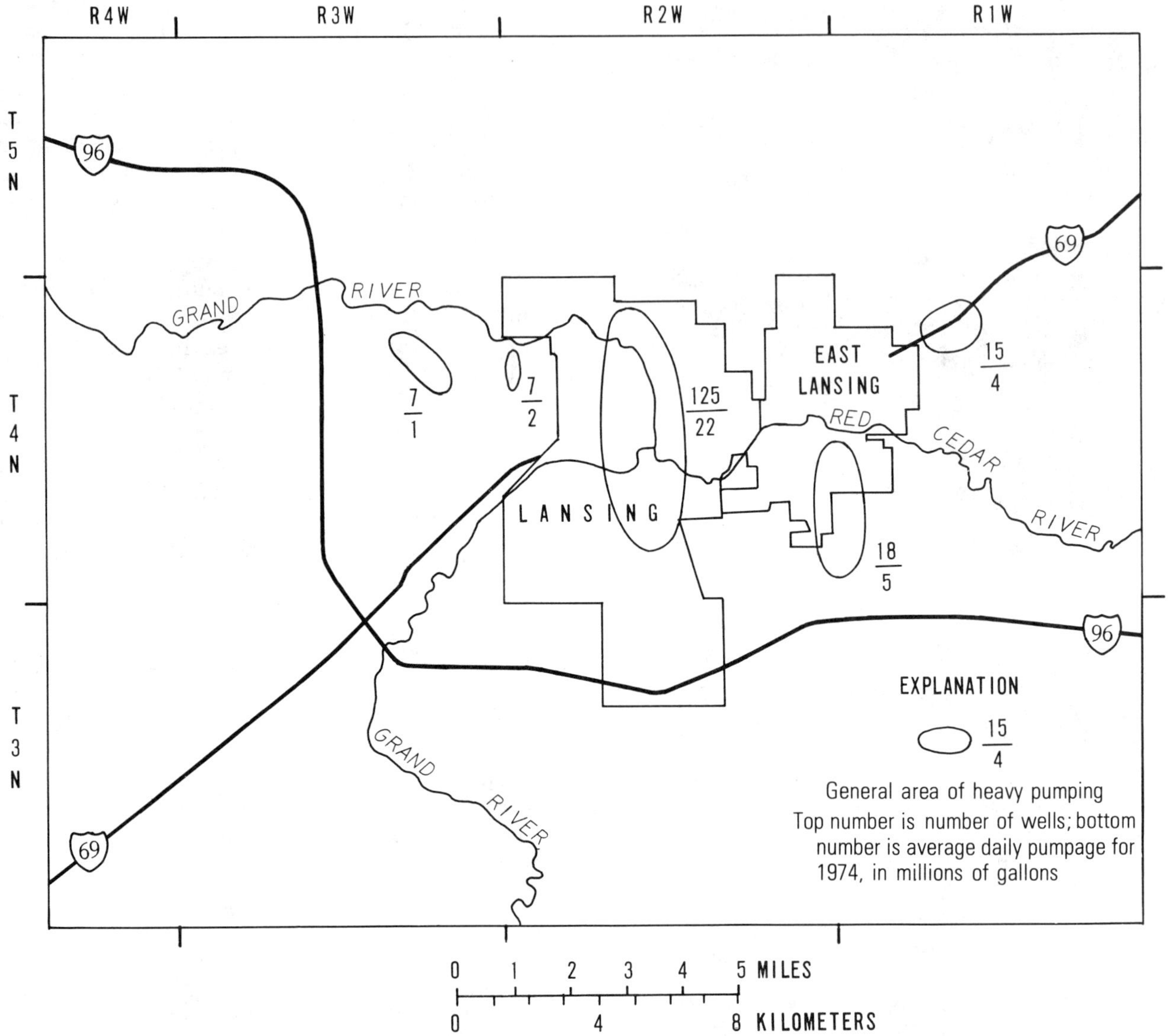

FIGURE 14.—Ground-water development in the Lansing, Mich., area.

some of the glacial deposits. Vanlier and others (1973) used an analog model to evaluate the hydrologic systems of the area. They discussed several different water-supply developments that included conjunctive use of ground water and surface water, increasing recharge to the aquifer, and additional development of the aquifer. They concluded that an intensive management program, with the cooperation of all the large water users in the area, would assure ample water to meet the needs of the area for many decades.

Recharge to the aquifers is an important consideration of water management. It is not possible to pump water from an aquifer without decreasing natural discharge, increasing natural recharge, removing water from storage, or some combination of these. In general, ground-water development is designed to take advantage of opportunities to increase recharge. In many places, wells can be located so that pumping will induce recharge from a nearby river.

Artificial recharge through wells may not be economically feasible in most of the Great Lakes Region. Augmentation of natural recharge by water spreading or diverting flow into gravel pits would probably be more effective.

MONITORING THE USE OF GROUND WATER

Managing ground-water resources does not end when the wells are put into service. Most ground-water development is planned using inadequate data. Although models of hydrologic systems are excellent tools for developing water resources, models are no better than the data used to construct them. Continued data collection during project operation, therefore, can be used for refining models. These data include, as a minimum, water-level measurements in observation wells and records of pumpage. If the aquifer is unconfined and is hydraulically connected to a river, flow upstream and downstream could be monitored. Predictions of well yields, water-level declines, and streamflow changes can be used to adjust pumping rates of wells and to locate additional wells for optimum development.

Water samples collected periodically from the pumping wells can be analyzed for chemical and bacteriological quality. Protecting the quality of the ground water might also involve monitoring land-use changes. As pumping increases, changes in the flow pattern of ground water can be determined. Landfills, sewage lagoons, storage pits, and similar sites in areas that are contributing to the ground-water supply constitute potential sources of pollution.

LEGAL CONSIDERATIONS

All the States in the Great Lakes Region have comprehensive laws and regulations to control the development and to protect the quality of their surface-water bodies. However, they do not have similar comprehensive laws and regulations for ground-water bodies.

Two common-law rules are applied to ground-water use in the Great Lakes Region: the rule of absolute ownership (English rule) and the rule of reasonable use (American rule). Under absolute ownership, the land owner can use all the ground water he can find; he is not liable, even if he drains his neighbor's supply. Only Wisconsin, among the Great Lakes States, follows this rule. The other States follow the reasonable-use rule, which allows a landowner to withdraw ground water, even by inducing flow from neighboring land, as long as the use of the water is reasonable. Nonexercise of the rights to use ground water does not result in forfeiture of those rights under either rule.

The Great Lakes Basin Commission (1975e) has summarized the State laws regarding ground-water resources. Within the Great Lakes Region, all the States except Ohio and New York require water-well drillers to obtain a license. The primary purpose of licensing are: (1) insures that the persons drilling a well are competent, (2) facilitates data collection, and (3) places control of drilling in a responsible agency. All the States except New York require that records be filed with the State on completion of a water well.

Michigan, Ohio, and Wisconsin have established standards for methods of constructing water wells and the materials used. Indiana, Michigan, and New York require a permit before a public-supply well can be installed. To protect ground-water resources from pollution or to protect scant supplies, Illinois, Indiana, and Minnesota can designate restrictive-use areas, specifying allowable pumping capacities under various geologic and hydrologic conditions. Wisconsin can do the same in areas where high-capacity wells (70 gal/min) can be developed. Michigan may specify the volume of water that may be used or allowed to flow in areas where a nuisance occurs owing to waste or unreasonable use. Indiana and Minnesota have statutes to conserve artesian pressure.

Illinois, Indiana, Minnesota, Ohio, and Pennsylvania control drilling, where wells penetrate several aquifers, in order to prevent interchange of fluids between the aquifers. Michigan has similar controls that apply to oil, gas, and brine wells but not to wells for private water supply. All the States except New York and Ohio require that abandoned wells and test holes be filled, capped, and(or) plugged, and New York and Ohio require these measures for oil and gas wells.

Regulation of waste disposal is also important in protecting ground-water resources. All the States in the Great Lakes Region require a license to operate a solid-waste disposal site and have adopted regulations governing the location of such sites. Michigan, Pennsylvania, and Wisconsin have regulations governing the size of lots in unsewered subdivisions. Indiana requires a minimum lot size of 15,000 ft^2 where 5 or more unsewered lots are developed along a public freshwater lake.

The National Water Commission (1973, p. 61–63) recommended enactment of laws outlining the establishment of water-management agencies that would have the power to manage surface water and ground water conjunctively, coordinate withdrawals from either or both sources, control the rate of depletion in an aquifer, and use artificial recharge to replenish aquifers. The commission does not believe that Federal regulation of ground water is desirable, but would let the States decide whether or not to regulate it.

In the framework study, the Great Lakes Basin Commission (1975e, p. ix) recognized nine legislative controls of ground water.

1. Regulation to control well spacing and withdrawal rates.
2. Plugging, filling, and capping of abandoned wells.
3. Determine rights of municipalities to withdraw water from land outside their limits.
4. Develop stricter laws to prevent ground-water pollution, and foster research and recordkeeping.
5. Require State certification of waste-treatment operators who are discharging wastes into ground-water bodies.
6. Regulation of solid-waste disposal sites to prevent pollution of ground water.
7. Regulation of abandoned mines.
8. Regulation of waste-disposal wells.
9. Establish ground-water rights.

Water-resources legislation need not be restrictive. As pointed out by Thomas (1961, p. 6), permissive legislation can encourage organization of public districts having the common interest of economic and effective development and use of water in a specified area. Laws recognizing potentialities as well as limitations of water resources can help rather than hinder development.

Because there are so many variations geographically, hydrologically, and temporally in water resources in the Great Lakes Region and even within a State, water laws, ideally, should be flexible enough to accommodate most any change. Thomas (1961, p. 6) suggested that one method to achieve flexibility is for such laws:

> to enunciate only the basic principles in statutes, to confer broad powers upon the administrator responsible for working out detailed solutions to specific problems, and to provide for prompt and effective action by the courts on appeals from the administrators' decisions.

In this manner, fundamental decisions on development and use of water resources would be made by those most knowledgeable of the local situation, subject to broad guidelines from the State.

FUTURE CONSIDERATIONS

Proper management of ground water and legislation to control its development and to protect its quality depend on a thorough knowledge of the ground-water system. Reports on ground-water resources are available for about 80 percent of the Great Lakes Region, but most are reconnaissance in scope. Michigan, in particular, is lacking in detailed ground-water studies. Waller and Allen (1975) include an extensive bibliography in their report on ground water in the Great Lakes Basin.

Many planners in the Region consider available ground-water data inadequate. The inadequacy is of several forms. In some areas, little or no basic ground-water data are available; in others, data are available but not in a form that planners can use. Elsewhere, available data are scant or not detailed for thorough planning. Several planners mentioned the need for a central clearing house for geologic and hydrologic data, perhaps on a statewide basis.

Many of the areas that have little or no basic data are also areas of little or no development. In such areas, wells are few and old, and little is known about them. The ground-water system in these areas, however, can be envisioned by interpolating from well records and geologic data in similar areas. As interest in these undeveloped areas increases, hydrologic studies would help evaluate the potential for ground-water development.

Data are generally available, but not in a usable form, in areas where most wells are stock and domestic wells. Numerous well records are available from drillers, State agencies, and the Geological Survey. Many of these records have been published as parts of reconnaissance reports. Of particular use are data organized to show areas of recharge, depth to water table, amount of water available, and rate of ground-water discharge into streams. This type of information can be interpreted by experienced ground-water hydrologists.

In areas of extensive ground-water development, or in areas having such a potential, some of which have been discussed earlier, quantitative ground-water studies would encourage fullest use of the resource. Digital or electric-analog models of the hydrologic system can be used as predictive tools to determine the best places for large withdrawals of ground water, the amount of water that can be safely pumped, and the long-term effects of pumping on water levels and streamflow.

Another concern to planners is the effect of urbanization on the local total hydrologic system. How much does paving increase runoff and decrease recharge? What is the effect on low flows? How is the quality of both surface water and ground water affected? Studies that answer these questions are useful to planners. Planners also seek information on the effects of filling in wetlands and the movement of pollutants in ground water. Few studies relating to the above have been made in the Great Lakes Region. Detailed ground-water studies have been made for some of the metropolitan areas, as previously mentioned (Chicago, Green Bay, Lansing and Milwaukee), and other studies will be made. Most of the studies, however, although suitable for initial planning, are not sufficiently detailed for designing.

Some planners were not fully aware of the studies that had been done in their area and indicated a need for both greater publicity and a clearing house on a local or State level where they could get information on geology, soils, water resources, and hydrology. The U.S. Geological Survey District offices in each State are usually informed on the availability of these types of data and serve as public-information offices. The State water-resources agencies are other sources of information.

The Geological Survey's NAWDEX (National Water Data Exchange) and RALI (Resources and Land Information) systems are clearing houses at the national level for sources of water and land data. The Survey's WATSTORE (Water Data Storage and Retrieval) system provides up-to-date basic water data collected by the Survey.

SUMMARY

Ground water available from storage in the Great Lakes Region is estimated to be 35,000 billion ft^3. Ground-water discharge to streams and lakes in the region is estimated to be 26 billion gallons per day. Despite this abundance of ground water, only 1.8 billion gallons per day was used in the region in 1970. This is about 11.6 percent of the total water used, excluding thermoelectric power use.

Ground water is available throughout the Great Lakes Region. However, the amount available locally varies widely. The most productive aquifers, those along the valleys of major streams, yield as much as 5,000 gal/min. Yields in excess of 1,000 gal/min have been obtained from the bedrock in some areas. Because well yields vary across the region, even within the same aquifer system, a detailed study of the hydrologic system is desirable before heavy development is started or continued.

In addition to being a potential source of water to meet increasing municipal and industrial demands, ground water can be used to augment low streamflow, to dilute water of poorer quality, and to maintain lake levels and wetlands, among other uses. In places, floodflows may be diminished by diverting them over areas where some of the water can infiltrate the ground and recharge the local aquifer.

Mineralized ground water has an economic potential for the region. Improved desalination processes make moderately saline (less than 10,000 mg/L dissolved solids) ground water a viable alternative source of water. Saline aquifers also could be used to store freshwater or natural gas, and the more highly saline aquifers have been used for liquid waste disposal.

Effective utilization of ground water depends on good management. This includes planning the development of the resource, monitoring pumping and modifying future development, and protecting ground water from pollution. Legislation that recognizes the variability and potential of ground water as well as its limitations can help in effective management.

Nature has endowed the Great Lakes Region with an abundance of water. Ground water, although only a small part of the total, is playing an increasingly important role in the Region's growth. Wise management of this resource will assure its orderly development and undiminished quality.

REFERENCES CITED

Born, S. M., and others, 1973, Dilutional pumping at Snake Lake, Wisconsin: Wisconsin Dept. Nat. Resources, Tech. Bull. 66, 32 p.

Braunstein, Jules, ed., 1973, Underground waste management and artificial recharge: Tulsa, Okla., Am. Assoc. Petroleum Geologists, v. 1 and 2, 931 p.

Coe, J. J., and others, 1972, Ground water management: Am. Soc. Civil Engineers, Manuals and Repts. on Engineering Practice—No. 40, 216 p.

Cook, T. D., ed., 1972, Underground waste management and environmental implications: Am. Assoc. Petroleum Geologists Mem. 18, 412 p.

Cotter, R. D., Young, H. L., Petri, L. R., and Prior, C. H., 1965, Ground and surface water in the Mesabi and Vermilion Iron Range area, northeastern Minnesota: U.S. Geol. Survey Water-Supply Paper 1759–A, p.A1–A36.

Cross, W. P., 1968, Flow duration of Ohio streams based on gaging-station records, through 1965: Ohio Div. Water Bull. 42, 68 p.

Environmental Data Service, 1968, Climatic atlas of the United States: U.S. Dept. Commerce, 80 p.

Feth, J. H., and others, 1965, Preliminary map of the conterminous United States showing depth to and quality of shallowest ground water containing more than 1,000 parts per million dissolved solids: U.S. Geol. Survey Hydrol. Inv. Atlas HA–199, 2 sheets.

Fidler, R. E., 1970, Potential development and recharge of ground water in Mill Creek valley, Butler and Hamilton Counties, Ohio, based on analog model analysis: U.S. Geol. Survey Water-Supply Paper 1893, 37 p.

Great Lakes Basin Commission, 1975a, Surface water hydrology: Great Lakes Basin Comm., Great Lakes Basin Framework Study, App. 2, 133 p.

———1975b, Water supply—municipal, industrial, rural: Great Lakes Basin Comm., Great Lakes Basin Framework Study, App. 6, 266 p.

———1975c, Irrigation: Great Lakes Basin Commission, Great Lakes Basin Framework Study, App. 15, 111 p.

———1975d, Economic and demographic studies: Great Lakes Basin Comm., Great Lakes Basin Framework Study, App. 19, 217 p.

———1975e, State laws, policies, and institutional arrangements: Great Lakes Basin Commission, Great Lakes Basin Framework Study, App. S20, 215 p.

Howard, J. F., 1974, The cost of ground water vs. surface water: Water Well Jour., v. 28, no. 8, p. 35–37.

Kimbler, O. K., Kazmann, R. G., and Whitehead, W. R., 1975, Cyclic storage of fresh water in saline aquifers: Louisiana Water Resources Research Inst. Bull. 10, 78 p.

Knowles, D. B., 1964, Ground-water conditions in the Green Bay area, Wisconsin, 1950–60: U.S. Geol. Survey Water-Supply Paper 1669–J, p. J1–J37.

Krieger, R. A., Hatchett, J. L., and Poole, J. L., 1957, Preliminary survey of the saline-water resources of the United States: U.S. Geol. Survey Water-Supply Paper 1374, 172 p.

Leonard, D. J., and Raoul, Joseph, Jr., 1975, Levels and flows: Great Lakes Basin Comm., Great Lakes Framework Study, App. 11, 206 p.

Murray, C. R., and Reeves, E. B., 1972, Estimated use of water in the United States in 1970: U.S. Geol. Survey Circ. 676, 37 p.

National Water Commission, 1973, New directions in U.S. water policy: Summary, conclusions and recommendations from the final report: U.S. Govt. Printing Office, 197 p.

Novitski, R. P., 1973, Improvement of trout streams in Wisconsin by augmenting low flows with ground water: U.S. Geol. Survey Water-Supply Paper 2017, 52 p.

Pettyjohn, W. A., 1972, Good coffee water needs body, *in* Pettyjohn, W. A., ed., Water quality in a stressed environment: Minneapolis, Minn., Burgess Publishing Co., p. 194–199.

Piper, A. M., 1969, Disposal of liquid wastes by injection underground—neither myth nor millennium: U.S. Geol. Survey Circ. 631, 15 p.

Richards, L. A., and Richards, S. J., 1957, Soil moisture, *in* Soil, the Yearbook of Agriculture 1957: Washington, D.C., U.S. Govt. Printing Office, p. 49–60.

Sherrill, M. G., 1975, Ground-water contamination in the Silurian dolomite of Door County, Wisconsin: Ground Water, v. 13, no. 2, p. 209–213.

Thomas, H. E., 1961, Ground water and the law: U.S. Geol. Survey Circ. 446, 6 p.

U.S. Environmental Protection Agency, 1974, Compilation of industrial and municipal injection wells in the United States: U.S. Environmental Protection Agency, 2 vols.

U.S. Geological Survey, 1974, Quality of public water supplies of New York, November 1970–April 1972: Albany, N.Y., U.S. Geol. Survey open-file rept., 198 p.

U.S. Public Health Service, 1962, Drinking water standards: Public Health Service Pamph. 956, 61 p.

Vanlier, K. E., Wood, W. W., and Brunett, J. O., 1973, Water-supply development and management alternatives for Clinton, Eaton, and Ingham Counties, Michigan: U.S. Geol. Survey Water-Supply Paper 1969, 111 p.

Vogt, J. E., 1972, Infectious hepatitis outbreak in Posen, Michigan, *in* Pettyjohn, W. A., ed., Water quality in a stressed environment, Minneapolis, Minn: Burgess Publishing Co., p. 188–193.

Waller, R. M., and Allen, W. B., 1975, Geology and ground water: Great Lakes Basin Comm., Great Lakes Basin Framework Study, App. 3, 152 p.

Waller, R. M., and Ayer, G. R., 1975, Water resources of the Black River basin, New York: N.Y. State Dept. Environmental Conserv. Basin Plan Rept. BRB-1, 205 p.

Winter, T. C., 1973, Hydrogeology of glacial drift, Mesabi Iron Range, northeastern Minnesota: U.S. Geol. Survey Water-Supply Paper 2029–A, p. A1–A23. [1974].

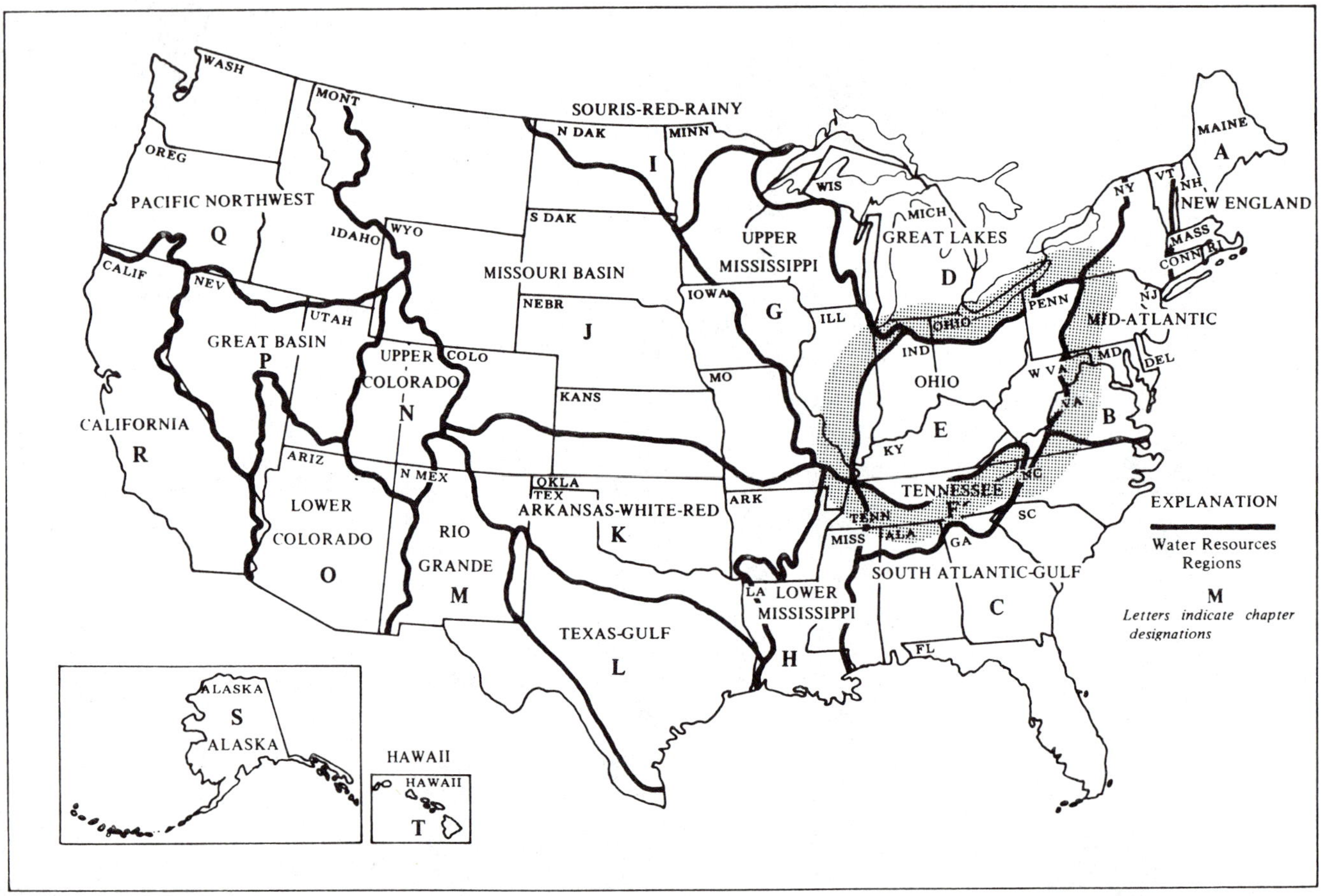

E – Location Map for the Ohio Region

E ....

OHIO REGION

By RICHARD M. BLOYD, JR.

ABSTRACT

Ground water in the Ohio Region is a large, important, and manageable resource that should have a significant role in regional water development.

On the basis of a comparison of ground-water withdrawals with estimated ground-water recharge, it appears that the ground-water resources of the Ohio Region probably will not be used at full potential under existing development plans. Annual ground-water use (1960) by municipalities and rural residents was about 1,000 million gallons per day. Average annual regional ground-water recharge is about 35,000 million gallons per day. Therefore, base-year (1960) municipal and rural ground-water use is only about 3 percent of recharge. Annual regionwide ground-water use (1965) by industry also is only about 3 percent of recharge.

Not all ground water in storage is recoverable for development, but estimates of the amounts that can be obtained from storage, under specified conditions, are calculated to show the magnitude of water that is available. Total potable ground water available from storage in the outwash and alluvial aquifers in the Ohio River valley and the subbasins is about 23,000 billion gallons. This about four times the flood-control storage of all Ohio Region Corps of Engineers reservoirs constructed, under construction, or in advance planning as of July 1965. Approximately 85,000 billion gallons of potable ground water is available from storage in the region in aquifers other than the outwash and alluvial aquifers. This is about 20 percent of estimated storage in Lake Ontario.

About 5 percent of the region has ground-water resources capable of supplying more than local needs. For example, under certain specified conditions the excess of ground-water recharge over base-year (1960) ground-water use is available for 22 million additional people in the Wabash subbasin; for 4, 1.5, and 12 million additional people, respectively, in the Miami, lower Scioto, and Allegheny subbasins; for 5 million additional people in the Ohio River valley; or for equivalent quantities of water supply for industrial or agricultural expansion or other use. A reasonable assumption is that much of the available ground water in these areas can be pumped and transported to reasonably distant points of need.

The Wabash and White subbasins probably have the highest potential of all Ohio River subbasins for additional ground-water development. About 30,000 billion gallons, or about 28 percent of the total potable ground water available from storage in the Ohio Region, is in storage in these subbasins. Estimated average annual ground-water recharge in the Wabash and White subbasins is 7,300 million gallons per day. Annual ground-water use (1960) by municipalities and rural residents of the subbasins of about 220 million gallons per day is only about 3 percent of estimated annual ground-water recharge and only about 0.3 percent of the potable ground water in storage in the subbasins. Also, many high-yield aquifers are present and offer excellent reservoir-manipulation possibilities in conjunction with existing and planned surface reservoirs.

Practically all areas of high population density in the Ohio Region have the potential for development of ground-water resources. The Indianapolis, Ind., area probably has the highest potential.

Assuming that future population growth will be heaviest in the areas paralleling the interstate highway system, much of the increased water demand associated with these growth areas can be supplied by ground water. The areas of population growth in Indiana and southwestern and south-central Ohio are especially well situated in terms of potential ground-water supplies.

Underground space in the Ohio Region, consisting of natural pore spaces and fractures in rocks and sediments, can be considered a regional resource in the sense that it can be included in regional water-pollution control or waste-disposal plans. Much of this space is already occupied by ground water or other fluids which must be displaced for any alternate use. There is a potential for underground waste storage in practically the entire Ohio Region.

Rapid advance of techniques in ground-water hydrology during recent years has provided methods which the hydrologist can use for evaluating planned ground-water development. Therefore, the manager can resolve the inherent problems that historically have bred caution when this part of our total water resource was considered for development.

INTRODUCTION

The purpose of this study was to make a broad-perspective analysis of the ground-water resources of the Ohio Region and thereby to demonstrate that the region's ground water is a large, important, and manageable resource that should have a significant role in regional water development. The analysis includes assessment of the significance of the ground-water resource in regional water supply, of quantities of ground water available, of quality of the water, of present and potential problems associated with its use, and of additional information needed for planning and efficient development.

This report emphasizes the role of ground water in water-resource planning and presents hydrologic and related information which will permit incorporation of the ground-water resource in the overall resource development and water-management scheme for the region.

The boundaries of the study area are the surface-water drainage boundaries of the Ohio River basin, exclusive of the Tennessee River drainage. For study purposes, the area was divided into 13 subbasins (fig. 1).

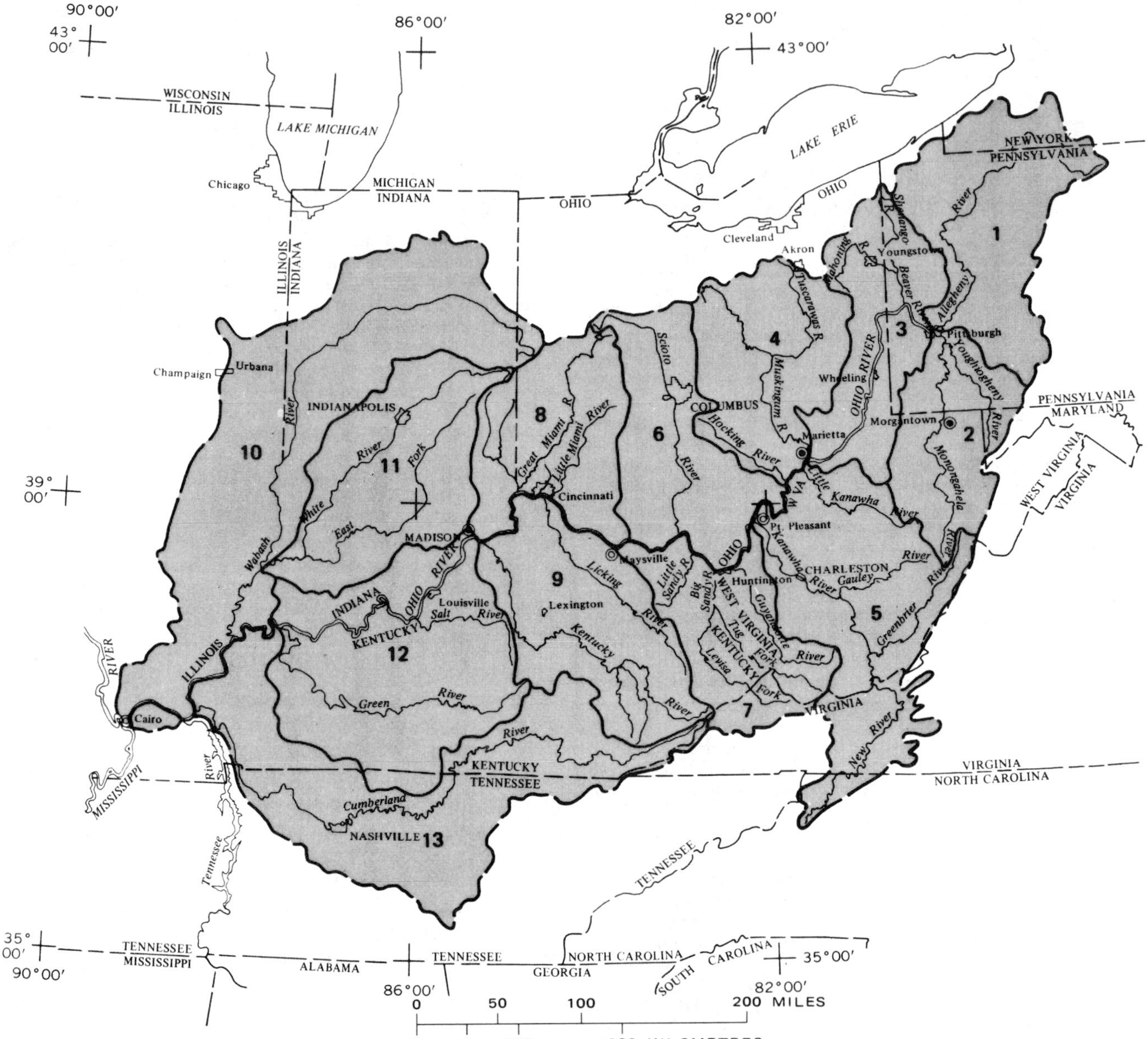

SUBBASINS IN THE OHIO REGION

1. Allegheny
2. Monongahela
3. Upper Ohio River
4. Muskingum
5. Kanawha-Little Kanawha
6. Scioto
7. Big and Little Sandy-Guyandotte
8. Great and Little Miami
9. Licking-Kentucky
10. Wabash
11. White
12. Green-Salt-Lower Ohio River
13. Cumberland

FIGURE 1. — Subbasins in the Ohio Region described in this report.

INTEGRATING GROUND WATER INTO WATER RESOURCE PLANNING

Historically, the tendency of planners has been to ignore or minimize the role of ground water, even though the subsurface system offers many advantages, including versatility. This tendency may have been justified in some situations because the ground-water resource could not be precisely evaluated in terms of availability, quality, cost of development, or effects of development on surface-water supplies. The water pumped from wells must be balanced by a change in water distribution within the total hydrologic system. Therefore, when ground water is used, one or more of the following conditions may arise and cause alarm: The ground-water supply is progressively depleted, the streamflow is reduced, and (or) the water quality is changed.

Rapid advance of techniques in ground-water hydrology during recent years has provided methods which the hydrologist can use for evaluating planned ground-water development. Therefore, the manager can resolve the inherent problems that historically have bred caution when this part of our total water resource was considered for development.

TECHNOLOGY AVAILABLE FOR PLANNING

An impressive array of tools is available for planning the development of ground water, for analyzing the alternative conjunctive or independent use of ground water and surface water, and for optimizing management objectives within specified physical, economic, and social constraints. Mathematical and statistical models have achieved greater utility since computers have become available. Models have been developed that simulate the ground-water regimen and predict the effects of stresses imposed by man.

Electric analog models have been used to describe ground-water hydrology in a variety of geologic environments in many places throughout the United States and in several foreign countries. The resistor-capacitor network of an analog permits display of a visual analogy of the ground-water system, which is especially useful in describing and understanding ground-water systems. Analogs also can predict the effects of stresses that may be applied to hydrologic systems.

Also, mathematical models have been programed in digital computers, providing yet another valuable tool to the hydrologist and water manager. Sufficient experience in digital modeling has been acquired to show that ground-water systems can be accurately simulated and the models manipulated to predict changes, both in quantity of flow and in water quality, resulting from natural phenomena and from man's activities. For example, it is possible to describe changes in water quality that might result from changes in water-management practices or to predict the fate of liquid contaminants intentionally or accidentally released to the environment. Thus, a pilot project now underway in the Arkansas River valley of Colorado will be used there to describe and predict on a monthly basis changes in salinity in the alluvial aquifer and in the adjacent stream.

The results of a model analysis are no more reliable than the data used to construct the model. However, with adequate geologic and hydrologic investigation, aided by models, ground water can be integrated into water-resource planning with a high degree of assurance and effectiveness.

DATA REQUIREMENTS FOR PLANNING

The data requirements for a planning study of a ground-water system depend upon the size and hydrologic complexity of the area and upon the types of water problems. Some planning studies need large-scale model analyses, whereas others require only a descriptive evaluation of the hydrologic data.

In determining the types of ground-water information to be presented in this report, it was assumed that water-resource planners in the Ohio Region will use systems engineering for establishing and implementing an optimal water-resource development scheme. What types of information are necessary for the application of the various methods, such as modeling and simulation, comprising the systems method?

The first type of information needed is that which helps the planner to determine whether the ground-water resource in the Ohio Region is available at points of need, or can be transported, in sufficient quantity at the right time and in acceptable quality to warrant further consideration.

The second type of information needed is estimates of the total resource available for allocation and of the individual demands among competing activities. Such estimates, along with cost estimates associated with allocation of the resource, are pertinent to the application of optimization techniques, such as linear programming. In the typical linear programming approach, available resources are allocated in an optimal manner among various competing activities.

This report provides an estimate of the total ground-water resource available. Initial estimates of water demand are available in a comprehensive survey study of the basin (U.S. Army Corps of Engineers, 1969). The present report includes no estimates of costs for supplying ground water because costs vary widely from place to place and from use to use. However, when specific requirements and a time frame are specified, cost estimates can be developed.

The third type of information needed is estimates of storage coefficients, transmissivities, aquifer boundaries, and the degree of connection between streams and aquifers. Such information is basic to the application of system modeling and simulation.

Many pertinent data, already available, are summarized in following sections of this report.

ASSESSING THE GROUND-WATER RESOURCE

GEOLOGIC FRAMEWORK

Alluvium, outwash, and glaciofluvial deposits constitute the most productive part of the ground-water system in the Ohio Region. Streams draining the glaciated part of the region (fig. 2), have redeposited size-sorted glacial sediments well beyond the southernmost encroachment of the glacial ice during the Ice Age and, thus, have helped to create highly permeable aquifers in widespread parts of the region.

Holocene alluvium, consisting of silt, sand, and gravel, is present in the lower reaches of the major tributary valleys south of the Ohio River. In general, these sediments are finer grained and less permeable than the glaciofluvial deposits.

Outwash, composed predominantly of sand and gravel deposited by melt-water streams, is generally highly permeable and is an excellent source of water. Outwash deposits are most extensive in valleys of major tributaries within the limits of glaciation. Where these deposits are present beyond the limits of glaciation, they occur as valley-train deposits.

Glaciofluvial deposits, which are a mixture of Pleistocene outwash deposits and Holocene alluvium, occur along many of the major tributary streams and in the Ohio River valley. These deposits are generally excellent sources of water.

Till is the principal type of glacial deposit present in the region. It is a heterogeneous mixture of clay, silt, sand, and gravel, has very low water-transmitting ability, and therefore is a relatively poor source of water. Moraines of unsorted till deposits form low ridges along former fronts of glacial ice sheets.

The Ohio Region is underlain by a series of bedrock units that vary greatly in thickness and in hydrologic characteristics and that range in age from Precambrian to Tertiary (fig. 3). The most conspicuous large features of the bedrock are the Appalachian and Illinois basins (fig. 3). These two basins are separated from each other and from the Michigan basin to the north by the Cincinnati, Findlay, and Kankakee arches and the Nashville dome (fig. 3). By envisioning a two-bowled candy dish with the bowls separated but molded together by rims, one can envision the general configuration of the surface of the basement rock in the Ohio Region.

The slope on the surface of the basement complex, from the arch areas toward the Appalachian and Illinois basins, is the central feature controlling the slope or dip of the younger bedrock unit overlying the basement. This general slope or dip is modified locally in areas of folding and faulting. The general dip on the bedrock aquifers east of the Nashville dome and Findlay arch is toward the low point of the Appalachian basement-rock depression; the dip on the bedrock aquifers west of the Findlay arch and south of the Kankakee arch is toward the low point of the Illinois basement-rock depression.

AQUIFER CHARACTERISTICS

Another important step in assessing the ground-water resource in the Ohio Region and in describing the aquifer system is the specification of values for aquifer characteristics, such as hydraulic conductivity and storage coefficient. Knowledge of aquifer characteristics makes it possible to delineate the various aquifers in the basin and to analyze the dynamic processes of ground-water movement.

Hydraulic conductivity, which is a unit measure of the ability of an aquifer to transmit water, depends primarily on the nature of the pore space in the aquifer and on the type of fluid occupying the space. In general, the greater the hydraulic conductivity, the greater is the ability of the aquifer to transmit water.

Transmissivity is a measure of the gross ability of the aquifer to transmit water. It is equal to the hydraulic conductivity multiplied by the saturated thickness of the aquifer.

The storage coefficient is the volume of water an aquifer releases or takes into storage per unit surface area of the aquifer per unit change in head. In a confined water body the water derived from storage with decline in head comes from expansion of the water and from compression of the aquifer; similarly, water added to storage with a rise in head is accommodated partly by compression of the water and partly by expansion of the aquifer. In an unconfined water body, the amount of water derived from storage by expansion of the water or by compression of the aquifer is generally negligible compared to that involved in gravity drainage or filling of pores. In an unconfined water body the storage coefficient corresponds to the specific yield and is generally greater than 0.01.

Estimated aquifer characteristics for aquifers in the Ohio Region are listed in table 1. The estimates are derived from previous studies and are considered to be sufficiently accurate for areal application.

STREAM-AQUIFER INTERRELATION

The purposes of this section of the report are to show that generally there is hydraulic connection between the ground-water and surface-water systems and to develop data needed in defining the ground-water resources of the region.

Hydraulic connection between the ground-water and surface-water systems and a vast supply of ground-water

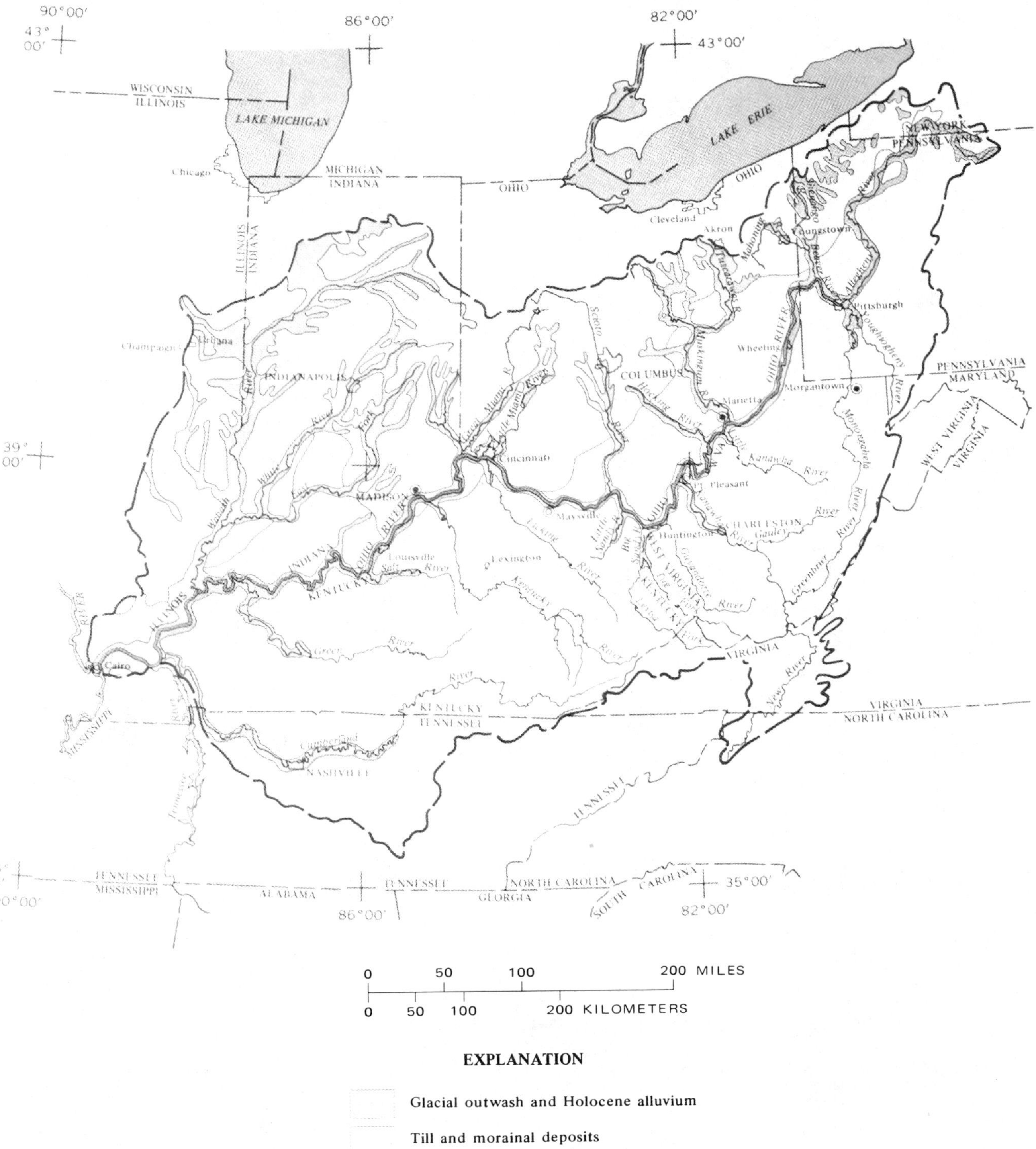

FIGURE 2. — Distribution of unconsolidated sediments.

in the region indicate a potential for large economic benefits. Nature has provided a natural subsurface multipurpose multireservoir system. Not only has nature supplied the system, she has tailored the system with a hydraulic link with the surface-water system.

Because of the interaction between ground water and surface water, streamflow data for nonregulated perennial streams can be used to estimate ground-water discharge to streams, average annual ground-water discharge and recharge, and the hydraulic conductivity of

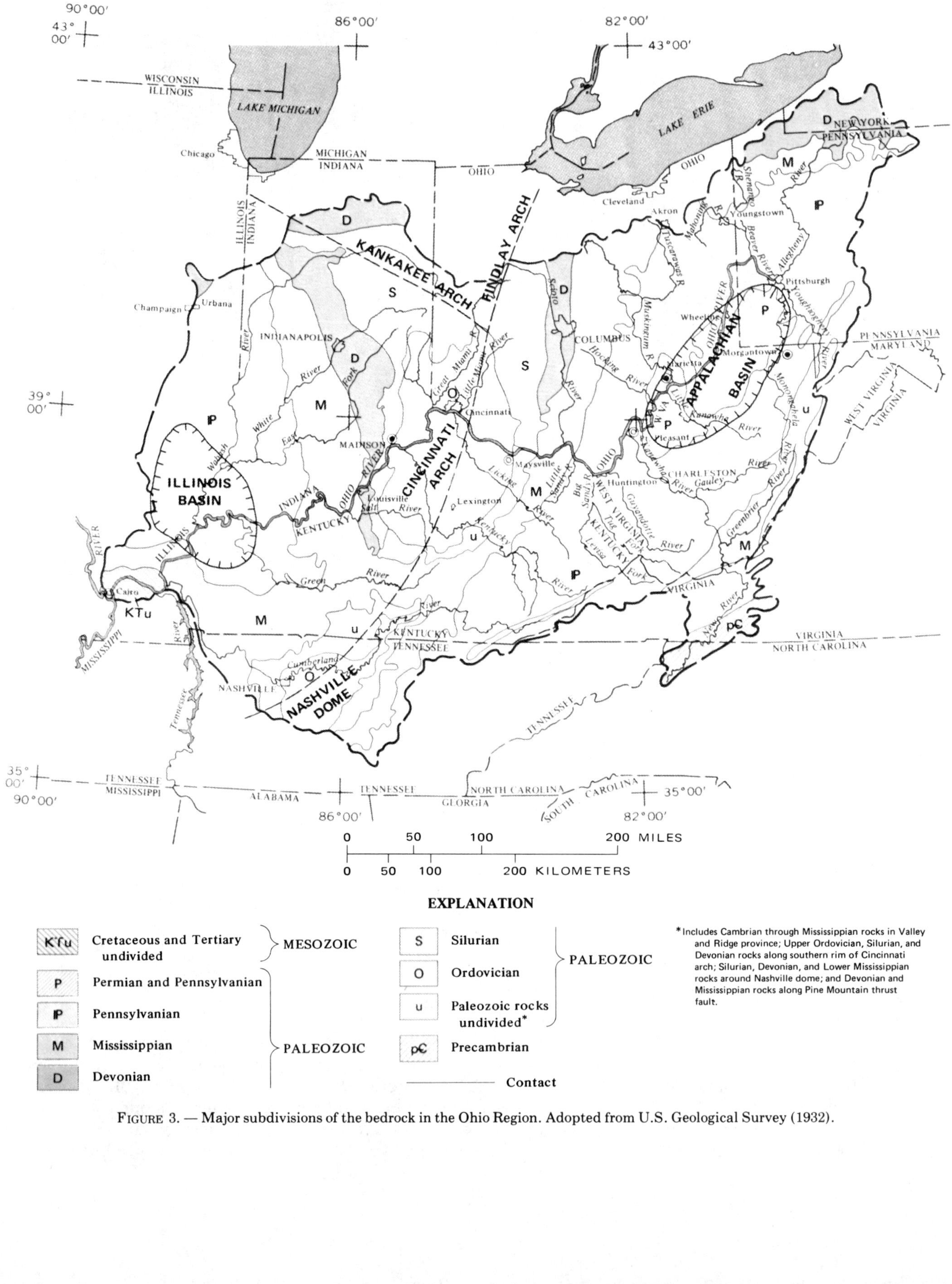

FIGURE 3. — Major subdivisions of the bedrock in the Ohio Region. Adopted from U.S. Geological Survey (1932).

TABLE 1. — *Characteristics of the consolidated and unconsolidated aquifers in the Ohio Region and ranking of aquifers in order of decreasing transmissivity*

Aquifer	Hydraulic conductivity (gpd/ft²)	Storage coefficient or specific yield
1. Mad River alluvial aquifer	4,000–4,500	0.25
2. Ohio River valley outwash and alluvial aquifer	400–8,400	0.05–0.20
3. Miami River, Scioto River, Upper Muskingum River, and Whitewater River alluvial aquifers	2,500–3,000	.15– .20
4. Allegheny, Lower Wabash, and White River alluvial aquifers	2,000	.15– .20
5. Hocking River, Lower Muskingum River, and Upper Wabash River alluvial aquifers	1,500–2,000	.15– .20
6. Beaver River alluvial aquifer and alluvial aquifers in the minor tributaries north of the Ohio River	500–1,000	.15
7. Alluvial aquifers in the major tributaries south of the Ohio River	500	.10
8. Mississippian bedrock aquifer (Green River Basin)	>20	.01– .05
9. Mississippian bedrock aquifer with glacial cover	>20	.01– .05
10. Pennsylvanian bedrock (Allegheny and Pottsville Formations)	>20	.01– .05
11. Pennsylvanian bedrock with glacial cover	>20	.01– .05
12. Silurian bedrock with glacial cover	>20	.01– .05
13. Pennsylvanian bedrock (Conemaugh Formation)	>20	.01– .05
14. Pennsylvanian and Permian bedrock (Dunkard Group)	>20	.01– .05
15. Ordovician bedrock with glacial cover	>20	.01– .05
16. Silurian bedrock	>20	.01– .05
17. Mississippian bedrock	>20	.01– .05
18. Pennsylvanian bedrock (Monongahela Formation)	>20	.01– .05
19. Devonian bedrock with glacial cover	>20	.01– .05
20. Ordovician and undifferentiated Paleozoic rocks	>20	.01– .05
21. Devonian bedrock	>20	.01– .05

aquifers that are in hydraulic connection with streams. Data from U.S. Geological Survey offices in the region were used for such purposes in the study., Most of the data were obtained from streamgage network evaluation studies made in each State. In these evaluation studies, a streamflow-frequency distribution was computed for each unregulated gaging site where at least 10 years of streamflow records were available. Except for the data on Ohio, which was adjusted to the standard period 1931–60 (Cross, 1968), the data used in this report were not adjusted to a common base period because of the short time allotted to the study. Even though the streamflow data were not adjusted to the standard base period, they can be used effectively in this general appraisal study as long as possible deficiencies in the data are kept in mind.

Variation in the overall regimen of streamflow can be depicted with a cumulative frequency curve constructed on the basis of the amount of time during which specific values of streamflow are equaled or exceeded. The water in streams during periods of high flow is derived from both direct surface runoff and ground-water discharge. As streamflow decreases, the percentage of streamflow derived from direct surface runoff also decreases.

The flow equaled or exceeded 90 percent of the time, or the 90-percent flow, generally is assumed to be all ground-water discharge. Because the choice is based on empirical studies and because ground-water discharge does vary seasonally, arguments have been presented for the choice of different percent flow values. Wyrick and Lloyd (1968, p. H19), mentioned data that indicated that the ground-water component of discharge to streamflow varies from the streamflow exceeded 60 percent of the time to that exceeded 90 percent of the time. Stuart, Schneider, and Crooks (1967, p. 42) showed that, for a stream in southeastern Pennsylvania, the flow exceeded more than about 75 percent of the time is all ground-water discharge.

The approach taken in this study is that two flow-duration parameters define the annual ground-water discharge to streams better than one. This approach is based upon the assumption that seasonal variation in base flow of a perennial stream is caused partly by seasonal variation in vapor discharge from aquifers. Because of the variation in vapor discharge, one parameter is necessary to define the ground-water discharge to streamflow during the season when vapor discharge is at a maximum, and another parameter is necessary for the season when vapor discharge is at a minimum.

Where ground water sustains streamflow in the Ohio Region, the water table in the vicinity of the streams is close to the land surface and is a ready source of moisture for plants. Therefore abundant vegetative growths generally line stream channels. Because vegetation on flood plains is situated between the ground-water-recharge areas and the stream channels or liquid-discharge areas, the vegetation gets "first crack" at the ground water that is enroute to discharge. During summer months, when transpiration is at a maximum, vegetation diverts more ground water to the vapor phase than during winter months when transpiration is at a minimum. This seasonal variation in magnitude of ground-water diversion by plants causes seasonal variation in base flow.

The assumption is made that the 90-percent flow-duration parameter is an indicator of base flow in a stream during that part of the year when ground-water vapor discharge is a maximum and that the 60-percent flow-duration parameter is an indicator during that part of the year when vapor discharge is at a minimum (fig. 4). Although the above choices may not be the "true" values, at least they seem reasonable.

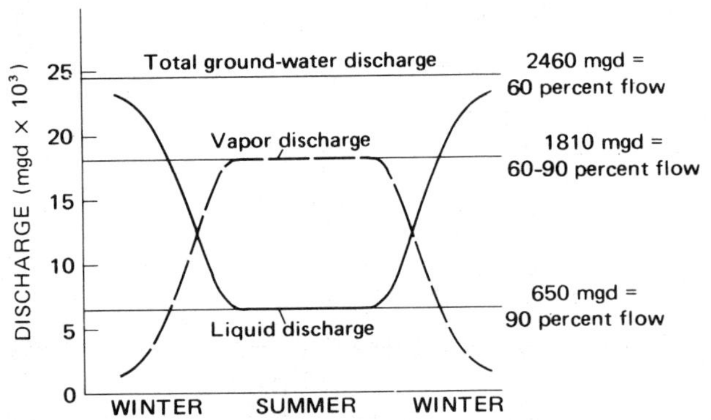

FIGURE 4. — The liquid- and vapor-discharge components and their relation to total ground-water discharge.

Where large reservoirs control flows, downstream flow records are not usable as indices of ground-water discharge. For the Allegheny, Upper Ohio River, Muskingum, Kentucky, and Cumberland subbasins, the 60- and 90-percent flow-duration parameters were determined by averaging values from nonregulated headwater and tributary flow records.

Further discussion on the use of the streamflow data to estimate ground-water recharge is presented in the section "Ground-Water Recharge."

PRECIPITATION — THE SOURCE

Gross water input to the Ohio Region is equal to total basin precipitation. Ground-water inflow to the region, if it occurs, is insignificant relative to total region precipitation.

Although precipitation is areally variable (fig. 5; table 2), average annual precipitation in all parts of the Ohio Region exceeds the national average. Monthly and seasonal variability of precipitation is greatest in the northwestern part of the region and least in the southern part.

On the basis of annual precipitation, average precipitation rates in the various subbasins range from 15,900 mgd (million gallons per day) in the Miami subbasin to about 47,200 mgd in the Cumberland subbasin.

EVAPOTRANSPIRATION—A MAJOR USE

The term "evapotranspiration" refers to the natural processes by which water on and beneath the land surface returns to the atmosphere as water vapor. Evapotranspiration is the largest single component of natural fresh-water discharge from the hydrologic system.

Mean annual lake evaporation is commonly accepted as potential evaporation or the maximum rate at which water can be taken into the atmosphere. It ranges from about 27 inches to about 37 inches in the region (fig. 6).

The type of vegetation and the depth of vegetative root systems in part determines the depth limit from which evapotranspiration can occur. This lower depth limit is not accurately known, but it probably is not more than 15 feet.

Maximum evapotranspiration probably occurs from alluvial aquifers adjacent to gaining reaches of many streams of the region, where water is readily available to the vegetation growing in the highly transmissive alluvial deposits. Evapotranspiration of ground water also occurs in marshlands and in the immediate vicinity of natural lakes fed by ground water.

THE SUPPLY

In simple terms, ground water in the region occurs in two zones: the supply of fresh water is at relatively shallow depths; the deeper zone contains a large supply of water of poorer quality. Hopkins (1966) mapped the interface between fresh and saline water in some parts of the region.

Although not all ground water in storage is recoverable by development, estimates of the amounts that can be obtained from storage, under specified conditions, are calculated to show the magnitude of water that is available.

FRESH WATER

For purposes of assessing the ground water available from storage, fresh water is defined as water containing less than 1,000 mg/l (milligrams per liter) dissolved solids. The definition is arbitrary because some waters with less than 1,000 mg/l dissolved solids are not potable. On the other hand, some municipal water systems in the Southwestern United States deliver water that has dissolved solids exceeding 2,000 mg/l. The Public Health Service specifies 500 mg/l as the desirable upper limit of dissolved solids in drinking water. Most waters can be treated to remove undesirable constituents at a cost. Therefore, in this report, all water is considered to be a resource, and it becomes a planning or management decision whether a given supply is potable or is usable only for other purposes.

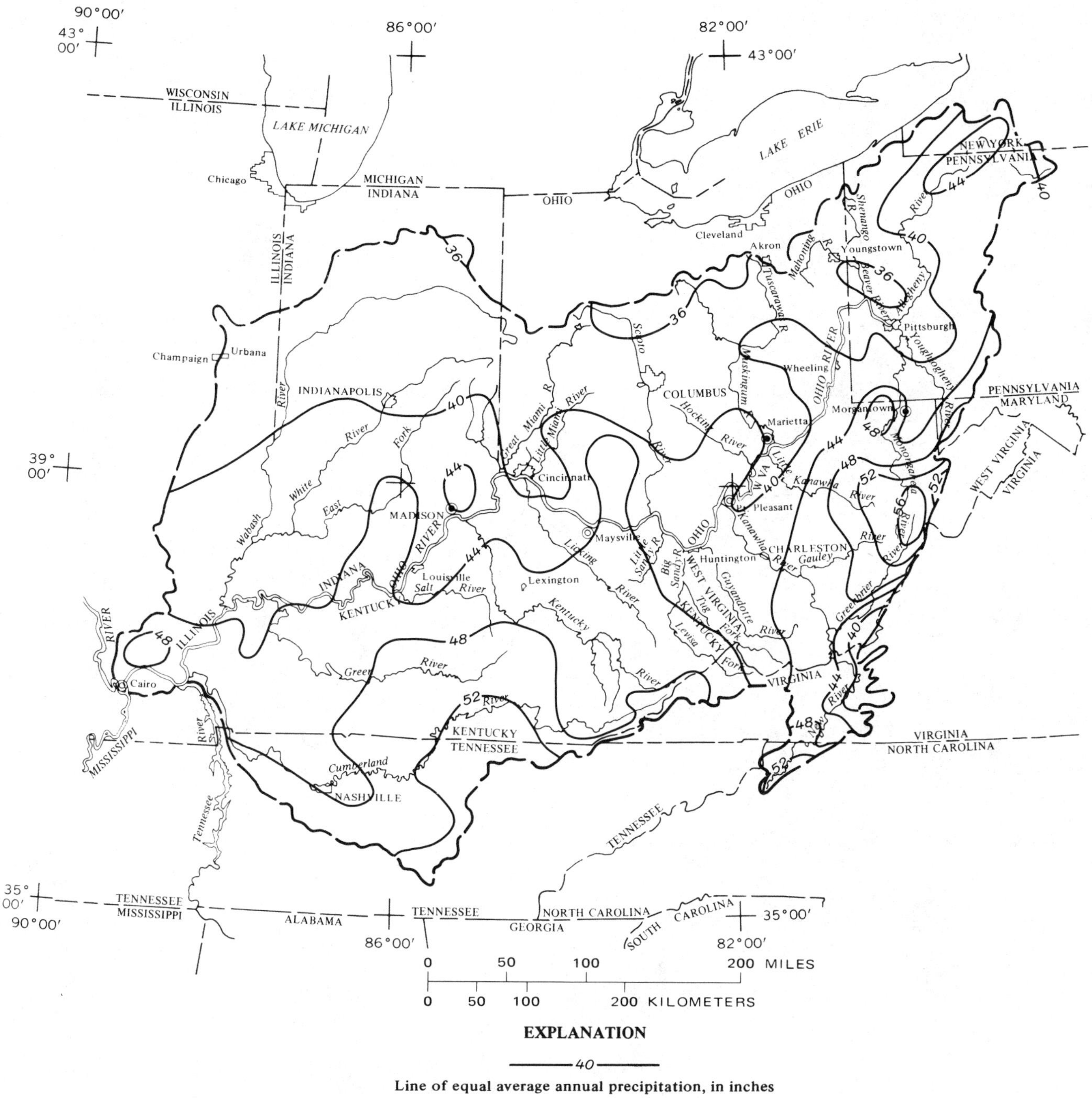

FIGURE 5. — Average annual precipitation, in inches (U.S. Department of Commerce, Environmental Science Services Administration, 1968).

Three categories of storage were specified for the purpose of computing the amount of potable ground water available from storage in the Ohio Region. They are (1) storage in the outwash and alluvial aquifer in the Ohio River valley, (2) storage in the outwash and alluvial aquifers in the various subbasins, and (3) storage in all other aquifers.

For making calculations of ground water available from storage, the simplifying assumption was made that all the outwash and alluvial aquifers in the Ohio Region are unconfined. In calculating the ground-water available from storage in an unconfined aquifer, the surface area of the aquifer, the saturated thickness of the aquifer, and the specific yield of the aquifer are needed. The areas of aquifers were estimated from geologic maps. Lithologic well logs and water levels in wells were used to

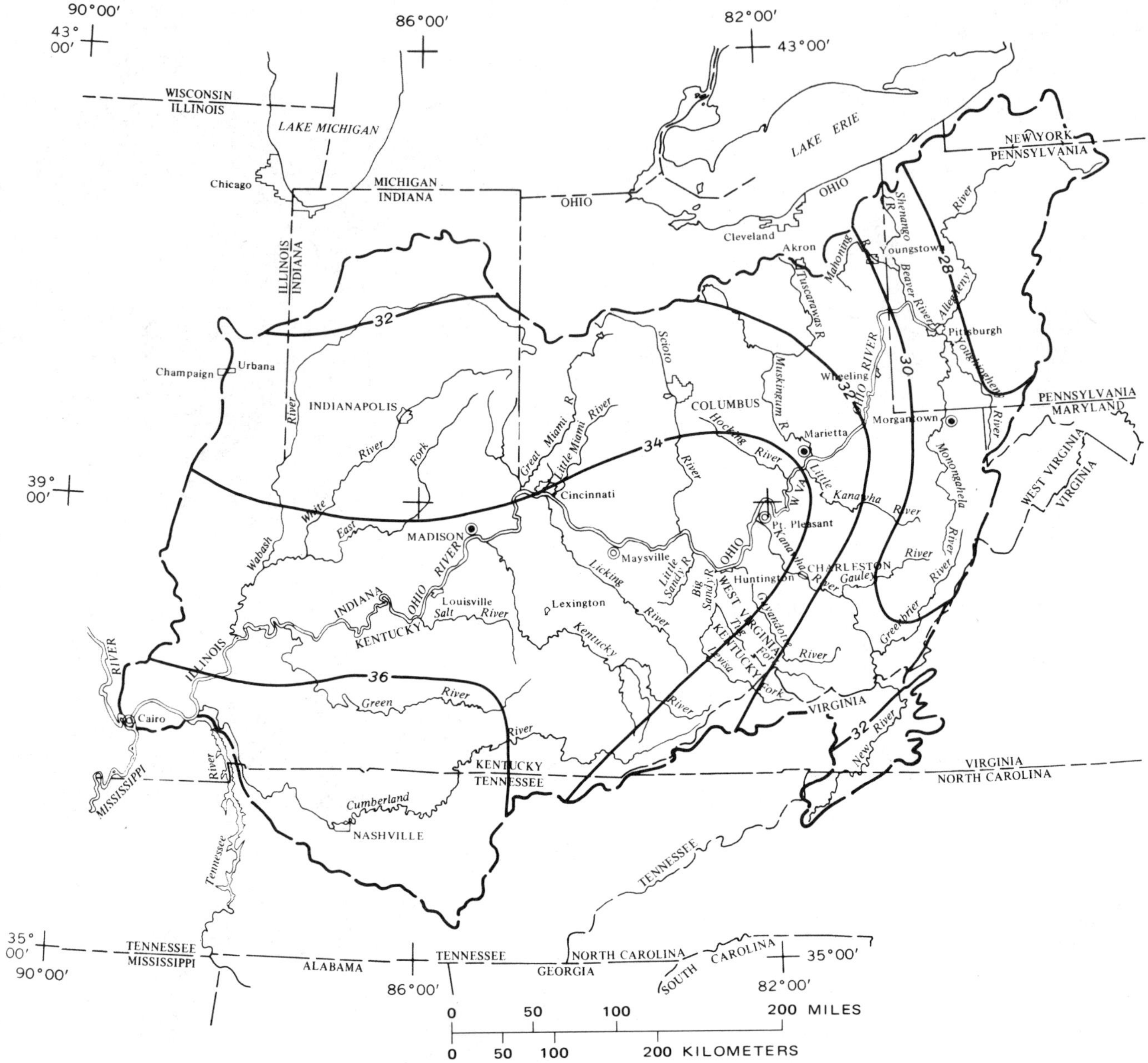

EXPLANATION

—32— Line of equal estimated mean annual lake evaporation, in inches

FIGURE 6. — Estimated mean annual lake evaporation, in inches (U.S. Dept. of Commerce, Environmental Science Services Admin., 1968).

estimate the saturated aquifer thickness — the lithologic logs to estimate formation thickness, and water levels in wells to define the zone of saturation. Much of the information on aquifer thickness is from Walker (1957) and Cross and Schemel (1956). Aquifer specific yields and storage coefficients are from aquifer tests.

Estimated amount of water available from storage in the outwash and alluvial aquifer in the Ohio River valley is about 4,500 billion gallons (table 3). About 60 percent of the water is in storage along the lower 200-mile reach of the river. The computed value of available stored water is crude but reasonable. The purpose of the com-

TABLE 2. — *Estimated average annual subbasin precipitation*

No. (in fig. 1)	Subbasin	Estimated annual precipitation (in/yr)	Subbasin area (sq mi)	Estimated precipitation (mgd)[1]
1.	Allegheny	42.0	11,500	23,000
2.	Monongahela	47.0	7,200	16,100
3.	Upper Ohio River	40.0	8,400	16,000
4.	Muskingum	39.5	9,250	17,400
5.	Kanawha-Little Kanawha	44.5	14,750	31,300
6.	Scioto	39.5	9,150	17,200
7.	Big and Little Sandy-Guyandotte	44.0	8,100	17,000
8.	Great and Little Miami	38.5	8,650	15,900
9.	Licking-Kentucky	45.0	11,250	24,100
10.	Wabash	40.0	23,000	43,800
11.	White	42.0	12,500	25,000
12.	Green-Salt-Lower Ohio River	48.0	16,350	37,400
13.	Cumberland	49.0	20,250	47,200

[1]mgd, million gallons per day.

TABLE 3. — *Calculation of potable water available from storage in the outwash and alluvial aquifer in the Ohio River valley*

Reach of river	Length of reach (miles)	Average aquifer width (ft)	Saturated aquifer thickness (ft)	Saturated aquifer volume (billion cu ft)	Specific yield	Stored ground water: Billions of cubic feet	Stored ground water: Billions of gallons
Pittsburgh to Beaver river	25.5	4,000	50	25	0.20	5	37
Beaver River to Ohio-Pennsylvania boundary	14.5	4,000	50	15	.20	3	22
Ohio-Pennsylvania boundary to Wellsburg	34.0	4,200	50	40	.20	8	60
Wellsburg to Captina	38.0	3,700	45	35	.20	7	52
Captina to Friendly	30.0	3,700	25	15	.20	3	22
Friendly to Muskingum River	30.0	4,700	35	25	.20	5	37
Muskingum River to Washington Bottom	22.5	6,300	35	25	.20	5	37
Washington Bottom to Mason	55.5	4,700	30	40	.15	6	45
Mason to Glenwood	36.5	7,400	35	50	.10	5	37
Glenwood to Big Sandy River	30.5	5,800	15	15	.05	1	7
Big Sandy River to Greenup	19.0	7,900	40	30	.05	2	15
Greenup to Portsmouth	20.5	8,400	40	35	.10	4	30
Portsmouth to Cincinnati	113.5	5,300	50	160	.20	32	239
Cincinnati to Warsaw	59.0	7,100	60	135	.10	14	105
Warsaw to Madison	28.5	6,800	60	60	.20	12	90
Madison to Charleston	31.5	5,300	60	55	.20	11	82
Charleston to NE. Louisville	13.0	8,400	70	40	.20	8	60
NE. Louisville to Valley Station	19.5	15,800	60	100	.20	20	150
Valley Station to West Point	8.5	11,600	70	35	.20	7	52
West Point to Brandenburg	16.0	7,400	70	45	.20	9	67
Brandenburg to dam 45	57.0	6,600	70	140	.20	28	209
Dam 45 to Hawesville	21.0	5,800	80	50	.20	10	75
Hawesville to Lewisport	14.0	10,000	70	50	.15	8	60
Lewisport to NE. Owensboro	15.0	18,200	70	100	.10	10	75
NE. Owensboro to dam 47	25.0	31,200	100	410	.10	41	307
Dam 47 to Henderson	26.0	35,400	70	340	.20	68	509
Henderson to Mississippi River	175.5	20,600	70	1,340	.20	270	2,020
Total						602	4,501

putations is to help planners assess the approximate magnitude of the ground-water resource, rather than to present precise figures.

About 18,300 billion gallons of water is available from storage in the outwash and alluvial aquifers in the subbasins (table 4). The amount ranges from 45 billion gallons in the Kanawha subbasin to 7,100 billion gallons in the Wabash subbasin. The estimates are considered to be conservative, especially those for some areas of higher potential yield. A conservative approach was taken because of the significant areal variations in the saturated thickness of the alluvial and outwash aquifers and because of the paucity of data on the thickness of alluvial sediments for some locations.

TABLE 4. — *Calculation of water available from storage in the outwash and alluvial aquifers in the subbasins*

No. (in fig. 1)	Subbasin	Surface area of outwash and alluvium (sq mi)	Assumed saturated thickness (ft)	Saturated volume (billion cu ft)	Assumed specific yield	Stored ground water	
						Billions of cubic feet	Billions of gallons
1.	Allegheny	1,610	25	1,120	0.15	170	1,270
2.	Monongahela	70	25	50	.15	8	60
3.	Upper Ohio River	370	50	515	.15	77	576
4.	Muskingum	1,050	50	1,460	.15	220	1,650
5.	Kanawha-Little Kanawha	90	25	65	.10	6	45
6.	Scioto	490	50	685	.15	100	748
7.	Big and Little Sandy-Guyandotte	310	25	215	.10	22	165
8.	Great and Little Miami	1,010	50	1,410	.20	280	2,090
9.	Licking-Kentucky	120	25	85	.10	8	60
10.	Wabash	3,410	50	4,750	.20	950	7,110
11.	White	1,980	50	2,760	.20	550	4,110
12.	Green-Salt-Lower Ohio River	560	25	390	.10	39	292
13.	Cumberland	290	25	200	.10	20	150
	Total					2,450	18,326

Total water available from storage in the outwash and alluvial aquifers in the Ohio River valley and the subbasins is about 23,000 billion gallons. This volume is about four times the flood-control storage of all Ohio Region Corps of Engineers reservoirs constructed, under construction, or in advance planning as of July 1965. By further comparison, the volume of ground water is about 5 percent of the volume of water in storage in Lake Ontario, based on Bue's estimate (1963, p. 11) of the volume of water in Lake Ontario.

Most of the ground water in the outwash and alluvial aquifers in the Ohio River valley and the subbasins is assumed to be potable. Exceptions are areas where acid mine drainage is prevalent. In such areas, even where the quality of the ground water is not now degraded, future degradation is possible. This is especially true where alluvial aquifers are hydraulically connected with polluted streams. This study does not delineate aquifers affected by acid mine drainage or other pollutants. Such a delineation should be made in subsequent, more intensive subregional studies. Without a delineation, satisfactory estimates of potable ground water available from storage are not possible.

For calculating the amount of water available from storage in the consolidated aquifers and the remaining unconsolidated aquifers of the region, the following simplifying assumptions were made:

1. That the aquifers are now confined but that lowering of water levels into the aquifers will produce unconfined conditions.
2. That a uniform vertical thickness of 250 feet of potable water is present throughout the Ohio Region in these aquifers.
3. That the amount of potable water available from storage in the aquifers is defined by the product of the aquifer surface area, a saturated thickness of 250 feet, and a storage coefficient or specific yield of 0.01.

The assumed 250 feet of saturated thickness is reasonable and probably conservative. Available data on water levels in wells in the region suggest that, in general, the potentiometric head in aquifers — other than the alluvial aquifers — is less than 100 feet below land surface. Feth and others (1965), showed that at depths greater than 500 feet in practically the entire Ohio region the dissolved solids in ground water is in excess of 1,000 mg/l. Using 100 feet and 500 feet below land surface as the upper and lower limits of potable water, the maximum saturated thickness in aquifers other than the alluvial aquifers is 400 feet. Such saturated thicknesses are not present everywhere. Actually, in many parts of the region, brine is present at shallow depths. Also, because practically all aquifers considered in this example are confined, the potentiometric head is closer to the land surface than the zone of saturation. The simplifying assumption was made that 250 feet is a reasonable value to use for the saturated thickness of potable water.

The assumed storage coefficient of 0.01 also is reasonable and probably conservative. As long as the aquifers are confined, the storage coefficient may range from 0.001 to 0.0001 or less. As the water levels are lowered into the aquifers to produce unconfined conditions, the storage coefficient probably increases and may range from 0.01 to 0.05 or more. In line with the other simplifying assumptions, the use of a constant storage coefficient of 0.01 seems justified.

On the basis of these assumptions, approximately 85,000 billion gallons of potable water is estimated to be available from storage in the Ohio Region in aquifers other than the outwash and alluvial aquifers. This amount is about four times the estimated water available from storage in the outwash and alluvial aquifers (tables 3, 4) and equivalent to about 20 percent of the volume of water in Lake Ontario.

SALINE WATER

Saline ground water is present throughout the Ohio Region. As shown by Feth and others (1965), the dissolved solids in ground water exceeds 1,000 mg/l at depths greater than 500 feet throughout the region. Salt concentrations in the saline zone range from 1,000 mg/l to considerably more than the 32,000–35,000 mg/l concentration in sea water.

The largest quantities of saline water, and therefore of equivalent storage volume in the saline zone, are in the Illinois and Appalachian basins. In these basins the depth to basement rock is the greatest in the Ohio Region. Assuming that the basement rock is the bottom of the saline zone, the bottom of the zone in the Illinois basin is about 9,000 feet below sea level; the bottom in the Appalachian basin is about 15,000–21,000 feet below sea level.

The saline zone can be used as a source of water for desalting, especially in areas where fresh-water supplies are small or as a storage reservoir for injection and recovery of fresh water.

UNDERGROUND SPACE

A simple definition of natural underground space might be expressed as the interconnected pore or fracture space into which a fluid can be emplaced. Underground space is not necessarily vacant. It is already occupied by a liquid or gas that must be compressed or displaced in order to use the space. That part of underground space occupied by potable water is evaluated in the section entitled "The Supply." In this section of the report, emphasis is upon unsaturated underground space at shallow depths. The presence of large volumes of unsaturated permeable deposits suggests the potential for additional temporary underground storage of potable water.

Geologic data suggest the presence of unsaturated outwash and alluvial sediment thicknesses of from 30 to 70 feet in the Ohio River valley adjacent to the main stem of the Ohio River. Seventy-foot thicknesses are present betweem dam 47 and Henderson (figs. 16, 17). Sixty-foot thicknesses are present between Cincinnati and Madison (figs. 13, 17) and between Brandenburg and Hawesville (fig. 17). Aquifer characteristics suggest that the sediments are permeable in all these reaches.

A gross estimate of the unsaturated volume in the above three reaches is 125 billion cubic feet; approximately 68 billion cubic feet is present between dam 47 and Henderson. Even if only one-half of the volume can be artificially recharged, temporary storage is available for approximately 450 billion gallons of water. Lesser volumes of unsaturated sediments and storage are present in other parts of the Ohio River valley.

Only a small potential exists for temporary storage water in other unsaturated geologic units in the region. The low relative permeability of the glacial till probably precludes the use of this unit. The small volumes of unsaturated alluvium in the various subbasins probably makes their use unfeasible. The consolidated aquifers are, for the most part, confined and already saturated, so that pressure effects of artificial recharge probably preclude their use.

DEFINING WATER FLOW THROUGH THE SYSTEM

Except for the almost stagnant brine in the deeper zone, most ground water in the Ohio Region flows through the porous rock stratum or aquifer system of the region from place of intake or recharge to place of discharge. The rate of flow under natural conditions generally ranges from a few tens of feet to a few hundreds of feet per year.

The recharge area for a consolidated or bedrock aquifer in the region is assumed to be the outcrop area of the aquifer; or, if the aquifer is covered with glacial or alluvial deposits, the recharge area is assumed to be the subcrop area underlying the glacial or alluvial deposits. The subcrop area underlying perennial streams generally is a discharge area. On the basis of these factors, the recharge areas for all bedrock aquifers can be inferred from a detailed geologic map.

Once water enters a consolidated aquifer there are two possible flow paths — the path through the intergranular, or primary, openings called pores and the path through secondary openings, such as rock fractures or solution channels. The intergranular openings generally were formed when the rocks were deposited as sediments. Although compaction and cementation generally occur with time and alter the original size and shape of the primary openings, some pore space generally is available in rocks. Rock fractures and solution channels, which occurred after the original deposition of the rocks, provide additional conduits for groundwater accumulation and movement. The secondary openings generally transmit greater quantities of water at higher rates than the primary openings.

The major discharge areas of the consolidated aquifers in the Ohio Region probably are the Illinois and Appalachian basins and the Ohio River valley. In these areas there is significant vertical ground-water discharge or interaquifer flow. Cartwright (1970, p. 917) used ground-water temperature anomalies in the Illinois basin to calculate the annual vertical ground-water discharge in an 1,800-square-mile area. He concluded that possibly as much as 95 percent of the discharge must be moving upward through vertical fractures or secondary openings.

Recharge to the unconsolidated alluvial and outwash aquifers comes from infiltration of precipitation, from upward interaquifer flow, and from percolation of streamflow from losing streams. In most areas the major source of recharge is precipitation. Under natural conditions practically all the exposed surface areas of the unconsolidated aquifers are recharge areas.

Under present manmade conditions of development, the recharge areas for unconsolidated aquifers encompass a smaller area than under former natural conditions. The unconsolidated alluvial aquifers generally are in the relatively flat stream valleys. The flat terrain favors economic development and habitation by man. As man develops and inhabits the area, some of the previously exposed land surface is covered by pavement and other impervious materials. Therefore, there is a decrease in the recharge area of the unconsolidated aquifers. In some areas of intensive urban and industrial development, almost all ground-water recharge is prevented.

The general direction of ground-water flow in the unconsolidated deposits of the region is toward the streams. As suggested in the section, "Stream-Aquifer Interrelation," the streams of the region are the "sinks" for ground-water discharge. Actually, there is interplay between the surface-water and ground-water systems. One reach of a stream may be a gaining reach, whereas a subsequent reach may be losing. Also, a gaining stream can become a losing stream and a losing stream can become a gaining one as the stream stage changes.

DELINEATION OF AQUIFERS

The aquifer system in the region was subdivided and ranked into 21 major units (table 1). The ranking of the units was made on the basis of the transmissivity of the aquifers in the units. The initial subdivision was made mainly on the basis of geologic age of the various units in the region (fig. 3). Modification of the initial subdivision was made on the basis of streamflow data, and the final refinement was made on the basis of available aquifer-test data.

The highest yielding aquifer in the Ohio Region is the outwash and alluvial aquifer occupying a valley that is incised into the Silurian bedrock system in the Mad River drainage area (fig. 15). Yields of individual wells in this area exceed 1,000 gpm (gallons per minute), and rather extensive development of the ground-water resource has already occurred. For example, in the Dayton, Ohio, area, 110 mgd of ground water was used in 1958. This was about one-fourth of the total ground-water use in the State at that time (Norris and Spieker, 1966).

The outwash and alluvial aquifer in the Ohio River valley is ranked as the second highest yielding aquifer in the region. Aquifer yields in some parts of the valley exceed those of the aquifer in the Mad River drainage, but as an overall unit the aquifer is less productive.

The Mississippian aquifer is the highest yielding bedrock unit in the region. Aquifer yields vary from place to place, but yields to wells are significant in the Green River subbasin, in Indiana and Ohio, and in Pennsylvania, where the unit is covered by glacial drift.

GROUND-WATER RECHARGE

Although various methods are used to determine available ground water in a region, most methods assume that average annual ground-water recharge is the upper limit of ground water perennially available. Assuming such a need for recharge estimates, they are presented in this report, but it should be recognized that not all recharge is necessarily available for use.

The 60-percent flow-duration data were used to make order-of-magnitude estimates of subbasin ground-water recharge (table 5). Assuming that the 60-percent flows are indicators of natural ground-water discharge, they also can be used as indicators of natural ground-water recharge. This assumes that an approximate hydrologic equilibrium must exist between ground water entering and leaving the hydrologic system. Actual ground-water discharge to streams probably is less than that indicated by the 60-percent flow parameter for the Monongahela, Allegheny, and Kanawha Rivers; therefore, ground-water recharge as determined by this parameter (table 5) for those rivers probably should be adjusted downward. The summary of recharge computations (table 5) suggests that recharge in the subbasins may be on the order of 15 percent of the subbasin precipitation. Although for the general purposes of this study no adjustments were made, detailed studies should include more precise estimates of recharge.

PRODUCTIVITY OF AQUIFERS IN THE OHIO REGION

Deutsch, Dove, Jordan, and Wallace (1969) estimated the rates at which aquifers in the Ohio Region can yield water to wells. These estimates, which were adopted in this study, were prepared on the basis of rock type, geologic structure, known well and spring production, low-flow streamflow records, drillers' records, and general knowledge of the local geohydrology.

Both the consolidated and unconsolidated aquifers of the region were classified on an areal basis into the categories of highest, high, intermediate, or low yield. The highest yield category applies to aquifers with a potential yield greater than 500 gpm; the high-yield category, 100–500 gpm; the intermediate-yield category, 25–100 gpm; and the low yield less than 25 gpm. In the Ohio Region the most productive wells are those situated adjacent to major streams and designed to induce flow from the stream to the alluvial ground-water reservoir. Areas ranked in the highest- and high-yield categories (fig. 7) have sufficient ground water for most municipal and industrial water demands. Areas ranked in the

TABLE 5. — *Summary of ground-water recharge computations*

No. (in fig. 1)	Subbasin	Subbasin area (sq mi)	60-percent flow (cfsm)[1]	Estimated ground-water recharge Cfs	Estimated ground-water recharge Mgd	Percent of subbasin precipitation
1.	Allegheny	11,500	0.55	6,300	4,100	18
2.	Monongahela	7,200	0.63	4,500	2,900	18
3.	Upper Ohio River	8,400	0.28	2,400	1,600	10
4.	Muskingum	9,250	0.27	2,500	1,600	9
5.	Kanawha–Little Kanawha	14,755	---	----	----	16
	Kanawha	12,461	0.58	7,200	4,700	--
	Little Kanawha	2,294	0.29	670	400	--
6.	Scioto	9,150	0.11	1,000	600	3
7.	Big and Little Sandy–Guyandotte	8,100	---	----	----	8
	Big Sandy	4,660	0.23	1,100	700	--
	Little Sandy	1,610	0.16	260	200	--
	Guyandotte	1,830	0.31	570	400	--
8.	Great and Little Miami	8,655	---	----	----	10
	Great Miami	6,276	0.29	1,800	1,200	--
	Little Miami	2,369	0.24	570	400	--
9.	Licking-Kentucky	11,231	---	----	----	5
	Licking	4,185	0.19	800	500	--
	Kentucky	7,046	0.19	1,300	800	--
10.	Wabash	23,000	0.32	7,400	4,800	11
11.	White	12,500	0.31	3,900	2,500	10
12.	Green–Salt–Lower Ohio River	16,350	0.31	5,100	3,300	9
13.	Cumberland	20,250	0.36	7,300	4,700	10
	Total			54,670	35,400	

[1]Cfsm, Cubic feet per second per square mile.

intermediate-yield category are capable of meeting the water demands of many small municipalities or industries. Areas ranked in the low-yield category (fig. 8) are capable in general of meeting only domestic water demands through individual wells.

Approximately 5 percent of the total region is classified in the highest yield category (table 6). Individual wells yield greater than 1,000 gpm in at least eight reaches of the Ohio River main stem and in the Wabash, Scioto, Muskingum, and Allegheny subbasins. Potential extraction rates in excess of 500 gpm are possible from practically the entire Ohio River valley outwash and alluvial aquifer and from parts of the unconsolidated aquifers in all subbasins north of the Ohio River.

WATER-SUPPLY NEEDS

Foregoing information in this report was presented to assist evaluation of the potential of the ground-water supply in the Ohio Region; the next logical step is to consider water demand.

CURRENT NEEDS

Partial estimates for ground-water withdrawals are presented in the "Ohio River Basin Comprehensive Survey." Estimates are presented by subbasin for the base-year (1960) municipal and industrial water use (U.S. Department of the Interior, Federal Water Pollution Control Administration, 1967), and for base-year rural water use (U.S. Army Corps of Engineers, 1968). The estimated base-year municipal water use is by water source so that estimated base-year municipal ground-water use is available. Industrial ground-water use by subbasin is not presented in the above reports. However, Murray (1968) presents basinwide industrial ground-water use for 1965. For the purposes of this report it was assumed that all base-year rural water use was from ground-water sources. This assumption probably is not far in error. The available base-year ground-water-use data were extracted from the reports of the comprehensive survey and from Murray's report and are presented in tabular form (table 7).

PROJECTED NEEDS

The gross demand for water and related functions in the region was estimated by the Corps of Engineers for the years 1980, 2000, and 2020 (table 8). By the year 2020, the total annual surface-water withdrawals in the region, including the nonconsumptive part returned to streams, are projected to be about 64 percent of the average annual volume of streamflow of the Ohio River at its mouth. Excluding the flow volumes from the Tennessee River, the Ohio River streamflow will, by the year 2020, about equal withdrawals in the Ohio Region. Reuse of water, of course, allows for greater use than total flow.

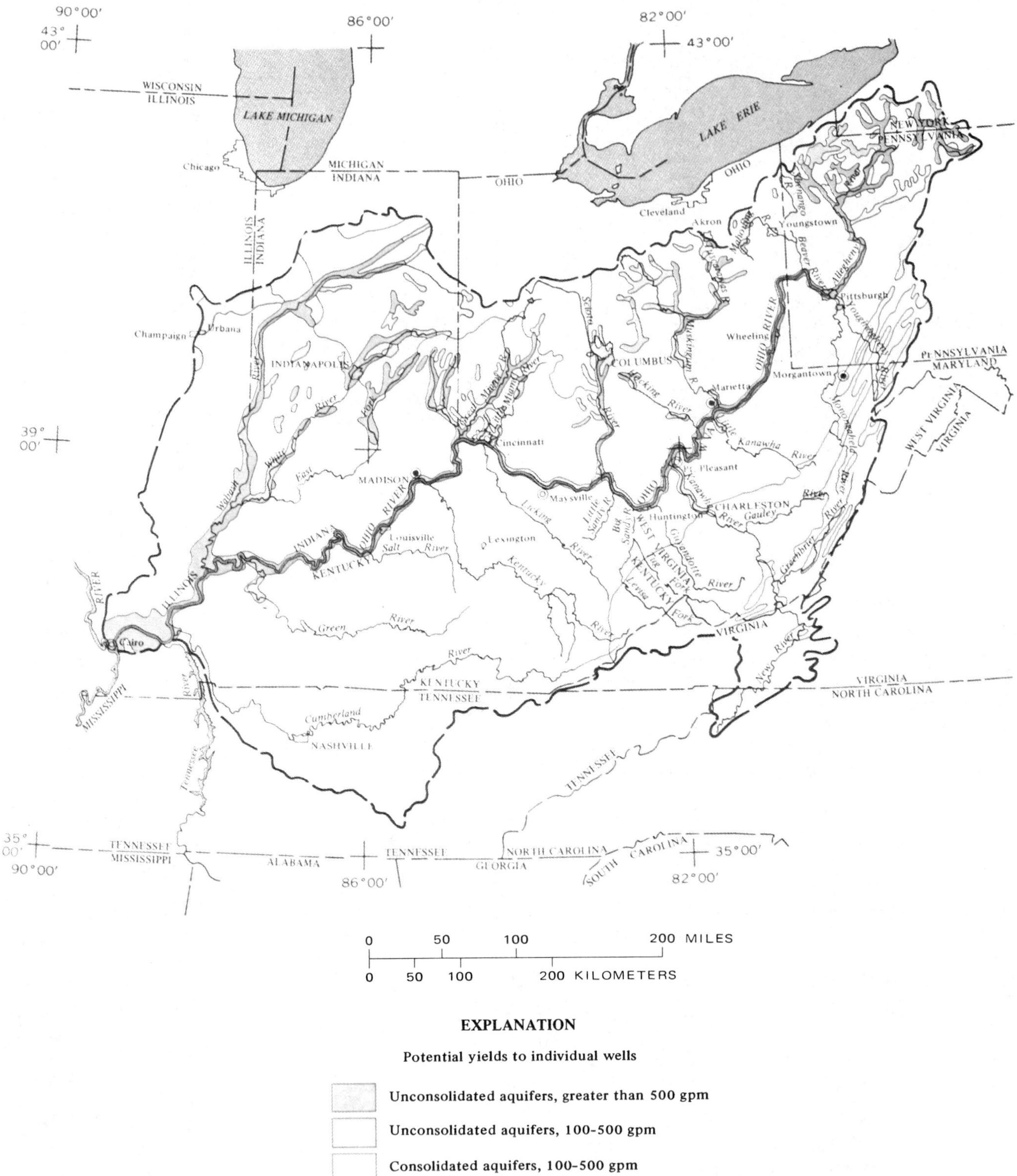

FIGURE 7. — High-yield sources of ground water.

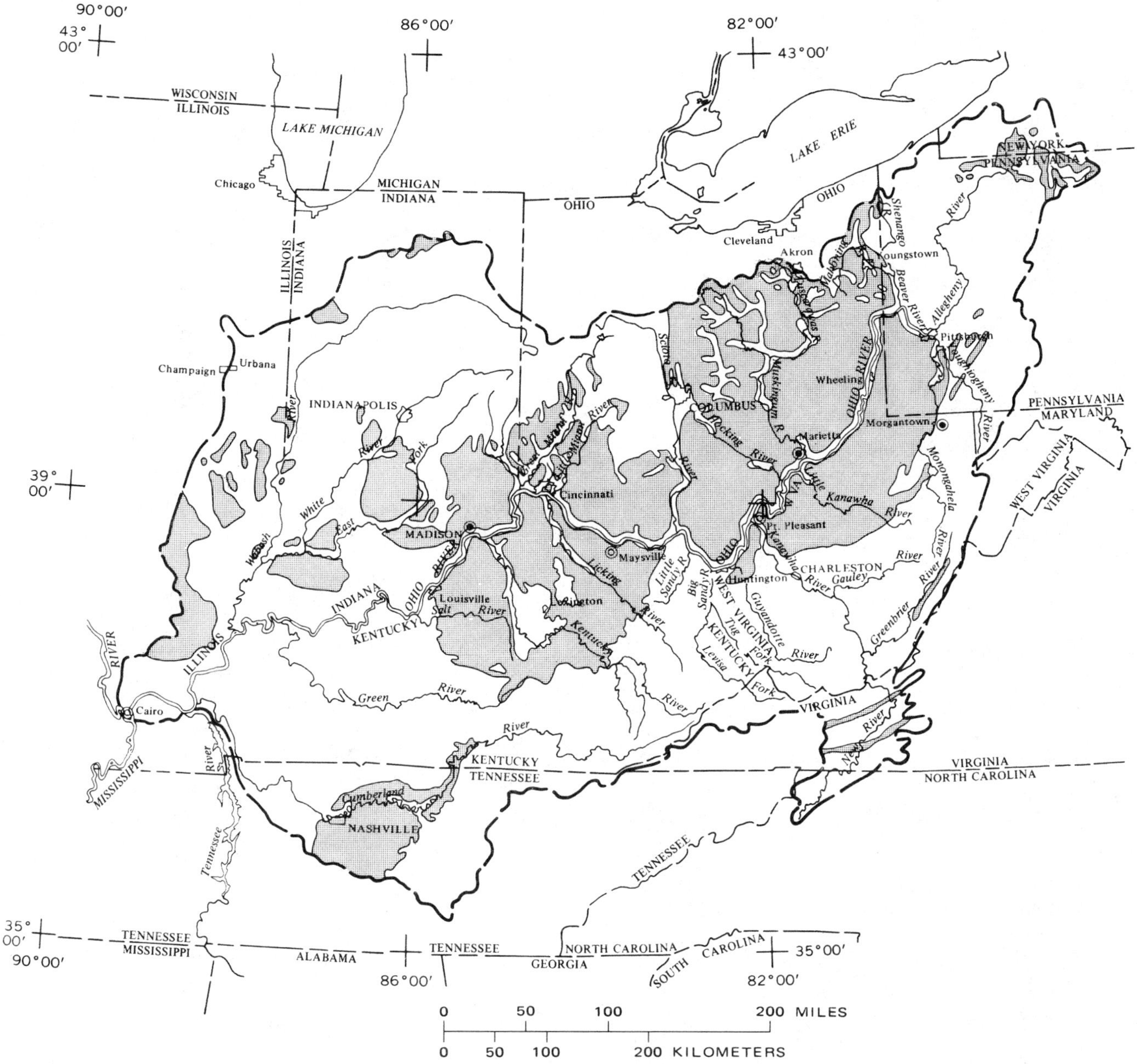

EXPLANATION

Areas where individual wells generally yield less than 25 gallons per minute

FIGURE 8. — Low-yield sources of ground water.

TABLE 6. — *Classification of subbasin area by potential amounts of ground-water withdrawal*

No. (in fig. 1) Subbasin	Percent of area			
	Highest yield	High yield	Intermediate yield	Low yield
1. Allegheny	8	12	50	30
2. Monongahela	0	26	49	25
3. Upper Ohio River	4	1	41	54
4. Muskingum	2	4	5	89
5. Kanawha–Little Kanawha	1	35	35	29
6. Scioto	3	26	6	65
7. Big and Little Sandy–Guyandotte	2	43	35	20
8. Great and Little Miami	5	19	19	57
9. Licking-Kentucky	1	3	42	54
10. Wabash	12	32	41	15
11. White	8	16	40	36
12. Green–Salt–Lower Ohio River	5	1	76	18
13. Cumberland	4	0	74	22
Total Ohio basin	5	16	44	35

TABLE 7. — *Base-year (1960) municipal, industrial,*[1] *and rural water use*

NOTE. — Total basin industrial ground-water use (1965) was 930 mgd (Murray, 1968).

Subbasin	Average municipal water use (mgd)			Average industrial use (mgd)	Average rural use (mgd)
	Total	Ground water	Surface water		
Allegheny	215	32	183	278	51
Monongahela	142	4	138	4,718	36
Upper Ohio River	172	60	112	1,608	28
Muskingum	86	50	36	72	53
Kanawha–Little Kanawha	45	10	35	1,477	53
Scioto	81	9	72	57	29
Big and Little Sandy–Guyandotte	16	8	8	51	31
Great and Little Miami	142	128	14	161	45
Licking-Kentucky	35	3	32	3	34
Wabash } White }	242	88	154	216	129
Green–Salt–Lower Ohio River	4	1	3	3	22
Cumberland	67	6	61	46	54
Ohio River Huntington	30	4	26	362	--
Ohio River Cincinnati	121	7	114	47	--
Ohio River Louisville	100	11	89	37	--
Ohio River Evansville	46	10	36	27	--
Total	1,544	431	1,113	9,163	565

[1]Industrial water use does not include water withdrawn for electric-power cooling purposes.

Estimates of future subbasin water withdrawals in addition to the base-year (1960) amounts are also presented in the "Ohio River Basin Comprehensive Survey" (U.S. Army Corps of Engineers, 1968). The values presented for the years 1980 and 2020 represent the water required to satisfy projected demands for withdrawal and use in addition to those provided in the base year. Such additional withdrawals are assumed by the Corps of Engineers to be a part of the total water withdrawals, both surface and ground water, required to satisfy municipal and industrial, electric-power cooling, rural community, rural domestic and livestock, and

TABLE 8. — *Estimated gross demand for water in the Ohio Region*

[Data are from the U.S. Army Corps of Engineers (1968). Values are in millions of gallons per day]

Use	Withdrawals in 1965	Total demand Year 1980	Year 2000	Year 2020
Municipal and industrial	11,553	14,035	19,357	28,251
Farm, domestic and livestock	162	168	231	294
Rural, nonfarm domestic	587	673	794	934
Irrigation	46	102	352	682
Electric-power cooling	19,200	29,000	46,000	63,000
Mining	289	511	974	1,894
Total	31,837	44,489	67,708	95,055

TABLE 9. — *Projected ground-water withdrawals for the year 2020, in millions of gallons per day*

No. (in fig. 1)	Subbasin	Ground-water withdrawals 1960 Municipal	Rural	Estimated additional withdrawals for the year 2020[1]	Total withdrawal for the year 2020[2]
1.	Allegheny	32	51	120	200
2.	Monongahela	4	36	150	190
3.	Upper Ohio River	60	28	200	290
4.	Muskingum	50	53	110	210
5.	Kanawha-Little Kanawha	10	53	260	320
6.	Scioto	9	29	110	150
7.	Big and Little Sandy-Guyandotte	8	31	60	100
8.	Great and Little Miami	128	45	510	680
9.	Licking-Kentucky	3	34	20	60
10. 11.	Wabash } White }	88	129	570	790
12.	Green-Salt-Lower Ohio River	1	22	20	40
13.	Cumberland	6	54	40	100
	Total	399	565	2,170	3,130

[1]Additional ground water required to partially satisfy municipal and industrial, electric-power cooling, rural community, rural domestic and livestock, and irrigation demands.

[2]Summation of estimated municipal and rural ground-water withdrawals in 1960 and estimated additional withdrawals above. Values have been rounded off.

irrigation demands. The estimated allowable ground-water-withdrawal rates for the year 2020 were extracted from the Corps of Engineers' report and are presented in tabular form (table 9) to permit further comparison and discussion.

THE SIGNIFICANCE OF THE SUBSURFACE SYSTEM IN THE OHIO REGION

The foregoing information in the report suggests that the region's ground water is a large natural resource and that there is a demand for such a resource. The next logical step is to show how the ground-water resource might be used and that ideally it can have a significant role in regional water development.

THE POTENTIAL FOR ADDITIONAL DEVELOPMENT

On the basis of a comparison of the projected ground-water withdrawals for the year 2020 with estimated ground-water recharge (table 10), it appears that the ground-water resources of the Ohio Region probably will not be used at full potential under existing development plans. Annual ground-water use (1960) by municipalities and rural residents is about 1,000 mgd (table 7). Average annual regional ground-water recharge is about 35,000 mgd (table 5). Therefore, base-year (1960) municipal and rural ground-water use is only about 3 percent of recharge. Annual regionwide ground-water use (1965) by industry, also, is only about 3 percent of recharge.

The Miami subbasin, which includes the Little Miami River drainage, is the only subbasin with projected total

TABLE 10. — *Comparison of projected ground-water withdrawals with estimated ground-water recharge for the year 2020*

No. (in fig. 1)	Subbasin	Total[1] withdrawal (mgd)	Estimated ground-water recharge (mgd)
1.	Allegheny	200	4,100
2.	Monongahela	190	2,900
3.	Upper Ohio River	290	1,600
4.	Muskingum	210	1,600
5.	Kanawha-Little Kanawha	320	5,100
6.	Scioto	150	600
7.	Big and Little Sandy-Guyandotte	100	1,300
8.	Great and Little Miami	680	1,600
9.	Licking-Kentucky	60	1,300
10.	Wabash }		4,800
11.	White }	790	2,500
12.	Green-Salt-Lower Ohio River	40	3,300
13.	Cumberland	100	4,700
14.	Basinwide industrial use	930	----
	Total	4,060	35,400

[1]Summation of estimated municipal and rural ground-water withdrawals in 1960 and estimated additional withdrawals (from table 9).

withdrawal rates greater than 25 percent of annual ground-water-recharge rates. A similar comparison for the Wabash subbasin suggests that plans for development are grossly underestimating the ground-water potential of the Wabash subbasin. This is true even if all the base-year industrial water use is from ground-water sources, which in fact it is not. The ground-water resources of the Allegheny, Monongahela, and Kanawha-Little Kanawha subbasins are also underdeveloped quantitatively, but water-quality considerations may be the controlling factor in these subbasins.

The Wabash and White subbasins probably have the highest potential of all Ohio Region subbasins for additional ground-water development. About 30,000 billion gallons, or about 28 percent of the total potable ground water available from storage in the Ohio Region, is in storage there. Estimated average annual ground-water recharge in the Wabash and White subbasins is about 7,300 mgd (table 5). Annual ground-water use (1960) by municipalities and rural residents of the subbasins of about 220 mgd (table 7) is only about 3 percent of ground-water recharge and about 0.3 percent of the potable ground water available from storage in the subbasins. Also, many high-yield aquifers are present and offer excellent reservoir-manipulation possibilities in conjunction with existing and planned surface reservoirs. Additional evaluation of the available hydrologic data is needed to design an optimal ground-water-development program for the subbasin.

ADVANTAGES FOR MANAGEMENT OF THE SUBSURFACE SYSTEM

Realizing that there is a potential for further ground-water development in the region, the planner ideally should consider utilizing several advantages associated with the subsurface (geologic) system and ground water. A discussion of some of these advantages follows.

One major advantage is the versatility of the subsurface system. The uses of the subsurface system already are diverse. Broadly speaking, however, the uses may be grouped into two categories — the temporary storage of gases and fluids (principally natural gas and fresh water) for later withdrawal and the injection of waste fluids for which withdrawal is not contemplated. These uses are likely to be competing in many parts of the Ohio Region, hence the need for proper planning and water-resource utilization.

Another advantage associated with ground water is that it generally is a renewable resource. In the section entitled "Ground-Water Recharge," estimates of average annual ground-water recharge are given. In addition to this natural recharge, artificial recharge of aquifers is also possible. Aquifer recharge can be induced from surface streams if the aquifer and stream are hydrologically connected. Surface spreading basins can be constructed adjacent to the stream to increase the area over which infiltration can occur. Recharge wells are used where hydrologic connection between aquifer and streams is poor or where land values prohibit land use for infiltration.

Flexibility is possible in the choice of a type of recharge facility. Depending upon the type of site available, reservoirs, ponds, lagoons, ditches, or pits may be utilized.

A very successful surface spreading facility has been in operation in Dayton, Ohio, for many years. The greater part of the Dayton municipal water supply is ground water. Ground-water pumpage in the major well field averages about 40 mgd; peak pumpage has been as high as 90 mgd (Norris and Spieker, 1966, p. 83). This large supply is maintained by artificial recharge of the aquifer with Mad River water. Infiltration ditches and lagoons covering a 20-acre area have been dredged and are flooded periodically when the turbidity of river water is low. The lagoons and ditches are dredged each year to remove bottom-clogging material and, thus, to maintain a high rate of infiltration.

Some of the records on the use of recharge wells are conflicting, and successes as well as failures have been reported. The injection of chlorinated water, free from silt, into a well casing at a rate sufficient to keep the casing full, has yielded the best results. Chlorine helps prevent the growth of soil-clogging micro-organisms. Silt introduced into a well can clog the well perforations or even penetrate the aquifer and reduce its hydraulic conductivity; and introduction of air in the aquifer can also decrease hydraulic conductivity of the aquifer. Incrustations by chemical action can also occur on metal casings where perforations are above the normal water table and exposed to air, where recharge waters are incompatible with the native water, or for other reasons.

Another advantage associated with ground water is that in much of the region the ground-water supply generally is equally available and dependable throughout the year. Because the volumes of the ground-water reservoir and of the water available from storage are vast, the ratio of the annual volume of ground water pumped from a particular reservoir or aquifer to the total volume present is small. Therefore, the supply is not depleted significantly in any one year or even in a series of years.

Another advantage is that the ground-water supply is widely distributed throughout the region. It is not equal everywhere in volume or rate of availability but is almost universally present in some quantity. This is in contrast to surface water which is readily available only in areas adjacent to perennial streams. The fact that ground water is almost universally present makes it an excellent reserve resource. It is available for emergency use during drought periods when surface-water resources are not available. An emergency also could arise from an accidental contamination of surface-water resources. Another emergency could arise in event of war. Therefore, even though the ground-water resource may not be the prime water resource in an area, a knowledge of the potential of the resource is warranted in planning for contingencies.

Ground water generally is not subject to large seasonal changes in temperature, a feature of importance to some industrial water users, such as chemical companies. Seasonal temperature fluctuations generally are greatest at shallow depths. Seasonal temperature fluctuations may amount to only a few degrees in deeper parts of an aquifer.

Of further importance to many water users, ground water generally is uniform in chemical quality, and the quality is not subject to sudden changes or even seasonal changes. Because the volume of ground water in storage is large and because of low velocities associated with ground-water movement, the replacement or turnover of water in storage is slow; therefore, chemical quality and temperature do not fluctuate considerably over short periods of time in most parts of the aquifer.

An advantage to be considered relative to the present-day concern for the environment is that the local environment or ecology is not permanently altered or scarred when ground water is developed properly. Visible evidence of development is generally limited to a pump, pumphouse, and pipeline. If development at the site ceases, the pump, pumphouse, and pipeline can be dismantled, and the environment can readily be restored to its original state.

Another important advantage is that the subsurface system generally is hydrologically connected with the surface-water system. Where the subsurface and surface-water systems are in good hydrologic connection, beneficial manipulation of the subsurface system is possible.

The beneficial manipulation of the ground-water system involves the lowering or raising of ground-water levels in order to regulate the available storage volume in response to time-varying storage requirements. Because the ground-water reservoir or aquifer, just as a surface reservoir, need not be maintained at full storage throughout the year, storage volume can be regulated. The value of the ground-water reservoir can be measured in part by the practicality or feasibility of such regulation.

The feasibility of any manipulation plan depends partly on the benefits derived as compared with detrimental effects. Just as with a surface reservoir, detrimental effects upon the rest of the hydrologic system and upon the local ecology can result because of induced reservoir water-level changes. In considering possible effects, a range of water-level changes for each manipulation scheme can be estimated. The total range of change can be subdivided into a number of intervals or increments. Then the effect on the ecology or environment can be estimated for each incremental change in water level.

The feasibility of any manipulation plan also depends upon the availability of answers to certain basic technical questions, such as:

1. What is the size of the ground-water reservoir or aquifer?
2. Where are the boundaries of the ground-water reservoir?
3. How much is the natural recharge to and natural discharge from the ground-water reservoir?
4. Where are the areas of natural recharge and discharge?
5. At what rate can water be taken from or added to the reservoir?
6. Are there any local geologic or hydrologic conditions that may hinder use of the reservoir?
7. What is the quality of water in storage and what quality changes may occur because of reservoir manipulation?
8. What is the project life of the reservoir?

Answers to many of these questions can be obtained using the information in this and other reports. More specific and definitive answers are possible when specific manipulation plans for specific areas are proposed and considered.

A limiting factor in practically every potential manipulation plan is the rate at which the ground-water reservoir can be artificially recharged. The potential ground-water extraction rates from wells are general indicators of the rates at which the wells can accept water. Theoretically, an aquifer can be recharged using injection wells at the same rate at which water can be withdrawn from the well, under a given head differential. However, the tendency for wells and infiltration facilities to become clogged causes a decrease in recharge rates with time. Therefore, average recharge rates with time probably will be much less than the designed rates.

The sites with the highest potential for beneficial manipulation of available ground-water-storage volume are in and adjacent to the flood plains of the major streams in the region (fig. 2). These areas have relatively thick and permeable aquifers underlying a surface stream or ready source of recharge. The benefits to be derived in utilizing the ground-water reservoirs underlying these areas as part of the proposed basin-wide reservoir system should be considered in any subsequent subregional studies.

With proper planning and proper water-resource utilization the natural advantages of the subsurface system and ground water commonly outweight inherent disadvantages, and ground water ideally should receive full consideration in water planning.

GROUND WATER TO SATISFY LOCAL WATER REQUIREMENTS

An individual well in almost any part of the region will yield enough water to satisfy a domestic water requirement. However, approximately 35 percent of the region (table 6, fig. 8) in general is capable of meeting only a domestic requirement through an individual well.

Disregarding water-quality considerations, approximately 65 percent of the region (table 6) can satisfy water demands of many small municipalities or industries through individual wells. In such areas individual wells generally yield greater than 25 gpm. Also, ground water can be used in meeting recreational demands where a large surface reservoir is not a specific requirement; for example, for small off-stream fishing ponds, swimming pools, or ice-skating rinks.

In order to consider some local problem-solving alternatives, locations of existing and potential water problems (U.S. Army Corps of Engineers, 1968, appendix K) were superimposed on subbasin maps showing locations of existing ground-water use and productivity of aquifers. In almost all these locations the use is for municipal water supply. For illustrative purposes, only the maps for subbasins with a significant number of projected problem areas are presented in this report (figs. 9–17).

As indicated by figures 9–17, many of the potential problem areas and some of the existing problem areas are underlain by high-yield ground-water reservoirs, which could have an important role in areal or local water-resource development plans.

The Miami subbasin, one with highly developed ground-water resources, is indicated as having many potential water-supply problem areas (fig. 13). Most of the problem areas are ones of existing ground-water use and development. The Miami subbasin is the only subbasin in the Ohio Region with projected total withdrawal rates in excess of the subbasin 90-percent flow parameters. Therefore over-development may become a reality. Principles outlined in Spieker's report (1968) should be used for management and development decisions, and other subbasin areas should be studied in like manner.

A comparative illustration shows that more than 140 potential outdoor recreational areas in the Ohio Region are outside the areas of poor sources of ground water (fig. 18). Therefore the ground-water resource is capable of meeting many of the recreational water demands, especially in areas remote from a surface-water source. The U.S. Department of the Interior, Bureau of Outdoor Recreation (1966), determined the potential outdoor recreational areas. An outdoor recreational area is one which meets all or part of demands for swimming, boating, water skiing, picnicking, camping, sightseeing, nature walks, and hiking. In nearly all places ground water can satisfy the water demand associated with the last five of the above demands because the demands are generally small.

The uniform temperature of ground water, especially if the temperature approaches that needed for trout hatcheries, makes the ground-water resource a potentially valuable resource of the sports-fishery industry. The potential value is there because cold-water streams capable of producing or sustaining trout fishing generally are limited to the forested mountainous areas of the Appalachian area. The use of ground water, although it may be expensive, could significantly expand the area in which trout fishing is possible.

GROUND WATER AS A REGIONAL RESOURCE

For ground water to be considered a regional resource, a large supply must be present, a large part of the supply

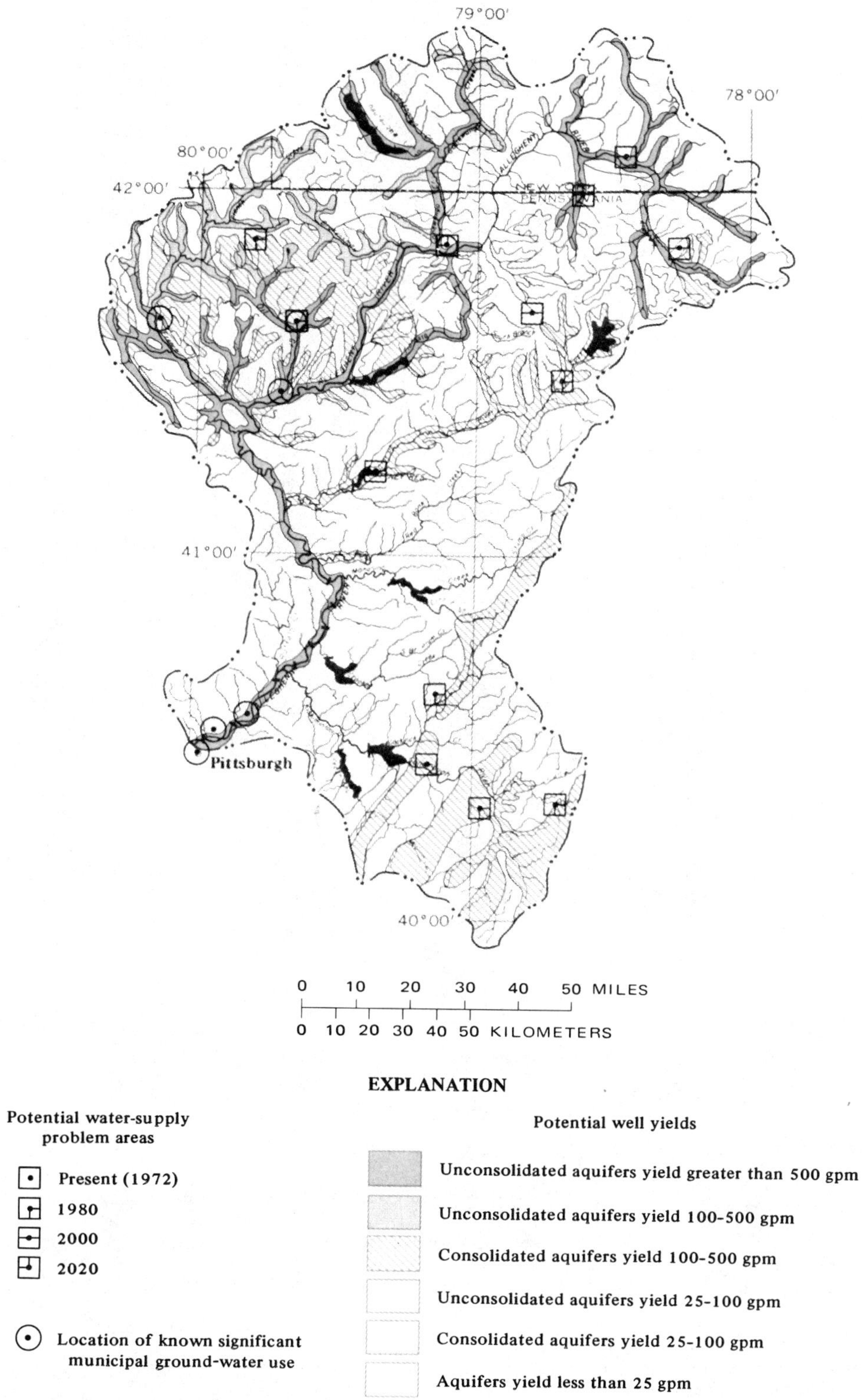

FIGURE 9. — Potential water-supply problem areas, potential well yields, and present ground-water use in the Allegheny subbasin.

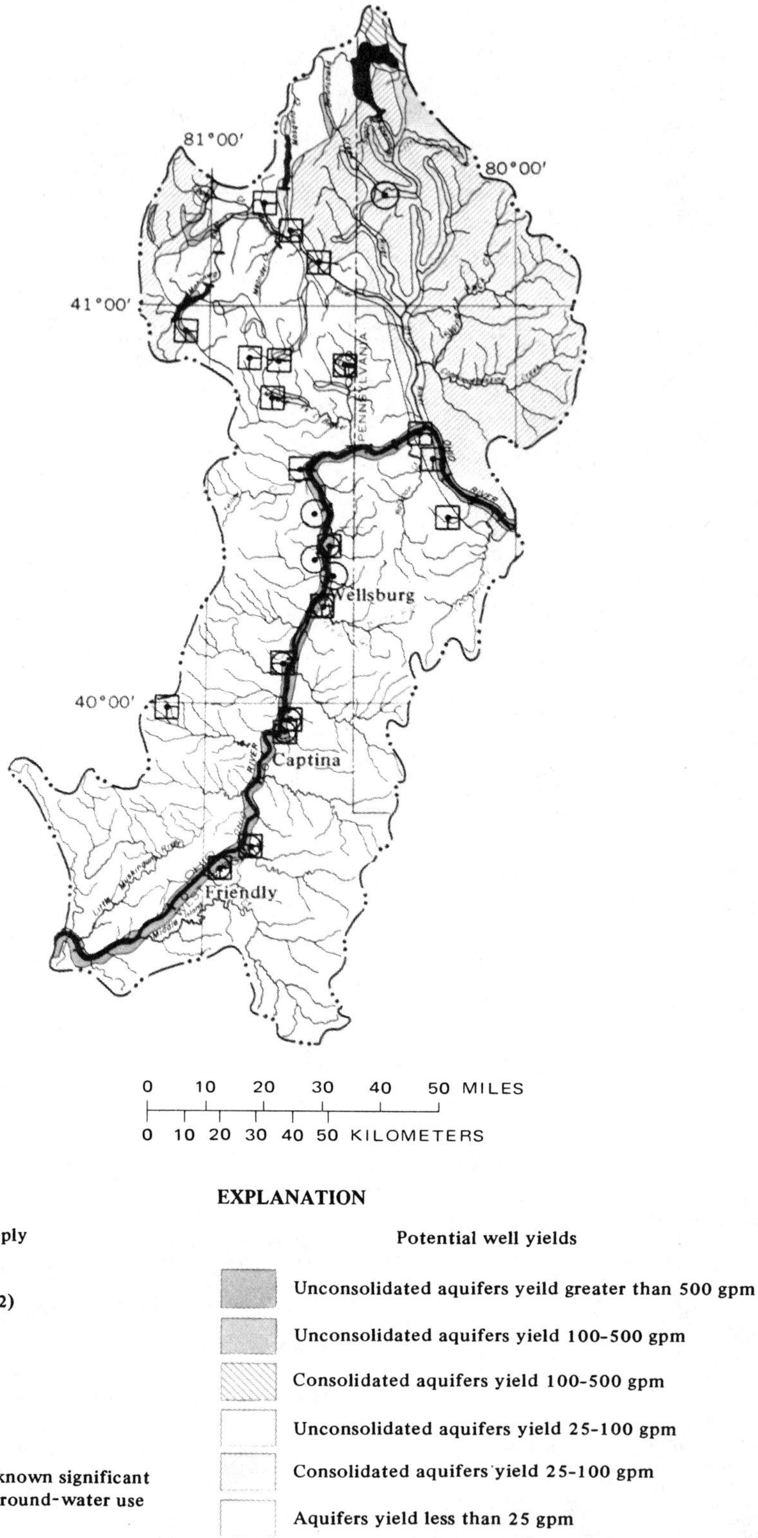

FIGURE 10. — Potential water-supply problem areas, potential well yields, and present ground-water use in the Upper Ohio River subbasin.

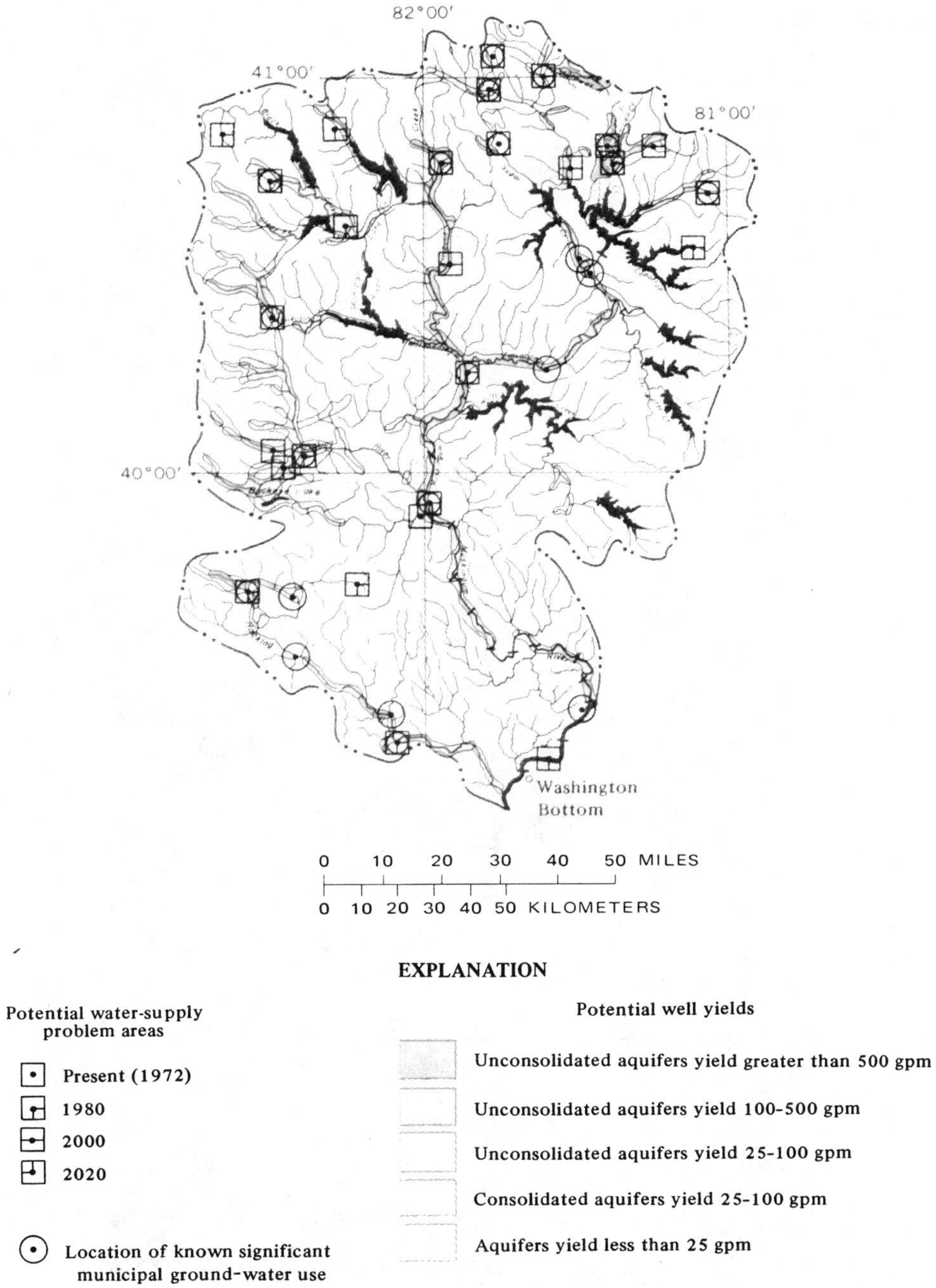

FIGURE 11. — Potential water-supply problem areas, potential well yields, and present ground-water use in the Muskingum subbasin.

must be available, and the supply must be renewable. Unless these criteria are met, large-scale regulation of the subsurface system probably is not warranted.

Specific criteria used in this report to classify the resource as regional are:

1. An unconsolidated aquifer(s) of significant areal extent and a saturated thickness of at least 50 feet must be present.
2. The aquifer(s) must have a potential to yield greater than 500 gpm to individual wells.
3. Surface water must be available for artificial recharge of the aquifer(s).

The ensuing discussion assumes that necessary legislation and land acquisition are possible if development of ground water is feasible. Also, all water referred to is assumed to be potable.

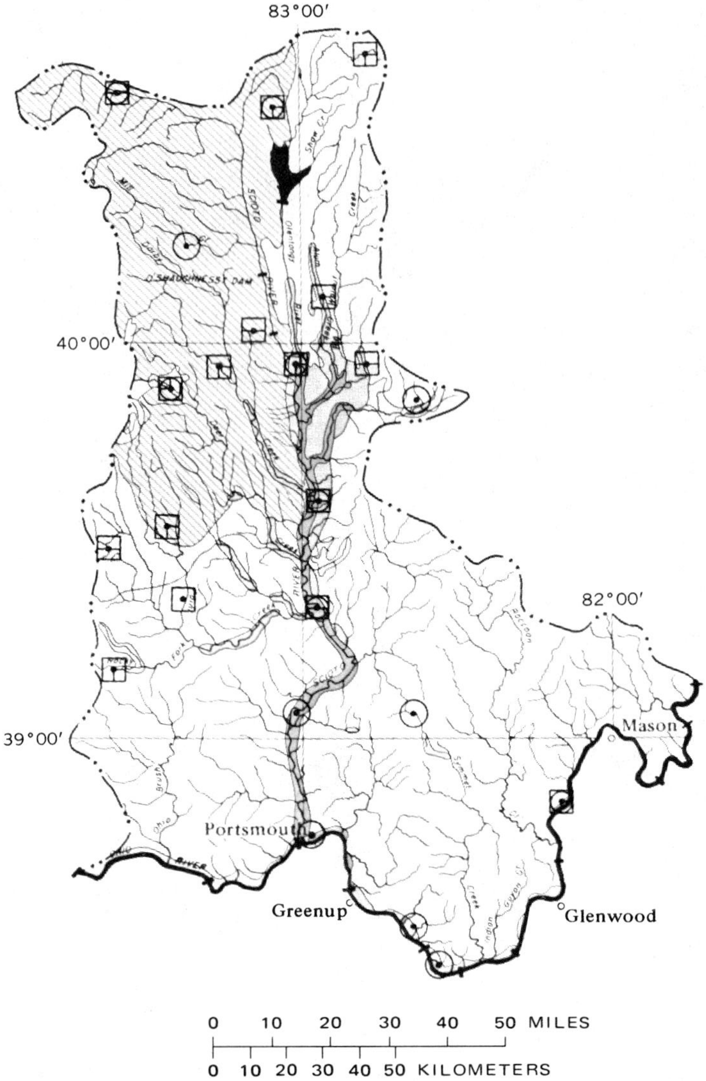

EXPLANATION

Potential water-supply problem areas

- Present (1972)
- 1980
- 2000
- 2020

Location of known significant municipal ground-water use

Potential well yields

- Unconsolidated aquifers yield greater than 500 gpm
- Unconsolidated aquifers yield 100-500 gpm
- Consolidated aquifers yield 100-500 gpm
- Unconsolidated aquifers yield 25-100 gpm
- Consolidated aquifers yield 25-100 gpm
- Aquifers yield less than 25 gpm

FIGURE 12. — Potential water-supply problem areas, potential well yields, and present ground-water use in the Scioto subbasin.

The above criteria are satisfied in approximately 5 percent of the region in that part underlain by unconsolidated aquifers with the potential to yield greater than 500 gpm to individual wells. The area in the Allegheny subbasin underlain by outwash and alluvial aquifers is included even though the available data

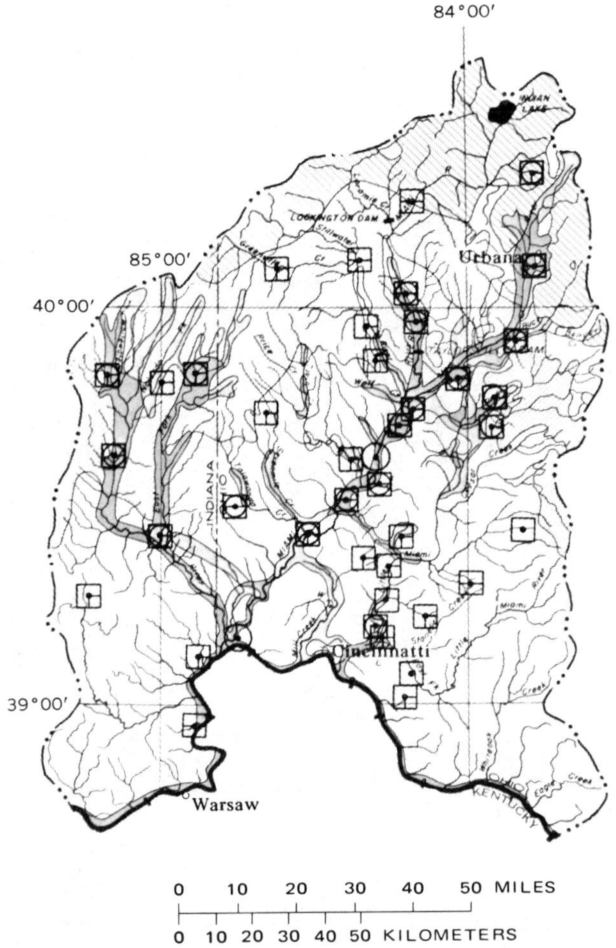

EXPLANATION

Potential water-supply problem areas

- Present (1972)
- 1980
- 2000
- 2020
- Location of known significant municipal ground-water use

Potential well yields

- Unconsolidated aquifers yield greater than 500 gpm
- Unconsolidated aquifers yield 100-500 gpm
- Consolidated aquifers yield 100-500 gpm
- Consolidated aquifers yield 25-100 gpm
- Aquifers yield less than 25 gpm

FIGURE 13. — Potential water-supply problem areas, potential well yields, and present ground-water use in the Great and Little Miami subbasin.

suggest a saturated thickness of only 25 feet (table 4). This area is included because the available data on aquifer thickness are extremely sparse.

Specifically, the ground water in the unconsolidated outwash and alluvial aquifers in much of the Ohio River valley, and in the Wabash, Miami, lower Scioto, and Allegheny subbasins can be considered a regional resource. In these areas ground water can satisfy more than local needs.

For example, the water requirements for a large population increase in the above areas can be satisfied with ground water. In order to estimate the actual population that can be satisfied, the following assumptions were made:

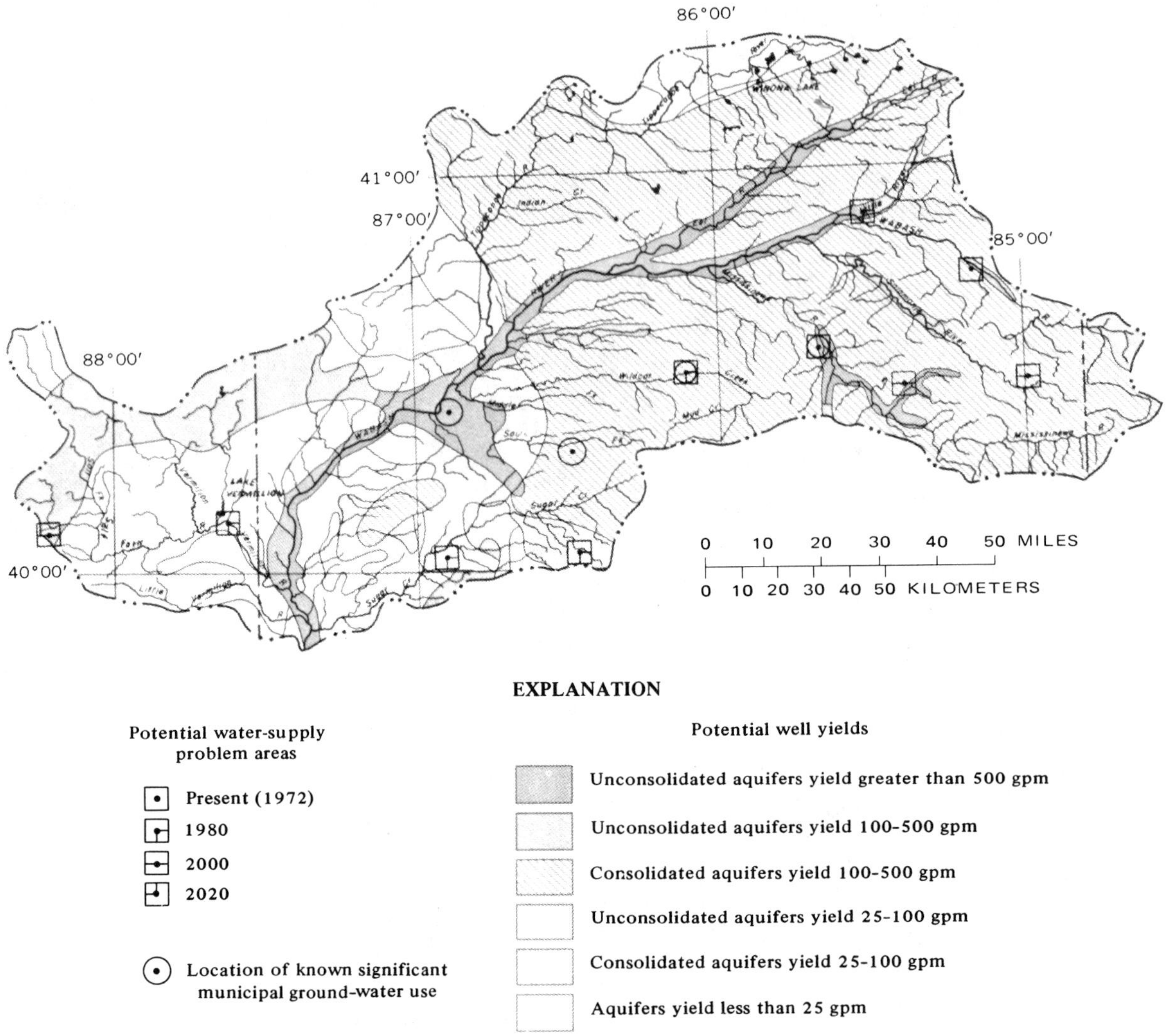

FIGURE 14. — Potential water-supply problem areas, potential well yields, and present ground-water use in the upper Wabash River drainage.

1. One-half the excess of annual ground-water recharge over base-year (1960) ground-water use is available. This should be a conservative estimate of available water.
2. All base-year industrial water use is from ground-water sources. This also causes the estimate of available water to be conservative.
3. The available water is used only once. Again, this causes the estimate of available water to be conservative.
4. Per-capita water use is 150 gpd (gallons per day). Therefore, a 1-mgd supply will satisfy the requirements of approximately 6,500 people.

On the basis of the above assumptions, ground water is available for 22 million additional people in the Wabash subbasin, and for 4, 1.5, and 12 million additional people, respectively, in the Miami, lower Scioto, and Allegheny subbasins.

The above estimates are given for illustration only. Instead of supporting a population increase, the available ground water in these subbasins could be utilized for industrial or agricultural expansion or for other uses. However, factors other than water supply constrain and determine development potential.

Ground water also is available in the Ohio River valley for approximately 5 million additional people. This estimate is based on the previously stated assumptions, except that a different method was used to estimate ground-water recharge. Because the Ohio River is regulated, streamflow data could not be used to estimate ground-water recharge. Instead, Rorabaugh's (1949) estimates of recharge were used.

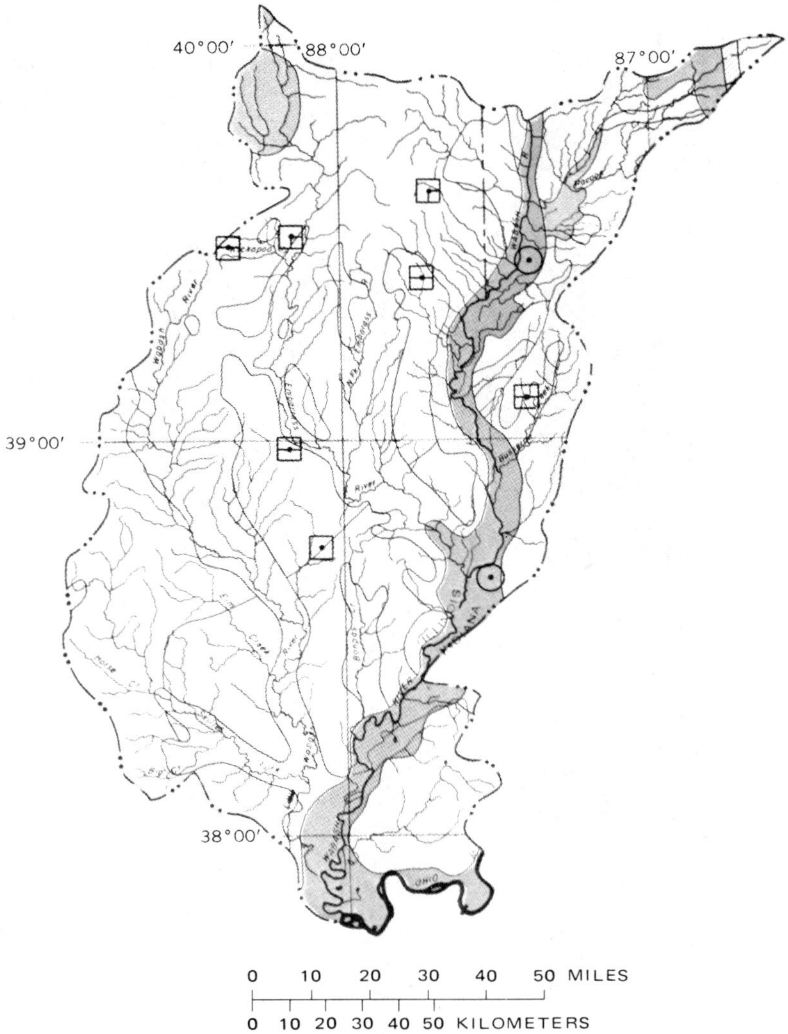

EXPLANATION

Potential water-supply problem areas

- Present (1972)
- 1980
- 2000
- 2020

Location of known significant municipal ground-water use

Potential well yields

- Unconsolidated aquifers yield greater than 500 gpm
- Unconsolidated aquifers yield 100-500 gpm
- Unconsolidated aquifers yield 25-100 gpm
- Consolidated aquifers yield 25-100 gpm
- Aquifers yield less than 25 gpm

FIGURE 15. — Potential water-supply problem areas, potential well yields, and present ground-water use in the lower Wabash River drainage.

Natural recharge to the outwash and alluvium in the Ohio River flood plain is partly by flow through the permeable rocks of the uplands adjacent to the outwash and alluvium and partly by direct downward seepage of precipitation.

For the purposes of this report, where a conservative estimate of recharge and available ground water is desired, zero recharge is assumed from the uplands. Sparse data on such recharge also make such an assumption necessary.

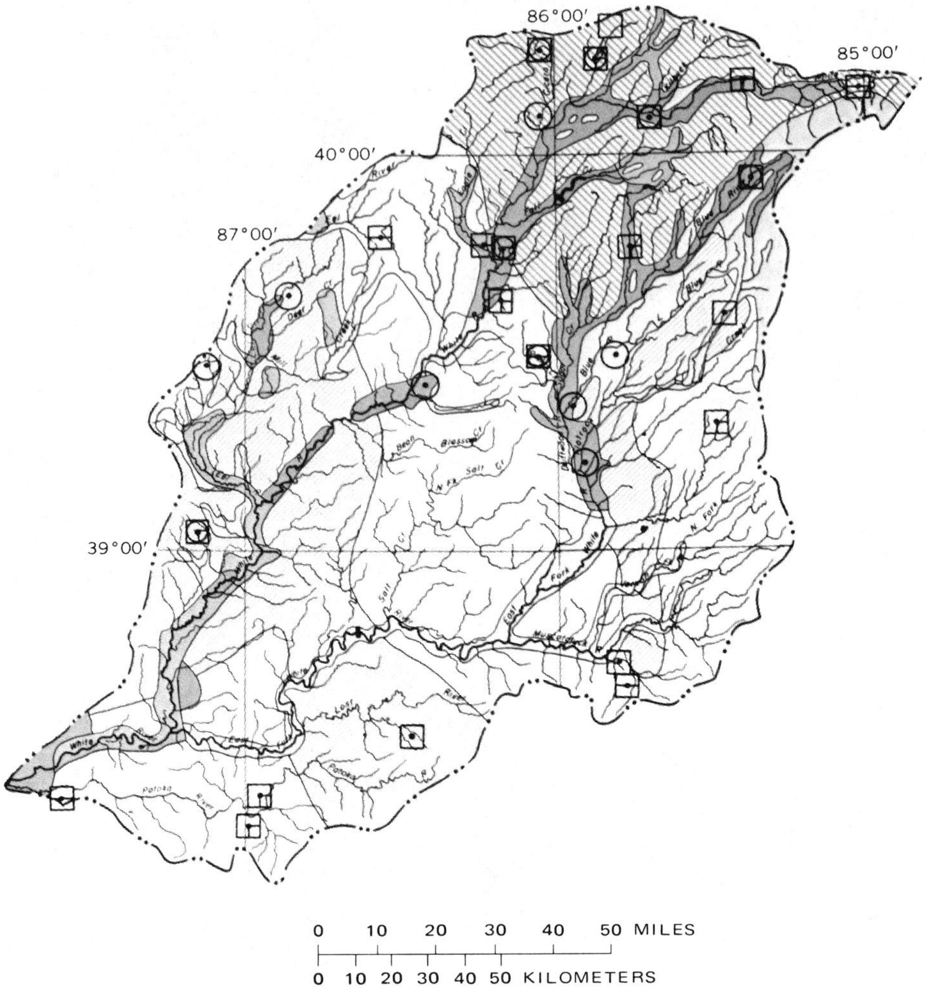

EXPLANATION

Potential water-supply problem areas

- Present (1972)
- 1980
- 2000
- 2020

Location of known significant municipal ground-water use

Potential well yields

- Unconsolidated aquifers yield greater than 500 gpm
- Unconsolidated aquifers yield 100-500 gpm
- Consolidated aquifers yield 100-500 gpm
- Unconsolidated aquifers yield 25-100 gpm
- Consolidated aquifers yield 25-100 gpm
- Aquifers yield less than 25 gpm

FIGURE 16. — Potential water-supply problem areas, potential well yields, and present ground-water use in the White River drainage.

Assumed recharge per square mile of outwash and alluvium from seepage of precipitation is 200,000 gpd. Rorabaugh (1949, p. 21) estimated seepage to the aquifer southwest of Louisville, Ky., in 1945 of about 250,000 gpd per square mile. However, because 1945 was a wet year, seepage would be less in dry years. Therefore, an average seepage of 200,000 gpd per square mile is assumed. The author also assumes that this estimate is applicable to the outwash and alluvium along the entire length of the Ohio River.

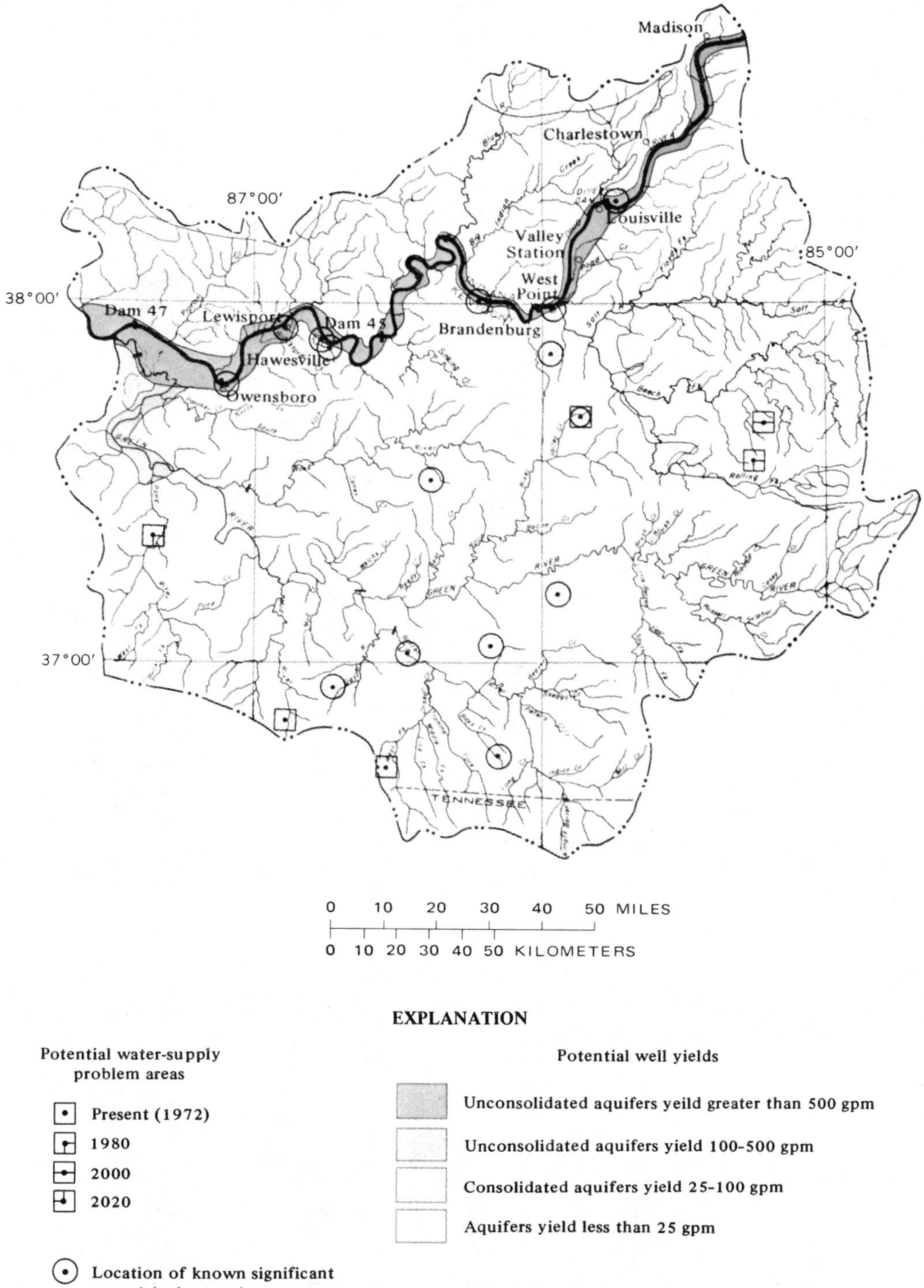

FIGURE 17. — Potential water-supply problem areas, potential well yields, and present ground-water use in the Green–Salt–Lower Ohio River subbasin.

Recharge from seepage of precipitation into the entire aquifer(s) in the Ohio River valley is about 2,000 mgd. This is the product of 10,000 square miles, or approximate aquifer area (table 3), and 200,000 gpd per square mile, or assumed recharge rate.

Finally, using the above estimate of recharge and

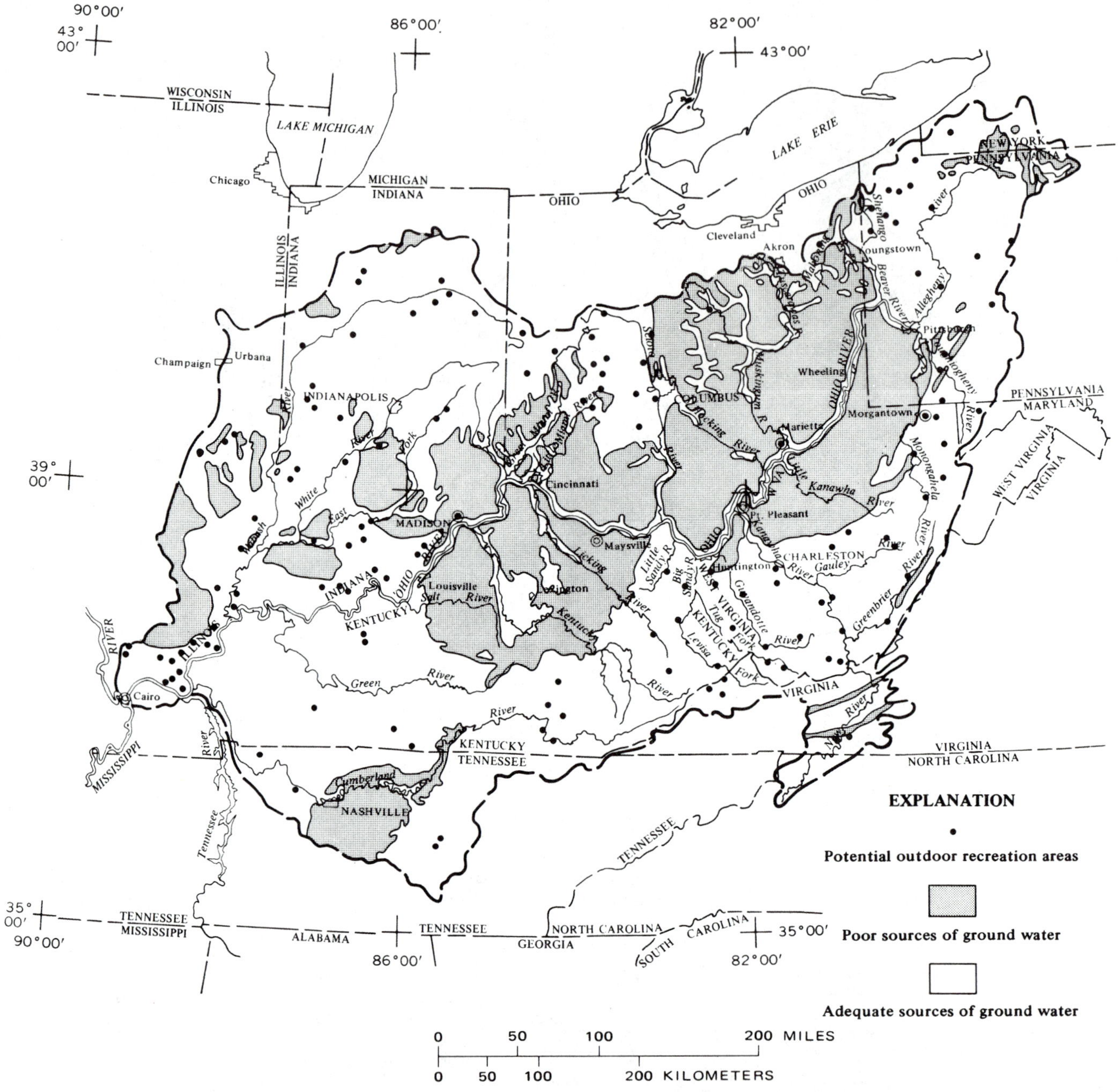

FIGURE 18. — Ground-water supply and potential outdoor recreation areas.

base-year water use (table 7), ground water is available in the Ohio River valley for approximately 5 million additional people.

In summary, ground water is available for 22 million additional people in the Wabash subbasin; for 4, 1.5, and 12 million additional people, respectively, in the Miami, lower Scioto, and Allegheny subbasins; and for 5 million additional people in the Ohio River valley.

Even though the above figures are gross, the obvious fact is that ground water is available in significant amounts in at least 5 percent of the region.

Of further importance to the planner, a reasonable assumption is that much of the available ground water in the above areas can be pumped and transported to relatively distant points of need. For example, approximately 55,000 people in the Fort Knox, Ky., area, southwest of Louisville, are served with ground water delivered from approximately 10 miles away (Mull and others, 1971). Several other communities in Kentucky also pump ground water from the outwash and alluvial aquifer in the Ohio River valley and transport the water to relatively distant points of use.

To make maximum use of the available ground-water resources of the region, the population-density pattern should be one of maximum population density adjacent to the Ohio River, and in the Wabash, Miami, lower Scioto, and possibly the Allegheny subbasins. An illustration, comparing ground-water supply, population density, and the interstate highway system (fig. 19), suggests that all areas of existing high population density have the potential for development of ground-water resources. The Indianapolis, Ind., area probably has the highest potential for ground-water development. Among the major population areas, the Youngstown, Ohio, area and the Nashville, Tenn., area probably have the poorest potential for ground-water development.

Assuming that future population growth will be heaviest in the areas paralleling the insterstate highway system, much of the increased water demand associated with these growth areas can be supplied by ground water. The areas of potential growth in Indiana and southwestern and south-central Ohio are especially well situated in terms of potential ground-water supplies. The development of the ground-water resources in western Pennsylvania and West Virginia probably depends primarily upon water-quality considerations, rather than upon availability of the resource.

The potential for additional industrial development also is enhanced in the areas where ground water is classified as a regional resource. Almost any type of industry can satisfy its water requirements with ground water in these areas.

THE POTENTIAL FOR UNDERGROUND WASTE DISPOSAL

Underground space in the Ohio Region also can be considered a regional resource in the sense that it can be included in regional water-pollution-control or waste-disposal plans. As previously suggested, a simple definition of natural underground space might be interconnected pore or fracture space into which a fluid can be emplaced. In this section of the report emphasis is upon underground space occupied by saline ground water. Before assessing the potential for underground waste storage in the region, a discussion is presented on the characteristics generally considered to be necessary for successful waste injection and underground storage.

Two geologic characteristics generally considered necessary for the use of waste-injection wells are (1) an injection zone with sufficient permeability, porosity, thickness, and areal extent to serve as a liquid-storage reservoir at safe injection pressures, and (2) an injection zone that is below the level of fresh-water circulation and is confined above by rocks that are, for practical purposes, impermeable to waste liquids.

Metamorphic and igneous crystalline rocks, which everywhere underlie the sedimentary rocks of the region, provide little potential for underground liquid-waste disposal because of lack of storage space and low permeability. Artificial tunnels or chambers in these rocks may facilitate storage under certain conditions. Warner (1969, p. B–28–A) listed characteristics of areas feasible for waste injection and underground storage.

On the basis of considering these characteristics, known conditions in the Ohio Region warrant investigation of waste-disposal potential in practically the entire region. In much of southern Illinois, southwestern Indiana, West Virginia, southwestern Pennsylvania, and extreme southeastern Ohio there is at least 1,500 to 2,000 feet of Mississippian and Pennsylvanian sedimentary rock present at considerable depth; in northwestern Pennsylvania, eastern Ohio, and south-central Indiana there are comparable thicknesses of Silurian and Devonian rock (Warner, 1969, fig. 9).

Although there is a great potential for underground waste storage in the Ohio Region, storing wastes underground is not a pollution panacea. The properties of the potential host rock and adjacent formations must be thoroughly understood. Waste injection undertaken without a knowledge of the limitations of the host formation may lead to undesirable or disastrous consequences. Possible consequences are the escape of waste to the surface or near-surface environment; the contamination of soils, surface water, ground water and other resources; and the denial of the use of other resources, such as oil and mineral ores. Evidently, effective underground waste management requires development of new technology (R. L. Nace, written commun., 1972).

Further discussion of waste management is beyond the scope of this report. However, the technical aspects of the use of salaquifers for waste storage is now under study by the U.S. Geological Survey.

PLANNING FOR GROUND-WATER UTILIZATION — AN EXAMPLE OF SIMULATION TECHNIQUES FOR GROUND-WATER SYSTEMS

The purpose of this section of the report is to demonstrate the relevancy of ground-water simulation techniques in solving a water-supply problem. As an example, a digital model of a small part of the stream-aquifer system south of Columbus, Ohio, was developed. S. E. Norris and other personnel of the U.S. Geological Survey office in Ohio contributed basic data and geohydrologic information that made development of the model possible. The model was developed in a very short time, for illustrative purposes. No verification of the

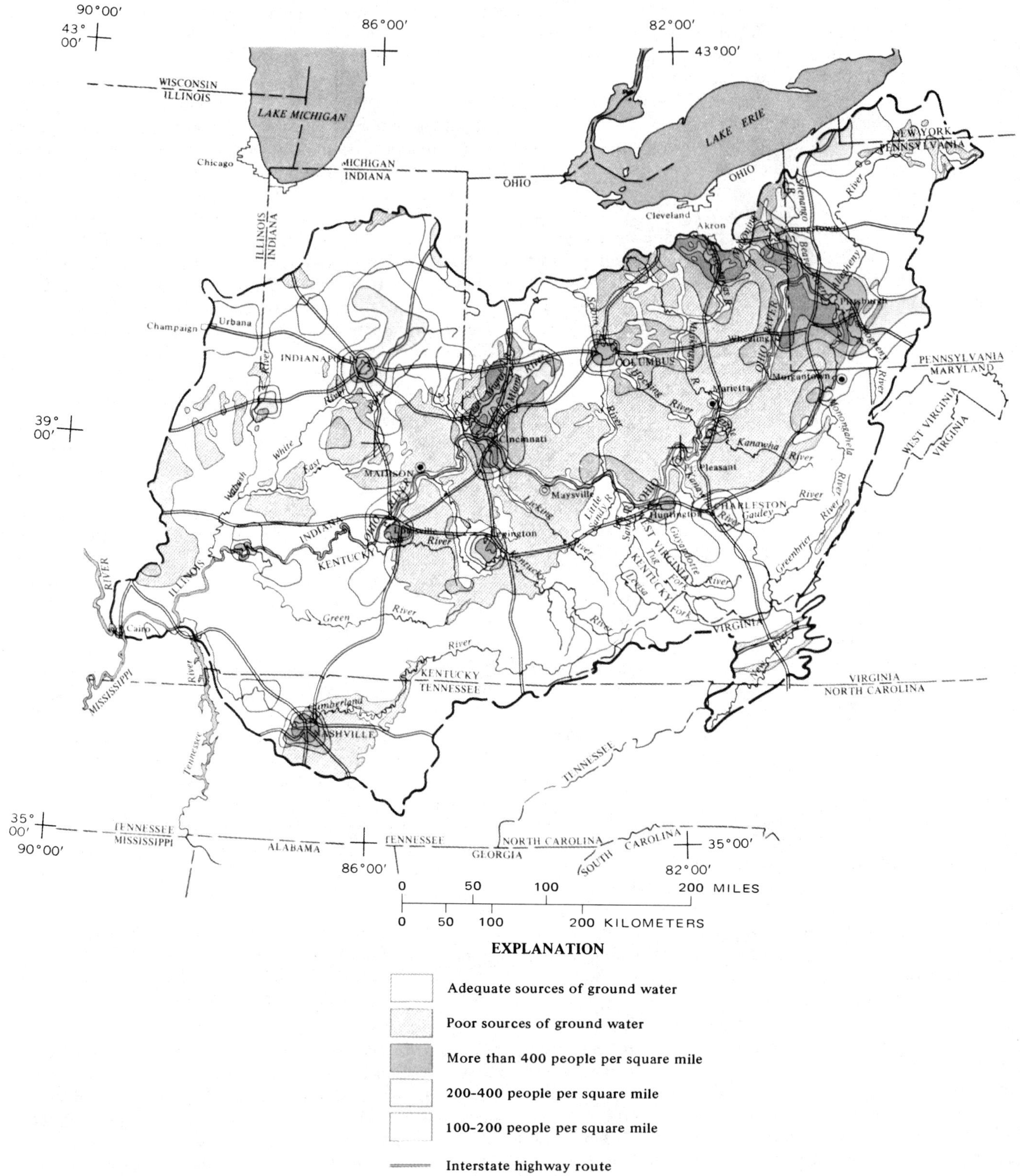

FIGURE 19. — Ground-water supply, population density, and interstate highway system.

model was attempted. In the study of an actual model the hydrologist generally verifies the model by simulating historical ground-water level changes caused by pumping, on the assumption that if the model simulates historical conditions it will also closely predict future conditions. The intent of the example is to demonstrate the methodology rather than to develop a verified model. Data related to the degree of stream-aquifer connection and data on hydraulic head differences between the stream and aquifer are scant. Such data are critical in determining actual drawdowns in the aquifer. Therefore, the model results should be considered as hypothetical.

The Columbus, Ohio, area was chosen for the example because large amounts of ground water are available in the aquifer system under the area and because the potential for additional surface storage in relation to need is small. Therefore, the Columbus area seemed a logical one in which to consider feasibility of conjunctive use of surface-water and ground-water resources.

A rather typical approach to solving a water-supply problem is used in the example. The surface-water supply is evaluated both in terms of average conditions and in terms of the critical period; the demand for water is compared with the surface-water supply; and, then, the ground-water supply is evaluated. Note, however, that the ground-water supply is evaluated for only a small part of the total Columbus area.

WATER RESOURCES VERSUS WATER DEMANDS IN THE COLUMBUS AREA

The major developed water resources of the Columbus area are the Scioto and Olentangy Rivers and Alum and Big Walnut Creeks. The city of Columbus has three surface reservoirs to store streamflow: O'Shaughnessy and Griggs Reservoirs on the Scioto River, with capacities of 5,520 and 1,440 million gallons, respectively, and Hoover Reservoir on Big Walnut Creek with a capacity of 19,660 million gallons. Because sediment-inflow estimates were not readily available, the above storage volumes for the surface reservoirs are assumed to be constant with time.

Streamflow records were used to estimate average monthly surface inflows to the three reservoirs operated by the city of Columbus (table 11). Streamflow records for the Scioto River near Prospect, Mill Creek near Bellepoint, and Big Walnut Creek at Central College were used in computing the monthly inflows. The amount of inflow to O'Shaugnessy Reservoir derived from an ungaged area of 235 square miles above the reservoir was estimated using the formula:

Ungaged inflow = 0.166 Prospect + 0.790 Bellepoint. In the above formula, Prospect refers to the Scioto River flow near Prospect and Bellepoint refers to the Mill Creek flow near Bellepoint. The formula is based on a simple areal relation. The first coefficient in the formula is the product of Prospect flow per unit drainage area and 40 percent of the ungaged drainage area. The second coefficient is the product of Bellepoint flow per unit drainage area and 60 percent of the ungaged drainage area. In other words, the assumption was made that 40 percent of the ungaged area had the same streamflow per unit drainage area as the Scioto drainage above Prospect and that 60 percent of the ungaged area had the same streamflow per unit of drainage area as the Mill Creek drainage. Much more accurate estimates of inflow to O'Shaughnessy and Griggs Reservoirs would be possible

TABLE 11. — *Estimated average monthly inflow to O'Shaughnessy, Griggs, and Hoover Reservoirs*

Month	Flow (mgd) Scioto River near Prospect	Mill Creek near Bellepoint	Big Walnut Creek at Central College	Estimated ungaged flow (mgd)	Estimated total inflow[1] (mgd)
October	115	11	15	19+ 12= 31	170
November	180	72	56	30+ 44= 74	380
December	377	194	118	63+ 93=156	850
January	791	362	288	131+228=359	1,800
February	746	378	291	124+230=354	1,770
March	1,016	400	363	169+287=456	2,240
April	859	326	285	143+225=368	1,840
May	422	161	149	70+118=188	920
June	364	202	130	60+103=163	860
July	178	95	35	30+ 28= 58	370
August	109	42	31	18+ 25= 43	230
September	106	17	18	18+ 14= 32	170

[1]Inflow values have been rounded off.

TABLE 12. — *Estimated critical-period inflow to O'Shaughnessy, Griggs, and Hoover Reservoirs*

Month	Flow (mgd)			Estimated ungaged flow (mgd)	Estimated total inflow (mgd)
	Scioto River near Prospect	Mill Creek near Bellepoint	Big Walnut Creek at Central College		
May	157	59	45	73	[1]330
June	85	63	35	64	[1]250
July	13	1	1	3	18
August	10	2	0	3	15
September	7	1	1	2	11
October	7	1	2	2	12
November	10	2	1	3	16
December	11	3	3	4	21
January	10	4	5	4	23
February	286	243	333	239	[1]1,100

[1]Inflow value has been rounded off.

if a stream gage were present on the Scioto River just above O'Shaughnessy Reservoir.

The estimated average yearly inflow to the three reservoirs, based on the above estimates of monthly inflows, is approximately 3.5×10^5 million gallons, or about 960 mgd.

Again using streamflow records and the above formula for estimating inflow from the ungaged drainage area, monthly values for critical-period inflow were estimated for the three surface reservoirs (table 12). The critical period of record for the Scioto River near Prospect is assumed as May 1944 to February 1945. This critical period is assumed to be applicable to the other records under consideration in this example.

Another major water resource in the Columbus area is the ground water available from storage in the unconsolidated glacial and alluvial deposits. Lesser volumes of water are available from the consolidated aquifers in much of the Scioto basin above Columbus, but this potential supply is not considered in this example. A 50-foot-thick layer of aquifer in an area of 70 square miles, just south of Columbus, is considered (fig. 20). Outwash deposits are present in about 19 square miles of the 70-square-mile area; till is present in the rest of the area. The use of the renewable ground-water resource from storage here is not new to the city of Columbus. During the severe drought of the early 1960's, the city planned and developed a well field to tap the outwash and alluvial aquifer adjacent to Alum Creek.

The average water demand for Columbus in the year 2000 is estimated to be 350 mgd (Ohio Division of Water, 1963, pl. 16), or almost 130,000 million gallons per year. Because demand is not constant throughout the year, average demand for each month was estimated. Monthly demand is assumed at a relative maximum during June, July, and August; demand is assumed at a relative minimum from November through February (table 13).

Even though the period of maximum water demand

TABLE 13. — *Comparison of estimated critical period inflow with estimated daily and monthly water demand for the year 2000*

Month	Percent of yearly water demand	Monthly water demand[1] (million gallons)	Daily water demand (mgd)	Surface-water supply (mgd)	Daily water deficit (mgd)	Monthly water deficit[1] (million gallons)
May	9	11,500	371	330	41	1,270
June	10	12,780	426	250	176	5,280
July	10	12,770	412	18	394	12,210
August	10	12,770	412	15	397	12,310
September	9	11,130	371	11	360	10,800
October	8	10,230	330	12	318	9,860
November	7	9,240	298	16	282	8,460
December	7	8,930	288	21	267	8,280
January	7	8,930	288	23	265	8,220
February	7	9,250	319	1,100	----	----
Total deficit	--	----	----	----	----	76,690

[1]Values have been rounded.

EXPLANATION

Transmissivity values, in gallons per day per foot, and storage coefficient

- T = 80,000 and S = 0.20
- T = 40,000 and S = 0.01
- T = 20,000 and S = 0.01
- T = 4,000 and S = 0.01
- T = 400 and S = 0.01
- T = 0

—— Model boundary

FIGURE 20. — Area of model and assumed values for transmissivity and storage coefficient.

does not coincide with the period of maximum surface-water supply, the available surface-water reservoirs have sufficient carryover storage volume to satisfy demands if average or greater than average streamflow occurs.

During the critical period, the carryover storage volume in the existing surface reservoirs is not sufficient to satisfy the estimated water demand in the year 2000. After considering dead-storage requirements, the total available surface-reservoir storage is 26,640 million gallons. During the critical period of 9 months there is a water deficit of about 77,000 million gallons, in terms of surface-water inflow versus demand (table 13). Even assuming considerable error in the above monthly breakdown in demand, there is a deficit in terms of critical-period surface-water supply and demand. Therefore, additional surface storage must be made available or the ground-water resource must be utilized.

Once a demand for ground water is evident, as in the example above, a hydrologic model can be used to evaluate the capability of an aquifer or aquifers to satisfy all or part of that demand.

The data input to the model is as follows:

1. Transmissivity: 40,000 to 80,000 gallons per day per foot (fig. 20).
2. Storage coefficient: 0.01 and 0.20.
3. Aquifer thickness: 50 feet.
4. Hydraulic conductivity of stream bottoms: 27 gallons per day per square foot.
5. Thickness of streambed: 5 feet.
6. Width of Sciotto River and Big Walnut Creek: 100 feet.
7. Nodal spacing: 1,056 feet.

USE OF A HYDROLOGIC MODEL TO EVALUATE AQUIFER CAPABILITY

A detailed theoretical development of the digital model utilized is given in a report by Pinder and Bredehoeft (1968), and is not repeated in this report. They also discuss the assumptions necessary in using the model. The digital model is designed to solve the equations of flow for the complex hydrologic systems encountered in nature, and is based upon finite-difference approximations to the flow equations. Solutions to the flow equations are calculated for each nodal point of an established finite-difference network. In the example model of the Columbus area, the established network is a rectangular 55-row by 45-column matrix with a total of 55x45 or 2,475 nodal points. The nodal points are equally spaced on ⅕-mile centers. Each of the nodel points is a point of data input and output.

The model was used to simulate the response of the aquifer system in the Columbus area to a hypothetical pumping stress during a 9-month critical period, with a 10-foot ground-water-level drawdown limitation. The 10-foot decline is 20 percent of the assumed aquifer thickness. Such a limitation or set of conditions must be stated prior to determining "available" water. Available water is the amount of ground water that can be withdrawn on some sort of an areal basis, and for a very long period of time, under the conditions stated.

Continuous pumpage of 22 mgd for 9 months was simulated. Each of 22 simulated wells located along lines parallel to the Scioto River and Big Walnut Creek was pumped at a rate of 1 mgd. The simulated wells were located approximately on 1-mile centers (fig. 21). By placing the wells adjacent to the stream channels, recharge is induced from the streams, and, therefore, not all the water pumped from the wells is from ground-water storage. Actually, about 80 percent of the water pumped is induced recharge from the two streams. In actual development the well-field configuration can be determined in such a way that some other percentage of the water pumped is river water induced to flow to the wells.

The output of the model — the computed ground-water-level decline values for each nodal point (fig. 22) — suggests that the modeled stream-aquifer system can supply 22 mgd under the conditions stated. Actually, much more water is available because nowhere in the modeled area is there a computed decline in excess of 7 feet after 9 months of pumping (fig. 22). However, the available water should not be overestimated from the small drawdowns indicated by the model. The drawdown, or head, calculated by the model at a node where pumping is simulated, should be related to a well with a diameter of about 440 feet. Using a formula derived by Prickett and Lonnquist (1971, p. 61), the drawdown in a 24-inch-diameter simulated well would be about 5.5 feet greater than was indicated by the model. In study of an actual model, more precise water-availability values could be estimated by designing an optimal well-field pattern. Again, however, the purpose of this example is simply to demonstrate the methodology.

SUMMARY

Ground water in the Ohio Region is a large, important, and manageable resource that should have a significant role in regional water development.

Total potable ground water available from storage in the outwash and alluvial aquifers in the Ohio River valley and the subbasins is about 23,000 billion gallons. This is about four times the flood-control storage of all Ohio region Corps of Engineers reservoirs constructed, under construction, or in advance planning as of July 1965.

Approximately 85,000 billion gallons of potable ground water is available from storage in the region in aquifers other than the outwash and alluvial aquifers.

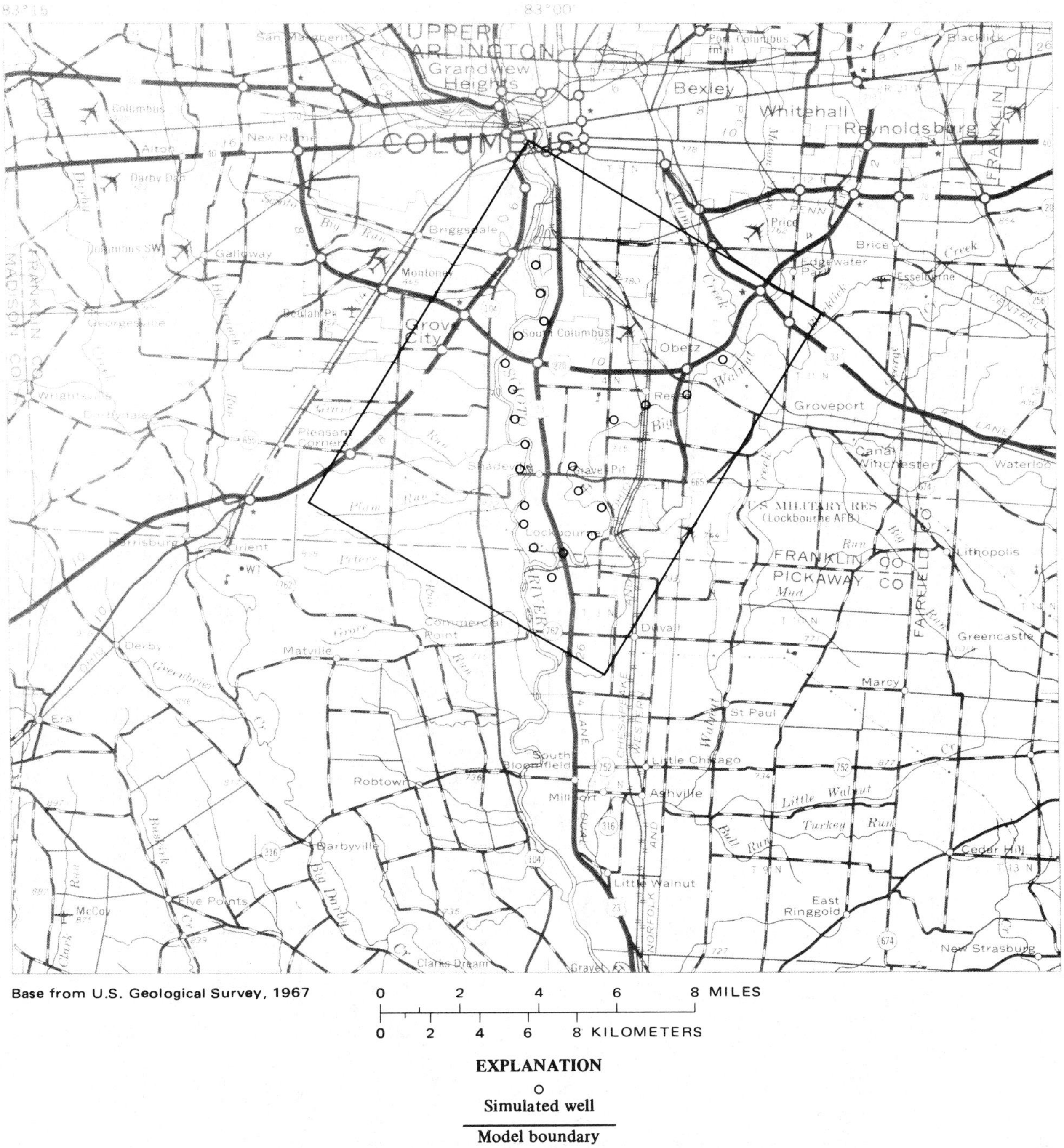

FIGURE 21. — Map showing placement of simulated pumping wells.

This is about 20 percent of estimated storage in Lake Ontario.

Estimated average annual ground-water recharge in the subbasins is about 35,000 mgd; estimated recharge from seepage of precipitation into the aquifer(s) in the Ohio River valley is about 2,000 mgd.

About 5 percent of the region has ground-water resources capable of satisfying more than local needs. For example, under certain specified conditions the excess of ground-water recharge over base-year (1960) ground-water use is available for 22 million additional people in the Wabash subbasin; for 4, 1.5, and 12 million

FIGURE 22. — Computed ground-water-level decline after pumping simulated wells for 9 months.

additional people, respectively, in the Miami, lower Scioto, and Allegheny subbasins; for 5 million additional people in the Ohio River valley; or for equivalent magnitudes of water supply for industry, agriculture, or other uses. A reasonable assumption is that much of the available ground water in these areas can be pumped and transported to reasonably distant points of need. However, factors other than water supply constrain and determine development potential.

Rapid advance of techniques in ground-water hydrology during recent years has provided methods which the hydrologist and planner can use for evaluating planned ground-water development in high potential areas such as those mentioned above. As an example of simulation techniques, a digital model of a small part of the stream-aquifer system south of Columbus, Ohio, was developed. Although the model results should be considered hypothetical because no model verification was attempted, the output of the model illustrates the technical feasibility of conjunctive development of a local ground-water surface-water system.

SELECTED REFERENCES

Bue, C. D., 1963, Principal lakes of the United States: U.S. Geol. Survey Circ. 476, 22 p.

Cartwright, Keros, 1970, Ground-water discharge in the Illinois basin as suggested by temperature anomalies: Water Resources Research, v. 6, no. 3, p. 912–918.

Cross, A. T., and Schemel, M. P., 1956, Geology of the Ohio River Valley in West Virginia, *in* v. XXII, Geology and economic resources of the Ohio River Valley in West Virginia: West Virginia Geol. Survey, 131 p.

Cross, W. P., 1968, Flow duration of Ohio streams: Ohio Div. Water Bull. BU-42, 68 p.

Deutsch, Morris, Dove, G. D., Jordan, P. R., and Wallace, J. C., 1969, Ground-water distribution and potential in the Ohio River basin, v. VI *of* Ohio River basin comprehensive survey: U.S. Army Corps of Engineers.

Feth, J. H., and others, 1965, Preliminary map of the conterminous United States showing depth to and quality of shallowest ground water containing more than 1,000 parts per million dissolved solids; U.S. Geol. Survey Hydrol. Inv. Atlas HA-199, 31 p. text.

Ohio Division of Water, 1963, Water inventory of the Scioto River basin: Ohio Div. Water Plan Inventory Rept. 17, 76 p.

Hopkins, H. T., 1966, Fresh-saline water interface map of Kentucky: Kentucky Geological Survey.

Mull, D. S., Cushman, R. V., and Lambert, T. W., 1971, Public and industrial water supplies of Kentucky, 1968–69: Kentucky Geol. Survey Information Circ. 20, 107 p.

Murray, C. R., 1968, Estimated use of water in the United States, 1965: U.S. Geol. Survey Circ. 556, 53 p.

Norris, S. E., and Spieker, A. M., 1966, Ground-water resources of the Dayton area, Ohio: U.S. Geol. Survey Water-Supply Paper 1808, 167 p.

Pinder, G. F., and Bredehoeft, J. D., 1968, Application of the digital computer for aquifer evaluation: Water Resources Research, v. 4, no. 5, 1069–1093.

Prickett, T. A., and Lonnquist, C. G., 1971, Selected digital computer techniques for ground-water resource evaluation: Illinois Water Survey Bull. 55, 62 p.

Rorabaugh, M. I., 1949, Progress report on the ground-water resources of the Louisville area, Kentucky: U.S. Geol. Survey open-file report, 64 p.

Spieker, A. M., 1968, Future development of the ground-water resource in the lower Great Miami River valley, Ohio — Problems and alternative solutions: U.S. Geol. Survey Prof. Paper 605-D, 15 p.

Stuart, W. T., Schneider, W. J., and Crooks, J. W., 1967, Swatara Creek basin of southeastern Pennsylvania — An evaluation of its hydrologic system: U.S. Geol. Survey Water-Supply Paper 1829, 79 p.

U.S. Army Corps of Engineers, 1968, Development program formulation, v. XII *of* Ohio River basin comprehensive survey: U.S. Army Corps of Engineers.

______1969, Ohio River basin comprehensive survey: v. I–XIV.

U.S. Department of Commerce, Environmental Science Services Administration, 1968, Climatic atlas of the United States: 80 p.

U.S. Department of the Interior, Bureau of Outdoor Recreation, 1966, Outdoor recreation study of the Ohio River basin, v. IX *of* Ohio River basin comprehensive survey: U.S. Army Corps of Engineers.

U.S. Department of the Interior, Federal Water Pollution Control Administration, 1967, Framework study of water-supply and water-pollution-control problem areas in the Ohio River basin, v. V *of* Ohio River basin comprehensive survey: U.S. Army Corps of Engineers.

U.S. Geological Survey, 1932, Geologic map of the United States.

Walker, E. H., 1957, The deep channel and alluvial deposits of the Ohio Valley in Kentucky: U.S. Geol. Survey Water-Supply Paper 1411, 25 p.

Warner, D. L., 1969, Administrative guidelines and evaluation criteria, pt. 2 *of* Ohio River Valley Water Sanitation Commission Perspective on the regulation of underground injection of wastewaters: 65 p.

Wyrick, G. G., and Lloyd, O. B., 1968, Ground-water resources, Appendix H *of* Report for development of water resources in Appalachia: U.S. Army Corps of Engineers, 122 p.

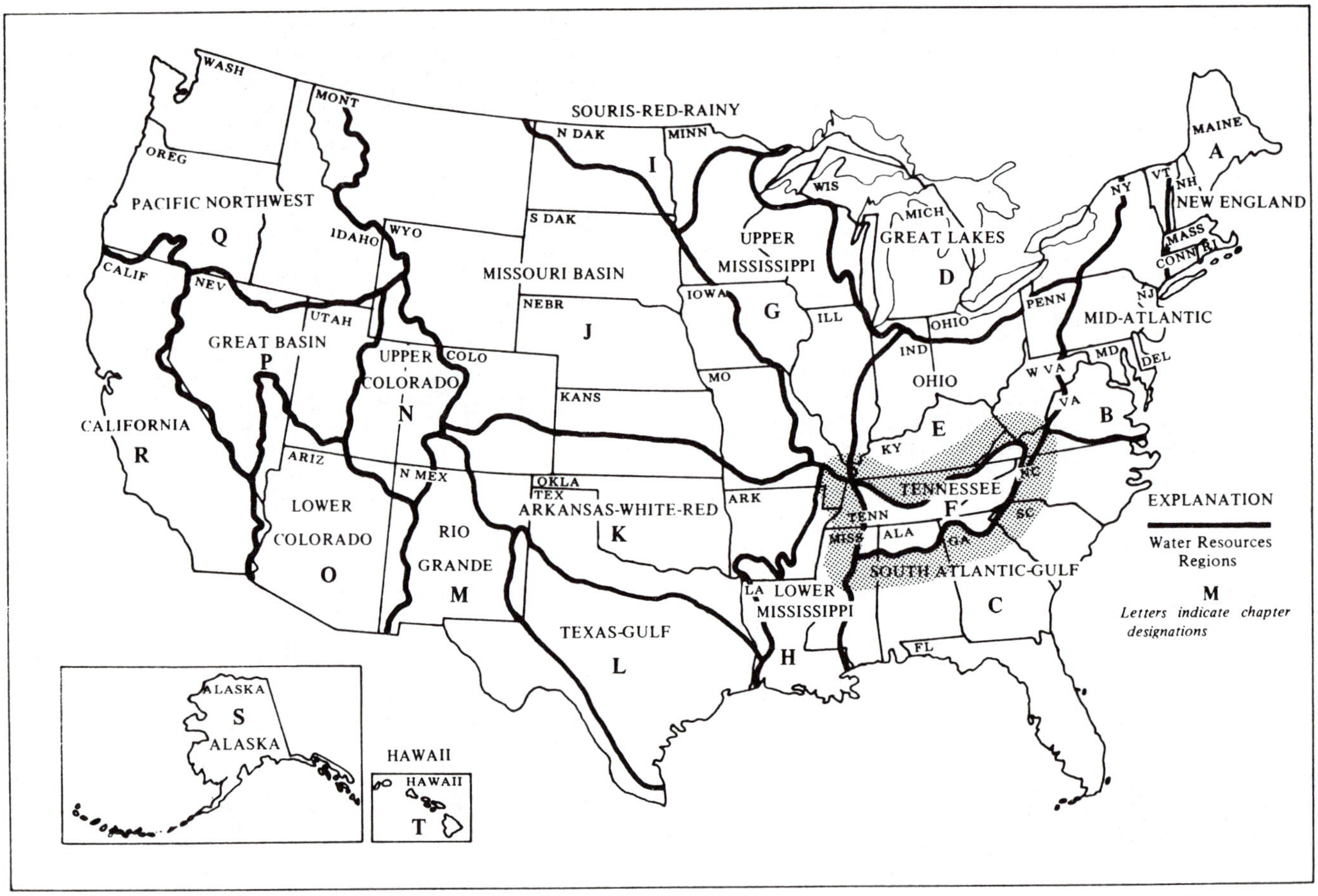

F – Location Map for the Tennessee Region

F

TENNESSEE REGION

By ANN ZURAWSKI

ABSTRACT

Ground water is an abundant and little-used resource in the Tennessee Region, a 41,000 square mile area dominated by the Tennessee River system and including parts of Alabama, Georgia, Kentucky, Mississippi, North Carolina, Tennessee, and Virginia. One-fifth to one-fourth of the precipitation that falls on the region enters the ground-water reservoirs. During the year approximately the same amount of water leaves the ground-water system, sustaining the dry-weather flow of streams. Recharge for the region is about 22,000 million gallons per day or 0.5 million gallons per day per square mile.

The major types of aquifers in the region are unconsolidated material (including sand and regolith), carbonate rocks, and fractured noncarbonate rocks. One or more of these aquifer types occurs in each of the six physiographic subdivisions of the region. The productivity of these aquifers depends on their hydraulic properties and on the distribution of these properties. The unconsolidated sand aquifers are the most homogeneous in composition and most predictable in occurrence. These aquifers commonly yield 200 to 600 gallons per minute per well depending on the thickness of sand penetrated.

The most difficult aquifers to predict in regard to depth and yield are the carbonate rocks. In these aquifers it is possible to drill dry holes within a few hundred feet of wells capable of producing several thousand gallons per minute. However, with an adequate reconnaissance study to determine the occurrence of ground water and a planned test drilling program, yields of up to 300 gallons per minute per well can be expected in the carbonate aquifers. Potential yields from the fractured noncarbonate aquifers are lower than in the carbonate rocks.

The chemical and physical properties of ground water in the Tennessee Region are usually within the limits recommended by the Environmental Protection Agency for drinking water, and the ground water in all but some very shallow aquifers tends to be free of pathogenic microorganisms. Saline water is not known to occur in significant quantities in the region.

In 1970, 173 million gallons per day of ground water were used in the Tennessee Region. This was less than 8 percent of the total quantity of water used in the region and only 0.8 percent of the estimated ground-water recharge. Ground water is used chiefly as a source of water supply for rural areas and small towns. A lesser amount is used by industries and commercial establishments located beyond the limits of municipal water-supply systems. However, there is potential for significantly increased use in order to augment surface-water supplies and to utilize the total water resource more efficiently.

Hydrologic studies and adequate test drilling would greatly increase the chances of locating large amounts of ground water, especially in the nine-tenths of the Tennessee Region that is underlain by either carbonate rocks or fractured noncarbonate rocks which have highly variable water-bearing properties. Collectively, such studies are useful in developing a concept of the hydrologic system which would permit the development of criteria for selecting well sites in other areas with a similar geological and hydrological setting. Hydrologic studies that include test drilling have been conducted in all parts of the region except the Cumberland Plateau.

Some of the basic data necessary for hydrologic studies, such as geologic maps, well records, and streamflow records are available throughout the region. However, detailed information on ground-water levels, ground-water quality, and aquifer characteristics are not equally available throughout the region. This type of information cannot be obtained quickly when it is needed; it must be the product of a continuing program of studies designed to evaluate the Tennessee Region's ground-water resource.

Because of the interdependence of ground water and surface water, water management efforts can be fully effective only if they involve the whole water resource. In the Tennessee Region, surface water is highly controlled, but there is at present no regionwide water-resources management plan that includes ground water.

INTRODUCTION

The significance of ground water as a resource is often not fully recognized in areas where surface water is abundant. Ground water is hidden from view, and requires special techniques to define its occurrence and availability. Planners and water managers who make decisions affecting development of water resources do not always have adequate information with which to evaluate ground water as an alternate or supplemental source of water.

In order to demonstrate that the Nation's ground water is a large and important resource, the U.S. Geological Survey has undertaken a broad-perspective appraisal of the ground-water resources in each of the 21 regions into which the United States has been divided by the Water Resources Council (fig. 1). The purpose of these regional appraisals is to show that in many parts of the Nation, ground water can play a significant role in regional water supply and that it warrants further study and consideration in regional developmental planning.

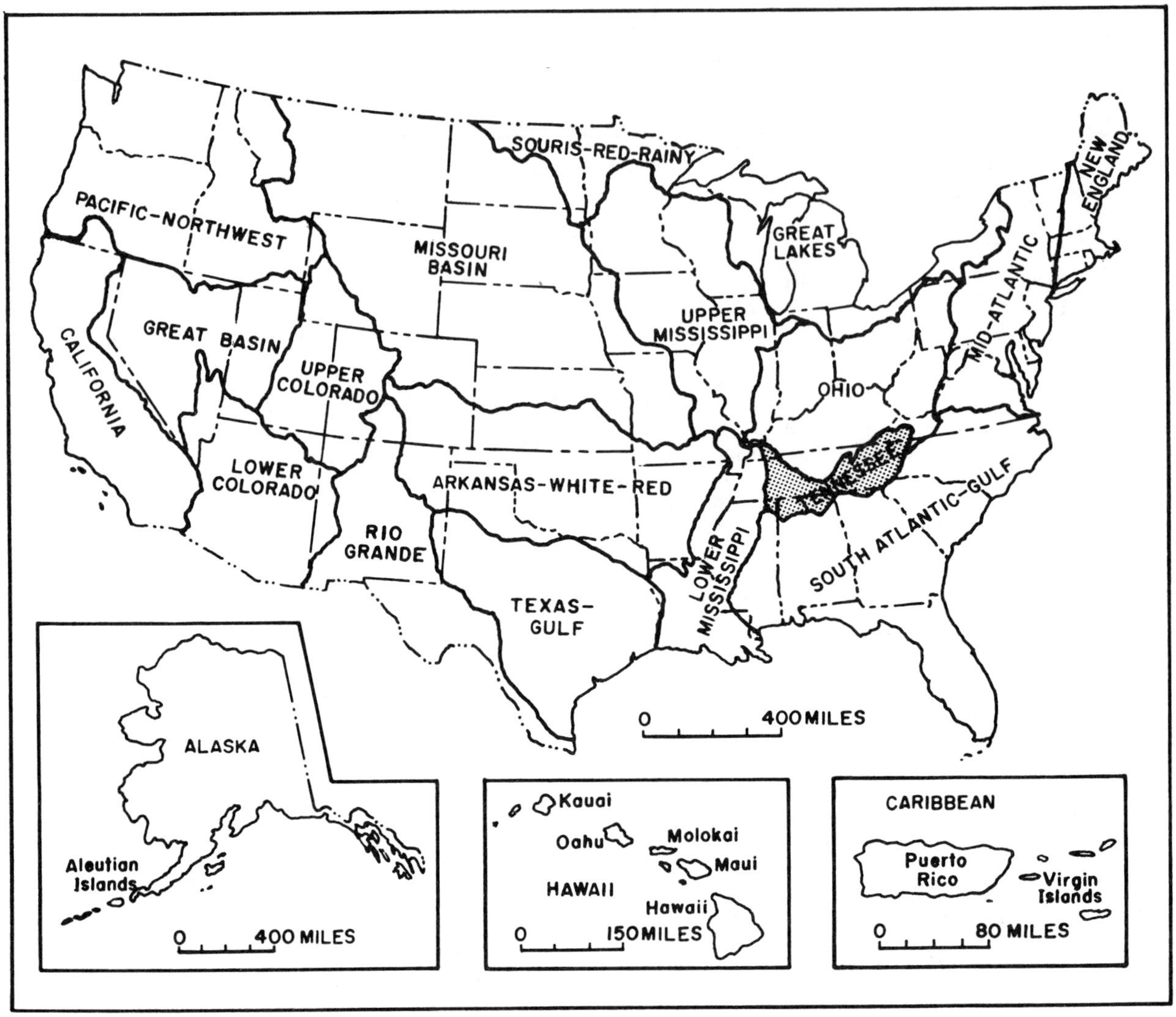

FIGURE 1.—Water Resources Council regions of the United States.

The Tennessee Region coincides with the Tennessee River basin, an area of 41,000 mi² which lies mainly in Tennessee, Alabama, and North Carolina but includes small parts of Virginia, Georgia, Kentucky, and Mississippi. This diverse area includes parts of six physiographic provinces each having distinctive topography and geology (fig. 2).

The Tennessee River system has played a major role in the region's development. Historically, severe flooding along the Tennessee River hindered industrialization of the region. The Tennessee Valley Authority, created by Congress in 1933 as a regional resource development agency, very early began to construct dams along the Tennessee to control flooding, promote navigation and produce electric power. The harnessing of the river was followed by industrialization, population growth, and a higher standard of living.

The flow of the Tennessee River and its tributaries, however, is only a small part of the region's water resources. In July of 1973, a record 10.3 million acre-ft of useable water was stored in the Tennessee River's reservoir system (Tennessee Valley Authority, 1973). This is 450 billion ft³ of water. M. I. Kaufman (written commun. 1975) estimated that 25,800 billion ft³ of ground water was available from storage in the Tennessee Region. While that approximated figure is difficult to verify, it is indicative of the magnitude of difference between ground-water and surface-water storage even for a fully regulated river system. In many areas where surface water is not available, large quantities of ground water are available for development. Ground water has been a neglected resource in the Tennessee Region, but because of increasing pressure to utilize all resources in the most efficient and

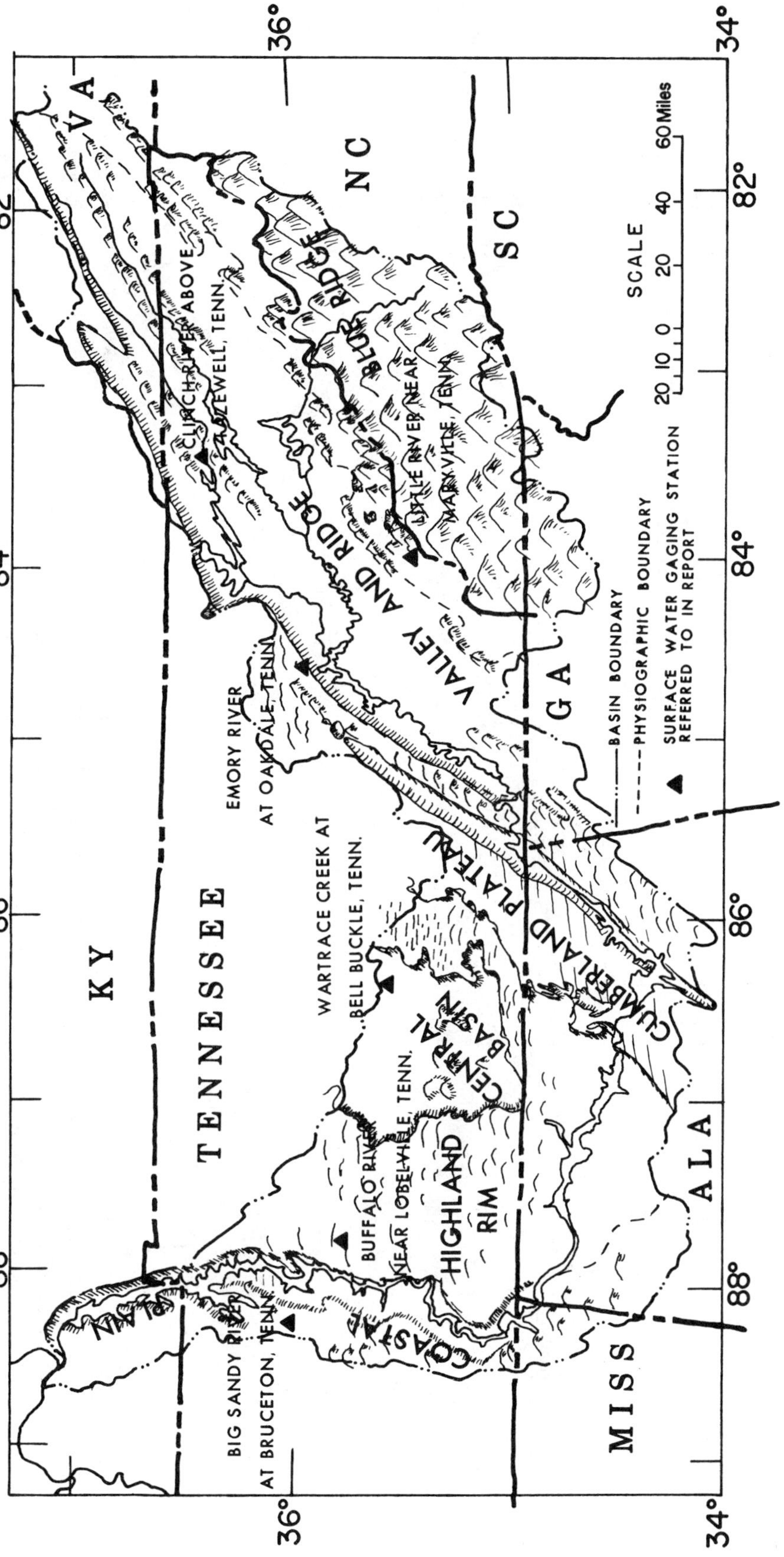

FIGURE 2.—Physiographic subdivisions of the Tennessee Region (modified from Hunt, 1967; Miller, 1974; Adams and others, 1926; Butts, 1933; and LaForge and others, 1925).

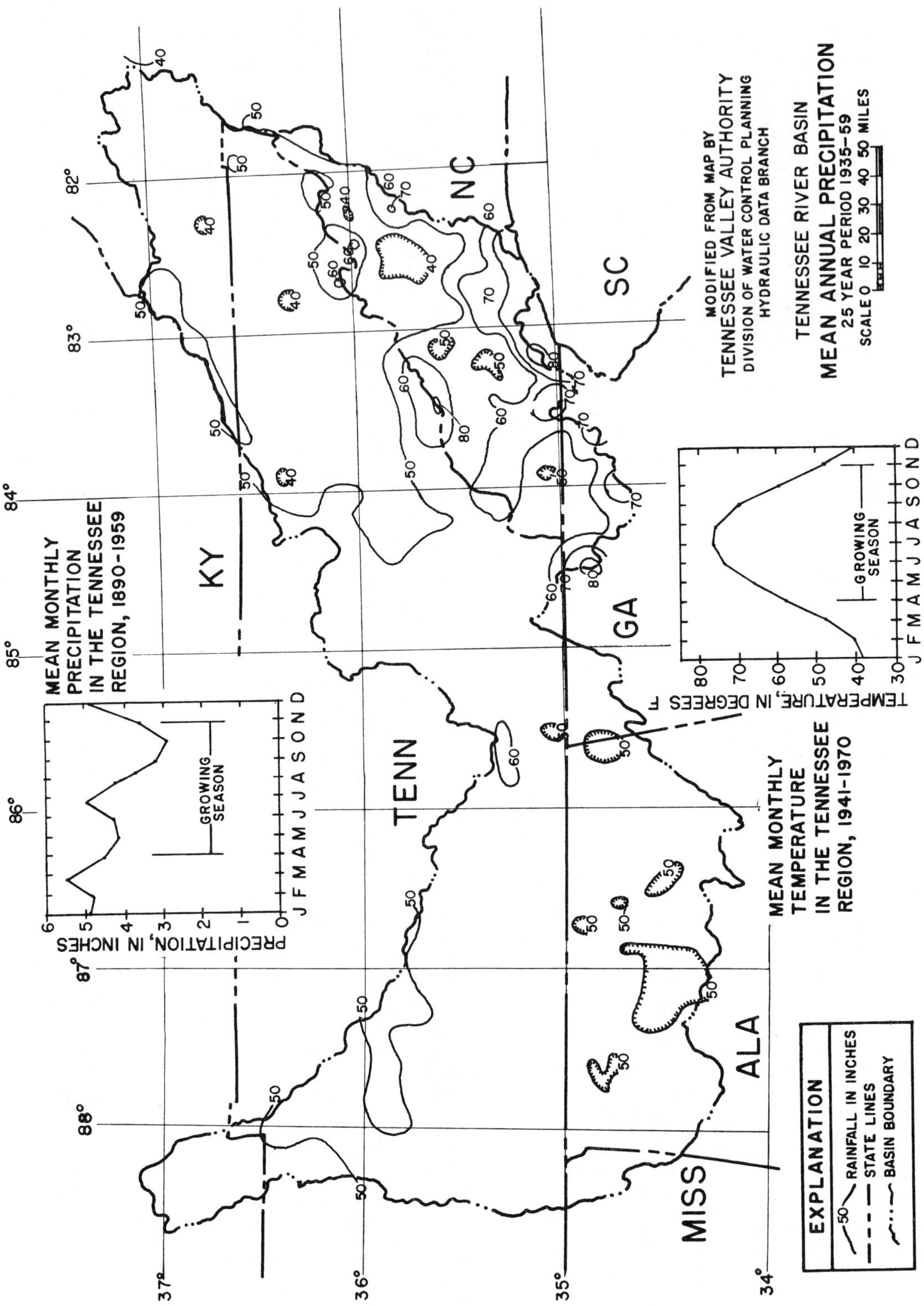

FIGURE 3.—Mean annual and monthly precipitation and mean monthly temperatures for the Tennessee Region.

productive manner, ground water is now being recognized as an integral part of the region's water resources.

This report synthesizes the results of previous studies in the Tennessee Region and appraises the role of ground water in the hydrologic systems of the Tennessee Region. It provides an overview of the occurrence of ground water, its present and potential use, and the need for management of ground water as an integral part of the water resources of the region.

THE GROUND-WATER RESOURCE

GROUND WATER AND STREAMFLOW

The source of both surface and ground water is precipitation. In an average year, precipitation in the Tennessee Region ranges from 40 in. in the northern Valley and Ridge province to 85 in. in the mountainous eastern part (fig. 3). The average for the region is 52 in. (Tennessee Valley Authority, 1975). In a normal year, most precipitation occurs in the winter and spring months, least in the summer and fall months. About 60 percent of the yearly precipitation returns to the atmosphere by evaporation and the transpiration of plants. Rates of evapotranspiration are highest during the hot, dry months when plants are growing. Of the remaining 40 percent of the precipitation, a part flows overland into streams and a part percolates into the ground to replenish, or recharge, the ground-water reservoirs (fig. 4).

Water entering the ground-water reservoirs is stored in the pore spaces in unconsolidated deposits and weathered rock, or in fractures and solution openings in the bedrock. Characteristically, all ground water in storage moves toward areas of discharge such as springs, streams, and wells. Water levels are related to the amount of water available to the base flow of streams and the volume of water that can be stored in the ground-water reservoir.

The effect that water-level fluctuation has on streamflow in areas underlain by a limestone aquifer with a thin soil cover is shown in figure 5. Ground water in these areas is discharged rapidly during periods of little or no rainfall, causing a decline in ground-water levels in the aquifer which results in the depletion of streamflow. Streams that cut deeply into thick, unconsolidated aquifers have sustained streamflow even through long periods of dry weather because of the slow release of ground water from these aquifers. In most aquifers, the replenishment of ground water at times exceeds the outflow or vice versa, but over a long period the recharge and discharge are about equal, so that ground-water discharge in a normal year can be approximately equated with the average annual base flow of streams.

The amount of ground-water recharge in the Tennessee Region was estimated using hydrographs of streamflow at six gaging stations during the 1968 water year, a year of nearly average streamflow across the region. Hydrograph separations were made to obtain maximum and minimum estimates of base flow (fig. 6). The values obtained ranged from 5 to 16 in./yr or 13 to 33 percent of the year's precipitation (fig. 7). The average rate of recharge for the region as a whole is estimated to be between 10 to 13 in./yr or 19 to 25 percent of the precipitation. This is about 0.5 (Mgal/d)/mi^2 or 22,000 Mgal/d for the entire region. The water budget shown in figure 4 is typical for the region. The numbers agree with previous water budget studies in the Pomperaug River basin in Connecticut, a basin underlain by fractured crystalline rocks and thin glacial drift (Meinzer and Stearns, 1929) and Beaverdam Creek basin in Maryland, underlain by coastal plain deposits (Rasmussen and Andreasen, 1957; table 1). Since the hydrologic properties of the aquifers in the Tennessee Region are intermediate between those of the other two basins, it is reasonable that the water budget figures for the Tennessee Region as a whole fall between those of the other two studies. Average recharge figures can serve only as a rough guide to the amounts of ground water available for withdrawal because they describe only the annual amount of water passing through the system, only part of which can be captured by wells.

Flow-duration curves of streams are, in part, indicators of the water-storing properties of aquifers. A flow-duration curve is constructed by plotting specified streamflows against the percentage of time they are equalled or exceeded at a gaging station. The streamflow at any time depends on the climate, drainage area, topography, overburden, and geology of the basin. Variations in these factors from one basin to another result in a variety of shapes of flow-duration curves (Burchett and Moore, 1971). The shape of the low-flow portions of duration curves is controlled chiefly by the geology of the basin (Searcy, 1959) and is indicative of the interaction of ground water and surface water in the basin.

Flow-duration curves of six long-term gaging stations on unregulated streams in the Tennessee Region are shown in figure 8. Each curve is representative of the flow characteristics of streams in a particular physiographic province. The effects of the difference in size of drainage areas have been minimized by plotting the streamflow in cubic feet per second per square mile.

The decreasing slope of four of the duration curves shows that those streams have well-sustained base flow indicating the ability of unconsolidated aquifer material to store and release water slowly. Big Sandy River at Bruceton, Tenn., traverses unconsolidated

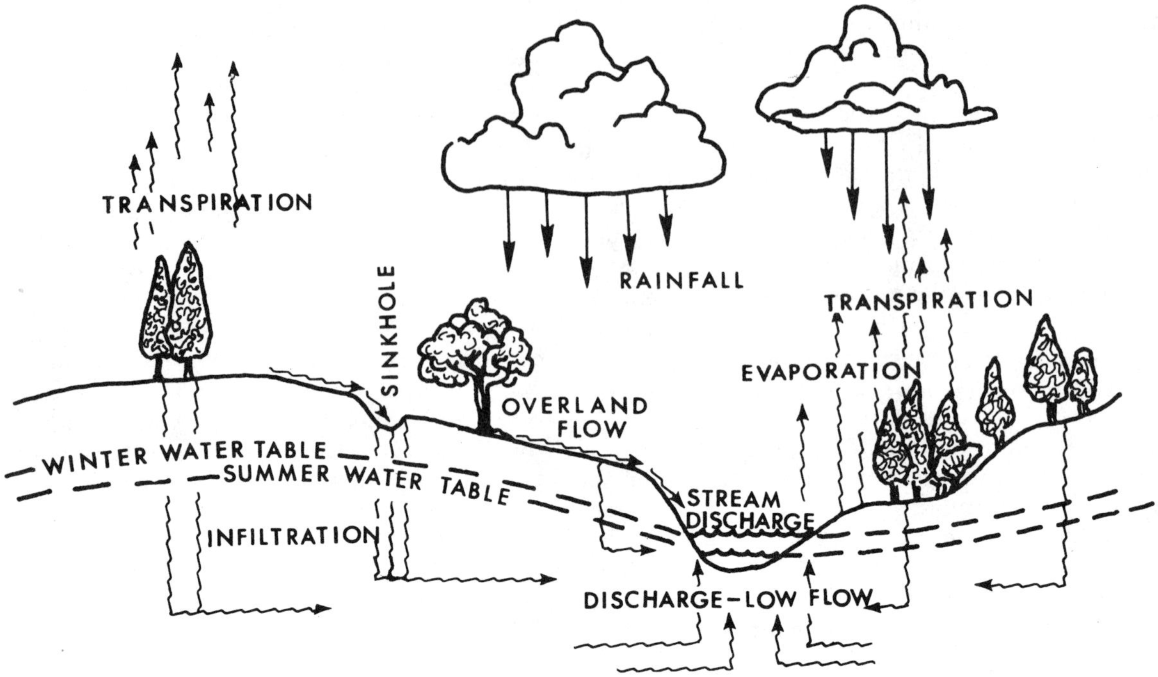

TYPICAL WATER BUDGET

	Inches
Average rainfall over drainage area	50.2
Stream discharge	21.5
Low flow	11.5
Overland flow	10.0
Evaporation and transpiration	28.7

FIGURE 4.—Hydrologic cycle and water budget for a typical stream in the Tennessee Region based on average of values for six drainage basins. (Hydrologic cycle from Burchett, 1977.)

sand. Buffalo River near Lobelville, Tenn., Clinch River above Tazewell, Tenn., and Little River near Maryville, Tenn., are in areas covered by thick regolith.

The curves for Wartrace Creek at Bell Buckle, Tenn., and Emory River at Oakdale, Tenn., are markedly different from the other curves, indicating the low water-storing properties of the rock underlying the basins and the very poorly sustained low flows. The curve for Emory River at Oakdale shows the low storage capacity of the fractured rocks capping the Cumberland Plateau. The curve for Wartrace Creek shows the low storage and rapid release of ground water in flat-lying limestone aquifers of the Central Basin. Interconnected solution openings in the limestone rapidly discharge the water that is temporarily stored above stream level. Low-flow frequency analyses for these gaging stations also show that base flow of Emory River and Wartrace Creek is much more poorly sustained than that of the other four streams (table 2).

MAJOR AQUIFERS

Three types of aquifers occur in the Tennessee Region: unconsolidated material with intergranular

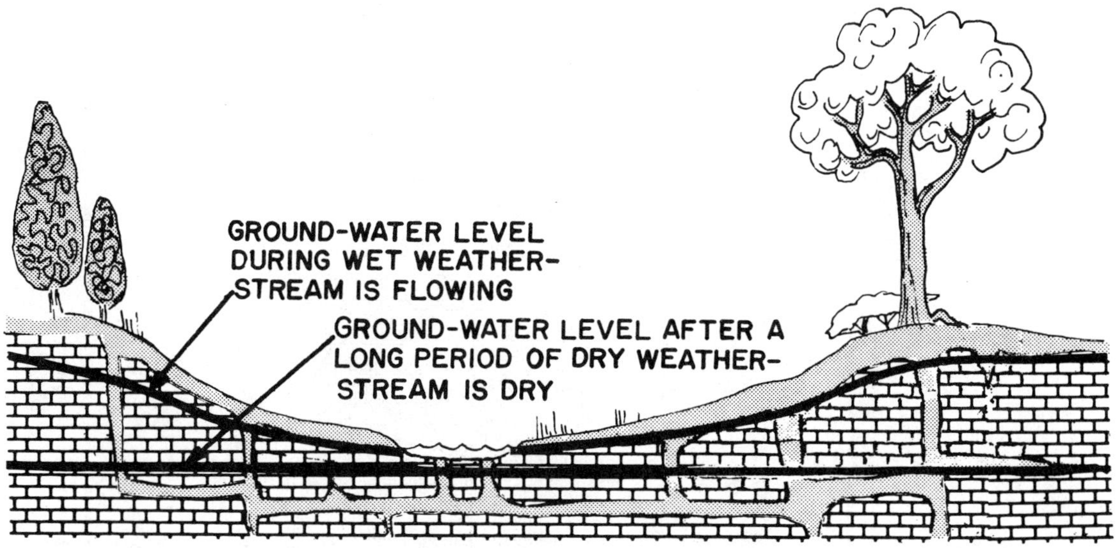

FIGURE 5.—Ground-water levels in an unconfined aquifer affect streamflow. In this limestone aquifer, typical of the Central Basin, recharge is stored temporarily above stream level. When there is a gradient toward the stream, ground water is discharged to the stream, sustaining its flow. As solution openings are drained, the water table becomes almost flat at or below the stream level, and streamflow ceases.

porosity, carbonate rocks with solution openings, and noncarbonate rocks with fractures (fig. 9). One or more of these aquifers is characteristic of each physiographic province (fig. 10).

UNCONSOLIDATED AQUIFERS

Uconsolidated materials are significant aquifers in about half of the Tennessee Region (fig. 10). In the Coastal Plain, an area of 4,000 mi² on the western edge of the region, the important aquifers are sand formations which dip to the west. These formations are the most uniformly productive aquifers of the region, commonly yielding 200 to 600 gal/min to single wells. Parts of another 18,000 mi² are covered by unconsolidated materials, referred to as regolith, which is a mantle of disintegrated rock that has accumulated over the bedrock. Grain size ranges from clay to coarse gravel and is a major factor in determining the regolith's water-bearing properties. The regolith is hydrologically significant in areas where it is thick and permeable, especially in the Highland Rim and parts of the Blue Ridge and Valley and Ridge. In these areas it acts as a sponge in absorbing and storing large amounts of ground water. Where saturated, the regolith yields dependable domestic supplies, but larger supplies can be obtained where fractures or solution openings in the underlying bedrock are hydraulically connected with the regolith.

CARBONATE AQUIFERS

Carbonate aquifers underlie about half of the Tennessee Region (fig. 10). The rocks in the Valley and Ridge section of the region are steeply tilted and west of the Valley and Ridge are essentially flat lying. Water occurs in openings along fractures, faults, and bedding planes which have become enlarged by circulating ground water. Solution openings occupy a small volume in the rock and generally occur within about 300 ft of land surface. Individual openings range from a fraction of an inch to several feet in height and can be laterally extensive. Wells that penetrate large openings may be able to produce several thousand gallons per minute, but such openings occupy a small proportion of the rock. As a result, well yields vary widely in carbonate terrane. For example, at Franklin, Tenn., in the Central Basin just north of the Tennessee Region, two test wells 100 ft apart produced 18 and about 180 gal/min respectively. Variability of solution openings makes well placement critical in efforts to develop large amounts of ground water from carbonate rocks.

FRACTURED NONCARBONATE AQUIFERS

Slightly more than a third of the Tennessee Region is underlain by noncarbonate rocks which, unlike the Coastal Plain deposits, have very little porosity aside from fractures.

These rocks range from sedimentary rocks such as shale and sandstone underlying the Cumberland Plateau to rocks that have been subjected to metamorphism and deformation in the Blue Ridge. Weathering along fractures and faults has created avenues for water movement. These fracture openings probably comprise less than 1 percent of the rock volume, and water-bearing fractures are uncommon at

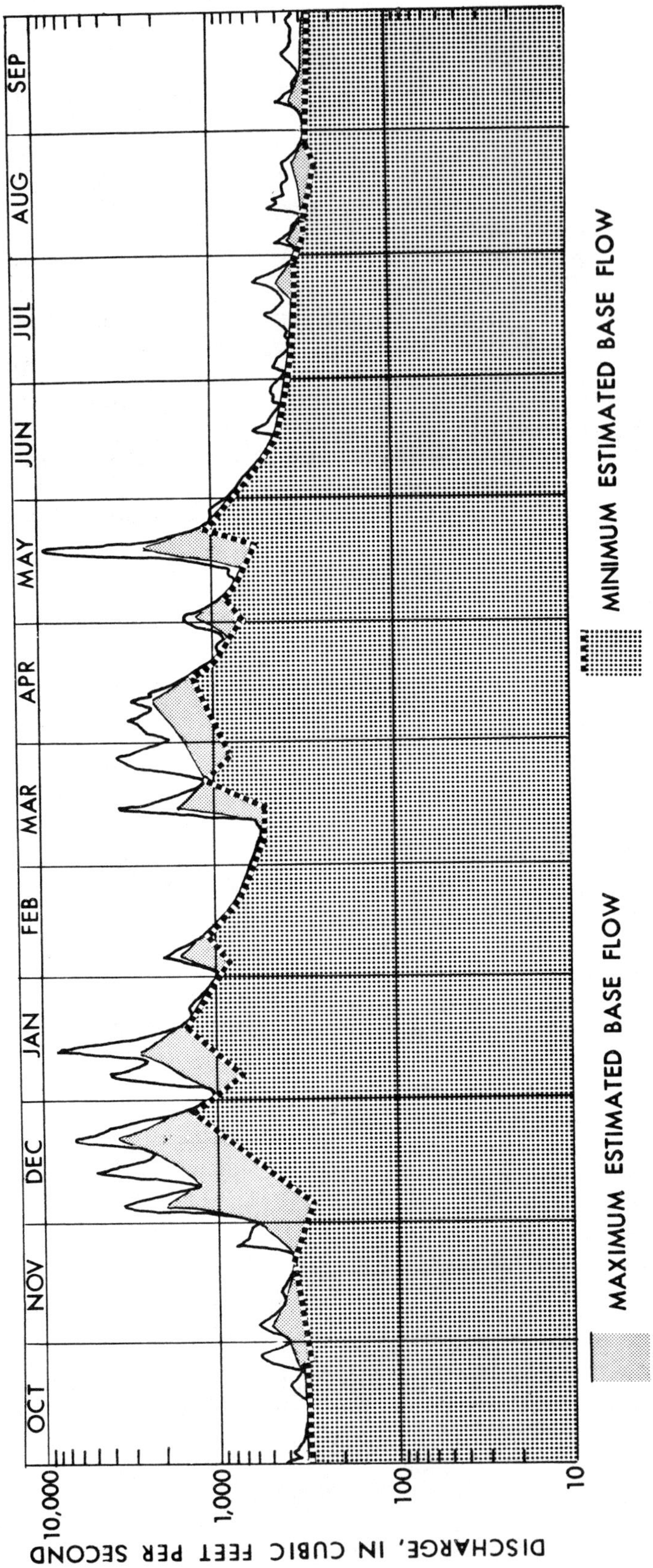

FIGURE 6.—Base flow of Buffalo River near Lobelville during the 1968 water year, a year of average streamflow. After about 6 days overland runoff has left the basin, and flow is supplied from the ground-water reservoir. Maximum and minimum ground-water contribution to streamflow was determined graphically for 1968 water year. For the maximum ground-water outflow, the recession curves from 6 days after each peak were projected back to the day of the peak. An arbitrary line was then drawn from the peak to the point where the stream began to rise. For the minimum, each recession curve was extended to the day of the following peak, then joined by a straight line to the point 6 days after that peak. This method is a modification of that described by Busby and Armentrout (1965) and Moore, Burchett, and Bingham (1969).

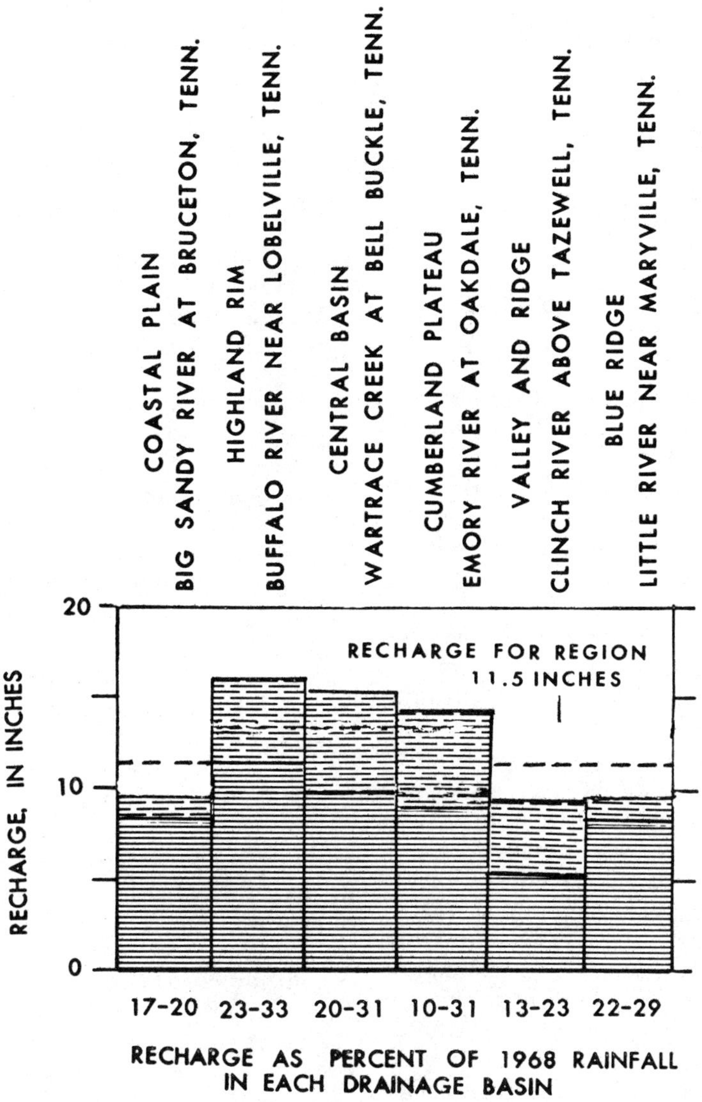

FIGURE 7.—Ground-water recharge (not including ground-water evapotranspiration) in 1968 in six drainage basins in the Tennessee region, and recharge for the region. This value is the average of the maximum and minimum values for the six basins.

depths over 300 ft (McMaster and Hubbard, 1970). The fractures at greater depths are likely to be unweathered and closed. Fractures are commonly more abundant near faults. Well production is determined by the depth, size, and degree of interconnection of fractures penetrated and by the thickness of overburden hydraulically connected to the fractures.

Fractured rocks are generally considered to be poor aquifers, but they have generally not been adequately tested to determine their water-yielding potential. Newcome and Smith (1958) report that wells producing 50 gal/min or more are rare on the Cumberland Plateau. However, the town of Wartburg, Tenn., has three wells which, at the time they were drilled, were reported to produce more than 100 gal/min each. Test drilling in the Great Smoky Mountains National Park (McMaster and Hubbard, 1970) showed that where fractured rock is hydraulically connected to thick regolith, properly located wells can produce 100 gal/min or more.

AQUIFER PRODUCTIVITY

Aquifer productivity as used in this report refers to the rate at which ground water can be withdrawn from an aquifer at a particular locality on a continuing basis by means of a well or group of wells. Well yields are determined by the hydraulic properties of the aquifer. In the Tennessee Region, however, the areal variability in properties of the aquifers, combined with lack of

TABLE 1.—*Comparison of water budget of a typical stream in the Tennessee Region with water budgets for Pomperaug River and Beaverdam Creek basins.*

[Values are given as percent of precipitation.]

	Pomperaug Basin	Beaverdam Creek Basin	Tennessee Region
Evapotranspiration	52.2	60.7	57.2
Total runoff (streamflow)	46.4	36.1	42.8
Ground-water runoff (base flow)	19.6	25.9	22.9
Change in storage	1.4	2.1	Not calculated

data for many areas make the definition of aquifer productivity difficult. Several approaches can be sed to quantify aquifer productivity.

The basic hydraulic properties of an aquifer are transmissivity (the rate at which water can be transmitted through the aquifer) and storage (related to the amount of water that is released by draining part of a water-table aquifer or lowering the pressure in an artesian aquifer). The more nearly an aquifer approaches uniformity in its water-bearing properties, the more applicable are these measures of hydraulic characteristics.

In the Tennessee Region, the unconsolidated aquifers of the Coastal Plain section, on the western margin of the region, are the lest variable in their hydraulic properties. Two aquifer tests indicate transmissivity of 3,300 and 4,300 ft^2/d and storage coefficients of 0.0008 and 0.0001 for the McNairy Sand, a confined unconsolidated aquifer (Boswell and others, 1965). However, nine-tenths of the region is underlain by carbonate or fractured noncarbonate aquifers which are far from uniform in their hydraulic properties. Five aquifer tests in Madison County, Ala., indicate transmissivity ranging from 650 to 130,000 ft^2/d and storage coefficients of 0.04 to 0.0004 for the Fort Payne Chert (Malmberg and Downing, 1957). A similar large variability in aquifer hydraulic properties is evident from the results of aquifer tests made elsewhere in the Tennessee Region and is typical of carbonate aquifers.

Another indicator of aquifer productivity is the specific capacity of wells, the rate of yield per unit of drawdown. Although specific capacity is influenced in part by conditions in and around the well, properly qualified specific capacity data can indicate the properties of the aquifer. The ranges of specific capacities shown in figure 11 illustrate the high degree of variability in the hydraulic properties of aquifers of the Tennessee Region, although some of the low values may be the result of poor well design rather than aquifer properties. This variability is of critical importance in developing a ground-water supply because the more variable the aquifer productivity, the greater the risk of failure to obtain the desired amount of ground water at a particular site. In the Tennessee Region, the least variability of yields is from wells that tap unconsolidated aquifers. The greatest variability is in carbonate-rock aquifers covered by thin or patchy mantles of regolith. In economic terms, greater variability in yields means that more preliminary study, more test wells, and more flexibility in well location will probably be needed to obtain a desired amount of ground water.

Another approach to quantifying aquifer productivity is to estimate probable yields that could be obtained from wells located on the basis of geologic and hydrologic information. Table 3 shows yields that might be expected from different types of aquifers in the Tennessee Region. These yields are for single wells constructed to obtain maximum yield having at least 50 ft of available drawdown. The numbers are based on figures given by Cederstrom (1973) for areas with geology similar to that of the Tennessee Region and on inferences drawn from records of well production in the region. Ground-water withdrawals are ultimately limited by the rate of recharge to the aquifer. In the Tennessee region, maximum withdrawals on a continuing basis are about 0.5 Mgal/d per square mile of area contributing recharge to the system.

Efficient and economical development of the ground-water resources of the Tennessee Region are strongly influenced by the variability of aquifer productivity. Predictability of well yields and the availability of ground-water supplies at particular sites is greatest in the Coastal Plain sand aquifers and the regolith-mantled carbonate rocks. Hence, it is probable that ground-water development will proceed most efficiently in these areas. The predictability of obtaining a ground-water supply is poorest in the noncarbonate rocks and in the carbonate rocks that have no regolith cover. In these areas, development will be less efficient until methods are developed to locate permeable zones and predict well yields with greater accuracy.

OCCURRENCE OF GROUND WATER

The distribution of ground water in the Tennessee Region is influenced by the difference in topography, geology, and hydrology among the six physiographic provinces (fig. 2). Though large amounts of ground water exist in each physiographic area, less preliminary study and exploratory drilling are usually needed to obtain a specified supply in areas where the ground water occurs in the intergranular pore spaces of unconsolidated aquifers, than in areas where water occurs in discrete fractures or openings in consolidated rocks. The following sections describe for each physiographic area the topographic, geologic, and hydrologic controls on the distribution of ground water.

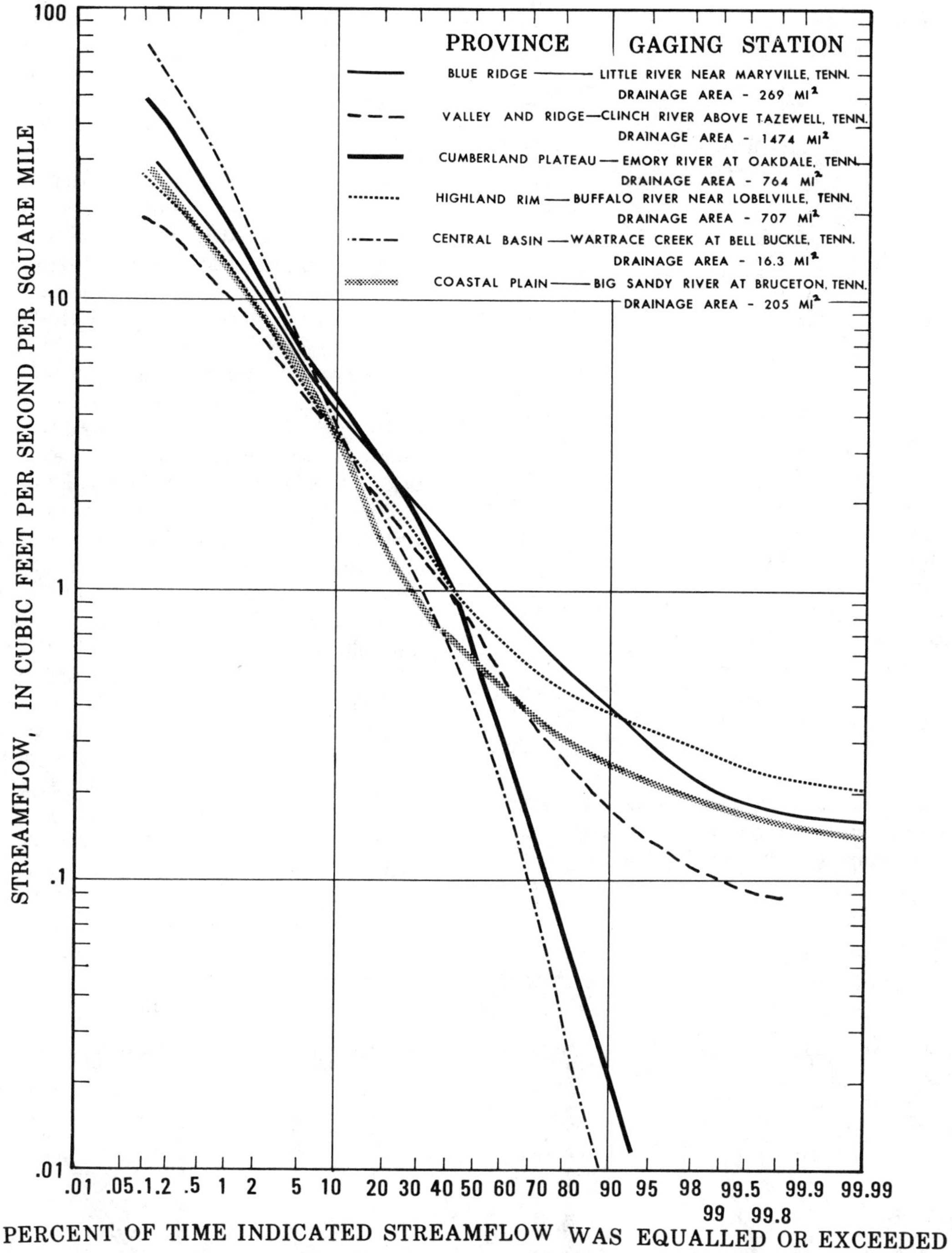

FIGURE 8.—Streamflow-duration curves for six gaging stations in the Tennessee Region.

BLUE RIDGE

The part of the Tennessee Region that lies within the Blue Ridge province is composed of remnants of an ancient mountain chain. The topography is rugged and relief is greater than in other parts of the region. In most places, the dense, massive bedrock contains little water except where faulted or fractured. Overlying the bedrock is a mantle of regolith that ranges in thickness from a few feet on the steeper slopes to more than 100 ft on the lower slopes of the mountains. The regolith is composed of sand, clay, and rock fragments (McMaster and Hubbard, 1970). It stores large amounts of water, releasing it slowly to the underlying fractures and to springs and streams. The fractures store only limited

TABLE 2.—*3-day 20-year low flows at six gaging stations in the Tennessee Region*

Province	Station	3-day 20-year low flow in cubic feet per second per square mile
Blue Ridge	Little River at Maryville, Tenn.	0.17
Valley and Ridge	Clinch River above Tazewell, Tenn.	.08
Cumberland Plateau	Emory River at Oakdale, Tenn.	.0005
Highland Rim	Buffalo River near Lobelville, Tenn.	.22
Central Basin	Wartrace Creek at Bell Buckle, Tenn.	.0014
Coastal Plain	Big Sandy River at Bruceton, Tenn.	.16

amounts of water, but they act as collectors which transmit water from the overlying regolith to points of discharge (fig. 12). Most of the bedrock in the Blue Ridge is noncarbonate, but a few areas are underlain by intensely weathered carbonate rocks that contain large amounts of ground water in solution openings and in porous zones.

In the Blue Ridge province, McMaster and Hubbard (1970) and LeGrand (1967) determined that the chances of drilling high-producing wells are increased at sites having a thick cover of regolith in relatively low topographic positions within a few hundred feet of a fault zone. The broad valleys underlain by carbonate rocks, such as Cades Cove in the Great Smoky Mountains National Park, may be favorable sites for locating ground-water supplies. However, these occur only in a few small areas of the Blue Ridge.

VALLEY AND RIDGE

The Valley and Ridge province is characterized by northeast-trending ridges underlain by resistant rock separated by valleys underlain by less resistant rock. The rock formations crop out in long, narrow belts parallel to the trend of ridges and valleys; some belts are bounded by faults. The linear ridges and valleys channel surface drainage into a "trellis" pattern of long streams flowing along valley floors fed by short lateral streams.

All three major aquifer types occur in the Valley and Ridge province. Because of the great differences in hydraulic characteristics among these aquifers, geologic maps are essential tools for well-site selection in this province. The shale and sandstone formations (fractured noncarbonate rocks) are the poorest aquifers. Limestone and dolomite of varying solubility occur with a cover of regolith ranging in thickness from a few feet to over 100 ft with extreme areal variability (DeBuchananne and Richardson, 1956). The largest ground-water supplies are in the soluble carbonate rocks, especially where they are associated with thick regolith (fig. 13).Water moves through enlarged fractures and solution openings in these carbonate rocks, emerging in places as large springs. Some of these flow at an average rate of 4,500 gal/min or more (Sun and others, 1963). The Knox Dolomite, which underlies about 60 percent of the province, is the most significant water-bearing formation. The most productive wells are located in areas of ground-water discharge such as stream valleys or else they penetrate fracture zones. These fracture zones are sometimes indicated by straight stream segments and aligned tributaries. Most of the larger water-bearing openings in the Knox occur at a depth of less than 300 ft. Swingle (1959) states that surface faults indicate areas with deep and numerous fractures which allow deep solution activity. Wells tapping water-filled solution openings in low areas are more dependable as a source of water supply than wells on ridges because the seasonal fluctuations of water level are small in the low areas.

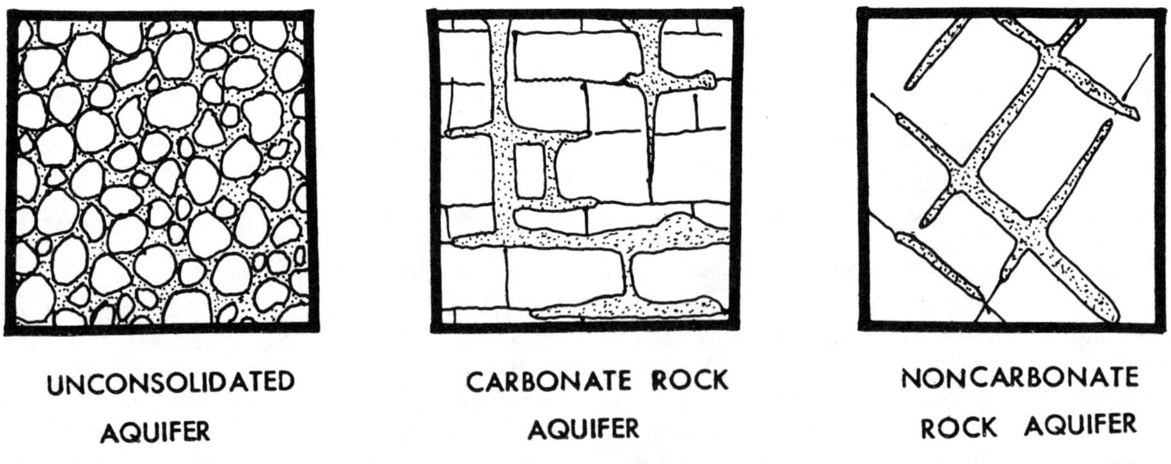

FIGURE 9.—The three major types of aquifers in the Tennessee Region, distinguished by the kind of water-bearing openings they contain (after Meinzer, 1923).

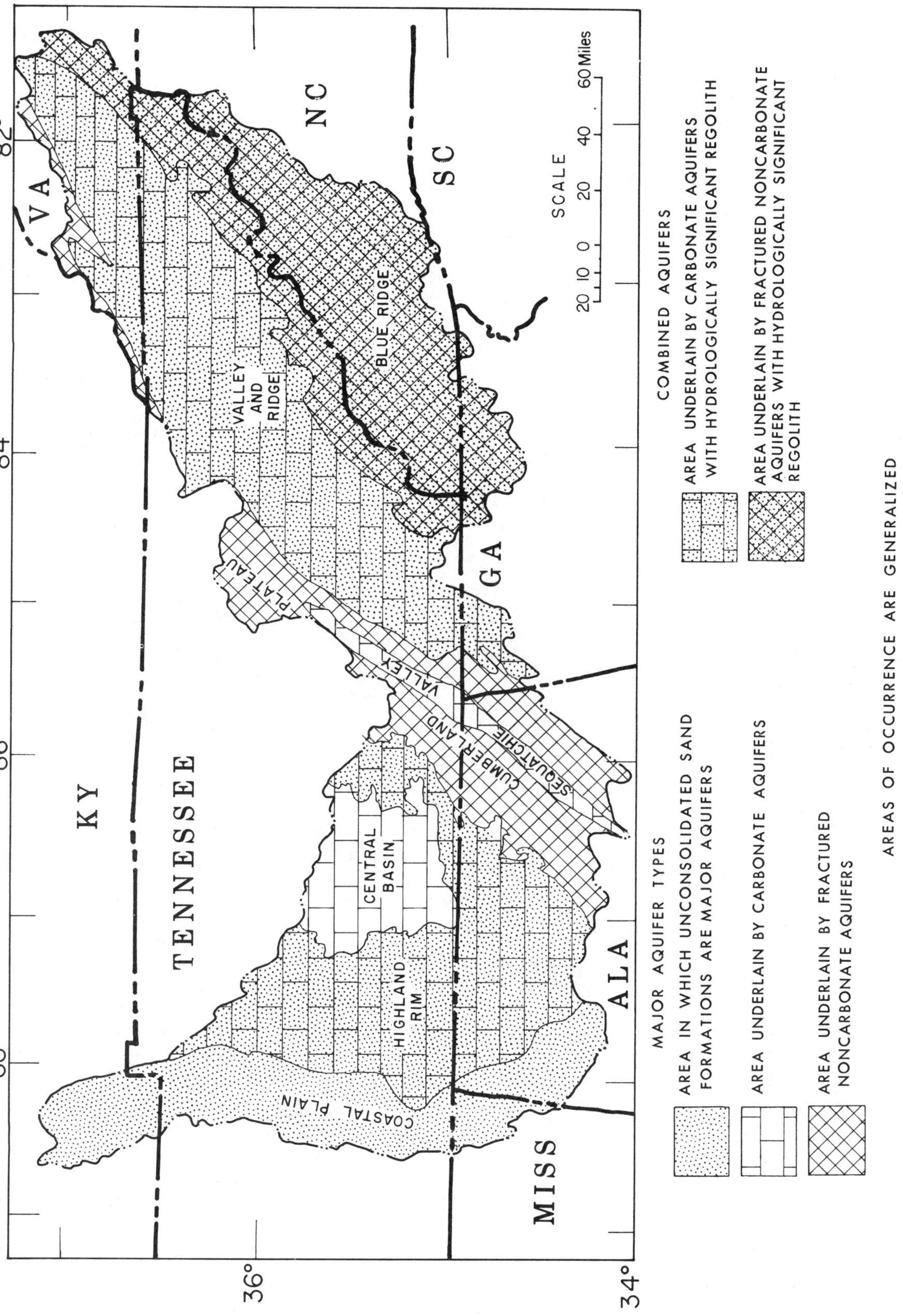

FIGURE 10.—Distribution of major aquifer types in the Tennessee Region.

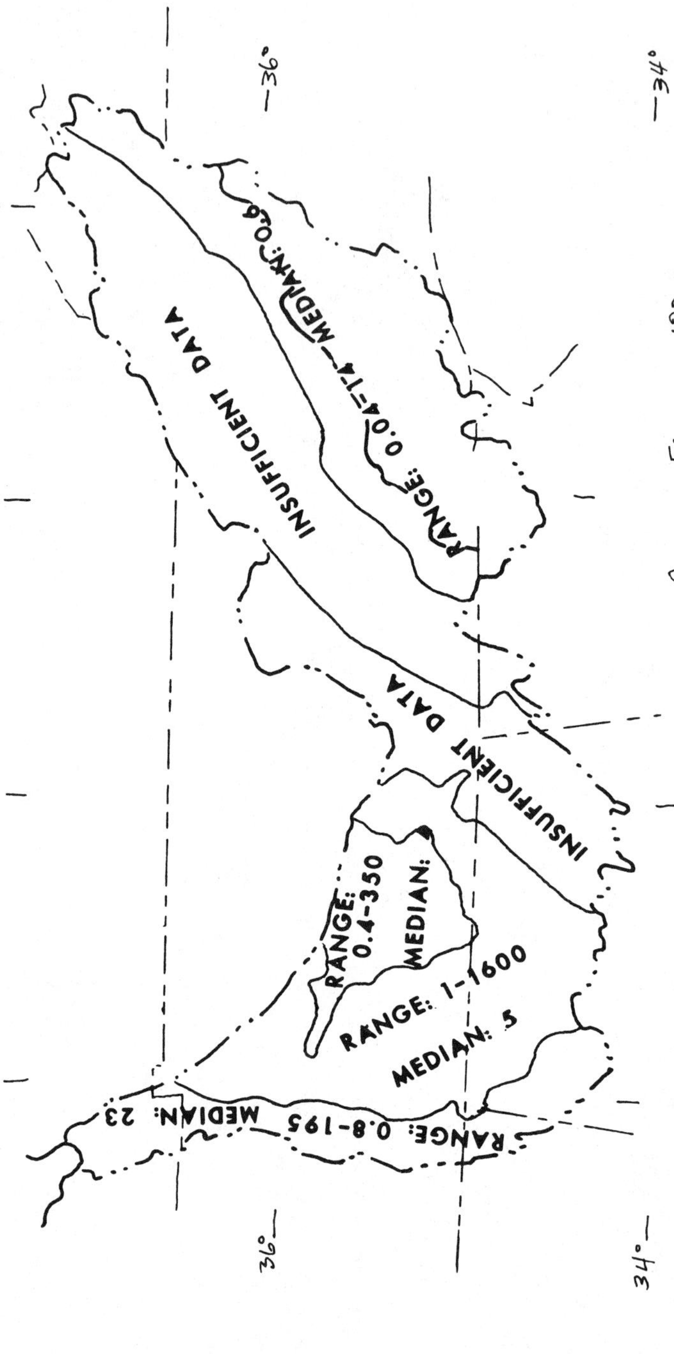

FIGURE 11.—Ranges and median values of specific capacities of selected wells in the Tennessee Region, in gallons per minute per foot.

TABLE 3.—*Probable yields from the major aquifer types of the Tennessee Region*

Type of aquifer	Yield per well, in gallons per minute
Unconsolidated sand	100–500
Carbonate rocks with regolith	100–300
Carbonate rocks without regolith	0–300
Noncarbonate rocks with regolith	25–100
Noncarbonate rocks without regolith	Unknown

CUMBERLAND PLATEAU

The Cumberland Plateau section of the Appalachian Plateaus province is about 1,000 to 1,500 ft higher than the adjoining Valley and Ridge province and Highland Rim. It extends about 175 miles northeast-south-west across the Tennessee Region. North from Anderson County, Tenn., the steep eastern escarpment of the Plateau forms the Tennessee River-Cumberland River divide.

The bedrock is a sequence of mostly horizontal Pennsylvanian sandstone, shale, conglomerate, and coal, underlain by Mississippian and older shale and carbonates. The Mississippian carbonates are exposed where Sequatchie Valley, a 130-mile long, linear valley, cuts deeply into the Plateau. Large springs emerge along the sides of the Plateau and from the Mississippian and older limestones exposed in Sequatchie Valley (fig. 14). An example of these large springs is Blue Spring which is used as a source of water supply by the city of Jasper, Tenn. An average of 170,000 gal/d was used by the city in 1970 (Tennessee Dept. of Public Health, 1975). Newcome and Smith (1958) reported the flow of this spring to be 900 gal/min but the seasonal variation in discharge has not been determined.

The Pennsylvanian rocks have low permeability except where fractured. The regolith is usually thin, providing little ground-water storage (DeBuchananne and Richardson, 1956). As a result of the uneven distribution of fractures, the average of reported well yields is less than 50 gal/min (Newcome and Smith, 1958). Even domestic supplies cannot always be obtained. However, since detailed studies involving test drilling have not been made, the full water-yielding potential of the Plateau is not known. Wilson (1965) states that in Cumberland County, Tenn., the Sewanee Conglomerate, 200 to 500 ft below the surface of the Plateau, is a relatively untested aquifer.

HIGHLAND RIM

The Highland Rim section of the Interior Low Plateaus province is a gently rolling plateau which occupies a large part of the center of the Tennessee Region. It is extensively dissected where it adjoins the Central Basin and Coastal Plain lowlands. The bedrock of the Highland Rim consists of flat-lying Mississippian carbonates. These formations, principally the Fort Payne Chert, a cherty dolomite, and the Tuscumbia Limestone (with its equivalents, the Warsaw Limestone and St. Louis Limestone), constitute the most areally extensive aquifer in the Tennessee Region.

These Mississippian formations weather to form a deep chert regolith typically having a "rubble zone" at the base. The residual chert grades from gravel-sized fragments at the base to a layer of clay-sized chert particles which partially seals the top of the aquifer creating artesian conditions in some areas. The regolith may rest directly on the Chattanooga Shale which retards downward movement of the water, but more commonly it rests on unweathered Fort Payne bedrock. In many places the carbonate bedrock contains solution openings which can transmit water rapidly.

In areas adjacent to the Highland Rim escarpment, water drains out of the bedrock openings as springs and seeps. These springs provide the well-sustained low flow of streams that dissect the edge of the Highland Rim. However, in the broad undissected areas of the Highland Rim, streams barely cut into the regolith and, therefore, have little effect on the discharge of water from the regolith and solution openings. The solution openings that supply water to wells in the undissected areas are hydraulically connected to the regolith (fig. 15).

The aquifer formed by the bedrock and regolith of the Fort Payne Chert has been named the "Manchester aquifer" (Burchett and Hollyday, 1974). Its areal extent has not been determined, especially on the western Highland Rim. This is basically the same aquifer that is present in the Tuscumbia Limestone and Fort Payne Chert in parts of northern Alabama. In places where the regolith is thin and little or no bedrock remains overlying the Chattanooga Shale, only small amounts of ground water are available. In areas where sections of Mississippian carbonate rocks remain beneath a thick regolith, however, extensive solution networks may develop that yield large quantities of water to wells and springs. For example, discharge measurements at Water Cress Spring in Madison County, Ala., indicate a sustained flow of about 5,000 gal/min (7½ Mgal/d) (Geol. Survey of Alabama, 1975). The Williams well, southwest of Huntsville, Ala., has been pumped continuously at 3,000 gal/min for 3 days with a maximum drawdown in water level of 2.5 ft. It was also noted during the period of pumping that water levels 0.8 mile away were unaffected (W. J. Powell, U.S. Geological Survey, oral commun., 1975).

Criteria that are important in selecting well sites on the Highland Rim are as follows: an area with at least

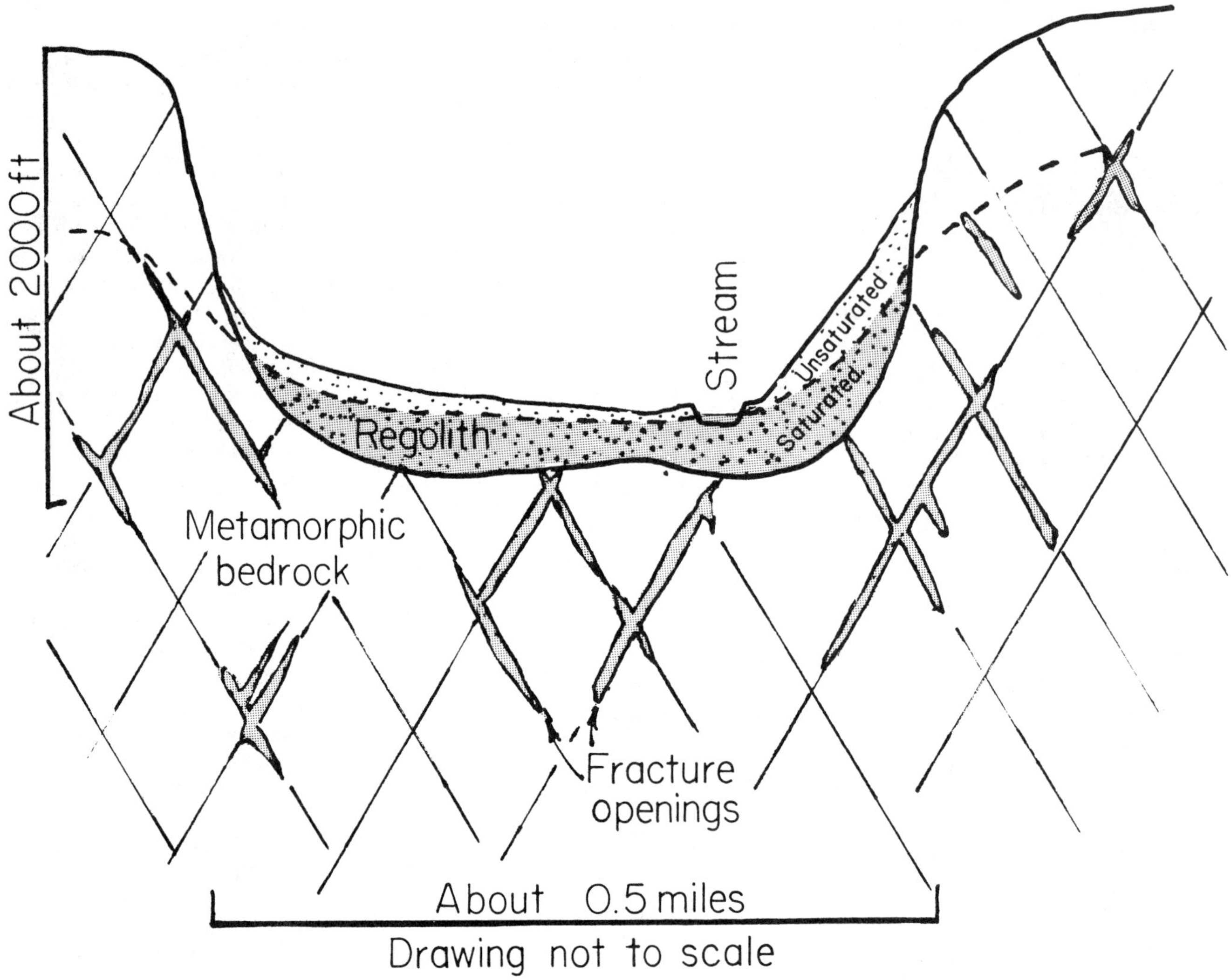

FIGURE 12.—Sketch of idealized geologic and hydrologic conditions in the Blue Ridge province.

40 ft of regolith overlying the Fort Payne Chert, amplitude of water level fluctuation less than 10 ft. (W. J. Powell, oral comm., 1975), and at least ½ to 1 mile from the Highland Rim escarpment in a topographic low area such as a long shallow depression (swale) parallel to the escarpment or other linear feature (C. R. Burchett and E. F. Hollyday, U.S. Geological Survey, oral commun., 1975). Seismic surveys, which indicate the depth to bedrock, may be a useful exploration tool in determining the thickness of overburden (Joiner and Scarbrough, 1969).

CENTRAL BASIN

The Tennessee Region includes the southern part of the Central Basin (Nashville Basin). The Central Basin, a section of the Interior Low Plateaus province, is an oval area in middle Tennessee lying about 200 ft lower than the surrounding Highland Rim. The bedrock is carbonate rocks that are generally flat lying but locally are folded. Ground water moves through solution-enlarged vertical joints and horizontal bedding-plane openings (fig. 16). Soil cover is usually thin and surface streams are cut into bedrock. The lack of regolith, along with open joints in the rock, allow rapid runoff and infiltration of precipitation. Water is relatively briefly stored above stream level and is rapidly discharged to streams through solution openings. As a result, small streams respond quickly to precipitation and have poorly sustained base flows.

Deeper solution openings are sheetlike and commonly range from 0.005 to 0.2 in. in height and about 100 to 2,500 ft in width (Moore, 1973). Most solution openings are within 300 feet of the surface. Burchett (1977) states that, in the upper Duck River Basin "46 percent of the water produced from wells in the Central Basin comes from a depth of 60 to 100 feet below land

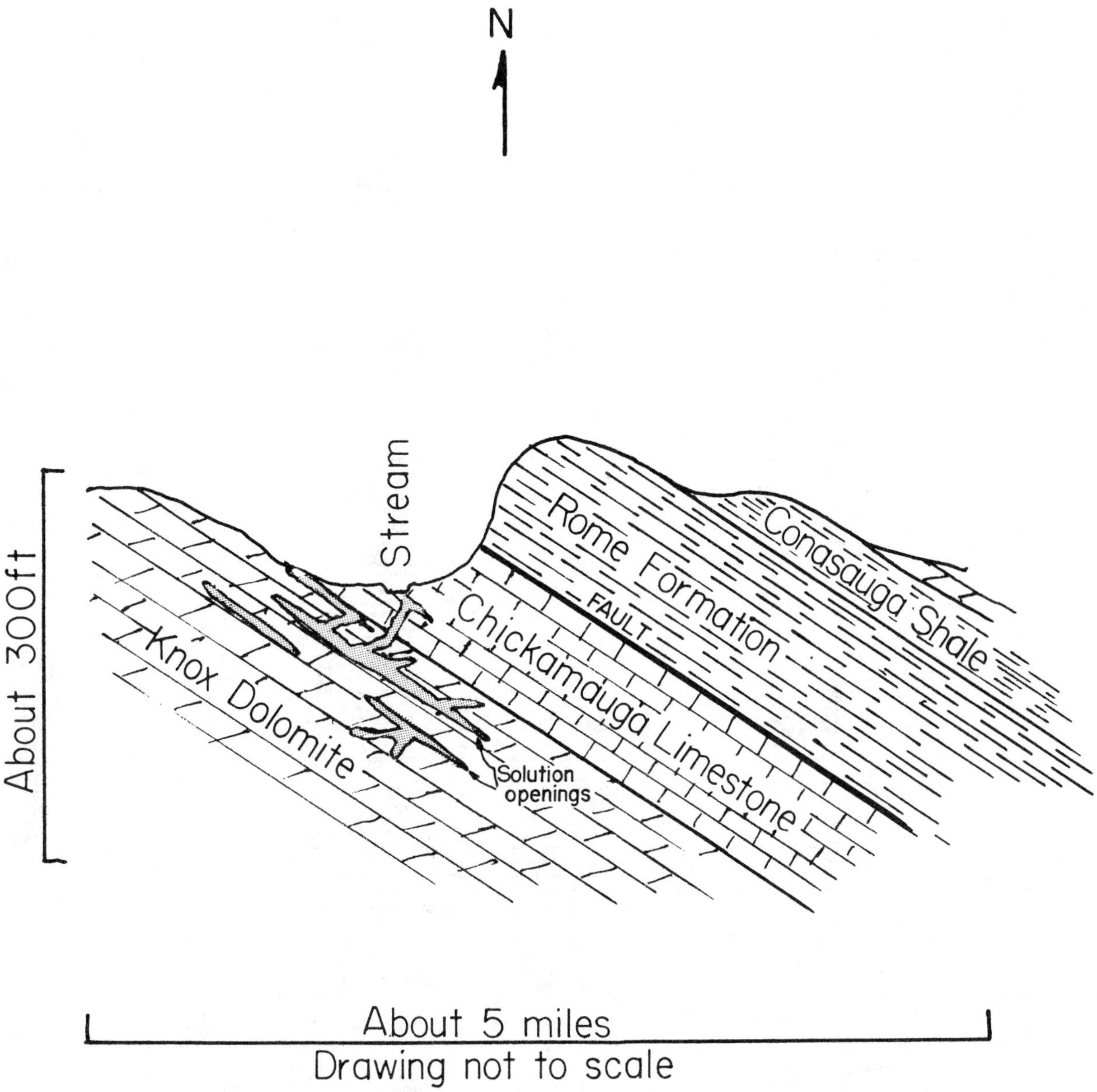

FIGURE 13.—Sketch of idealized geologic and hydrologic conditions in the Valley and Ridge province.

surface. Less than 1 percent . . . comes from a depth greater than 300 feet."

The distribution of ground water in the Central Basin is highly variable. Except for the secondary openings most of the limestone has extremely low porosity. Most wells produce at least some water from bedding-plane cracks as large solution openings are relatively uncommon. Of the wells reported by drillers near Center Hill Lake in Tennessee (north of the Tennessee Region), 46 percent of those in the Central Basin produced less than 4 gal/min and 89 percent produced less than 20 gal/min. Only 3 out of 74 wells produced over 50 gal/min (Moore and Wilson, 1972). In spite of the low reported well production, it is possible in many areas to drill wells capable of producing over 70 gal/min or 0.1 Mgal/d. In some areas it is possible to drill wells capable of producing several hundred gallons per minute. About 15 ft of solution openings were penetrated by a test well at Franklin, Tenn., (about 10 mi north of the Tennessee Region). The well was pumped at 200 gal/min for 8 hours with a drawdown in water level of 0.58 ft. Another test well about half a mile away penetrated a 5 ft opening. These drilling sites were selected on the basis of stratigraphic criteria, an investigation of streamflow gains and losses, and surficial evidence of solution activity. It ap-

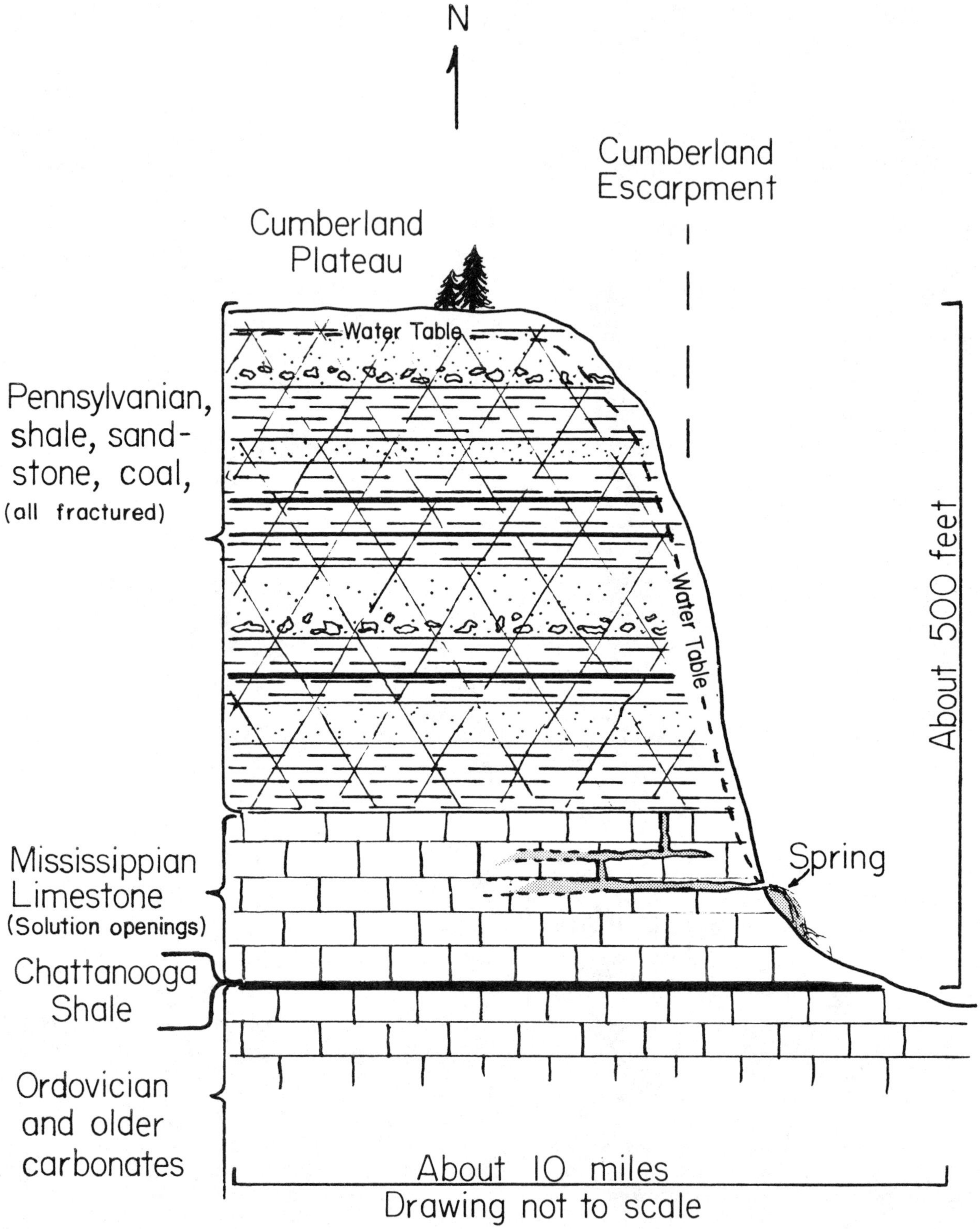

FIGURE 14.—Sketch of idealized geologic and hydrologic conditions in the Cumberland Plateau section. The prevalence of solution openings in the limestone underlying the Plateau is unknown.

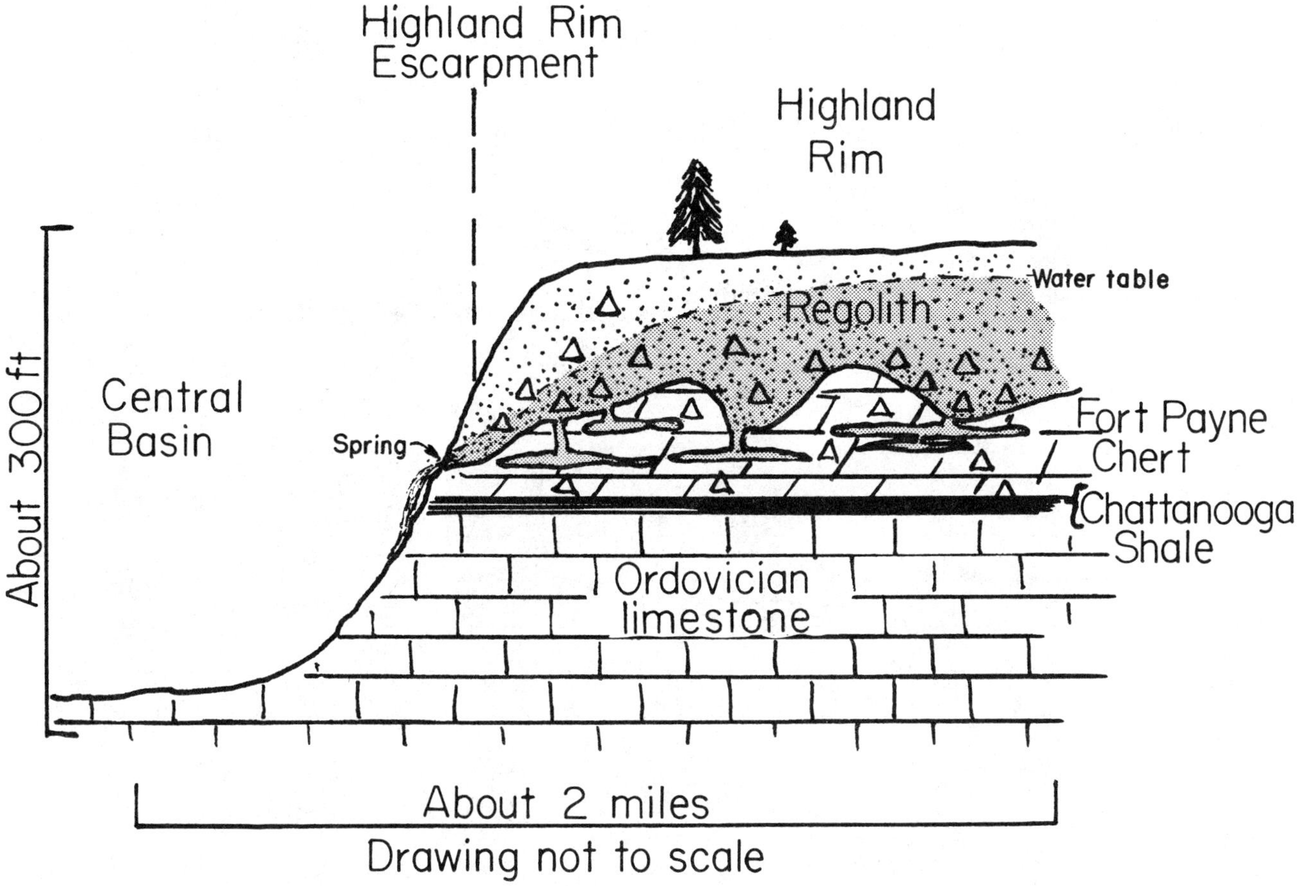

FIGURE 15.—Sketch of idealized geologic and hydrologic conditions in the Highland Rim section.

pears that the use of hydrologic and geologic studies in selecting drilling sites in the Central Basin greatly increases the chances of wells penetrating high-yielding solution openings.

The Knox Dolomite, at depths of 350 to 1,500 ft below the surface, comprises an areally extensive artesian aquifer throughout the Central Basin. It is a dependable source of domestic water supplies (usually 10 gal/min or less) of variable quality (Newcome and Smith, 1962). The largest known production from a well finished in the Knox in the Central Basin is 50 gal/min (Kernie Cothran, driller, oral commun., 1975). However, in eastern Tennessee, production from single wells in the Knox can exceed 1,000 gal/min. The potential of the Knox to yield large amounts of ground water in the central part of the region is unknown.

COASTAL PLAIN

A narrow strip of the Coastal Plain province extends along the western edge of the Tennessee Region. Its eastern boundary is approximately the edge of the Paleozoic rock outcrop and its western edge is the Tennessee River-Mississippi River drainage divide. The Tennessee River flows along the eastern boundary against the edge of the Highland Rim. The topography west of the river valley is of moderate relief, less than that of the dissected edge of the Highland Rim. The major aquifers in the Coastal Plain section of the Tennessee Region are two unconsolidated Cretaceous formations, the Coffee Sand and McNairy Formation (fig. 17).

According to Boswell and others (1965), both the McNairy Formation and the Coffee Sand are capable of yielding municipal and industrial supplies nearly everywhere to the west of the outcrop. The Coffee Sand yields up to about 300 gal/min to wells and is used as a source of water in the westernmost counties in the Tennessee Region as far north as Carroll County, Tenn. The McNairy Formation, however, is capable of much greater production, and good wells produce 500 to 1,000 gal/min. The McNairy Formation is used as a source of water in the Tennessee Region counties north

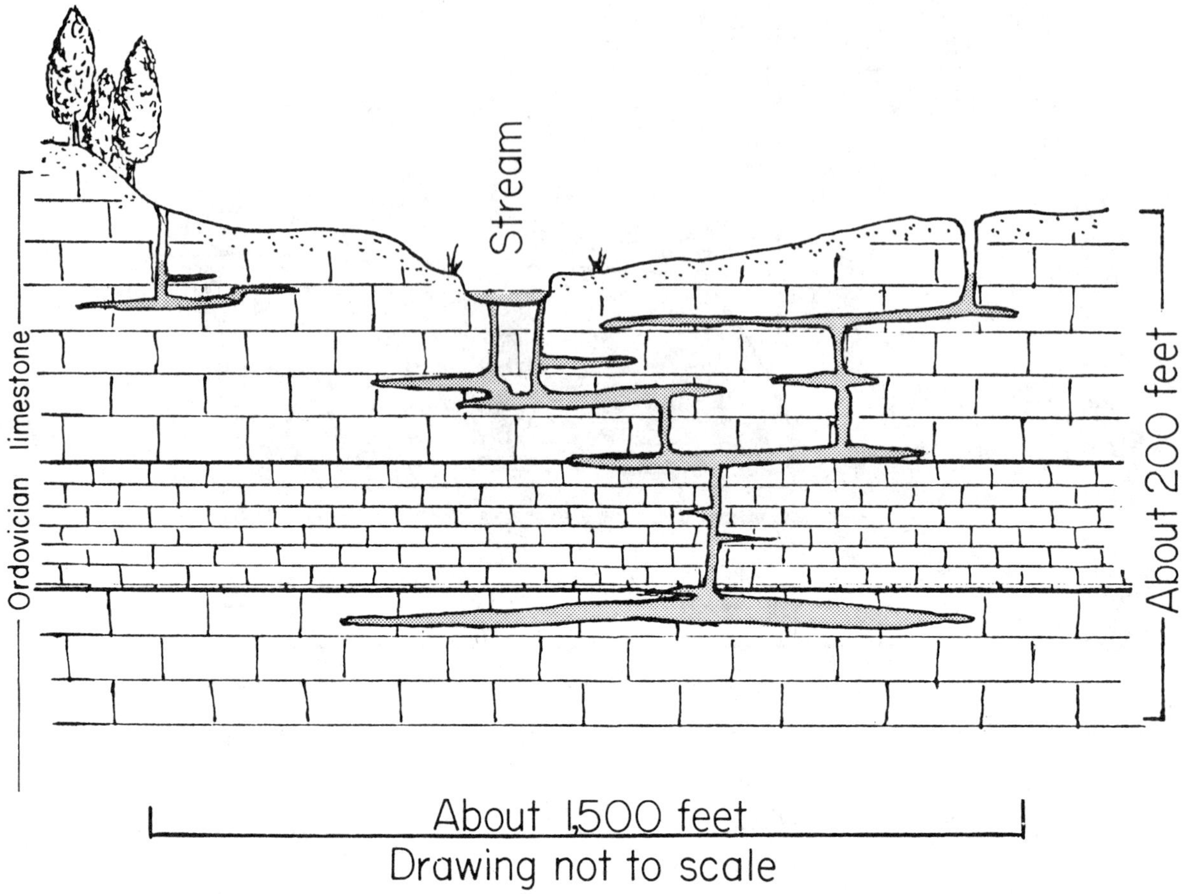

FIGURE 16.—Sketch howing idealized geologic and hydrologic conditions in the Central Basin (Nashville Basin section of the Interior Low Plateaus province).

of Carroll County, Tenn. The amount of water that can be withdrawn from these formations depends, in part, on the thickness of saturated sand layers and the construction of the wells that penetrate them.

In northern Mississippi, test wells in a shattered Paleozoic chert aquifer produced up to 550 gal/min with a specific capacity of 5 (gal/min)/ft (Newcome and Callahan, 1964). The Camden Chert and Fort Payne Chert comprise this fractured chert aquifer in Tennessee (Wells, 1933).

QUALITY OF GROUND WATER

The natural quality of ground water in the Tennessee Region depends on many factors, but mainly upon the chemical composition of the rock in which the water occurs. When water from precipitation enters the aquifer as recharge, it is generally low in dissolved solids, soft, and slightly acidic. As the water moves through the aquifer it acquires a greater concentration of dissolved constituents which change its chemical and physical properties. The least change occurs in the aquifers composed of regolith. The ground water in the regolith remains slightly acidic and low in dissolved solids. This type of ground water is common in the regolith of the Blue Ridge and Highland Rim. Water in the outcrop belt (recharge area) of the unconsolidated aquifers of the Coastal Plain is also fairly close to rainwater in composition, but becomes harder and higher in dissolved solids as it moves deeper below land surface.

Ground water that comes in contact with sandstone and shale containing pyrite remains soft but may become more acidic and high in concentrations of iron and hydrogen sulfide. This type of water occurs in some noncarbonate formations of the Valley and Ridge, in the Pennsylvanian shale and sandstone of the Cumberland Plateau, and immediately below the Chattanooga Shale of the Highland Rim.

A third kind of change occurs in water that contacts carbonate rocks. Because rainwater that has passed through the soil is somewhat acidic, it can dissolve limestone and dolomite, becoming enriched in bicarbonate, calcium, and magnesium. As the dissolved solids content increases, the water becomes harder and slightly alkaline. This type of chemical change occurs in carbonate aquifers such as those in the Valley and

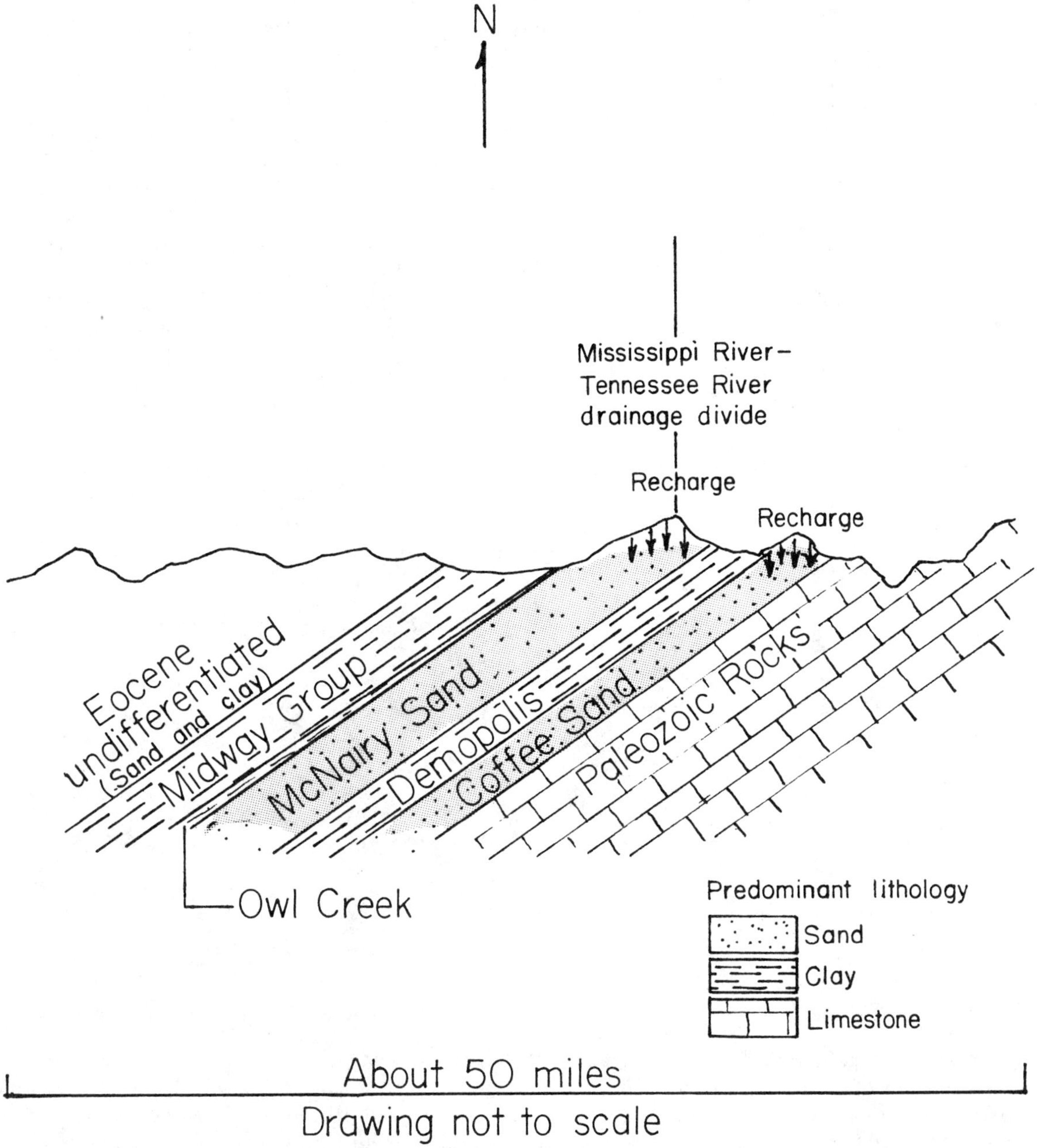

FIGURE 17.—Sketch showing idealized geologic and hydrologic conditions in the Coastal Plain province. The cross section extends east and west approximately at a latitude of 35°N.

Ridge, the Highland Rim, the Central Basin, Sequatchie Valley of the Cumberland Plateau, and in limestones underlying coves of the Blue Ridge.

The analyses shown in figure 18 are representative of the chemical quality of the ground water from the six physiographic areas of Tennessee. Dissolved constituents, consisting mainly of calcium and bicarbonate ions, are highest in the ground water of the Central Basin, with somewhat lower concentrations in the Highland Rim and Valley and Ridge. Unlike aquifers of the Central Basin, aquifers in the latter two areas do not consist entirely of carbonate rocks, and the influence of the regolith and other noncarbonate rocks is seen in the lower amounts of dissolved constituents, including calcium and bicarbonate.

The analyses for the Blue Ridge, Cumberland Plateau, and Coastal Plain, all areas with mainly noncarbonate aquifers, indicate considerably lower dissolved solids in the ground water of these areas than in carbonate terranes. In the Blue Ridge, where most of the ground water occurs in the highly weathered regolith, dissolved solids are lowest.

Physiographic Province	Silica	Iron	Calcium	Magnesium	Sodium	Potassium	Bicarbonate	Sulfate	Chloride	Fluoride	Nitrate
	Values are medians, in milligrams per liter										
1 Blue Ridge	16	0.05	4.6	0.5	3.3	0.8	29	1.6	0.9	0.1	0.1
2 Valley and Ridge	10	0.09	38	12	4.5		178	5	3.5	0.0	3.9
3 Cumberland Plateau	6.4	1.0	16	3.5	3.9	1.2	60	10	2.8	0.1	0.8
4 Highland Rim	11.5	0.00	39	3.8	3.4		146	4.2	4.0	0.1	1.9
5 Central Basin	7.3	0.08	79	9.7	4.4	1.5	256	26	5.0	0.3	0.5
6 Coastal Plain	14	0.5	16	4.6	22	3.4	95	12	4.7	0.2	0.7

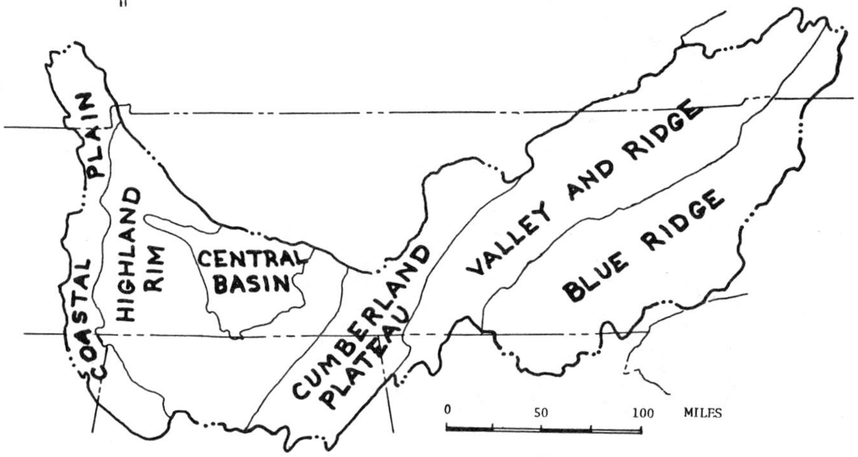

Source of analyses: 1, McMaster and Hubbard, 1970, 23 samples; 2, DeBuchananne and Richardson, 1950, 235 samples; 3, Newcome and Smith, 1958, 13 samples; 4, Geol. Survey of Alabama bulletins, 125 samples; 5, U.S.G.S. test wells at Columbia, Normandy, and Franklin, Tenn., 22 samples; 6, Boswell, Moore, MacCary, and others, 1965, median for Ripley Fm. (McNairy Sand), number of samples unknown.

FIGURE 18.—Results of chemical analyses of ground-water samples from each physiographic subdivision of the Tennessee Region.

Ground-water quality is reflected in the chemical character of stream water during periods of base flow. To some extent, streams can be used to obtain an integrated sample of discharge from the aquifers underlying the watershed. Betson and McMaster (1975) have developed, for the Tennessee River basin, a model to simulate mineral constituent concentrations in streamflow for watersheds underlain by different types of rock. Concentration values generated using their regression coefficients and a streamflow of 1 $(ft^3/s)/mi^2$ agree fairly closely with the analyses in figure 18 when the coefficients used are those for rock types most commonly found in each physiographic province.

The rate of water circulation within the aquifer has an important effect on ground-water quality. Where circulation has been rapid, aquifers may have been flushed of readily dissolved substances. Where the regolith is hydraulically connected to well-developed openings in the underlying bedrock, for example, water from the regolith, low in dissolved solids, can circulate rapidly through the openings without great increases in hardness. As a result of situations like these, water from wells that tap very permeable formations or highly developed solution or fracture systems tends to be lower in dissolved solids than water from poorly interconnected openings.

The quality of ground water from a particular aquifer at any one place tends to be relatively constant with time. This property is most evident where the regolith filters the water that replenishes the aquifer. In aquifers having direct connections with land surface (via sinkholes, for example), marked changes in quality may occur as storm runoff enters the system.

Well-developed openings and highly porous material, when less than about 100 ft below land surface are very susceptible to pollution, and strong protective measures are needed to ensure that the ground-water quality will remain unimpaired.

A study sponsored by the U.S. Public Health Service identified a high incidence of contamination of rural domestic water supplies in three counties in Tennessee, as shown by the presence of coliform and fecal coliform bacteria in water samples (Bureau of Water Hygiene, 1971). Fifty-nine percent of the water supplies examined failed to meet bacteriological standards. However, according to the report, nearly every one of the rural, individual systems examined had one or more facility deficiencies. Very few of these systems were constructed to prevent entrance of contamination. It is entirely probable that they represent contamination at the well site and not of the aquifers that furnish water to the wells. Water obtained from relatively deep aquifers penetrated by test wells has usually contained very few, if any, coliform bacteria, and it is not unusual for the water to be entirely free of any indications of contamination.

Most of the ground water in the Tennessee Region is of suitable chemical character for public drinking-water supplies. Figure 19 shows the medians and ranges of some of the chemical constituents for which maximum concentrations have been recommended by the Environmental Protection Agency (National Academy of Science-National Academy of Engineering, 1972; Environmental Protection Agency, 1975). In two areas the median values for iron exceed the recommended maximum concentration, but most of the samples are well within the recommendations.

The Tennessee Region has no known significant bodies of saline ground water. Of about 1,000 analyses (mostly published) of water from wells and springs throughout the region, only 40 indicate water with over 1,000 mg/L total dissolved solids. The high dissolved solids content is usually associated with stagnant ground water in poorly developed solution openings in flat-lying carbonate rocks. Most of the wells and springs with high dissolved solids are in the Central Basin or Highland Rim, and 17 of the wells tap the Knox Dolomite of central Tennessee. In almost every case they reportedly produce less than 20 gal/min (0.03 Mgal/d).

DEVELOPMENT OF THE GROUND-WATER RESOURCES

GROUND–WATER USE

In 1970, the use of ground water in the Tennessee Region totaled 173 Mgal/d (Murray and Reeves, 1972). This is less than 1 percent of the estimated 22,000 Mgal/d of ground water that is discharged annually to the streams of the Tennessee Region, which is an indication of the large amounts available for development. The ground water that is used amounts to slightly less than 8 percent of the total water use in the Region excluding that used for electric power generation (table 4). However, the percent of the population served by ground water is much larger than the total-use figures would indicate.

According to the 1970 census, the population of the Tennessee Region was about 3,300,000. One-third of the people were living in towns or cities with populations of 2,500 or more. The rest lived in small towns or rural areas (Delury, 1973).

At present, most of the large towns and cities in the Tennessee Region use surface water. A notable exception is Huntsville, Ala., a city of 138,000, which obtained all its water from Big Spring until 1950. Now it draws half its supply from five wells and one spring and the other half from the Tennessee River (Geol. Survey of Alabama, 1975). In that part of the State of Tennessee within the Tennessee River basin, the largest towns supplied entirely with ground water are Tullahoma, population 15,000 and Elizabethton, population 12,000. Both draw their water from springs (Tennessee Div. Water Resources, written commun., 1976).

Many of the smaller towns in the Tennessee Region use ground water. Seventy-nine percent of the small water-distribution systems (serving fewer than 2,500 people) use ground water for at least half their supply. As these towns and others that withdraw water from small streams grow, the present surface source may become inadequate, and ground water could play a major role in supplementing these supplies.

The rural population of the Tennessee Region obtains most of its water supply from wells and springs. Approximately 50 Mgal/d was used for rural domestic purposes in 1970. In contrast, only a third of water used for livestock and irrigation was ground water (Murray and Reeves, 1972).

Industries in the Tennessee Region that have their own source of water used 45 Mgal/d of ground water in 1970. This was only 3 percent of the water used by self-supplied industries other than power-generating plants.

In addition to industrial use, ground water supplies many businesses in isolated areas such as service stations and motels. Springs are used for raising fish because of the constant temperature and low turbidity of the ground water.

POTENTIAL FOR DEVELOPMENT

Ground water is generally overlooked as a water supply in the Tennessee Region because of the abundance of surface water. Only a small part of the available ground-water supply, about 0.8 percent of the average annual recharge, is being used. The advantages in

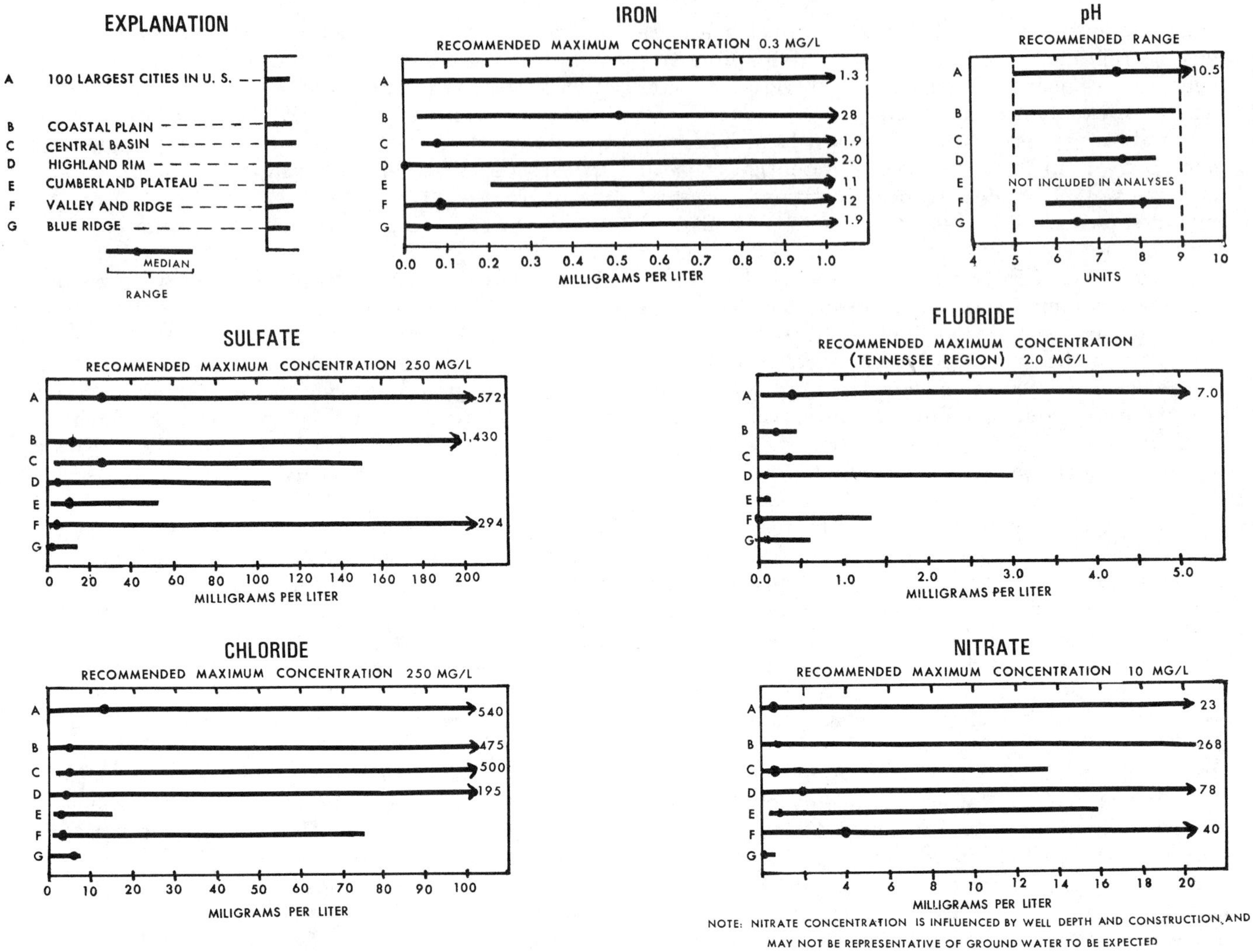

RECOMMENDATIONS FROM ENVIRONMENTAL PROTECTION AGENCY, 1975, AND ENVIRONMENTAL STUDIES BOARD, NAT. ACAD. SCI., NAT. ACAD. ENG., 1973.

VALUES FOR 100 LARGEST CITIES FROM DURFOR AND BECKER, 1965.

ANALYSES USED ARE THE SAME AS THOSE IN FIGURE 18.

FIGURE 19.—Medians and ranges of six chemical constituents in untreated ground water of the Tennessee Region and in the treated, finished water of the 100 largest cities in the United States. Also given are the recommended maximum or minimum values for these constituents in drinking water.

TABLE 4.—*Water use in the Tennessee Region in 1970*

[From Murray and Reeves (1972)]

Use of water	Water withdrawn (in Mgal/d) Ground water	Surface water
Public supplies	64	240
Rural domestic use	51	.9
Livestock and irrigation	13	25
Self-supplied industrial use (excluding power generation)	45	1,300
Power generation (thermoelectric)	0	6,100
Power generation (hydroelectric)	0	12,000
Total	173	19,700

using ground water as a water supply are as follows: (1) its widespread availability, (2) its general dependability, particularly at depth, (3) the minimal amount of treatment required, (4) the relatively low cost of developing a ground-water supply, and (5) its uniform temperature and chemical character. Ground water can be used very efficiently for small-scale developments such as water supplies for rural communities, industries, and small towns. In addition, ground water has potential for supplemental and conjunctive use with surface-water supplies.

There is growing interest in the feasibility and advisability of injecting fresh water into aquifers for storage. The use of underground space for storage of liquids has been practiced on only a limited scale in the region.

AVAILABILITY

In many parts of the Tennessee Region, ground water is available in amounts comparable to those that might be obtained from surface-water impoundments. In contrast to surface-water impoundments, however, the ground-water reservoir stores water with minimal evaporation, insulated to some degree from pollutants and available at diverse points. In addition, ground-water supplies can be developed more quickly and at lower cost compared to the time and cost of creating surface-water storage.

DEPENDABILITY

It is a common belief that ground-water levels are continually declining in all parts of the country. However, this is true only in places where pumpage of ground water greatly exceeds natural recharge. In the Tennessee Region, ground-water levels show only normal seasonal trends. Water levels usually decline during the growing season (April-November) and rise during the remainder of the year when most ground-water systems are being recharged from precipitation (fig. 20). Shallow dug wells are not dependable sources of supply during periods of prolonged drought as they are generally not dug deep enough to allow for any extreme decline in water level. However, production from drilled wells that penetrate deeper solution cavities or water-bearing formations is generally not seriously reduced by drought.

In areas where ground water is developed, local lowering of water levels can occur as a result of pumping. However, if pumpage is within the capacity of the aquifer to supply water, the rate of water-level decline will gradually diminish and the water level will stabilize. The response of the water level to various rates of pumping is shown in figure 21. Nowhere in the region has ground-water development caused significant depression of water levels, although there are instances where inadequate spacing between wells has caused excessive water-level drawdown in a localized area.

TREATMENT NEEDS

Ground water in the Tennessee Region usually needs less treatment than surface water to make it acceptable for most uses. It can often be used untreated for cooling and process water in industrial plants. Chlorination is the basic treatment needed for drinking-water supplies. In some cases aeration is needed to dissipate dissolved gases such as hydrogen sulfide or to precipitate dissolved iron. Buffering may be needed, especially for water systems in which low-pH ground water is to be mixed with slightly alkaline surface water. Ground water usually has low turbidity and does not generally require filtration to meet turbidity standards for drinking water. Ground water has the additional advantage of being uniform in chemical quality, and would therefore require less monitoring than a surface-water supply. For example, the turbidity of surface water is increased by storm runoff, but ground-water turbidity is usually constant, except in shallow carbonate aquifers that have direct connections with land surface, such as sinkholes.

As a result of the minimal treatment needed for most ground-water supplies, low-cost treatment facilities can be installed at the well field. This makes it possible to have multiple self-contained units located at points of use of the water rather than a single treatment plant with an extensive distribution system.

COST OF DEVELOPMENT

Another potential benefit of ground-water development is its relatively low cost. Cederstrom (1973) estimated that in the North Atlantic States the cost of large supplies of ground water at the wellhead, taking into account the costs of locating and developing a well or well field, ranged from 1.5 to 5 cents per thousand gallons in 1970 depending on the aquifer material. Many of the aquifer materials in this study area are similar to those of the Tennessee Region, ranging from Coastal Plain sediments from which ground water can be obtained most cheaply to carbonate rocks in which ground-water development is most costly. The investment required to construct and operate a ground-water distribution facility is further reduced at places where the water requires little treatment.

A study of the alternatives for water supply may reveal an economic incentive to use ground water. An example is a small utilities district in central Hamilton County, Tenn. Owing to the proximity of the surface-water intake for this system to the site of a nuclear powerplant on the Tennessee River, a study of alternative sources of water was made. The results of the investigation indicated that the use of ground water from the Knox Dolomite would lower the cost of providing finished water to the consumers by as much as 50 percent, and the utility district is now using wells.

TEMPERATURE STABILITY

A relatively constant annual temperature, about the same as the average annual air temperature of the area where it occurs, is characteristic of ground water. This characteristic makes it extremely useful for cooling and for industrial processes where constant temperature is required. In the Tennessee Region ground-water temperatures normally range from 50 to 65°F.

A growing use of ground water is as a heat exchange medium for heat pumps. Heating and cooling of build-

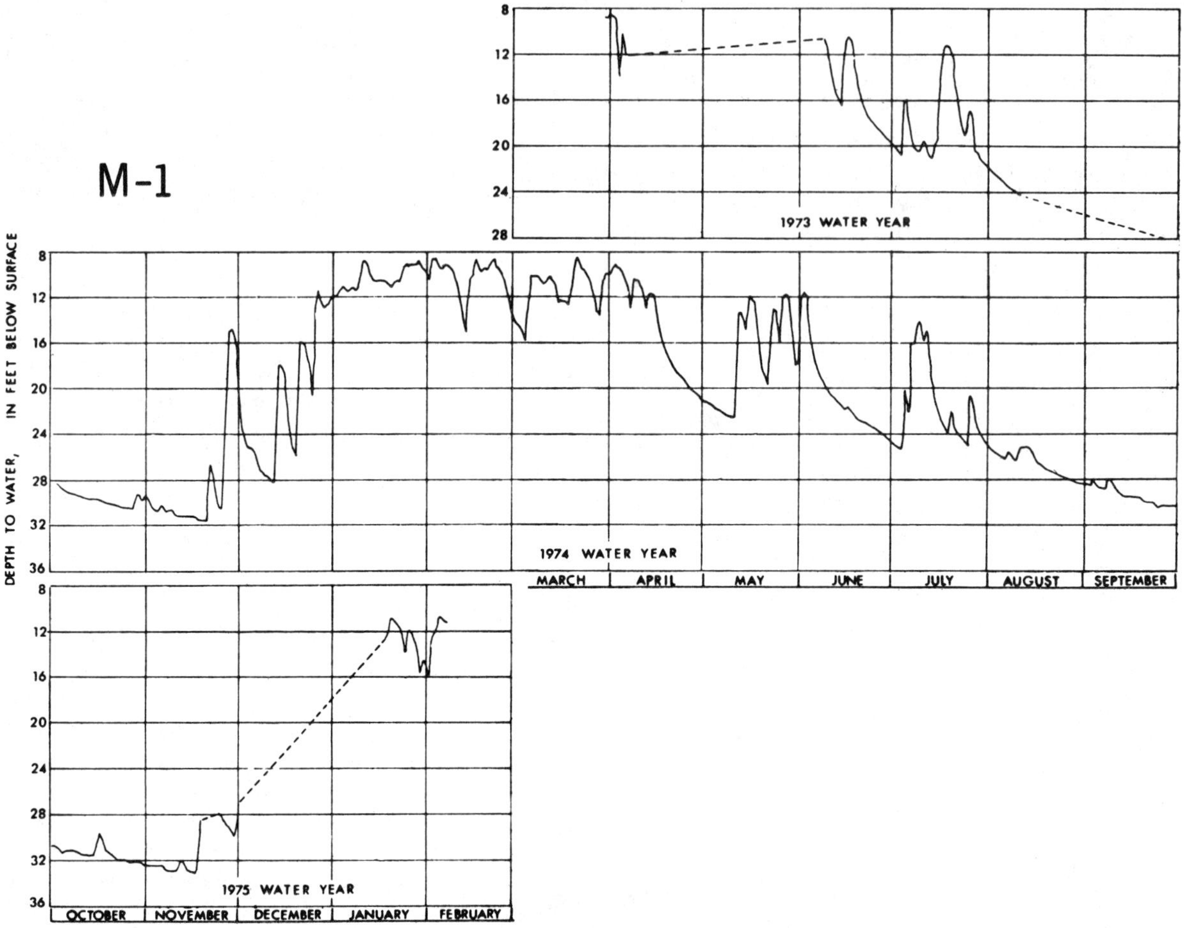

FIGURE 20.—Water-level fluctuations in test well M-1, Manchester, Tenn. showing seasonal trends (from Burchett, 1977).

ings is much more efficient with a ground-water-to-air rather than air-to-air interface, because of its constant temperature and the high specific heat of water.

EFFICIENCY FOR SMALL-SCALE DEVELOPMENT

Owing to the scale of ground-water developments, ground water in the Tennessee Region has great potential as a sole water source for small communities and industries that are remote from large reservoirs.

In sparsely populated areas and small rural communities a multifamily well may be a more efficient water supply than a connecting pipeline to a large centralized water distribution system (Lehr, 1976). Cederstrom (1973) concluded in his cost analysis of ground-water supplies in North Atlantic States (1973) that "where large water requirements consist of many small to moderate demands at distinctly separate points ground-water supplies may serve admirably from a cost point of view." This approach is being taken in Lincoln County, Tenn. where the U.S. Geological Survey has undertaken a cooperative study with the Lincoln County Public Utilities Commission to investigate the occurrence and availability of ground water as an aid in developing ground-water supplies for small communities throughout the county.

For industrial use, the cost of installing a well or developing a well field may compare favorably with the cost of a long pipeline connecting to a municipal system. Also, the initial cost would be defrayed by the low operation and maintenance cost of the ground-water facility. Where sufficient ground water cannot be obtained on site, water can be piped from wells at more favorable sites, as is being contemplated in Chattanooga, Tenn. (D. R. Rima, U.S. Geological Survey, oral commun., 1976).

SUPPLEMENTAL AND CONJUCTIVE USE

In areas where ground water alone is inadequate to supply a needed amount of water, it can play a supplemental role. There are several situations in which use of ground water would be an attractive alternative source of supply. For example, when a town grows gradually to the point where demand for water occasionally exceeds the supply available from ground-water sources, a surface-water impoundment might be needed. However, until the need was severe enough to

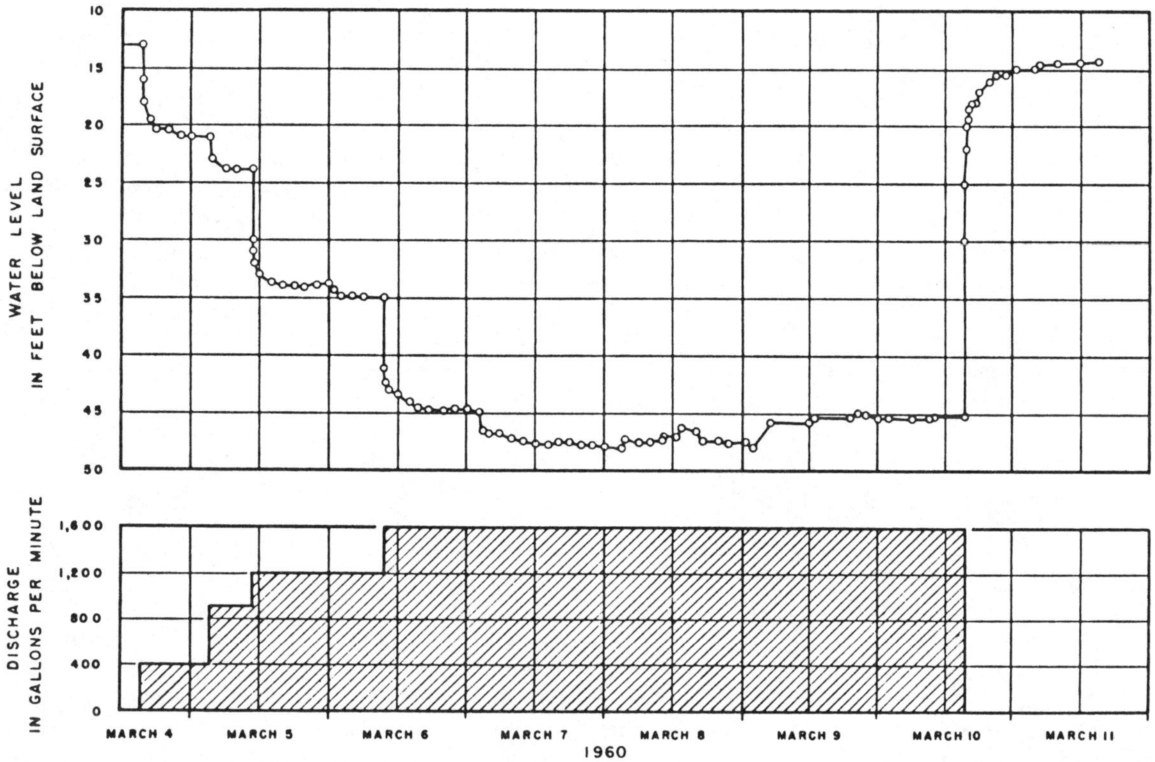

FIGURE 21.—Pumping rate and water levels in well M-157, Colbert County, Ala. during pumping test (from Harris and others, 1963).

justify major construction of this nature, ground water, even in relatively small amounts, could supply the peak water demand. In areas where surface water is not available, municipalities are faced with the alternatives of building a pipeline to connect to another water system or of developing ground-water supplies. In this situation, a thorough investigation of available ground-water resources is warranted because of the high cost of constructing a pipeline. Ground water can also play a role in urban development as the water supply for outlying areas of growth.

Conjuctive use of ground water and surface water has received little attention in the Tennessee Region. The following examples illustrate how knowledge of ground water and surface water interaction could increase the efficiency with which both are used. In Chattanooga, Tenn., there is a plan to modify a water system that normally uses water from a stream which is subjected to occasional chemical pollution. The stream will be monitored and, when pollution occurs, the system will draw water from nearby wells until the surface water is again usable. Ground water could also be used to augment the low flows of streams, both to supply water systems and to maintain sufficient flow to assimilate waste.

When a large surface-water impoundment is made, it could benefit ground-water users by reducing the ground-water level fluctuations in the vicinity of the lake. Surface water can be stored without the use of an impoundment where a suitable aquifer is available. Such an aquifer can be recharged with surface water during times of high flow for withdrawal during dry periods. Well fields can also be used to capture subsurface flow in stream beds by inducing flow from the stream toward the wells.

In an alluvial aquifer, of which there are few in the Tennessee Region that are not covered by reservoirs, there is a fluctuation of temperature in the aquifer caused by infiltration of river water. The ground-water temperature lags about 6 months behind the river temperature. It is warmest in winter and coolest in summer. This property could be useful for heating and cooling purposes.

UNDERGROUND STORAGE

Deep wells are being used in the Tennessee Region for disposal of liquid waste. Two industries, in New Johnsonville, Tenn., and Mt. Pleasant, Tenn., use wells for disposal of chemical waste. At Oak Ridge, Tenn., medium-level liquid radioactive wastes are injected into a shale formation in cement grout so that once they solidify they cannot move from the point of injection (de Laguna, 1968).

No aquifer in the Tennessee Region is capable of completely isolating injected liquid substances. Therefore, the possibility exists that injected liquids would displace poor-quality water and migrate upward into aquifers used for sources of water supplies.

The use of aquifers for storage of fresh water has not been strongly considered in the region. The inadequacy of information on local ground-water movement currently makes it difficult to evaluate the feasibility and environmental impact of injecting fresh water into a deep aquifer for later withdrawal.

LOCATING A GROUND-WATER SUPPLY

Ground water is not necessarily available in adequate quantities precisely where it is needed, especially in carbonate rocks. The chances of finding adequate supplies of ground water are greatly increased if drilling of production wells is preceded by a hydrologic study of the area to determine the most favorable areas for high-producing wells and by test drilling to verify these areas. In many areas the drilling and test pumping of more than one test well might be required before a satisfactory supply can be obtained.

In his cost analyis of ground-water supplies in the North Atlantic Region, Cederstrom (1973) states that the "average yield" of wells in any one area, as commonly given in the literature, is no guide to what might be obtained because most wells were constructed to supply water for domestic use and the potential yield of the aquifer was not determined. The "average yield," therefore, represents something a little greater than the average need and is not a measure of the full potential of wells in the rock type being studied.

In the Tennessee Region the generally low production from wells reported by drilling contractors has tended to discourage exploration for large ground-water supplies in almost all areas except the Coastal Plain. For example, in the part of the upper Duck River basin of Tennessee that is on the Highland Rim, 86 percent of wells reported to the Tennessee Division of Water Resources produced 20 gal/min or less. However, of 19 test holes drilled in the same area, only 16 percent produced 20 gal/min or less and 74 percent produced 100 gal/min or more (Burchett, 1977). The reason for this success in test drilling was the use of site-selection criteria based on hydrologic concepts of the occurrence and availability of water in the Fort Payne regolith and bedrock.

Site-selection criteria are the practical application of an understanding of the hydrologic system in an area. In some parts of the Tennessee Region the controls on ground-water occurrence are well defined. Two such areas are the Coastal Plain and northern Alabama. Test drilling in the Great Smoky Mountains National Park (McMaster and Hubbard, 1970) has helped to identify criteria for the Blue Ridge. However, in other parts of the region, criteria are either nonexistent or incomplete. The Cumberland Plateau is the only area where very little test drilling for water has been done. Considering the unreliability of the flow of surface streams on the Plateau and the reported difficulty in obtaining even domestic ground-water supplies in some areas, an intensive study of the Plateau's water resources would contribute information needed to fill the biggest gap in knowledge of Tennessee Region's ground-water hydrology.

In the Central Basin, Highland Rim, and Valley and Ridge Provinces, ongoing studies by the U.S. Geological Survey, including test drilling, are leading to development of site selection criteria which can greatly improve chances of obtaining large ground-water supplies in carbonate-rock terranes. These criteria are continually being tested and refined.

Some of the hydrologic controls in each province of the Tennessee Region have been discussed in the section on "Occurrence of Ground Water." Site selection criteria vary from one area to another depending on which factors have the greatest influence on ground-water distribution. Use of the criteria does not guarantee that large ground-water supplies will be located; it merely increases the chances of drilling successful wells, especially when a site is chosen on the basis of several different criteria. For example, test drilling, test pumping, definition of surface geology and subsurface structure, mapping of soil thickness, and correlation of the withdrawal of ground water with piezometric maps prepared during periods of high and low water levels were the basis for locating and developing a well field in a limestone terrane near Huntsville, Ala., capable of producing 14,000,000 gal/d at a fraction of the cost of a proposed surface-water supply (Lamoreaux and Powell, 1963). Figure 22 illustrates the kinds of information used in selecting drilling sites in central Tennessee. It should also be noted that in areas of great variability in ground-water occurrence, especially where carbonate rocks are the aquifers, a single test hole is not adequate to determine the maximum amount of ground water available at a particular site.

Once a ground-water supply has been located, the characteristics of the hydrologic system should be taken into account in designing the pumping facility. The optimum design is a balance of well yield and operating efficiency against undesirable impacts on the hydrologic system (for example, diminution of stream or spring flow, interference with nearby wells, and deterioration of water quality as a result of induced recharge). One of the most frequently neglected princi-

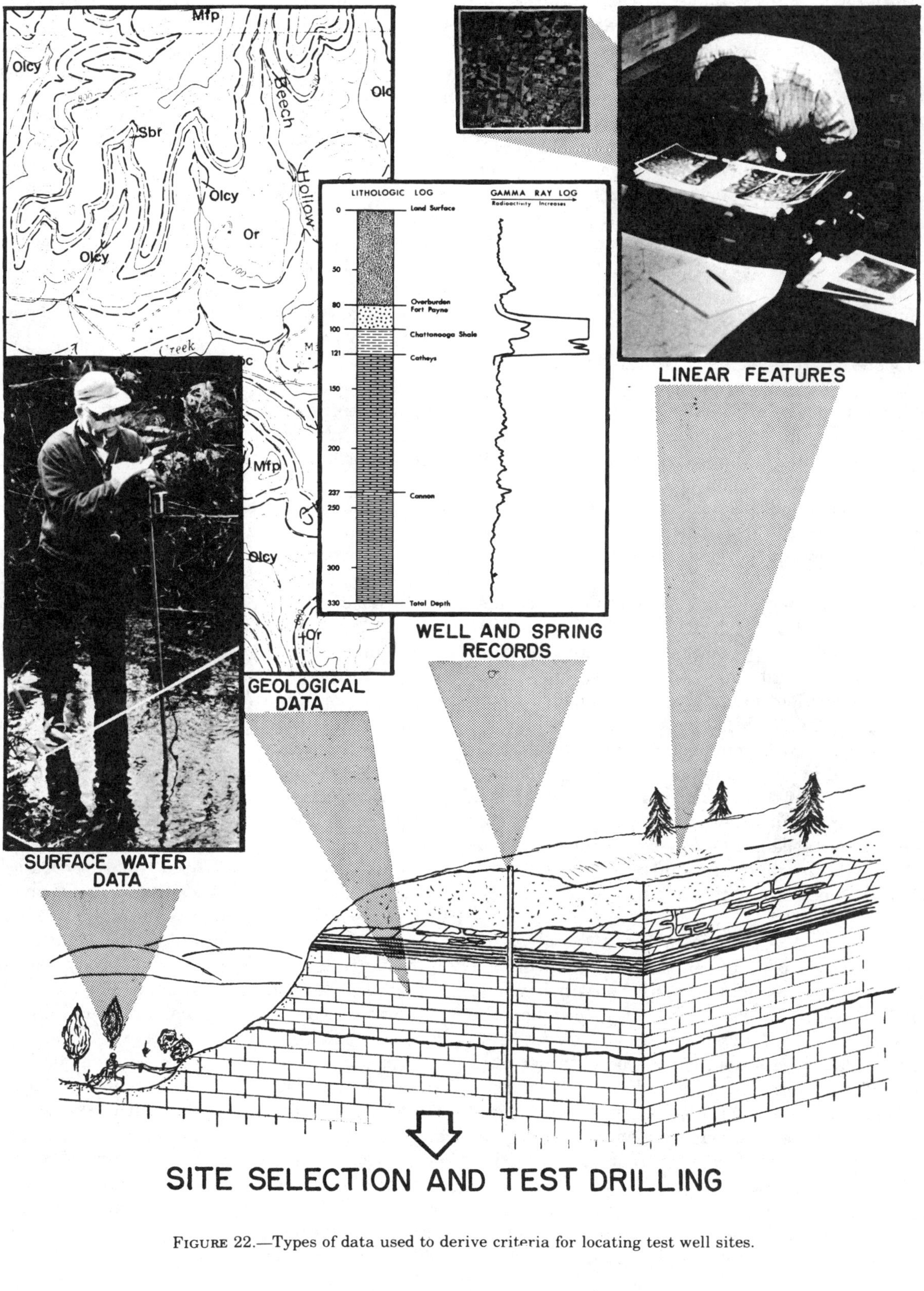

FIGURE 22.—Types of data used to derive criteria for locating test well sites.

ples of well-field design is adequate spacing of wells to avoid interference. For example, in 1960 the town of Waverly, Tenn., was provided with water from two wells less than 500 ft apart, each reportedly producing 300 gal/min. Pumpage from these wells had created a 25 ft deep cone of depression around the wells (Marcher and others, 1964). In 1960 a third well was drilled in the same city-owned lot as the other two municipal wells, but it could not supply sufficient water to be used as a production well. In effect the three wells were functioning as a single well, and withdrawals had exceeded the capacity of the aquifer to supply water to that small area. It is likely that had the well been placed sufficiently far away so as not to be influenced by pumping from the other two, it would have produced an adequate amount of water. The necessary spacing between wells could have been calculated from aquifer test data.

Land use can also be affected by ground-water development. For example, sinkhole development is a possible consequence of ground-water utilization in the parts of the Tennessee Region underlain by carbonate rocks. Where considerable lowering of ground-water levels is predicted, the likelihood of accelerated sinkhole formation should be investigated (Newton and others, 1973).

Wells withdraw water that would naturally discharge to streams or springs. Pumping wells will inevitably reduce the ground-water supply to these discharge points and may eventually alter the ground-water gradient enough to cause surface water to enter the aquifer. While these effects may not be serious or may even be beneficial, plans for ground-water use would ideally include an evaluation of their impact on both ground water and surface water. This is particularly important when ground water and surface water are to be used conjunctively because ground-water withdrawals affect streamflow at times of low flow when the need for sustained surface-water flow is greatest.

DATA NEEDS

Accurate assessment of an area's potential for ground-water development is only possible where adequate data exist or can be acquired to define the ground-water system. Much of the basic data required, even for a small-area study, cannot be obtained in a short period, but must be collected on a continuing basis. Geologic data, well records, water-level and water-quality data, and information on aquifer characteristics are typical of information required as a foundation for hydrogeologic studies.

Large-scale general purpose geologic mapping is available nearly everywhere in the Tennessee Region. In addition, mineral exploration has provided subsurface stratigraphic data in many areas. A large quantity of unpublished well records are available as a result of state laws requiring drillers to submit information for wells they drill. This information soon will exceed 50 wells per county in each state. Some of this information has been computer-listed by state water resources agencies. Federal and State water-resource agencies also keep well records and logs.

The greatest deficit in ground-water basic data is in records of water-levels, water-quality, and aquifer tests. Northern Alabama is unusual in the completeness of its basic records. For example, as shown by figure 23, the number of network observation wells operated by the U.S. Geological Survey in the northern Alabama part of the Tennessee Region is equal to the number in the remainder of the region even though only 17 percent of the region is in Alabama. Outside of Alabama, observation wells are sparse, and they do not give representative information for all the physiographic provinces in the region. Strengthening the observation well network before large-scale ground-water development takes place would provide necessary information on water levels, on "base line" ground-water quality, and on aquifer behavior under natural stresses such as drought.

Records of natural ground-water level fluctuations are particularly important, alone or in conjunction with stream seepage investigations, for identifying recharge and discharge areas. Identification of sources of contamination is most critical in recharge areas in order to manage ground-water quality. Large ground-water supplies that are subject to minimal water-level variation can often be located in or near ground-water discharge areas. Hence, the identification of these areas aids in exploration for ground water.

Aquifer characteristics, as determined by pumping tests, are not known in most of the Region. Many of the pumping tests that have been made served chiefly to test well performance rather than aquifer characteristics. Governmental regulation of public water supplies is increasing and more quantitative information from pumping tests will be required to satisfy the needs anticipated under these regulations. The results of these pumping tests would be useful in defining aquifer characteristics on a regional basis.

According to an assessment of the availability of ground-water data in the Tennessee Valley (W. M. McMaster, Tennessee Valley Authority, written commun., July 1975) the density, utility, and age of published reports of ground-water information are highly variable. Large-area reconnaissance reports, published between 1932 and 1962, exist for most of the Region. These reports are generally based on rather cursory well inventories and cover large areas. Only the parts

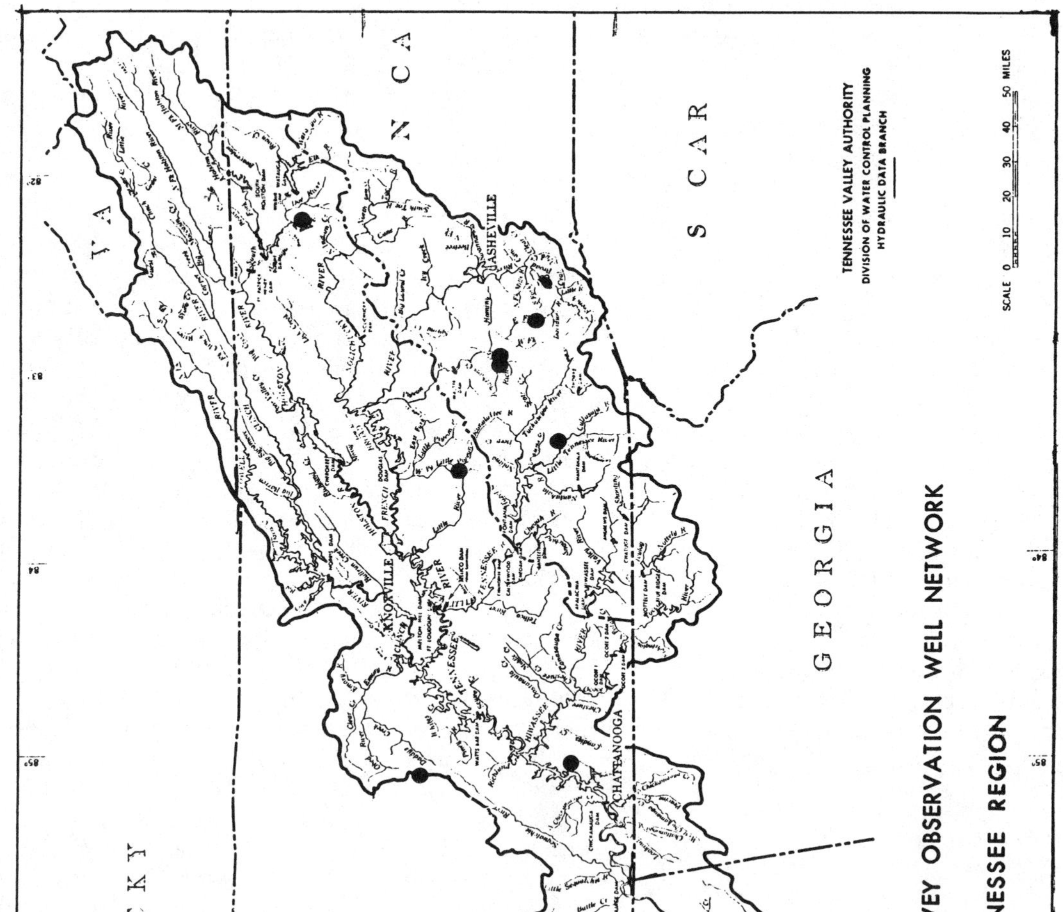

FIGURE 23.—Locations of wells in the U.S. Geological Survey observation well network.

of the Tennessee Region in northern Alabama, Kentucky, and North Carolina are adequately covered by reports on smaller areas that are useful for defining well-site selection criteria.

Detailed geohydrologic studies involving test drilling and aquifer testing are necessary for an understanding of ground-water hydrology that would allow the development of criteria for locating feasible areas in which to develop high-producing wells. The results of these studies should be considered in regional, metropolitan, and industrial water planning and management programs.

The study of hydrologic systems in the Tennessee Region requires considerable time. However, if present sources of supply are placed under a severe stress resulting from a major drought or population expansion, the degree to which water management can deal with problems will depend, in part, on how much information is available and on how well the hydrologic systems of critical areas are understood. Information required to deal with these problems might include such things as where untapped ground-water supplies can be located, which parts of the region have the least ground-water storage, how ground water and surface water can be used conjunctively, and what the effects of utilizing one will be on the other.

WATER-RESOURCES MANAGEMENT

Effective management of water resources cannot deal with water problems in isolation but as they relate to the hydrologic system as a whole. Stress imposed on one part of a hydrologic system is certain to have repercussions in other parts of the system. This is true whether the stress is applied to ground water or surface water. For example, ground-water withdrawals can reduce the low flow of streams by intercepting ground-water that is naturally discharged to the stream. Impoundment of surface water can increase the quantity of ground water in storage by altering head relationships in an aquifer. Although the surface-water resources in the Tennessee Region are heavily developed, there are no regionwide plans for managment of ground water or of the combined ground- and surface-water resources of the region.

The response of a hydrologic system to stress may be experienced at points distant from the place where the stress is applied. For this reason, study and management of water resources can be accomplished most effectively within hydrologic boundaries rather than political boundaries. Yet many of the organizations, including most government agencies, whose function is to study or manage water resources have programs which operate within political boundaries such as states or counties. To deal with the broad implications of hydrologic problems, either of two approaches could be considered: close coordination among the organizations, or the involvement of an organization whose jurisdiction includes the entire hydrologic system. In the past, conflicts as to matters such as funding and the means of implementing management policies have arisen in the course of some attempts to coordinate the efforts of several organizations concerned with water. Notable success at managing water resources of a basin-wide scale has been achieved by agencies such as the Delaware River Basin Commission and the Tennessee Valley Authority. However, because the Tennessee Valley Authority was established primarily to control the Tennessee River, it has not been called upon to direct a major effort toward developing the region's ground-water resources, though it has supported ground-water studies in some parts of the region.

The need for a regionwide ground-water management plan is not yet pressing in the present infancy of ground-water development in the Tennessee Region. In the absence of crisis, a regional ground-water management plan could serve several purposes: (1) to coordinate data collection and interpretive studies, (2) to indicate the most efficient and economical use of ground-water resources in the region, (3) to recommend measures to maintain the quality of present and potential ground-water supplies, and (4) to provide support for predicting the environmental impact of ground-water diversions by means of digital models which would simulate the conjunctive functioning of both ground-water and surface-water systems. Regardless of how a ground-water management plan would be administered, its formulation and implementation in conjunction with surface-water management plans should recognize the interdependence of ground water and surface water and provide for utilization of both aspects of the region's water resources to their fullest potential.

CONCLUSIONS

Ground water is an abundant resource in the Tennessee Region. As much as one-fifth to one-fourth of the precipitation that falls in the region enters the ground-water reservoirs each year. A significant part of the 22,000 Mgal/d contributed annually to the ground-water reservoir from precipitation is available for development.

In 1970, less than 1 percent of the estimated ground-water recharge, amounting to 173 Mgal/d, was used in the Tennessee Region. This was less than 8 percent of the total quantity of water used in the region. There is, therefore, a large potential for increased development of ground-water supplies.

At present, ground water is used chiefly in rural areas and small communities and by industries and commercial establishments beyond the limits of municipal water-supply systems. This use of ground water for small-scale developments is practical and economical because in many parts of the region, ground water can be obtained at or near the points of use, eliminating the need for water impoundments and extensive distribution systems. The cost is further reduced by the need for only minimal treatment facilities for most ground-water developments.

Ground water can also be used in larger communities to augment existing supplies. Use of ground water to provide for peak demand and as a standby or emergency water source is not uncommon in the Tennessee Region, but many more opportunities exist in which ground water could be used along with surface-water supplies to obtain maximum benefit from the water resources. Conjunctive use of ground water and surface water, such as the use of wells to obtain water to maintain a minimum streamflow, has not been given much consideration in the Tennessee Region. This kind of development as well as the use of aquifers for storage of surface water during periods of excess flow require a degree of knowledge of ground-water occurrence and movement which is at present unavailable in most of the region.

The amount of ground water available in the Tennessee Region, if all the recharge to an area were recoverable, would be about 0.5 $(Mgal/d)/mi^2$. The degree to which this quantity is recoverable depends on the hydraulic properties of the aquifers and their areal variability. A narrow strip along the western edge of the region, in the Coastal Plain province, is underlain by unconsolidated sand aquifers from which a large part of the ground-water recharge could be recovered with proper development. Aquifer yields at a given site are more predictable in the Coastal Plain than anywhere else in the region, and wells producing 500 to 1,000 gal/min are possible.

However, only one-tenth of the Tennessee Region lies in the Coastal Plain. The remainder is underlain by carbonate rocks or fracture noncarbonate rocks. The water-bearing properties of these rocks are variable resulting in high exploration costs in developing ground water. Carbonate rocks with little or no regolith, as they occur in the Central Basin and parts of the Valley and Ridge, are the most variable with well yields ranging from less than 1 gal/min to as much as several thousand gal/min within a short distance.

An important factor in aquifer productivity in the region is the occurrence of the regolith. A thick regolith stores ground water and releases it slowly to openings in the underlying rock. Carbonate aquifers with a thick regolith occur in the Highland Rim and parts of the Valley and Ridge. Because of their great areal extent and relatively uniform distribution of ground water, these areas have the greatest potential for ground water development.

Chemical constituents and physical properties of ground water in the Tennessee Region are usually within the limits recommended by the Environmental Protection Agency for drinking water. Water in unconsolidated aquifers and regolith tends to be soft, low in dissolved solids, and slightly acidic. In carbonate rocks the water is usually hard and somewhat alkaline. Water in noncarbonate rocks is generally soft and in some parts of the region, may contain certain undesirable amounts of iron and sulfate. Saline water is not known to occur in significant quantities in the region.

Because nine-tenths of the Tennessee Region is underlain by rocks that have highly variable water-bearing properties, one of the major problems with developing the ground-water resource is locating openings that would supply the quantity of water required. A hydrologic study that includes adequate test drilling and pumping tests would greatly increase the chances of obtaining the quantity of ground water needed. Information from such a study would be useful in developing a concept of the hydrologic systems in which ground water occurs, thus permitting the formulation of criteria for selecting well sites in geologically and hydrologically similar areas. Hydrologic studies, including test drilling, have been undertaken in each section of the region except the Cumberland Plateau. These studies can only be regarded as a beginning but they have already indicated that the ground-water resource in much of the region is significantly larger than had been anticipated. Coverage by detailed reports is lacking in most of the region and the basic data required to make adequate appraisals of the water resources are not always available. Some of the data for hydrologic studies, such as geologic maps, well records, and streamflow records are available throughout the region. However, adequate information on ground-water levels, ground-water quality, and aquifer characteristics is not available in all parts of the region.

Both basic data and interpretation derived from intensive studies are essential tools for managing the ground-water resources and predicting the results of developmental activities. The information needed for management cannot be obtained immediately when it is needed; it must be the product of a continuing program to understand and evaluate the Tennessee Region's water resources. Because ground-water development has been largely neglected in the region, there is opportunity to establish the data base and management capability before stress on the region's water resources increases to the point that management problems become difficult to solve.

At present the management of the region's water resources is unbalanced. Due to the establishment of the Tennessee Valley Authority, surface water is controlled to a high degree regionwide; however, there has been no comparable attempt to manage the ground water systematically. Any plans for fully developing the water resources should be based on hydrologic principles which recognize the interdependence of ground water and surface water and should provide for utilization of both aspects of the regions' water resources to their fullest potential.

REFERENCES

Adams, George I., Butts, Charles, Stephenson, L. W., and Cooke, Wythe, 1926, Geology of Alabama: Alabama Geol. Survey Special Rept. no. 14, 312 p.

Betson, Roger P., and McMaster, William M., 1975, Nonpoint source mineral water quality model: Water Pollution Control Federation Jour., v. 47, no. 10, p. 2461–2473.

Boswell, E. H., Moore, G. K., MacCary, L. M., and others, 1965, Cretaceous aquifers in the Mississippi embayment, U.S. Geol. Survey Prof. Paper 448–C, 37 p.

Burchett, Charles R., 1977, Water resources of the upper Duck River basin, central Tennessee: Tennessee Div. of Water Resources, Water Resources Series no. 12, 103 p.

Burchett, Charles R., and Hollyday, E. F., 1974, Tennessee's newest aquifer: Geol. Soc. America Abs. with Programs, v. 6, no. 4, p. 338.

Burchett, Charles R., and Moore, Gerald K., 1971, Water resources in the upper Stones River basin: Tennessee Div. of Water Resources, Water Resources Series no. 8, 62 p.

Bureau of Water Hygiene, 1971, Evaluation of the Tennessee water supply program summary: U.S. Environmental Protection Agency, 23 p.

Busby, M. W., and Armentrout, G. W., 1965, Kansas streamflow characteristics, part 6A, base flow data: Kansas Water Resources Board, Tech. Rept. no. 6A, 207 p.

Butts, Charles, 1933, Geologic map of the Appalachian Valley of Virginia with explanatory text: Virginia Geol. Survey Bull. 42, 56 p.

Cederstrom, D. J., 1973, Cost analysis of ground-water supplies in the North Atlantic Region, 1970: U.S. Geol. Survey Water-Supply Paper 2034, 48 p.

Cressler, Charles W., 1964, Geology and ground-water resources of Walker County, Georgia: Georgia Geol. Survey Inf. Circ. 29, 15 p.

Cushing, E. M., Boswell, E. H., Speer, P. R., Hosman, R. L., and others, 1970, Availability of water in the Mississippi embayment: U.S. Geol. Survey Prof. Paper 448–A, 13 p.

Davis, R. W., Lambert, T. Wm., and Hansen, Arnold J., Jr., 1973, Subsurface geology and ground-water resources of the Jackson Purchase region, Kentucky: U.S. Geol. Survey Water-Supply Paper 1987, 66 p. [1974].

DeBuchananne, G. D., and Richardson, R. M., 1956, Ground-water resources of East Tennessee: Tennessee Div. Geology Bull. 58, 393 p.

de Laguna, Wallace, 1968, Radioactive waste disposal by hydraulic fracturing [abs.]: Ground Water, v. 6, no. 6, p. 47.

Delury, George E., editor, 1973, The world almanac: New York, Newspaper Enterprise Assoc., 1040 p.

Dodson, Chester L., and Harris, Wiley F., 1963, Geology and ground-water resources of Morgan County, Alabama: Alabama Geol. Survey Bull. 76, 90 p.

Dufor, Charles N., and Becker, Edith, 1964, Public water supplies of the 100 largest cities in the United States, 1962: U.S. Geol. Survey Water-Supply Paper 1812, 364 p.

Environmental Protection Agency, 1975, Interim primary drinking water standards: Federal Register, v. 40, no. 51, Part II, March 14, 1975.

Feth, J. H., 1965, Preliminary map of the conterminous United States showing depth to and quality of shallowest ground water containing more than 1,000 parts per million dissolved solids: U.S. Geol. Survey Hydrol. Inv. Atlas HA–199.

Geological Survey of Alabama, 1975, Environmental geology and hydrology Huntsville and Madison County, Alabama: Alabama Geol. Survey Atlas Ser. 8, 118 p.

Grantham, Rodney G., and Stokes, William R., 1976, Ground-water quality data for Georgia: U.S. Geol. Survey open-file report, 216 p.

Harris, H. B., Moore, G. K., and West, L. R., 1963, Geology and ground-water resources of Colbert Co., Alabama: Alabama Geol. Survey County Rept. 10, 71 p.

Harris, Wiley F., Jr., and McMaster, W. M., 1965, Geology and ground-water resources of Lawrence County, Alabama: Alabama Geol. Survey Bull. 78, 70 p.

Hunt, Charles B., 1967, Physiography of the United States: San Francisco and London, W. H. Freeman and Company, 480 p.

Joiner, Thomas J., and Scarbrough, Leon W., 1969, Hydrology of limestone terranes, geophysical investigations: Alabama Geol. Survey Bull. 94, Part D, 43 p.

LeForge, Lawrence, Cooke, Wythe, Keith, Arthur, and Campbell, Marius R., 1925, Physical geography of Georgia: Georgia Geol. Survey Bull. 42, 189 p.

Lamoreaux, P. E., and Powell, W. J., 1963, Stratigraphic and structural guides to the development of water wells and well fields in a limestone terrane: Alabama Geol. Survey Rept. Ser. 6, 13 p.

LeGrand, H. E., 1967, Ground water of the Piedmont and Blue Ridge provinces in the Southeastern States: U.S. Geol. Survey Circ. 538, 11 p.

Lehr, Jay H., 1976, Let's promote individual and cluster wells in rural water systems: Water Well Jour., March 1976.

McMaster, W. M., 1963, Geology and ground-water resources of the Athens area, Alabama: Alabama Geol. Survey Bull. 71, 45 p.

McMaster, W. M., and Harris, W. F., Jr., 1963, General geology and ground-water resources of Limestone County, Alabama, a reconnaissance: Alabama Geol. Survey County Rept. 11, 43 p.

McMaster, W. M., and Hubbard, E. F., 1970, Water Resources of the Great Smoky Mountains National Park, Tennessee and North Carolina: U.S. Geol. Survey Hydrol. Inv. Atlas HA–420.

Malmberg, G. T., and Downing, H. T., 1957, Geology and ground-water resources of Madison County, Alabama: Alabama Geol. Survey County Rept. 3, 225 p.

Marcher, M. V., Bingham, R. H., and Lounsbury, R. E., 1964, Ground-water geology of the Dickson, Lawrenceburg, and Waverly areas in the western Highland Rim, Tennessee: U.S. Geol. Survey Water-Supply Paper 1764, 50 p.

Meinzer, O. E., 1923, Occurrence of ground water in the United States, with a discussion of principles: U.S. Geol. Survey Water-Supply Paper 489, 321 p.

Meinzer, O. E., and Stearns, N. D., 1929, A study of ground water in the Pomperaug Basin, Connecticut, with special reference to intake and discharge: U.S. Geol. Survey Water-Supply Paper 597–B, p. 73–146.

Miller, Robert A., 1974, The geologic history of Tennessee: Tennessee Div. Geol. Bull. 74, 63 p.

Moore, G. K., 1973, Hydraulics of sheetlike solution cavities: Ground Water, v. 11, no. 4, July-August 1973.

Moore, Gerald K., Burchett, Charles R., and Bingham, Roy H., 1969, Limestone hydrology in the upper Stones River basin, central Tennessee: Tennessee Div. Water Resources, 58 p.

Moore, Gerald K., and Wilson, John M., 1972, Water resources of the Center Hill Lake region, Tennessee: Tennessee Div. Water Resources, Water Resources Series no. 9, 77 p.

Murray, C. Richard, and Reeves, E. Bodette, 1972, Estimated use of water in the United States in 1970: U.S. Geol. Survey Circ. 676.

National Academy of Science—National Academy of Engineering, 1973, Water quality criteria 1972: U.S. Environmental Protection Agency Ecol. Research Ser., EPA–R3–033, 594 p.

Newcome, Roy, Jr., and Callahan, J. A., 1964, Water for industry in the Corinth area, Mississippi: Mississippi Board of Water Commissioners Bull. 64–2, 24 p.

Newcome, Roy, Jr., and Smith, Ollie, Jr., 1958, Ground-water resources of the Cumberland Plateau in Tennessee: Tennessee Div. Water Resources, 72 p.

———1962, Geology and ground-water resources of the Knox Dolomite in middle Tennessee: Tennessee Div. Water Resources, Water Resources Series no. 4, 43 p.

Newton, J. G., Copeland, C. W., and Scarbrough, L. W., 1973, Sinkhole problem along proposed route of interstate highway 459 near Greenwood, Alabama: Alabama Geol. Survey Circ. 83, 63 p.

Peace, Richard R., Jr., 1964, Geology and ground-water resources of the Russelville area, Alabama: Alabama Geol. Survey Bull. 77, 83 p.

Peace, Richard R., Jr., and Link, Donald R., 1971, Geology and ground-water resources of northwestern North Carolina: North Carolina Dept. of Water and Air Resources, Div. of Ground Water, Ground-Water Bull. 19, 135 p.

Rasmussen, W. C., and Andreasen, G. E., 1957, A hydrologic budget of the Beaverdam Creek basin, Maryland: U.S. Geol. Survey open-file report, 211 p.

Rima, D. R., Moran, Mary S., and Woods, Jean E., 1978, Ground-water supplies in the Murfreesboro area, Tennessee: U.S. Geol. Survey Water Resources Inv. 77–86, 73 p.

Sanford, T. H., Jr., 1964, Ground-water conditions in the Huntsville area, Alabama, Jan. 1960 thru June 1961: Alabama Geol. Survey Circ. 24, 46 p.

———1966, Ground water in Marshall County, Alabama, a reconnaissance: Alabama Geol. Survey Bull. 85, 66 p.

Searcy, James K., 1959, Flow-duration curves: U.S. Geol. Survey Water-Supply Paper 1542–A, 33 p.

Sun, P-C. P., Criner, J. H., and Poole, J. L., 1963, Large springs of East Tennessee: U.S. Geol. Survey Water-Supply Paper 1755, 52 p.

Swingle, George D., 1959, Geology, mineral resources, and ground water of the Cleveland area, Tennessee: Tennessee Div. Geol. Bull. 61, 125 p.

Tennessee Dept. of Public Health, Division of Water Quality Control, 1975, Water quality management plan for the Lower Tennessee River basin, 220 p.

Tennessee Valley Authority, 1973, Division of Water Control Planning, annual report for fiscal year 1973, 49 p.

———1975, Precipitation in Tennessee River basin, annual 1975: Tennessee Valley Auth., Div. of Water Management, Data Services Branch, Rept. no. O–243–A75.

Wells, Francis G., 1933, Ground-water resources of western Tennessee: U.S. Geol. Survey Water-Supply Paper 656, 319 p.

Wilson, John M., 1965, Ground-water resources and geology of Cumberland County, Tennessee: Tennessee Div. Water Resources, 56 p.

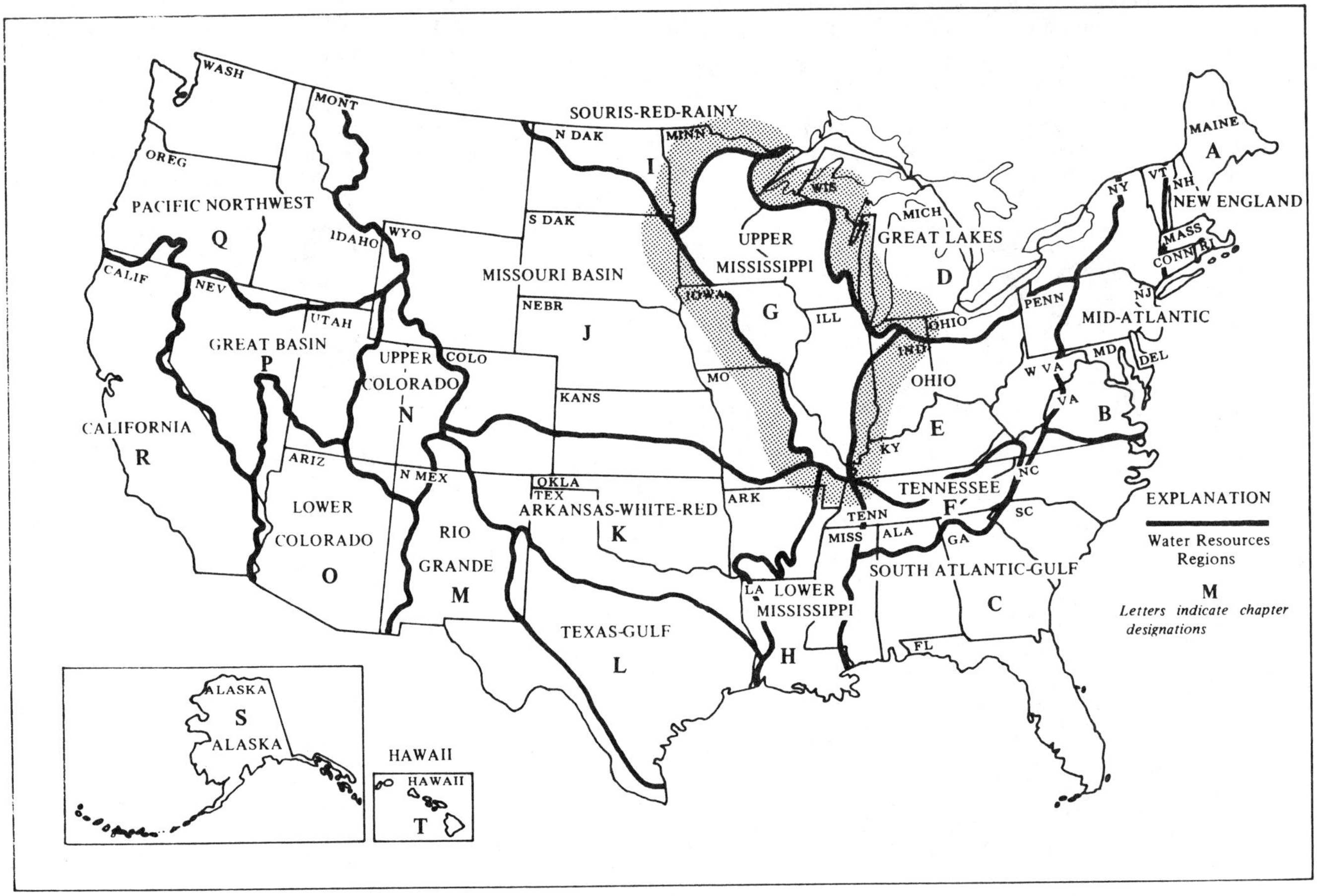

G – Location Map for the Upper Mississippi Region

G

UPPER MISSISSIPPI REGION

By R. M. BLOYD, JR.

ABSTRACT

The Upper Mississippi Region in general is rich in water—surface water is plentiful, and ground water is a large, important, and manageable resource. Total potable water in storage in the outwash and alluvial aquifers of the Mississippi River valley and the subbasins is about 45,000 billion gallons. This is about 10 percent of the water in storage in Lake Ontario. Water in storage in other aquifers of the region is probably at least several times that in the outwash and alluvial aquifers. Estimated ground-water recharge in the subbasins is 23,000 million gallons per day.

A comparison of ground-water withdrawals with estimated ground-water recharge suggests that the large ground-water resource of the region is not being fully utilized. Ground-water use by domestic, commercial, and rural interests is only 4 percent of recharge. Ground-water use (1965) by industry is only 3 percent of recharge.

Water in the outwash and alluvial aquifers of much of the valley of the Mississippi River in and south of St. Paul, Minn. of the Illinois River, the Lower Minnesota River, the Wisconsin River, the Lower Black River, the Wapsipinicon River, the Lower Rock River, and the Upper Des Moines River can be considered a regional resource. In these areas the ground-water resources are of sufficient magnitude to satisfy more than just local needs. For example, under certain specified conditions, ground water in the above areas can supply approximately 20 million additional people. Factors other than water supply, of course, will be constraints on development in the region.

Advances in techniques in ground-water hydrology during recent years have provided methods that the hydrologist and planner can use for planning and design of ground-water developments. Therefore, the planner can now resolve some of the development and management questions that historically have bred uncertainty when this part of the water resource was considered for development.

INTRODUCTION

This report emphasizes the role of ground water in water-resource planning for a region of expanding development and changing economic environment. The boundaries of the study area are the surface-water drainage boundaries of the Mississippi River basin above the mouth of the Ohio River, exclusive of the Missouri River drainage. For purposes of analysis, the area is divided into 16 subbasins (fig. 1). The subbasin boundaries coincide with the boundaries established in the Upper Mississippi River comprehensive basin study (U.S. Army Corps of Engineers, 1970a, Appendices A–P.)

According to projections in the Upper Mississippi River comprehensive basin study, increasing urbanization, with an attendant decline in agricultural employment, will characterize the future economy of the region. Such a development pattern is not unique to this region—the relative importance of farm-labor employment in crop production has declined and continues to decline in much of the United States.

Increasing urbanization will result in a greater population being served by municipal water-supply systems and waste-treatment facilities. Total regional municipal and industrial water requirements are expected to increase 4.7 times from 1960 to 2020 (table 1). The most rapidly growing economic subregion, historically and projected, is the Milwaukee subregion, Wis. (U.S. Army Corps of Engineers, 1970c, Appendix P, p. P–10), which includes the Madison and Milwaukee areas (U.S. Army Corps of Engineers, 1970, Appendix P, p. P–74). Other rapidly growing economic subregions are Minneapolis–St. Paul, Chicago, Davenport–Rock Island–Moline, Peoria, Des Moines–Fort Dodge, and St. Louis.

Fortunately, the region is rich in water, so that the increasing water demands are not cause for alarm. These rising demands do, however, signal the need for planned and orderly development of the region's water. Currently, water supplies are sufficient for municipal and industrial water needs throughout the region, with few exceptions (U.S. Dept. of the Interior, FWQA, 1970, Appendix H, p. H–31). With increasing water demands, analysis of supply and demand data suggests 26 water-supply problem areas in the region (U.S. Dept. of the Interior, FWQA, 1970, Appendix H, p. H–321–H–331) by the year 2020 (table 2). Water-supply problem areas are those where evaluation of local water resources indicates that the resource is insufficient in relation to anticipated water demands. Water quality, too, is a consideration in the identification of problem areas.

Because the region is rich in water, programs of action, ideally, would utilize that richness in the most effective manner. In order to determine what those

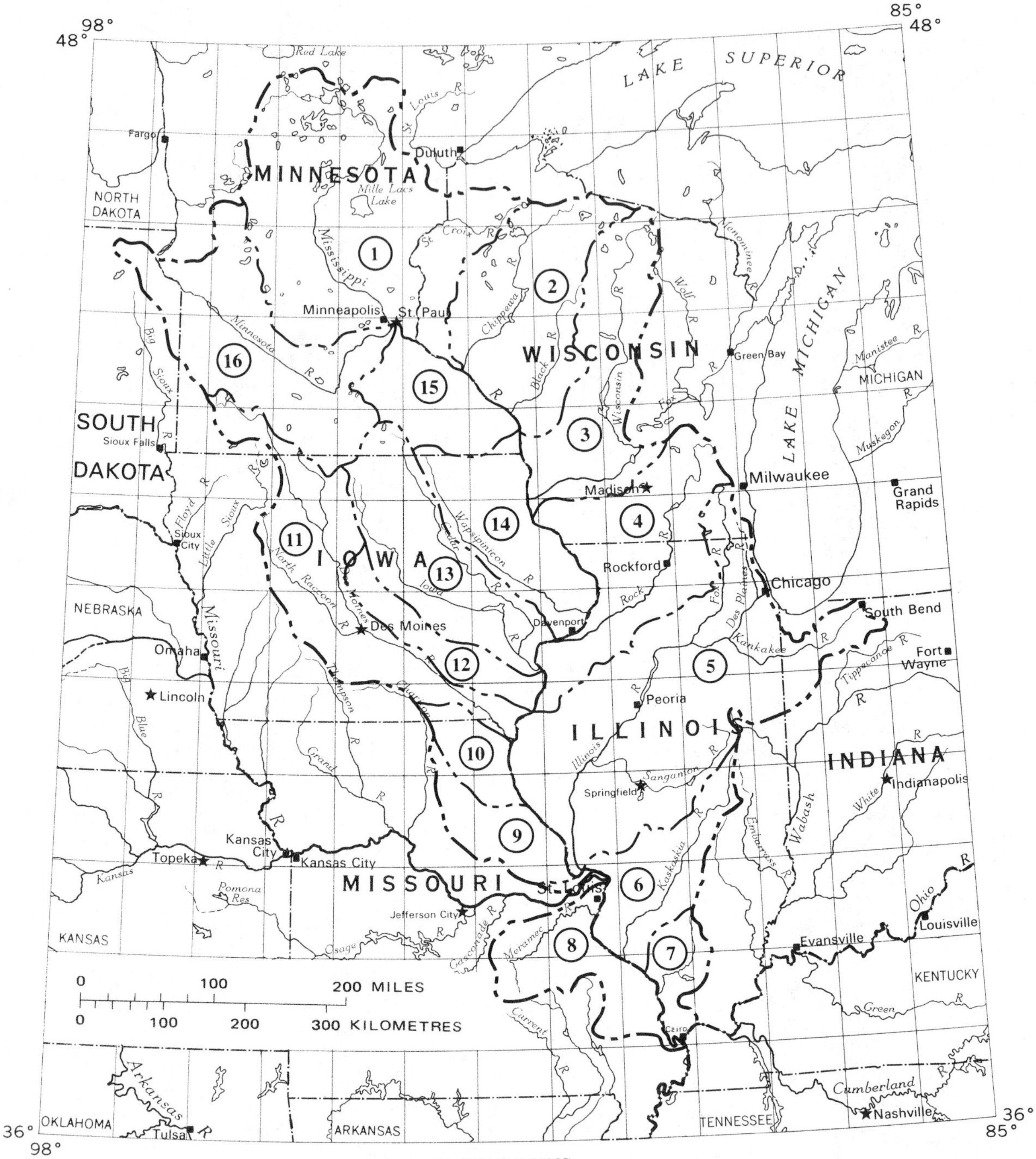

SUBBASINS

1. Mississippi headwaters
2. Chippewa-Black
3. Wisconsin
4. Rock
5. Illinois
6. Kaskaskia
7. Big Muddy
8. Meramec
9. Salt
10. Fox-Wyaconda-Fabius
11. Des Moines
12. Skunk
13. Iowa-Cedar
14. Turkey-Maquoketa-Wapsipinicon-Upper Iowa
15. Cannon-Zumbro-Root
16. Minnesota

FIGURE 1.—Subbasins in the Upper Mississippi Region described in this report.

TABLE 1.—*Year 1960 and projected municipal and industrial water use—Upper Mississippi Region*

[Table after U.S. Dept. of the Interior, FWQA, 1970, Appendix H, Table H-9. Units in million gallons per day]

Subbasin	Water use 1960	1980	2000	2020
Mississippi headwaters	291	501	717	1,054
Chippewa-Black	51	95	151	220
Wisconsin	196	340	582	946
Rock	236	432	736	1,208
Illinois	2,850	4,727	7,887	12,297
Kaskaskia	235	377	758	1,318
Big Muddy	19	28	41	51
Meramec	260	577	991	1,959
Salt	12	24	46	89
Fox-Wyaconda-Fabius	11	19	33	53
Des Moines	133	234	405	657
Skunk	25	48	85	143
Iowa-Cedar	152	273	498	800
Turkey-Maquoketa-Upper Iowa-Wapsipinicon	121	217	372	622
Cannon-Zumbro-Root	45	91	174	327
Minnesota	50	99	180	328
Total	4,687	8,082	13,656	22,072

TABLE 2.—*Potential water-supply problem areas*

[Table after U.S. Dept. of the Interior, FWQA, 1970, Appendix H]

Subbasin	Areas
Mississippi headwaters	Minneapolis-St. Paul metropolitan area.
Chippewa-Black	None.
Wisconsin	Marshfield, Wis.
Rock	None.
Illinois	Chicago suburbs, Decatur and Springfield, Ill.
Kaskaskia	None.
Big Muddy	None.
Meramec	None.
Salt	Mexico, Paris, Vandalia, Centralia, and Troy, Mo.
Fox-Wyaconda-Fabius	Kahoka and Palmyra, Mo.
Des Moines	Worthington, Minn.
Skunk	Ames, Iowa.
Iowa-Cedar	Austin and Albert Lea, Minnesota; Forest City, Mason City, Waterloo, and Cedar Rapids, Iowa.
Turkey-Maquoketa-Upper Iowa-Wapsipinicon	None.
Cannon-Zumbro-Root	Rochester, Minn.
Minnesota	Marshall, New Ulm, Mankato, and Fairmont, Minn.

programs would be, knowledge of the effects of man's activities on the hydrologic system in the region is necessary. Regardless of the plan of action chosen, development will alter the pattern of movement and discharge of water.

In this report, figures for measures are given in English units. The following table contains the factors for converting these English units to metric units:

English unit	*Symbol*	*Equivalent (multiply by)*	*Metric unit*	*Symbol*
Inches	(in)	25.4	Milimetres	(mm)
Feet	(ft)	0.3048	Metres	(m)
Miles	(mi)	1.609	Kilometres	(km)
Acres	----	0.4747	Hectares	(ha)
Square miles	(mi^2)	2.590	Square kilometres	(km^2)
U.S. gallons	(gal)	3.785	Litres	(l)
Acre-feet	(acre-ft)	1,233	Cubic metres	(m^3)
Cubic feet	(ft^3)	0.02832	Cubic metres	(m^3)
U.S. gallons per minute	(gal/min)	0.06309	Litres per second	(l/s)
Cubic feet per second	(ft^3/s)	0.02832	Cubic metres per second	(m^3/s)
Million gallons per day	(Mgal/d)	3,785	Cubic metres per day	(m^3/d)
Cubic feet per second per square mile	($ft^3/s/m^2$)	0.0109	Cubic metres per second per square kilometre	($m^3/s/km^2$)

Sophisticated tools are available to assist in the evaluation of alternative plans. The following section of the report considers some of the technology available and the data requirements for applying the technology.

INTEGRATING GROUND WATER INTO WATER-RESOURCE PLANNING

Advances in techniques in ground-water hydrology during recent years have provided methods that the hydrologist can use for planning and design of ground-water developments. Therefore, the manager can now resolve some of the development and management questions that historically have bred uncertainty when this part of the water resource was considered for development.

TECHNOLOGY AVAILABLE FOR PLANNING

An impressive array of tools is available for planning development of ground water and efficient conjunctive use of ground water and surface water. Models, in particular, have achieved greater utility as analytical tools in hydrology since computers have become available. A mathematical model of a ground-water basin or aquifer is designed to describe in mathematical language how the basin or aquifer functions under various conditions of development.

An advantage that a mathematical model has over a verbal description of a system is that the model describes the system in more concise, quantitative terms. Such a description facilitates considering a problem in its entirety and considering all interrelations simultaneously. For example, a mathematical model of a ground-water basin facilitates analysis of the mutual influence of the climatic, geologic, hydraulic, and man-made conditions that affect the basin.

A mathematical model of part of the multiple aquifer system in northeastern Illinois (Prickett and Lonnquist, 1971) is an example of a successful mathematical model in the Upper Mississippi Region. The purpose of the model was to improve on previous methods for predicting water-level declines in response to projected ground-water development. Also, Weeks and Stangland (1971) used a mathematical model to show the effects of irrigation on streamflow in a part of central Wisconsin.

Computer programs have been written that incorporate certain water-quality parameters into digital models. Though still in the early stages of development, these programs make it possible to describe changes in water quality that result from changes in water-management practices or to predict the fate of liquid contaminants released into the environment. Thus, a pilot project now underway in the Arkansas River valley of Colorado will attempt to describe and predict on a monthly basis changes in salinity in the alluvial aquifer and in the adjacent stream.

In the Upper Mississippi Region, excellent opportunities exist for the application of ground-water-quality modeling techniques in Minnesota and Wisconsin. For example in Minnesota, "the land of 10,000 lakes," real-estate development has resulted in many cases of lake-water degradation. Septic tanks, or other waste-disposal facilities, release nutrients to ground water in the vicinity of a lake. The nutrients migrate to the lake and degrade the quality of lake water. In many places little is known about the ground-water flow pattern near lakes or of the diffusion pattern and rate of movement of nutrients in the ground water. In order to set meaningful standards for waste-disposal facilities near lakes, such knowledge is essential. Model studies offer one method of gaining such an understanding.

The results of a model analysis are no more reliable than the data that are used to construct the model. The technical capability and experience of the hydrologist and his staff also markedly influence the accuracy of model calibration and of prediction analysis. Most model studies benefit from a team approach. The team may consist of a geologist, hydrologist, and geochemist, or other professionals. Some studies may require assistance of specialists, such as lawyers, economists, ecologists, and soil scientists.

Modeling is but one of the techniques that compose the systems engineering method. Hydrologists have successfully applied the systems method such that the literature is replete with useful applications of systems analysis to regional water-resource studies. The application of systems engineering to solving water-resource problems was in part pioneered by the Harvard Water Program staff (Maass and others, 1962). Although the Harvard group did not deal explicitly with ground water in their experimental systems, they stated that methods could be elaborated to account for ground water.

In summary, tools have been developed and are available for use in planning the development of ground-water supplies, analyzing alternative conjunctive uses of ground water and surface water, and for evaluating management objectives within specified physical, economic, and social constraints. With the aid of these tools, ground water can be integrated into water-resource planning with a high degree of assurance and effectiveness.

DATA REQUIREMENTS FOR PLANNING

The data requirements for a planning study of a ground-water system depend upon the size and hydrologic complexity of the area and upon the type of water problems. Some planning studies need intensive investigation and large-scale model analyses, whereas others require only descriptive evaluation of the hydrologic data.

In determining the types of ground-water information to be presented in this report, it was assumed that water-resource planners in the Upper Mississippi Region will use systems engineering for establishing and implementing an optimal water-resource development scheme. What types of information are necessary for the application of the various methods that compose the systems method, including modeling and simulation?

The first type of information needed is that which helps the planner determine if the ground-water resource in the Upper Mississippi Region is available where needed or if it can be transported in sufficient quantity and acceptable quality at the right time. Regarding water quality, a study by the U.S. Department of the Interior, Geological Survey (1970a) summarizes ground-water-quality characteristics in the Upper Mississippi Region. A more detailed analysis of water quality is beyond the scope of this report.

The second type of information needed is estimates of the total resource available for allocation and of the individual demands among competing activities. Such estimates, along with cost estimates associated with allocation of the resource, are pertinent to the application of optimization techniques, such as linear programming. In the typical linear programming approach, available resources are allocated in an optimal manner among various competing activities.

This report provides an estimate of the total ground-water resource available. Initial estimates of water demand are available in the Upper Mississippi River comprehensive basin study (U.S. Army Corps of Engineers, 1970a, Appendices A–P). This report includes no estimates of costs for developing and delivering ground water because costs vary widely from place to place and from use to use. However, when specific requirements and a time frame are specified, cost estimates can be developed for a given locality.

The third type of information needed is estimates of storage coefficients, transmissivities, aquifer boundaries, the degree of connection between streams and aquifers, and other relevant hydrologic parameters. Such information is basic to the application of system modeling and simulation.

Many pertinent data are already available; these are summarized in the following sections of this report.

ASSESSING THE GROUND-WATER RESOURCE

GEOLOGIC FRAMEWORK

Alluvium and outwash deposits (fig. 2) constitute the most productive part of the ground-water system in the Upper Mississippi Region. Streams draining the glaciated part of the region have redeposited size-sorted glacial sediments beyond the southernmost encroachment of the glacial ice during the Ice Age and thus have helped to create permeable aquifers throughout the region. Subsequent references to the valley of the Mississippi River refer to the area adjacent to the main stem of the river that is underlain by outwash and alluvial deposits (fig. 2).

Holocene-age alluvium consisting of silt, sand, and gravel is present in the valley of the Mississippi River and adjacent to many of the larger streams. These deposits generally are finer grained and are less permeable than the outwash deposits. However, locally, they are excellent sources of water.

Outwash, which is composed predominantly of sand and gravel deposited by melt-water streams, is generally highly permeable and is an excellent source of water. Outwash deposits are most prevalent in Minnesota and Wisconsin, where they are generally in hydrologic connection with and not differentiated from the alluvial deposits.

Another highly productive part of the ground-water system is the buried sand and gravel deposits in the Mahomet bedrock valley of east-central Illinois (Visocky and Schicht, 1969). The buried valley and its major tributaries, though not shown in figure 2, underlie about 3,700 square miles in Illinois. Other buried valleys are present in various parts of the region.

Till is the principal glacial deposit in the region. It is a heterogeneous mixture of clay, silt, sand, and gravel; generally it has low transmissivity and, therefore, is generally a poor source of water. Moraines of unsorted till form low ridges along former fronts of glacial ice sheets. Glacial deposits are present in most of the region except for a part of Wisconsin and most of Missouri.

The Upper Mississippi Region is underlain by a series of bedrock units (fig. 3) that vary greatly in thickness and in hydrologic characteristics (U.S. Dept. of the Interior, Geological Survey, 1970b, p. 74–75). Depth to the bottom of the bedrock system, or basement, is greatest near the center of the Illinois basin (fig. 3). The basement crops out in the region in two centers of uplift—the Superior Upland and the Ozark Dome (fig. 3). Around these centers of uplift, rocks of younger age crop out in successive belts.

The slope on the basement complex is the central feature controlling the slope or dip of the younger bedrock units overlying the basement. The basement gently slopes toward the regional geosyncline of the Great Plains, west of the region, and slopes less gently toward the Michigan basin, the center of which is in the State of Michigan, and slopes toward the Illinois basin.

Three major bedrock aquifers are present in the Upper Mississippi Region. They are the Mount Simon-Hinckley of Precambrian and Cambrian age, the Cambrian-Ordovician, and the Silurian-Devonian aquifers. Locally, other names may be applied to these aquifers.

The Mount Simon-Hinckley aquifer is mostly sandstone. It is a source of water in southeastern Minnesota, western and southern Wisconsin, northern Illinois, and easternmost Iowa.

The Cambrian-Ordovician aquifer consists of several sandstone and dolomite formations which act as a hydraulic unit. The aquifer is a source of water in southeastern Minnesota, southern Wisconsin, northern Illinois, Iowa, northwestern Indiana, and much of eastern Missouri.

The Silurian-Devonian aquifer, which consists of several limestone and dolomite formations, supplies water to northeastern Iowa, northern Illinois, southeastern Wisconsin, and northwestern Indiana.

PRECIPITATION—THE SOURCE

Gross water input to the Upper Mississippi Region consists of the total regional precipitation. Ground-water inflow, if present, is insignificant relative to the total regional precipitation.

On the basis of average annual precipitation (fig. 4), estimated average precipitation rates in the various subbasins range from 5,000 M gal/d (million gallons per day) in the Fox-Wyaconda-Fabius subbasin to 47,900 M gal/day in the Illinois subbasin (table 3).

EVAPOTRANSPIRATION—A MAJOR USE

Evapotranspiration is the largest single component of natural fresh-water discharge from the hydrologic system. Average annual potential evapotranspiration, or the maximum rate at which water can be taken into the atmosphere, ranges from about 22 inches to about 32 inches (fig. 5).

The type of vegetation and the depth of vegetal root systems, in part, determine the depth limit of evapotranspiration. The lower limit of depth from which

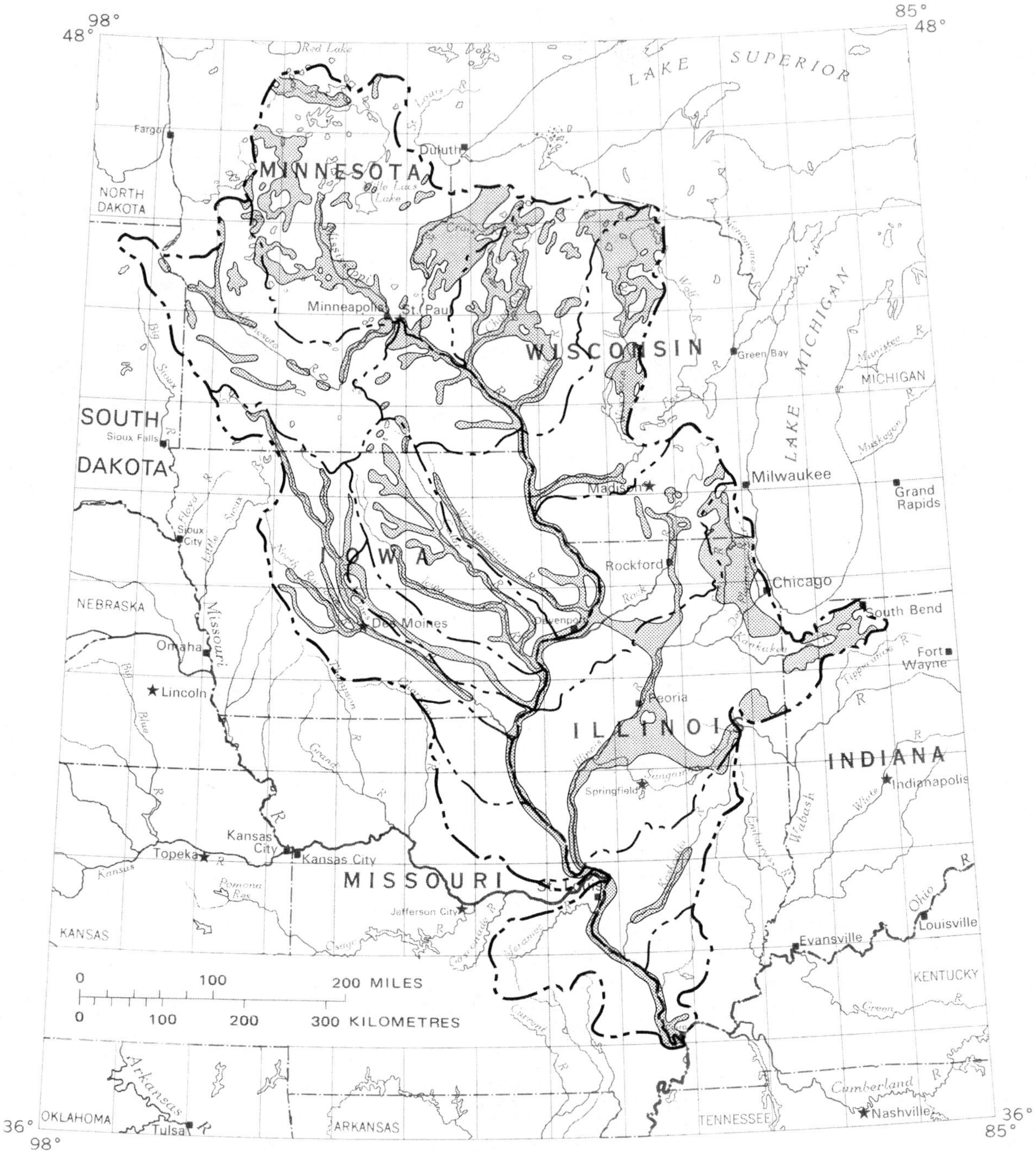

EXPLANATION

Outwash and alluvial deposits

FIGURE 2.—Distribution of outwash and alluvial deposits.

evapotranspiration occurs is not accurately known, but it is generally not more than 15 feet.

Evapotranspiration is probably at a maximum from alluvial aquifers adjacent to gaining reaches of many streams where water is readily available to streamside vegetation. Ground water also evaporates and is transpired in wetlands and in the immediate vicinity of natural lakes fed by ground water.

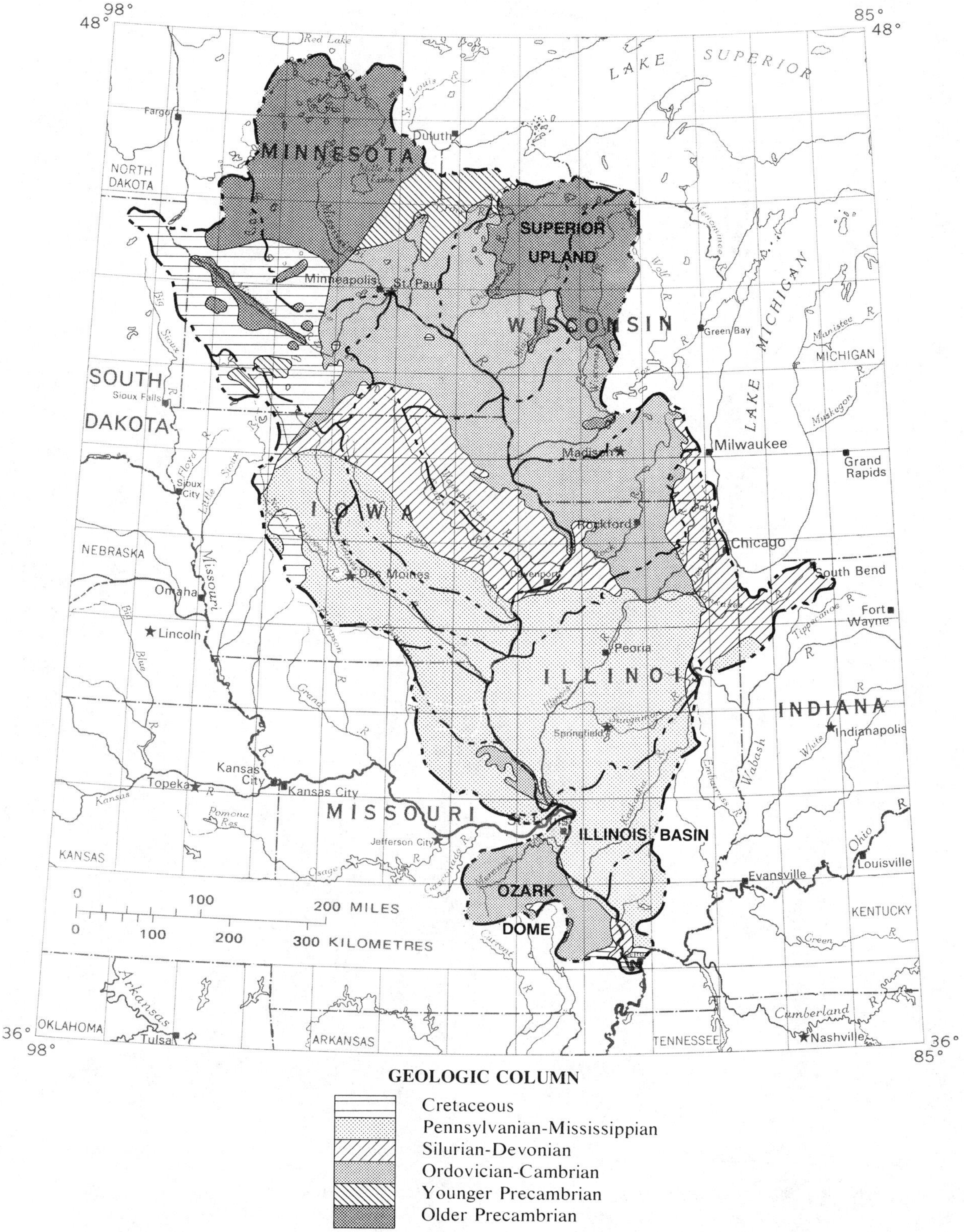

FIGURE 3.—Major geologic subdivisions of the bedrock in the Upper Mississippi Region. Adapted from U.S. Department of the Interior, Geological Survey (1970b).

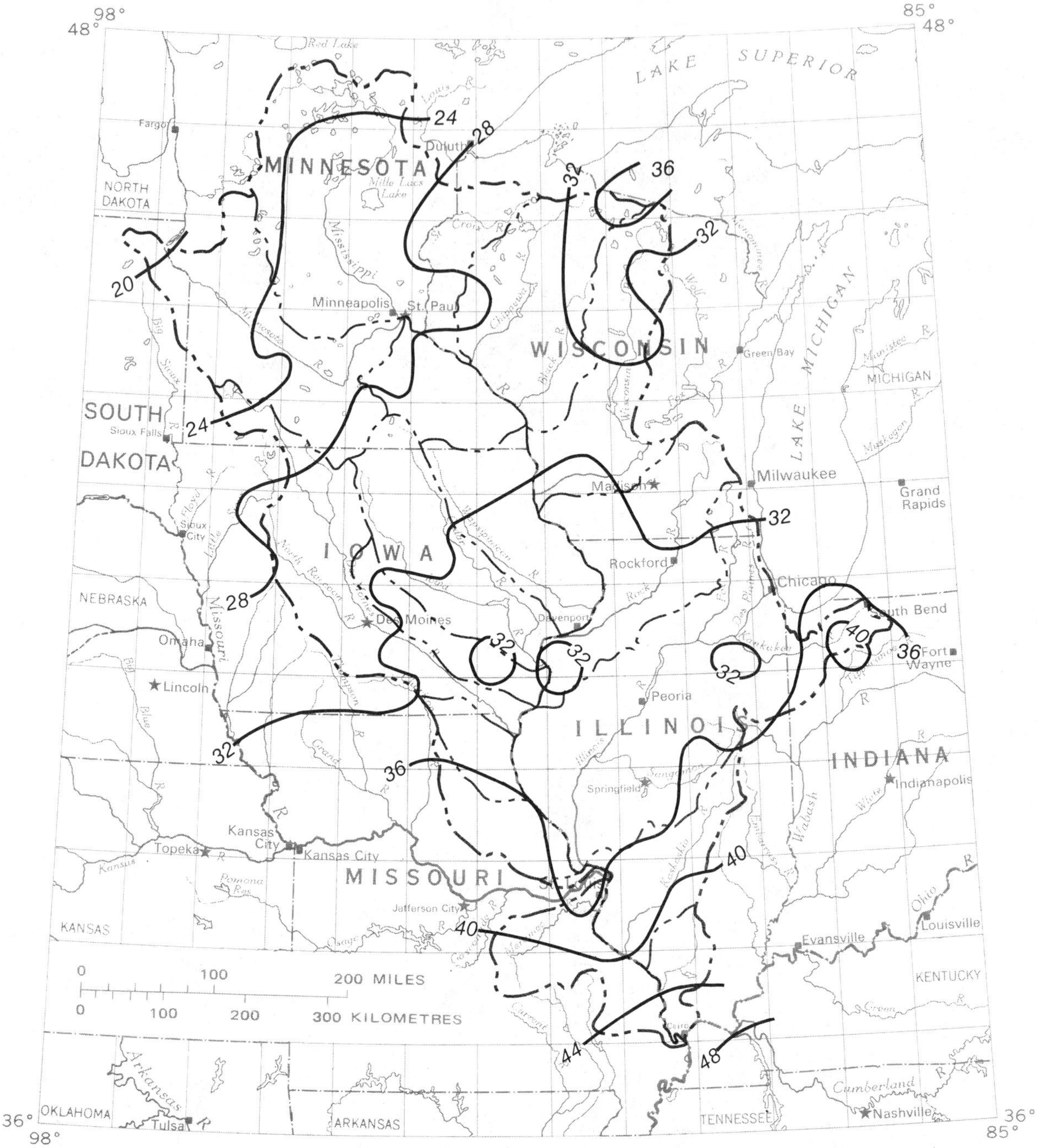

FIGURE 4.—Average annual precipitation, in inches. (Source is U.S. Dept. of Commerce, 1968.)

THE SUPPLY

Not all ground water in storage is available for development, but the following estimates of storage show the magnitude of water which, under specified conditions, can be considered available.

POTABLE WATER

For purposes of assessing the ground water in storage, potable water is defined as water containing less than 1,000 mg/l (milligrams per litre) dissolved solids. The definition is arbitrary because some water with less

TABLE 3.—*Estimated average annual subbasin precipitation*

Subbasin	Estimated annual precipitation (in/yr)[1]	Subbasin area (mi^2)	Estimated precipitation (M gal/d)
Mississippi headwaters	26.0	28,100	34,800
Chippewa-Black	30.0	13,100	18,700
Wisconsin	32.0	12,800	19,500
Rock	33.0	14,500	22,800
Illinois	34.0	29,600	47,900
Kaskaskia	38.0	7,000	12,700
Big Muddy	42.0	2,800	5,600
Meramec	40.0	7,000	13,300
Salt	37.0	4,500	7,900
Fox-Wyaconda-Fabius	35.0	3,000	5,000
Des Moines	31.0	14,800	21,800
Skunk	32.0	4,600	7,000
Iowa-Cedar	32.0	12,800	19,500
Turkey-Maquoketa-Upper Iowa-Wapsipinicon	32.0	8,800	13,400
Cannon-Zumbro-Root	29.0	5,700	7,900
Minnesota	24.0	16,900	19,300

[1]in/yr = inches per year.

than 1,000 mg/l dissolved solids is not potable. On the other hand, some municipal water-supply systems in the southwestern United States deliver water having dissolved solids exceeding 2,000 mg/l. In this report, all water is considered a resource because most waters can be treated to remove undesirable constituents. Whether a given supply is potable or is usable only for other purposes is a planning or management decision.

Three categories of storage were specified for the purpose of computing the amount of potable ground water in storage in the region: (1) storage in the outwash and alluvial aquifer in the valley of the Mississippi River, (2) storage in the outwash and alluvial aquifers in the various subbasins, and (3) storage in all other aquifers.

The areal variance in available basic data on saturated aquifer thickness and aquifer characteristics make simplifying assumptions necessary in computing ground water in storage. For example, Illinois has excellent descriptive records of its water resources (Illinois Technical Advisory Committee on Water Resources, 1967, p. 9). Conversely, only sparse hydrogeologic data are available for the outwash and alluvial aquifer in the valley of the Mississippi River.

For calculating water in storage in the outwash and alluvial aquifer in the valley of the Mississippi River, the following assumptions were made: (1) all stored water is potable; (2) the aquifer is unconfined and has an average specific yield of 0.15; (3) above St. Paul, Minn., the average aquifer width is one-eighth mile, or 660 feet, the average saturated aquifer thickness is 30 feet, and the aquifer length is 500 miles; (4) below St. Paul the average aquifer width is 1 mile, the average saturated aquifer thickness is 75 feet, and the aquifer length is 850 miles; and (5) the product of the surface area of the aquifer, the saturated aquifer thickness, and the aquifer specific yield is the approximate volume of water in storage in the aquifer.

The assumptions are consistent with available data and probably yield conservative results. For example, the total thickness of outwash and alluvial deposits along the Mississippi River in Iowa is from 100 to 160 feet at most places (Steinhilber and Horick, 1970, p. 35). Also, the American Bottoms of the Mississippi River at East St. Louis, Ill., an area 30 miles long and as much as 11 miles wide consisting of outwash and alluvial deposits, averages 120 feet in thickness (Bergstrom and others, 1968, p. 160–161).

On the basis of the above assumptions, approximately 2,050 billion gallons of potable ground water is available from storage in the outwash and alluvial aquifer in the valley of the Mississippi River.

For calculating water in storage in the outwash and alluvial aquifers in the subbasins, the following assumptions were made: (1) all stored water is potable, (2) the aquifers are unconfined, and (3) the specific yield of all aquifers is 0.15. There are some locales where the assumptions may not hold, but, in general, available geohydrologic data suggest that they are reasonable.

The surface areas and saturated thicknesses of the outwash and alluvial aquifers were estimated from available geologic maps and geohydrologic data. The assumed saturated thicknesses are probably conservative estimates. Steinhilber and Horick (1970, p. 35) state that the thickness of alluvial deposits along the principal interior streams in Iowa ranges from 30 to 70 feet. In this report a saturated thickness of 30 feet was used for all outwash and alluvial aquifers in the State. Holt, Young, and Cartwright (1964, p. 193) state that the Wisconsin, Chippewa, and Rock River valleys contain as much as 200 feet of sand and gravel. In this report a saturated thickness of 50 feet was used for all outwash and alluvial aquifers in these subbasins. The assumed saturated thicknesses of the aquifers in the Rock, Illinois, Kaskaskia, and Big Muddy subbasins are probably conservative also (Illinois Technical Advisory Committee on Water Resources, 1967, figs. 29 and 30, p. 69).

On the basis of the above assumptions, approximately 43,000 billion gallons of potable water is in storage in the outwash and alluvial aquifers in the subbasins (table 4). The amount ranges from 300 billion gallons in the Big Muddy subbasin to 12,600 billion gallons in the Illinois subbasin.

Estimated total water in storage in the outwash and alluvial aquifers in the valley of the Mississippi River and the subbasins is 45,000 billion gallons. This volume is about twice that in storage in the outwash and alluvial aquifers of the Ohio River region (Bloyd, 1972, p. 46) or about 10 percent of the volume of water in storage in Lake Ontario. (See Bue, 1963, p. 11.)

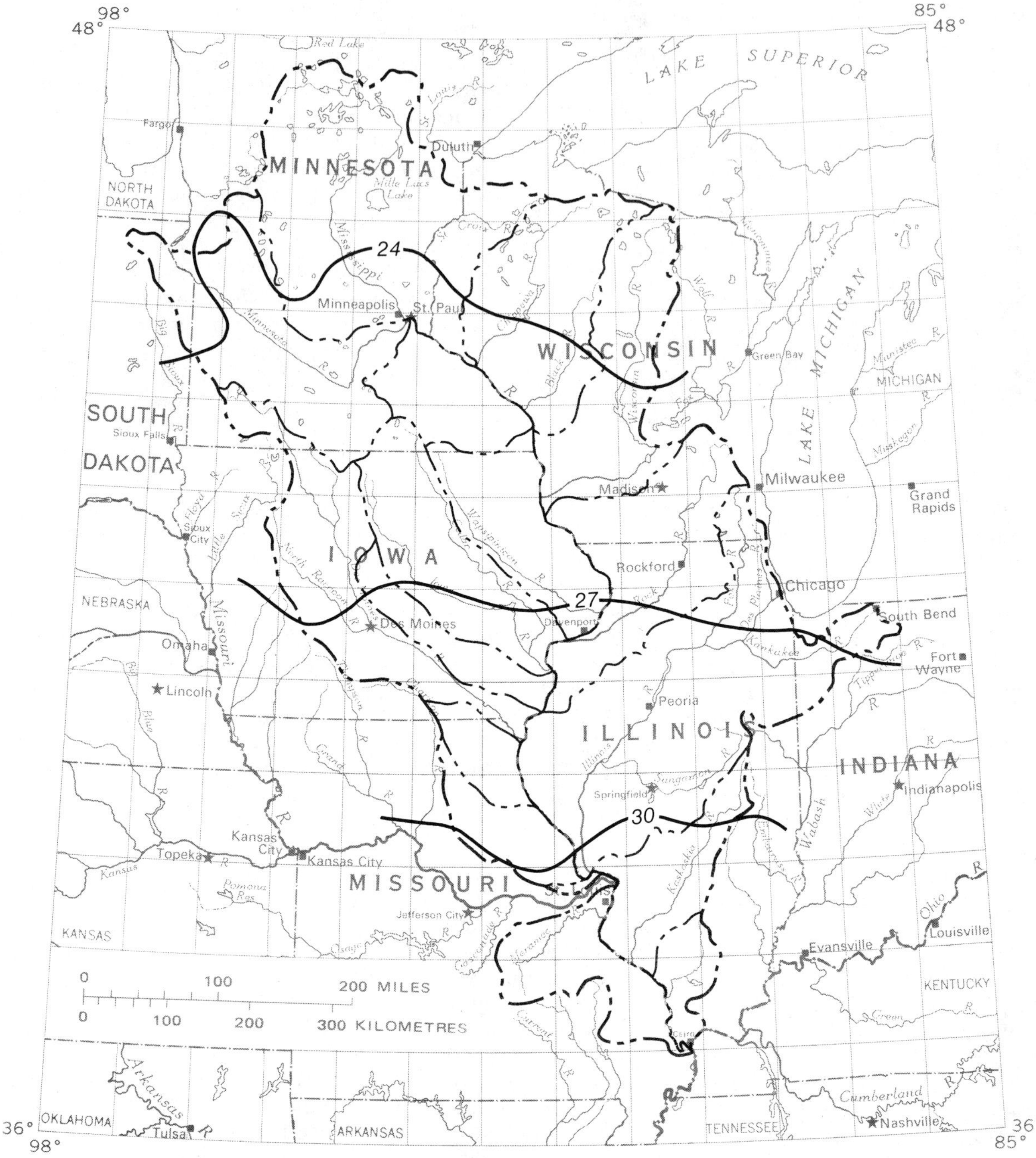

FIGURE 5.—Estimated average annual potential evapotranspiration, in inches. (Source is U.S. Army Corps of Engineers, 1970b.)

No attempt was made to calculate water in storage in the consolidated bedrock aquifers and the remaining unconsolidated surficial aquifers of the region. The sparse geohydrologic data make such a calculation unreliable. However, in the Ohio Region, where basic data are somewhat more abundant, potable water stored in the consolidated and minor unconsolidated aquifers is estimated to be four times that in the outwash and alluvial aquifers of the region (Bloyd, 1972). Possibly, such a ratio is applicable in the Upper Mississippi Region.

SALINE WATER

Saline ground water is most prevalent in the deep bedrock aquifers of central and southern Illinois and

TABLE 4.—*Calculation of water in storage in the outwash and alluvial aquifers of the subbasins*

Subbasin	Assumed surface area of outwash and alluvium (mi^2)	Assumed saturated thickness (ft)	Saturated volume (billion ft^3)	Assumed specific yield	Stored ground water	
					Billion ft^3	Billion gal
Mississippi headwaters	5,900	30	4,900	0.15	740	5,500
Chippewa-Black	3,600	50	5,000	.15	750	5,600
Wisconsin	3,800	50	5,300	.15	800	6,000
Rock	1,400	50	2,000	.15	300	2,200
Illinois	8,000	50	11,200	.15	1,680	12,600
Kaskaskia	600	50	800	.15	120	900
Big Muddy	200	50	300	.15	40	300
Meramec	------	----	------	----	------	------
Salt	------	----	------	----	------	------
Fox-Wyaconda-Fabius	------	----	------	----	------	------
Des Moines	3,400	30	2,800	.15	420	3,100
Skunk	900	30	800	.15	120	900
Iowa-Cedar	2,900	30	2,400	.15	360	2,700
Turkey-Maquoketa-Upper Iowa-Wapsipinicon	1,100	30	900	.15	140	1,000
Cannon-Zumbro-Root	700	25	500	.15	70	500
Minnesota	1,900	30	1,600	.15	240	1,800
Total	------	----	------	----	5,780	43,100

western Iowa. The largest quantity of saline water is present in the Illinois basin (fig. 3). Here the basement rock is deeper than elsewhere in the Upper Mississippi Region. Assuming that the basement rock forms the bottom boundary of the saline zone, the bottom of the zone is about 9,000 feet below sea level. Sparse geohydrologic data negates the possibility of computing the amount of saline water in storage in the region.

DEFINING MOVEMENT OF WATER THROUGH THE SYSTEM

Except for the almost stagnant brine in the deeper zone of the Illinois basin, almost all ground water in the Upper Mississippi Region flows through the aquifers from place of intake or recharge to place of discharge. The rate of movement, under natural conditions, generally ranges from a few feet to a few hundreds of feet per year.

The recharge area of a consolidated or bedrock aquifer in the region is assumed to be the outcrop area of the aquifer; or if the aquifer is covered with glacial or alluvial deposits the recharge area is assumed to be the subcrop area underlying the glacial or alluvial deposits. The outcrop or subcrop area underlying perennial streams is generally a discharge area. On the basis of these factors, the recharge areas for all bedrock aquifers can be inferred from a detailed geologic map.

Once water enters a consolidated aquifer, there are two probable flow paths—the path through the intergranular or primary openings, called pores, and the path through secondary openings, such as rock fractures or solution channels. The intergranular openings were usually formed when the rocks were deposited as sediments. Although compaction and cementation generally occur with time and alter the original size and shape of the primary openings, some pore space generally remains available in rocks. Rock fracture and solution channels, which occur after original deposition, provide additional conduits for ground-water accumulation and movement. In consolidated rocks secondary openings generally transmit greater quantities of ground water at higher rates than primary openings.

The major discharge areas of the consolidated aquifers of the region are probably the Illinois basin (fig. 3) and the Mississippi River valley. Also Cambrian and Ordovician age aquifers probably discharge water into Lake Michigan (Weidman and Schultz, 1915, p. 78–81). In these discharge areas, vertical ground-water discharge or interaquifer flow is significant. Cartwright (1970, p. 917) used ground-water temperature anomalies in the Illinois basin to calculate the annual vertical ground-water discharge in an 1,800-square-mile area. He concluded that possibly as much as 95 percent of the discharge must be moving upward through vertical fractures or secondary openings.

Unconsolidated alluvial and outwash aquifers are recharged by infiltration of precipitation, upward interaquifer flow, and percolation of streamflow from losing streams. In most places the major source of recharge is precipitation. Under natural conditions, practically all exposed upland areas of unconsolidated aquifers are recharge areas.

The general direction of ground-water flow in the unconsolidated deposits is toward the streams. As suggested in the section "Stream-aquifer Interrelation," the streams are primary "sinks" for ground-water discharge. Interchange between surface-water and ground-water systems, however, is common. One reach of a stream may be gaining and another losing. Also, streams may gain or lose according to stage changes.

STREAM-AQUIFER INTERRELATION

The purposes of this section of the report are: To demonstrate the general hydraulic connection between the shallow ground-water and surface-water systems, and to develop streamflow data needed in defining the ground-water resources.

Hydraulic connection between the shallow ground-water and surface-water systems is more evident in the Upper Mississippi Region than in many parts of the country. Thousands of natural ground-water fed lakes in Minnesota and Wisconsin are evidence of such hydraulic connection. The numerous perennial streams, which continue to flow even during drought, also testify to the connection between the two systems.

Hydraulic connection between the ground-water and surface-water systems and a vast supply of ground water indicate a potential for large economic benefits.

Because of the interchange between ground water and surface water, streamflow data for nonregulated perennial streams can be used to estimate ground-water discharge to streams, to estimate recharge as an approximate equivalent to discharge, and to estimate the hydraulic conductivity of aquifers that are in hydraulic connection with streams. Data from U.S. Geological Survey offices in the region were used for such purposes. Most of the data were obtained from studies of stream-gaging networks made in each State by the Geological Survey. In these studies, a streamflow-frequency distribution was computed for each unregulated gaging site, where at least 10 years of streamflow record was available. Records for most sites were for more than the minimum 10-year period. The streamflow data used in this report were not adjusted to a common base period but were considered sufficiently accurate for the estimates sought.

Variation in streamflow can be depicted statistically with a cumulative frequency curve constructed on the basis of the amount of time during which stated values of streamflow are equaled or exceeded. The water in streams during periods of high flow is from both surface runoff and ground-water discharge. As streamflow decreases, the percentage of streamflow derived from surface runoff commonly decreases also. For most streams, below some reduced rate of flow almost all water in the channel is from ground water.

The flow equaled or exceeded 90 percent of the time, or the 90-percent flow, is generally assumed to be all ground-water discharge. Because the above assumption is based on empirical studies and because ground-water discharge varies seasonally, the use of other percentage flow values has been urged. Wyrick and Lloyd (1968, p. H19) mention data that indicate that the ground-water component of discharge to streamflow varies from the streamflow that is exceeded 60 percent of the time to that exceeded 90 percent of the time. Stuart, Schneider, and Crooks (1967, p. 42) concluded that for a stream in southeastern Pennsylvania, the flow exceeded more than about 75 percent of the time is all ground-water discharge.

The approach taken in this study is that two flow-duration parameters define the annual ground-water discharge to streams better than one. This approach is based on the assumption that seasonal variation in base flow of a perennial stream is caused partly by seasonal variation in vapor discharge from aquifers. Because of the variation in vapor discharge, one parameter is necessary to define the ground-water discharge to streamflow during the season when vapor discharge is at a maximum, and another parameter is necessary for the season when vapor discharge is at a minimum.

Where ground water sustains streamflow in the Upper Mississippi Region, the water table in the vicinity of the streams is close to the land surface and is a ready source of moisture for plants; therefore, abundant vegetation generally grows along stream channels. Because vegetation on flood plains is situated between ground-water recharge areas and stream channels or ground-water discharge areas, the vegetation intercepts ground water enroute to discharge areas. During the summer, when transpiration is at a maximum, vegetation diverts more ground-water to the vapor phase than during the winter, when transpiration is at a minimum. This seasonal variation in magnitude of ground-water diversion by plants causes seasonal variation in base flow.

The assumption is made that the 90-percent flow-duration parameter is an indicator of base flow in a stream during that part of the year when ground-water vapor discharge is at a maximum and that the 60-percent flow-duration parameter is an indicator during that part of the year when vapor discharge is at a minimum (fig. 6). For the purposes of this study, these choices seem reasonable.

Further discussion on the use of streamflow data to

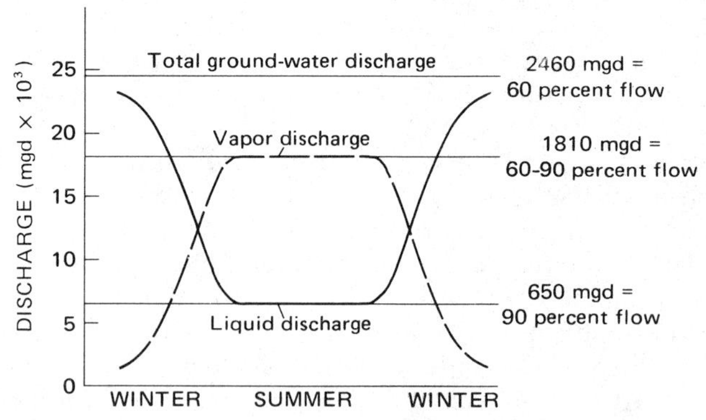

FIGURE 6.—The liquid- and vapor-discharge components and their relation to total ground-water discharge.

estimate ground-water recharge is presented in the section on "Ground-water Recharge."

The above approach of using two flow-duration parameters to define the annual ground-water discharge to streams is assumed applicable to the entire region even though the reason for the variation in base flow in the northern part of the region is complicated by frozen ground in winter that inhibits recharge and discharge from aquifers.

Where large reservoirs control flows, downstream flow records are not usable as indices of ground-water discharge. Also, water diverted from Lake Michigan is a major part of the annual flow of the Illinois River. For the Mississippi headwaters, Chippewa-Black, Wisconsin, Illinois, Des Moines, and Iowa-Cedar subbasins, the 60- and 90-percent flow duration parameters were determined by averaging values from nonregulated headwater and tributary flow records.

GROUND-WATER RECHARGE

Although various methods are used to determine "available" ground water or "safe yield" of a ground-water region, most methods assume that average annual ground-water recharge is the upper limit of ground water available.

The 60-percent flow-duration data were used to make order-of-magnitude estimates of subbasin ground-water recharge (table 5). Assuming that the 60-percent flows are indicators of natural ground-water discharge (fig. 6), they also can be used as indicators of natural ground-water recharge. This assumption implies that an approximate hydrologic equilibrium must exist between ground water entering and leaving the hydrologic system. Further discussion of this concept is presented in the section on "Principle of Continuity."

PRODUCTIVITY OF AQUIFERS IN THE UPPER MISSISSIPPI REGION

Estimates of the rates at which aquifers in the region can yield water to wells were made in the upper Mississippi River comprehensive basin study (U.S. Department of the Interior, Geol. Survey, 1970a, Appendix E). These estimates, which are modified slightly for this report, were made on the basis of rock type, geologic structure, known well and spring production, low-flow streamflow records, driller's records, and general knowledge of local geohydrology. For this report, only those areas having potential well yields greater than 500 gal/min (gallons per minute) or less than 25 gal/min are illustrated (figs. 7 and 8).

WATER-SUPPLY NEEDS

The foregoing information in this report is presented in order to define the geohydrologic system and to assist evaluation of the potential of the ground-water supply in the Upper Mississippi Region; the next logical step is to consider water demand in the region.

CURRENT

Partial estimates of ground-water withdrawals are presented in the upper Mississippi River comprehensive basin study (U.S. Department of the Interior, Federal Water Quality Administration, 1970, Appendix H). Estimates are given for domestic and commercial ground-water use and for rural domestic and livestock use by subbasin for base year (1960) (U.S. Department of Agriculture, 1970, Appendix N). For this report, it was assumed that all 1960 rural water use was from ground-water sources. This assumption is probably not

TABLE 5.—*Summary of ground-water recharge computations*

Subbasin	Subbasin area (mi²)	60-percent flow [1] (ft³/s/m²)	Estimated ground-water recharge [2] (ft³/s)	Estimated ground-water recharge Mgal/d	Percent of subbasin precipitation
Mississippi headwaters	28,100	0.26	7,300	4,700	14
Chippewa-Black	13,100	.26	3,400	2,200	12
Wisconsin	12,800	.35	4,500	2,900	15
Rock	14,500	.32	4,650	3,000	13
Illinois	29,600	.15	4,450	2,900	6
Kaskaskia	7,000	.16	1,100	700	6
Big Muddy	2,800	.07	200	150	3
Meramec	7,000	.27	1,900	1,250	9
Salt	4,500	.05	250	150	2
Fox-Wyaconda-Fabius	3,000	.05	150	100	2
Des Moines	14,800	.07	1,050	700	3
Skunk	4,600	.14	650	400	6
Iowa-Cedar	12,800	.15	1,900	1,250	6
Turkey-Maquoketa-Upper Iowa-Wapsipinicon	8,800	.25	2,200	1,400	10
Cannon-Zumbro-Root	5,700	.22	1,250	800	10
Minnesota	16,900	.05	850	550	3
Total				23,150	----

[1]ft³/s/m² = cubic feet per second per square mile. [2]ft³/s = cubic feet per second.

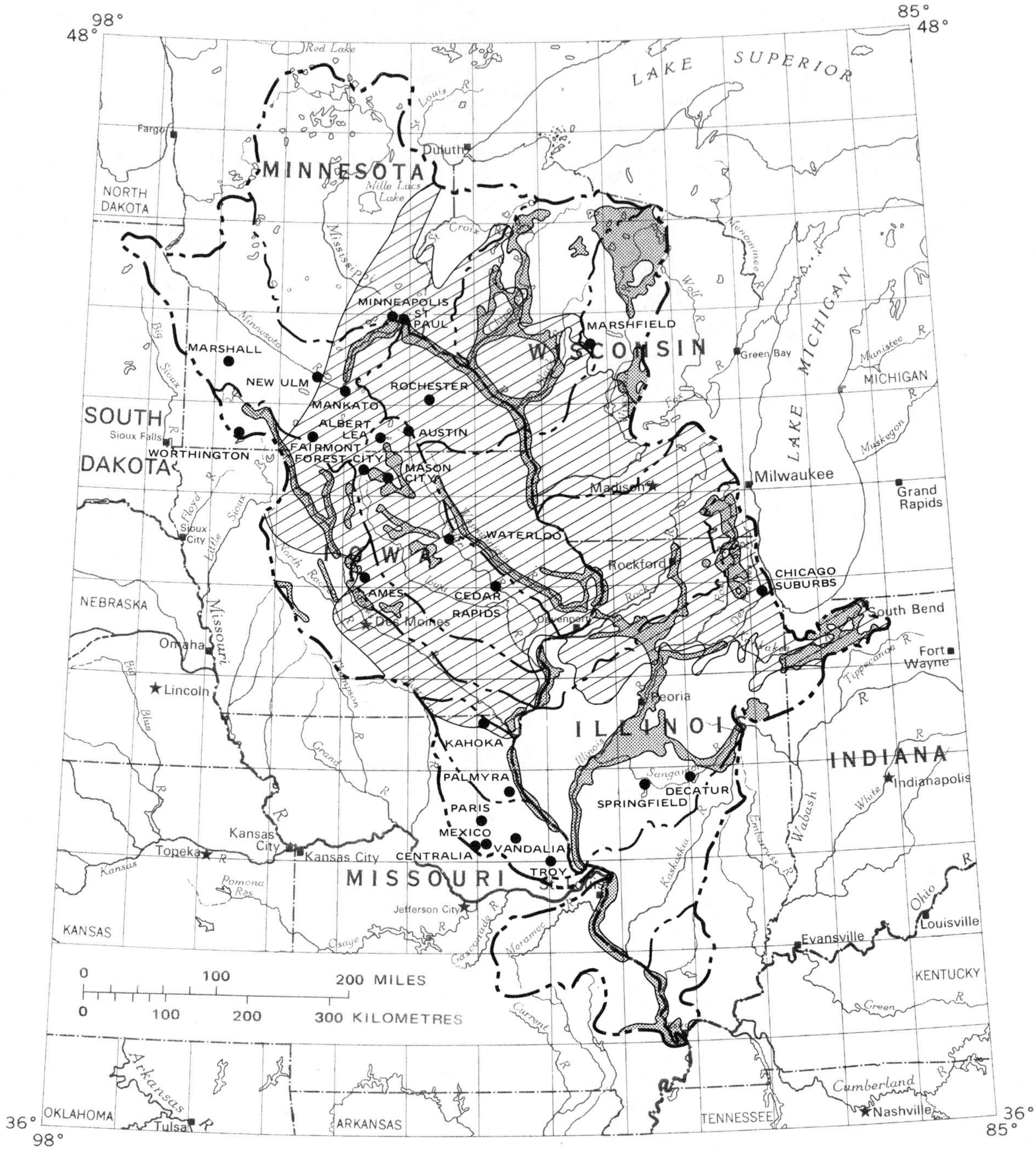

EXPLANATION

POTENTIAL YIELDS TO INDIVIDUAL WELLS

- Unconsolidated aquifers; generally more than 500 gpm
- Consolidated aquifers; generally more than 500 gpm
- ● Potential water-supply problem areas

FIGURE 7.—High-yield sources of ground water.

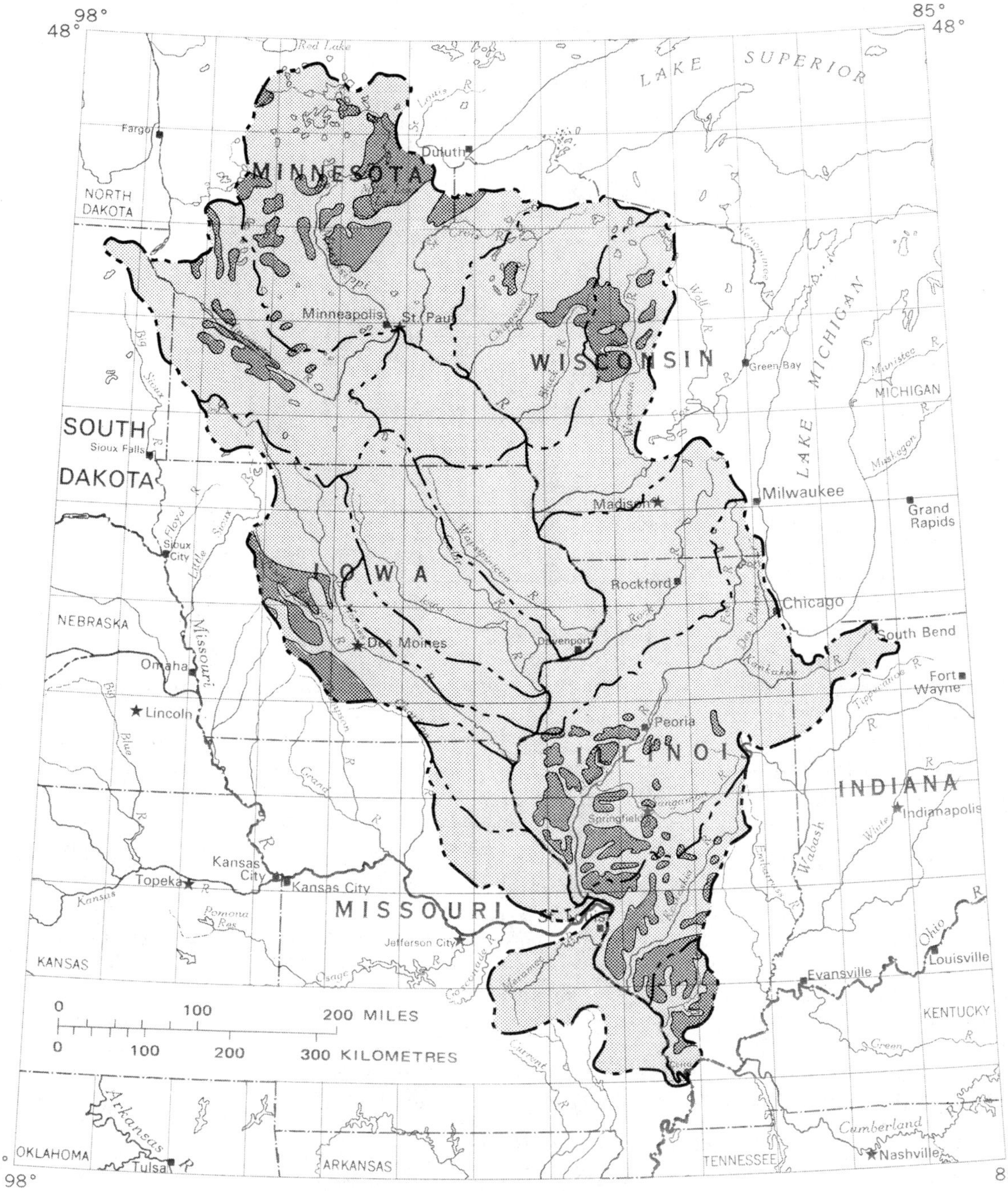

EXPLANATION

Areas where wells generally yield less than 25 gpm

Areas where wells generally yield more than 25 gpm

FIGURE 8.—Low-yield sources of ground water.

TABLE 6.—*Base year (1960) domestic, commercial, and rural ground-water use, Mgal/d*

Subbasin	Domestic and commercial	Rural domestic and livestock	Total
Mississippi headwaters	56	41	97
Chippewa-Black	11	22	33
Wisconsin	23	24	47
Rock	83	30	113
Illinois	154	75	229
Kaskaskia	9	11	20
Big Muddy	.4	5	5
Meramec	9	11	20
Salt	2	11	13
Fox-Wyaconda-Fabius	.6	5	6
Des Moines	23	48	71
Skunk	8	16	24
Iowa-Cedar	37	52	89
Turkey-Maquoketa-Upper Iowa-Wapsipinicon	15	36	51
Cannon-Zumbro-Root	15	17	32
Minnesota	16	35	51
Totals	462	439	901

Total basin industrial ground-water use (1965) was 629 Mgal/d (Murray, 1968). Industrial ground-water use does not include water withdrawn for electric power cooling purposes.

much in error. Industrial ground-water use by subbasin was not presented in the above reports. However, Murray (1968) presents basinwide industrial ground-water use for 1965. Ground-water use data from the comprehensive study report and from Murray's report are presented in table 6.

PROJECTED

Estimates of future subbasin water withdrawals, in addition to the base-year (1960) amounts, also are presented in the upper Mississippi River comprehensive basin study (U.S. Department of the Interior, Federal Water Quality Administration, 1970, Appendix H; U.S. Department of Agriculture, 1970, Appendix N). The estimated 2020 total water-use data were extracted from the reports and are presented in table 7 to permit further comparison and discussion.

THE SIGNIFICANCE OF THE SUBSURFACE SYSTEM IN THE UPPER MISSISSIPPI REGION

The foregoing information in the report suggests that the region's ground water is a large natural resource and that a demand for such a resource exists. The next logical step is to show how the ground-water resource may have a significant role in regional water development.

THE POTENTIAL FOR ADDITIONAL DEVELOPMENT

A comparison of ground-water withdrawals with estimated ground-water recharge (table 8) suggests that the ground-water resources of the Upper Mississippi Region are not being used at full potential. Annual domestic, commercial, and rural ground-water use (1960) is about 900 Mgal/d. Average annual regional ground-water recharge is about 23,000 Mgal/d; therefore, base-year use is only 4 percent of recharge. Ground-water use (1965) by industry is only 3 percent of recharge.

The Des Moines subbasin is the only subbasin with base-year withdrawal rates greater than 10 percent of annual subbasin ground-water recharge rates. A similar comparison for the Mississippi headwaters and the Chippewa-Black and the Wisconsin subbasins suggests that present ground-water development for those subbasins is but a small part of the recharge, even if all the base-year industrial water use were from ground-water sources, which is not true.

TABLE 7.—*Estimated total municipal and industrial water use for the year 2020, (in Mgal/d)*

Subbasin	Total water use
Mississippi headwaters	1,050
Chippewa-Black	220
Wisconsin	950
Rock	1,200
Illinois	12,300
Kaskaskia	1,320
Big Muddy	50
Meramec	1,960
Salt	90
Fox-Wyaconda-Fabius	50
Des Moines	660
Skunk	140
Iowa-Cedar	800
Turkey-Maquoketa-Upper Iowa-Wapsipinicon	620
Cannon-Zumbro-Root	330
Minnesota	330
Total	22,070

TABLE 8.—*Comparison of ground-water withdrawals with estimated ground-water recharge, in Mgal/d*

Subbasin	Total[1] withdrawal	Estimated ground-water recharge
Mississippi headwaters	97	4,700
Chippewa-Black	33	2,200
Wisconsin	47	2,900
Rock	113	3,000
Illinois	229	2,900
Kaskaskia	20	700
Big Muddy	5	150
Meramec	20	1,250
Salt	13	150
Fox-Wyaconda-Fabius	6	100
Des Moines	71	700
Skunk	24	400
Iowa-Cedar	89	1,250
Turkey-Maquoketa-Upper Iowa-Wapsipinicon	51	1,400
Cannon-Zumbro-Root	32	800
Minnesota	51	550
Regionwide industrial use	629	------
Totals	1,530	23,150

[1]Summation of base year (1960) domestic and commercial ground-water use, rural water use, and basinwide industrial ground-water use (1965).

GROUND WATER TO SATISFY LOCAL WATER REQUIREMENTS

A well in almost any part of the region will yield enough water to satisfy domestic water requirements. The ubiquitous windmill of the past testifies to this fact.

Disregarding the chemical quality of water, water demands of many small municipalities or industries can be satisfied by wells in areas where wells yield more than 25 gal/min. Also, ground water can be used to meet recreational demands, if a large surface reservoir is not required, for campsites, small off-stream fishing ponds, swimming pools, ice-skating rinks, or roadside rest stops.

In the Chicago suburban and Minneapolis–St. Paul metropolitan areas, large amounts of ground water are withdrawn locally. Total 1961 ground-water withdrawal in the Chicago area was 182 Mgal/d (U.S. Department of the Interior, Federal Water Quality Administration, 1970, Appendix H, p. H101). Ground-water withdrawal in the Minneapolis–St. Paul metropolitan area in 1960 was 136 Mgal/d (U.S. Department of the Interior, Federal Water Quality Administration, 1970, Appendix H, p. H41).

Rochester, Austin, and Albert Lea, Minn., three potential water-problem areas (table 2 and fig. 7), are underlain by the Prairie du Chien–Jordan aquifer, from which much water in the Minneapolis–St. Paul area is withdrawn. Detailed studies of this aquifer and of the Mount Simon–Hinkley aquifer would aid any future planning for water supplies for Rochester, Austin, and Albert Lea.

Ground water may partly satisfy future irrigation water requirements in the region. The potential for irrigation, and the water requirements for it have already been estimated (U.S. Department of Agriculture, 1970, Appendix N). Estimated water requirements range from 20,000 acre-feet per year (50 Mgal/d) in the Chippewa-Black subbasin to 1,756,000 acre-feet per year (3,370 Mgal/d) in the Illinois subbasin (table 9); these estimates are based on the assumption that the irrigation period coincides with the freeze-free period. The growing seasons for the subbasins were estimated from published data (U.S. Department of Agriculture, 1970, Appendix N, p. N10).

Although all irrigation water requirements cannot be satisfied with ground water, the comparison of base year (1960) ground-water withdrawals with estimated ground-water recharge (table 7) suggests that ground water is available in substantial amounts for irrigation use.

Ground water is potentially valuable to the sportsfishery industry. For example, part or all of the water requirements for a fish hatchery may be satisfied with ground water. Also ground water may be used to augment critically low streamflows during peak fishing seasons. Such augmentation may be extremely profitable in prime fishing areas. In both examples, ground water can be utilized as a secondary source during periods when surface water is not available in sufficient quantity or acceptable quality. Ground water also can be used as the primary water source for fish hatcheries at those places where surface water is in short supply and ground-water temperature is favorable for growing game fish.

TABLE 9.—*Estimated irrigation-water requirements for the year 2020*

Subbasin	Requirement (1,000 acre-feet per year)	Growing season (days)	Requirement (Mgal/d)
Mississippi headwaters	255	120	690
Chippewa-Black	20	120	50
Wisconsin	74	120	200
Rock	745	165	1,470
Illinois	1,756	170	3,370
Kaskaskia	318	190	550
Big Muddy	132	190	230
Meramec	248	190	430
Salt	189	190	320
Fox-Wyaconda-Fabius	84	180	150
Des Moines	653	165	1,290
Skunk	175	165	350
Iowa-Cedar	537	150	1,170
Turkey-Maquoketa-Upper Iowa-Wapsipinicon	381	150	830
Cannon-Zumbro-Root	217	150	470
Minnesota	515	150	1,120
Total			12,690

R. P. Novitzki (written commun., 1972) demonstrated that two trout streams in Wisconsin could be improved by augmenting streamflow with ground water. In the two streams, temperature extremes were moderated, and dissolved-oxygen concentrations increased during the augmentation. A model was used to predict potential temperature improvement during critical times. Improvement in living space was indicated by increases in stream volume and stage. Flow was also increased through the study reaches of both streams.

GROUND WATER AS A REGIONAL RESOURCE

Specific criteria used in this report to classify the ground water as a regional resource are:

1. Unconsolidated aquifer(s) of significant areal extent and saturated thickness of at least 30 feet must be present.
2. The aquifer(s) must be capable of yielding greater than 500 gal/min to individual wells.
3. Surface water must be available for artificial recharge of the aquifer(s).

The ensuing discussion assumes that necessary legislation and land acquisition are possible if development of ground water is feasible. Also all water referred to is assumed to be potable.

The above criteria for classifying ground water as a regional resource are satisfied in the outwash and alluvial aquifer along much of the main stem of the Mississippi River in and south of St. Paul, and in the outwash and alluvial aquifers of the Illinois, the Lower Minnesota, the Wisconsin, the Lower Black, the Wapsipinicon, the Lower Rock, and the Upper Des Moines Rivers. Essentially these areas are underlain by outwash and alluvial aquifers with the potential to yield greater than 500 gal/min to individual wells (fig. 7). In these areas ground-water resources are of sufficient magnitude to satisfy more than just local needs.

The water in some of the bedrock aquifers, such as that in the Prairie du Chien–Jordan aquifer, could be considered a regional resource. A large supply is present and is available. Because only scant data are available on the location and extent of recharge to the aquifer, more detailed geohydrologic studies seem warranted.

In the above-mentioned areas where ground water can be considered a regional resource, the water requirements for a large population increase or other major water demands can be satisfied with ground water. For example, to show the approximate magnitude of a population increase that could be adequately supplied, the following assumptions are made:

1. Half of the excess of annual ground-water recharge over base-year (1960) ground-water use can be used to satisfy a particular water demand.
2. Half of the base-year industrial water use is from ground-water sources.
3. Water withdrawn is used only once.
4. Per capita water use is 150 gal/d (gallons per day), therefore, a 1-Mgal/d supply will satisfy the water requirements of approximately 6,500 people.

On the basis of the above assumptions, ground water is available for about 9 million additional people in the Wisconsin subbasin, 6 million in the Illinois subbasin, 2 million in the Des Moines subbasin, 1½ million in the Minnesota subbasin, and 1 million in both the Wapsipinicon and Black River drainage areas.

The above calculations for "increased population" are made only as an example of the opportunities that ground-water development presents. Note that artificial recharge of aquifers was not considered nor were the effects of development upon streamflow evaluated. Factors other than water supply, of course, will constrain development in the region.

In order to make maximum use of the available ground-water resources without excessive transportation costs, the maximum population density should be in the areas where ground water is considered a regional resource. There is a high population density pattern in the upper Illinois River subbasin but not in the other subbasins. The upper Des Moines subbasin is a farming area. Only 4 percent of the total land area of the subbasin was urbanized or built up in 1958 (U.S. Department of Agriculture, 1970, Appendix N, p. N37). The land-use pattern is similar in the Wisconsin subbasin and in the lower Black River and Wapsipinicon drainage areas.

The potential for additional industrial development also is enhanced in areas where ground water is classified as a regional resource. Almost any type of industry can satisfy their water requirements in these areas through the use of ground water.

THE POTENTIAL FOR UNDERGROUND WASTE DISPOSAL

Underground space in the Upper Mississippi Region can be considered a regional resource in the sense that it could be included in regional water-pollution control or waste-disposal plans. A simple definition of natural underground space might be interconnected pore or fracture space into which a fluid can be emplaced. Underground space is not necessarily vacant. In this section of the report emphasis is upon underground space occupied by saline water. Before assessing the potential for underground waste storage in the region, a discussion is presented on the characteristics generally considered necessary for successful waste injection and underground storage.

Two geologic characteristics generally considered necessary for the use of waste-injection wells are: (1) An injection zone with sufficient permeability, porosity, thickness, and areal extent to serve as a liquid-storage reservoir at safe injection pressures, and (2) an injection zone that is below the level of fresh-water circulation and is confined above by rocks that are, for practical purposes, impermeable to waste liquids.

Three characteristics indicate that an area has potential for waste injection and underground storage. These characteristics are: Aquifers containing water having more than 1,000 mg/l dissolved solids are available, the sedimentary rock sequence is 1,500 feet thick below the minimum depth of ground water containing more than 1,000 mg/l dissolved solids, and subsurface hydrology can be predicted because of relatively simple geologic structure. These characteristics are the converse of three listed by Warner (1969, p. B–28–A), which indicate an area has little feasibility for waste injection and underground storage.

On the basis of the above characteristics, especially the requirement for aquifers containing water having more than 1,000 mg/l dissolved solids, there seems to be only slight potential for underground storage of waste in the Upper Mississippi Region. As stated in the section on "Saline Water," saline ground water is most prevalent in the deep aquifers of central and southern Illinois. These may be the areas with a potential for underground waste storage.

Further discussion of waste management is beyond the scope of this report. However, the technical aspects of the use of salaquifers for waste storage is now under study by the U.S. Geological Survey.

WATER-MANAGEMENT IMPLICATIONS OF THE HYDROLOGIC ANALYSIS

The effects of man's activities on the hydrologic system is a prime consideration in future planning. Basic to such consideration are an understanding of how the hydrologic system reacts to stresses and a knowledge of management alternatives. The purpose of this section is to discuss some basic hydrologic concepts and their relation to management and some specific water-management alternatives.

SOME CONCEPTS RELATED TO WATER MANAGEMENT

PRINCIPLE OF CONTINUITY

As a consequence of the principle of continuity, the quantities of water entering or leaving an arbitrary volume, such as the geohydrologic system of the Upper Mississippi Region, within a given time, are related by the hydrologic equation:

$$\text{Inflow} = \text{outflow} \pm \text{Change in storage}.$$

If the change in storage is positive, denoting an increase in storage, the change is added to the right side of the equation; if the change is negative, denoting a decrease in storage, the change is subtracted.

The hydrologic system of the Upper Mississippi Region will respond to any water-management program in a way that is consistent with the hydrologic equation. If ground water is used so that freshwater storage is not continually decreased, according to the hydrologic equation, ground-water inflow (recharge) and outflow (discharge) will be in balance. That is, increased consumptive use of ground water would have to be balanced by increased inflow (such as artificial recharge) or by a reduction of natural outflow (such as ground-water discharge to streams) to avoid a continual decrease in storage.

If inflow is less than outflow, fresh ground water in storage will be depleted. In Bellwood, Ill., a western suburb of Chicago, water levels in wells tapping bedrock aquifers declined 9 feet per year from 1913 to 1929 (Illinois Technical Advisory Committee on Water Resources, 1967, p. 107). By 1951 the water level had declined 495 feet, 305 feet since 1925, or 11.7 feet per year. By the mid-1960's the water level in the aquifer in the Bellwood area was 510 feet below the ground surface. If water withdrawal continues at the mid-1960 rate, the level is expected to decline another 300 feet by 1980, to about 810 feet below ground surface (Illinois Technical Advisory Committee on Water Resources, 1967, p. 108).

Such an overdraft of water in storage is not necessarily adverse. If temporary overdraft is part of a management program, management alternatives are increased and economic gains are possible.

YIELD OF THE SYSTEM

Discussions of development of the water resources of a region often focus on the "yield" or "safe yield" of the ground-water system. Todd (1959, p. 200) defines safe yield as the amount of water that can be withdrawn annually without producing any undesired result.

Safe yield is useful in evaluating alternatives in managing a hydrologic system (Franke and McClymonds, 1972, p. F52). In practice, however, the concept has been misused—usually because a single value of safe yield is adopted and comes to be regarded as an inflexible natural law.

Often discussions focus on "available ground water," which can be defined as the amount of ground water that can be withdrawn on an areal basis for a long time, with conditions and limitations stated. As with safe yield, there is no one realistic value for available ground water. Depending upon conditions stated, available water can be defined and is useful. However, if conditions change, available water changes.

CONSERVATION OF GROUND WATER

Conservation can be defined as preservation, protection, or wise use of a resource—especially its planned management to prevent its irreversible depletion or deterioration. A diversity of opinion exists regarding exploitation of water resources and what is desirable or undesirable development and management. Various management alternatives are, thus, bound to arise. Planning involves evaluation of alternatives and choice of the most desirable or least undesirable alternative for action. Before discussing alternatives in the Upper Mississippi Region, the advantages (for management) of the subsurface system are discussed. These advantages are common to all management alternatives.

ADVANTAGES (FOR MANAGEMENT) OF THE SUBSURFACE SYSTEM

Realizing the potential for further ground-water development in the region, the planner considers several advantages of using the subsurface system. A discussion of some of these advantages follows.

A major advantage is the versatility of the subsurface system. Uses of the subsurface system are diverse. Broadly speaking, however, uses may be grouped into two categories: Storage of gases and fluids—principally natural gas and fresh water—for withdrawal, and injection of waste fluids for which withdrawal is not contemplated. These uses are likely to compete in many parts of the Upper Mississippi Region; hence, proper planning and resource utilization are warranted.

Another advantage is that ground water is generally renewable in the Upper Mississippi Region. In the section on "Ground-water Recharge," average annual ground-water recharge is estimated. In addition to natural recharge, aquifers may be recharged artificially. Recharge can be induced from streams if an aquifer and stream are hydraulically connected. Surface-spreading basins may be constructed adjacent to a stream in order to increase the infiltration area. Recharge wells may be used where hydraulic connection is poor or where land values are prohibitive. The choice of the type of a recharge facility is flexible. Depending on the site available, wells, reservoirs, ponds, lagoons, ditches, or pits may be utilized for recharge.

A successful surface-spreading facility has been in operation in Peoria, Ill., for about 20 years. A large part of the water supply for the Peoria area is obtained from sand and gravel aquifers adjacent to the Illinois River. The supply is maintained by artificial recharge of the aquifer with Illinois River water.

Records on the use of some recharge wells are conflicting, and both successes and failures have been reported. The injection of chlorinated water free from silt into a full well has yielded the best results. Chlorine helps prevent the growth of soil-clogging microorganisms; silt can clog well perforations or can even penetrate the aquifer, reducing its hydraulic conductivity; and introduction of air into an aquifer can also decrease the aquifer hydraulic conductivity. Metal casings can become encrusted where perforations are exposed to air or if the injected water is incompatible with the native water.

Another advantage associated with ground water is that in much of the region ground water is generally dependable throughout the year. Because the amount of ground water in storage is vast compared with annual pumpage, storage is not depleted significantly over a pumping year or even a series of years.

Another advantage is that ground water is widely distributed throughout the region. It is not equal everywhere in volume or rate of availability, but it is almost everywhere present in some quantity—in contrast to surface water, which is readily available only adjacent to perennial streams. An excellent reserve resource is, thus, available for emergency use during drought when surface-water resources are not available. Emergencies also could arise from contamination of surface-water resources or through natural catastrophies or war. Although ground water may not be the prime water resource in an area, a knowledge of its potential is necessary in planning for contingencies.

Ground water is generally not subject to large seasonal changes in temperature, which is of importance to some industrial water users. Seasonal temperature fluctuations are generally greatest at shallow depths and least in deeper parts of an aquifer.

Of further importance to many water users, ground water is generally uniform in chemical quality, and the quality is not subject to sudden or even seasonal changes. Because the volume of ground water in storage is large and because it moves slowly, replacement of water in storage is slow and, therefore, chemical quality and temperature do not fluctuate much over short time periods in most parts of the aquifer.

Another advantage is that the local environment is not severely altered when ground water is developed properly. Visible evidence of development is generally limited to a well, pump, pumphouse, and pipeline, most of which could be dismantled on cessation of pumping.

A major advantage is that the subsurface system is generally hydraulically connected with the surface-water system. Where the connection is good, the combined system can be managed beneficially.

Beneficial management of a ground-water system involves lowering and raising of ground-water levels to regulate storage in response to time-varying requirements. Because a ground-water reservoir need not be maintained at full storage throughout the year, storage can be advantageously regulated.

The feasibility of a management plan depends partly on benefit compared with detriment caused by water-level changes. For simplification, ranges of water-level changes can be estimated for various pumpage rates. The ranges then can be divided into several increments. The probable effect on environment then can be estimated for each increment.

The feasibility of any management plan also depends upon the availability of answers to certain basic questions such as:

1. What is the areal extent, depth, and thickness of the ground-water reservoir?
2. Where are the hydrologic boundaries?
3. What is the amount of natural recharge and discharge?
4. Where are the areas of natural recharge and discharge?
5. At what rate can water be taken from or added to the reservoir?
6. Are there local geologic or hydrologic conditions

that may hinder use of the reservoir?

7. What is the quality of water in storage and how may quality of water change because of reservoir management?
8. What is the probable life of the reservoir?

Many of these questions can be answered through this and other reports. More specific and definitive answers become possible when specific management plans are given.

A factor to be considered in a management plan is the rate at which the ground-water reservoir can be artificially recharged. Extraction rates are indicators of the maximum possible rates of artificial recharge. Theoretically, an aquifer can be recharged by injection wells at the same rate that water can be withdrawn under a given head differential. However, the tendency for wells and infiltration facilities to become clogged causes a decrease in recharge rate with time. Therefore, average recharge rate, with time, will probably be much less than design rate.

Sites with the best potential for management of storage are in and adjacent to flood plains of the major streams (fig. 2). These sites are underlain by thick and permeable aquifers fed by streams, which are ready sources of recharge.

With proper planning and resource use, the natural advantages of ground water commonly outweigh disadvantages, and, ideally, ground water receives full consideration in water planning.

WATER MANAGEMENT ALTERNATIVES

The following alternatives for developing the water resources of the Upper Mississippi River are classified according to the terms of the hydrologic equation that are most affected by the particular alternative. In all the alternatives, consideration is given to changes in fresh-water recharge and discharge, changes in ground-water storage, changes in the flow pattern within the aquifer, and changes in the quality of water.

INCREASE INFLOW TO THE HYDROLOGIC SYSTEM

Assuming that precipitation is the only natural source of fresh water in the Upper Mississippi Region, increased precipitation and the importing of water from other regions are the only methods of increasing the total amount of fresh water available. It may be possible to increase precipitation by cloud seeding. Cloud seeding would probably be most beneficial during the summer when water demands are at a maximum. Potential outside sources of potable water include the Great Lakes and, possibly, the water resources of the Wabash River subbasin in the Ohio Region. The legal, political, economic, and technical aspects of transport of imported water may be formidable and are beyond the scope of this report, but importing of water is, of course, feasible physically.

SALVAGE OUTFLOW FROM THE HYDROLOGIC SYSTEM

Reduction of ground-water evapotranspiration would yield additional quantities of water without an attendant significant change in storage. Suppression of evapotranspiration would be at the expense of vegetal growth where evapotranspiration is at a maximum. Such areas are adjacent to gaining reaches of many streams, in marshlands, and in the immediate vicinity of natural lakes fed by ground water. However, the likelihood of salvaging much of such water over long periods of time is probably remote. Accordingly salvage of evapotranspiration in the region does not seem to be a significant water-management alternative at present.

USE OF ARTIFICIAL RECHARGE TO MAINTAIN BALANCE BETWEEN INFLOW AND OUTFLOW

Although it is difficult to recharge the aquifers of the region artificially, such recharge could be considered in areas of significant ground-water development.

Theoretically, if, after use, water pumped from an aquifer is returned to it, yield would be almost limitless. In practice, each time water is recirculated the quality of the water deteriorates. However, tertiary treatment of waste water before recharge might alleviate to some degree the quality-deterioration problem.

If much pumpage is from deeper artesian aquifers and if water is recharged to shallow unconsolidated aquifers, a local imbalance will result in the deeper aquifers. Because vertical hydraulic conductivity is generally less than horizontal conductivity, water recharged to shallow aquifers will probably not move downward into the deeper aquifers as rapidly as it is pumped from them. After detailed geohydrologic studies are made of the high-yielding bedrock aquifers, such as the Prairie du Chien–Jordan aquifer, natural recharge areas can be determined, and artificial recharge could be attempted.

SUMMARY

The water-resource planner in the Upper Mississippi Region in many ways is in an enviable position. The region is rich in both surface water and ground water, therefore, increasing water demands are not cause for alarm. Planning does not have to focus on programs for solving drastic problems of water supply. Rather, the planner can act to conserve the present water resources. Further, the planner is fortunate in that sophisticated tools are available to assist in evaluating alternative plans.

This report emphasizes the role of ground water in the

region. Total potable ground water in storage in the outwash and alluvial aquifers in the Mississippi River valley and the subbasins is about 45,000 billion gallons. This volume is about 10 percent of the water in Lake Ontario. Ground water in storage in other aquifers of the region is probably several times that in the outwash and alluvial aquifers. Estimated ground-water recharge in the subbasins is 23,000 M gal/d.

A comparison of ground-water withdrawals with estimated ground-water recharge suggests that ground water is not being used at full potential.

Ground-water use by domestic, commercial, and rural interests is only 4 percent of recharge. Ground-water use (1965) by industry is only 3 percent of recharge.

Part of the region has ground-water resources capable of satisfying more than local needs. Specifically, the water in the outwash and alluvial aquifers in much of the valley of the Mississippi River south of St. Paul, Minn., of the Illinois River, the lower Minnesota River, the upper Wisconsin River, the lower Black River, the Wapsipinicon River, the lower Rock River, and the upper Des Moines River can be considered a regional resource. Ground-water in the above areas could supply approximately 20 million additional people.

SELECTED REFERENCES

Bergstrom, R. E., and others, 1968, Ground-water resources of the Quaternary deposits of Illinois in the Quaternary of Illinois: Univ. of Ill. Coll. of Agriculture Spec. Pub. 14, p. 157–164.

Bloyd, R. M., Jr., 1972, Summary appraisals of the Nation's ground-water resources—Ohio Region: U.S. Geol. Survey open-file report, 136 p.

Bue, C. D., 1963, Principal lakes of the United States: U.S. Geol. Survey Circ. 476, 22 p.

Cartwright, Keros, 1970, Groundwater discharge in the Illinois basin as suggested by temperature anomalies: Water Resources Research, v. 6, no. 3, p. 912–918.

Franke, O. L., and McClymonds, N. E., 1972, Summary of the hydrologic situation on Long Island, New York, as a guide to water-management alternatives: U.S. Geol. Survey Prof. Paper 627–F, 59 p.

Holt, C. L. R., Jr., Young, K. B., and Cartwright, W. H., 1964, The water resources of Wisconsin, *in* Blue Book: Univ. of Wis. Geol. and Nat. History Survey, p. 178–198.

Illinois Technical Advisory Committee on Water Resources, 1967, Water for Illinois, a plan for action: 452 p.

Maass, Arthur, and others, 1962, Design of water-resource systems: Cambridge, Mass., Harvard Univ. Press, 620 p.

Murray, C. R., 1968, Estimated use of water in the United States, 1965: U.S. Geol. Survey Circ. 556, 53 p.

Prickett, T. A., and Lonnquist, C. G., 1971, Selected digital computer techniques for ground-water resource evaluation: Urbana, Illinois Water Survey Bull. 55, 62 p.

Steinhilber, W. L., and Horick, P. J., 1970, Ground-water resources of Iowa, *in* Water resources of Iowa—a symposium: Iowa Acad. Sci., p. 29–49.

Stuart, W. T., Schneider, W. J., and Crooks, J. W., 1967, Swatara Creek basin of southeastern Pennsylvania, an evaluation of its hydrologic system: U.S. Geol. Survey Water-Supply Paper 1829, 79 p.

Todd, D. K., 1959, Ground-water hydrology: New York, John Wiley & Sons, 336 p.

U.S. Army Corps of Engineers, 1970a, Upper Mississippi River comprehensive basin study, Washington, U.S. Govt. Printing Office. Volumes I–IX.

———1970b, Climatology and meteorology: Appendix C to Upper Mississippi River Comprehensive Basin Study, v. III: Washington, U.S. Govt. Printing Office.

———1970c, Economic base study and projections: Appendix P to Upper Mississippi River Comprehensive Basin Study, v. VIII: Washington, U.S. Govt. Printing Office.

U.S. Department of Agriculture, 1970, Agriculture: Appendix N to Upper Mississippi River Comprehensive Basin Study, v. VI: Washington, U.S. Govt. Printing Office.

U.S. Department of Commerce, Environmental Science Services Administration, 1968, Climatic atlas of the United States: Washington, U.S. Govt. Printing Office, 80 p.

U.S. Department of the Interior, Federal Water Quality Administration, 1970, Water supply and quality control, *in* Appendix H to Upper Mississippi River Comprehensive Basin Study, v. IV: Washington, U.S. Govt. Printing Office.

U.S. Department of the Interior, Geological Survey, 1970a, Ground water and geology, *in* Appendix E to Upper Mississippi River Comprehensive Basin Survey, v. III: Washington, U.S. Govt. Printing Office.

———1970b, National Atlas of the United States of America: Washington, U.S. Govt. Printing Office, 417 p.

Visocky, A. P., and Schicht, R. J., 1969, Groundwater resources of the buried Mahomet bedrock valley: Urbana, Illinois Water Survey, Rept. Inv. 62, 52 p.

Warner, D. L., 1969, Administrative guidelines and evaluation criteria, *in* Ohio River Valley Water Sanitation Commission Perspective on the regulation of underground injection of wastewaters, Cincinnati, Ohio: Part 2, 65 p.

Weeks, E. P., and Stangland, H. G., 1971, Effects of irrigation on streamflow in the central sand plain of Wisconsin: U.S. Geol. Survey open-file report, 109 p.

Weidman, S., and Schultz, A. R., 1915, The underground and surface water supplies of Wisconsin: Wisconsin Geol. and Nat. History Survey, Bull. 35, Econ. Ser. 17, 664 p.

Wyrick, G. G., and Lloyd, O. B., 1968, Ground water resources: Appendix H to report for development of water resources in Appalachia, U.S. Army Corps of Engineers, 122 p.

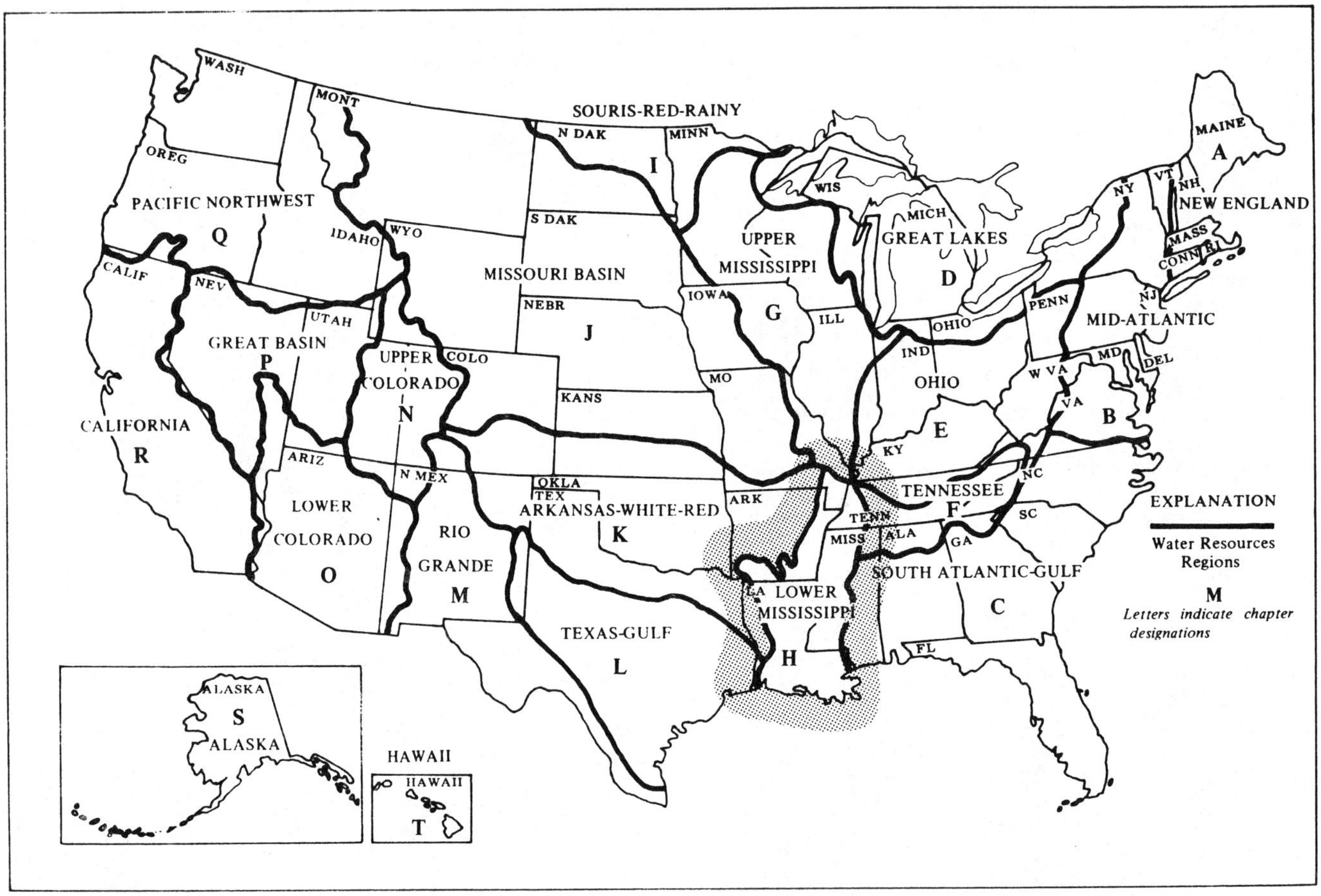

H – Location Map for the Lower Mississippi Region

H

LOWER MISSISSIPPI REGION

By J. E. TERRY, R. L. HOSMAN, and C. T. BRYANT

ABSTRACT

The Lower Mississippi Region comprises an area of 102,400 square miles (265,200 square kilometers). Almost all this area is in the physiographic province known as the Gulf Coastal Plain. Three small areas on the northwest boundary of the region are in the Interior Highlands.

The Lower Mississippi Region has an abundance of ground water. The geologic structure in that part of the region within the Coastal Plain is an elongated trough which has been filled with permeable materials, resulting in vast subsurface reservoirs. Except in local areas where continued large withdrawals have caused significant water-level declines, these reservoirs are full.

Recharge to the region's aquifers is primarily from rainfall. Annual rainfall in most of the region is well distributed throughout the year and is sufficient to satisfy evapotranspiration requirements and still provide recharge to the aquifers.

An estimated 844 billion cubic feet (24 billion cubic meters) of fresh ground water is available for withdrawal annually in the region. Only about one-third of this quantity is being utilized. Therefore, on this basis alone, the region still has much potential for ground-water development.

The Coastal Plain aquifers within the Lower Mississippi Region contain large reserves of saltwater in the downdip limits of the aquifers. The quantity of saltwater in the region is several times that of freshwater. As desalinization techniques are developed and as more uses are found for saltwater, this reserve could become an important source of water for the region.

At present (1976), the most productive and potentially productive aquifers or aquifer systems in the region are the Mississippi River valley alluvial aquifer of Quaternary age and the Sparta Sand and the Memphis aquifer (Memphis Sand in Tennessee) of Tertiary age. The Sparta Sand and the Memphis aquifer are heavily utilized and have shown significant water-level declines. However, selected well hydrographs indicate that water levels may be stabilizing under present pumping conditions. The Mississippi River valley alluvial aquifer is the most extensive high-yielding aquifer in the region; yields of several thousand gallons per minute may be obtained at depths of less than 200 feet (61 meters).

To obtain maximum benefit from the vast quantities of ground water in the region, adequate attention must be given to the effects of proposed development upon the ground-water regime. Knowledge of the geologic structure and hydraulic properties of the aquifer systems is essential to an evaluation of the effects of such development. Some studies have been made in sufficient detail to provide this knowledge, but additional studies are needed.

Activities that could cause significant changes in the ground-water regime should be undertaken only after all available information has been considered. Failure to seek out and use such information may result in inefficient development of the ground-water resource and, in some instances, degradation of the quality of the resource.

Some changes always result from ground-water development. The possible changes can be grouped into three categories: hydraulic, water quality, and those affecting the physical framework of the aquifers. Generally, they are small in magnitude and areal extent. Because these changes occur below the ground surface, they are unknown to the ground-water user unless they noticeably affect the quantity or quality of water produced or cause obvious physical effects, such as land subsidence.

Great advances have been made in hydrologic technology in recent years. Predictive models have been developed that make it possible for the hydrologist to simulate aquifer responses to proposed development or other stresses. These models would be invaluable tools in progressive water-resources planning and management.

INTRODUCTION

LOCATION AND SIZE OF AREA

The Lower Mississippi Region, as defined by the Water Resources Council, 1970, includes all the drainage basin of the Mississippi River downstream from its confluence with the Ohio River, except those parts of drainage basins of the Arkansas, Red, and White Rivers upstream from the backwater limits of the Mississippi River. It also includes the flood-protected areas at Cairo, Ill., and the Louisiana coastal area. The region comprises parts of Arkansas, Illinois, Kentucky, Louisiana, Mississippi, Missouri, and Tennessee, and encompasses about 102,400 mi^2 (265,200 km^2). Drainage from almost half of the conterminous United States culminates in the Lower Mississippi Region (fig. 1). The Mississippi River terminates at the lower end of the region, completing a river course totaling 2,348 mi (3,778 km) (U.S. Geological Survey, 1970).

THE NEW LAND

Explorers, traders, and hunters who sought adventure and fortune were the first to travel the valley of the Mississippi River. Following the trails they blazed, immigrants came to farm the river soil and to settle the wilderness. These early settlers in the Lower Mississippi Region found the Mississippi River to be both friend and foe. The river provided a mode of transportation and fertile soil for farming, yet it be-

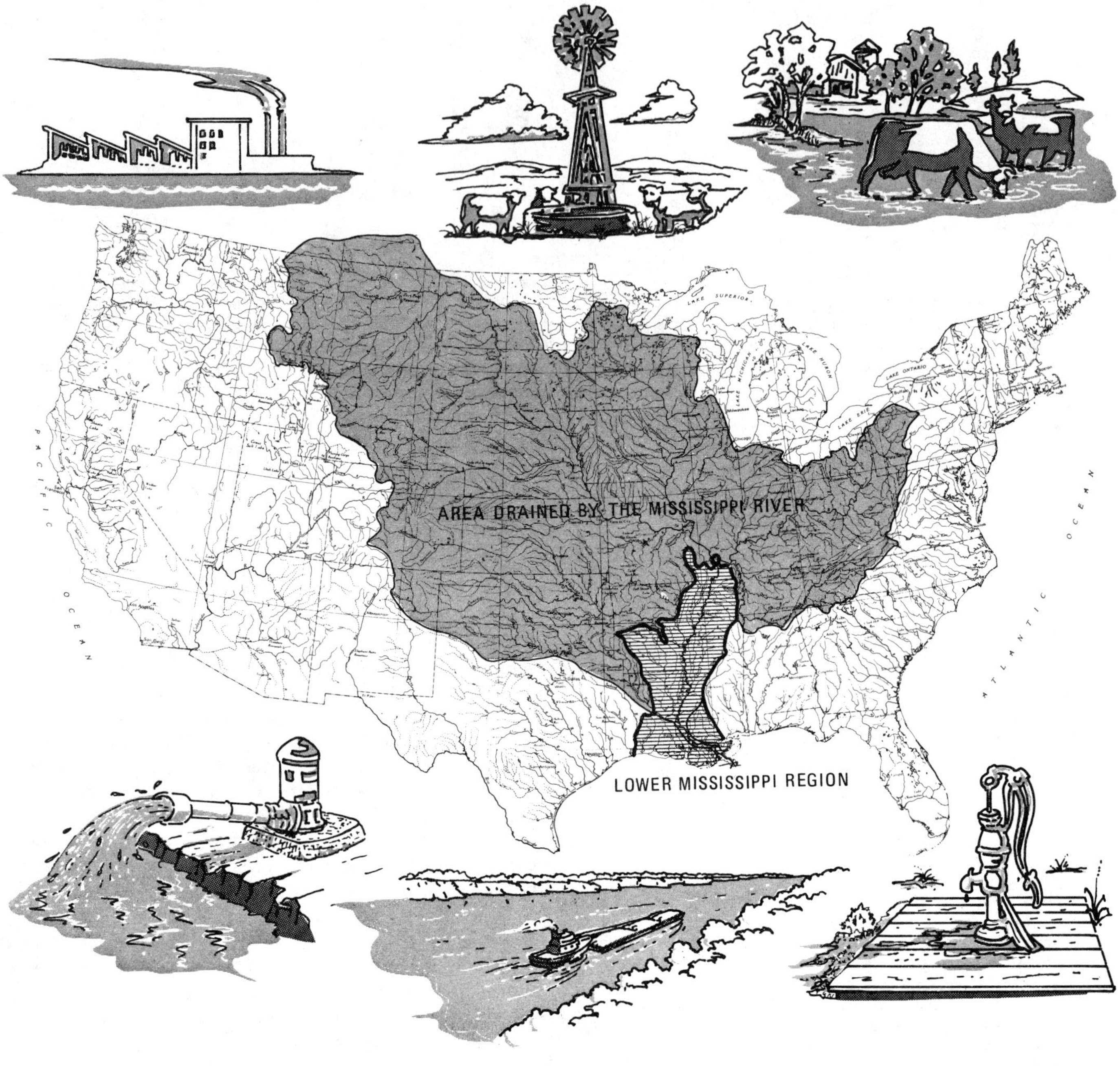

FIGURE 1.—The Lower Mississippi Region.

came a destructive force during floods. Since the time when the first plots of land were cleared for farming, agriculture has played a major role in both the economic and the cultural development of the region, and because the most productive land for farming lies along the Mississippi and its major tributaries, flood control has been the object of much concern and effort.

In more recent years, industry has greatly increased in the region. From Cairo, Ill., to the Gulf of Mexico, and between New Orleans and Lake Charles, La., are found petroleum refineries and related facilities, industrial and agricultural chemical plants, grain elevators, processing plants for food and kindred products, shipyards, textile mills, manufacturers of paper and related products, powerplants, cement plants, and aluminum-producing complexes. The reliance of many of these industries upon the river for movement of raw materials and finished products continues to focus regional attention upon the surface-water resource.

PURPOSE AND SCOPE

In contrast to the recognition given to surface water, only minor attention has been given to ground water in the Lower Mississippi Region since the early 1900's. The purpose of this report is to direct attention to the region's large ground-water resource so that it will not be overlooked when plans for water-resource-related changes within the region are devised and implemented. The pertinent questions about ground water that should be considered in developing and implementing water-management plans are:

1. Where is the ground water?
2. How much ground water is available?
3. What is the quality of the ground water?
4. What effects will development and use of ground water have on the total water resources and environment in the region?

And, inasmuch as ground water is such an important part of the total water resources of the region and should be considered in planning—

5. What kinds of data are needed, and where are data insufficient, to permit full consideration of ground water in water-resource planning?

This report supplies answers to these questions by presenting a regional assessment of the ground-water resource with emphasis on its significance. The scope is intended to be sufficient to permit evaluation of broad concepts of water planning for the region and to determine whether ground water has been adequately considered. This report is also intended to provide sufficient detail to serve as a basis for planning and program development.

All unreferenced quantitative values in this report were taken from either the "Lower Mississippi Region Comprehensive Study," by the Lower Mississippi Region Comprehensive Study Coordinating Committee (1974), or the "1975 National Water Assessment: Ground Water in the Lower Mississippi Region," by Boswell (1979).

Most numbers in this report are given in inch-pound units followed by metric units in parentheses. The conversions to metric units were made as follows:

Multiply		*By*	*To obtain*	
Inch-pound unit	*Inch-pound abbreviation*	*Conversion factor*	*Metric unit*	*Metric abbreviation*
Acre	acre	0.4047	Hectare	ha
Acre-foot	acre-ft	.0012335	Cubic hectometer	hm^3
Cubic foot	ft^3	.02832	Cubic meter	m^3
Cubic foot per second	ft^3/s	.02832	Cubic meter per second	m^3/s
Foot	ft	.3048	Meter	m
Gallon	gal	3.7854	Liter	L
Gallon	gal	.0037854	Cubic meter	m^3
Gallon per minute	gal/min	.06309	Liter per second	L/s
Inch	in.	25.4	Millimeter	mm
Mile	mi	1.6093	Kilometer	km
Square mile	mi^2	2.59	Square kilometer	km^2

Chemical concentrations are given only in metric units—milligrams per liter (mg/L). For concentrations less than 7,000 mg/L, the numerical value is about the same as for concentrations in the inch-pound unit, parts per million.

Throughout this report references to small, moderate, and large quantities of water in relation to aquifer yields have the following meaning: small, 0–50 gal/min (0–30 L/s); moderate, 50–500 gal/min (3–30 L/s); large, greater than 500 gal/min (30 L/s).

THE WATER-RESOURCES SCENE

The Lower Mississippi Region is indeed water rich. Precipitation throughout the region is generally abundant and well distributed areally. The normal annual precipitation ranges from 44 in. (1,100 mm) in the northern part of the region to 64 in. (1,600 mm) in the southeastern part (fig. 2). Seasonally, precipitation maximums occur in the winter in the northern part and in the summer in the southern part. A part of this precipitation is returned to the atmosphere by evapotranspiration, part infiltrates to the aquifers, and part becomes runoff.

Potential evaportranspiration is the combination of evaporation from the ground surface and transpiration from plants that would occur if there were complete vegetation coverage and adequate soil moisture. In the Lower Mississippi Region, average annual potential evapotranspiration ranges from 30 in. (760 mm) in the north to 44 in. (1,100 mm) in the south (fig. 3). Because periods of limited soil-moisture availability occur from time to time and vegetation coverage is not complete in many areas, actual evapotranspiration throughout a long period averages only about 70 to 90 percent of the potential value.

Runoff, including base runoff and direct runoff, combines with surface inflow to the region to make up streamflow and maintain water levels in surface-water impoundments. Runoff ranges from 18 in. (460 mm) in the north to 26 in. (660 mm) in the south (fig. 4).

Recharge to the confined aquifers occurs primarily in the outcrop areas. In the alluvial unconfined aquifers, recharge occurs in areas where the vertical hydraulic conductivity of the overlying material is sufficient to allow downward movement of water. When evapotranspiration requirements are met and soil-moisture storage maximums are exceeded, water infiltrates to the aquifers. An estimate of available recharge can be expressed by:

$$\text{Recharge} = \text{precip} - (\text{ET} + \text{direct runoff}),$$

where

precip = precipitation,
ET = evapotranspiration, and
direct runoff = direct surface runoff.

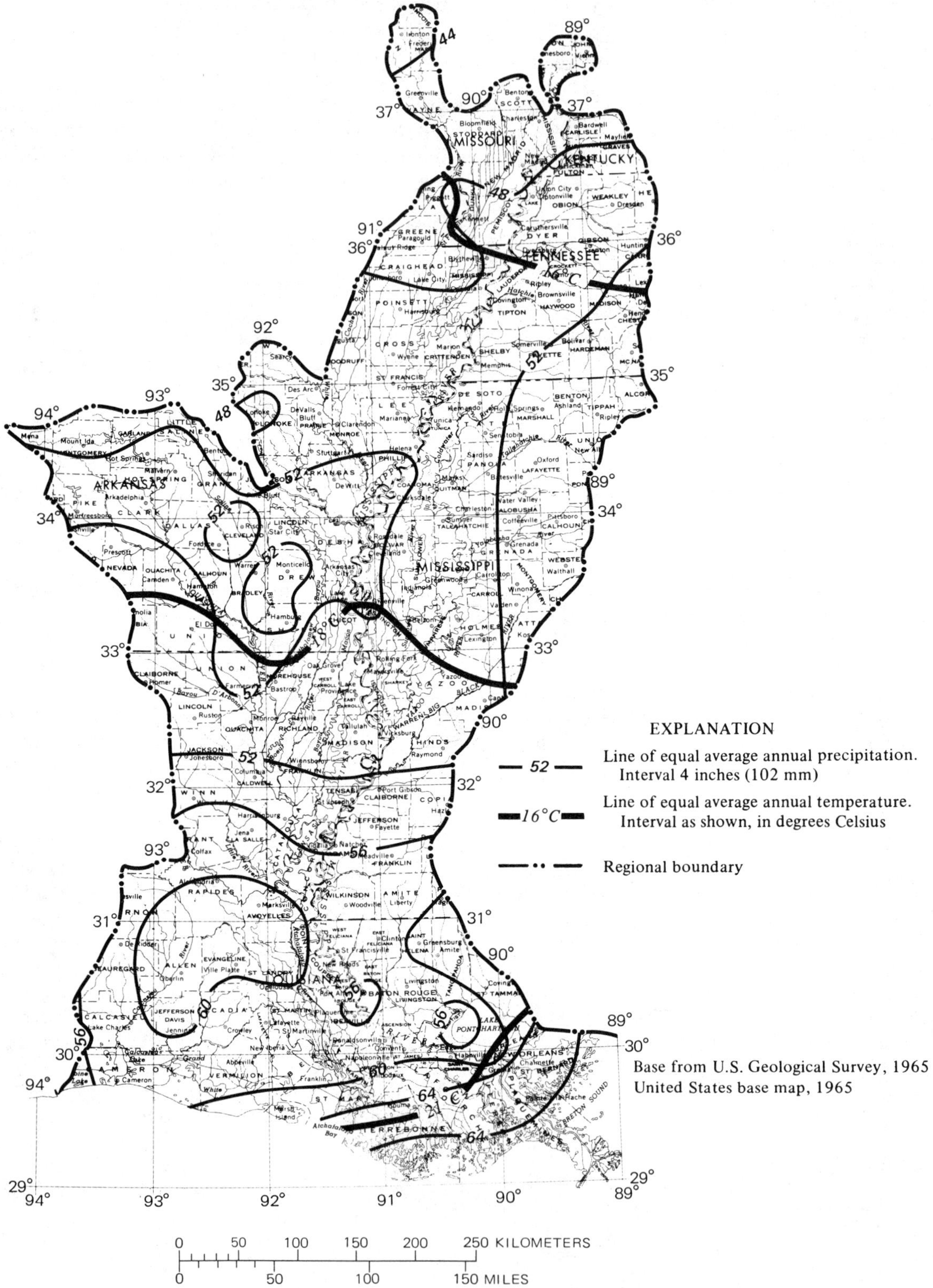

FIGURE 2.—Normal annual precipitation and temperature (modified from Lower Mississippi Region Comprehensive Study, 1974, v. 1, app. C).

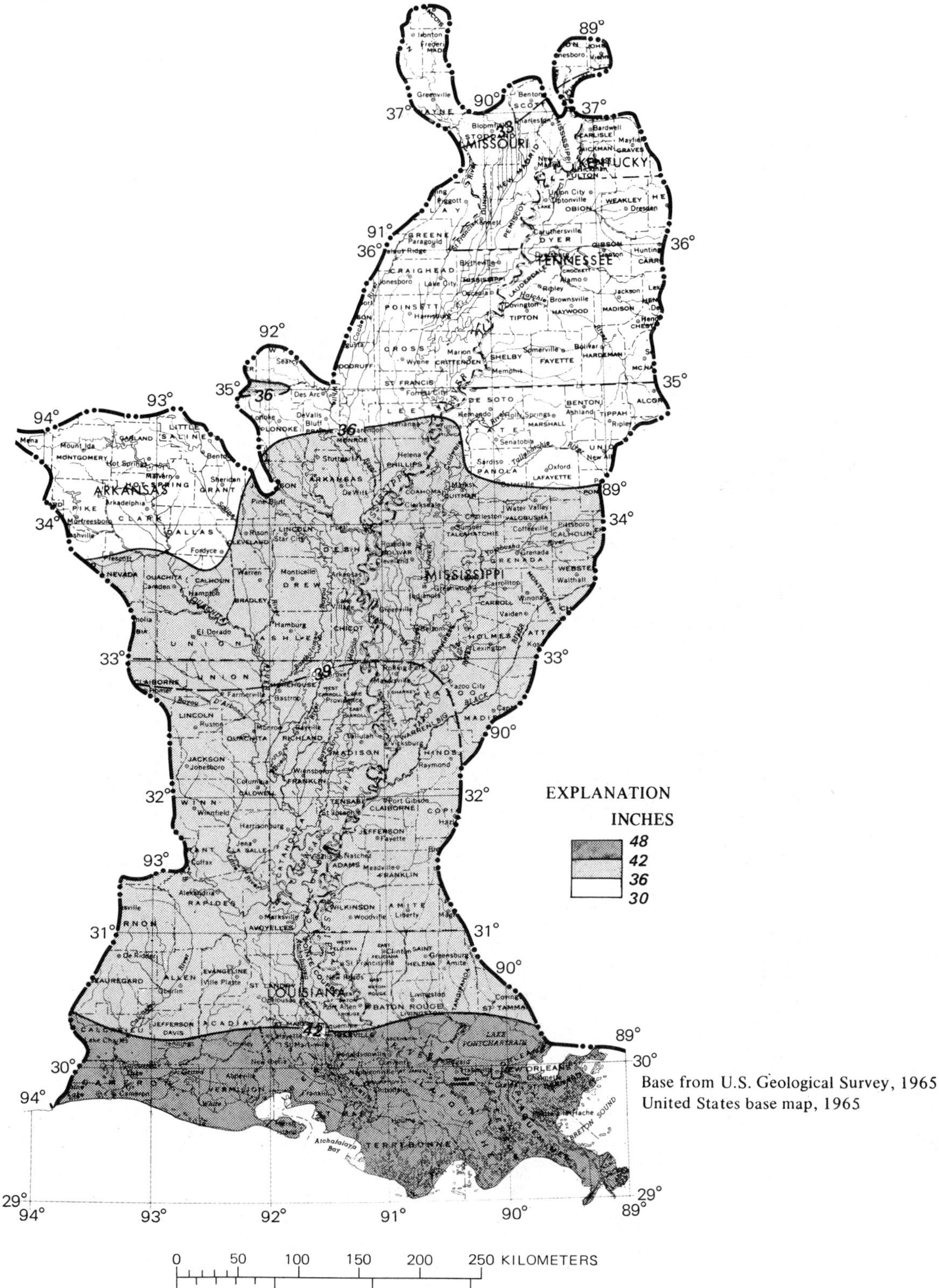

FIGURE 3.—Average annual potential evapotranspiration (from Thornthwaite, 1948).

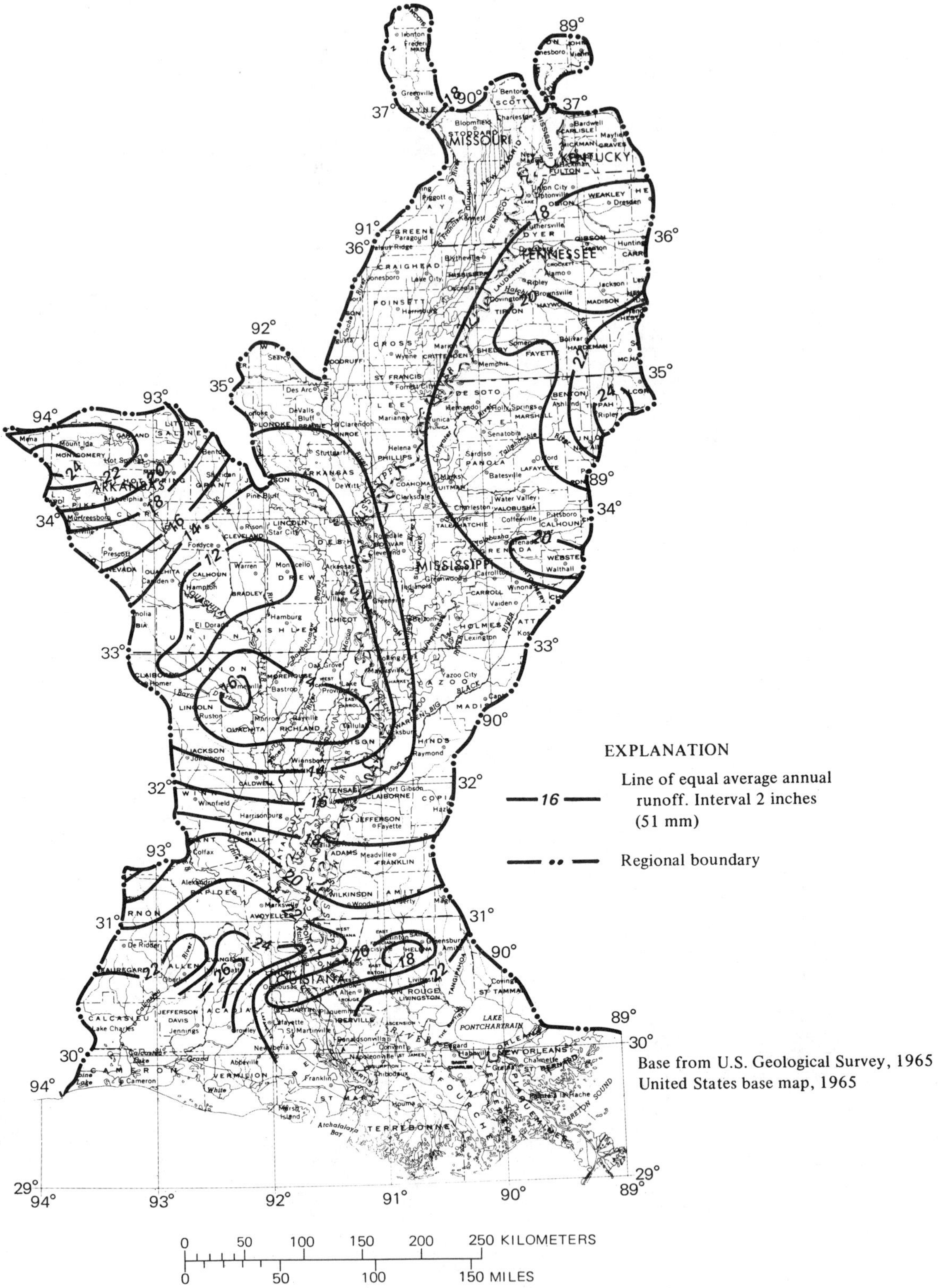

FIGURE 4.—Mean annual runoff, in inches.

HYDROLOGIC BOUNDARIES

The Lower Mississippi Region does not constitute a single or discrete hydrologic system. Although the regional boundaries are located primarily on drainage-basin divides, three major streams in addition to the Mississippi River bring substantial quantities of water across the regional boundary. In addition, and of even more significance in this region, underlying aquifers extend into adjacent areas and ground water moves into and out of the region. Because the regional boundaries are not completely hydrologically restrictive, the water resources of the region are affected by hydrologic events outside the region.

SURFACE WATER

The Lower Mississippi Region's surface-water supply is derived from precipitation and runoff within the region, streamflow including ground-water discharge entering the region from adjacent areas, and ground-water discharge to streams within the region.

The total mean annual inflow in major streams entering the region is nearly 550,000 ft^3/s (15,600 m^3/s). The mean annual stream discharge generated within the region is about 120,000 ft^3/s (3,400 m^3/s).

Each of the 29 controlled surface-water reservoirs within the region has a capacity of 5,000 acre-ft (6 hm^3) or more. The reservoirs have a combined storage of about 10 million acre-ft (12,300 hm^3).

Mean-annual stream outflow from the region is about 670,000 ft^3/s (19,000 m^3/s). Theroretically, this is the ultimate quantity of surface water available for use. However, because of the small number of available storage sites and the increased evaporative losses of surface water that occur with development, this quantity is not realistically obtainable. The dependable surface-water yield must therefore be defined on the basis of the percentage of time a given flow is available. For the Lower Mississippi Region, the 95-percent-duration flow (the flow that will be equaled or exceeded 95 percent of the time) is 216,400 ft^3/s (6,100 m^3/s), or 160 million acre-ft (197,000 hm^3) per year.

In 1970, regional surface-water withdrawals averaged 22,000 ft^3/s (620 m^3/s) or 16 million acre-ft (20,000 hm^3) per year. The total water returned to streams, including ground water withdrawn and not consumed, was 18,000 ft^3/s (510 m^3/s). The result was a net streamflow loss of 4,000 ft^3/s (113 m^3/s).

GROUND WATER

Ground water occurs in large quantities in the Lower Mississippi Region and is readily accessible because of the regional geological framework. Almost all the region is within the Gulf Coastal Plain physiographic province; three small areas along the northwest boundary are in the Interior Highlands (fig. 5). The Lower Mississippi Region includes most of the Mississippi embayment, a northeast-trending structural trough underlying part of the Coastal Plain. The Coastal Plain and the embayment received sediment during the Cretaceous, Tertiary, and Quaternary Periods. The older deposits generally consist of alternating layers of sand and clay; the Quaternary beds contain considerable gravel. The more permeable sand and gravel deposits now form the extensive and productive aquifers that underlie the Lower Mississippi Region.

The Cretaceous and older Tertiary units in the northern or embayment part of the area dip toward the axis of the Mississippi embayment, which coincides approximately with the present course of the Mississippi River. In the southern part of the area, the younger Tertiary deposits dip gulfward. Quaternary alluvium blankets most of the area and forms a gulfward-thickening wedge in the southern part (fig. 6).

Except in areas of outcrop and where local dewatering has taken place, water in the Cretaceous and Tertiary aquifers is confined under pressure; that is, water levels in wells tapping these aquifers rise above the top of the aquifer. Most of the Quaternary aquifers are also confined. In some areas, hydrostatic pressures are sufficient to produce natural flows from wells; in some areas, wells that once flowed no longer flow due to pressure declines. Although most of the declines have been caused by heavy pumping, a contributing and in places critical factor has been the practice of allowing uncapped wells to flow unregulated.

Recharge to the aquifers is primarily from rainfall. The movement of water in the confined aquifers generally is downdip unless affected by large withdrawals. In the alluvial aquifers, movement is toward points of discharge. The base of freshwater in the Coastal Plain aquifers is shown in figure 7. All aquifers locally contain some saltwater: Quaternary aquifers in coastal areas and the older aquifers in their downdip areas.

The vast reserves of saltwater in the Coastal Plain aquifers may prove to be an asset to the area as desalinization technology advances. The quantity of saltwater available is several times that of freshwater. Some saltwater is now used for purposes such as industrial cooling. Similar uses that can tolerate water of this quality will further enhance the value of the resource.

Aquifers in the parts of the Lower Mississippi Region that lie in the Interior Highlands do not repre-

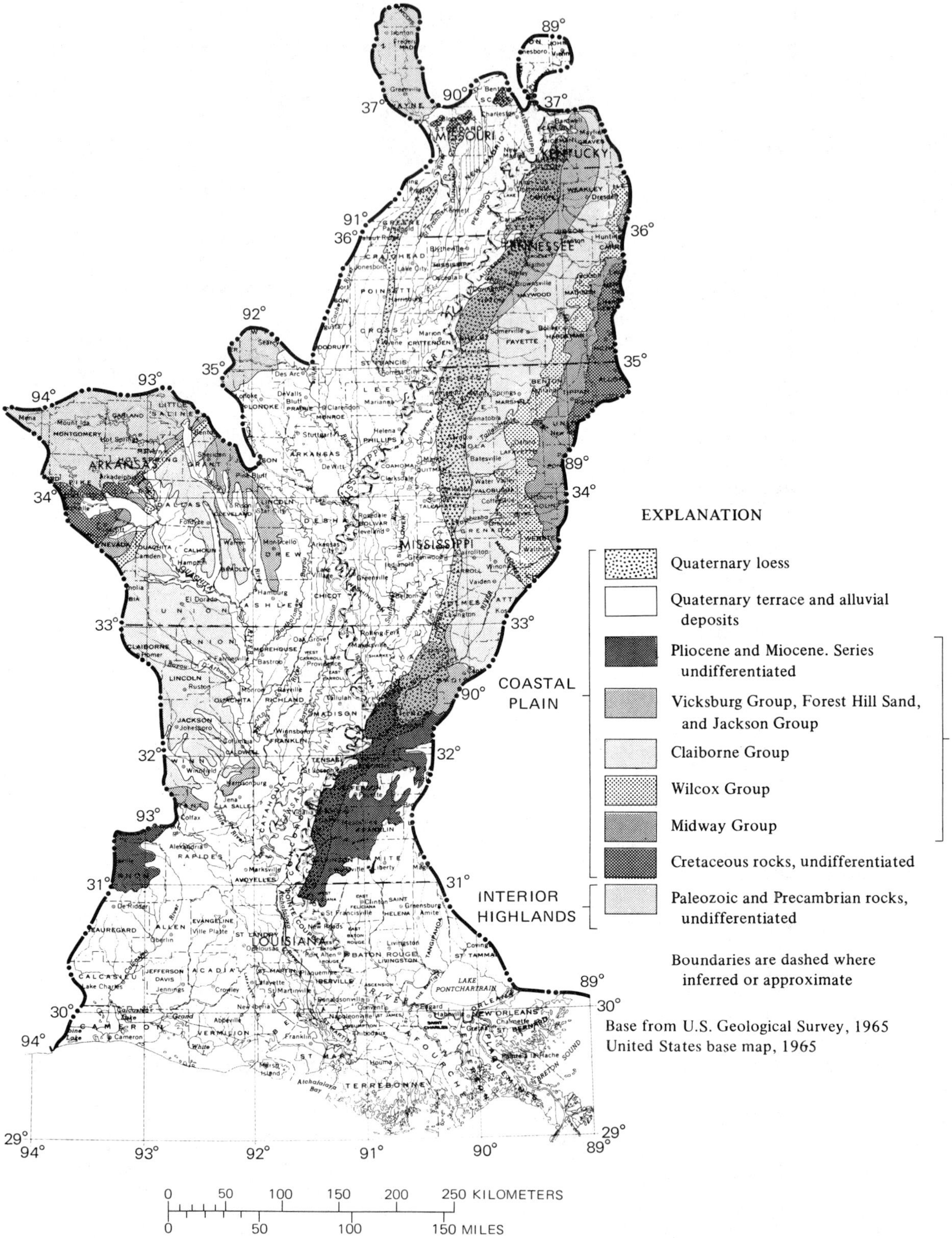

FIGURE 5.—Geologic map (modified from Lower Mississippi River Comprehensive Study, 1974, v. 1, app. C).

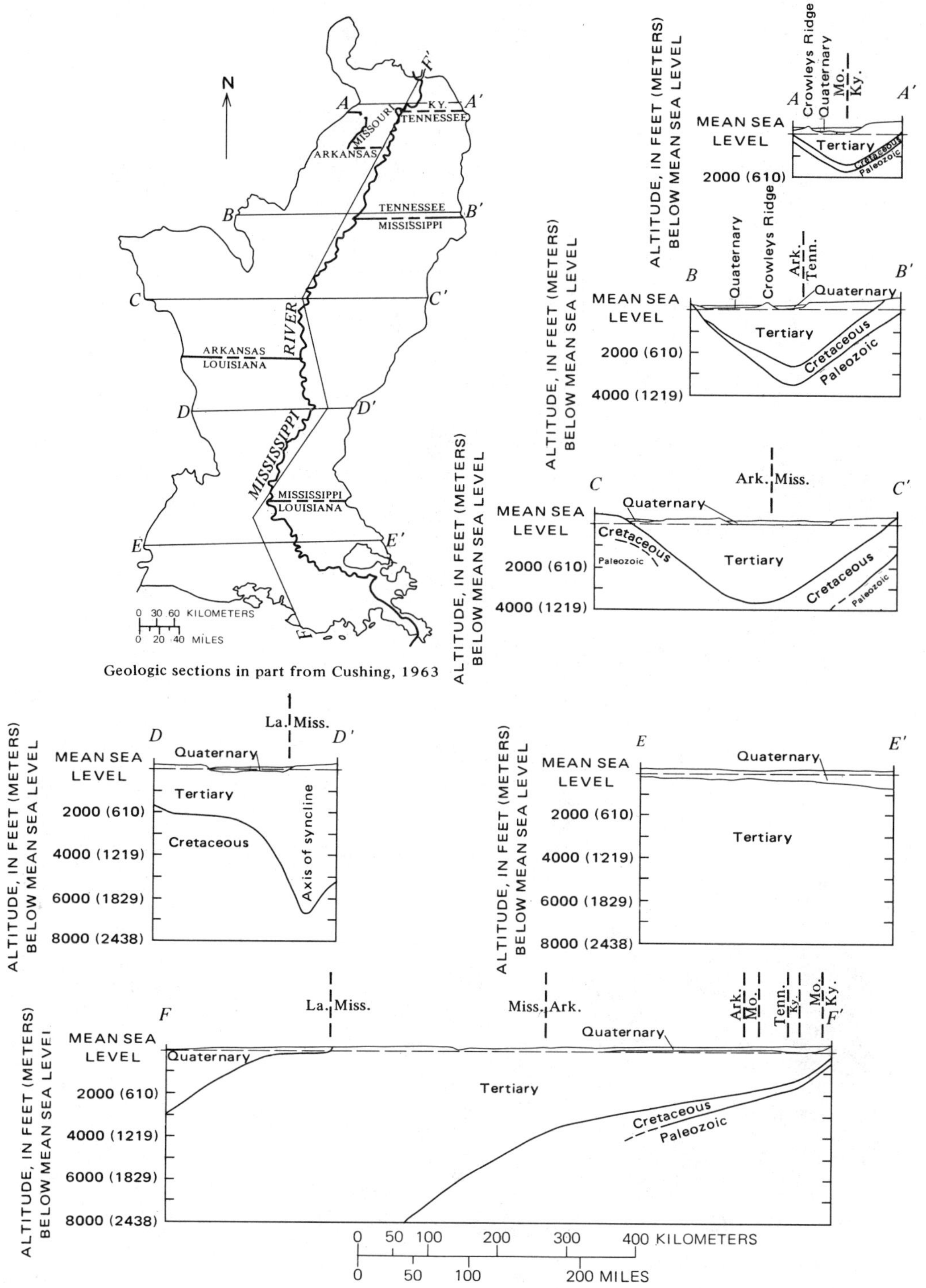

FIGURE 6.—Generalized geologic sections in the Lower Mississippi Region (modified from Lower Mississippi River Comprehensive Study, 1974, v. 1, app. C).

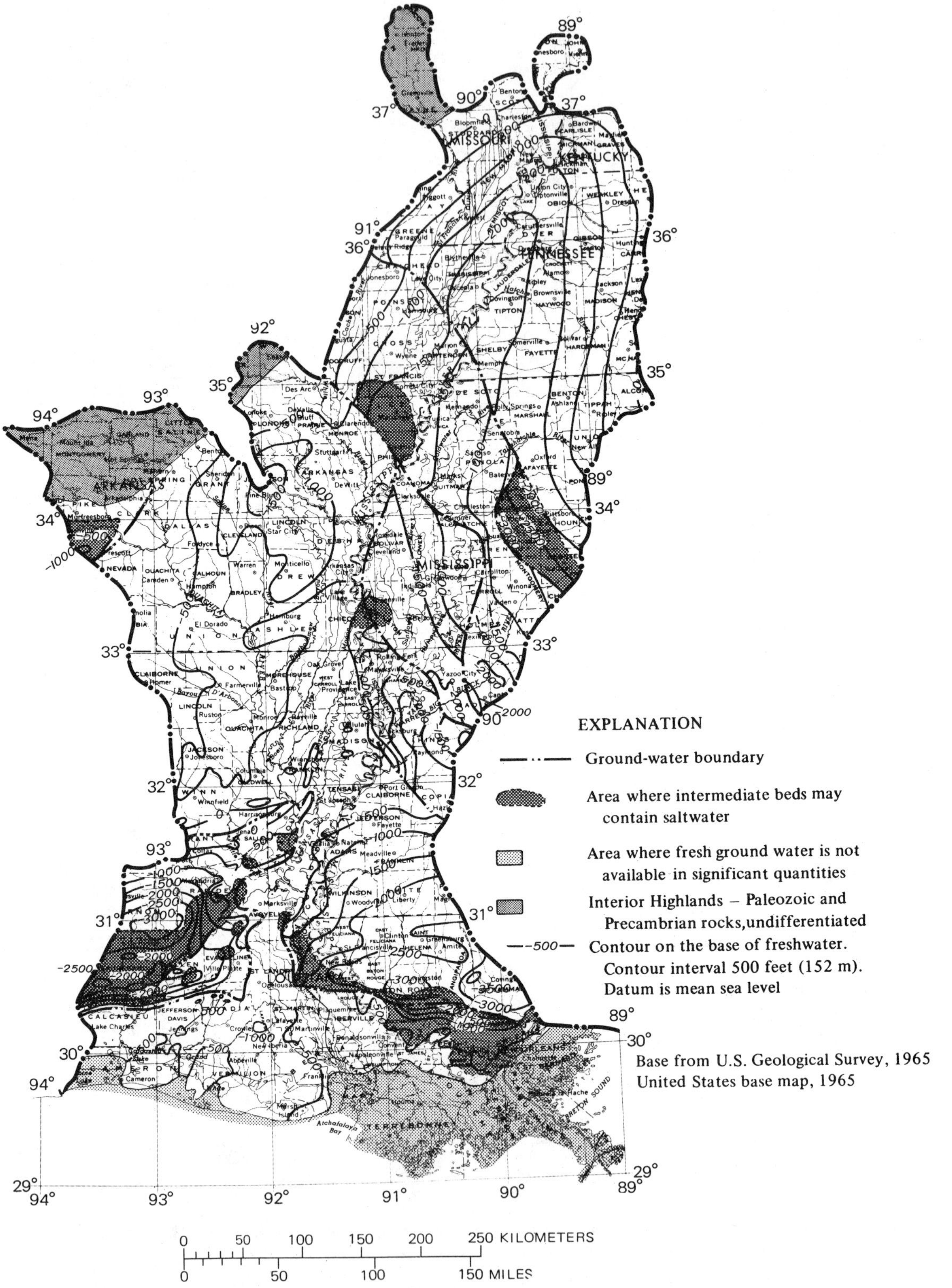

FIGURE 7.—Base of freshwater in Coastal Plain aquifers (modified from Lower Mississippi River Comprehensive Study, 1974, v. 1, app. C).

sent a significant resource to the region. Rocks in these areas are of Paleozoic age, mostly hard sandstone and shale, and the ground water they contain occurs in openings along fractures and bedding planes. The interstitial porosity and permeability that make the unconsolidated aquifers of the Coastal Plain so productive do not exist in the Interior Highlands.

An estimated 347,000 billion ft^3 (9,800 billion m^3) or 7,900 million acre-ft (10 million hm^3) of freshwater underlies the Lower Mississippi Region. Of this total, about 844 billion ft^3 (24 billion m^3) or 19 million acre-ft (23,900 hm^3) is available annually for development, using conventional methods. In 1970, regional ground-water withdrawals averaged about 8,300 ft^3/s (240 m^3/s) or 6 million acre-ft (7,400 hm^3) per year.

About 65 percent of the ground water withdrawn in the region is used for irrigation, about 15 percent by industry, and about 8 percent for municipal supply. The remaining 12 percent is used for domestic supply, livestock watering, and other uses.

WHY IS GROUND WATER OF IMPORTANCE IN THE LOWER MISSISSIPPI REGION?

WIDESPREAD ACCESSIBILITY

Ground water is available beneath the entire Lower Mississippi Region. Except for small areas in the Interior Highlands and the coastal area of Louisiana, one or more major aquifers (fig. 8) make moderate to large quantitites of freshwater available throughout the region (fig. 9, 10).

The three areas in the region that are in the Interior Highlands are the Arkansas Valley and the Ouachita Mountains in west-central Arkansas and the Ozark Plateaus in southeast Missouri. In these areas, small quantities of ground water are available from Paleozoic rocks (figs. 9, 10).

Aquifers of Cretaceous age underlie the northern part of the region. Except for relatively small areas in Arkansas, Tennessee, Mississippi, and Missouri, the Cretaceous material is overlain by Tertiary and (or) Quaternary aquifers (fig. 6) that can yield moderate to large quantitites of water to individual wells. For this reason, the deeper Cretaceous aquifers are not utilized extensively.

The major Cretaceous aquifer is the McNairy Sand Member of the Ripley Formation in Mississippi (equivalent to the McNairy Sand in Tennessee, Missouri, Illinois, and Kentucky, and the Nacatoch Sand in Arkansas). These aquifers are present throughout the northern one-fifth of the region within the Coastal Plain and in a small area in southwest Arkansas, a total area of nearly 20,000 mi^2 (52,000 km^2). Withdrawals have been restricted to areas where the

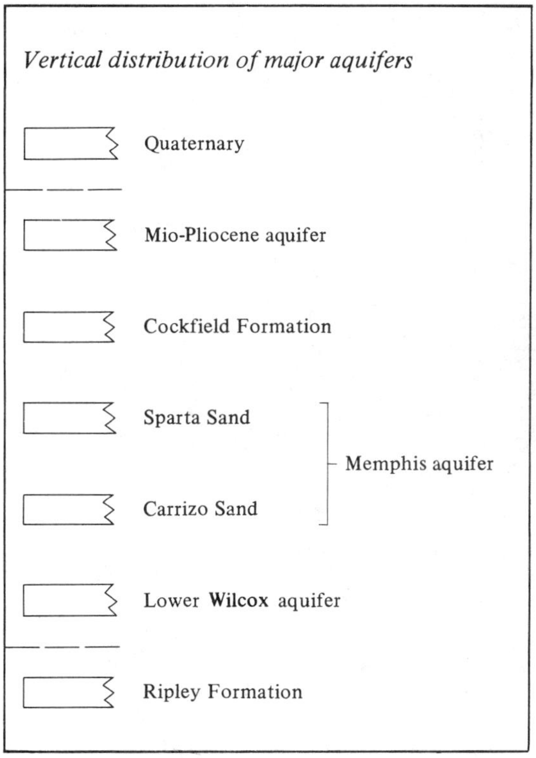

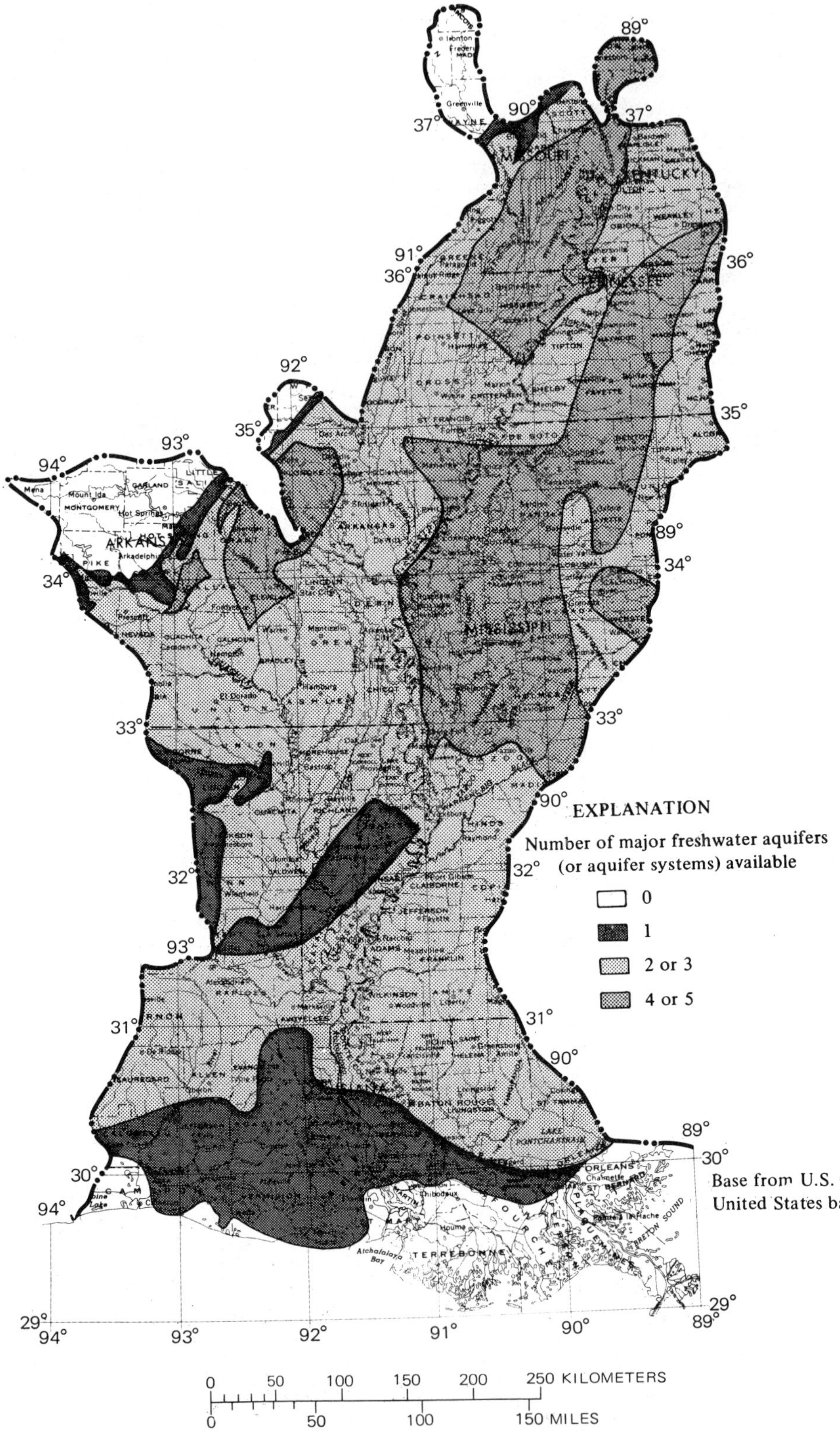

Base from U.S. Geological Survey, 1965
United States base map, 1965

FIGURE 8.—Multiple freshwater aquifers underlie 90 percent of the Lower Mississippi Region.

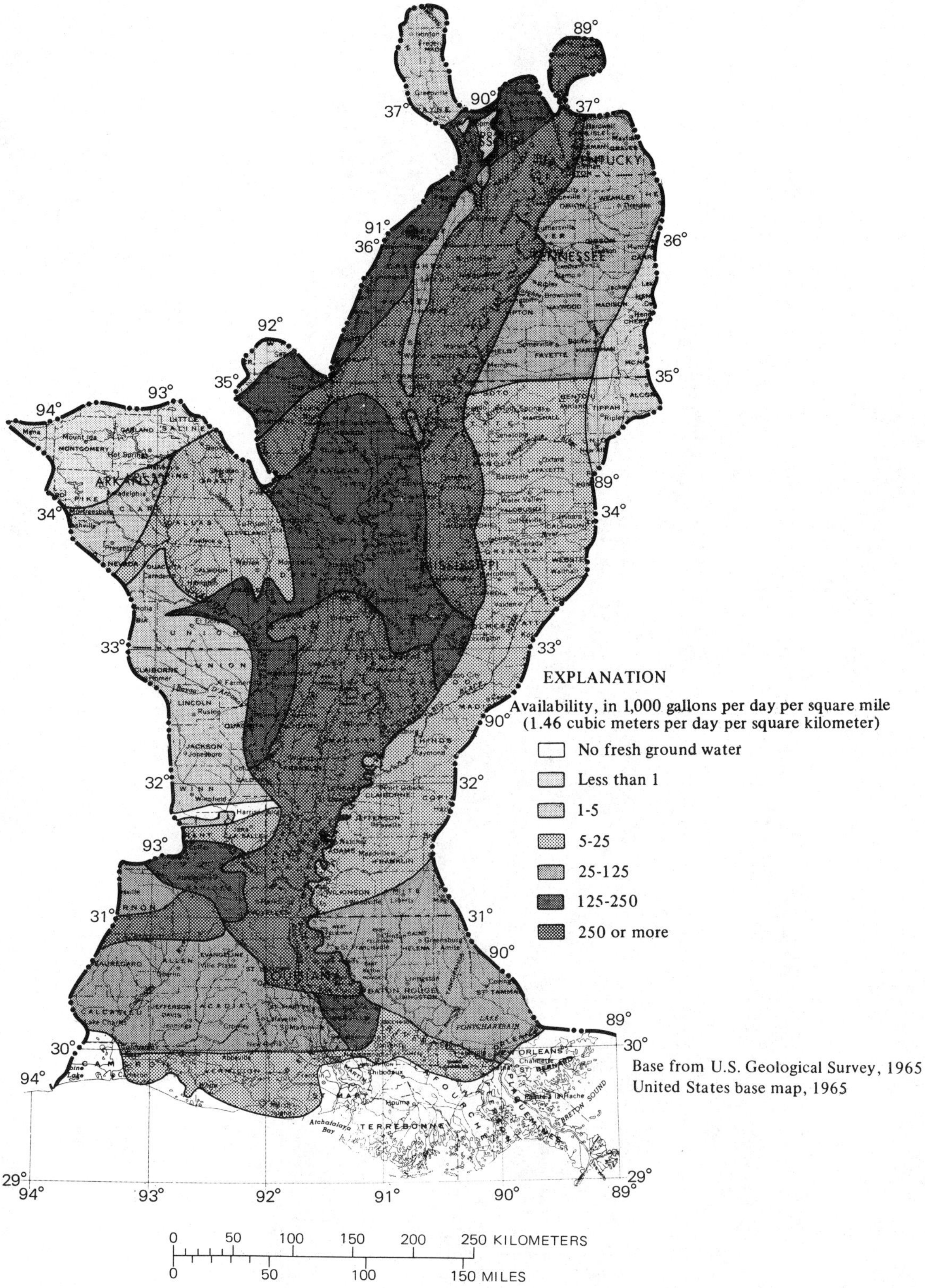

FIGURE 9.—Availability of fresh ground water.

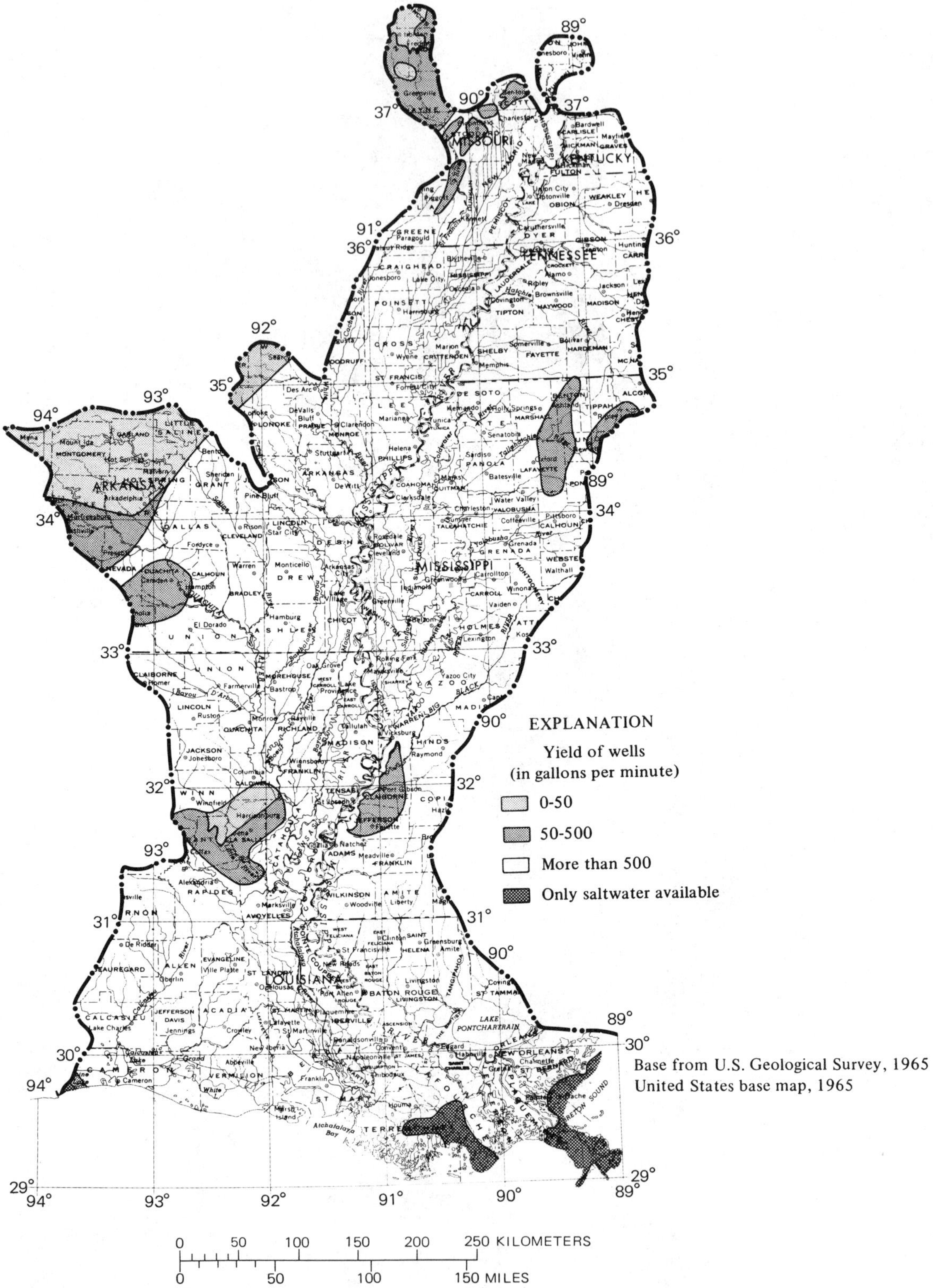

FIGURE 10.—Ranges in well yields throughout the region; only in small areas do wells yield less than 500 gallons per minute (modified from Boswell, 1975).

aquifers are present at moderate depths and water levels have been practically unaffected by pumping.

Other Cretaceous aquifers are important in some areas where they are the best or only sources of ground water available. In north-central Mississippi and parts of west Tennessee, the Coffee Sand is utilized. It is the best or only source of ground water in many places in this area where the underlying Eutaw Formation is too thin or yields highly mineralized water, or where Paleozoic rocks do not contain aquifers. The Gordo Formation underlies the Eutaw Formation in north-central Mississippi, primarily south of the downdip limit of the Coffee Sand. Water in the Gordo Formation generally is good quality and is the best available in this area.

Tertiary aquifers underlie virtually the entire Lower Mississippi Region, except for three small areas on the western border, two in Arkansas and one in Missouri, and a small strip on the eastern border in Tennessee and Mississippi. The Tertiary aquifers north of a line approximately through Vicksburg, Miss., to Colfax, La., are of Eocene age. South of this line, Miocene aquifers overlie the Eocene deposits. These beds dip southward and are, in turn, overlain by Pliocene deposits south of the 31st parallel.

The significant Eocene aquifers, in ascending order of their occurrence, are: the lower Wilcox aquifer, the Carrizo Sand (and its stratigraphic equivalent, the Meridian Sand Member of the Tallahatta Formation), the Sparta Sand, and the Cockfield Formation.

The lower Wilcox aquifer, the basal unit of the Wilcox Group, occurs throughout the northern one-third of the Lower Mississippi Region and in a strip across central Arkansas. It crops out in a narrow belt in north Mississippi and west Tennessee (Fort Pillow Sand in the subsurface of west Tennessee) and occurs as a subcrop beneath the Quaternary alluvium in Arkansas and Missouri. The lower Wilcox aquifer is a source of water for several cities in northeast Arkansas and northwest Mississippi.

Except for the lower Wilcox aquifer, the Tertiary aquifers mentioned previously are in the Claiborne Group. The basal unit of the Claiborne is the Carrizo Sand in Arkansas and Louisiana and its equivalent in Mississippi, the Meridian Sand Member of the Tallahatta Formation. This unit crops out in a narrow band in central Mississippi and southwest Arkansas and is a relatively minor aquifer in central Arkansas and in west Mississippi north of the latitude of Vicksburg. Separating the Carrizo from the overlying Sparta Sand is the Cane River Formation and its equivalents, composed mostly of clay and a few thin beds of fine, almost impermeable sand. The Cane River and its equivalents contain only very minor aquifers in south-central and southwest Arkansas and in west-central Mississippi. In northwest Mississippi, the Winona Sand (equivalent to the Cane River Formation) becomes more significant and merges into the Memphis aquifer. The Sparta Sand underlies the entire central part of the region. It crops out on the eastern side in a wide belt, from southwest Kentucky through Tennessee and Mississippi, and on the west side in northeast and south-central Arkansas. The Sparta occurs as subcrops beneath the Quaternary alluvium in some areas in Arkansas and Mississippi. The Sparta is a very productive aquifer throughout the northern three-fourths of the region. The Cook Mountain Formation, which is not an aquifer, overlies the Sparta Sand and separates it from the uppermost unit of the Claiborne Group, the Cockfield Formation. The Cockfield Formation directly underlies the Quaternary alluvium in most of the central part of the region. It includes productive aquifers in Arkansas, Louisiana, and Mississippi. The Cockfield crops out in small areas in southeast Arkansas, along the Arkansas-Louisiana State boundary, in northwest Louisiana, and in central Mississippi.

North of approximately the 35th parallel, the Memphis aquifer (Memphis Sand in Tennessee) comprises all Claiborne units from the base of the Carrizo Sand to the top of the Sparta Sand. This part of the Claiborne section is a massive sand several hundred feet thick which constitutes a vast ground-water reservoir. As such, the Memphis aquifer is second only to the Mississippi River valley alluvial aquifer as a potential source of large quantities of ground water.

Aquifers of Oligocene age in the Forest Hill Sand and the overlying Vicksburg Group occur in a small area in the southern half of the Lower Mississippi Region. These aquifers, although they are not extensive, are the only sources of fresh ground water in the 500 ft (150 m) or more of sediments between the top of the Claiborne Group and the base of the Miocene Series.

Aquifers of Miocene age occur south of a line approximately through Vicksburg, Miss., to Colfax, La. South of approximately the 31st parallel, the Miocene deposits are overlain by the Pliocene Series. These two series are lithologically similar and are referred to in Louisiana, where they occur together, as the Mio-Pliocene aquifer. In Mississippi, the Miocene deposits are divided into the undifferentiated upper Miocene aquifer and the Catahoula Sandstone. Both the Miocene and Pliocene Series are overlain by Quaternary deposits and, except for Miocene outcrops in west-central and southwest Louisiana, are not exposed at the surface. Both of these series are good present and potential sources of moderate to large quantities of fresh ground water. Near the Louisiana coast, water in both series becomes salty gulfward.

Deposits of Quaternary age cover most of the Lower Mississippi Region. Sediments of Pleistocene and Holocene age compose the Mississippi River valley alluvial aquifer, the most extensive source of ground water in the region. The Pleistocene deposits contain gravel at the base, grading upward to finer sand, and are the most productive parts of the aquifer. The overlying Holocene material is composed of very fine sand, silt, and clay. In many areas it forms a confining layer, although it is permeable to varying degrees.

Large quantities of water are available from the Mississippi River valley alluvial aquifer throughout most of the region. The aquifer's value is enhanced in that, generally, only shallow well depths are required, pumping lifts are small, and recharge conditions are favorable. Throughout most of the Lower Mississippi Region, recharge to the alluvial aquifer is by precipitation. In some areas, where overlying fine-grained materials are nearly impermeable, the aquifers are recharged by underflow. Along the coast in Louisiana, especially in the southeastern part of the State, the alluvial aquifer contains saltwater.

Ground water can be obtained with relative ease almost everywhere in the Lower Mississippi Region. For this reason, most public, industrial, and agricultural supplies are from wells. In most areas within the region, obtaining adequate quantities of water from surface-water supplies would be economically unfeasible. Of the total 102,400 mi^2 (265,200 km^2) within the Lower Mississippi Region, about 5 percent is covered by surface water. In contrast, 90 percent of tbe region is underlain by two or more aquifers that can yield 100 gal/min (6L/s) or more to individual wells.

In contrast to surface reservoirs that inundate many acres of land, large quantities of water are stored in subsurface reservoirs without loss of surface area. Some disadvantages of surface reservoirs are high construction costs, cost of land purchase, loss of land use, and maintenance. Another consideration is that some terrain is not suited for large reservoir construction. Much of the area within the southern half of the Lower Mississippi Region could be classified as unsuitable because of the absence of deep, broad stream valleys that could be dammed. For these reasons, much dependence is placed on the region's subsurface reservoirs.

LARGE AMOUNT AVAILABLE

Approximately 347,000 billion ft^3 (9,800 billion m^3) of water, containing less than 3,000 mg/L dissolved solids, is stored in the subsurface of the Lower Mississippi Region. This quantity of water is more than 16 times the average annual surface-water outflow from the region, and if contained in a reservoir it would cover an area the size of the entire region to a depth of 120 ft (37 m). Of this total, about 844 billion ft^3 (24 billion m^3) is available annually for development, based on withdrawals consistent with economically and environmentally acceptable water-level declines.

The primary containers of ground water in the lower Mississippi Region are the extensive unconsolidated sand-and-gravel aquifers of Tertiary and Quaternary age. The ability of these aquifers to store and transmit water varies due to differences in thickness and hydraulic conductivity. However, single aquifers that can yield 500 gal/min (32 L/s) or more to individual wells underlie about 90 percent of the region (fig. 10). The highest yields, often several thousand gallons per minute, are from wells screened in sand and gravel of the Quaternary alluvial-terrace deposits. The alluvial and terrace deposits account for two-thirds of the potential ground-water supply in the region.

Within most of the Lower Mississippi Region, water-table aquifers commonly discharge water to streams that are connected with them (fig. 11A). Also, some streams that have sufficient hydraulic connection with a confined aquifer may receive contributions from the aquifer. Such a condition occurs when the altitude of the water surface in the stream is less than the head (potentiometric surface) in the aquifer (fig. 11B). Under dry, low-flow conditions, perennial streams are sustained completely by discharged ground water. The lowest flow that occurs in a stream for 7 consecutive days once every 10 years is commonly accepted to be composed of discharged ground water. U.S. Geological Survey stream-gaging stations, within the region, where the 7-day, 10-year low-flow has been defined, are shown in figure 12. Table 1 contains the 7-day, 10-year low-flows for the stations plotted in figure 12.

Where there is good connection between a major stream and an aquifer, advantage can be taken of the relationship by locating wells near the stream. When a well is pumped, one of two things will occur. If movement has been from the aquifer to the stream, gradients will flatten and may reverse in the vicinity of the well, utilizing water normally discharged to the stream and even taking water from the stream if the gradient reverses. If movement has been from the stream to the aquifer, the gradient will become steeper as the head in the aquifer near the well is reduced by pumping. In either instance, some water is diverted from the stream, thereby reducing the stress on the aquifer (fig. 13).

Most aquifers in the Lower Mississippi Region are full; consequently much potential recharge is rejected

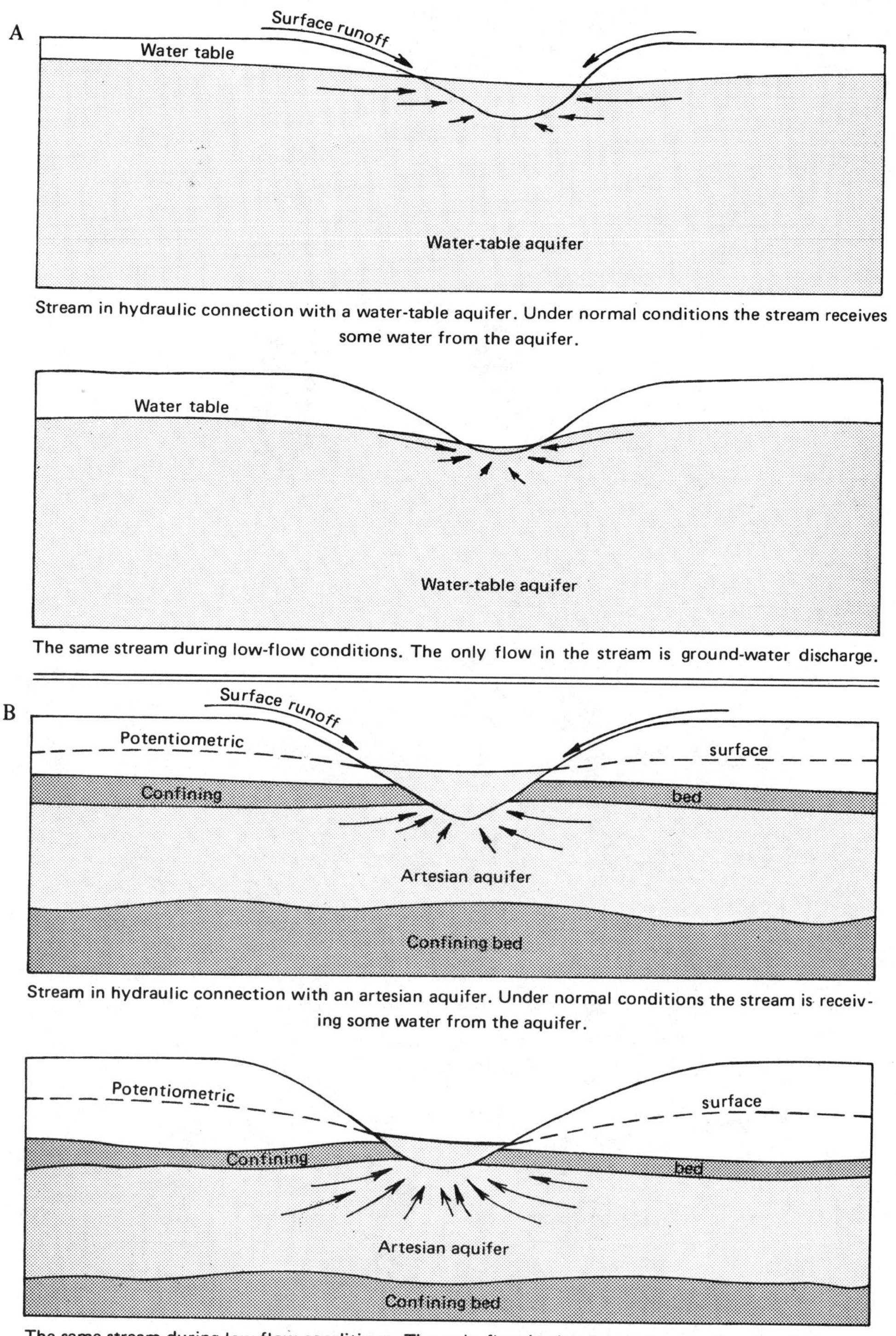

FIGURE 11.—Ground water-surface water relations during normal and dry conditions.

and evapotranspired or discharged to streams. When an aquifer is tapped and water is withdrawn, changing head relations can cause the aquifer to become receptive to recharge (fig. 14). Under the most favorable

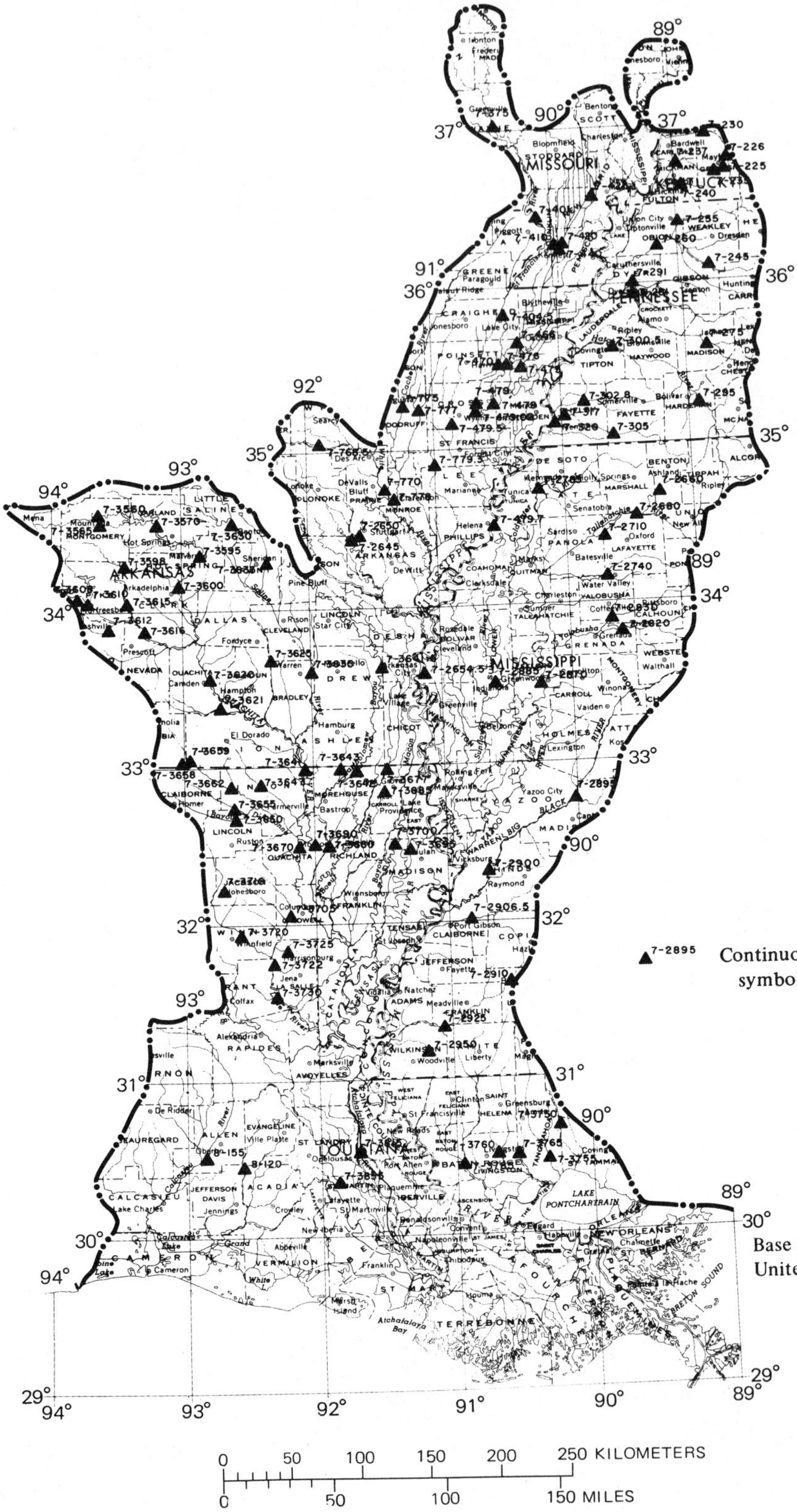

EXPLANATION

▲7-2895 Continuous-record gaging station. Numeral beside symbol is abbreviated station number

Base from U.S. Geological Survey, 1965
United States base map, 1965

FIGURE 12.—Selected 7-day, 10-year low-flow sites.

TABLE 1.—*The 7-day, 10-year low flow at selected stream-gaging stations plotted in figure 12*

Station Number	Station Name	7-day, 10-yr low flow (ft^3/s)
07022500	Perry Creek near Mayfield, Ky	0.0
07022600	Mayfield Creek at Mayfield, Ky	0
07023000	Mayfield Creek at Lovelaceville, Ky	7.7
07023500	Obion Creek at Pryorsburg, Ky	0
07023700	Obion Creek near Arlington, Ky	3.3
07024000	Bayou du (de) Chien near Clinton, Ky	6.3
07024500	South Fork Obion River near Como, Tenn	79
07025500	North Fork Obion River near Union City, Tenn	90
07026000	Obion River at Obion, Tenn	260
07027500	South Fork Forked Deer River at Jackson, Tenn	80
07028100	South Fork Forked Deer River near Halls, Tenn	145
07029100	North Fork Forked Deer River at Dyersburg, Tenn	93
07029500	Hatchie River at Bolivar, Tenn	122
07030050	Hatchie River at Rialto, Tenn	284
07030280	Loosahatchie River at Brunswick, Tenn	58
07030500	Wolf River at Rossville, Tenn	124
07031700	Wolf River at Raleigh, Tenn	158
07032000	Mississippi River at Memphis, Tenn	99,000
07037500	St. Francis River near Patterson, Tenn	14.7
07040100	St. Francis River at St. Francis, Ark	76
07040450	St. Francis River at Lake City, Ark	97
07041000	Little River Ditch 81 near Kennett, Mo	15
07042000	Little River Ditch 1 near Kennett, Mo	17
07042500	Little River Ditch 251 near Lilbourn, Mo	30
07044000	Little River Ditch 251 near Kennett, Mo	69
07046600	Right Hand Chute of Little River at Rivervale, Ark	146
07047000	St. Francis River floodway near Marked Tree, Ark	0
07047500	St. Francis River at Marked Tree, Ark	97
07047600	Tyronza River near Tyronza, Ark	27
07047800	St. Francis River at Parkin, Ark	284
07047900	St. Francis Bay at Riverfront, Ark	38
07047902	St. Francis River at latitude of Wittsburg, Ark	405
07047950	L'Anguille River at Palestine, Ark	<.1
07047970	Mississippi River at Helena, Ark	102,000
07076850	Cypress Bayou near Beebe, Ark	0
07077000	White River at DeValls Bluff, Ark	4,830
07077500	Cache River at Patterson, Ark	33
07077700	Bayou DeView at Morton, Ark	<.1
07077800	White River at Clarendon, Ark	3,530
07077930	Big Creek near Moro, Ark	0
07264500	Bayou Meto near Stuttgart, Ark	0
07265000	Crooked Creek near Humphrey, Ark	0
07265450	Mississippi River near Arkansas City, Ark	115,000
07266000	Cane Creek near New Albany, Miss	.4
07268000	Tallahatchie River at Etta, Miss	8.6
07271000	Clear Creek near Oxford, Miss	4.2
07274000	Yocona River near Oxford, Miss	7.2
07282000	Yalobusha River at Calhoun City, Miss	<.1
07283000	Skuna River at Bruce, Miss	2.0
07287000	Yazoo River at Greenwood, Miss	720
07288500	Sunflower River at Sunflower, Miss	94
07289500	Big Black River at Pickens, Miss	41
07290000	Big Black River near Bovina, Miss	75
07290650	Bayou Pierre near Willows, Miss	19
07291000	Homochitto River at Eddiceton, Miss	31
07292500	Homochitto River at Rosetta, Miss	140
07295000	Buffalo River (Bayou) near Woodville, Miss	20
07356000	Ouachita River near Mount Ida, Ark	6.0
07356500	South Fork Ouachita River at Mount Ida, Ark	0
07357000	Ouachita River near Mountain Pine, Ark	19
07359500	Ouachita River near Malvern, Ark	73
07359800	Caddo River near Alpine, Ark	13
07360000	Ouachita River at Arkadelphia, Ark	110
07360800	Muddy Fork Creek near Murfreesboro, Ark	0
07361000	Little Missouri River near Murfreesboro, Ark	3.6
07361200	Ozan Creek near McCaskill, Ark	0
07361500	Antoine River at Antoine, Ark	0
07361600	Little Missouri River near Boughton, Ark	27
07362000	Ouachita River at Camden, Ark	175
07362100	Smackover Creek near Smackover, Ark	.1
07362500	Moro Creek near Fordyce, Ark	0
07363000	Saline River at Benton, Ark	1.1
07363300	Hurricane Creek near Sheridan, Ark	.1
07363500	Saline River near Rye, Ark	11
07364100	Ouachita River near Arkansas-Louisiana State Line	780
07364150	Bayou Bartholomew near McGee, Ark	4.5
07364200	Bayou Bartholomew near Jones, La	39
07364300	Chemin-a-Haut Bayou near Beekman, La	<.1
07364700	Bayou de Loutre near Laran, La	2.4
07365000	Bayou D'Arbonne near Dubach, La	.1
07365500	Middle Fork Bayou D'Arbonne near Bernice, La	.1
07365800	Cornie Bayou near Three Creeks, Ark	<.1
07365900	Three Creeks near Three Creeks, Ark	<.1
07366200	Little Corney Bayou near Lillie, La	.1
07367000	Ouachita River at Monroe, La	780
07367700	Boeuf River near Arkansas-Louisiana State line	21
07368000	Boeuf River near Girard, La	17
07368500	Big Colewa Bayou near Oak Grove, La	0
07369000	Bayou Lafourche near Crew Lake, La	4.6
07369500	Tensas Bayou at Tendal, La	4.0
07370000	Bayou Macon near Delhi, La	33
07370500	Castor Creek near Grayson, La	<.5
07371000	Garrett Creek at Jonesboro, La	0
07372000	Dugdemona River near Winnfield, La	.2
07372200	Little River near Rochelle, La	16
07372500	Bayou Funny Louis near Trout, La	<.1
07373000	Big Creek at Pollock, La	6.9
07375000	Tchefuncta River near Folsom, La	35
07375500	Tangipahoa River at Robert, La	270
07376000	Tickfaw River at Holden, La	23
07376500	Natalbany River at Baptist, La	2.5
07378500	Amite River near Denham Springs, La	290
07381500	Atchafalaya River at Krotz Springs, La	24,000
07385500	Bayou Teche at Arnaudville, La	90
08012000	Bayou Nezpique near Basile, La	.4
08015500	Calcasieu River near Kinder, La	200

management conditions, withdrawal should not be greater than average annual recharge.

Ground water has a low evaporation loss compared with surface water. Evaporation loss varies with the hydraulic conductivities of materials overlying the water-bearing zone and approaches zero as water levels deepen.

The same protective covering that minimizes evaporation losses from ground water also tends to filter the water and protect it from contamination under natural conditions. However, this attribute should not be taken too much for granted. For example, the dumping of concentrated pollutants in or near recharge areas, or the discharge of such pollutants into streams that may be recharging an aquifer, could have drastic effects on the quality of water in the aquifer.

Water-well development, operation, and maintenance are relatively economical in most of the Lower Mississippi Region. With two or more freshwater aquifers underlying most of the region, good-quality water in adequate amounts generally can be obtained at moderate depths. Thus, construction and pumping costs tend to be moderate.

ACCEPTABLE FOR MANY USES

The quality of ground water in the Lower Mississippi Region varies from aquifer to aquifer and in some instances with geographic or vertical location within the aquifer. However, at almost any location within the region, ground water that has a total dissolved-solids concentration of less than 1,000 mg/L can be obtained (fig. 15). The exceptions are areas along the Louisiana coast and a small area in west-central Louisiana, where all the ground water is saline to varying degrees. The prevalent chemical types and dissolved-solids concentrations of water in the shallow aquifers control the quality of water in streams at low flow (figs. 16, 17).

Under most conditions, the quality of water in an aquifer will remain unchanged. If the ground water being withdrawn is of good quality, with proper man-

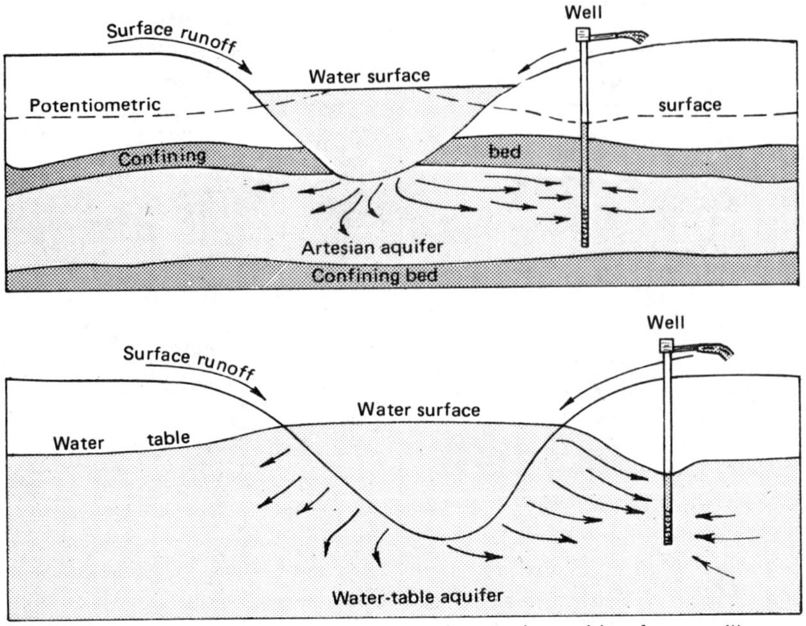

If a pumping well is located near a losing stream, increased quantities of water will move from the stream to the aquifer in the vicinity of the well.

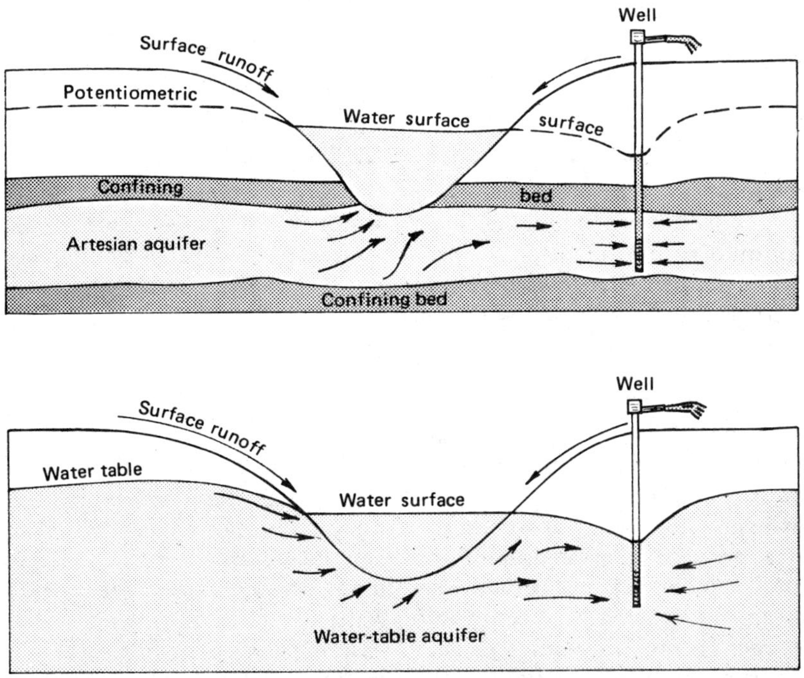

If a pumping well is located near a gaining stream, the direction of movement of water in the vicinity of the well may change and water normally discharged to the stream will be utilized by the well.

FIGURE 13.—Utilization of streamflow by nearby wells.

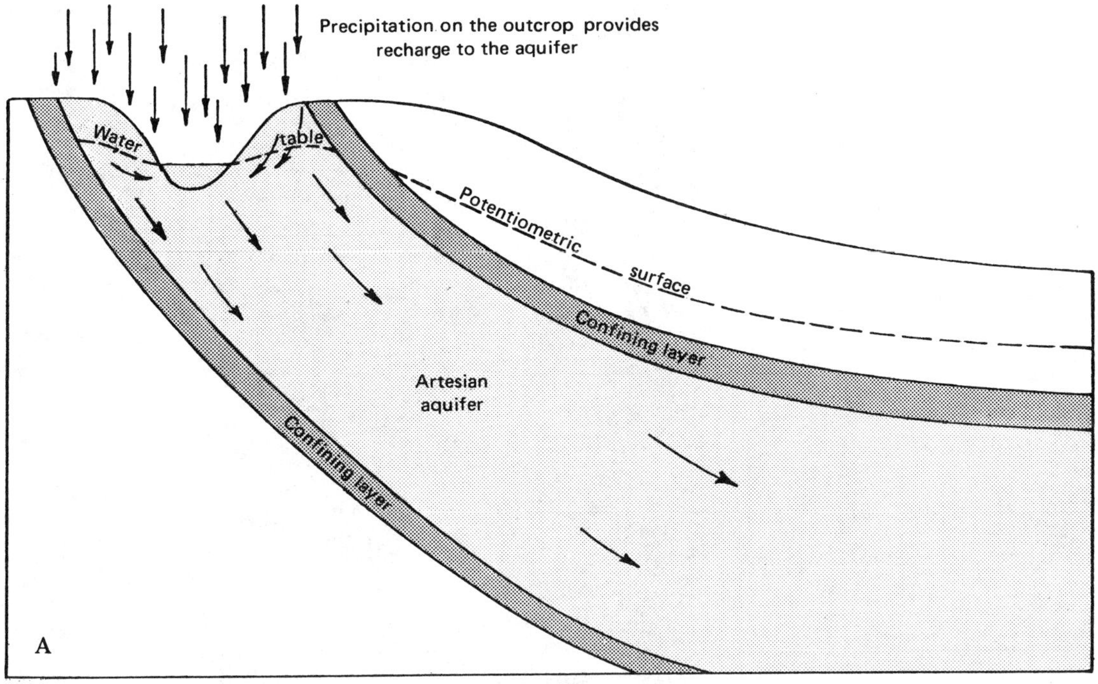

Pressure in the aquifer is great enough to maintain a head above the water-surface altitude in streams connected with the aquifer; therefore most potential recharge is discharged to the streams before it can move downdip.

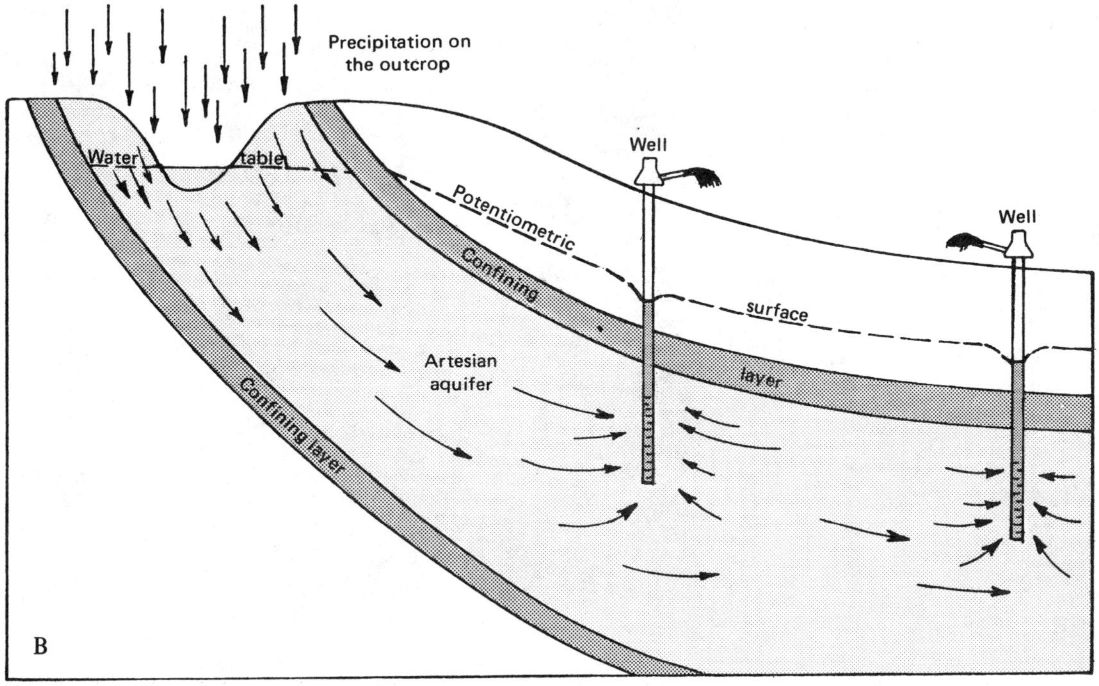

When water is withdrawn from the aquifer, pressure (and, thus, head) is reduced, allowing previously rejected potential recharge to move downdip.

FIGURE 14.—Full aquifers reject recharge; movement of water downdip increases with utilization of the aquifer.

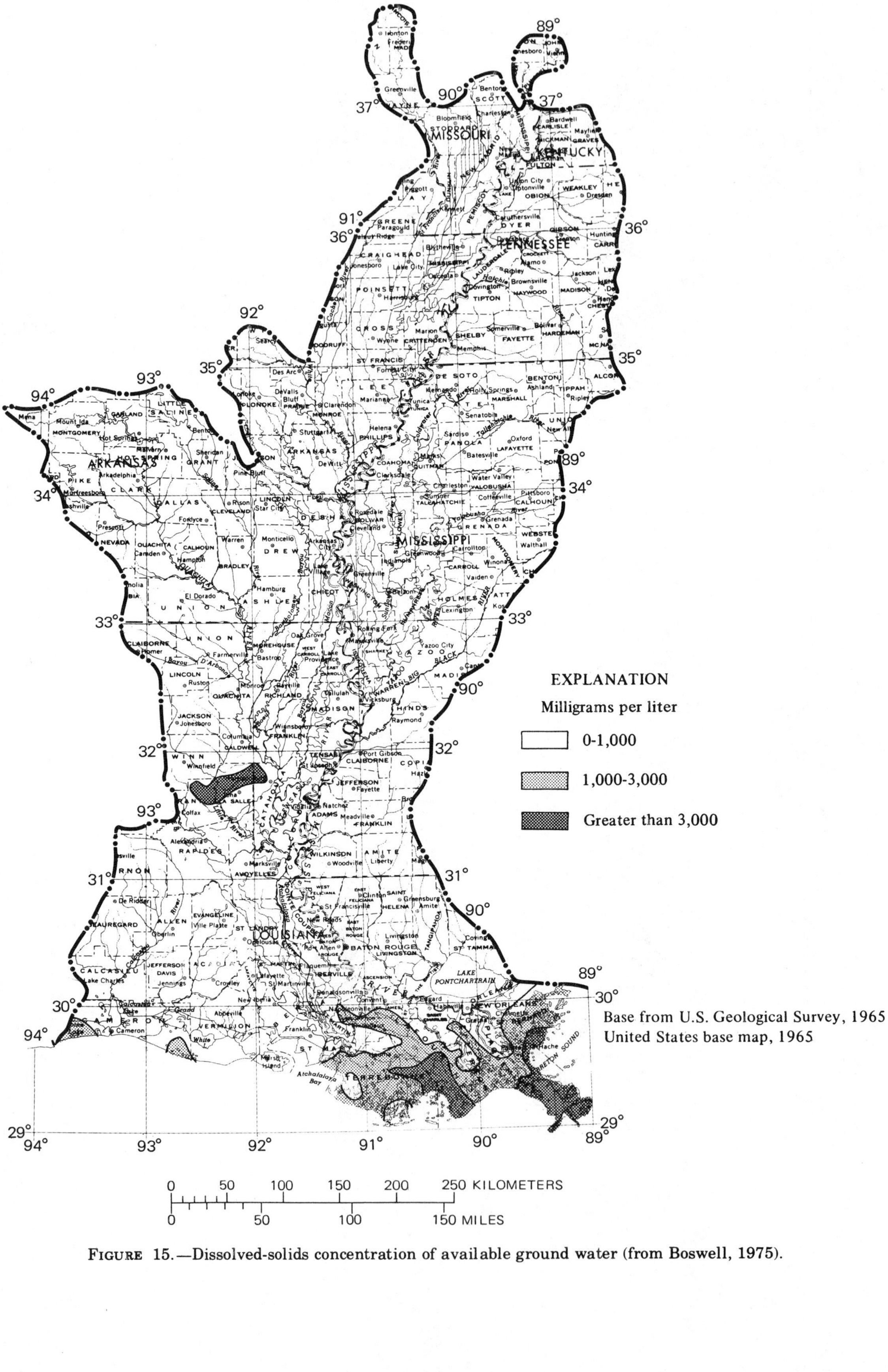

FIGURE 15.—Dissolved-solids concentration of available ground water (from Boswell, 1975).

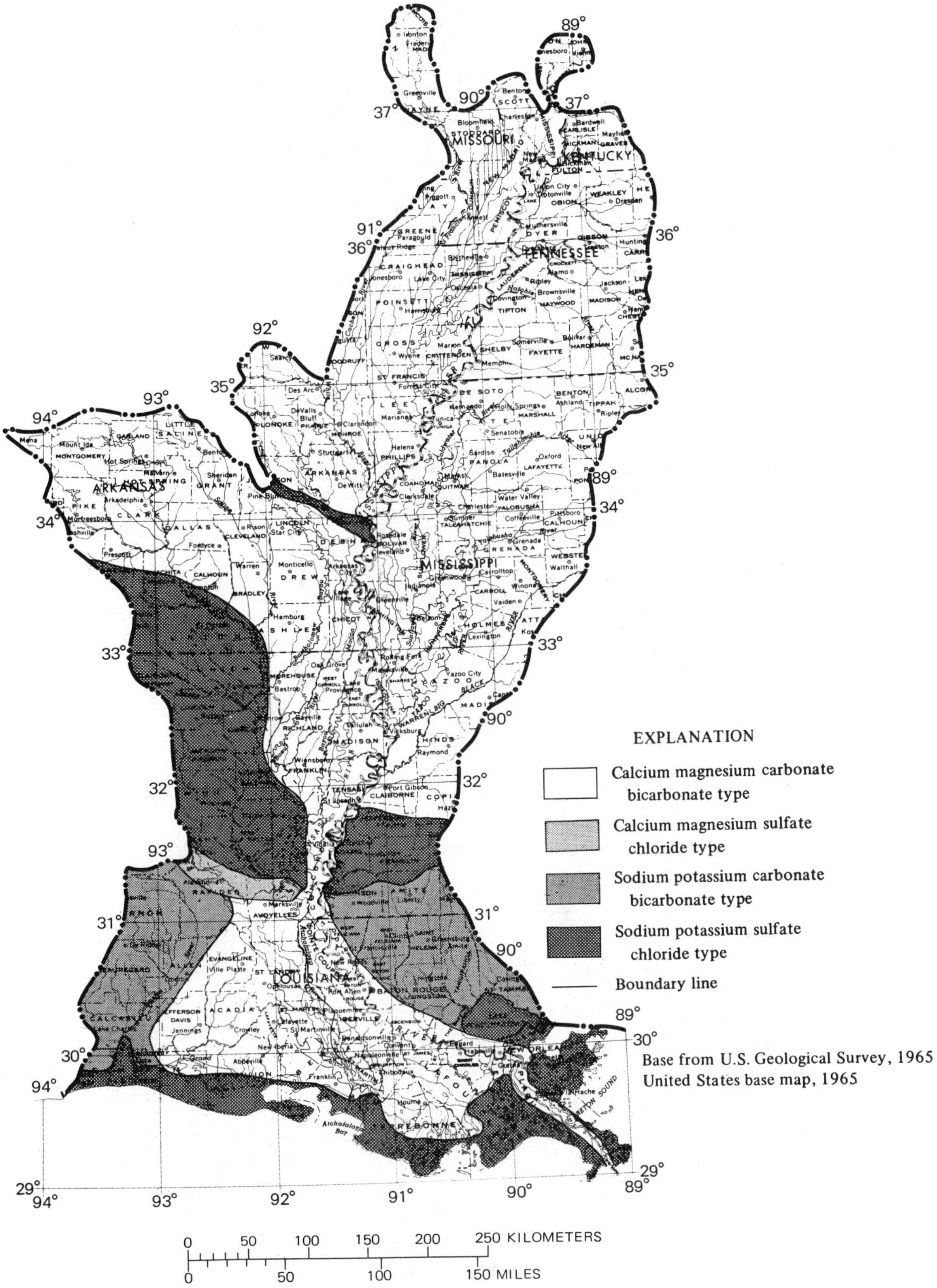

FIGURE 16.—Prevalent chemical types of water in streams at low flow (modified from Rainwater, 1962).

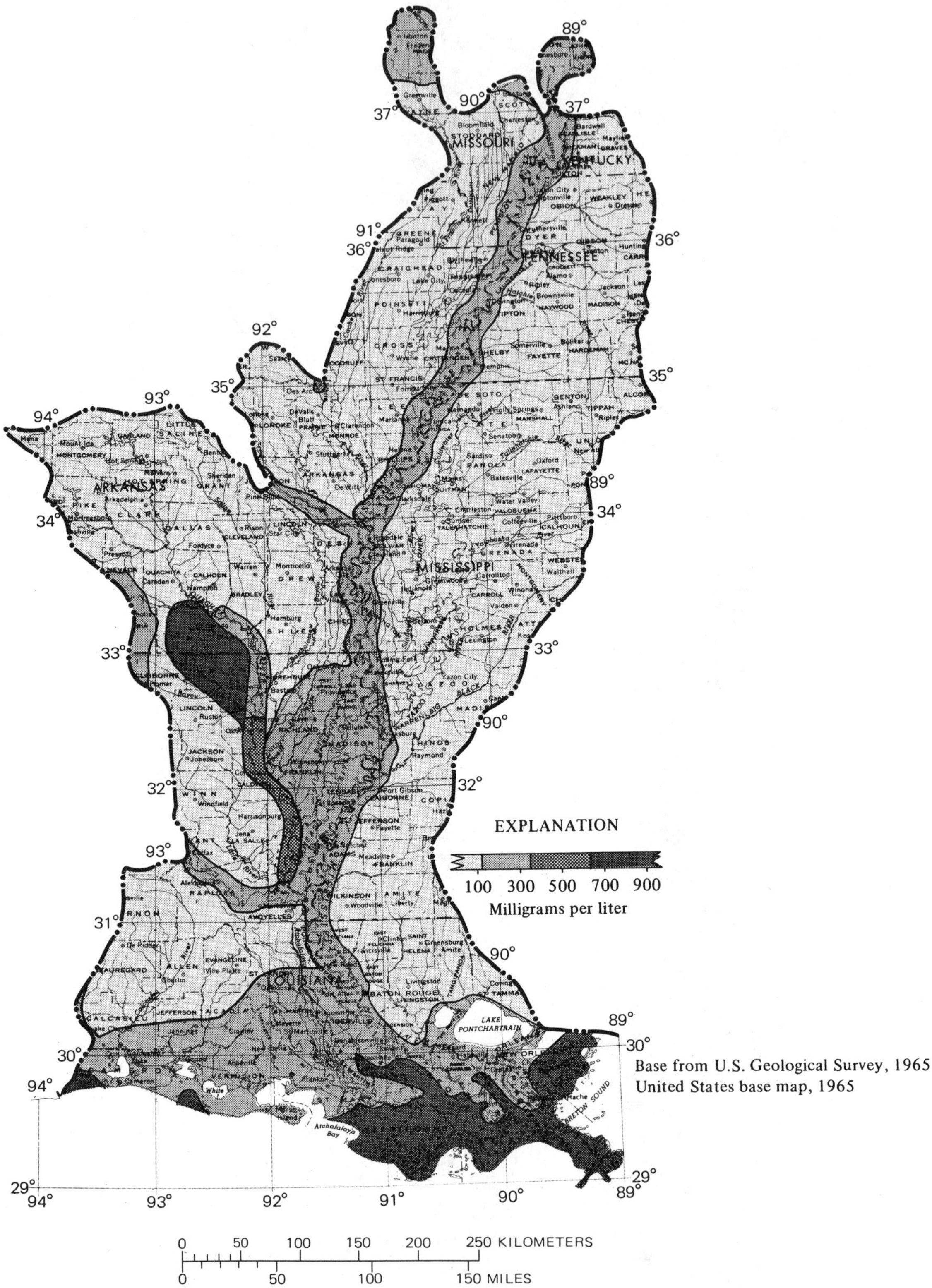

FIGURE 17.—Prevalent dissolved-solids concentration of water in streams at low flow (modified from Rainwater, 1962).

agement it is quite likely to remain so; if the ground water being withdrawn requires treatment, it is not likely that, with proper management, treatment facilities will have to be changed significantly due to changes in the ground-water quality.

Ground water is believed to be generally free of bacteria and chemical pollution. This belief is generally valid because ground water moves through natural soil and rock filtering media which can reduce natural bacterial pollution to almost zero. No widespread occurrence of bacterial pollution of ground water in the Lower Mississippi Region has been observed. However, locally, individual wells may yield bacterially contaminated water due to faulty well construction or location.

By definition, ground water is considered to be naturally "polluted" when natural mineral concentrations exceed established criteria for various uses; so, whether water from a particular aquifer is considered to be polluted depends upon the intended use. If a use problem arises, it can generally be solved by treatment of the water or, in some places, by tapping another available aquifer containing water that is more suitable.

WHY ARE WE NOT GETTING MAXIMUM BENEFIT FROM GROUND WATER?

INADEQUATE CONSIDERATION

Although ground water is being widely utilized, regionally much of its potential is not fully realized. Because of flooding problems and the need for navigation, high priority has generally been given to detailed studies of the region's surface-water systems. As a result, steps have been taken to alleviate most of the severe flooding problems, and good water-transportation systems have been developed. The characteristics of most of the major streams in the region are well defined, and consequently the behavior of these streams during periods of flood and drought is reasonably predictable.

Knowledge of the behavior of the region's subsurface water systems (aquifers) under natural or imposed stresses is also important in water resources management. The ability to predict with some precision the effects of additional ground-water development is needed. Data such as aquifer characteristics, interaquifer relations, and stream-aquifer relations must be available in order to make such predictions. At least as much effort should be made to define and control our subsurface waters as has been made to control our surface waters. Detailed ground-water studies, whose end results are predictive models, cover only small parts of the Lower Mississippi Region (fig. 18).

There is a general lack of public awareness of the overall significance of ground water and the possible widespread effects of ground-water development. Many times the proper information is not sought or is not properly analyzed before a ground-water-related development is started. There is also a lack of public awareness concerning specific problems, such as the possible impact upon the local ground-water regime of certain seemingly unrelated activities, such as land clearing, excavations, and the proximity of sewage facilities to shallow wells.

IMPROPER DEVELOPMENT

Improper ground-water development may entail one or any combination of the following:

1. Drilling below the base of freshwater.
2. Finishing a well above the most suitable aquifer.
3. Locating a well too near other pumping wells.
4. Locating a well too near a source of contamination.
5. Overdevelopment.

Some of these development problems may be due to an information deficit, as mentioned in the preceding section, but often available information is adequate but it is not given due consideration.

Sometimes when a new well is drilled, available information is not considered to determine the depth of the water best suited to the need. Drilling below the base of freshwater, for example, is generally a waste of time and money (fig. 19A) and will increase the chances of well contamination, as head is reduced by pumping (fig. 19B). Of course, the reverse is also true; when a driller does not have adequate knowledge of the section he is drilling, he may stop short of the best water-bearing zone.

Deep test holes, such as those drilled by oil companies, commonly pass through both freshwater aquifers and aquifers containing undesirable water (fig. 20A). If these holes are not properly plugged, they may become conduits through which undesirable water may leak into the freshwater zones that have lower hydrostatic heads (fig. 20B). Such leakage has caused saltwater pollution in local areas in the Lower Mississippi Region. The Mississippi River valley alluvial aquifer in northeast Louisiana has experienced some saltwater pollution due to leaky abandoned wells (Whitfield, 1975).

Interference between wells occurs when they are located too near each other and are screened in the same aquifer. The cones of depression, sometimes termed "zones of influence," created by continued withdrawal from the wells, may coalesce. This condi-

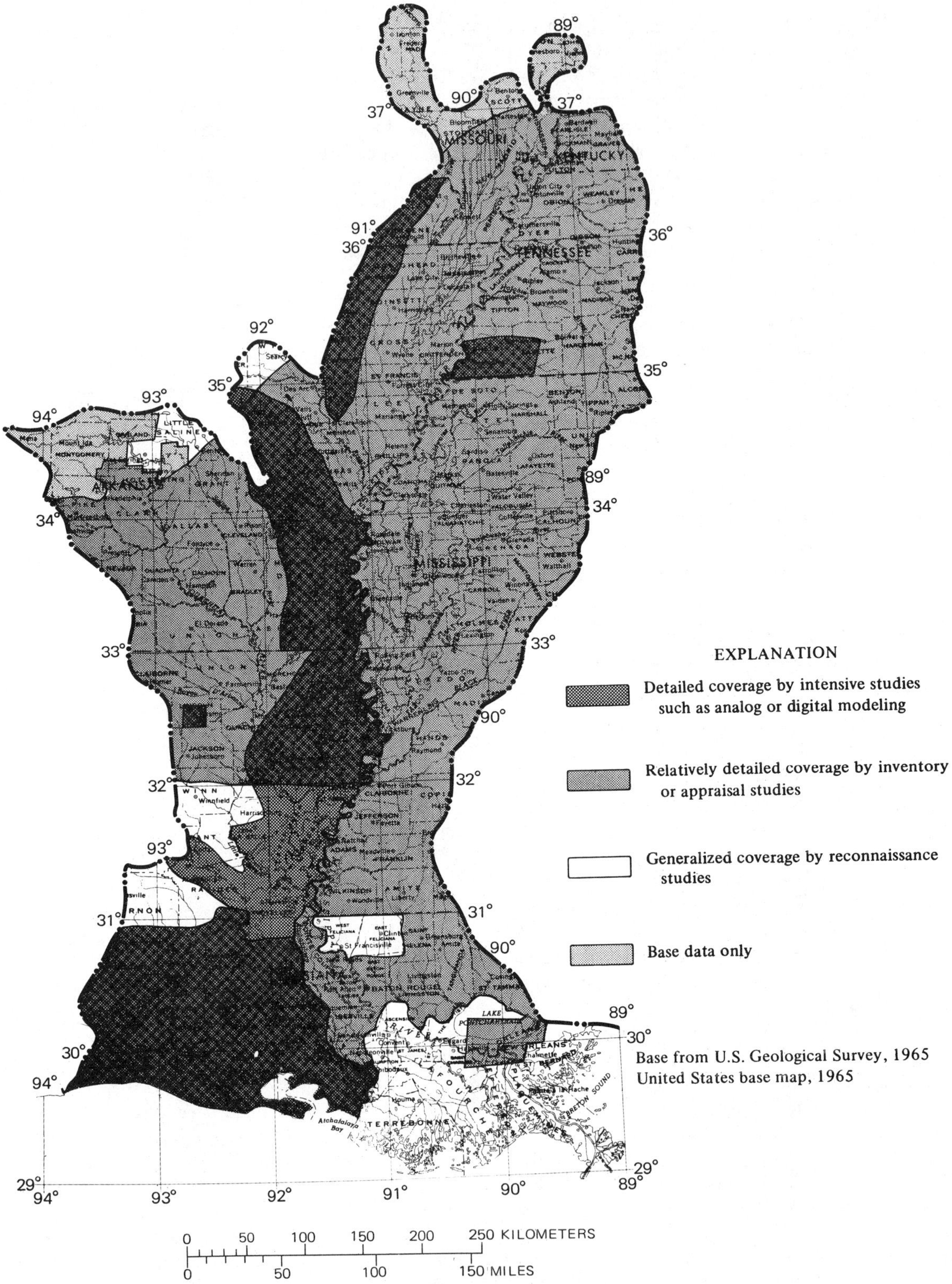

FIGURE 18.—Delineation of areas covered in ground-water-related reports.

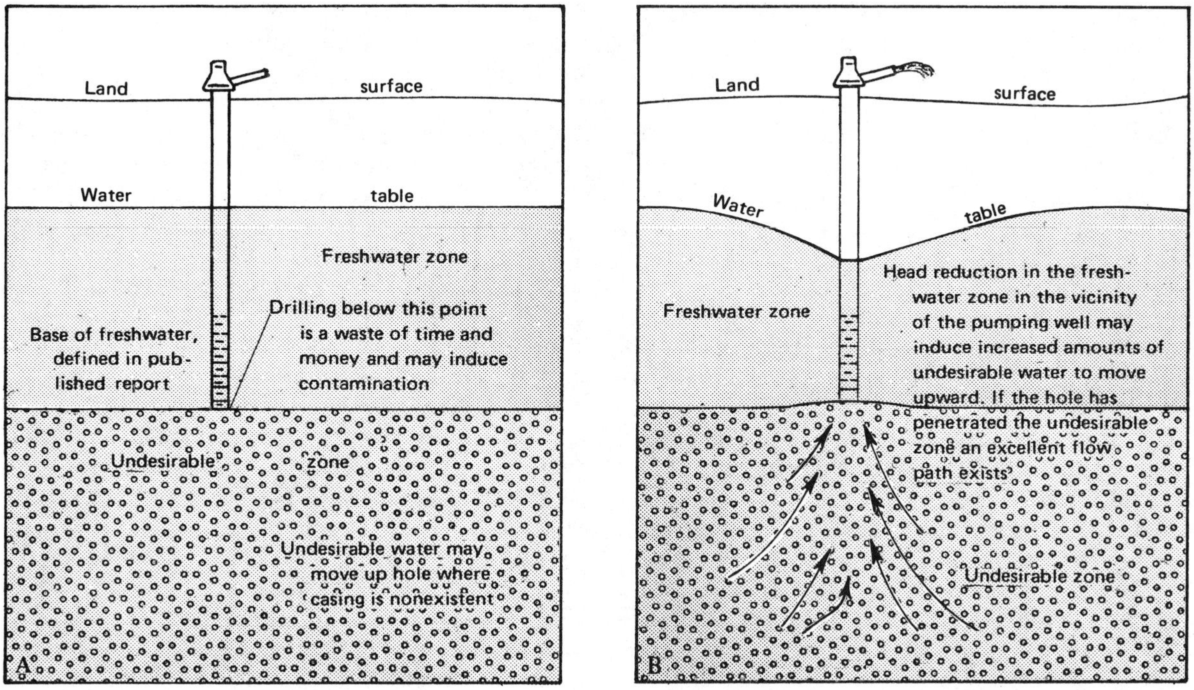

FIGURE 19.—Disregard of available information can be expensive economically and ecologically.

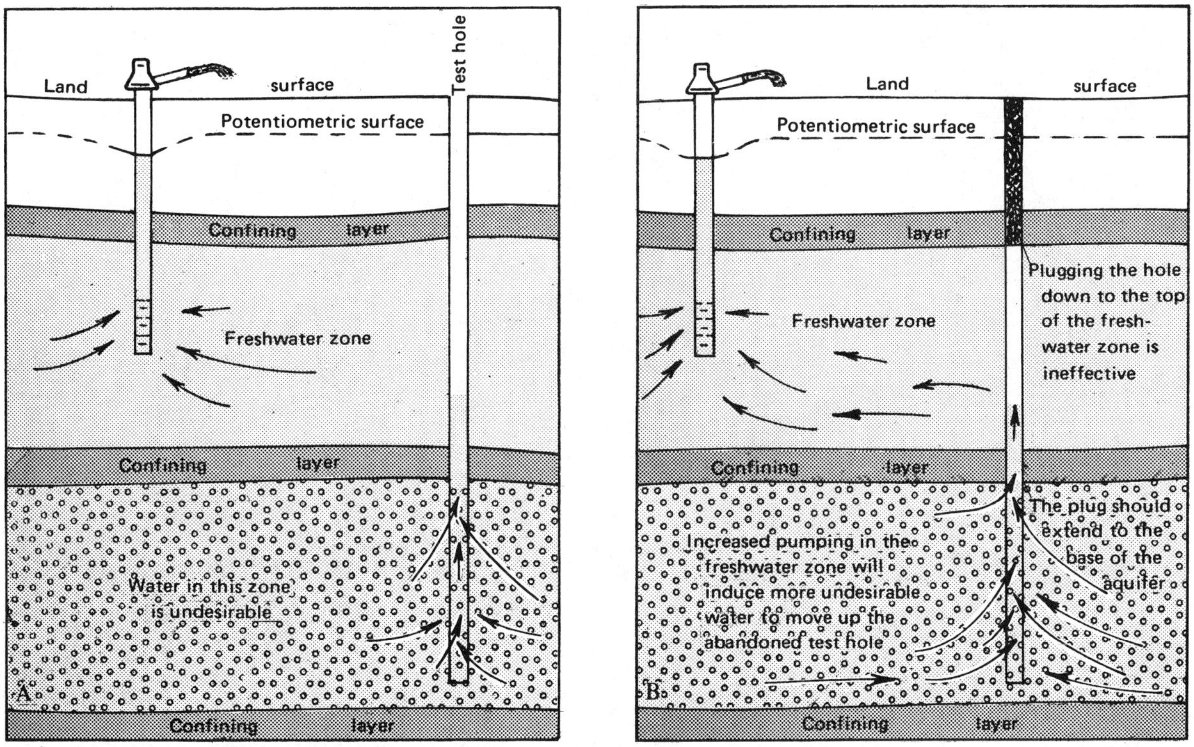

FIGURE 20.—Abandoned, deep test holes cause contamination by interaquifer water exchange if not properly plugged.

tion will significantly reduce the amount of water the aquifer will yield to these wells. Observe wells A, B, and C shown in figure 21. Well A depicts a very poor location for a new well; well B depicts a fair location, if pumping is not increased significantly from either the new or the existing well; well C depicts the best location of the three for the conditions indicated.

The development of a new water-supply well for a city within the region provides an example of both poor well spacing and disregard of information about the depth of the best water-bearing zone available. The city's existing water supply is from two wells screened at depth intervals of 365 ft (111 m) to 405 ft (123 m) and 370 ft (113 m) to 410 ft (125 m), in the upper part of the Claiborne Group. These wells each produce about 600 gal/min (38 L/s). About 300 ft (91 m) from one of these existing wells, the new supply well has been drilled to a depth of 575 ft (175 m). This well is screened in the 364-ft (111-m) to 398-ft (121-m) interval, the same water-bearing zone as the existing wells. It is intended that the well produce 1,000 gal/min (63 L/s), which is the capacity of a new water-treatment plant that the city is now building. Before the well was screened, a gamma-ray log, run by the Geological Survey on the pilot hole, indicated the presence of a possible aquifer 120 ft (37 m) thick below the lower confining layer of the thin (35 ft to 45 ft or 11 m to 14 m) aquifer now being used. Based on the piezometric surface in other wells tapping this aquifer, the pressure in this lower aquifer would have been sufficient to cause flow at ground surface. Further testing could have determined the potential of this lower sand.

Some wells are located too near freshwater-saltwater interfaces. When withdrawals are made from such wells, head(pressure) in the vicinity of the well is reduced, allowing saltwater to move toward the well. This problem can be avoided or possibly solved if adequate information about the aquifer is available. Although saltwater is the most common pollutant associated with ground water in the Lower Mississippi Region, the proximity of proposed well sites to other sources of contamination should be carefully considered before locating a well.

"Overdevelopment" of an aquifer is a relative term. In local areas where water levels have been lowered enough to significantly increase pumping lifts and costs, the term "overdevelopment" is often applied (fig.

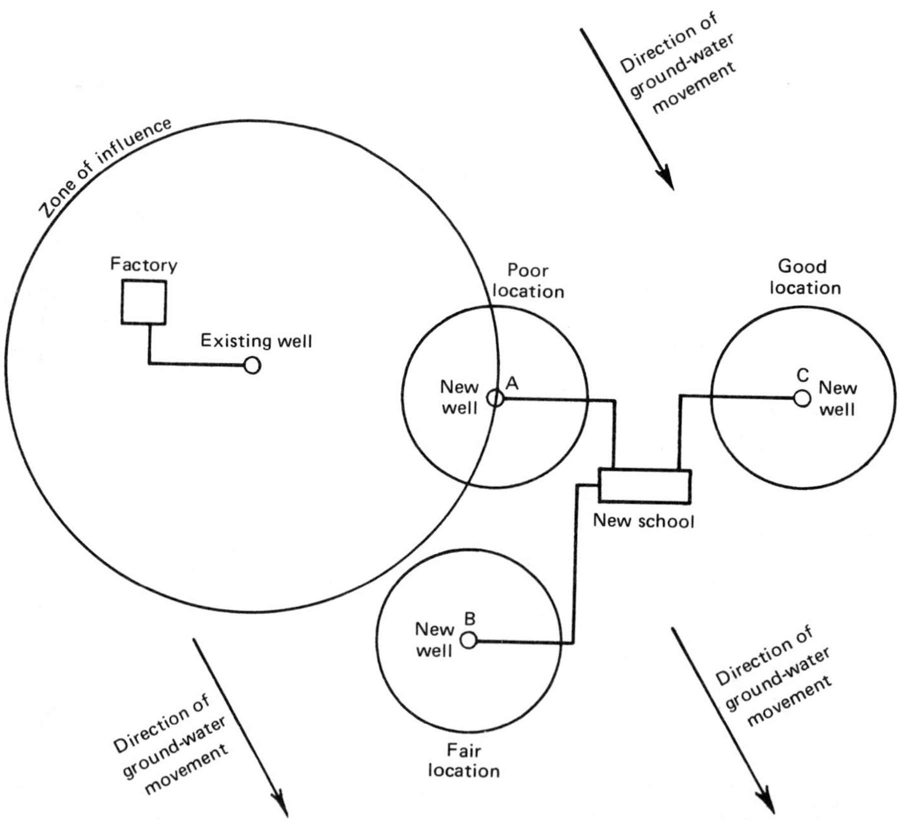

FIGURE 21.—A simplified sketch of how a well location should be selected.

22). If water levels around such pumping centers continue to decline, the zone of influence will spread, resulting in increased pumping costs for users in the area even though they do not significantly contribute to the cause of the decline.

Another cause of some loss in artesian pressure in aquifers within the region is the practice of allowing some naturally flowing wells to flow to waste. In southwest Arkansas, many wells that tap Cretaceous aquifers originally flowed at and above the land surface. Very few of these wells were capped and today, due at least in part to this waste, heads in these wells have dropped considerably and many of them have ceased to flow.

FRACTIONAL USE OF AVAILABLE SUPPLY

The availability of fresh ground water varies throughout the Lower Mississippi Region (fig. 9). Because annual ground-water withdrawal in the region is about one-third of the annual amount available (fig. 23), rejected potential recharge constitutes a loss of water that would have entered the subsurface (fig. 14) if storage space were available. A maximum continuous withdrawal that could be sustained from an aquifer would be that quantity of water necessary to reduce the hydrostatic pressure and maintain it at such a level that the controlling factor on recharge to the aquifer becomes either the amount of recharge available or the ability of the aquifer to transmit water.

As desalinization technology advances and as more uses are found for saltwater, the amount of usable ground water that is available in the region will increase several times. Utilization of saltwater will have a twofold effect—a reduction in demands placed upon freshwater aquifers, making more freshwater available for uses requiring better quality water, and a reduction of pressure in saltwater zones, thus allowing more freshwater to be withdrawn near saltwater interfaces without inducing coning or encroachment problems.

HOW CAN WE OBTAIN MAXIMUM BENEFIT FROM GROUND WATER?

Merely increasing the usage of ground water does not assure that the most benefit will be obtained from the available supply; in fact, unwise development could result in the opposite effect. The determination to protect, as well as utilize, ground water can prevent the waste that has resulted from misuse and mismanagement of some of our other natural resources.

EXPAND INFORMATION BASE, AND USE IT

Ground-water management in the future should consider not only the local aspects of a planned development but also the regional framework into which the development must fit. To do this there should be an adequate information base containing at least the following kinds of data: physical and hydraulic conditions within the aquifers, the quality of water in the aquifers, and interaquifer hydraulic relations.

Comprehensive planning or the development of proper planning tools cannot be done without knowing the size and properties of the ground-water container involved. The vital statistics of an aquifer that must be determined are its thickness, areal extent, configuration, and texture. Also, the quality of water, and any areal or vertical changes in quality within an aquifer, should be known. In addition to the physical features, the hydraulic conditions within the aquifer must be understood, that is, the movement of water as influenced by recharge and discharge.

The hydraulic characteristics of an aquifer are those properties that determine its ability to store and transmit water. Aquifer characteristics, such as hydraulic conductivity and transmissivity, control the well yields, the amount of drawdown incurred to produce a specified yield, and the magnitude of water-level declines produced by pumping. With this information, the probable effects of a planned development can be predetermined, and well fields can be designed to minimize well interference. In addition, these aquifer characteristics can be used to estimate the gross yield available from an aquifer throughout a large area under a prescribed set of conditions.

Interaquifer relations—the way aquifers interact hydraulically with one another—must also be determined regionally and locally. Some aquifers receive recharge through or from other aquifers. Confining beds may be sufficiently permeable to permit exchange of water between aquifers. The movement of water into an aquifer that is being pumped can significantly affect water levels and water quality. In fact, in some areas where such conditions exist, the quality of water in an aquifer may be manipulated by carefully planned pumping patterns and schedules.

The extent of hydraulic connection between aquifers must be known for other reasons, too. For example, if poor-quality water or waste were to be injected into an aquifer for storage, a hydraulic connection with another aquifer could induce contamination of the second aquifer. Because of interaquifer hydraulic connections, aquifers considered to be separate entities locally may actually be part of a large system when considered regionally.

Much of the information base needed for planning can be provided by regional aquifer studies. Such studies have been made in most of the Lower Missis-

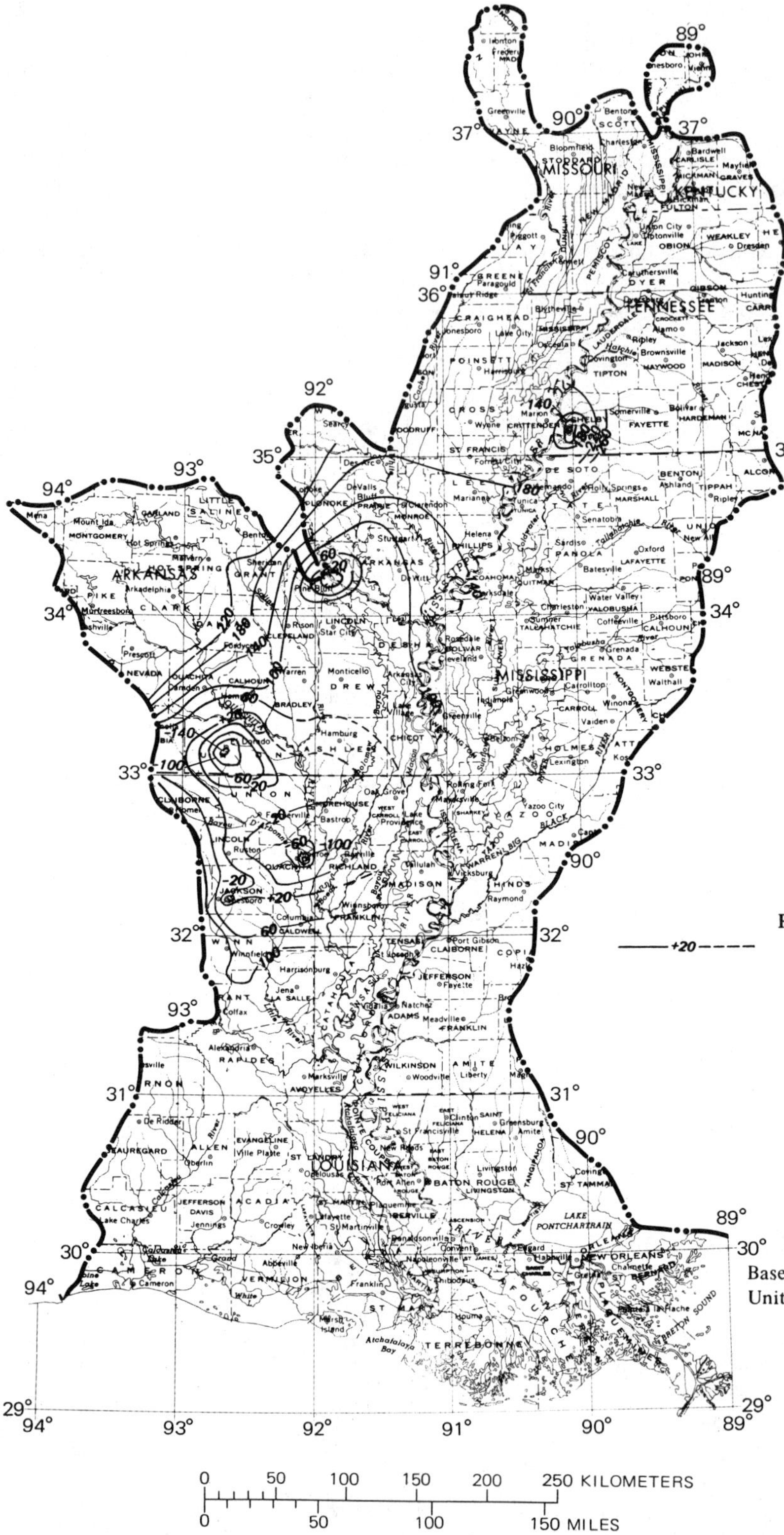

EXPLANATION

POTENTIOMETRIC CONTOUR – Dashed where approximately located. Contour interval 40 feet (12 m). Datum is mean sea level.

——— +20 – – – –

Base from U.S. Geological Survey, 1965
United States base map, 1965

FIGURE 22.—Areas where extensive lowering of water levels has occurred in the Sparta Sand in Arkansas and Louisiana (contours based on 1974–75 data) and in the Memphis Sand in the area of Memphis, Tenn. (contours based on 1973 data).

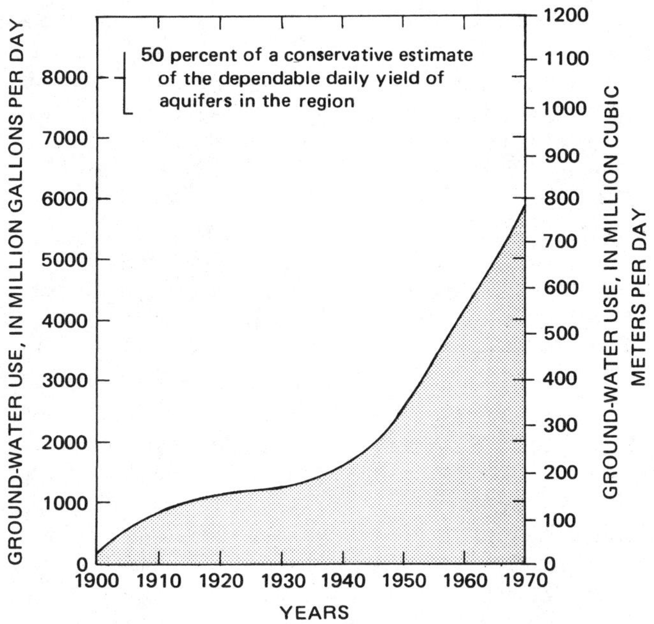

FIGURE 23.—Estimated ground-water use.

sippi Region. Most of the area north of the 32d parallel is included in studies of the water resources of the Mississippi embayment (Cushing and others, 1963, 1964; Boswell and others, 1965, 1968; Hosman and others, 1968). The reports describe the Cretaceous, Tertiary, and Quaternary aquifer systems. Further studies by Payne (1968, 1970, 1972, 1975) describe the hydrologic significance of lithofacies of aquifers of the Claiborne Group in Arkansas, Louisiana, Mississippi, and Texas. Much of the information necessary for the development of predictive aquifer models is provided by these studies. The only major aquifer systems in the Lower Mississippi Region that have not been studied regionally are the vast coastal aquifers of Miocene, Pliocene, and Pleistocene age.

A large amount of water-quality information is available in the Lower Mississippi Region, and regionally the water quality in the different aquifers is well known. Consideration of this information by water managers would help them in choosing a suitable source of water. However, the quality of the water in some aquifers varies in short distances, and substantial additional testing may be necessary at these localities. Water-quality testing is also advisable in the proximity of saltwater interfaces. Monitoring networks can be used to detect anticipated water-quality changes, such as an advancing saltwater front, before the change affects the pumping center. Information on water quality will become increasingly important as concerned management organizations seek measures to conserve and protect the resource.

In addition to the regional aquifer studies mentioned previously, local ground-water studies have been, and are being, made throughout the region. The areas investigated range in size from a few square miles to one or more counties (or parishes) or a river basin. Most of these studies are conducted by the Geological Survey in cooperation with State agencies. Reports based on these studies, plus abundant data in Geological Survey and State agency files, represent a sizable background of ground-water information for the region. Additional investigations should be undertaken in areas within the region where information is insufficient, especially where only reconnaissance studies or basic data are available (fig. 18). These investigations should provide information concerning aquifer properties sufficient for the development of predictive aquifer models.

USE MODERN TECHNOLOGY

In the past, ground-water technology was almost entirely oriented toward locating and developing supplies of potable water. Little attention was given to the possible consequences of such developments. As the science evolved, techniques were developed (and are still being developed) that meant to our ground-water resource and its users what reforestation technology meant to our timberlands and the lumber industry. With proper use and coordination of these techniques and the continued development of new techniques, benefits from use of ground water will approach a maximum, and our ground-water resource will be protected for future users.

AQUIFER MODELING

Aquifer models are the best-available predictive tool for ground-water planners and managers. Among the early models was the analog type: using a network of resistors and capacitors to simulate hydraulic properties, it simulated electrically the effects of pumping stresses on an aquifer. The early analog model was the predecessor of the digital model. Construction of the digital model has been made possible by the sophistication of the digital computer and the development of numerical methods for the solution of the equations of ground-water flow. Use of the digital model requires adequate knowledge of the physical and hydraulic properties of the aquifers, pumping information, and a history of water-level fluctuation for calibration of the model. After the model has been calibrated to reproduce verifiable results, the planner and manager can use it to predict the effect of proposed development upon the ground-water system. Models can also be used to predict the movement of a saltwater front or fluids injected into the aquifer.

Aquifer modeling is still in a relatively early stage of development, and the degree of sophistication of the technology is steadily increasing.

Within the Lower Mississippi Region, four ground-water studies that utilized either analog or digital modeling techniques have been completed. Two additional studies are currently underway. Each of these studies was, or is being, conducted by the Geological Survey in cooperation with another Federal, or a State, agency. Two of the completed studies used only analog models. One of these completed studies predicted the effects that the imposition of navigation structures on the Arkansas River would have on the ground-water regime (Bedinger, 1970), and the other simulated water-level declines in the Sparta Sand in the Mississippi embayment (Reed, 1972). Two of the completed studies used only digital models. In the Ruston, La., area a digital model was used to predict the effects of projected pumping upon water levels in the Sparta Sand (Sanford, 1973). Digital models were used to predict the effects that the construction of locks and dams on the Red River in Louisiana would have upon ground-water levels in the Red River alluvial aquifer (A. H. Ludwig, oral comm., 1976). One of the ongoing studies will model the hydrology of the Bayou Bartholomew alluvial aquifer-stream system in Arkansas. Originally, an analog model was constructed (Broom and Reed, 1973); however, adequate controlling parameters could not be incorporated into the analysis by analog methods. Today, development of a more versatile digital model is underway. The other ongoing study will determine, with the aid of a digital model, the effects of pumping stress upon ground-water levels in the Memphis Sand in the area of Memphis, Tenn.

ARTIFICIAL RECHARGE

Artificial recharge may be used to augment natural recharge. It has been used to salvage excess streamflow and has also been applied to problems associated with ground-water development, such as overdevelopment and subsidence. Artificial recharge is done by two basic methods: (1) impounding surface water where it can infiltrate a permeable part of the aquifer that is exposed at land surface and (2) injecting water into the aquifer through a well.

A method that combines injection with surface-water impoundment has been tried in some areas outside the Lower Mississippi Region, primarily in the West, with only limited success. In such experiments, playa lakes with relatively impermeable beds were used as catch basins. Wells drilled through the lake bottom and into the aquifer were intended to allow the accumulated water that otherwise would be lost to evaporation to drain by gravity into the aquifer. Problems with well plugging, both at the intake and in the screened interval, reduced the efficiency of this method to an unacceptable level. Whether or not this hybrid technique could be made feasible in the Lower Mississippi Region might be worth investigating.

Extensive experiments in which treated surface water was injected into the Quaternary aquifer in the Grand Prairie region of Arkansas were conducted by Sniegocki and others (1963). Heavy pumping for rice irrigation had depleted the ground-water supply in the shallow aquifer, and natural recharge to the aquifer was impeded by an extensive overlying clay layer. The conclusions of the experimental study were that the costs of the extensive treatment necessary to render raw surface water suitable for injection, without plugging the well screen or the aquifer, are economically prohibitive with present technology. However, well injection may become feasible in the Lower Mississippi Region, either through some technological advance or by locating a compatible combination of aquifer conditions and a surface-water recharge source of such quality that the cost of treatment to prevent well-screen and aquifer plugging would be acceptable.

At present, water spreading seems to offer the best hope for artificial recharge at places where water levels have been drawn down in the recharge area and where excess surface water is readily available. Aside from possible technical problems, a major consideration may be the high cost or unavailability of adequate land for water spreading. However, recharge by water spreading should be given consideration as a management option, wherever additional replenishment to the aquifer would be beneficial.

BLENDING OF WATER

The blending of waters from more than one source can enable the use of water that would require treatment if used by itself. The waters to be blended could come from different aquifers or from ground and surface sources if the waters are chemically compatible. The resulting concentration of chemical constituents in the blend will be in direct proportion to the quantity of each of the contributed waters. For example, a blend containing equal amounts of two types of water would have an average chemical composition of the two.

Considering the types of ground water in the Lower Mississippi Region, water blending could prove to be an extremely beneficial practice. The hard, high-iron, low-chloride water in the Quaternary aquifers is plentiful in most of the region; however, the water generally requires treatment for iron removal and softening. A soft, high-chloride, low-iron water occurs at depth in most aquifers; this water generally remains unused because of the difficulty in lowering the chloride concentration to acceptable levels. A blending of these two waters would dilute the iron concentration of one, the chloride concentration of the other, and produce an intermediate hardness. Blends could be designed that would require little or no treatment for most uses. Excesses in iron, chloride, and hardness represent the most common chemical-quality problems with ground water in the Lower Mississippi Region. Before considering a water blend, competent professional advice should be sought to determine the chemical compatibility of the different waters.

CONTROL AND USE OF SALTWATER

The potential for saltwater encroachment exists where withdrawals are large in the proximity of saltwater interfaces. However, if it is necessary to plan a large development of wells near an interface, techniques are available for maintaining a dependable supply of freshwater at the pumping center.

Barrier wells, either discharge or recharge, have been the most successful means of arresting saltwater encroachment. In a discharge-well system, a line of pumping wells between the saltwater front and the pumping center intercepts the migrating saltwater and forms a low-pressure trough in the potentiometric surface beyond which the saltwater cannot pass. The water discharged by the barrier wells will become increasingly salty. If the saltwater cannot be used, it must be disposed of, possibly by injection into a deeper, saltwater-bearing aquifer. Recharge barrier

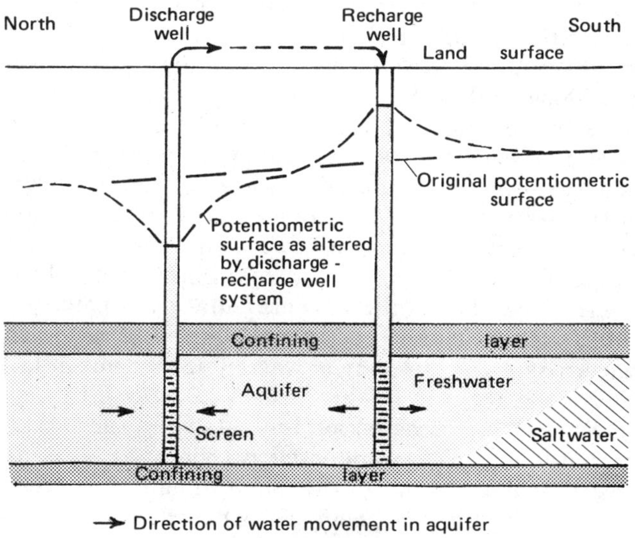

FIGURE 24.—A discharge-recharge barrier well system to control saltwater encroachment (modified from Rollo, 1969.)

wells, also located between the pumping center and the saltwater, require a supply of freshwater that is chemically compatible with water in the receiving aquifer. The injected water creates a high-pressure ridge in the potentiometric surface and retards the advancement of the saltwater.

A barrier-well system, combining both discharge and recharge wells, has many attractive features and solves most of the problems created by either of the separate systems. In such a system, the discharge wells pump into the recharge wells, which are located between the discharge wells and the interface (fig. 24). Thus, both a high-pressure ridge and a low-pressure trough are formed as an effective barrier to the saltwater. The problems of disposal of water from the discharge wells and a source of compatible water for the recharge wells solve each other, and there is no net loss of water from the aquifer. Although this combined discharge-recharge barrier system is not known to have been tried, its practical features may make it the most economic long-term technique for arresting saltwater encroachment.

Aside from functioning as a control tool, the combined barrier-disposal well system offers an additional application. Water from the discharge wells can be used elsewhere until it becomes too saline; then it can be diverted to the recharge wells. If the pressure ridge formed by the recharge wells effectively stops the progressive saltwater movement, water in the aquifer between the two lines of wells will gradually become fresh. When the chloride concentration in water from the discharge wells decreases to an acceptable level, the water can again be used elsewhere until it becomes too saline. The possibility that the combined barrier system could produce usable water cyclically is a distinct advantage over the discharge-well system that produces only saltwater.

Saltwater also presents problems in areas where it underlies freshwater in an aquifer. This condition exists mostly in coastal aquifers, as a result of the gentle gulfward dip of the aquifers and the low angle of the freshwater-saltwater interface. In a thick sand, the area in which freshwater overlies saltwater can be large. A well screened in the upper (freshwater-bearing) part of the aquifer will produce freshwater. The length of time it will continue to produce freshwater depends upon the rate and duration of pumping, the hydraulic characteristics of the aquifer, and the proximity of the saltwater. When the well is pumped, water in the aquifer moves toward the well. The underlying saltwater moves upward and can eventually reach the well if it is pumped hard enough and long enough. One method of producing freshwater under these conditions is to use several low-producing wells screened at the top of the aquifer. Although this method can be successful, it can also be expensive because of the number of wells and the amount of land required for proper well spacing.

A more economical way to withdraw large quantities of freshwater from a zone that is underlain by saltwater is through the use of a scavenger well (Long, 1965). The scavenger well is installed next to the supply well, but it is screened lower in the aquifer. The supply well can be pumped until the saltwater reaches it. At this time, the scavenger pump is turned on to intercept the saltwater, and the supply well will then produce freshwater. Of course, the saltwater from the scavenger well must be disposed of. It could be injected into a deeper aquifer or possibly into the basal part of the same aquifer. The basic scavenger-well technique offers interesting opportunities for innovative approaches.

RECYCLING

The possibilities for recycling water, or returning it to the subsurface for reuse, must be considered in order to derive maximum benefit from the ground-water resource. The water used for many processes is only slightly altered, sometimes only thermally, and could be reused, reclaimed for reuse, put to other uses, or returned to the same or another aquifer. Recycling should become a standard practice as far as is feasible for the sake of conservation. The advantages of recycling should be emphasized now, rather than at some future time when a shortage may force conservation measures.

GEOTHERMAL ENERGY

Harnessing geothermal energy is a newly developing technology, insofar as the United States is concerned. Whether or not development of geothermal-energy resources is feasible in the Lower Mississippi Region has not been determined. In this area, extremely deep strata contain hot water as a result of the natural geothermal gradient. Experiments with hot water from an abandoned, deep gas well in southern Louisiana are being planned to determine the feasibility of geothermal development. If results of the study are favorable, this would indicate similar possibilities in approximately the southern one-third of the region. The only other known source of hot ground water in the region is in the Hot Springs, Ark., area. The temperature of the water issuing from the hot springs is 145°F (62°C), which is sufficient only for space heating. The heat source for the springs does not seem to be large, and hotter water probably would not be available by drilling. Tapping the source of the water at depth probably would diminish the natural flow of the springs (Bedinger and others, 1974), and whether or not the energy that might be available would offset the loss to the tourist industry is questionable. The deep-seated, superheated water in the southern part of the region seems to offer the best potential for geothermal energy sources and may be more economically accessible where abandoned, deep oil and gas wells are available.

PLAN PROPERLY

Proper planning will be the key to maximizing benefits from the ground-water supply in the lower Mississippi Region. Planning must include provisions for management so as to prevent waste and assure conservation and protection of the resource. Plans must be coordinated within the region, as well as with adjacent regions, because most of the aquifers cross regional boundaries.

To coordinate work, planners should establish communications with all local, State, and Federal agencies that manage or investigate water resources (see following list). The main function of water-oriented agencies has been investigation. However, because of the need for planning and management, regulatory bodies are being formed and others are planned. Regulation should be based on cooperative planning by regions, States, and local districts in order to form workable management plans that account for hydrologic reality.

The underlying philosophy for planning and management efforts should be to promote wise use, conservation, and protection of the ground-water resource. Conservation measures, such as reuse or recycling, pumping controls, and blending, should be combined with measures to protect the ground-water supply, primarily from man's activities. One potentially harmful activity is the underground injection of liquid wastes. Saltwater-bearing aquifers are generally considered to be suitable receptacles for waste. However, as the result of advances that are being made in developing economical desalinization techniques, saline aquifers may eventually become important sources of water. Therefore, in addition to the other indirect hazards associated with waste injection, such as leakage through wells or confining beds, there is also the risk of directly contaminating a potential water supply. These practices should be of as much concern as is the dumping of waste into streams or the Gulf of Mexico.

Strip mining in the region is a possible future activity that should be evaluated for its potential effect on ground water. Most of the environmental concern with strip mining is about disturbance of the landscape; however, the geohydrologic effects could be far more serious. Mining companies now contemplate the

Selected agencies within the Lower Mississippi Region that manage or investigate ground-water resources

Arkansas
Arkansas Department of Pollution Control and Ecology
Arkansas Division of Soil and Water Resources
Arkansas Geological Commission
U.S. Geological Survey, Water Resources Division
U.S. Army Corps of Engineers

Louisiana
Capitol Area Water Conservation Commission
Capitol Region Planning Commission
Louisiana Department of Public Works
Louisiana Geological Survey
U.S. Geological Survey, Water Resources Division
U.S. Army Corps of Engineers

Mississippi
Mississippi Air and Water Pollution Commission
Mississippi Board of Water Commissioners
Mississippi Geological Economic and Topographical Survey
Mississippi Research and Development Center
Mississippi State Board of Health
Environmental Protection Agency
U.S. Geological Survey, Water Resources Division
U.S. Army Corps of Engineers

Tennessee
Chickasaw Basin Authority
Memphis, Arkansas, Tennessee Council of Governments, Memphis Delta Development District (MATCOG MDDD)
Memphis Department of Public Works
Memphis Light, Gas, and Water Development
Memphis Planning Commission
Memphis-Shelby County Health Commission
Shelby County Conservation Board
Utility Districts in cities in west Tennessee
Tennessee Department of Health
  Division of Water Quality Department
  Division of Solid Waste Management
  County Health Superintendents
Tennessee Division of Water Resources
U.S. Geological Survey, Water Resources Division
U.S. Army Corps of Engineers

Kentucky
Department for Natural Resources and Environmental Protection
Kentucky Geological Survey
U.S. Geological Survey, Water Resources Division
U.S. Army Corps of Engineers

Missouri
Missouri Department of Natural Resources
U.S. Geological Survey, Water Resources Division
U.S. Army Corps of Engineers

removal of as much as 200 ft (61 m) of overburden in order to mine small lignite layers. The current emphasis on locating new sources of energy has awakened interest in mining coal from deposits that heretofore have been considered uneconomical to develop. The Wilcox Group is such a deposit; it consists of complexly interbedded sands, silts, clays, and thin lignite beds. The sands are water bearing and are interconnected to varying degrees. Large, deep excavations in Wilcox deposits would act as sumps and would drain the sands. Such draining could have widespread effects, which would be difficult to predict, depending upon the size of the excavation and the extent and degree of hydraulic connection between the sands. The effects of such drainage would range from lowering the water level in aquifers to the removal of ground water from a sizable part of the aquifer system, thus endangering many existing wells, as well as future well development. Drainage of freshwater-bearing sands into excavation sites could also adversely affect the quality of the ground water downdip from the excavation by introducing contaminated water into the aquifer or by inducing the advance of saltwater into freshwater zones. Much of the damage to the ground-water regime caused by strip mining may be irreversible. Although refilling the excavations would help to restore the esthetic quality of the area, the continuity and interconnection of aquifers necessary to conduct recharge downdip probably could not be reestablished.

Conservation of ground water should go beyond eliminating the more obvious forms of waste, such as disposing of usable water once it has served a purpose. Water is also wasted when a superior-quality water is used where an inferior-quality water would suffice—for example, using potable water for industrial cooling. Obviously, if only one or two aquifers are available for supply in an area, there may be little or no choice as to which aquifer is to be used for which purpose. However, at most locations in the Lower Mississippi Region, several aquifers (fig. 8) that contain water of differing chemical quality are available. Under such conditions, water-use priorities keyed to water quality should be established for different-use categories. Each category would be assigned water of the lowest acceptable quality; however, drinking water would be assigned the top priority.

It is apparent that there is a need to plan carefully for the use of ground water. The resource is of no value unless used, and it should receive its due consideration. However, while use of ground water is to be encouraged, care must be taken to prevent its abuse. Development should be guided by sound precepts. The consequences of other planned activities that could affect the ground-water regime should be weighed carefully as are the more commonly recognized environmental consequences. Ground water is an unseen part of the environment but, like the visible part, it must not be neglected.

WHAT ARE THE POSSIBLE CONSEQUENCES OF INCREASED GROUND-WATER USE?

When considering the development of ground-water supplies, managers should not overlook possible consequences to the resource. Development of large-producing wells without an adequate knowledge of the aquifer system could result in undesirable changes in the system. These changes may be in the hydraulics of the system, in the quality of water in the aquifer, and (or) in the framework of the water-bearing unit.

HYDRAULIC CHANGES

When a well is pumped, the water level in the vicinity of the well is lowered. The lowering of the water level may be temporary if the pumping is cyclical or seasonal, or it may be permanent if the pumping is continuous. If recharge to the aquifer is increased or discharge is decreased to compensate for the water discharged from the well, the water level in the vicinity of the well should stabilize. Otherwise, the water level will continue to decline, and pumping lifts and pumping costs will increase.

Water levels in several aquifers have been lowered in sizable areas of the Lower Mississippi Region. The aquifers most affected are the Sparta Sand and the Memphis aquifer in Louisiana, Arkansas, and Tennessee (fig. 25). Other affected aquifers include the upper Wilcox in northwest Mississippi, the lower Wilcox at Memphis, Tenn., the Chicot-Atchafalaya aquifer near Lake Charles, La., and alluvial deposits of Quaternary age in the Grand Prairie region in Arkansas.

Normally, aquifers in the Lower Mississippi Region are recharged by streams only during high flows. Ground water discharges to streams during normal and low flows. However, continuous pumping from an aquifer in the vicinity of a stream may reverse the hydraulic gradient so that water is demanded from the stream (fig. 13). Inducing the flow of water from a stream to the aquifer can be of benefit if the stream carries sufficient quantities of water to meet both the normal surface-water demands and the demands of aquifer recharge.

Where an aquifer overlies another and the two are separated by a confining layer, movement of water from one aquifer to the other is minimal if a near balance of pressure (head) exists in the aquifers. When water is pumped from one of these aquifers, the head imbalance between the aquifers may induce flow through the confining layer. After continuous pumping, a stable pumping level will be reached in the receiving aquifer at a higher level than would be reached in the absence of recharge from the adjacent aquifer. Also, the head will be reduced in the contributing aquifer even if this aquifer is not being pumped directly. Reports by Hosman and others (1968), Whitfield (1975), and Boswell (1979) have pointed out several areas in the Lower Mississippi Region where water moves from one aquifer to another through a confining layer.

WATER QUALITY CHANGES

Where aquifers are hydraulically connected, as discussed previously, and where pumping of one aquifer has induced a change in the flow system, water-quality changes may occur. Inducing flow from a stream to an aquifer may also cause changes in the quality of water in the aquifer. Because the ground water generally will be of better quality than water in the stream, any significant change will probably be detrimental. However, an improvement in quality could result if water in the stream is of exceptionally good quality. Changes in quality should be carefully considered before inducing recharge from streams or from other aquifers.

Saltwater encroachment could occur in many places in the Lower Mississippi Region. Northward saltwater migration has been occurring for several years in the Chicot aquifer in southwest Louisiana due to pumpage in the Lake Charles area (Zack, 1971). However, the movement of saltwater thus far observed is small.

Many aquifers in the interior of the Lower Mississippi Region contain saltwater downdip. Under natural conditions, the saltwater is either virtually static or is moving very slowly downdip. However, heavy pumping of these aquifers near the saltwater-freshwater interface can reverse the hydraulic gradient and cause saltwater to move toward the point of withdrawal. This condition has occurred in the vicinity of Baton Rouge, where heavy industrial pumping has caused northward movement of saltwater in several aquifers (Rollo, 1969). In at least one place, the saltwater is moving across a fault south of the area of pumping. Conversely, the same fault acts as a barrier and protects some other aquifers from saltwater encroachment (fig. 26). Although saltwater has been moving toward the withdrawal point, no saltwater has yet migrated into the zone affected by pumping. The dense saltwater in the basal part of the aquifer appears to take a circuitous route because of irregularities in the base of the aquifer. Because the nature of these irregularities is not known, the time required for the saltwater to reach pumping centers is not predictable with any degree of accuracy.

AQUIFER-FRAMEWORK CHANGES

Large water-level declines caused by withdrawals of ground water may result in major changes in the framework of an aquifer. Major aquifers in the Lower Mississippi Region are composed principally of sand and gravel interbedded with clays and silts, and when large amounts of water are withdrawn, the fine-grained materials can become compacted.

Land-surface subsidence is the most visible effect of compaction due to pumping. In areas outside the region, land surfaces have been lowered in amounts ranging from a few feet to a few tens of feet. In the

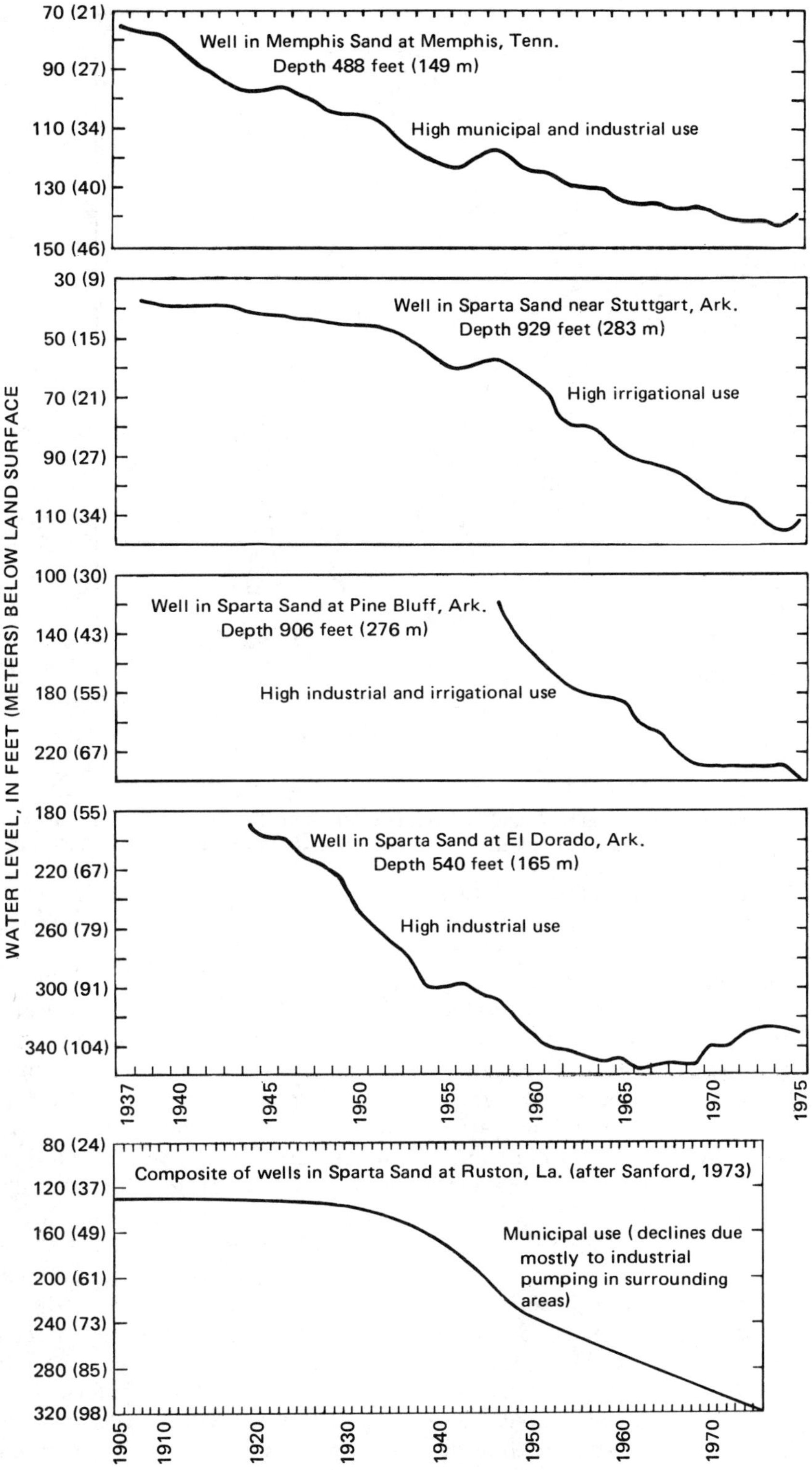

FIGURE 25.—Water-level declines in observation wells in the Memphis Sand and Sparta Sand.

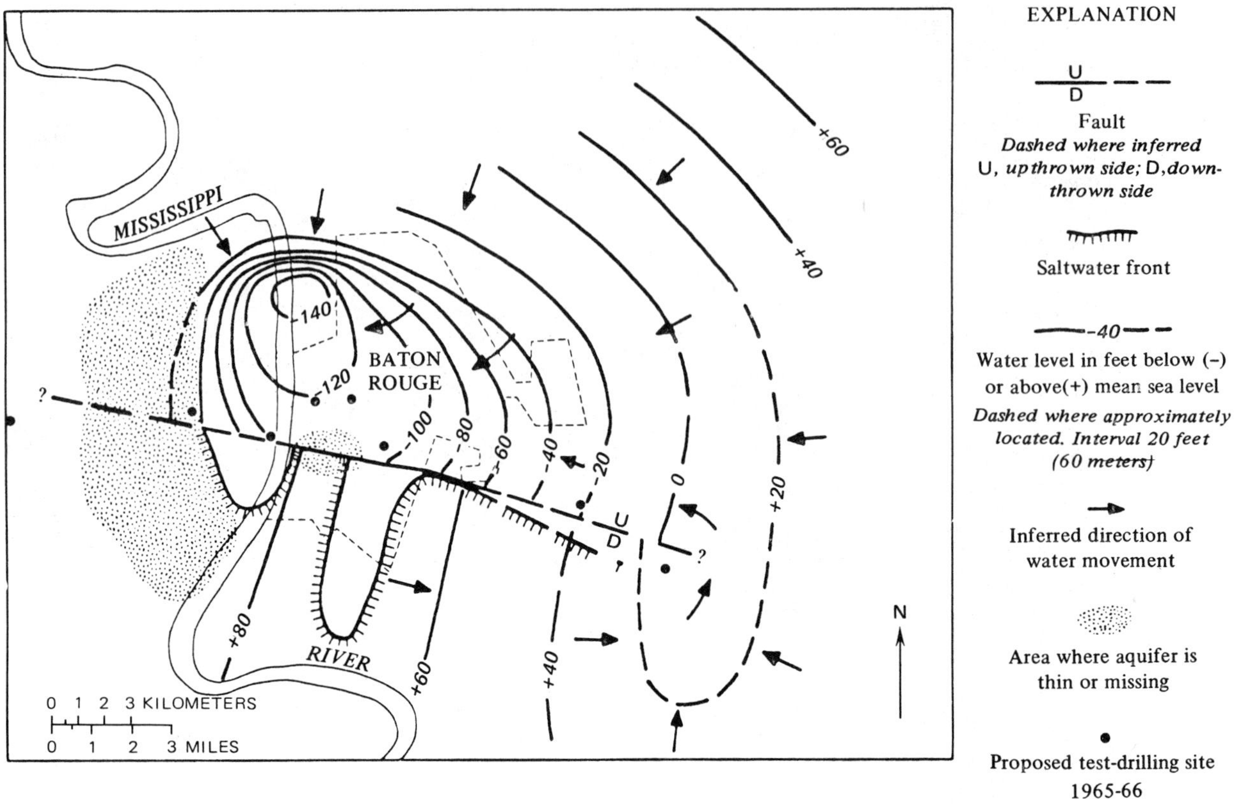

FIGURE 26.—Saltwater front, water-level contours, and location of fault in the "2,000-foot" sand as determined during the 1965 test-drilling program in the Baton Rouge, La., area (from Meyer and Rollo, 1965).

San Joaquin Valley of California, for example, withdrawals of water for irrigation have allowed compaction of fine-grained materials to the extent that by 1972 the land surface had subsided as much as 29 ft (9 m). More than 5,000 mi^2 (13,000 km^2) in the valley have been affected (Poland and others, 1975). Subsidence can cause damage to roadways and structures and allow innundation of low-lying coastal areas.

Known subsidence in the Lower Mississippi Region has been confined to Baton Rouge, La., where large withdrawals of ground water have caused the land surface to subside a maximum of about 1.5 ft (0.5 m). In other areas in the Lower Mississippi Region, where hydrostatic pressures have been significantly lowered in sand-and-gravel aquifers, compaction probably has occurred as a result of drainage from underlying, overlying, and interbedded fine-grained materials. Although noticeable subsidence may not have been produced, deformation generally is accompanied by a permanent reduction in storage of water in the fine-grained materials. Therefore, although subsidence is not a major problem in the Lower Mississippi Region at present, it should be regarded as a serious potential problem.

IN SUMMARY— IS THE OUTLOOK OPTIMISTIC?

The outlook for additional ground-water use in the Lower Mississippi Region is bright. Multiple freshwater aquifers are available throughout most of the region. By conservative estimate, 844 billion ft^3 (24 billion m^3) of fresh ground water is available annually in the region. Today, nearly two-thirds of this available supply is not being utilized, and an additional large quantity of water is being lost—water that might have infiltrated the surface to become ground water if storage space were available. In other words, increased ground-water usage would not only allow ground-water benefits to more nearly approach a maximum but would also save some water that today cannot recharge the aquifers.

Much of the ground water withdrawn and commonly termed "ground water used" could be used more than once or put to other uses. As existing water-oriented conservation techniques become more widely implemented and as new ones are developed, 1 gallon of water withdrawn may become several gallons used. It is quite conceivable that in the future, with the implementation of such practices are recycling, reuse, injection of unaltered ground water back into the aquifer, and induced aquifer recharge from streams, annual ground-water use (including reuse) could exceed the quantity now estimated to be available annually.

The quantity of subsurface saltwater available in the Lower Mississippi Region is several times that of freshwater. Further development and utilization of desalinization techniques would multiply the region's fresh ground-water supply several times. Furthermore, untreated saltwater may be used in some applications instead of freshwater; for example, at places, such as some areas of Louisiana, where saltwater occurs at shallow depths and is not too warm, it can be used for cooling purposes.

To make maximum use of this vast quantity of ground water, adequate information about the subsurface reservoirs and the hydraulics of the system must be available. This information, as well as adequate legislation, is necessary in order to properly protect the quality of the ground water and the physical integrity of the aquifers.

Today, there is an increasing awareness of the vital role that ground water plays in the total water-resources scene in the Lower Mississippi Region. Planning and management efforts are beginning to reflect this awareness. Interpretive and predictive investigations required by the water manager in order to make sound decisions have been, and are being, made in the region. Each State in the region maintains active water-resources-study programs of its own, as well as cooperative programs with other water-oriented agencies, including support of programs with the Geological Survey. All these agencies give considerable attention to ground-water resources. Increased cooperation between water-oriented agencies not only broadens the information base and helps prevent duplication of effort, but also stimulates new approaches to old problems and the recognition of areas where attention is needed.

The technology and methodology for creating useful management tools for the water-resource manager have greatly advanced in the last 10 years. The sophistication of the digital computer and the development of digital-modeling techniques have made it possible, with adequate data, to make long-term stress-response predictions for a large area in a very short time. Also, methods are available for dealing with problems that may be encountered when utilizing the ground-water resource; for example, selective use based on water quality where multiple aquifers are available, barrier-well systems, scavenger wells, and artificial recharge.

In the past, ground water and surface water were considered by many people to be separate entities that act and react independently. Today, such false ideas have been discarded, and the close and dependent relationship of the two resources is accepted. Total water-resource planning considers this relationship and the ways in which changes in one regime may affect the other.

Each State in the Lower Mississippi Region has water laws that directly and indirectly concern ground water. These laws are changed and improved as legislators are made more aware of the immediate need for enforceable guidelines in ground-water development and protection. Each State has an agency that is responsible for the identification and control of environmental pollution, including ground-water pollution. Some water-resources groups or agencies function on the local, county, or parish level. For example, in Louisiana, the Capital Area Ground-Water Conservation District encompasses a multiparish area in the Baton Rouge vicinity. The findings of the Louisiana Legislature that precipitated the organization of this conservation district are expressed in Article A of Section 3071, Part XIII, Chapter 13, Title 38, of the Louisiana Revised Statutes. The article states: "The orderly utilization of groundwater resources is hereby found and declared to be a matter of public interest." The purpose of the conservation district is, as stated in Article B of the same section "to provide for the efficient administration, conservation, orderly development and supplementation of groundwater resources***." The existence and resolve of such local groups is good; the extension of these concepts to a regional conservation organization concerned with ground-water planning and management would be even more compatible with the regional nature of ground-water occurrence.

SELECTED REFERENCES

Bedinger, M. S., 1970, Methods and applications of electrical simulation in ground-water studies in the lower Arkansas and Verdigris River valleys, Arkansas and Oklahoma: U.S. Geol. Survey Water-Supply Paper 1971, 69 p.

Bedinger, M. S., Emmett, L. F., and Jeffery, H. G., 1963, Ground-water potential of the alluvium of the Arkansas River between Little Rock and Fort Smith, Arkansas: U.S. Geol. Survey Water-Supply Paper 1669–L, 29 p.

Bedinger, M. S., Pearson, F. J., Jr., Reed, J. E., Sniegocki, R. T., and Stone, C. G., 1974, The waters of Hot Springs National Park Arkansas—Their origin, nature, and management: U.S. Geol. Survey open-file report, 102 p.

Boswell, E. H., 1979, 1975 National Water Assessment: Ground water in the Lower Mississippi Region: Natl. Water Resources Council Rept. (in press).

Boswell, E. H., Cushing, E. M., and Hosman, R. L., 1968, Quaternary aquifers in the Mississippi embayment: U.S. Geol. Survey Prof. Paper 448–E, 15 p.

Boswell, E. H., Moore, G. K., MacCary, L. M., and others, 1965, Cretaceous aquifers in the Mississippi embayment: U.S. Geol. Survey Prof. Paper 448–C, 37 p.

Broom, M. E., and Reed, J. E., 1973, Hydrology of the Bayou Bartholomew alluvial aquifer-stream system, Arkansas: U.S. Geol. Survey open-file report, 91 p.

Bull, W. B., 1964, Alluvial fans and near surface subsidence in western Fresno County, California: U.S. Geol. Survey Prof. Paper, 437–A, 71 p.

Bull, W. B., and Miller, R. E., 1975, Land subsidence due to ground-water withdrawal in the Los Banos-Kettleman city area, California, Part 1. Changes in the hydrologic environment conducive to subsidence: U.S. Geol. Survey Prof. Paper 437–E, 67 p.

Committee on Geological Sciences, 1972, The earth and human affairs: San Francisco, Canfield Press, 138 p.

Committee on Resources and Man, 1969, Resources and man: San Francisco, W. H. Freeman and Company, 239 p.

Cushing, E. M., 1963, Water Resources and the Mississippi embayment: U.S. Geol. Survey Circ. 471, 8 p.

Cushing, E. M., Boswell, E. H., and Hosman, R. L., 1964, General geology of the Mississippi embayment: U.S. Geol. Survey Prof. Paper 448–B, 28 p.

Davis, R. W., Lambert, T. W., and Hanson, A. J., Jr., 1973, Subsurface geology and ground-water resources of the Jackson Purchase region, Kentucky: U.S. Geol. Survey Water-Supply Paper 1987, 66 p.

Dial, D. C., 1968, Water-level trends in southeastern Louisiana: Louisiana Geol. Survey and Dept. Public Works Water Resource Pamph. 22, 11 p.

Eaton, W. R., 1958, Summary of flow duration and low flow at stream gaging stations in Tennessee: Tennessee Div. Geology and Div. Water Resources open-file report, 137 p.

Flawn, P. T., 1970, Environmental geology: New York, Harper and Row, 251 p.

Gabrysch, R. K., and Bonnet, C. W., 1975, Land-surface subsidence in the Houston-Galveston region, Texas: Texas Water Devel. Board Rept. 188, 19 p.

Halberg, H. N., 1972, Use of water in Arkansas, 1970: Arkansas Geol. Comm. Water Resources Summ. 7, 17 p.

Harrison, R. W., 1961, Alluvial empire: Little Rock, Ark., Pioneer Press, v. 1, 344 p.

Hines, M. S., 1965, Water-supply characteristics of selected Arkansas streams: Arkansas Geol. Comm. Water Resources Circ. 9, 43 p.

———1975, Flow-duration and low-flow frequency determinations of selected Arkansas streams: U.S. Geol. Survey Water Resources Circ. 12, 75 p.

Hosman, R. L., 1969, Geohydrology of the Coastal Plain aquifers of Arkansas: U.S. Geol. Survey Hydrol. Inv. Atlas HA–309.

Hosman, R. L., Long, A. T., Lambert, T. W., and others, 1968, Tertiary aquifers in the Mississippi embayment: U.S. Geol. Survey Prof. Paper 448–D, 29 p.

Long, R. A., 1965, Ground water in the Geismar-Gonzales area, Ascension Parish, Louisiana: Louisiana Geol. Survey and Dept. Public Works Water Resources Bull. 7, 67 p.

Lower Mississippi Region Comprehensive Study Coordinating Committee, 1974: Lower Mississippi Region Comprehensive Study.

Meyer, R. R., and Rollo, J. R., 1965, Salt-water encroachment, Baton Rouge area, Louisiana: Water Resources Pamph. 17, 9 p.

Moore, G. K., 1965, Geology and hydrology of the Claiborne Group in western Tennessee: U.S. Geol. Survey Water-Supply Paper 1809–F, 44 p.

Morgan, C. O., and Winner, M. D., Jr., 1964, Salt-water encroachment in aquifers of the Baton Rouge area—preliminary report and proposal: Louisiana Dept. Public Works pub., 37 p.

Newcome, Roy, Jr., 1975, The Miocene aquifer system in Mississippi: U.S. Geol. Survey open-file [map] report.

Page, L. V., 1963, Water supply characteristics of Louisiana streams: Louisiana Dept. Public Works Techn. Rept. 1, April, 109 p.

Payne, J. N., 1968, Hydrologic significance of the lithofacies of the Sparta Sand in Arkansas, Louisiana, Mississippi, and Texas: U.S. Geol. Survey Prof. Paper 569–A, 17 p.

———1970, Geohydrologic significance of lithofacies of the Cockfield Formation of Louisiana and Mississippi and the Yeyua Formation of Texas: U.S. Geol. Survey Prof. Paper 569–B, 14 p.

———1972, Hydrologic significance of lithofacies of the Cane River Formation or equivalents of Arkansas, Louisiana, Mississippi, and Texas: U.S. Geol. Survey Prof. Paper 569–C, 17 p.

———1975, Geohydrologic significance of lithofacies of the Carrizo Sand in Arkansas, Louisiana, and Texas, and the Meridian Sand of Mississippi: U.S. Geol. Survey Prof. Paper 569–D, 11 p.

Poland, J. F., Lofgren, B. E., Ireland, R. L., and Pugh, R. G., 1975, Land subsidence in the San Joaquin Valley, California: U.S. Geol. Survey Prof. Paper 437–H, 77 p.

Rainwater, F. H., 1962, Stream composition of the conterminous United States: U.S. Geol. Survey Hydrol. Inv. Atlas HA–61.

Reed, J. E., 1972, Analog simulation of water-level declines in the Sparta Sand, Mississippi embayment: U.S. Geol. Survey Hydrol. Inv. Atlas HA–434.

Rollo, J. R., 1966, Ground-water resources of the greater New Orleans area, Louisiana: Louisiana Geol. Survey and Dept. Public Works Water Resources Bull. 9, 69 p.

———1969, Salt-water encroachment in aquifers of the Baton Rouge area, Louisiana: Louisiana Geol. Survey and Dept. Public Works Water Resources Bull. 13, 43 p.

Sanford, T. H., Jr., 1973, Water resources of the Ruston area, Louisiana: Louisiana Dept. Public Works Water Resources Tech. Rept. 8, 30 p.

Skelton, John, 1966, Low-flow characteristics of Missouri streams: Missouri Geol. Survey and Water Resources Rept. 20, 95 p.

Sniegocki, R. T., 1963, Problems in artificial recharge through wells in the Grand Prairie region, Arkansas: U.S. Geol. Survey Water-Supply Paper 1615–F, 25 p.

———1964, Hydrology of a part of the Grand Prairie region, Arkansas: U.S. Geol. Survey Water-Supply Paper 1615–B, 72 p.

Sniegocki, R. T., Bayley, F. H., III, and Engler, Kyle, 1963, Equipment and controls used in studies of artificial recharge in the Grand Prairie region Arkansas: U.S. Geol. Survey Water-Supply Paper 1615–C, 39 p.

Sniegocki, R. T., and Reed, J. E., 1963, Principles of siphons with respect to the artificial recharge studies in the Grand Prairie region, Arkansas: U.S. Geol. Survey Water-Supply Paper 1615–D, 19 p.

Speer, P. R., Hines, M. S., Janson, M. E., and others, 1966, Low-flow characteristics of streams in the Mississippi embayment in northern Arkansas and Missouri: U.S. Geol. Survey Prof. Paper 448–F, 25 p.

Swisshelm, R. V., Jr., 1974, Low-flow characteristics of Kentucky streams: U.S. Geol. Survey open-file [map] report.

Tharpe, E. J., 1975, Low-flow characteristics of Mississippi streams:

Mississippi Board of Water Commissioners Bull. 75–1, 60 p.
Thornthwaite, C. W., 1948, An approach to a rational classification of climate: The Geol. Rev., p. 55.
U.S. Geological Survey, 1970, The National atlas of the United States of America: U.S. Geol. Survey, 417 p.
U.S. Water Resources Council, 1970, Water resources regions and subregions for the national assessment of water and related land resources: Washington, U.S. Govt. Printing Office, 187 p.
Whitfield, M. S., Jr., 1975, Geohydrology and water quality of the Mississippi River alluvial aquifer, northeastern Louisiana: Louisiana Dept. Public Works, Water Resources Tech. Rept. 10, 26 p.
Zack, A. L., 1971, Ground-water pumpage and related effects, southwestern Louisiana, 1970, *with a section on* surface-water withdrawals: Louisiana Dept. Public Works Water Resources Pamph. 27, 33 p.

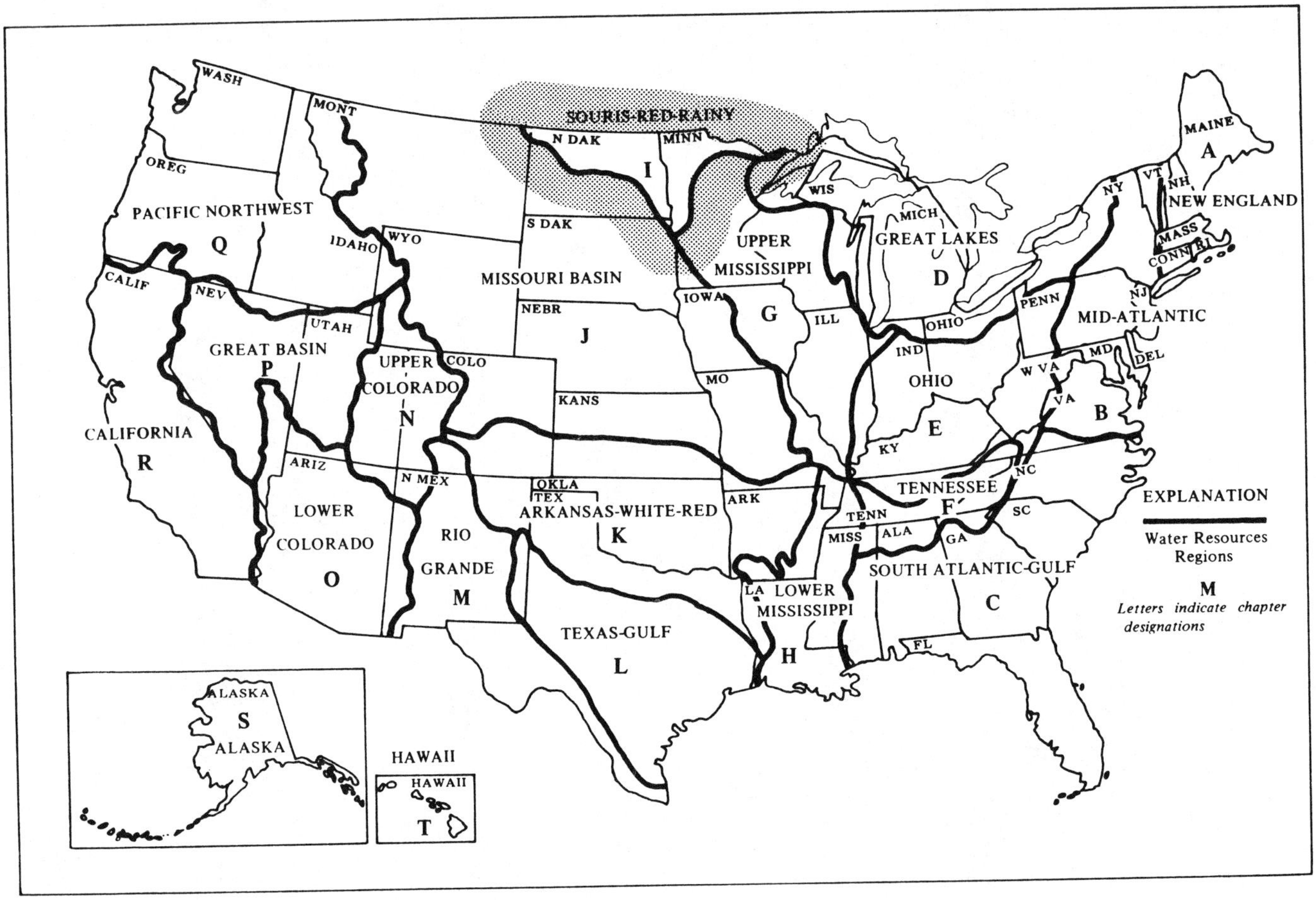

I — Location Map for the Souris-Red-Rainy Region

I

SOURIS-RED-RAINY REGION

By HAROLD O. REEDER

ABSTRACT

A broad-perspective analysis of the ground-water resources and present and possible future water development and management in the Souris-Red-Rainy Region is presented. The region includes the basins of the Souris River within Montana and North Dakota; the Red River of the North in South Dakota, North Dakota, and Minnesota; and the Rainy River within Minnesota. The region includes 59,645 square miles, mostly in North Dakota and Minnesota.

The terrain is relatively flat, but ranges in altitude from 2,541 to 750 feet. Annual average precipitation ranges from 14 inches in the west to 28 inches in the east and about 75 percent of it is rain. The mean annual snowfall ranges from 32 inches in the west to 64 inches in the east. Temperatures range from -55° to 118° F (-48.3° to 47.8° C). Irrigation is needed at least part of the time to assure crop production, particularly in the western part of the region.

Sand and gravel deposits in the drift form the most important freshwater aquifers. Other aquifers are found in at least parts of the region in the Precambrian, Paleozoic, Cretaceous, and Tertiary rocks. The potentiometric surface in the bedrock aquifers generally decreases in altitude toward the Red River of the North, indicating that the general direction of ground-water movement is toward the river.

Ground water with less than 3,000 milligrams per liter dissolved solids is available throughout the region. Ground water with less than 1,000 milligrams per liter occurs in most of the region east of the Red River of the North and in most of the shallow aquifers west of the river. The total volume of water available from storage having less than 3,000 milligrams per liter dissolved solids is estimated to be $5x10^8$ acre-feet. In addition to the fresh and slightly saline water, the region has abundant highly mineralized water that can be considered as a resource. Yields of wells in individual bedrock aquifers are generally less than 100 gallons per minute but locally yields may be as much as 500 gallons per minute and more. Yields in drift aquifers are frequently less than 100 gallons per minute but range from 5 to 1,000 gallons per minute. In a few places outwash yields more than 1,000 gallons per minute.

Ground water is the sole or a primary source of water supply in much of the region, including supplies for irrigation, domestic and livestock, municipal, and industrial needs. Reportedly, the potential irrigation development is 1,550,000 acres, as compared with 50,200 acres in 1975. Both ground- and surface-water supplies would be required to meet these demands. Rural domestic and livestock water supplies are derived almost entirely from ground-water sources. Smaller communities and towns generally rely on ground water, and the cities and industries use ground water, surface water, or both. The municipalities using surface water generally depend upon reservoir storage. Water quality rather than quantity is the greater water-supply problem for many communities in the region.

Increased demands on both ground-water and surface-water supplies likely will be made in the future. Storage of surface water in the ground-water reservoirs during times of surplus for withdrawal during times of scarcity would aid in meeting these demands. The surplus (flood) water is of better chemical quality than underlying ground water in parts of the western half of the region. Freshwater could be stored in saline- or freshwater aquifers, and pumped out later, as needed. Thus, the ground-water reservoirs have a definite present and potential role in water management.

To understand the hydrologic system for management purposes there is a need to determine more adequately the geologic and hydrologic characteristics of existing aquifers and the location of new aquifers. Also, as pumping and other stresses on any part of the hydrologic system affect other parts of the system, monitoring programs ideally should be started and maintained to detect changes and determine effects of the stresses.

Many alternatives are available for managing water in the region. Some of these are operational and others are undergoing research. Adequate hydrologic information is needed to aid in solving problems of water supply, use, and pollution.

INTRODUCTION

PURPOSE AND SCOPE OF THIS REPORT

The purpose of this report is to present a broad-perspective analysis of ground-water resources in the Souris-Red-Rainy Region. The region's ground water is a large and manageable resource that could have a more significant role in regional water development. This report is one of a U.S. Geological Survey series that summarizes information on the Nation's ground water for the guidance of planners. New data were not collected for this appraisal, but information from many sources has been utilized.

In addition to summarizing the knowledge of ground-water resources of the region, the report points out deficiencies in knowledge. The primary objective of evaluating information deficiencies is to direct attention to types of studies and information that will lead to fuller understanding and description of ground-water reservoirs for better evaluation, planning, and management of the region's water resources. With proper knowledge, utilization, and conjunctive management of all water resources, ground water can assume greater significance in the region's development.

PHYSICAL SETTING

The region is located along the northern boundary of the United States in North Dakota and Minnesota and extends short distances into Montana and South Dakota (fig. 1). The region discussed in this report includes that part of the three river basins within the United States. The region, which includes 59,645 square miles in the United States, drains northward into Hudson Bay. The Red River of the North, hereafter referred to as the Red River, drains 39,199 square miles in the central part of the region and flows into Lake Winnipeg and Nelson River to Hudson Bay (all in Canada and not shown on map). The Souris River drains 9,142 square miles in the western part of the region, joins the Assiniboine River in Manitoba, Canada, and flows into the Red River. The Rainy River drains 11,304 square miles in the eastern part of the region, flows through Lake of the Woods to the Winnipeg River in Canada, and eventually joins the Nelson River in Canada. The three river basins in the United States are in the Western Lake section of the Central Lowland physiographic province (Fenneman, 1931).

The terrain is relatively flat. One of the most prominent features is the plain along the Red River from 30 to 50 miles wide and 315 miles long. During the glacial epoch the Red River was occupied by glacial Lake Agassiz (fig. 2). Its outlet was southward into Big Stone Lake (outside the region) and through the present valley of the Minnesota River. Altitude ranges from 2,541 feet in north central North Dakota to 750 feet where the Red River crosses the Canadian boundary. Figure 2 Also shows the location of other former glacial lakes in the region.

REGIONAL WATER SUPPLY

Precipitation is the ultimate source of water supply. Average annual precipitation in the region ranges from about 14 inches in the west to 28 inches in the east (fig. 3). About 75 percent of the annual precipitation is rain. The mean annual snowfall ranges from about 32 inches in the west to 64 inches in the east (fig. 4). Precipitation is adequate for crop production during normal years, although the western half of the region has occasional droughts. The average annual natural runoff originating within the region ranges from less than 0.2 inch in the western part of the Souris River basin to about 15 inches in the eastern part of the Rainy River basin (fig. 5) owing to less precipitation and a higher evapotranspiration rate in the west than in the east. Generally, minimum flows occur during the winter under ice cover 2 to 4 feet thick. Maximum flows generally occur during the spring breakup, and about 50 percent of the annual flow occurs during April and May.

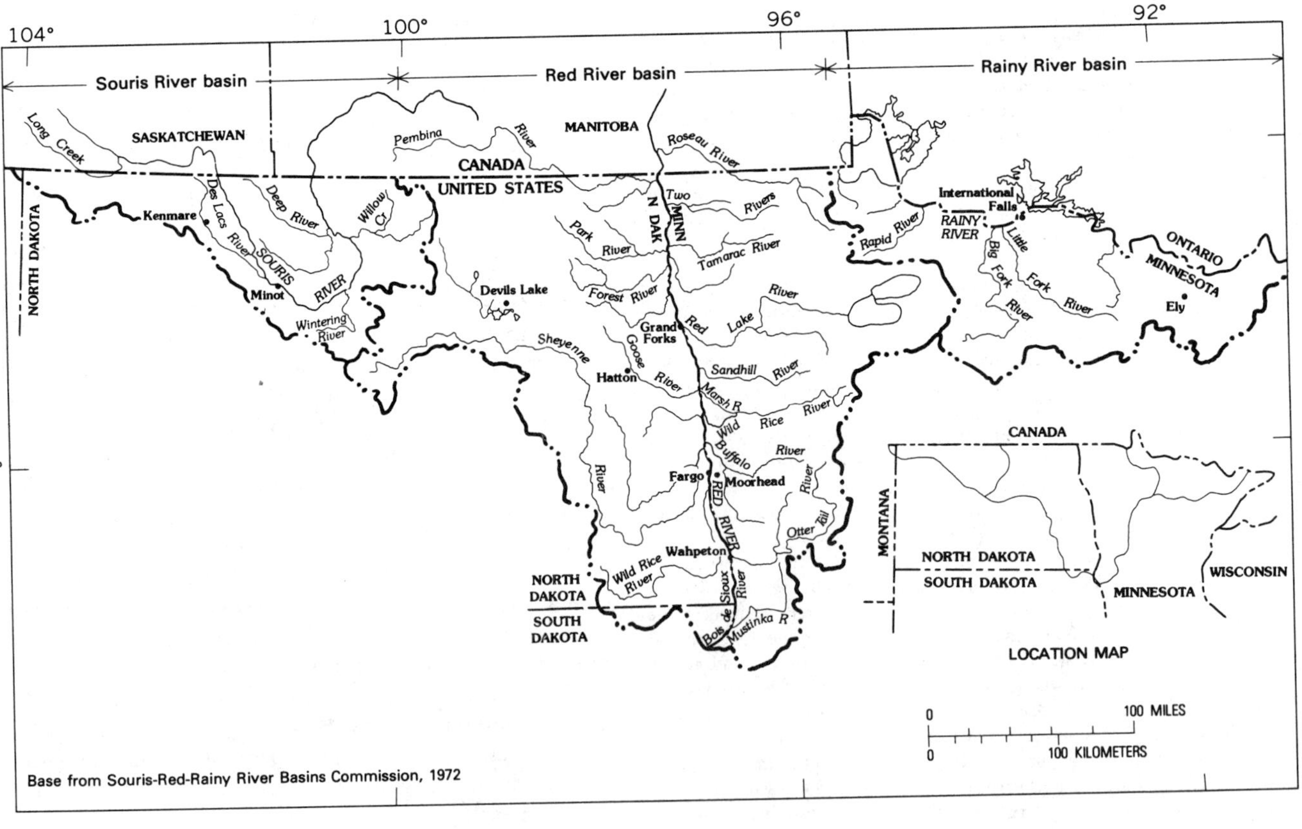

FIGURE 1.—Souris-Red-Rainy Region, showing subbasins.

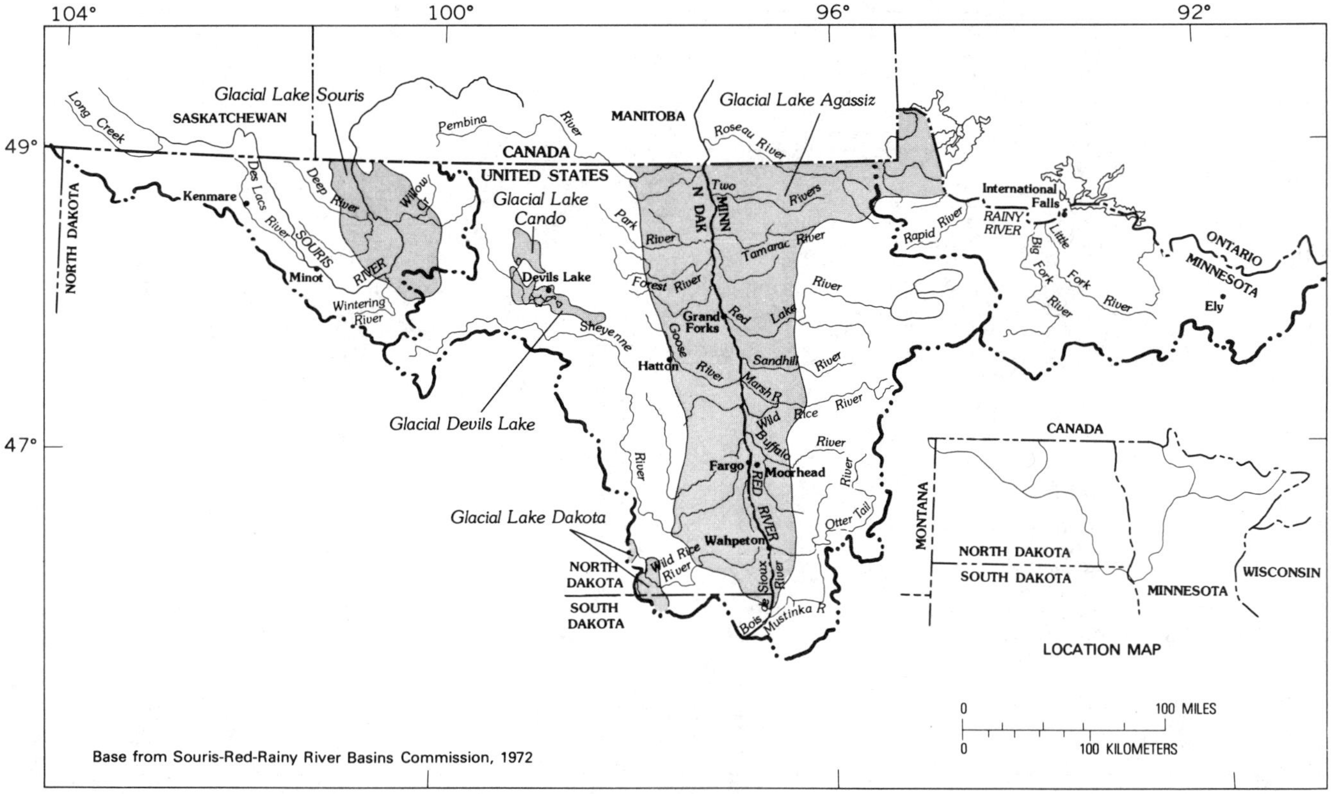

FIGURE 2.—Location of former glacial lakes in the Souris-Red-Rainy Region.

The quantity and quality of water resources differ greatly from place to place, season to season, and year to year. In the Souris River basin, streamflow is normally inadequate to satisfy water needs, and the quality is marginally acceptable for most present uses. During normal years, streamflow in most of the Red River basin is adequate, and the quality is generally satisfactory for the predominantly agricultural demands of the basin. However, water quality in the main stem of the Red River is poor due to heavy sediment loads; periodic low flows with attendant increases in dissolved solids; and increasing municipal, industrial, and agricultural polution. Water is abundant for the extent of present development in the Rainy River basin, and, except below International Falls and Shagawa Lake at Ely, its quality is excellent for most uses (U.S. Water Resources Council, 1968, p. 6-8-3).

The glacial aquifers and much of the bedrock are recharged by water from precipitation within the region. Some ground-water recharge is from surface-water bodies and from bedrock aquifers of adjacent regions. Recharge to the ground-water reservoirs from precipitation can occur only after the soil moisture deficiency is satisfied. Water requirements of plants and the intensity and duration of a rain, and certain additional factors such as soil type, also affect the amount of recharge. Consequently, recharge is not necessarily proportional to rates of precipitation, although when the precipitation rate is large, recharge is generally greater than during relatively dry periods.

Recorded temperature extremes range from -55° to 118° F (-48.3° to 47.8° C); however, mean monthly temperatures range from a monthly minimum of about -10° F (-23.3° C) in January to a monthly maximum of about 85° F (29.4° C) in July (U.S. Geological Survey, 1970, p. 104-107). The occasional droughts, hot winds, and prolonged high temperatures that occur, particularly in the western half of the region, cause crop failure and create the need for irrigation at least part of the time. Large quantities of ground water evaporate through the swamps, marshes, lakes, and streams. The average annual lake evaporation rate ranges from 36 inches in the west to less than 24 inches in the east (fig. 6).

Discharge from aquifers occurs naturally and by pumping. Natural discharge occurs by flow into adjacent rocks having lower hydraulic head, by seepage into streams, as springs, and by evapotranspiration. Trends in water levels reflect the balance or imbalance of ground-water recharge and discharge. Current pumping is not large enough to cause a significant general impact. Significant water-level declines are noted, however, in several areas (Kenmare, Minot, Fargo-West Fargo-Moorhead, and Hatton areas).

Ground water constitutes a major element of the region's water supply, as discussed in detail in the following section.

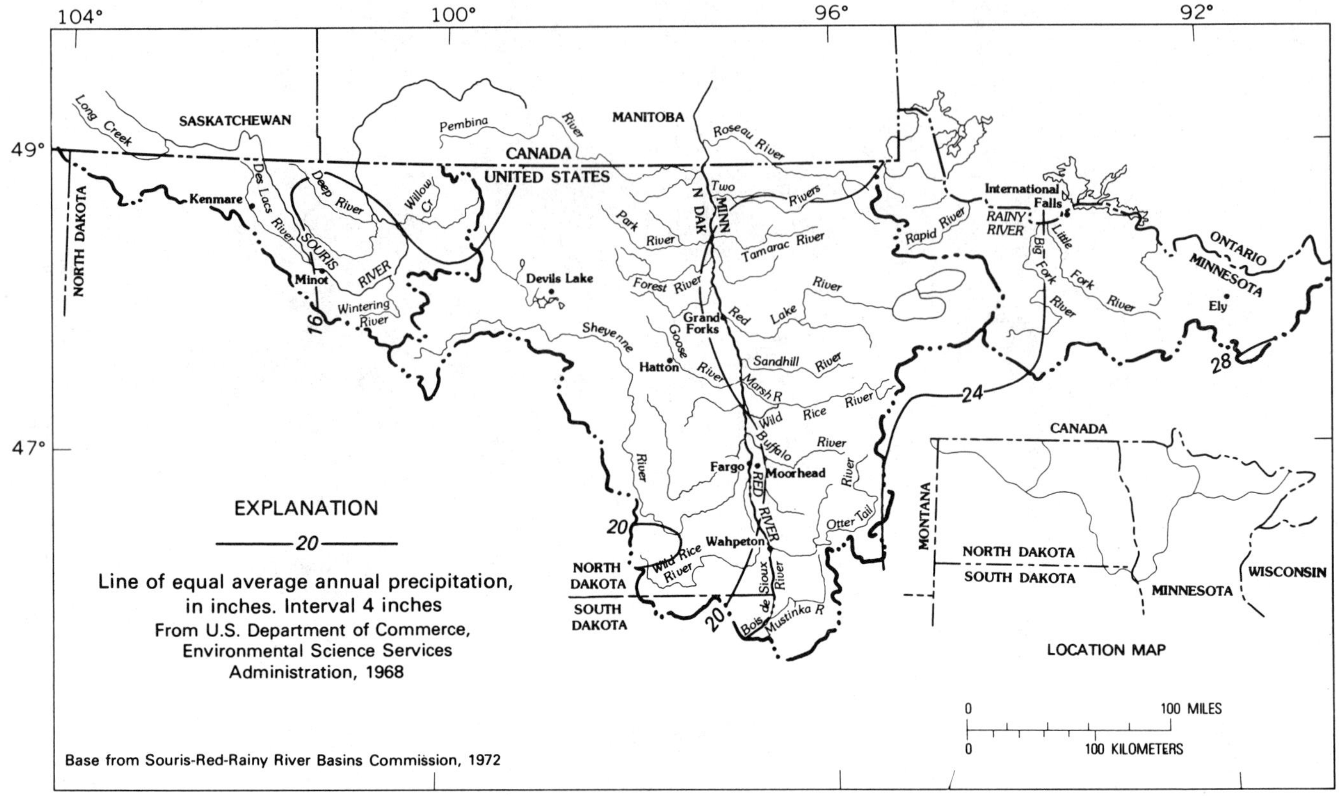

FIGURE 3.—Average annual precipitation.

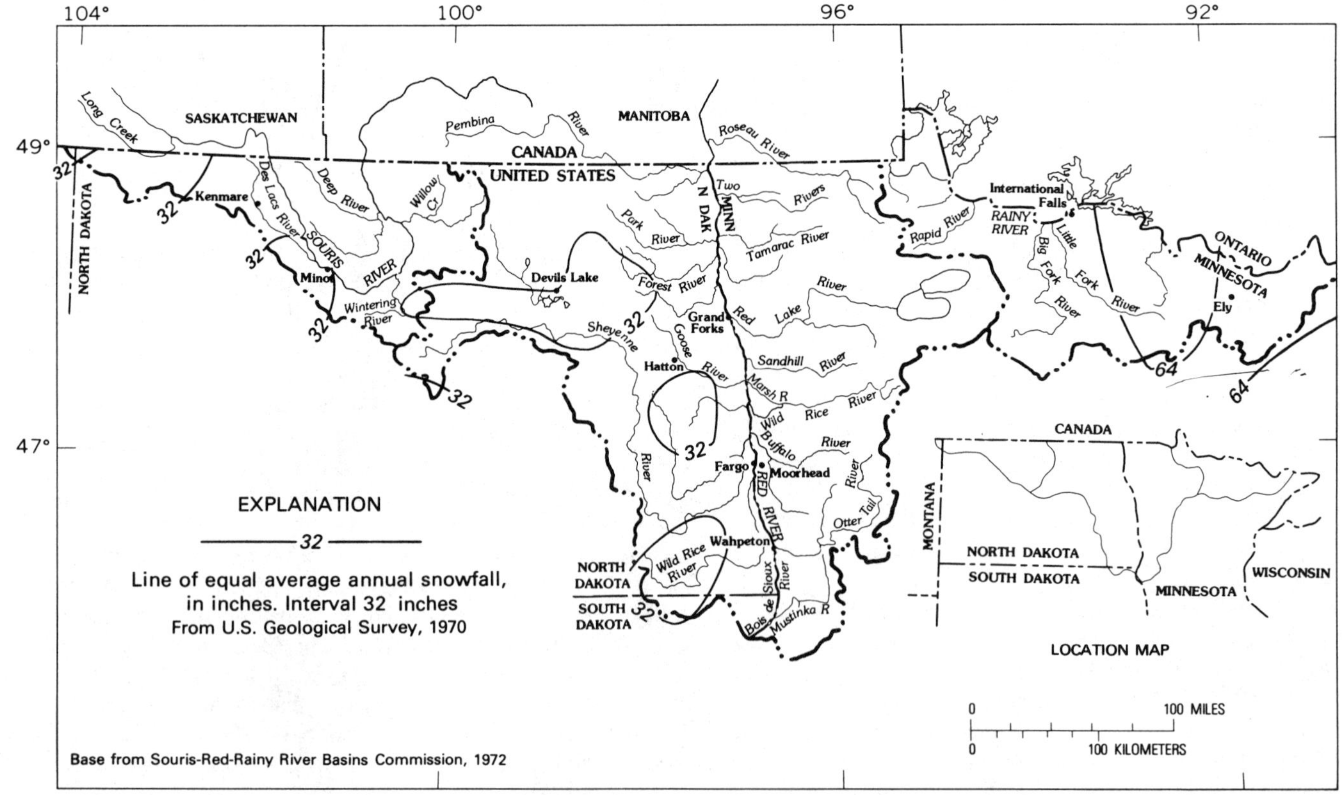

FIGURE 4.—Average annual snowfall.

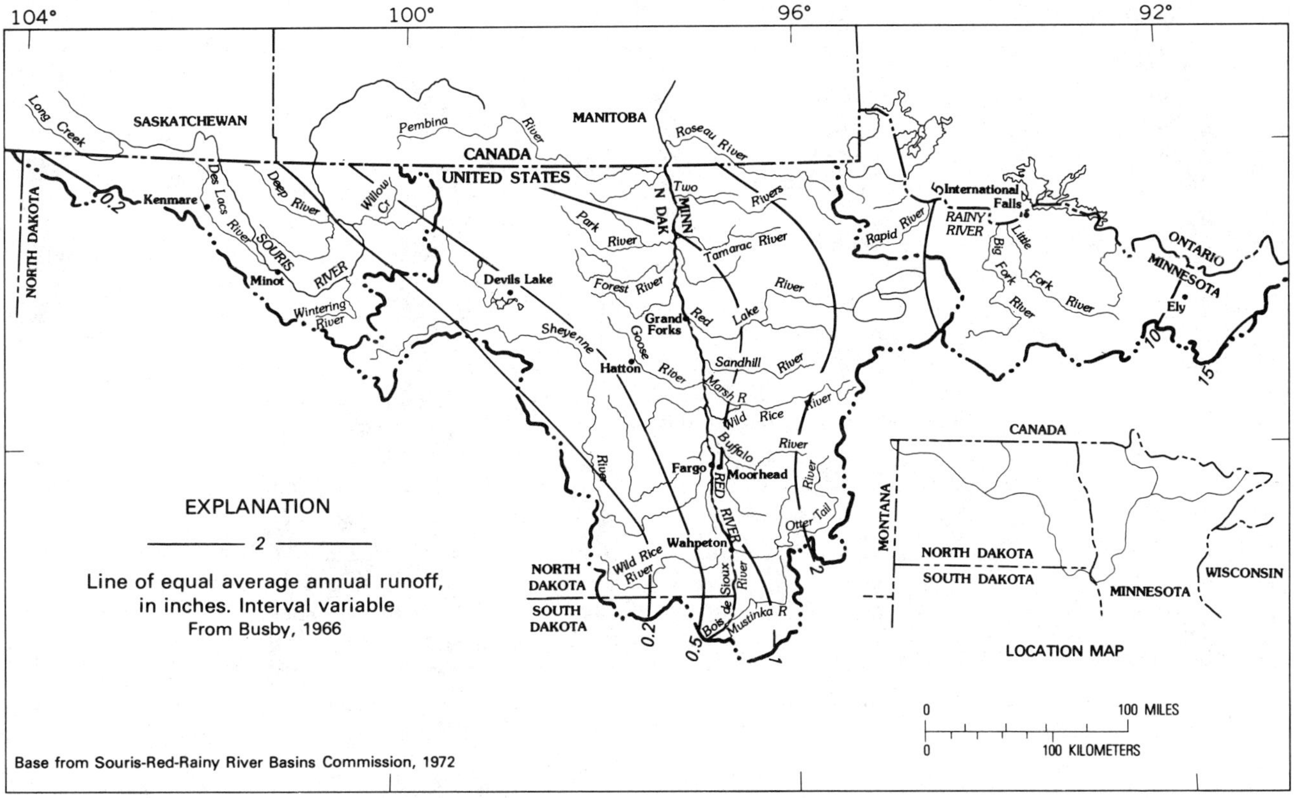

FIGURE 5.—Average annual runnoff.

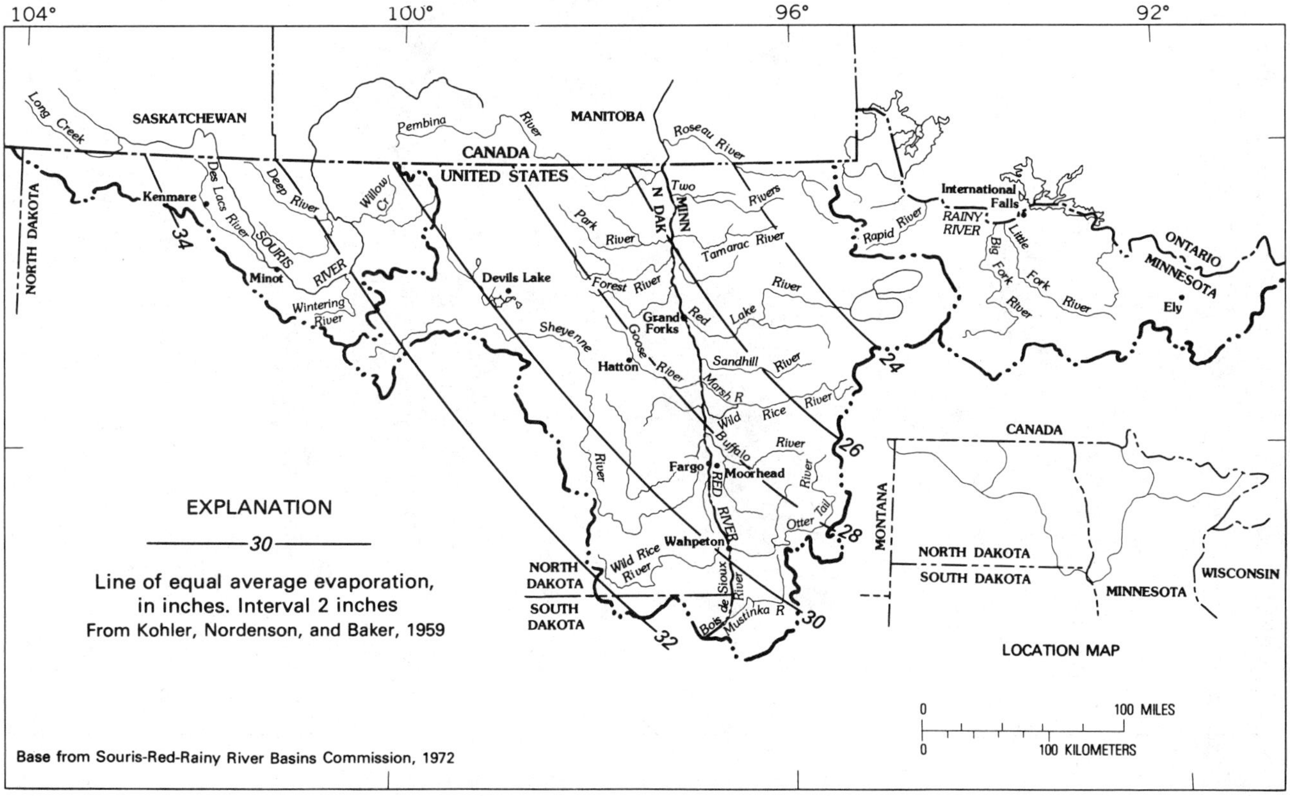

FIGURE 6.—Average annual lake evaporation.

GROUND-WATER RESOURCES

Ground water in the Souris-Red-Rainy Region is obtained mainly from aquifers in Pleistocene drift such as drainage-channel deposits, lake deltas, beach deposits, outwash deposits, and small bodies of sand and gravel interbedded with till. In addition, the Souris and Red River basins in North Dakota have aquifers of Precambrian and Paleozoic age; the Dakota, Pierre, and Fox Hills-Hell Creek aquifers of Cretaceous age; and the Fort Union aquifer of Tertiary age (Crosby and others, 1973, p. 176). In the Red and Rainy River basins in Minnesota, aquifers are of Precambrian, Paleozoic, and Cretaceous age. The total volume of water having less than 3,000 mg/L of dissolved solids available from storage in the region is estimated to be $5x10^8$ acre-feet. The estimate is based on areas, estimated and known saturated thicknesses, and estimated and known values of specific yields of the various aquifer materials.

The region is underlain by a series of bedrock units that differ greatly in thickness and in hydrologic characteristics and that range in age from Precambrian to Quaternary. Precambrian crystalline rocks are at or near the land surface (about 1,300 feet above mean sea level) locally in the eastern part of the region and about 15,000 feet below the surface (about 12,000 feet below mean sea level) in the center of the Williston Basin (figs. 7 and 8). The central and deepest part of this basin is in the westernmost part of North Dakota, southwest of and beyond the limits of the Souris River basin (fig. 7). Most of North Dakota, including the Souris and the western part of the Red River basins, is in the Williston Basin, which extends northward into Canada, southward into South Dakota, and westward into Montana. Sedimentary rocks of Paleozoic, Cretaceous, and Tertiary age occupy the Williston Basin. These strata gradually thin toward the east and are missing in the southern and eastern parts of the Red River basin and in the Rainy River basin, where Precambrian rocks directly underlie the glacial deposits (fig. 8).

The oldest Paleozoic beds dip westward at an average slope of nearly 50 feet per mile across the eastern three-fourths of North Dakota, but the Tertiary beds dip westward at only a few feet per mile (Crosby and others, 1973, p. 176). Sedimentary rocks of Ordovician age overlie the crystalline rocks in most of the Red River and Souris

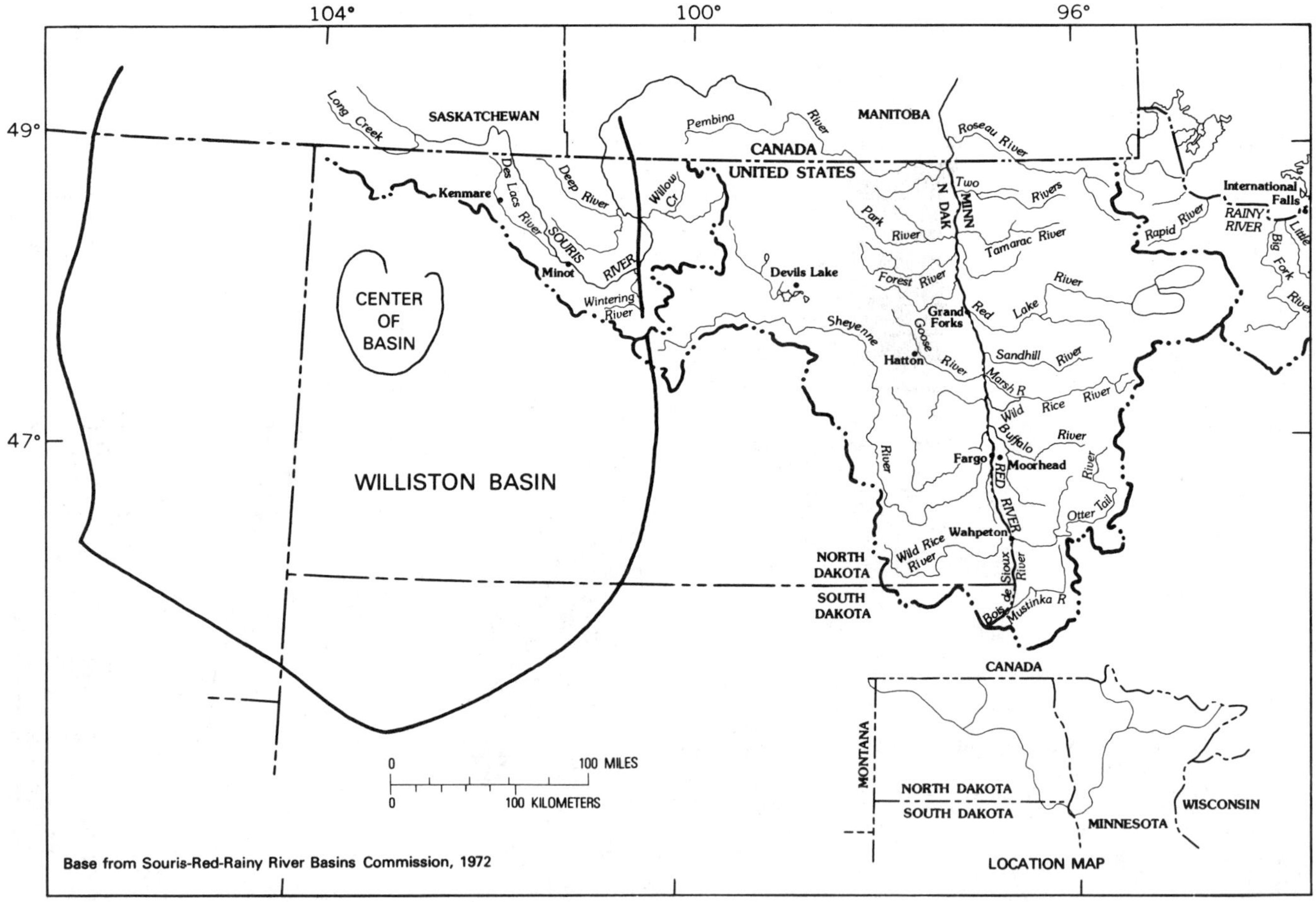

FIGURE 7.—Location of the Williston Basin.

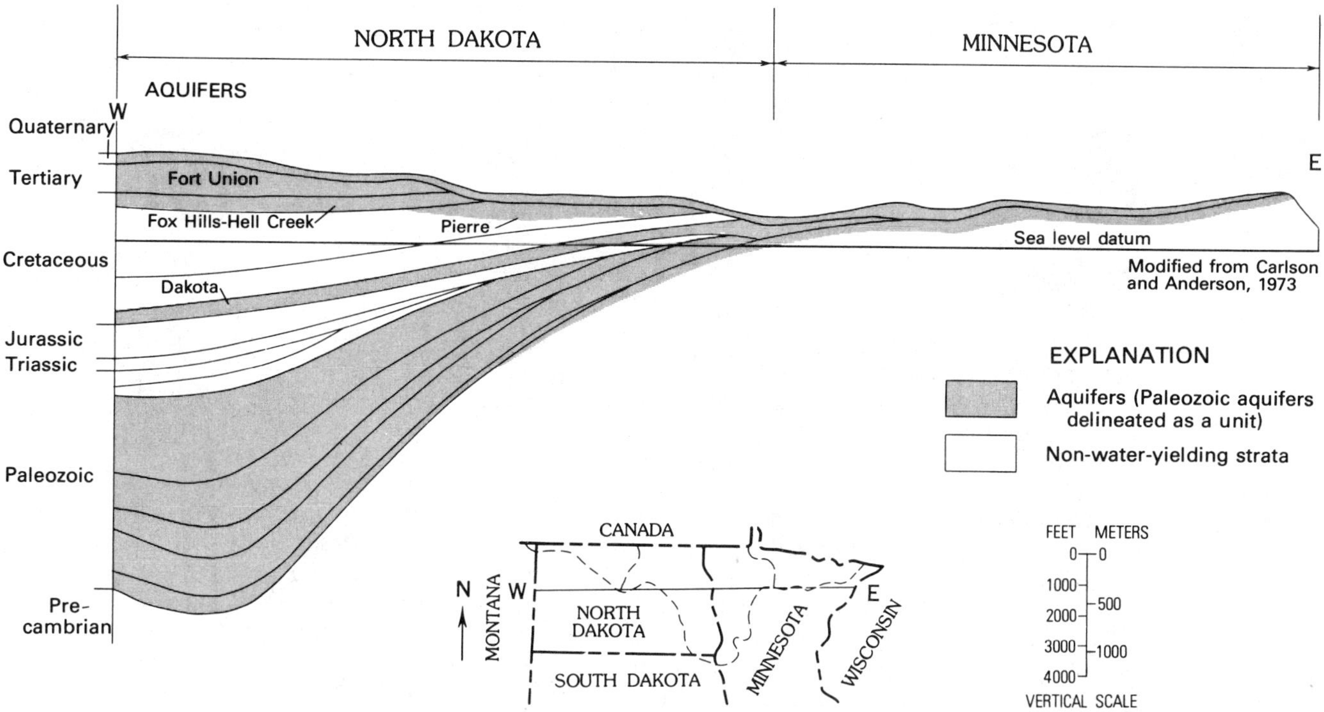

FIGURE 8.—Diagrammatic section of North Dakota and Minnesota showing location of aquifers.

River basins, sloping and thickening westward. Rocks of Jurassic and Cretaceous age, which thicken westward from the west border of Minnesota overlie the Paleozoic rocks. Tertiary rocks occur on the west edge of the Red River basin and in much of the Souris River basin (Glover and others, 1972).

The entire region has been glaciated, and most of it is covered with drift that ranges in thickness from less than a foot to several hundred feet. The drift is largely till, a heterogeneous mixture of clay, silt, sand, and gravel, but it also contains buried lenses of stratified sand, gravel, silt, and clay, which were deposited along the margin of the glacier or near it. The buried sand units, which range in thickness from a few inches to many feet, cover areas of many square miles.

Lake Agassiz sediments significantly affect the hydrologic system in the Red River basin because they are extremely fine grained, thick, and widespread. The thickness of these sediments in the Minnesota part of the lake deposits ranges from less than 1 foot to more than 140 feet (Maclay and others, 1972, p. 29).

Small well yields are obtained from bedrock aquifers underlying the drift in the Souris and Red River basins. In parts of the Rainy River basin, where productive glacial deposits are thin or absent, small yields of water are obtained locally from fractures in the crystalline bedrock (Glover and others, 1972, p. B-27, B-52, B-77).

Yields of wells in individual bedrock aquifers are generally less than about 100 gal/min but may be as much as 500 gal/min locally, and more than 500 gal/min in a few places. Yields in drift aquifers average less than 100 gal/min but range from 5 to 1,000 gal/min. A few wells in outwash yield more than 1,000 gal/min. Table 1 is a list of aquifers and range of yields to wells.

The potentiometric surface in the bedrock formations generally decreases in altitude toward the Red River, indicating that the regional direction of water movement is toward the river. In the Rainy River basin, some ground water probably migrates westward into the Red River basin; but most ground water moves northwestward and is discharged from the area through the Rainy River and Lake of the Woods (Glover and others, 1972, p. B-53, B-77). These regional patterns of movements of the ground water indicate that recharge from precipitation on upland areas, from surface water, and from ground water entering from adjacent regions moves regionally to discharge areas along the Red River Valley. The general ground-water flow system is much more complex because of the many local flow patterns within the regional system. Much of the ground-water discharge occurs near recharge areas within the region (Maclay and others, 1972, p. 104-113).

PRECAMBRIAN AQUIFER

Precambrian crystalline rocks, which underlie all of the region, generally are poor aquifers, but locally are sources of small supplies. Along the southeastern edge of North Dakota, and in the eastern part of the Red River basin in Minnesota and in most of the Rainy River basin (fig. 9), small supplies of water can be obtained from the fractures or from weathered zones in the upper part of the crystalline rocks. Generally, wells in the Precambrian aquifer will not yield more than a few gallons per minute (Crosby and others, 1973, p. 176). The Precambrian rock surface also generally defines the base of the water-yielding zone of the hydrologic system. The general slope of the crystalline rock surface is from east to west (Maclay and others, 1972, p. 23). (See figure 8.)

PALEOZOIC AQUIFER

Water occurs in the Paleozoic rocks in the Souris River basin and in most of the Red River basin (fig. 9), but in places it is highly saline. Although there are several water-bearing units within the Paleozoic rocks, data are insufficient to determine the areal extent, thickness (except locally), and degree of interconnection of the individual units; therefore, all water-bearing Paleozoic rocks are treated here as a single aquifer. The top of the aquifer occurs at depths of about 150 feet in eastern North Dakota and deepens westward to more than 13,500 feet at the bottom of the aquifer in western North Dakota, beyond the limits of the region (figs. 7 and 8). It is composed of fine-grained sandstone which yields small dependable supplies of water; and porous, cavernous limestone, which yields large supplies of water. The salinity of the water, however, severely limits its usefulness. Production of water in eastern North Dakota is limited to one industrial well that flows, when not closed, at a rate of about 700 gal/min, and a few domestic wells that are used for sanitation. Water from the Paleozoic aquifer also is produced with oil in the western part of the region (Crosby and others, 1973, p. 177). Because of the highly saline water, the Paleozoic aquifers in Minnesota have never been used. Test holes into the Paleozoic rocks have flowed as much as 60 gal/min (Maclay and others, 1972, p. 60).

TABLE 1.—*Aquifers and well yields in the Souris-Red-Rainy Region*

[Bedrock aquifers adapted from Crosby and others (1973, p. 176-185) and Glover and others (1972, tables B-10, B-16, B-24). Glacial drift aquifers adapted from Glover and others (1972, tables B-11, B-12, B-17, B-18, B-19, B-24) and Q. F. Paulson (written commun., 1976)]

Aquifer	Well yields (gal/min)		
	Common rate pumped	General range	Largest known
Drift	<100	5-1,000	>1,000
Fort Union	2-4	<1-50	100
Fox Hills-Hell Creek	<5	<1-30	150
Pierre	<5	<1-6	100
Dakota	2-3	<1-350	500
Paleozoic	<5	<1-60	700
Precambrian	<5	<1-10	

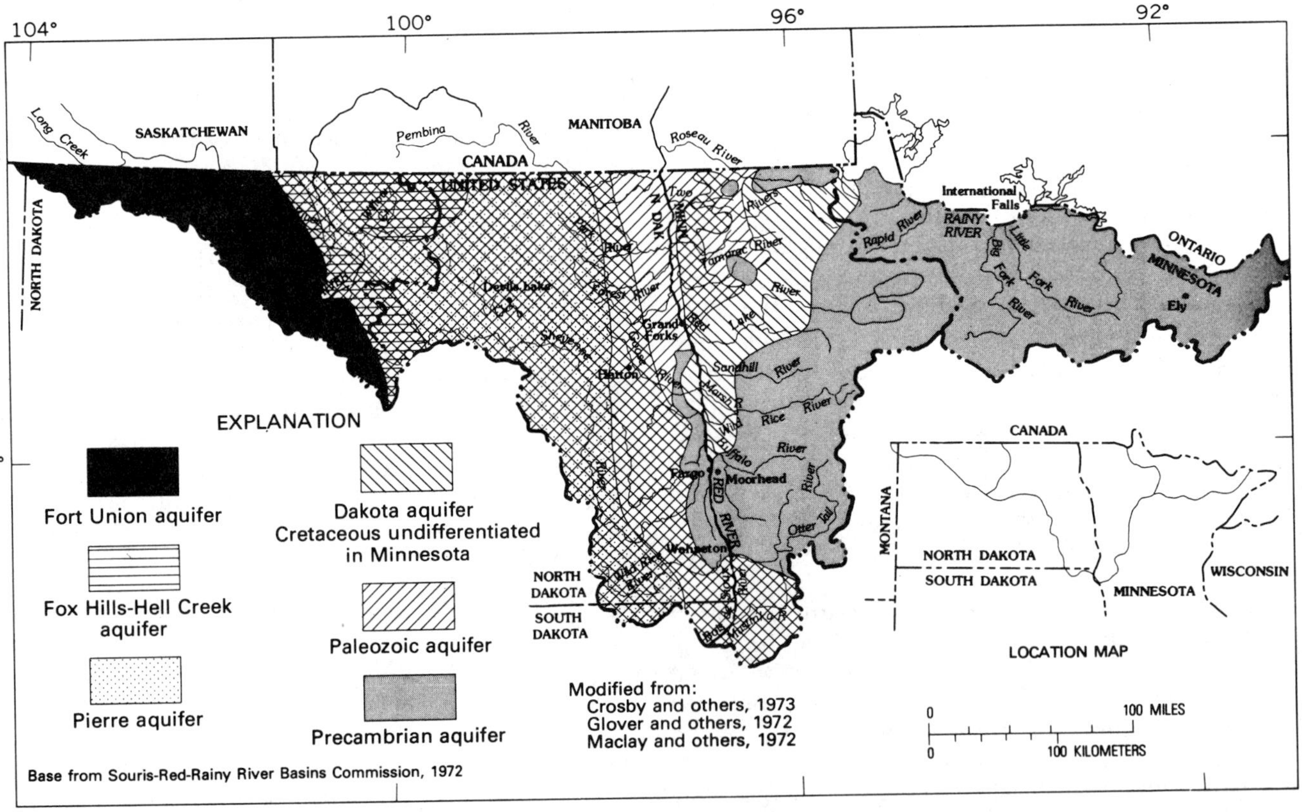

FIGURE 9.—Areal extent of bedrock aquifers in the Souris-Red-Rainy Region.

DAKOTA AQUIFER

The Dakota aquifer, which is composed of basal Cretaceous sandstone and shale, underlies all of the Souris River basin and the western half of the Red River basin (fig. 9). Aquifers in Cretaceous rocks, which probably are equivalent to the Dakota aquifer, underly the northwestern Red River basin in Minnesota. The depth to the top of the aquifer gradually increases from 100 feet along the eastern limit of the aquifer in the Red River basin to more than 5,600 feet in the deepest part of the Williston Basin southwest of the Souris River basin (fig. 8). The aquifer materials differ from place to place, but generally consist of interbedded quartzose sand and shale. Sand predominates in the eastern part of North Dakota and shale predominates in the western part.

In the eastern part of North Dakota, the individual sand beds generally are less than 30 feet thick, but locally are as much as 100 feet thick. They generally are composed of fine, medium, or coarse sand that has very little interstitial silt or clay. Interstitial silt and clay decreases permeability of the aquifer and, thus, decreases well yields and the rate of flow of water through the aquifer. In the western part, individual beds generally are composed of fine sand and minor amounts of medium sand, and interstitial silt and clay are more common (Crosby and others, 1973, p. 178).

Numerous wells tap the Dakota aquifer in the eastern part of North Dakota. Most of these wells are for domestic and stock use but a few are for municipal and industrial purposes. Flowing wells in the Dakota have discharges that range from less than 1 to 100 gal/min and average 2 to 3 gal/min in most of the Souris-Red-Rainy Region. Pumping rates in excess of 500 gal/min are obtained from some wells (Crosby and others, 1973, p. 179). In the Souris basin part of the region potential well yields range from 50 to 350 gal/min. This water is used mostly by the oil industry for maintaining pressure in oil reseservoirs (Glover and others, 1972, p. B-31).

PIERRE AQUIFER

The Pierre aquifer is in the upper part of the Pierre Shale of Cretaceous age in the eastern part of the Souris River basin and the western part of the Red River basin (fig. 9). The aquifer may extend westward for a short distance beneath the Fox Hills-Hell Creek aquifer, but no data are available to determine the western limit accurately.

The Pierre is composed of light-gray to black siliceous shale, marlstone, and claystone, and is locally fractured in the upper part. Unfractured shale, marlstone, and claystone are nearly impermeable and will not yield a significant amount of water to wells; therefore, the fractured zones are the only sources of water. The fractures, where present, may extend several hundred feet below the land surface, but they generally are too small to yield significant quantities of water below depths of about 100 feet (Crosby and others, 1973, p. 180).

Although the Pierre is not a major aquifer, it is the only source of water for many farms and a few municipal supplies. Most farm wells yield less than 6 gal/min. Locally, where the fracture zone is exceptionally thick or the fractures are unusually large, pumping rates range from 50 to 100 gal/min (Crosby and others, 1973, p. 180).

FOX HILLS-HELL CREEK AQUIFER

The Fox Hills-Hell Creek aquifer of Cretaceous age underlies most of the Souris River basin (fig. 9). The depth to the top of the aquifer gradually increases from a few feet in topographically low areas near the eastern boundary of the aquifer to more than 1,400 feet in topographically high areas near the center of the Williston Basin southwest of the Souris River basin (figs. 7 and 8). The aquifer is composed of interbedded sand, clay, silt, and lignite, but the lithology differs considerably from place to place. Many of the individual sandy layers in the aquifer are thin and lenticular and do not extend for more than a few miles. However, at least one, and commonly more than one, sandy layer more than 20 feet thick is present. Generally, the individual sandy layers are composed of fine to medium sand with interbedded silt and clay lenses, but locally the sandy layers may be either very fine sand or include considerable amounts of interstitial silt (Crosby and others, 1973, p. 180).

Wells in the Fox Hills-Hell Creek aquifer generally yield less than 30 gal/min; but locally, where the sandy layers are unusually thick or contain very little interstitial or interbedded silt or clay, yields may be as much as 150 gal/min. Most of the water is used for rural domestic and stock purposes (Crosby and others, 1973, p. 181

FORT UNION AQUIFER

Sand and lignite beds in the Fort Union Formation of Tertiary age form the Fort Union aquifer in the western two-thirds of the Souris River basin (fig. 9). The Fort Union, which may be as thick as 1,100 feet near the center of the Williston Basin, is composed of beds of silty clay, clay, sand, and lignite. Individual beds in the formation generally cannot be traced as much as a mile; however, a few exceptionally thick and extensive beds extend many miles. Most sand beds are less than 10 feet thick, but some are as much as 150 feet thick. Lignite beds generally are 2 to 5 feet thick, but thicknesses of as much as 40 feet have been reported. The sand is generally fine, with medium sand reported locally. The sand also may contain considerable amounts of interstitial clay (Crosby and others, 1973, p. 181-182).

The nonmarine Tongue River Member of the Fort Union Formation, stratigraphically in the upper part of the formation, is the principal bedrock aquifer in the Souris River basin. The underlying marine Cannonball Member yields water that is used for watering livestock but that is generally too salty for human consumption, according to Glover and others (1972, p. B-27). Although there may be several aquifers in the Fort Union Formation, data generally are insufficient to determine the areal extent, thickness, or degree of interconnection of the units; therefore, all water-bearing units in the Fort Union are considered here as a single aquifer.

The quantity of water the Fort Union aquifer will yield to wells depends on (1) the thickness, sorting, and grain size of the sand beds, and (2) the quantity of interstitial or interbedded clay near each well. Properly constructed wells finished in sand beds as thick as 100 feet will yield as much as 50 gal/min with 20 feet of draw-down. A few wells will yield as much as 100 gal/min, but drawdowns are greater than 20 feet. Most wells that tap the Fort Union aquifer are used for rural domestic and stock supplies. Some wells, however, are used for municipal and industrial supplies. Most farm wells are completed in the uppermost saturated sand lens. The wells commonly are equipped with cylinder pumps generally having capacities of only 2 to 4 gal/min (Crosby and others, 1973, p. 182).

DRIFT AQUIFERS

Drift aquifers of Quaternary age, which are the most important sources of ground water in the region, are distributed throughout most of the Souris and Red River basins and part of the Rainy River basin. Although most of the drift in the region is composed of several tens to several hundred feet of till, which yields little or no water to wells, some of the drift consists of stratified sand and gravel which form important sources of water supply (Crosby and others, 1973, p. 183). In the Rainy River basin, however, glacial erosion rather than deposition was the primary process shaping the surface of the land, and the deposits are thin or absent, except in the south and west parts of the basin where the glacial deposits thicken to nearly 250 feet (Glover and others, 1972, p. B-77). Figure 10 shows the areal distribution of known major drift aquifers and the expected yields to wells of water having less than 3,000 mg/L of dissolved solids. Ranges in yields to wells in the various aquifers delineated in figure 10 have been simplified from those tabulated by Glover and others (1972).

Most of the aquifers are either valley fill or outwash deposits. Melt water flowing from and along the ice front deposited outwash and ice-contact deposits in a network of channels. During subsequent periods of glacial advance and retreat, old channels were blocked and new ones formed. Each successive advance of glacial ice smoothed the terrain by filling the valleys and eroding the hills. Parts of these channels were subsequently filled with sand, gravel, and till, such as in the New Rockford aquifer in east-central North Dakota. Some of these glacial channels are now occupied by modern streams, as illustrated by the Sheyenne River valley in southeastern North Dakota. Deposits in these channels

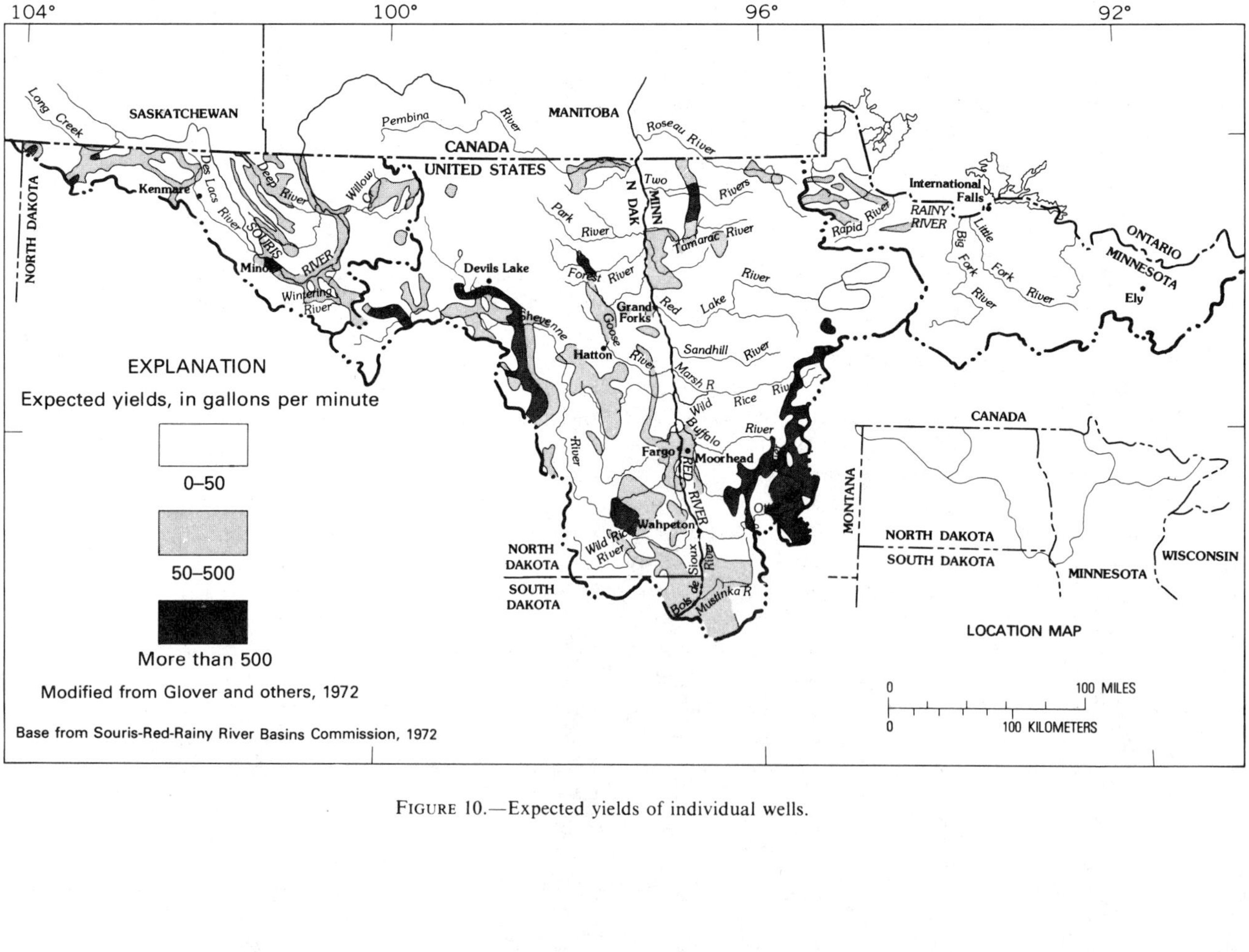

FIGURE 10.—Expected yields of individual wells.

form numerous aquifers scattered throughout the Red and Souris River basins and the western part of the Rainy River basin as shown by the long narrow patterns in figure 10. Many of the earlier channels were buried by subsequent glacial deposits, and most of the glacial channels exposed at the surface were formed during the most recent advance of glacial ice. Outwash deposits in these youngest channels are generally thin, and only locally form significant aquifers.

The deposits in glacial Lake Agassiz, in the center of the Red River basin, occupy almost one-third of the area of the basin (fig. 2). These deposits are generally too fine grained to be important as a source of ground water. However, some lakeshore deposits, such as deltas and beach ridges, are coarse grained and form important aquifers (Glover and others, 1972, p. B-29, B-53). The Sheyenne delta aquifer in the southeast corner of North Dakota is a notable example (Glover and others, 1972, p. B-62). Fine-grained sediments also were deposited in glacial Lakes Cando and Dakota, and glacial Devils Lake in the western part of the Red River basin and in glacial Lake Souris in the Red and Souris River basins (fig. 2).

Alluvium deposited after the glacial periods consists of silt, sand, and gravel in the valleys of some of the larger streams. These deposits generally are finer grained and less permeable than the outwash deposits, but locally they are sources of water.

The ability of the deposits to yield water to wells depends on the grain size and thickness of the materials, but in broad view and ultimately the quantity of water that an aquifer will yield depends on the amount of water in storage and on the amount of recharge. If a sand or gravel deposit is small and enclosed in till or other fine-grained material, it receives little recharge; consequently, such deposits will not yield large quantities of water for sustained periods. Large deposits of saturated sand and gravel, however, contain large quantities of water in storage and may be capable of receiving sustained quantities of recharge over the large area of contact with the clay layers even if they are enclosed within fine-grained deposits; consequently, such deposits form major aquifers that will yield substantial quantities of water for sustained periods (Crosby and others, 1973, p. 183).

QUALITY OF GROUND WATER

The chemical quality of water can be classified according to its dissolved-solids concentration. The classification of Winslow and Kister (1956, p. 5) used by the U.S. Geological Survey is as follows:

Class	Dissolved solids (mg/L)
Fresh	less than 1,000
Slightly saline	1,000- 3,000
Moderately saline	3,000-10,000
Very saline	10,000-35,000
Brine	more than 35,000

Figure 11 shows areas in the Souris-Red-Rainy Region where ground water occurs with less than 1,000 mg/L and with 1,000 to 3,000 mg/L dissolved solids. Ground water with less than 3,000 mg/L dissolved solids is available throughout the region; however, yields may be small.

In addition to the fresh and slightly saline water, the region has abundant highly mineralized water. Robinove and others (1958, p. 9) stated that many of the formations in North Dakota are capable of yielding only very small supplies of water. The Pierre Shale, for example, yields only minor amounts of saline water to many wells in the eastern part of North Dakota. Table 2 lists and describes water-yielding formations and their water-quality and water-supply potential.

The quality of ground water in the Souris-Red-Rainy Region differs from place to place, even within an aquifer. Generally, water with the best quality is in or near recharge areas or areas with flow sufficient to flush out saline water. Water in surficial outwash in or near recharge areas almost everywhere contains less than 1,000 mg/L dissolved solids and in many places less than 500 mg/L. Highly mineralized water is found in older, deeper rocks in much of the region and in shallow aquifers where water has migrated from mineralized aquifers. For example, in the north-central part of the Red River basin, highly mineralized water migrates upward into the drift from the underlying Paleozoic and Dakota aquifers. Water in the drift in these areas generally is a sodium chloride type, with many of the characteristics of the water in the underlying bedrock (Crosby and others, 1973, p. 195).

In the Pierre aquifer, water is a sodium chloride or sodium sulfate type or a combination of the two. The dissolved-solids concentration ranges from 700 to 12,500 mg/L, and the shallow fractured zone generally yields water with 2,000 to 4,000 mg/L dissolved solids. Most of the water in the Pierre aquifer contains greater quantities of chloride or sulfate, or both, than the limits for drinking water recommended by the U.S. Public Health Service (1962). The Pierre is not a highly productive aquifer, but it is the only source of water for many farms and a few municipalities (Crosby and others, 1973, p. 194).

Water in the Fox Hills-Hell Creek aquifer generally is a sodium bicarbonate type, but locally the water is a sodium sulfate type. Although water from the Fox Hills-Hell Creek aquifer ranges from about 300 to 3,700 mg/L dissolved solids, water containing less than 1,000 mg/L dissolved solids occurs near outcrop areas, where little drift overlies the aquifer. Most of the water that contains more than 2,500 mg/L of dissolved solids occurs where significant thicknesses of drift cover the Fox Hills-Hell Creek aquifer. The drift probably contributes much of the dissolved-solids concentration to recharge water that reaches the Fox Hills-Hell Creek aquifer (Crosby and others, 1973, p. 194).

Water in the Fort Union aquifer is a sodium bicarbonate or sodium sulfate type in different places. However, some of

TABLE 2.—*Description of aquifers, water quality, and water-supply potential in the Souris-Red-Rainy Region*

[Compiled from Crosby and others, 1973; Glover and others, 1972; and Robinove and others, 1958. Small yields are 5 to 150 gal/min, moderate yields 150 to 350 gal/min, and large yields more than 350 gal/min]

Aquifer	Character of rocks	Water quality	Water-supply potential and aquifer occurrence
Drift and alluvium .	Drift: till, outwash, and lacustrine deposits. Alluvium: gravel, sand, silt, and clay.	Extremely variable from place to place Dissolved solids generally 250 to 1,000 mg/L, but may be as high as 3,000 to 10,000 mg/L in a few places.	Extremely varible from place to place. Large potential for large supplies in sand and gravel aquifers. Large potential for small supplies almost everywhere in region.
Fort Union: Upper part of Fort Union (includes Tongue River Member).	Sandstone, shale and lignite.	Sodium bicarbonate type and locally high sulfate. Dissolved solids 200 to 7,000 mg/L.	Small to moderate supplies. Occurs in western two-thirds of Souris basin.
Fort Union: Cannonball Member of Fort Union.	Marine sandstone and sandy shale.	Sodium chloride type. Dissolved solids less than 3,000 mg/L, most water ranges from 1,000 to 2,500 mg/L.	
Fox Hills-Hell Creek	Sandstone; sandstone and shale.	Sodium bicarbonate type generally, sodium sulfate type locally. Dissolved solids 300 to 3,700 mg/L, but most water contains 1,000 to 2,000 mg/L.	Small to moderate supplies. Occurs in most of Souris basin and extreme western parts of Red basin.
Pierre	Shale, brittle, fissile; upper part fractured.	Extremely variable from place to place. Sodium chloride, sodium sulfate type, or combination. Dissolved solids 700 to 12,500 mg/L but most is less than 3,000 mg/L.	Widely used for domestic and stock supplies. Occurs in eastern part of Souris basin and western two-thirds of Red basin.
Dakota	Sandstone and shale.	Souris basin: sodium chloride type, dissolved solids 4,000 to 15,000 mg/L. Red basin: sodium sulfate type, dissolved solids 2,000 to 8,000 mg/L in North Dakota, less than 2,000 mg/L in Minnesota.	Moderate to large supplies. Occurs in western two-thirds of region.
Paleozoic	Shale, sandstone, limestone, and dolomite.	Highly saline. Dissolved solids 14,000 to 54,000 mg/L in central part of Red basin, and as high as 330,000 mg/L in western North Dakota.	Potential for development of small to large supplies. Occurs in North Dakots and northwestern corner of Minnesota.
Precambrian	Crystalline rocks, upper part fractured and weathered.	Sodium bicarbonate sulfate type, depending largely on type in the adjacent aquifer. Dissolved solids about 1,000 mg/L or more in Red basin, less than 500 mg/L in Rainy basin.	Small supplies, low yields. Occurs in Minnesota and eastern North Dakota.

the water from the deeper parts of the aquifer in the northern part of North Dakota is a sodium chloride type, with chloride concentrations as high as 3,830 mg/L. Elsewhere, the water in the Fort Union aquifer generally contains less than the U.S. Public Health Service recommended limit for chloride of 250 mg/L.

The quality of water differs greatly among shallower (300 feet or less) wells in the Fort Union aquifer. Dissolved-solids concentrations range from about 200 to 6,700 mg/L in North Dakota. Most water in the Fort Union aquifer that contains more than 3,000 mg/L dissolved solids is in areas covered with drift. The dissolved-solids concentrations in most of the deeper part of the aquifer is in the 1,000- to 2,500-mg/L range. The high sodium and dissolved-solids concentrations generally make the water from the Fort Union aquifer unsuitable for irrigation (Crosby and others, 1973, p. 194).

In most of the Red River basin in Minnesota the dissolved-solids concentrations in the drift are between 300 and 600 mg/L (Maclay and others, 1972, p. 62, 63). In the Rainy River basin and in the northeastern part of the Red River basin, the dissolved-solids concentrations in the drift are less than 300 mg/L. The dissolved solids in drift aquifers in the Red and Souris River basins in North Dakota are generally greater than 600 mg/L. Dissolved solids in the drift in North Dakota are as high as 26,200 mg/L, although most water in areas where flow is upward from bedrock aquifers contains less than 10,000 mg/L (Crosby and others, 1973, p. 195).

ROLE OF GROUND WATER IN WATER-RESOURCES MANAGEMENT

Increased demands on ground-water and surface-water supplies of the Souris-Red-Rainy Region will be made in the future (Souris-Red-Rainy River Basins Commission, 1972, p. 15; and Ferris and others, 1972, p. G-3, G-5). The population is expected to increase 20 percent between 1960 and 2020 but the municipal and rural domestic water uses are expected to increase 90 percent. Increased irrigation and industrial uses will cause even greater demands on the water supply. Municipal and industrial growth in an area generally depend directly upon the availability of adequate supplies of water of suitable quality. Anticipated increased use may

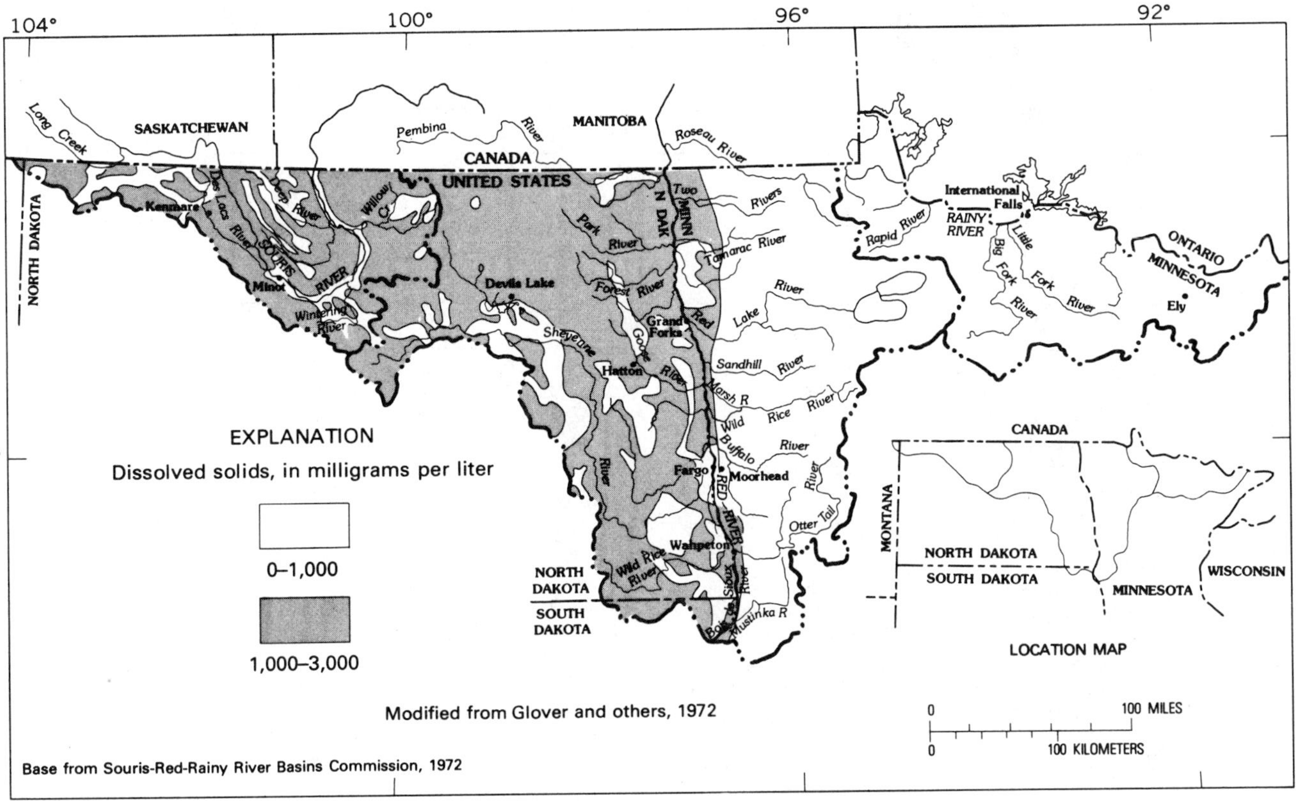

FIGURE 11.—Dissolved solids in ground water.

result in ground-water shortages. Little difficulty is expected in meeting this need in the eastern part of the Souris-Red-Rainy Region, but in the western part, particularly in the Souris basin, good quality water from known sources is in short supply.

Most of the anticipated additional needs for water in the western half of the region probably can be met from the Garrison Diversion Unit and through greater use of ground- and surface-water supplies. The Garrison Diversion Unit is a large-scale water redistribution project. It is planned to divert initially 2,000 ft^3/s and eventually 8,850 ft^3/s from the Missouri River at Garrison Reservoir to the Red and Souris River drainages and the James River (tributary to the Missouri) drainage for irrigation and municipal-industrial uses. Also included as part of the Garrison Diversion Unit plan are measures for flood control, drainage of nonirrigable lands, pollution abatement, and recreation.

Maclay and others (1972, p. 47-84) discussed water-resource management in the Red River basin, including ground water. A summary of their discussion of ground water in the Red River basin applies as well to the Souris-Red-Rainy Region.

Water management before 1967 included major flood-control projects, local flood-protection projects, watershed-protection projects, fish and wildlife developments, land drainage, and irrigation. Many man-made reservoirs have been constructed, particularly in North Dakota. The largest of these is Lake Ashtabula on the Sheyenne River, which has about 69,000 acre-feet of usable storage.

Primary considerations in the management of water supply are: (1) location and amount of the supply, (2) accessibility, (3) dependability, and (4) quality. Table 3 shows the advantages and disadvantages of ground water and surface water in regard to these four considerations. Ground water is important to the region's future development because of its large total quantity and general availability in most of the region; however, careful planning will be required in developing ground-water supplies because of its different quality from place to place and with depth, and the limited geographic extent of many of the Quaternary aquifers. Adapting McGuinness' (1963, p. 119) discussion of the role of ground water in the Nation's water situation to the Souris-Red-Rainy Region:

1. Ground-water reservoirs are important and indispensable in securing the region's future water supply.
2. Existing knowledge is grossly inadequate to form a basis for effective development and management of the ground-water reservoirs.
3. The region must overcome informational inadequacies in ground-water hydrology as well as in techniques of planning and water management.

TABLE 3.—*Advantages and disadvantages of ground water and surface water as sources of water supply in the Souris-Red-Rainy Region*

[After Maclay, Winter, and Bidwell, 1972]

Surface water	Ground water
Location and amount	
Surface water is easily located.	Ground water can be easily located where surficial deposits are water bearing.
Amount of surface water available is not large in some areas, even with storage.	In most places exploration by test drilling or other methods is generally necessary to locate sources capable of supplying high-yield wells. Locally, even obtaining a small supply is difficult.
Accessibility	
Surface water is readily accessible to riparian land owners in Minnesota.	Ground water is generally accessible by drilling wells.
Surface water is not accessible to riparian land owners in North Dakota. Users must obtain State permits in North Dakota and Minnesota. For the large number of people not living along streams and lakes, surface water is generally not available. The cost of physically storing and transporting water is an important economic factor.	Locally, an aquifer capable of supplying high-yield wells may not be in the immediate vicinity and a municipality may have to go several miles to obtain a supply. A State permit is required for supplies larger than for domestic purposes.
Dependability	
Because streamflow can be observed and measured it is possible to calculate and project the amount of water that will probably be available in the future. Data are already available for some locations on many of the streams.	Ground-water sources are ususally very dependable. If discharge does not exceed recharge, ground water can normally be used virtually indefinitely, including periods of extended droughts.
Stream flow is highly variable in most streams in the region. Dependable water supplies require storage reservoirs as a rule.	Determining the yield of an aquifer is difficult. Data are not available for many areas and they are costly to obtain. Shallow surficial aquifers are generally not dependable during drought years.
Quality	
Surface water is generally low in dissolved solids and the concentration is fairly uniform east of the Red River but variable west of the Red River.	Ground-water quality differs over the region but generally does not vary with time at any location. Temperature is constant throughout the year.
Treatment is generally necessary to remove suspended sediment and organic matter. Quality, temperature, and color vary widely over the year at any location.	Certain chemical characteristics, such as hardness and iron content, are high in many places.

FRESHWATER USE

Ground water is the sole or a primary source of supply in much of the Souris-Red-Rainy Region. In areas distant from streams and in uplands, where surface water is not available physically, legally, or in the quality required for a particular use, ground water is the sole source of supply.

Ground water and surface water are used conjunctively in part of the region, but conjunctive use is not fully exploited to meet quantity and quality requirements of water supplies. For example, water of good quality from one source could be mixed with mineralized water from another source in some areas to get a greater quantity within specified quality requirements. Also during times of low or no streamflow, ground water could supplement or replace surface water. Distribution systems could be used for surface water or ground water.

Estimates of water use in the United States according to categories of use were listed by State and region by MacKichan and Kammerer (1961), Murray (1968), and Murray and Reeves (1972). Water-use projections to 2020 were given for the Souris-Red-Rainy Region by Ferris and others (1972) and Weber and others (1972) in the Souris-Red-Rainy River Basins Comprehensive Study. The U.S. Bureau of Reclamation (1973) summarized water-resources development in North Dakota, with emphasis on surface-water supplies, and included a brief description of the Garrison Diversion Unit which was mentioned earlier.

IRRIGATION

About 15,700 acres of land was irrigated in the region in 1967 and 1968, including about 12,000 acres in the Souris River basin, about 3,500 acres in the Red River basin, and about 200 acres in the Rainy River basin. Irrigated land in the Red River basin increased to about 38,000 acres in 1975 but remained about the same in the Souris and Rainy River basins. The total land irrigated in 1975 was 50,200 acres.

The largest potential use of water in the Souris-Red-Rainy Region is for irrigation. Forecasts of irrigation development were described by Weber and others (1972) in the Souris-Red-Rainy River Basins Comprehensive Study for optimum utilization of land and water resources. Potentially irrigable lands were considered to be suitable for development by either private (small-scale) or project (large-scale) methods, depending on availability of water. Private methods considered were farm-size systems irrigated primarily with ground water. Project methods considered would use surface-water storage and distribution works, which probably could be constructed only with public funds.

Sources of water in quantity for irrigation include ground water in the Souris and Red River basins and water from the Missouri River (Garrison Diversion Unit) and the Rainy River. The Souris and Red River basins have large areas of potentially irrigable land (fig. 12), but water is not available from the Souris River for irrigation and the Red River can supply only small parts of the potentially irrigable area in the basin, such as the irrigated area east of Moorhead where several irrigators tap tributaries to the Red River. Surface water for potential irrigation development in the Souris and Red basins must be supplied from other sources, such as the Missouri and Rainy Rivers. In the Rainy River basin the abundance of surface water of good quality and apparent absence of large aquifers preclude the consideration of ground water for major irrigation.

The ultimate potential irrigation development in the region totals 1,550,000 acres, table 4. It is predicted that the amount of irrigated lands will reach 530,000 acres by the year 2000 and 1,404, 000 acres by 2020. Ground-water development is projected to decrease between 2000 and 2020, and

TABLE 4.—*Projected irrigation in Souris, Red, and Rainy River basins*

[From Souris-Red-Rainy River Basins Comprehensive Study by Weber and others, 1972, tables F-17, F-20]

River basin	Water source	Acreage (thousand acres) Year 2000	Year 2020	Ultimate
Souris	Ground water	20	16	16
Do	Surface water	134	369	369
Red	Ground water	200	180	180
Do	Surface water	169	824	970
Rainy	Surface water	7	15	15
Total ground water		220	196	196
Total surface water		310	1,208	1,354
Grand total irrigated		530	1,404	1,550

Note.—Does not include 190,700 acres as initial stage of Garrison Diversion Unit.

surface-water (or project) development is projected to increase nearly 400 percent in the same period.

The source of water and the irrigable area for the ultimate irrigation development within each State in the region is shown in table 5. It is estimated that 579,000 acres could be irrigated from the Missouri River (Garrison Diversion Unit), 775,000 acres could be irrigated from the Rainy River, and 196,000 acres from ground-water supplies. The estimate of potential irrigation development is 1,203,500 acres in North Dakota, 343,000 acres in Minnesota, and 3,500 acres in South Dakota.

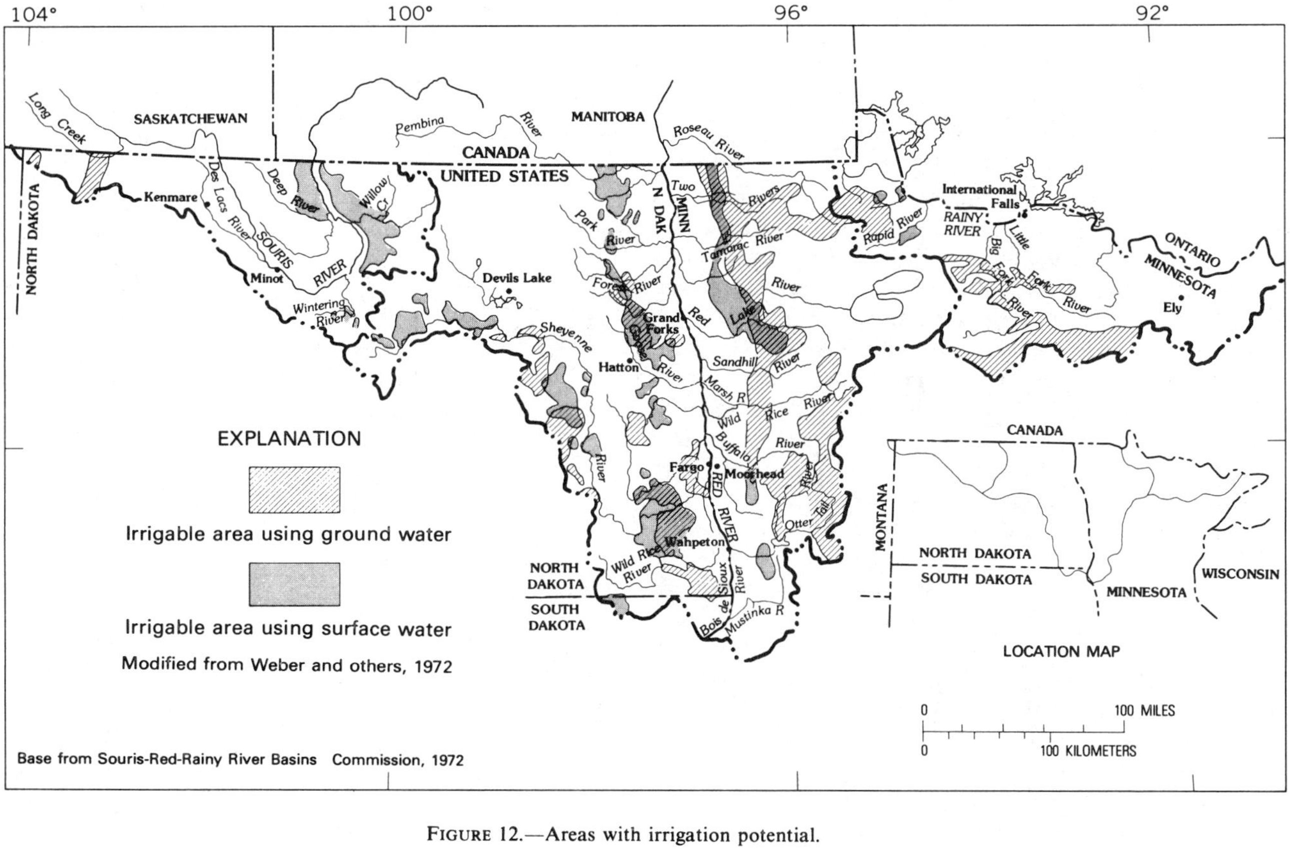

FIGURE 12.—Areas with irrigation potential.

TABLE 5.—*Ultimate irrigation in each State in the Souris-Red-Rainy Region*

[From Souris-Red-Rainy River Basins Comprehensive Study by Weber and others, 1972, p. F-51 to F-60]

Development plan	Potential irrigation development (thousand acres)			
	North Dakota	Minnesota	South Dakota	Total
Souris River basin:				
From ground water[1]	16	...	...	16
From Missouri River[2]	369	...	...	369
Red River basin:				
From ground water[1]	122	58	...	180
From Missouri River[2]	206.5	...	3.5	210
From Rainy River[2]	490	270	...	760
Rainy River basin:				
From Rainy River[1]	...	15	...	15
Total	1,203.5	343	3.5	1,550

[1]Private development
[2]Project development

The projections in the Souris-Red-Rainy River Basins Comprehensive Study (Weber and others, 1972) show that most of the near-future irrigation will be with ground water. By 2000, 41 percent of the acreage will be irrigated from ground water, but by 2020 surface-water development will account for 86 percent of the irrigated area. These projections are based on present knowledge of the availability of ground water and do not allow for methods of water management such as artificial recharge, salvage of water for beneficial use, and use of saline aquifers. Also, with continued ground-water investigations, additional aquifers probably will be found. When water-management possibilities are fully utilized, ground water may become increasingly important to irrigation.

MUNICIPAL

The region is primarily rural, with few population centers of more than a few thousand persons. Only 11 cities are projected to have a population of greater than 10,000 by 2020 (Ferris and others, 1972, p. G-7). Ten are in the Red River basin and one in the Souris basin.

Of the 208 cities or towns in the region, 178 have ground-water supplies, 21 have surface-water supplies, and 9 have combined supplies. Table 6 lists the number of cities and towns and population served in each of the river basins. The smaller communities and towns of the region generally rely on ground water, and the larger cities and industries rely on surface water, ground water, or both. The municipalities using surface water generally depend on reservoir storage.

Municipalities with water-distribution systems have a per capita use of about 100 gallons of water per day.The per capita use is 108 gallons per day in the Souris basin, 98 gallons in the Red basin, and 79 gallons per day in the Rainy basin. Projections indicate a per capita use of about 140 gallons per day by 2020 for the Souris-Red-Rainy Region. Projection for the Souris, Red, and Rainy basins are 152, 138, and 119 gallons per day per person, respectively, in 2020. Table 7 lists the 1960 population and water withdrawal, and projections of population and withdrawal to 2020 for each of the basins.

TABLE 6.—*Municipal water supplies in the Souris-Red-Rainy Region*

[From Souris-Red-Rainy River Basins Comprehensive Study by Ferris and others, 1972, p. G-19, G-36, G-75. Populations are from 1960 census]

Source of water	Ground	Surface	Combined
Souris River basin:			
Number of cities or towns	31	0	2
Population served	23,019	0	35,800
Red River basin:			
Number of cities or towns:			
North and South Dakota	69	8	5
Minnesota	69	8	1
Population served:			
North and South Dakota	62,620	104,317	13,903
Minnesota	50,355	42,590	22,935
Rainy River basin:			
Number of cities or towns	9	5	1
Population served	9,915	14,705	465
Total:			
Number of cities or towns	178	21	9
Population served	145,909	161,612	73,103

According to Ferris and others (1972, p. G-2) in the Souris-Red-Rainy River Basins Comprehensive Study, for a few communities the threat of water shortages persists or may develop as requirements increase. These communities include Minot, in the Souris River basin, and several other cities (such as Neche, Pembina, Grafton, and Mayville, N. Dak., and Crookston, Minn.) that depend on tributaries of the Red River. The largest municipal water-supply demands of the region are along the main stem of the Red River. The

TABLE 7.—*Municipal, domestic, livestock, and industrial water supply in the Souris-Red-Rainy Region*

[From Souris-Red-Rainy River Basins Comprehensive Study by Ferris and others, 1972, tables G-12, G-15, G-36, G-39, G-49, G-52]

	Population served		Withdrawal	
	1960	2020	1960 Mgal/d	2020 Mgal/d
Municipalities:				
Souris	58,819	121,899	6,36	18.50
Red	296,720	507,709	29.11	69.87
Rainy	25,085	36,511	1.98	4.36
Total	380,624	666,119	37.45	92.73
Domestic:				
Souris	44,852	20,701	2.11	1.15
Red	244,393	135,791	13.12	7.25
Rainy	15,323	8,152	.65	.36
Total	304,568	164,644	15.88	8.76
Livestock:				
Souris			4.50	8.18
Red			52.63	100.06
Rainy			2.35	4.25
Total			59.48	112.49
Industrial:				
Souris			2.33	4.75
Red			151.38	92.78
Rainy			54.23	157.62
Total			207.94	255.15

Red River and its tributary storage can meet the present municipal and industrial requirements. Shortages in some areas, however, have occurred during prolonged drought. Communities that depend on ground water have solved such problems by deepening wells or constructing additional wells. As water use along the main stem of the Red River increases, additional storage for water supply may be required, or ground water may be needed as a supplement. Water supplies in the Rainy River basin exceed present and foreseeable requirements.

DOMESTIC AND LIVESTOCK

Domestic and livestock water supplies are almost entirely from ground-water sources. Withdrawals are small and well distributed and shortages are not expected. About two-thirds of the 305,000 rural population are served by pressurized water systems, and per capita use is 60 gallons per day. By 2020, the rural population is expected to be about half the present number and to have the same per capita use, according to Ferris and others (1972) in the Souris-Red-Rainy River Basins Comprehensive Study. The 2020 water-use projections (table 7) are intended to indicate the trend of water requirements based on population projections and the general water-use criteria developed for the basinwide projections. These projections do not reflect the individual circumstances for a given community.

In the Souris River basin, about half the domestic and livestock wells tap bedrock aquifers. Drift aquifers supply the other half. Surface water, where available, is used for livestock. Some water, which is too mineralized for human consumption, can be tolerated by livestock. In the Red River basin drift aquifers, where available, are generally tapped by domestic and livestock wells. Well yields and water quality from the bedrock aquifers are generally poorer than from the drift. Bedrock aquifers can be tapped nearly anywhere on the Dakota side of the Red River basin, but are not present in parts of northwestern Minnesota. As a result, local water shortages exist where drift aquifers are also absent. Livestock production has been inhibited due to lack of adequate water resources in parts of northwestern Minnesota. Rural domestic water in the Rainy River basin is supplied by wells. Livestock supplies are mostly from surface-water sources.

In the last several years, water districts have been formed in North Dakota to supply water to rural and small-city residents for domestic and stock uses. In general, the rural water districts are formed as a consequence of county-wide ground-water studies which delineated aquifers and identified ground-water resources available for development. Water supplies of about 100 gal/min are developed from glacial outwash, buried sand and gravel channels, and delta aquifers such as are found along the east edge of North Dakota in the Red River valley. The water districts are formed to utilize good-quality water from these aquifers and to supply nearby rural areas where water quality and quantity are inadequate.

INDUSTRIAL

The economy of the Souris and Red River basins is based primarily on agriculture and related food-processing industries. The food-processing industries, which include sugar-beet refining, potato processing, meat packing, and creameries, account for a large part of the industrial water demands. Most soils in the Rainy River basin and in the northern Minnesota part of the Red River basin, however, are unsuitable for cultivation but valuable for forestry. In these areas forestry resources and recreation are the economic foundation, and the pulp and paper industry is the primary water user. Manufacturing is increasing in the Souris basin. The large decrease in water requirements indicated in the Red River basin from 1960 to 2020 (table 7) is due to a shift in the power industry from flow-through to cooling-tower operation. Table 7 reflects only water use by self-supplied industry. Some industries obtain water from municipal supplies. A rough guideline is that municipal water use in excess of 80 to 90 gallons per day per person is attributable to commercial and industrial use.

Ferris and others (1972) in the Souris-Red-Rainy River Basins Comprehensive Study assumed that future industrial food processing in the region would increase at about the same rate as agricultural production. Figures in table 7 reflect the expected increase in the Souris and Rainy River basins. A similar increase in the Red River basin is overshadowed by the expected decrease in water requirements of the power industry. That is, future water use was derived by applying the projected production of power to current water use. Where appropriate, adjustments were made to reflect changing trends, such as the anticipated increase of water recycling or reuse.

SALINE-WATER USE

Highly mineralized ground water is plentiful in the western part of the region. Feth and others (1965) showed the availability of water containing more than 1,000 mg/L dissolved solids at depths less than 500 feet throughout the western half. In the northeastern and northwestern corners of North Dakota, water containing 10,000 to 35,000 mg/L dissolved solids lies at depths of less than 500 feet. At greater depths the water is generally more mineralized. Such water can be considered a resource or asset rather than a liability. The water can be used for a number of purposes, including cooling and oil-field repressuring. Possibly this highly mineralized water could be utilized by the chemical industry as a source of minerals and chemicals. Mineralized water from the Dakota aquifer is being used for washing sugar beets and potatoes.

In the western part of the region highly mineralized water is being used to some extent to repressure oil fields to increase production. Saltwater produced with the oil must be disposed of so that it does not damage the environment and contaminate surface water and fresh ground water. About 94 percent of the oil-field brine produced in North

Dakota is returned to saline ground-water reservoirs. Most of the brine is returned to the reservoirs from which it was withdrawn (Folsom, 1973, p. 102). Where oil-field brines are insufficient for repressuring oil fields, highly mineralized water available from other ground-water sources ideally should be used to conserve freshwater supplies.

A large part of the available water supply of the western part of the region is saline. Robinove and others (1958) stated that a large proportion of the available water supply of North Dakota is saline. Water-resources investigations in the region have been principally oriented to municipal, industrial, domestic, and irrigation uses requiring freshwater. However, saline water can be treated to increase its usability for these purposes. Congress, in 1952, passed the Saline Water Act that provided for research into and development of practicable and economic means to produce freshwater from saline water to conserve and increase the water resources of the Nation. Considerable progress has been made in developing desalinization processes. The former Office of Saline Water (now Office of Water Research and Technology), Department of the Interior, once operated an electrodialysis desalting plant at Webster, S. Dak.. Although only a small number of desalinization plants are in operation throughout the Nation, they may be used more extensively in the future.

In some areas of the region water of acceptable quality for a particular purpose may be inadequate in quantity. Where saline water is available, it could be diluted with the freshwater to create an adequate supply of usable water.

Feth (1965) prepared a map of the United States showing distribution of saline ground water. His report includes a large number of references. Several reports for North Dakota and Minnesota are listed and the reader is referred to these reports for a more detailed discussion of saline ground water in the region.

ARTIFICIAL RECHARGE

Artificial recharge is the process of increasing the storage of water in an aquifer system by artificial means. Storage in the ground has several advantages over surface storage: (1) the sand and gravel would filter the water; (2) the water would be kept at a relatively constant temperature; and (3) there would be no evaporation. Care is necessary, however, to avoid pollution or physical impairment of the underground system. For example, recharge waters must be chemically compatible with the native ground water to prevent formation of precipitates that would reduce hydraulic conductivity. Suspended solids must be removed before injection, and pollution by bacteria must be prevented. Storage of surface water in aquifers during times of surplus for withdrawal during times of scarcity would aid in meeting increased demands. Floodwater is chemically better in most respects than underlying ground water in parts of the western half of the region. Freshwater could be stored in saline-water aquifers, as well as freshwater aquifers, and pumped out as needed.

Artificial recharge may be accomplished by direct and indirect methods. Direct methods of recharge include water spreading by means of ponds, check dams, pits, furrows, or ditches to increase the amount of water infiltrating from the surface into the ground-water reservoir, and injection of water directly into the ground-water body by means of wells or shafts (fig. 13). Indirect methods of recharge include inducing movement of water from streams or lakes into underground formations (fig. 14A), and preventing the natural flow of ground water to the land surface by lowering ground-water levels (fig. 14B). Further information is available in the published literature on methods, case studies, and other aspects of artificial recharge. See Todd (1959), Signor, Growitz, and Kam (1970), and Knapp (1973) for listings of literature to 1973.

REPLENISHING FRESHWATER AQUIFERS

Surplus water is available during winter and spring from surface-water sources, principally the larger rivers and tributaries, which could be used to recharge ground water.

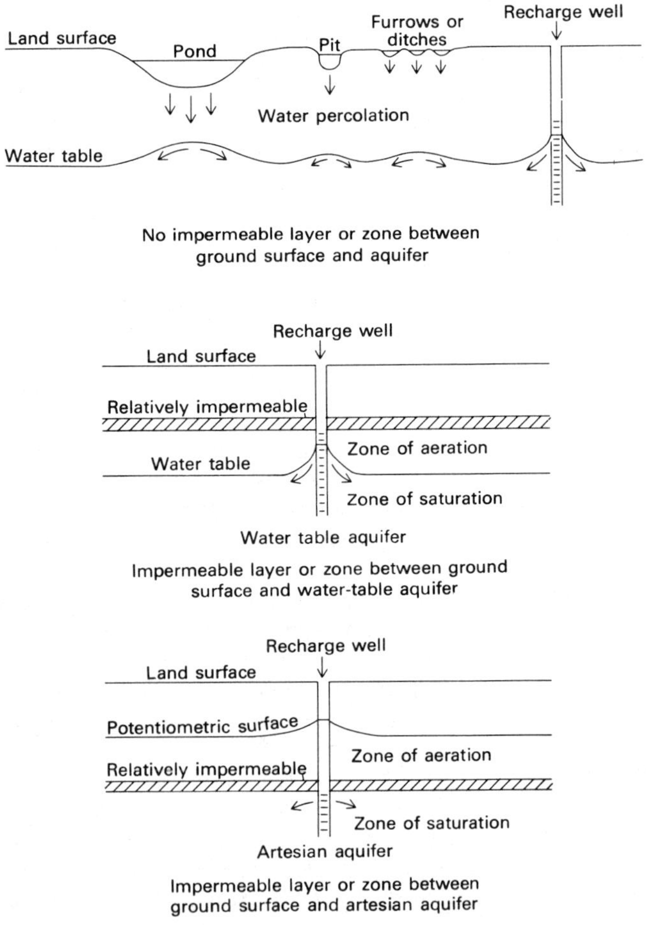

FIGURE 13.—Diagrams of direct methods of recharge.

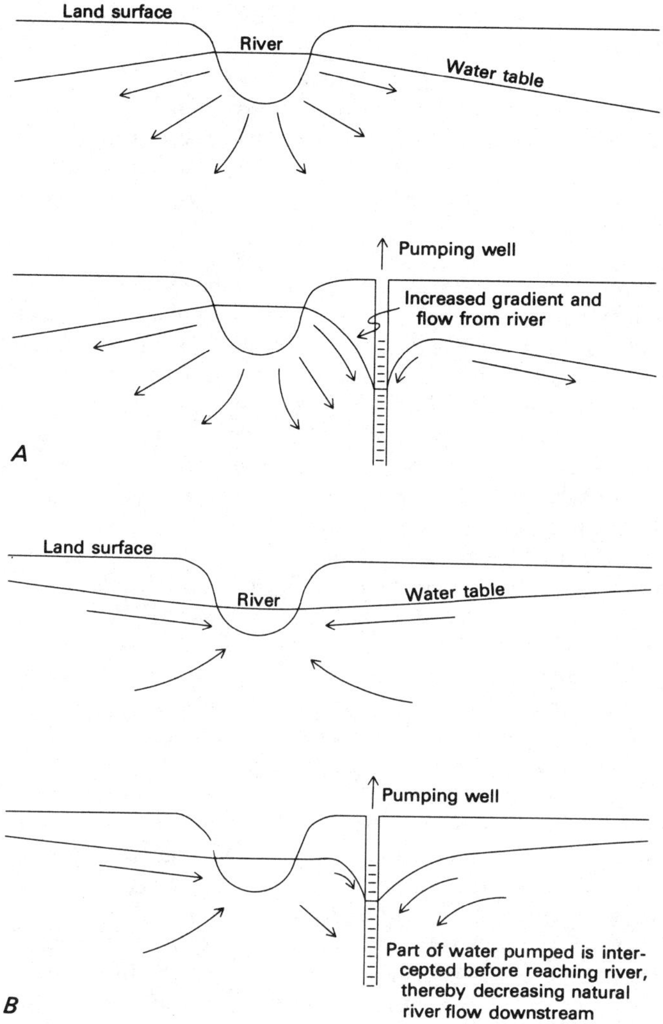

FIGURE 14.—Diagrams of indirect methods of recharge.

Utilization of floodwater to replenish ground-water supplies also would diminish flooding to some extent. Additionally, water injected into the ground in winter or early spring would be cold and well suited for air conditioning and as a process coolant. This technique probably could be used near Moorhead where high flows from the Buffalo River could be injected into the aquifer and pumped out later in the year. Artificial recharge is not new to the region, as shown by the following two examples.

Artificial recharge has been in operation at Minot, N. Dak. since 1965. The Souris River has been a source of part of Minot's municipal water supply for many years. In 1961, Minot constructed eight wells to supplement the stream supply. Excessive water-level declines in the following years indicated that a ground-water shortage was developing. Although the Souris River is not a reliable source of direct supply, relatively large peak flows in the spring indicated that the river was a potential source of water for recharge to the aquifer. A ground-water recharge facility was designed and constructed in 1965 to use both direct and indirect recharge methods, (Ferris and others, 1972, p. G-22).

Kelly (1967, p. 20) reported that Valley City, N. Dak., had an average daily water use of 750,000 gallons in 1966, which was obtained from wells tapping partly confined gravel deposits in the Sheyenne River valley. These deposits at Valley City have a maximum thickness of more than 50 feet and an areal extent of about 1 square mile. The aquifer has been artificially recharged successfully since 1932 by diversion of water from the Sheyenne River to an abandoned gravel pit. During this time the potentiometric surface in the aquifer has risen more than 22 feet.

Before 1958, the recharge system was operated annually from January until June; however, when the potentiometric surface rose to within 8 feet of the surface, the recharge operation was discontinued. Between June and January the potentiometric surface declined as ground water was withdrawn. During the recharge-discharge cycle, the average annual fluctuation of the potentiometric surface was 10 feet, amounting to a change in storage of about 1,000 acre-feet of water. Since 1958, the recharge system has been operated throughout the year. The quality of water in the aquifer has gradually improved since the installation of the recharge system.

FRESHWATER STORAGE IN SALINE AQUIFERS

Saline aquifers can be used to store freshwater during times of abundant surface supply and pumped out as needed. Tests and experiments by the U.S. Geological Survey at Norfolk, Va., recovered about 85 percent of the injected freshwater (Brown and Silvey, 1973). The freshwater displaces the saline water and forms a fresh-water body, most of which can be recovered. The success of such projects would depend on knowledge of the hydrologic system in the vicinity of the projects, favorable site conditions, and on the operation of adequate monitoring systems.

This possibility might be investigated in the Minot area, N. Dak., the Fargo-Moorhead area, N. Dak. and Minn., the Hallock area, Minn., and other population centers or areas of central rural supplies where spring floodwater might be retained temporarily in upland reservoirs until it can be treated and stored underground. In the Red River basin, the Dakota Sandstone or the Pierre Shale might be used in North Dakota for this purpose; in Minnesota aquifers in Cretaceous or Ordovician rocks might be so used. In the Souris River basin, the Fox Hills-Hell Creek aquifer or the Fort Union aquifer might be so used. The Dakota Sandstone in the Souris basin probably is too deep to be used economically for this purpose. To use these aquifers for storage of surplus surface water, suitable site conditions would have to be located.

LIQUID-WASTE DISPOSAL IN SALINE AQUIFERS

Saline aquifers can be used for liquid waste disposal. Much has been written concerning underground waste disposal (Rima and others, 1971). A knowledge of the geology, ground-water hydraulics, and geochemistry would aid in determining whether or not injected wastes could be contained and isolated, or recovered later if it is warranted. (Injection of waste underground can present serious hazards to water supplies and the local environment.)

GROUND-WATER DEVELOPMENT

Water use has had a rapid upward trend in the Souris-Red-Rainy Region. As the gross water supply is constant and water demands are rising, the answer to problems of supply is the development of untapped sources of water and the possible reuse of water.

PROBLEMS OF DEVELOPMENT

The problems of development differ widely with the intended use of the water. Problems inherent in development of most small wells are chiefly the availability, quantity, and quality of the water. Water quality is a serious problem in parts of the Souris and Red River basins where supplies of potable water are scarce. The development of large supplies has additional problems related to proper planning and administration of the development pattern.

Initial development generally involves haphazard drilling for water that is readily available. In some areas where the ground and surface waters are closely related, the development of ground-water supplies has depleted surface flow (Theis, 1941). Also, unplanned development may result in aquifer overdraft in one area and surpluses at another.

Ideally, the development of an aquifer would be designed to achieve the optimum economic recovery of the water. To achieve this, detailed information would be needed on the depth, quality, quantity, and annual recharge and discharge of the water in the various aquifers. Thus it is seen that water problems during initial development in much of the region are related to hydrologic and geologic conditions.

In general, investigations that would provide information about the geologic and hydrologic conditions of the aquifer are not undertaken until the degree of development has created problems that threaten the continued use of the aquifer. In this region, these problems are: changes in ground-water levels, depletion of streamflow, and deterioration of water quality. Also there are management problems relating to ground-water replenishment, ground-water conservation, and effects of the development on the environment.

CHANGES IN WATER LEVELS

Pumping necessarily causes water levels to decline near the pumped well, creating a gradient and flow toward the well. Water-level declines, although not yet alarming in most of the region, are increasing the cost of pumping. Declines of ground-water levels near West Fargo, due to municipal and industrial pumpage, indicate possible future supply problems from this aquifer. Large water-level declines in the Minot well field in the early 1960's indicated that a critical water shortage was developing rapidly. The problem was alleviated by artificial recharge of the aquifer. (See the section "Replenishing Freshwater Aquifers.")

If the extraction of water continues until withdrawal exceeds recharge, significant depletion of water storage may occur in the area of overdraft. With time, the continued or increased water use may cause water-level declines in the entire aquifer.

In the late 19th century and early 20th century near the Red River, wells flowed that tapped the glacial drift below the lake deposits. Since then water levels in the western part of the lake plain have declined and wells have ceased to flow. Serious water-level declines have occurred near West Fargo because of municipal and industrial pumpage of about 500 million gallons annually. Water levels have declined as much as 150 feet at Moorhead where buried sand and gravel aquifers have been tapped for a municipal water supply. Because of the declines in water levels in this area, the ground-water gradient has been reversed locally and the underlying drift aquifers are being recharged rather than discharging upward toward the land surface (Maclay and others, 1972, p. 113).

Water-level rises can be caused by irrigation of land, leakage from canals, flooding, and construction of surface reservoirs. Percolation of irrigation water to the water table in time may cause the water table to rise. Infiltration of irrigation water additionally could cause deterioration of ground-water quality. Excessive water-level rises may cause other problems such as flooding of basements, damage to structures, and detrimental changes to vegetation.

An example of potential rising water levels is near the planned Kindred Dam and lake on the Sheyenne River in southeastern North Dakota. Downey and Paulson (1974) have predicted that the maximum rise in water levels will occur about 50 years after filling of the lake.

DEPLETION OF STREAMFLOW

The withdrawal of ground water can decrease streamflow by decreasing ground-water discharge to the stream or by inducing infiltration of water from the stream to the aquifer. The pumping effect on a stream depends on the pumping rate, distance of well from the stream, and the aquifer characteristics. The amount of ground-water pumpage derived from a nearby stream or drain can be computed by methods developed by Theis (1941).

Decreased streamflow caused by pumping of ground water is generally considered an adverse effect. However, benefits may outweigh the disadvantages, as in the Minot area, where a dam on the Souris River augments the surface supply to the city and increases infiltration of ground water to the city's well field (Ferris and others, 1972, p. G-22).

DETERIORATION OF WATER QUALITY

Deterioration in chemical quality of the ground water may result when the natural equilibrium of the aquifer is disturbed by development. Intensive development can cause large-scale deterioration of quality by recirculation of water, interformational leakage, and pulling poor water from distant areas.

Where good quality water is being pumped from glacial aquifers, upward migration of deeper mineralized water may be caused by the lowered head due to pumping. The Paleozoic and Cretaceous rocks in the northwestern part of the Red River basin contain highly saline water that is discharging where these rocks pinch out against crystalline rocks in Minnesota. Upward migration of saline water has resulted in contamination of ground water in the overlying drift in northwestern Minnesota and northeastern North Dakota (Maclay and others, 1972, p. 113; Crosby and others, 1973, p. 195). To the west of the area, wells tapping the better water in the overlying drift aquifers could induce upward migration of the saline water if the heads are lowered sufficiently by the pumping.

Water may leak vertically between aquifers if the seal between them is broken by improperly constructed wells. If the quality of water is poorer in the aquifer having the higher head, the quality of water in the receiving aquifer will deteriorate.

Municipalities and industry have to dispose of a variety of wastes. The important types of waste in the region are chemicals from food and industrial processing, sewage, concentrated chemicals in cooling water, oil-field brines, mine discharge, and return flow of irrigation water. Any of these wastes may cause deterioration of ground-water quality if they enter the aquifer. Other types of waste, such as the effluent from saline-water conversion plants may be added in the future.

The use and reuse of water for irrigation results in the increase of dissolved mineral concentrations in the water. Each time water is applied to the land, some returns to the surface stream or to the ground-water reservoir. This return flow is wastewater, and its utility has been decreased because of the increase in dissolved-solids concentration.

GROUND-WATER CONSERVATION

When intensive development reaches advanced stages, the conservation of water becomes important as a solution to several types of problems. Farm conservation could include reduction of irrigation losses between well and field; more efficient irrigation, planting crops with a smaller water requirement, and eradication of phreatophytes. Leaky artesian wells could be repaired and artesian flow controlled by valves at the wellhead. Municipalities could repair leaky distribution systems, and limit lawn watering. Commercial users could return cooling water to the aquifer. Industry could process and return used water underground through injection wells or filter beds. All of these conservation measures could be used to various degrees in the Souris-Red-Rainy Region.

Elimination of phreatophytes would reduce consumptive use of the ground water. Phreatophytes, where found, are in areas where ground water is less than about 30 feet deep. Phreatophytes may intercept and waste ground water moving toward streams or lakes. They also may use enough water to induce recharge from surface-water bodies into shallow aquifers. Phreatophytes in some parts of the Nation are undesirable. However, in this region phreatophytes are desirable as wind breaks and ground cover to prevent erosion.

Water may be salvaged by use of evaporation suppressants on surface reservoirs in the western half of the region where evaporation rates are high. Experiments with evaporation suppressants (Magin and Randall, 1960) indicate suppression may not be practical or feasible on large reservoirs, but is promising for use on small ponds. Evaporation losses could be eliminated by storing water underground and pumping it out as needed. The underground storage could use known aquifers or permeable, unsaturated materials.

Another method of water conservation that could be used advantageously in the region is the salvage and recycling of water. Such recycling includes air conditioning or industrial water where the water can be treated to restore its usefulness. The possibility of reuse should be considered for as many different situations as possible. Under some circumstances ground water used for air conditioning could be returned to the reservoir without detrimental effects, thereby maintaining the ground-water supply.

EFFECTS ON THE ENVIRONMENT

A Souris-Red-Rainy River Basins Commission report (1972, p. 215) states:

> Although this Region is relatively small in comparison to other water resource planning regions in the Nation, its natural environment is characterized by a wide variety of physiographic and vegetative features. Over the years, public concern for maintaining and/or enhancing the quality of this natural environment has been increasing rapidly. Proper use and management of the water and land resources, which make up the Region's natural environment, will preserve and maintain a high quality environment.

Development of water supplies may have a detrimental effect on the environment. However, adequate monitoring programs can detect adverse effects in most situations. Declining water levels caused by pumping can cause migration of highly mineralized water, depletion of streamflow, or a change in natural vegetation in the case of near-surface ground-water levels. Irrigation can cause deterioration of water quality as natural minerals and fertilizers are dissolved in recycled irrigation water.

Long distance transportation of water may cause environmental changes. Such changes depend on (1) quality of the transported water in relation to the native water, (2) depth to the ground water along the path of transportation and in the area of use, and (3) effects of removing water from one area for use in another.

Along much of the Red River in North Dakota and Minnesota, natural upward migration of highly mineralized water has reduced soil productivity. Pumping could lower heads of the mineralized water and improve the quality of the soils, but provision would have to be made to dispose of the mineralized water. Lowering the artesian heads is a drainage project that could be detrimental to wildlife and natural vegetation.

A broad-scale analysis of land resources and watershed management as related to economic development and certain aspects of the environment is presented for the region in the Souris-Red-Rainy River Basins Comprehensive Study (McClure and others, 1972).

Currently more than 2 million acres of pasture and range land is suitable for crop production. Conversely, one-half million acres of land currently used for crop production could be converted to more economical uses such as forest, pasture or range land. The projected land development depends largely on availability of water, and particularly on availability of ground water.

INFORMATION AND GROUND-WATER MANAGEMENT NEEDS

Few aquifers are so continuous and uniform that high-production wells may be assured without preliminary test drilling. In some places the aquifers are defined by data from scattered test holes; future work may modify interpretations. For example, very little is known about the Paleozoic rocks in the Souris-Red-Rainy Region owing to a lack of drill-hole data and information on depth to the rocks. Exploration, testing, and study may help to find water at depth in quantity and quality suitable for a particular purpose. Study and planning also are needed to determine the best use of the water and land resources.

A properly designed and operated monitoring program can alert managers to approaching problems of diminishing supply or other detrimental effects. Monitoring programs ideally should be started before development, and continued to detect changes and determine effects of the development. For example, pumping ground water causes water-level declines near the pumped wells. As pumping continues or is increased, these declines generally spread. Return irrigation water, a new canal or reservoir, or excess precipitation can cause water-level rises. Monitoring water levels near pumped areas could give useful information on possible migration of inferior water toward the well field.

A ground-water quality monitoring program would detect changes in water quality such as migration of highly mineralized water or other pollution within or into the aquifer. Data from such a program could be used to suggest corrective action such as altering the pumping rates or patterns.

Reconnaissance and qualitative studies have been made for most of the region. Quantitative studies, including the use of models, have been made for small parts of the region to define the hydrologic and geochemical aspects of the system. Where sufficient data are available to develop digital models, these would benefit the planned development of the ground- and surface-water supplies. All of the water in storage, both surface water and ground water, is not usable, and the amount of each fluctuates with changing conditions. Regulation of surface- and ground-water storage can increase the usable supply. Adequate surface storage possibly could eliminate the loss of floodwaters and reduce surface discharge to that amount required for downstream water users.

Some surface water is lost during floods when streams overflow onto adjacent lands and the water is lost to evapotranspiration. Additional surface storage could reduce these losses. These surplus waters could be conserved by artificial recharge through wells and infiltration beds. Data are needed to determine the amount of surface water that could be salvaged by artificial recharge. Also, data are needed in all areas to locate suitable recharge sites.

Some water in storage is not usable because of high mineral content. Information is needed to determine where desalinization could be effective to increase the usable water supply.

Information and inventories are needed to identify existing and potential sites of pollution and means of waste disposal. In a water-short area such as the western half of the region, the reduction of waste through intensive use and conservation would reduce problems of waste disposal. The increased concentration of chemical constituents that results from intensive reuse of water, however, may aggravate problems in other areas, such as restrictions on, or lack of, suitable disposal sites or methods. Wise and efficient use of water requires that waste problems be solved at the stage of prevention, rather than when the problem becomes one of correction. Further research may lead to better means of reducing agricultural processing wastes. Also, research may determine the efficiency and suitability of various types of soils for the land disposal of sewage sludges and effluent.

Information gained by continued monitoring could be effective in preventing pollution from oil-field waste. Water produced with oil in the western part of the Souris River basin is an important quality problem in North Dakota. The water is highly mineralized and is produced in relatively large volume. Pollution from petroleum production, however, does not appear to be a serious problem at this time. Surveillance of these activities is conducted by the State of North Dakota and prevention of pollution rather than curative measures are emphasized in the Souris-Red-Rainy River Basins Comprehensive Study (Young and others, 1972, p. H-16). Most of this water is disposed of by injection into deep wells, either for disposal, or for repressuring oil producing formations.

Data are required on water-level fluctuations in connection with the inventory of water supplies, and to minimize the possibility of waterlogging of the land, and the invasion of inferior water. Data also are needed on the quantity of return flow from existing irrigation areas and on the relation among crop type, water application practices, fertilizer use, and other factors causing increase in dissolved solids in the return flow.

Maclay, Winter, and Bidwell (1972, p. 71-73) discussed water-resource management alternatives and related information needs. Their discussion is summarized here with respect to ground water in the Souris-Red-Rainy Region. Many alternatives for managing water are available to planners, some of which could be applied to water problems in the region. Some of these are operational and others are currently undergoing research. Many of them require adaptation to each local situation.

Improved methods to locate ground-water supplies are greatly needed. This would assure that the best and nearest possible source is being used. Better methods of well drilling and construction could lead to better development of water supplies. More efficient well development is needed, particularly in small aquifers. Pumping water from the lower yielding aquifers by using several low-yield wells rather than one high-yield well might be desirable. This is particularly applicable to beach-ridge aquifers associated with deposits of former glacial lakes. Also, special types of well construction such as infiltration galleries may be useful for developing thin but widespread aquifers, for example, those composed of beach deposits.

Special uses of saline or other poor quality water could be beneficial in part of the region where this type of water is abundant. Desalinization might be considered for these areas when it becomes economical.

Joint use of ground water and surface water also offers many management possibilities, particularly in the lake plain. Mixing of good quality water with poor quality water may provide an intermediate type that is acceptable for a particular use. This might be considered where poor quality water is abundant.

The storage capacity of aquifers would determine the possibility of storing surface water in them. Ground-water pumpage could greatly exceed natural recharge if stream water could be injected into ground-water reservoirs during periods of high flow. This technique probably could be used near Moorhead, where high flows from the Buffalo River could be injected into the aquifer and pumped out later.

More efficient and less costly methods are needed for transporting water from source to central distribution system, and then to individual users. Better metering and more realistic water billing in some of the communities could reduce water demands.

Many alternatives in the management of water supplies are possible. The selection of methods would depend upon their economic and hydrologic feasibility. In selecting any one or a set of these alternatives, the development of a water resource should be considered in the context of the total hydrologic system, with the realization that hydrologic changes in one part of a system will cause changes at other places.

McGuinness (1969, p. 1) stated, "...management of aquifers requires vast amounts of data plus a much better understanding of aquifer-system behavior than now exists. Implicit in this deficiency of knowledge is a need for much new research, lest aquifers be managed according to ineffective rule-of-thumb standards, or even abandoned as unmanageable."

SUMMARY

The Souris-Red-Rainy Region is underlain by a series of bedrock units that differ greatly in thickness and hydraulic characteristics and that range in age from Precambrian to Quaternary. Precambrian rocks are at or near the surface locally in the eastern part of the region and more than 15,000 feet below the surface in the center of the Williston Basin in western North Dakota. The Paleozoic, Cretaceous, and Tertiary sedimentary rocks of the Williston Basin gradually thin eastward and are missing in the Rainy River basin and in the southern and eastern parts of the Red River basin. The entire region has been glaciated, and most of the region is covered with glacial deposits that range in thickness from less than a foot to several hundred feet. Sand and gravel deposits in the drift form the most important freshwater aquifers. Other aquifers in the region are in Precambrian Paleozoic, Cretaceous, and Tertiary rocks. The potentiometric surface in the bedrock aquifers generally decreases in altitude toward the Red River, indicating that the general direction of ground-water movement is toward the Red River.

Ground water with less than 3,000 mg/L dissolved solids is available throughout the Souris-Red-Rainy Region; however, yields are small in places because some aquifers contain highly mineralized water. Ground-water quality in the Rainy River basin generally is better than in the Souris or Red River basins. Ground water with less than 1,000 mg/L of dissolved solids occurs in most of the region east of the Red River and in most of the shallow aquifers west of the Red River. The dissolved-solids concentration generally is less than 500 mg/L in water from the fractured crystalline rock and outwash-delta aquifers and less than 1,000 mg/L in other aquifers in the glacial deposits. The total volume of water available from storage having less than 3,000 mg/L of dissolved solids is estimated to be 5×10^8 acre-feet. In addition to the fresh and slightly saline water, the region has an abundance of highly mineralized water that also can be considered as a resource.

Yields of wells in individual bedrock aquifers are generally less than 100 gal/min but locally yields may be as much as 500 gal/min and more. Yields in drift aquifers are commonly less than 100 gal/min but range from 5 to 1,000 gal/min. In a few places the outwash yields more than 1,000 gal/min.

Water quality rather than quantity is the most important problem for many communities in the region. Excessive natural concentrations of dissolved solids, sulfate, and chloride are common in the western parts. Iron and manganese concentrations also are excessive in many supplies throughout the region. However, at the natural concentration levels none of these constituents are hazardous to health.

Ground water is the sole or a primary source of water supply in much of the Souris-Red-Rainy Region. In areas distant from streams, and in upland areas where surface water is not available physically, legally, or in the quality required for a particular use, ground water is the sole source of supply.

Reportedly, the potential irrigation development is 1,550,000 acres as compared with 50,200 acres irrigated in 1975. Both ground- and surface-water supplies will be required to meet these potential demands.

Rural domestic and livestock water supplies are derived almost entirely from ground-water sources. Small communities and towns generally rely on ground water, and the cities and industries use ground water, surface water, or both. The municipalities using surface water generally depend upon reservoir storage.

The quantity of water available to most municipalities has been adequate. Shortages in some areas, however, have occurred during prolonged drought. Communities depending upon ground water have solved such problems by deepening wells or constructing additional ones. For a few communities, potential water shortages persist or may develop in the future as requirements increase. Included are Minot in the Souris River basin, and several cities (such as Neche, Pembina, Grafton, and Mayville, N. Dak., and Crookston, Minn.) that depend on tributaries of the Red River. The largest municipal and industrial water-supply demands are in areas along the main stem of the Red River. As water use increases, however, additional reservoir storage or ground water may be required. Careful analysis and management of this complex water-supply system is warranted. Water supplies in the Rainy River basin are abundant in terms of present and foreseeable requirements.

Increased demands on both ground-water and surface-water supplies will be made in the future. Storage of water in the ground-water reservoir during times of surplus for withdrawal during times of scarcity would aid in meeting these demands. Similarly the surplus (flood) water is of better chemical quality than underlying ground water in parts of the western half of the region. Freshwater could be stored in saline-water aquifers as well as in freshwater aquifers, and pumped out as needed. The ground-water reservoir has a definite potential in water management. This reservoir volume is larger than all of the surface reservoirs in the region and should be more fully used.

Supplies of water adequate for recharging the ground-water reservoir are available in the region during the spring from several rivers and tributaries. The management of floodwater for replenishing ground-water supplies also could diminish flooding to some extent. Examples of artificial recharge projects that are operational in the region are at Minot and Valley City, N. Dak.

To understand the hydrologic system for management purposes there is a need to determine more adequately the hydrologic and geologic characteristics of existing aquifers, to determine the feasibility of providing treatment for improvement of the quality of water contained in those aquifers, and to locate new, undeveloped aquifers. As pumping and other stresses on one part of the hydrologic system affect other parts, monitoring programs ideally should be started before development and continued to detect changes and determine effects of the stresses. Monitoring water-level changes, withdrawals, and ground-water quality can alert managers to approaching problems of diminishing supply, waterlogging of land, or other detrimental effects. A water-quality monitoring program could detect migration of highly mineralized water or of other pollutants in the aquifer and aid in determining corrective action. Information is needed to locate and identify pollution sources. Further research may lead to better means of reducing wastes derived from product processing.

Many alternatives are available to planners for managing water. Some of these are operational and others are undergoing research. Many of them require adaptation to specific local situations. Adequate hydrologic information is needed by the water manager or planner to aid in solving problems of water supply, use, and pollution.

REFERENCES CITED

Brown, D. L., and Silvey, W. D., 1973, Underground storage and retrieval of fresh water from a brackish-water aquifer *in* Underground waste management and artificial recharge: Second Internat. Sym. on Underground Waste Management and Artificial Recharge, New Orleans, p. 379-406.

Busby, M. W., 1966, Annual runoff in the conterminous United States: U.S. Geol. Survey Hydrol. Inv. Atlas HA-212.

Carlson, C. G., and Anderson, S. B., 1973, Stratigraphy *in* Mineral and water resources of North Dakota: North Dakota Geol. Survey Bull. 63, p. 30-41.

Crosby, O. A., Armstrong, C. A., and Paulson, Q. F., 1973, Water resources of North Dakota *in* Mineral and water resources of North Dakota: North Dakota Geol. Survey Bull. 63, p. 161-198.

Downey, J. S., and Paulson, Q. F., 1974, Predictive modeling of effects of the planned Kindred Lake on ground-water levels and discharge, southeastern North Dakota: U.S. Geol. Survey Water-Resources Inv. 30-74, 22 p.

Fenneman, N. M., 1931, Physiography of the Western United States: New York, McGraw-Hill Book Co., 534 p.

Ferris, Ted, (chm.), and others, 1972, Water supply and health aspects, Appendix G *of* Souris-Red-Rainy River basins comprehensive study: Souris-Red-Rainy River Basins Commission, v. 4, 110 p.

Feth, J. H., 1965, Selected references on saline ground-water resources of the United States: U.S. Geol. Survey Circ. 499, 30 p.

Feth, J. H., and others, 1965, Preliminary map of the conterminous United States showing depth to and quality of shallowest ground water containing more than 1,000 parts per million dissolved solids: U.S. Geol. Survey Hydrol. Inv. Atlas HA-199.

Folsom, C. B., Jr., 1973 Petroleum and natural gas *in* Mineral and water resources of North Dakota: North Dakota Geol. Survey Bull. 63, p. 99-119.

Glover, D. H. (chm.), and others, 1972, Water resources, Appendix B *of* Souris-Red-Rainy River basins comprehensive study: Souris-Red-Rainy River Basins Commission, v. 2, 80 p.

Kelly, T. E., 1967, Artificial recharge at Valley City, North Dakota, 1932 to 1965: Jour. of Ground Water, v. 5, no. 2, April, p. 20-25.

Knapp, G. L., 1973, Artificial recharge of ground water — A bibliography: Water Resources Scientific Information Center 73-202, 309 p.

Kohler, M. A., Nordenson, T. J., and Baker, D. R., 1959, Evaporation maps of the United States: U.S. Dept. of Commerce Tech. Paper 37, 13 p.

McClure, N. A. (chm), and others, 1972, Land resources and watershed management, Appendix C *of* Souris-Red-Rainy River basins comprehensive study: Souris-Red-Rainy River Basins Commission, v. 2, 122 p.

McGuinness, C. L., 1963, The role of ground water in the National water situation: U.S. Geol. Survey Water-Supply Paper 1800, 1,121 p.

McGuinness, C. L., 1969, Scientific or rule-of-thumb techniques of ground-water management—Which will prevail?: U.S. Geol. Survey Circ. 608, 8 p.

MacKichan, K. A., and Kammerer, J. C., 1961, Estimated use of water in the United States, 1960, U.S. Geol. Survey Circ. 456, 26 p.

Maclay, R. W., Winter, T. C., and Bidwell, L. E., 1972, Water resources of the Red River of the North drainage basin in Minnesota: U.S. Geol. Survey Water-Resources Inv. 1-72, 129 p.

Magin, G. B., Jr., and Randall, L. E., 1960, Review of literature on evaporation suppression: U.S. Geol. Survey Prof. Paper 272-C, p. 53-69.

Murray, C. R., 1968, Estimated use of water in the United States, 1965: U.S. Geol. Survey Circ. 556. 53 p.

Murray, C. R., and Reeves, E. B., 1972, Estimated use of water in the United States in 1970: U.S. Geol. Survey Circ. 676, 37 p.

Rima, D. R., Chase, E. B., and Meyers, B. M., 1971, Subsurface waste disposal by means of wells — A selective annotated bibliography: U.S. Geol. Survey Water-Supply Paper 2020, 305 p.

Robinove, C. J., Langford, R. H., and Brookhart, J. W., 1958, Saline water resources of North Dakota: U.S. Geol. Survey Water-Supply Paper 1428, 72 p.

Signor, D. C., Growitz, D. J., and Kam, William, 1970, Annotated bibliography on artificial recharge of ground water, 1955-67: U.S. Geol. Survey Water-Supply Paper 1990, 141 p.

Souris-Red-Rainy River Basins Commission, 1972, Souris-Red-Rainy River basins comprehensive study: Souris-Red-Rainy River Basins Commission, v. 1, Type I Framework study, 216 p. and Type II study of selected basins, 128 p.

Theis, C. V., 1941, The effect of a well on the flow of a nearby stream: Am. Geophys. Union Trans., v. 22, pt. 3, p. 734-738.

Todd, D. K., 1959, Annotated bibliography on artificial recharge of ground water through 1954: U.S. Geol. Survey Water-Supply Paper 1477, 115 p.

U.S. Bureau of Reclamation, 1973, Water resources development in North Dakota *in* Mineral and water resources of North Dakota: North Dakota Geol. Survey Bull. 63, p. 199-252.

U.S. Department of Commerce, Environmental Science Services Administration, 1968, Climatic atlas of the United States: 80 p.

U.S. Geological Survey, 1970, The National atlas of the U nited States of America: U.S. Geol. Survey, Washington, D. C., 417 p.

U.S. Public Health Service, 1962, Drinking water standards — 1962: U.S. Public Health Service Pub. No. 956, 61 p.

U.S. Water Resources Council, 1968, The Nation's water resources: Washington, U.S. Govt. Printing Office.

Weber, W. R. (chm.), and others, 1972, Irrigation, Appendix F *of* Souris-Red-Rainy River basins comprehensive study: Souris-Red-Rainy River Basins Commission, v. 4, 69 p.

Winslow, A. G., and Kister, L. R., 1956, Saline-water resources of Texas: U.S. Geol. Survey Water-Supply Paper 1365, 105 p.

Young, L. A. (chm.), and others, 1972, Water quality and pollution control, Appendix H *of* Souris-Red-Rainy River basins comprehensive study: Souris-Red-Rainy River Basins Commission, v. 4, 86 p.

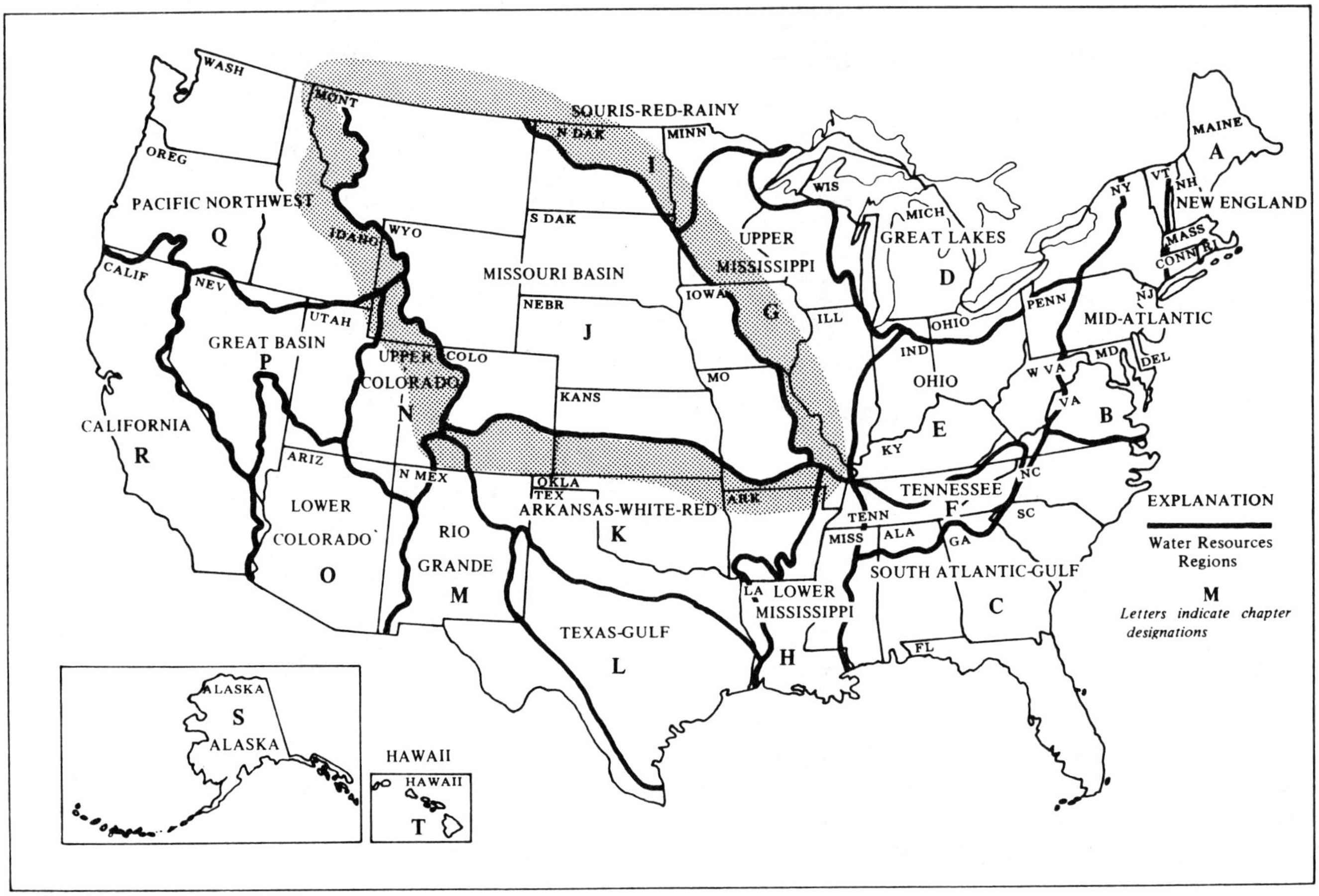

J – Location Map for the Missouri Basin Region

J

MISSOURI BASIN REGION

By O. JAMES TAYLOR

ABSTRACT

The Missouri Basin Region lies in the north-central part of the United States and southern Canada. It includes parts of Alberta and Saskatchewan in Canada; parts of Montana, Wyoming, North Dakota, South Dakota, Minnesota, Iowa, Colorado, Kansas, and Missouri, and all of Nebraska in the United States. The region includes about one-sixth of the contiguous United States and requires large water supplies for irrigation, industrial, public, and rural uses. Climate ranges from semiarid to subhumid. Normal annual precipitation increases generally eastward in the downstream direction, but precipitation is not a dependable source of supply. The Missouri River and its tributaries furnish water to many users, but surface water is often inadequate to meet large demands. Numerous surface reservoirs help to regulate streamflow and provide storage, but they also allow an increase in evapotranspiration, which in some areas exceeds normal precipitation. Ground water occurs in aquifers classified as alluvial deposits of sand and gravel, glacial deposits, dune-sand deposits, basin-fill deposits of sand and gravel, sandstone, siltstone, fractured sandy clay, limestone, and dolomite. Ground water can be developed and managed in an orderly manner provided adequate geologic and hydrologic data are available to determine aquifer characteristics and response to pumping and other hydraulic stresses. These data and determinations are essential to design, testing, and implementation of water-management plans.

Unconsolidated and semiconsolidated aquifers include valley-fill alluvium, areally extensive alluvium, glacial deposits, and basin-fill deposits. The aquifers normally consist of alluvial sand and gravel that contain unconfined ground water that lies near the land surface. Many wells completed in the alluvial aquifers have high yields of good-quality water because most alluvial aquifers are highly transmissive and hydraulically connected to streams. Aquifers in glacial deposits may be difficult to locate and in some areas contain saline water; nevertheless, these aquifers are sources of water supply for many users. Basin-fill aquifers are as much as several thousand feet thick, and many contain large ground-water supplies. Ground-water mining has occurred in semiconsolidated aquifers because of withdrawals from wells. Unconsolidated and semiconsolidated aquifers have potential for conjunctive use with surface water, recycling to reuse available supplies, artificial recharge, and salvage of evapotranspired water.

Sandstone aquifers lie near the land surface and in structural basins. Interbasin movement of ground water occurs in the Virgelle (Milk River aquifer), Fox Hills-basal Hell Creek, and Dakota aquifers. Sandstone aquifers are less transmissive than unconsolidated and semiconsolidated aquifers in general. Confined sandstone aquifers are common, and flowing wells are obtained in many areas. However, flowing wells may cause large declines in water levels if uncontrolled. Water quality is variable in sandstone aquifers but is adequate for most needs. Sandstone aquifers have potential for artificial recharge, induced interaquifer leakage, conjunctive use with surface water, and mining of ground water.

Limestone and dolomite aquifers are extensive in the region, but in some areas they lie deep below the land surface. The occurrence of ground water in small pores, fractures, or large caverns causes yields from wells to range widely. Large flows through cavern systems in the aquifers are indicated by large springs in some areas. Water quality is extremely variable and must be considered in any water-development plan. Limestone and dolomite aquifers have potential for development of large water supplies in some areas. The development may be aided by induced recharge and inter-aquifer leakage.

Saline ground water occurs throughout the Missouri Basin Region. Dissolved-solids concentration as much as 30,000 milligrams per liter has been measured in aquifers in glacial deposits in Montana. Saline water is common in sandstone aquifers in Wyoming, North Dakota, and South Dakota; maximum reported concentration is 280,000 milligrams per liter in water from the Tensleep Sandstone in Wyoming. Limestone contains saline water in many areas; maximum dissolved-solids concentration is about 350,000 milligrams per liter for the Madison Group in the Williston Basin in North Dakota.

Comprehensive water-management planning in the Missouri Basin Region will require periodic or continuing inventory of precipitation, streamflow, surface-water storage, and ground water. Water demands for irrigation, industrial, public supply, and rural use are increasing rapidly. Reliance on ground-water supplies is increasing even though in many areas the ground water is still mostly undeveloped. Optimal use of water supplies will require the establishment of realistic goals and carefully conceived water-management plans, each of which will necessarily be based on an adequate baseline of hydrologic data and knowledge of the highly variable hydrologic systems in the region.

INTRODUCTION

LOCATION AND EXTENT OF BASIN

The Missouri Basin Region lies in the north-central part of the contiguous United States and southern Canada; it includes small parts of Alberta and Saskatchewan Provinces; parts of Montana, Wyoming, North Dakota, South Dakota, Minnesota, Iowa, Colorado, Kansas, Missouri, and all of Nebrska (fig. 1). The area includes about 519,000 mi^2, (about one-sixth) of the United States and about 9,700 mi^2 of Canada. Resources vary widely over the region and include corps, timber, livestock, fossil fuels, economic minerals, and water. The Missouri Basin Inter-agency Committee (1969a) prepared an extensive description of the resources of the Missouri Basin Region.

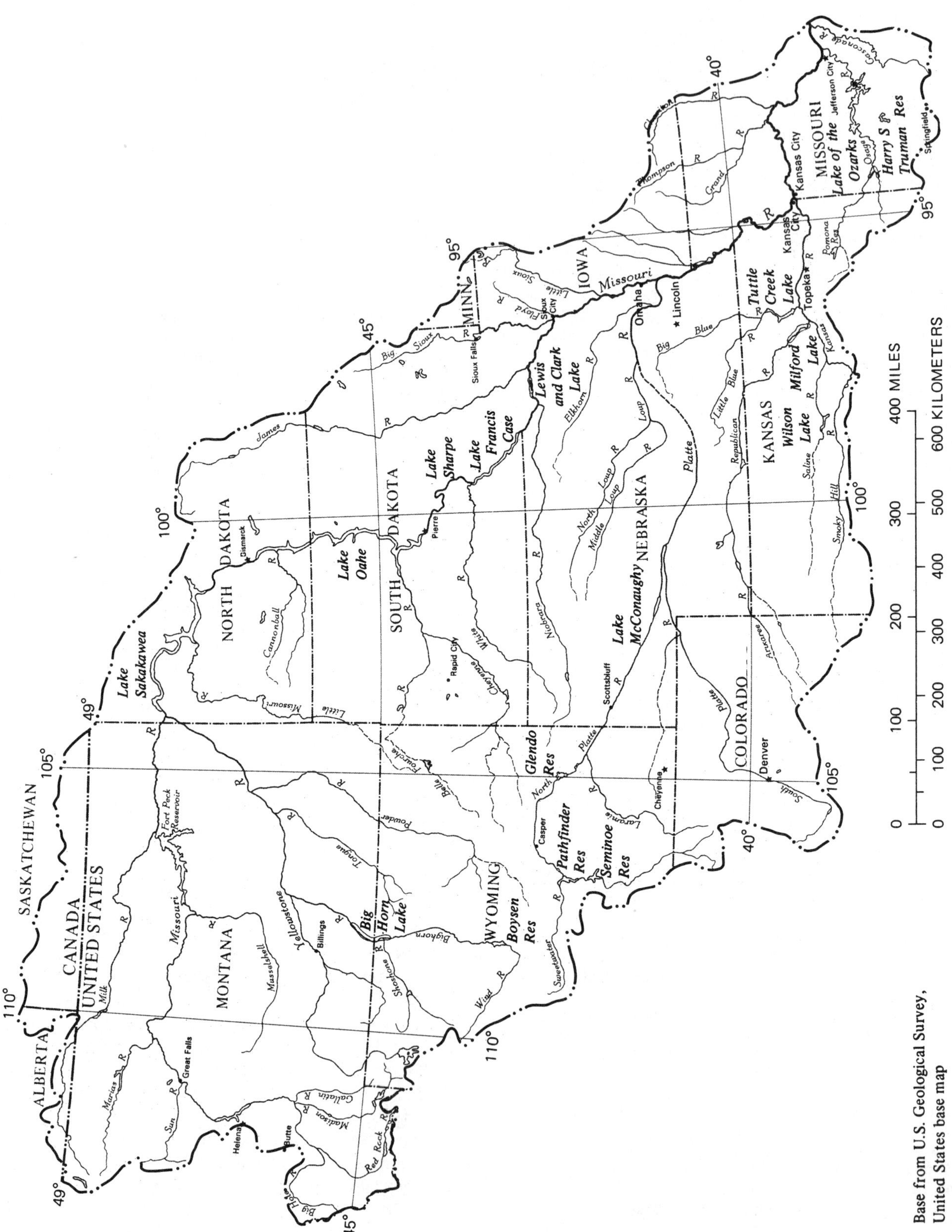

Base from U.S. Geological Survey, United States base map

FIGURE 1.—The Missouri Basin Region, showing surface drainage.

Physiography of the Missouri Basin Region is extremely variable. The western part of the region includes mountainous and foothill areas of Montana, Wyoming, and Colorado where igneous, metamorphic, and sedimentary rocks are severely folded and faulted. Headwaters of the Missouri River and many of its tributaries lie in the western part of the region where surface relief is large. East of the mountains lie glaciated areas in the northern part of the region and the plains of Montana, Wyoming, western Dakotas, and Colorado. The plains are underlain by extensive structural basins and interrupted by domes and uplifts, including the Black Hills Uplift. Sedimentary rocks crop out over the basins, and most rock types are exposed in uplifted areas. Surface relief is moderate in the plains, and streams drain toward the east and northeast.

Topographic and structural features of the lower part of the region are subdued. The plains become more gentle in south-central North Dakota, eastern South Dakota, Nebraska, southwestern Iowa, and northern Kansas. Sedimentary rocks are exposed and streams drain to the south and east. In the plains of northwestern Missouri, limestone is exposed. Hilly uplands and rolling plateaus are interrupted by the broad gentle uplift of the Ozark Region, where karst topography is developed. Streams drain generally toward the east into the Mississippi River.

PURPOSE AND SCOPE

Ground-water resources in the Missouri Basin Region constitute a large part of the total water resources. In parts of the region, ground-water resources are barely used; in other parts of the region the use of ground water contributes greatly to overall water supply. Full utilization of ground water in water management certainly has not been reached. The purpose of this report is to encourage further consideration of the use of ground water in regional water planning and management by:

1. Describing the known occurrence, characteristics, and behavior of ground-water resources in the Missouri Basin Region.
2. Indicating areas of the region in which hydrologic data and interpretive investigations of ground water are needed.
3. Suggesting preliminary ways to broaden water management to include fuller use of ground water.
4. Explaining proven methods for investigating and utilizing ground-water supplies.

Various State and Province officials supplied valuable hydrologic information and made helpful review comments. Their assistance is greatly appreciated.

WATER SUPPLY OF THE BASIN

HYDROLOGIC SETTING

CLIMATE

Normal annual precipitation (Environmental Science Services Administration, 1968, p. 43–44) varies greatly in the region as shown on figure 2. Range is from about 48 in. in northwestern Montana to as little as 8 in. in central Wyoming. A broad area of relatively high precipitation lies in eastern Nebraska, western Iowa, eastern Kansas, and Missouri. Normal annual precipitation does not indicate the wide variation in annual or seasonal precipitation observed in the region, which is noted for wide climatic variations. Annual precipitation commonly ranges from less than half to more than double the normal annual precipitation. Seasonal precipitation is more variable than annual precipitation. Below- and above-normal precipitation may occur concurrently in different parts of the region. Therefore precipitation as a source of surface-water replenishment and ground-water recharge is not as dependable as the figures for normal annual precipitation would seem to indicate.

Consumption of surface and ground water by evaporation and transpiration is termed evapotranspiration. Evaporation is large during periods of intense solar radiation, low relative humidity, and rapid wind movement. Distribution of lake evaporation (Environmental Science Services Administration, 1968, p. 63) is shown in figure 3. Annual evaporation ranges from about 28 in. in western Montana and northern Wyoming to about 62 in. in western Kansas. Over most of the region annual lake evaporation greatly exceeds annual precipitation. Therefore, little precipitation remains to replenish surface-water and ground-water supplies. Transpiration consumes ground water in areas in which dense growths of phreatophytes are maintained by a high water table. Transpiration consumes surface water in areas of phreatophytes near streams or lakes. Transpiration losses depend on local hydrologic conditions and cannot be generalized accurately over the region.

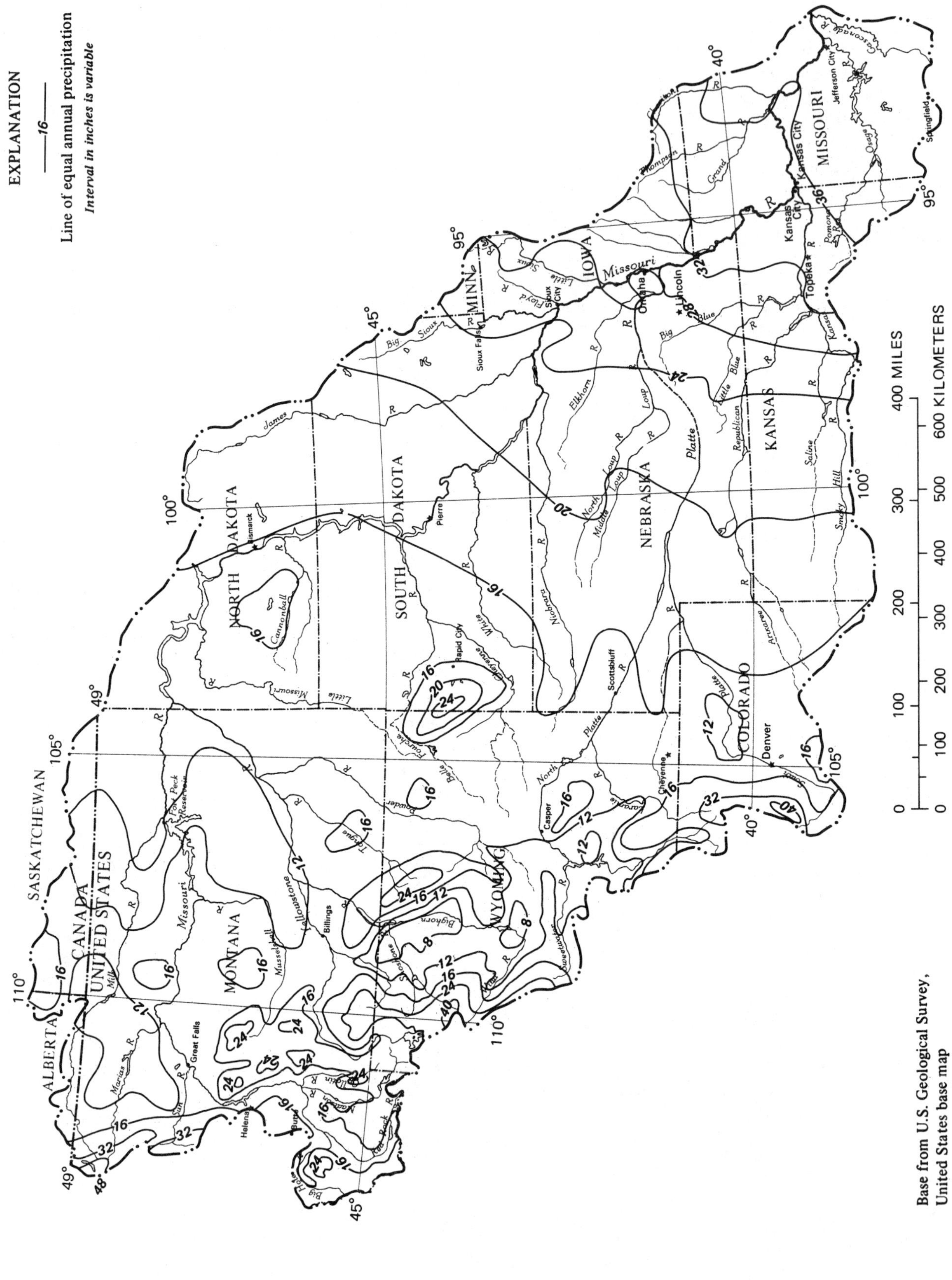

FIGURE 2.—Normal annual precipitation 1931–60 in the Missouri Basin Region.

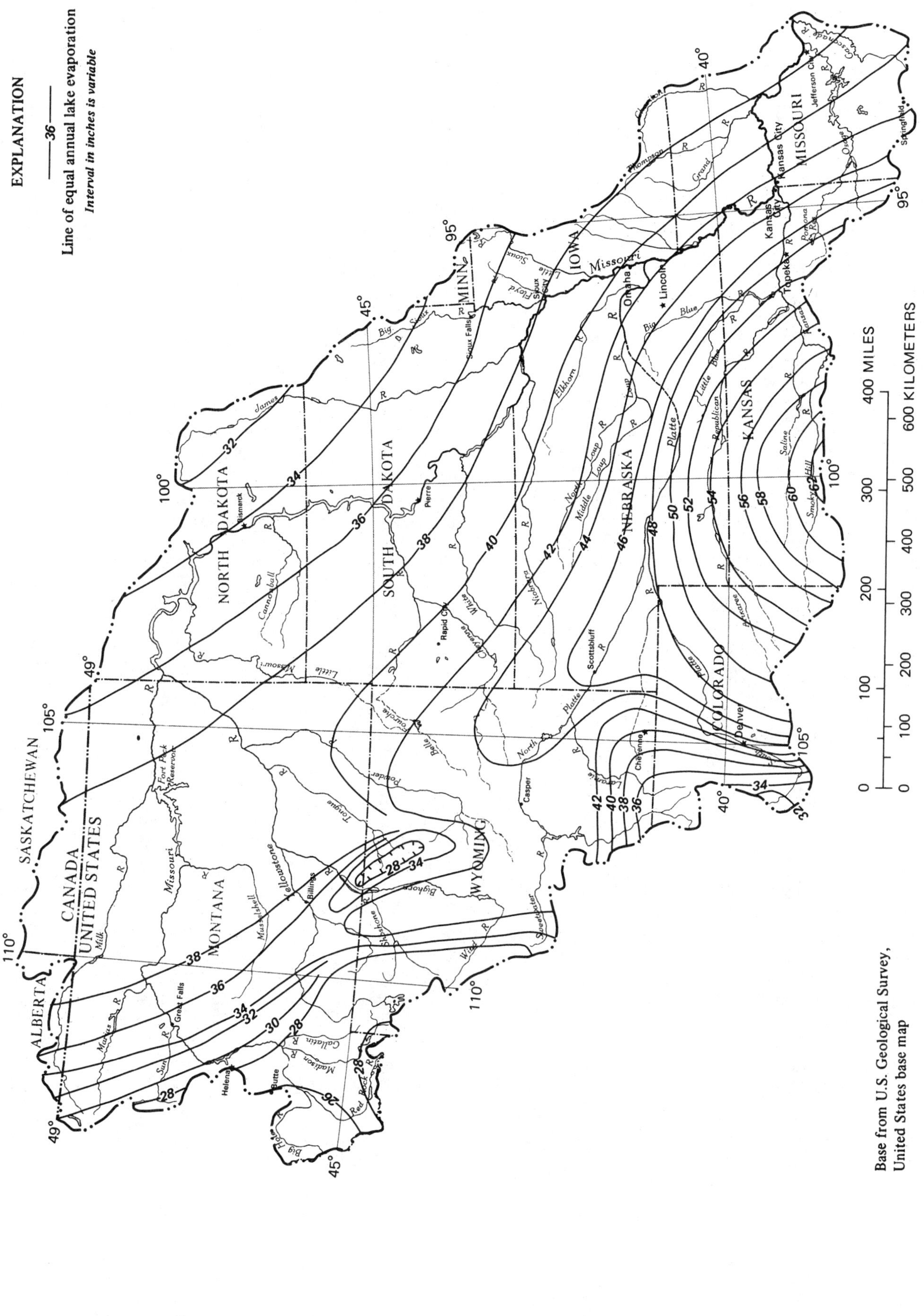

FIGURE 3.—Average annual lake evaporation 1946–55 in the Missouri Basin Region. (Environmental Sciences Administration, 1968, p. 63.)

SURFACE WATER

The Missouri is the longest river in the United States. Headwaters of the Missouri River are in southwestern Montana, 2,315 mi from its confluence with the Mississippi River at St. Louis, Missouri. Surface drainage in the Missouri Basin Region is shown on figure 1. Major tributaries to the Missouri River in downstream order include the Milk, Yellowstone, Little Missouri, Cheyenne, Niobrara, James, Big Sioux, Little Sioux, Platte, Kansas, Grand, Chariton, Osage, and Gasconade Rivers. Surface runoff within the region is approximately proportional to the precipitation and drainage area. Tributaries in Missouri and small parts of Kansas and Iowa drain about 8 percent of the area of the region but contribute 30 percent of the average annual flow of the region (Missouri Basin Inter-agency Committee, 1969b, p. 58).

Surface reservoirs constructed for varied and multiple purposes are numerous in the region. Major reservoirs along the Missouri River include Fort Peck Reservoir in northeastern Montana; Lake Sakakawea in western North Dakota; Lake Oahe in north-central South Dakota; Lake Sharpe, Lake Francis Case, and Lewis and Clark Lake in southern South Dakota. Numerous reservoirs along the tributaries contribute to the estimated total surface storage capacity of 112 million acre-ft in the region (Missouri Basin Inter-agency Committee, 1969b, p. 68).

Water quality of streams is suitable for most purposes. Over most of the region the average dissolved-solids concentration is less than 1,000 mg/L. Concentrations range from 1,000 to 4,000 mg/L in most streams of western North Dakota and western South Dakota, the lower South Platte River Valley in Colorado and Nebraska, and the upper Powder River Valley in central Wyoming (Missouri Basin Inter-agency Committee, 1969c, p. 108).

GROUND WATER

Ground water of the Missouri Basin Region occurs in a variety of aquifers that have been classified as alluvial deposits of sand and gravel, glacial deposits, dune-sand deposits, basin-fill deposits of sand and gravel; and as formations of sandstone, siltstone, fractured sandy clay, limestone, and dolomite. The alluvium of Pleistocene and Holocene age is generally located along stream valleys, except in central Nebraska, where the alluvium of Pleistocene age is areally extensive. Glacial deposits of Pleistocene age in the region are mostly in a broad band along the Missouri River. A major dune-sand aquifer of Pleistocene and Holocene age occurs in north-central Nebraska. Unconsolidated basin-fill deposits of Tertiary age that are thousands of feet thick occur in intermontane valleys of Montana. The Brule aquifer, siltstone and fractured sandy clay of Oligocene age, underlies parts of Wyoming, South Dakota, Colorado, and Nebraska. Sandstone aquifers ranging in age from Cambrian to Cretaceous lie near the land surface and in basins of various depths. The Madison Limestone aquifer of Early Mississippian age lies at depth in much of the western part of the Missouri Basin Region but remains mostly untested. Limestone and dolomite aquifers of Permian and Pennsylvanian age are present also in Nebraska, Iowa, Kansas, and Missouri. In addition the region may contain undiscovered aquifers capable of yielding large water supplies. Recharge to the aquifers is derived from precipitation, irrigation water, leaky reservoirs, losing streams, and leakage from other aquifers. Natural ground-water movement is generally in the direction of surface drainage; however, movement in deep aquifers may be unrelated to surface drainage because ground-water movement between surface drainage basins has been identified. The aquifers discharge to streams, springs, phreatophytes, wells, and to other aquifers. General yields to wells and depth to water for the region are summarized by LaRocque (1966).

DEVELOPMENT OF WATER SUPPLIES

AVAILABLE SOURCES OF WATER SUPPLY

Sources of water supply in the Missouri Basin Region are listed below with a brief description of each. All supplies are not available at all places and times, but normally the water user can choose from several alternative sources.

Precipitation.—Snowfall and rainfall replenish all sources of supply in the Missouri Basin Region. Precipitation provides for agricultural needs and contributes to surface runoff, surface storage, and ground-water recharge throughout the Missouri Basin Region. Precipitation is not predictable over extended periods of time, and water-management plans must take into account departures from normal precipitation patterns. However, data are available on the normal monthly precipitation and, for some areas, on the probability of specified rates of precipitation. Snowmelt in mountainous areas and resulting surface runoff are predictable to a degree of accuracy proportional to the frequency of field measurements of the snowpack.

Streamflow.—Precipitation, snowmelt, surface runoff, releases from surface reservoirs, base flow, and underflow contribute to streamflow. Resulting surface-water supplies are available in the extensive stream network of the Missouri Basin Region. Numerous long-term streamflow records are available that describe the average streamflow and observed or calculated variations at stream-gaging sites. However, most streamflow is derived from unpredictable and undependable precipitation. Streamflow is most abundant in the lower part of the region, where it serves a large part of the population but only a small part of the area of the Missouri Basin Region.

Stored surface water.—Water supplies are stored throughout the Missouri Basin Region in on-channel reservoirs, off-channel reservoirs, and lakes. The surface storage facilities help to regulate streamflow and provide carryover storage. Average lake evaporation is so high in the Missouri Basin Region that surface storage for extended periods of time will result in large losses of water that are accompanied by increases in the dissolved-solids concentration. Evaporation occurs throughout the year. For example, surface storage from November through April will result in evaporation of 16 to 38 percent of the average annual lake evaporation shown on figure 3 (Environmental Science Services Administration, 1968, p. 63).

Ground water.—As described in this report, aquifers of many different types that contain large supplies of water are numerous and widespread in the Missouri Basin Region. Ground water is a dependable and predictable resource if the hydrologic characteristics of the aquifers are well known. Ground water in storage may be increased by salvage of rejected recharge to, or reduction of discharge from, the ground-water reservoir. Sound water-management practices may require depletion of ground-water reserves or use of the ground-water reservoirs to store surplus amounts of surface water. Changes in ground-water storage may be determined by means of periodic measurements of water levels in an observation well network. Ground water in storage is not normally subject to the high evaporative losses associated with stored surface water. Estimated volumes of ground water stored in the Missouri Basin Region are listed in table 1. Volumes in each area include ground water in all aquifers, provided that dissolved-solids concentration is less than 3,000 mg/L. Estimated total ground-water storage of 150,000 billion ft^3 is about 64 times the average annual flow of the Missouri River in the lower part of the region near Hermann, Mo., and about 30 times larger than the total capacity of all surface-water reservoirs.

TABLE 1.—*Estimated ground-water in storage in the Missouri Basin Region*

[Adapted from written communication, C. F. Keech, 1976]

Area	Ground water in storage (billion ft^3)
Northeastern Montana	17
Western Montana	4
Central Montana	8
Southeastern Montana and northeastern Wyoming	22,000
Western North Dakota and western South Dakota	20,000
Eastern North Dakota and eastern South Dakota	13,000
Southeastern Wyoming and northeastern Colorado	2,000
Northern and central Nebraska	54,000
Western Iowa, eastern Nebraska and northeastern Kansas	3,300
Northern Kansas and southern Nebraska	20,000
Northwestern Missouri	15,000
Total	150,000

DEMANDS FOR WATER—PAST, PRESENT, AND FUTURE

Recent trends in water use in the United States have been summarized by Murray (1968) and Murray and Reeves (1972). Ground- and surface-water use in the Missouri Basin Region increased 20 percent from 1965 to 1970. Total ground-water use in the region increased by more than 67 percent from 1965 to 1970, as shown in table 2. Ground-water use also increased for each use category. Irrigation use was increased by more acreage brought under cultivation and efforts to increase crop yield by applying more water. The use of sprinklers for irrigation purposes is a major cause of the increased irrigated acreage and increased use of ground water for irrigation. In Nebraska, the number of registered irrigation wells increased from about 7,000 in 1950 to 55,078 in 1976 (Ellis and Pederson, 1977). Water withdrawn for irrigation probably will continue to increase throughout the region because under present economic conditions, irrigated-land farming is more profitable than dry-land farming. In the 1960's the region produced one-third of the Nation's wheat, 21–31 percent of the feed grains, and a large part of the rye, sugar beets, and flaxseed.

Industrial use of ground water in the region increased by 153 percent from 1965 to 1970. Water requirements for energy development are likely to cause

TABLE 2.—*Water use, in million gallons per day, Missouri Basin Region, 1965 and 1970*

Use	Water withdrawn, 1965			Water withdrawn, 1970		
	Ground water	Surface water	Ground and surface water	Ground water	Surface water	Ground and surface water
Irrigation	2,700	13,000	16,000	4,500	14,000	19,000
Industrial	340	2,400	2,700	860	3,200	4,100
Public supply	340	630	970	430	590	1,000
Rural	300	170	470	380	180	560
Total (rounded)	3,700	16,000	20,000	6,200	18,000	24,000

a further increase in the use of ground water by industry. Surface-minable coal deposits occur in Alberta, Saskatchewan, Montana, Wyoming, North Dakota, South Dakota, Colorado, Nebraska, Iowa, Kansas, and Missouri in the Missouri Basin Region. The coal may occur within aquifers, and in some areas the coal itself is an aquifer. In other areas the coal is part of the confining beds that may overlie or underlie other aquifers. Coal mining, coal transportation, and coal utilization affect ground water in various ways. Mining may disrupt natural recharge, movement, storage, and discharge of ground water. Water-quality changes are common due to leaching associated with changes in the flow system. The Surface Mining Control and Reclamation Act, Public Law 9587, was passed in 1977, partly to insure that future mining practices preserve ground- and surface-water resources and to restore previously mined areas to more natural hydrologic conditions.

Expected high demand for coal in the future will affect ground-water supplies and quality in the region. A coal slurry pipeline has been proposed to transport coal from Wyoming to steam generating plants in Arkansas, and water from the Madison aquifer is proposed as the transport medium for the coal. The estimated maximum withdrawal is 15,000 acre-ft/yr for 20 yr. Water withdrawn would never be returned to the aquifer. The use of ground water for steam generating plants near coal fields is also common. Water is either consumed or returned to the aquifer or streams at high temperatures that cause thermal pollution.

Ground water is pumped in the region by the petroleum industry for waterflooding, a process in which water is used to flush petroleum out of reservoir rocks. Waterflooding using ground water is currently practiced in Saskatchewan, Montana, Wyoming, North Dakota, Colorado, Nebraska, and Kansas. Production wells are completed in aquifers that range from shallow alluvial aquifers that contain freshwater to deeply buried bedrock aquifers that contain saline water.

Increases in public supply and rural use of ground water are also given in table 2. The rates of increase indicate an increasing reliance on ground water.

Demand for water, including ground water, is influenced by economic factors, especially the price for water that users are willing to pay. Normally, users of water for irrigation or rural purposes require low-cost supplies; users of water for industrial or public-supply purposes can afford to pay higher prices for supplies. Competition for water supplies may cause a shift in water demand and use toward those users that are willing to pay higher prices, provided that a change in use is permitted by regulations. The market price of water also varies with current quantity available. The quantity of water available for use may be increased by effective water management. Increased quantities will lower prices if accompanied by favorable production conditions. The quantity available for use may be decreased by overdrafts or by seasonal or extended droughts. Decreased quantities will raise prices if accompanied by unfavorable production conditions such as increased withdrawal costs due to declining water levels or increased treatment costs due to impaired water quality.

LEGAL AND REGULATORY CONSTRAINTS

Legal and administrative regulations are used to apportion available water supplies. Regulations are imposed by special purpose districts, Provinces, States, treaties, compacts, and decrees. Local special purpose districts may impose regulations that apply within the district boundaries. The Provinces of Alberta and Saskatchewan require licensing to use ground water for most purposes. States commonly select riparian or appropriation regulations that apply within their boundaries, although special rules may be designated for certain basins, areas, streams, or aquifers. Riparian rights allow a landowner to divert water from streams or lakes adjacent to his property or from ground water beneath his property, provided that he does not appreciably reduce the flow or impair the quality for downstream uses. Appropriation rights allow the owner of a right to divert or store water at a given rate based on a priority system. The amount of the right is based on need, and the priority is based on the beneficial use in time. The senior right is assigned to the user who first benefically used the water, and more junior rights are assigned to those who beneficially used the water later in time. Both the riparian and appropriation systems may be constrained to encourage conservation. In the Missouri Basin Region the riparian rights doctrine is used in Minnesota, Iowa, and Missouri. The appropriation rights doctrine is used in Montana, Wyoming, North Dakota, South Dakota, Colorado, Nebraska, and Kansas.

An international treaty, interstate compacts, and court decrees regulate streams that cross political boundaries in the region. The treaty apportions streamflow of the Milk and St. Mary Rivers and their tributaries between the United States and Canada. Interstate compact agreements have been ratified for division of streamflow among States drained by the Yellowstone River and tributaries, Belle Fourche River, Upper Niobrara River, South Platte River and certain tributaries, Republican River, Big Blue River, and Little Blue River. The court decrees affect the apportionment of streamflow among States drained by the Laramie River and the North Platte River. Some of the regulations limit storage, diversion, and irrigated acreage in affected areas. Only the Upper Niobrara Compact and the Big Blue River Compact (Big and Little Blue Rivers) regulate ground-water use in the drainage basins. Most international and interstate agreements do not consider ground water because of the lack of a data base for determining, and the means of controlling, ground-water flow across political boundaries. However, apportioned streamflow and regional water supplies are affected by ground water because:

1. Streamflow may be augmented by natural drainage of aquifers.
2. Streamflow may be augmented by return flow from irrigated areas.
3. Streamflow may be depleted by natural recharge of aquifers.
4. Streamflow may be depleted by discharging wells near streams.
5. Ground-water underflow and regional flow commonly cross political boundaries.

Constraints to basinwide management of water resources will be numerous and varied. Lack of ground-water information in the Missouri Basin Region is a constraint that limits development. Proper planning will improve the level of information and permit more utilization of ground water. Water apportionment rules should lead to efficient use of supplies through modern water-management techniques. Apportionment rules should be based on knowledge of all sources of water supply and on knowledge of the behavior of water-resource systems in order to facilitate efficient use. Costly projects such as water transfer, reservoir construction, deep drilling, or water treatment may not be necessary if all alternatives in water management are considered.

GROUND-WATER RESOURCES

For discussion purposes the ground-water resources of the Missouri Basin Region are grouped into three general types: unconsolidated and semiconsolidated aquifers, sandstone aquifers, and limestone and dolomite aquifers. All principal aquifers of the region in each State and Province are given in table 7 at the end of the report.

UNCONSOLIDATED AND SEMICONSOLIDATED AQUIFERS

Unconsolidated aquifers include valley-fill alluvium, alluvium, dune sand, glacial deposits, and basin-fill alluvium. Semiconsolidated aquifers include deposits of gravel, sand, sandstone, and fractured sandy clay. Exposures of unconsolidated and semiconsolidated aquifers are shown on plate 1. In many areas the unconsolidated and semiconsolidated aquifers overlie each other or other aquifers that are discussed later.

VALLEY-FILL ALLUVIAL AQUIFERS

The valley-fill alluvial aquifers consist of unconsolidated gravel, sand, silt, and clay, and are located along, and hydraulically connected to, the principal streams and their tributaries. Therefore the aquifers tend to be elongate. The maximum thickness of valley-fill aquifers is about 240 ft in Colorado along the South Platte River valley but probably averages about 50 ft over the Missouri Basin Region. In general, aquifers in main stream valleys are more extensive and thicker than aquifers in tributary valleys. Valley-fill aquifers are absent in some reaches of the main streams and in the upper reaches of most tributaries. In parts of North Dakota and South Dakota the aquifers are not distinguishable from glacial aquifers and are inundated in some areas by surface-water reservoirs. Data describing the areal extent and saturated thickness of the valley-fill aquifers are poorly known in many parts of the region, and information is needed to design plans for improved use of the aquifers.

Valley-fill aquifers are recharged by precipitation, irrigation water, leakage from other aquifers, and losing streams. Ground-water movement normally is in the downstream direction, so underflow in tributary valleys contributes to the ground water in the main valleys. In general the ground water in the valley-fill aquifers is in transit toward the lower part of the basin. Discharge from the aquifers is by drainage to streams, evapotranspiration, wells, and leakage to other aquifers. Evapotranspiration occurs where phreatophytes grow along stream valleys or anywhere the water table lies close to the land surface, as in many valleys in the Missouri Basin Region. Large-capacity wells completed in valley-fill alluvium yield as much as 4,000 gal/min along the North Platte River in Wyoming. Generally the maximum yields from wells in tributary valleys are smaller because aquifer materials are finer grained and aquifers are less extensive than those in the main stream valleys. More water-table maps and gain-and-loss studies in streams are needed to define the flow systems of the valley-fill aquifers.

Aquifer testing of valley-fill aquifers indicates that the transmissivity and specific yield are normally large. In many areas additional aquifer testing is needed in order to prepare maps showing the areal distribution of transmissivity and specific yield for use in models and planning. Aquifer testing can be conducted either by pumping large-capacity wells and observing the effect on the water table or by measuring fluctuations of the water table that are caused by changes in the stage of the stream. For example, the annual stage change, normally about 12 ft for the Missouri River in Missouri, is propagated through the alluvial aquifer and could be analyzed to determine the areal distribution of transmissivity and specific yield of the aquifer (Ferris, 1950).

Water quality in the valley-fill aquifers is suitable for most intended purposes. The dissolved-solids concentration ranges from about 100 mg/L to about 4,000 mg/L; however, the concentration is normally less than 1,000 mg/L.

Valley-fill alluvial aquifers and the associated streams in the Missouri Basin Region form stream-aquifer systems in which pumping of water wells in various locations may be used to control the timing of depletion of ground water in storage and streams, the two sources that supply the wells. Examples of well and stream relationships are shown in figure 4. Typically, the width of the valley is 10 mi, and the thickness of aquifer is 50 ft. Wells discharging on both sides of the stream are shown in figure 4*A*. Wells discharging on one side of the stream have little or no effect on the aquifer or wells on the other side of the stream, provided that streamflow is maintained and good hydraulic connection exists between the aquifer and the stream. However, all wells compete for water in the stream. The discharging well A, close to the stream, depletes streamflow more rapidly than ground water. The discharging well B, distant from the stream, depletes ground water more rapidly than streamflow. However, eventually all water withdrawn by the wells will deplete the stream. Ground-water levels will stabilize during prolonged withdrawals, and the levels will recover after the cessation of withdrawals. Depletion of streamflow also can be delayed by pumping wells that are completed in an aquifer that has a low transmissivity and (or) a high specific yield. The effects of natural or artificial recharge or discharge on streamflow and ground water are similarly controlled by distance, aquifer transmissivity, and specific yield.

The value of water management for the stream-aquifer system is apparent because the timing of some of the effects on the stream and aquifer can be selected. During periods of surplus or adequate surface runoff, it may be desirable to pump wells near the stream in order to deplete streamflow rapidly. When surface runoff is small or when all streamflow is allocated among the other users, it may be desirable to pump wells distant from the stream and delay depletion of the stream until a later period when streamflow is more abundant. Withdrawals from wells in the valley-fill aquifers of the Missouri Basin Region deplete the streamflow and affect downstream users. However, part of the streamflow probably is derived from ground water, so the surface and ground water in the stream-aquifer system actually constitute a single water supply.

Effects of an aquifer boundary and well location on the timing of depletion are shown in figure 4B. The location of a discharging well C on the west side of the valley in a narrow zone between the stream and the aquifer boundary causes the stream depletion to be extremely rapid because the stream is near and no ground water can cross the aquifer boundary. Well D is very distant from the stream, and the percentage of stream depletion is small, even after a pumping period as long as 400 days. The situation portrayed in figure 4B exists along many streams in the Missouri Basin Region, especially along the Missouri River in Missouri. Because the river lies close to one boundary in many reaches of the alluviated valleys, the water manager has many choices to design beneficial timing of stream depletion due to withdrawals from wells, or beneficial return flow due to recharge operations.

Additional water-management techniques are shown in figure 5. Reuse of the available water supply is shown in a valley irrigated from canals and wells (fig. 5A). Part of the canal water is consumed by evapotranspiration, and part recharges the alluvial aquifer. On the west side of the valley a well field withdraws the recharged water and reapplies it, causing additional recharge which in turn is withdrawn again. The well field allows timely and localized reuse of the water supply. About 50 percent of each application of irrigation water in the South Platte River valley recharges the valley-fill aquifer (Hurr and others, 1975). The recharge resulting from each application indicates that a water supply is only partly consumed with each application and the supply can be partly reused with successive applications. On the east side of the valley, without wells to withdraw water from the aquifer, the return flow can be reused only after the delay of return to the stream. Return flow also migrates downstream and therefore may not be available for use in the reach of the valley where it is needed. The withdrawal and recharge scheme described above is not possible near the stream, because pumped wells rapidly deplete the stream rather than stored ground water and recharged water flows rapidly to the nearby stream rather than into storage in the aquifer. Water-quality problems may develop from reuse schemes. If dissolved solids remain as the water is consumed, the dissolved-solids concentration may become large.

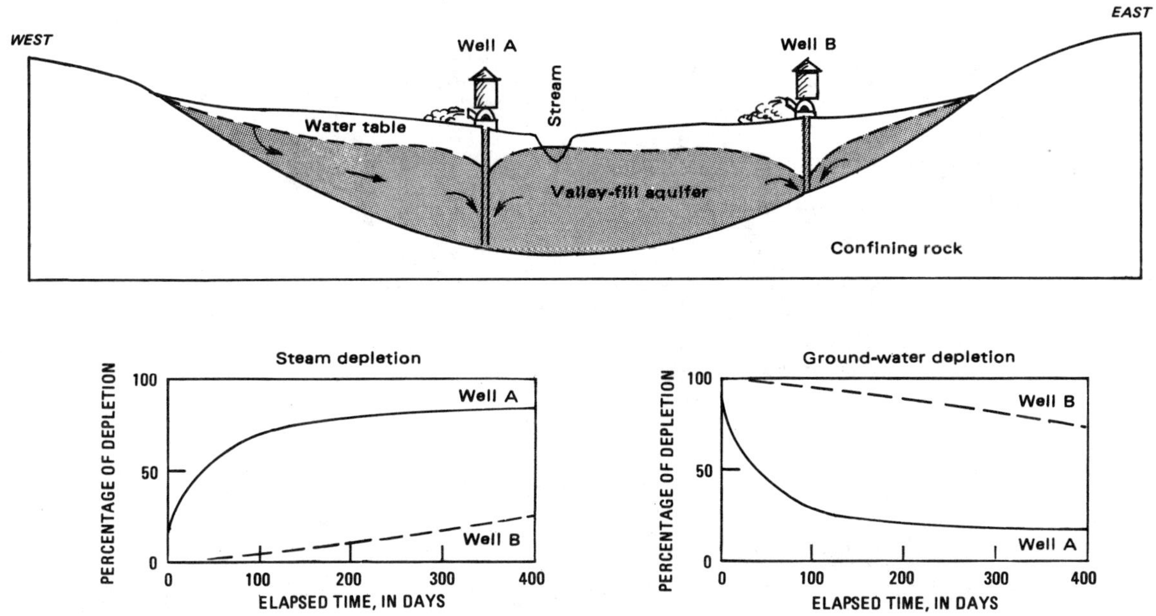

A. **Effect of well location on timing of depletion**

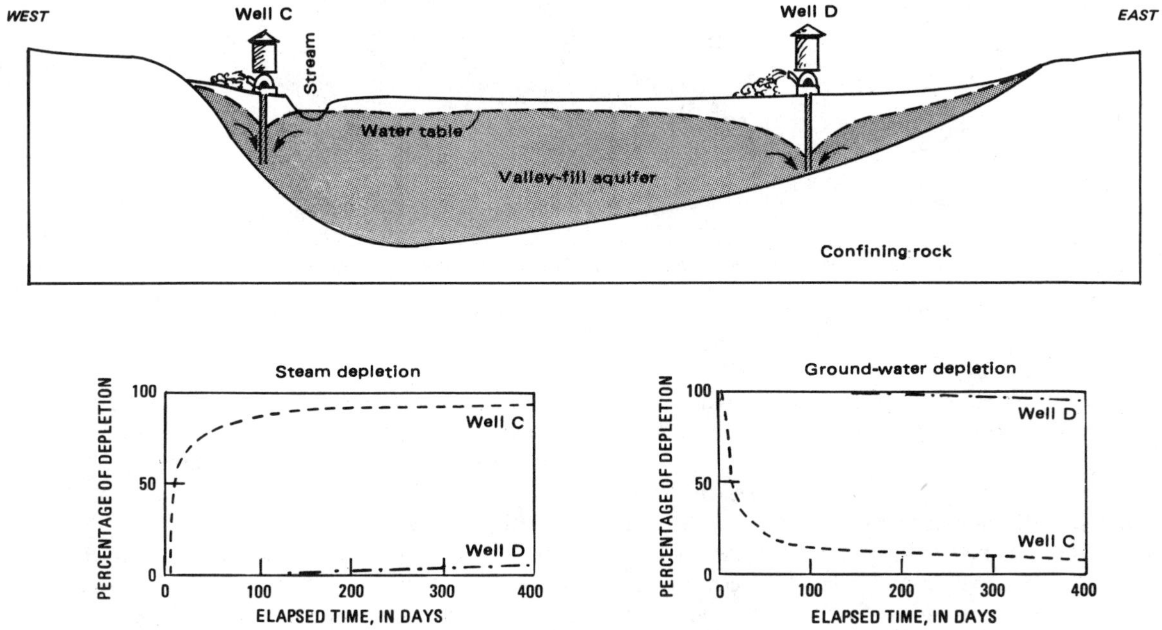

B. **Effect of aquifer boundary and well location on timing of depletion**

FIGURE 4—Effects of well location and aquifer boundary on depletion in a stream-aquifer system.

An urbanized area is sketched in figure 5B. Water supply is obtained from wells completed in the valley-fill aquifer. Yields to wells are maintained by natural recharge to the aquifer from precipitation and by depletion of the stream. However, as impervious cover is added to land in the urbanized area on the west side of the valley, natural recharge to the aquifer is decreased and overland flow to the stream is increased. The decrease in recharge reduces the yield of wells and the stored ground water. A change in well placement permits the water supply to be restored. Existing wells are abandoned, and new wells are drilled closer to the stream. The new wells deplete the stream more rapidly and derive less water from aquifer recharge and

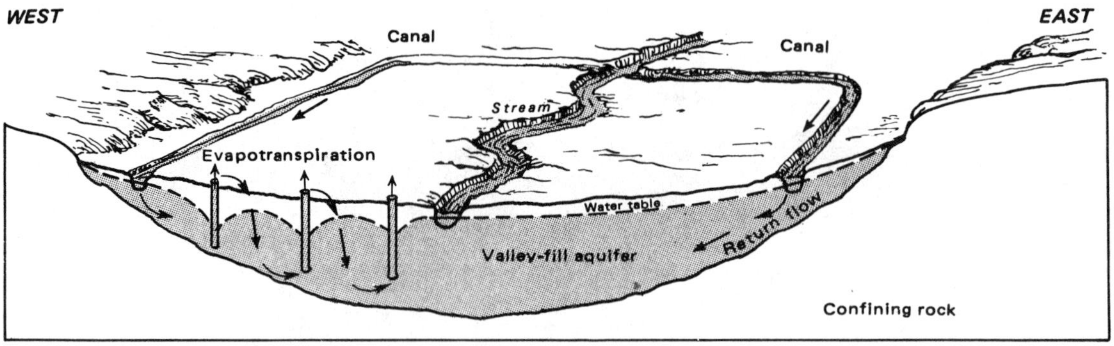

A. **Reuse of water supply by successive recharge and discharge**

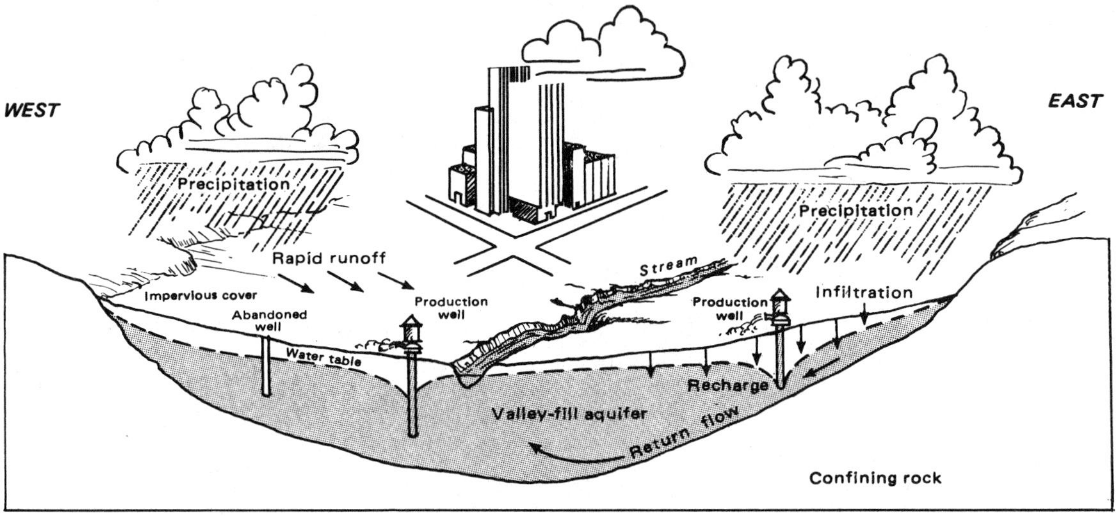

B. **Capture of rapid runoff from urbanized area by pumping well near stream**

FIGURE 5.—Reuse of water supply and capture of runoff from urbanized area.

storage. More rapid stream depletion captures the additional streamflow resulting from increased overland flow.

A specific water-management plan may be designed for any area in which water data and interpretive studies are available. For example, water-use regulations in the South Platte River Valley in Colorado are based on the known connection between surface and ground water. Conjunctive use of ground and surface water in the South Platte River Valley is related to the history of the development of water use in the valley. Prior to the beginning of irrigation the saturated thickness of the valley-fill deposits was small and the South Platte probably was not a gaining stream. After surface-water irrigation began in the middle 1800's, the recharge from irrigation raised the water table and increased the saturated thickness of the valley-fill aquifer. Resulting return flow to the South Platte has made it a gaining stream. However, the surface-water supply was extremely variable and undependable even when supplemented with imported surface water from the Colorado River and Laramie River basins. By 1970 more than 3,200 large capacity wells had been drilled for irrigation use (Hurr and others, 1975). The South Platte remains a gaining stream, in spite of the decrease in return flow that results from ground-water withdrawals that average 420,000 acre-ft/yr. An increase in withdrawal rate in the main valley would reduce further the return flow to the stream. However, flow of the South Platte River at the Nebraska State line averages 335,000 acre-ft/yr, and a reduction in streamflow may not harm downstream users because of normally adequate surface supplies. Present regulations of the State Engineer of Colorado require that a volume of water equal to 5 percent of streamflow depletion, caused by ground-water pumping during the year, be added to the stream or to surface-water storage from a supplemental source for use in augmenting streamflow the following year. The regulation was designed to protect senior water rights and eliminate water shortages in the overappropriated stream that occur about 5 percent of the time.

Management of water use in the South Platte River Valley might be improved by integrating forecasts of seasonal water supply with the known behavior of the hydrologic system. Predictions of surface runoff during the irrigation season are prepared monthly by the U.S. Soil Conservation Service and the National Oceanic and Atmospheric Administration, beginning in January. Magnitude of ground-water withdrawals depends on surface runoff because together they constitute the water supply. In years of above-normal runoff and recharge, ground-water withdrawals are usually small, and surplus supplies should be stored; in years of below-normal runoff and recharge, ground-water withdrawals are usually large, and stored supplies should be used. Therefore, the magnitude of ground-water withdrawals and the use of storage facilities can be anticipated from predictions of surface runoff.

Specially constructed artificial-recharge pits in the upper reaches of the tributary valleys could be used during periods of above-normal runoff in the South Platte River Basin. Minor surface-water irrigation in the tributary valleys provides little recharge. Mining of ground water in many tributary valleys has occurred because withdrawals from wells deplete ground water in storage rather than the flow of ephemeral tributaries. Surplus streamflow in tributaries could be diverted to recharge depleted valley-fill aquifers in the tributary valleys, rather than allowed to pass downstream to the main valley where it also may be surplus. Ground water could be withdrawn from the aquifers in tributary valleys and transmitted to the main valley through the tributary channels during periods of below-normal runoff (fig. 6). Aquifers in tributary valleys would function as off-channel reservoirs with the advantage of smaller evaporative losses than surface reservoirs, particularly if the water table were maintained at least 10 ft below the land surface.

The use of tributary valleys as off-channel reservoirs for the South Platte River Valley is permissible under the regulations of the South Platte River Compact between Colorado and Nebraska, and the Doctrine of Prior Appropriation. Briefly, the regulations state that during the irrigation season in the lower valley in Colorado, diversions from the river are forbidden for appropriators having adjudicated dates of priority after 1897 if Nebraska users diverting water from the river put their supply to beneficial use. However, during the winter season in the lower valley, Colorado may divert and store as much as 35,000 acre-ft of water in either surface or ground-water reservoirs.

ALLUVIAL AND DUNE-SAND AQUIFERS

Other unconsolidated aquifers in the Missouri Basin Region include alluvial and dune-sand aquifers. A transmissive and extensive sand and gravel alluvial aquifer occurs in south-central Nebraska and part of northeastern Kansas. The aquifer attains a thickness of several hundred feet and is hydraulically connected to valley-fill aquifers, dune-sand aquifers, the underlying Ogallala aquifer, and adjacent streams (pl. 1). Large-capacity wells in the aquifer yield as much as 2,000 gal/min for irrigation. Dissolved-solids concentration is normally less than 500 mg/L.

An extensive dune-sand aquifer in central Nebraska and southern South Dakota consists of fine sand and clay and has a maximum thickness of several hundred feet. Precipitation recharges the permeable aquifer at high rates, and surface runoff does not occur. Dissolved-solids concentration is only several hundred milligrams per liter. Streams that originate in the dune-sand area have steady flows because they consist of base flow from the ground-water reservoir. The water table in the dune-sand aquifer lies close to the land surface in many areas; evapotranspiration is about 10 million acre-ft/yr from natural lakes and from the shallow ground water. Increased use of ground water may salvage water that is normally evapotranspired. However, implementation of an incautiously designed plan for additional withdrawal of ground water might reduce streamflow, eliminate some natural vegetation, and allow wind erosion to increase (Keech and Bentall, 1971).

Salvage of water normally evapotranspired is portrayed schematically in figure 7. Lowering the water table by pumping reduces the losses due to transpiration by plants and evaporation from the soil and ground-water reservoir. Some pheatrophytes have roots that grow deeper as the water table is lowered, necessitating physical or chemical eradication of the plants to stop transpiration. The scheme shown in figure 7 does not work well near the stream in a stream-aquifer system because the wells mostly deplete the stream and do not lower the water table appreciably.

Several model studies of unconsolidated aquifers in Nebraska have been completed. An analog model of the Big Blue River Basin and Little Blue River Basin in southeastern Nebraska was prepared by Emery (1966). Results from the model study indicated that withdrawals from wells in Nebraska reduce base flow of the Big Blue and Little Blue Rivers by less than 5 percent. Knowledge of the simulated reduction in base flow is important to Nebraska and Kansas because the streams drain parts of both States. Huntoon (1974) prepared a mathematical model of the upper

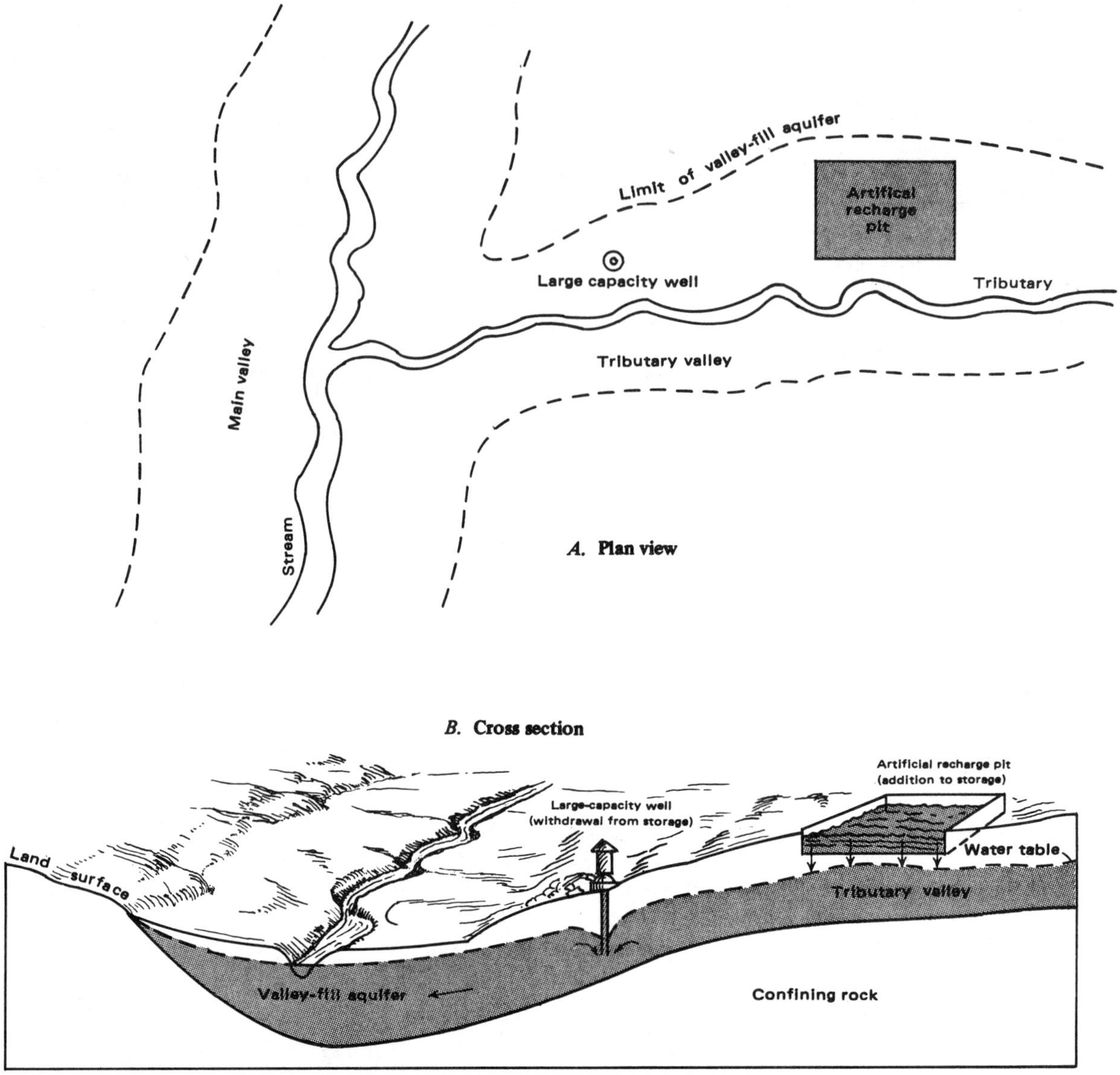

FIGURE 6.—Use of alluvial aquifer in tributary valley as off channel reservoir.

Big Blue River Basin to consider alternative methods to halt the continuous decline in the water table caused by ground-water withdrawals for irrigation. Artificial recharge was the only simulated alternative that would halt the decline. An areally extensive mathematical model that covers the Platte, Loup, and Elkhorn River Basins was prepared by the Missouri River Basin Commission (1975). The model was formulated for the streams and hydraulically connected alluvial, dune-sand, and Ogallala aquifers which have a maximum saturated thickness of 1,100 ft. Effects of recharge and discharge on the stream-aquifer system were simulated to predict the effect of additional ground-water use for irrigation. Results indicate that additional ground-water use will have one or more of the following effects: increased depletion of ground water, increased depletion of streamflow, or increased salvage of water normally evapotranspired. Of the three effects, salvage of water normally evapotranspired is preferred because water is obtained by intercepting a normal loss from the stream-aquifer system, rather than by depleting existing supplies. Additional model study is needed to determine optimal withdrawal of available supplies for improved water use. A study of optimal operation would determine the best sources of irrigation water obtainable under stated

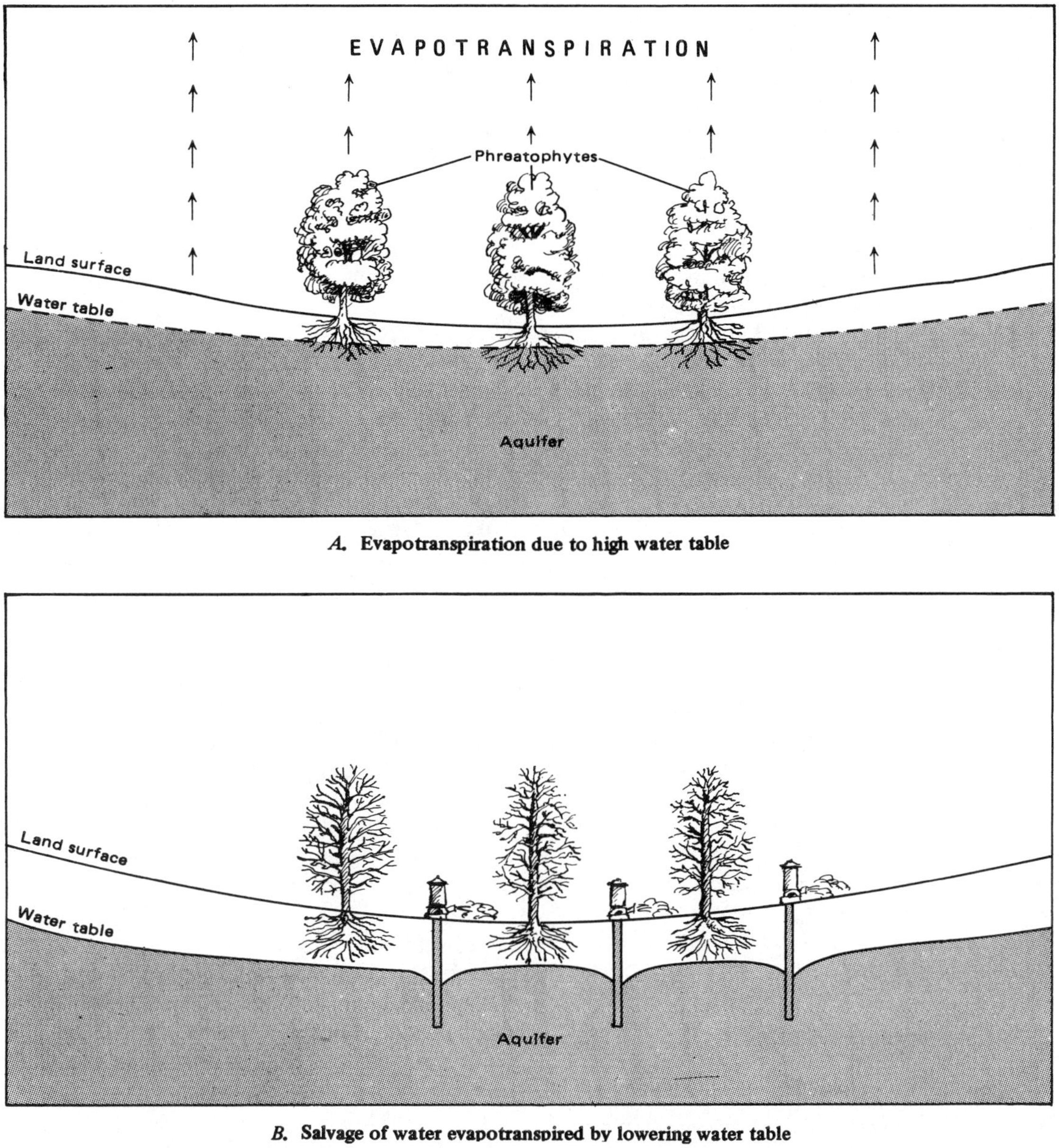

FIGURE 7—Salvage of evapotranspired water.

conservation and production goals and within legal and economic constraints.

GLACIAL AQUIFERS

Unconsolidated aquifers also occur in glacial deposits that extend in a belt as much as 200 mi wide, as shown on plate 1. The deposits reach a maximum thickness of about 800 ft and contain sand and gravel aquifers in the form of buried channel deposits, extensive outwash deposits, and lenticular outwash deposits. Aquifers in glacial deposits may be formed during various stages of glaciation, complicating the geologic interpretation of their origin and distribution. Glacial aquifers are recharged by precipitation, streams, or leakage from bedrock aquifers below the glacial deposits. Ground water in the glacial aquifers is discharged to streams, springs, underlying bedrock, evapotranspiration, or wells. Well yields are extremely variable and are as much as 2,000 gal/min. Flowing wells derive water from aquifers confined by clay beds. Withdrawals from wells are used for domestic, livestock, municipal, and irrigation purposes.

Water quality in the glacial aquifers is extremely variable, and the dissolved-solids concentration ranges from 300 to 30,000 mg/L. Saline seeps from glacial aquifers in northern Montana have affected a growing area that currently is about 170,000 acres of land; the seeps make the land too saline for most uses (Marvin Miller, Montana Bureau of Mines and Geology, oral commun., 1976). Water-quality sampling of ground water and streamflow under low-flow conditions helps to delineate reaches of the streams affected by the seeps. Water-table maps and depth-to-water maps help to delineate areas that may be subject to future damage and help to develop schemes to reduce or halt the spread to the seep areas. Reducing recharge to the glacial aquifers by planting crops that transpire more water helps to moderate the problem. In some cases it may be possible to remove the saline water by wells or drains and evaporate the water from ponds. The residue could be processed to separate trace elements that are known to be present in unusually high concentrations. The elements include selenium, lead, cadmium, copper, zinc, nickel, tin, chromium, aluminum, silver, molybdenum, and vanadium.

Many aquifers in the glacial deposits are not mappable at the surface and must be located by test drilling, geologic interpretation, or geophysical exploration. Several States in the Missouri Basin Region have drilling programs to define aquifers in glacial deposits, but more extensive and ordered exploration is needed. Exploration procedures for aquifers in glacial deposits have been discussed by Winter (1975). Statistical analyses of data describing drillers' logs, rock types, well yields, well dependability, and aquifer transmissivity are needed to provide information useful in locating the aquifers and to describe areal variations in the hydraulic properties of the aquifers.

BASIN-FILL AQUIFERS

Basin-fill aquifers in southwestern Montana consist mostly of poorly consolidated deposits of gravel, sand, silt, and clay of Tertiary age and attain a maximum thickness of about 6,000 ft in intermontane valleys. Basin-fill aquifers are overlain by valley-fill alluvial aquifers in many areas. Basin-fill aquifers are recharged by precipitation, irrigation water, and streams. Ground water moves in the downstream direction. Aquifers discharge to streams, to the atmosphere by evapotranspiration where the water table is close to the land surface, and to wells drilled for irrigation, domestic, livestock, and industrial purposes. Although basin-fill aquifers are not as transmissive as the overlying valley-fill alluvium, wells yield as much as 1,000 gal/min. Dissolved-solids concentration is normally less than 1,000 mg/L, making the water suitable for most purposes.

An extensive analysis of the ground-water resources in the basin-fill aquifers of the Gallatin Valley near Bozeman, Montana, was prepared by Hackett and others (1960). The authors proposed greater conjunctive use of ground water and surface water to intercept more of the water that passes downstream. Artificial recharge is proposed by the authors as a means to store the occasional surface-water surplus in the ground-water reservoir. Similar studies should be made for each of the basin-fill aquifers shown on plate 1.

A comprehensive water-management plan could be designed so that each year a water supply could be obtained efficiently from a combination of available sources. The combination would depend on the current available supply in each source, the known behavior of the hydrologic system, snowmelt forecasts, and runoff forecasts. The large thickness of the basin-fill deposits indicates that a large volume of ground water is available, which would be particularly useful if temporary depletion of stored supplies became necessary during a drought of several years duration.

OGALLALA, ARIKAREE, AND BRULE AQUIFERS

The Ogallala aquifer is widespread over parts of Colorado, Nebraska, and Kansas and present in small areas of Wyoming and South Dakota. The Ogallala Formation consists of semiconsolidated gravel, sand, silt, caliche, and clay and attains a maximum thickness of about 400 ft. Recharge to the Ogallala occurs directly from precipitation and by leakage through the overlying alluvial and dune-sand aquifers. Average annual recharge ranges from about ½ to about 3 in. where the aquifer is exposed, according to numerous analyses. Jenkins and Hofstra (1970) concluded that part of the recharge to the Ogallala is from ephemeral streams. The water table slopes toward the east, indicating that water in the Ogallala moves slowly from Wyoming and Colorado into Nebraska and Kansas. Discharge from the Ogallala is to streams or to wells. Large-capacity wells yield as much as 3,600 gal/min for irrigation, municipal, domestic, and livestock purposes. Average well yield is about 1,000 gal/min. Dissolved-solids concentration ranges from about 200 to 600 mg/L.

Continual declines of the water table of the Ogallala aquifer in Colorado, Kansas, Nebraska, and Wyoming indicate that ground water is being depleted as the result of discharge exceeding the long-term recharge. In Colorado the water-table decline averages about 3 ft/yr; in Kansas the decline averages about 0.7 ft/yr. A small decline of the water table in extreme southeastern Wyoming is due to withdrawals from the aquifer that began in 1968. Decline in the water table at some places in southwestern Nebraska is about 0.2 ft/yr. Withdrawal from the Ogallala in South Dakota is small. The Colorado Ground Water Commission has established standards that limit well spacing and storage depletion for any new wells completed in the Ogallala in Colorado. However, model studies of the Ogallala in Colorado indicate that the water table will continue to decline even if the withdrawal from wells is reduced gradually as the water table declines (Luckey and Hofstra, 1973, 1974; Kapple and others, 1976).

A needed regional water-management and economic study of the Ogallala aquifer covering Colorado, Kansas, Nebraska, and Wyoming would require data on the areal distribution of the following:

1. Specific yield of the aquifer.
2. Saturated thickness of the aquifer.
3. Hydraulic conductivity of the aquifer.
4. Depth to water below land surface.
5. Ground-water withdrawal.
6. Recharge from precipitation and applied irrigation water.
7. Artificial-recharge capability.
8. Cost of pumping from the aquifer.
9. Expected yield from, and value of, various crops.

The study could search for methods of operation that would reduce ground-water mining, increase the beneficial use of the supply, and extend the life of the aquifer. A regional aquifer system analysis for the Ogallala aquifer was begun by the U.S. Geological Survey in 1977.

The Arikaree aquifer occurs in parts of Wyoming, South Dakota, Colorado, and Nebraska and underlies the Ogallala aquifer. The Arikaree Formation consists of semiconsolidated sandstone, siltstone, and claystone and attains a maximum thickness of about 600 ft. Aquifer recharge is largely from precipitation and irrigation. The Arikaree aquifer may be recharged also by leakage from the overlying Ogallala aquifer in some areas. Discharge is to streams and to irrigation, municipal, stock, and domestic wells. Some wells yield as much as 2,000 gal/min. Ground-water mining has occurred in the Arikaree in northwestern Nebraska, as evidenced by a progressive decline in the water table. Continued decline in the water table will gradually increase the cost of lifting water from the aquifer to the land surface and may discourage use of the aquifer. The dissolved-solids concentration in the Arikaree aquifer is only several hundred milligrams per liter. A mathematical model study (Lines, 1976) of the Arikaree aquifer along the Laramie River in Wyoming indicated that part of the natural discharge from the aquifer to the river could be captured with wells. Additional water-table maps, aquifer testing, and gain-and-loss studies in streams would help in the design of plans to capture more natural discharge from the aquifer and reduce ground-water depletion.

The Brule Formation of the White River Group underlies the Arikaree Formation and is exposed over parts of Wyoming, South Dakota, Colorado, and Nebraska. The Brule Formation consists of siltstone and fractured sandy clay with a maximum thickness of about 450 ft. Precipitation and streamflow recharge the Brule aquifer. Ground-water movement occurs in the siltstone and in the fractures in the sandy clay. Therefore, the Brule aquifer is transmissive even though the aquifer material is fine grained. Extensive vertical and horizontal fractures in the Brule are described by Rapp and others (1953) and Babcock and Bjorklund (1956). Discharge is to streams and to wells that yield as much as 2,000 gal/min. Withdrawals from wells have caused minor ground-water depletion in Wyoming and Nebraska. The dissolved-solids concentration is normally only several hundred milligrams per liter. Artificial-recharge experiments in the Brule aquifer would determine if the fracture system in the aquifer could facilitate rapid recharge. Conceivably the vertical fractures could transmit the recharged water rapidly downward, and the horizontal fractures could transmit the water rapidly laterally through the extent of the aquifer. Successful artificial-recharge operations, using occasionally surplus streamflow, could result in a greatly improved water supply in the Brule aquifer.

Water supplies in unconsolidated and semiconsolidated aquifers of the Missouri Basin Region are summarized in table 3. Included in the table are summaries of water-management suggestions for each aquifer.

SANDSTONE AQUIFERS

Sandstone aquifers are widespread in the Missouri Basin Region except in southwestern Montana and most of Nebraska, Kansas, Missouri, and parts of Wyoming and Colorado (pl. 2). Some sandstone aquifers are exposed at the land surface; others are in structural basins of various depths and alternate with relatively impervious beds of shale and limestone.

AQUIFERS IN STRUCTURAL BASINS

Structural basins are depressions that result from folding, faulting, subsidence, or nearby uplifting. Many structural basins in the region contain thick accumulations of sediment. Extensive accumulations of

TABLE 3.—*Summary of water supplies in unconsolidated and semiconsolidated aquifers, Missouri Basin Region*

Source of supply	Location	Yield to wells (gal/min)	Suggestions for improved hydrologic data, analyses, and water management	Outlook for future use if suggestions are followed
Valley-fill alluvial aquifers.	Montana, Wyoming, North Dakota, South Dakota, Colorado, Nebraska, Minnesota, Iowa, Kansas, Missouri.	10–4,000	Mapping and test drilling to determine location and type of boundaries, aquifer thickness and extent. Aquifer testing to prepare maps of transmissivity and specific yield. Preparation of water-table maps. Gain-and-loss studies in streams. Water-management analysis of stream-aquifer system.	Improved water supply by controlled changes in streamflow and ground-water storage. Additional reuse of available supply. Storage benefits from artificial recharge.
Alluvial aquifer, dune-sand aquifer.	Nebraska, Kansas.	50–2,000	Mathematical model studies designed to determine optimal development of available supplies.	Increased water supply by additional salvage of water evapotranspired. Reduced shortages during periods of drought.
Glacial-deposit aquifers.	Alberta, Saskatchewan, Montana, North Dakota, South Dakota, Nebraska, Minnesota, Iowa, Kansas, Missouri.	10–2,000	Test drilling to locate aquifers. Statistical analyses of drillers' logs, aquifer transmissivity, well yield, well dependability, and water quality to define major aquifers and saline ground-water resources. In northern Montana, ground water and streamflow sampling under low-flow conditions. Preparation of water-table maps and depth-to-water maps. Changes in agricultural practices to reduce recharge. Withdrawal and disposal of saline water from areas where conditions permit.	Increased and more dependable water supply. Reclamation of land in northern Montana affected by saline seeps. Use of saline ground water.
Basin-fill Aquifers.	Southwestern Montana.	15–1,000	Incorporation of ground water into water-management plans that include all water supplies, current status of each, and known behavior of hydrologic system. Exploration for additional basin-fill aquifers.	Reduced shortages, even during drought of several years' duration. Possible discovery of additional ground-water supplies.
Ogallala aquifer.	Wyoming, South Dakota, Colorado.	10–3,600	Regional water-management and economic studies to determine availability and best use of supply.	Reduced ground-water mining and increased benefit from available supply. Increased aquifer life.
Arikaree aquifer.	Wyoming, South Dakota, Colorado, Nebraska.	10–2,000	Preparation of additional water-table maps. Aquifer testing to determine transmissivity and specific yield. Gain-and-loss studies along streams hydraulically connected to the aquifer.	Improved understanding of aquifer potential. Conjunctive use of ground and surface water. Reduced ground-water mining.
Brule aquifer.	Wyoming, South Dakota, Colorado, Nebraska.	10–2,000	Artificial-recharge experiments to determine if fractures in aquifer will facilitate rapid recharge.	Increased ground-water storage and decreased shortages during extensive drought. Reduction of pumping lift in wells.

fine sediment form confining beds; accumlations of coarser sediment form aquifers. Numerous freshwater aquifers that alternate with confining beds occur in structural basins in Montana, Wyoming, North Dakota, South Dakota, and Colorado. In some cases, a shallow sandstone aquifer may be exposed over the areal extent of the structural basin. Normally, the sandstone aquifers and confining formations are exposed at the land surface at the margin of the basin. Confined conditions are common, and in many cases flowing wells may be obtained in the sandstone aquifers. Principal structural basins that contain freshwater aquifers in the Missouri Basin Region are shown in figure 8 (Rocky Mountain Association of Geologists, 1972). Intervening areas between basins are uplifted domes or arches and volcanic rocks. Margins are not shown for most basins on plate 2 and figure 8 because they tend to merge and exact basin margins are arbitrary. Local sandstone aquifers that occur in western Montana and northwestern Wyoming are not shown because they lie in an area of complex geologic structure and have not been mapped thoroughly. All principal aquifers in structural basins of the region are shown on plate 2 and tabulated in table 8 at the end of the report.

The Virgelle Sandstone occurs in northwestern Montana in the southern part of the Alberta Basin. In other parts of Montana the Virgelle is the basal member of the Eagle Formation; in southern Alberta the virgelle is known as the Milk River Sandstone (Meyboom, 1960). The Virgelle consists of sandstone and shale and attains a maximum thickness of about 180 ft. Precipitation and leakage from overlying glacial deposits recharge the Virgelle aquifer (Zimmerman, 1967). A potentiometric map of the Virgelle indicates a ground-water divide about 12 mi south of the Canadian border. Ground water moves north and south from the divide. The ground water moving north is leaving the Missouri River Basin and entering the Saskatchewan River Basin in Alberta. Discharge in Montana and Alberta is to streams and springs and to wells that yield as much as 250 gal/min for industrial, municipal, domestic, and livestock uses. The aquifer is confined in parts of Alberta, and a large number of flowing wells have been drilled. Withdrawals from pumped and flowing wells in the aquifer have caused continuing declines in water levels in wells. Meyboom (1960) suggested the implementation of conservation practices to protect the ground-water supply, even though the current riparian laws discourage conservation. Location and plugging of flowing wells would help to preserve the supply in the aquifer, but the wells are not located easily. Zimmerman (1967) suggested that industrial demands might be supplied

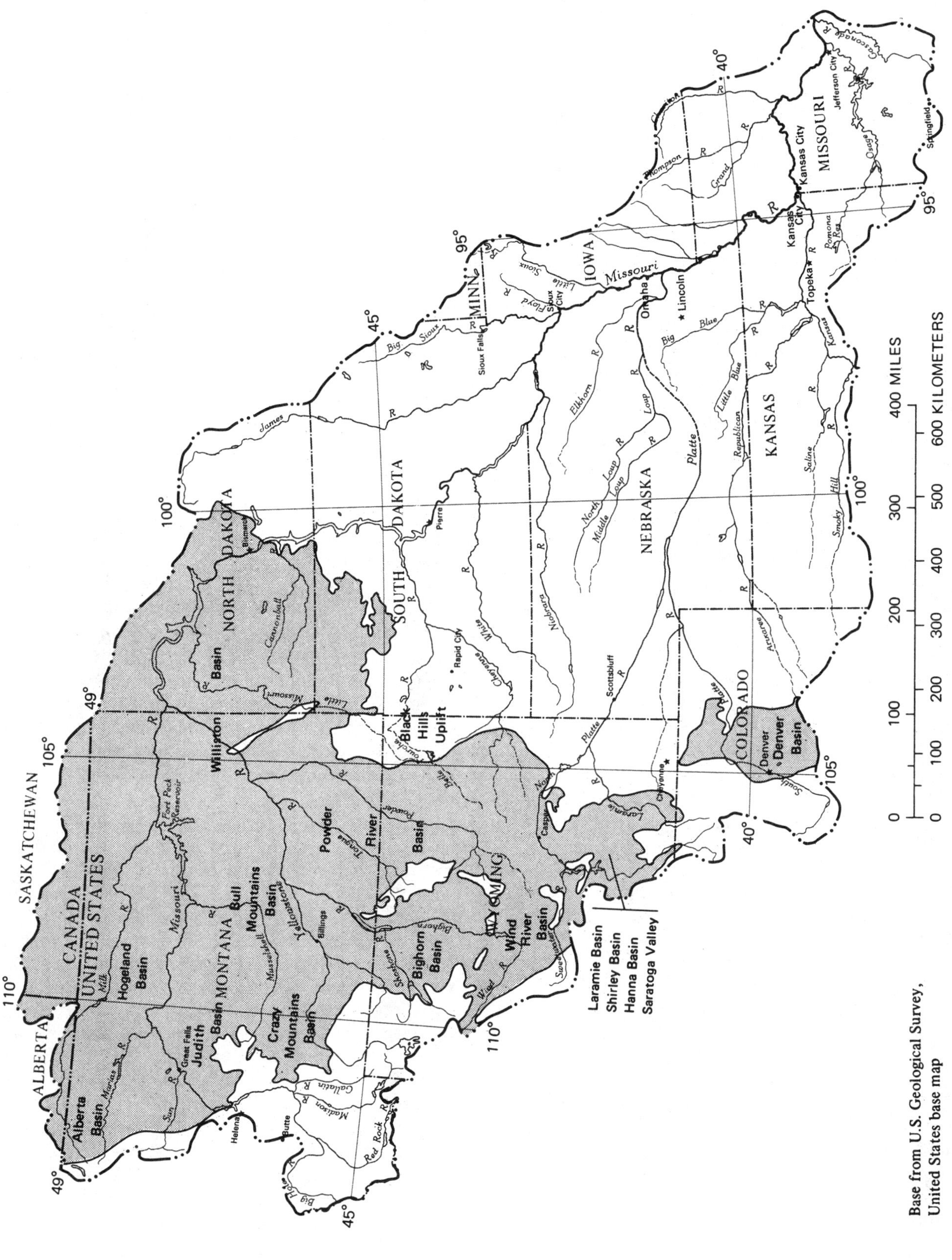

Base from U.S. Geological Survey,
United States base map

FIGURE 8.—Structural basins that contain freshwater aquifers in the Missouri Basin Region.

from Cut Bank Creek or aquifers in the Madison Group. Dissolved-solids concentration in the aquifer reaches a maximum of about 2,500 mg/L and varies widely over the aquifer.

Maximum transmissivity of the Virgelle aquifer in Montana (Zimmerman, 1967) is about 25 times greater than the maximum transmissivity of the Milk River aquifer in Alberta (Meyboom, 1960). The greater maximum transmissivity of the Virgelle aquifer in Montana is probably due to extensive fracturing of the aquifer, which facilitates water movement. Laboratory tests of horizontal and vertical hydraulic conductivity of samples of the Milk River aquifer from Alberta indicated that the vertical conductivity of the sandstone is nearly as great as the horizontal conductivity. Therefore, the feasibility of artificial recharge of the aquifer should be investigated. The reported small storage coefficient of the confined zone of the Milk River aquifer in Alberta indicates that recharge will cause a large rise in water levels. Uncontrolled flowing wells would have to be plugged before beginning artificial-recharge operations. The water supply for artificial-recharge operations could be obtained from the Milk River when streamflow exceeds demands.

The Blood Reserve and Judith River Formations are sandstone aquifers that overlie the Milk River Sandstone in the Alberta Basin in southern Alberta (McLean, 1971; Tokarsky, 1974). The dissolved-solids concentration in the Blood Reserve and Judith River aquifers ranges from about 1,000 to 4,000 mg/L. Additional potentiometric maps and aquifer testing are needed in order to understand the flow system in each aquifer, including the possibility of interaquifer leakage, and to design plans for water management.

The Judith Basin lies in central Montana near the headwaters of the Missouri River and many of its upstream tributaries. Reconnaissance geologic and hydrologic studies of the basin were made by Zimmerman (1966) and Feltis (1973). Sandstone aquifers include the Judith River Formation, Eagle Sandstone, Dakota Sandstone, Kootenai Formation, Swift Formation, Amsden Group, and Kibbey Formation. The sandstone aquifers are recharged by precipitation, streams, and irrigation water. Direction of ground-water movement is poorly understood because not enough wells exist to allow preparation of potentiometric maps. Discharge is to wells, streams, or by evapotranspiration in areas where the water table is near the land surface. Wells completed in the Judith River, Eagle, and Dakota aquifers yield as much as 70 gal/min; wells completed in the Kootenai and Swift aquifers yield as much as 300 gal/min. Flowing wells are obtained where the aquifers are confined. Few wells have been completed in the Amsden or Kibbey aquifers.

Additional potentiometric maps, aquifer testing, and water-quality sampling are needed for the Judith Basin. Improved understanding of the ground-water resources could lead to comprehensive conjunctive use of ground and surface water.

The Hogeland Basin lies in southern Saskatchewan and north-central Montana. However, the sandstone aquifers present in the basin extend over a wide area (pl. 2) beyond the limit of the basin (fig. 8). Sandstone aquifers described by Whitaker (1967, 1974) and McLean (1971) include the Frenchman, Whitemud, Eastend, and Judith River Formations. Sandstone aquifers of the Hogeland Basin in north-central Montana described by Alverson (1965) and Osterkamp (1968) include the Judith River Formation, Eagle Sandstone, Dakota Sandstone, Kootenai Formation, and Swift Formation. Undoubtedly formations in Montana have stratigraphic equivalents in Saskatchewan, but such correlation is beyond the scope of this report. Toward the east in northeastern Montana, the Eagle Sandstone and the Swift Formation become less permeable and are not aquifers. Where exposed, the sandstone aquifers are recharged by precipitation and elsewhere probably by leakage from the overlying glacial aquifers. Discharge from the aquifers is to springs, streams, wells, possibly to the atmosphere by evapotranspiration, and to other aquifers. Flowing wells are obtained in some places where the aquifers are confined. Wells completed in the Judith River aquifer yield as much as 200 gal/min; wells in the Eagle yield as much as 75 gal/min. A potentiometric map prepared by Osterkamp (1968) indicates discharge from the Judith River aquifer to Fort Peck Reservoir and the Missouri River.

Additional hydrologic information is needed for the Hogeland Basin and surrounding areas in order to conserve or develop the water resources. Aquifers that are continuous across the United States-Canadian border need to be identified because effects of water use may be transmitted across the border. Potentiometric maps and aquifer testing of extensive areas are needed to provide information on direction and rate of ground-water movement, as well as areas of recharge and discharge. Additional water-quality sampling and interpretation are needed. Currently (1977) the vast ground-water resources are barely used because of the small population and small demand.

The Crazy Mountains Basin and the Bull Mountains Basin contain numerous sandstone formations that are known to be aquifers in other areas of the Missouri Basin Region. Hydrologic properties of the basins are poorly known, and hydrologic investigations are needed in both basins to obtain the information required for planning any development of the largely unused ground-water supplies.

The Bighorn Basin lies mostly in Wyoming but extends slightly into Montana. Principal sandstone formations in Wyoming are described by Lowry and others (1975); principal sandstone formations in Montana are described by Hamilton and Paulson (1968). Some of the sandstone formations in Montana correlate with formations in Wyoming. The sandstone aquifers are recharged by precipitation, streamflow, and leakage from other aquifers. Discharge is to springs, streamflow, to the atmosphere by evapotranspiration, other aquifers, and wells. Flowing wells completed in the Tensleep and Amsden aquifers have yielded 2,000 gal/min. Uses of the ground water include municipal, industrial, irrigation, and fish culture.

Potentiometric maps and aquifer testing are needed for each extensive sandstone aquifer in the Bighorn Basin. Collection and interpretation of water-quality data are needed also. Improved understanding of the ground-water resources in the Bighorn Basin is needed in order to plan for conjunctive use of ground water with surface water in the adjacent reaches of Clarks Fork of the Yellowstone River, the Shoshone, and the Bighorn Rivers.

Ground-water resources of the Wind River Basin were summarized by Whitcomb and Lowry (1968). The aquifers probably are recharged by precipitation or by interaquifer leakage. Measured gains in streamflow indicate that the aquifers discharge to streams, as well as to springs and wells. Some of the sandstone aquifers are fractured, increasing aquifer transmissivity and well yield. Maximum well yield is about 1,100 gal/min (an industrial well completed in the Arikaree aquifer). The concentration of dissolved solids or of a particular chemical constituent is very high in several aquifers. Water-quality mapping is needed to locate saline water zones so that they may be avoided or developed as saline water resources.

Numerous extensive sandstone aquifers in the Wind River Basin contain large volumes of ground water. Additional data on the hydrologic properties of the aquifers are necessary before additional ground water can be developed properly. The Wind River Basin might serve as a storage reservoir common to the two major streams that drain the basin, the Wind River and the Sweetwater River. Conceivably, the ground-water reservoir could be used to reduce shortages in water supply in each of the surface drainage areas through water-management procedures for storage, transfer, and withdrawal. During periods of below-normal runoff, ground-water withdrawals could be increased. During periods of above-normal runoff, surplus streamflow could be artificially recharged to the ground-water reservoirs.

Ground-water resources of the Laramie, Shirley, and Hanna Basins and Saratoga Valley were appraised by Lowry and others (1973). The basins contain large number of formations that consist partly or completely of sandstone. All of the sandstone aquifers are not present in each basin or valley, but the presence of several aquifers in any basin may represent a large thickness of sandstone and a correspondingly large volume of stored ground water. Most of the sandstone aquifers are recharged by precipitation where exposed. Interaquifer leakage is likely but remains untested. Measured gains in streamflow indicate that some of the aquifers discharge to streams. Wells yield as much as 1,000 gal/min especially where fracturing has increased the permeability of the aquifers. Water quality is variable, although the water is generally suitable for most purposes. Potentiometric maps, aquifer testing, and gain-and-loss studies along streams are needed.

The Laramie, Shirley, and Hanna Basins and Saratoga Valley lie near the headwaters of the North Platte River in northern Colorado. Ground-water reservoirs in the basins could be used in total water management in the main valley downstream. For example, in some areas a water surplus might be artificially recharged to the basins for later withdrawal and use. However, careful planning would be required to insure that the recharged water would be available when needed and not discharged from the aquifer or transmitted beyond the area influenced by recovery wells. Ground water in storage is subject to virtually no evaporative losses compared to lake evaporation of 36 to 42 in./yr for surface reservoirs in this area (see fig. 3). However, full use of the ground water would include conjunctive use with streamflow and surface-water reservoirs. Because of the large reserves, mining of ground water in storage may be feasible.

The Powder River Basin contains sandstone aquifers that extend over a large area in northeastern Wyoming and southeastern Montana (Hodson and others, 1973). The Wasatch Formation is an aquifer that yields as much as 500 gal/min to wells in Wyoming, but it is mostly absent in Montana. The dissolved-solids concentration commonly ranges from 500 to 1,500 mg/L. Aquifers in the Fort Union Formation are sandstone and coal beds. Wells in Wyoming in the Fort Union yield as much as 150 gal/min of water with a normal dissolved-solids concentration of 500 to 1,500 mg/L. The Fox Hills Sandstone and overlying sandstone beds of the Lance or Hell Creek Formations form an areally extensive and persistent aquifer. The aquifer is probably recharged where exposed at the edge of the basin and by leakage from other aquifers. In southeastern Montana ground water in the Fox Hills-basal Hell Creek aquifer moves toward the northwest, away from the Black Hills Uplift. Discharge is to the Yellowstone River and the Powder River. The ground-water movement is very slow; the transit time from the outcrop of the Fox Hills-basal Hell Creek aquifer in the Black Hills to the Yellowstone River is about 400,000 yr (Taylor, 1968). In contrast, the transit time for surface water over approximately the same distance is only several days. In northern Wyoming, water in the Fox Hills moves north into Montana. Industrial, municipal, and domestic wells yield as much as 200 gal/min. Dissolved-solids concentration is normally less than 2,000 mg/L. More aquifer testing and potentiometric maps are needed for the Fox Hills and overlying sandstone beds in order to predict the hydrologic response of the aquifer to future development.

Additional sandstone aquifers lie below the Fox Hills in the Powder River Basin. The Mesaverde and Frontier aquifers yield as much as 50 gal/min to wells; the Inyan Kara aquifer may yield 200 gal/min. Dissolved-solids concentration in the Mesaverde, Frontier, and Inyan Kara aquifers is normally less than 3,000 mg/L. Wells completed in the Sundance aquifer yield a maximum of 50 gal/min. Dissolved-solids concentration ranges from about 500 to 2,000 mg/L in the Sundance. The Tensleep Sandstone is partly equivalent to the Minnelusa Formation in the Black Hills. Wells completed in the Tensleep aquifer yield from 20 to 1,200 gal/min. Higher yields are obtained where the Tensleep is fractured, because of the resulting increase in transmissivity. Dissolved solids in water from the Tensleep aquifer commonly range from 200 to 500 mg/L and consist mostly of calcium and bicarbonate. A hydrologic model of the Powder River Basin would help to indicate ways of developing the ground-water resources. Potentiometric maps and aquifer testing are lacking for all aquifers. Data describing the vertical hydraulic conductivity of intervening layers between aquifers would help to predict the rate of interaquifer leakage.

The Williston Basin extends over western North Dakota, eastern Montana, northwestern South Dakota,and southern Saskatchewan. Sandstone aquifers occur in the Fort Union Formation, Fox Hills Sandstone, Hell Creek Formation, and Dakota Sandstone. The Fort Union Formation consists of shale, sandstone, siltstone, lignite, and clinker beds. Sandstone aquifers in the Fort Union are recharged principally by precipitation. Ground water in southwestern North Dakota generally moves northeast, in the direction of surface drainage. Discharge is to streams and to domestic and stock wells that yield as much as 200 gal/min, but generally much less. Dissolved-solids concentration normally is less than 2,000 mg/L but may be as high as 7,000 mg/L. Dependable sandstone aquifers in the Fort Union are difficult to locate because of their discontinuous and irregular nature. A statistical compilation and analysis of well logs, aquifer properties, well characteristics, and water quality is needed to improve the level of understanding of ground-water occurrence in the Fort Union.

The Fox Hills-basal Hell Creek aquifer is an extensive aquifer in the Williston Basin. Recharge occurs through outcrops of the aquifer near the Black Hills Uplift and seepage from overlying aquifers in part of the area. Ground-water movement is to the north, northeast, and east. Natural discharge is by upward leakage through confining beds or by evapotranspiration where the aquifer crops out, as in central North Dakota. Apparently the water moving to the north and northeast passes from the Missouri Basin Region into the Souris-Red-Rainy Region where it is discharged in north-central North Dakota and possibly in Saskatchewan and Manitoba. In south-central North Dakota east of the Missouri River the potentiometric surface slopes westward, indicating that ground water moves from east to west. Wells in the aquifer yield as much as 700 gal/min. Uncontrolled or partly controlled flow from wells in the aquifer has caused water-level declines of about 30 ft in Mercer County, N. Dak. Dissolved-solids concentration generally ranges from 1,000 to 2,000 mg/L. An areally extensive mathematical model is needed for evaluation of the Fox Hills-basal Hell Creek aquifer. The model could provide information on suspected leakage from aquifers in the overlying Fort Union Formation and aid in the design of plans for optimal use of the supply.

The Ravenscrag, Frenchman, Whitemud, and Eastend Formations of southern Saskatchewan in the Williston Basin probably are stratigraphically equivalent to the Fort Union Formation, Fox Hills Sandstone, and Hell Creek Formation. The Ravenscrag Formation consists of 600 ft of sand, silt, clay, and lignite. Sandstone and coal aquifers of the Ravenscrag Formation yield as much as 500 gal/min to wells. Extensive dewatering of a Ravenscrag aquifer for coal mining in southern Saskatchewan may affect water levels in aquifers in northern Montana. Dissolved-solids concentration ranges from 260 to 3,770 mg/L. The Frenchman, Whitemud, and Eastend aquifers together consist of 280 ft of sand, silt, and clay. Dissolved-solids concentration ranges from 560 to 5,350 mg/L. Any areally extensive hydrologic analysis of the Williston Basin should include equivalent aquifers in Saskatchewan.

The Dakota Sandstone is a principal aquifer in the Williston Basin. At the deepest part of the basin near Williston, N. Dak., the top of the Dakota aquifer is more than 5,600 ft below the land surface. In western North Dakota, shale predominates over sandstone in the aquifer; in eastern North Dakota the sandstone predominates over shale. Natural recharge, movement, and natural discharge for the Dakota are not well understood in the Williston Basin. Large artesian pressures in the Dakota led to the drilling of numerous flowing wells, many of which were allowed to flow continuously and wasted ground water from the Dakota aquifer. Water levels declined an estimated 200 feet from 1902 to 1923 in LaMoure and Dickey Counties of North Dakota (North Dakota Geological Survey, 1973). The Dakota aquifer may be hydraulically connected to other aquifers whose supply is also being depleted by uncontrolled discharges from wells completed in the Dakota. Dissolved-solids concentration is as much as 11,000 mg/L in western North Dakota but generally ranges from 2,000 to 3,000 mg/L in areas of greatest use. Principal ions include sodium, chloride, and sulfate.

A detailed potentiometric map and additional aquifer tests are needed for the Dakota aquifer. A mathematical model of the aquifer would help to define areas of natural recharge and provide information necessary to extend the life of the water supply. Delineations of saline zones in the aquifer by water-quality mapping would be useful in avoiding withdrawals of saline water or in developing saline-water resources.

The Denver Basin lies in northeastern Colorado and extends over parts of the Missouri River Basin and the Arkansas River Basin. However, most of the Denver Basin is included in the Missouri Basin Region. Potentiometric maps of the sandstone aquifers in the Denver Basin indicate that the potentiometric surfaces are similar and slope to the north in the direction of surface drainage, indicating ground-water movement toward the north. However, the general pattern of natural recharge to and discharge from the aquifers is unknown. Long-term measurements indicate a decline in the water levels in wells completed in the aquifers of the Denver Basin. The greatest declines are in areas where withdrawals from wells are concentrated. Wells yield a maximum of about 300 gal/min. Withdrawals are for domestic, municipal, and industrial uses. Discharge from naturally flowing wells has been reduced, and in many areas the flow has ceased and wells must be pumped to obtain a water supply. Dissolved-solids concentration ranges from about 170 to 3,700 mg/L.

Ground-water supplies in the Denver Basin lie in the area east of the Front Range in Colorado where population growth is rapid, water demands are large, and much of the surface-water supply is imported. The lack of hydrologic data and interpretive hydrologic studies of the aquifers in the Denver Basin limits the design of water-management plans for developing the ground-water supply. Potentiometric maps and aquifer testing are needed for the sandstone aquifers, as well as information on the recharge to, movement through, and discharge from each aquifer. Data describing the hydraulic properties of confining layers between aquifers are also needed to analyze possible flow from one aquifer to another. A hydrologic model of the aquifer system could simulate the response of the system to expected withdrawals and provide information that could lead to better utilization of the existing ground-water resources.

Development of a water supply from sandstone aquifers in structural basins may induce ground-water movement that helps to sustain the yield of wells, as portrayed in figure 9. Aquifer A is unconfined; aquifers B and C are confined. However, the confined aquifers have unconfined zones wherever the aquifers are exposed, often ring-shaped zones near the periphery of the basin. The confined and unconfined zones have different storage characteristics that affect withdrawals by wells. For example, in the confined zone of a typical sandstone aquifer, a relatively small volume of water may be derived from storage by lowering the water level a few feet. In the unconfined zone of the same aquifer, a relatively large volume of water may be derived from storage by lowering the water level a few feet. Induced ground-water movement from the unconfined region to the confined region may help to sustain well yields as water supplies are developed in the structural basins.

Well 1 in figure 9A is completed in the confined zone of aquifer B but near the unconfined zone of the aquifer. Withdrawal from well 1 will initially yield water by expansion of water in the aquifer and by compression of the aquifer skeleton. In time drawdown will induce ground-water movement from the unconfined zone to the confined zone of aquifer B, as shown in figure 9B. The relatively large volume of water released from the unconfined zone helps to maintain the water levels and yield of well 1 in the confined zone. The decline of the water table in the unconfined zone of aquifer B causes the unconfined zone to migrate toward the well and helps to sustain the yield of well 1. Well 2 is completed in aquifer B at a greater distance from the unconfined zone than well 1. Withdrawals from well 2 and well 1 with time may induce leakage from the confining shale beds, and with greater time

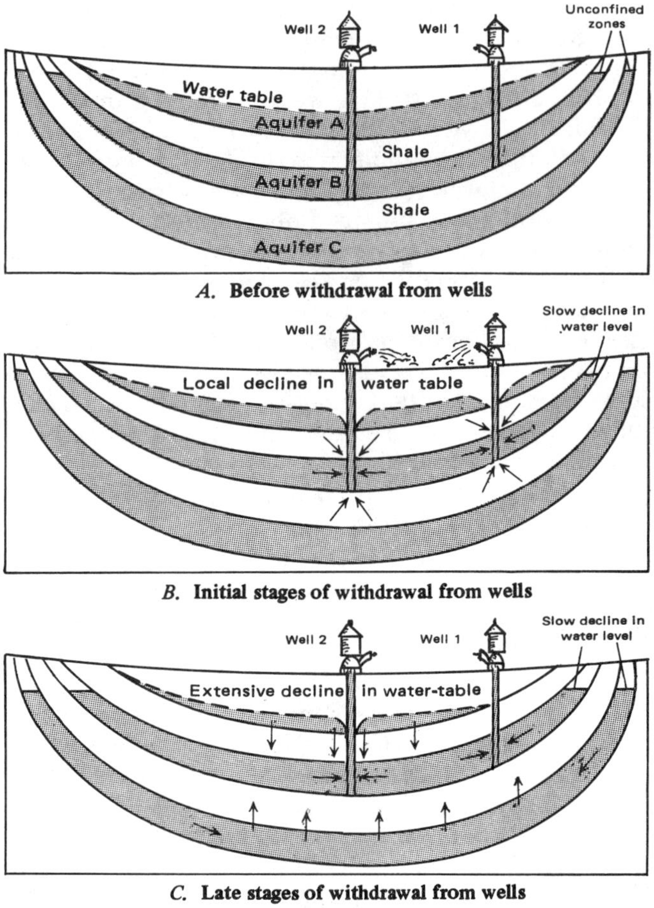

FIGURE 9.—Possible effects of withdrawal from aquifer in structural basin.

water may be induced through shale beds from aquifer A above and aquifer C below, into aquifer B. Leakage from and through shale beds may be substantial over the large areas covered by the structural basins. Eventually the leakage into aquifer B may replenish some of the ground water being lost from storage by pumping and will help to sustain the water level in the unconfined zone of aquifer B as shown in figure 9C. However, a continued withdrawal from wells will exhaust the supply to the wells.

Increased withdrawal of water from sandstone aquifers in the basins is likely because of expected increase in demands for energy development, agriculture, and municipal use throughout the Missouri Basin Region. The hypothetical example portrayed in figure 9 illustrates how the properties and hydrologic response of a basin may help to sustain the yield of wells.

DAKOTA, INYAN KARA, SUNDANCE, MINNELUSA, AND DEADWOOD AQUIFERS

Important sandstone aquifers in the Missouri Basin Region include the Dakota Sandstone, Inyan Kara Group, Sundance Formation, Minnelusa Formation, and Deadwood Formation. The Dakota Sandstone crops out in eastern Nebraska, western Iowa, central Kansas, possibly southwestern Minnesota, and near uplifted areas. Maximum aquifer thickness, where the aquifer crops out, is about 400 ft. At depth, west of the outcrop area, the maximum thickness is about 700 ft. Recharge is from precipitation and streams on the exposed sandstone and by leakage from shallow aquifers to the underlying Dakota. Direction of ground-water movement is not well known, because in most areas insufficient data are available to construct potentiometric maps. However, a potentiometric map prepared from available data indicates that in southwestern Kansas the water in the Dakota aquifer passes from the Arkansas-White-Red Region into the Missouri Basin Region. Discharge is to streams, springs, or into valley-fill aquifers in local areas. Wells completed in the exposed Dakota Sandstone yield as much as 1,500 gal/min. Dissolved-solids concentration is variable and ranges from about 300 to 30,000 mg/L. The concentration is low at shallow depths where the aquifer is exposed but high at depth and toward the west. Large concentrations of chloride and sulfate are reported in some areas. In parts of Kansas the water quality of some streams has been impaired by base flow from the Dakota. Additional potentiometric maps and water-quality maps are needed for the Dakota aquifer. Mapping of saline zones in the Dakota aquifer will delineate areas in which wells should not be drilled. Streamflow sampling under low-flow conditions will help to locate reaches where saline base flow is degrading the quality of streams that cross the Dakota aquifer. Well fields carefully designed to withdraw the saline ground water may reduce the saline base flow and improve water quality in the stream. Saline ground water pumped from the Dakota could be evaporated or flushed downstream during periods of above-normal runoff. Conjunctive use of ground water from the Dakota and other aquifers could lead to increased efficiency in use of water supplies. For example, near Lincoln, Nebr., ground water is pumped from wells in alluvium, treated, and recharged through injection wells into the Dakota aquifer during the winter. The ground water is withdrawn from the Dakota in the summer demand period.

Principal sandstone units that underlie the Dakota Sandstone primarily in central South Dakota and south-central North Dakota include the Inyan Kara Group and Sundance, Minnelusa, and Deadwood Formations. Inyan Kara is not recognized in North Dakota, but the Dakota Sandstone of North Dakota is partly equivalent to the Inyan Kara Group of South Dakota. The Sundance Formation is not recognized in North Dakota (Sandberg, 1962), although an equivalent unit may be present.

The geology and hydrology of the Dakota aquifer in South Dakota have been described by Schoon (1971). The potentiometric surface of the Dakota slopes from the Black Hills Uplift toward southeastern and northeastern South Dakota. Recharge is from precipitation and streamflow on outcrops near the Black Hills and from leakage from other aquifers. Ground water moves toward the east and partly into the Souris-Red-Rainy Region of northern and eastern North Dakota and Upper Mississippi Region of Minnesota. Natural discharge from the Dakota is to overlying glacial deposits or possibly other aquifers. Numerous flowing wells have been drilled in the Dakota since the late 1800's. Wells yield as much as 1,500 gal/min but average about 15 gal/min. An estimated 25 percent of the population of South Dakota obtains water supplies from the Dakota. Uncontrolled flowing wells have lowered the potentiometric surface, especially along the Missouri River Valley where wells are relatively shallow and artesian pressures are high. Dissolved-solids concentration ranges from 1,700 to 8,000 mg/L.

Casings of uncontrolled flowing wells should be sealed if necessary and the flow of the wells regulated with valves to preserve the ground-water supply. For example, consider the possible effect of withdrawals on the water levels of a typical artesian aquifer. Assume that an uncontrolled flowing well discharges only 1 gal/min for a period of 1 yr and that discharge depletes ground water in storage over an area of 1 mi^2. The average decline in water levels in wells over the area may be about 13 ft at the end of 1 yr, and the rate of decline will continue in successive years if the discharge is allowed to continue. The decline represents a decrease in ground water in storage and eventually will cause a reduction in the yield of flowing wells. The flows will eventually cease and continual lowering of water levels will require continually increasing energy to lift water to the land surface. The well owner that permits his well to flow uncontrolled is depleting not only his own water supply but the supplies of others nearby.

The Inyan Kara aquifer and Sundance aquifer lie directly below the Dakota aquifer in South Dakota. Flowing wells are obtained in both aquifers, but few wells have been drilled. Water quality is variable, and dissolved-solids concentrations range from less than 1,000 mg/L to more than 10,000 mg/L. Combined use of water from the Dakota, Inyan Kara, and Sundance aquifers should be investigated because wells completed in all three aquifers may have large yields.

The Minnelusa and Deadwood aquifers are mostly undeveloped. A well completed in the Minnelusa aquifer yielded 4,000 gal/min when drilled. Water in the Deadwood aquifer may be saline, but the aquifer is mostly untested. Additional drilling to the Minnelusa and Deadwood aquifers is needed to determine well yields and to prepare potentiometric maps. Aquifer testing and water-quality sampling should be conducted to appraise the potential water supply.

PRAIRIE DU CHIEN AND JORDAN AQUIFERS

The Jordan Sandstone is the most productive sandstone and dolomite aquifer in the part of Iowa within the Missouri Basin Region. The Jordan aquifer is hydraulically connected to an overlying dolomite, the Prairie du Chien Formation, and the two units form a single aquifer with a combined thickness that ranges from 300 to 600 ft. Recharge, movement, and discharge for the combined aquifers are poorly understood. Municipal wells yield as much as 1,000 gal/min. Dissolved-solids concentration ranges from about 1,500 to 2,000 mg/L. Care should be taken when constructing wells in southwestern Iowa to insure that saline water from the overlying limestone does not leak into the Jordan or Prairie du Chien aquifers. A potentiometric map is needed for the combined aquifers, as well as additional aquifer testing, to improve understanding of recharge, movement, and discharge and help design plans to sustain the water supply.

Water supplies in sandstone aquifers of the Missouri Basin Region are summarized in table 4. The sandstone aquifers are numerous and of diverse types, but water supplies can be improved greatly if the hydrologic characteristics of each aquifer are known and incorporated into management plans.

LIMESTONE AND DOLOMITE AQUIFERS

The few limestone and dolomite aquifers that occur in the Missouri Basin Region are areally extensive, as shown on plate 3, and capable of large yields where cavernous. Ground water in limestone or dolomite aquifers occurs in openings that range in size from

TABLE 4.—*Summary of water supplies in sandstone aquifers, Missouri Basin Region*

Source of supply	Location	Yield to wells (gal/min)	Suggestions for improved hydrologic data, analyses, and water management	Outlook for future use if suggestions are followed
Virgelle aquifer. Milk River aquifer.	Montana Alberta	5–250	Control of flowing wells and implementation of other conservation practices. Use of other supplies for industry. Artificial-recharge experiments to determine if fractures and vertical hydraulic conductivity of aquifer will facilitate rapid recharge.	Reduction of ground-water mining and pumping lift in wells.
Blood Reserve, Judith River, and Milk River aquifers.	Alberta	5–120(?)	Preparation of potentiometric maps and aquifer testing for each aquifer.	Development of ground-water supply designed to take advantage of possible interaquifer leakage.
Judith Basin	Montana	5–300	Additional potentiometric maps, aquifer testing, and water-quality sampling for sandstone aquifers of Judith Basin and adjacent areas to northwest and southwest.	Conjunctive use of ground and surface water in upper reach of the Missouri Basin Region.
Hogeland Basin	Saskatchewan	10–200	Identification of aquifers that are continuous across the United States-Canadian border. Extensive aquifer testing, preparation of potentiometric maps, and water-quality appraisal.	Development of water supplies in vast area in which water resources are barely used.
Crazy Mountains and Bull Mountains Basins.	Montana	?	Complete hydrologic investigation.	Development of unused ground-water supplies.
Bighorn Basin	Wyoming, Montana.	5–2,000	Aquifer testing and preparation of potentiometric maps for each aquifer. Water-quality data and interpretation.	Conjunctive use with surface water in adjacent reaches of Clarks Fork, Shoshone River, and Bighorn River.
Wind River Basin.	Wyoming	5–1,100	Preparation of potentiometric maps. Aquifer testing to prepare maps of transmissivity and storage. Preparation of water-quality maps.	Conjunctive use of ground water and surface water from Wind River and Sweetwater River. Improved quality of water supplies and use of saline water.
Laramie, Shirley, and Hanna Basins and Saratoga Valley.	Wyoming	5–1,000	Conjunctive use of ground water, streamflow, and surface-water reservoirs. Artificial recharge to, and mining from, the large ground-water supplies.	Improved water management North Platte River valley. Reduction of evaporative losses.
Powder River Basin.	Montana, Wyoming.	5–1,200	Preparation of potentiometric maps and aquifer testing of all aquifers. Determination of vertical hydraulic conductivity of intervening layers between aquifers. Construction of mathematical model of entire basin.	Development of water-management plan for optimal development of ground-water resources.
Williston Basin.	Montana, North Dakota, South Dakota.	5–700	Statistical compilation of well logs, aquifer properties, well characteristics, and water quality for aquifers in Fort Union Formation. Preparation of areally extensive model for Fox Hills-basal Hell Creek aquifer. Control of flowing wells that discharge from Fox Hills-basal Hell Creek and Dakota aquifers. Potentiometric maps, aquifer testing, mathematical modeling, and water-quality mapping for Dakota aquifer.	Increased success in drilling dependable wells in Fort Union. Improved water supply in Fox Hills-basal Hell Creek aquifer. Extended life and improved quality of water supplies for Dakota aquifer. Development of saline-water resources in Dakota aquifer.
Denver Basin	Colorado	10–300	Preparation of potentiometric maps and aquifer testing of all aquifers. Determination of vertical hydraulic conductivity of intervening layers between aquifers. Construction of mathematical model of entire basin.	Better utilization of existing ground-water resources in area of large demands and frequent shortages.
Dakota aquifer.	Nebraska, Iowa, Kansas, Minnesota(?).	5–1,500	Preparation of potentiometric maps and water-quality maps. Streamflow sampling under low-flow conditions to locate saline base flow. Conjunctive use with water supplies in other aquifers.	Improved quality of ground- and surface-water supplies. Increased water supplies.
Dakota, Inyan Kara, Sundance, Minnelusa, and Deadwood aquifers.	North Dakota, South Dakota.	5–4,000	Preparation of up-to-date potentiometric map and additional aquifer testing of Dakota aquifer. Preparation of mathematical model of Dakota aquifer. Preparation of water-quality maps. Control of flowing wells that discharge from Dakota aquifer. Combined use of water from Dakota, Inyan Kara, and Sundance aquifers. Additional drilling, aquifer testing, and water-quality sampling of Minnelusa and Deadwood aquifers.	Improved understanding of natural recharge to Dakota aquifer. Improved quality of water supplies and development of saline water. Extended life of Dakota aquifer. Development of large yields from combined use of water from several aquifers. Development of additional ground-water supplies.
Prairie du Chien and Jordan aquifers.	Iowa	100–1,000	Preparation of potentiometric map and additional aquifer testing. Identification of recharge and discharge areas.	Sustained municipal supply.

small pores, joints, and other fractures to large caverns formed by the solution and removal of rock by ground water. Fractures also may increase the storage and transmissive properties of limestone and dolomite aquifers.

MADISON AQUIFER

The Madison Group (also Madison Limestone) is an extensive limestone aquifer that occurs in parts of Alberta, Saskatchewan, Montana, Wyoming, North Dakota, South Dakota, Nebraska, and Colorado. The

Madison is partly equivalent to the Pahasapa Limestone. A regional description of the Madison aquifer was prepared by Swenson and others (1978). Thickness of the Madison Group increases toward the center of the Williston Basin in northwestern North Dakota, where it is about 2,300 ft thick. The Madison is exposed near major structural uplifts in the Missouri Basin Region, but at the deepest location, about 40 mi northeast of Casper, Wyo., it is about 16,000 ft below land surface. The areal extent of the Madison aquifer, as shown on plate 3, is only approximate because the geologic structure in Montana and Wyoming is complex and few drillers' logs are available.

Streamflow and precipitation recharge the Madison aquifer where it is exposed; leakage from adjacent aquifers also may recharge the aquifer. Ground-water movement is generally to the north and northeast in northern Wyoming and Montana. Movement is generally toward the east in North Dakota and South Dakota. Discharge is to streams, springs, other aquifers, or wells. Numerous springs discharge from the Madison in Montana, and Giant Spring near Great Falls, Mont., flows about 300 ft^3/s. Wells yield from several hundred to several thousand gallons per minute. Industrial wells that withdraw water for use in oil fields have caused declines of water levels of as much as several hundred feet at places in Wyoming. Water temperature in the Madison aquifer increases with depth below land surface and ranges from about 50°F to 250°F. Large concentrations of dissolved solids are often associated with high water temperatures because of increased mineral solubilities. Dissolved-solids concentration in the Madison aquifer ranges from about 300 mg/L near Lewistown, Mont., to 350,000 mg/L in the Williston Basin.

A preliminary model of the Madison aquifer in the Powder River Basin and adjacent areas was prepared by Konikow (1976). The model simulated the effect of proposed withdrawals of water from wells at the rate of 20 ft^3/s for 100 yr, for use in coal transportation. An estimated 200 ft^3/s flows through the Madison, mostly from recharge along outcrops in the Bighorn Mountains and the Black Hills; the presence of springs in the Black Hills suggests that part of the recharge is rejected because the water table is shallow. Sources of withdrawn water in the model included salvaged rejected recharge, ground water in storage, and water otherwise lost by natural discharge. The model was used to predict declines in water levels in the Madison after 100 years of hypothetical pumping. Maximum predicted declines are 300 ft at a distance of approximately 5 mi from the pumped wells, presuming leakage from other aquifers into the Madison, and 600 ft, presuming no leakage from other aquifers.

More hydrologic data are needed if the Madison aquifer is to be developed properly. Additional observation wells are required for a more accurate potentiometric map. Aquifer testing is needed to prepare maps showing the areal distribution of transmissivity and storage. Inventory and discharge measurements of springs are necessary to provide information on natural discharge. Identification of aquifers that are hydraulically connected to the Madison and measurement of water levels in observation wells in the aquifers would provide information regarding leakage to and from the Madison aquifer. A regional aquifer system analysis for the Madison aquifer was begun by the U.S. Geological Survey in 1975. The analysis should supply much of the needed data. Development of water in the Madison aquifer will be more efficient if development plans are based on the known behavior of the aquifer, as determined from field measurements and hydrologic models. The high water temperatures at depth in the Madison aquifer may indicate a potential source of thermal energy. Figure 10 shows how recharge to the aquifer may be increased. Natural recharge is normally low where the water table is high, reducing space available for recharge and permitting evapotranspiration from the water table (fig. 10A).The use of wells to lower the water table increases space for recharge and captures water formerly occurring as surface runoff or evapotranspiration (fig. 10B). Normally only part of the rejected recharge can be converted to recharge.

AQUIFERS IN ROCKS OF PERMIAN AND PENNSYLVANIAN AGE

Limestone and sandstone aquifers occur in rocks of Permian and Pennsylvanian age in Nebraska and Kansas. Similar aquifers occur in rocks of Pennsylvanian age in Iowa and Missouri. Total thickness of the limestone and sandstone ranges widely; maximum thickness is several thousand feet. Ground water occurs in pores, caverns, and fractures of the limestone and in interbedded sandstone. Aquifers are recharged by precipitation and are discharged to streams, springs, and wells. Direction of ground-water movement is poorly understood because few potentiometric maps have been prepared. Reported well yields range from 3 to 500 gal/min. Water quality is variable. In Iowa and Missouri the dissolved-solids concentration generally increases with depth below land surface. In Kansas the dissolved-solids concentration in the sandstone aquifers increases toward the west. Saline water discharges from mineralized shale, included with aquifers of Permian age, to streams in central Kansas and impairs the quality of surface water. In a

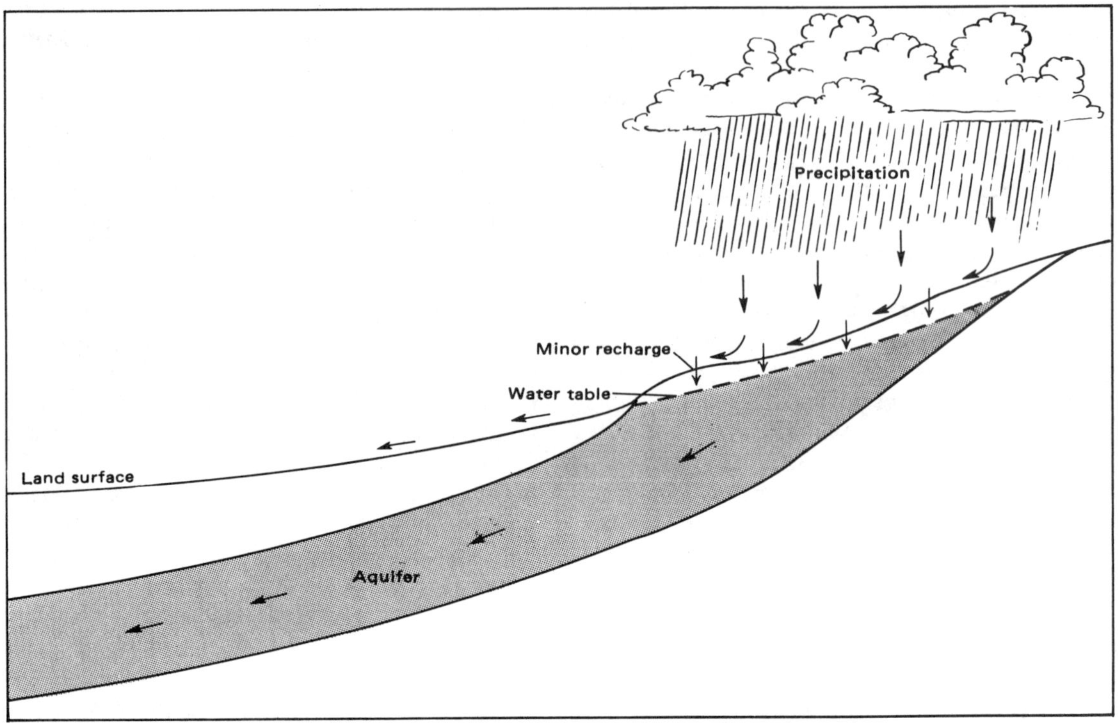

A. **Minor recharge due to high water table**

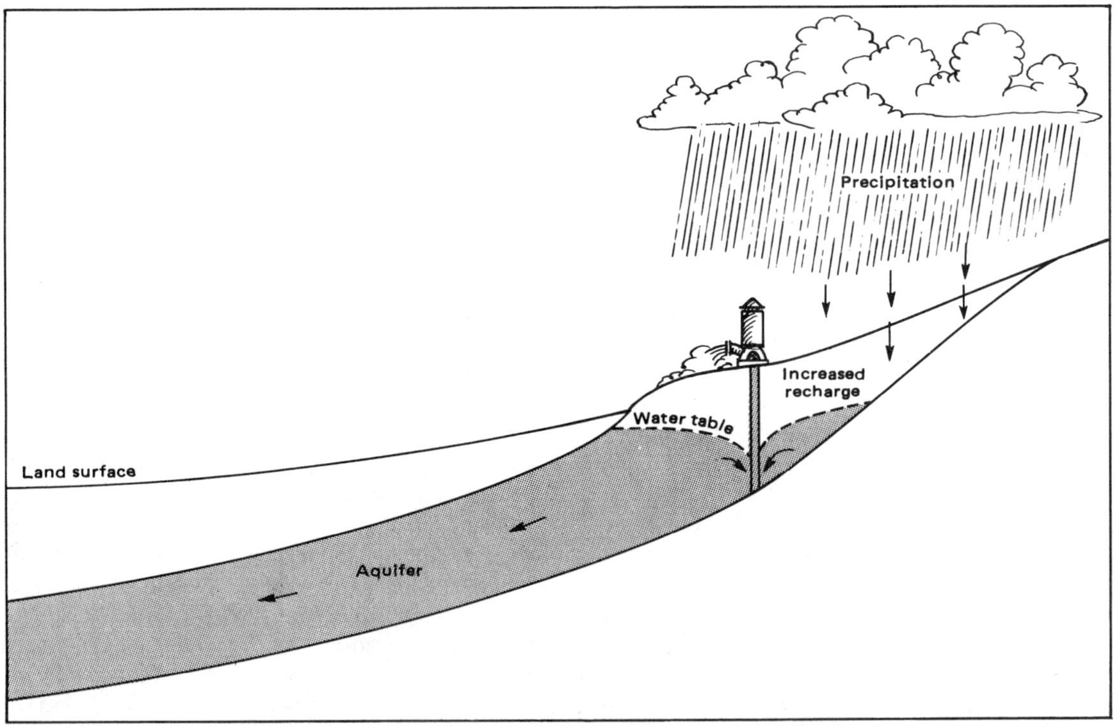

B. **Increase in recharge induced by lowering water table**

FIGURE 10—Induced increase in recharge.

few places in Kansas, high concentrations of sulfate, chloride, and nitrate have been detected in water from the aquifers in Permian and Pennsylvanian rocks.

Zones in the limestone aquifers that transmit and store water readily are difficult to locate because of their random occurrence. Therefore, a statistical compilation of data describing well logs, aquifer performance, and water quality is needed to search for trends and improve our understanding of the location of the most productive zones in the limestone. A water-quality investigation would help to determine the cause of saline zones in the aquifers. Water-management plans based on improved understanding could be designed to either (1) eliminate the saline ground water by withdrawing it through wells and using it or disposing of it or (2) avoid the withdrawal of saline water through carefully designed well fields.

BIGHORN, RED RIVER, AND WHITEWOOD AQUIFER

An areally extensive dolomite aquifer lies beneath the Madison aquifer in eastern Montana, northern Wyoming, and the Williston Basin. The aquifer lies at great depths except near uplifts; therefore, it is not shown on plate 3. The dolomite aquifer is known as the Bighorn Dolomite in northern Wyoming and southern Montana, the Red River Formation in the Williston Basin, and the Whitewood Dolomite near the Black Hills. However, the Bighorn, Red River, and Whitewood are only approximately equivalent. Maximum thickness of the aquifer is about 550 ft. Water temperatures as much as 250°F have been measured, and high artesian pressures are reported in the aquifer in South Dakota. Locally, the aquifer may yield several hundred gallons per minute to wells. The aquifer might furnish water to the overlying Madison aquifer through leakage if large amounts of water were withdrawn from the Madison. An observation-well network would provide data on suspected leakage. A potentiometric map is needed to define the recharge, movement, and discharge for the aquifer.

ROUBIDOUX, GASCONADE, EMINENCE, AND POTOSI AQUIFERS

The Roubidoux Formation, Gasconade Dolomite, Eminence Dolomite, and Potosi Dolomite in south-central Missouri and southeastern Kansas are utilized for water supplies where shown in plate 3. The formations are primarily dolomite and represent the principal aquifers of the Ozarks area (Missouri Geological Survey and Water Resources, 1963). The formations constitute a series of adjacent dolomite aquifers that are recharged by precipitation and streamflow. Normal monthly precipitation in Missouri ranges from 1.6 in. during February to 4.4 in. during June (Environmental Science Services Administration, 1968), assuring adequate recharge to the dolomite aquifers on a long-term basis. Ground-water movement is toward streams where the aquifers discharge naturally. Ground water is transmitted as rapidly as 1 mi/d through an extensive system of caverns. Relatively large and rapid ground-water movement is also indicated by the uniform cool temperatures of 56° to 65°F of ground water over a wide range in depth below land surface and the low dissolved-solids concentration (Feder, 1973). Most of the many large springs in Missouri issue from the Gasconade, Eminence, and Potosi aquifers (Vineyard and Feder, 1974). Bennett Spring in Missouri within the Missouri Basin Region discharges about 150 ft^3/s from the dolomite aquifers. Maximum yield of wells completed in the Roubidoux, Gasconade, Eminence, and Potosi aquifers is about 2,000 gal/min. The dissolved-solids concentration in the aquifers is normally less than 350 mg/L. Potentiometric maps and aquifer testing are needed to provide improved definition of the vertical and horizontal flow system for the aquifers. Tracer studies would be valuable in defining ground-water movement, transit time, and water mixing. An improved understanding of the flow system also would help to avoid pollution of the ground-water supply.

Limestone and dolomite aquifers are very productive as indicated by large spring flows and well yields. Transmission of ground water in the caverns is more like streamflow than like flow in a fine-grained aquifer. The presence of cavern systems gives the water developer the option of obtaining a large supply in some areas from sources other than the streamflow network. The advantages of the ground-water supply over a surface-water supply are greater protection from pollution and greater dependability because ground water in storage fluctuates less than streamflow. However, withdrawals from wells may deplete spring discharge and streamflow derived from springs. Water supplies in limestone and dolomite aquifers are summarized in table 5.

SALINE GROUND WATER

Many possible uses of saline ground water are summarized by Kohout (1970). Desalinization may be used to obtain usable water of any quality, selected according to costs and needs. Minerals recovered as byproducts of the desalinization process may have economic value. Saline ground water may be used sometimes without treatment for cooling purposes in industry or for waterflooding in oil fields to increase

TABLE 5.—*Summary of water supplies in limestone and dolomite aquifers, Missouri Basin Region*

Source of supply	Location	Yield to wells (gal/min)	Suggestions for improved hydrologic data, analyses, and water management	Outlook for future use if suggestions are followed
Madison aquifer.	Alberta, Saskatchewan, Montana, Wyoming, North Dakota, South Dakota, Nebraska, Colorado.	20–9,000	Preparation of improved potentiometric map and additional aquifer testing to define areal distribution of transmissivity and storage. Inventory of springs and measurement of spring discharges. Identification of aquifers hydraulically connected to Madison aquifer and measurement of observation wells completed in the aquifers. Preparation of maps showing areal distribution of dissolved-solids concentration. Continued use of simulation models to determine optimal development plans.	Development of large supply derived from increased recharge, reduced discharge, storage, and leakage to the Madison from other aquifers. Development of saline water resources.
Limestone and sandstone aquifers in rocks of Permian and Pennsylvanian age.	Iowa, Nebraska, Kansas, Missouri.	3–500	Statistical compilation of well logs, aquifer performance, and water-quality data. Investigations to determine the cause of deterioration of water quality at depth and laterally. Design of water-management plans to eliminate or avoid saline ground water.	Improved success in drilling wells and maintaining well performance. Improvement of water quality.
Bighorn, Red River, and Whitewood aquifer.	Montana, Wyoming, North Dakota, South Dakota.	20–300	Establishment of observation-well network. Preparation of potentiometric map.	Development of local supplies where aquifer is shallow. Development of large supplies maintained by induced leakage from adjacent aquifers.
Roubidoux, Gasconade, Eminence, and Potosi aquifers.	Missouri, Kansas.	20–2,000	Preparation of potentiometric maps and aquifer testing. Tracer studies to define cavern systems in all aquifers.	Sustained municipal, industrial, and domestic supply. Reduced hazard of pollution of supply.

oil production. Hot saline water in deep aquifers may also be utilized as a geothermal resource from which energy may be extracted. Saline ground-water reservoirs may be used for storage of several types of materials. Toxic wastes may be stored in reservoirs if no other use of the ground water is planned. Natural gas may be stored in saline ground-water reservoirs for later withdrawal. Finally, freshwater may be stored in saline ground-water reservoirs for later use, probably with very little deterioration in quality (Moulder, 1970).

Saline ground water in this report is defined arbitrarily as ground water in which the maximum dissolved-solids concentration exceeds 4,000 mg/L. Saline ground-water resources of the Missouri Basin Region are listed in table 6. Data in table 6 are based on a relatively small amount of reported information.

TABLE 6.—*Saline ground-water resources of the Missouri Basin Region*

State or Province	Aquifer (listed in order of increasing depth below land surface in each area)	Maximim dissolved - solids concentration (mg/L)
Saskatchewan	Glacial deposits	5,600
	Frenchman, Whitemud, Eastend	5,400
Montana	Glacial deposits	30,000
	Virgelle	5,100
Wyoming	Wind River	6,500
	Mesaverde	16,300
	Frontier	20,000
	Cloverly	24,000
	Sundance	4,700
	Nugget	4,900
	Casper	9,600
	Tensleep	280,000
North Dakota	Glacial deposits	5,000
	Fort Union	7,000
	Dakota	11,000
	Madison	350,000
South Dakota	Glacial deposits	10,000
	Dakota	8,000
	Inyan Kara	10,000
	Sundance	7,600
	Minnelusa	4,300
	Madison	120,000
	Red River	130,000
	Deadwood	40,000
Nebraska	Dakota	30,000
Iowa	do	5,000
Kansas	do	30,000
Missouri	Limestone of Pennsylvanian age	20,000

Additional drilling and water-quality sampling in the region may indicate other saline ground-water resources. The most saline ground water in the Missouri Basin Region occurs in Wyoming, North Dakota, and South Dakota in aquifers that lie deep below the land surface. However, data in table 6 indicate that dissolved-solids concentration does not necessarily increase with the depth of aquifer below land surface and ground-water temperature. Salinity of ground water is controlled by a combination of mineral solubility, ground-water temperature, and rate of ground-water movement. Water-management plans developed for the Missouri Basin Region should include saline ground water as a valuable resource. Possible uses of the resource may expand in the future and are not limited to the desalinization, cooling, waterflooding, geothermal, and storage uses summarized in this report.

BASIS FOR WATER MANAGEMENT IN MISSOURI BASIN REGION

Although several water-management plans have been implemented in the Missouri Basin Region, numerous alternative plans for improvement in water supply may be designed. Management plans can be derived from information on all available sources of water supply, expected demands for current and future types of water use, constraints on desired use, and the effects of implemented use.

DEVELOPMENT OF GROUND-WATER SUPPLIES

Development of a water supply is accomplished by *intercepting* water from the hydrologic cycle, the natural circulation of water among the oceans, atmosphere, and continents. Examples of interception include diversion of streamflow using a canal or pumping ground water using wells. Development of a ground-water supply or incorporation of ground water into basinwide management of water procedures. A typical sequence of events is described below, although administrative, hydrologic, or legal constraints may dictate different sequences.

COLLECTION AND INTERPRETATION OF HYDROLOGIC DATA

Knowledge of location and type of lateral and vertical aquifer boundaries is needed to define the geometry of the flow systems. Types of boundaries include a stream or lake that is hydraulically connected to an aquifer, confining rock that completely or partially confines flow in an aquifer, or an areal change in aquifer properties. Boundaries are identified by data collection and interpretation. For example, boundaries of aquifers can be determined from geologic mapping, well logs, or the observed response of the aquifer to pumping or to other hydraulic stresses. If a stream forms a boundary of an aquifer (fig. 11*A*), the water interchange influences the stream and aquifer. Measurements of streamflow and the slope of the water table may indicate a gaining or losing stream. Valley-fill aquifers in the region commonly have stream boundaries. If tight confining rock forms a boundary of an aquifer (fig. 11*B*), as it does in most places in the region, little or no flow may cross the boundary. In some cases, however, confining layers of clay or shale between aquifers leak (fig. 11*C*) and allow water interchange between the aquifers. Leaky confining rock boundaries may occur in deep structural basins of the region. The contact between unconfined and confined zones of an aquifer (fig. 11*D*) also represents a type of boundary. The unconfined zone of an aquifer is subject to the draining of and filling with ground water; the confined zone is subject to pressure changes. A relatively large amount of ground water per unit volume of aquifer may be released from or taken into storage in the unconfined zone, compared to the amount that may be released from or taken into storage by pressure changes in the confined zone. Unconfined zone boundaries are common near uplifts such as the Black Hills.

Inventory of existing wells and test holes followed by aquifer testing is needed to provide information on the geologic and hydraulic properties of the flow systems and on changes in ground-water storage with time. For example, logs of wells and test holes may be useful to determine the depth to the contacts between aquifers and confining beds. Water levels in observation wells may be used to prepare water-table or potentiometric maps that indicate direction of ground-water movement. Water-level changes measured in observation wells are useful to calculate changes in ground-water storage. Aquifer tests determine the hydraulic properties of the aquifers, such as the transmissivity and specific yield or storage coefficient. Data describing the hydraulic properties are necessary for calculations designed to predict the behavior of the aquifers in response to water use. Chemical analysis of water samples from the aquifer helps to define relations between the flow system and water quality.

An example of the interpretation of hydrologic data is shown schematically in figure 12. A sand and gravel aquifer lies beneath a stream and lake (fig. 12*A*). Additional wells are drilled into the aquifer to provide data for preparing the potentiometric map, shown in figure 12*B*. The potentiometric surface slopes toward the north, as shown by the altitudes on the contours. Arrows indicate the direction of ground-water movement toward the north, perpendicular to the contours. Flow from the lake to the stream boundary is indicated in the upper reach of the stream. Flow along the granite outcrop indicates the granite is a confined rock boundary. Aquifer testing provides data to subdivide the aquifer into three zones of transmissivity, as shown in figure 12*C*. Flow through the aquifer may be calculated using data from the potentiometric surface and the transmissivity value for each zone. The calculated flow from the aquifer to the stream may be checked by measurements of gain in the stream. The two zones of specific yield for the aquifer (fig. 12*D*) are also based on data from aquifer testing. The value of specific yield for each zone may be used to calculate changes in ground-water storage due to changes in the position of the potentiometric surface, as measured in wells.

After basic information is obtained, a water budget for the aquifers can be prepared from general data. Water budgets serve to indicate average recharge, changes in storage, and discharge, and they help to emphasize relative magnitudes of water sources, water uses, and water losses. Budgets may also provide information on the general long-term behavior of the aquifers and contribute to an understanding that may lead to better use of the aquifers.

DETERMINATION OF AQUIFER RESPONSE TO USE

Response of the aquifers to use may be determined using data describing the geologic and hydraulic properties of the aquifers. A simple problem may be solved with a few calculations by a skillful hydrologist. A

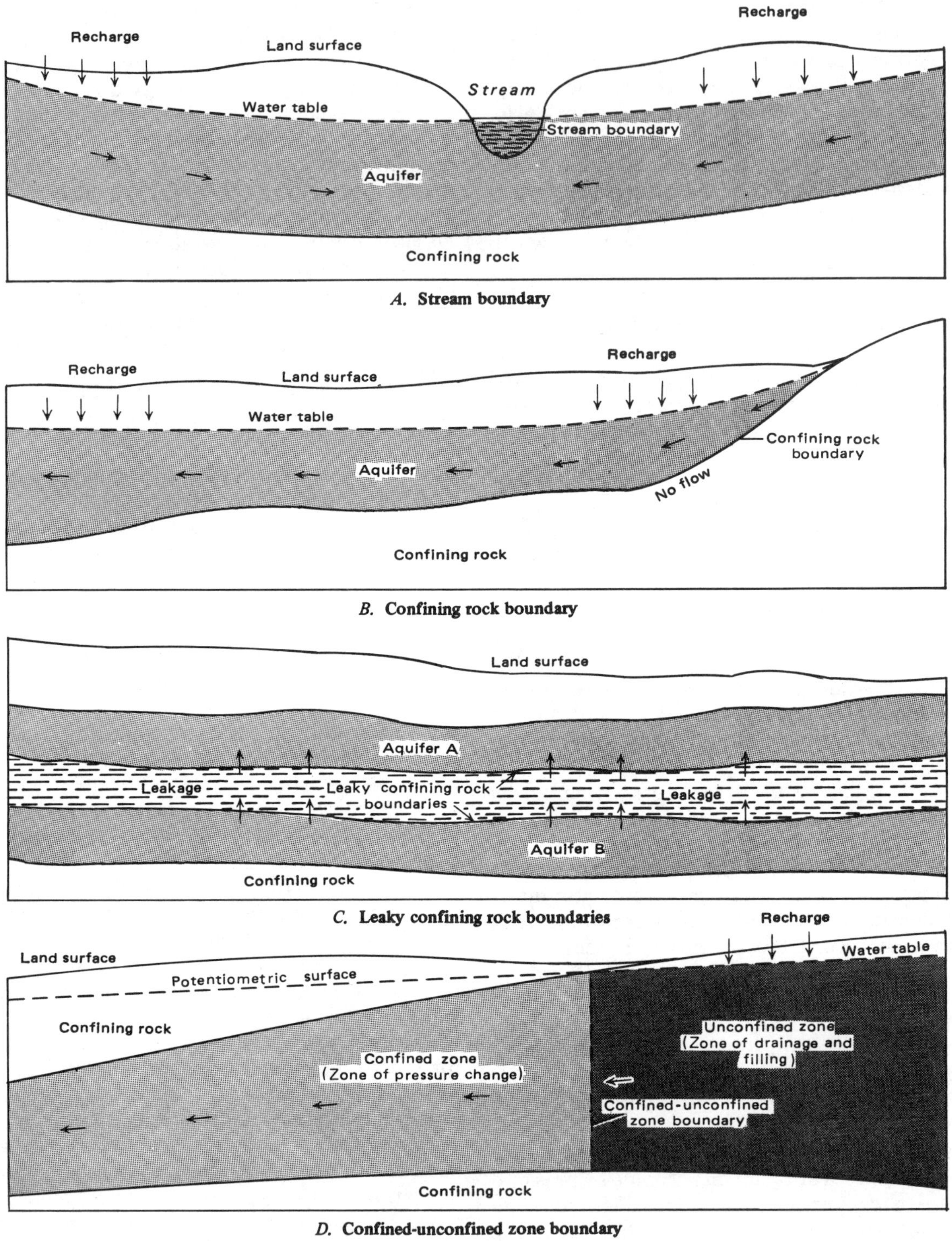

FIGURE 11.—Types of aquifer boundaries.

complex problem may require a simulation model in order to incorporate and integrate all data, including aquifer properties, necessary for an accurate solution. Various types of analog or mathematical simulation

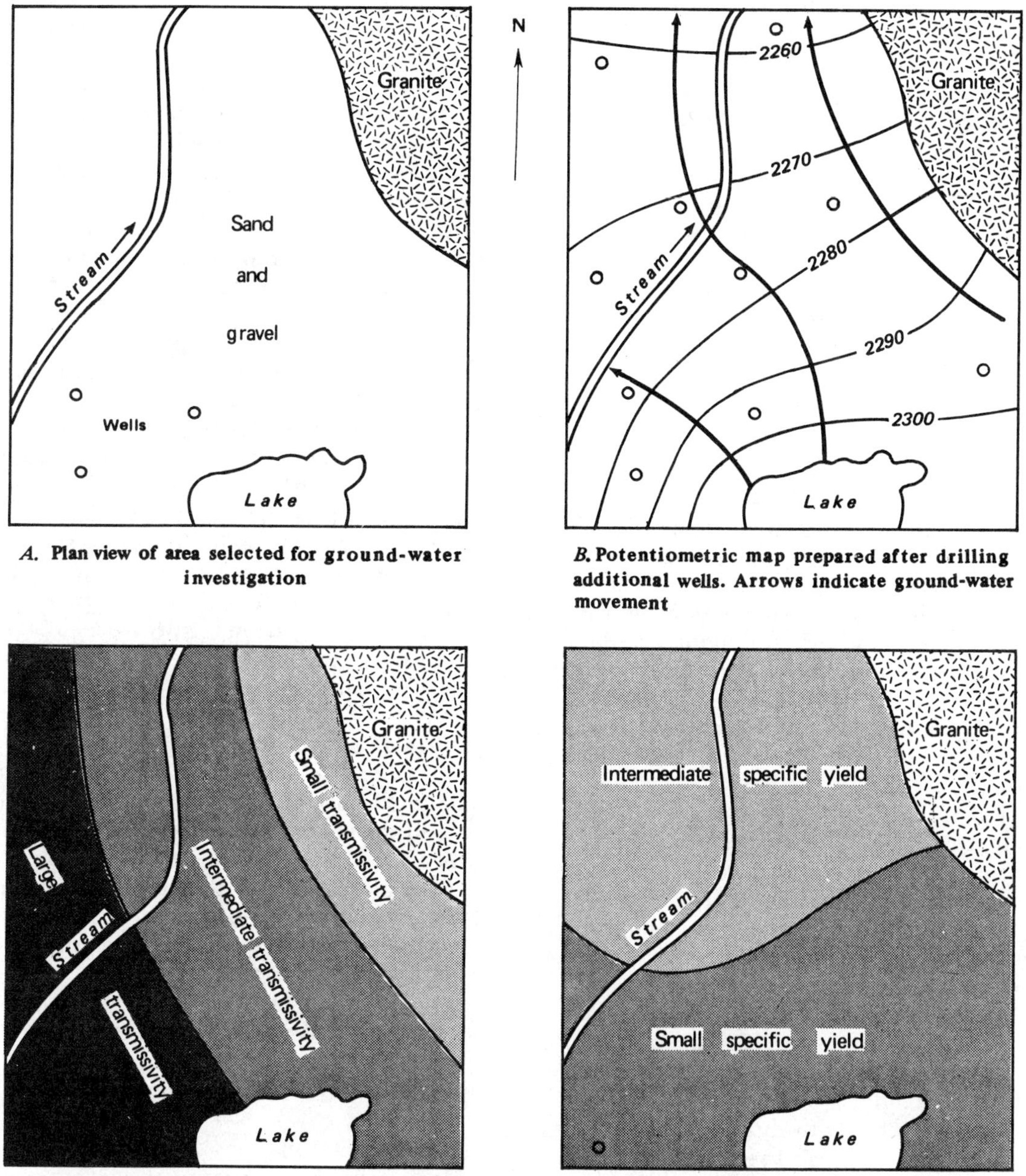

A. Plan view of area selected for ground-water investigation

B. Potentiometric map prepared after drilling additional wells. Arrows indicate ground-water movement

C. Distribution of aquifer transmissivity obtained from aquifer testing

D. Distribution of specific yield obtained from aquifer testing

FIGURE 12.—Interpretation of collected hydrologic data.

models are used routinely to help understand and predict behavior of aquifers under hypothetical conditions.

WATER-MANAGEMENT PLANS

Hydrologic data and interpretation provide hydrogeologic understanding and predictive capability to aid the design of water-management plans that may lead to improved water use. A simple development plan may involve artificial recharge, withdrawals from wells, or ground-water mining. A comprehensive plan might include the use of ground water based on information on snowpack, surface runoff and storage, stream-aquifer interrelations, water quality, and expected fluctuations in water supply and demand. Design of water-management plans offers the challenge of integrating information on the known behavior of hydrologic systems with goals designed to improve water use. Variety of possible plans is limited only by the imagination and ingenuity of the analyst.

Most water-management plans are designed with the goals of alleviating shortages in water supply and reducing the harmful effects of saline water. The ground-water reservoir can contribute toward these goals in many ways, as discussed in this report. Most plans involve a redistribution of water supplies in time and location in order to improve availability. Water-management plans that utilize the ground-water reservoir are listed below to indicate the variety possible.

1. Withdrawal from wells in selected locations to control sources of water depleted with time.
2. Recharge to aquifers in selected locations to control movement and storage of water with time.
3. Reuse of available water supply through schemes that successively capture the part of the water supply that is not consumed with each use.
4. Use of valley-fill aquifers in tributary valleys as off-channel reservoirs by recharging surplus streamflow in the tributaries to the aquifers.
5. Salvage of water normally evapotranspired by plants, using discharging wells to lower the water table.
6. Reduction of saline discharge water from aquifers by planting crops that reduce recharge to the aquifers or by wells and surface drains that control movement of the saline water.
7. Plugging of flowing wells in confined aquifers to conserve ground water and to reduce the lowering of water levels.
8. Use of an aquifer that underlies two adjacent surface-drainage areas as a reservoir common to both areas by means of storage, withdrawal, and transfer techniques.
9. Mining of ground water to obtain a temporary water supply, even though the ground water may not be replenished for a long time.
10. Withdrawal from wells in an aquifer in a multi-aquifer system so that the aquifer boundaries and interaquifer leakage help to sustain the withdrawal rate.
11. Use of wells to lower a high water table in a recharge area of an aquifer, in order to increase the rate of recharge.

Normally a plan would not be implemented until tested in the field. Testing does not indicate lack of confidence in hydrologic data or analyses, only a desire to insure that the selected plan is sound, as evidenced by verification. Field testing of a selected water-management plan may proceed by trial in the area of analysis for a limited time period. For example, withdrawals from industrial wells in the Fox Hills Sandstone in eastern Montana were planned for 1964–84 and began on schedule in 1964. A hydrologic analysis was made, and the magnitude of declines in water levels was predicted (Taylor, 1965). However, periodic measurements in observation wells were made in order to compare predicted and observed declines before State agencies permitted the continuation of scheduled withdrawals. Alternatively a plan may be tried over part of the area for a limited time period. Field-measurement procedures must be designed and data collected carefully to insure that collected data are sensitive to the expected results of the plan. Data of the required accuracy must be obtainable with currently available field instruments and methods of procedure. Field verification of the expected response of the hydrologic system may prove the value of the selected plan. However, the decision to implement a water-management plan also requires economic justification, expressed as a proven net benefit from the plan (Howe, 1971). Net benefit is the sum of all benefits from the plan, less the sum of all costs of the plan and damages caused by the plan.

Optimal use of water supplies to meet water demands is defined as the best plan of use under stated goals and constraints. Frequently goals are conflicting, and the water manager may be tempted to make a decision that solves only part of the problem at the expense of the rest of the problem or that benefits part of a region at the expense of the rest of the region. Nevertheless, a plan can be designed that will be beneficial to water users as a group.

Establishment of goals is the first step in implementing optimal use of water resources. Typical hydrologic objectives toward the goal of adequate water supply might include:

1. Development of a new water supply.
2. Reduction of shortages in an existing water supply.
3. Reduction of surpluses in an existing water supply.
4. Improvement or maintenance of water quality.

Use of ground water might help meet any of the above objectives. A new water supply could be derived from ground water or conjunctive use of ground water and surface water (fig. 4). shortages may be reduced through ground-water salvage (figs. 7 and 10), reuse schemes that utilize ground water (fig. 5*A*), or ground-water storage for later withdrawal (fig. 6). Depletion of ground-water reservoirs may also help to relieve shortages (fig. 9). Reduction of water surpluses might be accomplished by artificial recharge to utilize the storage capacity of the ground-water reservoir (fig. 6) or by a comprehensive plan that utilizes stored ground water, stored surface water, and streamflow. The flow of saline water from the ground-water reservoir to streams may be eliminated by hydrologically planned withdrawals from wells and carefully designed use or disposal of the saline water withdrawn.

Water-management plans are the key to meeting goals set to improve a water supply. Effective plans of operation require inventories of currently available water supplies derived from precipitation, streamflow, stored surface water, and stored ground water. Hydrologic models can be prepared that will simulate the use of all available water supplies in the basin and the best overall use of the supplies at various times and in various parts of the basin (Taylor and Luckey, 1974). Alternatives in water management can be appraised through use of models, and contingency plans can be devised to cope with variations in water supply and water demand. Basinwide management of water resources must consider uncertain supplies and uncertain demands. Nevertheless, through statistical descriptions of the uncertainties and continual upgrading of information, sound decisions can be made, even though the decisions may be selected on probability rather than certainty of outcome.

The final step for incorporation of ground water into basinwide management of water resources is full implementation of a selected plan. However, the plan must conform to State or interstate regulations for water administration. Other possible constraints include availability of required capital and cooperation among participants. After implementation of a selected plan, a field monitoring program should be maintained to determine if the plan is producing the expected response in the hydrologic system and contributing benefits to the water users and managers.

SOURCES OF ADDITIONAL WATER INFORMATION

Additional specific information on water supplies, other than discussed in this report, is available for the Missouri Basin Region. Hydrologic data include historical, current, and forecasted data. Interpretive reports, which are available for some areas, provide detailed descriptions of water supplies. General sources of water-resource information include:

Forest Service of U.S. Department of Agriculture
National Oceanic and Atmospheric Administration
Soil Conservation Service of U.S. Department of Agriculture
State universities and colleges
State water-resource agencies
U.S. Army Corps of Engineers
U.S. Bureau of Reclamation
U.S. Department of Housing and Urban Development
U.S. Environmental Protection Agency
U.S. Geological Survey, Water Resources Division
Water-resource consultants

Specific sources of water-resource information by State and Province include:

Alberta

Alberta Research Council
Groundwater Division
3rd Floor, Campus Tower
8625 112th Street
Edmonton, Alberta
T6G 1K8

Alberta Research Council
Transportation and Surface Water
Engineering Division
301 Civil Electrical Engineering Building
University of Alberta
Edmonton, Alberta
T6G 2J2

Alberta Research Council
Department of the Environment
Oxbridge Place
9820 106th Street
Edmonton, Alberta
T6K 2J6

Alberta Research Council
Federal Government
Department of the Environment
Inland Waters Branch
4616 Valiant Drive N.W.
Calgary, Alberta
T3A 0X9

Federal Government
Department of the Environment
Northern Forest Research Center
5320 122d Street
Edmonton, Alberta
T6H 3S5

Saskatchewan

Chemistry Division
Saskatchewan Research Council
30 Campus Drive
Saskatoon, Saskatchewan
S7N 0X1

Family Farm Improvement Branch
Saskatchewan Department of Agriculture
1318 Winnipeg Street
Regina, Saskatchewan
S4R 1J6

Geology Division
Saskatchewan Research Council
30 Campus Drive
Saskatoon, Saskatchewan
S7N 0X1

Hydrology Branch
Saskatchewan Department of the Environment
1855 Victoria Avenue

Regina, Saskatchewan
S4P 3T1

Saskatchewan District Office
Water Survey of Canada
210-1102 8th Avenue
Regina, Saskatchewan
S4R 1C9

Water Pollution Control Branch
Saskatchewan Department of the Environment
1855 Victoria Avenue
Regina, Saskatchewan
S4P 3T1

Water Quality Branch
Inland Waters Directorate
Environment Canada
1901 Victoria Avenue
Regina, Saskatchewan
S4P 0R3

Water Rights Branch
Saskatchewan Department of the Environment
1855 Victoria Avenue
Regina, Saskatchewan
S4P 3T1

Montana

Montana Bureau of Mines and Geology
Room 203-B Main Hall
Montana College of Mineral Science and Technology
Butte, Montana 59701

Montana State Department of Health and Environmental Sciences
1400 11th Avenue
Helena, Montana 59601

Montana State Department of Natural Resources and Conservation
32 South Ewing
Helena, Montana 59601

U.S. Geological Survey
Water Resources Division
Montana District Office
P.O. Box 1696
421 Federal Building
Helena, Montana 59601

Wyoming

U.S. Geological Survey
Water Resources Division
Wyoming District Office
2120 Capitol Avenue
P.O. Box 1125
Cheyenne, Wyoming 82001

Wyoming State Engineer
Barrett Building
Central Avenue
Cheyenne, Wyoming 82002

North Dakota

North Dakota State Health Department
Capital Building
Bismarck, North Dakota 58501

North Dakota State Water Commission
900 E. Blvd. Avenue
Bismarck, North Dakota 58501

U.S. Geological Survey
Water Resources Division
North Dakota District Office
Room 332 New Federal Building
Third Street and Rosser Avenue
Bismarck, North Dakota 58501

South Dakota

City of Sioux Falls
Sioux Falls, South Dakota 57101

Division of Water Rights
Department of Natural Resource Development
Joe Foss Building
Pierre, South Dakota 57501

East Dakota Conservancy Sub-District
P.O. Box 68
Brookings, South Dakota 57006

South Dakota Department of Environmental Protection
Joe Foss Building
Pierre, South Dakota 57501

State Geologist
Division of Geological Survey
State University
Science Center
Vermillion, South Dakota 57069

U.S. Bureau of Reclamation
Missouri-Oahe Project Office
Federal Building
200 Fourth St. SW
Huron, South Dakota 57350

U.S. Geological Survey
Water Resources Division
South Dakota District Office
Federal Building Room 308
200 4th St. SW
Huron, South Dakota 57350

Colorado

Colorado Division of Water Resources
Office of the State Engineer
818 State Centennial Building
1313 Sherman Street
Denver, Colorado 80203

Colorado Geological Survey
715 State Centennial Building
1313 Sherman Street
Denver, Colorado 80203

Colorado Water Conservation Board
823 State Centennial Building
1313 Sherman Street
Denver, Colorado 80203

U.S. Geological Survey
Water Resources Division
Colorado District Office
Mail Stop 415, Box 25046
Denver Federal Center
Denver, Colorado 80225

Nebraska

Conservation and Survey Division
University of Nebraska at Lincoln
Room 113 Nebraska Hall
Lincoln, Nebraska 68588

Nebraska Department of Water Resources
301 Centennial Mall, South
P.O. Box 94676
Lincoln, Nebraska 68509

Nebraska Natural Resources Commission
301 Centennial Mall, South
P.O. Box 94876
Lincoln, Nebraska 68509

U.S. Geological Survey
Water Resources Division
Nebraska District Office
Room 406, Federal Building and U.S. Courthouse
100 Centennial Mall North
Lincoln, Nebraska 68508

Minnesota

Minnesota Department of Natural Resources
Division of Waters
Space Center Building, 444 Lafayette Road
St. Paul, Minnesota 55101

U.S. Geological Survey
Water Resources Division
Minnesota District Office
1033 Post Office Building
St. Paul, Minnesota 55101

Iowa

Department of Environmental Quality
Henry Wallace Office Building
Des Moines, Iowa 50319

Iowa Conservation Commission
Henry Wallace Office Building
Des Moines, Iowa 50319

Iowa Geological Survey
123 N. Capitol Street
Iowa City, Iowa 52240

Iowa Natural Resources Council
Grimes State Office Building
E. 14th Street and Grand Avenue
Des Moines, Iowa 50319

Iowa State Hygienic Laboratory
Medical Laboratories
University Hospital
University of Iowa
Iowa City, Iowa 52242

U.S. Geological Survey
Water Resources Division
Iowa District Office
Room 269 Federal Building
400 South Clinton Street
Iowa City, Iowa 52240

Kansas

Kansas Geological Survey
1930 Avenue A
Camp West, University of Kansas
Lawrence, Kansas 66045

Kansas State Department of Health and Environment
Building 740
Forbes Air Force Base
Topeka, Kansas 66620

Kansas Water Resources Board
Suite 303
503 Kansas Avenue
Topeka, Kansas 66603

U.S. Geological Survey
Water Resources Division
Kansas District Office
1950 Avenue A – Campus West
University of Kansas
Lawrence, Kansas 66045

Kansas State Board of Agriculture
Division of Water Resources
1720 South Topeka Avenue
Topeka, Kansas 66612

Missouri

Missouri Department of Conservation
P.O. Box 180
Jefferson City, Missouri 65101

Missouri Department of Natural Resources
Division of Environmental Quality
P.O. Box 1368
Jefferson City, Missouri 65101

Missouri Department of Natural Resources
Division of Geology and Land Survey
P.O. Box 250
Rolla, Missouri 65401

U.S. Corps of Engineers
Kansas City District
700 Federal Building
601 E. 12th Street
Kansas City, Missouri 64106

U.S. Geological Survey
Water Resources Division
Missouri District Office
1400 Independence Road
Rolla, Missouri 65401

CONCLUSIONS

The Missouri Basin Region is a large area that requires large water supplies to meet increasing demands for irrigation, industrial, public supply, and rural uses. Development of petroleum, coal, and oil shale reserves in the region also will require larger water supplies. Available supplies are obtained from precipitation, streamflow, stored surface water, and ground water. Although ground-water use is increasing, large supplies available in numerous aquifers are still not completely defined or completely developed. Effective and efficient use of ground water requires additional hydrologic information so that withdrawals from aquifers can be designed to furnish beneficial supplies without creating problems. A better understanding of ground-water resources will aid in the design of alternative water-management plans that consider all available supplies.

Ground-water reservoirs have unique storage properties that enable them to contribute to sound water-management practices in the region. Large quantities of water are stored in many aquifers because of the storage capacity of the aquifer material and the great areal extent of the aquifers. The stored ground water is not rapidly affected by fluctuations in seasonal precipitation, because recharge and movement of ground water are slow processes. Evaporation also affects ground water much less than surface water.

Ground water in storage can be withdrawn, replaced, or increased if the aquifer is not hydraulically connected to streams or surface-water reservoirs at numerous locations. For example, during a drought or any period in which water demand increases, water supplies can be obtained by pumping wells and mining ground water from storage. Mining eventually lowers the potentiometric surface throughout the extent of the aquifer, rather than locally. A large volume of water may result from a small lowering of the potentiometric surface over a large area. Surplus water supplies can be added to the ground water in storage by using artificial-recharge techniques. Water added to storage also eventually raises the potentiometric surface throughout the areal extent of the aquifer so that the increased quantity in storage is areally distributed.

The recharge, movement, and discharge of ground water also may facilitate water-management practices. Any technique that increases natural recharge to or decreases natural discharge from an aquifer will increase ground water in storage that can be withdrawn by pumping wells. Improved water supplies can be obtained by designed and induced changes in ground-water movement, especially if the aquifer is hydraulically connected to nearby streams. Production wells can be located according to aquifer properties and boundaries and pumped according to calculated time schedules. The resulting withdrawals induce changes in ground-water movement, control the distribution of ground water in storage, and affect the nearby streamflow. The resulting changes in ground water and surface water can result in an improved supply that is available at the desired time and location.

REFERENCES CITED

Alverson, D. C., 1965, Geology and hydrology of the Fort Belknap Indian Reservation Montana: U.S. Geol. Survey Water-Supply Paper 1576–F, 59 p.

Babcock, H. M., and Bjorklund, L. J., 1956, Ground-water geology of parts of Laramie and Albany Counties, Wyoming, and Weld County, Colorado: U.S. Geol. Survey Water-Supply Paper 1367, p. 14–15.

Ellis, M. J., and Pederson, D. T., 1977, Groundwater levels in Nebraska, 1976: Nebraska Water Survey Paper 44, 96 p.

Emery, P. A., 1966, Use of analog model to predict streamflow depletion, Big and Little Blue River basin, Nebraska: Ground Water, v. 4, no. 4, p. 13–19.

Environmental Science Services Administration, 1968, Climatic atlas of the United States: Environmental Data Service (repr. by Natl. Oceanic and Atmospheric Adm., 1974), 80 p.

Feder, G. L., 1973, A conceptual model of the hydrologic system supplying the large springs in the Ozarks: U.S. Geol. Survey Open-File Rept., 148 p.

Feltis, R. D., 1973, Geology and water resources of eastern part of Judith Basin, Montana: Montana Bur. Mines and Geology Bull. 87, 51 p.

Ferris, J. G., 1950, Quantitative method for determining ground-water characteristics for drainage design: Agr. Eng., v. 31, no. 6, p. 285–291.

Hackett, O. M., Visher, F. N., McMurtrey, R. G., and Steinhilber, W. L., 1960, Geology and ground-water resources of the Gallatin Valley, Gallatin County, Montana: U.S. Geol. Survey Water-Supply Paper 1482, 282 p.

Hamilton, L. J., and Paulson, Q. F., 1968, Geology and ground-water resources of the lower Bighorn valley, Montana: U.S. Geol. Survey Water-Supply Paper 1876, 39 p.

Hodson, W. G., Pearl, R. H., and Druse, S. A., 1973, Water resources of the Powder River Basin and adjacent areas, northeastern Wyoming: U.S. Geol. Survey Hydrol. Inv. Atlas HA–465.

Howe, C. W., 1971, Benefit-cost analysis for water system planning: American Geophys. Union Water Resources Mon. 2, 144 p.

Huntoon, P. W., 1974, Predicted water-level declines for alternative groundwater developments in the upper Big Blue River basin, Nebraska: Conserv. and Survey Div., Inst. Agr. and Nat. Resources, Univ. Nebraska, Lincoln, Resource Rept. 6, 105 p.

Hurr, R. T., Schneider, D. A., and Minges, D. R., 1975, Hydrology of the South Platte River valley, northeastern Colorado: Colorado Water Conserv. Board Circ. 28, 24 p.

Jenkins, C. T., and Hofstra, W. E., 1970, Problems of artificially recharging the Ogallala Formation in Colorado, *in* The Ogallala aquifer, a symposium: Lubbock, Tex., Tech. Univ. Spec. Rept. 39, p. 154–164.

Kapple, G. W., Luckey, R. R., and Hofstra, W. E., 1976, Digital ground-water model of the Ogallala aquifer in parts of Washington and Yuma Counties, northern High Plains of Colorado: Colorado Water Resources Circ. 34, 25 p.

Keech, C. F., and Bentall, Ray, 1971, Dunes on the plains, the sand hills region of Nebraska: Univ. Nebraska Conserv. and Survey Div., Resource Rept. 4, Feb. 1971, 18 p.

Kohout, F. A., 1970, Reorientation of our saline water resources thinking: Water Resources Research, v. 6, no. 5, p. 1442–1448.

Konikow, L. F., 1976, Preliminary digital model of ground-water flow in the Madison Group, Powder River Basin and adjacent areas, Wyoming, Montana, South Dakota, North Dakota, and Nebraska: U.S. Geol. Survey Water-Resources Inv. 63–75.

La Rocque, G. A., Jr., 1966, General availability of ground water and depth to water level in the Missouri River basin: U.S. Geol. Survey Hydrol. Inv. Atlas HA–217.

Lines, G. C., 1976, Digital model to predict effects of pumping from Arikaree aquifer in the Dwyer area, southeastern Wyoming: U.S. Geol. Survey Water Resources Inv. 8–76, 24 p.

Lowry, M. E., Lowham, H. W., and Lines, G. C., 1975, Water resources of the Bighorn Basin, northwestern Wyoming: U.S. Geol. Survey Hydrol. Inv. Atlas HA–512.

Lowry, M. E., Rucker, S. J., IV, and Wahl, K. L., 1973, Water resources of the Laramie, Shirley, Hanna Basins and adjacent areas, southeastern Wyoming: U.S. Geol. Survey Hydrol. Inv. Atlas HA–471.

Luckey, R. R., and Hofstra, W. E., 1973, Digital model of the hydrologic system, northern High Plains of Colorado: Colorado

Water Resources Circ. 19, 25 p.
——— 1974, Digital model of the Ogallala aquifer of the northern part of the northern High Plains of Colorado: Colorado Water Resources Circ. 24, 22 p.
McLean, J. R., 1971, Stratigraphy of the upper Cretaceous Judith River Formation in the Canadian Great Plains: Research Council Saskatchewan, Geol. Div., Rept. 11, 96 p.
Meyboom, Peter, 1960, Geology and groundwater resources of the Milk River Sandstone in southern Alberta: Alberta Research Council Mem. 2, 89 p.
Missouri Basin Inter-agency Committee, 1969a, Comprehensive framework study Missouri River basin: v. 1–7.
——— 1969b, Comprehensive framework study Missouri River basin: v. 1, 274 p.
——— 1969c, Comprehensive framework study Missouri River basin, *in* Hydrologic Analyses and Projections: v. 6, 143 p.
Missouri Geological Survey and Water Resources, 1963, Groundwater maps of Missouri.
Missouri River Basin Commission, 1975, Stream-aquifer hydrology, technical paper, Platte River basin, Nebraska Level B Study.
Moulder, E. A., 1970, Freshwater bubbles: A possibility for using saline aquifers to store water: Water Resources Research, v. 6, no. 5, p. 1528–1531.
Murray, C. R., 1968, Estimated use of water in the United States, 1965: U.S. Geol. Survey Circ. 556, 53 p.
Murray, C. R., and Reeves, E. B., 1972, Estimated use of water in the United States in 1970: U.S. Geol. Survey Circ. 676, 37 p.
North Dakota Geological Survey, 1973, Mineral and water resources of North Dakota: North Dakota Geol. Survey Bull. 63, 252 p.
Osterkamp, W. R., 1968, Occurrence of ground water in the Judith River Formation north-central Montana: U.S. Geol. Survey Hydrol. Inv. Atlas HA–308.
Rapp, J. R., Warner, D. A., and Morgan, A. M., 1953, Geology and ground-water resources of the Egbert-Pine Bluffs Carpenter area Laramie County, Wyoming: U.S. Geol. Survey Water-Supply Paper 1140, p. 40–42.
Rocky Mountain Association of Geologists, 1972, Geologic Atlas of the Rocky Mountain region: Denver, Colo., A. B. Hirschfeld Press, 331 p.
Sandberg, C. A., 1962, Geology of the Williston Basin, North Dakota, Montana, and South Dakota, with reference to subsurface disposal of radioactive wastes: U.S. Geol. Survey Rept. TEI-809, 148 p.
Schoon, R. A., 1971, Geology and hydrology of the Dakota Formation in South Dakota: South Dakota Geol. Survey Rept. Inv. 104, 55 p.
Swenson, F. A., Miller, W. R., Hodson, W. G., and Visher, F. H., 1978, Water in the Madison Group, Powder River Basin, Wyoming and Montana: U.S. Geol. Survey Inv. Map I–847–C (in press).
Taylor, O. J., 1965, Ground-water resources along Cedar Creek anticline in eastern Montana: Montana Bur. Mines and Geol. Mem. 40, 99 p.
——— 1968, Ground water resources of the northern Powder River valley, southeastern Montana: Montana Bur. Mines and Geol. Bull. 66, 34 p.
Taylor, O. J., and Luckey, R. R., 1974, Water-management studies of a stream-aquifer system, Arkansas River valley, Colorado: Ground Water, v. 12, no. 1, p. 22–38.
Tokarsky, O., 1974, Hydrogeology of the Lethbridge-Fernie area, Alberta: Alberta Research Council Rept. 74-1, 18 p.
Vineyard, J. D., and Feder, G. L., 1974, Springs of Missouri: Missouri Geol. Survey and Water Resources, Water Resources Rept. 29, 267 p.
Whitaker, S. H., 1967, Geology and groundwater resources of the Wood Mountain Area (72–G): Saskatchewan Research Council, Geology Div. Map 5.
——— 1974, Geology and groundwater resources of the Willow Bunch area (72–H): Saskatchewan Research Council, Geol. Div. Map 20.
Whitcomb, H. A., and Lowry, M. E., 1968, Ground-water resources and geology of the Wind River Basin area, central Wyoming: U.S. Geol. Survey Hydrol. Inv. Atlas HA–270, 13 p.
Winter, T. G., 1975, Delineation of buried glacial-drift aquifers: U.S. Geol. Survey Jour. Research, v. 3, no. 2, p. 137–148.
Zimmerman, E. A., 1966, Geology and ground-water resources of western and southern parts of Judith Basin, Montana: Montana Bur. Mines and Geol. Bull. 50–A, 33 p.
——— 1967, Water resources of the Cut Bank area, Glacier and Toole Counties, Montana: Montana Bur. Mines and Geol. Bull. 60, 37 p.

TABLE 7.—*Principal aquifers in States and Provinces and their geologic ages in the Missouri Basin Region*

Alberta:
- Glacial deposits of Pleistocene age
- Blood Reserve Formation—Cretaceous
- Judith River Formation—Late Cretaceous
- Milk River Sandstone—Late Cretaceous
- Madison Group—Early Mississippian

Saskatchewan:
- Glacial deposits of Pleistocene age
- Ravenscrag Formation—Paleocene
- Frenchman Formation—Late Cretaceous
- Whitemud Formation—Late Cretaceous
- Eastend Formation—Late Cretaceous
- Judith River Formation—Late Cretaceous
- Madison Group—Early Mississippian

Montana:
- Alluvium—Pleistocene and Holocene
- Glacial deposits of Pleistocene age
- Basin-fill deposits of Tertiary age
- Fort Union Formation—Late Cretaceous and Paleocene
- Hell Creek Formation—Late Cretaceous
- Fox Hills Sandstone—Late Cretaceous
- Judith River Formation—Late Cretaceous
- Eagle Sandstone—Late Cretaceous
- Virgelle Member of Eagle Sandstone—Late Cretaceous
- Dakota Sandstone—Cretaceous
- Kootenai Formation—Early Cretaceous
- Swift Formation—Middle and Late Jurassic
- Tensleep Sandstone—Pennsylvanian
- Amsden Group or Formation—Mississippian and Pennsylvanian
- Kibbey Formation—Late Mississippian
- Madison Group—Early Mississippian
- Bighorn Dolomite—Middle and Late Ordovician

Wyoming:
- Alluvium—Pleistocene and Holocene
- Ogallala Formation—Miocene
- North Park Formation—Late Miocene and Pliocene (?)
- Arikaree Formation—Miocene
- White River Formation—Oligocene
- Brule Formation—Oligocene
- Wagon Bed Formation—Eocene
- Battle Spring Formation—Eocene
- Wind River Formation—Eocene
- Indian Meadows Formation—Eocene
- Wasatch Formation—Eocene
- Hanna Formation—Paleocene
- Fort Union Formation—Paleocene
- Ferris Formation—Late Cretaceous and Paleocene
- Medicine Bow Formation—Late Cretaceous
- Lance Formation—Late Cretaceous
- Fox Hills Sandstone—Late Cretaceous
- Lewis Shale—Late Cretaceous
- Mesaverde Group or Formation—Late Cretaceous
- Frontier Formation—Late Cretaceous
- Inyan Kara Group—Early Cretaceous
- Cloverly Formation—Early Cretaceous
- Sundance Formation—Middle and Late Jurassic
- Nugget Sandstone—Late Triassic (?) and Jurassic (?)
- Jelm Formation of Chugwater Group—Late Triassic
- Casper Formation—Pennsylvanian and Permian
- Tensleep Sandstone—Pennsylvanian and Early Permian
- Amsden Formation—Mississippian and Pennsylvanian
- Madison Limestone—Early Mississippian
- Bighorn Dolomite—Middle and Late Ordovician

North Dakota:
- Alluvium—Pleistocene and Holocene
- Glacial deposits of Pleistocene age
- Fort Union Formation—Paleocene
- Hell Creek Formation—Late Cretaceous
- Fox Hills Sandstone—Late Cretaceous
- Dakota Sandstone—Cretaceous

North Dakota—Continued:
- Minnelusa Formation—Pennsylvanian and Permian
- Madison Group—Early Mississippian
- Red River Formation—Late Ordovician
- Deadwood Formation—Late Cambrian and Early Ordovician

South Dakota:
- Alluvium—Pleistocene and Holocene
- Glacial deposits of Pleistocene age
- Ogallala Formation—Miocene
- Arikaree Formation—Miocene
- Brule Formation—Oligocene
- Hell Creek Formation—Late Cretaceous
- Fox Hills Sandstone—Late Cretacous
- Dakota Sandstone—Cretaceous
- Inyan Kara Group—Early Cretaceous
- Sundance Formation—Middle and Late Jurassic
- Minnelusa Formation—Pennsylvanian and Permian
- Madison Group—Early Mississippian
- Red River Formation and Whitewood Dolomite—Late Ordovician
- Deadwood Formation—Late Cambrian and Early Ordovician

Colorado:
- Alluvium—Pleistocene and Holocene
- Ogallala Formation—Miocene
- Arikaree Formation—Miocene
- Brule Clay—Oligocene
- Dawson Formation—Late Cretaceous and Paleocene
- Denver Formation—Late Cretaceous and Paleocene
- Arapahoe Formation—Late Cretaceous
- Laramie Formation—Late Cretaceous
- Fox Hills Sandstone—Late Cretaceous
- Madison Limestone—Early Mississippian

Nebraska:
- Alluvium—Pleistocene and Holocene
- Dune-sand deposits of Pleistocene and Holocene age
- Glacial deposits of Pleistocene age
- Ogallala Formation—Miocene
- Arikaree Formation—Miocene
- Brule Formation—Oligocene
- Dakota Sandstone—Cretaceous
- Limestone of Pennsylvanian and Permian age
- Madison Limestone equivalent—Early Mississippian

Minnesota:
- Alluvium—Pleistocene and Holocene
- Glacial deposits of Pleistocene age

Iowa:
- Alluvium—Pleistocene and Holocene
- Glacial deposits of Pleistocene age
- Dakota Sandstone—Cretaceous
- Limestone of Pennsylvanian age
- Prairie du Chien Formation—Early Ordovician
- Jordan Sandstone—Late Cambrian

Kansas:
- Alluvium—Pleistocene and Holocene
- Glacial deposits of Pleistocene age
- Ogallala Formation—Miocene
- Dakota Sandstone—Cretaceous
- Limestone and sandstone of Pennsylvanian and Permian age

Missouri:
- Alluvium—Pleistocene and Holocene
- Glacial deposits of Pleistocene age
- Limestone of Pennsylvanian age
- Roubidoux Formation—Early Ordovician
- Gasconade Dolomite—Early Ordovician
- Eminence Dolomite—Late Cambrian
- Potosi Dolomite—Late Cambrian

TABLE 8.—*Principal aquifers in structural basins, their geologic ages and maximum thicknesses, Missouri Basin Region*

Formation and age	Maximum thickness (ft)	(m)
Alberta Basin, Alberta and Mont.:		
Blood Reserve Formation—Cretaceous	100	30
Judith River Formation—Late Cretaceous	1,200	370
Milk River Sandstone—Late Cretaceous	220	67
Bighorn Basin, Mont.:		
Hell Creek Formation—Late Cretaceous	600	180
Judith River Formation including Parkman Sandstone Member—Late Cretaceous	250	76
Kootenai Formation including Cloverly Formation—Early Cretaceous	400	120
Tensleep Sandstone—Pennsylvanian	100	30
Amsden Formation—Mississippian and Pennsylvanian	300	91
Bighorn Basin, Wyo.:		
Fort Union Formation—Paleocene	5,000	1,500
Lance Formation—Late Cretaceous	1,800	550
Mesaverde Formation—Late Cretaceous	1,800	550
Frontier Formation—Late Cretaceous	700	210
Cloverly Formation—Early Cretaceous	400	120
Tensleep Sandstone—Pennsylvanian and Early Permian	400	120
Amsden Formation—Mississippian and Pennsylvanian	350	110
Denver Basin, Colo.:		
Dawson Formation—Late Cretaceous and Paleocene	2,000	610
Denver Formation—Late Cretaceous and Paleocene	800	240
Arapahoe Formation—Late Cretaceous	600	180
Fox Hills Sandstone and basal Laramie Formation—Late Cretaceous	200	60
Hogeland Basin, Mont.:		
Judith River Formation—Late Cretaceous	460	140
Eagle Sandstone—Late Cretaceous	270	82
Dakota Sandstone—Cretaceous	60	18
Kootenai Formation—Early Cretaceous	190	58
Swift Formation—Middle and Late Jurassic	240	73
Hogeland Basin, Saskatchewan:		
Frenchman Formation—Late Cretaceous	130	40
Whitemud Formation—Late Cretaceous	60	18
Eastend Formation—Late Cretaceous	100	30
Judith River Formation—Late Cretaceous	650	200
Judith Basin, Mont.:		
Judith River Formation—Late Cretaceous	400	120
Eagle Sandstone—Late Cretaceous	300	91
Dakota Sandstone—Cretaceous	80	24
Kootenai Formation—Early Cretaceous	660	200
Swift Formation—Middle and Late Jurassic	185	56
Amsden Group—Mississippian and Pennsylvanian	920	280
Kibbey Formation—Late Mississippian	300	91
Laramie, Shirley, Hanna Basins, and Saratoga Valley, Wyo.:		
North Park Formation—Late Miocene and Pliocene (?)	1,500	460
Arikaree Formation—Miocene	1,400	430
White River Formation—Oligocene	400	120
Wagon Bed Formation—Eocene	150	46
Wind River Formation—Eocene	500	150
Hanna Formation—Paleocene	13,500	4,110
Ferris Formation—Late Cretaceous and Paleocene	6,500	2,000
Medicine Bow Formation—Late Cretaceous	6,200	1,900
Lewis Shale, including Fox Hills Sandstone—Late Cretaceous	3,900	1,200
Mesaverde Group—Late Cretaceous	4,000	1,200
Cloverly Formation—Early Cretaceous	100	30
Sundance Formation—Middle and Late Jurassic	230	70
Nugget Sandstone—Late Triassic (?) and Jurassic (?)	50	15
Jelm Formation—Late Jurassic	360	110
Casper Formation—Pennsylvanian and Permian	800	240
Powder River Basin, Wyo. and Mont.:		
Wasatch Formation—Eocene	2,000	610
Fort Union Formation—Late Cretaceous and Paleocene	3,450	1,050
Fox Hills Sandstone and Lance Formation, Wyo.—Late Cretaceous	700	210
Fox Hills Sandstone and basal Hell Creek Formation, Mont.—Late Cretaceous	250	76
Mesaverde Formation—Late Cretaceous	900	270
Frontier Formation—Late Cretaceous	830	250
Inyan Kara Group—Early Cretaceous	450	140
Sundance Formation—Middle and Late Jurassic	400	120
Tensleep Sandstone—Pennsylvanian and Early Permian	500	150
Williston Basin, N. Dak., Mont., S. Dak., and Saskatchewan:		
Fort Union Formation—Paleocene	1,100	340
Fox Hills Sandstone and basal Hell Creek Formation—Late Cretaceous	400	120
Dakota Sandstone—Cretaceous	460	140
Inyan Kara Group—Early Cretaceous	700	210
Sundance Formation—Middle and Late Jurassic	740	230
Minnelusa Formation—Pennsylvanian and Permian	1,400	430
Deadwood Formation—Late Cambrian and Early Ordovician	600	180
Wind River Basin, Wyo.:		
Arikaree Formation—Miocene	2,700	820
White River Formation—Oligocene	650	200
Wagon Bed Formation—Eocene	700	210
Battle Spring, Wind River, and Indian Meadows Formations—Eocene	8,000	2,400
Casper Formation—Pennsylvanian and Permian; Tensleep Formation—Pennsylvanian and Early Permian and Amsden Formation—Mississippian and Pennsylvanian	900	270

PLATE 1A. SURFACE EXTENT OF PRINCIPAL UNCONSOLIDATED AND SEMICONSOLIDATED AQUIFERS IN THE MISSOURI BASIN REGION (WESTERN AREA)

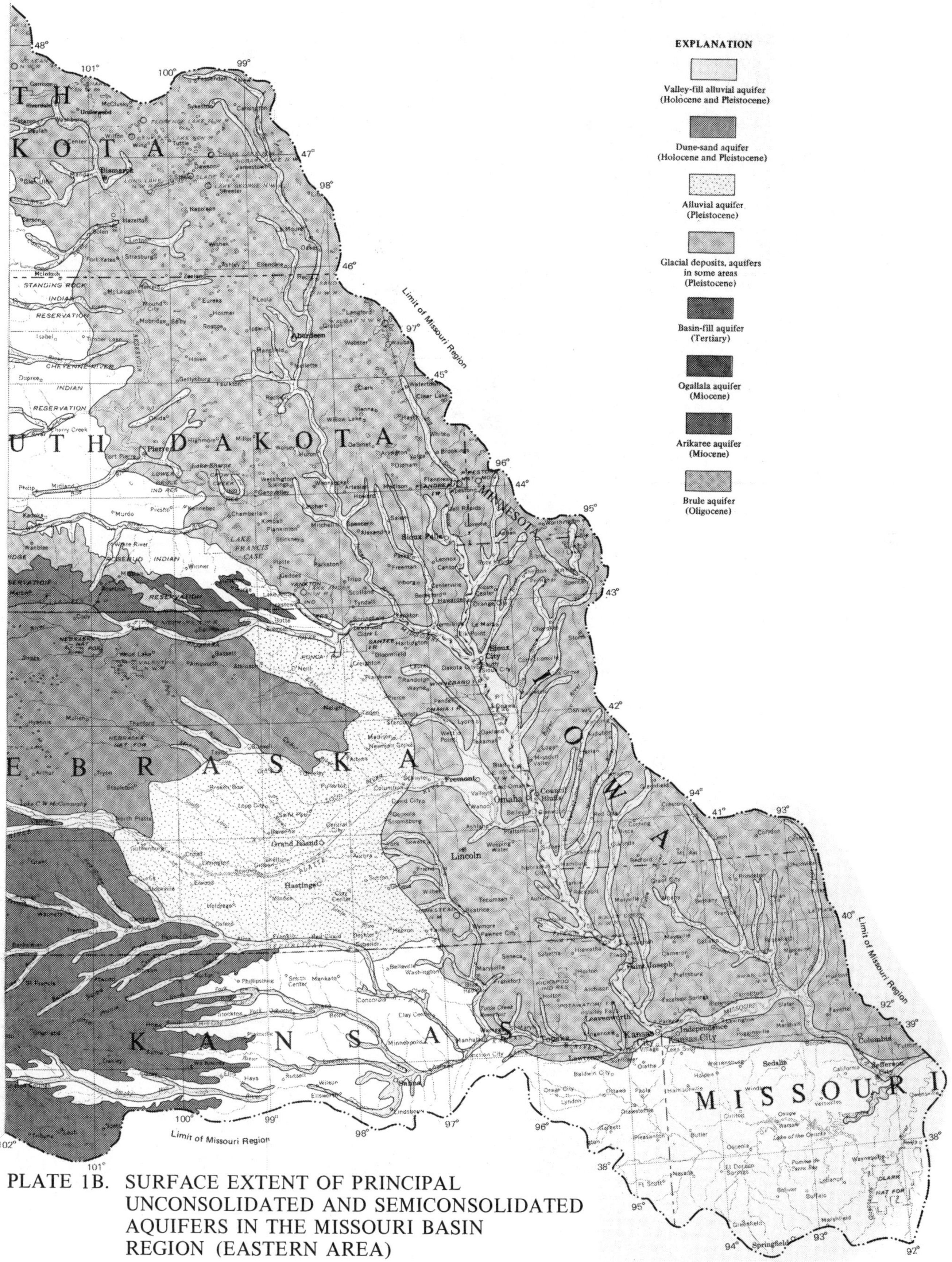

PLATE 1B. SURFACE EXTENT OF PRINCIPAL UNCONSOLIDATED AND SEMICONSOLIDATED AQUIFERS IN THE MISSOURI BASIN REGION (EASTERN AREA)

PLATE 2A. SURFACE AND SUBSURFACE EXTENT OF PRINCIPAL SANDSTONE AQUIFERS IN THE MISSOURI BASIN REGION (WESTERN AREA)

PLATE 2B. SURFACE AND SUBSURFACE EXTENT OF PRINCIPAL SANDSTONE AQUIFERS IN THE MISSOURI BASIN REGION (EASTERN AREA)

PLATE 3A. SURFACE AND SUBSURFACE EXTENT OF PRINCIPAL LIMESTONE AND DOLOMITE AQUIFERS IN THE MISSOURI BASIN REGION (WESTERN AREA)

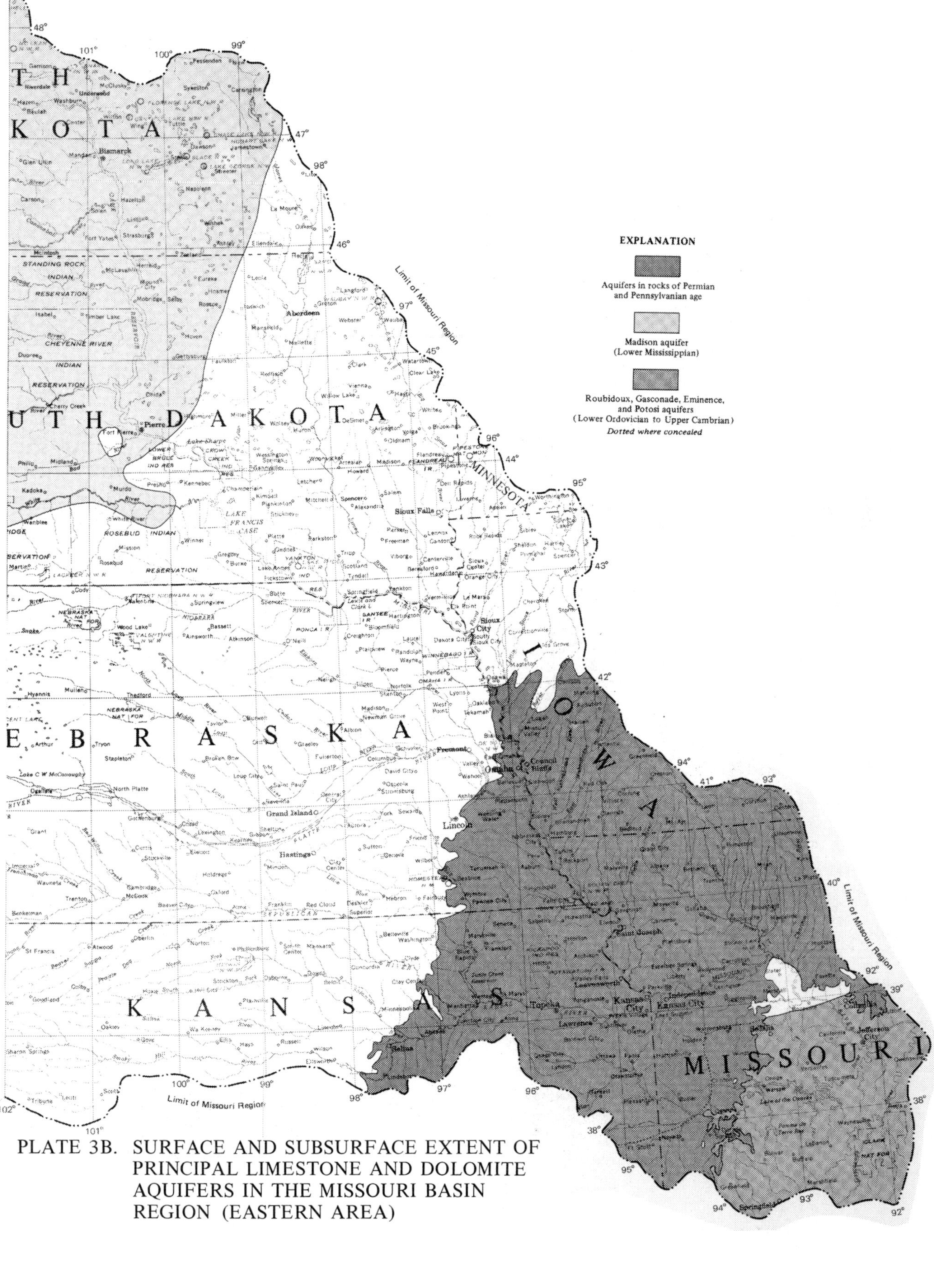

PLATE 3B. SURFACE AND SUBSURFACE EXTENT OF PRINCIPAL LIMESTONE AND DOLOMITE AQUIFERS IN THE MISSOURI BASIN REGION (EASTERN AREA)

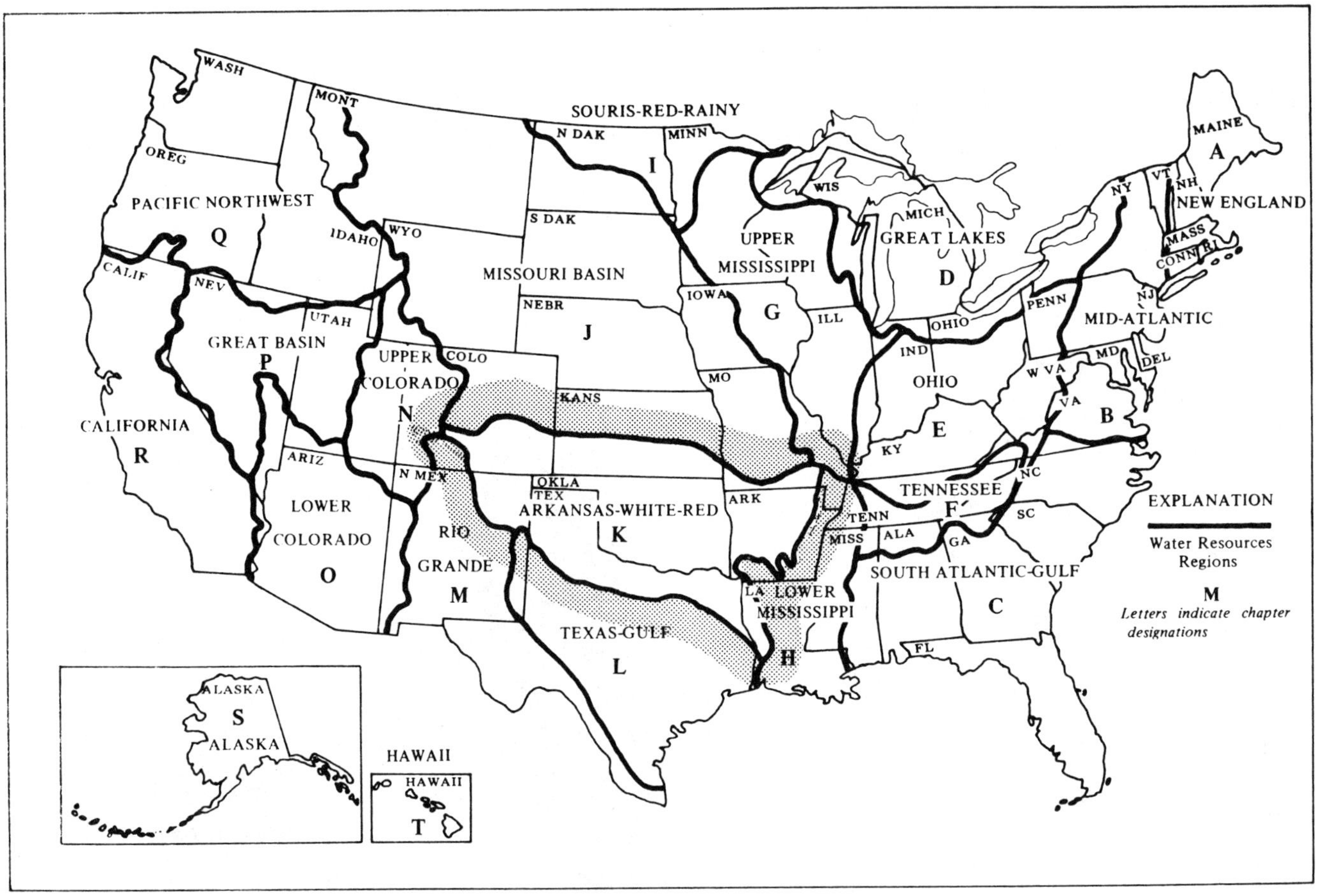

K – Location Map for the Arkansas-White-Red Region

K

ARKANSAS-WHITE-RED REGION

By M. S. BEDINGER and R. T. SNIEGOCKI

ABSTRACT

The Arkansas-White-Red Region, an area of 265,000 square miles (6.86×10^{11} square metres), is characterized by diversity in geography, climate, and geology and, in turn, by diversity in water resources and water problems. The western semiarid part of the region is water deficient, that is, potential evapotranspiration exceeds precipitation. The eastern, humid part has a surplus. Water use in the region in 1970 averaged 10 billion gallons per day (438 cubic metres per second), of which more than 65 percent was ground water. The largest use of ground water was for irrigation of crops, mostly in the water-deficient areas of Texas, Oklahoma, Kansas, and Colorado. Because of its ready availability and widespread occurrence, ground water is used throughout the region to supply municipal and rural water needs. The most productive aquifers, capable of yielding more than 50 gallons per minute (0.0032 cubic metres per second) to individual wells, are alluvium, carbonate rocks, gypsum, and sandstone. Fresh water in storage in aquifers in the region is estimated to be 2 billion acre-feet (2.5×10^{12} cubic metres). In addition, a large, unmeasured volume of saline water (containing more than 1,000 milligrams per litre of dissolved solids) underlies the fresh water at depths generally less than 500 feet (150 metres).

The flow of water in each aquifer depends upon the physical and hydrologic characteristics of the aquifer, the climate, and the relation to, and the character of, adjacent rocks and streams. These factors also determine the effect of water-supply development or other man-induced stresses on the flow and the quality of water in the aquifers. Analog and digital models of aquifers can be used to evaluate stresses on aquifers and thereby provide water managers and planners with efficient tools for planning the development and continued use of aquifers.

INTRODUCTION

The Arkansas-White-Red Region, as defined by the Water Resources Council, includes about 265,000 square miles (6.86×10^{11} m^2) in the South-Central United States (fig. 1). The three largest rivers—the Arkansas, White, and Red—drain approximately one-tenth of the Nation's conterminous land area, including all of Oklahoma and parts of Colorado, New Mexico, Kansas, Texas, Missouri, Arkansas, and Louisiana.

The Arkansas-White-Red Region is characterized by diversity in geography, climate, and geology, and, in turn, by diversity in water resources and water problems. The principal surface features of the region (Fenneman, 1931) consist of the high Southern Rocky Mountains in the west; the low mountains of the Ozark Plateaus and the Ouachita province, which rise abruptly from the Coastal Plain in the east; and between the two mountain areas, a broad expanse of the Great Plains and the Central Lowland, sloping gradually from west to east, broken in places by escarpments, hills, and a few old, eroded mountains (fig. 2). The climate varies from humid in the east to semiarid in the west. The western half of the region experiences temperature extremes and moisture deficiencies associated with its interior continental location. The eastern part is influenced by the warm moist air from the Gulf of Mexico.

THE WATER RESOURCE

Precipitation (mean annual) is about 50 inches (1,270 mm) in Arkansas and Louisiana, decreases rather uniformly to about 12 inches (300 mm) in the western Great Plains (fig. 3), and increases to 32 inches (810 mm) in the mountains of Colorado and New Mexico. In the Great Plains, annual precipitation is low and highly variable, and serious deficiencies occur during the growing season. The eastern section is subject to climatic extremes—to severe rainstorms and flash flooding and to droughts. Localized floods in the central and western sections result from infrequent but intense rainstorms of short duration. High wind velocities and high evaporation rates, associated with the dry climate of most of the western half of the region, cause annual lake evaporation to exceed precipitation.

The eastern half of the region is characterized by a surplus of water—that is, annual precipitation exceeds potential evapotranspiration. Areas of natural water surplus and of natural water deficiency within the Arkansas-White-Red Region are shown in figure 4. The water deficiency was computed by subtracting values of potential evapotranspiration from the average precipitation. However, even the areas of annual water surplus have seasonal or short-term periods of deficiency.

The occurrence of ground and surface water is entirely different in flow rate, quantity in storage, variability in quality, variability in quantity with time, and

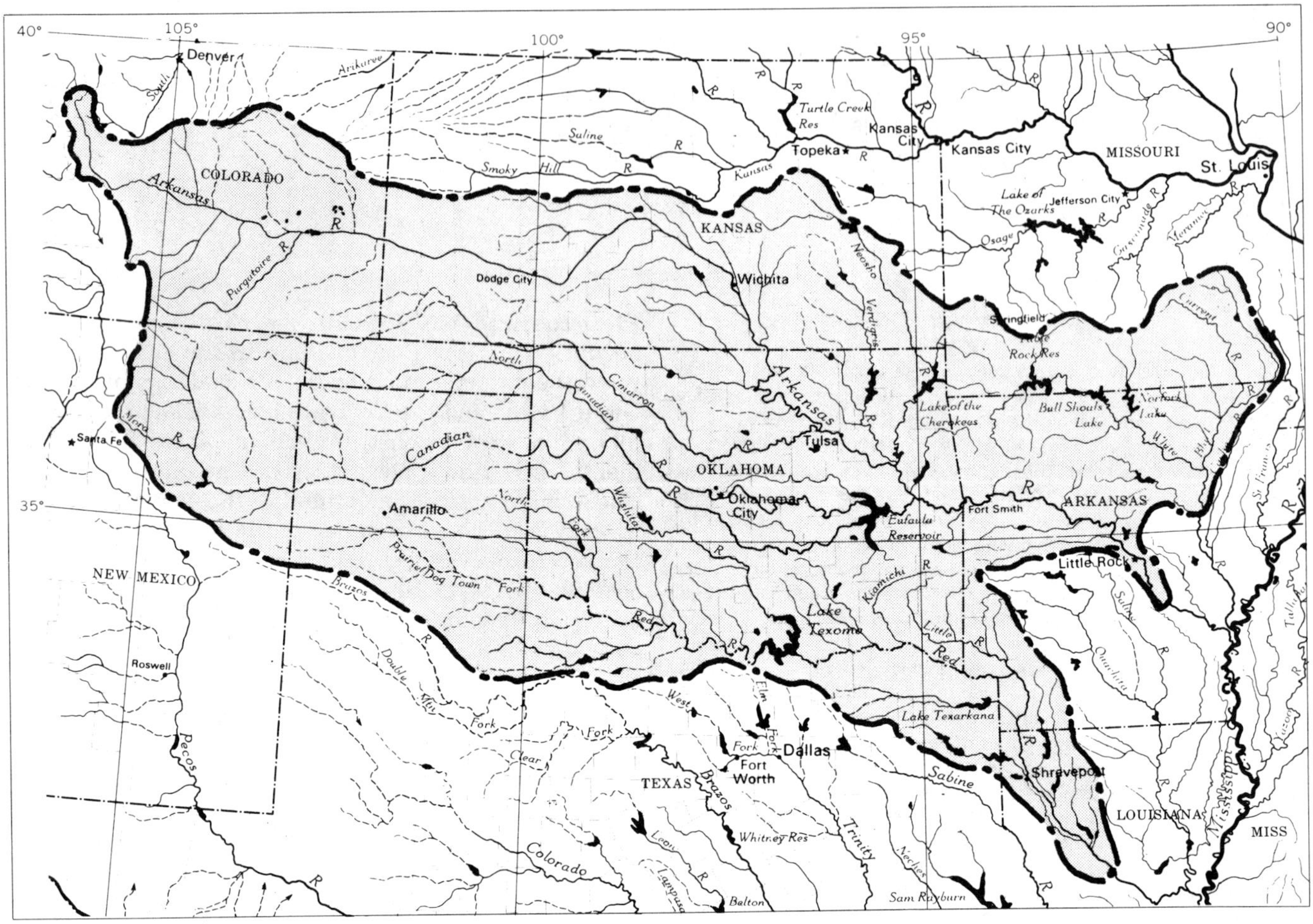

Base from U.S. Geological Survey,
The National Atlas of the United States of America

0 50 100 200 300 MILES

0 50 100 200 300 400 KILOMETRES

EXPLANATION

Area of this report

FIGURE 1.—Location of the Arkansas-White-Red Region.

areal distribution. These differences can be used to advantage in managing ground and surface water conjunctively to achieve the most efficient use of the total supply.

The natural streamflow, shown in figure 5, is the average annual flow in streams if there were no upstream development. In areas of direct hydraulic connection between surface water and ground water, the natural runoff may include natural base flow from aquifers.

Water in the Arkansas-White-Red Region is derived from precipitation in the region and interbasin transfers. The water resource fluctuates with variations in precipitation. Fluctuations are moderated by storage in surface-water and ground-water reservoirs. Regionally, the water resource consists of an average annual runoff of about 6 inches (150 mm) from a 265,000-square-mile (6.86×10^{11} m^2) area and an underground storage of about 2 billion acre-feet (2.5×10^{12} m^3). From 1961 through 1973, an average of 82,440 acre-feet ($1{,}016.0 \times 10^8$ m^3) of water was transferred from the Colorado River basin to the Arkansas River basin. Interbasin transfers during this period increased. Plans for diversion call for an eventual average of more than 200,000 acre-feet (2.47×10^8 m^3) to be imported annually from the Colorado River basin to the Arkansas River basin.

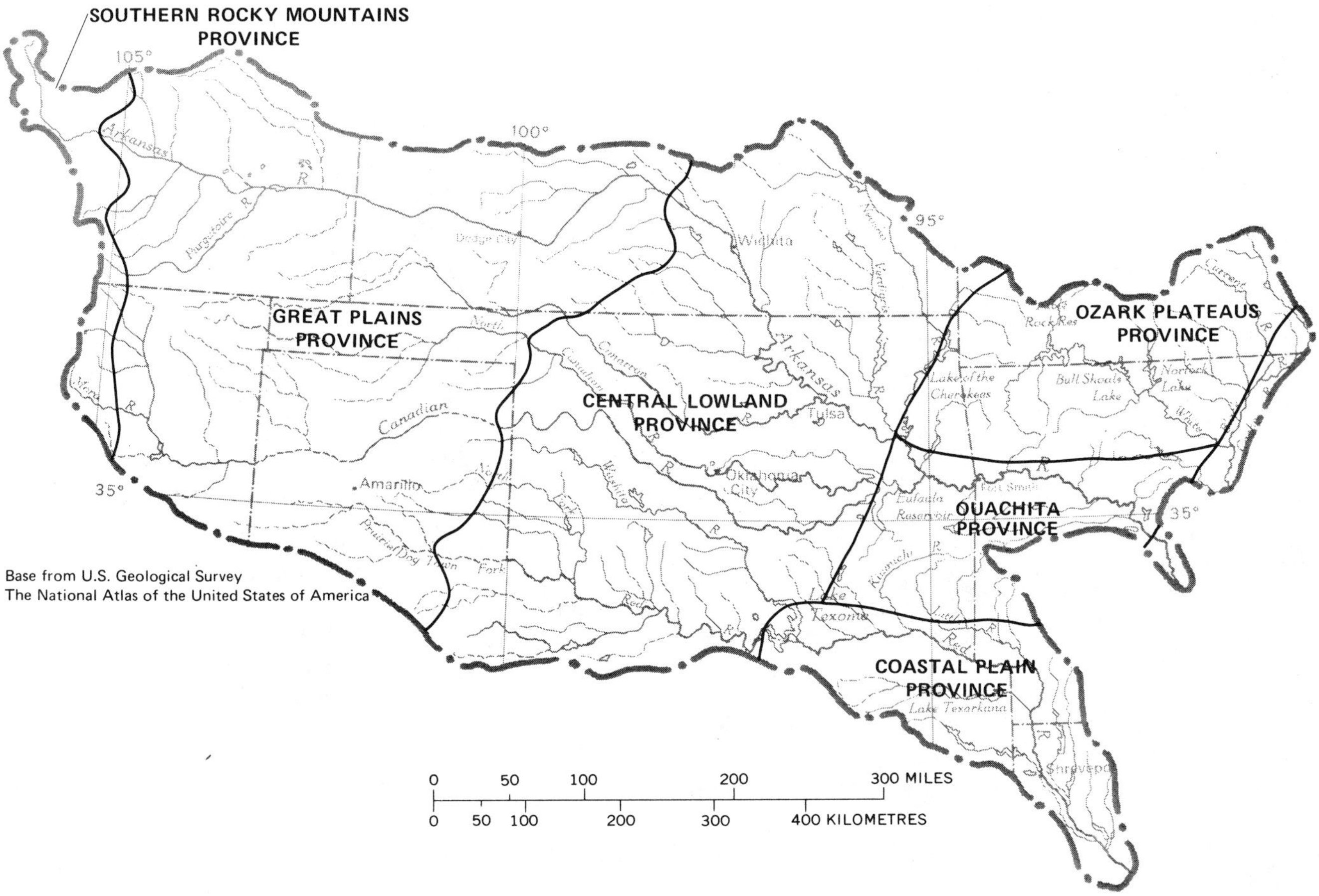

FIGURE 2.—Physiographic subdivisions of the Arkansas-White-Red Region.

The year-to-year difference between extremes of runoff is called the variability, which greatly affects the amount of water that can be developed for use. The natural runoff in the Arkansas-White-Red Region is 73 Bgal/d (3.2×10^3 m^3/s). The natural runoff varies widely—in 90 years out of 100 years, the flow exceeds 36 Bgal/d (1.6×10^3 m^3/s; Murray and Reeves, 1972), and in 95 of 100 years, the flow exceeds 33.4 Bgal/d (1.5×10^3 m^3/s). In addition, great variations in natural streamflow occur seasonally. Annual runoff is more variable in the west-central water-deficient part of the Arkansas-White-Red Region (fig. 6). Storage is required to meet the water demand in dry years. Even at such levels of development as those indicated possible by the available runoff in 90 percent and 95 percent of the years, storage would be required to make the indicated runoff available for use. Thus, the percentage of the mean runoff that is available for development depends on the variability of the annual runoff, the storage in the reservoirs, and evapotranspiration from the reservoirs. Woodward (1957) estimated that in 1955 the dependable water supply was 10 Bgal/d (4.4×10^2 m^3/s) and projected that in 1980 the dependable supply will be 20 Bgal/d (8.8×10^2 m^3/s), principally because of increased surface-water storage.

ACCENT ON GROUND WATER

Approximately two-thirds of the 10 Bgal/d (4.4×10^2 m^3/s) of water withdrawn in the region for public, rural, industrial, and irrigational use in 1970 was from ground water (table 1). Of the 6.6 Bgal/d (2.89×10^2 m^3/s) withdrawn from ground-water sources, about 90 percent was for irrigation. The general distribution of ground-water withdrawal and the major areas irrigated by ground water are shown in figure 7. Irrigation is practiced primarily in the major stream valleys and terraces of the Great Plains and Central Lowland, where aquifers provide large well yields. Of the remaining 10 percent of ground water withdrawn, about 4 percent is for industrial supplies, 4 percent is for public supplies, and 2 percent is for rural supplies.

An estimated 2 billion acre-feet (2.5×10^{12} m^3) of fresh water is in storage in aquifers in the region, about 60,000 times more than the 1970 water use from both ground- and surface-water sources. Further, the

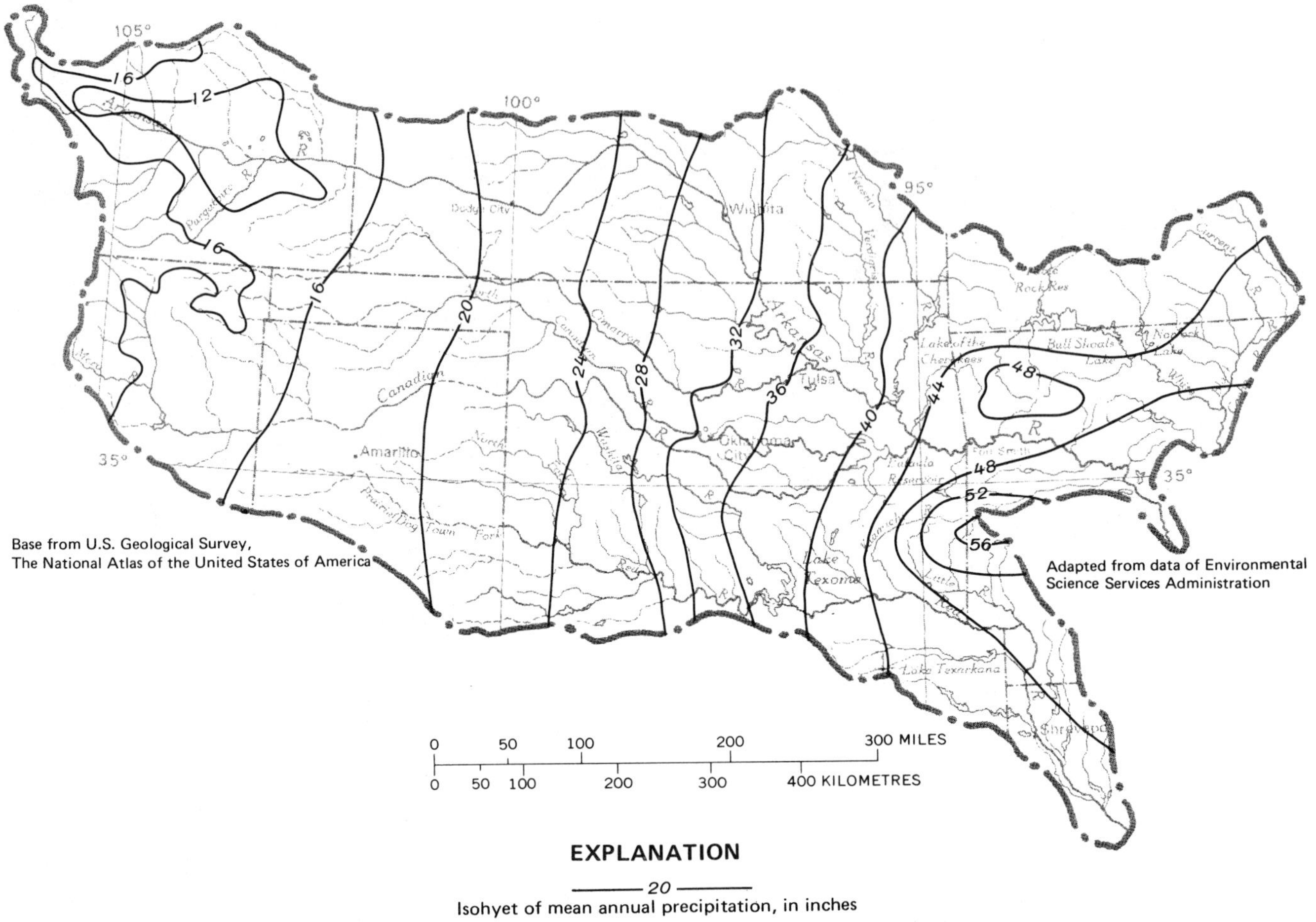

FIGURE 3.—Mean annual precipitation.

amount of fresh water stored in aquifers is more than 30,000 times that estimated to be perennially available from both ground- and surface-water sources in the region.

TABLE 1.—*Summary of ground water withdrawn in the Arkansas-White-Red Region in 1970, except for hydroelectric-power and thermoelectric-power generation*

[After Murray and Reeves (1972), and Murray, written commun. (1947)]

State	*Withdrawal (Bgal/d)*
Arkansas	0.150
Colorado	.220
Kansas	2.500
Louisiana	.027
Missouri	.058
New Mexico	.074
Oklahoma	.880
Texas	2.700
Total (rounded)	6.600

HYDROLOGIC ASPECTS OF WATER-RESOURCE PLANNING

The quantity of water entering and leaving a hydrologic system, such as that of the Arkansas-White-Red Region, within a given time can be related by the following continuity equation:

$$\text{Inflow} = \text{outflow} \pm \text{change in storage.}$$

If inflow is greater than outflow, the change in storage is positive; if inflow is less than outflow, the change in storage is negative.

Prior to development by wells, aquifers are in a state of dynamic equilibrium in which inflow and outflow are equal. Ground-water withdrawal upsets this balance by producing a loss from storage. A new state of equilibrium can be reached only by an increase in inflow and a decrease in outflow. Inflow can be increased by a lowering of water level, resulting in:

1. Storage space in the aquifer for recharge that would otherwise be rejected.
2. Induced leakage to the aquifer from confining beds.
3. Induced inflow from surface water.

Outflow can be decreased by a lowering of water level, resulting in:

1. Decrease in evapotranspiration from the aquifer.
2. Decrease in discharge from the aquifer by springs and flow to surface water.

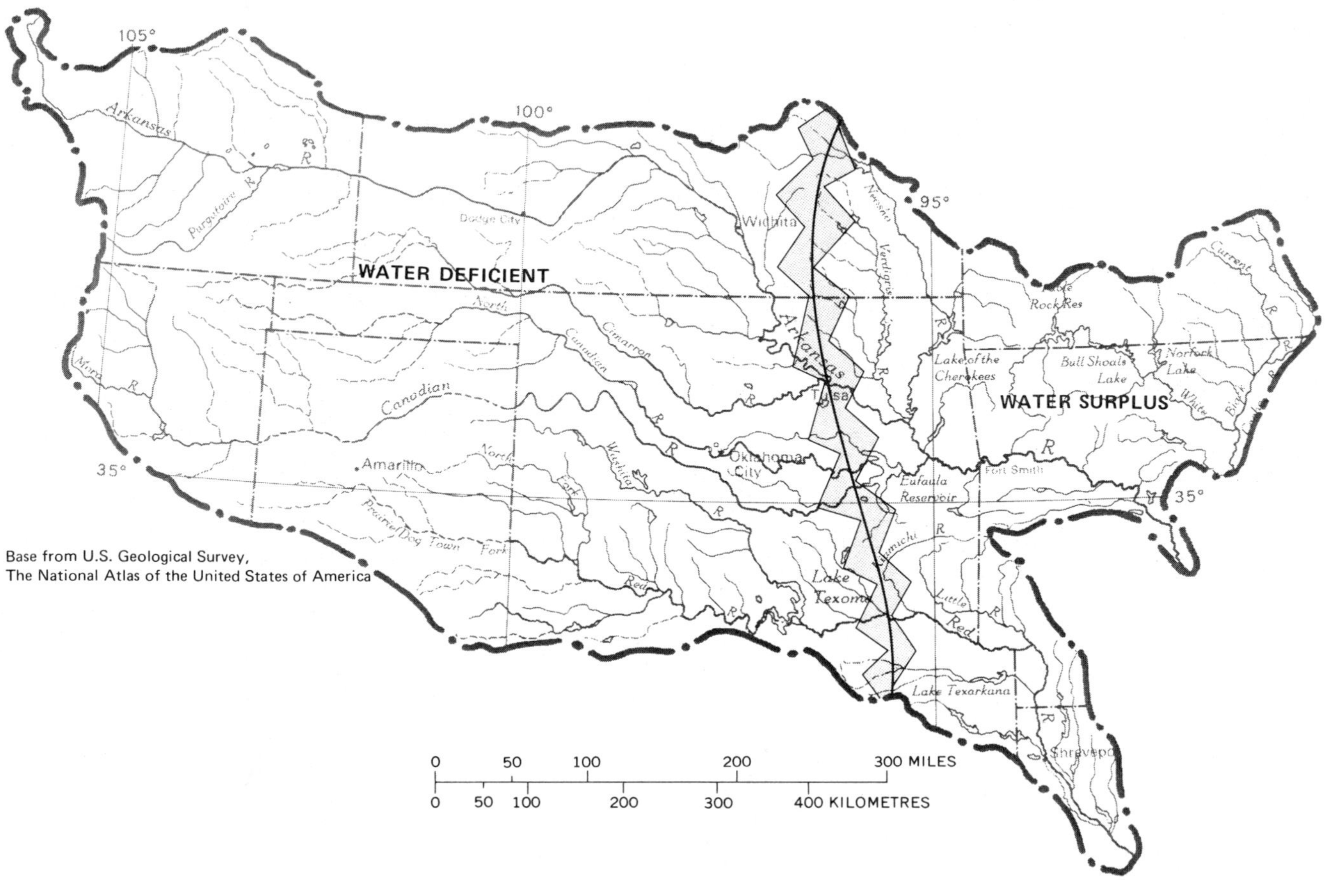

FIGURE 4.—Areas of natural water deficiency and natural water surplus (U.S. Water Resources Council, 1968).

3. Decrease in leakage from the aquifer through confining beds.

Use of ground water (as opposed to the exclusive use of surface water) increases the potential amount of water available for development. The increase includes the amount of water permanently withdrawn from storage in excess of recharge, and the water salvaged by reduction in evapotranspiration. Reduction in exapotranspiration from the aquifer represents an increase in the water available for use in the basin. The water in storage in an aquifer can be managed as a reserve to be withdrawn during seasons of peak demand or periods of drought. The depletion from storage can be replenished during seasons or periods when water is available for recharge.

To express the hydraulic response of an aquifer system to development, the equation of continuity may be written:

$$I + \Delta I = O + \Delta O + Q + \Delta S,$$

in which

I = inflow rate,
ΔI = change in inflow rate,
O = natural outflow rate,
ΔO = change in outflow rate,
Q = rate of withdrawal from wells,

and

ΔS = rate of change in storage.

It is implicit from this equation that an aquifer flow system is a complete functional unit. Application of the equation requires an understanding of the nature of the flow between the aquifer and contiguous streams and confining beds.

A desirable management objective in a groundwater development is to use the aquifer such that a continued decrease in subsurface storage is minimized. To do this, the withdrawal from storage must be balanced by an increase in inflow to the aquifer or a decrease in outflow.

Another management objective is to prevent or retard deterioration of the physical or chemical quality of the water or the environment. Under natural conditions, a state of balance is reached between hydraulic, thermal, and geochemical stresses in the flow system. An imposed hydraulic stress upsets not only the natural hydraulic balance but also the thermal and geochemical balances.

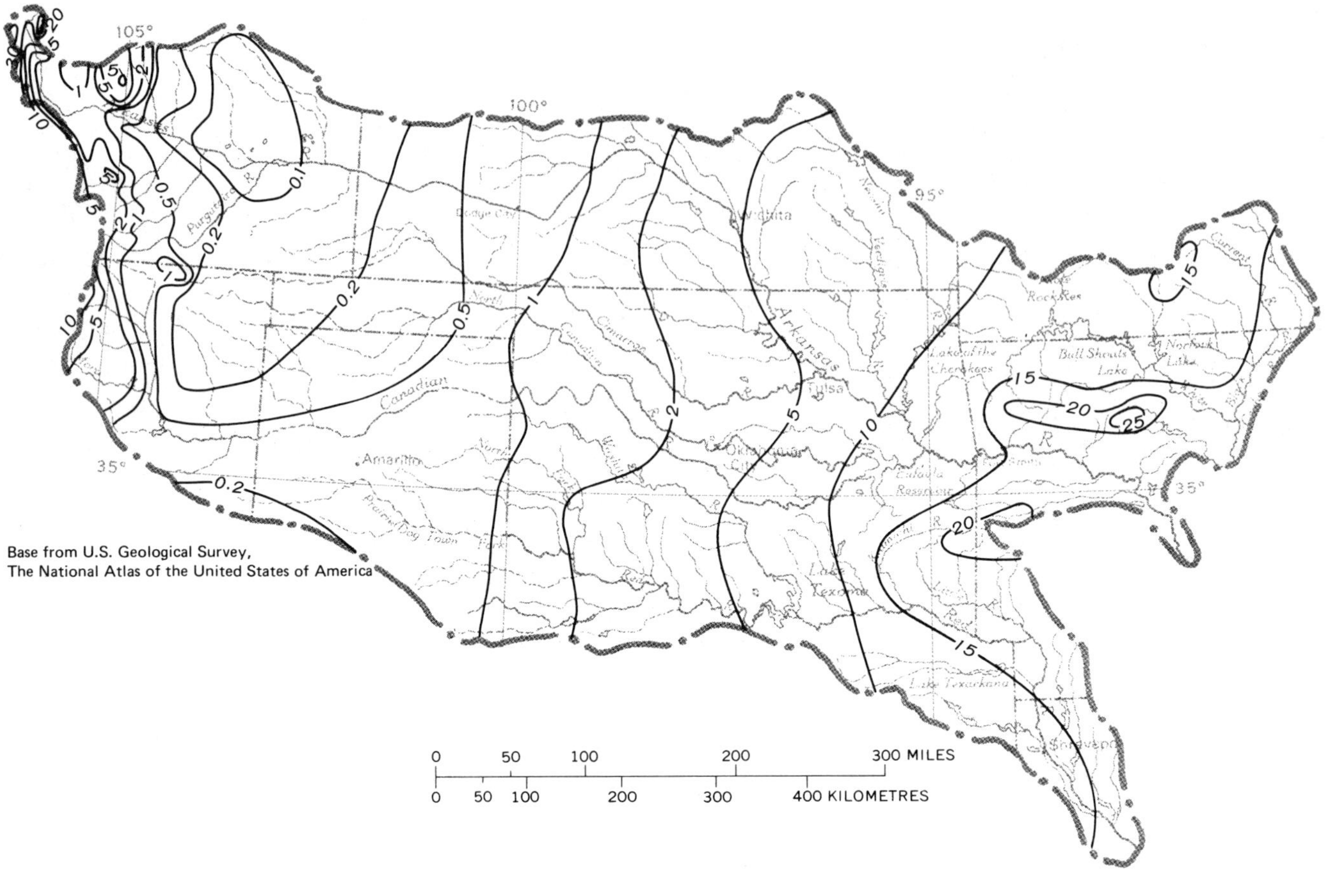

EXPLANATION

——— 20 ———
Line of equal average annual runoff, in inches

FIGURE 5.—Average annual runoff (after Busby, 1966).

The rationale for response of the system to changes in stress may be stated as follows:

1. The system reacts to each stress in accordance with the laws of conservation of energy and mass.
2. Any given stress or combination of stresses on the flow system will result in a trifold response, that is, hydraulic, thermal, and chemical responses in the system.
3. The response of the system to the stress is uniquely dependent upon the geometry and the physical and chemical characteristics of the system. In addition, a stress or response may involve a change in the physical or chemical character of the system framework.

The breadth of impact of a given stress on the aquifer may be illustrated by the use of table 2. The object of this table is to point out that a wide range of actions can affect an aquifer system. Actions that can affect the ground water include not only actions directed toward management of ground water and surface water but also to those related to agriculture, urbanization, manufacturing, and many others that are not directed toward water management.

Neither the list of actions nor the responses to stress shown in table 2 are intended to be complete. The table is intended as a checklist and an aid to the planner or water manager in considering systematically the full ranges of stresses and responses that an action will produce. The table may be considered to be an expansion of a part of an environmental-impact matrix, as presented by Leopold, Clarke, Hanshaw, and Balsley (1971). Also, it may be expanded and redesigned to aid in evaluation of the impact of a particular action on ground water.

The principal nature of stress on the aquifer for each action can be indicated in the table. Framework stresses include physical changes in the system framework, such as dredging, resulting in greater connection between the aquifer and streams, or channelization, resulting in new boundary conditions on part of the aquifer. A hydraulic stress is one that produces a

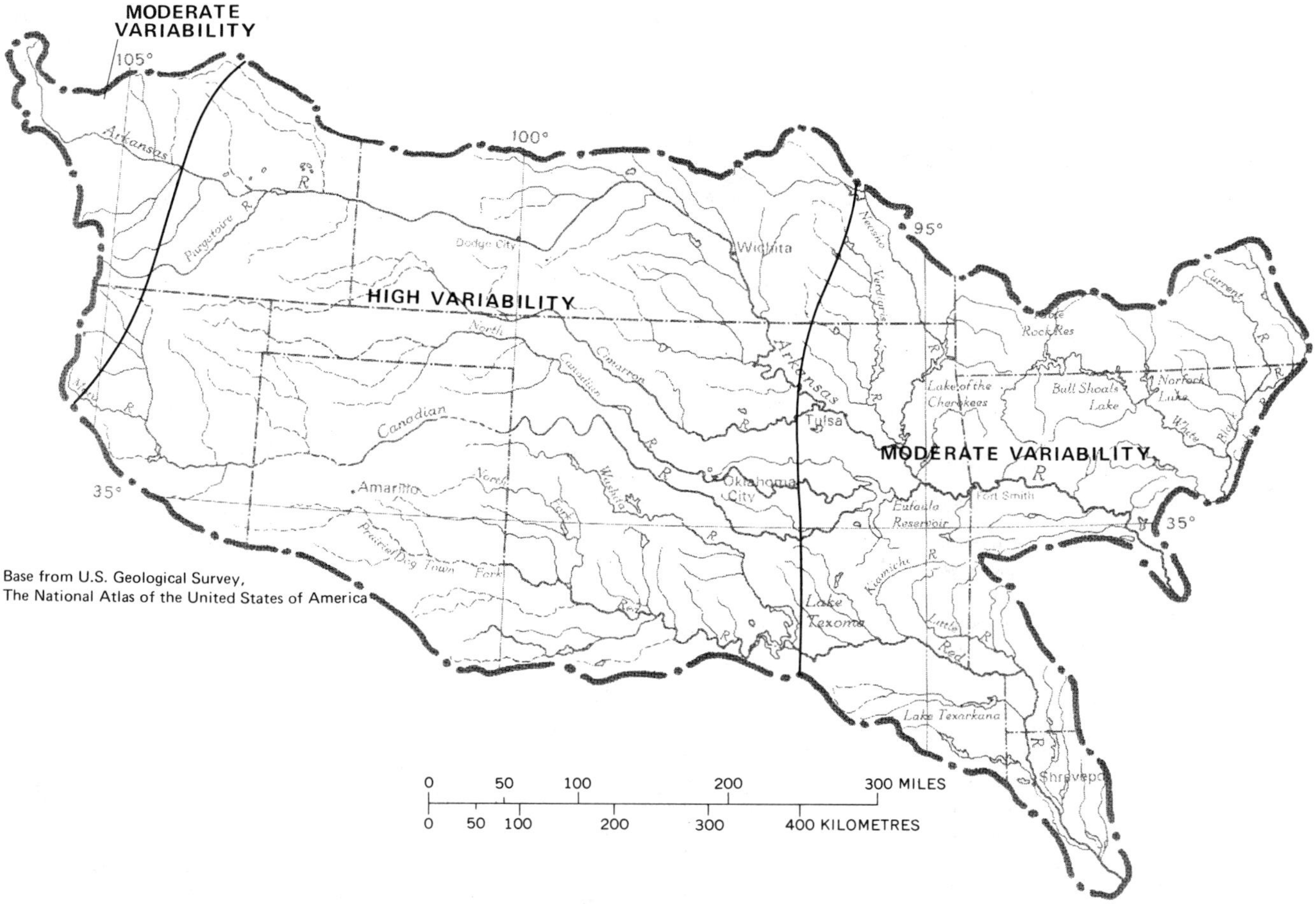

FIGURE 6.—Variations of the annual runoff (after Busby, 1966).

change in head or flow on a hydrologic boundary. Likewide, a thermal stress is one that produces a change in heat flow or heat gradient across a hydrologic boundary. A chemical stress is one that results in a direct change in chemical character of the water flowing into the aquifer or a chemical change in the system framework itself.

Responses of the aquifer are classified by the same groupings as those used for stress; however, the nature of the response of the aquifer is not limited to the same nature as the stress. Large changes in the aquifer framework in response to stress are not common but can be very significant, such as the extensive subsidence caused by large withdrawals in certain areas of California, at Mexico City, Mexico, and in the Houston, Tex., area. Earth movements in Denver, Colo., and vicinity have been attributed to injection of waste liquids. Subsidence has not been documented in the Arkansas-White-Red Region. The areas susceptible to subsidence are thick unconsolidated deposits, such as occur in the Gulf Coastal Plain and in deep basins in Colorado.

Hydraulic response in the aquifer is the most common response, particularly to water-management stresses. Hydraulic response has been divided into several changes in head and flow conditions in the aquifer and directly related effects, such as changes in well yield and specific capacity. The divisions of hydraulic response, as well as the other major response categories, are not intended to be comprehensive nor exclusive. Overlap between some of the divisions occurs and is sometimes desirable. Change in saturated thickness of the aquifer has been included in this table as a hydraulic response rather than a change in the aquifer framework.

Thermal response may be significant, particularly in applications where cooling water is required. The spectrum of chemical response is vast and complex. Only a few broad subdivisions of chemical response are given in table 2.

Table 2 can be used in the manner similar to that described by Leopold and others (1971) as follows: First, identify all actions that are a part of the proposed project. Second, in the rows of the proposed ac-

TABLE 2.—*Impact matrix relating*

Actions of potential impact on ground water	Principal nature of stress				Response of aquifer				
						Hydraulic			
	Framework	Hydraulic	Thermal	Chemical	Framework	Change in seepage to or from streams	Increase in leakage from confining beds and interbedded clays	Rise in water level	Decline in water level
Direct modification of hydrologic regime:									
Weather modification									
Desalinization									
Surface-water management:									
Reservoirs									
Flow regulation									
Water-retarding structures									
Locks and dams									
Revetments									
Levees									
Dredging									
Ditches and canals									
Channelization									
Interbasin diversions									
Pumping plants									
Surface-waste disposal									
Storm drains									
Irrigation									
Ground-water management:									
Relief wells									
Supply wells									
Dewatering wells									
Artificial-recharge									
Underground-waste disposal									
Subsurface drains									
Indirect modification of hydrologic regime:									
Agriculture:									
Fish farming									
Cultivation									
Land clearing									
Land leveling									
Terracing									
Fertilizer application									
Pesticide application									
Feedlots									
Pastures and grazing									
Timber management									
Urbanization:									
Homes and buildings									
Streets, parking lots, and airports									
Sewage lines									
Sanitary landfill									
Sewage spreading									
Septic tanks									
Mining and mineral production:									
Production									
Surface excavation									
Subsurface excavation									
Well drilling and fluid removal									
Manufacturing and processing									
Tourism and recreation									

stress and response in aquifer systems

Response of aquifer—Continued												
Hydraulic—Continued							Chemical					
Change in rate or direction of ground-water movement	Change from nonartesian to artesian, or vice versa	Change in saturated thickness of aquifer	Change in well yield	Change in specific capacity	Change in evapotranspiration or rejected recharge	Thermal	Change in dissolved-solids content	Dispersion and diffusion of introduced fluid of exogenous quality	Salinization of soils	Chemical reaction between introduced fluid and natural ground water	Chemical reaction between introduced fluid and system framework

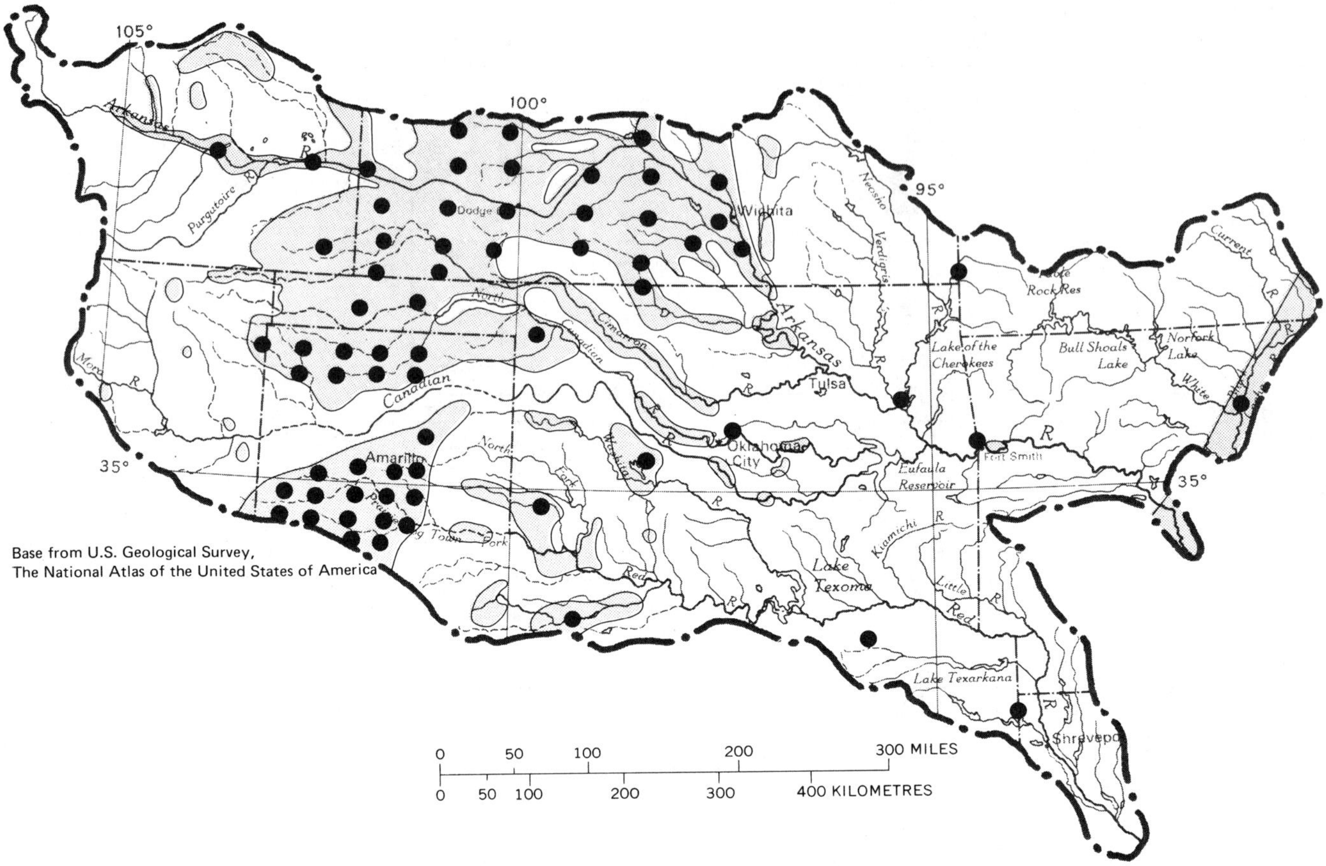

EXPLANATION

Areas irrigated with ground water

Represents withdrawal of 100 million gallons per day of ground water for all purposes

FIGURE 7.—Ground-water withdrawal and principal irrigated areas.

tions, place an "X" at the intersection with each applicable stress category. Third, consider the response to the combined stresses. Place a slash at the intersection of the action row with the response column. At this stage, an expansion of the response categories may be desired, depending upon the nature of the stress and the characteristics of the system being evaluated. Fourth, in the upper left corner of each box containing a slash, place a number from 0 to 3, which indicates the magnitude of the response; 3 represents the greatest magnitude of the response and 0, the least. In the lower right corner of the box, place a number from 0 to 3, which also indicates the significance of the possible response.

As examples, the matrix is used to illustrate the responses of aquifers to three projects (table 3). First, consider a navigation project involving construction of locks and dams, dredging of the channel, ditching to redirect interior drainage, and building revetments to stabilize the channel. These actions will produce changes in the framework and hydraulic, thermal, and chemical stresses. The greatest and most significant change will be a rise in water level in the aquifer. Small water-level declines will occur immediately downstream from the navigation dams. The resulting change in saturated thickness will be large but not significant in affecting transmissivity or the specific capacity of wells. In areas where water levels are raised, there will be significant increases in evapotranspiration from the aquifer. Increased evapotranspiration and seepage around dams will produce small thermal and chemical responses in water in the aquifer, and possibly salinization of the soils. The revetments and dredging will produce changes in the degree of connection between the aquifer and the stream and consequential changes in seepage. Ditching will cause water-level changes and attendant changes in the rate and direction of flow.

An irrigation project, in a semiarid climate, that derives water by ground-water withdrawals and surface-water diversion, will cause hydraulic and chemical stresses and small thermal changes. Declines in water levels near pumping wells, rises in water levels near leaky canals, and changes in seepage to the river will be the most significant hydraulic responses. Dissolved-solids concentration of the water in the aquifer will increase because of reuse and evaporation of the applied irrigation water. Salinization of soils will occur where applied water is insufficient to flush salts from the soil.

A water supply developed in a large artesian sand aquifer imposes a hydraulic stress. The response is largely hydraulic—lowering of water levels and related changes and the changes in the flow regime in the outcrop area. Responses may include a framework response, such as subsidence, and small thermal and chemical responses.

GROUND WATER IN WATER-RESOURCE PLANNING

CONJUNCTIVE USE WITH SURFACE WATER

Few projects have been deliberately planned to use ground water and surface water conjunctively. However, as discussed, many aquifer systems include hydrologically connected streams. This natural aquifer-stream connection results in an unplanned form of conjunctive use. The consequence may be deleterious or beneficial. But, even if beneficial, the benefits can be increased by sound hydrologically based planning. Planned conjunctive use of ground water and surface water unfolds a new dimension of water planning in which practices such as interbasin transfers of water, artificial recharge, and streamflow augmentation are employed.

The most readily amenable types of hydrologic systems for conjunctive management are the alluvial aquifer-stream systems along most of the length of the Arkansas and Red Rivers and along the lower White River. Pumping from the aquifer induces flow from the river to the aquifer or intercepts water that is moving toward the river; thus, drawdown in the aquifer is less than if there were no connected river. Poor-quality water in the middle reaches of the Arkansas River (Kansas and Oklahoma) is mixed with good-quality ground water, resulting in a quality that is satisfactory for irrigation and suitable for municipal supply after minimal treatment. The upper Arkansas River valley is intensively irrigated by ground and surface water. Conjunctive use of ground water and surface water, as practiced in the Arkansas River valley in Colorado, increases the usable water supply by 30 percent over the supply that could be developed by using surface water alone. The dependable supply could be further increased 23 percent by water-management practices that more fully utilize conjunctive-use techniques (Taylor and Luckey, 1974).

ARTIFICIAL RECHARGE

Conjunctive-use potential, involving utilization of the vast underground space in the western part of the Arkansas-White-Red River basins, calls for the technique of artificial recharge for storage of seasonal excess water or imported water. Artificial recharge has been the topic of much discussion and study. Although artificial recharge is frequently cited as a panacea for water-shortage problems, in reality its feasibility for application depends upon several factors.

Artificial recharge through wells is costly because of technical problems, maintenance costs, and large capital expenditure requirements. One of the inherent advantages of use of ground water over the use of surface water is the wide distribution and availability of ground water throughout large areas. In artificial recharge through wells, the reverse of withdrawal, the wide areal distribution is a disadvantage, requiring a surface-distribution system to many artificial-recharge wells. Generally, water recharged through wells and later withdrawn will cost at least twice as much as withdrawing native ground water. Studies by Sniegocki (1963) showed the cost of recharging through wells to be prohibitive for supplying water for irrigation in the Grand Prairie Region of Arkansas.

Artificial recharge through water spreading may be feasible on parts of the Arkansas-White-Red Region, including the High Plains. The feasibility of recharge by spreading is dependent upon the local conditions, requiring a large surface area of highly pervious earth materials above the aquifer.

REDUCTION OF CONSUMPTIVE WASTE

The salvage of evapotranspiration losses is the only practical means at present by which the amount of water available for use could be increased. A large part of the evapotranspiration loss is by transpiration by nonbeneficial plants such as phreatophytes.

TABLE 3.—*Impact matrices of water-resource*

Actions of potential impact on ground water	Principal nature of stress: Framework	Hydraulic	Thermal	Chemical	Response of aquifer: Framework	Hydraulic: Change in seepage to or from streams	Increase in leakage from confining beds and interbedded clays	Rise in water level	Decline in water level
Navigation project in alluvial-stream aquifers:									
Locks and dams		X	X	X	1/1	3/1		3/3	2/1
Revetments	X					1/0		0/0	
Dredging	X	X	X			2/1			2/1
Ditching	X	X	X			3/3			3/3
Irrigation project in alluvial-stream aquifers:									
Reservoirs		X	X	X		2/1		2/1	
Diversion canals	X	X	X	X	3/2	3/3		3/2	
Irrigation		X	X	X	3/2	2/1		2/2	
Supply wells		X				3/3	1/0		3/3
Industrial supply from artesian-sand aquifers:									
Supply wells		X			2/1	2/1	3/3		3/3

Phreatophytes send their roots down to the water table or capillary fringe. They consume large quantities of water in arid and semiarid regions, where they infest large acreages, particularly along those streams where the water table is shallow. They may cause serious siltation problems by clogging stream channels, thereby retarding streamflow.

In the eastern part of the Arkansas-White-Red Region, where water is more plentiful, as well as in the western part, many species of plants either of low economic value or of no value consume large quantities of water that would otherwise be available for beneficial use. Transpiration by noneconomic plants in the Arkansas River valley, downstream from Pueblo, Colo., averages 140,000 acre-feet (1.7×10^8 m^3) per year (J. E. Moore, oral commun., 1974).

Reduction of evapotranspiration losses may require removing noneconomic plants or depriving them of their source of water. The efficacy of a given procedure depends upon proper planning, based on an understanding of the conditions. Perhaps the most direct method of removing the water supply from the phreatophyte would be the lowering of the water level below the root depth of the plant. This procedure may also reduce loss of water by evaporation from the aquifer through the capillary fringe. Phreatophyte control by pumping from wells may not be feasible near perennial streams.

Control of plants of no economic value by removal could be augmented by replacement with economically valuable plants. Such practices have been successful in restoring rangeland in the western part of the Arkansas-White-Red Region. There, deep-rooted nonbeneficial plants and shrubs intercept water before it reaches the water table. Lowering of the water level would be of minimal value. In rangelands it has been shown (Rechenthin and Smith, 1967, p. 11), that many of the nonbeneficial plants consume up to four times as much water per pound of dry matter produced than the beneficial grasses consume.

LEGAL AND ORGANIZATIONAL CONSIDERATIONS

Water-resource management includes a comprehension of legal theories and procedures involving water. The laws are relatively few and simple in the eastern water-surplus part of the Arkansas-White-Red Region but are more complex in the western water-deficient *development actions on aquifer system*

Response of aquifer—Continued											
Hydraulic—Continued							Chemical				
Change in rate or direction of ground-water movement	Change from water table to artesian or vice versa	Change in saturated thickness of aquifer	Change in well yield	Change in specific capacity	Change in evapotranspiration or rejected recharge	Thermal	Change in dissolved-solids content	Dispersion and diffusion of introduced fluid of exogenous quality	Salinization of soils	Chemical reaction between introduced fluid and natural ground water	Chemical reaction between introduced fluid and system framework
3/2		3/0	1/0	0/0	3/3	2/1	2/2	2/2	1/1	1/0	1/0
2/1		0/0				1/0					
3/2		1/0	1/0	0/0	2/2	2/1					
1/1		2/1	1/0	1/0	1/1	0/0	1/0				
2/2		2/1	1/1	1/0	2/1	1/0	2/1			1/1	1/1
2/1		2/1	2/1	1/0	2/2	1/0				1/1	1 1
3/1		3/2	3/3	2/2	2/2	0/0					
3/2		0/0	3/2	0/0	2/1	1/0	1/0	1/0		1/0	1/0

part (fig. 4.). Obviously, a resource in surplus supply has little need of legal tools for its local allocation. Regional management by methods such as water transport to water-deficient areas from water-surplus areas may necessitate allocation procedures.

In most of the eight States in the Arkansas-White-Red Region, the laws are now in the process of modification or development. Consequently, only the principal terms presently used in water law in the Arkansas-White-Red Region are considered.

These terms are riparian, overlying, and appropriative rights. Under riparian rights, the land adjacent to a stream or lake is riparian to that body of water. Under common law, the waters of the stream are available on a correlative basis for the use of all riparian owners. In overlying rights, the overlying landowner either possesses rights that are analogous to riparian rights on a stream or asserts absolute title to water under his land. Under appropriative rights, the right to water use is based upon priority of diversion and application of the water to beneficial use.

The riparian doctrine of reasonable use has been accepted by the Arkansas Supreme Court. This general rule applies to surface water, as well as to ground water. In addition to the riparian doctrine, the Arkansas Legislature has adopted some aspects of the appropriative doctrine, whereby a State agency may allocate a fair share of water to persons where there is a shortage.

Colorado's basic principle of water law is that of appropriation and includes ground water that is tributary to a watercourse. The State recognizes the need for clarification of the question regarding ground water not adjacent to a watercourse and is in the process (1974) of modification of the water laws.

The water law in Kansas is based on the principle of prior appropriation and applies to surface and ground water.

In a general way, Louisiana follows the riparian rule. In 1910 the State declared that all surface water then not in private ownership was the property of the State, subject only to the jurisdiction of the United States on navigable waters. Almost no attention has been given to water laws concerning ground water.

Missouri adheres to the riparian doctrine in the administration of water rights and includes ground water and surface water in this category. Modifications in the State's water laws are being considered.

In New Mexico, both surface water and ground water are managed under the doctrine of prior appropriation. The State engineer recognizes the relation between surface water and ground water and requires that new pumping of ground water be offset by retirement of existing surface-water rights equivalent to the anticipated depletion of streamflow.

Oklahoma is operating under the appropriative theory of water rights. Riparian rights are recognized only to the extent of domestic, household, and stock water. The State's ground-water law provides for adjudication of ground-water rights in designated critical areas. Ground water and surface water are treated as separate entities.

Texas enacted a ground-water law in 1949 that authorized the formation of local districts having the power to make and enforce regulations governing the withdrawal of percolating ground water. By virtue of this law, Texas has given the power of regulation to local groups, thus placing the responsibility for regulation at the lowest possible governmental level capable of performing the desired functions. The districts in Texas are unique in that their boundaries are required to be coterminous with the underground reservoir or subdivision thereof.

A compact, in the general sense, is a type of water law that deals with streamflow across State boundaries. Noteworthy examples are compacts between Colorado and Kansas, Kansas and Oklahoma, and Oklahoma and Arkansas, on the flow of the Arkansas River and the proposed compact on the flow of the Red River. For the most part, these compacts do not deal with ground-water flow across State boundaries. Considerable quantities of ground water are exchanged between Colorado and Kansas, and Kansas and Oklahoma, in the vicinity of the Arkansas River. Only negligible amounts of ground water move across the boundary between Oklahoma and Arkansas.

This brief review of the water laws in the Arkansas-White-Red River basins shows that variations exist in the application of the three principal doctrines. As indicated, most of the States are in the process of modifying these laws.

MODELS FOR WATER-RESOURCE PLANNING

The development of sophisticated and accurate digital and analog models for representation of ground-water-flow systems has provided efficient tools for water-resource planning and for water-management decisions. The scope of modern-modeling techniques provides a basis for environmental evaluations of quantity and quality of water and for considering the effects of economic and legal constraints on water use. This section discusses modern techniques of ground-water modeling, with special reference to model applications in the Arkansas-White-Red Region.

Analog and digital models simulate the hydrologic properties and boundaries of the hydrologic system. The scientific basis for both models is the finite-difference approximations of the equations that define the flow system. The main difference between the two types of models is that in the digital model flow equations are solved mathematically and in an electric analog model the equations are simulated with a resistor-capacitor network. Each type has certain advantages, depending on the system to be modeled.

The analog model can simulate very large, complex hydrologic problems, such as one that involves two or more aquifers with fine-grained interbeds and confining beds. It provides a visual display of the aquifer characteristics (transmissivity and storage) and boundaries (impermeable and stream contacts). The analog is also programed and read by the use of visual-display units. The visual nature of the analog and its operation, and the readiness with which stress acting on the aquifer, such as pumping or river stage, can be changed and the effects observed, make the analog useful in visibly modeling the operation of the hydrologic system.

The digital model can solve the more complex problems, such as heat flow, movement of contaminants, chemical quality changes, and transient changes in transmissivity with saturated thickness. Digital models can also incorporate legal, economic, and environmental constraints on the system and can be programed to seek the best water-use conditions. Data input, data output, and program changes can be managed more efficiently by using digital models.

Analog models of the lower Arkansas and Verdigris River alluvial aquifers were used to analyze the effects of rocks and dams on the potentiometric surface (fig. 8). The Arkansas and Verdigris Rivers are made navigable by a series of 17 locks and dams for a distance of 450 miles (7.2×10^5 m) from the Mississippi River through Arkansas to near Tulsa, Okla. The analog models represent 2,350 square miles (6.1×10^9 m^2) of alluvial aquifer adjacent to the river (Bedinger and others, 1970). They were made up of resistors and capacitors in a rectangular network to represent the transmissivity and storage characteristics of the aquifer. Resistor-capacitor nodes in the model were spaced to represent distances of ⅛ – ½ mile (201 to 805 m). Resistors connected at each node of the aquifer network were used to model evapotranspiration as a function of the depth to water level.

A combination of an electric analog and a digital model was used in the Arkansas River valley of Colorado to define the operation of an aquifer-stream sys-

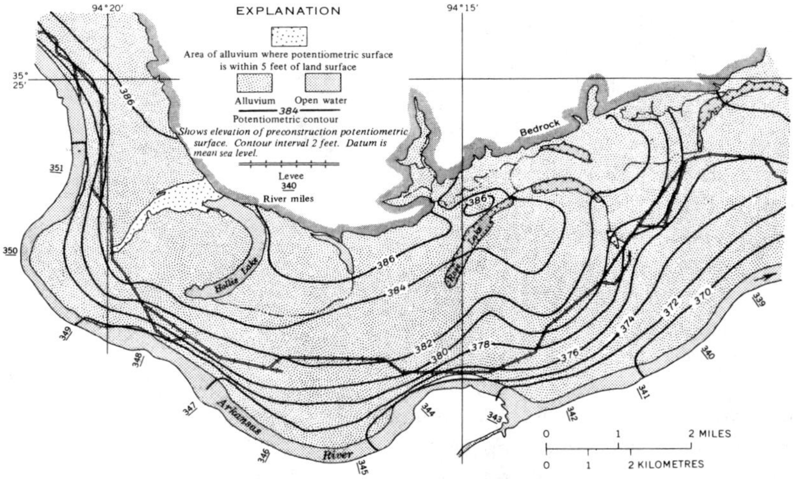

A. Area studied in selection of a site for lock and dam 13, near Fort Smith, Ark.

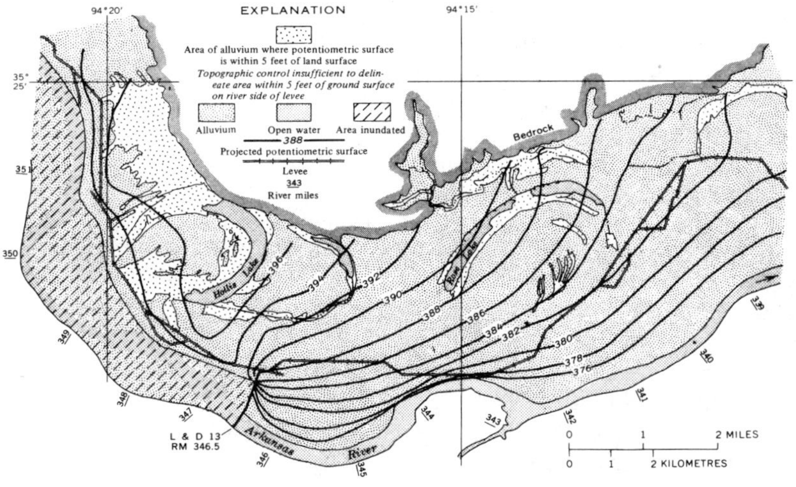

B. Projected potentiometric surface for damsite (lock and dam 13) at river-mil

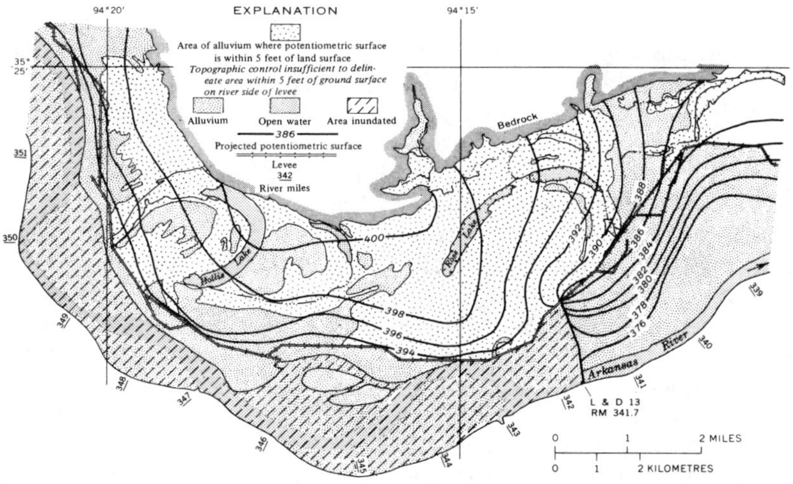

C. Projected potentiometric surface for damsite (lock and dam 13) at river-mile 341.7.

FIGURE 8.—Preconstruction ground-water conditions and projected ground-water conditions in a part of the Arkansas River valley, Arkansas (from Bedinger and others, 1970, p. 63–65).

tem and to predict effects of changes in water management. The area modeled, extending about 150 miles (2.4×10^5 m) from Pueblo, Colo., to the Kansas boundary (Moore and Wood, 1967), is underlain by an alluvial aquifer that is hydraulically connected to the river. The ground water and surface water constitute one supply. The development of ground water for irrigation has caused legal disputes between ground-water and surface-water users because withdrawals by wells have reduced the flow of the Arkansas River. An electric analog model was developed to simulate the stream and the aquifer. Applied irrigation water, precipitation, evapotranspiration, and well withdrawal were the hydrologic stresses incorporated in the model.

The analog model was stressed at 266 points to obtain reponse curves showing the effect of aquifer stress on streamflow. The response curves, summarizing differences in aquifer transmissivity, specific yield, and boundaries, were then prepared as a stream-depletion factor map (Moulder and Jenkins, 1969). The lines of equal-stream-depletion factors connect points where the hydrologic stresses on the aquifer have the same effects on the streamflow (fig. 9). The stream-depletion-factor map was the basis for constructing a digital model of the aquifer-stream system, to analyze and optimize water-management plans for any given set of management objectives. The model was designed to predict the availability of surface water at successive diversion points downstream, on a month-to-month schedule, and to show change in ground-water storage. Output data from the model of the Arkansas River valley have been used for planning maximum water use within legal, economic, environmental, and hydrogeologic constraints. The results have been used for modifying water law, for developing ground-water supplies, and for distributing water.

Digital models have been developed to couple the computer solutions of ground-water-flow equations with mass-transport and dispersion equations of a dissolved-chemical constituent in an aquifer (Bredehoeft and Pinder, 1973). This coupling permits the modeling of the spread of saline water or other contaminants in an aquifer. The model has been used to simulate changes in salinity related to irrigational practices in the Arkansas River valley of southeast Colorado (fig. 10) (Konikow and Bredehoeft, 1973), where the ground water and surface water are interrelated in an aquifer-stream system. Crops are irrigated by both diverted surface water and pumped ground water from the alluvial aquifer. Much of the irrigation water is lost by evapotranspiration, but some of it infiltrates the alluvial aquifer and provides return flow to the stream. Dissolved solids in the return flow become concentrated because of the evapotranspiration. The downvalley reuse of water causes a buildup of salts

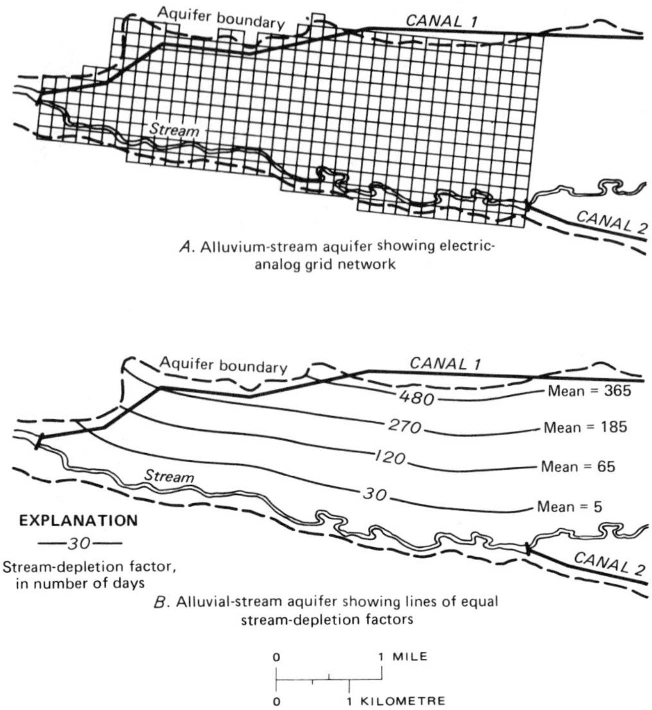

FIGURE 9.—An alluvial-stream aquifer in part of the Arkansas River valley, Colorado, showing (A) the electric analog grid network and (B) lines of equal-stream-depletion factors (from Moulder and Jenkins, 1969).

approaching levels that could restrict the use of the water.

The model can be used to predict the effects of changes in management or irrigational practices on the quality and quantity of ground and surface water. For example, the model could be used to evaluate the impact of increased pumping from irrigation wells, increased surface-water diversions, increased irrigated acreage, floods, droughts, and increased salinity in the river due to upstream activities. The model can aid in determining the feasibility of plans to improve the quality of ground water or to limit salinity increases in the river during critical low-flow periods.

One application that illustrates the versatility of digital models for use by the planner is the development of models that simulate the physical and chemical responses of a flow system to changes in environmental and managerial stresses. The digital models may also incorporate features to perfect management of the system under environmental, social, legal, and economic constraints. A model of the aquifer-stream system in southeastern Colorado was used to simulate the dependable supply under various water-distribution schemes. Included in analysis are flow components, such as salvage of water from phreatophytes, evapotranspiration, different

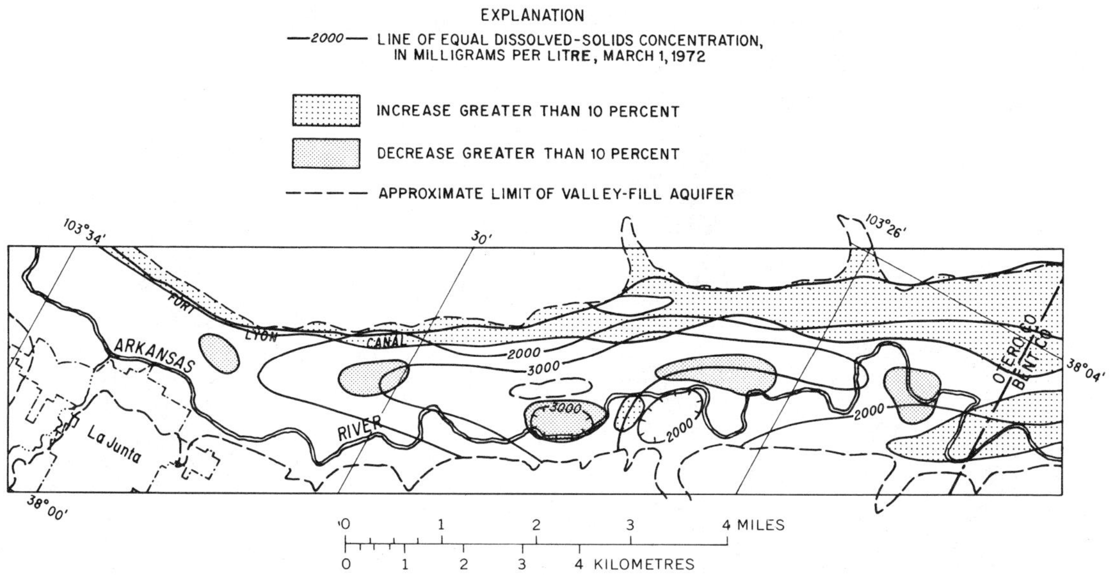

A. Observed dissolved-solids concentration on March 1, 1972, and observed changes since March 1, 1971

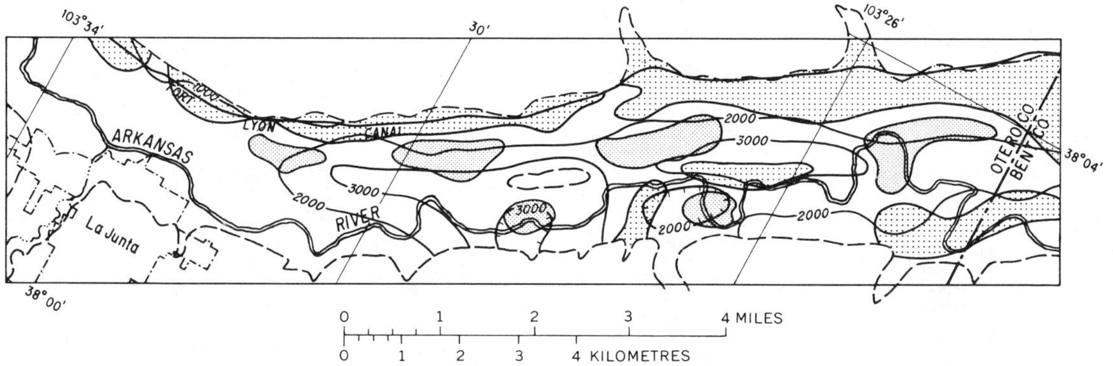

B. Calculated dissolved-solids concentration on March 1, 1972, and calculated changes since March 1, 1971

FIGURE 10.—Observed and calculated dissolved-solids concentrations in ground water in a part of the Arkansas River valley, Colorado (from Konikow and Bredehoeft, 1973).

reservoir-operation regulations, use of imported ground water and surface water, additional reservoirs, additional ground-water use, and application of excess streamflow (Taylor and Luckey, 1974).

GROUND WATER IN THE ARKANSAS-WHITE-RED REGION

The Arkansas-White-Red Region comprises the following six aquifer types: (1) Stream-valley alluvium, (2) terrace alluvium, (3) alluvium of intermontane valleys and buried alluvial valleys, (4) carbonate and gypsum, (5) sand and sandstone, and (6) undifferentiated sandstone, carbonate rock, shale, and (or) basalt. Their distribution is shown in figure 11.

Information is available about the aquifers from offices of the U.S. Geological Survey and from State agencies in the respective States in which the aquifers occur; references to published information and unpublished data can be obtained from the agencies listed later in this report. Included are data on thickness, permeability, transmissivity, storage coefficient, recharge rates, water in storage, depth to water, direction and rate of ground-water movement, water use, well yields, water quality, and other information, such as grain size of aquifer materials. A summary of data is given in table 4. Geologic and geographic distribution of formations that are principal aquifers are given in table 5.

The quality of ground water throughout the Arkansas-White-Red River basin varies considerably, depending upon the original character of water entering the aquifer, the nature of the rocks through which

TABLE 4.—*Principal aquifers in the Arkansas-White-Red Region*

Aquifer type	Nature of rock	Thickness (ft)	Areal extent	Depth to water (ft)	Hydraulic conductivity (ft/d)	Well yields (gal/min)	Development and use	Ground water in storage (acre-ft × 10^8)
Stream valley alluvium	Sand and gravel	50-200	Along large streams in flood plains. Extensive in Coastal Plain of Arkansas and Louisiana.	0-30	100-1,500	300-5,000	Extensive; principal source of ground water; frequently overdeveloped. Not used in some areas.	2.8
Terrace alluvium		50-600	Plains of Texas, New Mexico, Colorado, Kansas, and Oklahoma.	50-300	10-700	50-1,000	Extensive; subject to overdevelopment and water mining, particularly in High Plains of Texas.	4.1
Alluvium of intermontane valleys and buried alluvial valleys		100-5,000	Arkansas River basin in Colorado.	0-50				0.2
Carbonate and gypsum	Limestone and dolomite, and gypsum beds. Generally a dense rock, but subject to solution along fracture and bedding planes.	50-1,500	Limestone and dolomite in southern Missouri, northern Arkansas, southeastern Kansas, and Oklahoma. Gypsum in Oklahoma and Texas.	30-450	50-1,500	50-1,000	Moderately to heavily developed; overlooked as a source of water in some areas. More subject to pollution than other aquifers because of cavernous nature.	3.2
Sand and sandstone	Sand grains ranging from very fine to coarse. Generally cemented with siliceous material or carbonate. Unconsolidated in the Coastal Plain.	100-500	Sandstone principally in Kansas, New Mexico, and Oklahoma. Sand in Coastal Plain of Arkansas, Texas, and Louisiana.	20-300	([1])	10-1,000	Extensive; subject to overdevelopment and water mining. Loss of artesian head in many areas ranging from 2 to 300 feet.	7.9
Undifferentiated sandstone, carbonate, shale, or basalt	Consolidated rocks, including sandstone, interbedded shale, carbonate, and crystalline igneous rocks.	100-5,000	Sandstone, carbonate, and shale locally throughout region; basalt in parts of New Mexico, Colorado, and northwestern Oklahoma.	1,200	([2])	5-50	Mainly domestic use, not heavy, concentrated use, because of low permeability and low well yields. Difficult to predict well yields.	2.2

[1] Generally less than 100 ft/d.
[2] Generally less than 10 ft/d.

the water has moved, the contaminants that may have been introduced, and (or) the concentration of mineralization by evaporation or transpiration. The quality ranges from fresh to saline. Fresh water (containing less than 1,000 mg/l of dissolved solids) is generally present at shallow depths, and saline water, at greater depths.

The prevalent dissolved-solids concentrations and chemical types in streams at low flow are shown in figures 12 and 13. At low flow the streamflow is largely ground-water discharge. The quality characteristics of water in streams at low flow are therefore indicative of the quality of the water in the aquifers in contact with the streams. The dissolved-solids concentration (fig. 12) is generally low in the eastern part of the region and at the higher elevations along the west margin of the region. The areas having high dissolved-solids concentrations coincide with the distribution of shale beds

TABLE 5.—*Names and geographic distribution of major geologic units that form aquifers*

Aquifer type	Major geologic unit names	Distribution
Stream valley alluvium		Valleys of major streams.
Terrace alluvium	Seymour Formation	Texas.
	Equus zone of McPherson Formation	Kansas.
	Ogallala Formation	Texas, New Mexico, Colorado, Kansas, and Oklahoma.
Alluvium of intermontane valleys and buried alluvial valleys		Upper Arkansas River valley and Wet Mountain valley, Colorado.
Carbonate	Reeds Spring Limestone	Oklahoma.
	Boone Formation	Arkansas.
	Roubidoux and Gasconade Formations and the Eminence and Potosi Dolomites of the Arbuckle Group	Arkansas, Missouri, Kansas, and Oklahoma.
	Arbuckle Group undivided	Oklahoma and Kansas.
Gypsum	Dog Creek Shale and Blaine Gypsum	Oklahoma and Texas.
Sand and sandstone	Sparta Sand	Arkansas and Louisiana.
	Wilcox Group, Carrizo Sand, and Trinity Group	Oklahoma, Texas, and Arkansas.
	Purgatoire Formation, Cheyenne Sandstone, and Dakota Sandstone	Colorado, Kansas, New Mexico, and Oklahoma.
	Vamoosa Formation, Rush Springs Sandstone, Garber Sandstone, Wellington Formation, Member of Gasconade Formation, Simpson Group, and Gunter Sandstone	Oklahoma.

in southern Colorado and northern New Mexico, with salt springs and the pollution by oil-field brine in northern and central Oklahoma; and with the occurrence of gypsum beds in Texas through central Oklahoma to southern Kansas. The chemical types (fig. 13) reflect the nature of water in the shallow aquifers.

Saline ground water (containing more than 1,000 mg/l dissolved solids) is at depth beneath fresh water in most of the Arkansas-White-Red Region (Feth and others, 1965). Saline water is at depths less than 500 feet (150 m) throughout most of the area. The rate of use of saline ground water in the region in 1970 was estimated to be 38 Mgal/d (1.66 m^3/s; Murray and Reeves, 1972)—a very low rate compared with the rate of use of fresh water and compared with the amount of saline ground water that is available. When considered as a resource, the potential of saline water falls mainly into two categories—(1) direct use in industrial processes, such as cooling, or for irrigation where a moderate mineral concentration may not be a disadvantage, and (2) use after dilution or demineralization to whatever degree may be required by the intended user. Aquifers containing saline water may be considered as a resource potential for temporary storage of fresh water (Moulder, 1970) and disposal of wastes.

The areas of potential or existing ground-water pollution (fig. 14) are chiefly associated with areas of naturally high dissolved solids in streams and aquifers and the limestone terrane of the Ozark region. Water of naturally high dissolved solids may be further concentrated by evaporation and reuse for irrigation. Infiltration of irrigation water to the aquifer increases the salinity of the aquifer. The highly permeable limestones of the Ozark region are subject to pollution by entry of polluted wastes. Saline water may be induced to enter freshwater aquifers from adjacent saline-water aquifers and streams.

AQUIFER SYSTEMS IN THE REGION

The flow operation of each aquifer system depends on the characteristics of the aquifer, the climate, and the character of adjacent rocks and streams. These are the principal factors that also determine the response of aquifer flow systems to development or other man-induced stresses.

STREAM-VALLEY ALLUVIAL AQUIFERS

Alluvial deposits in stream valleys cover more than 28,000 square miles (7.25×10^{10} m^2) of surface area and constitute some of the most productive aquifers. The alluvium generally ranges in thickness from 50 to 200 feet (15.2 to 61.0 m) and ranges in width from 1 to 10 miles (1.6×10^3 to 16.0×10^3 m), except in the Coastal Plain, where the width is several tens of miles. The aquifers are generally composed of sand and gravel in the lower part of the alluvium.

The alluvial aquifers are in connection with the predominantly perennial streams. Ground water in the alluvium generally is at shallow depths, within reach of roots of plants. The alluvial material above the sand and gravel varies from fine sand to silt and clay; consequently, the water in the aquifer may be under water-table or semiconfined conditions.

The alluvium commonly lies in a valley cut in less permeable rock. Locally, the alluvium is in contact with rocks of equal or greater permeability.

OPERATION OF THE FLOW SYSTEM IN THE WATER-DEFICIENT AREA

Recharge by rainfall is low, ¼–½ inch (6.4×10^{-3} to 1.3×10^{-2} m) per year. Streams may be perennial or intermittent. Discharge is by evapotranspiration and by flow to streams. In response to withdrawal of ground water, water levels are lowered, reducing the amount of both evapotranspiration and rejected recharge. The amount of water salvaged from rejected recharge is relatively small because of low precipitation and high loss of precipitation by evaporation. Water-level declines also reduce streamflow, either by decreasing flow to the stream or by inducing flow from the stream into the aquifer. Operation of a stream-valley alluvial aquifer in a water-deficient area is illustrated in figure 15.

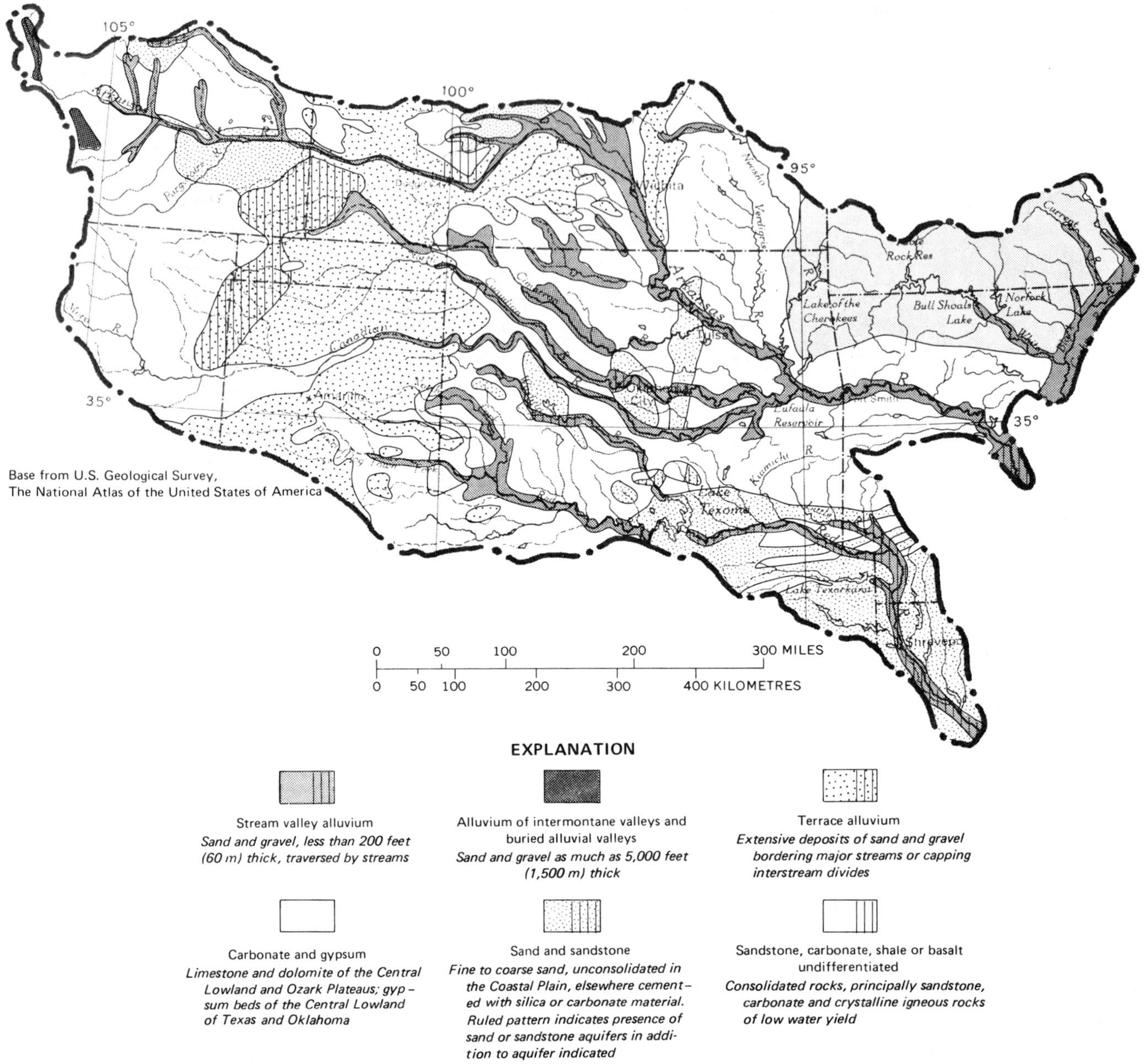

FIGURE 11.—Principal aquifers in the Arkansas-White-Red Region (modified from Lohman and others, 1953).

The amount of water in storage in the aquifer is small, about 10 times the annual withdrawal. Perennial overdrafts in excess of recharge cannot occur. Large developments, such as those in which a large part of the valley is irrigated by ground water, temporarily draw water from storage, but most of the water withdrawn must be restored by induced infiltration from rivers. Thus, the water supply is limited by available streamflow. Excess salts will build up in soil unless water in excess of evapotranspiration is applied.

CASE HISTORY—ARKANSAS RIVER VALLEY, COLORADO

The 150-mile (2.41×10^5-m) reach of the Arkansas River valley from Pueblo to the Colorado-Kansas State boundary is intensively irrigated by ground and surface waters. The annual delivery by wells and canals is 725,000 acre-ft (8.95×10^8 m^3) of water, of which about 150,000 acre-ft (1.85×10^8 m^3) is from wells. Storage in the aquifer is small, about 10 times the annual with-

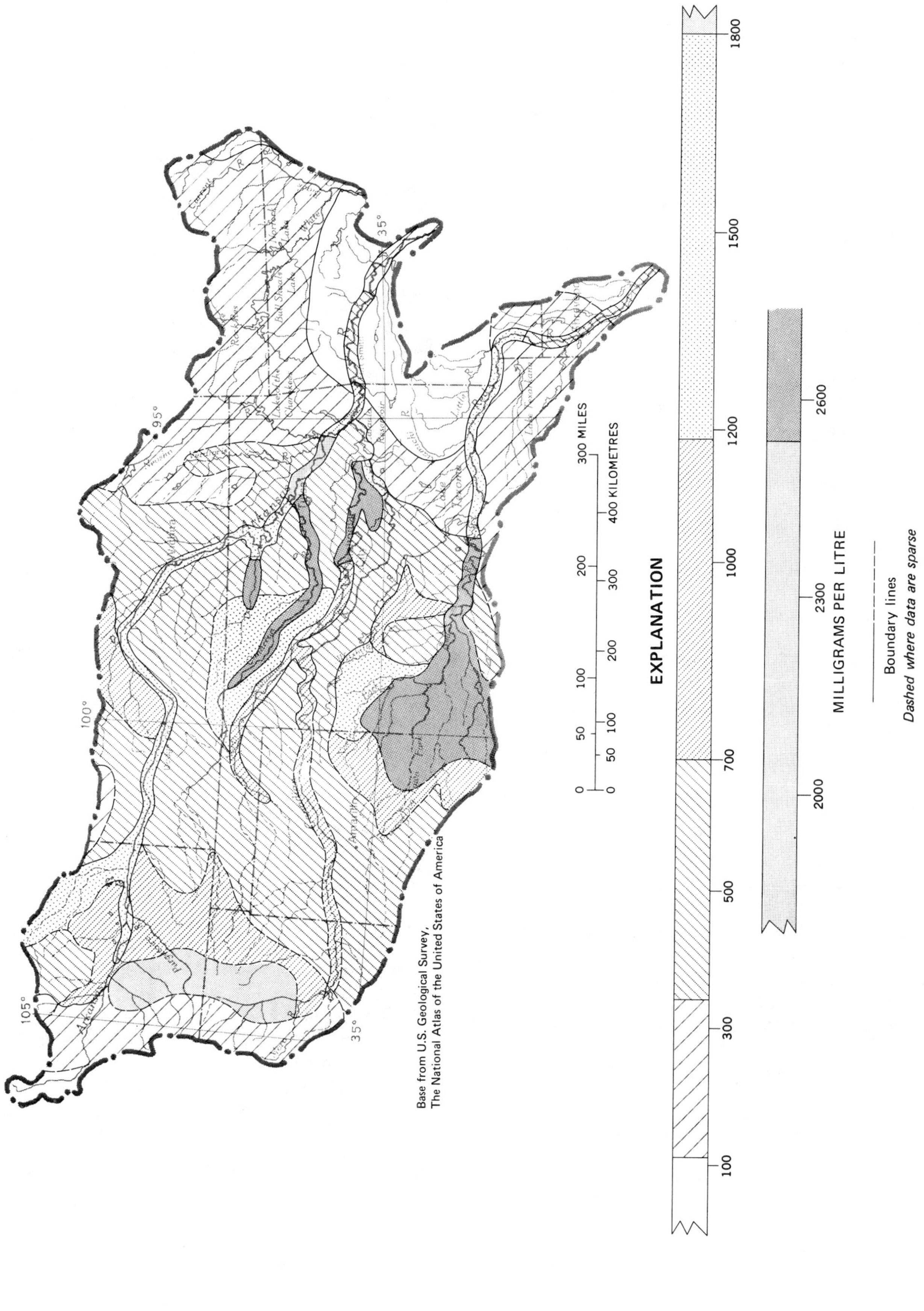

FIGURE 12.—Prevalent dissolved-solids concentration of water in the rivers at low flow (modified after Rainwater, 1962).

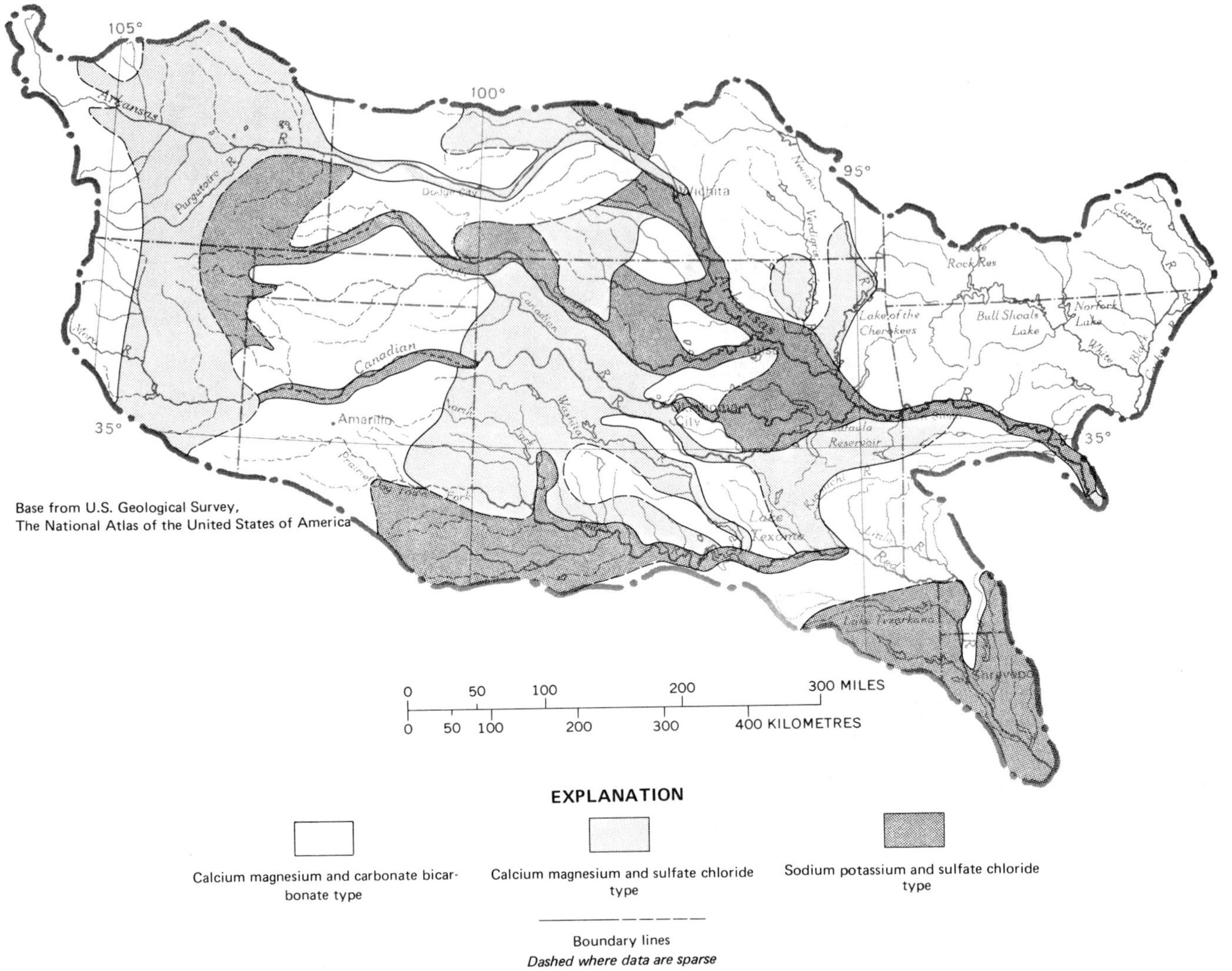

FIGURE 13.—Prevalent chemical types of water in the rivers at low flow (after Rainwater, 1962).

drawal. Water is withdrawn from storage during the irrigation season and is then replenished by rainfall, by infiltration of surface applications for irrigation, and influent seepage from the rivers and canals during and after the irrigation season. Because little water is salvaged from evapotranspiration by lowering the water level, the water supply in the aquifer is limited by the streamflow. Because of the heavy demand and reuse of water, the water quality in the aquifer progressively deteriorates downstream in the valley. Studies by Taylor and Luckey (1974) show that under present conditions use of ground water from storage during the irrigation season increases the water available during the irrigation season by as much as 113,000 acre-ft (1.39×10^8 m^3).

OPERATION OF THE FLOW SYSTEM IN THE WATER-EXCESS AREA

Recharge by infiltration of rainfall is about 3–10 inches (0.076–0.25 m) per year. Discharge is by evapotranspiration and seepage to the stream (fig. 16). In response to withdrawal, water levels are lowered, thereby reducing losses by evapotranspiration, reducing rejected recharge, and reducing base flow to the stream. Drawdown near the stream will induce flow from the stream to the aquifer. Large ground-water developments can be supported by water salvaged from evapotranspiration and by increased net infiltration of rainfall and reduced ground-water flow to streams. Streamflow is large and does not limit development. The quality of water in the aquifer is generally good.

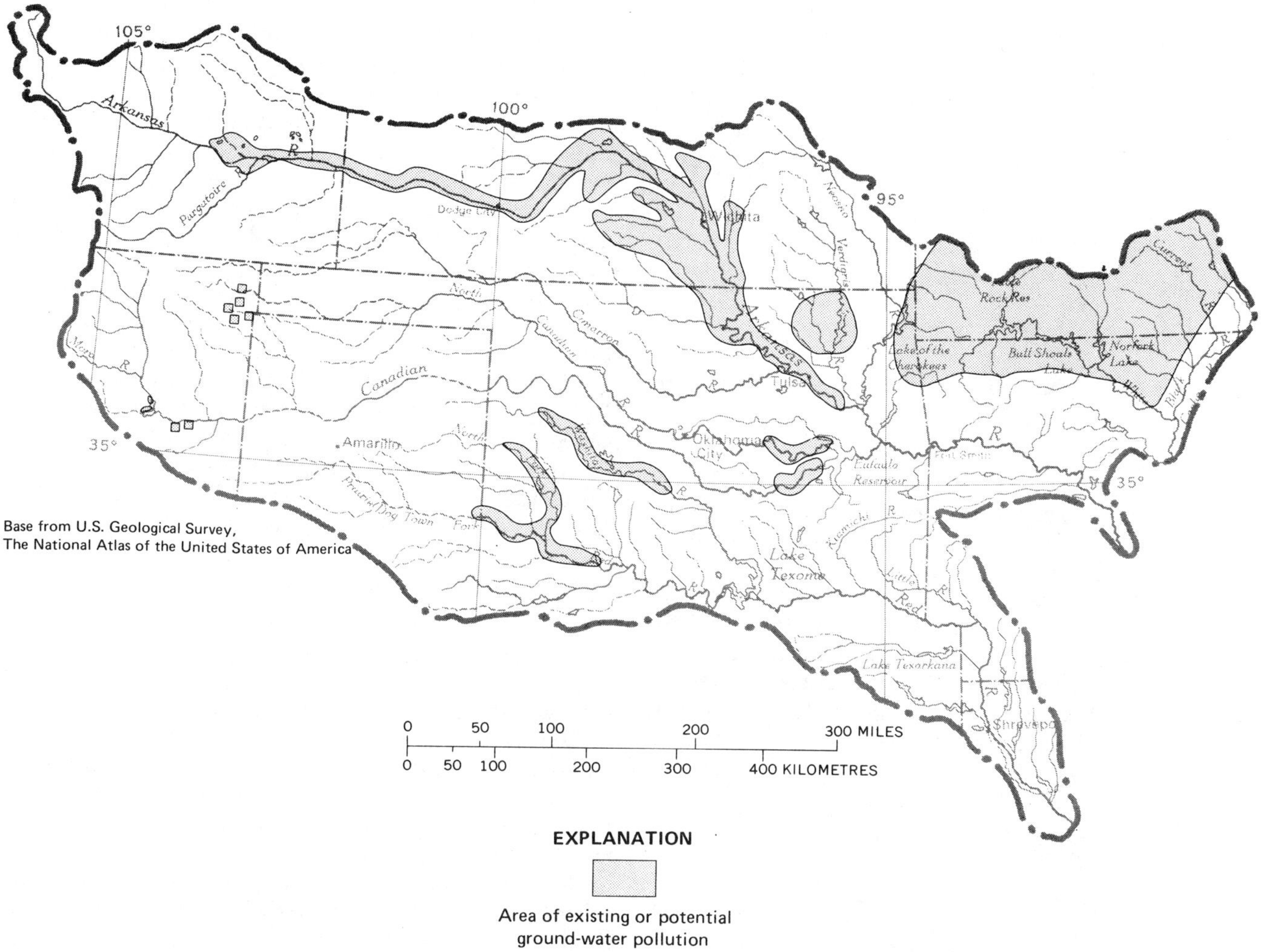

FIGURE 14.—Areas of potential or existing ground-water pollution in the Arkansas-White-Red Region.

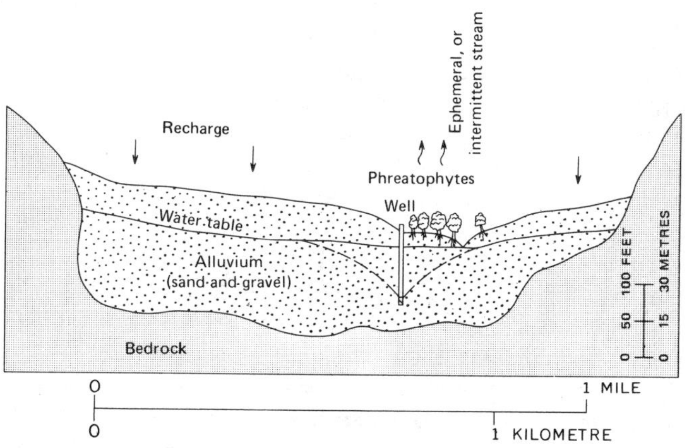

FIGURE 15.—Stream-valley alluvial aquifer with an ephemeral, or intermittent, stream in the water-deficient area.

TERRACE-ALLUVIAL AQUIFERS

Terrace alluvium is extensive in the plains of Texas, Colorado, Oklahoma, Kansas, and New Mexico. The sand-and-gravel section of the terrace alluvium is as much as 600 feet (1.8×10^2 m) thick and averages 300 feet (91 m) thick. Perennial streams generally are incised below the base of the terrace deposits. Terrace-alluvial aquifers contain large amounts of water in storage—as much as 75,000 acre-feet (9.2×10^7 m^3) per square mile (2.59×10^6 m^2) of surface area. Water levels are 50 feet (15 m) or more below the land surface.

OPERATION OF THE FLOW SYSTEM

Recharge to the aquifer, derived solely from sparse precipitation, is estimated to be $1/20$–$1/2$ inch (1.3×10^{-3} to

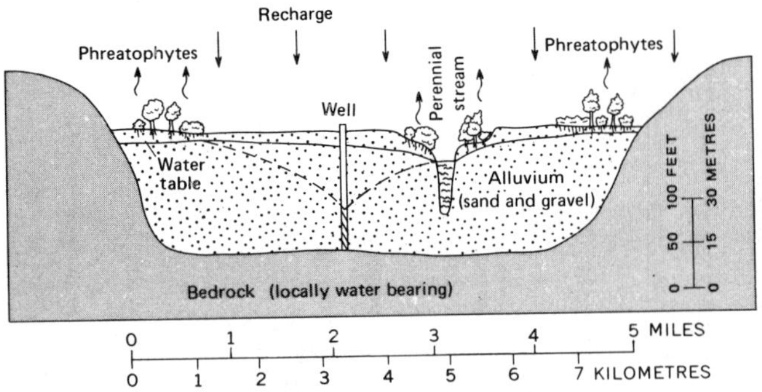

FIGURE 16.—Stream-valley alluvial aquifer with a perennial stream in the water-excess area.

1.3×10^{-2} m) per year. The natural discharge is to the adjacent valley alluvium or by seeps and springs along the terrace escarpment (fig. 17).

In response to withdrawal, there is little or no salvage of evapotranspiration from the water table because of generally low initial water levels. Also because streamflow is generally intermittent, there is little water induced from the stream to the aquifer. Withdrawals in excess of the meager recharge must come from storage.

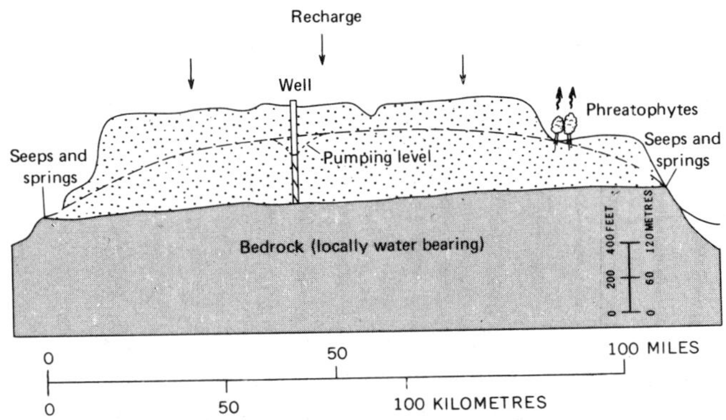

FIGURE 17.—Terrace alluvial aquifer.

CASE HISTORY—OGALLALA FORMATION IN THE SOUTHERN HIGH PLAINS OF TEXAS AND NEW MEXICO

Ground-water withdrawals increased from about 92,000 acre-feet (1.1×10^8 m^3) per year, in 1934, to more than 4,600,000 acre-feet (5.7×10^9 m^3) per year and are used mainly for irrigation. The storage of ground water prior to pumping was very large, approximately 460,000,000 acre-feet (5.7×10^{11} m^3).

Recharge from precipitation is approximately 69,000 acre-feet (8.5×10^7 m^3) per year. Rechenthin and Smith (1967) estimated that a program in the High Plains to remove wasteful nonbeneficial plants and restore grass vegetation would save approximately 290,000 acre-feet (3.59×10^8 m^3) of water annually. Much of the area of nonbeneficial plants does not overlie the aquifer, but, even if a large part of this salvaged water infiltrated to the water table, the increased annual accretion to the aquifer would amount to only about 5 percent of the annual withdrawal.

Salvaged rejected recharge and natural discharge are very low. The withdrawal is depleting water in storage, and the ground water is being mined.

A form of conjunctive use has been suggested for the High Plains in which surface water would be transported to the area from the Mississippi River valley. The large volume of unsaturated sand and gravel in the terrace deposits is a potential reservoir for the imported water. Increasing population and demand for food and fiber production may shift the plan into economic reality.

SAND AND SANDSTONE AQUIFERS

Unconsolidated sand in the Coastal Plain and sandstone partly cemented with calcium carbonate or silica in the High Plains and Central Lowland constitute the sand and sandstone aquifers. The sand and sandstone range in thickness from 100 to 500 feet (30 to 150 m). The aquifer is nonartesian (under water-table conditions) in the outcrop area. Where the sand or sandstone in the subsurface is confined by shale, clay, or siltstone of low permeability, the aquifer is artesian (under confined conditions).

OPERATION OF THE AQUIFERS

Recharge is by infiltration of rainfall in the outcrop area (fig. 18). The recharge rate is highly dependent upon climatic conditions; it ranges from less than 1 inch (2.5×10^{-2} m) per year in the western, water-deficient part of the region to 4–6 inches (0.1 to 0.15 m) per year in the eastern part of the region. Recharge to the sand also occurs where it is overlain by extensive, saturated sand and gravel in the Coastal Plain. Natural discharge is by seepage to streams incised into the aquifer in the outcrop area and by seepage to confining beds in the subsurface.

Pumping in the area where the aquifer is artesian produces large, widespread declines in head. Water is initially pumped from storage; as pumping continues, water is induced from interbedded fine-grained beds and confining beds. As the cone of depression reaches outcrop areas, water is induced from the streams and from the overlying alluvium into the sand aquifer.

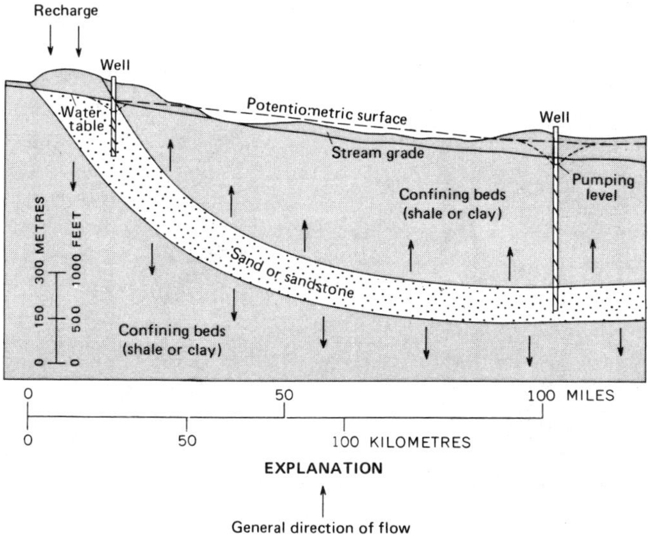

FIGURE 18.—Sand or sandstone aquifer.

CASE HISTORY—THE SPARTA SAND OF THE COASTAL PLAIN

The Sparta Sand and the contiguous sand beds above and below it form an aquifer that extends over thousands of square miles in Arkansas, Louisiana, Missouri, Tennessee, Kentucky, and Mississippi. Only a small part of the Sparta's extent is in the Arkansas-White-Red Region. But the aquifer cannot be dissected along arbitrary lines for analysis, such as those bounding States or Water Resources Council regions. Significant water withdrawal from the Sparta Sand began in 1886 at Memphis, Tenn. Since then, more than 3½ trillion gallons (10,740,000 acre-ft) (1.32×10^{10} m^3) of water has been pumped from the aquifer. Analog-model analysis of flow in the Sparta Sand by Reed (1972) showed that in 1965 only about 20 percent of the 350 million gallons (1,074 acre-ft) (1.32×10^6 m^3) of water pumped was from storage in the aquifer. Sixty percent of the water was derived from leakage from confining beds and 20 percent of the water was induced recharge or captured discharge in the outcrop area.

Pumping from the aquifer is not evenly distributed areally. Consequently, large drawdowns in artesian head have occurred near centers of large withdrawals. Analog-model studies have shown that the Sparta Sand can provide large sustained yields. Projections to the year 1990 of the drawdown in artesian head, in response to a pumping rate of 630 Mgal/d (27.6 m^3/s), show that at some pumping centers water levels will decline below the top of the aquifer. Local excessive drawdowns can be alleviated by more even distribution of pumping.

CARBONATE-ROCK AND GYPSUM AQUIFERS

Carbonate rocks crop out extensively over the Ozark Plateaus region, principally in Missouri and Arkansas. Gypsum beds form productive aquifers in southwest Oklahoma and north Texas. Limestone, dolomite, or gypsum aquifers range in thickness from 50 to 1,500 feet (15 to 460 m). Porosity and permeability are due to solution of the rock. Water in storage ranges in amount from moderate to very large.

OPERATION OF THE FLOW SYSTEM

Recharge is by infiltration of rainfall in the outcrop area (fig. 19). Natural discharge is to streams in the outcrop area and by seepage to confining beds where the formation is in the subsurface. The rates of water movement and well yields are variable but are highest in the outcrop area. Recharge may be as much as 20 inches (0.51 m) a year. The depth to water is 30–450 feet (9–137 m). Water pumped from the aquifer in the outcrop area is derived from storage and results in a decrease in natural discharge. Where these aquifers are confined in the subsurface, they operate similarly to the confined sandstone aquifers.

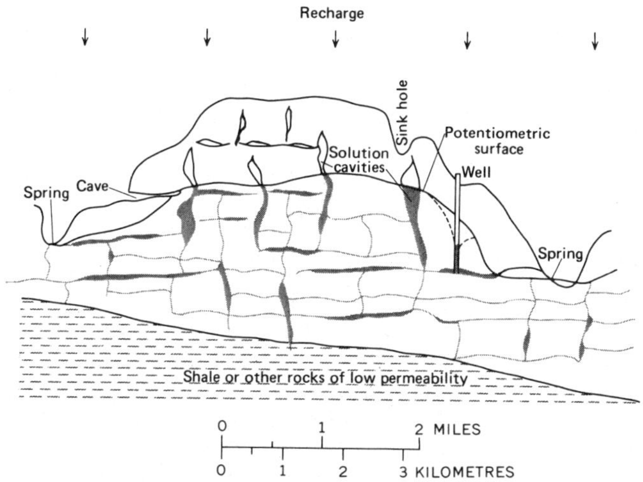

FIGURE 19.—Carbonate-rock and gypsum aquifer.

CASE HISTORY—LIMESTONE AREA OF THE OZARK REGION IN MISSOURI AND ARKANSAS

Thick, dense limestone and dolomite of Paleozoic age underlie a large area. Permeability of the limestone is due to joints, bedding planes and solution openings. Recharge is high in some areas where there is no surface runoff. Recharge and ground-water flow are rapid, soon appearing as increased discharge in springs.

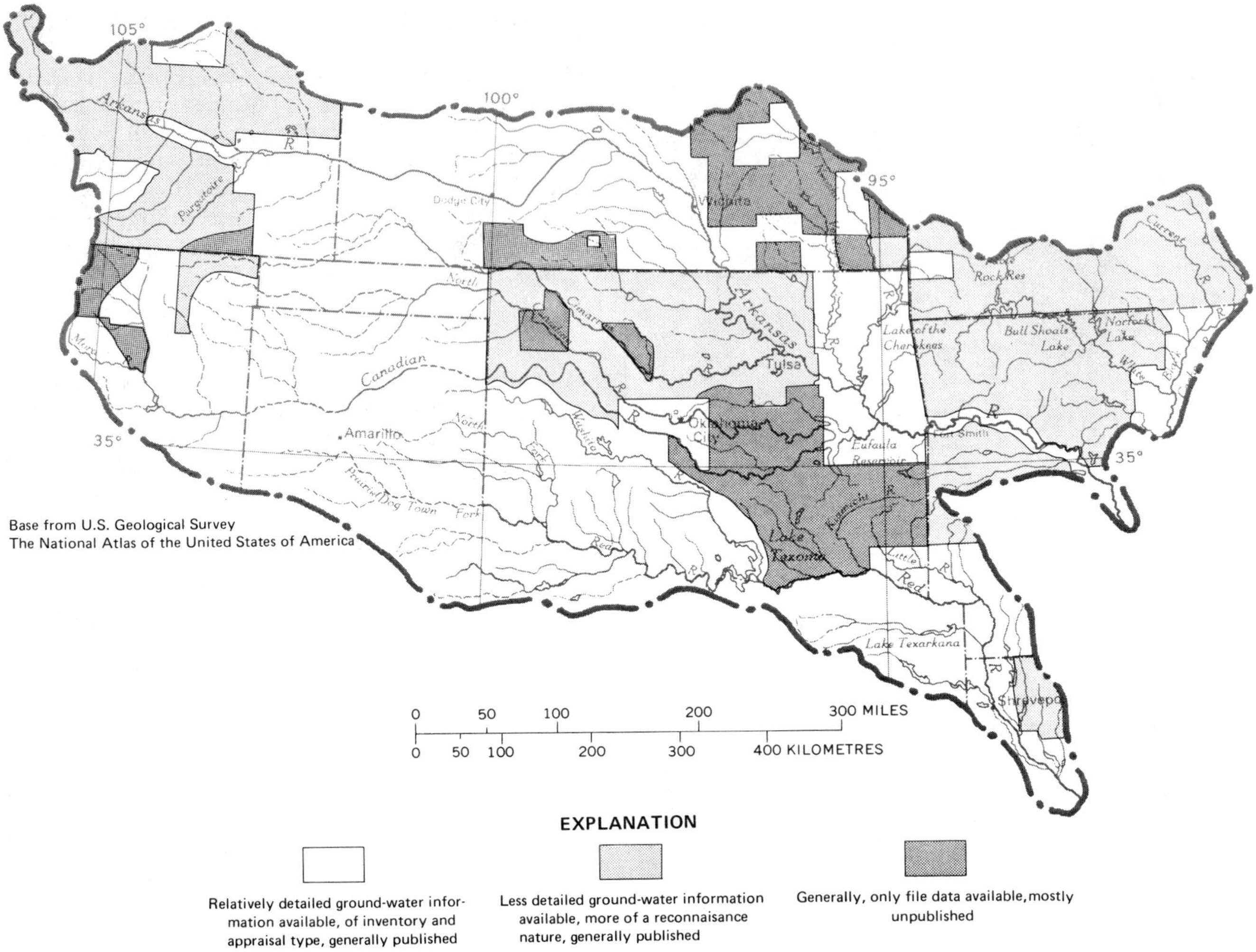

FIGURE 20.—Areas of principal ground-water investigations in the Arkansas-White-Red Region.

There are many large springs and caves, and sinkholes are common. Water travels long distances rapidly underground. Wells yield as much as 500 gal/min (3.2×10^{-2} m³/s), depending on openings encountered by the well.

Pollution is a problem; pollutants travel rapidly and can enter in large quantities. Pollution by mining wastes and by collapse of sewage ponds has occurred in the carbonate area of Missouri.

PREVIOUS STUDIES AND NEEDS IN THE ARKANSAS-WHITE-RED REGION

Ground-water information is available for nearly all the Arkansas-White-Red Region. In those areas not covered by reports, file data are available from appropriate water-oriented State and Federal agencies. A list of these agencies in each State is given immediately preceding "Selected References" in this report. Areas of principal ground-water investigations for which reports are available are shown in figure 20. Reports on the area are classified as detailed or reconnaissance. Both types of reports generally contain information on the most productive aquifers, including well yields, thickness, extent, potentiometric surface, water-table configuration, top of the aquifer, storage coefficient, transmissivity, water quality, and the nature of overlying and underlying beds. The detailed reports generally contain more data at more points than the reconnaissance reports and also contain information on recharge, discharge, water use, and relation of aquifers to streams.

To date, few areas in the region have been modeled by analog or digital models. The few model studies of ground-water systems reflect a prevailing lack of quantitative consideration of ground water in water-resource planning in the Arkansas-White-Red Region.

Ground water now can be adequately evaluated in terms of quantity, availability, quality, cost of development, and impact on the environment through the use of models. Greater consideration should be given to use of ground water in making planning decisions. The inclusion of ground water in water planning affords many more alternatives for development and control of water supply. If ground-water reservoirs underlying an area can yield water to wells at rates exceeding 50 gallons per minute (3.2×10^{-3} m³/s), planners can be certain that these sources are significant and that they should not be ignored. If these reservoirs also contain good-quality water in amounts equal to the storage capacity of existing surface-water reservoirs and potential surface-water reservoir sites, planners can be certain that ground water should play a principal role in any water-development plans (Moore, 1971). Future studies in the region should incorporate the use of models in providing planners with quantitative analyses required for incorporating ground water into water-management plans.

General studies provide a valuable basis for conceptual models or even initial analog or digital models useful in designing data-collection and analysis programs. The appraisal and reconnaissance studies (fig. 20) provide some data for all three modeling needs—definition of the system, stress on the system, and response of the system. The reports generally lack definition of aquifer hydrologic characteristics in quantitative terms required for detailed modeling, as well as lack quantitative hydrologic data on confining beds or the degree of connection of the aquifer with streams. Data on stress and response of the system to stress are generally incomplete for adequate model analysis. For example, many studies lack sufficient data on the history and the distribution of withdrawals from the aquifer. In many model studies, withdrawals are historically the major stresses on the aquifer and the pumping history should be duplicated in testing the model.

Other critical information on aquifer stresses that are not commonly collected or analyzed as a part of an appraisal study are recharge to the aquifer and its relation to the depth to water. In addition to the natural recharge from precipitation, information is required on the recharge and return surface flow of applied irrigation water.

Reconnaissance and appraisal studies generally are limited by political boundaries. Areas for model studies are best delineated by the natural boundaries of the hydrologic system (Bedinger and Sniegocki, 1972). Ideally, these are boundaries across which there is no flow, or boundaries at which the hydrologic head or flow conditions can be defined—that is, streams, lakes, or wells. Natural system boundaries should be used in planning future studies in the area. Exceptions may include the very extensive water-table aquifers in which intensive development is limited to a small part of the aquifer. Nevertheless, ultimate needs for systemwide water planning should be considered in continuing data-collection programs and in comprehensive ground-water studies.

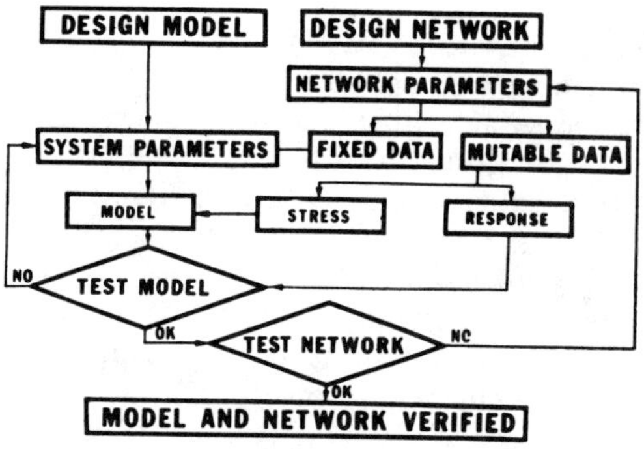

FIGURE 21.—Model and network development (after Bedinger and Sniegocki, 1972).

The relation between data networks and the use of data in modeling is shown by the flow diagram in figure 21. Data on response of the system to stress are used conjunctively with stress data in testing the model. Response data are thus needed for adequate historic and spatial coverage to assess the validity of the model. These data, or the lack thereof, may be more critical than stress data. In some instances much of the stress history, such as pumpage and recharge, can be reconstructed from indirect records of agricultural or industrial production or from population and climatic records. Response data cannot be so reconstructed and therefore must be available before a model study is initiated. It follows that areas of potential model studies should be identified and that continuing records be kept of stress and response. These data-collection programs should be designed to provide adequate sampling in frequency and areal distribution.

Regionally, continuing ground-water data-collection programs (exclusive of studies in designated areas) by the U.S. Geological Survey and by State and private agencies consist principally of observations of water levels in wells. In some areas periodic measurements of spring flow and analyses of ground-water quality are made every 5 years by the Survey. These data do not provide a sufficient history of stress and response for adequate model analysis. The most serious deficiency in data collection in the Arkansas-White-Red Region concerns water-use information. Not only is the coverage inadequate, but the accuracy of some of the information is variable and questionable. Inadequate data collection constitutes a major problem and causes lengthy delays in model analysis.

Outstanding needs for detailed quantitative definition of several ground-water systems exist in the western, water-deficient part of the Arkansas-White-Red Region. The systems needing study include the large, basin-fill aquifers in Colorado and the limestone aquifer of the Arbuckle Group of Oklahoma. These aquifers are virtually unused and could potentially play a significant part in meeting future water needs in the Arkansas-White-Red Region. The large basin-fill ground-water reservoirs in the Arkansas River basin in Colorado are estimated to hold 20 million acre-feet (2.46×10^{10} m^3) of water (P. A. Emery, L. A. Hershey, and J. M. Klein, written commun., 1974). Presently, they are virtually unused. The reservoirs could be utilized in conjunctive-use planning for irrigation requirements in the Arkansas River valley and may be used for storing water from interbasin transfers during times of excess. This water could then be used during times of demand for irrigation or other use, or, as an alternative, the water in these basins could be mined to supply water during water-deficient periods. Water-development plans in Oklahoma include interbasin transport of water from Broken Bow, Hugo, Pine Creek, and several to-be-constructed reservoirs in the vicinity of Oklahoma City. Part of the water would be used for municipal supplies and the rest moved to southwest Oklahoma for other uses. These plans call for transporting surface water across a very productive aquifer in the limestone of the Arbuckle Group. This aquifer is not included in the water-transport plan, but it could potentially supply a significant part of the water required, and it would be relatively near the area of use.

The aquifer systems mentioned could play significant parts in regional and local water-management plans, but further study would be required before ground water can be fully incorporated in these plans. Studies are needed to define the extent and characteristics of the aquifer and to define the flow system quantitatively, so that the response of the system to stress could be predicted. These studies would also provide information needed to determine well spacing, well yields, cost of pumping, and cost of construction of wells and collection systems.

SELECTED STATE AND FEDERAL WATER-ORIENTED AGENCIES IN THE ARKANSAS-WHITE-RED REGION

Arkansas:

U.S. Geological Survey
Water Resources Division
2301 Federal Office Building
Little Rock, Arkansas 72201

Arkansas Geological Commission
State Capitol Building
Little Rock, Arkansas 72201

Arkansas Department of Pollution Control and Ecology
8801 National Drive
Little Rock, Arkansas 72209

Colorado:

U.S. Geological Survey
Water Resources Division
Denver Federal Center, Building 25
Lakewood, Colorado 80225

Colorado Water Conservation Board
102 Columbine Building
1845 Sherman Street
Denver, Colorado 80203

Colorado Division of Water Resources
Office of the State Engineer
303 Columbine Building
1845 Sherman Street
Denver, Colorado 80203

Southeastern Colorado Water Conservancy District
905 Highway 50 West
Pueblo, Colorado 81008

Kansas:

U.S. Geological Survey
Water Resources Division
1950 Avenue A, Campus West
University of Kansas
Lawrence, Kansas 66045

Kansas Water Resources Board
4th Floor, Mills Building
109 West 9th Street
Topeka, Kansas 66612

Kansas Geological Survey
1930 Avenue A, Campus West
University of Kansas
Lawrence, Kansas 66045

Kansas State Board of Agriculture
Division of Water Resources
10th Floor, State Office Building
Topeka, Kansas 66612

Louisiana:

U.S. Geological Survey
Water Resources Division
6554 Florida Boulevard
Baton Rouge, Louisiana 70806

Louisiana Geological Survey
Department of Conservation
Post Office Box G, University Station
Baton Rouge, Louisiana 70803

Louisiana Department of Public Works
Post Office Box 4155, Capitol Station
Baton Rouge, Louisiana 70804

Louisiana Department of Highways
Post Office Box 4245, Capitol Station
Baton Rouge, Louisiana 70804

Missouri:

U.S. Geological Survey
Water Resources Division
Post Office Box 340
Rolla, Missouri 65401

Missouri Department of Natural Resources
Division of Research and Technical Information
Post Office Box 250
Rolla, Missouri 65401

New Mexico:

U.S. Geological Survey
Water Resources Division
Post Office Box 4369
Albuquerque, New Mexico 87106

New Mexico State Engineer
State Capitol Building
Santa Fe, New Mexico 87501

New Mexico Interstate Stream Commission
State Capitol Building
Santa Fe, New Mexico 87501

Oklahoma:

U.S. Geological Survey
Water Resources Division
200 Northwest 4th Street, Room 4301
Oklahoma City, Oklahoma 73102

Oklahoma Water Resources Board
2241 Northwest 40th Street
Oklahoma City, Oklahoma 73112

Oklahoma Geological Survey
The University of Oklahoma
Norman, Oklahoma 73069

Texas:

U.S. Geological Survey
Water Resources Division
Federal Building
300 East 8th Street
Austin, Texas 78701

Texas Water Development Board
Post Office Box 13087
Austin, Texas 78711

Texas Water Quality Board
Post Office Box 13246
Austin, Texas 78711

Texas Water Rights Commission
Post Office Box 13207
Austin, Texas 78711

SELECTED REFERENCES

Alexander, W. H., Jr., 1961, Geology and ground-water resources of the northern High Plains of Texas, Progress Report No. 1: Texas Board Water Engineers Bull. 6109, 39 p.

Aley, T. J., and others, 1976, Ground-water contamination and sinkhole collapse in soluble rock terrain: Missouri Geol. Survey and Water Resources Eng. Geology Ser. 5 (in press).

Arkansas-White-Red Basins Inter-Agency Committee, 1955, A report on the conservation and development of the water and land resources, Part 1, 252 p.

Baker, E. T., Jr., 1960, Geology and ground-water resources of Grayson County, Texas: Texas Board Water Engineers Bull. 6013, 152 p.

Bayne, C. K., 1956, Geology and ground-water resources of Reno County, Kansas, with a geologic map, by O. S. Fent and C. K. Bayne: Kansas Geol. Survey Bull. 120, 130 p.

Bedinger, M. S., Reed, J. E., Wells, C. J., and Swafford, B. F., 1970, Methods and applications of electrical simulation in ground-water studies in the lower Arkansas and Verdigris River valleys, Arkansas and Oklahoma: U.S. Geol. Survey Water-Supply Paper 1971, 71 p.

Bedinger, M. S., and Sniegocki, R. T., 1972, Definition of hydrologic units for water studies in Arkansas: Water Resources Research, v. 8, no. 1, p. 217–221.

Berkstresser, C. F., Jr., and Mourant, W. A., 1966, Ground-water resources and geology of Quay County, New Mexico: New Mexico Bur. Mines and Mineral Resources, Ground-Water Rept. 9, 115 p.

Bredehoeft, J. D., and Pinder, G. F., 1973, Mass transport in flowing ground water: Water Resources Research, v. 9., no. 1, p. 194-210.

Broadhurst, W. L., and Follett, C. R., 1944, Ground-water resources in the vicinity of Nocona, Montague County, Texas: Texas Board Water Engineers Misc. Pub., 15 p.

Busby, M. W., 1963, Yearly variations in runoff for the conterminous United States, 1931–60: U.S. Geol. Survey Water-Supply Paper 1669–S, 49 p.

——— 1966, Annual runoff in the conterminous United States: U.S. Geol. Survey Hydrol. Inv. Atlas HA–212.

Cooper, J. B., and Davis, L. V., 1967, General occurrence and quality of ground water in Union County, New Mexico: New Mexico Bur. Mines and Mineral Resources, Ground-Water Rept. 8, 168 p.

Cronin, J. G., 1964, A summary of the occurrence and development of ground water in the southern High Plains of Texas, *with a section on* Artificial recharge studies, by B. N. Myers: U.S. Geol. Survey Water-Supply Paper 1693, 88 p.

———1969, Ground water in the Ogallala Formation of the High Plains of Texas and New Mexico: U.S. Geol. Survey Hydrol. Inv. Atlas HA–330.

Davis, L. V., 1955, Geology and water resources of Grady and northern Stephens Counties, Oklahoma: Oklahoma Geol. Survey Bull. 73, 184 p.

———1960, Geology and ground-water resources of southern McCurtain County, Oklahoma: Oklahoma Geol. Survey Bull. 86, 108 p.

Dole, R. B., and Stabler, Herman, 1909, Denudation, *in* Papers on the conservation of water resources: U.S. Geol. Survey Water-Supply Paper 234, p. 78–93.

Feder, G. L., Skelton, John, Jeffery, H. G., and Harvey, E. J., 1969, Water resources of the Joplin area, Missouri: Missouri Geol. Survey and Water Resources, Water Resources Rept. 24, 97 p.

Fenneman, N. M., 1931, Physiography of Western United States: New York and London, McGraw-Hill Book Co., 534 p.

Feth, J. H., and others, 1965, Preliminary map of the conterminous United States showing depth to and quality of shallowest ground water containing more than 1,000 parts per million dissolved solids: U.S. Geol. Survey Hydrol. Inv. Atlas HA-199.

Gutentag, E. D., Lobmeyer, D. H., and McGovern, H. E., 1972, Ground water in Kearny County, southwestern Kansas: U.S. Geol. Survey Hydrol. Inv. Atlas HA-416.

Hart, D. L., Jr., 1965, Ground water in the alluvial deposits of the Washita River between Clinton and Anadarko, Oklahoma: Oklahoma Water Resources Board Bull. 26, 25 p.

Hart, D. L., Jr., 1965, Ground water in the alluvial deposits of the Washita River between Clinton and Anadarko, Oklahoma: Oklahoma Water Resources Board Bull. 26, 25 p.

Hurr, R. T., and Moore, J. E., 1972, Hydrogeologic characteristics of the valley-fill aquifer in the Arkansas River valley, Bent County, Colorado: U.S. Geol. Survey Hydrol. Inv. Atlas HA-461.

Irwin, J. H., 1971, Ground-water investigations in Oklahoma, *in* Environmental aspects of geology and engineering in Oklahoma, Annals of The Oklahoma Academy of Science Pub. No. 2: Oklahoma Geol. Survey, p. 58–70.

Jenkins, E. D., 1964, Ground water in Fountain and Jimmy Camp Valleys, El Paso County, Colorado, *with a section on* Computations of drawdowns caused by the pumping of wells in Fountain Valley, by R. E. Glover and E. D. Jenkins: U.S. Geol. Survey Water-Supply Paper 1583, 66 p.

Johnston, J. R., 1957, The significance of a water budget for Texas: Texas A & M Univ., 3d Ann. Water for Texas Conf.

Knight, R. D., 1962, Groundwater areas in Missouri, *in* Groundwater maps of Missouri: Missouri Geol. Survey and Water Resources [1963].

Konikow, L. F., and Bredehoeft, J. D., 1973, Simulation of hydrologic and chemical-quality variations in an irrigated stream-aquifer system—A preliminary report: Colorado Water Conserv. Board Water Resources Circ. 17, 43 p.

Latta, B. F., 1950, Geology and ground-water resources of Barton and Stafford Counties, Kansas: Kansas Geol. Survey Bull. 88, 228 p.

Leonard, R. B., 1972, Chemical quality of water in the Walnut River basin, south-central Kansas: U.S. Geol. Survey Water-Supply Paper 1982, 113 p.

Leopold, L. B., Clarke, F. E., Hanshaw, B. B., and Balsley, J. R., 1971, A procedure for evaluating environmental impact: U.S. Geol. Survey Circ. 645, 13 p.

Lohman, S. W., Burtis, V. M., and others, 1953, General availability of ground water and depth to water level in the Arkansas, White, and Red River basins: U.S. Geol. Survey Hydrol. Inv. Atlas HA-3.

Lohman, S. W., and others, 1972, Definitions of selected ground-water terms—Revisions and conceptual refinements: U.S. Geol. Survey Water-Supply Paper 1988, 21 p.

McGovern, H. E., and Jenkins, E. D., 1966, Ground water in Black Squirrel Creek valley, El Paso County, Colorado: U.S. Geol. Survey Hydrol. Inv. Atlas HA-236.

McGuinness, C. L., 1963, The role of ground water in the national water situation: U.S. Geol. Survey Water-Supply Paper 1800, 1121 p.

McLaughlin, T. G., 1943, Geology and ground-water resources of Hamilton and Kearny Counties, Kansas: Kansas State Geol. Survey Bull. 49, 220 p.

_____1949, Geology and ground-water resources of Pawnee and Edwards Counties, Kansas: Kansas State Geol. Survey Bull. 80, 189 p.

_____1954, Geology and ground-water resources of Baca County, Colorado: U.S. Geol. Survey Water-Supply Paper 1256, 232 p.

Maher, J. C., 1939, Fluoride in the ground water of Avoyelles and Rapides Parishes, Louisiana: Louisiana Dept. Conserv. Geol. Pamph. 1, 23 p.

Major, T. J., Hurr, R. T., and Moore, J. E., 1970, Hydrogeologic data for the lower Arkansas River valley, Colorado: Colorado Water Conserv. Board Ground-Water Series, Basic-Data Release 21, 125 p.

Marcher, M. V., 1969, Reconnaissance of the water resources of the Fort Smith quadrangle, east-central Oklahoma: Oklahoma Geol. Survey HA-1.

_____1972, Major sources of water in Oklahoma, *in* Johnson, K. S., and others, Geology and earth resources of Oklahoma, an atlas of maps and cross sections: Oklahoma Geol. Survey Educ. Pub. 1, p. 8.

Marcher, M. V., and Bergman, D. L., 1976, Reconnaissance of the water resources of the McAlester and Texarkana quadrangles, southeastern Oklahoma: Oklahoma Geol. Survey HA (in press).

Marcher, M. V., and Bingham, R. H., 1971, Reconnaissance of the water resources of the Tulsa quadrangle, northeastern Oklahoma: Oklahoma Geol. Survey HA-2.

Meyer, W. R., Gutentag; E. D., and Lobmeyer, D. H., 1970, Geohydrology of Finney County, southwestern Kansas: U.S. Geol. Survey Water-Supply Paper 1891, 117 p.

Moore, J. E., 1971, Integrating the use of ground water into water-resource planning, *in* Minutes of 165th Meeting: Missouri Basin Inter-Agency Comm., June 23, 1971, 9 p.

Moore, J. E., and Wood, L. A., 1967, Data requirements and preliminary results of an analog-model evaluation, Arkansas River valley in eastern Colorado: Ground Water, v. 5, no. 1, p. 20–23.

Morton, R. B., and Goemaat, 1973, Reconnaissance of the water resources of Beaver County, Oklahoma: U.S. Geol. Survey Hydrol. Inv. Atlas HA-450.

Moulder, E. A., 1970, Freshwater bubbles, a possibility for using saline aquifers to store water: Water Resources Research, v. 6, no. 5, p. 1528–1531.

Moulder, E. A., and Jenkins, C. T., 1969, Analog-digital models of stream-aquifer systems: Ground Water, v. 7, no. 5, p. 19–24.

Moulder, E. A., Jenkins, C. T., Moore, J. E., and Coffin, D. L., 1963, Effects of water management on a reach of the Arkansas valley, La Junta to Las Animas, Colorado: Colorado Water Conserv. Board Ground-Water Series, Circ. 10, 20 p.

Murray, C. R., and Reeves, E. B., 1972, Estimated use of water in the United States in 1970: U.S. Geol. Survey Circ. 676, 37 p.

Newcome, Roy, Jr., and Page, L. V., 1962, Water resources of Red River Parish, Louisiana: U.S. Geol. Survey Water-Supply Paper 1614, 133 p. [1963].

Newcome, Roy, Jr., Page, L. V., Sloss, Raymond, 1963, Water resources of Natchitoches Parish, Louisiana: Louisiana Dept. Conserv. Geol. Survey and Louisiana Dept. Public Works, Water Resources Bull. 4, 189 p.

Newcome, Roy, Jr., and Sloss, Raymond, 1965, Water resources of Rapides Parish, Louisiana: Louisiana Dept. Conserv. Geol. Survey and Louisiana Dept. Public Works, Water Resources Bull. 8.

Page, L. V., and May, H. G., 1964, Water resources of Bossier and Caddo Parishes, Louisiana: Louisiana Dept. Conserv. Geol. Survey and Louisiana Dept. Public Works, Water Resources Bull. 5.

Page, L. V., and Preé, H. L., Jr., 1964, Water resources of De Soto Parish, Louisiana: U.S. Geol. Survey Water-Supply Paper 1774, 152 p. [1965].

Petri, L. R., Lane, C. W., and Furness, L. W., 1964, Water resources of the Wichita area, Kansas: U.S. Geol. Survey Water-Supply Paper 1499-I, 69 p.

Rainwater, F. H., 1962, Stream composition of the conterminous United States: U.S. Geol. Survey Hydrol. Inv. Atlas HA-61.

Rechenthin, C. A., and Smith, H. M., 1967, Effect on water yield and supply, pt. 5 *of* Grassland restoration: Temple, Tex., U.S. Dept. Agriculture, Soil Conserv. Service, 46 p.

Reed, J. E., 1972, Analog simulation of water-level declines in the Sparta Sand, Mississippi embayment: U.S. Geol. Survey Hydrol. Inv. Atlas HA–434.

Robertson, C. E., 1962, Water well yield map of Missouri, *in* Ground water maps of Missouri: Missouri Geol. Survey and Water Resources [1963].

Sapik, D. B., and Goemaat, R. L., 1973, Reconnaissance of the ground-water resources of Cimarron County, Oklahoma: U.S. Geol. Survey Hydrol. Inv. Atlas HA–373.

Sniegocki, R. T., 1963, Problems in artificial recharge through wells in the Grand Prairie region, Arkansas: U.S. Geol. Survey Water-Supply Paper 1615–F, 25 p.

Stramel, G. J., Lane, C. W., and Hodson, W. G., 1958, Geology and ground-water hydrology of the Ingalls area, Kansas: Kansas Geol. Survey Bull. 132, 154 p.

Tanaka, H. H., 1972, Geohydrology of the lower Verdigris River valley between Muskogee and Catoosa, Oklahoma: U.S. Geol. Survey Water-Supply Paper 1999–A, 23 p.

Tanaka, H. H., and Hollowell, J. R., 1966, Hydrology of the alluvium of the Arkansas River, Muskogee, Oklahoma, to Fort Smith, Arkansas, *with a section on* Chemical quality of the water, by J. J. Murphy: U.S. Geol. Survey Water-Supply Paper 1809–T, 42 p.

Taylor, O. J., and Luckey, R. R., 1974, Water management studies of a stream-aquifer system, Arkansas River valley, Colorado: Ground Water, v. 12, no. 1, p. 22–38.

Voegeli, P. T., Sr., and Hershey, L. A., 1965, Geology and ground-water resources of Prowers County, Colorado: U.S. Geol. Survey Water-Supply Paper 1772, 101 p.

Waite, H. A., 1942, Geology and ground-water resources of Ford County, Kansas: Kansas Geol. Survey Bull. 43.

———1961, Geology and ground-water resources of Sumner County, Kansas: Kansas Geol. Survey Bull. 151.

Water Resources Council, 1968, The Nation's water resources—The first national assessment of the Water Resources Council: 403 p.

Williams, C. C., and Lohman, S. W., 1949, Geology and ground-water resources of a part of south-central Kansas, *with special reference to* the Wichita municipal water supply: Kansas Geol. Survey Bull. 79, 455 p.

Wood, P. R., and Hart, D. L., Jr., 1967, Availability of ground water in Texas County, Oklahoma: U.S. Geol. Survey Hydrol. Inv. Atlas HA–250.

Woodward, D. R., 1957, Availability of water in the United States, with special reference to industrial needs by 1980: Washington, [U.S.] Armed Forces Indus. Coll., 74 p.

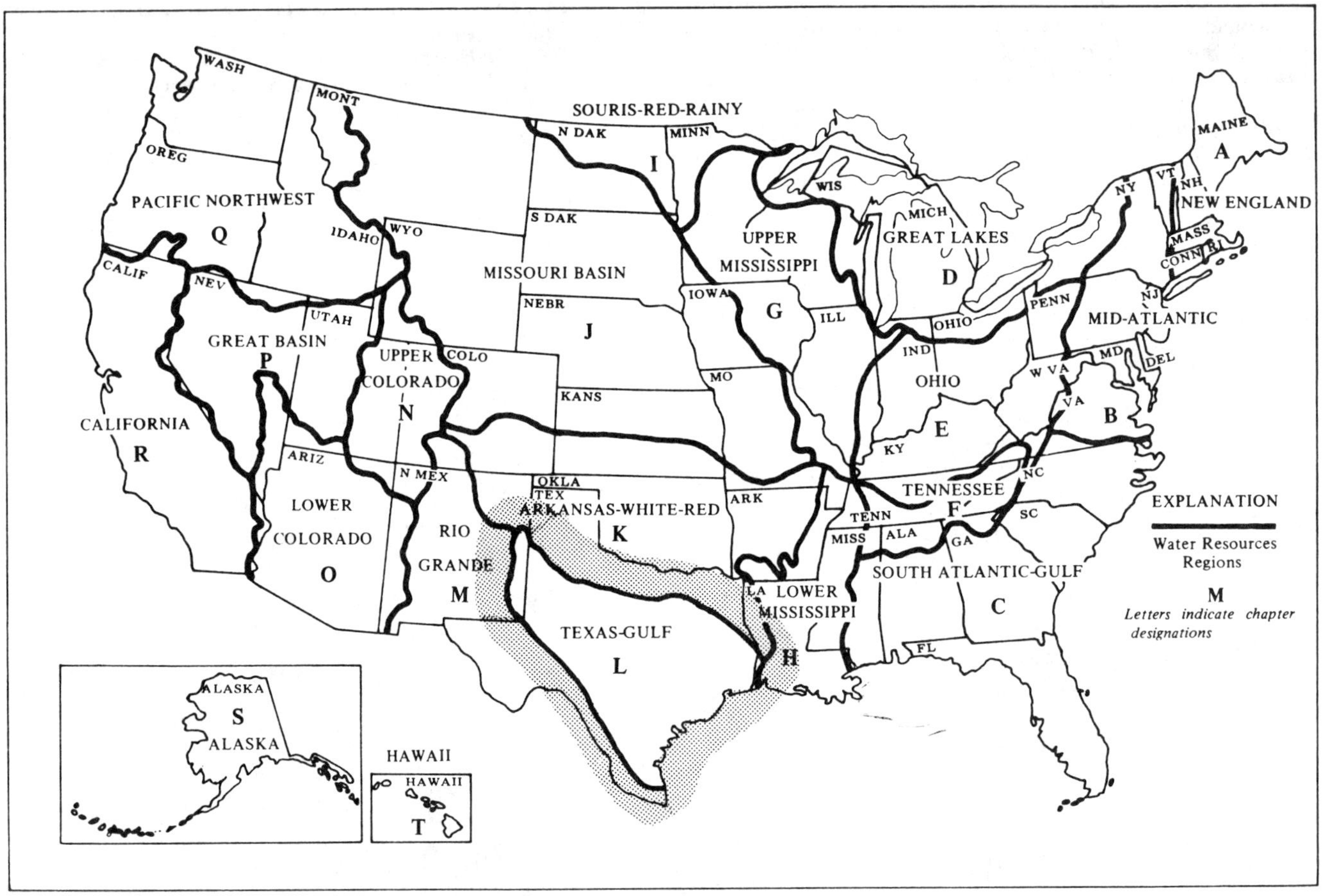

L — Location Map for the Texas-Gulf Region

L

TEXAS-GULF REGION

By E. T. BAKER, JR., and JAMES R. WALL

ABSTRACT

Ground water in the Texas-Gulf Region is a large and important resource that can provide a more significant percentage of the total water supply of the region. Total water requirements within the region are projected to rise sharply from 14 million acre-feet (17 cubic kilometres) in 1970 to nearly 26 million acre-feet (32 cubic kilometres) in 2020. About half of the water used in 1970 was ground water.

An estimated total of 1.04 billion acre-feet (1,280 cubic kilometres) of recoverable water containing less than 3,000 milligrams per litre dissolved solids is stored above a depth of 400 feet (122 metres) in the aquifers of the region. In addition, part of an estimated 3.28 billion acre-feet (4,040 cubic kilometres) of water in storage below 400 feet (122 metres) is recoverable. Although not all of the ground water in storage is recoverable, a significant amount is available for development; and an enormous quantity is accessible should occasions prompt its use on a time-limited basis.

The total steady-state yield (amount of water that approximates the maximum perennial replenishment from precipitation) of the region's aquifers is about 4.6 million acre-feet (5.7 cubic kilometres) annually, or about 2.3 times the ground-water usage in 1970 if the large mining draft on the High Plains is not considered as part of the total steady-state yield. Because of the large quantity of recoverable ground-water in storage, the steady-state yield can be augmented for a very long time on a "deferred" basis, whereby the economic use of the water that is withdrawn from storage in excess of the steady-state yield may result in a strengthened economy.

An important goal in programs for meeting future water needs should be to identify the full potential of the available ground-water resources. The subsurface reservoirs may be utilized not only as sources of fresh and treatable water, but as storage facilities for other freshwater supplies and as possible sources of geothermal energy. Some saline-water reservoirs may be suitable for liquid-waste storage or disposal.

Large-scale and unregulated ground-water pumping may result in hydrologic problems such as declining water levels, streamflow depletion, and land-surface subsidence; but studies on proper development of the ground-water reservoirs could provide solutions to many of the problems. Water rights and other legal concepts should be based on sound hydrologic principles to assist development of water resources in an orderly and efficient manner.

Because significant amounts of ground water are available, the opportunities for expanded and conjunctive use of ground water and surface water should be considered in regional plans for water development and conservation. The complexities of water management and the difficulties of achieving an integrated system of total-water management will require additional technical information.

INTRODUCTION

PURPOSE OF THIS REPORT

The purpose of this report is to provide information on the importance of subsurface reservoirs in water-development and conservation plans in the Texas-Gulf Region (fig. 1). The report emphasizes the desirability of considering the ground-water and surface-water subsystems as closely related and equal parts of the water-supply system as a whole.

Most regional water-development plans are based on the use of surface water, which is highly visible and easily obtained, and for which legal procedures for allocation and management have been established for many years. The total water supply of a region, however, is contained in a complex system, and ground water in the Texas-Gulf Region is a large and important resource that could provide a greater percentage of the water used. The subsurface reservoirs may be utilized as sources of fresh and treatable water, as possible storage facilities for other freshwater supplies, and as possible sources of geothermal energy. Some saline aquifers may be suitable for storage of liquid waste.

Large-scale and unregulated ground-water pumping may result in hydrologic problems such as declining water levels, streamflow depletion, and land-surface subsidence. However, because significant amounts of water are available, the opportunities for expanded and conjunctive use of the ground-water resources should be considered in regional plans for water development and conservation.

The complexities of total water management and the difficulties of achieving a workable system are understandably numerous. Although a discussion of these complexities and difficulties is beyond the scope

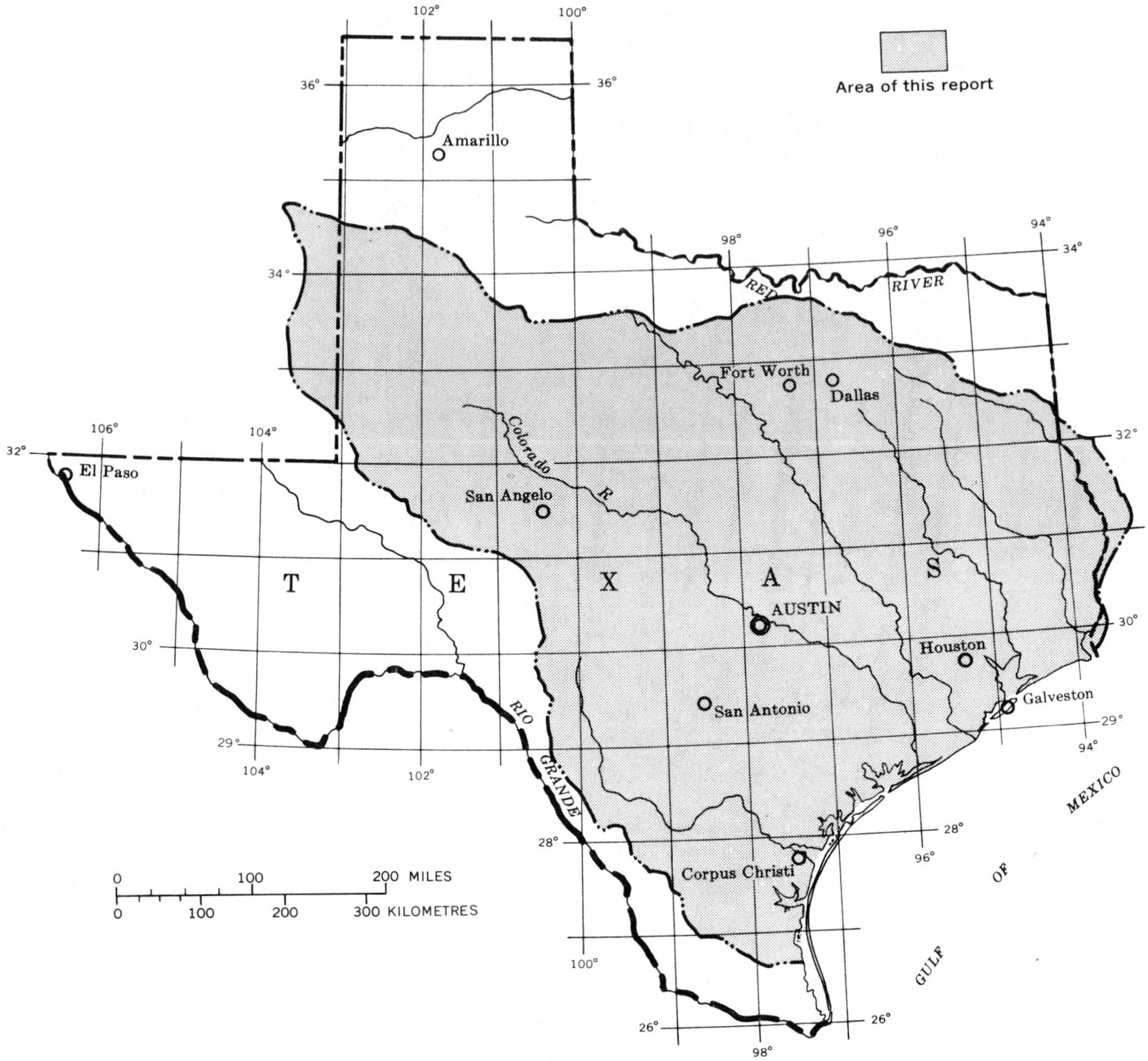

FIGURE 1.—Index map showing location of the Texas-Gulf Region.

of this report, such constraints affecting total water management should not be viewed as insurmountable barriers to future achievement. Nevertheless, efforts to accomplish an integrated system of ground- and surface-water management will require extensive technical and economic feasibility studies.

GEOGRAPHIC SETTING OF THE TEXAS-GULF REGION

The Texas-Gulf Region comprises an area of 173,000 mi^2 (448,000 km^2) in Texas, New Mexico, and Louisiana. About 95 percent of the area is in Texas. The region includes 10 major river basins and 8 intervening coastal basins (Texas Water Development Board, 1968) within the physiographic provinces of the Great Plains, the Central Lowland, and the Coastal Plain (pl. 1A).

The climate of the Texas-Gulf Region is marked by extremes in precipitation and evaporation—climatic elements that to a large degree determine the availability of both ground water and surface water. Average annual precipitation ranges from slightly more than 56 in (1,420 mm) in extreme southeastern Texas and southwestern Louisiana to slightly less than 16 in (410 mm) on the High Plains of New Mexico (pl. 1B).

Average annual net water-surface evaporation (effective evaporation loss obtained as total evaporation minus precipitation) ranges from zero in

Louisiana, where precipitation completely offsets total evaporation, to between 60 and 80 in. (152–203 cm) in west Texas and New Mexico (pl. 1B).

Streamflow, which is influenced primarily by precipitation, varies considerably. The average annual streamflow is about 30 million ac-ft (37 km^3), about two-thirds of which originates in the eastern one-fourth of the region. During a wet year (1941), total runoff was about 65 million ac-ft (80 km^3). During a dry year (1956), total runoff was only about 8 million ac-ft (10 km^3). Low flow of the streams, however, is controlled to a large degree by ground-water outflow, which tends to stabilize streamflow during periods of little or no runoff.

The economy of the Texas-Gulf Region is in a period of transition and rapid expansion marked by increasing population, urbanization, and industrialization. The economy has evolved from a largely noncommercial base through periods dominated first by commercialization of agriculture and later by development of mineral resources, principally oil and gas. Agriculture, though proportionately a smaller part of the economy than previously, continues to contribute a large share of the region's total income.

Increased opportunities for industrial employment, coupled with decreased personnel requirements in agriculture, has radically shifted the distribution of the population. Just prior to World War II, about half the population of the Texas-Gulf Region was in urban areas. During the next 30 years, the population shifted so strongly to urban areas that about four-fifths of the 1970 population of 9.7 million was urban. About one-third of the total population in 1970 was in the four largest cities—Houston, Dallas, San Antonio, and Fort Worth.

The concentrations of the population along the interior and coastal margins of the Coastal Plains, together with projections of large increases in the populations of these areas, are particularly significant aspects of the problems of water and related land-resource planning.

METRIC CONVERSIONS

For those readers interested in using the metric system, metric equivalents of English units of measurements are given in parentheses. The English units used in this report may be converted to metric units by the conversion factors given in table 1.

THE CHALLENGE OF MEETING FUTURE WATER REQUIREMENTS

The demands for water in the Texas-Gulf Region have increased rapidly over the past several decades and are expected to increase more rapidly in the future. By 2020, the population of the region is projected to be 22 million, a 125-percent increase over 1970 (Texas Water Development Board, written commun., 1974). The effects of the population growth upon demands for water will be intensified by urbanization and by industrial and agricultural expansion. The total-water requirements projected to 2020 for municipal supply, industrial use, and irrigation will rise sharply from 14 million ac-ft (17 km^3) in 1970 to nearly 26 million ac-ft (32 km^3) in 2020 (fig. 2). The needs for irrigation are pro-

TABLE 1.—*Factors for conversion from English to metric units used in this report*

From			To obtain	
Unit	Abbreviation	Multiply by	Unit	Abbreviation
acres	ac	4,047	square metres	m^2
		0.004047	square kilometres	km^2
acre-feet	ac-ft	1,233	cubic metres	m^3
		.000001233	cubic kilometres	km^3
barrels	bbl	.159	cubic metres	m^3
		.000159	cubic kilometres	km^3
cubic feet	ft^3	.02832	cubic metres	m^3
feet	ft	.3048	metres	m
gallons	--------	.003785	cubic metres	m^3
		3.785×10^{-12}	cubic kilometres	km^3
gallons per minute	gal/min	.06309	litres per second	l/s
inches	in	2.540	centimetres	cm
		25.4	millimetres	mm
miles	mi	1.609	kilometres	km
pounds per square inch	lb/in^2	.07031	kilograms per square centimetre	kg/cm^2
square miles	mi^2	2.59	square kilometres	km^2

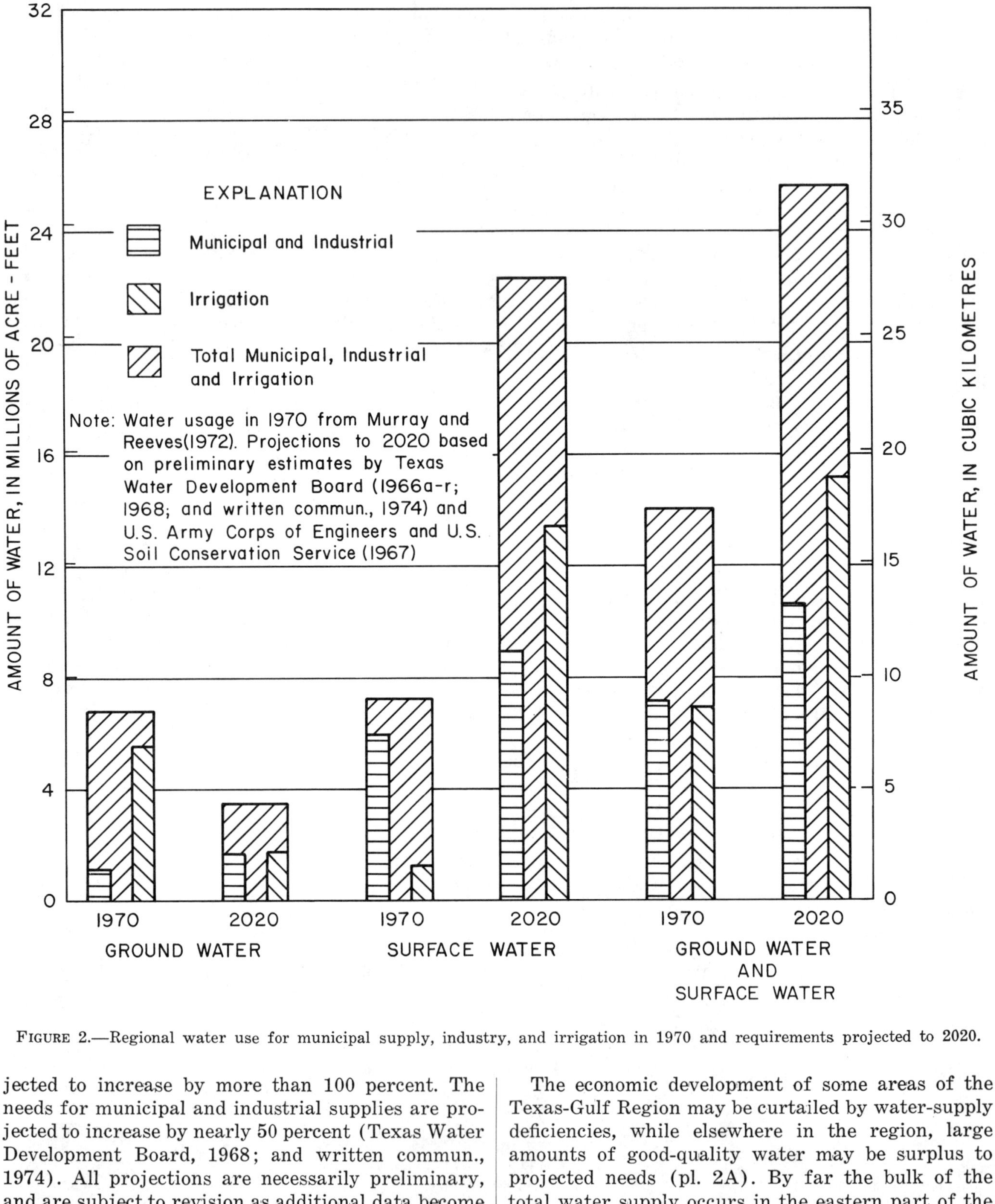

FIGURE 2.—Regional water use for municipal supply, industry, and irrigation in 1970 and requirements projected to 2020.

jected to increase by more than 100 percent. The needs for municipal and industrial supplies are projected to increase by nearly 50 percent (Texas Water Development Board, 1968; and written commun., 1974). All projections are necessarily preliminary, and are subject to revision as additional data become available.

The economic development of some areas of the Texas-Gulf Region may be curtailed by water-supply deficiencies, while elsewhere in the region, large amounts of good-quality water may be surplus to projected needs (pl. 2A). By far the bulk of the total water supply occurs in the eastern part of the region, where projected increases in water requirements are in most places moderate. In contrast, future requirements for water are very large in the western part of the region, where the available supplies are inadequate to meet the demands.

A projected decrease in total ground-water usage is attributed to a reduction in the availability of ground water for irrigation on the High Plains, where the reserves will largely be depleted by 2020. Ground-water requirements are projected to be 3.4 million ac-ft (4.2 km^3) in 2020, or about half of the 1970 ground-water usage of 6.8 million ac-ft (8.4 km^3). Of the 6.8 million ac-ft, 83 percent was used for irrigation, seven-eighths of which was used for irrigation on the High Plains. Ground-water demands for municipal supply and industrial use are projected to increase by nearly 50 percent from almost 1.2 million ac-ft (1.5 km^3) in 1970 to 1.7 million ac-ft (2.1 km^3) in 2020 (Texas Water Development Board, 1968; and written commun., 1974).

Surface-water use is projected to triple from 7.3 million ac-ft (9.0 km^3) in 1970 to 22.3 million ac-ft (27.5 km^3) in 2020. Although municipal and industrial use of surface water will increase by about one-third from 1970 to 2020, irrigation use will increase 10-fold by the end of this 50-year span. Most of this increase for irrigation will be required on the High Plains to replace the projected depletion of the ground-water supply (Texas Water Development Board, 1968; and written commun., 1974). Because of the geographic pattern of projected surpluses and deficiencies, planning and development should be carried out on a regional basis, and all sources of water should be utilized in the most beneficial way. Although the region will require about 26 million ac-ft (32 km^3) of water per year by 2020 for municipal supply, industrial use, and irrigation, the regional supply is ample for these uses, although other needs, such as freshwater inflow to the estuaries, impose an added strain on the supply.

The average annual discharge of surface water to the Gulf of Mexico is 30 million ac-ft (37 km^3); the potential steady-state yield of the ground-water reservoirs is 4.6 million ac-ft (5.7 km^3) per year; and the great reserve of recoverable ground water in storage above a depth of 400 feet (122 m) is about 1.0 billion ac-ft (1,233 km^3). The challenge, therefore, of meeting future water requirements is not so much in coping with a regional shortage, but in planning for regional and conjunctive development of the total resource and in devising the means for overcoming the problems of unequal geographic distribution and demand. The relative scarcity of water as compared to long-term needs on the High Plains and elsewhere in the Texas-Gulf Region have counterparts in adjoining regions, so that most proposals for large-scale transfer of water supplies extend well beyond the region's boundaries. Examples of such proposed transfers are contained in the Texas Water Plan (Texas Water Development Board, 1968) which is currently being updated and refined.

SIGNIFICANCE OF GROUND-WATER RESOURCES IN THE REGIONAL WATER SUPPLY

During the past several decades, ground-water resources have played a sizable role in satisfying the demands for freshwater in the Texas-Gulf Region. Many cities, industries, and irrigators, as well as most rural inhabitants have depended upon this resource because of its accessibility, good quality, and low cost of utilization.

In 1970, ground-water reservoirs supplied about 45 percent of the total freshwater used in the region. In that year, an estimated 7 million ac-ft (8.6 km^3) of freshwater was withdrawn for municipal supply, rural-domestic and livestock supply, industrial use, and irrigation (Murray and Reeves, 1972). This large ground-water use, relative to the use of surface water, contrasts sharply with practices in other Water-Resources Council regions, where the average use of fresh surface water exceeds the use of fresh ground water by about 4 to 1.

Although the ground-water reservoirs provided an estimated 7 million ac-ft (8.6 km^3) on the water used in 1970, about 5 million ac-ft (6 km^3) was used on the High Plains and about 1 million ac-ft (1.2 km^3) was used in the Houston and San Antonio areas. Because of this geographic concentration of ground-water usage, the potential significance of the ground-water supply in the rest of the region is greatly increased. The ground-water reservoirs, excluding the Ogallala aquifer that provided the 5 million ac-ft (6 km^3) used on the High Plains, could provide 4.6 million ac-ft (5.7 km^3) per year, or about 2.3 times the amount used outside the High Plains in 1970.

The unequal hydrologic distribution of the total-water supply in the Texas-Gulf Region is striking. Considering only water having less than 3,000 mg/l dissolved solids, the data summarized in table 2 show that the total regional supply is an estimated 8.6 billion ac-ft (10,600 km^3). Of this amount, 99.6 percent is ground water in storage. Surface water in artificial lakes at capacity and in natural lakes and ponds, and the average amount of water in stream

TABLE 2.—*Distribution of the total regional water supply having less than 3,000 mg/l dissolved solids*

Location	Volume of water (thousands of acre-feet)	Percentage of total water
Ground water		
Above depth of 400 feet	1,975,000	23.08
Below depth of 400 feet	6,545,000	76.49
Total ground water	8,520,000	99.57
Surface water		
Natural and artificial lakes and ponds.	36,000	0.42
Average in stream channels	640	.01
Total surface water	36,640	0.43
Total regional water supply	8,556,640	100

channels at any one time is about 37 million ac-ft (46 km³), or only 0.4 percent.

Most of the 8.5 billion ac-ft (10,600 km³) of ground water in storage cannot be recovered feasibly. Nevertheless, a large amount of the water in storage is recoverable and should be considered in plans for water management. It is reassuring to know that this reserve can accommodate overdrafts when necessary, but it is important to remember that this supply of water has been accumulating for generations and that the annual recharge adds only a small increment to the total. The extent to which this principal reserve of water can be conserved and managed may be a decisive factor in determining whether or not an adequate supply will be available in the future.

GEOGRAPHIC AND GEOLOGIC DISTRIBUTION OF GROUND-WATER SUPPLIES

Although some water-bearing formations are important sources of water in local areas, only those of regional significance are considered in this report. The geographic distribution of the aquifers, in those areas where they will yield water having less than 3,000 mg/l dissolved solids, is shown on plate 2B. The descriptions of the aquifers apply only to the sections containing water having less than 3,000 mg/l dissolved solids. The stated capacities of wells should not be construed to imply that such yields can be obtained everywhere in the aquifer or that such yields represent the maximum yields possible.

The Hickory aquifer underlies about 5,000 mi² (12,950 km²) of the Edwards Plateau and Llano Uplift of central Texas, where the strata are highly faulted and steeply dipping. The aquifer, which consists principally of sand and sandstone, is more than 400 ft (122 m) thick and extends to maximum depths of almost 5,000 ft (1,520 m). The yields of the larger-capacity wells generally range between 200 and 500 gal/min (13 to 32 l/s), but the yields of a few wells exceed 1,000 gal/min (63 l/s). The aquifer is only slightly developed and is capable of supporting a sizable increase in draft if declines in water levels are acceptable to water planners and users.

The Ellenburger-San Saba aquifer underlies an area of about 4,000 mi² (10,360 km²) and almost completely surrounds the Precambrian core of the Llano Uplift. The aquifer, which consists of more than 1,000 ft (305 m) of limestone and dolomite beds that are characterized by substantial secondary permeability, extends to depths of about 3,000 ft (914 m). A few large springs issue from the aquifer and help sustain the base flow of several streams in the Colorado River basin. Although the yields of large-capacity wells reach 1,000 gal/min (63 l/s), the aquifer is not extensively developed. Additional development will reduce the base flow of some streams in the Colorado River basin.

The Santa Rosa aquifer underlies an area of about 1,000 mi² (2,590 km²) east of the High Plains, but its usefulness as a source of potable water is restricted to parts of a few counties. The aquifer, which is composed of gravel, sand, and finer materials, yields water to wells as deep as 450 ft (137 m). Yields of the larger capacity wells average about 250 gal/min (16 l/s), but some wells yield as much as 1,000 gal/min (63 l/s). The aquifer is capable of sustaining slight to moderate increases in the volume of water pumped.

The Trinity aquifer, which underlies about 20,000 mi² (51,800 km²), constitutes an important ground-water reservoir because of its areal extent and its ability to yield more than 1,000 gal/min (63 l/s) of water to large-capacity wells in some places. The aquifer is composed of interbedded sand, shale, and limestone and reaches a maximum thickness of about 1,200 ft (366 m) at its eastern limit (of water of less than 3,000 mg/l dissolved solids) where it extends to depths of about 3,500 ft (1,060 m). The Trinity aquifer is intensively developed in the Dallas-Fort Worth area where the water is pumped for municipal supply and industrial use.

The Edwards-Trinity (Plateau) aquifer underlies about 15,000 mi² (38,200 km²) of the Edwards Plateau. Freshwater (containing less than 1,000 mg/l dissolved solids) is contained in the sand and sandstone layers of the lower part of the aquifer and in the honeycombed limestone of the upper part. The maximum thickness of the aquifer exceeds 1,000 ft (305 m). Although the yields of wells in most places are relatively small, they may exceed 3,000 gal/min (190 l/s) in places where secondary permeability of the limestone is well developed. Varying degrees of secondary permeability, however, tend to govern the rate of production of individual wells. The base flow of several streams in the Colorado and Nueces River basins is substantially augmented by natural discharge from numerous springs issuing from solution channels in the limestone. The aquifer is not extensively developed and is capable of supporting a sizable increase in pumpage, but additional development will reduce base flows.

The Edwards (Balcones Fault Zone) aquifer occupies a relatively small area of about 2,500 mi^2 (6,470 km^2) along the Balcones Escarpment, which separates the Edwards Plateau from the Gulf Coastal Plain. The aquifer consists of about 500 ft (152 m) of extensively faulted limestone and dolomite. The faulting has greatly facilitated the development of secondary permeability, which in turn has created storage space for large quantities of ground water.

The aquifer is extensively developed in much of its extent for municipal supply, industrial use, and irrigation. Ground-water drafts are particularly heavy at San Antonio, where the aquifer provides the entire water supply for the metropolitan area. The capacities of the wells operated by the city of San Antonio are among the world's largest; some exceptional wells yield more than 16,000 gal/min (1,000 l/s). Springs issuing under artesian pressure from the aquifer largely sustain the low flow of several streams in the Nueces, San Antonio, Guadalupe, and Colorado River basins. Prolonged droughts together with heavy ground-water withdrawals may seriously reduce or even stop some of the springflow.

The Woodbine aquifer underlies about 6,000 mi^2 (15,500 km^2) near the inland extent of the Coastal Plain. It consists of as much as 600 ft (183 m) of sand, sandstone, and shale and extends to maximum depths of about 2,000 ft (610 m). Yields of wells completed in the aquifer approach 700 gal/min (44 l/s) in areas of thick sand accumulations. Although the aquifer is tapped by numerous municipal and industrial wells, it is capable of supporting additional development if water-level declines are acceptable.

The Carrizo-Wilcox aquifer underlies about 30,000 mi^2 (77,700 km^2) near the inland margin of the Coastal Plain and extends across the region in a band from 30 to 80 mi (48 to 129 km) wide. The Carrizo-Wilcox aquifer, which consists of interbedded sand and clay, is one of the most productive aquifers in the region. The maximum thickness of the aquifer is about 3,000 ft (914 m) at its southern (downdip) limits. Yields of wells vary widely, but some large-capacity wells discharge as much as 3,000 gal/min (190 l/s).

In the Carrizo-Wilcox aquifer, water having less than 3,000 mg/l dissolved solids occurs at depths of more than 5,000 ft (1,520 m), which is deeper than in any other aquifer in the region. Artesian pressure is sufficient to cause many of the deep wells to flow. The aquifer is intensively developed by irrigation wells in the Winter Garden area southwest of San Antonio, where the annual ground-water draft exceeds the annual recharge.

The Queen City aquifer, which underlies about 14,000 mi^2 (36,300 km^2) of the Coastal Plain, consists of interbedded sand and clay as much as 500 ft (152 m) thick. The maximum depth of water having less than 3,000 mg/l dissolved solids is about 2,000 ft (610 m). Yields of wells are relatively low, but a few large-capacity wells exceed 400 gal/min (25 l/s). The aquifer is only slightly developed throughout its extent.

The Sparta aquifer underlies about 9,000 mi^2 (23,300 km^2) of the Coastal Plain and mostly overlies the Queen City and Carrizo-Wilcox aquifers. The Sparta consists of as much as 350 ft (107 m) of interbedded sand and clay, and water containing less than 3,000 mg/l dissolved solids extends to depths of more than 2,000 ft (610 m) in places along the southeastern boundary of the aquifer. Large-capacity wells generally yield about 500 gal/min (32 l/s) although the yields of some wells exceed 1,000 gal/min (63 l/s) Except for relatively heavy localized development, the aquifer is only slightly developed throughout most of its extent.

The Gulf Coast aquifer, which includes the Catahoula, Jasper, Evangeline, and Chicot aquifers, underlies about 35,000 mi^2 (90,600 km^2) of the Coastal Plain and extends from the coastline inland for about 90 to 120 mi (145 to 193 km). This aquifer, which is the most extensive of the region's ground-water reservoirs, consists of more than 3,500 ft (1,060 m) of interbedded sand, clay, and gravel. Huge quantities of water are pumped from the aquifer for municipal supply, industrial use, and irrigation. The most intensive and concentrated development is in the Houston area, where large-capacity wells yield from 1,000 to more than 3,000 gal/min (63-189 l/s), and average about 2,000 gal/min (126 l/s). Even though very large quantities of water are being pumped, the Gulf Coast aquifer could support substantial additional development, especially in the area northeast of Houston.

The Ogallala aquifer underlies about 19,000 mi^2 (49,200 km^2) of the High Plains of Texas and New Mexico. It is one of the most intensively developed aquifers in the Nation. Within the Texas-Gulf Region, the Ogallala supplies water to an estimated 50,000 irrigation wells.

The Ogallala consists of as much as 500 ft (152 m) of lenticular beds of gravel, sand, and finer material, and the zone of saturation ranges in thickness from a few feet to more than 250 ft (76 m). The depths to water range from less than 50 ft (15 m) to more than 300 ft (91 m). The yields of irrigation wells range from less than 100 gal/min (6.3 l/s) to more than 1,000 gal/min (63 l/s), depending largely upon the saturated thickness of the aquifer. Large-scale pumping, which greatly exceeds recharge, is drastically depleting the ground-water reserve. Recharge was estimated by Lohman (1972, p. 66) to range from 0.05 to 0.5 in (0.13 to 1.3 cm) per year. In some areas, recharge may exceed 0.5 in (1.3 cm) and perhaps approach 1 in (2.5 cm) (A. W. Wyatt, Texas Water Development Board, oral commun., 1974).

The various alluvial aquifers in the region occupy an aggregate area of about 1,800 mi^2 (4,660 km^2). Although alluvium occurs as flood-plain deposits along the Colorado and Trinity Rivers and other streams, the principal deposits are in the Central Lowland east of the High Plains and along the Brazos River. The Central Lowland aquifers are scattered and isolated terrace and flood-plain deposits that consist of as much as 150 ft (46 m) of gravel, sand, and finer sediments. The yields of large-capacity wells in these deposits range from less than 100 gal/min (6.3 l/s) to more than 1,000 gal/min (63 l/s). The Central Lowland aquifers are not capable of furnishing additional large quantities of water for future development.

The Brazos River alluvial aquifer underlies a flood plain from 1 to 7 mi (1.6 to 11.3 km) wide and follows the long sinuous course of the Brazos River for 350 mi (563 km). The aquifer consists of as much as 100 ft (30.5 m) of gravel, sand, and finer sediments. More than 1,000 irrigation wells tap the aquifer, which yields from 250 to 500 gal/min (16 to 32 l/s) to 50 percent of the wells. The maximum yields exceed 1,000 gal/min (63 l/s). The Brazos River alluvial aquifer is capable of sustaining additional ground-water withdrawals, but the discharge of the Brazos River will be reduced.

The geographic distribution of the aquifers, taking into consideration the overlap that occurs in places, provides coverage of about four-fifths of the Texas-Gulf Region. Only about one-fifth of the region is not underlain by one or more significant ground-water reservoirs, and even these areas are not totally devoid of ground water, but must rely on smaller or less dependable supplies from relatively minor water-bearing formations.

QUALITY OF THE GROUND WATER

In any appraisal of a water resource, the quality of the water must be known before the value of the supply can be judged. A knowledge of the quality of the available water is, therefore, in the same order of importance as a knowledge of the quantity of available water. To cite an obvious example, the oceans contain an estimated 97 percent of the world's total supply of water, but, under present technology and the limits of economic feasibility, it is of little value as a water supply for most uses.

The chemical, physical, biological, and radiological characteristics of water determine its suitability for various uses. Physical problems (turbidity, odor, color, taste, and temperature) and biological problems (bacteria) usually can be alleviated economically, but the removal or neutralization of undesirable chemical constituents can be difficult and expensive.

The type and amount of dissolved constituents in both ground water and surface water are different from place to place, but the chemical quality and physical properties of ground water at any one place generally remain constant. Ground water is also less susceptible than surface water to biological and radiological contamination.

Analyses of water from wells and springs in the Texas-Gulf Region show that there are great differences in the chemical quality of the ground water, both geographically and with depth of occurrence. Such differences are too numerous to depict on a map, but the lowest concentrations of dissolved solids usually occur in the areas of greatest rainfall and least evaporation—in the aquifers of the eastern part of the region where recharge is abundant. The exceptions to this general rule are the occurrences of water with low concentrations of dissolved solids in the limestone aquifers of the central and western parts of the region where recharge is rapid.

Widely ranging levels of salinity are characteristic of the water in storage in the ground-water reservoirs of the region, and this fact, in turn, largely governs the suitability of the water for various uses. For the purpose of this report the degrees of chemical quality, in general terms of total mineralization, are differentiated according to the following classification of Winslow and Kister (1956):

Description	Dissolved solids (mg/l)
Fresh	Less than 1,000
Slightly saline	1,000 to 3,000
Moderately saline	3,000 to 10,000
Very saline	10,000 to 35,000
Brine	More than 35,000

The fresh and slightly saline ground-water supplies have been sufficiently identified in previous ground-water studies (Alexander and others, 1964; Baker and others, 1963a, b; Cronin and others, 1963; Mount and others, 1967; Peckham and others, 1963; and Wood and others, 1963) to permit a delineation of their combined extent (modified from Texas Water Development Board, 1968, figs. II–2 and II–7) in the various ground-water reservoirs of the region (pl. 2B).

Ground water having concentrations of dissolved solids in excess of 3,000 mg/l commonly occurs in the deep subsurface below the slightly saline water and downdip in most of the aquifers of the region. Only the Edwards-Trinity (Plateau), Ogallala, and alluvial aquifers, most of which do not have deep subsurface extensions, contain water that is almost entirely fresh to slightly saline. Largely because of the natural increase in the salinity of ground water with depth, saline water occurs throughout most of the region. The typical range in dissolved solids in water that is being used from the various aquifers is given in table 3.

TABLE 3.—*Typical range of dissolved solids in water used from each aquifer*

Aquifer	Typical range in dissolved solids (mg/l)
Alluvium	500–2,000
Ogallala	400–1,200
Gulf Coast	300–1,000
Sparta	200– 800
Queen City	200– 800
Carrizo-Wilcox	200–1,500
Woodbine	500–1,200
Edwards (Balcones Fault Zone)	300–1,200
Edwards-Trinity (Plateau)	400–1,000
Trinity	500–1,500
Santa Rosa	400–2,500
Ellenburger-San Saba	400–2,000
Hickory	300– 700

QUANTITIES OF GROUND WATER AVAILABLE FOR DEVELOPMENT

Because of the hydrologic relationship of ground-water recharge to precipitation, runoff, and evapotranspiration, potential recharge in the Texas-Gulf Region increases from northwest to southeast and ranges from a miniscule amount on the High Plains to several inches in the Sabine River basin. As a result of this difference in potential recharge, development of the ground-water resources may vary from unavoidable "mining" (pumping in excess of the potential replenishment rate) of ground water in storage in the western part of the region to a plan of limiting withdrawals to the steady-state yield of the aquifers in the eastern part of the region.

The quantities of ground water available for development have been estimated largely on the basis of the amount of natural discharge subject to capture by pumping which is an index of the maximum or near maximum recharge rate from precipitation. The methods used to obtain the data given in this report were: (1) the U.S. Study Commission method, which uses theoretical lines of recharge and discharge and maximum pumping lifts of 400 ft (122 m) to determine the yield of clastic artesian aquifers by increasing natural recharge at the outcrop from increased hydraulic gradients (Alexander and others, 1964; Baker and others, 1963a, b; Cronin and others, 1963; Mount and others, 1967; Peckham and others, 1963; and Wood and others, 1963); (2) the base-flow method for determining recharge to the Edwards-Trinity (Plateau) aquifer, in which recharge approximates base flow of the streams draining the aquifer (Alexander and Patman, 1969; Long, 1958; and Reeves, 1969); (3) the stream inflow-outflow method used for Edwards (Balcones Fault Zone) aquifer, in which recharge is computed on the basis of historical streamflow records (Petitt and George, 1956; and Garza, 1962); and (4) the method of determining the amount of water moving through the aquifers (Ogallala and alluvium) under existing hydraulic gradients (Theis, 1937; Cronin and Wilson, 1967; and Brown and Signor, 1973).

Availability is given also in this report as the amount of water recoverable and partly recoverable from storage. These values have their greatest significance in areas where ground-water mining is unavoidable, as on the High Plains.

The quantities of ground water available for development (table 4) are expressed as the steady-state yield. Under this concept, the yield approximates the maximum perennial replenishment from precipitation. The retention in the aquifer of at least a part of the recharge that would have been released as base flow is inherent in this concept, and development of an aquifer to this extent necessarily results in diminution of surface runoff. The concept also presupposes an idealized pattern of aquifer development whereby construction and spacing of individual wells are suitable for recovering much of the water added to the zone of saturation and for stabilizing water levels.

As shown in table 4, the total steady-state yield of the aquifers of the region is 4.6 million ac-ft (5.7 km^3) annually. The largest single yield is 2.5 million ac-ft (3.1 km^3) annually from the Gulf Coast aquifer. The quantities of water given as steady-state yields are, at best, only approximations and are to be regarded as preliminary estimates that should be revised and kept current as development progresses and more basic data become available.

It is not suggested that the aquifers should be developed to the indicated extent of the yields in any plan of total water management. As appropriately expressed by Lohman (1972, p. 62), where the framework of prevailing laws and regulations permits the total water resource to be operated for maximum efficiency, water planners, given the knowledge of the combined ground- and surface-water systems and their interrelationship, should then be able to select the most desirable or equitable distribution of available water.

An estimated total of 4.32 billion ac-ft (5,330 km^3) of recoverable and partly recoverable water containing less than 3,000 mg/l dissolved solids is stored in the aquifers of the region, and of this amount, about 1.04 billion ac-ft (1,280 km^3) of recoverable water, or 24 percent of the total, is estimated to be stored within 400 ft (122 m) of the land surface. The Gulf Coast aquifer, which contains an estimated 450 million ac-ft (555 km^3) of recoverable water in the top 400 ft (122 m), has the largest storage capacity of the region's aquifers.

The 400-ft (122-m) depth as used here is an arbitrary level that does not necessarily imply an economic limit of recovery, as economics of pumping can and usually do change for various reasons. Obviously in the case of the Ogallala aquifer, the 5 million ac-ft (6 km^3) of water below a depth of 400 ft is economically recoverable. Nevertheless, for the purpose of this report, the available water above a depth of 400 ft (table 4) is considered to be recoverable and implies dewatering (mining) to that depth, which would necessitate pumping lifts in excess of 400 ft. On the other hand, much of the water below 400 ft, especially in those aquifers where freshwater extends to great depths and is in lateral and vertical connection with saltwater, may not be economically recoverable at present costs and may never be practical to recover. Largely for this reason, the volume of water below a depth of 400 ft is referred to in table 4 as partly recoverable.

TABLE 4.—*Quantities of ground water available for development* [1]

Aquifer	Steady-state yield (thousands of acre feet per year)	Recoverable water in storage above depth of 400 ft (thousands of acre-feet)	Partly recoverable water in storage below depth of 400 ft (thousands of acre-feet)
Alluvium	130	5,000	0
Ogallala	90	135,000	5,000
Gulf Coast	2,500	450,000	1,150,000
Sparta	130	20,000	65,000
Queen City	120	70,000	200,000
Carrizo-Wilcox	560	150,000	1,150,000
Woodbine	10	10,000	70,000
Edwards (Balcones Fault Zone).	410	2,000	13,000
Edwards-Trinity (Plateau).	540	70,000	70,000
Trinity	70	100,000	450,000
Santa Rosa	30	8,000	0
Ellenburger-San Saba	20	8,000	12,000
Hickory	40	10,000	100,000
Total (Rounded).	4,650	1,038,000	3,285,000

[1] Steady-state yield and recoverable water in storage from data in Alexander and others (1964), Baker and others (1963a,b), Brown and Signor (1973), Cronin and others (1963), Mount and others (1967), Peckham and others (1963), Petitt and George (1956), Texas Water Development Board (1966a–r), Wood (1956), Wood and others (1963), and other computations by U.S. Geological Survey. Specific yield of 2 percent used in computing recoverable water in storage in Edwards (Balcones Fault Zone), limestone part of Edwards-Trinity (Plateau), and Ellenburger-San Saba aquifers; specific yield of 15 percent used for other aquifers. Ogallala storage as of 1967 after Cronin (1969).

Because of the enormous amount of recoverable water in storage, each aquifer contains a stock of water that can be mined if necessary (as is being done with the Ogallala aquifer on the High Plains), whether the rate of recharge is large or small. The steady-state yield can be augmented for a very long time on a "deferred" basis, whereby the use of the water that is withdrawn from storage may result in a strengthened economy, which is turn can support programs for developing additional sources of supply. (See American Society of Civil Engineers, 1972, p. 21.)

The surpluses and deficiencies in the ground-water supplies in the region result from three factors: (1) the aquifers are not uniformly distributed within the region; (2) the yields of the aquifers are different; and (3) the demands for water vary from one part of the region to another. Indices of areal potential for future development can be established on the basis of the geographical variance in the steady-state yield and the demands projected to 2020. The areas of greatest potential for additional development of ground-water supplies within the Texas-Gulf Region, as shown on plate 2C, are areas of ground-water surpluses where the locally available supplies exceed the projected local demands.

Projected ground-water requirements exceed the available ground-water supply in only three areas—the High Plains, the Dallas-Fort Worth area, and the San Antonio area. The aquifers in these areas are already heavily pumped, and projected withdrawals to satisfy the 2020 demands will result in the initiation or continuance of ground-water mining.

The steady-state supply of ground water exceeds the projected ground-water requirements by different amounts in the rest of the region. Ground-water surpluses of as much as 300,000 ac-ft (0.37 km^3) annually are available in the Sabine and Neches River basins, and the occurrence of highly productive aquifers and relatively low requirements in the Nueces River and Middle Colorado River basins indicate that these areas in the generally water-deficient western part of the region have the potential for additional development (pl. 2C).

In many parts of the region, however, the steady-state supply only slightly exceeds the projected ground-water requirements, and surpluses in these areas are projected to be less than 100,000 ac-ft (0.12 km^3) annually by 2020. In the heavily pumped Houston area, for example, ground-water withdrawals slightly in excess of the projected demand would result in mining of the water in storage in the sand and gravel beds. Water is being mined already from the clay beds by compaction and accompanying land-surface subsidence.

The foregoing discussion of the availability of ground water and of the potential for future development by area does not consider the possible impact of such development. It should be reiterated that full development of ground-water resources to the extent indicated, or even less in some areas, could be detrimental to the adequacy of surface-water supplies by reducing basin runoff. Also, part of the ground-water surpluses in some aquifers augments streamflow that in turn provides recharge to other aquifers.

OPPORTUNITIES FOR MEETING WATER REQUIREMENTS BY GREATER USE OF GROUND-WATER RESERVOIRS

The large ground-water resources of the Texas-Gulf Region, with some notable exceptions, are still in an incipient stage of development. This situation exists despite the fact that development has continued almost unabated for the past half century and that ground water now constitutes the source of about half the water used in the region.

Provisions should be made in programs for meeting future water requirements to develop the available ground-water supply to the most reasonable extent and to use the underground resources, as appropriate, in integrated systems of importing and exporting water, storage, artificial recharge, desalinization, reclamation, and waste disposal.

USE OF GROUND WATER AS THE PRIMARY SOURCE OF WATER SUPPLY

The selection of ground water as the primary source of water supply has generally been based on one or more of the following advantages (Thomas, 1951, p. 223–224):

1. *Accessibility.*—Ground water usually may be reached within a few hundred feet of where it is to be used, and on the same property, whereas surface water may require pipelines and rights-of-way over stretches of several miles.
2. *Availability.*—Ground water may be available for use in areas where the water in streams and lakes has already been appropriated by other users.
3. *Dependability.*—Yield from wells generally fluctuates less than streamflow in alternating wet and dry climatic cycles.
4. *Uniformity in quality.*—Ground water is more uniform in temperature and soluble-mineral load than surface water, and is generally free of turbidity and bacterial pollution.

The use of ground water also has inherent disadvantages, some of which are: (1) a potential for land-surface subsidence in some areas; (2) the absence of recreation facilities associated with surface-water reservoirs; and (3) possible interference with surface-water rights if streamflow is reduced.

Many of the advantages, however, are applicable to ground-water supplies in the Texas-Gulf Region, and in many places the potential for greater utilization of ground-water resources should warrant investigation. In some areas, the ground-water resources can provide for the needs of additional urban development, industrial expansion, and increased irrigation. Consequently, if ground water is utilized as the primary source of water supply, excess surface water in amounts larger than otherwise anticipated, might then become available for export to water-deficient areas. Because most of the eastern part of the region contains such large ground-water resources, much of the proposed development and local use of surface-water supplies might be reconsidered on the basis of regional needs.

CONJUNCTIVE USE OF GROUND WATER AND SURFACE WATER

The problem of meeting demands for water in the future can be solved more efficiently if the resources are developed in accordance with hydrologic principles. One basic principle is that within the dimensions of a region, which include the subsurface, surface, and atmosphere, there is a given amount of water to be managed, regardless of where it may be in storage or in transit at any particular time. The concept of "conjunctive use" means, therefore, in its broadest sense, that ground water, surface water, and atmospheric water are treated as a single resource.

Conjunctive use provides the opportunity to develop the inherent advantages of both surface-water and ground-water reservoirs. The storage capacity of ground-water reservoirs, for example, is far greater than the storage capacity of surface-water reservoirs; and the water stored underground is relatively safe from evaporation and contamination. On the other hand, surface-water supplies are more easily renewed and are usually more desirable for providing large amounts of water to areas of concentrated demand.

The problems of water-resources management would be less difficult if a fully integrated natural and man-made plumbing system for the entire region could be modeled. However, the legal and technological constraints imposed on regional planning cannot be ignored, and the existing facilities cannot be eliminated, because they were not designed as part of a theoretical system. In some places, however, it may be possible to use existing and proposed facilities more efficiently. The following definitions demonstrate the point:

1. Surface-water reservoirs that are constucted for the primary purpose of providing water for artificial recharge of ground-water reservoirs, can be designated as "recharge reservoirs" and could be considered as a proper addition to such primary classifications of surface-water reservoirs as "flood control," "conservation," and "hydroelectric-power generation."

2. Ground-water well fields that are developed for the primary purpose of maintaining conservation storage in surface-water reservoirs, can be designated as "reservoir-maintenance well fields" and could be considered as a proper addition to the primary classification of well fields as "public supply," "industrial use," and "irrigation."

In the Texas-Gulf Region, conjunctive-use programs might be applicable to the following problems: (1) water shortages during droughts, such as the severe Texas drought of the 1950's and the drought in the Concho River basin duing 1962–71; (2) the overdevelopment of the Carrizo-Wilcox aquifer for irrigation in the Winter Garden area; (3) the ever increasing and hardly sustainable demands that are being imposed upon the Edwards aquifer in the San Antonio area; (4) the declining water levels, land-surface subsidence, and saltwater intrusion that have resulted from pumping some 400 million gallons (0.0015 km^3) of water per day from the Gulf Coast aquifer in the Houston area; and (5) the depletion of water in storage in the Ogallala aquifer of the High Plains to the extent that the economy may suffer.

A prolonged drought imposes a severe strain upon the water resources of a region, because it results in a reduction in the total amount of water available for management. The most immediate effect, however, is the reduction in streamflow and the need to conserve storage in surface-water reservoirs. In some places, the local ground-water resources could be employed to help maintain storage in the reservoirs; in other places, it is realistic to consider piping ground water from other areas. Where ground water and underground space are available, well fields consisting of both recharge and discharge (or dual-purpose) wells could be considered as a possibility for management of excesses and deficiencies in streamflow.

During the drought of the 1950's, surface-water storage in Lake Corpus Christi, which is the source of water for public supply in the area, was greatly reduced. The problem was alleviated by the use of four flowing wells that had been drilled in the Carrizo-Wilcox aquifer in Atascosa County by the Lower Nueces Water Supply District in 1951. The water from these wells was directed into the Nueces River, via the Atascosa River, where it eventually reached and was impounded in Lake Corpus Christi. Although 65 to 75 percent of the water was lost by evaporation, transpiration, and infiltration during the 118-mi (190-km) trip from the wells to the lake, the rate of flow of 10 million gallons (37,800 m^3) per day at the wellheads was sufficient to increase the supply in Lake Corpus Christi.

During the drought of 1962–71 in the Concho River basin, the city of San Angelo, which obtains its municipal supply from Lake Nasworthy, contended with the lack of adequate water in storage by purchasing irrigation rights on the South Concho River, constructing a pipeline to E. V. Spence Reservoir, and drilling wells for ground-water supplies. The wells would have served to ease the critical situation, but the drought ended, and the reservoir filled before the well field was developed.

The important fact in these examples is not that ground water was used or was planned to be used, but that it was used after water shortages had occurred rather than as a planned contingency for drought conditions. The situations in the Winter Garden, San Antonio, and Houston areas provide examples in which heavily pumped subsurface reservoirs could be used advantageously in conjunction with the relatively abundant surface-water resources.

In the Winter Garden area, in Dimmit and Zavala Counties north of Laredo, about 300,000 ac (1,200 km^2) are irrigated by an annual pumpage of about 350,000 ac-ft (0.43 km^3) of ground water from the Carrizo-Wilcox aquifer. The amount of water being pumped greatly exceeds recharge to the aquifer; therefore, the potential problems are declining water levels and possibly economically prohibitive pumping costs.

The solution to the Winter Garden problem as proposed by the Texas Water Plan is to import 200,000 ac-ft (0.25 km^3) of water annually from Amistad Reservoir in the Rio Grande basin. The Water Plan states that this imported water offers the possibility for "conjunctive operation." In addition, the productive capacity of the aquifer can be prolonged by recharging the ground-water reservoir with surface water intercepted within the Nueces River basin.

The Edwards (Balcones Fault Zone) aquifer provides water for public supply, industrial use, and irrigation in an extensive area of south-central Texas. The city of San Antonio, which has a population of about 800,000, depends upon the Edwards for its entire supply. If the demands for water from the Edwards continue to increase, as they are likely to do in this rapidly growing area, the predictable results will be: (1) severe declines in water levels during periods of drought; (2) large reductions in the amount of water available to all users; and (3) reductions in flow, or even periods of no-flow, from the important springs such as those at San Antonio, New Braunfels, and San Marcos.

These problems might be avoided or attenuated by using surface water from the Guadalupe, Nueces, and San Antonio River basins in conjunction with the storage capacity and transmissivity of the Edwards reservoir. Surface-water reservoirs could be constructed on or near the recharge area of the Edwards for the primary purpose of providing recharge to the aquifer.

In the Houston area, about 400 million gallons (0.0015 km^3) of water per day are pumped from the Gulf Coast aquifer for municipal supply, industrial use, and irrigation. Although the potential ground-water recharge is very large, maximum development is limited by the capability of the aquifer to transmit water from areas of recharge to areas of discharge. The Houston area is not short of water, but the sustained heavy pumping in local areas has already resulted in hydrologic and economic problems.

Expanding the use of abundant surface-water supplies or the development of well fields nearer to the recharge areas could be helpful in the Houston area because of the serious problem of land-surface subsidence, which is subjecting the highly developed near-shore areas to progressive inundation by normal tides and to catastrophic flooding by hurricane tides. The lost ground cannot be regained, but subsidence can be arrested by decreasing the pumping and recharging the aquifer.

The problem on the High Plains presents a special situation, where a finite supply of water is being consumed. The water in storage in the Ogallala aquifer has been depleted in recent years at the rate of about 3.5 million ac-ft (4.3 km^3) per year, based on about 3.5 million irrigated acres (14,200 km^2) in 1969 and on an assumed average depletion rate of 1 ac-ft per irrigated acre per year (1,230 m^3 per 4,050 m^2). Surface-water supplies, except from the playa lakes, are practically nonexistent, and recharge to the aquifer is extremely small.

The consequences of depleting the water in storage in the Ogallala can be delayed, however, by modifying the land-use practices and by judicious use of all available water. The amount of water available annually from the playa lakes within the region is estimated (Schwiesow, 1965) at about 2 million ac-ft (2.5 km^3), and the amount of recoverable water remaining in the Ogallala aquifer in 1967 (Cronin, 1969) was calculated to be 140 million ac-ft (173 km^3). If the playa-lake water is used to supplement supplies from the Ogallala, if the water remaining in the Ogallala is used with great care, and if land-use practices can be modified to require less water, the available supplies may last for several more decades.

In the meantime, higher food prices because of worldwide shortages, together with improved technology, may change the cost-benefit ratio so that importation of water will be economically feasible, and the dewatered part of the Ogallala aquifer may be the best reservoir available for storing imported water.

ARTIFICIAL RECHARGE

Although an enormous total amount of water is annually recharged to the aquifers by natural means, it may be desirable for a variety of reasons to increase the amount or rate of recharge in some areas. Artificial recharge may be useful in coping with present or potential problems, but under certain circumstances and in certain areas, the practice is probably necessary in any program to manage the ground-water subsystem in conjunction with and in a manner similar to the management of the surface-water subsystem.

Artificial recharge is best known for its application in alleviating water-supply problems, but the practice has important applications in liquid-waste disposal, secondary oil recovery, maintenance of reservoir pressures, and in arresting land-surface subsidence. As applied to water-supply problems, artificial-recharge programs are designed to: (1) supplement the quantity of available ground water; (2) retard or prevent the depletion of ground-water reservoirs; (3) prolong acceptable ground-water quality by retarding or preventing salt-water encroachment; (4) conserve and dispose of runoff, including floodwaters; (5) store water to reduce costs of pumping and piping; and (6) store water in times of low demand for use during periods of high demand. Although applications relating to liquid-waste disposal and to petroleum-industry operations are currently being practiced in the Texas-Gulf Region, only a few diverse programs relative to water supply are in operation.

The principal area where artificial recharge may be useful is on the High Plains, where progressive depletion of the Ogallala aquifer will require adjustments in irrigation practices. Water supplies can be increased by importing water, but any plan that presupposes an increase in available water must consider the problem of storage facilities, particularly during periods of low use. Adequate storage space is available in the Ogallala if the water can be placed underground by injection wells or by infiltration methods. The method or combination of methods used will be determined by the quality of the recharge water and by the hydraulic characteristics and geochemistry of the aquifer (Brown and Signor, 1973).

Artificial recharge operations and experiments have been conducted on a small scale in several places on the Southern High Plains. Successful recharge by injection wells was achieved by intra-aquifer transfer of Ogallala water (Moulder and Frazer, 1957; Reed, unpub., rept. 1959) and by recharge of water from Lake Meredith (in the Arkansas-White-Red Region) and Lake J. B. Thomas in well fields at Lubbock and in Martin County, Texas, respectively (Brown and Signor, 1973). Although these operations were not carried on long enough to preclude the possibility of failure because of bacterial contamination or slowly occurring chemical changes, the success indicates that there are no insurmountable short-term technical difficulties in recharging high-quality water. However, the pump-back cost of the water, including the expense of the recharge operation, may exceed that which the agricultural users now pay.

The largest available local supply of surface water for artificial recharge on the High Plains is the 2 million ac-ft (2.5 km^3) estimated to be available annually in the playa lakes. Recharge of this water through injection wells, however, has generally been unsuccessful because of clogging of the aquifer by suspended material. The most successful effort in recharging playa-lake water through wells was that of Hauser and Lotspeich (1967), whereby surface treatment of the highly turbid water removed 90 percent or more of the suspended material. Thus far, artificial recharge using playa-lake water has been tried experimentally and is currently being attempted at numerous sites on the High Plains of Texas and New Mexico. (See Broadhurst, 1960; Valliant, 1961; Cronin, 1964; Clyma, 1964; Havens, 1966; Meyer, 1972.)

Artificial recharge by surface spreading, which is ineffective in many places on the High Plains because of the occurrence of layers of low permeability between the land surface and the water table, has been accomplished near Lubbock, Texas, where treated sewage effluent has been used for flood irrigation of 4,000 ac (16.2 km^2) since 1938. After this operation began, the water table in the Ogallala rose from about 120 ft (37 m) below land surface to within a few feet of the surface by 1968 (Brown and Signor, 1973). Artificial recharge by surface spreading has not been attempted elsewhere in the region.

Because of the potential value of artificial recharge in coping with present and future problems, intensive research studies of various aspects of this process are continuing. Such studies will provide information on artificial recharge of the Ogallala aquifer, and the results of these studies may have application in other areas where artificial recharge is contemplated.

Artificial-recharge methods could be used advantageously in segments of two significant aquifers in the southwestern part of the region—in the fully-developed Carrizo-Wilcox aquifer in the heavily irrigated Winter Garden area southwest of San Antonio and in the heavily-pumped Edwards (Balcones Fault Zone) aquifer in the San Antonio area. Imported water is envisioned by the Texas Water Plan (Texas Water Development Board, 1968) to supplement ground water pumped in the Winter Garden area. Excess water available after the growing season could be used to recharge the ground-water reservoir. In addition, the diversion of floodwaters within the Nueces River basin might provide a local source of water for recharging the Carrizo-Wilcox aquifer by surface spreading on its sandy outcrop.

In the San Antonio area, artificial-recharge facilities could include a series of surface-water reservoirs on or near the recharge area of the Edwards (Balcones Fault Zone) aquifer. Such reservoirs could be designed to impound flood waters from the upper Nueces, San Antonio, and Guadalupe River basins for controlled release into the highly permeable aquifer.

In the north-central and west-central parts of the region, pumping from the alluvial aquifers for municipal supply and irrigation has already resulted in water-level declines that will become more serious if demands increase. Imported water envisioned by the Texas Water Plan for this area would sustain agricultural production and would meet the needs for municipal supply. In addition to direct use of the surface water, a program of artificial recharge could be used to store part of the imported water in the ground-water reservoirs, especially when such water is in excess of seasonal needs.

Various applications of artificial recharge in the Houston area and nearby coastal areas of heavy ground-water pumping could be beneficial although reduced pumping and increased surface-water use may be more practical. In this water-rich part of the Coastal Plain, generally adequate supplies of freshwater are available, but the continually expanding and deepening cone of depression in a part of the Gulf Coast aquifer has caused land-surface subsidence and saltwater encroachment.

Repressuring the heavily pumped part of the aquifer by artificial recharge would decrease the rate of subsidence, and if enough water is recharged, subsidence could be halted. Reduced ground-water pumping would have the same effect, however, and would probably be a more feasible method. It is not likely, however, that any measurable recovery in the altitude of the land surface could be effected (Gabrysch, 1972).

Artificial recharge might also be used in the Houston area to halt or retard the advancement of saline water toward the freshwater well fields. This might be achieved by injecting freshwater along the interface of fresh and saline water to create and maintain a high-pressure ridge that would function as a barrier to the saline water. Because of the availability of surface water in this area, reduced ground-water pumping would, again, probably be a more feasible method.

Applications of artificial-recharge methods to liquid-waste disposal, secondary oil recovery, and maintenance of reservoir pressures are discussed in a separate section of this report.

SALVAGE OF WATER FOR BENEFICIAL USE

The process of salvaging ground water may be viewed as a means of effecting a prolongation or an increase in supply, and for this reason salvage becomes a factor in aquifer management. Salvage of ground water for beneficial use can be exercised primarily by: (1) reducing consumption of water by evaporation and phreatophytes; (2) practicing conservation; and (3) reclaiming and recycling waste water that has already served some beneficial use.

Consumption of ground water by most phreatophytes generally is conceded to be excessive in relation to the benefit of these plants to man. Although some phreatophytes furnish cover and browse for game, provide erosion control, serve as windbreaks, or have esthetic value, most of them have low economic value. Salvage of ground water by reducing consumptive use by phreatophytes may be accomplished by eradication or by substituting plants of high economic value such as alfalfa.

The most common phreatophytes in the Texas-Gulf Region are saltcedar and mesquite. Cottonwood, willow, baccharis, and other species occur where the accessibility and quality of water are favorable to their existence. The phreatophyte problem is more acute and has received more attention in the semiarid western part of the region, where the phreatophytes are mostly confined to the flood-plain alluvium along stream channels where ground water is within reach of the root systems.

An example of the density, spread, and water consumption of woody phreatophytes in the region is afforded by an intensive study along a reach of the Brazos River and selected tributaries above Possum Kingdom Reservoir (Busby and Schuster, 1973). More than 60 percent of about 100,000 ac (405 km^2) of flood plain in the reach studied was occupied by various phreatophytes. For a part of the area where a succession of aerial photographs was studied, saltcedar showed a progressive invasion of the area and doubled in extent over a period of about 30 years. Mesquite and saltcedar alone in the total area consume approximately 90,000 ac-ft (0.11 km^3) of water annually.

With different land-use practices, a part of this water could be salvaged for greater economic benefits in one form or another. The total amount of ground water that might be salvaged by the control of saltcedar throughout the Brazos River basin has been preliminarily estimated by the Soil Conservation Service to be 625,000 ac-ft (0.77 km^3) annually (Texas Water Development Board, 1968, p. III–30). This amount, when placed in perspective, equals the amount of ground water pumped for all purposes in 1970 in the Houston area. Not all of the 625,000 ac-ft (0.77 km^3), however, would be available for beneficial use, because a part of the potential salvage includes saline water.

Another extensive area of saltcedar growth is along the Colorado River and its tributaries in the west-central part of the region, where dense stands of phreatophytes occupy parts of the flood plain over a distance of approximately 300 river miles (483 km) (Smith and Rechenthin, 1964). A study by Larner and others (1974) in this reach showed that if 1 ac-ft (1,230 m^3) of water per acre (4,047 m^2) could be salvaged through a control program on the flood plain, and 3 ac-ft (3,700 m^3) per acre along the river banks and along the shoreline of Lake J. B. Thomas, a total of at least 10,200 ac-ft (0.01 km^3) of water could be salvaged annually and made available for other uses.

Reduction of consumption of water by phreatophytes in the humid eastern part of the region obviously could salvage a large quantity of ground water, but in this area much of the consumption of water is by dense stands of timber. The relatively high economic value of the timber justifies consumption of the water.

Water-conservation practices, such as reclaiming or recycling water that has already served some beneficial use, are a means of increasing the available ground-water supply by reducing the demand for new water. Reuse allows more service per unit of water withdrawn.

Studies conducted by the Texas Water Development Board to update projected water requirements for agriculture envision a highly efficient use of water from conservation and reclamation practices in the future. For example, application of water per acre on the High Plains of Texas by 1990 is preliminarily projected by the Texas Water Development Board to be about 0.8 ft (0.24 m), whereas the historical average application of water per acre is estimated to be about 1.6 ft (0.49 m) in Texas and about 2.0 ft (0.61 m) in New Mexico (U.S. Bureau of Reclamation, 1973, p. 41). Such a large reduction in use of water would stretch the supply of water in storage in the Ogallala and extend the productive life of the aquifer.

The relative scarcity of water in some areas of the region, coupled with Federal and State laws regulating the quality of waste water before discharge, have spurred many industries to develop techniques that permit multiple use of small amounts of the same water rather than larger amounts of new water that is run only once through the system. Thomas (1951) has shown that some large industries have been able to reduce their demand for new water from underground or surface resrvoirs to as little as 5 or 10 percent of the quantities used in similar industrial operations in other areas.

Barring some breakthrough in weather modification or saline-water conversion, the actual water supply of the region cannot be increased. The supply can be used more effectively, however, by a reduction in unit requirements by all consumers. The result then of conservation, recycling, and reclamation is that more water would be immediately available where it is needed to meet deficiencies or to support economic expansion.

USE OF SALINE AQUIFERS

An evolution in attitudes about saline aquifers has taken place in an accelerating manner and within a relatively short time. The change has been from considering saline aquifers as a nuisance to recognizing them as a potentially valuable resource, not only for their content of usable or treatable water, but also for their potential use for the disposal of liquid waste or for storage of fresh water. The importance of the saline aquifers in the Texas-Gulf Region is enhanced by their nearly ubiquitous areal extent. About 170,000 mi^2 (440,300 km^2) or about 98 percent of the region is underlain by saline aquifers (Winslow and Kister, 1956; Hood and Kister, 1962; and Core Laboratories, 1972).

SALINE AQUIFERS AS SOURCES OF TREATABLE WATER

The demand for desalted water to supplement naturally occurring freshwater did not develop until the beginning of the second half of this century. While the demands for freshwater were steadily increasing over the past several decades, the freshwater supply in some areas was decreasing. Inevitably the requirement to tap sources of saline water became compelling.

Under the Saline Water Act, which was passed by Congress in 1952, provisions were made for the development of practical means for the economical production of water, from saline sources, of a quality suitable for municipal supply, industrial use, irrigation, and other beneficial uses. Since 1952, progress has been made in developing desalination processes, and under four basic classifications—distillation, membrane, crystallization, and chemical—about a dozen processes are currently under study. Some processes are being used commercially while others have progressed only to the pilot-plant stage.

The progress in technology increases the potential value of the saline ground-water resources of the Texas-Gulf Region. Although the supply of saline water in underground reservoirs has never been determined quantitatively, it is accessible almost everywhere. The depth to saline water ranges from a few tens of feet to several thousand feet, and many of the aquifers in the region that yield fresh water also contain saline water in their downdip extensions. Some of these aquifers are capable of yielding from 100 gal/min (6.3 l/s) to more than 5,000 gal/min (315 l/s) of saline water to wells (Winslow and Kister, 1956; Winslow, Hillier, and Turcan, 1968) either by pumping or by normal flow.

In 1965 and 1967, the Office of Saline Water investigated the potential contribution of saline-water conversion to future water supply in several areas of the Texas-Gulf Region. The study in 1965 (Prehn and Sigafoos, 1966) showed that, at a few of the sites, the unit cost of desalted water, including disposal of the effluent, is less than or about the same as the cost of developing the most feasible alternative fresh-water supply. At the remaining sites, the calculated unit cost of desalted water exceeded the cost of developing conventional supplies.

The study in 1967 (Ralph M. Parsons Co., 1968) in several counties in the western part of the region concentrated on the feasibility of producing 2 to 20 million gallons (7.6 to 76 km^3) per day of water for municipal supply and industrial use. The results indicated that saline ground water in a few of these areas can be desalted at reasonable cost to partly fulfill future needs. A major cost factor was the disposal of the effluent resulting from the conversion process.

The studies concluded that the costs of producing desalted water for municipal and industrial use at some sites may be less expensive than or at least economically competitive with alternative sources. The costs of desalting were considered too high for producing irrigation water.

As of January 1, 1972, at least 14 desalting plants, each with a capacity of 25,000 gallons (95 m^3) or more freshwater daily, were operating in the region for production of freshwater for industrial and municipal use (pl. 3A). Most of these plants were owned by industry. The total daily capacity was 5.66 million gallons (21.400 m^3), and the largest single plant capacity was 2.16 million gallons (8,200 m^3) (O'Shaughnessy, 1973).

USE OF SALINE AQUIFERS FOR LIQUID-WASTE DISPOSAL

The urgency being given to programs for reducing stream pollution and pollution of fresh ground-water supplies is stimulating interest in saline aquifers for use in waste disposal. V. E. McKelvey, Director of the U.S. Geological Survey, in a keynote address delivered in Houston, Texas, December 6, 1971, to a Symposium on Underground Waste Management and Environmental Implications (McKelvey, 1972, p. 1) remarked:

> * * * all our growing activities and interest in underground storage and waste management directly involve the increased utilization of what is becoming an extremely important resource—namely, underground space. * * * I believe there is considerable value in thinking about underground space as a natural resource that needs to be appraised in order that we may understand the role it can play in our economy and thus plan for its wise use.

To this end, successful utilization of underground space is predicated upon developing the capability to predict, with reasonable confidence, the consequences of using this resource (Meyer, 1972). "Proper" disposal underground, which here implies a prior knowledge of all interactions within the injected zone, requires that: (1) the fate of the waste is, in general, known and understood; (2) the waste is contained and can be isolated from man's food, water, and activity; and (3) the waste can be recovered if desired or if the need arises (Hill, 1972a, p. 384).

On the other hand, in the absence of technical ability to conduct underground waste disposal in a safe manner, which may be due largely to deficiencies in knowledege of ground-water hydraulics, geology, and geochemistry, placement of liquid wastes in underground reservoirs can create serious and irreparable hazards (Stallman, 1972, p. 6–10; Piper, 1969). In this case, any one or all of the following consequences may occur (Stallman, 1972, p. 6): (1) pollution of groundwater supplies; (2) pollution of surface-water supplies because waste fluids are discharged by upward leakage into streams; (3) changes in permeability caused by thermal, chemical, and mechanical alterations of the rocks (such changes make it difficult to predict future subsurface flow regions); (4) land-surface subsidence and earthquakes; and (5) possible contamination of mineral resources, including fossil fuel.

In the Texas-Gulf Region, emplacement of liquid wastes in subsurface reservoirs is regulated by State agencies. Each application for disposal is handled individually and conditions are evaluated at each proposed disposal site. Because the Texas-Gulf Region is underlain extensively by relatively thick sequences of saline aquifers, the potential of the subsurface strata for liquid-waste disposal is regarded by Hill (1972b, fig. 1) as generally favorable in most of the region.

Numerous liquid-waste disposal wells are in operation throughout the region. Excluding the subsurface disposal of saltwater from the production of oil and gas, 56 municipal and industrial waste-disposal wells were in operation in 1973, and by that year, a cumulative total volume of about 18 billion gallons (0.068 km^3) of liquid wastes, some containing toxic substances, had been injected into saline aquifers (Texas Water Quality Board, written commun., 1973). About 85 percent of the waste-disposal wells are in the heavily industrialized areas of southeast Texas (pl. 3A).

The petroleum industry pioneered the techniques of waste disposal in underground reservoirs. The East Texas Salt Water Disposal Co., operating in the huge East Texas oil field in the northeast part of the region, began the systematic disposal of saltwater in 1942, after an experimental well drilled in 1936 demonstrated the feasibility of returning saltwater to the oil and gas reservoir. By 1972, about 4.5 billion bbl (0.7 km^3) of saltwater had been returned to the underground reservoir. This operation has been successful both in eliminating saltwater pollution of land areas and freshwater sources and in maintaining reservoir pressure to prolong the productive life of the field (East Texas Salt Water Disposal Co., 1958, p. 10).

Spurred by State regulations banning evaporation pits, the oil industry injects the major part of the saltwater produced in the Texas-Gulf Region. The degree of effectiveness achieved by the industry in disposing of these wastes underground will become increasingly more important in terms of using underground space to protect the surface environment.

USE OF SALINE AQUIFERS FOR FRESHWATER STORAGE

The proposed use of saline aquifers for storing freshwater has promise, under certain conditions, as a method of water management. Basically the process involves artificial recharge, where freshwater is emplaced in saline aquifers for later retrieval.

The utilization of saline aquifers for this purpose has been hindered, until recently, by the lack of a sound basis for predicting possible quality attenuation of the freshwater from repetitive cycles of injection and withdrawal. Mathematical and laboratory models constructed a few years ago showed conclusively that the efficiency of the process—ratio of freshwater recovered to freshwater injected—exceeded 90 percent and improved with the number of cycles of injection and withdrawal (Kimbler and others, 1973, p. 203). Kimbler and others (1973, p. 198) state that potentially, there is an overwhelming economic advantage in using underground-storage space as compared to the construction and use of surface space.

Freshwater was injected and retrieved through a well open to a saline aquifer in 1971 and 1972 by the U.S. Geological Survey at Norfolk, Virginia, in cooperation with the city of Norfolk. The purpose of this experiment was to determine if it would be feasible to use surface water that is surplus to immediate needs, process it in existing treatment plants, and store it underground in saline aquifers. The freshwater would be retrieved when peak demands place severe strains on the water systems.

The results of the study showed that although the permeability of the aquifer was reduced by clay dispersion, chemical treatment of the aquifer could overcome this problem, and that a well field could be constructed and operated economically (Brown and Silvey, 1973, p. 405).

This process could be useful in the flat coastal and estuarial area of the Texas-Gulf Region where the ground water is saline, such as the Galveston area and the Beaumont-Orange area. Old well fields in these areas that are no longer used because of saltwater encroachment might be used for storage and retrieval of highly treated fresh surface water.

Saline aquifers used for injection of toxic wastes would not be usable for freshwater storage at the same locations and depths. Disposal of such wastes generally would be in deeper zones that are not likely to be used as reservoirs for freshwater storage.

PROBLEMS ASSOCIATED WITH GROUND-WATER DEVELOPMENT

Ground-water problems usually result from developing the resources with inadequate knowledge of the geologic and hydrologic framework of the reservoirs. Solution of some of the problems will require the physical means of coordinating surface-water and ground-water storage and use and the removal of legal and economic constraints that prevent the required coordination. Development of ground-water reservoirs in conjunction with surface-water supplies may provide solutions to some problems, mitigate the harmful effects of others, and in some instances prevent their occurrence.

The following discussion of four typical problems of varying degrees of concern in the Texas-Gulf Region indicate the kind of difficulties that may be anticipated where conjunctive planning and development practices are lacking.

DECLINING WATER LEVELS

Declining water levels are inevitable until a steady-state gradient is established. Depending upon the hydraulic characteristics of the aquifer and the rates of recharge and pumping, the declines may be small and temporary or they may be severe and continual.

Severe water-level declines result from two basically different conditions: (1) sustained pumping in excess of captured discharge; or (2) water being withdrawn at a rate greater than the rate of flow through the aquifer at the existing gradient, even though potential recharge in the area may be more than adequate. Examples of these two conditions in the Texas-Gulf Region are the declining water levels on the High Plains, where pumpage exceeds captured discharge from the Ogallala aquifer, and the declining water levels in the Houston area, where potential recharge is adequate but where changing withdrawal rates have not permitted the gradient to stabilize, or enough time has not elapsed at any withdrawal rate for the gradient to adjust throughout the aquifer.

The most serious water-level declines in the region are occurring on the High Plains of Texas and New Mexico, where ground water is being mined from the Ogallala aquifer. The water-level decline in the aquifer represents a decline in the altitude of the water table rather than a decline in pressure, because all of the ground water is being taken from storage by dewatering the reservoir.

The decline of the water table in the Ogallala began shortly after irrigation reached a substantial state of development in the 1930's, and the decline has continued at a rapid rate ever since. The rate of depletion of the reservoir from the middle 1930's to 1973 averaged about 2.5 ft (0.76 m) per year at a site near Plainview, Texas, (fig. 3) and averaged about 1.5 ft (0.46 m) per year at a site in Roosevelt County near Portales, New Mexico. Since 1941 the rates of depletion have increased sharply and the yields of wells have decreased (Cronin, 1964). Conservation measures, in advance of long-range plans for possibly importing water into the area, are already being applied to slow the rate of depletion, but imported water probably will not be available soon enough to prevent large reductions in the irrigated acreage.

The water-level declines in the Houston area, unlike the declines on the High Plains, represent diminishing artesian pressure and not dewatering of the aquifer. In a sense, the artesian head is being mined, as expressed by Lohman (1972, p. 67). Most of the water is being pumped from a part of the Gulf Coast aquifer (Evangeline and Chicot aquifers) where the ground water is under artesian conditions. Although the aquifer is saturated at the outcrop, and is even rejecting potential recharge, the transmissivity of the aquifer is not sufficient at existing gradients to convey the water to the wells rapidly enough to stabilize water levels. Most of the water pumped is coming from and through the clays in the area of the cone of depression.

Although there is no water shortage, the water-level declines at Houston have created serious problems. A representative well in a heavily industrialized part of the Houston area shows that the rate of decline averaged about 8 ft (2.4 m) per year from 1939 to 1973. During this period, depths to water in the well increased from about 90 to 370 ft (27 to 113 m) below land surface (fig. 3). The declines of water levels in other places, especially in the less heavily industrialized sector of the city, have been slightly less severe.

The problems of water-level declines commonly afflict areas where pumping of very large amounts of ground water is concentrated within small areas. The problem may develop inauspiciously at first, but

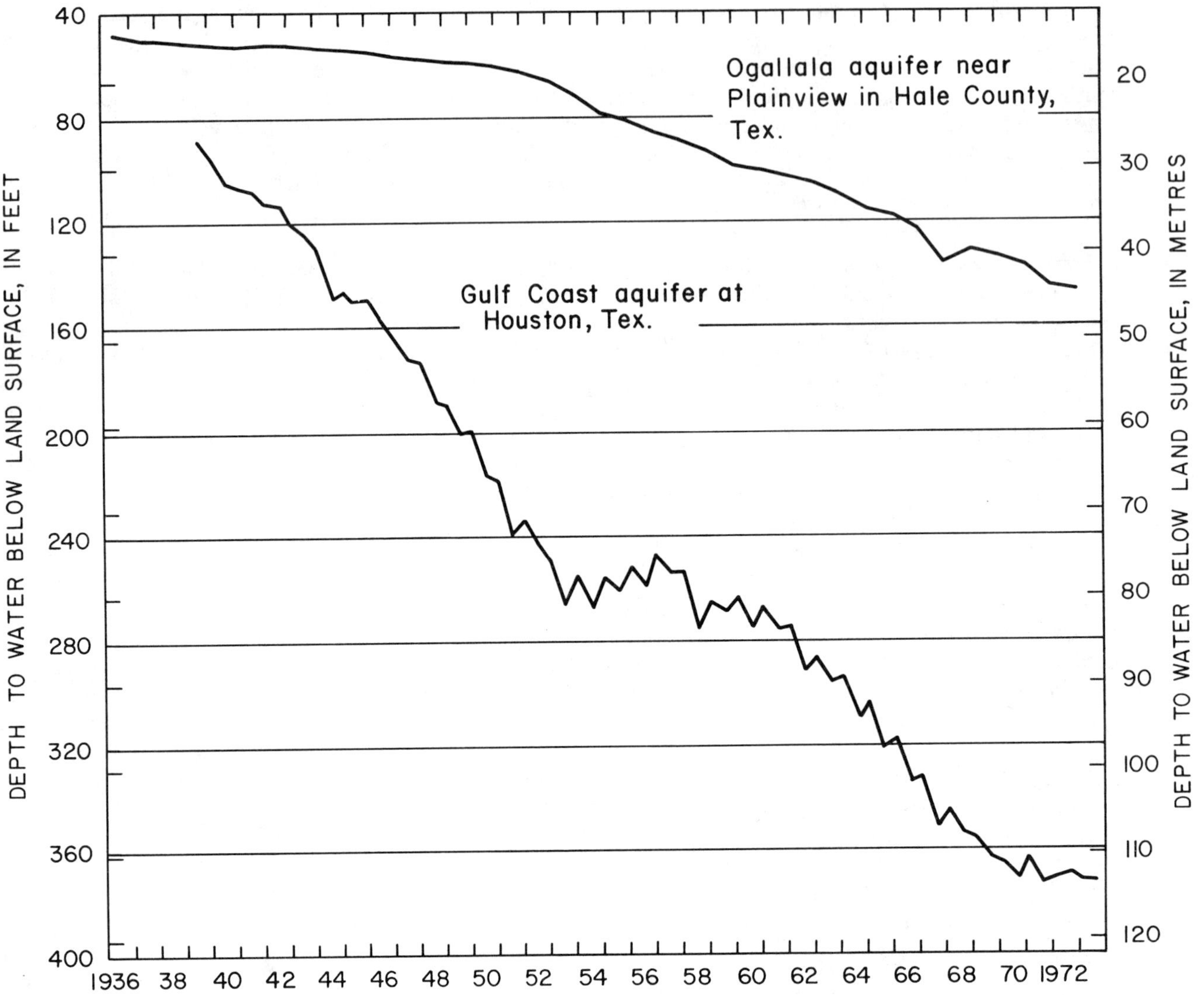

FIGURE 3.—Hydrographs showing water-level declines in the Ogallala aquifer near Plainview, Texas, and in the Gulf Coast aquifer at Houston, Texas.

later becomes significant, even when water supplies are adequate, because the difficulties may include prohibitive pumping costs, saltwater intrusion, and land-surface subsidence.

Distributing the pumping over a comparatively larger area, particularly over an area that includes the principal region of recharge or area of more rapid recharge, would mitigate the problem considerably. In addition to well-field dispersion, a water-management program of conjunctively utilizing ground water and surface water might be beneficial.

DEPLETION OF STREAMFLOW

Ground water and surface water are so interrelater that in most places the withdrawal of ground water will modify the natural regimen of streamflow. In any area where there is hydraulic continuity between the surface-water sources and the ground-water reservoirs, pumping from wells will result in depletion of streamflow; and the surface water that is diverted or intercepted may be required by or belong to downstream users.

The interrelationship of ground water and surface water in the San Antonio area serves to illustrate how ground-water development affects the streamflow. The Edwards (Balcones Fault Zone) aquifer, which is heavily tapped by wells, receives recharge from the streams that flow across the outcrops. Water that is recharged from the streamflow moves rapidly through the cavernous, fractured, and honeycombed framework of the aquifer and then discharges in large springs that sustain the flow of the Frio, San Antonio, Guadalupe, and San Marcos Rivers. The rate of springflow changes rapidly in response to changes in the volume of water in storage in the aquifer, and the volume of water in storage changes rapidly in response to rainfall and pumping rates. Flow of Comal Springs in Comal County, the largest of the springs, fluctuates chiefly from precipitation trends and from changes in pumping rates in San Antonio and Bexar County (Garza, 1962, p. 20).

A continued increase in ground-water development in the San Antonio area, without provisions for increasing the recharge, could result in a reduction of ground water in storage to a degree that could eliminate the flow of the springs, at least part of the time. Cessation of the springflow would substantially reduce the downstream surface-water supply and destroy one of the major esthetic values of the area.

Depletion of streamflow as a result of ground-water development is more subtle elsewhere in the region, and has been detected (or suspected) largely by inference. One such area is on the Gulf Coastal Plain between Houston and Corpus Christi. Heavy ground-water drafts for rice irrigation have caused a gradual lowering of the water table until it is below the stream level of the Lavaca and Navidad Rivers in some places. Although a complete loss of the base flow of the rivers seems remote, a reduction in streamflow probably has occurred (Baker and Follett, 1973, p. 41).

SALTWATER INTRUSION

The movement of poor-quality water into freshwater supplies can be as serious a problem as that of ground-water depletion. A deterioration of water quality may necessitate abandoning wells while there are ample supplies of water in the reservoir. A drawn-down reservoir may fill again by natural processes if pumping is reduced, but dissolved contaminants may be difficult or impossible to remove. Because natural dilution is a slow process, the effects of deterioration of ground-water quality may be long-lasting.

Saltwater intrusion into fresh ground water is the principal cause of ground-water deterioration in the Texas-Gulf Region. This problem is not only currently troublesome, but poses the greatest threat to future ground-water development in some parts of the region. The problem manifests itself to a far greater degree in the coastal areas than elsewhere in the region, and is most acute in southeast Texas from Houston to Orange. In this area, fresh ground water is in sharp contact with saltwater, which may be positioned above, below, or seaward of the freshwater.

Intrusion of saltwater in the Houston-Galveston area is occurring along several fronts in two different zones (Evangeline and Chicot aquifers) in the Gulf Coast aquifer. Regional depressions in the potentiometric surface in the Evangeline aquifer imply that the interface of the fresh and saline water in Harris County is transient and that the saltwater front is moving updip toward the centers of pumping. In the Chicot aquifer, the saltwater front in Galveston County has already invaded the city of Galveston's "old" well field by either lateral or upward movement of the water (Petitt and Winslow, 1955, p. 81). Farther inland, saltwater contamination is occurring in shallow wells in the Chicot aquifer along the Houston Ship Channel, presumably from the intrusion of saline water in the channel.

In the Beaumont-Port Arthur-Orange area, heavy pumping of ground water initiated the deterioration of ground-water quality prior to 1939, and the deterioration is continuing. The pattern of saltwater intrusion, at least in Orange County, indicates that the saltwater front in the Chicot aquifer is advancing laterally toward the center of pumping (Gabrysch and McAdoo, 1972, p. 10).

These examples show that the problems of saltwater intrusion result from development of the water resources. With foresight, the problems might have been prevented or reduced in severity. For example, the creation of pressure ridges, by artificial recharge, during the early stages of development would have provided barriers to saltwater intrusion and would have protected the freshwater supply. Ground-water quality, as well as quantity, must be managed if the supply is to remain suitable, as well as available, for its intended use.

SUBSIDENCE OF THE LAND SURFACE

Conclusive evidence from studies in the Houston area (Winslow and Doyel, 1954; Winslow and Wood, 1959; and Gabrysch, 1969) shows that a decrease in artesian pressure resulting from large withdrawals of ground water creates stresses in the aquifer that cause water to move slowly out of the water-saturated clays and into the interbedded sands. The clays, which are highly compressible, become increasingly compacted, and this in turn results in subsidence of the land surface.

The areas of major subsidence in the Texas-Gulf Region are centered in and around Houston, but subsidence is either known or suspected in other coastal areas of heavy pumping from Louisiana to possibly Kingsville, south of Corpus Christi.

In Houston, the increasing draft of ground water which causes the progressive lowering of artesian pressures is enlarging the subsidence bowl in depth and lateral extent. Between 1943 and 1973, the land surface subsided a maximum of about 7.5 ft (2.3 m) near the Houston Ship Channel, while in parts of eight counties in southeast Texas, subsidence was at least 0.5 ft (0.15 m) (fig. 4). Subsidence prior to 1943 in the area of the Houston Ship Channel was about 2 ft (0.6 m), which added to the 7.5 ft (2.3 m) that occurred between 1943 and 1973 makes a total of 9.5 ft (2.9 m) of subsidence in that area (Gabrysch and Bonnet, 1974). Subsidence will continue if the decline in artesian pressures continues, and even if the pressure could be maintained at its present level, the land surface would nevertheless subside a few additional feet near the center of the cone of depression.

Some of the effects of subsidence in Houston and surrounding areas are severe. Some buildings have been cracked, roadways have been damaged, well casings and pipelines have been damaged, and water-logging has increased the cost of construction and repairs. Some waterfront properties are now flooded by normal high tides and are subject to inundation and destruction by hurricane tides. Efforts to minimize the subsidence problem will necessitate a decrease in the rate of artesian-pressure decline, which can be accomplished only by reducing the ground-water draft or by recharging the aquifers.

Subsidence in the other coastal areas is much less severe. In the urbanized Beaumont-Port Arthur-Orange area, subsidence has been generally less than 0.5 ft (0.15 m) between 1918 and 1973, but it is expected to continue if there is a continued decline in water levels. In rural areas between Houston and Corpus Christi, the land surface is subsiding from large ground-water withdrawals for rice irrigation. In Jackson County, land-surface subsidence to 1972 has been as much as 1 ft (0.3 m), and is expected to continue at annual rates estimated to range from 0.012 to 0.026 ft (0.004 to 0.008 m) (Baker and Follett, 1973, p. 32, 34). In these rural areas, the subsidence is fairly well distributed, which minimizes the undesirable effects. Changes in the land slope, however, will affect the drainage patterns and stream gradients to some extent.

POTENTIAL FOR OBTAINING ELECTRICAL ENERGY AND FRESHWATER FROM THE GEOPRESSURED RESERVOIRS

Studies by Jones (1969) and others and information obtained from drilling operations for oil and gas have shown that thousands of square miles of the Coastal Plain are underlain, at depths of 6,000 to 20,000 ft (1,800 to 6,100 m), by ground-water reservoirs containing zones of extremely high pressures and temperatures that have been termed "geopressured reservoirs" (Dickinson, 1953).

Geopressured reservoirs occur in the Texas-Gulf Region in part of the Coastal Plain in an area that extends from near the edge of the continental shelf to 50 to 150 mi (80 to 240 km) inland from the coastline (pl. 3B). Within this area, the principal ground-water reservoirs that are geopressured are the downdip extensions of the Carrizo-Wilcox aquifer and the Gulf Coast aquifer. Geopressured reservoirs may underlie other areas in the Texas-Gulf Region, but their presence is mostly a matter of speculation at this time.

The most significant aspects of these geopressured reservoirs are the relatively low salinity of the water, the unusually high temperature and water pressure, and the content of natural gas in solution in the water. Below depths of several thousands of feet, a progressive freshening of the ground water with depth in the geopressured reservoirs is typical. The dissolved-solids content commonly decreases vertically downward from more than 100,000 mg/l to less than 10,000 mg/l, and is less than 5,000 mg/l in some places. In some areas at depths greater than 6,000 ft (1,800 m), the water is reported to contain less than 1,000 mg/l dissolved solids, while highly saline water occurs both above and below the freshwater (Jones, 1969, oral commun., 1974).

Abnormally high geothermal gradients and water pressures accompany the relatively low salinities. Drainage restrictions imposed on the water-bearing deposits by growth faults (faults that have been more or less continuously active) and chemical changes are probably responsible for the high pressures and temperatures in the geopressured reservoirs.

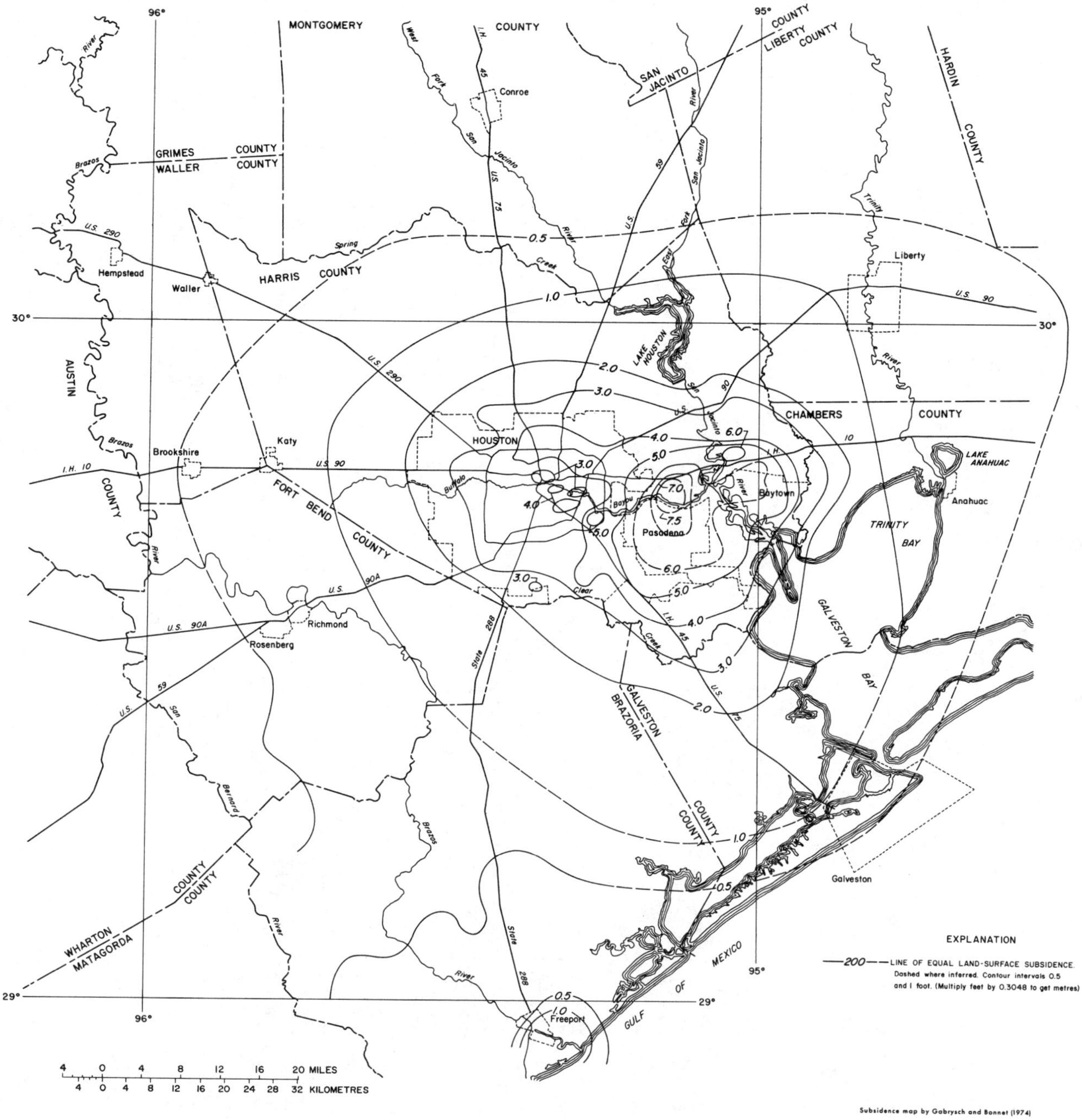

FIGURE 4.—Map showing land-surface subsidence in the Houston area, 1943–73.

A typical increase in the geothermal gradient in a well pentrating a geopressured reservoir in south Texas is shown on figure 5. Geothermal gradients within the reservoirs are as much as 3.4°C (6.2°F) per 100 ft (30.5 m) (Jones, 1969). Minimum temperatures of water in the geopressured reservoirs generally range between 80° and 120°C (176°–248°F) near the tops of the reservoirs; maximum temperatures are a matter of speculation because the reservoirs may be geopressured to the zone of metamorphism or to the ultimate depth at which fluids are free to move (Jones, 1970). The highest temperature recorded in the Texas-Gulf Region was 273.6°C (525°F) at a depth of 19,222 ft (5,859 m)

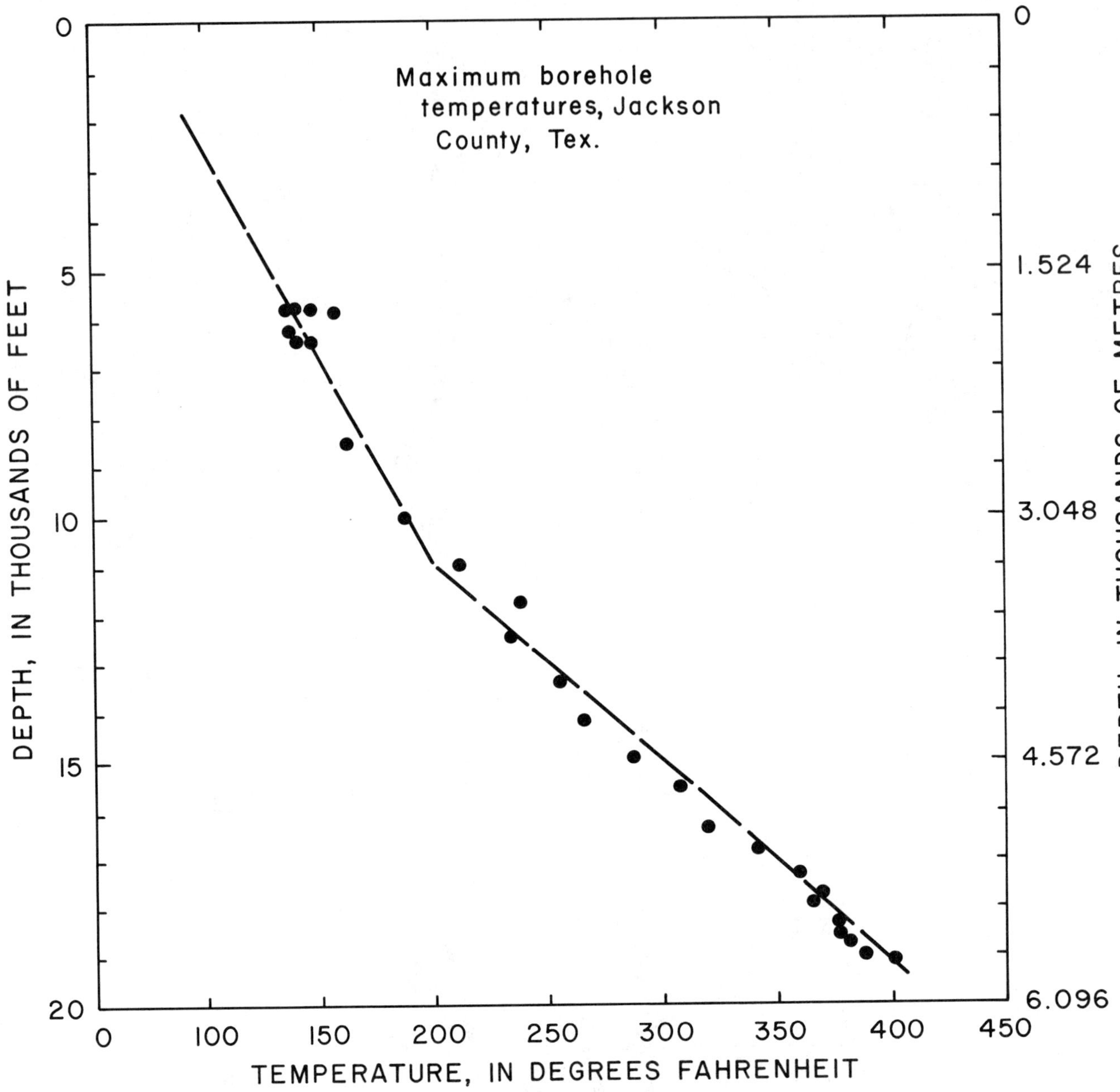

FIGURE 5.—Graph showing thermal gradient in the geopressured zone.

in a well in Matagorda County between Houston and Corpus Christi (Hough and Couvillion, 1966).

Water pressures in wells tapping the geopressured reservoirs range from a few hundred to as much as 500 atmospheres (7,350 pounds per square inch or 517 kilograms per square centimetre) or more at the land surface (Stuart, 1970). This is equivalent to static heads of more than 15,000 ft (4,570 m) above land surface.

The geopressured aquifers may be potentially important for the generation of electrical power and the production of freshwater by desalting. Electrical power might be generated by exploiting the three forms of energy inherent in the water: geohydraulic energy from high-pressured flow at the surface, thermal energy from the heat content of the water, and natural gas in solution in the water. However, research is needed in geothermal-power production and generation equipment as well as detailed studies of the geopressured reservoirs before the economic feasibility can be evaluated and any development of the resource can take place within the region.

CONSTRAINTS ASSOCIATED WITH THE DEVELOPMENT OF GROUND-WATER RESOURCES

WATER RIGHTS

The two major doctrines of water law in the United States—the common-law doctrine of riparian rights and the doctrine of prior appropriation—are observed in the Texas-Gulf Region. These doctrines, which evolved in respect to surface water but have subsequently been extended to ground water by case law or by statute, are based on the conflicting but fundamental principles of private versus public ownership of the water.

In Texas and Louisiana, which adhere to the common-law doctrine, ground waters are private and are not subject to appropriation by the State. The waters are subject to capture and use by the owners of the surface rights, their agents, or assignees. Louisiana exercises no regulation in any form, and Texas exercises no regulation except as provided by the establishment of ground-water conservation districts, which are local and completely autonomous after establishment. New Mexico invokes the doctrine of prior appropriation. Under this concept, the ground waters are public and subject to appropriation in accordance with the law.

The adherence to one or the other of these water laws and regulations was basically influenced by the climatic environment of the State. In the arid to semiarid parts of the West, the dependence of agriculture and other businesses upon the limited supply gave water a high value, and led to the necessity of imposing appropriate constraints on water development. Conversely, in the humid or subhumid East, the needs of most water users were readily satisfied within a legal framework of little or no constraint. Despite the fact that Texas has both climatic environments, the one doctrine of common-law rights applies Statewide to the ground-water resources.

To cope with excessive competition for limited supplies of ground water in the semiarid parts of Texas, or to resolve other ground-water controversies, the State currently resorts to limited regulation of ground-water use in some areas. It authorizes groups of water users to form underground-water conservation districts for the regulation of well spacing and production and for the preservation, protection, recharging, and prevention of waste of the ground water from an aquifer.

Under Texas law, an aquifer that is proposed to be regulated must have ascertainable boundaries and must be capable of yielding at least 150,000 gallons (568 m^3) per day to a well. Wells producing less than 100,000 gallons (378 m^3) per day are not regulated, so far as withdrawal is concerned. The act creating the underground water districts was amended in 1973 (House Bill 935) primarily to allow for the control of land-surface subsidence caused by withdrawal of ground water.

By 1974, two underground-water conservation districts were operating in the Texas part of the Texas-Gulf Region, one to regulate development of a major part of the Ogallala aquifer on the High Plains and the other to regulate development of the Edwards (Balcones Fault Zone) aquifer in the San Antonio area.

In New Mexico, the doctrine of prior appropriation allows the State's waters to be appropriated for beneficial use within the State. Although the ground water is public, no permit or license to appropriate ground water is required except in designated or declared basins. The declaration of an area by the State Engineer as an "underground water basin" thus enables the State to control orderly development of ground water in order to protect prior water rights—for example, to protect senior ground-water rights in areas of overdraft. By 1974, two underground-water basins had been declared by the State in the New Mexico part of the Texas-Gulf Region and were operating to control development of the Ogallala aquifer on the High Plains.

TECHNOLOGICAL PROBLEMS

Various deficiencies in scientific knowledge, procedures, and equipment, or in many instances failure to use present technological capabilities, continue to impede progress in utilizing the ground-water reservoirs to their full potential. Advances have been made, especially within the past 2 decades, in water-related research, but the increasing requirements for water and the complex problems that are caused by fulfilling these needs reveal the desirability of continued technological improvement.

Innovative methods and techniques of combating the problems related to full use of both the fresh-water and saline-water resources should be developed and perfected. The lack of adequate funding, however, probably constitutes the basic constraint to technological advancement.

One field of research that has experienced technological problems for years is that of artificial recharge, especially recharge by wells. Laboratory research and experimental efforts have shown that the process is complicated by problems ranging from the quality of the recharge water to permeability reductions in the aquifer. Scientific progress may eventually solve these problems, but until they are resolved, the benefits to be derived from artificial recharge will necessarily be delayed.

Technological and economic problems in desalinization continue to restrict this process from being used to good advantage where freshwater is scarce. Research is continuing, and improvements are being made both in procedure and equipment; but until significant technological advancements are achieved, the economic feasibility of the process prevents widespread use of the large supplies of saline water.

NEEDS FOR ADDITIONAL INFORMATION ON GROUND-WATER RESOURCES

To preserve the economy of the Texas-Gulf Region and to provide for future water demands, it will be necessary to conduct water-resources investigations that yield information geared to present and anticipated water development.

A full presentation of the deficiencies in information is beyond the scope of this report, but some of the most important needs, as they apply to the Texas-Gulf Region, are as follows:

1. *Evaluation of aquifers and aquifer systems employing digital or electric-analog models.*—Such models would be used to analyze the effects of future conditions of pumping and would serve as tools for gathering the information needed to formulate long-term plans for ground-water management. In addition to these applications, the use of highly sophisticated models, as stated by Moore (1971), could optimize management objectives within specified physical, economic, and social constraints, which makes it possible to integrate use of ground water into water-resource planning with a high degree of certainty. The effectiveness of models in local areas has already been demonstrated. (See Wood and Gabrysch, 1965; Jorgensen, 1974).
2. *Determination of the amount of natural discharge from the aquifers that is subject to capture.*—Computations of long-term availability of ground water are generally being made with inexact knowledge of the rate of discharge. In some areas, the rates of withdrawal should be no more than the rates of natural discharge subject to capture. As an approach to this problem, such studies should relate the base flow of streams that drain the recharge areas of the aquifers to fluctuations of water levels in shallow wells. These studies will require a more detailed knowledge of aquifer coefficients in the recharge areas and a better understanding of the movement of water in these areas.
3. *Information on diffusion and dispersion in relation to transport of pollutants from surface disposal of wastes.*—This is an inincreasingly important field of study as development increases and wastes are discharged onto the land surface or into surface-water bodies from which the pollutants can enter the ground-water subsystem. A good example is the study of the subsurface effects of sewage disposal by septic tanks. Such studies will provide a basis for effective water-quality management.
4. *Determination of potential contamination and other problems associated with underground storage of water.*—Liquid wastes in unprecedented quantities are being injected into saline aquifers because of stringent controls on disposal to streams. Knowledge is required regarding the geologic, geochemical, and hydrologic factors related to injection processes; waste mobility; quality attenuation; and degree of containment.
5. *Intensified research and testing to determine the potential of the geopressured reservoirs to produce electrical energy.*—Regional studies of the geopressured aquifers will permit an appraisal of the extent and magnitude of the geothermal energy that might be developed, and will identify possible detrimental effects of development. The geologic properties of the aquifer materials, especially the clays, need to be determined for predicting land-surface subsidence; observation wells need to be drilled; and possibly a prototype production well could be designed and drilled.
6. *Additional information on saltwater encroachment in the Gulf Coastal Plain.*—Studies of saltwater encroachment will be necessary before measures can be taken to protect fresh ground-water supplies in the heavily industrialized and populated cen-

ters along the Gulf Coast. To map accurately the saltwater fronts and to monitor their advancement toward centers of pumping, a network of properly located observation wells of various depths should be established.

7. *Additional information on land-surface subsidence along the Gulf Coast.*—Additional research on the mechanics of land-surface subsidence must be made to allow more accurate predictions of future subsidence. Significant technological progress has already been made through studies in the Houston area, but this research needs to be refined and expanded to accommodate different geologic and hydrologic conditions in other areas of known or suspected subsidence.
8. *Development of techniques for artificial recharge.*—The importance of this field of research should be emphasized because its potential for application in the Texas-Gulf Region is widespread. It is possible that many problems relating to overdraft of ground-water supplies, such as saltwater encroachment, land-surface subsidence, underground storage of surplus freshwater, and disposal of liquid wastes could be reduced by development of practical and economical methods of artificial recharge. Research, with an interdisciplinary approach, directed toward descriptions of the geologic environment, hydraulics, geochemistry, bacteriology, engineering details for an entire management scheme, and economics of the recharge system would have the potential for answering many questions.
9. *Determination of the amount of water in storage in the ground-water reservoirs, especially the Edwards (Balcones Fault Zone) aquifer.*—Complete and accurate appraisals of the quantity of water in storage in ground-water reservoirs are needed by water planners concerned with eventual conjunctive management of the ground-water supply with water from other sources.
10. *Studies of the relationship of ground water and surface water to determine the effects of aquifer development on surface flow.*—Water planners are becoming increasingly aware of the significance of the relationship between ground water and surface water, and the extent to which one affects the other must be determined. A water plan that presupposes total development of the ground-water resources in a manner that captures all discharge should include reassessment of the surface-water resources in terms of possible reductions in streamflow.

REFERENCES CITED

Alexander, W. H., Jr., and others, 1964, Reconnaissance investigation of the ground-water resources of the Guadalupe, San Antonio, and Nueces River basins, Texas: Texas Water Commission Bull. 6409, 106 p.

Alexander, W. H., Jr., and Patman, J. H., 1969, Ground-water resources of Kimble County, Texas: Texas Water Devel. Board Rept. 95, 93 p.

American Society of Civil Engineers, 1972, Ground-water management: New York, Am. Soc. Civil Eng., Manuals and reports on engineering practice no. 40, 216 p.

Baker, B. B., and others, 1963a, Reconnaissance investigation of the ground-water resources of the Neches River basin, Texas: Texas Water Commission Bull. 6308, 67 p.

——— 1963b, Reconnaissance investigation of the ground-water resources of the Sabine River basin, Texas: Texas Water Commission Bull. 6307, 57 p.

Baker, E. T., Jr., and Follett, C. R., 1973, Effects of ground-water development on the proposed Palmetto Bend Dam and Reservoir in southeast Texas: U.S. Geol. Survey Water-Resources Investigations 18–73, 70 p.

Broadhurst, W. L., 1960, Recharge well used in conducting flocculating chemical test: The Cross Section, High Plains Underground Water Dist. no. 1, Lubbock, Tex. V. 7, no. 2, p. 1.

Brown, D. L., and Silvey, W. D., 1973, Underground storage and retrieval of fresh water from a brackish water aquifer *in* Underground waste management and artificial recharge, v. 1, second international symposium on underground waste management and artificial recharge, Sept. 26-30, 1973, New Orleans, Louisiana: Am. Assoc. Petroleum Geologists, p. 379–419.

Brown, R. F., and Signor, D. C., 1973, Artificial recharge experiments and operations on the Southern High Plains of Texas and New Mexico: U.S. Geol. Survey Water-Resources Investigations 10–73, 54 p.

Busby, F. E., and Schuster, J. L., 1973, Woody phreatophytes along the Brazos River and selected tributaries above Possom Kingdom Lake: Texas Water Devel. Board Rept. 168, 41 p.

Clyma, Wayne, 1964 ,Artificial ground-water recharge by a multiple-purpose well: Texas Agr. Expt. Sta. Misc. Pub. 712, 7 p.

Core Laboratories, Inc., 1972, A survey of the subsurface saline water of Texas: Texas Water Devel. Board Rept. 157, v. 1, 113 p.

Cronin, J. G., 1964, A summary of the occurrence and development of ground water in the Southern High Plains of Texas *with a section on* Artificial recharge studies by B. N. Myers: U.S. Geol. Survey Water-Supply Paper 1693, 88 p.

——— 1969, Ground-water in the Ogallala Formation in the Southern High Plains of Texas and New Mexico: U.S. Geol. Survey Hydrol. Inv. Atlas HA–330, 9 p., 4 maps.

Cronin, J. G., and others, 1963, Reconnaissance investigation of the ground-water resources of the Brazos River basin, Texas: Texas Water Commission Bull. 6310, 152 p.

Cronin, J. G., and Wilson, C. A., 1967, Ground water in the flood-plain alluvium of the Brazos River, Whitney Dam to vicinity of Richmond, Texas: Texas Water Devel. Board Rept. 41, 206 p.

Dickinson, George, 1953, Reservoir pressures in Gulf Coast Louisiana: Am. Assoc. Petroleum Geologists Bull., v. 37, no. 2, p. 410–432.

East Texas Salt Water Disposal Company, 1958, Salt-water disposal—East Texas oil field: Austin, Texas, Petroleum Extension Service, Texas Univ. Div. of Extension, 131 p.

Fenneman, N. M., 1931, Physiography of Western United States: New York, McGraw-Hill Book Co., Inc. 534 p.

——— 1938, Physiography of Eastern United States: New York, McGraw-Hill Book Co., Inc., 691 p.

——— 1946, Physical divisions of the United States: U.S. Geol. Survey map, scale, 1 :7,000,000.

Gabrysch, R. K., 1969, Land-surface subsidence in the Houston-Galveston region, Texas, *in* Land subsidence, Tokyo Symposium: Internat. Assoc. Sci. Hydrology Pub., v. 1, no. 88, p. 43–54, 1971.

——— 1972, Development of ground water in the Houston district, Texas, 1966–69: Texas Water Devel. Board Rept. 152, 24 p.

Gabrysch, R. K., and Bonnet, C. W., 1974, Land-surface subsidence in the Houston-Galveston region, Texas: Texas Water Devel. Board Rept. [in press].

Gabrysch, R. K., and McAdoo, G. D., 1972, Development of ground-water resources in the Orange County Area, Texas and Louisiana, 1963–71: Texas Water Devel. Board Rept. 156, 47 p.

Garza, Sergio, 1962, Recharge, discharge, and changes in ground-water storage in the Edwards and associated limestones, San Antonio area, Texas—A progress report on studies, 1955-59: Texas Board of Water Engineers Bull. 6201, 42 p.

Hauser, V. L., and Lotspeich, F. B., 1967, Artificial ground-water recharge through wells: Jour. Soil and Water Conserv., v. 22, no. 1, p. 11–15.

Havens, J. S., 1966, Recharge studies on the High Plains in northern Lea County, New Mexico: U.S. Geol. Survey Water-Supply Paper 1819–F, 52 p.

Hilchie, D. W., 1968, Maximum temperatures recorded in wellbores: Log Analyst, v. 9, No. 5, p. 21–24.

Hill, R. B., 1972a, Regulation of subsurface disposal in Texas *in* Underground waste management and environmental implications: Am. Assoc. Petroleum Geologists Memoir 18, p. 381–385.

———1972b, Subsurface waste disposal in Texas: Texas Water Quality Pub. 72–05, 40 p.

Hood, J. W., and Kister, L. R., 1962, Saline-water resources of New Mexico: U.S. Geol. Survey Water-Supply Paper 1601, 70 p.

Hough, H., and Couvillion, J. A., 1966, Deep hot drilling marks Gulf Frio play: Oil and Gas Jour., v. 64, no. 19, p. 142–158.

Jones, P. H., 1969, Hydrology of Neogene deposits in the northern Gulf of Mexico basin: Louisiana Water Resour. Res. Inst., Bull GT–2, 105 p.

——— 1970, Geothermal resources of the northern Gulf of Mexico basin *in* Geothermics, Special Issue 2, v. 2, pt. 1, Pisa, Italy, p. 14–26.

Jorgensen, D. G., 1974, Analog-model studies of ground-water hydrology in the Houston district, Texas: U.S. Geol. Survey open-file rept. 89 p.

Kane, J. W., 1967, Monthly evaporation rates for Texas, 1940 through 1965: Texas Water Devel. Board Rept. 64, 111 p.

Kimbler, O. K., and others, 1973, Saline aquifers—future storage reservoirs for fresh water? *in* Underground waste management and artificial recharge, v. 1, second international symposium on underground waste management and artificial recharge, Sept. 26–30, 1973, New Orleans, Louisiana: Am. Assoc. Petroleum Geologists, p. 192–206.

Larner, D. C., and others, 1974, Woody phreatophytes along the Colorado River from southeast Runnels County to the headwaters in Borden County, Texas: Texas Water Devel. Board Rept. 182, 19 p.

Lohman, S. W., 1972, Ground-water hydraulics: U.S. Geol. Survey Prof. Paper 708, 70 p.

Long, A. T., 1958, Ground-water geology of Real County, Texas: Texas Board Water Engineers Bull. 5803, 46 p.

McKelvey, V. E., 1972, Underground space—an unappraised resource *in* Underground waste management and environmental implications: The Am. Assoc. Petroleum Geologists Memoir 18, p. 1–5.

Meyer, Gerald, 1972, New trends in the usage of natural underground reservoirs: Am. Inst. Chem. Eng. Symposium Series, v. 69, no. 129, p. 326–328.

Moore, J. E., 1971, Integrating the use of ground water into waste-resources planning: U.S. Geol. Survey open-file rept., 9 p.

Moulder, E. A., and Frazer, D. R., 1957, Artificial recharge experiment at McDonald well field, Amarillo, Texas: Texas Board Water Engineers Bull. 5701, 34 p.

Mount, J. R., and others, 1967, Reconnaissance investigation of the ground-water resources of the Colorado River basin, Texas: Texas Water Devel. Board Rept. 51, 107 p.

Murray, C. R., and Reeves, E. B., 1972, Estimated use of water in the United States in 1970: U.S. Geol. Survey Circ. 676, 37 p.

O'Shaughnessy, Frank, 1973, Desalting plants inventory report no. 4: U.S. Office of Saline Water, 27 p.

Peckham, R. C., and others, 1963, Reconnaissance investigation of the ground-water resources of the Trinity River basin, Texas: Texas Water Commission Bull. 6309, 110 p.

Petitt, B. M., Jr., and George, W. O., 1956, Ground-water resources of the San Antonio area, Texas: Texas Board Water Engineers Bull 5608, v. 1, 80 p.

Petitt, B. M., Jr., and Winslow, A. G., 1955, Geology and ground-water resources of Galveston County, Texas: Texas Board Water Engineers Bull. 5502, 219 p.

Piper, A. M., 1969, Disposal of liquid wastes by injection underground neither myth nor millennium: U.S. Geol. Survey Cir. 631, 15 p.

Prehn, W. L., and Sigafoos, R. A., 1966, The potential contribution of desalting to future water supply in Texas: Houston, Texas, Southwest Research Institute, 262 p.

Ralph M. Parsons Co., 1968, Preliminary possibility and economics of desalting brackish waters in candidate areas of West Texas: Ralph M. Parsons Co., 262 p.

Reeves, R. D., 1969, Ground-water resources of Kerr County, Texas: Texas Water Devel. Board Rept. 102, 58 p.

Schwiesow, W. F., 1965, Playa lake use and modification in the High Plains *in* Studies of playa lakes in the High Plains of Texas: Texas Water Devel. Board Rept. 10, 17 p.

Smith, H. N., and Rechenthin, C. A., 1964, Grassland restoration—the Texas brush problem: U.S. Soil Conservation Service open-file rept., 17 p.

Stallman, R. W., 1972, Subsurface waste storage—the earth scientists' dilemma *in* Underground waste management and environmental implications: Am. Assoc. Petroleum Geologists Memoir 18, p. 6–10.

Stuart, C. A., 1970, Geopressure: Proceedings of second symposium on abnormal subsurface pressures, La. State Univ., Baton Rouge, La., Jan. 1970, 12 p.

Texas Water Development Board, 1966a, A summary of the preliminary plan for proposed water resources development in the Brazos River basin: Austin, Texas, Texas Water Devel. Board, 87 p.

——— 1966b, A summary of the preliminary plan for proposed water resources development in the Brazos-Colorado coastal basin: Austin, Texas, Texas Water Devel. Board, 22 p.

——— 1966c, A summary of the preliminary plan for proposed water resources development in the Colorado River basin: Austin, Texas, Texas Water Devel. Board, 21 p.

——— 1966d, A summary of the preliminary plan for proposed water resources development in the Colorado-Lavaca coastal basin: Austin, Texas, Texas Water Devel. Board, 21 p.

——— 1966e, A summary of the preliminary plan for proposed water resources development in the Guadalupe River basin: Austin, Texas, Texas Water Devel. Board, 52 p.

——— 1966f, A summary of the preliminary plan for proposed water resources development in the Lavaca River basin: Austin, Texas, Texas Water Devel. Board, 25 p.

——— 1966g, A summary of the preliminary plan for proposed water resources development in the Lavaca-Guadalupe coastal basin: Austin, Texas, Texas Water Devel. Board, 24 p.

——— 1966h, A summary of the preliminary plan for proposed water resources development in the Neches River basin: Austin, Texas, Texas Water Devel. Board, 45 p.

——— 1966i, A summary of the preliminary plan for proposed water resources development in the Neches-Trinity coastal basin: Austin Texas, Texas Water Devel. Board, 26 p.

——— 1966j, A summary of the preliminary plan for proposed water resources development in the Nueces River basin: Austin, Texas, Texas Water Devel. Board, 50 p.

——— 1966k, A summary of the preliminary plan for proposed water resources development in the Nueces-Rio Grande coastal basin: Austin, Texas, Texas Water Devel. Board, 44 p.

———1966l, A summary of the preliminary plan for proposed water resources development in the Sabine River basin: Austin, Texas, Texas Water Devel. Board, 40 p.

——— 1966m, A summary of the preliminary plan for proposed water resources development in the San Antonio River basin: Austin, Texas, Texas Water Devel. Board, 46 p.

——— 1966n, A summary of the preliminary plan for proposed water resources development in the San Antonio-Nueces coastal basin, Austin, Texas, Texas Water Devel. Board, 26 p.

——— 1966o, A summary of the preliminary plan for proposed water resources development in the San Jacinto River basin: Austin, Texas, Texas Water Devel. Board, 35 p.

——— 1966p, A summary of the preliminary plan for proposed water resources development in the San Jacinto-Brazos coastal basin: Austin, Texas, Texas Water Devel. Board, 26 p.

——— 1966q, A summary of the preliminary plan for proposed water resources development in the Trinity River basin: Austin, Texas, Texas Water Devel. Board, 66 p.

——— 1966r, A summary of the preliminary plan for proposed water resources development in the Trinity-San Jacinto coastal basin: Austin, Texas, Texas Water Devel. Board, 19 p.

——— 1968, The Texas Water Plan: Austin, Texas, Texas Water Devel. Board, 215 p.

Theis, C. V., 1937, Amount of ground-water recharge in the Southern High Plains: Am. Geophys. Union Trans., v. 18, p. 564–568.

Thomas, H. E., 1951, The conservation of ground water: New York, McGraw-Hill Book Co., Inc., 327 p.

U.S. Army Corps of Engineers and U.S. Soil Conservation Service, 1967, Report on comprehensive basin study, Sabine River and tributaries, Texas and Louisiana, v. 2: U.S. Army Corps of Engineers and U.S. Soil Conservation Service, 778 p.

U.S. Bureau of Reclamation, 1973, West Texas and eastern New Mexico import project: U.S. Bureau of Reclamation open-file report, 201 p.

Valliant, J. C., 1961, Artificial ground-water recharge: Halfway, Tex., High Plains Research Foundation Rept. 18, 8 p.

Winslow, A. G., and Doyel, W. W., 1954, Land-surface subsidence and its relation to the withdrawal of ground water in the Houston-Galveston region, Texas: Econ. Geol., v. 49, no. 4, p. 413–422.

Winslow, A. G., Hillier, D. E., and Turcan, A. N., Jr., 1968, Saline ground water in Louisiana: U.S. Geol. Survey Hydrol. Inv. Atlas HA–310, 4 maps.

Winslow, A. G., and Kister, L. R., 1956, Saline-water resources of Texas: U.S. Geol. Survey Water-Supply Paper 1365, 105 p.

Winslow, A. G., and Wood, L. A., 1959, Relation of land subsidence to ground-water withdrawals in the upper Gulf Coast region, Texas: Mining Eng., no. 10, p. 1030–1034, Am. Inst. Mining Metall. and Petroleum Engineers Trans., v. 214.

Wood, L. A., 1956, Availability of ground water in the Gulf Coast region of Texas: U.S. Geol. Survey open-file rept., 55 p.

Wood, L. A., and Gabrysch, R. K., 1965, Analog model study of ground water in the Houston district, Texas, with a section on design, construction, and use of electric analog models by E. P. Patten, Jr.: Texas Water Commission Bull. 6508, 103 p.

Wood, L. A., and others, 1963, Reconnaissance investigation of the ground-water resources of the Gulf Coast region, Texas: Texas Water Commission Bull. 6305, 114 p.

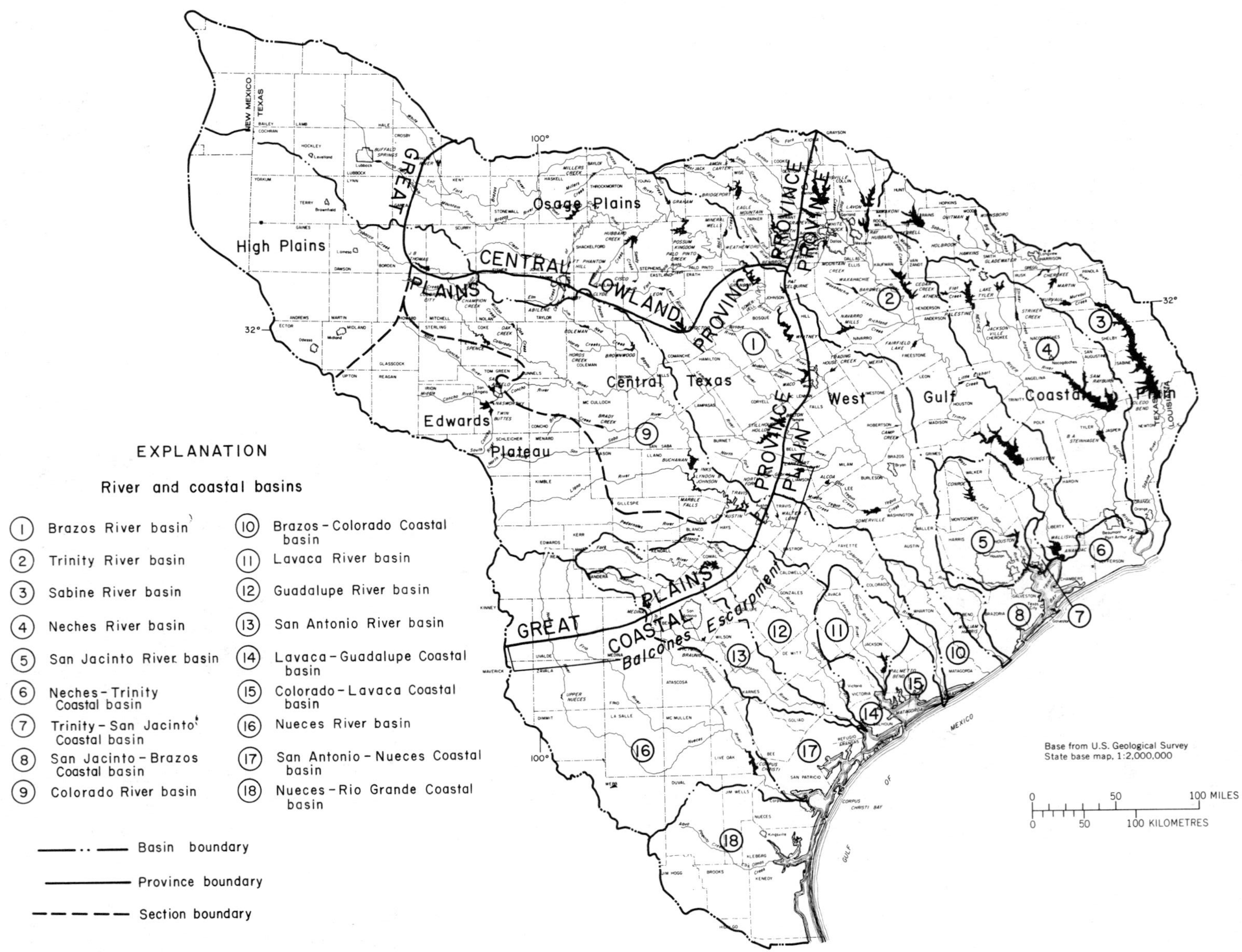

PLATE 1A. PHYSIOGRAPHIC REGIONS AND DRAINAGE BASINS, TEXAS-GULF REGION

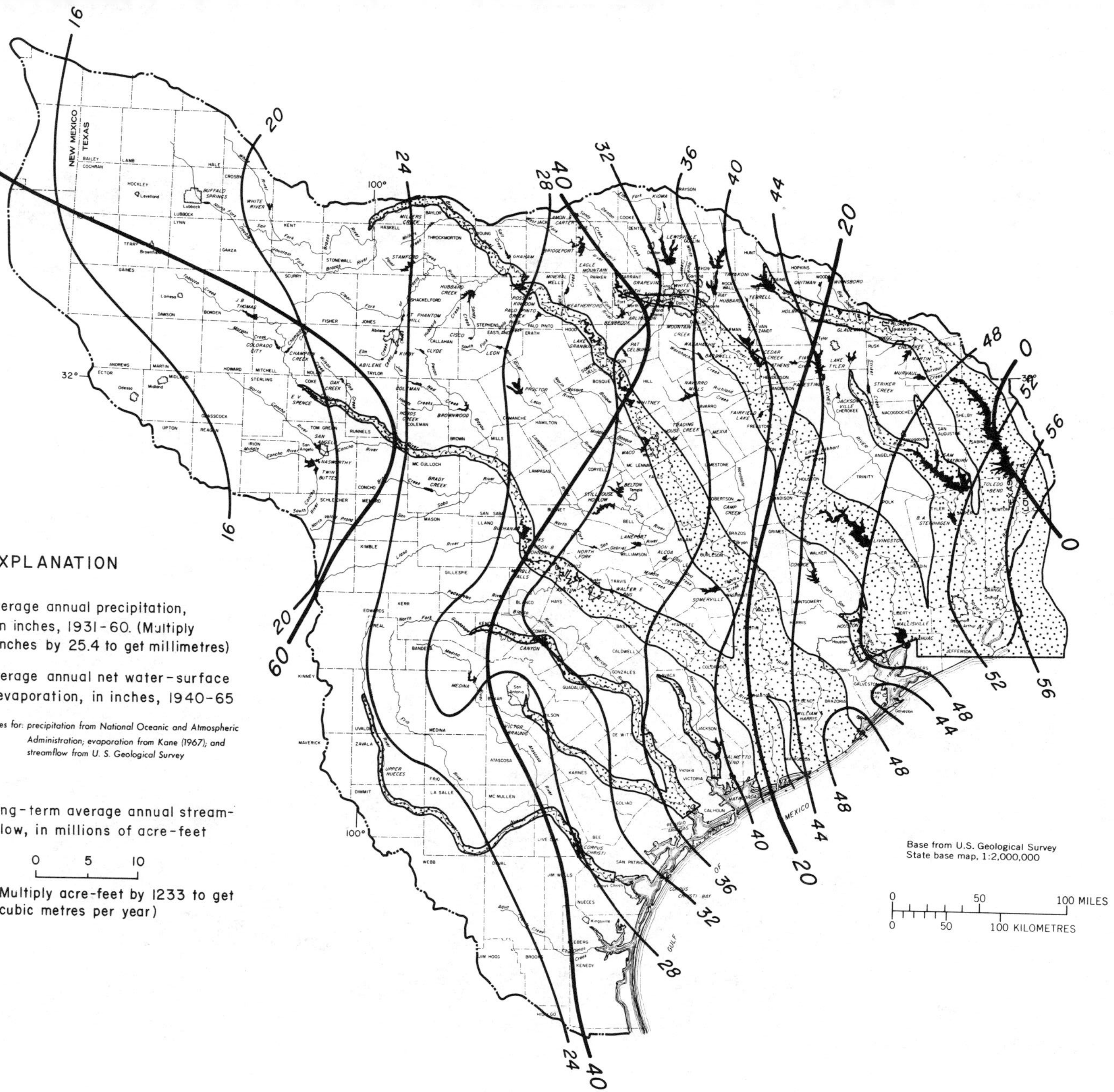

EXPLANATION

—24— Average annual precipitation, in inches, 1931-60. (Multiply inches by 25.4 to get millimetres)

—40— Average annual net water-surface evaporation, in inches, 1940-65

Values for: precipitation from National Oceanic and Atmospheric Administration; evaporation from Kane (1967); and streamflow from U. S. Geological Survey

Long-term average annual streamflow, in millions of acre-feet

0 5 10

(Multiply acre-feet by 1233 to get cubic metres per year)

Base from U.S. Geological Survey State base map, 1:2,000,000

0 50 100 MILES

0 50 100 KILOMETRES

PLATE 1B. ANNUAL PRECIPITATION, STREAMFLOW, AND NET WATER-SURFACE EVAPORATION, TEXAS-GULF REGION

PLATE 2A. GEOGRAPHIC DISTRIBUTION OF TOTAL ANNUAL WATER REQUIREMENTS AND SUPPLIES PROJECTED TO THE YEAR 2020, TEXAS-GULF REGION

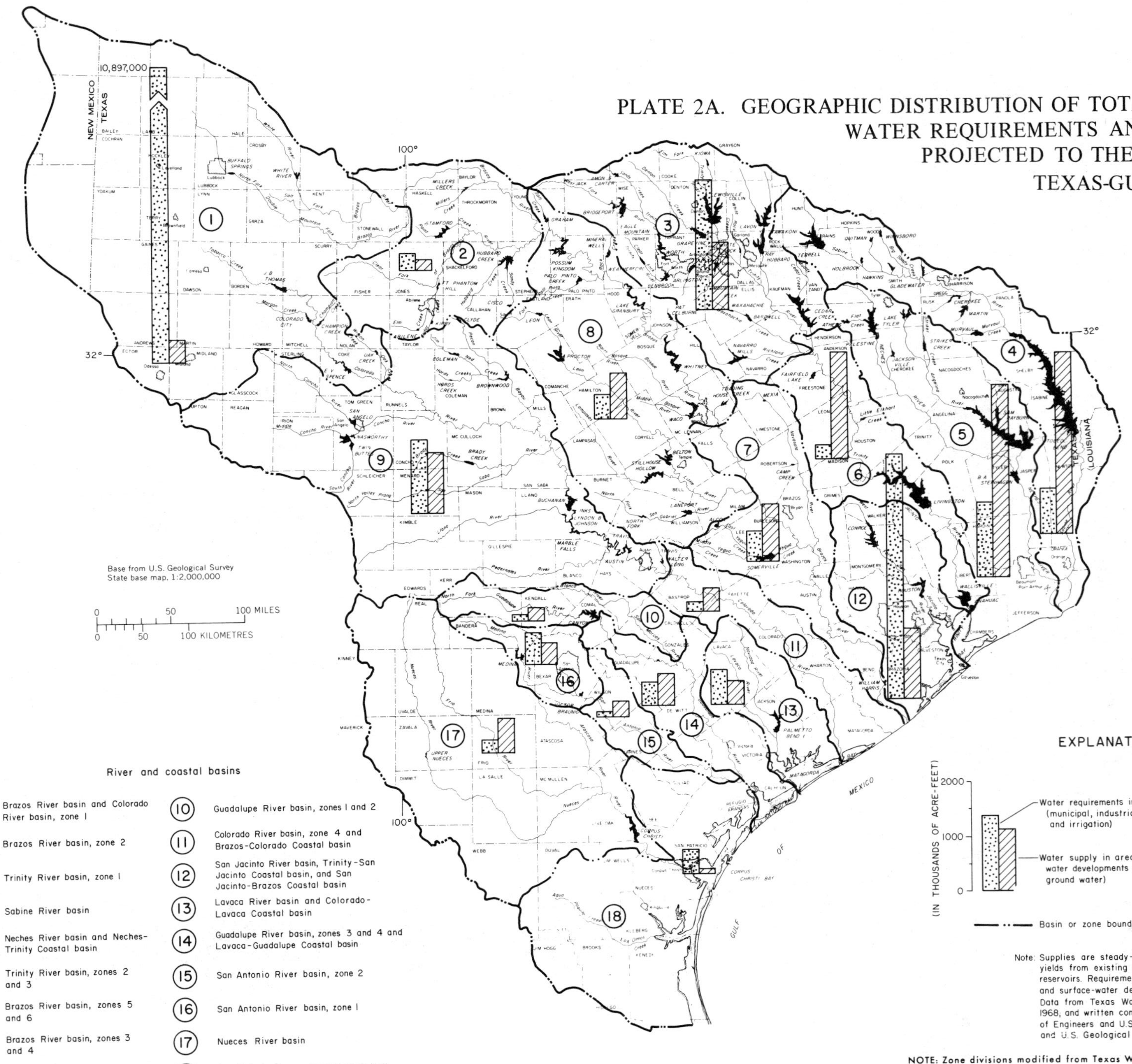

Base from U.S. Geological Survey State base map. 1:2,000,000

0 50 100 MILES

0 50 100 KILOMETRES

River and coastal basins

1 Brazos River basin and Colorado River basin, zone 1

2 Brazos River basin, zone 2

3 Trinity River basin, zone 1

4 Sabine River basin

5 Neches River basin and Neches-Trinity Coastal basin

6 Trinity River basin, zones 2 and 3

7 Brazos River basin, zones 5 and 6

8 Brazos River basin, zones 3 and 4

9 Colorado River basin, zones 2 and 3

10 Guadalupe River basin, zones 1 and 2

11 Colorado River basin, zone 4 and Brazos-Colorado Coastal basin

12 San Jacinto River basin, Trinity-San Jacinto Coastal basin, and San Jacinto-Brazos Coastal basin

13 Lavaca River basin and Colorado-Lavaca Coastal basin

14 Guadalupe River basin, zones 3 and 4 and Lavaca-Guadalupe Coastal basin

15 San Antonio River basin, zone 2

16 San Antonio River basin, zone 1

17 Nueces River basin

18 San Antonio-Nueces Coastal basin and Nueces-Rio Grande Coastal basin

EXPLANATION

Basin or zone boundary

Note: Supplies are steady-state ground-water yields plus yields from existing and proposed surface-water reservoirs. Requirements are ground-water and surface-water demands projected to 2020. Data from Texas Water Development Board (1966a-r; 1968; and written commun., 1974), U.S. Army Corps of Engineers and U.S. Soil Conservation Service (1967), and U.S. Geological Survey

NOTE: Zone divisions modified from Texas Water Development Board (1968)

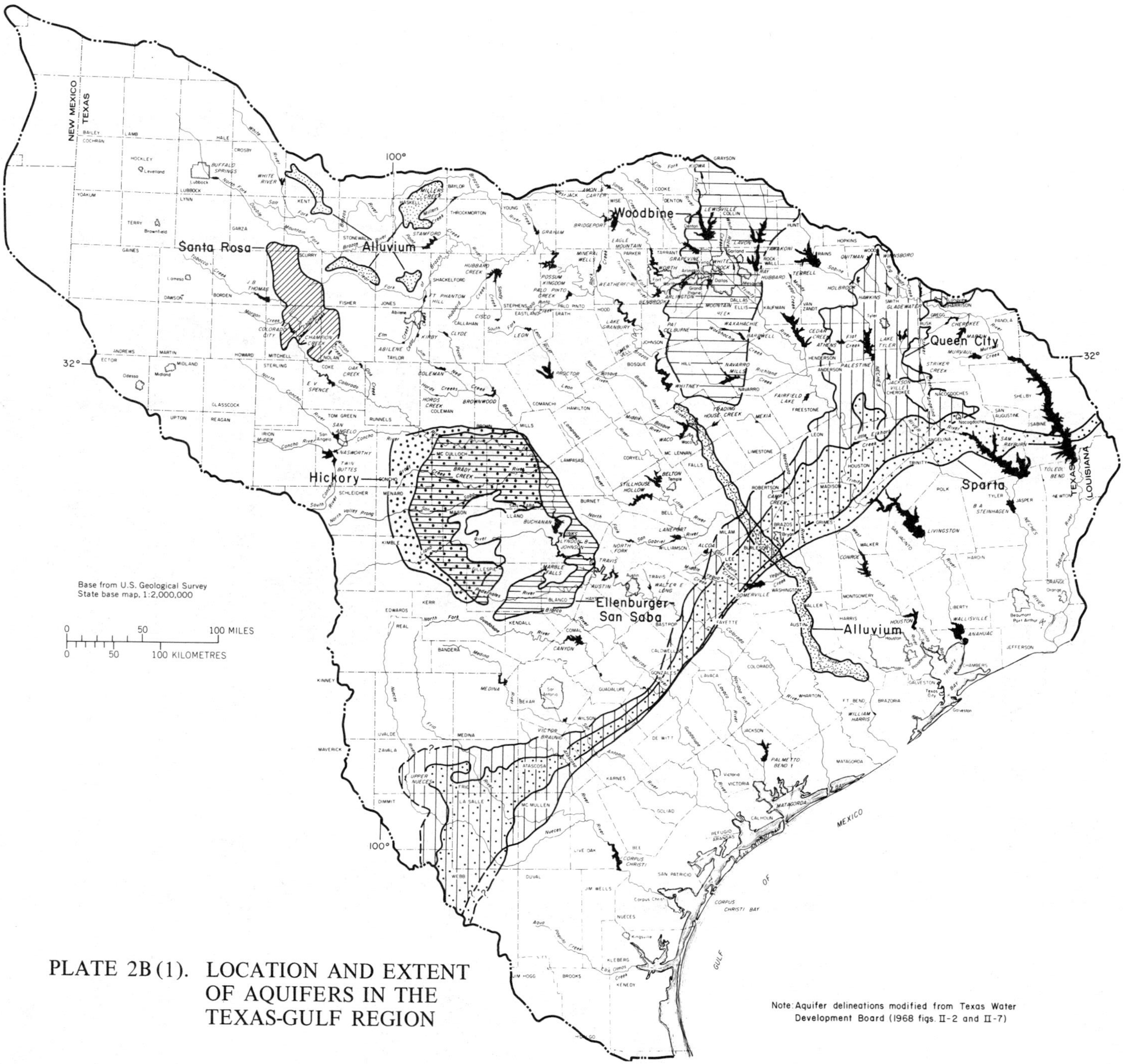

PLATE 2B (1). LOCATION AND EXTENT OF AQUIFERS IN THE TEXAS-GULF REGION

Note: Aquifer delineations modified from Texas Water Development Board (1968 figs. II-2 and II-7)

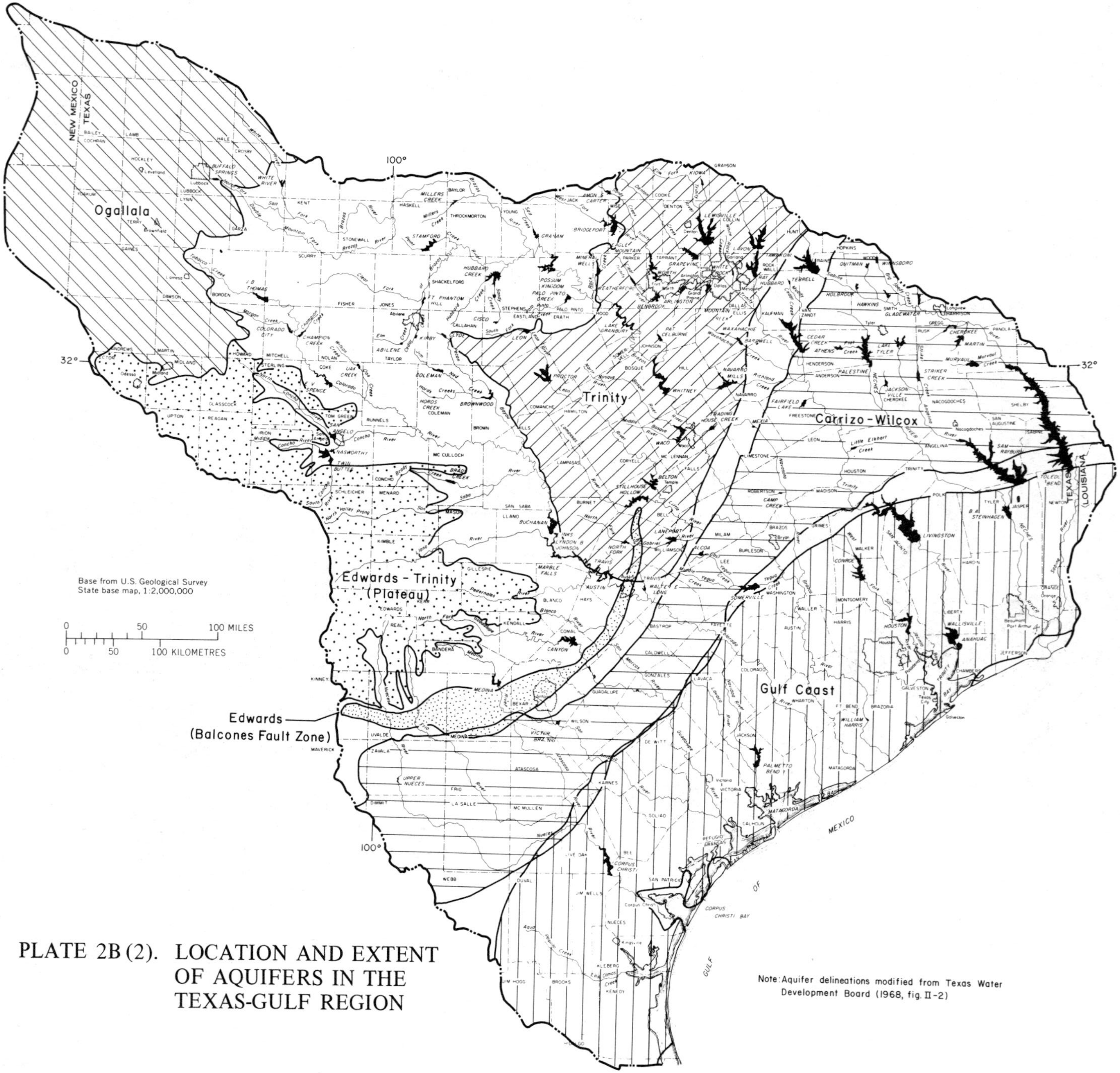

PLATE 2B (2). LOCATION AND EXTENT OF AQUIFERS IN THE TEXAS-GULF REGION

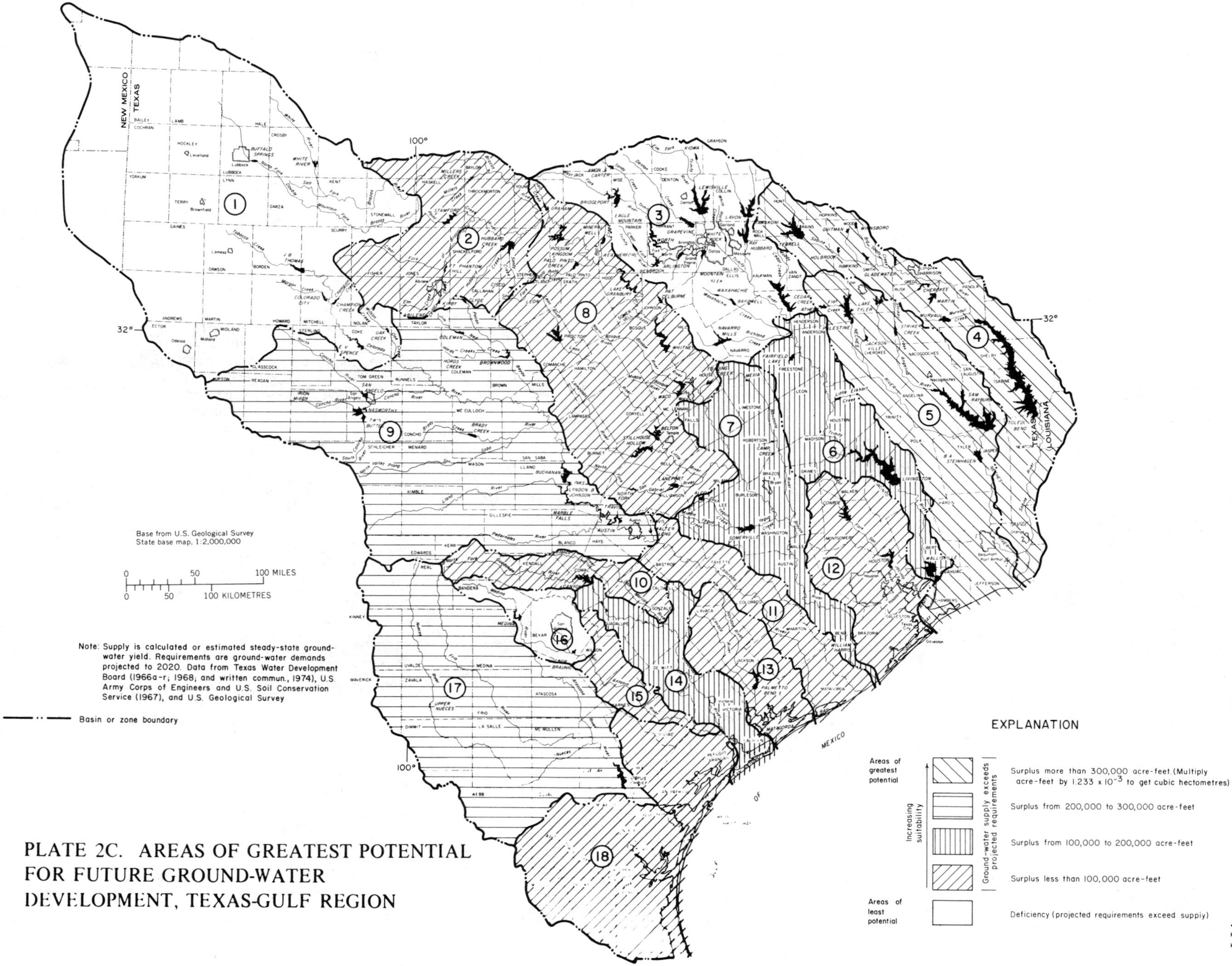

PLATE 2C. AREAS OF GREATEST POTENTIAL FOR FUTURE GROUND-WATER DEVELOPMENT, TEXAS-GULF REGION

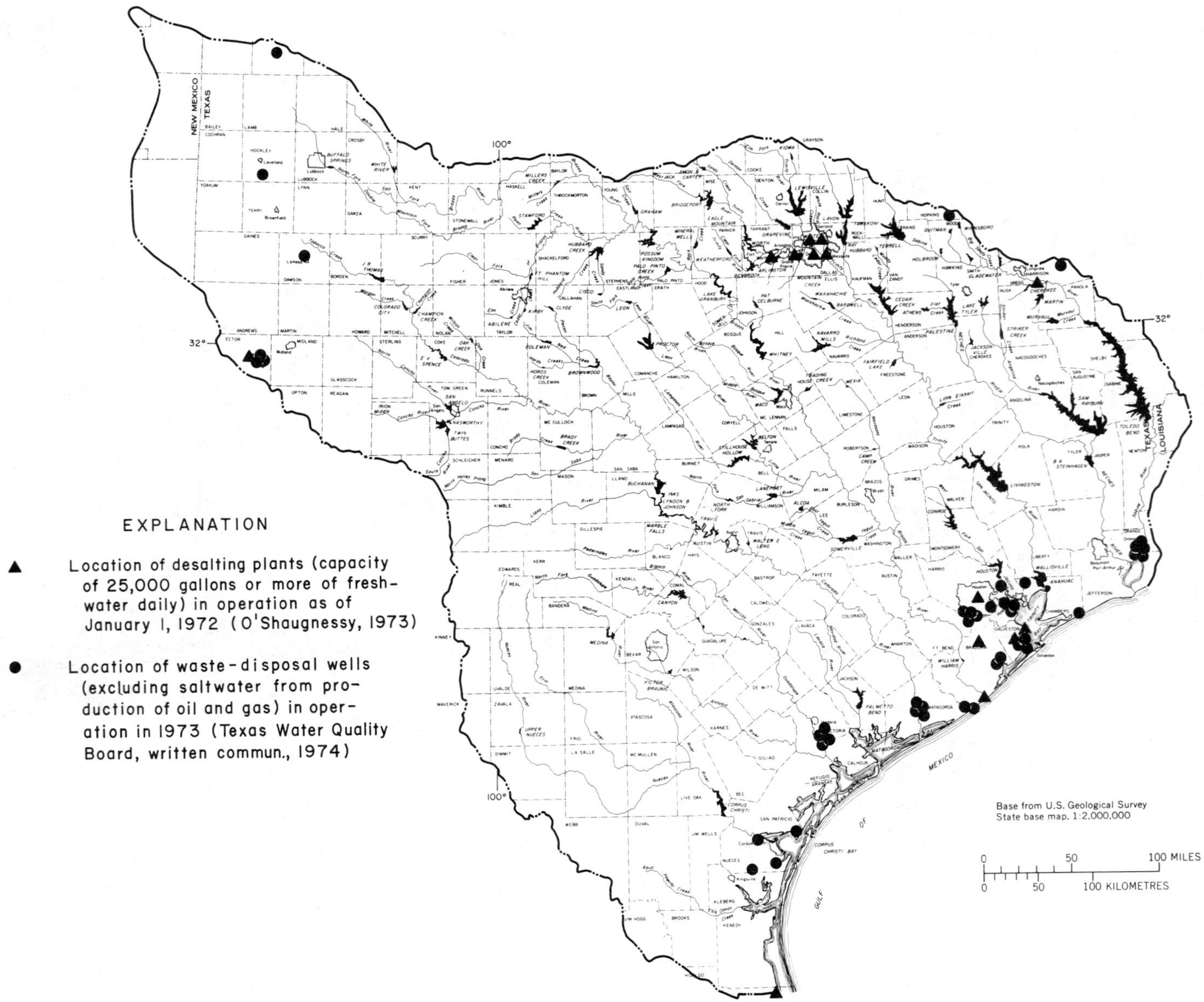

PLATE 3A. LOCATION OF DESALTING PLANTS AND LIQUID-WASTE DISPOSAL SITES, TEXAS-GULF REGION

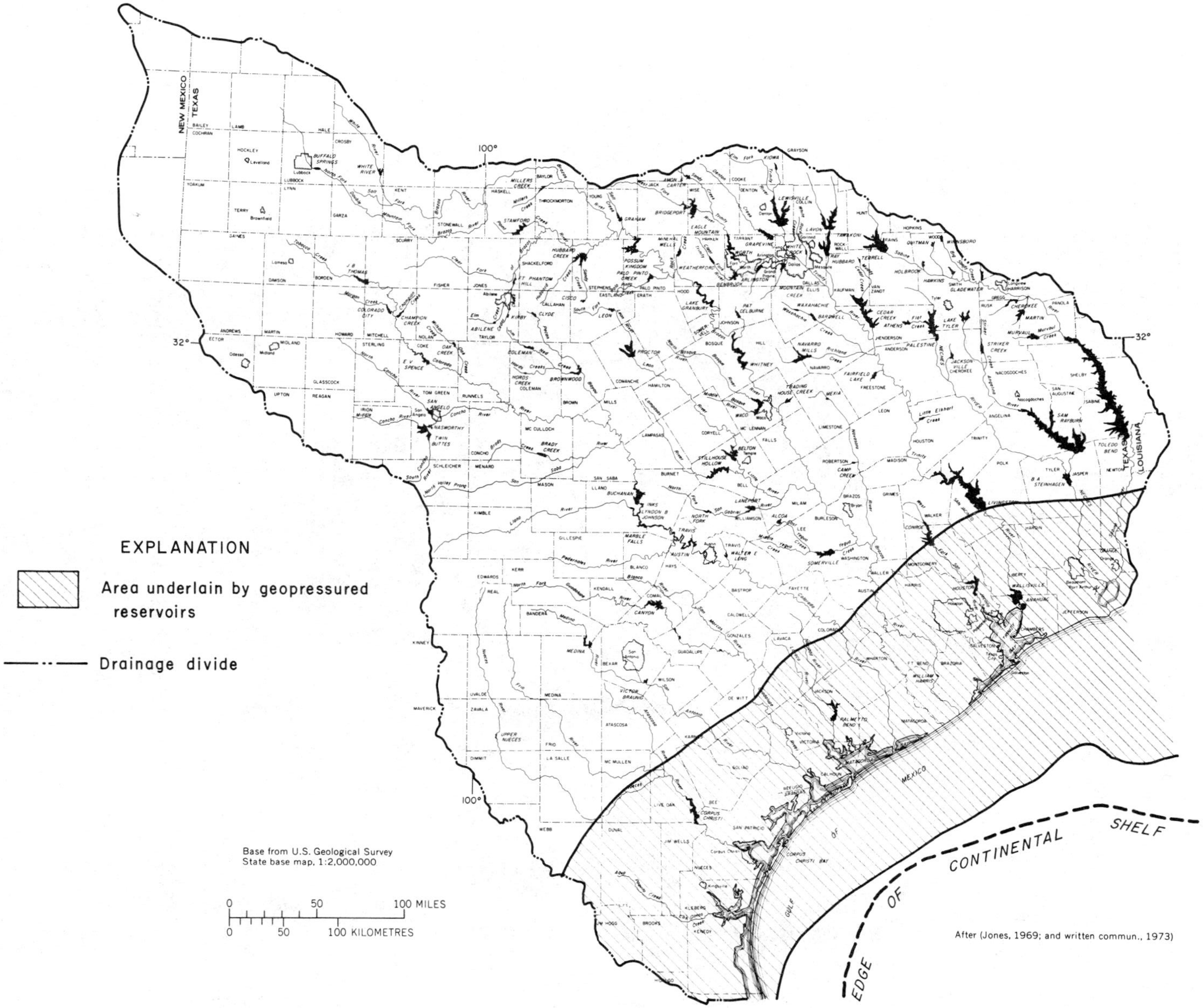

PLATE 3B. AREAS IN WHICH GEOPRESSURED RESERVOIRS OCCUR IN THE TEXAS-GULF REGION

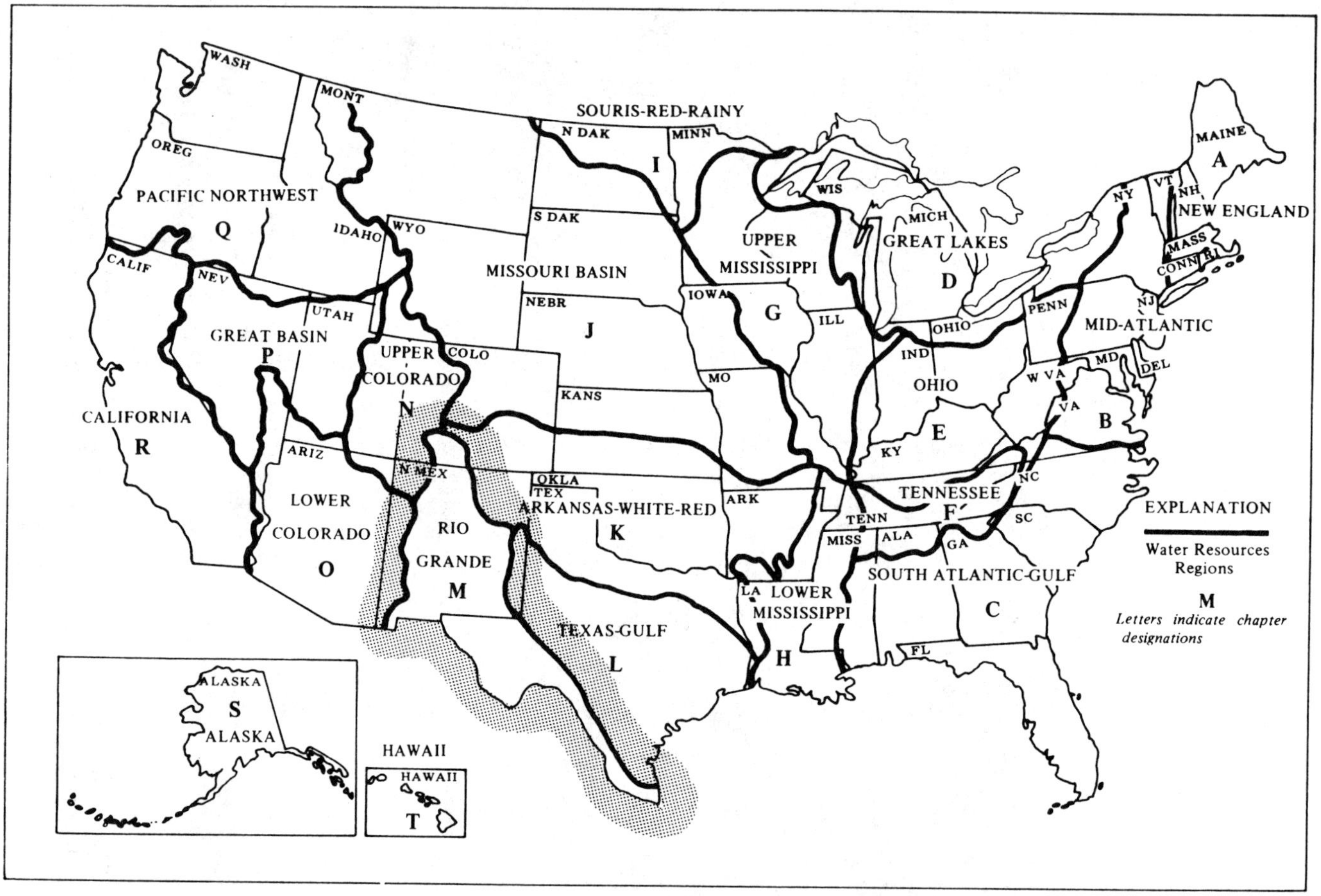

M – Location Map for the Rio Grande Region

M ....

RIO GRANDE REGION

By S. W. WEST and W. L. BROADHURST

ABSTRACT

The Rio Grande is an interstate and international stream which begins in high mountains of Colorado, flows across New Mexico, and forms the boundary between Texas and Mexico. Precipitation ranges from 8 inches (20 cm) to more than 30 inches (76 cm), but irrigation is required for growing crops throughout the region.

The population of the region has been increasing rapidly, from 750,000 in 1929 to 1,700,000 in 1970, and it is expected to increase to 2,500,000 by 2020. The basic economy of the region was agricultural until recent years. Since 1950, the mining and petroleum industries have increased much more rapidly than agriculture.

Annual precipitation on the region is about 86 million acre-feet (110,000 hm^3); however, all but 4 million acre-feet (4,900 hm^3) is returned to the atmosphere by evapotranspiration. The ground-water reservoirs contain an aggregate of 5,800 million acre-feet (7,200,000 hm^3) of fresh and slightly saline water in storage, which could be withdrawn through wells. In contrast, the surface reservoirs have a combined storage capacity of only 18 million acre-feet (22,000 hm^3).

Thick deposits of valley fill in stream and intermontane valleys comprise the principal ground-water reservoirs. In most areas they are capable of yielding large supplies of water to wells. In some areas, limestone constitutes major aquifers.

Withdrawal of ground water in the region in 1970 was 2.7 million acre-feet (3,300 hm^3), of which 88 percent was used for irrigation. About 53 percent of the water withdrawn was consumed. Ground water has been "mined" in some areas, and severe declines in water levels have resulted.

The loss of water by evapotranspiration in wetlands and phreatophyte areas is 2.5 million acre-feet (3,100 hm^3) per year. In comparison, about 3.7 million acre-feet (4,600 hm^3) per year of surface water and ground water is consumed by man's activities.

Salvage of water lost to noneconomic evapotranspiration in wet and phreatophyte-infested areas offers the greatest possibility of improving the effective water supply in the region. Salvage of half the water lost would increase the effective supply by 1.2 million acre-feet (1,500 hm^3) per year. The usable water supply could be increased tremendously by drawing on the large reserve of ground water in storage, but this withdrawal could affect the flow of streams in some areas.

The region appears to offer several possibilities for utilizing underground space for purposes other than the withdrawal of water, such as waste disposal, artificial recharge, water-quality control, and development of geothermal energy.

Planners for ground-water management should have detailed information on the physical parameters that affect ground water, so improved management would be possible.

INTRODUCTION

Water resources are neatly catalogued, for convenience, as surface water and ground water. According to common usage, surface water is defined as any water on the land surface, regardless of whether it was derived directly from precipitation or from discharge of ground water. Ground water, in contrast, is defined as water below the land surface in the zone of saturation, regardless of whether it was derived from direct infiltration of precipitation or from infiltration of water flowing across or standing on the land surface. Traditionally, surface water has received more attention than ground water, because surface water is visible, is easily measured, and commonly can be diverted for use by gravity flow. Although ground water is out of sight, modern technology has provided tools for its measurement and utilization. Technology has also shown such a close relationship between surface water and ground water that they cannot be treated as separate sources of water. A change in the regimen of either will generally affect the other. In this report, the ground water is emphasized, because surface water has been fully appropriated and numerous structures exist for its regulation. In contrast, vast supplies of ground water lie beneath the surface.

In the past, much of the systematic planning for economic development or management of water resources has been straightforward and simple. The flows of principal streams were measured, and plans were drawn for diverting water from the streams to points of use. If variations in streamflow presented problems in meeting water demands, the situation was improved by building dams to regulate the streamflow by surface storage and controlled release.

Ground water has not been entirely ignored in water-resources development, but, in general, it has not been considered in systematic planning for total water management. Although ground water comprised 21 percent of the water used in the United States in 1970, planning for ground-water development has been limited to the municipal, water district, or private level, except for a few instances, and the impact of ground-water development on surface supplies has been largely neglected. This report attempts to place ground water in its proper perspective in regard to the total water resources.

This report is one of a series that will constitute a national compendium on ground water for the guidance of planners. New data for this appraisal were not collected. The many excellent reports dealing with the occurrence, development, and use of ground water for selected areas of the Rio Grande Region, from the headwaters to the mouth of the river, have been utilized. Also, unpublished data in the files of Federal, State, and other agencies and statements from many individuals concerned with the water problems of the region added to the information.

This report summarizes the knowledge of the ground-water resources of the region and evaluates deficiencies in our knowledge. The primary objectives are to direct attention to the locations and storage capacities of the principal ground-water reservoirs, to delineate the types of information needed for fuller evaluation of the opportunities for ground-water management, and to describe the role of ground-water reservoirs in meeting the region's water needs.

Although some of the water that falls as precipitation in the United States is used in Mexico and vice versa, that topic is not within the scope of this report. Of primary concern for this study are the water problems of the Rio Grande Region within the United States, including the drainage basin of the Pecos River and certain closed basins in Colorado, New Mexico, and Texas.

Numerous individuals in the district offices of the U.S. Geological Survey in Colorado, New Mexico, and Texas furnished many publications, permitted use of material being prepared for publication, and gave freely from their store of knowledge. Individuals in the New Mexico State Engineer Office provided unpublished records and reviewed the report.

Special thanks are extended to State officials and their staffs, including H. P. Burleigh, Executive Director of Texas Water Development Board; S. E. Reynolds, State Engineer of New Mexico; and C. J. Kuiper, State Engineer of Colorado. Data, assistance, and encouragement were received from many others. Grateful acknowledgement is extended to Dr. Gerald Thomas, President of New Mexico State University, Dr. John Clark, Director of New Mexico Water Resources Research Institute, and Jesse Gilmer, Texas member of the Rio Grande Compact Commission.

Most numbers in this report are given in English units followed by metric units in parentheses. The conversions to metric units were made as follows:

English			Metric	
Unit	Abbreviation	Multiplied by	Unit	Abbreviation
Acre	acre	0.4047	Hectare	ha
Acre-foot	acre-ft	.0012335	Cubic hectometre	hm^3
Foot	ft	.3048	Metre	m
Gallons per minute	gpm	5.45	Cubic metres per day	m^3/d
Inch	in.	2.54	Centemetre	cm
Mile	mi	1.6093	Kilometre	km
Square mile	mi^2	2.59	Square kilometre	km^2

Chemical concentrations are given only in metric units — milligrams per liter (mg/l). For concentrations less than 7,000 mg/l, the numerical value is about the same as for concentrations in the English unit, parts per million.

THE PHYSICAL SETTING

The Rio Grande is an interstate and international stream. It rises in southern Colorado, flows southward more than 400 miles (640 km) across New Mexico, and then forms the boundary between Mexico and Texas for about 1,250 miles (2,000 km) from El Paso to its mouth at the Gulf of Mexico (fig. 1). The total length of the river is about 1,800 miles (2,900 km).

The region ranges in altitude from sea level at the Gulf to a little more than 14,000 feet (4,300 m) on the higher peaks near the headwaters. The upper half of the region is characterized by disconnected mountain ranges and intervening intermontane valleys, except for the Pecos Valley, which lies mostly east of the principal mountain ranges. High hills and mountains or tablelands predominate locally in the northern half of the region. The southern half of the region is characterized by tablelands, canyons, and plains (fig. 2). The mountains and tablelands slope steeply, almost precipitously in some areas, toward the valley floors. Coalescing alluvial fans lie at the foot of many mountains and form intermediate slopes between the mountains and valley floors.

The altitude of irrigated lands averages about 7,700 feet (2,300 m) in the San Luis Valley, 5,000 feet (1,500 m) in the Albuquerque area, 3,800 feet (1,200 m) in the Las Cruces–El Paso area, 3,400 feet (1,000 m) in the Roswell-Carlsbad area, 2,600 feet (790 m) in the Pecos area, and only a few tens of feet in the coastal area.

Because of its geographic position and wide range in altitude, the region has a variety of climatic zones, including semitropic, arid, and arctic. As shown by available records, annual precipitation averages more than 30 inches (76 cm) in the mountainous headwaters

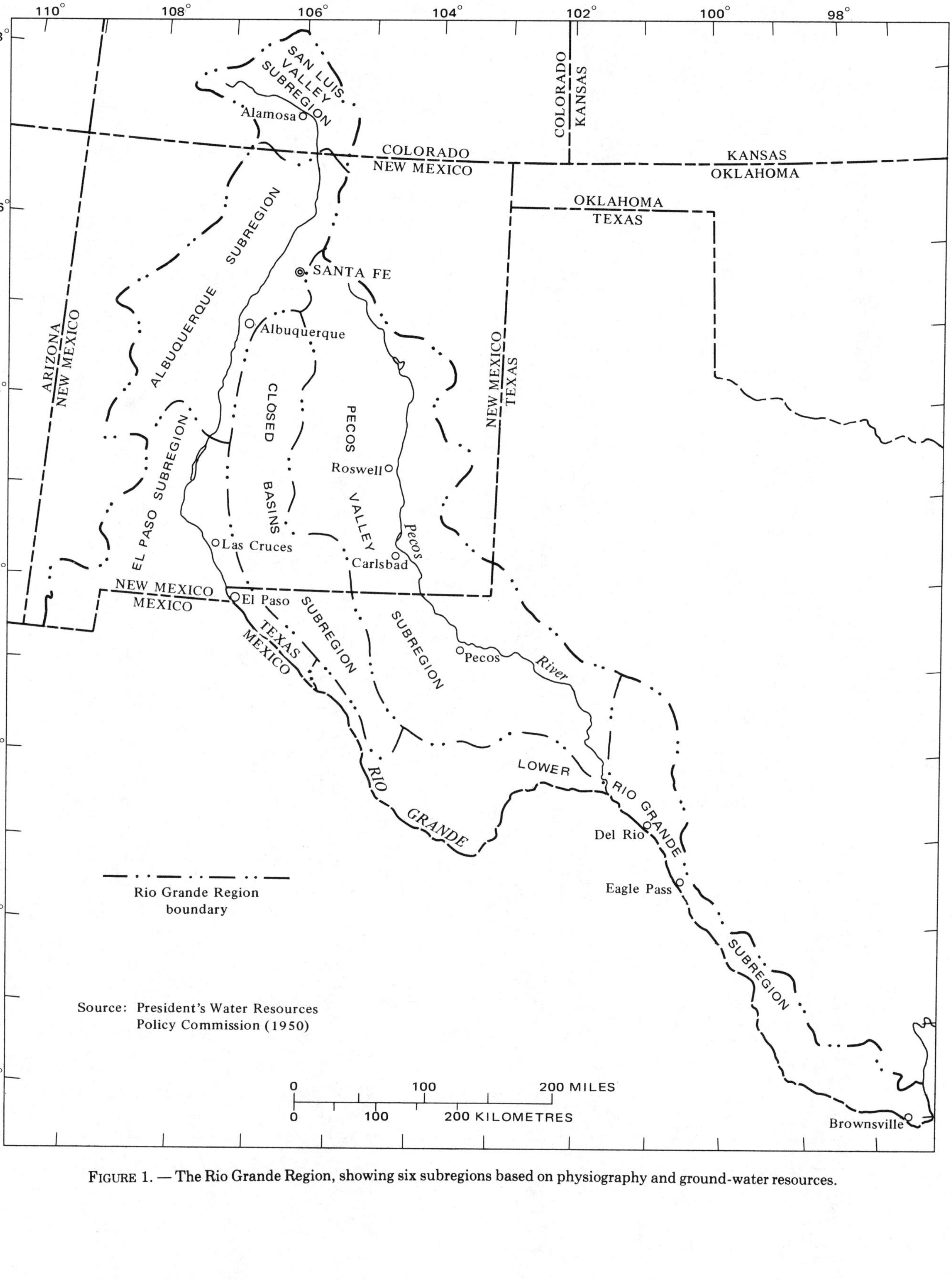

FIGURE 1. — The Rio Grande Region, showing six subregions based on physiography and ground-water resources.

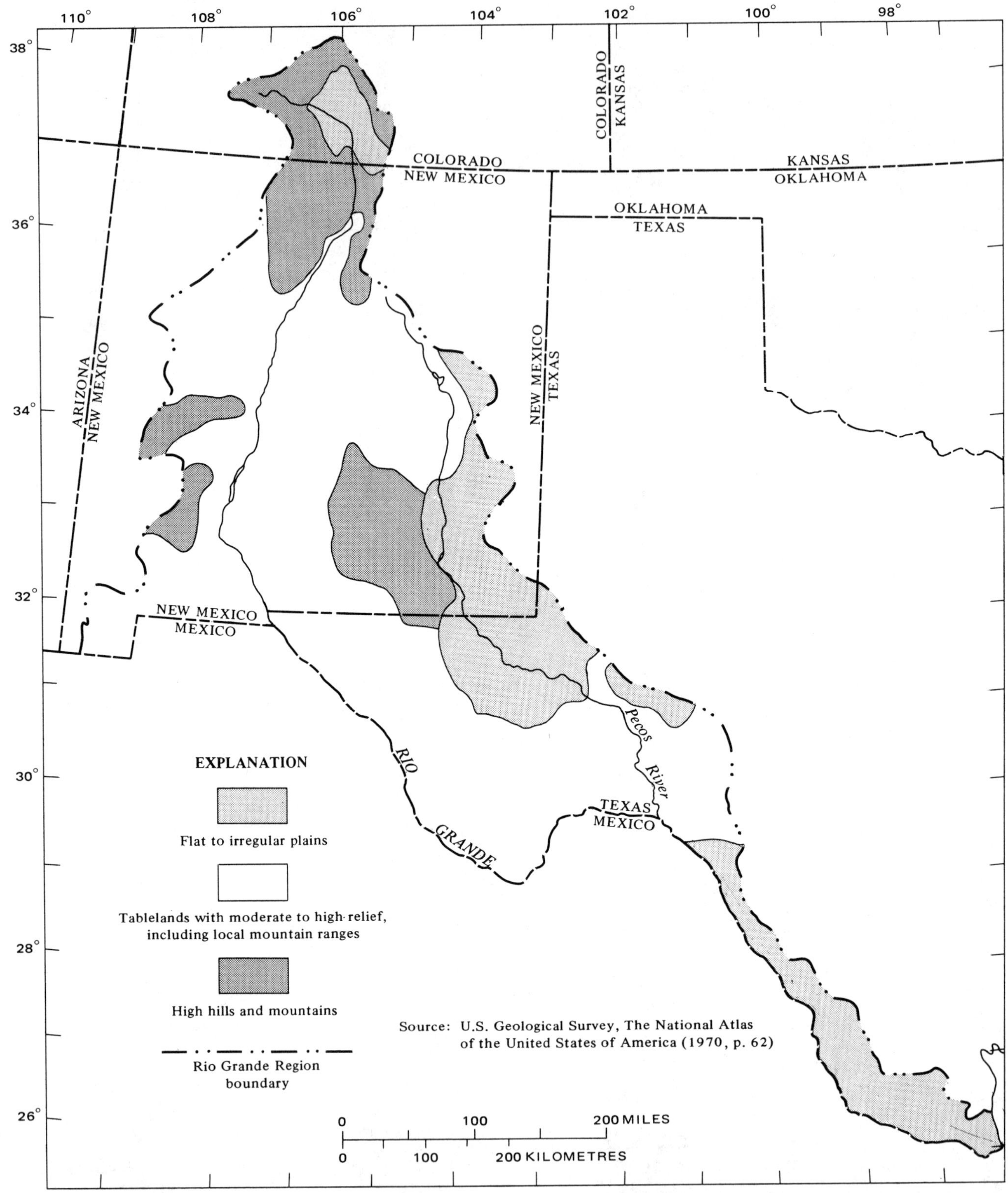

FIGURE 2. — Generalized terrain. Most of the region consists of tablelands with moderate to high relief. Mountains are limited mostly to the northern half of the region.

in Colorado and New Mexico, less than 8 inches (20 cm) in the intermontane valleys in Colorado, southern New Mexico, and western Texas, and about 25 inches (64 cm) along the Gulf Coast. Rainstorms of high intensity during summer are common throughout most of the region. Average precipitation for the entire basin is about 12 inches (30 cm) a year. That part of the basin within the United States has an area of approximately 135,000 square miles (350,000 km^2) — about 86 million acres (35 million ha). Therefore, total annual precipitation on the region averages about 86 million acre-feet (110,000 hm^3). But the considerable variability of precipitation from year to year and the recurrence of prolonged dry periods between wet periods create the major problems of water supply. Irrigation is required for growing crops throughout the region.

Temperatures vary widely, depending on latitude and altitude, but the entire region experiences a high percentage of bright sunny days. Both Alburquerque and El Paso have at times boasted of more than 365 consecutive days during which the sun made an appearance, and each city has about 75 percent of maximum possible sunshine.

Evaporation of water from lakes and reservoirs in the region ranges from 42 inches (110 cm) in the San Luis Valley to 80 inches (200 cm) near Big Bend, Tex., depending on temperature, humidity, wind velocity, and depth of water. Water loss by evaporation from the Elephant Butte and Caballo Reservoirs in southern New Mexico is estimated to be 255,000 acre-feet (310 hm^3) per year (Sorensen and Linford, 1967), and the loss from all the storage reservoirs, including small ponds, and from streams is about 950,000 acre-feet (1,200 hm^3) per year (Meyers, 1962).

The average growing season ranges from 90-120 days in the San Luis Valley, Colo., to 365 days in the lowermost part of the region near the Gulf of Mexico. The length of growing season, the types of crops grown, and the amount and timeliness of precipitation significantly affect the amount of water required for growing irrigated crops.

Much of the region is sparsely vegetated, as would be expected in areas of low rainfall. The prevalent types of vegetation are grasses, desert shrubs, juniper, pinon, forest trees, and alpine shrubs, depending on the altitude of the land and the amount of precipitation.

The steep slopes, the high-intensity rainstorms in summer, and the sparse vegetation combine to cause rapid runoff and erosion. The resultant sediment loads of tributary streams tend to overload the Rio Grande, causing aggradation along a large part of its course in New Mexico and rapid filling of surface storage reservoirs.

THE ECONOMIC SETTING

Human occupation of the Rio Grande Region has a long and varied history. Parts of the region were inhabited by nomadic Indians for thousands of years before the first white men arrived. These nomads were replaced by pastoral Pueblo Indians, who built villages in the valleys and tilled nearby irrigated farms. The first Spanish colonists arrived in 1598, establishing the first white settlement in the region near the present San Juan Indian Pueblo, at the confluence of the Rio Chama and the Rio Grande. Many Spanish colonists followed, and numerous villages were established along the Rio Grande and its tributaries. Descendants of the Pueblo Indians and Spanish colonists still constitute a significant part of the population in the northern part of the region. Since the region became a part of the United States, it has experienced a large influx of people from all parts of the United States, as well as from several foreign countries.

For centuries the water supply in the Rio Grande and its tributaries was adequate to meet the small demands of the pastoral Indians and Spanish colonists. In the present century, the demands for water have been increasing dramatically, owing to extensive irrigation, new industries, rapid increase in population, and construction of numerous military facilities. The population of the region increased from 750,000 in 1929 to 1,700,000 in 1970 and is expected to increase to 2,500,000 by 2020. In addition to the resident population, thousands of tourists visit the region each year. Of particular interest to tourists are national parks (Carlsbad Caverns, N. Mex.; Big Bend, Tex.; and Guadalupe Mountains, Tex.), national monuments, state monuments, prehistoric Indian sites, other areas of historic interest, and areas of great scenic beauty. The principal population centers are Albuquerque, N. Mex., and El Paso, Tex. The distribution of population is shown in figure 3, and population trends are shown in figure 4.

From early Pueblo Indian times to recent years, the basic economy of the region was related to agriculture. Successful production of crops depended on irrigation, and most of the irrigated lands are on or adjacent to the flood plains of perennial streams (fig. 5). In some areas, irrigation with ground water has been pursued for many years. Water is a limiting factor in agricultural expansion, so the earnings from agriculture have not increased significantly during the last 20-30 years, because the surface water has been fully appropriated and ground-water development has been limited by small supplies or has been restricted to protect surface-water rights.

The earnings from economic activities that require less water per dollar of earnings than irrigation agriculture increased rapidly during 1950-70. These activities are expected to continue their growth, as shown in table 1. The

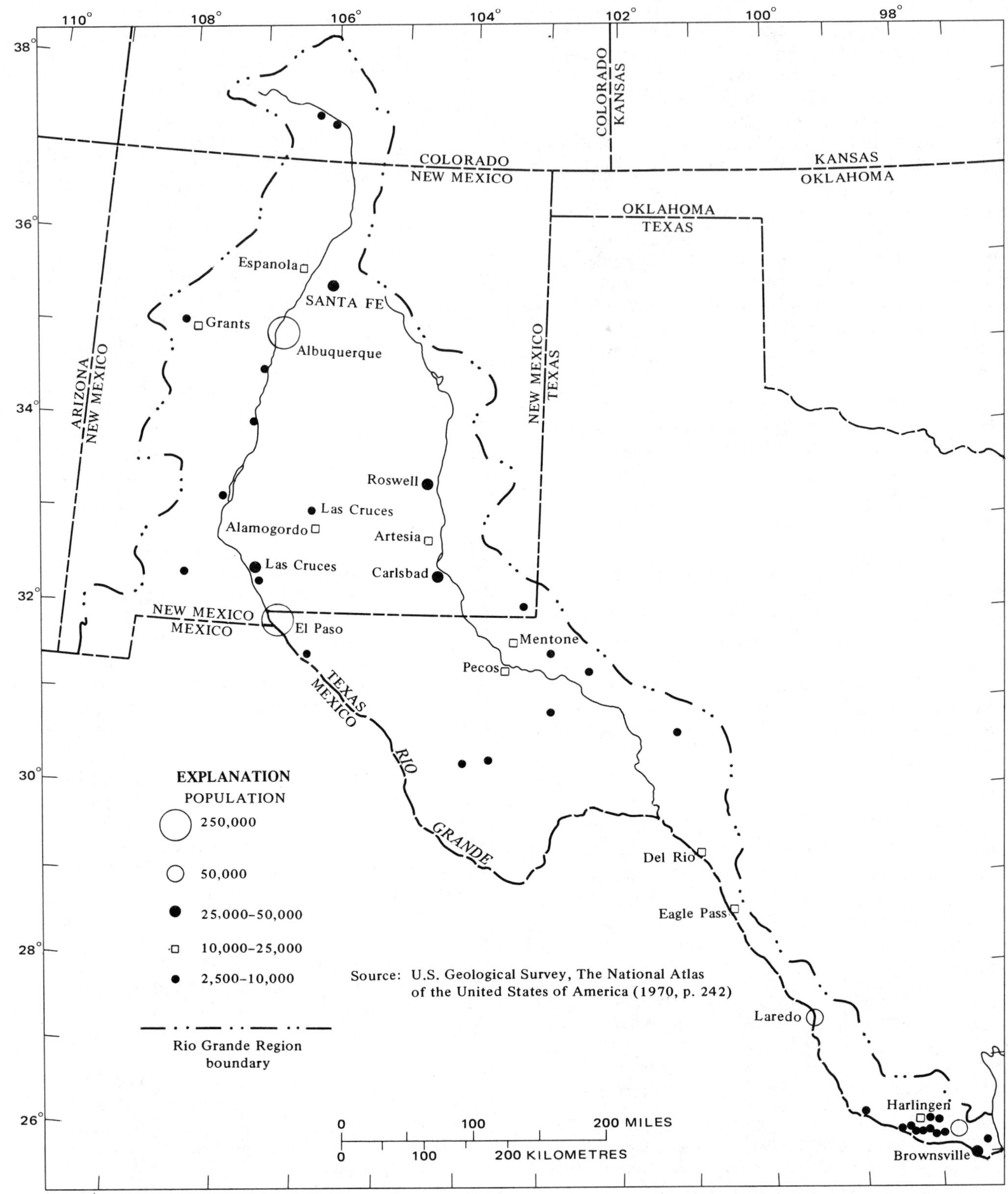

FIGURE 3. — Population distribution. Most of the population is concentrated along the major rivers — the Rio Grande and the Pecos River. The largest cities are Albuquerque, N. Mex., and El Paso, Tex.

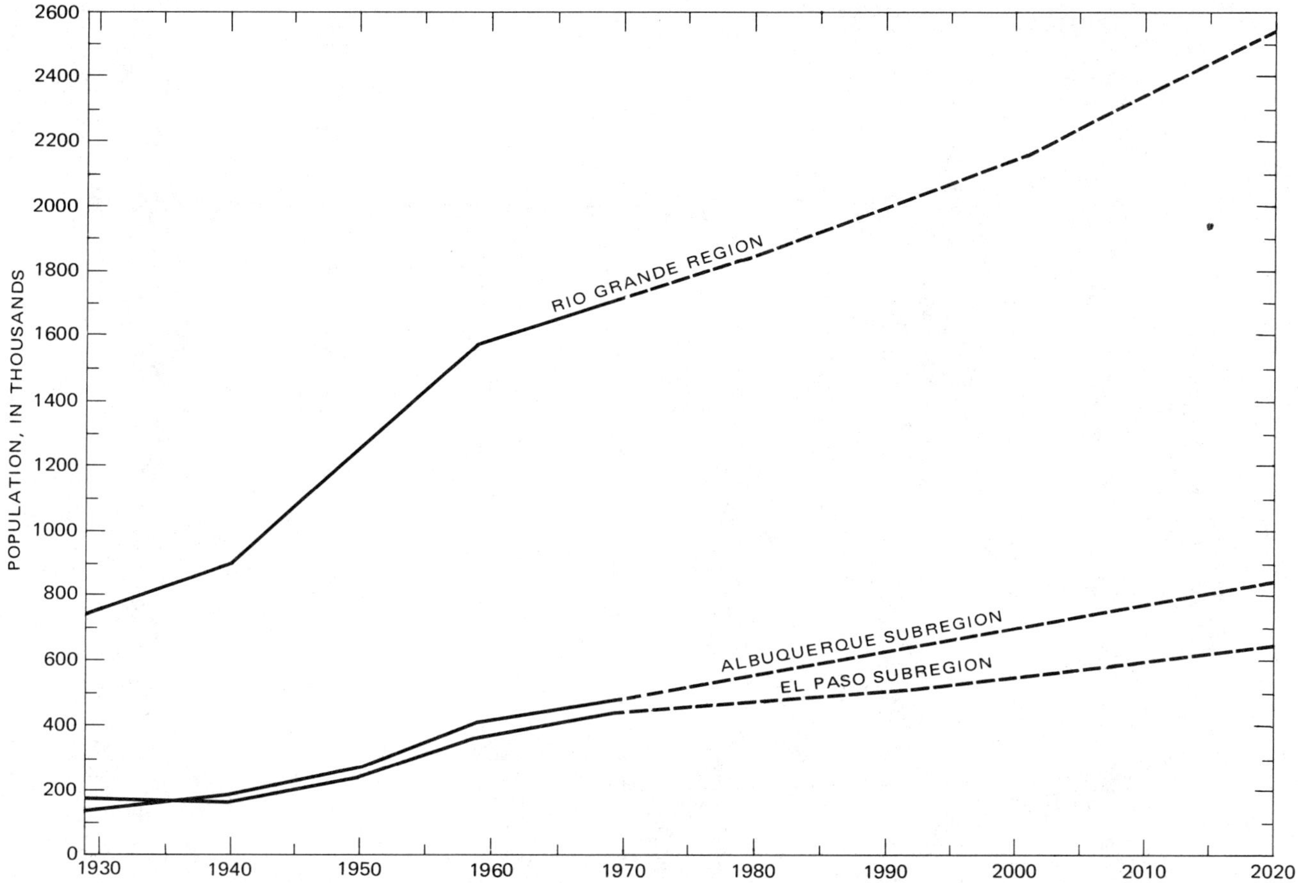

Source: U.S. Water Resources Council (1972a)

FIGURE 4. — Population trends. The population of the Rio Grande Region has been increasing rapidly, and the trend of the last decade is expected to continue for the next 50 years. Dashed lines indicate projected trends.

TABLE 1. — *Population, personal income, and industrial earnings: Historical and projected*

[Adapted from U.S. Water Resources Council (1972a). Values are based on the value of the dollar in 1967]

Year	1950	1969	1980	2000	2020
Population (midyear)	1,238,201	1,684,853	1,845,500	2,154,100	2,536,000
Per capita income	$1,503	$2,456	$3,597	$6,717	$12,274
Earnings per worker	$3,882	$5,665	$8,253	$14,639	$25,599
Industrial earnings (in thousands of dollars):					
Agriculture, forestry, fisheries	$286,783	$282,653	$287,200	$358,500	$628,300
Mining, oil, and gas	58,603	112,375	131,200	189,000	287,200
Contract construction	129,595	196,828	291,700	628,500	1,329,800
Manufacturing	107,846	276,003	413,800	843,900	1,761,600
Transportation and public utilities	142,256	229,654	330,900	638,800	1,256,900
Wholesale and retail trade	290,476	545,115	866,300	1,856,700	3,927,100
Finance, insurance, and real estate	47,158	131,629	208,700	455,900	980,500
Services	162,963	539,388	941,400	2,230,100	4,986,100
Government	334,698	1,089,227	1,790,000	4,039,700	8,649,600
Total	$1,560,382	$3,402,870	$5,261,600	$11,241,400	$23,807,600

distribution of minerals and petroleum is shown in figure 6. Innumerable sand and gravel quarries constitute a significant part of the mineral industry of the region, but they could not be shown in figure 6.

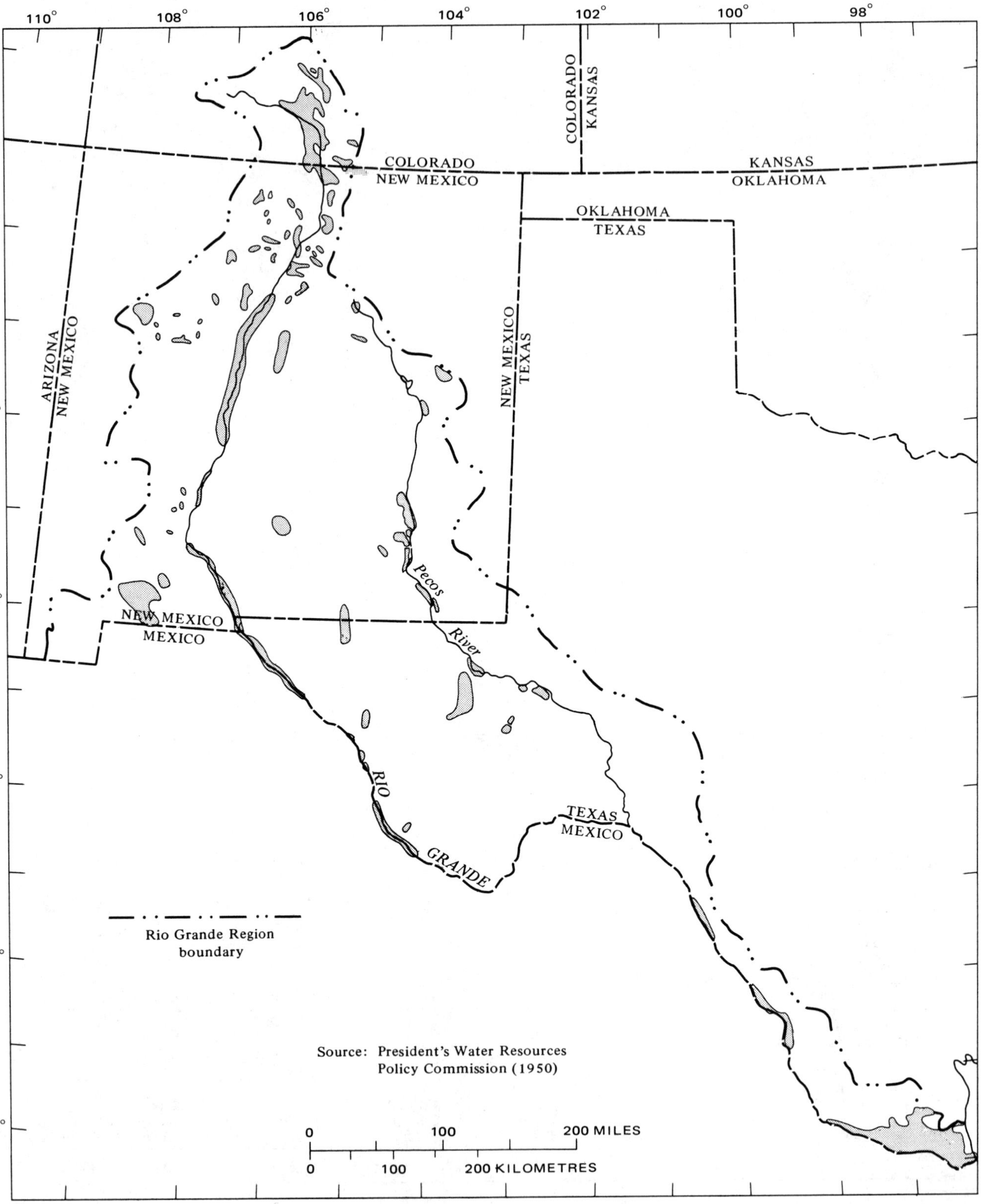

FIGURE 5. — Areas of irrigated land. Most of the irrigated land is alongside the Rio Grande and the Pecos River.

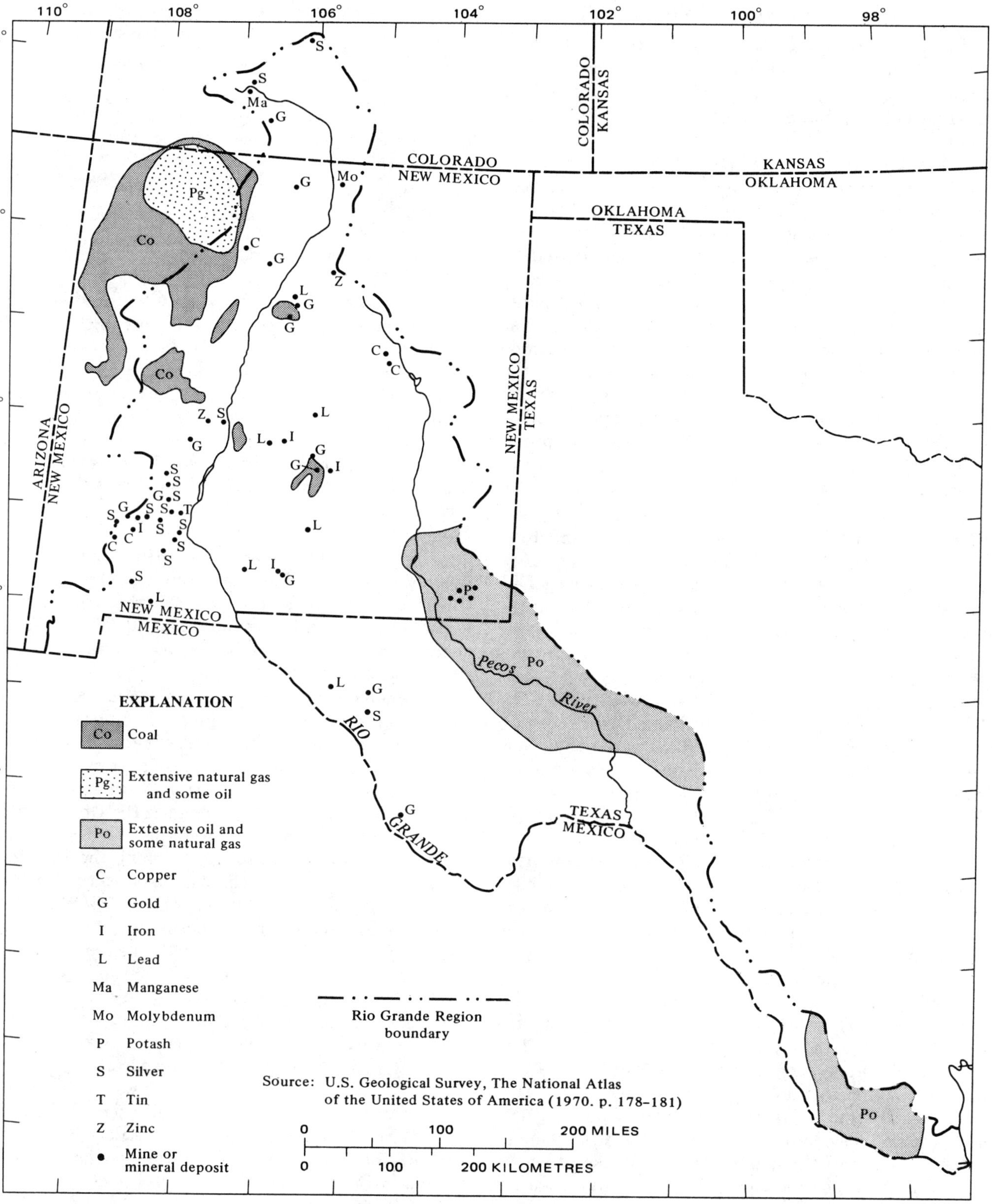

FIGURE 6. — Mineral and petroleum deposits. Extensive mineral deposits are found in the northern half of the region. Innumerable sand and gravel quarries are not shown.

THE LEGAL SETTING

Management of the surface waters of the Rio Grande Region is subject to international treaty, interstate compacts, and laws of three States. Management of ground water within each State is subject to regulation by that State.

In the late 1880's, water shortages experienced by Mexican users near Ciudad Juarez resulted in protests to the United States by Mexico, alleging that shortages resulted from increasing diversions for irrigation in New Mexico and Colorado. The International Boundary Commission, United States and Mexico, was directed to investigate the water situation in the upper Rio Grande, and as a result of that investigation, an embargo was placed on further surface-water development in the two States until the problem could be resolved. In 1906, the United States and Mexico negotiated a treaty whereby Mexico was guaranteed an annual delivery of 60,000 acre-feet (74 hm^3) of water in perpetuity, with the provision that the two Nations would share shortages in times of drought (Sorensen and Linford, 1967, p. 148).

Both the Rio Grande main stem and the Pecos River are subject to apportionment of surface water under interstate compacts. The Rio Grande Compact specifies that Colorado must deliver to New Mexico a specified quantity of water in proportion to water available at index stations in Colorado, and that New Mexico must deliver to Texas a specified quantity of water in proportion to water available at index stations in New Mexico. At times, both Colorado and New Mexico have been delinquent in water deliveries. The Pecos River Compact specifies the proportion of Pecos River water that New Mexico must deliver to Texas. Uncontrolled development of ground water in either the Rio Grande or the Pecos River valleys could make it impossible to deliver the specified quantities of surface water down the natural river channels; however, orderly development and management of the ground-water reservoirs in conjunction with management of the rivers could assure delivery of the specified quantities annually for the foreseeable future.

When the Colorado Constitution was adopted in 1876, it provided for appropriation of the water of natural streams not previously appropriated. All natural streams were declared to be the property of the public. The courts in Colorado have, in effect, defined "waters of a natural stream" by ruling that if the waters in question would reach a natural watercourse or were in a natural watercourse, they were waters of a natural stream. In cases involving water pumped from wells, some courts have adjudicated on the theory that underground waters are waters of a natural stream. Other courts have held that for adjudication purposes waters produced from wells are not waters of a natural stream (Sparks, 1970).

In 1965 the Colorado Legislature passed a new ground-water law. Section 148-18-1 of the law states:

> It is hereby declared that the traditional policy of the state of Colorado requiring the water resources of this state be devoted to beneficial use in reasonable amounts through appropriation, is affirmed with respect to the designated ground waters of this state, as said waters are hereinafter defined. While the doctrine of prior appropriation is recognized, such doctrine should be modified to permit the full economic development of designated ground-water resources. Prior appropriations of ground water should be protected and reasonable ground water pumping levels maintained, but not to include the maintenance of historical water levels. All ground waters in this state are therefore declared to be subject to appropriation in the manner herein defined.

The law also specifically prohibits the diversion of ground water outside the State for use within another State.

The State Engineer of Colorado administers the laws relative to the distribution of surface waters, including ground waters tributary thereto.

In 1969 the Colorado Legislature enacted a bill that revised the entire water code and allowed for the use of ground water in conjunction with surface water. It is now possible to change the point of diversion from a surface diversion to a well diversion.

By 1970, about 30,000 wells had been drilled in Colorado, none of which had adjudicated water rights. Under existing laws, if rights on the wells are adjudicated, they will be subordinate to all the surface decrees and generally will have a priority date later than 1950 (Sparks, 1970).

The Pueblo Indians in the Rio Grande Valley of New Mexico were operating acequias (community ditches) along perennial streams to support subsistence farming when Coronado led the first Spanish exploration through the area in 1540. From 1609 until 1680, Spanish colonists established several villages along the Rio Grande and its perennial tributaries and began the same type of subsistence farming that prevailed among the Pueblo Indians. When Mexico gained its independence from Spain in 1821, the territory including the Rio Grande Valley passed from Spanish to Mexican rule, but the water rights of both the Indians and the colonists were recognized by Mexico. The territory was ceded to the United States in 1848 by the Treaty of Guadalupe Hidalgo, and again the water rights of residents were recognized by the new government.

The Rio Grande Valley was still sparsely settled when ceded to the United States, so there was little competition for water, and the local acequias served the people well. With the influx of American settlers in the latter part of the 19th century, the competition for water became serious, so the Territorial Legislature of New Mexico passed the first legislation to regulate the appropriation of surface water in 1905, and the legislation was rewritten in 1907. Also in 1905, the legislature adopted an act to regulate the use of artesian wells and to prevent the waste of subterranean flows of water (Mechem, 1961).

When New Mexico became a State in 1912, the constitution contained the following sections pertaining to water rights:

1. All existing rights to the use of any waters in this state for any useful or beneficial purpose are hereby recognized and confirmed.

2. The unappropriated water of every natural stream, perennial or torrential, within the state of New Mexico, is hereby declared to belong to the public and to be subject to appropriation for beneficial use, in accordance with the laws of the state. Priority of appropriation shall give the better right.

3. Beneficial use shall be the basis, the measure and the limit of the right to the use of water.

The first ground-water statute was passed by the State in 1927. It provided that

All waters in the State found in underground streams, channels, artesian basins, reservoirs, or lakes, the boundaries of which may be reasonably ascertained by scientific investigations of surface indications, are hereby declared to be public waters and to belong to the public, and subject to appropriation for beneficial uses under the existing laws of this state relating to appropriation and beneficial use of waters from surface streams.

This law was declared invalid by the New Mexico Supreme Court, because it attempted to extend existing legislation by reference (Mechem, 1961). In 1931 the statute was reenacted in a form the court outlined in its opinion on the 1927 act. The 1931 statute was not challenged in the courts until 1949, and in 1950 the New Mexico Supreme Court upheld the statute (Mechem, 1961). It is still the law governing ground-water appropriation and use in New Mexico.

Under the New Mexico law, the State Engineer, after adequate evaluation, may declare an "underground water basin" and control further development of ground water in order to protect prior water rights. The State Engineer has declared 14 such basins in the Rio Grande Region, the largest of which extends on each side of the Rio Grande from Elephant Butte Reservoir to the Colorado–New Mexico line. The State Engineer defines and declares such basins whenever it becomes apparent that regulation is necessary to (1) prevent impairment of existing rights, (2) insure beneficial use of water, and (3) provide for an orderly development of ground-water reservoirs. The New Mexico Supreme Court has found that it is unreasonable to attempt a legal distinction between ground water and surface water, because of their interrelationship.

Texas does not have provisions for statewide administration and control of ground water in its legal code. The courts, in implementing statutes on the books, have in general applied the principle of the law of capture (riparian rights) in ground-water disputes (Dixon, 1961). In 1949 the Texas Legislature enacted the "Underground Water District Act" (Art. 7880-3c), which authorized water districts to take action to promulgate rules regarding the conservation and use of percolating ground water. However, the districts were not authorized to regulate the use of underground water in defined channels or the underflow of rivers, both of which are subject to appropriation the same as surface water. Districts can be created by the legislature, the Texas Water Rights Commission, and county commissioners' courts. Prior to creation of an underground-water conservation district, the Water Rights Commission must determine through studies conducted by the Texas Water Development Board that an underground-water reservoir, or subdivision thereof, having definable boundaries and meeting other predetermined requirements actually exists. The district is a corporate unit which can own property and act in all ways as an entity having financial and legal responsibilities. The initial legislation was amended in 1955 to strengthen the power of the districts with regard to well spacing, regulation of production, and prevention of waste (Dixon, 1961). This legislation was amended again in 1972.

THE GROUND-WATER SUPPLY

The total water supply of the Rio Grande Region is the sum of the surface-water runoff, the ground-water outflow (which is small), and the natural evapotranspiration. Annual precipitation averages about 12 inches (30 cm) on the 86 million acres (35 million ha) within the region; however, all but 4 million acre-feet (4,900 hm^3) of the precipitation returns to the atmosphere through natural evapotranspiration. The runoff, or water yield, of the region and part of the natural evapotranspiration represent the water that can be controlled or modified to a degree by man. For example, man may be able to intercept for economic use (agricultural, municipal, and industrial supplies) water that normally is lost to noneconomic consumption by native vegetation. Also, he may import more water or intercept for beneficial use the water that runs off to the Gulf of Mexico. A small amount of surface water is imported into the region by transmountain diversion.

The ground-water reservoirs of the region contain an aggregate of more than 5,800 million acre-feet (7,200,000 hm^3) of fresh (less than 1,000 mg/l dissolved solids) to slightly saline (1,000-3,000 mg/l dissolved solids) water in storage, which under the laws of nature is available to wells. In contrast, the surface reservoirs of the region have a combined storage capacity of only 18 million acre-feet (22,000 hm^3).

The ground-water reservoirs of the region can be classified on the basis of rock characteristics, physiography, and geographic distribution. For convenience of description, the rocks have been divided into four basic types: (1) valley fill — unconsolidated to poorly consolidated sand and gravel interbedded or in-

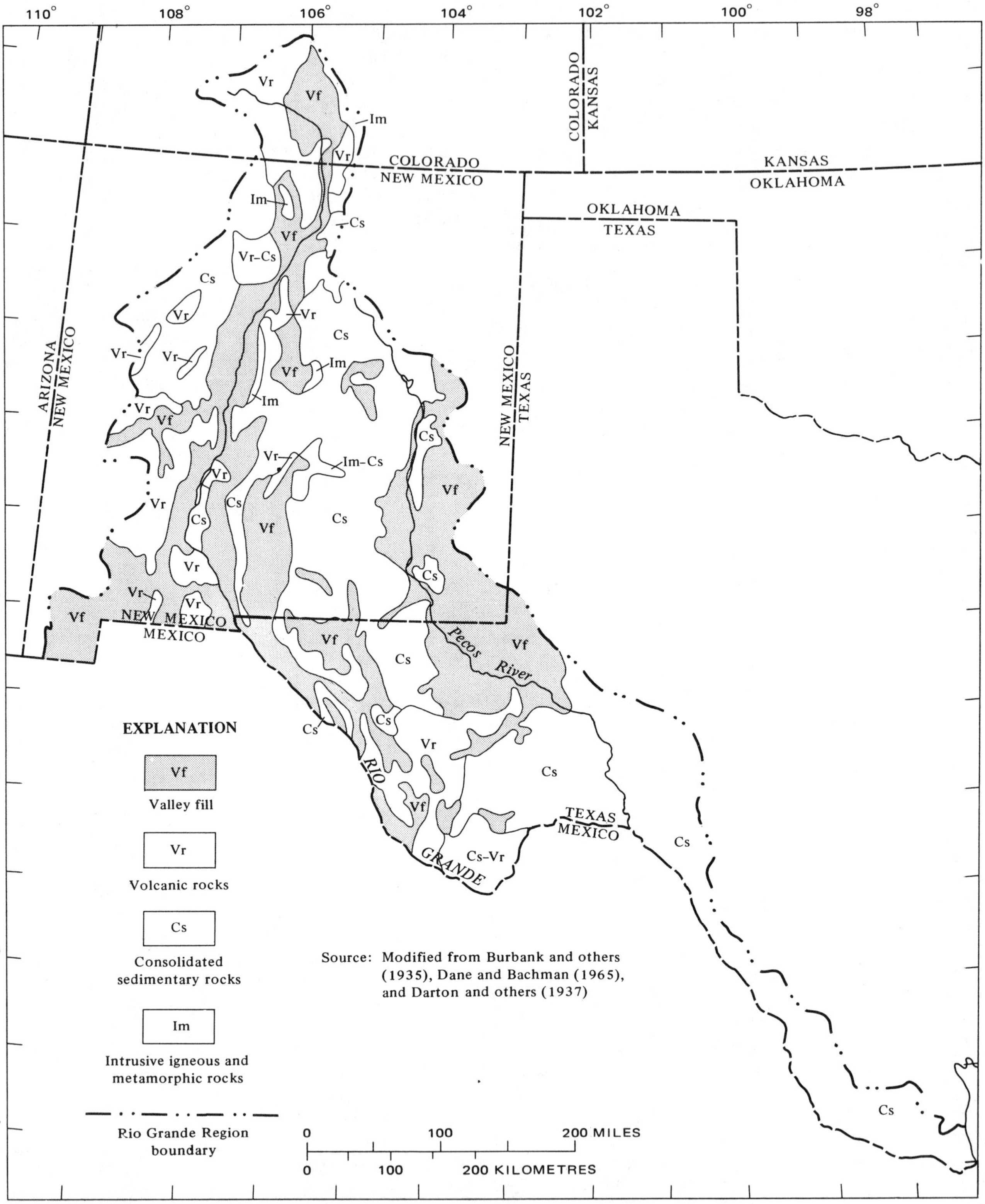

FIGURE 7. — Generalized distribution of geologic units. Valley fill comprises the principal ground-water reservoirs.

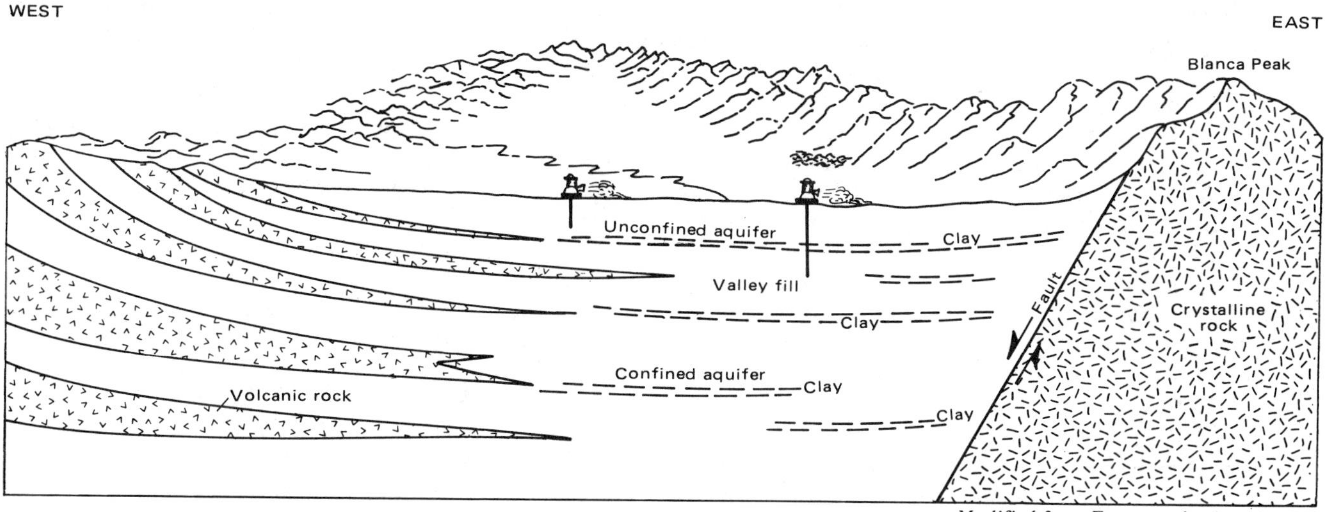

FIGURE 8. — The San Luis Valley. The valley is bounded by high mountains and is underlain by valley fill and volcanic rocks; the water table is within 12 feet (3.6 m) of land surface in most areas.

termixed with clay and silt; (2) volcanic rocks — primarily basalt but including other flow rocks, tuff, and small intrusive bodies; (3) consolidated sedimentary rocks — primarily shale and sandstone but including limestone, gypsum, and salt; and (4) crystalline rocks — intrusive igneous rocks and metamorphic rocks.

The intermontane valleys of the region have been partly filled with unconsolidated sand, gravel, and clay — termed "valley fill" — derived by weathering and erosion from the rocks in the adjacent mountains and from similar rocks in the mountains upstream (fig. 7). Alluvial deposits in the Pecos Valley are included in the valley-fill category, but the valley fill shown east of the Pecos River (fig. 7) is thin and is not a significant ground-water reservoir.

Locally, the valley fill is interbedded with basalt and other flow rocks derived from volcanic centers in and bordering the valleys. Large masses of basalt and andesite are common, especially in the northern part of the basin. The valley fill, including the interbedded volcanic rocks, is as much as 9,000 feet (2,700 m) thick in New Mexico and is reported to be more than 30,000 feet (9,100 m) thick in the north-central part of the San Luis Valley, Colo. (Gaca and Karig, 1966). The valley fill comprises the principal ground-water reservoir in the region. In most areas it is capable of yielding a few hundred to a few thousand gallons of water per minute to individual wells.

The mountains consist of volcanic rocks, consolidated sedimentary rocks, and crystalline rocks (fig. 7). The volcanic rocks commonly cap plateaus and low mountain ranges and lie above the regional water table. These rocks generally are not significant aquifers except where they are interbedded with or overlie valley fill. The consolidated sedimentary rocks form most of the hills and low mountains. Generally,these rocks are poor aquifers, but locally, beds of limestone containing extensive fractures and solution channels lie below the water table and yield large supplies of water to wells. In many areas the consolidated sedimentary rocks contain soluble minerals, such as halite and gypsum — the principal sources of dissolved solids in ground water. The crystalline rocks generally are dense and yield insignificant quantities of water to wells.

SAN LUIS VALLEY SUBREGION

The valley-fill ground-water reservoir in the San Luis Valley contains both unconfined and confined aquifers, which are separated in places by a clay series or by layers of volcanic rocks (fig. 8). These confining beds are discontinuous and lenticular, so it is difficult to differentiate between unconfined and confined aquifers except locally. This discontinuity in the clay series permits varying degrees of hydraulic connection between the aquifers; therefore, all the aquifers in the valley north of the San Luis Hills should be considered a single ground-water reservoir (Emery and others, 1971).

Recharge to the unconfined aquifer is mainly by infiltration of irrigation water from canals, ditches, and fields and by upward leakage from the confined aquifer. Some water percolates from the many streams flanking the valley, but very little precipitation on the valley floor recharges the unconfined aquifer. Natural discharge from this aquifer is by evapotranspiration and seepage to streams.

The principal source of recharge to the confined aquifer is seepage from mountain streams that flow across the alluvial fans flanking the valley floor. At the edge of the valley the clay series is absent, permitting recharge to beds that constitute the confined aquifer in the main part of the valley. The mountain streams show significant losses as they cross the porous surface of the fans. The confined aquifer underlies most of the valley, and the water has sufficient head to flow at the land surface. The natural discharge from the confined aquifer is by springs and by upward leakage through the confining beds into the unconfined aquifer. A small amount discharges as underflow into New Mexico.

The ground-water reservoir in the San Luis Valley underlies about 2 million acres (809,000 ha). It ranges in thickness from a featheredge around the rim of the valley to as much as 7,000-9,000 feet (2,100-2,700 m) beneath much of the area and to a reported thickness of more than 30,000 feet (9,100 m) locally (Gaca and Karig, 1966). If we assume the average thickness of the permeable deposits to be 5,000 feet (1,500 m), the volume of reservoir material is 2 million × 5,000, or about 10 billion acre-feet (12 million hm^3), and if the average specific yield is 0.2 (20 percent), the amount of water in storage in the reservoir available to wells is about 2 billion acre-feet (2.5 million hm^3).

In 1967 about 2,800 wells in the San Luis Valley yielded more than 300 gpm (1,635 m^3/d) each. Of this total, 2,160 were completed in the unconfined aquifer. In addition to the large-capacity wells, there are more than 7,000 small-capacity flowing wells. The annual water income to the San Luis Valley averages about 2.5 million acre-feet (3,100 hm^3); about 1.5 million acre-feet (1,800 hm^3) is streamflow derived chiefly from snowmelt in the surrounding mountains, and about 1 million acre-feet (1,200 hm^3) is from precipitation on the valley floor. Annual discharge of water from the valley also averages 2.5 million acre-feet (3,100 hm^3) — about 2 million acre-feet (2,500 hm^3) by evapotranspiration and about 500,000 acre-feet (620 hm^3) as flow across the State line (table 2). The streamflow at the State line averages 445,000 acre-feet (550 hm^3), and ground-water underflow is 55,000 acre-feet (68 hm^3). About half the evapotranspiration is noneconomic; that is, it does not contribute to the growth of plants having economic or commercial value. Much of the noneconomic consumption is by phreatophytes in areas where the depth to water is less than 12 feet (3.6 m).

According to Powell (1958), the quality of water in the confined aquifer is generally better than that in the unconfined aquifer. The concentration of dissolved solids in 41 samples from the confined aquifer ranged from 70 to 437 mg/l, and the concentration in 271 samples from the unconfined aquifer ranged from 52 to 13,800 mg/l. The least mineralized water in the unconfined aquifer occurs on the west side of the valley. The mineral concentration increases toward the sump area of the closed basin, probably because of solution from the rocks and concentration by evapotranspiration in areas having a shallow water table.

ALBUQUERQUE SUBREGION

The valley fill in the Rio Grande depression, which underlies about 5,000 square miles (13,000 km^2), or 3,200,000 acres (1,300,000 ha), comprises the principal ground-water reservoir in the Albuquerque subregion (fig. 9). The thickness of the fill is not well known, but it may average about 4,000 feet (1,200 m). The maximum thickness is probably about 9,000 feet (2,700 m) (Dinwiddie, 1967). The estimated volume of recoverable fresh ground water in storage in this and other ground-water reservoirs is 2,300 million acre-feet (2,800,000 hm^3); an additional 540 million acre-feet (670,000 hm^3) of recoverable slightly saline water is in storage, making a total of 2,800 million acre-feet (3,400,000 hm^3) of fresh

TABLE 2. — *The water budget*

Subregion	Income		Outgo			Storage capacity	
				Evapotranspiration			
	Surface-water inflow (ac-ft per yr)	Surface-water yield (ac-ft per yr)	Surface-water outflow[1] (ac-ft per yr)	Man's activities (ac-ft per yr)	Wetlands and phreatophytes (ac-ft per yr)	Surface water (ac-ft)	Ground water[2] (ac-ft)
San Luis Valley	0	1,500,000	500,000	1,000,000	1,000,000	400,000	2,000,000,000
Albuquerque	500,000	1,500,000	[3]990,000	300,000	400,000	230,000	2,800,000,000
El Paso	[4]730,000	200,000	260,000	500,000	500,000	2,500,000	230,000,000
Lower Rio Grande	[5]620,000	700,000	0	1,000,000	200,000	14,000,000	20,000,000
Pecos Valley	0	1,700,000	360,000	800,000	280,000	650,000	410,000,000
Closed Basins	0	[6]0	0	130,000	100,000	0	290,000,000
Total (rounded)		5,600,000		3,700,000	2,500,000	18,000,000	5,800,000,000

[1]Includes some ground-water underflow.
[2]Recoverable fresh and slightly saline ground water in storage.
[3]Surface flow at San Marcial above Elephant Butte Reservoir.
[4]Surface flow below Elephant Butte Reservoir.
[5]The sum of surface-water outflow from both the El Paso and the Pecos Valley subregions.
[6]Some runoff from mountainous areas, which infiltrates the ground or evaporates in playas.

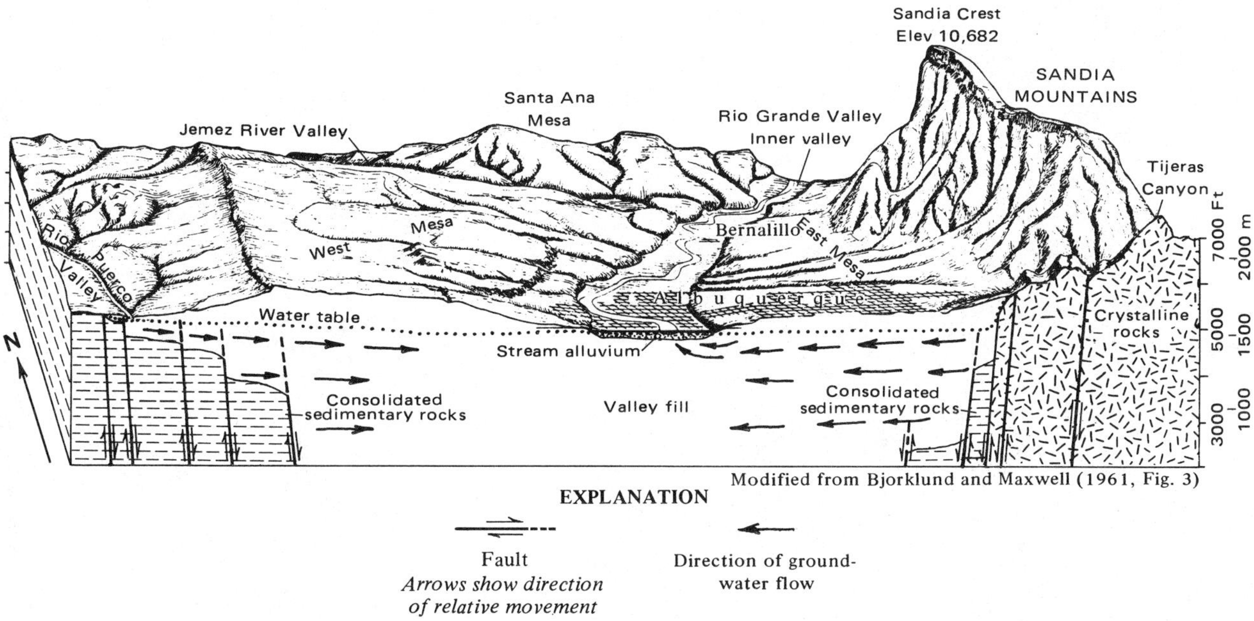

Modified from Bjorklund and Maxwell (1961, Fig. 3)

FIGURE 9. — The Rio Grande Valley at Albuquerque, N. Mex. High mountains bound the valley on the east; agricultural lands and the older part of the city are in the inner valley, where the water table and the river level coincide.

and slightly saline water in storage (based on a specific yield ranging from 5 to 15 percent) (table 2).

The yield of 83 large-discharge wells that tap the valley fill in the Albuquerque area ranges from 240 to 2,000 gpm (1,300 to 11,000 m^3/d) and averaged 860 gpm (4,700 m^3/d) (Bjorklund and Maxwell, 1961). Volcanic rocks in the northern part of the subregion locally yield water to wells. Consolidated sedimentary rocks are not significant aquifers, except in the Rio San Jose valley near Grants, N. Mex., where yields of more than 2,000 gpm (1,300 m^3/d) per well have been obtained from limestone (Gordon, 1962).

Recharge in the subregion is from precipitation on the valley fill and associated volcanic rocks, from infiltration of surface water diverted for irrigation, and from intermittent runoff during intense rainstorms. In the northern part of the subregion, some recharge is from runoff of snowmelt.

The chemical quality of ground water varies widely in this subregion. A study of municipal water supplies in New Mexico (Dinwiddie and others, 1966a, b) revealed that the concentration of dissolved solids in municipal or community water supplies in this subregion ranges from 125 to 2,620 mg/l and that the water supply for 23 communities exceeds the concentration of 500 mg/l recommended for public water supplies by the U.S. Public Health Service (1962). The quality of municipal water supplies is probably an indication of ground-water quality in general, although even higher concentrations can be expected locally, because public supplies are generally drawn from the best water available. The water in the valley fill of the Rio Grande Valley is better in quality than in most of the tributary valleys along the west side of the subregion.

EL PASO SUBREGION

A bedrock high in the vicinity of Elephant Butte Reservoir tends to separate the ground-water reservoirs in the Albuquerque subregion from those in the El Paso subregion. Some ground water is forced to the land surface where the valley fill becomes narrower and thinner in this area. A section across the Rio Grande Valley and adjacent areas near Elephant Butte Reservoir is shown in figure 10.

As in the San Luis Valley and the Albuquerque subregions, the major ground-water reservoir in the El Paso subregion is valley fill. The thickness of the fill and interbedded volcanic rocks in the Las Cruces area is locally more than 5,000 feet (1,500 m) and probably averages at least 3,000 feet (910 m) (fig. 11). The valley fill narrows in the vicinity of El Paso but widens again into the Hueco bolson east of the Franklin Mountains and in broad band alongside the Rio Grande down to Fort Quitman. The fresh and slightly saline ground water stored in these reservoirs is about 230 million acre-feet (280,000 hm^3). (table 2).

Individual wells in the Las Cruces–El Paso area yield as much as 3,000 gpm (16,400 m^3/d) and average about 1,000 gpm (5,400 m^3/d). Farther down the valley the

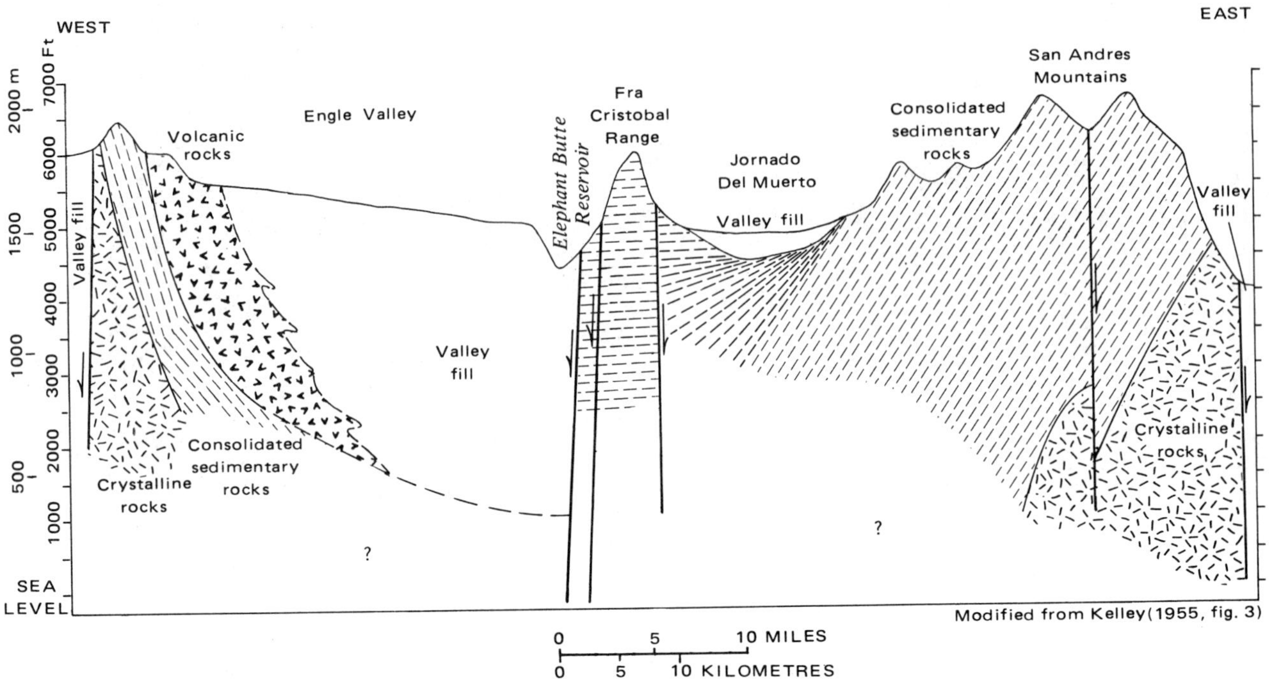

FIGURE 10. — The ground-water reservoir in valley fill at Elephant Butte Reservoir in New Mexico. The ground-water reservoir is thinner and narrower here than at most places in New Mexico.

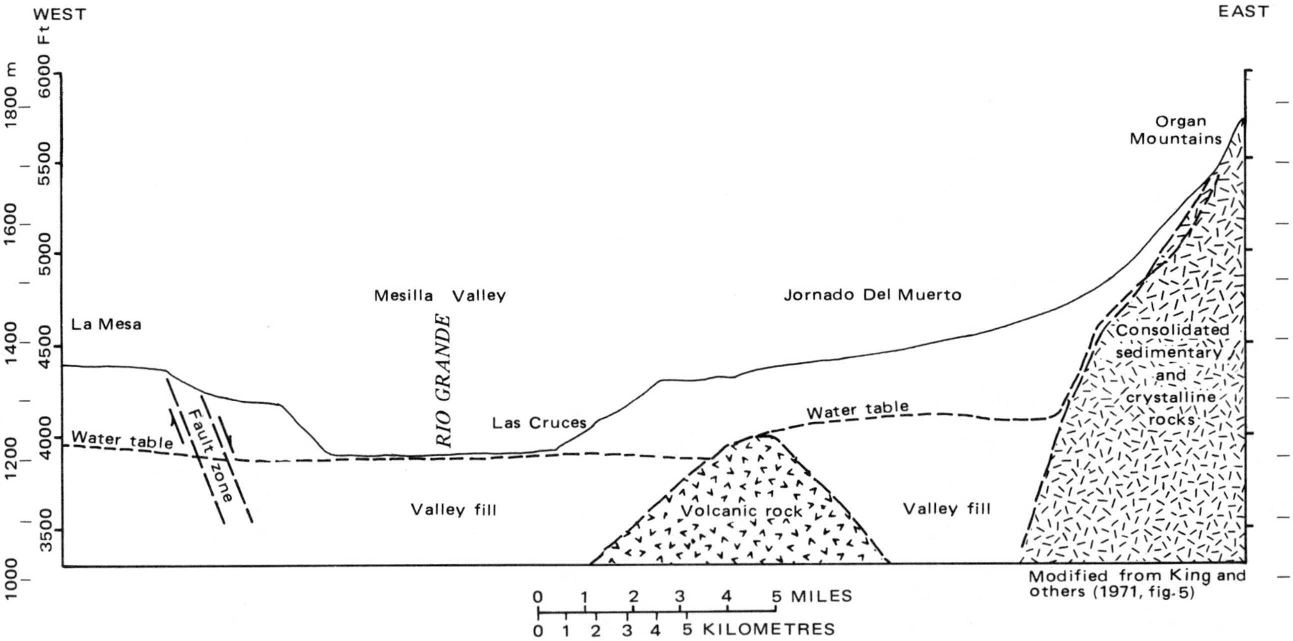

FIGURE 11. — The ground-water reservoir in valley fill at Las Cruces, N. Mex. A buried ridge of volcanic rock divides the broad ground-water reservoir.

yields are much smaller, ranging from about 100 to 500 gpm (54 to 2,700 m^3/d) (Davis and Leggat, 1965).

Recharge is relatively small, because the subregion is in the minimum rainfall belt. Infiltration from surface water diverted for irrigation locally provides significant recharge.

The chemical quality of water in the valley-fill aquifers in the El Paso subregion varies widely, both laterally and vertically. The water in the shallow alluvial deposits along the river generally contains higher concentrations of dissolved solids than the water in the underlying older fill, but the salinity may increase at depths of a few thousand feet in the older fill. The dissolved solids in water in the river alluvium have been concentrated by evapotranspiration in areas where the water table is shallow and by return of irrigation water, which contains dissolved soil salts and fertilizers. The concentration of dissolved solids in ground water of the Las Cruces–El Paso area ranges from about 200 to more than 6,000 mg/l (Davis and Leggat, 1965).

PECOS VALLEY SUBREGION

Consolidated sedimentary rocks, which consist of shale, sandstone, limestone, gypsum, and salt, predominate in the Pecos Valley subregion. Beds of limestone and, locally, gypsum are excellent aquifers in some areas. However, extensive deposits of gypsum and salt contribute large quantities of dissolved solids to the water in the subregion. The valley fill constitutes productive aquifers where it is thickest, such as in the Roswell basin (fig. 12) and in several areas between the Texas–New Mexico line and Girvin, Tex.

The quantity of fresh and slightly saline ground water in storage in the subregion is about 410 million acre-feet (510,000 hm^3) (table 2). Most of the water is stored in valley-fill and sandstone reservoirs; but in the Roswell and Carlsbad areas, some 10 million acre-feet (12,000 hm^3) is stored in limestone and gypsum reservoirs, and an equal amount is stored in limestone and sandstone in the Texas part of the subregion.

Wells completed in either the limestone or the valley fill in the Roswell basin generally yield more than 300 gpm (1,600 m^3/d) each, and yields of 1,000-3,500 gpm (5,400-19,000 m^3/d) are common. The limetone aquifer has outcrops west of the Pecos River near Roswell, N. Mex., that accept recharge readily from direct precipitation and from surface flow. The overlying valley-fill aquifer is recharged by upward leakage of artesian water from the limestone aquifer and by infiltration of a part of the water used for irrigation.

The chemical quality of ground water varies widely in the subregion, ranging from fresh to briny (less than 1,000 to more than 35,000 mg/l). All or part of the water supply for 16 communities in the New Mexico part of the subregion contains more than 500 mg/l dissolved solids. The concentration of dissolved solids in municipal supplies ranges from 150 mg/l in the northern part to 2,410 mg/l for one community in Chaves County. The chemical quality of public water supplies in the Texas part of the subregion is comparable.

In some areas, as fresh ground water is withdrawn it is replaced by more saline water, which mixes with the fresh water in storage and causes a general deterioration in water quality.

CLOSED BASINS SUBREGION

The closed basins subregion comprises the Estancia, the Jornada del Muerto, and the Tularosa basins in New Mexico and the Salt basin in New Mexico and Texas. All these basins have internal surface drainage, but some water may move underground to adjacent basins or subregions.

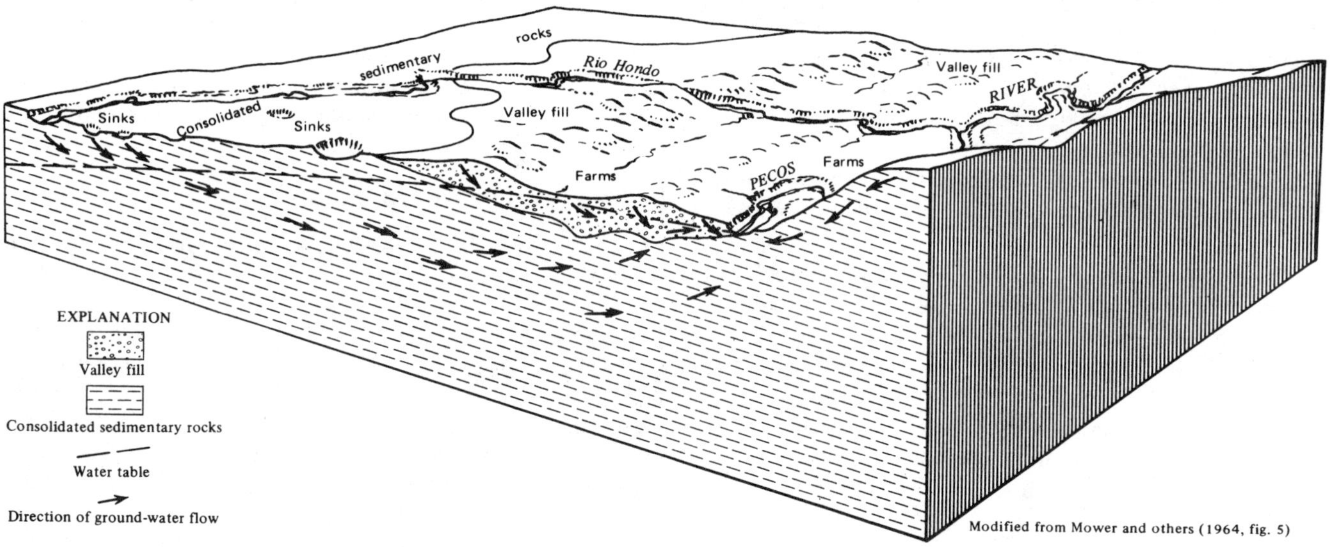

FIGURE 12. — The complex relationship between the ground-water flow system and the surface water in the Pecos Valley at Roswell, N. Mex.

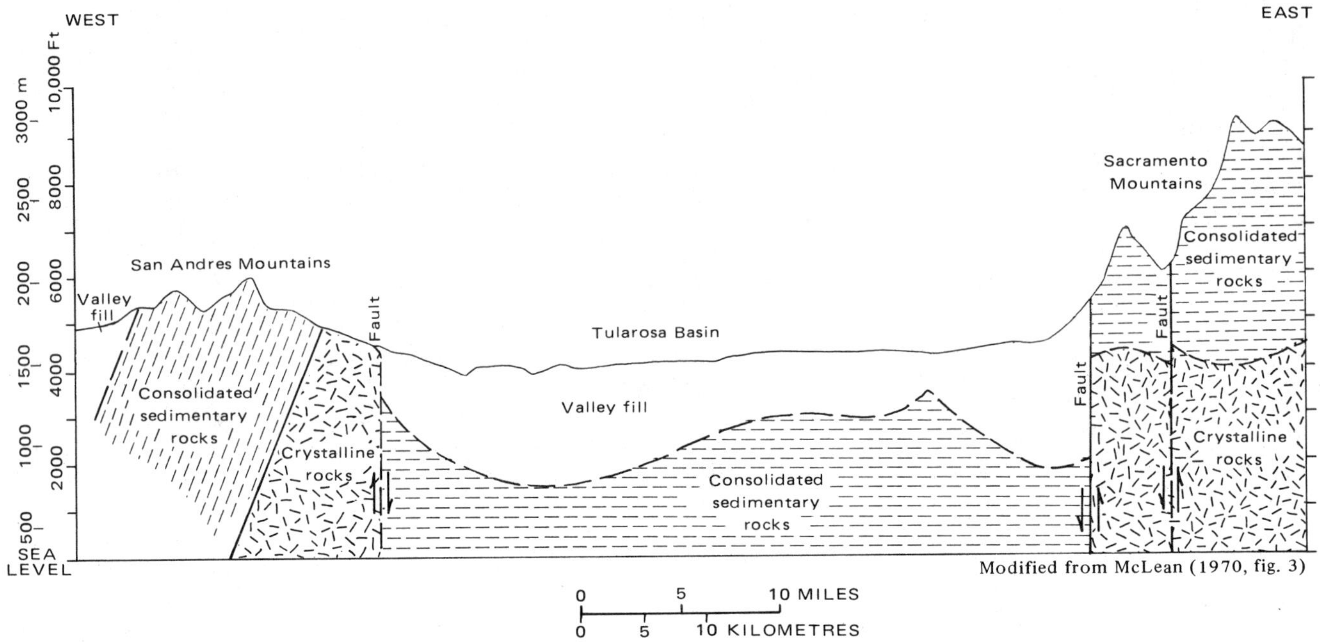

FIGURE 13. — The ground-water reservoir in valley fill of the Tularosa basin in New Mexico. The reservoir is large, but fresh water is found only in narrow bands adjacent to the mountains.

Valley fill predominates in the closed basins and constitutes the principal ground-water reservoirs. The fill consists of both alluvial deposits at the foot of the mountains and lake deposits in the central part of each basin, except in the Jornada del Muerto basin. The lake deposits consist of silt, clay, and evaporites, especially gypsum. These evaporites are a common source of high salinity in ground water in the central parts of the basins. The valley fill commonly is thin around the margins but may be a few thousand feet thick in the deeper parts of the basins. A generalized section across the Tularosa basin is shown in figure 13. Yields as large as 2,400 gpm (13,000 m^3/d) have been obtained from individual wells in the fill, and yields of several hundred gallons per minute are common.

Limestone constitutes an important aquifer in the northern part of the Salt basin, where it is a source of water for irrigation.

The volume of fresh and slightly saline ground water stored in the closed basins is summarized below:

Basin	Volume (ac-ft)
Estancia	10,000,000
Jornada del Muerto	110,000,000
Tularosa	150,000,000
Salt	20,000,000
Total	290,000,000

In addition to the fresh and slightly saline water in storage, large volumes of more saline water are stored in each of the basins. McLean (1970) estimated that about 60 million acre-feet (74,000 hm^3) of water with a dissolved-solids content of more than 3,000 mg/l is in storage in the Tularosa basin. Similar estimates for the other basins have not been made. The fresh water (less than 1,000 mg/l) is limited to narrow bands in the valley fill next to mountain ranges of most basins. The concentration of dissolved solids in ground water ranges from about 200 mg/l to more than 100,000 mg/l.

A study of public water supplies in southeastern New Mexico by Dinwiddie (1963) showed that the supplies for 16 communities in the New Mexico part of the subregion contain concentrations of dissolved solids higher than 500 mg/l. Public supplies have not been developed in the Jornada del Muerto.

LOWER RIO GRANDE SUBREGION

The ground-water reservoirs in the lower Rio Grande subregion are insignificant in comparison with those in the other subregions. Most of the aquifers are consolidated sedimentary rocks consisting of sandstone and limestone (Brown and others, 1965). In the northernmost and the southernmost parts of the subregion, stream alluvium along the Rio Grande is the principal aquifer (Baker, 1965). The estimated quantity of fresh and slightly saline water in the subregion is only 20 million acre-feet (25,000 hm^3)(table 2).

The yields of wells in this subregion are highly variable depending on the type of aquifer. Yields range from a few tens of gallons per minute to as much as 3,000 gpm

(16,000 m^3/d). Beds of limestone are the most productive.

The water generally contains from a few hundred to a few thousand milligrams per liter dissolved solids.

WITHDRAWAL AND CONSUMPTION OF GROUND WATER

Withdrawal of ground water in the Rio Grande Region in 1970 was 2,700,000 acre-feet (3,300 hm^3), in comparison to a withdrawal of 4,300,000 acre-feet (5,300 hm^3) of surface water (fig. 14). The largest part of the ground water (88 percent) was used for irrigation (fig. 15). About 180,000 acre-feet (220 hm^3) (5 percent) of the ground water withdrawn was used for public supplies (ground water furnished 58 percent of all the public supply requirements). The largest metropolitan areas (Albuquerque and El Paso) depend entirely on ground water as a source of public supplies. The average per capita use of public supplies in the region is 228 gallons per day, or 0.86 m^3/d (Murray and Reeves, 1972). Ground water withdrawn for all uses other than irrigation and public supplies was only 7 percent (fig. 15). Of the total water withdrawn from all sources for all uses, 53 percent was consumed and 47 percent returned to the streams or ground-water reservoirs (fig. 16).

About 680,000 acre-feet (840 hm^3) of ground water was withdrawn for use in the San Luis Valley in 1970, mostly for irrigation (fig. 14). Of this amount, about half (340,000 acre-feet, or 420 hm^3) was consumed, and half was returned to the ground-water reservoir. In comparison, about 1 million acre-feet (1,200 hm^3) of ground water was lost by noneconomic evapotranspiration from wetlands and phreatophyte areas (Emery and others, 1971) (table 2).

Withdrawal of ground water in the San Luis Valley has not significantly affected the quantity of water in storage. In fact, the amount of ground water in storage has been greatly increased, owing to extensive use of surface water for irrigation.

The amount of ground water withdrawn in the Albuquerque subregion in 1970 was 160,000 acre-feet (200 hm^3), but only 67,000 acre-feet (83 hm^3) of the water withdrawn was consumed (figs. 14 and 16). The remainder (93,000 acre-feet, or 110 hm^3) returned to the river or the underground reservoir. About 400,000 acre-feet (490 hm^3) of water per year is lost to noneconomic evapotranspiration in this region (table 2). Overall changes in ground-water storage have been insignificant. In some areas the quantity of ground water in storage has increased, owing to infiltration of surface water diverted for irrigation, and in other areas the quantity in storage has decreased, owing to extensive withdrawal of ground water.

The amount of ground water withdrawn in the El Paso subregion in 1970 was 390,000 acre-feet (480 hm^3) (fig. 14), of which about one-fourth was for municipal, industrial, and military uses and the remainder was for irrigation. Pumpage exceeds recharge in the Mesa and artesian well fields, and water levels have declined more than 60 feet (18 m) (fig. 17). However, in both the upper and the lower valleys, where most of the pumping is for irrigation and the amount of ground water pumped annually varies inversely with the amount of surface water available, water levels have fluctuated a few feet but show no overall decline.

The amount of ground water withdrawn for use in the Pecos Valley subregion in 1970 was about 480,000 acre-feet (590 hm^3) in New Mexico and 680,000 acre-feet (840 hm^3) in Texas (fig. 14). About 55 percent, or 638,000 acre-feet (790 hm^3), was consumed, and the remainder was returned to the ground-water reservoir.

The estimated loss of water by noneconomic evapotranspiration along the main stem of the Pecos River in New Mexico is 185,000 acre-feet (230 hm^3) per year (Sorenson and Borton, 1967a), and the loss along the main stem in Texas is about 90,000 acre-feet (110 hm^3) per year (table 2). An additional 5,000 acre feet is lost from tributaries and closed depressions.

Ground water has been extensively "mined" in several areas, especially near Roswell and Carlsbad, N. Mex., and in Pecos and Reeves Counties, Tex. (fig. 17). The rate of "mining" in the Roswell basin, is estimated to be 120,000 acre-feet (150 hm^3) annually (Sorensen and Borton, 1967a). Water levels have declined as much as 225 feet (68 m) in the artesian aquifer in the Roswell basin and more than 300 feet (91 m) in Reeves and Pecos Counties, Tex., since ground-water development began.

The amount of ground water withdrawn for use in the closed basins in 1970 was 200,000 acre-feet (250 hm^3) (fig. 14). More than 50 percent of this was consumed, and the remainder was returned to the ground-water reservoir. Most of the water was used for irrigation, but about 9,000 acre-feet (11 hm^3) was for public supplies, including military supplies.

Evapotranspiration losses in the New Mexico part of the closed basins subregion is only 50,000 acre-feet (62 hm^3) per year — mainly as direct evaporation from playas in the Estancia basin (Sorensen and Borton, 1967b). An equivalent amount probably is evaporated from playas in the Texas part of Salt basin.

The amount of ground water withdrawn in the lower Rio Grande subregion in 1970 was 85,000 acre-feet (100 hm^3) (fig. 14), mostly for supplemental irrigation in the coastal area. Data are not available to show percentage consumed or water-level changes effected.

Table 2 summarizes, by subregions, the water supply, the water consumed by man's activities and by evapotranspiration from wetlands and phreatophyte

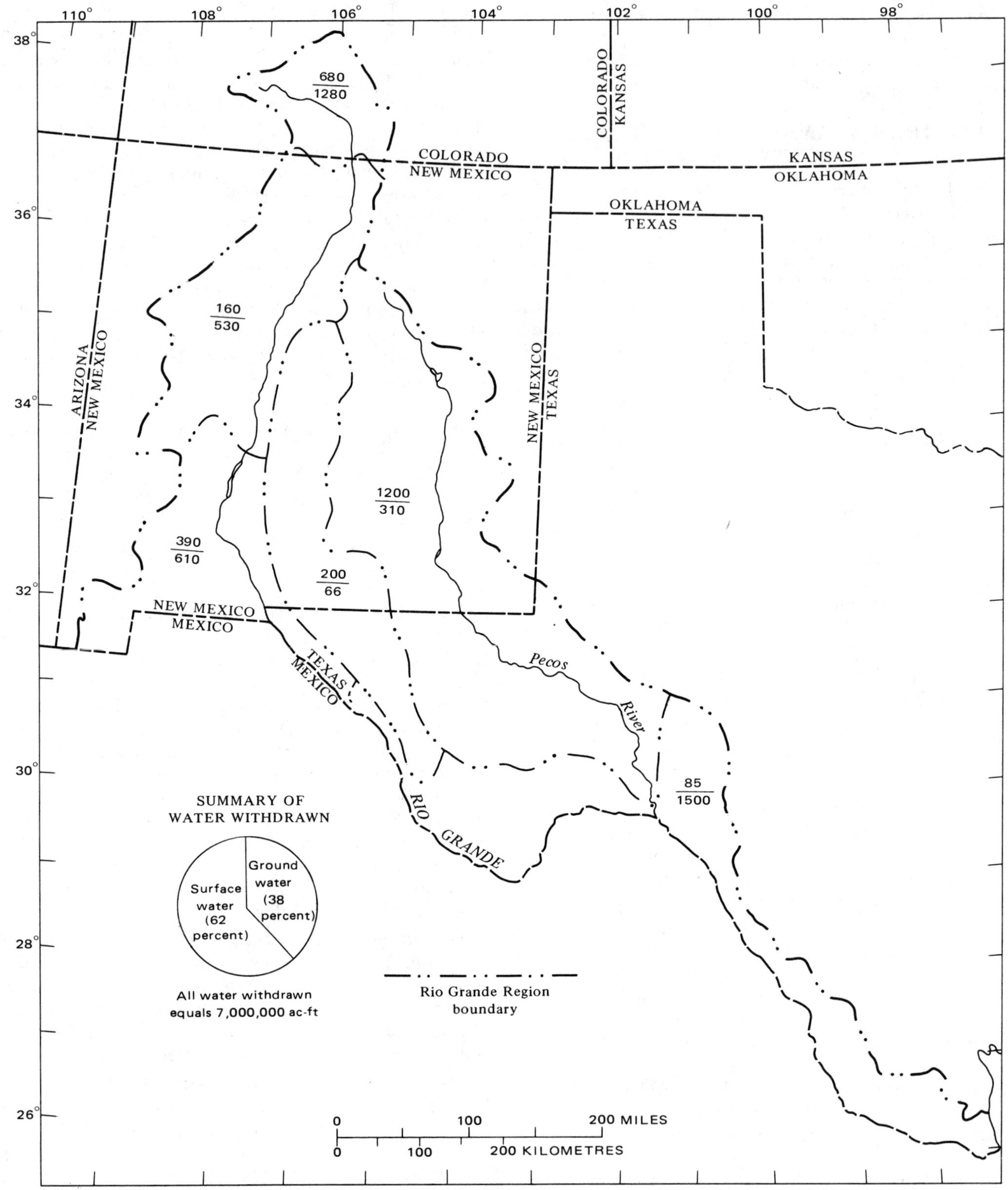

FIGURE 14. — Ratio between ground water withdrawn (upper number) and surface water withdrawn (lower number). The ratio varies widely from one subregion to another. Units are thousands of acre-feet.

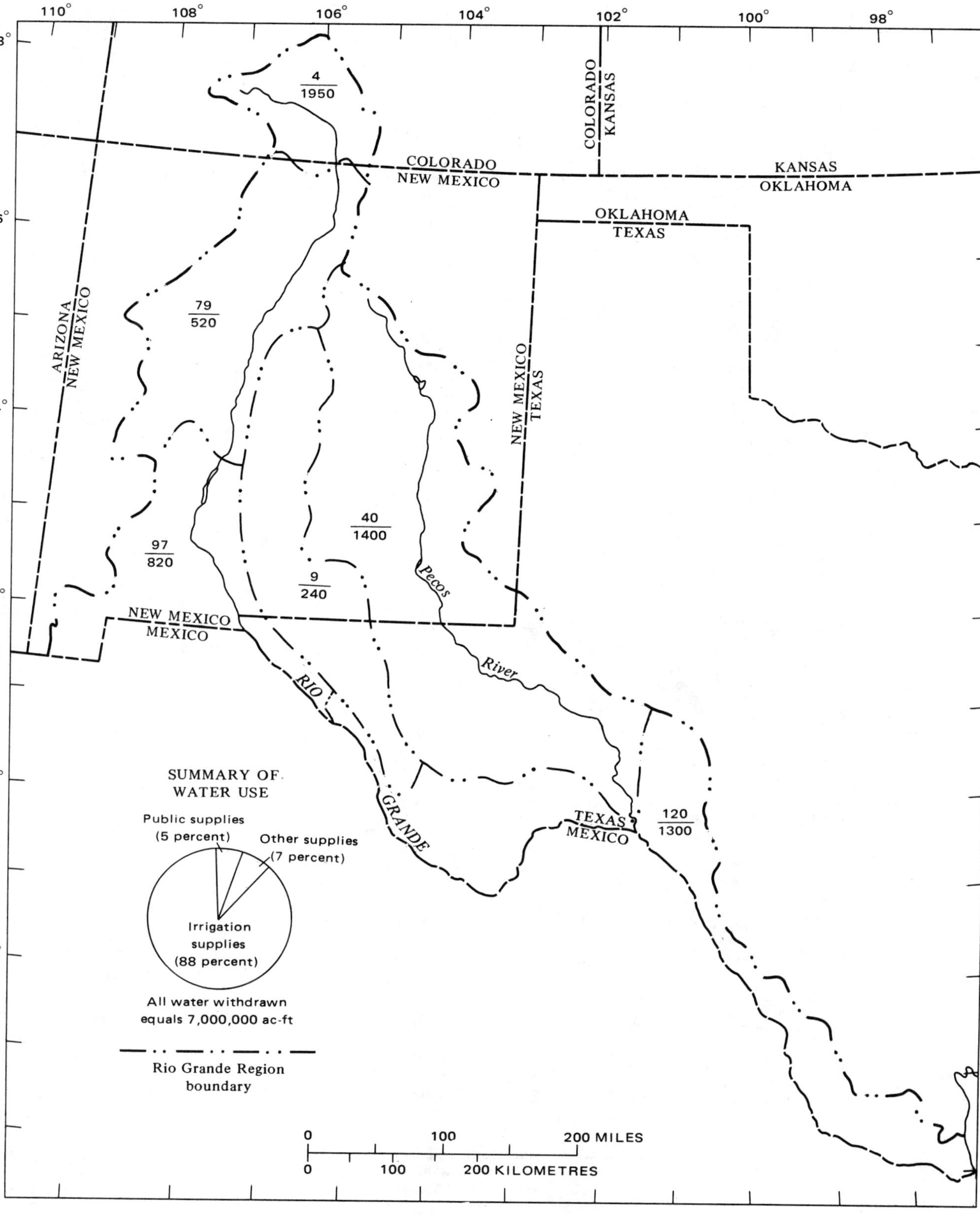

FIGURE 15. — Ratio between water withdrawn for public supplies (upper number) and water withdrawn for irrigation (lower number). The ratio is larger in the Albuquerque and El Paso subregions than in the others. Units are thousands of acre-feet.

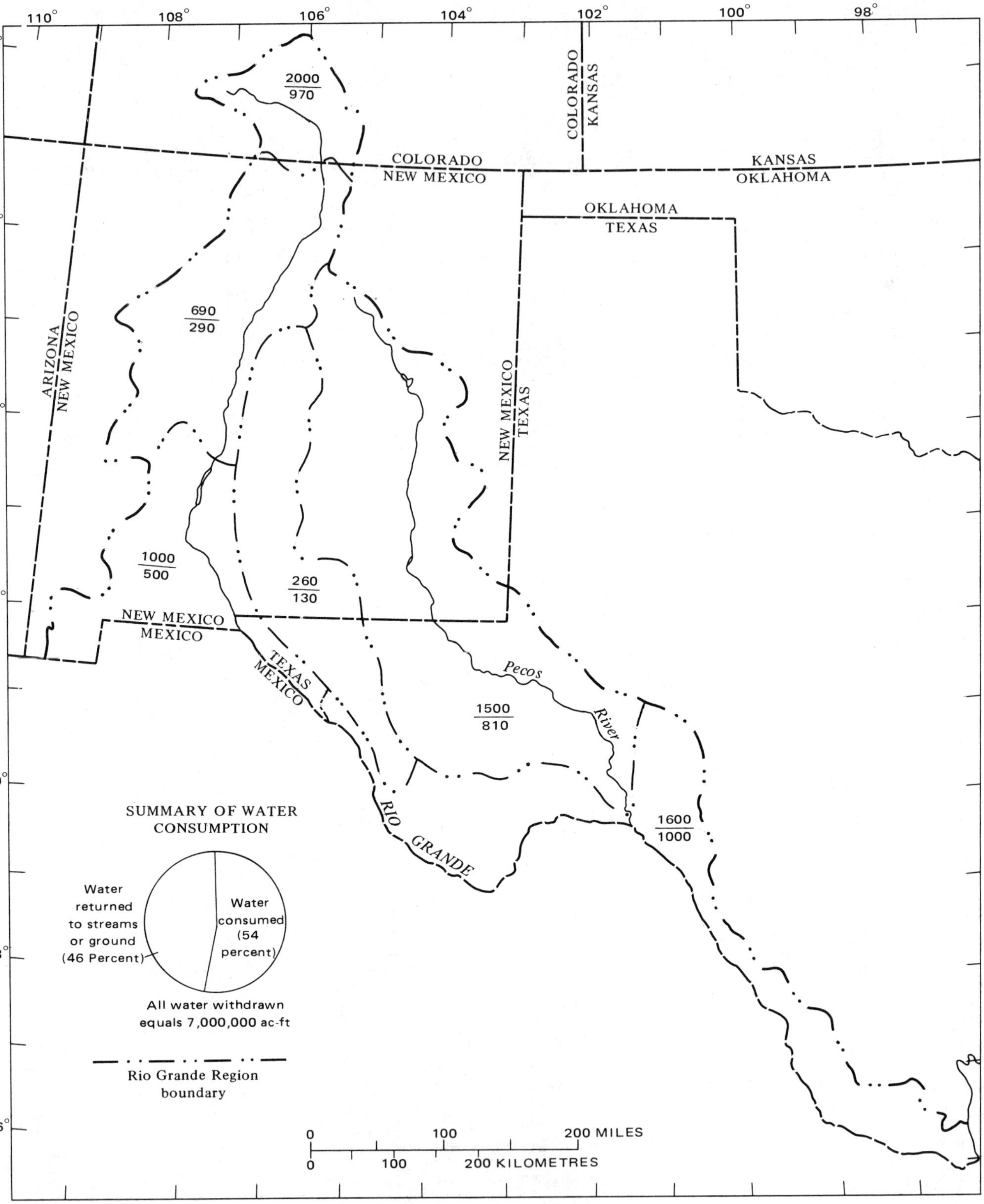

FIGURE 16. — Ratio between water withdrawn (upper number) and water consumed (lower number). About half the water withdrawn is consumed. Units are thousands of acre-feet.

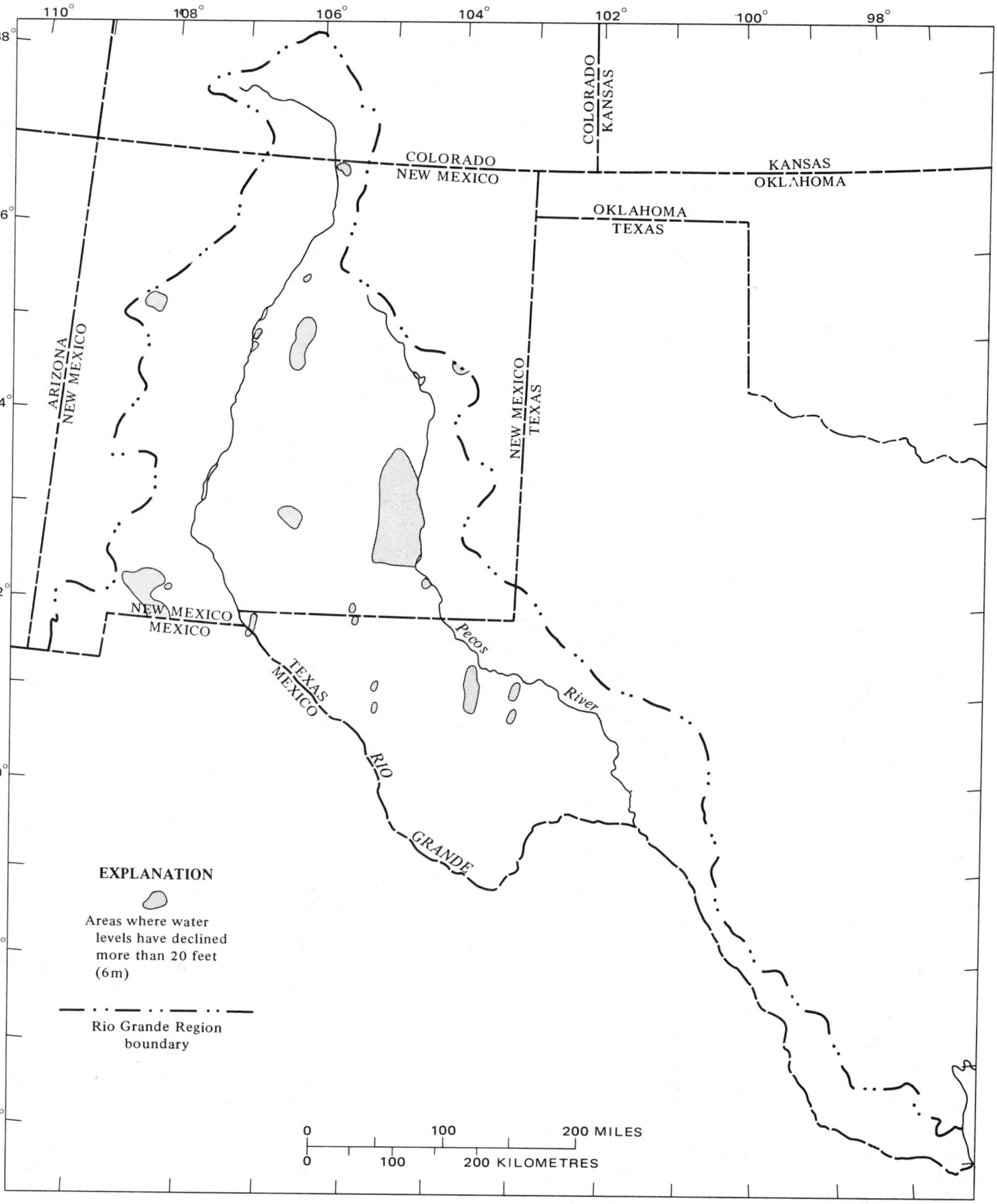

FIGURE 17. — Areas of water-level decline. Water levels have declined more than 20 feet (6 m) in many areas, owing to overdraft of ground water.

areas, and the surface and subsurface storage capacity. This tabulation shows that more than half as much water is consumed by evapotranspiration in wetlands and areas infested with phreatophytes (2.5 million acre-feet, or 3,100 hm^3, per year) as is consumed by man's activities (3.7 million acre-feet, or 4,600 hm^3, per year).

Man's use of water commonly creates potential sources for ground-water pollution. The areas of highest potential for pollution due to man's activities are in irrigated areas, in areas of mining and petroleum production, and in the large metropolitan areas.

In general, the water underground is less susceptible to pollution than the water in streams and lakes. If ground water does become polluted, the pollution is likely to persist much longer than a similar pollution of surface water.

The most widespread pollution from man's activities in the region results from irrigation. When fields are irrigated, part of the water is evaporated from the soil or transpired by the crops, leaving a greater concentration of salts in the remaining water. Also, part of the fertilizer added to fields is dissolved and carried down to the ground-water reservoirs. Another major source of pollution from man's activities is sewage effluent from cities and individual home sewage systems. The sewage effluent may cause bacterial pollution as well as chemical pollution. Effluents from industrial plants, mines, and oil fields commonly contain high concentrations of dissolved solids and require special management to avoid severe pollution of ground water.

Soluble minerals in the host rocks are the principal sources of dissolved solids in ground water in the Rio Grande Region. Large areas in the region are underlain by consolidated sedimentary rocks that contain gypsum and salt or by valley fill derived by erosion of those rocks. These minerals have caused large quantities of ground water in the region to become unfit for many uses, especially in the Pecos Valley and closed basins subregions. In ground-water discharge areas, the salts in the water are further concentrated by evapotranspiration. In some areas, saline ground water is discharged to streams and transported to other areas, where it reenters the ground and causes a deterioration in the chemical quality of the water in those areas.

POSSIBLE ALTERNATIVES IN GROUND-WATER MANAGEMENT

Management and use of water under any alternative has both beneficial and adverse effects. In planning water management, the effects and the related cost of each alternative should be considered.

The use of ground water for meeting the water demands of a region is a possible alternative in water-management planning that should be considered, keeping in mind that ground water and surface water are actually a single resource. Ground-water reservoirs can be used for input, storage, and withdrawal of water the same as surface reservoirs. General alternatives that should be considered in planning ground-water management are described below.

Under natural conditions, water from precipitation and surface flow infiltrates the ground and moves slowly underground toward points of discharge at lower elevations. In general, underground storage space is available in recharge areas, but the underground reservoirs are completely filled and are spilling in the discharge areas, generally along major stream valleys or in natural lakes or playas. Ground-water discharge provides the base flow of perennial streams, and the ground-water contribution throughout the year may exceed the direct overland flow of water.

The principal beneficial effect of maintaining full ground-water reservoirs relates directly to the maintenance of surface streams or base flow. Wet areas and associated phreatophytes, sustained by ground-water discharge, provide unique habitats for many species of wildlife. Free-flowing springs and spring pools are common in many ground-water discharge areas. The circulating ground water also flushes soluble salts from the underground reservoirs and soil zones.

Adverse effects of maintaining full ground-water reservoirs include benefits foregone from lack of potential economic development, inability of many areas to support a human population at desirable places to live, and limited space for storage of additional water during periods of excess precipitation and streamflow. The closed basins subregion is dependent almost entirely upon withdrawal of ground water for human habitation. On the other hand, much water is lost by evaporation in the closed basins, and the concentration of salts in the water in and near discharge areas is increased by evaporation, causing a general degradation of water quality.

The principal aquifers in the Rio Grande valley and its major tributaries are directly connected with the streams. The dominant beneficial effect of utilizing ground water from a stream-connected aquifer is assurance of a water supply whenever the need is greatest and, thus, the maximum potential economic return from water use. Ground water use can also assure a water supply for large human populations at desirable places to live, such as at Albuquerque and El Paso. Lowering of the water table through extensive ground-water withdrawals in areas of ground-water discharge can reduce significantly the quantity of water lost by noneconomic evapotranspiration and provide additional space for underground storage of water during periods of excess precipitation and streamflow. Ground water requires little treatment for human consumption, as it is free of sediments and generally free of bacteria.

The principal adverse effect of ground-water withdrawal from stream-connected aquifers is reduction in streamflow due to interception of natural ground-water discharge and to induced infiltration from streams. Lowering of water levels by extensive ground-water withdrawal can adversely affect wildlife habitats that are dependent on wet areas and phreatophytes. However, water levels could be lowered several feet in many areas without destroying all the phreatophytes, because of the ability of some to grow roots to tens of feet to obtain their water supply. Lower ground-water levels would dry up some springs and spring pools. A deterioration in water quality would result from extensive withdrawals of ground water in some parts of the Rio Grande Region.

The induced infiltration from streams could be controlled by constructing lined canals for transport of surface water past areas of ground-water pumping.

Aquifers in the closed basins subregion generally are isolated from perennial streams. Ground-water withdrawal from these aquifers can be limited to an amount equal to or less than the average discharge that can be intercepted, thus assuring a relatively small yield of water indefinitely; or ground-water withdrawal can exceed the average discharge (commonly termed "mining" of ground water), thus providing a larger supply of water during a finite period of time.

The principal beneficial effect of developing ground-water supplies from aquifers that are isolated from streams is provision of water for economic development and for human populations where no other source of water exists. Lowering of ground-water levels in areas of natural discharge can salvage water otherwise lost to noneconomic evapotranspiration. Lowering of water levels will also provide additional storage space, which can be refilled by recharge during periods of excess precipitation and runoff. Otherwise, the excess runoff would enter playa lakes and be lost from the basin by evaporation.

Continuous withdrawal of ground water in excess of the natural discharge that can be intercepted will have the adverse effects of increasing depths to water and increasing pumping lifts. The lower water levels may dry up natural wet areas in some basins. Dewatering of unconsolidated sediments may cause subsidence of the land surface due to accelerated compaction of the sediments. Deterioration in water quality in closed basins, due to migration of saline water from discharge areas to areas of extensive ground-water withdrawal, is common. Continuous pumping of ground water in excess of the replenishment rate will eventually deplete the water supply, as happens when any resource is mined.

Because ground water and surface water are so closely related in most of the Rio Grande Region, any alternative for ground-water development must recognize the effects of the development on the surface-water supplies, and ideally, should integrate the two sources of supply into a plan for conjunctive use of ground water and surface water. The following descriptions of potential management of ground water are based primarily on physical factors and do not fully consider the legal constraints.

Once the policy for governing the utilization of ground-water reservoirs is established, all possibilities for management of the water within that policy should be considered. The rate of economic return per unit of water used varies widely from one type of use to another, but water supplies for some uses are critical, regardless of cost. After water supplies to meet critical needs have been allocated, should further allocations be based on the maximum rate of economic return per unit of water consumed until all the demands have been met in order of highest return? Broadhurst (1964) prepared the following rates of economic returns for the specified uses on the plains of west Texas:

Type of use	*Gross return (dollars per ac-ft of water used)*
Irrigaton of crops:	
Grain sorghum	$50- 100
Cotton	100- 200
Vegetables	1,500-2,000
Secondary recovery of oil	[1]20,000

[1]Based on the use of one barrel (42 gal) of water to recover one barrel of oil and on a market value of $3.00 per barrel for oil.

Obviously, the uses that provide the highest economic returns have relatively small demands. Regardless of how the water supply is allocated, maximum efficiency in water use should be achieved to stretch the supply as far as possible. The largest potential for increasing the effective water supply by improving efficiency is related to irrigation of croplands. Better efficiency in irrigation could salvage considerable quantities of water commonly lost by evaporation and runoff from irrigated fields.

The salvage of water lost to noneconomic evapotranspiration in wet and phreatophyte-infested areas (table 2) offers the greatest possibility of improving the water-supply situation in the region.

SAN LUIS VALLEY SUBREGION

The Closed Basin Division of the San Luis Valley Project, proposed by the U.S. Bureau of Reclamation (1963), contemplates salvaging about 101,000 acre-feet (124 hm^3) of water annually, of which 86,000 acre-feet (106 hm^3) would be pumped ground water (mostly from a salvage area of 109,00 acres, or 44,000 ha) and 15,000 acre-feet (18 hm^3) would be surface water. According to the Bureau of Reclamation plan, "Under the assumptions of schedule of construction (1965) and continuation of the average rate of debit accrual and the estimate of annual water salvage, the accrued debit of Colorado would be offset in about 35 years."

If 86,000 acre-feet (110 hm^3) of water can be salvaged annually from the salvage area of 109,000 acres (44,000 ha), what are the prospects of salvaging 1 million acre-feet (1,200 hm^3) a year from the 1.5 million acres (610,000 ha) where the water table is less than 8 feet (2 m) below ground surface? This question seems reasonable when 1 million acre-feet (1,200 hm^3) of water per year is being lost from the valley by noneconomic evapotranspiration. Under such a salvage program, the accrued debit of Colorado could be offset in a year or two, or additional economic development based on use of ground water would be possible. An electric-analog model of the valley has been constructed, and several alternatives of ground-water management should be evaluated with the model. The sharp decrease in the rate of evapotranspiration in the San Luis Valley as the water table is lowered is shown in figure 18.

In addition to the saving of water through increased efficiency in irrigation and through salvage operations, consideration should be given to withdrawal from ground-water storage. The ground-water reservoir contains about 2 billion acre-feet (2.5 million hm^3) of water in storage. Withdrawal of 2 million acre-feet (2,500 hm^3) a year would cause only a 10 percent decrease in storage in 100 years. Most large surface reservoirs in the Rio Grande region are being filled with sediment at rates near 50 percent in 100 years. (Some surface reservoirs are currently being filled at the rate of 100 percent in 100 years.) Although the capacity of a surface reservoir being filled with sediment is continually decreasing, ground-water reservoirs retain indefinitely a nearly constant storage capacity and can be used again and again to store and recover water.

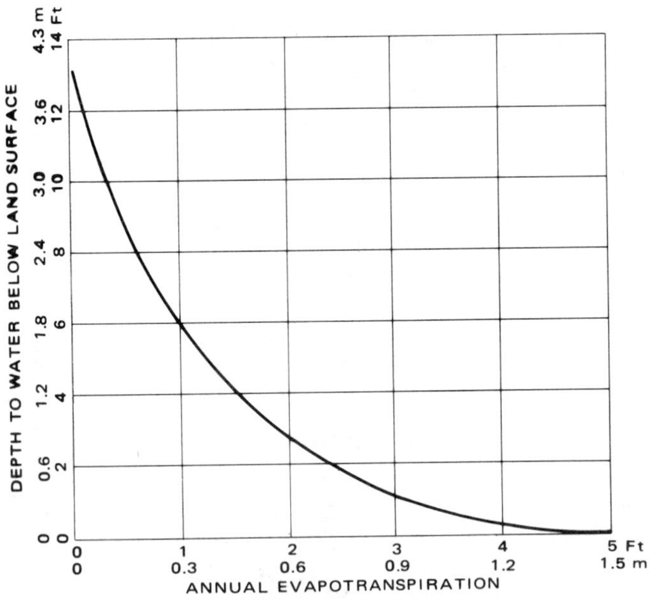

Source: Emery and others (1971)

FIGURE 18. — Relationship between the rate of water loss by evapotranspiration and the depth to the water table in the San Luis Valley. The rate decreases dramatically as the water table gets deeper.

ALBUQUERQUE SUBREGION

An approach to improving the water-supply situation in the Albuquerque subregion could be an expansion of conjunctive use of ground water and surface water, which is already being practiced to some extent. Because of the physical connection between the river and the ground-water reservoir, withdrawal of large quantities of ground water in some reaches would cause a reduction in streamflow. On the other hand, a large part of the 400,000 acre-feet (490 hm^3) of water now being lost annually to noneconomic evapotranspiration probably could be salvaged by lowering ground-water levels in the wetlands and phreatophyte areas of the subregion.

Lowering the ground-water level below the bed of the river would cause losses from the river, but once the ground-water level is below the river, additional lowering would not cause a further increase in loss of water from the river. However, ground water that normally would discharge into the river would be intercepted before it reaches the river. Water losses from the river could be controlled in areas where the water table is lowered by diverting into an artificial, lined channel the amount of streamflow needed to satisfy downstream rights. Excess flow could follow the natural channel to recharge the ground-water reservoir. The economics of this approach have not been analyzed.

The increasing demand for water in the Albuquerque subregion could be met readily by drawing on the large volume of water stored in the underground reservoirs. About 65 million acre-feet (80,000 hm^3) of ground water could be withdrawn in the subregion by uniformly lowering the water level 100 feet (30 m) in the valley fill. This approach could more than double the water supply for the next 90 years, even if all the water withdrawn is consumed, which rarely is true. An additional 400,000 acre-feet (490 hm^3) per year would be salvaged in the process because evapotranspiration in wetlands and phreatophyte areas would be eliminated.

Large reserves of coal and natural gas have been mapped in and adjacent to the Albuquerque subregion (fig. 6). Development of these resources for electrical power generation to help meet the rapidly growing demand for electrical energy in the Southwest could be enhanced by drawing on salvaged water or on the large reserves of ground water in storage.

Some, and possibly all, of the wetlands and phreatophyte areas are considered by many to be ecologically desirable. Therefore, a uniform lowering of water level in the ground-water reservoir would not be reasonable. However, a program of ground-water withdrawal could be designed to lower the water levels significantly more than 100 feet (30 m) in some areas and significantly less, or none, in selected wet and phreatophyte areas. This approach would not permit as much salvage of water normally lost to noneconomic evapotranspiration. Some areas would be conducive to maintaining wildlife habitat by surface irrigation, which would use much less water than is now lost by evapotranspiration.

A program could be designed also to utilize the ground-water reservoir to replace part of the surface storage. Water levels could be lowered by pumping ground water into lined canals or pipes for delivery to points of use, and excess surface water could be used for recharging the underground reservoirs by infiltration, either directly from the streambed or indirectly from spreading areas or recharge ponds in areas favorable for infiltration. This approach has the advantage of salvaging water normally lost to evaporation from surface reservoirs (255,000 acre-feet, or 310 hm^3, per year from Elephant Butte and Caballo Reservoirs), but it increases the available supply only in the amount of this salvage. Surface reservoirs have recreational values which may justify large evaporation losses. However, many recreationists prefer a small reservoir of constant size to a larger reservoir of widely varying size, and at some places surface reservoirs could be regulated for nearly constant size by utilizing underground storage for part of the water.

Sewage effluent from municipalities and individual homes in the Rio Grande Valley and its principal tributaries contributes to the salt load in streams and underground reservoirs. The principal chemical constituents added to the water are phosphates and nitrates. Methods of removing these constituents are being studied by several organizations at the Federal, State, and local level. One method being investigated is the spraying of treated effluent on irrigated fields, where much of the phosphates and nitrates are adsorbed in the soil and used by the crops. An excess of water is applied to the land so that a large part returns to the ground-water reservoir after the phosphates and nitrates have been adsorbed. This process might work very well on the types of soils in much of the subregion.

The water supply of the subregion possibly could be improved to some extent with better watershed management, that is, with control of the types and density of vegetation growing on the watershed.

EL PASO SUBREGION

Conjunctive use of ground water and surface water has been practiced by individuals in the Elephant Butte irrigation district for many years. Irrigation water from Elephant Butte Reservoir is used when the supply is adequate, but when surface water is in short supply, more than 90 percent of the land receives supplemental ground water from privately owned wells (Sorensen and Linford, 1967). Part of the water applied to irrigated fields returns to the ground-water reservoir carrying soluble soil salts and fertilizers. Consequently, the chemical quality of shallow ground water has deteriorated extensively, some now containing as much as 6,000 mg/l dissolved solids. A systematic approach to total water management, including mixing of the shallow ground water with deeper water of better quality or with surface water possibly would improve the general quality of irrigation and municipal water.

Evaporation loss from Elephant Butte and Caballo Reservoirs averages about 255,000 acre-feet (310 hm^3) per year (Sorensen and Linford, 1967). This loss could be reduced significantly by storing more of the water underground and reducing the surface area of these reservoirs. Elephant Butte Reservoir is used extensively for recreation, which must be considered in any alternative plan for water management. The recreational value of the reservoir possibly would be improved by maintaining a smaller surface area of a constant level.

Large amounts of water are lost by evapotranspiration from wetlands and phreatophyte areas. Much of this water could possibly be salvaged by lowering the water table in selected areas.

Population has been increasing rapidly in the Las Cruces and El Paso areas, creating problems in both municipal water supply and sewage disposal. The water-supply problem at Las Cruces primarily involves poor quality of the shallow ground water. This problem can be solved by drilling deeper wells to get below the zone influenced by return of poor quality water from irrigated fields. The water supply for El Paso is placing a heavy demand on local ground-water supplies, but large quantities of fresh ground water are available in surrounding areas. The long-term effects of ground-water withdrawal have been evaluated through use of an electric-analog model as part of a cooperative study of the City of El Paso, the Texas Water Development Board, and the U.S. Geological Survey (Leggat and Davis, 1966). A new model to incorporate recently obtained data is being constructed to further refine the analysis of ground-water management possibilities and resultant effects in the area.

Treated sewage effluent in the subregion probably could be used for irrigation — the nitrates and phosphates being used by plants, and the excess effluent returning to the ground-water reservoir.

Because the chemical quality of ground water in the subregion is highly variable, consideration should be given to mixing the water of best quality with water of inferior quality to stretch the supply of usable water.

An interagency study of the subregion, involving both State and Federal agencies, is currently (1973) in process. This study, termed the "Rio Grande Regional Environmental Project," is considering a full range of potential resource-development plans, including the water resources. Various alternatives in water management will be a part of this study.

PECOS VALLEY SUBREGION

The effective supply of water in the Pecos Valley subregion could be increased, and the chemical quality of the water improved, by reduction of evapotranspiration from wetlands and phreatophyte areas and by reduction of evaporation from surface storage reservoirs. Salvage of 50 percent of the water now lost in these processes would increase the effective supply by 140,000 acre-feet (170 hm^3) per year.

The loss of water to evapotranspiration possibly could be reduced by lowering the water table in the wetlands and phreatophyte areas. However, the salinity of the water in some of the phreatophyte areas is too high for direct use of the water, and desalination might be required before it could be used beneficially. In some areas of intensive pumping, withdrawal of saline water and subsequent desalination would reduce the encroachment of saline water into fresh-water zones.

If it were decided to salvage evapotranspiration loss, some wetlands possibly should be preserved, such as those at the Bitter Lakes National Wildlife Refuge and the Bottomless Lakes State Park. A sparse stand of salt cedars and cottonwoods could be preserved in selected areas at small loss of water by evapotranspiration by lowering the water table slowly to an optimum level. Control of ground-water levels could prevent the spread and revegetation of salt cedars.

Water salvaged by control of wetlands and phreatophytes could be used to meet the increasing demand for public water supplies, to supplement short supplies of irrigation water, or to irrigate new lands. Water salvaged by reduction of evaporation losses from surface storage reservoirs also could be used for these purposes.

The volume of water stored in surface reservoirs, and thus the quantity of water lost by evaporation, could be reduced by controlled recharge to and withdrawal from the ground-water reservoirs. The limestone aquifer in the Roswell basin has a large capacity for receiving, storing, and transmitting water. This approach to water storage would necessarily have to guarantee delivery of water to satisfy prior rights to diversion of surface water.

The overdraft, or "mining," of ground water has resulted in significant decreases in ground water in storage and has caused increased pumping lifts. This overdraft could be alleviated by sufficient salvage of water now lost through evapotranspiration, by artificial recharge to the ground-water reservoir, or by reduction of irrigation. Salvaged water, as described above, could replace some of the ground water being pumped and reduce the ovedraft. Artificial recharge could alleviate the overdraft, but the supply of water that might be available for this purpose is limited. Eventually, reducing the irrigated acreage may become necessary, but that would have a severe adverse impact on the economy of the area.

Local floods have caused serious problems in parts of the subregion. Under favorable circumstances, the floodwaters could possibly be diverted to recharge areas before they reached populated areas and irrigated farms. However, the floodwaters of the region have been appropriated, and these rights would have to be protected. The floodwaters generally carry heavy loads of sediment, and the storage capacity of surface reservoirs is reduced by accumulation of sediment trapped during floods. In contrast, the capacity of underground reservoirs is not reduced significantly by deposition of sediments on the land surface.

The high salinity of much of the water in the Pecos Valley subregion is one of the most serious water problems. It is difficult to manage the fresh water so it does not become mixed with the abundant saline water, which in effect reduces the supply of fresh water. One approach to improving the water quality without reducing the total water supply significantly is pumping and desalting in areas of saline-water discharge or in areas where saline water is encroaching into fresh-water zones of underground reservoirs. An experimental desalting plant to obtain data on the feasibility of desalting water of a chemical character prevalent in the subregion has been operated successfully at Roswell. Another possible approach is pumping the more saline water into evaporation ponds. At one place, near Malaga Bend, brine that normally discharges into the Pecos River has been intercepted by pumping the water from a well and discharging it into a natural depression. The effect of this experiment is still being evaluated.

Elimination of evaporation from wetlands and transpiration by phreatophytes would improve the quality of water. The salts are concentrated by these processes, and the salts eventually return to the principal water systems, contaminating more of the freshwater supply. Evapotranspiration could be reduced in many areas by lowering the water table.

Management of the scarce water supplies of the subregion possibly could be improved by regarding all the sources of supply and storage facilities as a single system, managed for the maximum benefit of all users. However, this approach would require considerable reorganization of the complex managerial structures for water that are now in operation.

CLOSED BASINS SUBREGION

The large supply of saline water in the closed basins subregion offers an opportunity for economic development that could utilize the saline water. One possibility might be utilization of the water for cooling in powerplants. The unconsumed water from the cooling facilities could be reinjected into the underground reservoirs at adequate distances from the withdrawal points to permit heat dissipation before the water returns to the production wells, thus minimizing the consumption of water. The amount of fresh water required for the operation could be provided by desalting. Large supplies of fresh water for other uses also could be obtained by desalting the saline water.

Ground water has been "mined" in a large area of the Estancia basin. The water table has been lowered as much as 50 feet (15 m) in areas of maximum withdrawals. The overdraft could be alleviated to some extent by inducing recharge to the ground-water reservoir during the infrequent periods of surface runoff and by dispersing some of the pumping to areas of natural discharge. The quantity of water being lost by evaporation from the playas is more than the quantity of water being consumed by beneficial uses, and most of the water lost is from natural ground-water discharge.

Ground water also has been "mined" in parts of Salt basin. Water levels have declined as much as 20 feet (6 m) in the New Mexico part and as much as 75 feet (23 m) in the Texas part. Additional artificial recharge and dispersal of pumping to discharge areas in this basin could possibly reduce the overdraft of ground water. Some natural ground-water discharge has already been intercepted by pumping.

LOWER RIO GRANDE SUBREGION

Because of the limited quantity of fresh ground water and the generally adequate supply of surface water in the lower Rio Grande subregion, no further consideration is given to possible alternatives in this subregion.

SPECIAL UTILIZATION OF UNDERGROUND SPACE

Underground space can be used for more than providing a normal water supply. In many parts of the region, thick accumulations of valley fill lie above the water table, and the average porosity of this unsaturated material is equivalent to, or higher than, that of the saturated material. By constructing spreading ponds, excess surface flow from rainstorms could be diverted in favorable areas for artificial recharge into the unsaturated fill. Thus, the volume of water in storage could be increased significantly in some localities. Also, air could be pumped from thick unsaturated material for air conditioning buildings, because the air at a given depth below land surface has an almost constant temperature throughout the year.

Diversion of excess surface water to recharge facilities can regulate the flow of streams without the use of large surface storage facilities. The water recharged into the ground during periods of excess surface flow will eventually finds its way to a nearby stream as extra ground-water discharge.

Introduction of excess surface water of good chemical quality into aquifers containing water of inferior quality can improve the general quality of the ground water. This approach would have the greatest potential for quality improvement in the Pecos Valley and closed basins subregions.

As the war against pollution mounts, the interest in underground storage of wastes is growing. Underground space above the zone of saturation or in impermeable materials can be used for storage of solid wastes in favorable areas, where vaults can be excavated and kept dry. If deemed necessary, wastes stored in this manner can be recovered at any time. Liquid wastes can be stored in saturated materials, where adequate safeguards, such as enclosing impermeable beds, are present. These conditions exist in many localities in the region where the consolidated sedimentary rocks or the valley fill are thick and highly variable in permeability. Underground disposal of wastes requires thorough testing and analysis of the receiving environment before the safety of disposal can be assured.

The Anaconda Co. has operated an injection well for disposal of uranium-mill effluent near Grants, N. Mex., since December 1960 (West, 1972). The injection interval is sandstone, separated from all fresh-water aquifers by relatively impermeable, thick beds of mudstone and anhydrite. Large areas in the Rio Grande region should have similar features which would permit safe disposal of liquid chemical wastes.

Release of cooling water from thermal electric plants and some types of industrial plants has caused severe problems of thermal pollution of nearby streams in many areas of the United States. The ground-water reservoirs in the valley fill of the Rio Grande Region could be used as receptacles for thermal waters from these types of plants. If the thermal water were injected into the valley fill at considerable distances from discharge points, either natural discharge points or wells, the heat would be dissipated to the rocks and the atmosphere before the water could reach the surface again.

Development of geothermal energy is another special utilization of underground space. In favorable geothermal areas, natural steam can be withdrawn through wells and used to drive turbines for generation of electricity. Experiments to evaluate the injection of cool water into dry geothermal areas for conversion of the water to steam and recovery of the steam for driving turbines are underway, but the outcome is open to speculation. In some parts of the world, geothermal waters are used directly for space heating of homes, offices, and greenhouses.

Several geothermal areas have been identified along the margins of the Rio Grande depression, but they have not been adequately explored to evaluate their potential for energy development. The largest and best known geothermal area in the region is the Jemez Mountains area of northern New Mexico. Hydrologic studies of that area were begun recently (1972).

Experiments indicate that in favorable areas both heating and cooling of buildings can be accomplished by direct use of ground water from different depths. Shallow ground water, which commonly is cool, can be obtained in summer for cooling buildings and can be reinjected at greater depths for conservation of the water. Warm water from greater depths can be obtained for heating buildings in winter and can be reinjected at shallow depths to maintain the balance between withdrawal and injection. The thick permeable valley fill in the Rio Grande Valley should be favorable for this application.

INFORMATION NEEDED FOR PLANNING GROUND-WATER MANAGEMENT

Much work has been done toward evaluating the ground-water resources of the Rio Grande Region, as indicated by the references at the end of the report and by figure 19, but much more remains to be done before systematic planning for ground-water management can be accomplished. The types of information needed for systematic planning are summarized on the following pages.

Planners for ground-water management must recognize the close relationship of ground water and surface water. The planner must have detailed information on: (1) the flow characteristics of streams; (2) the position of the streambed in relation to the water table; (3) infiltration rates from streams, canals, ditches, and irrigated fields; (4) potential infiltration rates from recharge ponds or spreading areas; (5) the contribution of ground-water discharge to streams; (6) the chemical quality of ground water and surface water; and (7) the changes in ground-water storage due to past management of the water resources. Typically, use of surface water causes an increase in ground-water storage, because some water is lost to infiltration, and use of ground water causes a decrease in storage, which varies as a function of ground water withdrawn and consumed.

The quantity of recoverable ground water in storage to different depths should be determined, as it represents the reserves available for development. Rough estimates of ground-water storage in the major aquifers of the Rio Grande Region have been made to show the water resources in general perspective. However, more accurate information is needed in areas to be considered for systematic ground-water management.

The physical properties of aquifers control the quantity of water that can be stored or yielded, the rate at which water can be added to or withdrawn from the underground reservoirs, and the change in water levels that will result from withdrawal of a given volume of water. The mineral content of aquifer materials largely controls the chemical quality of the ground water. The physical properties of the aquifers should be determined by imposing a hydrologic stress to the system (pumping) and measuring the response of the system (water-level changes). Useful information has already been obtained by pumping from individual wells or from well fields and measuring water-level changes. Additional information could be obtained from existing wells. In some areas the depth and spacing of wells are inadequate for acceptable tests, and special test wells are needed.

The depth to water and the pumping lifts that will be required must be known in order to estimate pumping costs.

Valley-fill aquifers, the predominant type in the Rio Grande Region, commonly contain extensive beds or lenses of clay and silt. When water is withdrawn from these aquifers, slow drainage of water from the clay may permit its compaction. If the clay beds comprise a significant part of the valley fill, the compaction due to withdrawal of water may result in subsidence of the land surface. Therefore, a determination of total clay thickness and laboratory determination of hydraulic and mechanical properties of the clays would be needed.

Chemical analyses of ground water from many areas are needed to define the variations in quality, both laterally and vertically, within the aquifers. Variations in chemical quality of water within an aquifer can lead to intermixing of fresh and saline water as fresh water is withdrawn, because saline water moves into space previously occupied by fresh water. Withdrawal of saline water reverses the situation but has the same net effect.

Wetlands and phreatophyte areas should be mapped in detail, and the quantity of water lost by evapotranspiration should be determined as accurately as possible. The significance of these areas as wildlife habitats should be evaluated. Eradication of phreatophytes by mechanical means and selective elimination by controlling ground-water levels should be appraised in relation to wildlife habitats and potential for water salvage.

The responses of water-resource systems to the many potential hydrologic stresses can best be analyzed by employing an electric-analog or digital model, or a com-

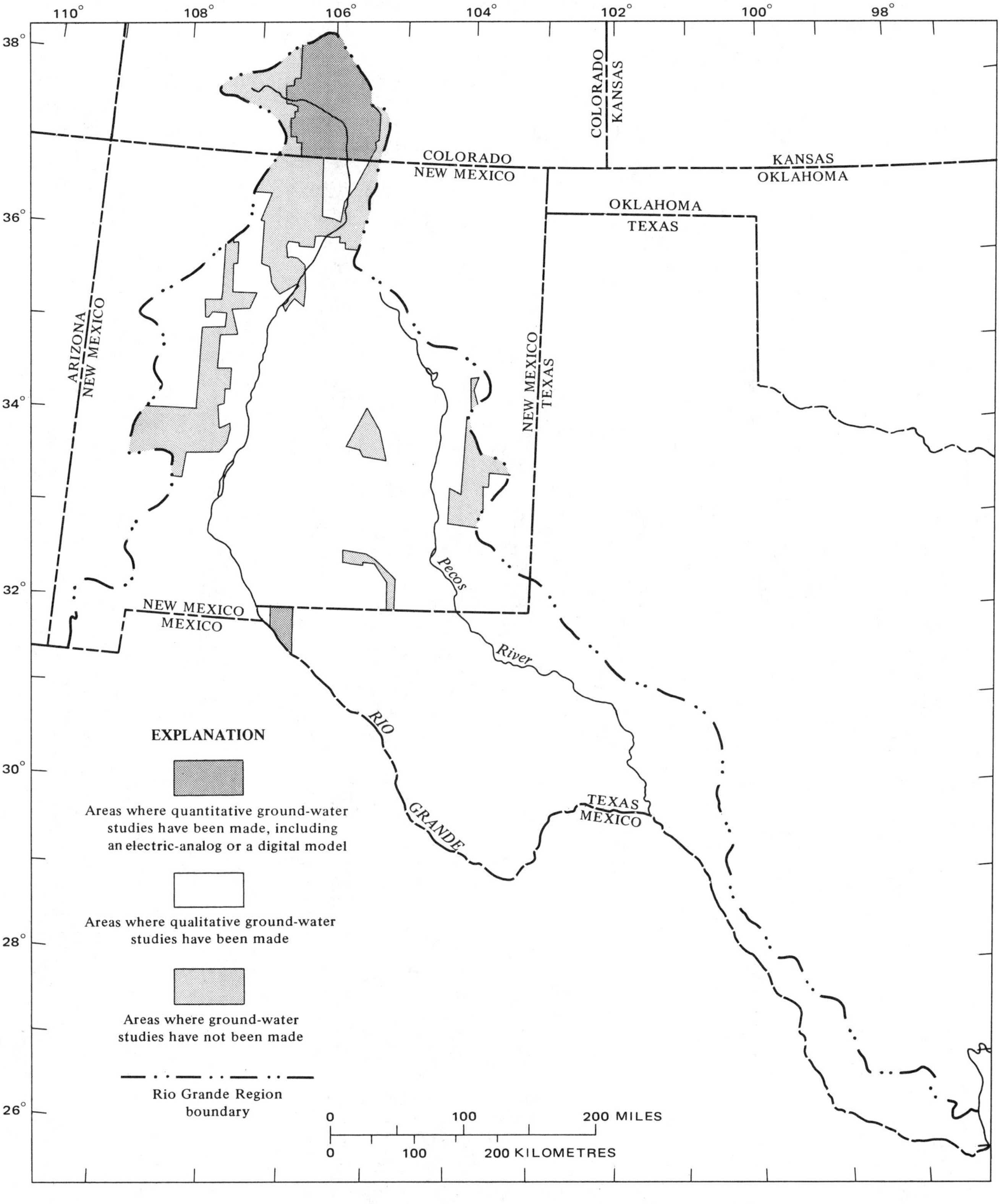

FIGURE 19. — Areas in which ground-water studies have been made. Quantitative studies have been made in only two areas, and several areas have not had even qualitative studies.

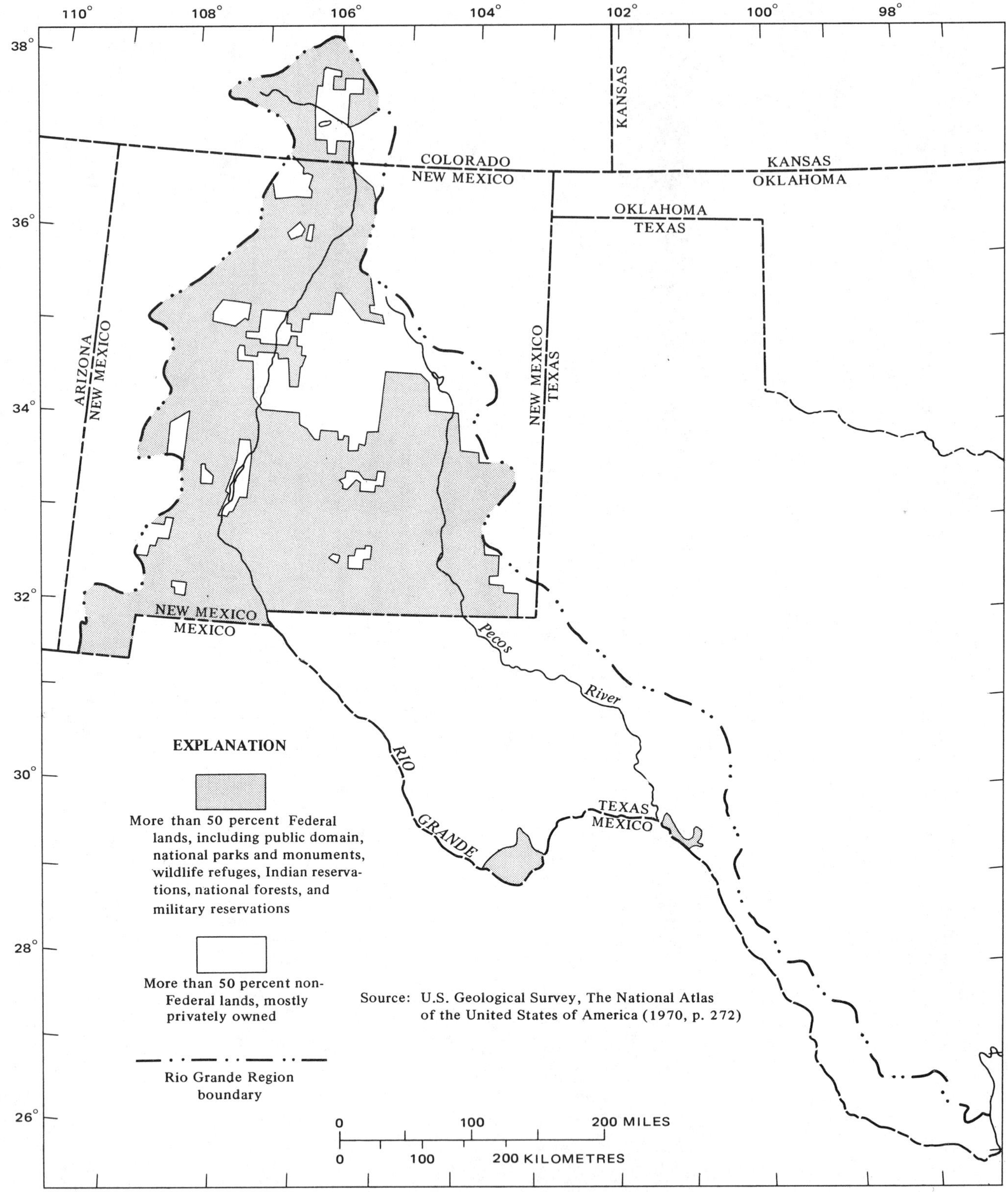

FIGURE 20. — Land ownership or control. Most of the land in New Mexico is owned or controlled by the Federal Government, and most in Texas is privately owned.

bination of the two, to simulate the ground-water reservoir. Once the models are completed and verified with historic records, the effects of alternative plans for water management and optimum locations, spacing, and depths of wells can be readily analyzed.

Information on the benefits and costs, both economic and environmental, is needed for evaluation of alternative plans for water management. The planners and the public should be aware of the benefits and costs of water use and the value of benefits foregone if the water is not used. The cost analysis of a proposed project should include: (1) the cost of the investigative program; (2) the cost of the construction program; and (3) the cost of the operational program, including the cost of monitoring the response of the water-resource system. Any proposed project, regardless of benefits and costs, must be acceptable within legal, social, and ecological constraints.

Once a water-resource project becomes operational, the response of the system to the new stresses should be monitored. The monitoring data very likely will show that modifications of the models are necessary to improve the predictive capability of the models. Depending on the actual response of the water-resource system, minor modifications in operations, as well as modifications of the models, may be necessary.

In some instances it might be possible to increase the effective supply of water by improved watershed management or by weather modification. These possibilities should be thoroughly evaluated as part of comprehensive water-resources planning.

All facets of data acquisition and analysis should be thoroughly documented with written reports for future use and current transfer of knowledge.

The Federal Government should have a strong interest in more detailed evaluation of the ground-water resources and possible alternatives in water use, as more than 50 percent of the land in the New Mexico part of the region is Federally owned or administered (fig. 20).

The types of studies needed in various parts of the region to provide critical information for planning systematic ground-water management are outlined in figure 21.

SUMMARY AND CONCLUSIONS

The Rio Grande is an interstate and international stream which begins in the mountains of Colorado, flows across New Mexico from north to south, and forms the boundary between Mexico and Texas for 1,250 miles (2,000 km). The region ranges in altitude from sea level at the Gulf of Mexico to more than 14,000 feet (4,300 m) in the headwaters area. Annual precipitation ranges from 8 to more than 30 inches (20 to more than 76 cm), depending on altitude and latitude. Irrigation is required for growing crops throughout the region. Evaporation rates are generally high, causing an annual water loss of 950,000 acre-feet (1,200 hm^3) from reservoirs, ponds, and streams.

The Rio Grande Valley in New Mexico has been inhabited for thousands of years, first by Indians, then Spaniards, and, later, Americans. Crops have been irrigated since the early part of this millenium. From earliest times, most of the Indian and Spanish agriculture in northern New Mexico has been based on subsistence farming, and the farms have become smaller as the land was divided amongst each new generation. The American settlers established larger farms or ranches, and, subsequently, many have been combined to create even larger ones.

The population of the region has been increasing rapidly in this century, from 750,000 in 1929 to 1,700,000 in 1970. It is expected to increase to 2,500,000 by 2020. The basin economy of the region was traditionally agricultural until recent years, but agricultural development is now increasing very slowly. Since 1950 the mining and petroleum industries have increased much more rapidly than agricultural development.

Management of surface waters in the region is subject to international treaty, interstate compacts, and the laws of three separate States. Management of ground water is subject to the laws of the State in which the ground water occurs, and the laws of each State are quite different.

The renewable water supply of the region is the sum of the surface-water runoff, the ground-water outflow (which is small), and the natural evapotranspiration. Annual precipitation on the region is 86 million acre-feet (110,000 hm^3); however, all but 4 million acre-feet (4,900 hm^3) is returned to the atmosphere by evapotranspiration.

The ground-water reservoirs of the region contain an aggregate of about 5,800 million acre-feet (7,200,000 hm^3) of fresh and slightly saline water in storage, which could be withdrawn through wells. In contrast, the surface reservoirs have a combined storage capacity of only 18 million acre-feet (22,000 hm^3).

Thick deposits of valley fill in stream and intermontane valleys of the region comprise the principal ground-water reservoirs. In most areas they are capable of yielding large supplies of water to wells. In some areas, consolidated sedimentary rocks, particularly limestone, yield a few hundred to a few thousand gallons per minute of water to wells. The largest ground-water reservoirs are in the San Luis Valley and Albuquerque subregions, each of which contains about 2 billion acre-feet (2,500,000 hm^3) of fresh and slightly saline water.

The chemical quality of water in the valley-fill aquifers varies widely. The dissolved solids in water in

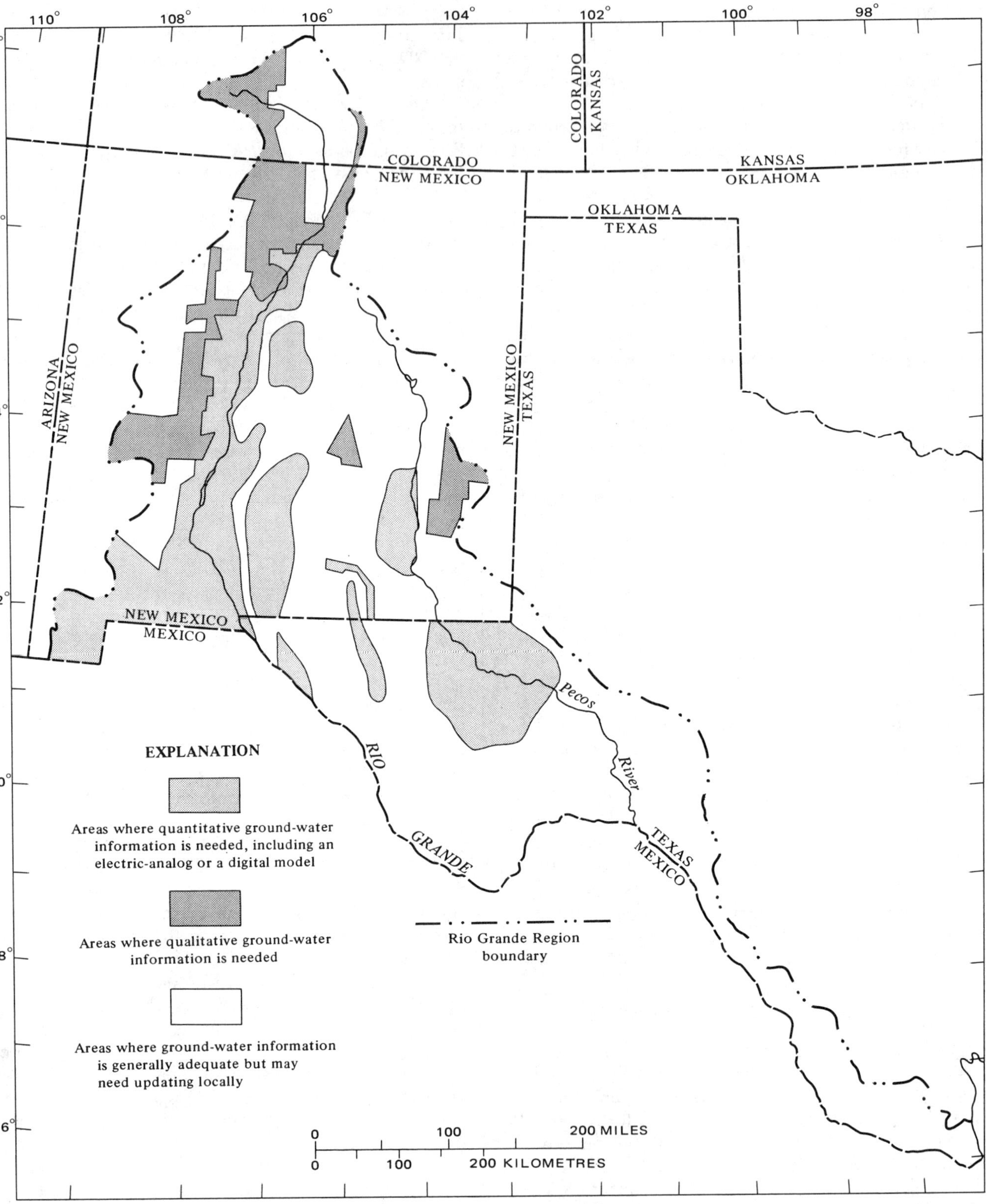

FIGURE 21. — Types of studies needed.

the shallow aquifers have been concentrated by water loss due to evapotranspiration and by return of water containing fertilizers from irrigated fields to the ground-water reservoir. The water at great depths in the valley fill also may have high concentrations of dissolved solids. The water of best quality generally is at intermediate depths. In the Tularosa basin, most of the water contains 3,000 to more than 35,000 mg/l.

Withdrawal of ground water in the region in 1970 was 2.7 million acre-feet (3,300 hm^3). The major part (88 percent) of this water was used for irrigation. Of the water withdrawn, 53 percent was consumed, and 47 percent was returned to the streams or ground-water reservoirs.

Ground water has been extensively overdrawn, or "mined," in several areas, especially near Roswell, N. Mex., and in Pecos and Reeves Counties, Tex., and severe decline in ground-water levels have resulted.

The loss of water by evapotranspiration in wetlands and phreatophyte areas is about 2.5 million acre feet (3,100 hm^3) per year in the region. In comparison, 3.7 million acre-feet (4,600 hm^3) of water is consumed by man's activities.

Ground water and surface water are really a single resource — water — although they commonly are treated separately. Ground-water reservoirs can be used for input, storage, and withdrawal of water the same as surface reservoirs, but ground water in storage constitutes a large reserve that can be withdrawn at any time. Alternatives that should be considered in ground-water management include: (1) maintenance of full ground-water reservoirs, (2) withdrawal of ground water from stream-connected aquifers, and (3) withdrawal of ground water from aquifers isolated from streams. Each alternative has definite beneficial and adverse effects that must be considered.

Because ground water and surface water are so closely related in most of the region, any alternative plan for ground-water development must recognize the effects of this development on the surface-water supplies and preferably should integrate the two sources of supply into a plan for conjunctive use of both the ground water and the surface water.

The salvage of water lost to noneconomic evapotranspiration in wet and phreatophyte-infested areas offers the greatest possibility of improving the effective water supply in the region. Salvage of half the water lost would increase the effective supply by 1.2 million acre-feet (1,500 hm^3) per year. However, salvage of this water could adversely affect the wildlife habitat.

The usable water supply for the region could be increased tremendously by drawing on the large reserve of ground water in storage. Pumping of ground water in many reaches of the streams could cause seepage losses from the streams. However, if the economics were favorable, these losses could be prevented by diverting the streams into lined channels.

Withdrawal of ground water, thus provision of more storage space underground, could help to regulate streamflow at times of excess flow and in some places could prevent floods. Water that enters the ground-water reservoirs during periods of excess surface flow would be available for later use. Storage of excess water underground has the added advantage of leaving the sediment behind on the land surface rather than letting it accumulate in the surface reservoirs.

The large supplies of saline water in parts of the region should have a market for some uses. Removal of the saline water would prevent its mixing with fresh water.

Development of geothermal energy appears to offer possibility for meeting part of the energy requirements of the future. Preliminary studies indicate geothermal areas at several places in the region.

Planners for ground-water management must recognize the close relationship between ground water and surface water and must have detailed information on the physical parameters that control or affect each supply. Large-scale development without adequate evaluation, planning, and wise management can be disastrous.

REFERENCES

Armstrong, C. A., and McMillion, L. G., 1961, Geology and ground-water resources of Pecos County, Texas: Texas Board Water Engineers Bull. 6106, v. 1, 241 p., v. 2, 295 p.

Baker, R. C., 1965, Reconnaissance investigation of the ground-water resources of the lower Rio Grande basin, Texas, *in* Reconnaissance investigation of ground-water resources of the Rio Grande basin, Texas: Texas Water Comm. Bull. 6502, p. L1-L34.

Baker, R. C., and Dale, O. C., 1961, Ground-water resources of the lower Rio Grande Valley area, Texas: Texas Board Water Engineers Bull. 6014, v. 1, 81 p., v. 2, 336 p.

———1964, Ground-water resources of the Lower Rio Grande Valley area, Texas: U.S. Geol. Survey Water-Supply Paper 1653, 56 p.

Baltz, E. H., Jr., and West, S. W., 1967, Geology and ground-water resources of the southern part of the Jicarilla Apache Indian Reservation and adjacent areas, New Mexico: U.S. Geol. Survey Water-Supply Paper 1576-H, 89 p.

Bean, R. T., 1949, Geology of the Roswell artesian basin, New Mexico, and its relation to the Hondo Reservoir: New Mexico State Engineer Tech. Rept. 9, p. 1-31.

Bennett, R. R., and Sayre, A. N., 1962, Geology and ground-water resources of Kinney County, Texas: Texas Water Comm. Bull. 6216, 163 p.

Bjorklund, L. J., 1957, Reconnaissance of ground-water conditions in the Crow Flats area, Otero County, New Mexico: New Mexico State Engineer Tech. Rept. 8, 19 p.

Bjorklund, L. J., and Maxwell, B. W., 1961, Availability of ground water in the Albuquerque area, Bernalillo and Sandoval Counties, New Mexico: New Mexico State Engineer Tech. Rept. 21, 117 p.

Bjorklund, L. J., and Motts, W. S., 1959, Geology and water resources of the Carlsbad area, Eddy County, New Mexico: U.S. Geol. Survey open-file report, 454 p.

Brackbill, R. M., and Gaines, J. C., 1964, El Capitan source water system: Jour. Petroleum Technology, December, p. 135-156.

Broadhurst, W. L., 1964, Utilization of ground water in the southwest: Am. Railway Devel. Assoc., 55th Ann. Mtg., St. Louis, Mo., May 3-6, 1964, Proc., p. 64-68.

Broadhurst, W. L., Sundstrom, R. W., and Rowley, J. H., 1950, Public water supplies in southern Texas: U.S. Geol. Survey Water-Supply Paper 1070, 114 p.

Broadhurst, W. L., Sundstrom, R. W., and Weaver, D. E., 1951, Public water supplies in western Texas: U.S. Geol. Survey Water-Supply Paper 1106, 168 p.

Brown, J. B., Rogers, L. T., and Baker, B. B., 1965, Reconnaissance investigation of the ground-water resources of the middle Rio Grande basin, Texas, *in* Reconnaissance investigation of the Rio Grande basin, Texas: Texas Water Comm. Bull. 6502, p. M1-M80.

Bryan, Kirk, 1938, Geology and ground-water conditions of the Rio Grande depression in Colorado and New Mexico, *in* [U.S.] Natl. Resources Comm., Regional planning — Part 6, Rio Grande Joint Investigation in the upper Rio Grande basin in Colorado, New Mexico, and Texas, 1936-37: Washington, U.S. Govt. Printing Office, p. 197-225.

Burbank, W. S., Lovering, T. S., Goddard, E. N., and Echel, E. B., 1935, Geologic map of Colorado: U.S. Geol. Survey.

Burkham, D. E., 1966, Hydrology of Cornfield Wash area and effects of land-treatment practices, Sandoval County, New Mexico: U.S. Geol. Survey Water-Supply Paper 1831, 87 p.

Code, W. E., 1952, Underground water resources of Colorado, *in* A hundred years of irrigation in Colorado: Colorado Water Conserv. Board and Colorado Agr. Mech. Coll., p. 49-55.

Conover, C. S., 1954, Ground-water conditions in the Rincon and Mesilla Valleys and adjacent areas in New Mexico: U.S. Geol. Survey Water-Supply Paper 1230, 200 p.

Conover, C. S., Herrick, E. H., Hood, J. W., and Weir, J. E., Jr., 1955, The occurrence of ground water in south-central New Mexico, *in* New Mexico Geol. Soc. Guidebook, 6th Field Conf., south-central New Mexico, 1955: p. 108-120.

Conover, C. S., Theis, C. V., and Griggs, R. L., 1963, Geology and hydrology of Valle Grande and Valle Toledo, Sandoval County, New Mexico: U.S. Geol. Survey Water-Supply Paper 1619-Y, 37 p.

Cooper, J. B., 1965, Ground-water resources of the northern Tularosa basin near Carrizozo, Lincoln County, New Mexico: U.S. Geol Survey Hydrol. Inv. Atlas HA-193.

Cooper, J. B., and John, E. C., 1968, Geology and ground-water resources in southeastern McKinley County, New Mexico: New Mexico State Engineer Tech. Rept. 35, 108 p.

Cooper, J. B., and West, S. W., 1967, Principal aquifers and uses of ground water between Laguna Pueblo and Gallup, Valencia and McKinley Counties, New Mexico, *in* New Mexico Geol. Soc. Guidebook, 18th Field Conf., 1967: p. 145-149.

Cox, E. R., 1957, Preliminary results of test drilling between Lake McMillan and Major Johnson Springs, Eddy County, New Mexico: U.S. Geol. Survey open-file report, 28 p.

———1967, Geology and hydrology between Lake McMillan and Carlsbad Springs, Eddy County, New Mexico: U.S. Geol. Survey Water-Supply Paper 1828, 55 p.

Cox, E. R., and Reeder, H. O., 1962, Ground-water conditions in the Rio Grande Valley between Truth or Consequences and Las Palomas, Sierra County, New Mexico: New Mexico State Engineer Tech. Rept. 25, 47 p.

Cushman, R. L., 1965, An evaluation of aquifer and well characteristics of municipal well fields in Los Alamos and Guaje Canyons, near Los Alamos, New Mexico: U.S. Geol. Survey Water-Supply Paper 1809, 50 p.

Dale, O. C., 1952, Ground-water resources of Starr County, Texas: Texas Board Water Engineers Bull. 5209, 47 p.

Dale, O. C., and George, W. O., 1954, Ground-water resources of Cameron County, Texas: Texas Board Water Engineers Bull. 5403, 63 p.

Dane, C. H., and Bachman, G. O., 1965, Geologic map of New Mexico: U.S. Geol. Survey.

Darton, N. H., Stephenson, L. W., and Gardner, J. A., 1937, Geologic map of Texas: U.S. Geol. Survey.

Davis, M. E., 1961, Ground-water reconnaissance of the Marfa area, Presidio County, Texas: Texas Board Water Engineers Bull. 6110, 41 p.

———1965, Development of ground water in the El Paso district, Texas, 1960-63, Progress report 9: Texas Water Comm. Bull. 6514, 34 p.

Davis, M. E., and Leggat, E. R., 1965, Reconnaissance investigation of the ground-water resources of the upper Rio Grande basin, Texas, *in* Reconnaissance investigations of the ground-water resources of the Rio Grande basin, Texas: Texas Water Comm. Bull. 6502, p. U1-U99.

Dillard, J. W., and Muse, R. W., 1964, Water levels and chemical analysis from observation wells in the Dell City area, Hudspeth and Culberson Counties, Texas, 1948 through January 1964: Texas Water Comm. Circ. 6401, 22 p.

Dinwiddie, G. A., 1963, Municipal water supplies and uses, southeastern New Mexico: New Mexico State Engineer Tech. Rept. 29A, 140 p.

———1964, Municipal water supplies and uses, northeastern New Mexico: New Mexico State Engineer Tech. Rept. 29B, 64 p.

———1967, Rio Grande Basin — Geography, geology, and hydrology, *in* Water resources of New Mexico: New Mexico State Plan. Office, p. 127-142.

Dinwiddie, G. A., and Clebsch, Alfred, Jr., 1973, Water resources of Gaudalupe County, New Mexico: New Mexico Bur. Mines and Mineral Resources Hydrol. Rept. 3, 43 p.

Dinwiddie, G. A., Mourant, W. A., and Basler, J. A., 1966a, Municipal water supplies and uses, northwestern New Mexico: New Mexico State Engineer Tech. Rept. 29C, 197 p.

———1966b, Municipal water supplies and uses, southwestern New Mexico: New Mexico State Engineer Tech. Rept. 29D, 98 p.

Dixon, R. M., 1961, General principals of ground water administration in Texas, *in* Sixth Annual New Mexico Water Conference: New Mexico State Univ., Nov. 1-2, 1961, p. 90-95.

Emery, P. A., 1970, Electric analog model evaluation of a water-salvage plan, San Luis Valley, south-central Colorado: Colorado Water Conserv. Board Ground-Water Series Circ. 14, 11 p.

Emery, P. A., Boettcher, A. J., Snipes, R. J., and McIntyre, H. J., Jr., 1971, Hydrology of the San Luis Valley, South-central Colorado: U.S. Geol. Survey Hydrol. Inv. Atlas HA-381.

Fiedler, A. G., and Nye, S. S., 1933, Geology and ground-water resources of the Roswell artesian basin, New Mexico: U.S. Geol. Survey Water-Supply Paper 639, 372 p.

Follett, C. R., 1954a, Records of water-level measurements in Loving and Ward Counties, Texas: Texas Board Water Engineers Bull. 5408, 24 p.

———1954b, Records of water-level measurements in Culberson, Hudspeth, and Jeff Davis Counties, Texas: Texas Board Water Engineers Bull. 5415, 31 p.

———1954c, Water-level measurements in El Paso County, Texas: Texas Board Water Engineers Bull. 5417, 49 p.

———1956a, Records of water-level measurements in Dimmit, Maverick, and Zavala Counties, Texas, 1920, 1928 to September 1956: Texas Board Water Engineers Bull. 5617, 74 p.

———1956b, Records of water-level measurements in Kinney, Uvalde, and Val Verde Counties, Texas: Texas Board Water Engineers Bull. 5611, 70 p.

Follett, C. R., White, W. N., and Irelan, Burdge, 1949, Occurrence and development of ground water in the Linn-Faysville area, Hidalgo County, Texas: Texas Board Water Engineers duplicated rept., 50 p.

Gaca, J. R., and Karig, D. E., 1966, Gravity survey in the San Luis Valley area, Colorado: U.S. Geol. Survey open-file report, 21 p.

Garza, Sergio, and Wesselman, J. B., 1959, Geology and ground-water resources of Winkler County, Texas: Texas Board Water Engineers Bull. 5916, 200 p.

——1962, Geology and ground-water resources of Winkler County, Texas: U.S. Geol. Survey Water-Supply Paper 1582, 162 p.

Gordon, E. D., 1962, Geology and ground-water resources of the Grants-Bluewater area, Valencia County, New Mexico: New Mexico State Engineer Tech. Rept. 20, 109 p.

Griggs, R. L., 1964, Geology and ground-water resources of the Los Alamos area, New Mexico: U.S. Geol. Survey Water-Supply Paper 1753, 107 p.

Hale, W. E., 1955, Ground-water conditions in the vicinity of Rattlesnake Springs, Eddy County, New Mexico: New Mexico State Engineer Tech. Rept. 3, 54 p.

——1961, Availability of ground water in New Mexico, *in* Sixth Annual New Mexico Water Conference: New Mexico State Univ., Nov. 1-2, 1961, p. 11-22.

Hale, W. E., Reiland, L. J., and Beverage, J. P., 1965, Characteristics of the water supply in New Mexico: New Mexico State Engineer Tech. Rept. 31, 131 p.

Hantush, M. S., 1955, Preliminary quantitative study of the Roswell ground-water reservoir, New Mexico: New Mexico Bur. Mines and Mineral Resources, 113 p.

Hendrickson, G. E., and Jones, R. S., 1952, Geology and ground-water resources of Eddy County, New Mexico: New Mexico Bur. Mines and Mineral Resources Ground-Water Rept. 3, 169 p.

Herrick, E. H., and Davis, L. V., 1965, Availability of ground water in the Tularosa basin and adjoining areas, New Mexico and Texas: U.S. Geol. Survey Hydrol. Inv. Atlas HA-191, 2 p.

Hiss, W. L., Peterson, J. B., and Ramsey, T. R., 1969, Saline water in southeastern New Mexico: Chem. Geology, v. 4, p. 341-360.

Hood, J. W., 1956, Summary of results of ground-water investigations in the vicinity of Boles well field, Otero County, New Mexico: U.S. Geol. Survey open-file report, 17 p.

——1959, Ground water in the Tularosa basin, New Mexico, *in* Guidebook for Joint Field Conf., Permian Basin Sec., Soc. Econ. Paleotologists and Mineralogists and Roswell Geol. Soc., Sacramento Mountains of Otero County, New Mexico: p. 236-250.

——1963, Saline ground water in the Roswell basin, Chaves and Eddy Counties, New Mexico, 1958-59: U.S. Geol. Survey Water Supply Paper 1539-M, 46 p.

Hood, J. W., and Herrick, E. H., 1965, Water resources of the Three Rivers area, Otero and Lincoln Counties, New Mexico: U.S. Geol. Survey Hydrol. Inv. Atlas HA-192.

Hood, J. W., and Kister, L. R., 1962, Saline-water resources of New Mexico: U.S. Geol. Survey Water-Supply Paper 1601, 67 p.

Hood, J. W., and Knowles, D. B., 1952, Summary of ground-water development in the Pecos area, Reeves and Ward Counties, Texas, 1947-51: Texas Board Water Engineers Bull. 5202, 11 p.

Hood, J. W., Mower, R. W., and Grogin, M. J., 1960, The occurrence of saline water near Roswell, Chaves County, New Mexico: New Mexico State Engineer Tech. Rept. 17, 93 p.

Hood, J. W., and Scalapino, R. A., 1951, Summary of the development of ground water for irrigation in the Lobo Flats area, Culberson and Jeff Davis Counties, Texas: Texas Board Water Engineers Bull. 5102, 25 p.

Hughes, P. W., and Kiersch, G. A., 1952, Structural localization of ground water in limestones, Big Bend District, Texas-Mexico: Geol. Soc. America Bull., v. 63, p. 1334-1335.

Hutchins, W. A., 1955, The New Mexico law of water rights: New Mexico State Engineer Tech. Rept. 4, 61 p.

Inglehart, H. H., 1967, Occurrence and quality of ground water in Crockett County, Texas: Texas Water Devel. Board Rept. 47, 150 p.

Kelley, V. C., 1955, Regional tectonics of south-central New Mexico, *in* New Mexico Geol. Soc. Guidebook, 6th Field Conf., south-central New Mexico, 1955: p. 96-104.

Kelly, T. E., Myers, B. N., and Hershey, L. A., 1970, Saline groundwater resources of the Rio Grande drainage basin — A pilot study: Office of Saline Water Research and Devel. Prog. Rept. 560, 71 p.

Kennon, F. W., and Peterson, H. V., 1960, Hydrology of Cornfield Wash, Sandoval County, New Mexico: U.S. Geol. Survey Water-Supply Paper 1475-B, p. 45-103.

King, W. E., Hawley, J. W., Taylor, A. M., and Wilson, B. P., 1971, Geology and ground-water resources of central and western Dona Ana County, New Mexico: New Mexico Bur. Mines and Mineral Resources Hydrol. Rept. 1, 64 p.

Knowles, D. B., 1952, Ground-water resources of Ector County, Texas: Texas Board Water Engineers Bull. 5210, 112 p.

Knowles, D. B., and Kennedy, R. A., 1956, Ground-water resources of the Hueco bolson northeast of El Paso, Texas: Texas Board Water Engineers Bull. 5615, 265 p.

——1958, Ground-water resources of the Hueco bolson northeast of El Paso, Texas: U.S. Geol. Survey Water-Supply Paper 1426, 186 p.

Lee, W. T., 1907, Water resources of the Rio Grande Valley in New Mexico and their development: U.S. Geol. Survey Water-Supply Paper 188, 59 p.

Leggat, E. R., 1962, Development of ground water in the El Paso district, Texas, 1955-60, Progress report 8: Texas Water Comm. Bull. 6204, 56 p.

Leggat, E. R, and Davis, M. E., 1966, Analog model study of the Hueco bolson near El Paso, Texas: Texas Water Devel. Board Rept. 28, 26 p.

Leggat, E. R., Lowry, M. E., and Hood, J. W., 1962, Ground-water resources of the lower Mesilla Valley, Texas and New Mexico: Texas Water Comm. Bull. 6203, 191 p.

Lingle, R. T., and Linford, Dee, 1961, The Pecos River Commisson of New Mexico and Texas — A report of a decade of progress, 1950-1960: Pecos River Comm., 284 p.

Littleton, R. T., and Audsley, G. L., 1957, Ground-water geology of the Alpine area, Brewster, Jeff Davis, and Presidio Counties, Texas: Texas Board Water Engineers Bull. 5712, 87 p.

McLean, J. S., 1970, Saline ground-water resources of the Tularosa Basin, New Mexico: Office of Saline Water Research and Devel. Prog. Rept. 561, 128 p.

Mechem, E. L., 1961, Ground water and ground water law in New Mexico, *in* Sixth Annual New Mexico Water Conference: New Mexico State Univ., Nov. 1-2, 1961, p. 70-78.

Meinzer, O. E., 1911, Geology and water resources of Estancia Valley, New Mexico, with notes on ground-water conditions in adjacent parts of central New Mexico: U.S. Geol. Survey Water-Supply Paper 275, 89 p.

Meyer, W. R., and Gordon, J. D., 1972, Development of ground water in the El Paso district, Texas, 1963-70: Texas Water Devel. Board Rept. 153, 51 p.

Meyers, J. S., 1962, Evaporation from the 17 Western States: U.S. Geol. Survey Prof. Paper 272-D, 100 p.

Motts, W. S., and Cushman, R. L., 1964, An appraisal of the possibilities of artificial recharge to ground-water supplies in part of the Roswell basin, New Mexico: U.S. Geol. Survey Water-Supply Paper 1785, 85 p.

Mourant, W. A., 1957, Reconnaissance of water resources in the upper part of Sacramento River Canyon, Otero County, New Mexico: U.S. Geol. Survey open-file report, 25 p.

______1963, Water resources and geology of the Rio Hondo drainage basin, Chaves, Lincoln, and Otero Counties, New Mexico: New Mexico State Engineer Tech. Rept. 28, 85 p.

______1967, Pecos River Basin — Geography, geology, and hydrology, *in* Water resources of New Mexico: New Mexico State Plan. Office, p. 60-73.

Mourant, W. A., and Shomaker, J. W., 1970, Water resources and geology of De Baca County, New Mexico: New Mexico Bur. Mines and Mineral Resources Ground-Water Rept. 10, 87 p.

Mower, R. W., Hood, J. W., Cushman, R. L., Borton, R. L., and Galloway, S. E., 1964, An appraisal of potential ground-water salvage along the Pecos River between Acme and Artesia, New Mexico: U.S. Geol. Survey Water-Supply Paper 1659, 98 p.

Murray, C. R., 1959, Ground-water conditions in the nonthermal artesian-water basin south of Hot Springs, Sierra County, New Mexico: New Mexico State Engineer Tech Rept. 10, 33 p.

Murray, C. R., and Reeves, E. B., 1972, Estimated use of water in the United States in 1970: U.S. Geol. Survey Circ. 676, 37 p.

Muse, W. R., 1966, Water-level data from observation wells in Culberson, Jeff Davis, Presidio, and Brewster Counties, Texas: Texas Water Devel. Board Rept. 16, 61 p.

Nash, J. P., 1918, Water resources of Val Verde County: Texas Univ. Bull. 1803, p. 44-45.

Ogilbee, William, Wesselman, J. B., and Irelan, Burdge, 1962, Geology and ground-water resources of Reeves County, Texas: Texas Water Comm. Bull. 6214, v. 1, 193 p., v. 2, 245 p.

Perkins, R. D., Buckner, A. W., and Henry, J. M., 1972, Availability and quality of ground water in the Cenozoic alluvium aquifer in Reeves, Pecos, Loving, and Ward Counties, Texas: Texas Water Devel. Board open-file rept., 28 p.

Powell, W. J., 1958, Ground-water resources of the San Luis Valley, Colorado: U.S. Geol. Survey Water-Supply Paper 1379, 284 p.

President's Water Resources Policy Commission, 1950, The Rio Grande, *in* Ten rivers in America's future: President's Water Resources Policy Comm. Rept., v. 2, no. 4, 67 p.

Renick, B. C., 1931, Geology and ground-water resources of Sandoval County, New Mexico: U.S. Geol. Survey Water-Supply Paper 620, 117 p.

Robinson, T. W., and Lang, W. B., 1938, Geology and ground-water conditions of the Pecos River valley in the vicinity of Laguna Grande de la Sal, New Mexico: New Mexico State Engineer 12th and 13th Bienn. Repts., 1934-38, p. 77-100; New Mexico State Engineer Bull. 4, 24 p.

Robinson, T. W., and Waite, H. A., 1938, Ground water in the San Luis Valley, Colorado *in* [U.S.] Natl. Resources Comm.. Regional planning — Part 6, Rio Grande Joint Investigation in the upper Rio Grande basin in Colorado, New Mexico, and Texas, 1936-37: Washington, U.S. Govt. Printing Office, p. 226-267.

Sayre, A. N., and Livingston, Penn, 1945, Ground-water resources of the El Paso area, Texas: U.S. Geol. Survey Water-Supply Paper 919, 190 p.

Scalapino, R. A., 1950, Development of ground water for irrigation in the Dell City area, Hudspeth County, Texas: Texas Board Water Engineers Bull. 5004, p. 1-38.

Scofield, C. S., 1938, Quality of water of the Rio Grande basin above Ft. Quitman, Texas: U.S. Geol. Survey Water-Supply Paper 839, 294 p.

Siebenthal, C. E., 1910, Geology and water resources of the San Luis Valley, Colorado: U.S. Geol. Survey Water-Supply Paper 240, 228 p.

Smith, R. E., 1956, Ground water of the El Paso district, Texas: Texas Board Water Engineers Bull. 5603, 33 p.

______1957, Geology and ground-water resources of Torrance County, New Mexico: New Mexico Bur. Mines and Mineral Resources Ground-Water Rept. 5, 186 p.

Sorensen, E. F., and Borton, R. L., 1967a, Pecos River Basin — Settlement, development, and water use, *in* Water resources of New Mexico: New Mexico State Plan. Office, p. 74-96.

______1967b, Central closed basins — Settlement, development, and water use, *in* Water resources of New Mexico: New Mexico State Plan. Office, p. 112-126.

Sorensen, E. F., and Linford, Dee, 1967, Rio Grande Basin — Settlement, development, and water use, *in* Water resources of New Mexico: New Mexico State Plan. Office, p. 143-168.

Sparks, F. L., 1970, The development of Colorado water law — A paper presented before a joint meeting of the U.S. Interagency Committee on the Missouri River Basin and the U.S. Interagency Committee on the Arkansas-White-Red River Basins: Fort Collins, Colo., July 8, 1970.

Spiegel, Zane, 1955, Geology and ground-water resources of northeastern Socorro County, New Mexico: New Mexico Bur. Mines and Mineral Resources Ground-Water Rept. 4, 99 p.

______1962, Hydraulics of certain stream-connected aquifer systems: New Mexico State Engineer Spec. Rept., 105 p.

Spiegel, Zane, and Baldwin, Brewster, 1958, Geology and water resources of the Sante Fe area, New Mexico, *with contributions by* F. E. Kottlowski and E. L. Barrows *and a section on* Geophysics by H. A. Winkler: U.S. Geol. Survey Water-Supply Paper 1525, 258 p.

Stearman, Jack, 1960a, Water-level measurements in Culberson, Hudspeth, and Jeff Davis Counties, Texas: Texas Board Water Engineers Bull. 6005, 7 p.

______1960b, Water levels in observation wells in Cameron, Hidalgo, and Starr Counties, Texas: Texas Board Water Engineers Bull. 6008, 7 p.

Texas Water Development Board, 1968, The Texas water plan: Austin, Tex., Texas Water Devel. Board.

Theis, C. V., and Conover, C. S., 1962, Pumping tests in the Los Alamos Canyon well field near Los Alamos, New Mexico: U.S. Geol. Survey Water-Supply Paper 1619-I, 24 p.

Theis, C. V., and Sayre, A. N., 1942, Geology and ground water, *in* [U.S.] Natl. Resources Plan. Board, Pecos River Joint Investigation — Reports of the participating agencies: Washington, U.S. Govt. Printing Office, p. 27-75.

Titus, F. B., Jr., 1963, Geology and ground-water conditions in eastern Valencia County, New Mexico: New Mexico Bur. Mines and Mineral Resources Ground-Water Rept. 7, 113 p.

______1967, Central closed basins — Geography, geology, and hydrology, *in* Water resources of New Mexico: New Mexico State Plan. Office, p. 97-111.

U.S. Bureau of Reclamation, 1963, Plan for development of Closed Basin Division, San Luis Valley project, Colorado: Amarillo, Tex., U.S. Bur. Reclamation, Region 5, Project Plan. Rept., 87 p.

U.S. Department of State, 1946, Flow of the Rio Grande and tributary contributions from San Marchial, New Mexico, to the Gulf of Mexico: International Boundary and Water Comm. Water Bull. 16, 88 p.

______1965, Flow of the Rio Grande and related data, from Elephant Butte Dam, New Mexico, to the Gulf of Mexico: International Boundary and Water Comm. Water Bull. 35, 153 p.

U.S. Geological Survey, 1961-71, Water resources data for New Mexico — Part 1, Surface-water records, 1961-71; Part 2, Water quality records, 1964-69 [ann. repts.].

______1970, The National atlas of the United States of America: U.S. Geol. Survey, Washington, D. C., 417 p.

[U.S.] National Resources Committee, 1938, Regional Planning —

Part 6, Rio Grande Joint Investigation in the upper Rio Grande basin in Colorado, New Mexico, and Texas, 1936-37: Washington, U.S. Govt. Printing Office, 566 p.

[U.S.] National Resources Planning Board, 1942a, Pecos River Joint Investigation — reports of the participating agencies: Washington, U.S. Govt. Printing Office, 407 p.

——1942b, Regional planning — Part 10, Pecos River Joint Investigation in the Pecos River basin in New Mexico and Texas: Washington, U.S. Govt. Printing Office, 207 p.

U.S. Public Health Service, 1962, Drinking water standards — 1962: U.S. Public Health Service Pub. 956, 61 p.

U.S. Water Resources Council, 1968, The Nation's water resources: Washington, U.S. Govt. Printing Office.

——1972a, Concepts, methodology, and summary data, v. 1 *of* 1972 OBERS projections regional economic activity in the U.S.: Washington, U.S. Govt. Printing Office, 109 p.

——1972b, Water resources regions, v. 4 *of* 1972 OBERS projections, regional economic activity in the U.S.: Washington, U.S. Govt. Printing Office, 283 p.

Weir, J. E., Jr., 1965, Geology and availability of ground water in the northern part of the White Sands Missile Range and vicinity, New Mexico: U.S. Geol. Survey Water-Supply Paper 1801, 78 p.

West, S. W., 1972, Disposal of uranium-mill effluent by well injection in the Grants area, Valencia County, New Mexico: U.S. Geol. Survey Prof. Paper 386-D, 28 p.

West, S. W., Cushman, R. L., Heckler, W. L., and Stow, J. M., 1965, Water resources, *in* Mineral and water resources of New Mexico: U.S. 89th Cong., 1st sess., Senate Comm. Interior and Insular Affairs, Comm. Print, p. 387-437 [New Mexico Bur. Mines and Mineral Resources Bull. 87].

White, D. E., 1968, Ground-water resources of Upton County, Texas: Texas Water Devel. Board Rept. 78, 132 p.

White, W. N., 1931, Preliminary report on the ground-water supply of Mimbres Valley, New Mexico: U.S. Geol. Survey Water-Supply Paper 637-B, p. 69-90.

White, W. N., Gale, H. S., and Nye, S. S., 1941, Geology and ground-water resources of the Balmorhea area, western Texas: U.S. Geol. Survey Water-Supply Paper 849-C, p. 83-146.

Winograd, I. J., 1959, Ground-water conditions and geology of Sunshine Valley and western Taos County, New Mexico: New Mexico State Engineer Tech. Rept. 12, 70 p.

Winslow, A. G., and Kister, L. R., 1956, Saline water resources of Texas: U.S. Geol. Survey Water-Supply Paper 1365, 105 p.

Wollman, Nathaniel, and others, 1962, The value of water in alternative uses: Albuquerque, New Mexico Univ. Press, 426 p.

Wood, L. A., 1956, Availability of ground water in the Gulf Coast region of Texas: U.S. Geol. Survey open-file report, 55 p.

Yates, J. C., 1961, Changes in quantity of ground water in New Mexico, *in* Sixth Annual New Mexico Water Conference: New Mexico State Univ., Nov. 1-2, 1961, p. 23-30.

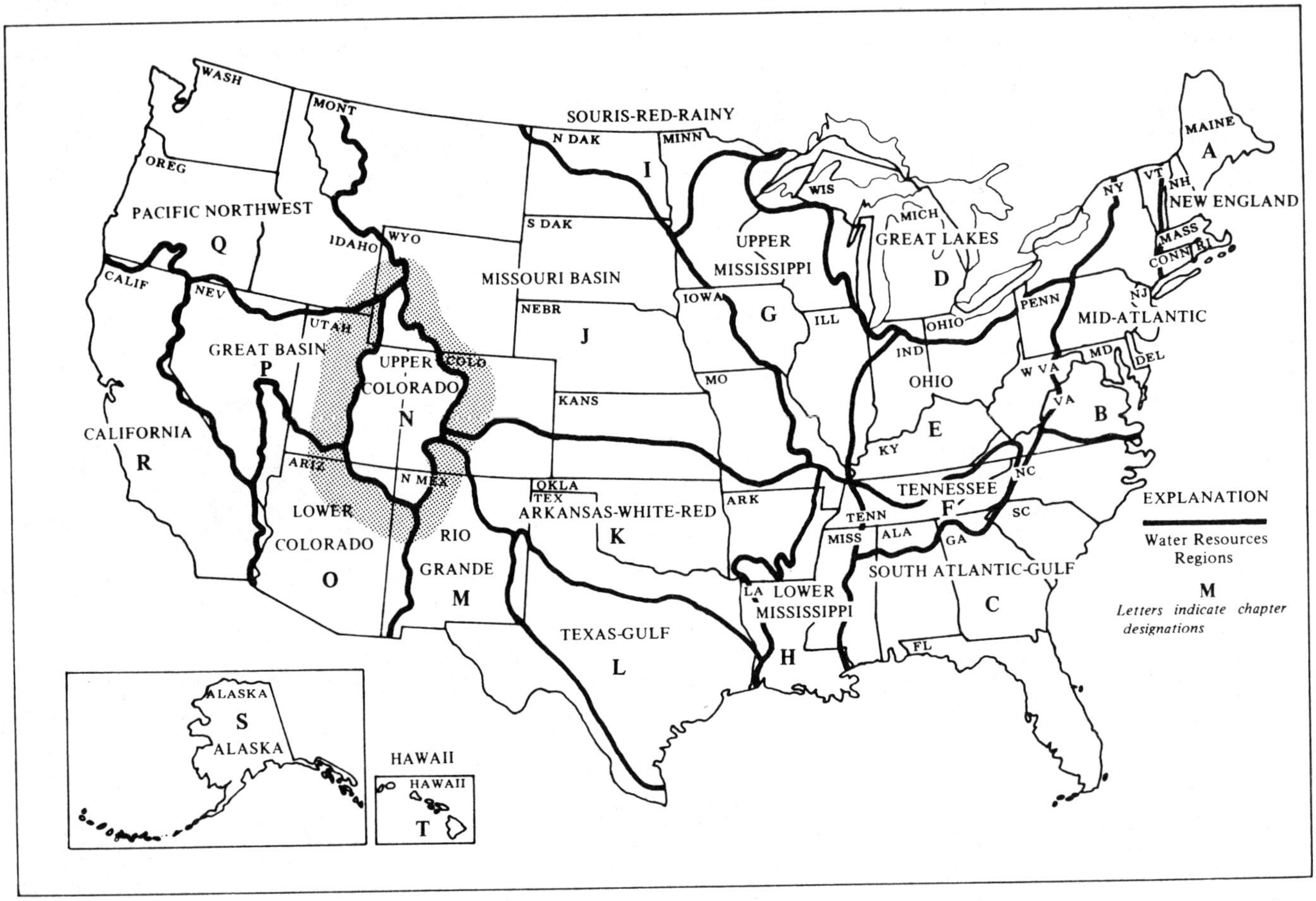

N – Location Map for the Upper Colorado Region

N

UPPER COLORADO REGION

By DON PRICE and TED ARNOW

ABSTRACT

The Upper Colorado Region covers about 113,500 square miles (293,965 km²) in parts of Arizona, Colorado, New Mexico, Utah, and Wyoming. Drainage from about 97 percent of the region is to the Colorado River. About 60 percent of the land is owned or administered by the Federal Government, and another 15 percent is in Indian trust. The predominantly arid to semiarid region is sparsely populated (averaging about three persons per square mile, or about two and one-half persons per km²) and is used chiefly for grazing, recreation, and mineral development.

The water supply for the region comes from precipitation within the region, which averages about 95 million acre-feet (117,182.5 hm³) per year. Development of the region's water supply has been limited almost entirely to surface water. Only about 2 percent of the total estimated volume of water withdrawn (about 5.7 million acre-ft, or 7,030.9 hm³) and consumed (about 3.6 million acre-ft, or 4,440.6 hm³) in the region in 1970 came directly from ground-water sources.

By the year 2020 consumptive use of water within the region and water exports to adjacent regions are expected to total more than 6.5 million acre-feet (8,017.8 hm³) per year. Use of the ground-water resources of the Upper Colorado Region in water-resources management can help to meet these water needs.

A tremendous amount of water is stored in the rocks (ground-water reservoirs) of the Upper Colorado Region. Recoverable water in just the upper 100 feet (30.5 m) of saturated rocks is estimated to be as much as 115 million acre-feet (141,852.5 hm³). That amount is nearly four times the total active storage capacity of all surface-water reservoirs in the region. The average annual replenishable supply of the ground-water reservoir is about 4 million acre-feet (4,934 hm³). This amount of water could irrigate about 1.3 million acres (526,110 ha) of crops having an annual water requirement of 3 feet per acre (0.9 m/ha), or it could provide about 3,600 million gallons (13,627,440 m³) per day for industrial use.

Most of the ground water is in consolidated rocks, which generally yield water to wells slowly. Much of the ground water is saline and, in some places, occurs at great depths. Nevertheless, the ground water is more uniformly distributed than is surface water, both areally and with time; therefore, it can be used advantageously in overall water-resources management. Recent advancements in the field of demineralization and in evaluation and development of ground water make this possible.

Options available for use of ground water in water-resources management in the region include conjunctive use with surface water or development of ground water as an independent supply. The latter option could be for a perennial supply or for a time-limited supply (mining ground water), depending on the need and the existing ground-water conditions. All options can be carried out so as to meet the requirements of the Colorado River Compact. The options could be implemented to optimally develop the Upper Colorado River Basin's allocation of Colorado River water while meeting the Compact commitments to the Lower Basin.

INTRODUCTION

PURPOSE AND SCOPE

The history of the Western United States is replete with stories of settlers and travelers who suffered grievously or abandoned their homes when streams, springs, and waterholes went dry. Agony for many of them would have been even greater if they had realized that beneath their feet, under the parched land, were huge reservoirs of cool clear water.

The early settlers were unaware of the underground reservoirs, but, even if they had known of them, they would not have been able to reach much of the water because of lack of drilling equipment. Today's occupants of the West know that the underground reservoirs exist and have at their command the equipment to tap the hidden water supplies, but still, for the most part, the underground water remains untouched even when the streams and waterholes cannot supply all the water needed. Why is this so?

Water on the surface is obvious — it can be seen, tasted, readily measured, and diverted for man's use. But ground water is hidden in the earth, and knowledge about ground water and the development of suitable techniques for its measurement and withdrawal have lagged for hundreds of years behind similar techniques for surface water.

Even today many water planners believe that ground water cannot be evaluated adequately in terms of availability, quantity, quality, cost of development, or the effect of development on surface-water sources. Such belief is no longer valid because advances in the science of ground-water hydrology in the last half century have provided the knowledge and techniques necessary for planning, developing, and managing water supplies using ground water. This same knowledge also can be applied to the joint management of interrelated surface water and ground water to make full use of the total water resource.

Unless the total water resource of the Upper Colorado Region is fully used, it may not be possible to satisfy the projected water needs of the region and the exports from the region (fig. 1) because of constraints on the full

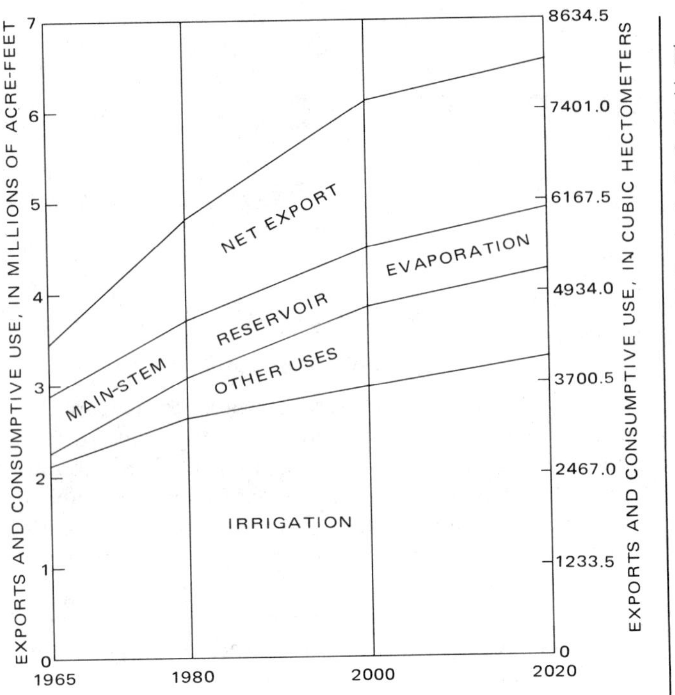

FIGURE 1. — Estimated annual exports and consumptive use of water within the region by the year 2020. Water needs and exports by then are expected to total more than 6.5 million acre-feet (8,017.8 hm^3). (After Hedlund and others, 1971, p. 47, 54.)

development of the surface-water sources. Even if the needs could be met entirely from surface water, the full development of surface water alone may not be possible because of the increasing social pressure to preserve and enhance the environment, which continues to place greater constraints on construction of surface-water storage facilities. More emphasis, therefore, must be placed on evaluation and development of the region's ground-water supply because ground water is the region's "unseen waterhole."

Optimal use of the total water resource of the Upper Colorado Region can best be achieved by considering all the alternatives — surface water alone in some areas, ground water alone in other areas, either as a sustained supply or an exhaustible supply, and conjunctive use of surface and ground water in still other areas.

The following questions about ground water are most commonly asked by water planners and managers in the Upper Colorado Region:

1. How much ground water is available for development?
2. What is the quality of the ground water?
3. What effect will large-scale ground-water development have on streamflow in general and on the Colorado River in particular?
4. What effect will large-scale ground-water development have on the environment?

The purpose of this report is to supply the best possible answers to those questions by presenting a gross regional assessment of the ground-water resources in a manner that will show the availability and significance of ground water in the region, and how it might be used alone or conjunctively with surface water to meet future water needs. The report shows what additional ground-water information is needed for project planning, design, and management, and what can be done to obtain this information. It also discusses possible uses of underground space as related to water and environmental problems.

The maps and discussions in this report (one of a series of similar reports that describe the entire United States by water-resources regions) are in sufficient detail to evaluate broad concepts of water planning, development, and management. The detail is not intended to be suitable for the design of new projects or implementation of specific proposals in terms of quantity, quality, or economic feasibility.

Most numbers are given in this report in English units followed by metric units in parentheses. The conversion factors used are as follows:

English unit (multiply)	Conversion factor (by)	Metric unit (to obtain)	Metric abbreviation
Acre	0.4047	Hectare	ha
Acre-foot	.0012335	Cubic hectometer	hm^3
Cubic foot	.02832	Cubic meter	m^3
Foot	.3048	Meter	m
Gallon	3.7854	Liter	l
Gallon	.0037854	Cubic meter	m^3
Inch	25.4	Millimeter	mm
Inch	2.54	Centimeter	cm
Mile	1.6093	Kilometer	km
Square mile	2.59	Square kilometer	km^2

Chemical concentration and water temperature are given only in metric units. Chemical concentration is given in milligrams per liter (mg/l). For concentrations less than 7,000 mg/l, the numerical value is about the same as for concentrations given in parts per million.

Water temperature is given in degrees Celsius (°C), which can be converted to degrees Fahrenheit by the following equation: $°F = 1.8\,(°C)+32$.

SETTING

The Upper Colorado Region includes the area drained by the Colorado River and its tributaries upstream from Lee Ferry, Ariz., and the area of the Great Divide Basin, a closed basin in Wyoming. The region covers about 113,500 square miles (293,965 km^2) in parts of Arizona, Colorado, New Mexico, Utah, and Wyoming. Physiographic subdivisions that lie wholly or partly within the region are shown in figure 2.

The Upper Colorado Region is characterized by high rugged mountains, broad basins, and high plateaus that

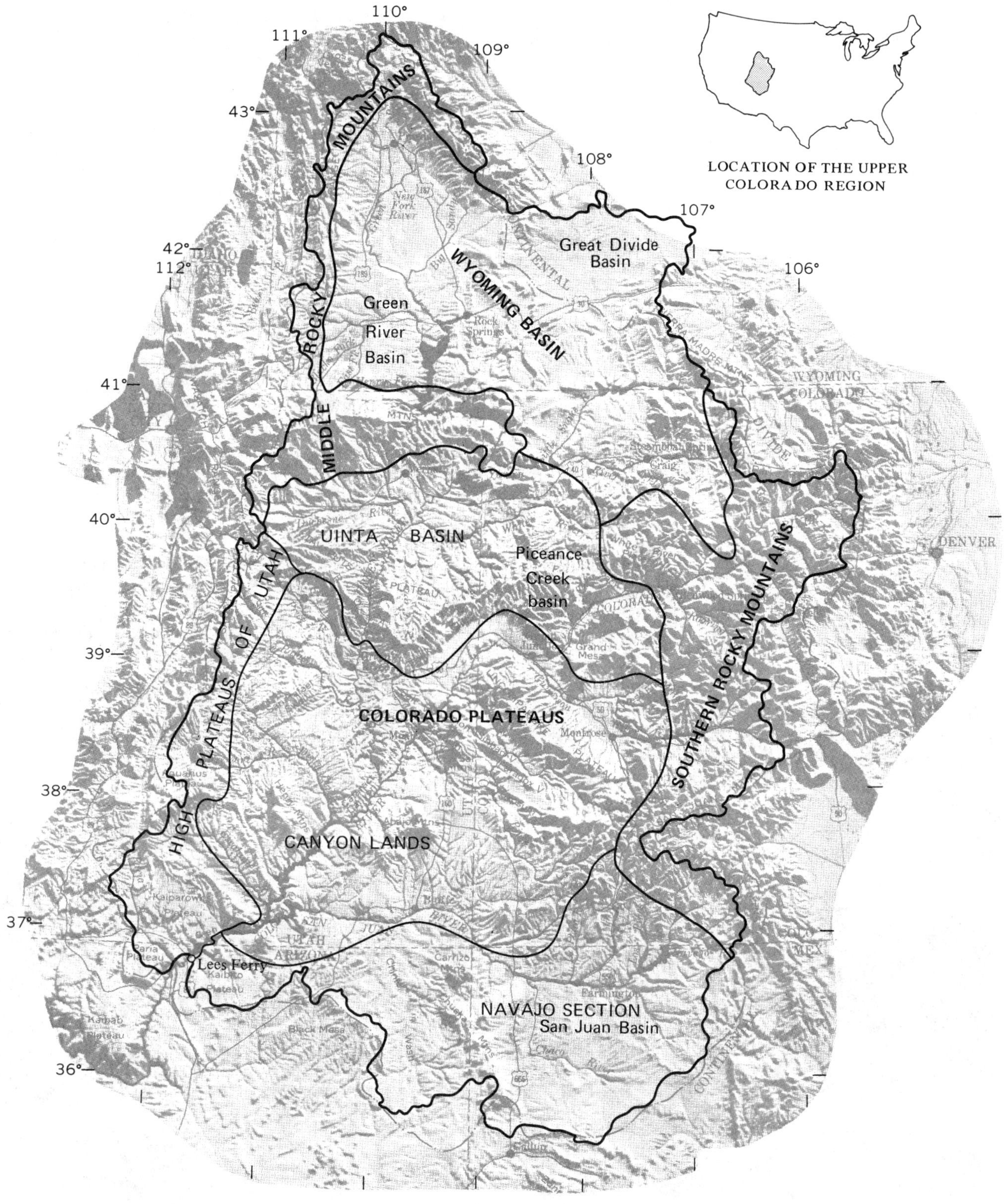

FIGURE 2. — The Upper Colorado Region, showing drainage and principal physiographic subdivisions. After Iorns and others (1965, fig. 1). Physiographic boundaries from Fenneman (1946).

TABLE 1. — *Average annual discharge of the three largest tributaries of the Colorado River — the Green, San Juan, and Gunnison Rivers*

Stream	Years of record	Period	Average annual discharge[1] cfs	m³/sec	Thousands of acre-feet	Thousands of hm³
Green River (at Green River, Utah)	72	Oct. 1894-Oct. 1899 Oct. 1904-Sept. 1971	6,360	180.12	4,610	5,686.4
San Juan River (near Bluff, Utah)	57	Oct. 1914-Sept. 1971	2,600	73.63	1,880	2,319.0
Gunnison River (near Grand Junction, Colo.)	62	1896-99, 1901-06, 1916-71	2,580	73.07	1,870	2,306.6

[1]Values have been rounded.

have been deeply entrenched and dissected by the Colorado River and many of its tributaries. Perhaps the most striking, unique physiographic feature of the region is the deep narrow intricate canyons that have been carved by streams (many of which are intermittent and ephemeral) in the varicolored rocks that underlie broad basins and plateaus (fig. 3). Altitudes range from about 3,100 feet (944.9 m) above mean sea level near Lee Ferry to more than 14,000 feet (4,267.2 m) in the Central and Southern Rocky Mountains. Most of the region has an arid to semiarid climate, and some areas receive less than 5 inches (127 mm) of precipitation a year. The higher plateaus and mountains have subhumid to alpine climatic zones, and more than 40 inches (1,016 mm) of precipitation a year falls on the highest peaks (pl. 1*B*).

Nearly 97 percent of the region drains to the Colorado River; the rest drains to the Great Divide Basin. Average annual discharge of the Colorado River near Lees Ferry was 17,760 cfs (cubic feet per second) (502.96 m^3/sec), or 12,860,000 acre-feet (15,862.8 hm^3) per year, for 49 years of record prior to completion of Glen Canyon Dam in 1963. The river and its three largest tributaries — the Green, San Juan, and Gunnison Rivers — all head in the Southern and Central Rocky Mountains (fig. 4), and the average annual discharge of each of these tributaries exceeds 2,000 cfs (56.64 m^3/sec) (table 1).

About 60 percent of the land in the region is owned or administered by the Federal Government, and another 15 percent is in Indian trust. (See fig. 5.) The region is sparsely populated, averaging about three persons per square mile (2.6 persons per km^2). Grand Junction, Colo., and Farmington, N. Mex., were the only communities with populations of more than 20,000 in 1970. (See fig. 6.) Because of the growing popularity of the region for recreation, however, many of the communities have large seasonal influxes of population. Most of the land is used for grazing, recreation, and mineral development (mostly fossil fuels).

FIGURE 3. — Deep narrow intricate canyon, such as those in the Canyon Lands, characteristic of much of the region. (Photograph by U.S. Bureau of Reclamation.)

FIGURE 4. — High rugged peaks of the San Juan Mountains, one of the headwater areas in the Upper Colorado River system. (Photograph by U.S. Bureau of Reclamation.)

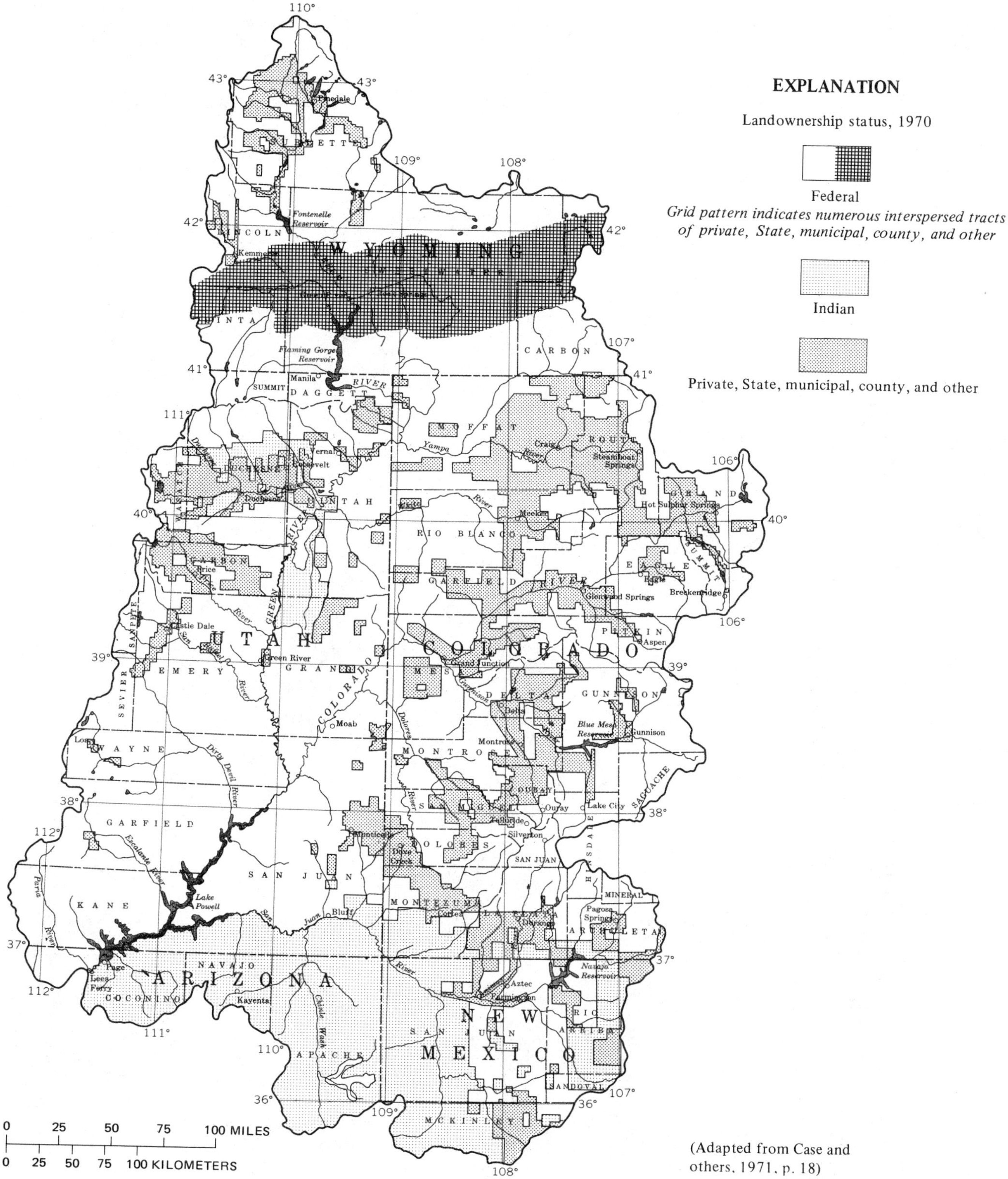

(Adapted from Case and others, 1971, p. 18)

FIGURE 5. — Most of the land in the region is owned or administered by the Federal Government.

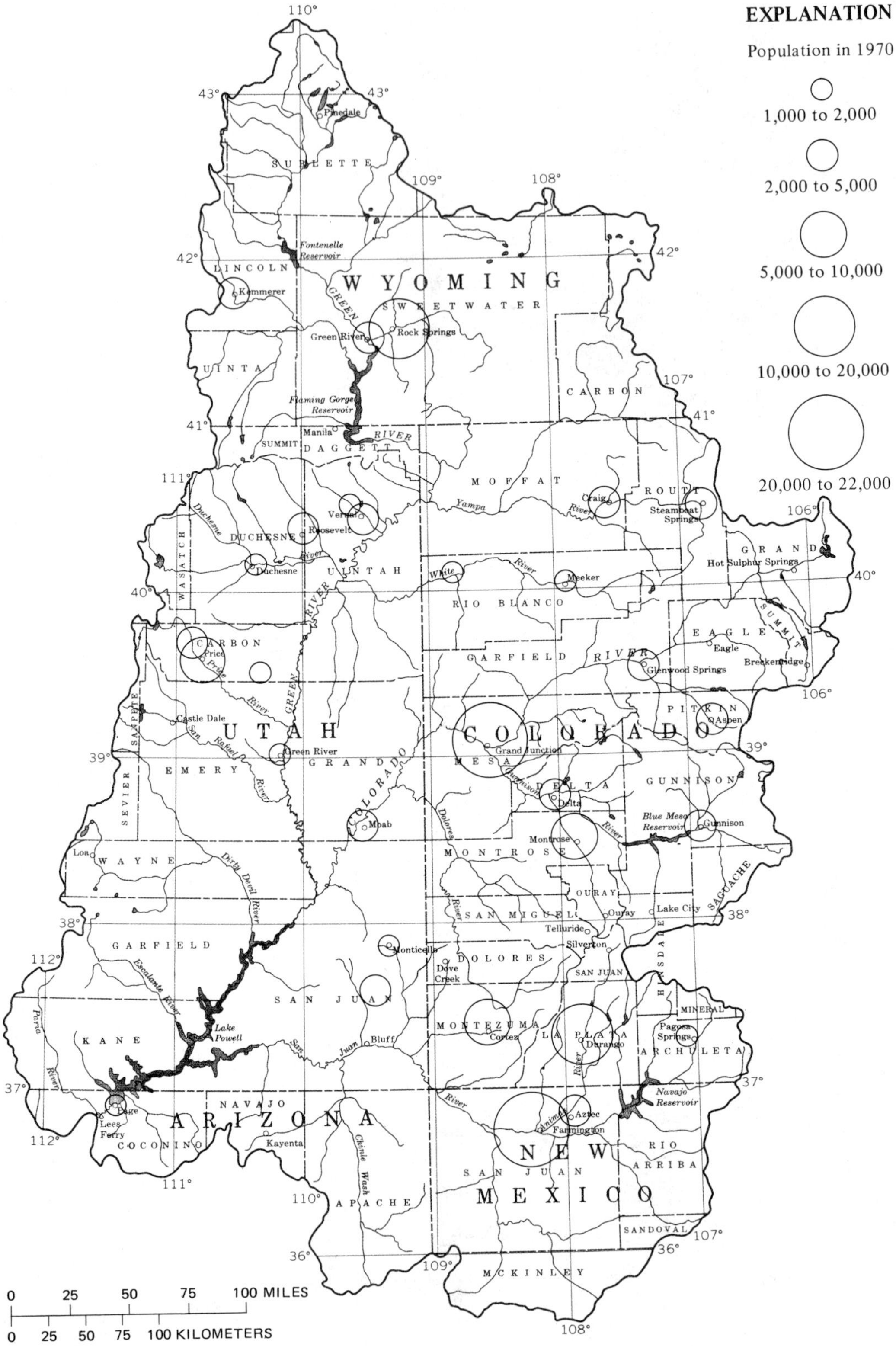

FIGURE 6. — Principal population centers in the region in 1970.

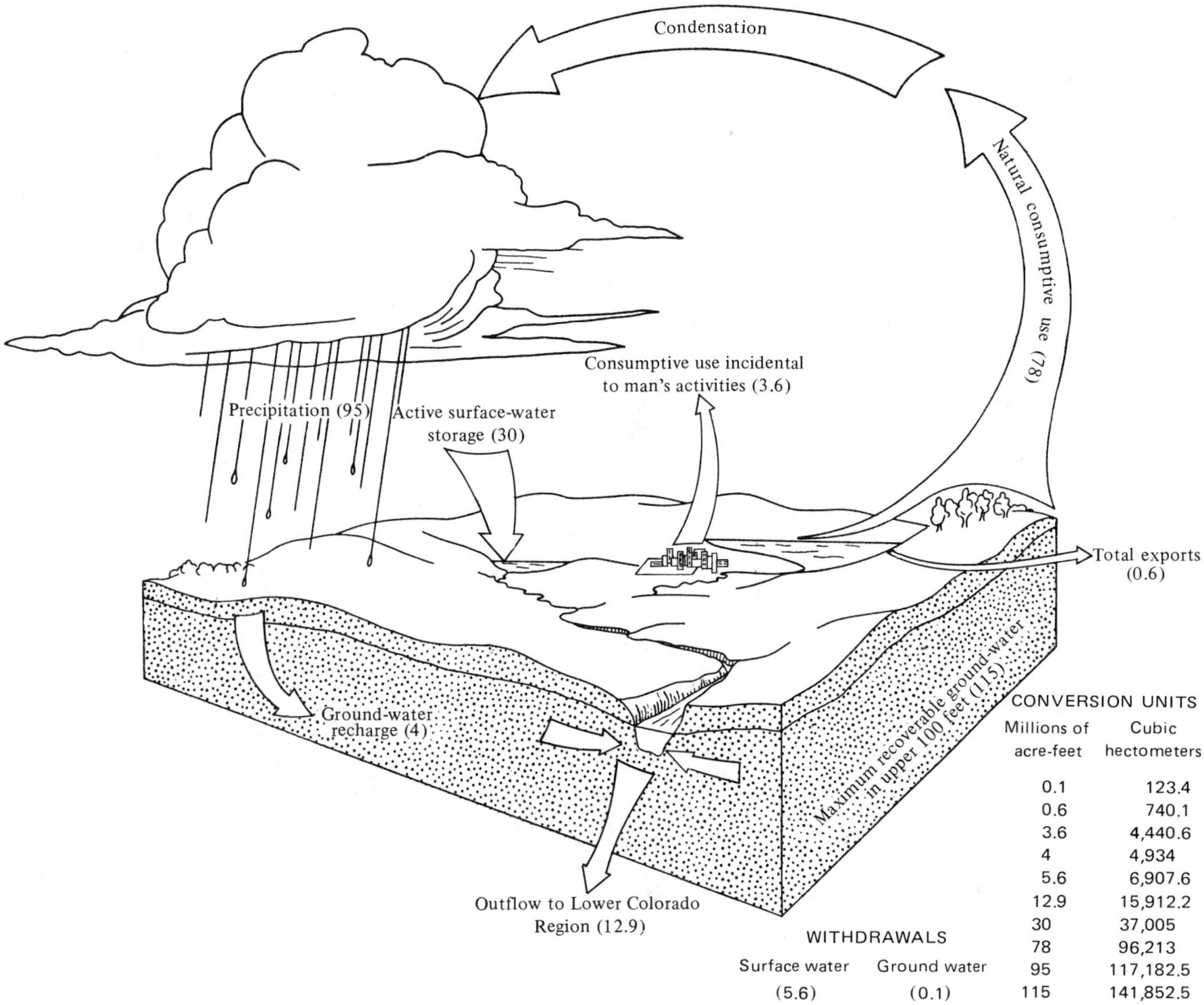

FIGURE 7. — The hydrologic system — the region's "water bank." Most of the supply is in ground-water storage, yet direct withdrawals of ground water are comparatively small. (Numbers in parentheses are in millions of acre-ft per year except those for storage, which are in millions of acre-ft.) Values are for the 1970 level of development, in part after Hedlund and others (1971) and Murray and Reeves (1972).

Development of the water resources of the region has been limited almost entirely to the development of surface water — particularly in relation to the Colorado River Storage Project of the U.S. Bureau of Reclamation and its participating agencies. The completion of such multipurpose dams as Flaming Gorge (1964) and Glen Canyon (1963) under this project has greatly stimulated the economy of the region. Development of ground water to date (1973) has been on a small scale and is limited in most places to withdrawals for domestic and stock supplies.

GROUND WATER IN OVERALL WATER-RESOURCES MANAGEMENT

THE WATER BANK

The hydrologic system in the Upper Colorado Region is the "water bank" for the region. (See fig. 7.) Deposits are made from the precipitation that falls within the region; withdrawals are the returns to the atmosphere by consumptive use, outflow in the Colorado River near Lees Ferry, and exports to other regions. The largest amount of water by far in the bank is in the ground-water reservoirs of the region. Under natural conditions these reservoirs help provide a more uniform annual and long-term water supply for the region. The reservoirs accept wet-season deposits for release during dry seasons, and provide large carryover storage for release during periods of drought. Proper development and management of these underground reservoirs by man can provide an efficient utilization of the "water bank."

UTILITY OF GROUND WATER

Because it is generally expedient to use what can be seen and to shy away from that which is hidden and not fully understood, surface water, if available, has been used as the source of water supply in many parts of the country without consideration being given to the alternative sources for development of a supply. Advances in techniques in ground-water hydrology during recent years now provide methods for resolving some of the development and management problems that historically have bred reluctance to consider ground-water development.

In the Upper Colorado Region, as in many other parts of the country, ground water has certain advantages, which in many areas make it a logical source of supply. It is widely distributed throughout the region and is generally dependable throughout the year. It is not everywhere equal in volume or rate of availability, but it is almost everywhere present in some quantity and quality, so that in many places the point of withdrawal can be at the point of use, and the effect upon the environment is minimal. Its temperature and chemical quality are fairly uniform, making it desirable for use where these characteristics are necessary. It is generally silt free and is stored in a spacious reservoir that is not subject to excessive losses by evaporation. Its reservoirs are ready made and generally self-regulated, yet many can be manipulated for more advantageous and beneficial use. The reservoirs are not easily polluted and are not subject to damage or destruction by natural or man-caused disasters.

ASSESSMENT OF THE GROUND-WATER RESOURCE

QUANTITY AND AVAILABILITY OF GROUND WATER

Because of the varying hydrologic properties of rocks, the geology of the Upper Colorado Region is the principal factor controlling the quantity and availability of ground water for development. The rocks that underlie the region consist mostly of consolidated and semiconsolidated sedimentary strata. Igneous and metamorphic rocks underlie parts of the mountains, and unconsolidated alluvial deposits underlie reaches of major stream valleys.

Iorns and others (1965, p. 4-8) grouped the rocks into eight hydrologic units on the basis of their age and general hydrologic properties. These units are herein regrouped into five geohydrologic units on the basis chiefly of relative hydrologic properties of the rocks (pl. 1*A*). The better known water-bearing formations in the units are listed in table 2.

TABLE 2. — *Principal water-bearing formations*

[Unconsolidated deposits of Quaternary age are the most permeable water-bearing formations in most parts of the region; sandstone strata of Jurassic, Cretaceous, and Tertiary age contain the most extensive bedrock aquifers. Letters correspond to areas shown on pl. 1*A*]

Age	Formation
WYOMING BASIN PROVINCE	
A — West-central part of Green River basin[1]	
Quaternary	Unconsolidated sedimentary deposits (mostly alluvial and lacustrine in origin)
Tertiary	Bridger Formation
	Green River Formation
	Wasatch Formation
Jurassic	Nugget Sandstone
Ordovician	Bighorn Dolomite
B — Great Divide and Washakie Basins[2]	
Quaternary	Unconsolidated sedimentary deposits (mostly alluvial and lacustrine in origin)
Tertiary	Browns Park Formation
	Green River Formation
	Wasatch Formation-Battle Spring Formation
Cretaceous	Ericson Formation
	Rock Springs Formation
	Mesaverde Formation (east part of area)
Mississippian	Madison Limestone
MIDDLE ROCKY MOUNTAINS PROVINCE	
C — South flank of Uinta Mountains[3]	
Quaternary	Unconsolidated sedimentary deposits (mostly alluvium)
Cretaceous	Dakota Sandstone
Jurassic	Nugget Sandstone
Permian	Park City Formation
Pennsylvanian	Weber Quartzite
	Morgan Formation
UINTA BASIN SECTION	
D — Eastern and central parts[4]	
Quaternary	Unconsolidated sedimentary deposits (mostly alluvium)
Tertiary	Duchesne River Formation
	Uinta Formation
	Green River Formation
	Wasatch Formation
Cretaceous	Frontier Sandstone Member of Mancos Shale
	Mowry Shale
	Dakota Sandstone
Triassic(?) and Jurassic	Glen Canyon Sandstone
Permian	Park City Formation
	Phosphoria Formation
Pennsylvanian	Weber Sandstone (Quartzite)
	Morgan Formation
	Round Valley Limestone

TABLE 2. — *Principal water-bearing formations* — Continued

HIGH PLATEAUS OF UTAH
E — Northern part[3]

- Quaternary
  - Unconsolidated sedimentary deposits (mostly alluvium)
- Tertiary
  - Crazy Hollow Formation
  - Flagstaff Formation
- Cretaceous and Tertiary
  - North Horn Formation
- Cretaceous
  - Emery Sandstone Member of Mancos Shale
  - Ferron Sandstone Member of Mancos Shale
- Jurassic
  - Carmel Formation
- Triassic(?) and Jurassic
  - Navajo Sandstone (Glen Canyon Group)
- Triassic
  - Wingate Sandstone (Glen Canyon Group)

F — Southern part[3]

- Quaternary
  - Unconsolidated alluvial and lacustrine deposits (probably some extrusive igneous rocks)
- Tertiary
  - Igneous rocks
  - Brian Head Formation
  - Wasatch Formation
- Cretaceous
  - Kaiparowits Formation
  - Wahweap Sandstone
  - Straight Cliffs Sandstone
- Jurassic
  - Carmel Formation
- Triassic(?) and Jurassic
  - Navajo Sandstone (Glen Canyon Group)
- Triassic
  - Wingate Sandstone (Glen Canyon Group)

CANYON LANDS
G — Henry Mountains vicinity[4]

- Quaternary
  - Unconsolidated sedimentary deposits (mostly alluvium and dune sand)
- Cretaceous
  - Dakota Sandstone
- Triassic(?) and Jurassic
  - Navajo Sandstone (Glen Canyon Group)
- Triassic
  - Wingate Sandstone (Glen Canyon Group)

H — La Sal Mountains vicinity[5]

- Quaternary
  - Unconsolidated sedimentary deposits (mostly alluvium and dune sand)
- Cretaceous
  - Dakota Sandstone
  - Burro Canyon Formation
- Jurassic
  - Entrada Sandstone
- Triassic(?) and Jurassic
  - Navajo Sandstone (Glen Canyon Group)
- Triassic
  - Wingate Sandstone (Glen Canyon Group)
- Permian
  - Cutler Formation

NAVAJO SECTION
I — North-central part[6]

- Quaternary
  - Unconsolidated alluvium and dune sand (some igneous rocks)
- Tertiary
  - Chuska Sandstone
- Cretaceous
  - Dakota Sandstone

TABLE 2. — *Principal water-bearing formations* — Continued

NAVAJO SECTION — Continued
I — North-central part[6] — Continued

- Jurassic
  - Recapture Shale Member of Morrison Formation
  - Salt Wash Sandstone Member of Morrison Formation
  - Summerville Formation
  - Cow Springs Sandstone
  - Bluff Sandstone
- Triassic(?) and Jurassic
  - Navajo Sandstone (Glen Canyon Group)
- Triassic(?)
  - Moenave Formation
- Triassic
  - Owl Rock Member of Chinle Formation
  - Shinarump Member of Chinle Formation
- Permian
  - Cedar Mesa Sandstone Member of Cutler Formation

J — Northeast part[7]

- Quaternary
  - Unconsolidated sedimentary deposits (mostly alluvium)
- Tertiary
  - San Jose Formation
- Cretaceous
  - Dakota Sandstone
- Jurassic
  - Entrada Sandstone

SOUTHERN ROCKY MOUNTAIN PROVINCE
K — North Park and Middle Park vicinity[8]

- Quaternary
  - Unconsolidated sedimentary deposits (mostly alluvium and glacial deposits)

L — Glenwood Springs-McCoy vicinity[9]

- Quaternary
  - Unconsolidated alluvium
- Mississippian
  - Leadville Limestone

From geologic columns and sections in:
[1]Oriel (1963).
[2]Welder and McGreevy (1966).
[3]Feltis (1966).
[4]Hunt and others (1953).
[5]Carter and Gualtieri (1965).
[6]Feltis (1966) and McGavock and others (1966).
[7]Baltz (1967) and McGavock and others (1966).
[8]Voegeli (1965) and Hall (1968).
[9]Boettcher (1972).

The source of virtually all ground water in transient storage in the Upper Colorado Region is the precipitation that falls within the region. Total annual precipitation in the region averages about 95 million acre-feet (117,182.5 hm^3) (Hedlund and others, 1971, p. i). Practically all this water is consumed at or near the place of fall by sublimation and evapotranspiration or becomes overland runoff. Only about 4 percent, or about 4 million acre-feet (4,934 hm^3), is estimated to become ground-water recharge. This includes percolation through the soil zone as well as seepage from streams and lands irrigated by streams. It is regarded as the peren-

TABLE 3. — *Estimated recoverable ground water in storage,*[1] *Upper Colorado Region*

Geohydrologic unit (pl. 1*A*)	Rock type	Area (thousands of acres)	Estimated specific yield[2] (percent)	Saturated thickness[3] (ft)	Estimated amount of water in storage (thousands of acre-ft)	
					Minimum	Maximum
1	Unconsolidated deposits	800	5-15	50	2,000	6,000
2	Volcanic rocks	2,200	2-5	100	4,400	11,000
3	Sedimentary rocks	40,000	1-2	100	40,300	80,600
4	do	24,300	0.2-0.7	100	4,900	17,000
5	Igneous and metamorphic rocks	5,100	0-0.3	100	0	1,500
	Total (rounded)				50,000	115,000

[1]About 85 percent of the recoverable ground water stored in the upper 100 feet of saturated rocks occurs in sedimentary rocks which have relatively low permeability and yield water slowly.

[2]The ratio of the volume of water that a saturated rock will yield by gravity to the volume of the rock.

[3]Ground-water storage is estimated for only 50 feet of the unconsolidated deposits because in many places the saturated thickness of this unit does not greatly exceed 50 feet. Ground-water storage is estimated for only 100 feet of the other geohydrologic units because the saturated thickness of the units is not known. In many places throughout the region the saturated thickness is much more than 100 feet; thus, total recoverable ground-water storage in the region greatly exceeds the maximum amount shown in this table.

nial yield[1] of the ground-water reservoirs or perennial ground-water supply for the region under the existing climatic and hydrologic conditions.

A perennial water supply of 4 million acre-feet (4,934 hm^3) could support a population of about 24 million people having an average daily per capita consumption of 150 gallons (568 l). It could irrigate about 1.3 million acres (526,110 ha) of crops having an average annual water requirement of 3 feet per acre (0.9 m/ha), or it could provide about 3,600 mgd (million gallons per day) (13,627,440 m^3) for industrial use. It should be noted, however, that, over a long period of years, the natural ground-water discharge balances the natural ground-water recharge. Consequently, any diversion by wells and consumptive use of ground water will cause some form of natural ground-water discharge, such as seepage to streams, to decrease proportionately.

The principal areas of natural ground-water recharge are in the higher mountains and plateaus, which receive the largest amount of annual precipitation (pl. 1*B*) and produce most of the runoff (pl. 1*C*). The ground water moves from the areas of recharge to areas of natural discharge, which include numerous widely scattered springs, gaining reaches of streams, and areas of phreatophyte growth. The principal areas of ground-water recharge and natural ground-water discharge are shown on plate 1*D*.

Ground water occurs under both water-table (unconfined) and artesian (confined) conditions, as illustrated in figure 8. Water-table conditions commonly exist in shallow alluvial aquifers along the larger streams, in principal recharge areas, and in the relatively flat-lying rocks that prevail in the Canyon Lands and Navajo sections (fig. 2) of the region. Artesian conditions occur locally throughout the region but are prevalent in the bedrock aquifers of the major structural basins, such as the Green River, Uinta, Piceance Creek, and San Juan Basins.

The volume of recoverable ground water in storage in the upper 100 feet (30.5 m) of the saturated rocks is estimated to be between about 50 and 115 million acre-feet (61,675 and 141,852.5 hm^3) (table 3). The maximum figure is nearly four times the total active storage capacity of all surface-water impoundments in the region, including Lake Powell and Flaming Gorge and Navajo Reservoirs. Total recoverable ground-water storage in the region — that is, in the complete section of saturated rocks — is many times the amount stored in the upper 100 feet (30.5 m) of saturated rocks. Plate 1*E* shows that the amount of recoverable storage per square mile (2.6 km^2) in the upper 100 feet (30.5 m) of saturated rocks varies considerably from place to place.

Although the total volume of recoverable ground water in storage is great, the water cannot be recovered in large quantities from wells at all places. For example, about 85 percent of the estimated maximum recoverable water in storage occurs in sedimentary rocks (table 3), which have relatively low permeability and yield water slowly. Only about 5 percent of the estimated maximum recoverable water in storage occurs in unconsolidated deposits, which includes permeable alluvium. Wells that yield more than 50 gpm (gallons per minute) (189 l/min) generally can be expected only along reaches of larger streams and in small widely scattered alluvium-filled basins (pl. 1*F*).

The depth to ground water and the depth from which it must be pumped have a great economic effect on the availability of ground water. In much of the Canyon Lands and Navajo sections of the region, ground-water levels are several hundred to more than 1,000 feet (several tens to more than 304.8 m) below the land surface,

[1]Perennial yield of a ground-water reservoir, as used in this report, is the maximum amount of water of suitable chemical quality that can be withdrawn from the reservoir each year for an indefinite period of years without causing a continuing depletion of storage. The perennial yield cannot exceed the average annual natural recharge to, or discharge from, the reservoir.

whereas ground water in alluvium of the larger perennial streams generally is within 50 feet (15.2 m) of the land surface (pl. 1*G*). As illustrated in figure 8 and shown on plate 1*G*, wells will flow under artesian pressure in a number of areas, but in most, several hundred feet of drilling through unsaturated rocks is required before the artesian aquifers are reached.

QUALITY OF THE GROUND WATER

Plate 2*A* shows the general chemical quality of the water with regard to its dissolved-solids content. Water containing less than 1,000 mg/l (milligrams per liter) of dissolved solids is fresh; water containing more than 1,000 mg/l is saline according to the following classification used by the U.S. Geological Survey:

Class	*Dissolved solids (mg/l)*
Fresh	0- 1,000
Slightly saline	1,000- 3,000
Moderately saline	3,000-10,000
Very saline	10,000-35,000
Briny	>35,000

Freshwater is generally available from shallow aquifers in most rock units in areas above an altitude of about 7,000 feet (2,133.6 m). At lower altitudes, fresh water most commonly occurs in the relatively permeable sandstone, such as the Navajo and Dakota Sandstones. Freshwater is also found in carbonate rocks, such as the Madison Limestone and Morgan Formation, which have good hydrologic connection with the principal recharge areas in the mountains.

Saline water commonly occurs in shale and siltstone strata that underlie most of the Green River, Great Divide, Washakie, Uinta, Piceance Creek, and San Juan Basins, but even in those areas some of the aquifers locally contain freshwater.

The quality of ground water in many parts of the region may be greatly altered by man's activities. Areas where fresh ground water might be contaminated by man's activities are shown on plate 2*B*.

Temperature is an important property of ground water, particularly with regard to artificial recharge, to use of the water for certain industries, and to possible development of geothermal energy. The degree of success of most artificial-recharge operations (or fluid-waste disposal) depends largely on the thermal compatability of the recharge water and the natural ground water. The success of industries that require cooling water (such as large thermal electric powerplants) depends at least in part on the initial temperature of the cooling water. Finally, the areal distribution of thermal water (water whose temperature is at least 5°C higher than the mean annual air temperature of the area in which the water occurs) is a guide in locating potential sites for development of geothermal energy.

The temperature of water from most water wells less than 1,000 feet (304.8 m) deep and from nonthermal springs is about 5°-10°C in the mountainous parts of the region and about 10°-20°C in the lower basin and plateau areas (pl. 2*C*). Discharge of thermal ground water is relatively rare in the region and is confined chiefly to a few widely scattered thermal springs (pl. 2*C*).

RELATION OF SURFACE WATER AND GROUND WATER

A close relation exists between surface water and ground water. This relation is shown diagrammatically in figure 9 and is known to exist throughout the Upper Colorado Region. Factors that control this relation include the degree of hydraulic connection between the aquifers and the stream channels, lakes, or reservoirs, the permeability of the aquifers, and the slope or gradient of the water table. As noted earlier, some recharge to aquifers occurs as seepage from certain reaches of streams in the region (fig. 9*A*); conversely, discharge from aquifers occurs as seepage to other reaches of the stream (fig. 9*B*). Aquifers in the permeable alluvial deposits along perennial streams receive recharge (bank storage) during peak runoff periods (fig. 9*D*) and release the water back to the streams as streamflow subsides. Ground-water withdrawal by wells or vegetation along streams depletes streamflow (fig. 9*C*).

Bank storage at manmade reservoirs can involve considerable quantities of water. For example, Madison and Waddell (1973, p. C16) estimated that about 240,000 acre-feet (296 hm^3) of water went into bank storage, mostly into bedrock aquifers, at Flaming Gorge Reservoir during the period 1963-68; also, it has been estimated that as much as 6.7 million acre-feet (8,264.4 hm^3) has gone into bank storage (mostly into the Navajo Sandstone) at Lake Powell since filling of that reservoir began in 1963. This increased bank storage is reflected by a rise of water levels in wells near Lake Powell. (See fig. 10.)

The interchange of ground water and surface water also has an effect on water quality. For example, the relatively large discharge-weighted-average concentration of dissolved solids in the streamflow in various parts of the region (pl. 2*D*) results partly from inflow of saline ground water from such formations as the Mancos Shale and Paradox Formation to streams. In some areas, seepage of highly mineralized irrigation return flow from surface sources into aquifers can deteriorate the quality of water in those aquifers. (See pl. 2*B*.) Conversely, the chemical quality of ground water in some aquifers (as the Uinta and Duchesne River Formations in the Uinta Basin) has improved where irrigation developments have augmented natural ground-water recharge with fresher water diverted from surface sources.

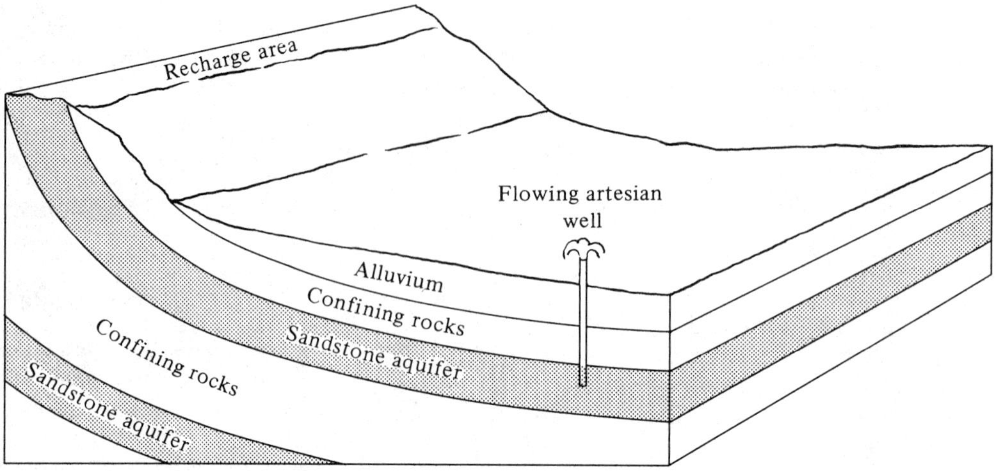

A. CONFINED GROUND WATER IN FOLDED ROCKS
Examples: Green River, Washakie, Uinta, Piceance Creek, and San Juan Basins

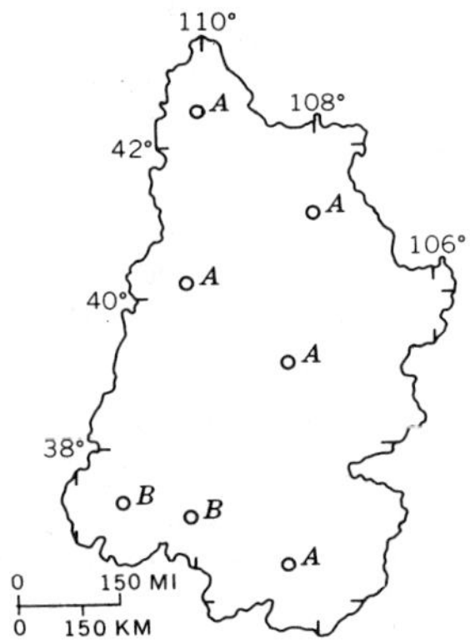

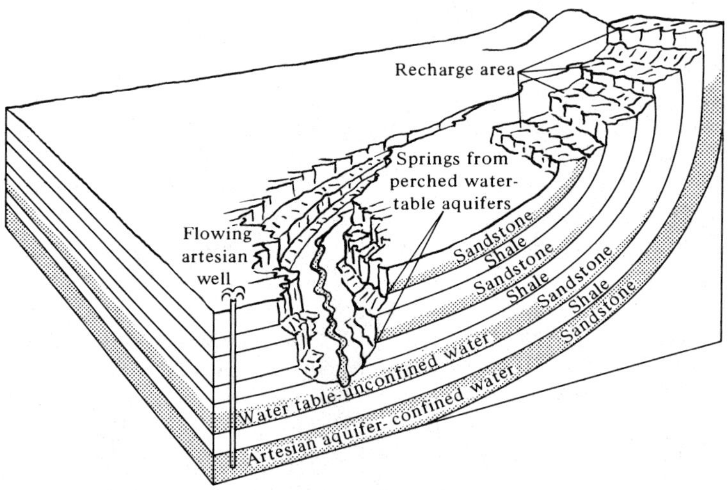

B. CONFINED AND UNCONFINED GROUND WATER
IN DEEPLY DISSECTED ROCKS
Example: Canyon Lands area

FIGURE 8 (above and facing page). — Examples of how ground water occurs under water-table (unconfined) or artesian (confined) conditions in the region.

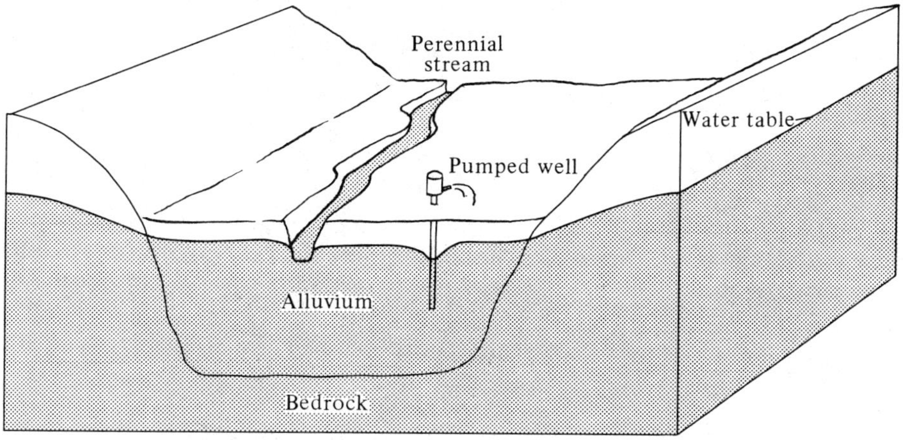

C. UNCONFINED GROUND WATER IN STREAM-VALLEY ALLUVIUM
Examples: Valley segments of the Green, Yampa, White, and Gunnison Rivers

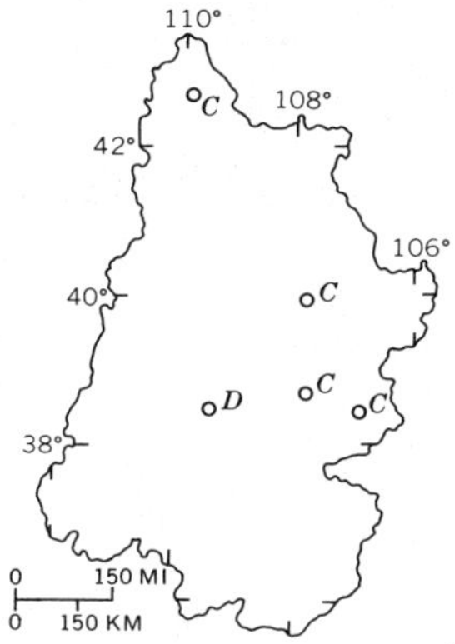

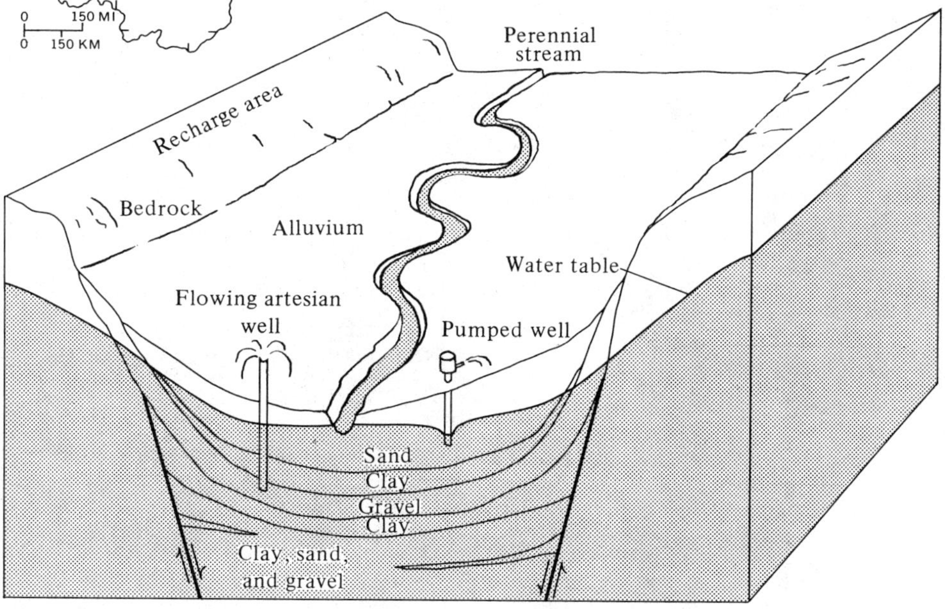

D. UNCONFINED AND LOCALLY CONFINED GROUND WATER
IN A DEEP ALLUVIUM-FILLED BASIN
Example: Spanish Valley

FIGURE 8. — Continued.

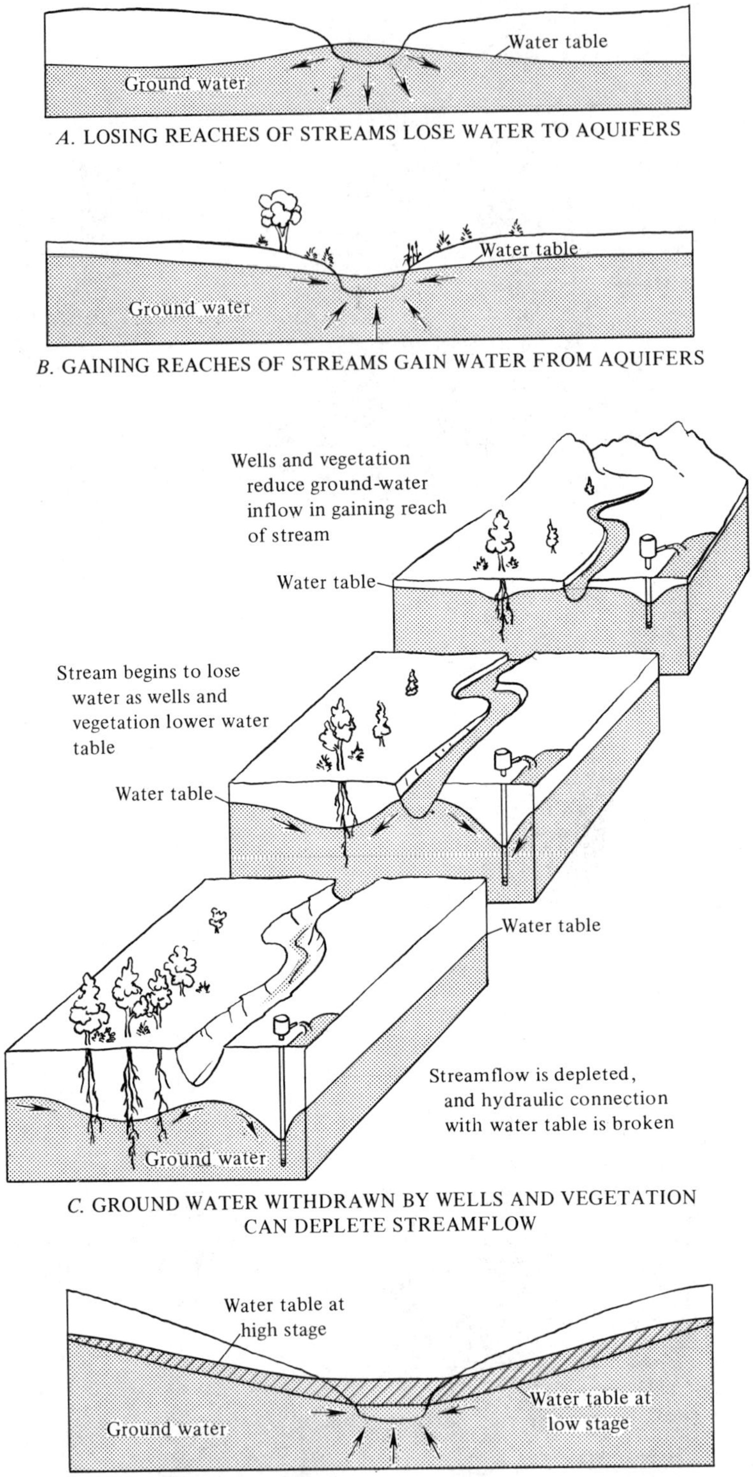

A. LOSING REACHES OF STREAMS LOSE WATER TO AQUIFERS

B. GAINING REACHES OF STREAMS GAIN WATER FROM AQUIFERS

C. GROUND WATER WITHDRAWN BY WELLS AND VEGETATION CAN DEPLETE STREAMFLOW

D. GROUND-WATER STORAGE IS INCREASED OR DEPLETED WITH THE INCREASE OR DEPLETION OF STORAGE IN THE BANKS OF STREAMS, LAKES, AND RESERVOIRS

FIGURE 9. — Surface water and ground water are closely related.

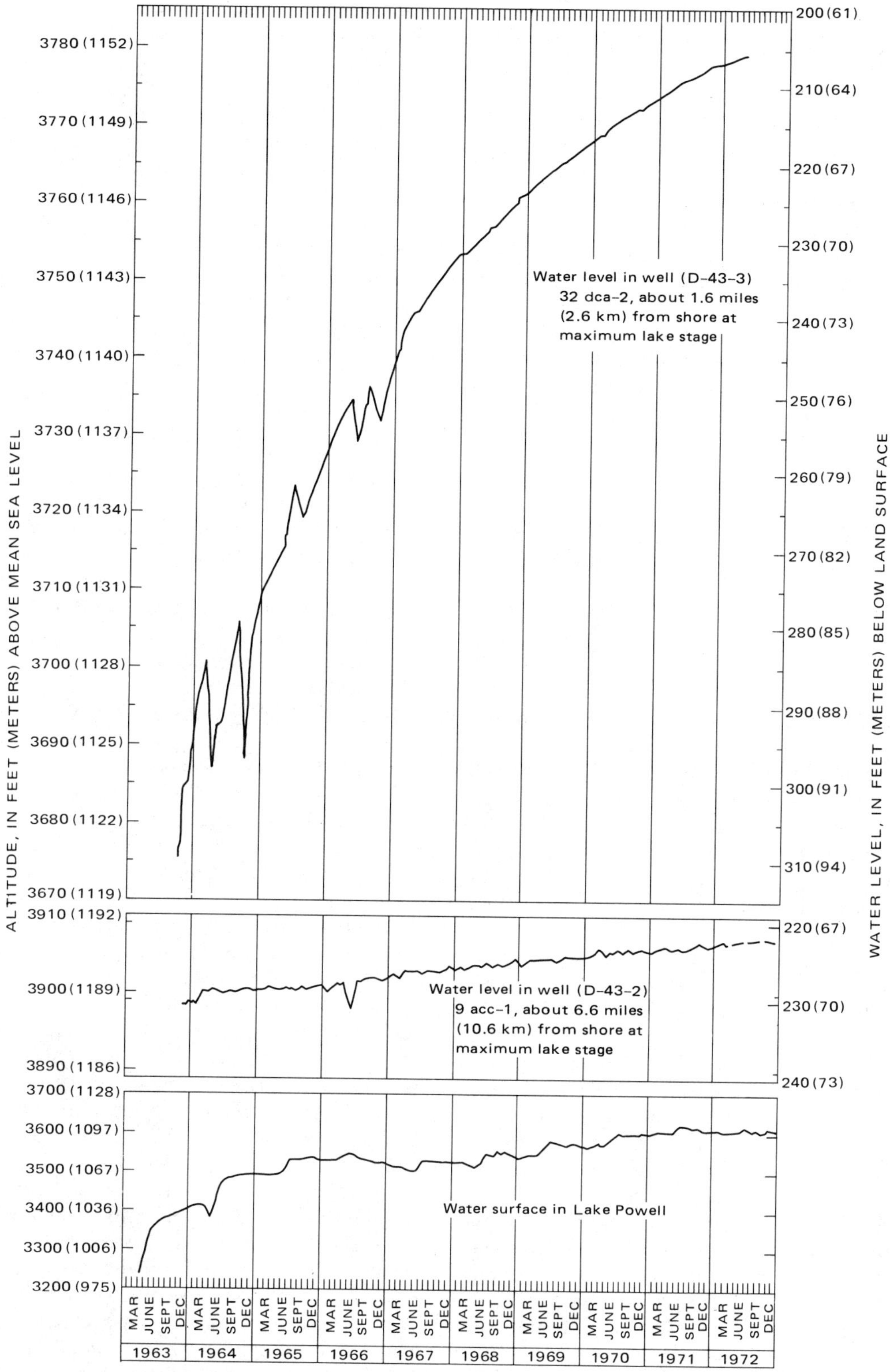

FIGURE 10. — Rising ground-water levels near Lake Powell reflect the increased bank storage resulting from filling of the lake. Note change in scale for bottom graph. Well locations are shown on plate 1*F*.

USE OF GROUND WATER

Use of ground water in the Upper Colorado Region is negligible compared with use of surface water. For example, estimated withdrawals and consumptive use of ground water in 1970 amounted to only about 122,000 and 63,000 acre-feet (150.5 and 77.7 hm^3), respectively (fig. 11). Total withdrawals and consumptive use of water from all sources (including main-stem reservoir evaporation) were about 5.7 million and about 3.6 million acre-feet (7,030.9 and 4,440.6 hm^3), respectively. Most of this water was from the Colorado River system, including the ground-water component to streamflow.

The following tabulation of estimated total annual withdrawal of ground water by States in 1970 was compiled from unpublished data used by Murray and Reeves (1972) to estimate the use of water in the United States:

State	Withdrawal	
	Acre-feet	hm^3
Arizona	12,000	14.8
Colorado	27,000	33.3
New Mexico	9,000	11.1
Utah	38,000	46.9
Wyoming	36,000	44.4

There are few areas of concentrated ground-water withdrawals from wells in the region. Local areas in which total annual withdrawals exceed 500 acre-feet (0.6 hm^3) are shown in figure 12. The areas of largest withdrawals are Ashley Valley and upper Fremont River valley, Utah, where the water is used chiefly to supplement surface water for irrigation, and the Gunnison, Colo., area, where the water is used chiefly for public supply.

Withdrawals of ground water by wells apparently have not had widespread significant effect on ground-water levels. The few long-term water-level records available for various parts of the region indicate that the changes in ground-water levels (which reflect changes in ground-water storage) are caused chiefly by changes in the annual supply of natural recharge from precipitation. (See fig. 13.) Local depletions of ground-water storage by pumping, as in the Uinta Basin (indicated by well U(D-1-1)14bbc-1 in fig. 13) and in the Grand Junction area (Lohman, 1965, p. 113), are probably more than offset regionally by the increased storage resulting from bank storage around new reservoirs, such as Lake Powell and Flaming Gorge and Navajo Reservoirs. Because of the close relation between surface water and ground water (fig. 9*C*), however, large-scale ground-water withdrawal over a long period of time would intercept water that naturally enters streams, and this would ultimately reduce the flow of the Colorado River.

INCREASING THE USABLE SUPPLY OF GROUND WATER

In any given area of the region, the available supply of ground water is not necessarily the usable supply. In many places, for example, a large percentage of the available supply is too saline for most uses. In other places, much of the available supply is consumed by vegetation that generally is considered to be nonbeneficial economically. Even in areas where all the available supply is of usable quality, it may not be in sufficient quantity to sustain withdrawals for an intended use.

A number of water-management practices are available to increase the usable ground-water supply. They include (1) artificial recharge, (2) reduction of natural discharge by nonbeneficial vegetation, (3) demineralization, and (4) recycling. (See fig. 14.)

ARTIFICIAL RECHARGE

Artificial recharge is the augmentation of natural ground-water recharge by the activities of man. It may be used either to replace ground water where water levels have been drawn down or to temporarily increase local ground-water storage for later recovery. Artificial recharge can also serve to help even out streamflow — that is, aquifers can be used to absorb flood runoff, which will be released slowly back to the stream as runoff subsides.

Artificial recharge can be accomplished by two basic methods — surface spreading and subsurface injection. (See fig. 15.) Both methods are described by Price and others (1965) for parts of the Columbia North Pacific Region, with specific examples and associated problems. Both methods are also being practiced in the Upper Colorado Region, but neither is part of a water-resources development project. Recharge by surface spreading is incidental with most irrigation projects in the region, and subsurface injection is used to repressurize oil fields and for brine disposal at oil fields.

Successful artificial recharge requires an adequate supply of recharge water and suitable reservoir space. Water generally is available during storm runoff or peak seasonal runoff periods. Reservoir space, if not naturally present, can be created by dewatering an aquifer. Areas in the region that have the greatest potential for the use of artificial recharge are designated in figure 20 as having promise for water-resource development involving the conjunctive use of ground and surface waters.

Natural reservoir space exists in permeable unsaturated terrace deposits and extensive strata of unsaturated rocks, such as the sandstones in the Canyon Lands and Navajo sections. In these areas, perched ground-water bodies might be created by artificial recharge (fig. 16). The success of this type of artificial recharge would depend on the percentage of recharge water that could be recovered, which in turn depends largely on the retaining

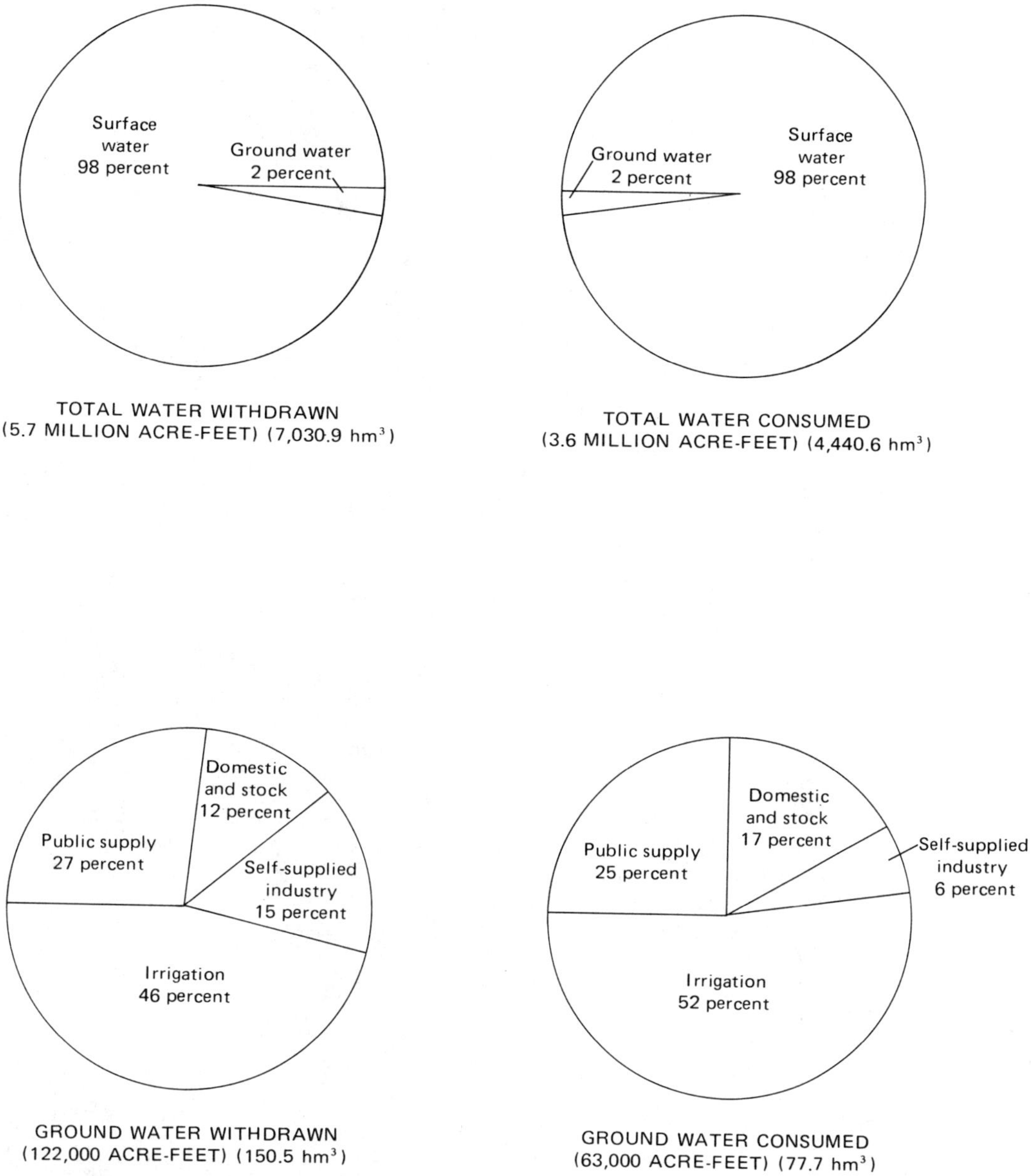

FIGURE 11. — Total withdrawal and consumptive use of ground water compared with the total withdrawal and consumptive use of water from all sources in 1970.

time of the reservoir rock. Some of the water will be lost by leakage from the artificial-recharge area, and some will be retained in the finer grained parts of the newly formed aquifer.

Deeper saline-water aquifers in the Canyon Lands and Navajo sections might also be used for recharge and later recovery from the freshwater bubbles (fig. 16). The freshwater source might be from natural runoff or it might be treated effluent. Injection and recovery of freshwater in saline aquifers is somewhat similar to injection and recovery of natural gas in aquifers, and it holds promise of utilizing some of the extensive saline-water aquifers in the region.

Reservoir space for artificial recharge can also be created by overpumping — that is, withdrawing ground water at a rate faster than it is replaced by natural recharge over an extended period of time. Overpumping is not uncommon in highly developed areas, such as many of the ground-water basins in the California and Lower Colorado Regions. It is practically nonexistent in the Upper Colorado Region, but it might be a desirable practice in some areas of the region where surface water

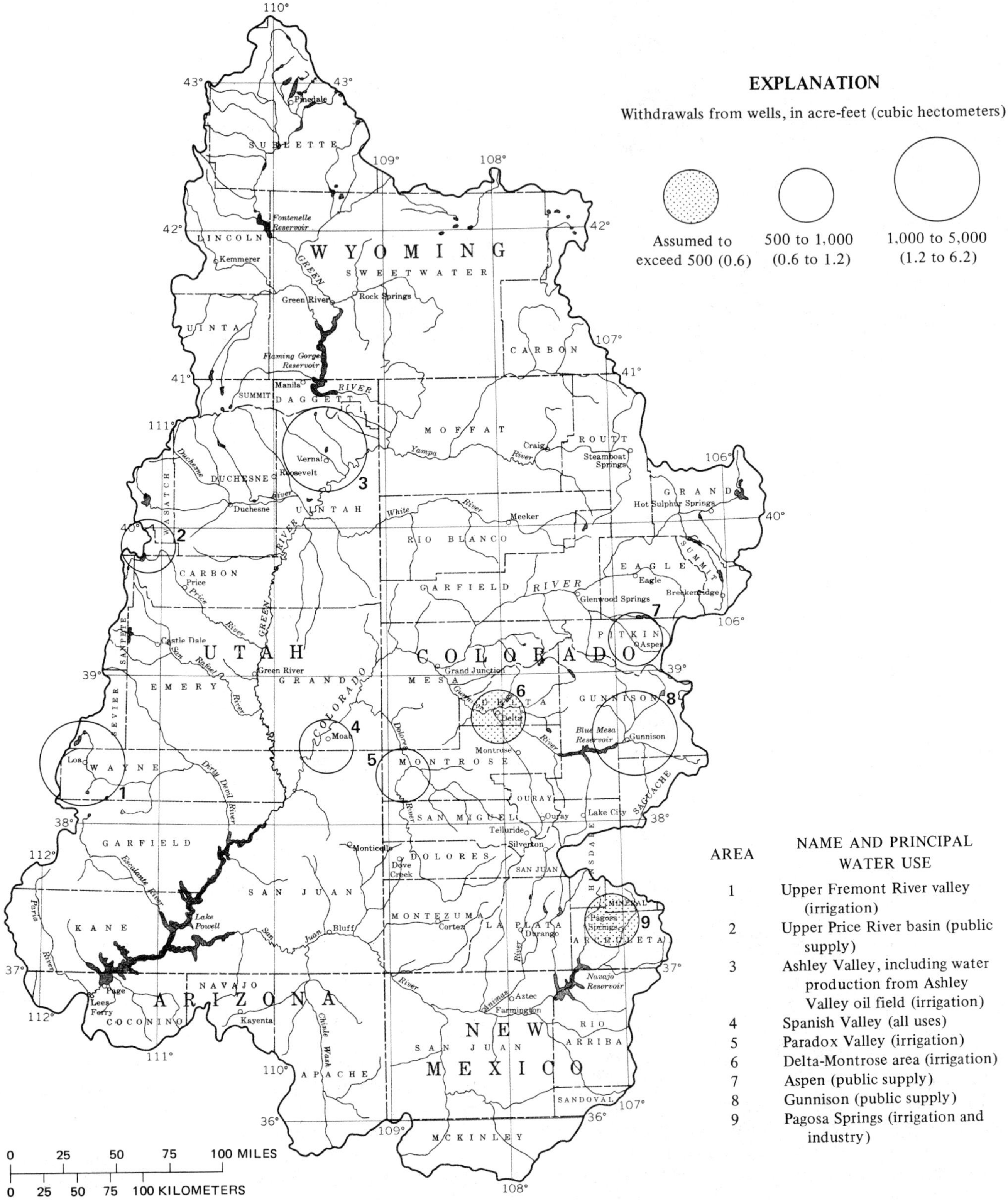

AREA	NAME AND PRINCIPAL WATER USE
1	Upper Fremont River valley (irrigation)
2	Upper Price River basin (public supply)
3	Ashley Valley, including water production from Ashley Valley oil field (irrigation)
4	Spanish Valley (all uses)
5	Paradox Valley (irrigation)
6	Delta-Montrose area (irrigation)
7	Aspen (public supply)
8	Gunnison (public supply)
9	Pagosa Springs (irrigation and industry)

FIGURE 12. — Annual withdrawals from wells exceeded 500 acre-feet (0.6 hm^3) in 1970 in only a few scattered areas in the region.

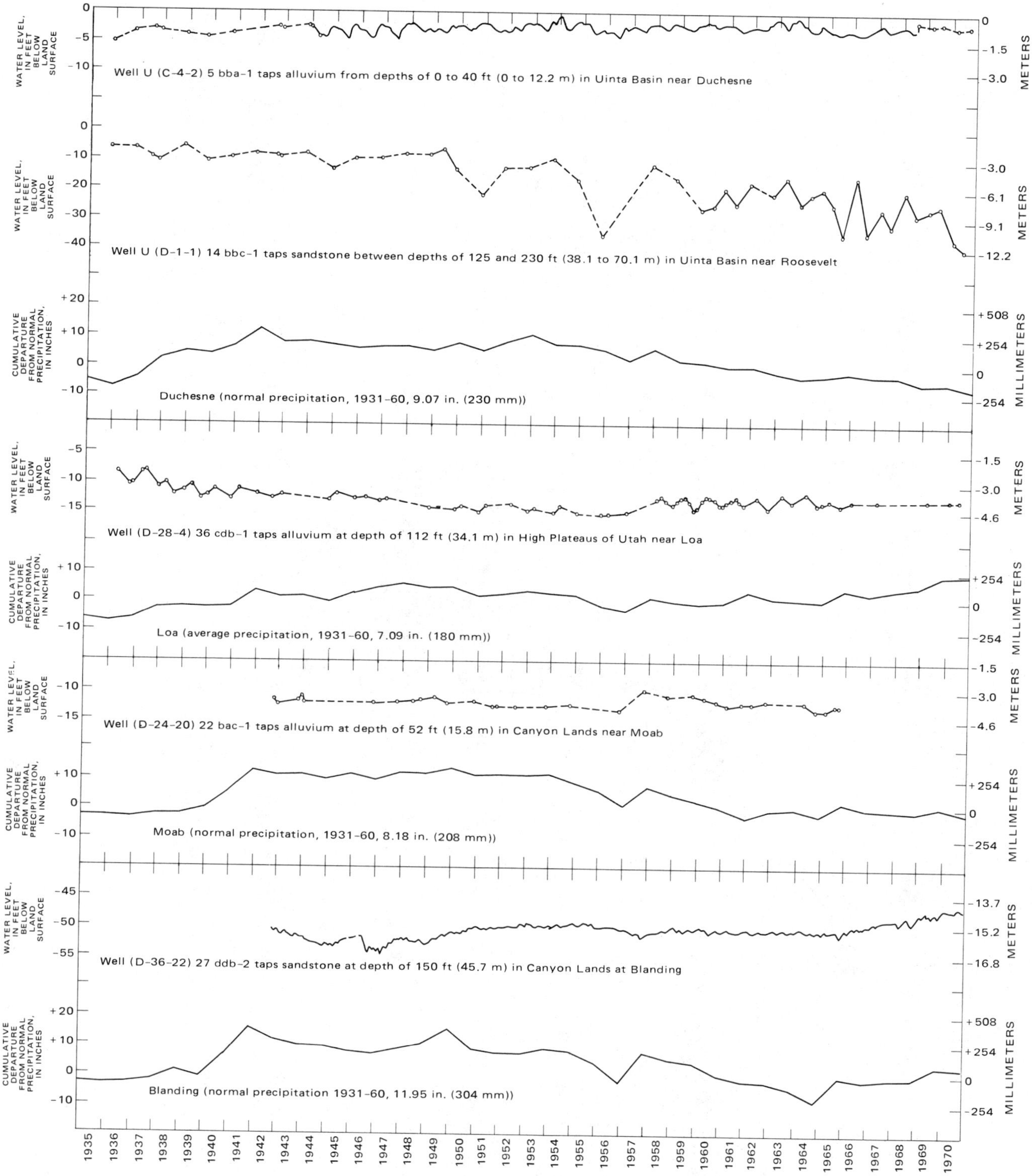

FIGURE 13. — Ground-water developments has had little effect on ground-water levels; changes in water levels reflect above- or below-average (normal) precipitation. Well locations are shown on plate 1F. Precipitation data are from National Weather Service.

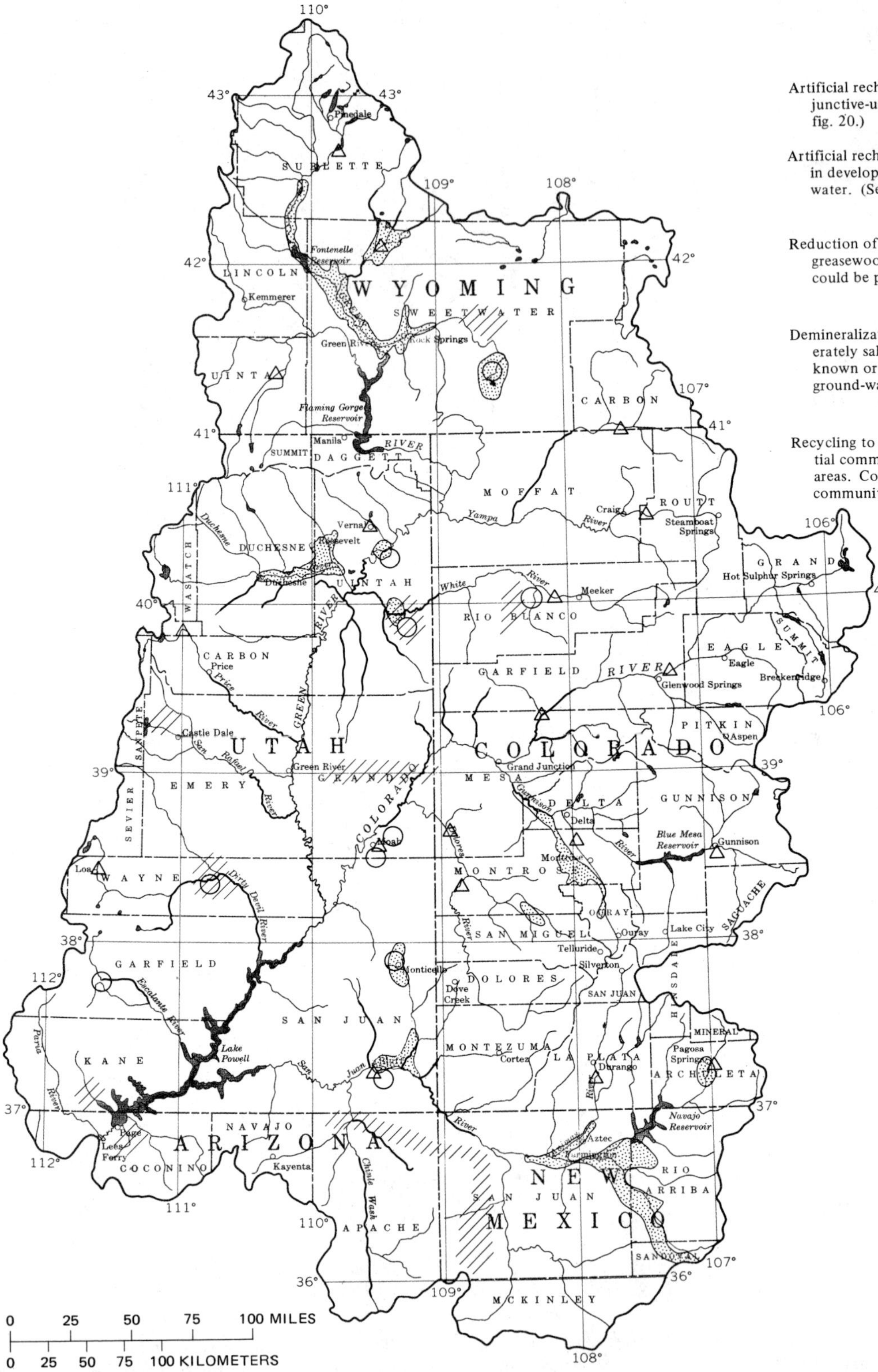

EXPLANATION

Water-management practices

△ Artificial recharge to regulate total supply in conjunctive-use water-development options. (See fig. 20.)

○ Artificial recharge to avoid problems of overdraft in development options that utilize only ground water. (See fig. 21.)

Reduction of natural discharge by eradication of greasewood and saltcedar so that salvaged water could be put to more beneficial use

Demineralization in areas of shallow slightly to moderately saline ground water where aquifers are known or assumed to support moderate to large ground-water withdrawals

Recycling to salvage effluents from existing or potential communities and industries in water-deficient areas. Could also be practiced by any of the larger communities shown in figure 6

FIGURE 14. — Areas where water-management practices can be used singly or in combination to increase the usable ground-water supply. Map shows selected areas of existing or potential supply problems where these practices might be feasible.

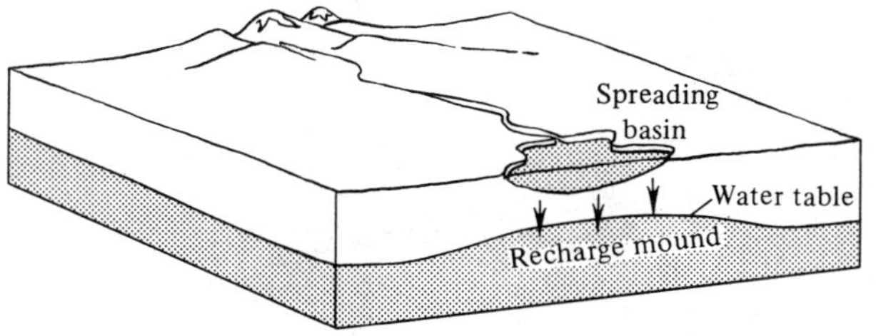

ARTIFICIAL RECHARGE BY SURFACE SPREADING

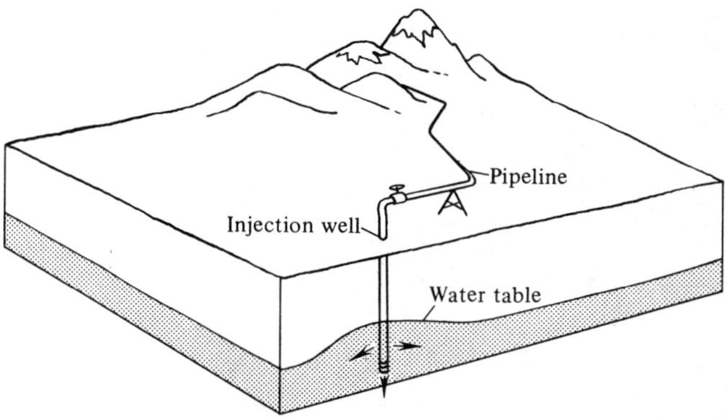

ARTIFICIAL RECHARGE BY SUBSURFACE INJECTION

FIGURE 15. — Artificial recharge is accomplished by surface spreading or by subsurface injection.

might be available for recharge during periods of high runoff. For example, large quantities of water could be pumped from alluvium of the larger perennial streams for irrigation during critical periods of low flow in the late summer. Water removed by the increased late summer irrigation withdrawals could be replaced by artificial recharge using winter runoff, a time of low streamflow but of low seasonal demand, and spring and early summer runoff. Perhaps the wells used for withdrawal could also be used for injection — a recharge operation similar to that practiced near Klamath Falls, Oreg. (Price and others, 1965), or the recharge could be accomplished by surface spreading.

Artificial recharge not only is a means of increasing the usable water supply, but it also may prevent deterioration of ground-water quality which results from overpumping aquifers that are hydraulically connected with saline-water aquifers. This condition probably exists in Spanish and Castle Valleys, Utah, Paradox Valley, Colo., and in many other places throughout the region. The mound created by artificial recharge forms a barrier to encroachment of the saline water in the same fashion that artificial-recharge mounds successfully form barriers to the encroachment of sea water in the Long Beach area of California (Johnson and Lundeen, 1967).

REDUCTION OF NATURAL DISCHARGE

Consumptive use of ground water by phreatophytes (plants that rely directly on ground water for their moisture) and hydrophytes (plants that live in water, such as cattails and rushes) accounts for a major portion of the total natural ground-water discharge in the Upper Colorado Region. The phreatophytes, which are natural ground-water pumps, not only deplete ground-water storage but also have marked effect on streamflow. (See fig. 9.) The effect of evapotranspiration, largely by phreatophytes, on streamflow in part of the Green River basin is shown in figure 17. Consumption of water by phreatophytes also concentrates the mineral content of the remaining ground water and streamflow.

Phreatophytes commonly are found growing along the alluvial plains of both perennial and intermittent streams in the Upper Colorado Region. The most common ones in the region are cottonwood (*Populus* sp.), willow (*Salix* sp.), greasewood (*Sarcobatus vermiculatus*), rabbitbrush (*Chrysothamnus* sp.), saltcedar (*Tamarix* sp.), and saltgrass (*Distichlis stricta*). Greasewood and saltcedar (figs. 18, 19) probably are the most aggressive and least beneficial of the phreatophytes. They both are salt tolerant, and, although saltcedar is generally most adaptable to the southern part of the region, it has been observed growing in thick stands along many streams as far north as the Uinta Basin.

Consumptive use of water by phreatophytes varies depending on many factors, such as depth to water, water quality, and plant density. Under ideal growing conditions and 100-percent plant density, greasewood might consume on the order of 2 feet (0.6 m) of water and saltcedar might consume as much as 9 feet (2.7 m) annually. (See Mower and Nace, 1957, p. 21; Robinson, 1958, p. 75.)

The total volume of ground water consumed by phreatophytes in the region has not been estimated. Robinson (1968) estimated the consumptive use by phreatophytes and hydrophytes in the Colorado and Utah parts of the region alone to be more than 2 million acre-feet (2,467 hm^3) per year. Total consumptive use in the entire region probably approaches 2.5 million acre-feet (3,083.8 hm^3) per year. Part of this consumption, however, is from streamflow, as indicated by figure 9.

Some phreatophytes and hydrophytes are considered beneficial vegetation. The grasses provide forage for livestock, and the larger plants provide shade or add to the scenic value of the area in which they grow. Nevertheless, eradication of the least beneficial phreatophytes or their replacement with vegetation requiring less water

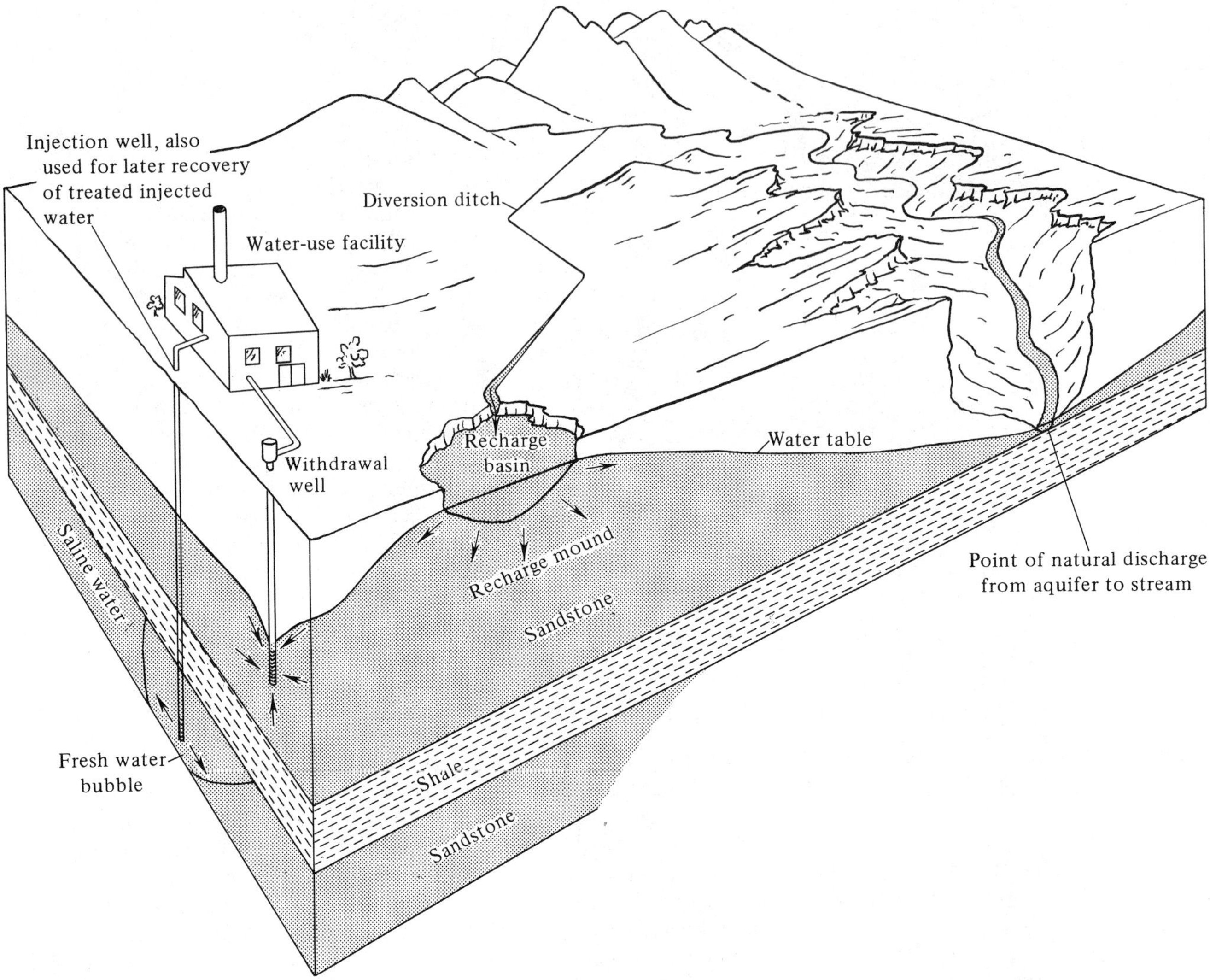

FIGURE 16. — Artificial recharge can increase the ground-water supply in water-deficient areas.

might add considerably to the total usable ground-water supply in some areas. In Spanish Valley, Utah, for example, an estimated 3,000 acre-feet (3.7 hm^3) per year of ground water is consumed by phreatophytes (Sumsion, 1971a, p. 24). A large percentage of this water might be salvaged if the phreatophytes were eradicated or replaced. Even if there were no need for the water in the area in which phreatophytes had been removed or replaced, the salvaged ground water would eventually add to the total streamflow and could be used downstream.

DEMINERALIZATION

One of the principal limiting factors in the development of ground water in the Upper Colorado Region has been the widespread occurrence of generally undesirable saline ground water. (See pl. 2*A*.) Perhaps as much as 70 percent of the estimated maximum recoverable ground water stored in the upper 100 feet (30.5 m) of saturated sedimentary rocks is saline; this amounts to about 80 million acre-feet (98,680 hm^3). If only 50 percent of this amount of water could be withdrawn and demineralized for beneficial uses, the total usable fresh-water supply for the region could be increased by about 40 million acre-feet (49,340 hm^3). This is considered a minimum figure, because considerably more recoverable saline water is stored at depths greater than 100 feet in the sedimentary rocks and in the other geohydrologic units shown on plate 1*A*.

Any development of saline water on a sustained basis, however, would be limited to the perennial yield of the saline-water aquifers. This might be only a few thousand

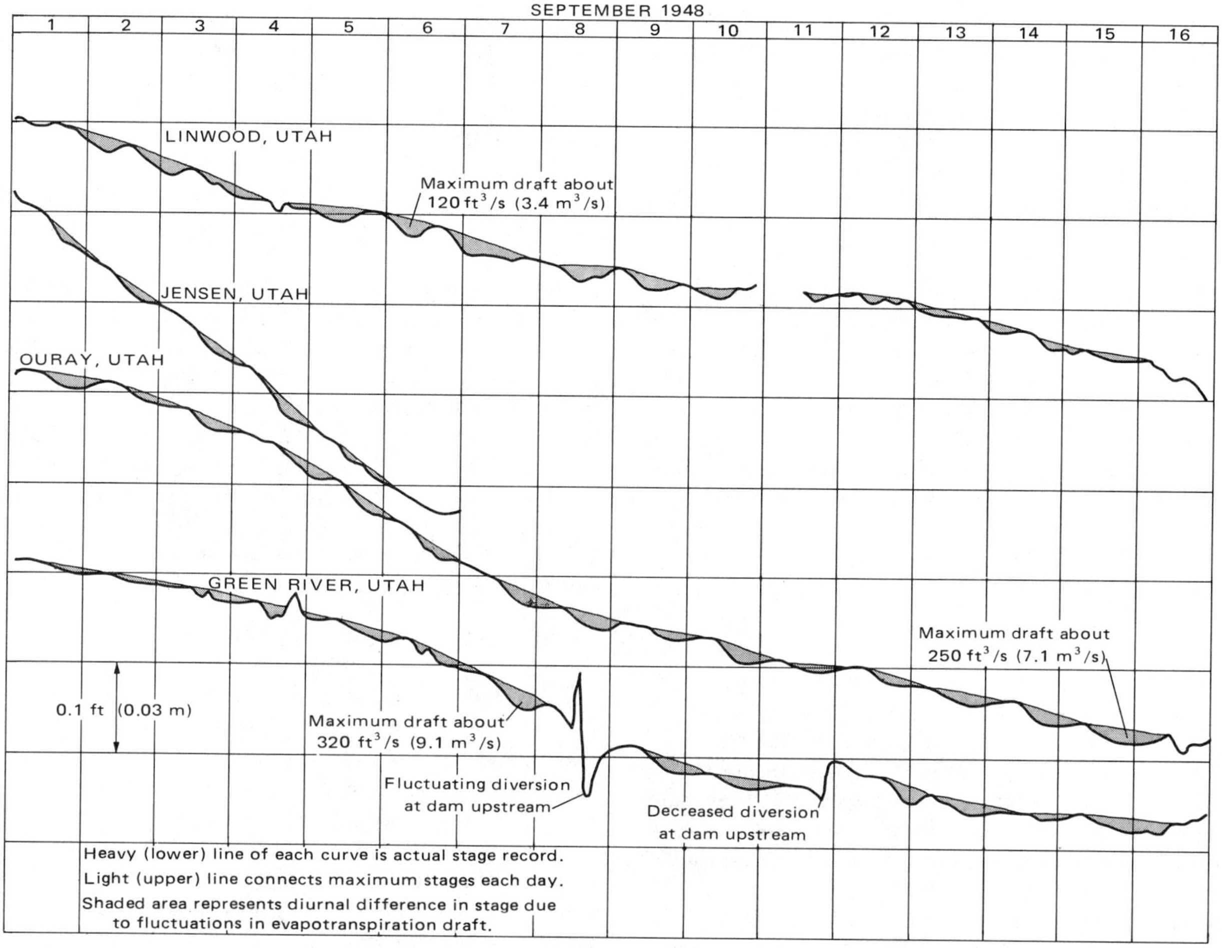

FIGURE 17. — The effect of evapotranspiration is reflected in fluctuations of the stage of the Green River (after Thomas, 1952, p. 20).

FIGURE 18. — Greasewood, a common phreatophyte which covers many square miles of the region (from Iorns and others, 1965).

FIGURE 19. — Saltcedar-lined perennial stream, one of many in the region.

acre-feet a year in many parts of the region, but it would provide an adequate supply for small communities and industries that use little water, and it probably would be the most feasible source to develop.

Demineralization is becoming an increasingly common practice throughout the world, and it should be adaptable to the Upper Colorado Region. According to the Office of Saline Water, more than 680 demineralization plants, with capacities ranging from as little as 25,000 gpd (gallons per day) to more than 7.5 mgd (94.6 to more than 28,390.5 m^3/day), were in operation in 1970. Existing commercial plants (in 1970), having capacities of 1 to 3 mgd (3,785 to 11,355 m^3/day), produce water in the general cost range of about $1 per 1,000 gallons (3.8 m^3). However, the status of large nuclear-powered dual-purpose plants, which produce both freshwater and electrical energy, indicates that a cost of 20 to 40 cents per 1,000 gallons (3.8 m^3) may be achieved before the year 2000 (Oak Ridge National Laboratory and Office of Saline Water, 1970).

Of the three basic demineralization processes — distillation, membrane, and vacuum freezing — distillation is most commonly used. This process can produce water with less than 25 mg/l of dissolved solids from natural supplies having a dissolved-solids content of 35,000 mg/l. The process might be well adapted to the Upper Colorado Region because it requires steam as its energy source, which might be produced from the vast fossil-fuel resources of the region. Because much of the recoverable saline water occurs near or in conjunction with major fossil-fuel reserves (as in the Green River, Uinta, Piceance Creek, and San Juan Basins), demineralization of saline water might be an ideal method of increasing the freshwater supply, if needed.

RECYCLING

Recycling of water is the reuse of water that has already been through a given supply system. The water might be recycled one or several times through the same system, or it may be taken from one system (such as a municipal supply) and put through another system (such as an industrial supply) where little or no treatment is required. Recycling does not physically increase the total usable water supply, but it does, in effect, "stretch" the supply by reducing nonbeneficial consumption of the effluent. It is becoming a common practice in many water-deficient areas of the world.

With recycling, it might be possible to develop sustained water supplies exceeding the natural sustained yields of certain aquifers. For example, an ore-processing plant might require a sustained water supply of 20,000 gpd (75.7 m^3/day), but the sustained perennial yield of the aquifers might be only 15,000 gpd (56.8 m^3/day). Assuming that the effluent from the plant is 10,000 gpd (37.8 m^3/day) and 50 percent could be recycled through the system, then the sustained yield of the aquifers plus the recycled water could support the industry on a perennial basis after the initial 20,000 gallons (75.7 m^3) has been withdrawn.

The amount of water that could be recycled depends on the consumptive use of the primary supply and on the percentage of the effluent that can be treated and reused. Many industries consume only a small percentage of the water withdrawn, but the effluent may be contaminated beyond treatment. For other industries, all the effluent may be economically treated and reused. The latter example is also true for effluents from municipal water supplies. In 1970 the difference between total withdrawals and total depletions of ground water for municipal use and self-supplied industry was about 32,000 acre-feet (39.5 hm^3). This, theoretically, is the volume of ground water regionally that might be recycled at least once at 100-percent recovery at the 1970 level of development.

AVAILABILITY OF GROUND WATER TO MEET FUTURE WATER NEEDS

As noted earlier, the perennial ground-water supply for the Upper Colorado Region is about 4 million acre-feet (4,934 hm^3); in addition, the recoverable reserve of ground water in storage in just the upper 100 feet (30.5 m) of saturated rocks may be as much as 115 million acre-feet (141,852.5 hm^3). Yet, in 1970, only about 2 percent of the estimated withdrawal and consumptive use of water for all uses in the region came directly from ground-water sources.

By the year 2020 the projected consumptive use of water for various uses in the region will be about 4.1 million acre-feet (5,057.4 hm^3) per year, as indicated in the following tabulation. Part of these needs could be met from ground-water sources.

Supply	Consumptive use[1]	
	Acre-feet	hm^3
Irrigation	[2]3,294,000	4,063.1
Thermal electric power	627,000	773.4
Municipal and industry	110,000	135.7
Minerals development	[3]85,000	104.8
Domestic and stock	25,000	30.8
Recreation	5,000	6.2

[1]All estimates except for domestic and stock supply and minerals development are after Hedlund and others (1971, p. 54). The estimate for domestic and stock supply assumes an increase of about 10 percent over the 1970 use.
[2]Includes reservoir and incidental evaporation.
[3]Assumes minimum additional requirement of 32,000 acre-feet (39.5 hm^3) for mining and processing oil shale (U.S. Dept. Interior, written commun., 1973).

The projected consumptive water need for irrigation annually by the year 2020 is estimated to be about 3.3

million acre-feet (4,070.6 hm^3). Although most of this need will be met with surface water, ground water is available in many of the irrigated areas to supplement surface water, particularly during periods of low flow and peak crop requirements. A minimum of 3,200 acre-feet (3.9 hm^3) of recoverable ground water per square mile is stored in alluvium in the irrigated areas along the Colorado and Green Rivers and other large streams. Most of these areas cover at least 10 square miles (25.9 km^2), indicating that at least 32,000 acre-feet (39.5 hm^3) of ground water is available to supplement the surface-water supply in each of the areas. If only half of this water were withdrawn in any given 10-square-mile (25.9 km^2) area, it could be used to irrigate about 16,000 acres (6,475.2 ha) during periods of low flow at a crop requirement of 1 foot per acre (0.3 m/ha).

An estimated 627,000 acre-feet (773.4 hm^3) of water will be needed annually for production of thermal-electric power by the year 2020 (Hanley and others, 1971, p. 65). The individual plants that will be producing this energy will require from about 6.6 to more than 130 million gallons (24,983.6 to 492,102 m^3) of water per day. Most will be located where ground water is plentiful (sites of plants are shown on pl. *2B*) but where the aquifers yield water too slowly to meet the daily requirements on a sustained basis. However, ground water could be developed at some sites as supplemental or emergency supply.

The use of ground water for municipal and industrial supply has been inhibited in many areas by undesirable chemical characteristics of the water. Recent advancements in the field of demineralization and progressive decline in cost per unit volume of demineralized water, however, now make the cost such that ground water could be used to supply a large part of the projected water needs of 110,000 acre-feet (135.7 hm^3) per year for municipal and industrial use. The larger communities, such as Farmington, N. Mex., and Grand Junction, Colo., may require between 15,000 and 20,000 acre-feet (18.5 and 24.7 hm^3) per year for municipal and industrial supply by the year 2020. Most of the other communities probably will require between 100 and 5,000 acre-feet (0.1 and 6.2 hm^3) per year. Ground water (with optional demineralization) is available in most places to meet these requirements completely or, in some places, as a supplemental supply. Communities near the larger streams could obtain water from alluvium through shallow wells or infiltration galleries. Such communities as Price and Vernal, Utah, Green River, Wyo., and Grand Junction, Colo., are underlain by aquifers that contain slightly to moderately saline ground water that probably could be economically withdrawn and demineralized.

Mountain resort communities such as Aspen and Steamboat Springs, Colo., have a seasonal influx in population. For example, the permanent population served by the Aspen municipal water supply is about 4,000 (1970), but during the winter ski season the population increases to more than 20,000. With a trend to development of such communities for seasonal leisure and recreation, there will be a growing need for adequate seasonal public water supply. Most of the existing or potential mountain resort communities are on or near mountains streams. A seasonal water supply for those communities could be developed from the permeable alluvium of the streams, and the ground water withdrawn during the tourist influx could be replaced by natural or artificial recharge during the off-season.

Most of the projected annual water need of 85,000 acre-feet (104.8 hm^3) by the year 2020 for minerals production could be obtained from ground-water sources. Most of the mineral production of the region is in areas where surface water is absent, inadequate, or used for other purposes. The ground water in the fossil-fuel-producing areas generally is saline, but it generally is suitable for such uses as repressuring oil and gas fields.

The amount of water needed to develop the vast oil-shale reserves in the Uinta, Piceance Creek, and Green River basins will depend on whether the oil is released by surface or in situ retorting. Mining and surface retorting of the shale may require large supplies of water. The actual volume required is not known, and it will depend on such factors as methods used to mine, crush, upgrade, and retort the shale, and the availability of electrical energy and the methods used to dispose of the spent shale. Preliminary estimates of the total volume of water needed range from as much as 189,000 to as little as 32,000 acre-feet (233.1 to as little as 39.5 hm^3) per year for an oil production capacity of 1 million barrels (158,987 m^3) per day. The estimated maximum need probably will have to be met largely from surface sources, but the minimum need probably can be met with ground water — part of which will be produced from the shale and part can be developed from local unaffected aquifers. In the Piceance Creek basin, however, water produced by mine dewatering activities might be sufficient to meet the entire estimated maximum need for the first few years of operation. In all areas, ground water could be used to supplement a surface-water supply for the life of the development. In situ retorting of the shale could reduce water requirements by more than 60 percent by eliminating water needs for mining, dust control, surface retorts, and disposal and revegetation of spent shale.

Only about 30,000 acre-feet (37.0 hm^3) of water will be needed annually for domestic, stock, and recreation supply by the year 2020. These needs can be met from ground water in most parts of the region.

SELECTED DEVELOPMENT OPTIONS

Various options for use of ground water in overall water management are possible in the Upper Colorado Region. In some areas, ground water might be used conjunctively with surface water. In other areas, ground water might be developed as the sole supply on a sustained basis or a time-limited basis. In order to approach optimal development of the total water resource of the region, the various options would include one or more of the practices for increasing the usable ground-water supply discussed in a preceding section of this report.

CONJUNCTIVE USE OF SURFACE WATER AND GROUND WATER

The opportunity to use ground water in conjunction with surface water exists locally along several of the larger perennial streams that are underlain by permeable alluvial aquifers, as illustrated in figures 8*C* and *D*. The areas are shown in figure 20. In these areas, ground water could be used to augment surface water during periods of low flow, which commonly coincide with periods of peak seasonal demand. Should the withdrawal of ground water exceed natural recharge during these periods, then artificial recharge could be implemented during peak runoff periods to facilitate ground-water replenishment.

Such a coordinated system would provide a more uniform year-round water supply without surface-reservoir construction. The well fields, regardless of their primary purpose, would also provide an emergency public water supply in populated areas. For example, the public water supply for Delta, Colo., is diverted from a stream and a spring which are 17 and 11 miles (27.4 and 17.7 km) north of the city, respectively. Should that supply be disrupted by a man-caused accident or a natural disaster, local well fields used to augment surface water for irrigation might serve as a temporary public supply for the city while the primary system is being restored.

Few areas in the region have been investigated in sufficient detail to determine their adaptability to a program of conjunctive development of ground and surface water. Selected potential areas for such development are shown in figure 20. One of these areas is the upper Fremont River valley of Utah, which might serve as a specific example of how ground water could be used to increase the total usable water supply for irrigation during seasonal low-flow periods. The ground-water resources of the upper Fremont River valley are described by Bjorklund (1969). Although only about 40 square miles (103.6 km^2) in areal extent, the valley contains one of the most productive ground-water reservoirs in the Upper Colorado Region. Yields to irrigation wells in the valley range from several hundred to more than 2,000 gpm (7.57 m^3/min). Total estimated recoverable storage in the upper 100 feet (30.5 m) of the ground-water reservoir in the valley is between 130,000 and 380,000 acre-feet (160.4 and 468.7 hm^3), and the perennial yield of the reservoir is on the order of 90,000 acre-feet (111 hm^3). Most of the water is fresh and suitable for irrigation and most other uses.

The surface-water supply for the valley is the Fremont River, and it is used primarily for irrigation. Flow of the river into the valley is regulated and has been measured during the period 1949-58, including the complete irrigation seasons during 1950-57. During those irrigation seasons, average monthly flow of the river was about 4,400 acre-feet (5.4 hm^3). However, the flow in September and October of some years was as low as 1,300 acre-feet (1.6 hm^3). This is an assumed deficiency of about 3,000 acre-feet (3.7 hm^3). Use of ground water in a management program for conjunctive use of water resources could make up this deficiency and perhaps provide enough water for irrigation of additional acreage.

Well fields could be located to deliver the water directly to the irrigated fields or to canals and distribution ditches. An inherent problem associated with large-scale ground-water development — well interference — could be minimized by proper well design and spacing. Any temporary overdraft could be replaced by artificial recharge during peak runoff periods.

A planned development that makes optimum conjunctive use of the ground- and surface-water resources in the upper Fremont River valley might provide a combined annual supply of more than 100,000 acre-feet (123.4 hm^3). This would consist of the perennial ground-water supply, plus surface-water inflow, plus irrigation-return flow, which could be as much as 25 percent of the total withdrawal. However, this type of development might virtually deplete the flow of the Fremont River.

USE OF GROUND WATER ONLY

Extensive areas of the Upper Colorado Region are far removed from the principal river systems. The supply from local intermittent and ephemeral streams supports some livestock but it is not sufficient to meet large-scale demands. Yet these areas are underlain by extensive bedrock aquifers that may be capable of supplying sufficient water for industry, public supply, or to irrigate small areas. Plans for development of this water might consider use on a sustained basis where the annual withdrawal would not exceed the perennial yield of the aquifers, or on a time-limited basis where the total recoverable ground-water storage would be withdrawn (mined) at a given annual rate for a given number of years. Because a large percentage of the water is saline, a demineralization process would be an integral part of most developments.

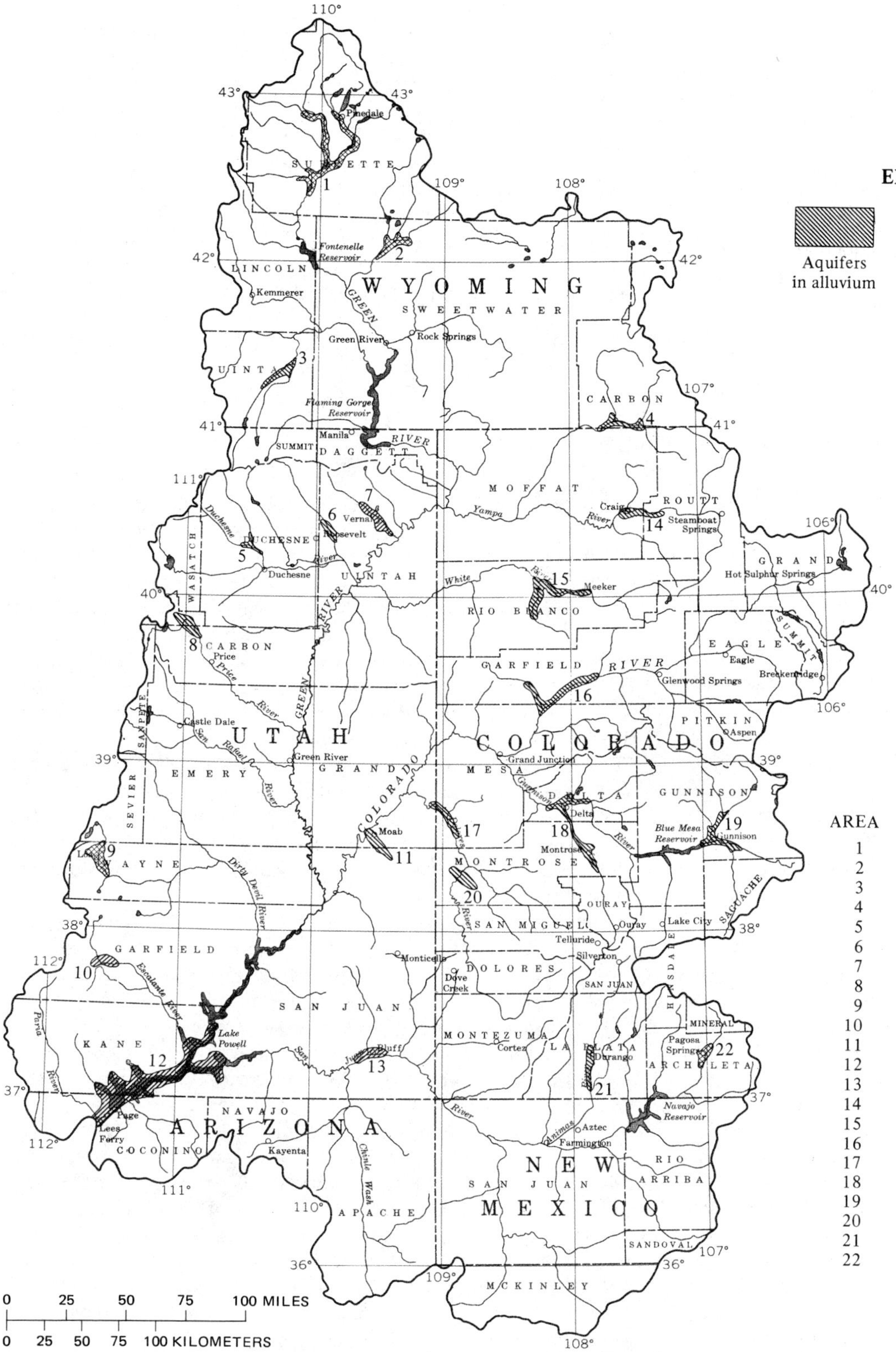

EXPLANATION

Aquifers in alluvium — Aquifers in bedrock — Aquifers in alluvium and bedrock

AREA	NAME
1	Pinedale–Big Piney
2	Farson
3	Fort Bridger
4	Baggs-Savery
5	Hanna-Tabiona
6	Roosevelt
7	Ashley Valley
8	Colton
9	Upper Fremont River valley
10	Escalante
11	Spanish Valley
12	Glen Canyon
13	Bluff
14	Craig
15	Meeker–Piceance Creek
16	DeBeque–Grant Valley
17	Gateway
18	Delta
19	Gunnison
20	Paradox Valley
21	Durango
22	Pagosa Springs

FIGURE 20. — Areas where aquifers are known or assumed to be sufficiently productive for conjunctive use with surface water in water-resources development.

SUSTAINED GROUND-WATER SUPPLY

A sustained ground-water supply is one where the planned annual withdrawal over a long time does not exceed the annual natural discharge of the aquifers. Successful development of ground water on a sustained basis requires that the size and character of the ground-water reservoirs be fairly accurately known and that the annual withdrawals be regulated to prevent excessive drawdowns and stream depletions. Short-term withdrawals exceeding the annual discharge rate might be tolerated if recycling or artificial recharge were a part of the development.

Areas that have some potential for moderate to large development (0.5 to 1.5 mgd or 1,892.7 to 5,678.1 m^3/day) of ground water as a sole supply are shown in figure 21. These areas appear to be underlain by aquifers that seem to have good sources of recharge. They have not been investigated in detail, however, so it is not possible to do more than speculate on the size or even the feasibility of such development in these areas.

A hypothetical development using ground water on a sustained basis as a sole source of supply is herein described using Spanish Valley, Utah, as an example. The water resources of Spanish Valley have been described by Sumsion (1971a). The valley covers about 18 square miles (46.6 km^2) and is underlain by permeable valley-fill deposits to depths of at least 360 feet (109.7 m). It is estimated that the ground-water reservoir will yield about 14,000 acre-feet (17.3 hm^3) per year and that about 58,000 to 173,000 acre-feet (71.5 to 213.4 hm^3) of water is recoverable from storage in the upper 100 feet (30.5 m) of the reservoir.

At the 1970 level of development, about 3,300 acre-feet (4.1 hm^3) of ground water was withdrawn from Spanish Valley for irrigation and public and domestic supply. About 8,000 acre-feet (9.9 hm^3) of ground water drained to the Colorado River, and another 3,000 acre-feet (3.7 hm^3) was consumed by phreatophytes and hydrophytes. Most of this 11,000 acre-feet (13.6 hm^3) of water could be put to beneficial use directly in Spanish Valley if it were withdrawn there from wells. An annual supply of about 11,000 acre-feet (13.6 hm^3) could be used by industry, or to supply a population of about 65,000 additional people at a per capita consumption rate of 150 gallons (0.57 m^3) per day, or to irrigate about 3,700 additional acres (1,497 ha) at an application rate of 3 feet per acre (0.9 m/ha) per year, or any combination of these. The water thus used would be of better chemical quality than it would be after seepage through the area of phreatophytes in the downstream part of the valley, where the dissolved-solids content increases as a result of evapotranspiration.

Sustained development of 14,000 acre-feet (17.3 hm^3) of ground water in Spanish Valley need not necessarily denude the area presently occupied by phreatophytes or completely terminate underflow to the Colorado River. The phreatophytes could be replaced by nonphreatophytic vegetation that uses less water and provides the same forage, shade, and scenic benefits. If the development were entirely for municipal and industrial supply, only about 54 percent (regional average in 1970) of the withdrawal would be consumed; thus, more than 7,500 acre-feet (9.2 hm^3) of water might still be allowed to reach the Colorado River. Withdrawals for irrigation or certain industries, which consume larger percentages of the diversions, would decrease proportionately the amount of water reaching the river; however, this could be offset to some extent by incorporating recycling systems in the development program.

TIME-LIMITED GROUND-WATER SUPPLY

A time-limited ground-water supply is one where the planned annual withdrawal will exceed the annual natural discharge, the excess amount being obtained (mined) from the recoverable water in storage in the aquifers. This supply is limited in time by the number of years required to deplete the water in storage or to lower water levels to depths at which it is no longer economically feasible to withdraw the water.

Large-scale sustained ground-water developments generally are not possible in those parts of the Upper Colorado Region underlain by bedrock aquifers and far removed from principal areas of recharge. The rate of ground-water movement to the area of withdrawal would be too slow, even if a highly efficient recycling system were incorporated in the development. Nevertheless, large quantities of ground water in transient storage in these areas are either being wasted or becoming chemically degraded as they move to areas of natural discharge. Diversion of this water for beneficial use on a time-limited basis (ground-water mining) might be the best solution to water-supply problems in the arid Canyon Lands and Navajo sections of the region (fig. 21).

Little quantitative data are available for any of the areas shown in figure 21, but provisional estimates are made herein to illustrate the development of a time-limited ground-water supply in the Escalante area of Utah. The estimates are based on a reconnaissance by Goode (1969) of the ground-water resources of this area.

The ground water to be developed is under artesian pressure in the Navajo Sandstone, which is about 1,500 feet (457.2 m) thick in the area. Goode (1969, p. 31) estimated that if the sandstone were completely saturated, the total quantity of water in storage might be as much as 60,000 acre-feet per square mile (28.6 hm^3/km^2). Assuming that 60 percent of this water could be recovered by wells, then recoverable water in storage

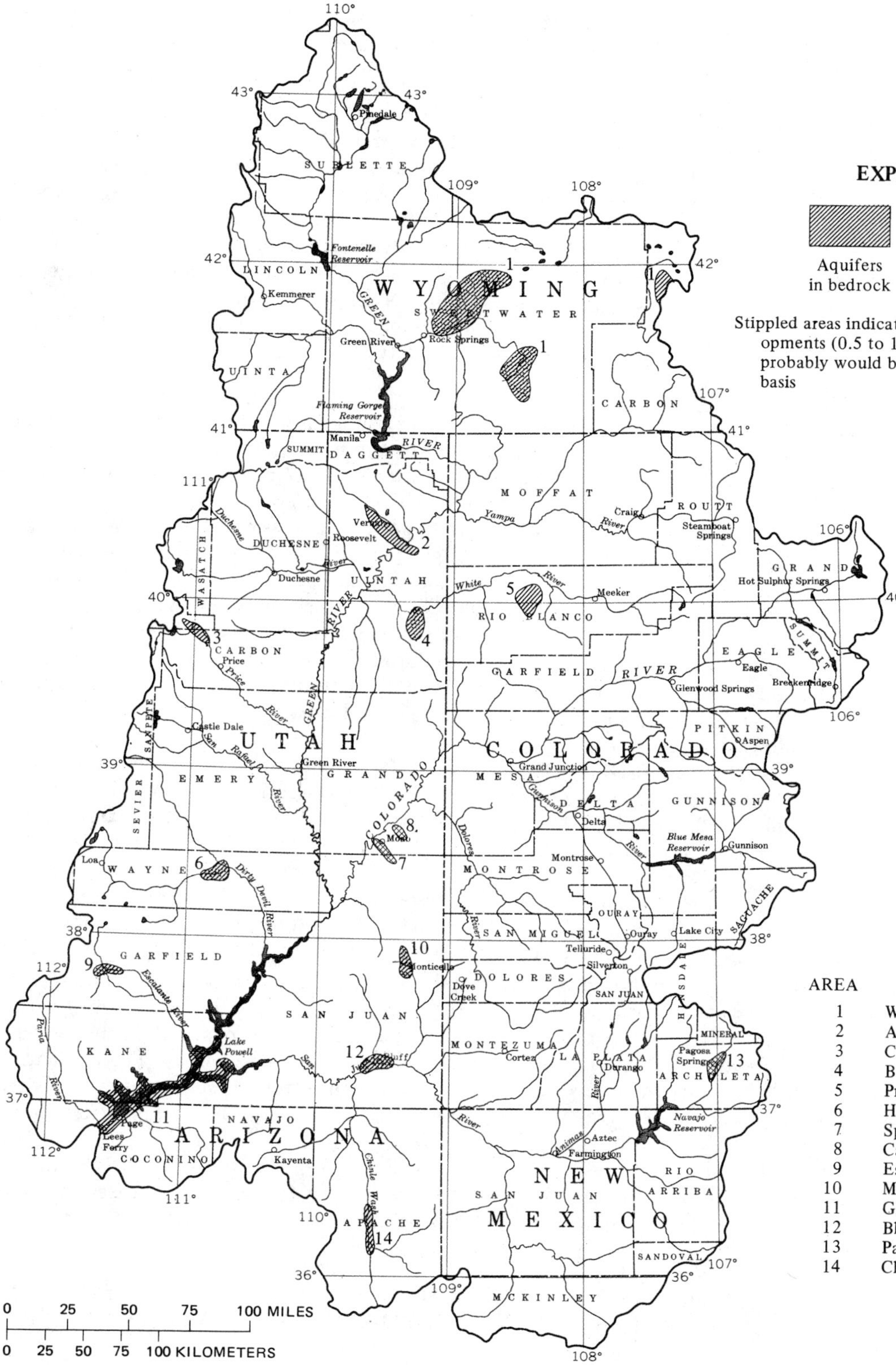

EXPLANATION

Aquifers in bedrock

Aquifers in alluvium and bedrock

Stippled areas indicate that moderate to large developments (0.5 to 1.5 mgd)(1,892.7 to 5,678.1 m^3/d) probably would be possible only on a time-limited basis

AREA	NAME
1	Washakie-Great Divide Basin
2	Ashley Valley
3	Colton
4	Bonanza (oil-shale area)
5	Piceance Creek (oil-shale area)
6	Hanksville
7	Spanish Valley
8	Castle Valley
9	Escalante
10	Monticello
11	Glen Canyon
12	Bluff
13	Pagosa Springs
14	Chinle-Many Farms

FIGURE 21. — Areas where aquifers are known or assumed to be sufficiently productive for use as a sole water supply in water-resources development.

would be about 36,000 acre-feet per square mile (17.1 hm^3/km^2) and would total about 360,000 acre-feet (444.1 hm^3) in a 10-square-mile (25.9 km^2) area from which the wells would most likely draw water. Average annual recharge to the Navajo Sandstone in this area probably is less than 2,000 acre-feet (2.5 hm^3) and, therefore, could be considered negligible. If the development is planned for a lifespan of 30 years, then the annual withdrawal would be nearly 12,000 acre-feet (14.8 hm^3). This much water could irrigate 4,000 acres (1,619 ha) of land at an annual crop requirement of 3 feet per acre (0.9 m/ha). If larger annual withdrawals are required, the lifespan of the development could be shortened. If smaller annual withdrawals are required, the lifespan of the development could be lengthened.

OTHER USE OF UNDERGROUND SPACE

Utilization of underground space for artificial recharge of ground water has already been discussed as a means of increasing the usable ground-water supply. Other uses of underground space include underground disposal of fluid waste, flood control, and underground storage of gas, liquid, and solid materials for later recovery and use.

FLUID-WASTE DISPOSAL

With increasing social pressures to reduce pollution of the surface environment, there has been an accelerated growth in underground disposal of fluid waste. According to Kohout (1970, p. 1445) the number of wells constructed for underground disposal of fluid industrial wastes in the country increased from 10 in 1950 to nearly 150 in 1969 and might approach 500 by 1980. The technique is used in the Upper Colorado Region almost exclusively for disposal of oil-field brines; it might also be used to dispose of blowdown water from the cooling towers of thermal-electric powerplants, waste brines from desalting plants, and other fluid industrial wastes.

Important requirements for successful underground disposal of fluid waste are that the reservoir rock be capable of accepting and retaining the waste indefinitely and that it be hydrologically isolated by impermeable strata from freshwater aquifers. Many areas in the Upper Colorado Region seem to meet these requirements. Deep aquifers in the Green River, Uinta, Piceance Creek, and San Juan Basins contain brines that are probably more highly mineralized than many untreated fluid wastes; therefore, they might not have a better use than as reservoirs for the fluid waste. They apparently transmit water very slowly and seem to be virtually isolated hydrologically from freshwater aquifers and streams. Areas in the region where underground disposal of fluid waste might be feasible are shown in figure 24. Before any underground fluid-waste disposal project is implemented, however, it should be preceded by detailed and careful study to insure that the waste remains out of the surface environment indefinitely and does not contaminate usable ground-water supplies.

FLOOD CONTROL

Underground space in the unsaturated zone can be used for the control of flash floods, which are a serious problem throughout the region. Floodwaters could be diverted over permeable unsaturated terrace deposits, which generally exist near mouths of canyons, to be absorbed and released gradually back to the stream channel. The cost of required diversion facilities and spreading basins might be minimal compared with the high cost of repeated repair or replacement of bridges and other structures generally damaged by floods.

UNDERGROUND STORAGE

Underground storage as used here is the emplacement underground of any material with the intent of recovering that material for later use. Underground storage of natural gas is a common practice in the Gulf Coast area and in other parts of the world. The space utilized for storage of gas generally is in depleted oil or gas reservoirs, but aquifers are also used. In aquifers, the gas is stored as a bubble below impermeable rock and above the heavier ground water. Such a reservoir has been created in the Paradox(?) Formation near Moab, Utah (C. T. Sumsion, oral commun., 1972), and others seem feasible in that general area.

A unique use of underground space for storage and recovery of water is the multiple-aquifer reverse-cycle heating and cooling system. This system has had considerable success in some large office buildings in Portland, Oreg., and Tacoma, Wash. It might be adaptable to the Upper Colorado Region because there are multiple-aquifer systems throughout the region (as illustrated in figs. 8*A*, *B*), and this method of heating and air conditioning consumes very little water.

A special use of the unsaturated zone in underground space is the storage in underground chambers of dangerous chemical and strategic war materials that must be isolated from densely populated areas. The great thicknesses of relatively unsaturated rock in the sparsely populated Canyon Lands section of the region might provide such space with a minimum of construction and security costs. Because of the widespread distribution of radioactive minerals in the rocks in the section, however, storage of materials that are affected by radiation might not be advisable.

PROBLEMS AND CONSTRAINTS ASSOCIATED WITH GROUND-WATER DEVELOPMENT

PHYSICAL PROBLEMS

Theis (1940, p. 280) listed some of the basic principles regarding ground-water development, and a summary of them follows:

1. All water discharged by a well is balanced by a loss of water somewhere in the system. The loss may be depletion of aquifer storage, reduction in streamflow, reduction of evapotranspiration, or reduced normal water flow from the region.
2. The loss is in many cases largely from aquifer storage. Some ground water is always mined. The ground-water reservoir is in effect bounded by time, by structural conditions, and by material boundaries. The amount of water removed is proportional to the drawdown, which, in turn, is proportional to the rate of pumping.
3. If the drawdown cone reaches an area of recharge, the well discharge will be made up, at least in part, by an increase in the recharge. If any recharge was previously rejected through nonbeneficial transpiration, no economic loss will probably occur. If the recharge was rejected through springs or by refusal of the aquifer to absorb surface water, an economic loss may result owing to resulting decrease in spring or stream discharge.
4. If the drawdown cone reaches areas of natural discharge, further discharge by wells will be made up, in part, by a reduction in natural discharge. If this natural discharge fed surface streams, prior rights to the surface water may be injured.
5. In artesian aquifers, because the cones of depression spread with great rapidity, each well in a short time has its maximum effect on the whole aquifer and obtains most of its water by increase of recharge or decrease of natural discharge.

DECLINES IN WATER LEVELS

Water levels in the vicinity of a discharging well or well field will decline as long as the rate of the well discharge exceeds the rate of recharge to the aquifers supplying the water. Decline in the water levels is in the form of an inverted cone (cone of depression), as shown in figure 16. As the cone expands, the amount of ground-water moving toward the well or well field will increase until it equals the amount of discharge from the wells; then, the water levels should cease to decline. The water levels will begin to rise once pumping ceases. In areas of closely spaced wells, however, even temporary water-level declines may create legal and economic problems, owing to the interference caused by a discharging well on the water levels in nearby wells.

If withdrawals exceed natural recharge over many years, water levels might decline until a level is reached where further drilling or pumping would be uneconomical. This possibility is readily apparent in areas of deep natural water levels (pl. 1*G*), where the cost of drilling and pumping might be greater than the benefits to be gained.

DEPLETION OF STREAMFLOW

The relation of ground water and surface water was discussed on page 00 and is illustrated in figure 9. As shown in figure 9*C*, pumping of ground water can deplete streamflow. This is true whether the well field is adjacent to a stream or some distance from a stream, as long as there is hydraulic connection between that stream and aquifers tapped by the wells.

The amount of depletion would be proportional to the amount of ground water withdrawn provided that there is no return flow. Therefore, the effect of small ground-water withdrawals probably would be minuscule and could not be detected by current methods of streamflow measurement. In contrast, long periods of large-scale pumping, such as several tens of cubic feet per second during an irrigation season, from the alluvium of relatively small streams (average discharges of several hundred cubic feet per second during the irrigation season) would unquestionably cause a significant reduction of streamflow. This could result in legal problems related to local water rights and the Colorado River Compact, and it might also have an effect on the ecology of the stream and the water-related land resources.

DETERIORATION OF WATER QUALITY

Plate 2*B* shows how ground water is subject to contamination or deterioration by activities that are not directly related to ground-water withdrawals. Deterioration of ground-water quality as a direct result of large-scale withdrawal can occur in two ways — inducement of saline water from adjacent aquifers and recycling of water whose quality has been deteriorated by use.

Some freshwater aquifers in the region are either overlain by, underlain by, or sandwiched between saline-water aquifers. Reduction of the hydrostatic pressure in the freshwater aquifer by pumping could allow saline water from an adjacent aquifer to move into the pumped aquifer, as illustrated in figure 22*A*. The possibility of this problem exists throughout the region because of the widespread occurrence of saline ground water. It might have its greatest impact in areas such as Spanish Valley, Utah, and Paradox Valley, Colo., where the Paradox Formation, which is composed largely of salt, gypsum, and other salines, seems to be in hydraulic connection with overlying unconsolidated valley deposits. The Paradox Formation also underlies sandstone aquifers throughout much of the Canyon Lands section.

Deterioration of water quality by recycling of previously used water can occur where withdrawal exceeds natural recharge. (See Handy and others, 1969.) Excessive pumping for irrigation, for example, can cause a reversal of the natural ground-water gradient, so that water seeping downward from irrigated fields no longer can move out of the area but instead percolates back

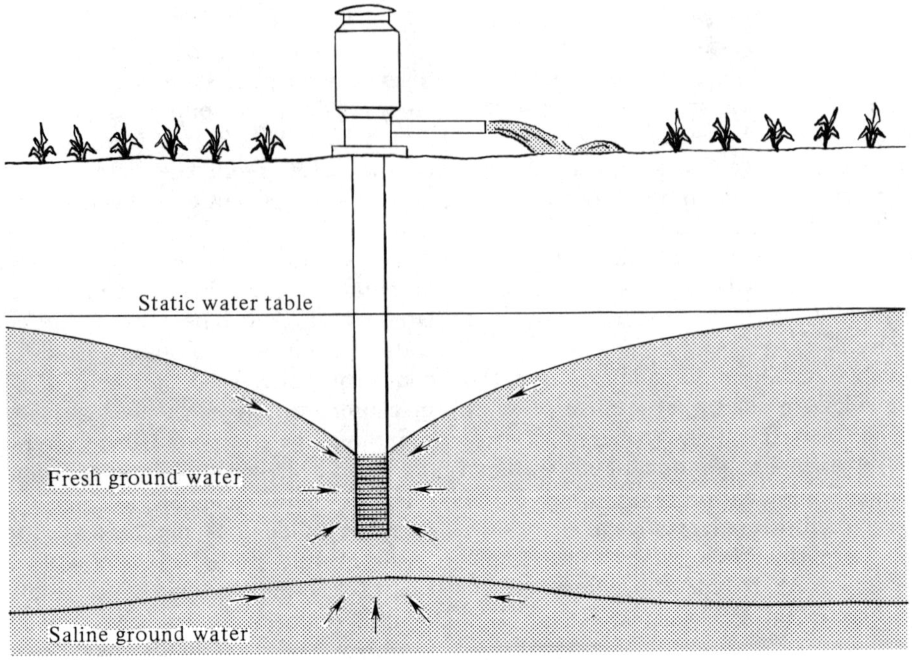

A. REDUCTION OF HEAD IN FRESH WATER AQUIFER CAN INDUCE ENCROACHMENT OF UNDERLYING SALINE WATER INTO THAT AQUIFER

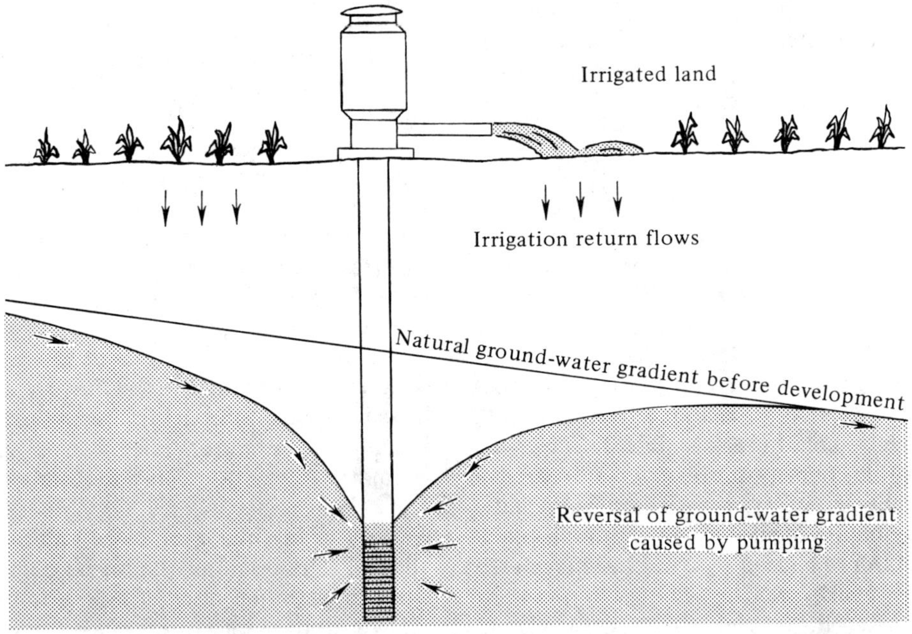

B. PERMANENT CONE OF DEPRESSION AROUND WELL FIELD PREVENTS OUTFLOW OF IRRIGATION RETURN FLOWS AND THEREFORE CONCENTRATES MINERAL CONTENT OF WATER IN PUMPED AREA BY RECIRCULATION

FIGURE 22. — Large-scale withdrawals from freshwater aquifers can result in deterioration of water quality.

toward the well fields. (See fig. 22*B*.) This return flow from irrigation is concentrated by evapotranspiration, and it carries salts leached from the soil and chemicals from fertilizers and soil amendments. The return of this water to the well field, therefore, serves to increase the dissolved-solids content of the water. This problem is most likely to occur in the shallow unconfined aquifers of the region, such as the broader alluvial plains of the larger streams, where natural ground-water gradients are fairly flat.

LAND SUBSIDENCE

Land subsidence associated with ground-water development is well known in Arizona, California, and other areas where withdrawals of ground water have exceeded natural recharge. In many places, it has resulted in considerable inconvenience and damage. (See Poland and Davis, 1969.)

Subsidence results principally from the compaction of lenses or layers of fine-grained materials within unconsolidated aquifers that contain water under artesian pressure. Aquifers meeting these conditions are rare in the Upper Colorado Region; therefore, subsidence may not be a major problem in the region.

EFFECT ON THE ENVIRONMENT

Development of a ground-water supply directly within the desired area of use probably would cause less damage to the environment than the impoundment of a stream a great distance from the area and the construction of long pipelines or canals. Some damage to the environment might occur, however, from a ground-water development that is on a time-limited basis in which water levels are allowed to decline indefinitely. For example, the ecological balance of an area might be upset if ground-water withdrawals drastically depleted streamflow or dried up a natural wetland habitat. Similarly, the esthetic beauty of a natural spring area would be destroyed if the development dried up the spring.

Other ecological problems might result from developments that require demineralization of ground water. If distillation were used in the demineralization process, the burning of fossil fuel to create the necessary energy from steam may result in a source of air pollution. If a membrane process were used to demineralize the water, fuel might also have to be burned to produce the required electrical energy.

CONSTRAINTS

WATER RIGHTS

Each State in the Upper Colorado Region administers its water resources according to the appropriation doctrine. The basic concept of this doctrine is that ownership of land provides no inherent right to water from sources upon, contiguous to, or underlying his land, but that rights to these sources are based on priority in time of beneficial use. The doctrine originally applied to surface water, but it was later extended to include ground water. The general knowledge that a proposed ground-water development in a given area might affect existing water rights in the area, therefore, is a major constraint to large-scale ground-water development.

Utah, for example, has restricted Spanish Valley and has closed or restricted parts of the Uinta Basin to further ground-water withdrawal until more information can be obtained regarding the effects of such withdrawal on existing water rights. The State of Colorado generally will not issue permits for wells that divert more than 15 gpm (57 l/min) in areas where streamflow is fully appropriated and ground-water withdrawals would reduce the streamflow, unless the permit is to replace or change the point of diversion of an existing water right.

Nevertheless, large-scale surface-water development has successfully been implemented within the framework of the appropriation doctrine. It seems reasonable to believe, therefore, that planned large-scale ground-water development, such as a conjunctive-use development, might also be successfully implemented without major legal complications.

Differences in interpretation of various State water laws and the Federal "reserved-water doctrine" might impose some constraints to ground-water development in the region. The State water laws imply that all the water in the State is subject to appropriation, whereas the Federal reserved-water doctrine implies that when land is reserved by the Federal government (fig. 5), the water necessary for use of the land is simultaneously reserved.

THE COLORADO RIVER COMPACT

The Colorado River Compact, which was approved August 19, 1921, states: "There is hereby apportioned from the Colorado River system in perpetuity to the upper basin and to the lower basin, respectively, the exclusive beneficial consumptive use of 7,500,000 acre-feet [9,251.2 hm^3] of water per annum, which shall include all water necessary for the supply of any rights which may now exist." (Article III a.) The Compact further states: "The States of the upper division will not cause the flow of the river at Lee Ferry to be depleted below an aggregate of 75,000,000 acre-feet [92,512.5 hm^3] for any period of 10 consecutive years reckoned in continuing progressive series beginning with the 1st day of October next succeeding the ratification of this Compact." (Article III d.)

These statements could be construed by some to imply that the compact is a major constraint to large-scale ground-water development in the Upper Colorado Region, but this is not true. For example, no provision in the compact specifies that the allotted consumptive use of 7.5 million acre-feet (9,251.2 hm^3) per year must come directly from streams. The allotted consumptive use could come from streams, aquifers, or the combined sources in a conjunctive-use development. In fact, large-scale ground-water development in the region, particularly using saline water with demineralization, would reduce mineral inflow to the Colorado River. This would help to maintain better chemical quality of outflows to the Lower Colorado River Basin.

ECOLOGICAL CONSIDERATIONS

Growing social pressure to preserve the environment in its natural state is now a significant constraining factor to surface-water development. This type of constraint has not been felt in ground-water development, partly because ground water is an unseen resource and the effects of its development on the environment are not as obvious as a major surface-water development. Also, ground-water developments have grown in a piecemeal fashion, and any effects on the environment have been gradual. This is especially true in the Upper Colorado Region, where there is little large-scale withdrawal of ground water. Nevertheless, some of the problems associated with large-scale ground-water development, such as the land-subsidence problems of California, are well known. Thus, in addition to considering the benefits of a large-scale ground-water development, groups concerned with the environment quite surely will be evaluating the impact on the environment and will be prepared to contest the development if adverse effects on the ecology and environment are believed to outweigh the benefits of the development.

LIMIT OF SCIENTIFIC KNOWLEDGE

Little detailed information about ground water is available for the Upper Colorado Region. Reconnaissance studies of the ground-water resources have been made in most parts of the region, but only in a few widely scattered areas have detailed studies been made (fig. 23). Few of these studies contain quantitative information, and only one (Sumsion, 1971a) has involved the use of a model to define the ground-water system and predict the effects of withdrawals on water levels.

Ironically, the major emphasis on ground-water development in all five of the Upper Colorado Region States has been in parts of those States that lie outside the region — the High Plains of Colorado, New Mexico, and Wyoming and the Basin and Range segments of Utah and Arizona. Consequently, very little of the States' available funds for ground-water study have been expended for detailed or comprehensive studies to evaluate the relatively undeveloped ground-water resources of the Upper Colorado Region.

If large-scale ground-water development is to be implemented in the Upper Colorado Region, studies will be required that are similar in scope, magnitude, and cost to those that precede the planning and implementation of large-scale surface-water developments. These would involve feasibility studies, environmental-impact studies, and project design and implementation studies. Figure 24 shows selected types of studies that are needed in various parts of the region that have promise for use of ground water in overall water-resources management.

INFORMATION NEEDED FOR PROJECT DESIGN

Planned ground-water development, like planned surface-water development, requires considerable background information about the geohydrologic system in which the development is to function. The limitations or even the feasibility of the development cannot be determined until reasonable answers are obtained for such questions as:

1. How permeable are the aquifers? Will they support the required well yields?
2. How deep is the water? Will drilling and pumping costs be within the economic limit of the development?
3. What is the perennial yield of the ground-water reservoirs? Can the reservoirs support anticipated perennial withdrawals including peak seasonal or periodic requirements?
4. How much water is recoverable from storage within economic pumping limits? Is the storage adequate to support the anticipated perennial withdrawals for the life of the project, or until a supply can be imported?
5. How good is the water? Is it chemically and physically suitable for the anticipated needs, or will it require desalinization or treatment?
6. What effects will the development have on streamflow and the environment in general?

Because of the small amount of ground-water development to date (1973) in the region, there is little background data on which to base detailed studies. Such data must be acquired in the earliest possible stages of an investigation; thus, in essentially all parts of the region, detailed geologic mapping in conjunction with test and drilling and geophysical mapping is needed to determine the location, thickness, and extent of aquifers. Static water levels in the test holes would provide information about direction of ground-water movement and location of areas of natural recharge and discharge. Pumping tests, using the test wells as production wells

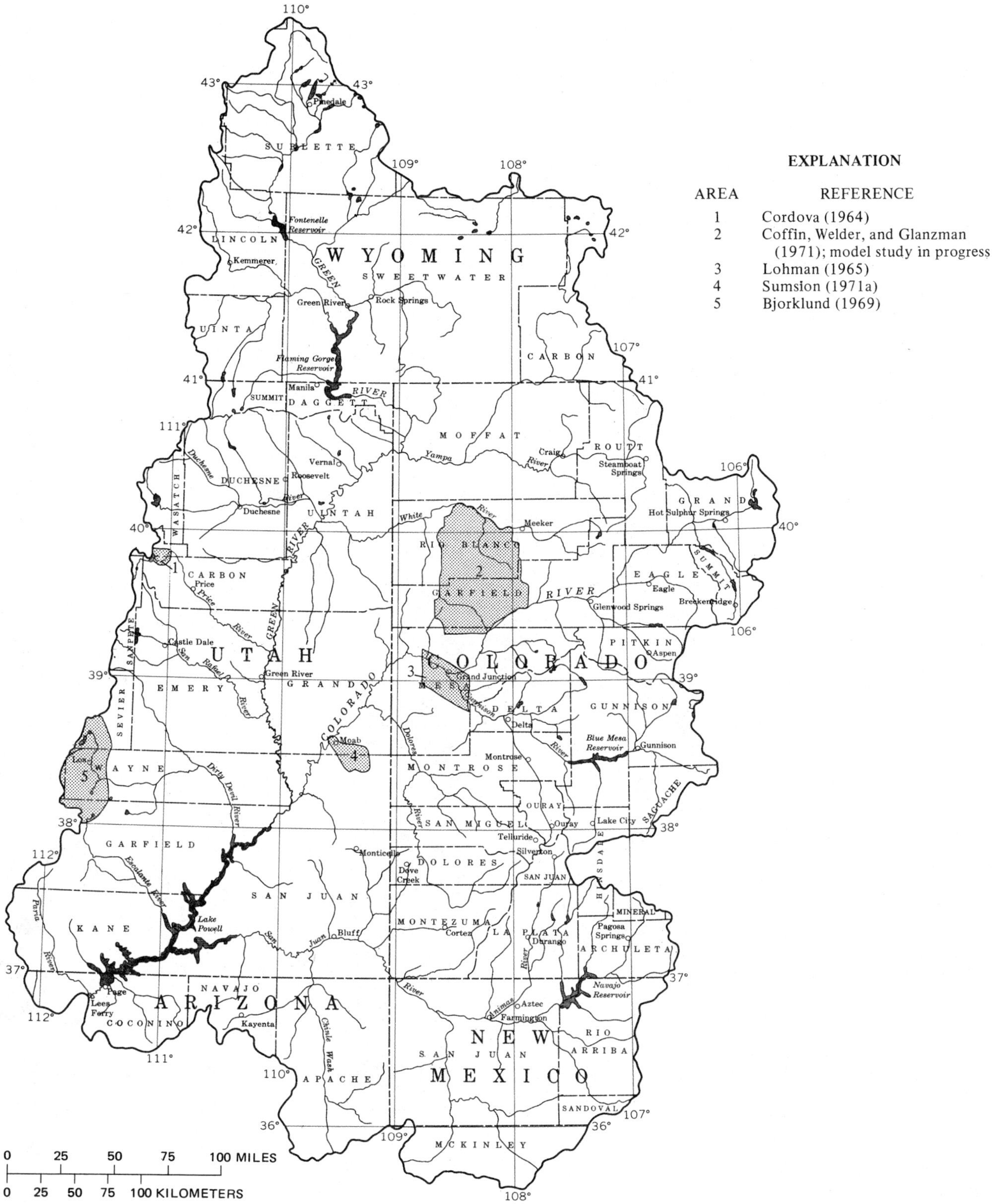

FIGURE 23. — The ground-water resources have been studied in detail in only a few widely scattered areas in the region.

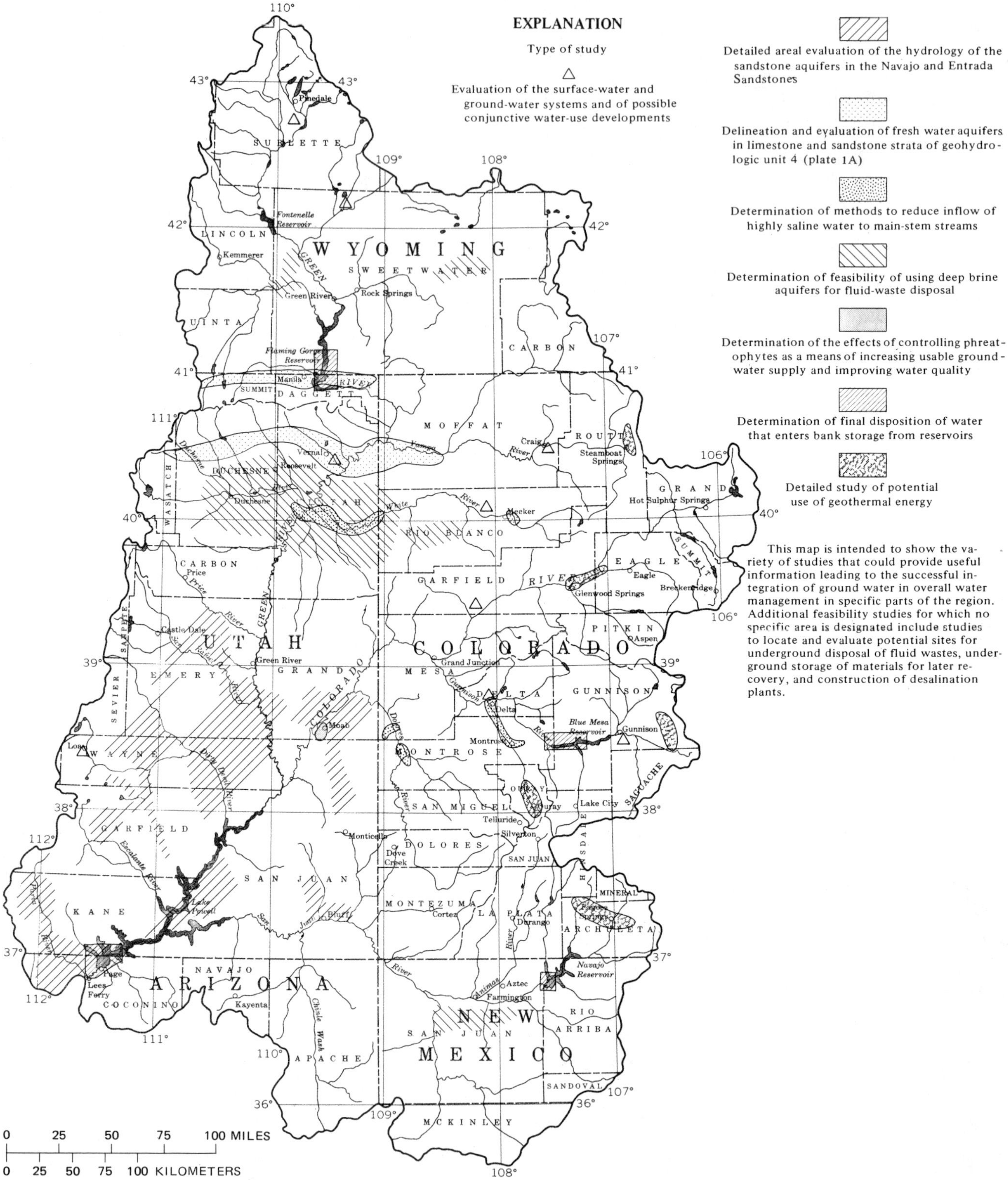

FIGURE 24. — Areas of the region that have potential for implementation of ground water and use of subsurface space in overall water-resources management.

and observation wells, would provide information about the hydraulic properties of the aquifers. Analyses of water samples collected from the test wells would provide information about the ground-water quality both areally and with depth.

By use of mathematical, statistical, and electric-analog models the field data can be evaluated. Successful models can accurately portray the ground-water regimen and predict the effects and changes imposed by various types and stages of ground-water development. A mathematical model of the hydrologic system in the Piceance Creek basin is currently being developed. It will help planners evaluate the ground-water supply and to predict effects of oil-shale development on the water resources of the basin.

Basic information needed for accurate and successful model studies is listed in table 4, and some applications of models to specific problems are listed in table 5.

TABLE 4. — *Data requirements for prediction model studies*

[After Moore, 1971]

Physical framework

1. Geologic map showing areal extent and boundaries of all aquifers.
2. Topographic map showing surface-water bodies.
3. Water-table, bedrock-configuration, and saturated-thickness maps.
4. Transmissivity map showing aquifer and its boundaries.
5. Map showing variation in storage coefficient of aquifer.
6. Relation of saturated thickness to transmissivity.
7. Relation of stream and aquifer (hydraulic connection).

Hydrologic stress

1. Recharge areas (irrigated areas, recharge basins, recharge wells, and so forth).
2. Surface-water diversions.
3. Ground-water pumpage distributed in time and space.
4. Depth-to-water map (with evapotranspiration rate).
5. Tributary inflow (distribution in time and space).
6. Ground-water inflow and outflow.
7. Precipitation.
8. Distribution of water quality in aquifer.
9. Streamflow quality (distribution in time and space).

Model calibration

1. Water-level-change maps and hydrographs.
2. Streamflow (including gain and loss measurements).

Prediction and optimization analysis

1. Economic information on water supply.
2. Legal and administrative rules.
3. Environmental factors.

The model can be used for planning the project and for evaluation of development, use, and management plans. Some of the possible uses of the model are shown in table 5.

TABLE 5. — *Summary of problems that could be analyzed with a model*

[After Moore, 1971]

1. Assist in identifying deficiencies in hydrologic field data.
2. Quantitative evaluation of stream-aquifer relations.
3. Evaluation of the efficiency of the surface-water distribution system, including surface storage.
4. Define areas where additional pumping of ground water would be beneficial, such as salvage of ground water now consumed non-beneficially.
5. Study the effect of increased pumpage of ground water on return flow to the river, on evapotranspiration, and on aquifer storage.
6. Aid in the selection of sites where large-capacity wells can be developed for use in satisfying surface-water rights.
7. Evaluation of the most efficient spacing of irrigation wells in different areas.
8. Outline areas where growth of phreatophytic plants can be controlled by lowering the water table by pumping from wells, determine the number and spacing of salvage wells that would be required, and evaluate the effect of these wells on seepage to and from the river.
9. Measure the effects of the importation of additional surface water.
10. Aid in the formulation of effective management criteria for the control and maintenance of favorable water quality.
11. Identify areas where additional water might be stored underground.
12. Provide a means of predicting water availability in different areas under varying schemes of surface-water delivery and ground-water pumpage.
13. Optimize plans for utilizing both ground water and surface water based on economic, legal, and social constraints.
14. Use of additional water in the tributary valleys.
15. Study the effect of lining canals.

SUMMARY

Ground water often has been neglected in water-resources planning and management. The reasons most commonly given are that ground water cannot be adequately evaluated in terms of quantity, quality, availability, and cost of development. These reasons are no longer valid. Advancements in the field of ground-water hydrology in the past half century have provided the knowledge and techniques to evaluate ground-water resources with considerable accuracy and reliability. Therefore, the resource should be considered in every phase of water-resources planning and included in management options. This is particularly true in the Upper Colorado Region, where limitations on surface-water development are imposed by the Colorado River Compact and by a growing demand for preservation and enhancement of the environment.

Tremendous quantities of ground water are available in the region. The volume of recoverable water stored in the upper 100 feet (30.5 m) of saturated rocks alone may be as much as 115 million acre-feet (141,852.5 hm^3), and many times this amount is available from greater depths. In addition, the annual replenishable supply of ground water of about 4 million acre-feet (4,934 hm^3) is enough water to support industries requiring 3,600 million gallons (13,627,440 m^3) of water a day, or to irrigate about 1.3 million acres (526,110 ha) of crops having an annual water requirement of 3 feet per acre (0.9 m/ha). Yet, in 1970 only about 2 percent of the water used in the region came directly from ground-water sources.

It is true that most of the ground water is in consolidated rocks which generally yield water to wells slowly, and it is also true that much of the ground water is saline and, in some places, occurs at great depths. Nevertheless, the ground water in the region has certain characteristics, such as its uniform areal and time distribution, that make it highly useful in overall water-resources management.

Under various development options, ground water can be used in conjunction with surface water or it can be developed as an independent supply where surface water is not available. The various development options all are possible within the framework of the Colorado River Compact; in fact, the full use of the ground-water resource could help the water users in the Upper Colorado River Basin to meet projected needs for water in the region and still satisfy the various provisions of the compact.

Benefits that could be gained by the optimal use of ground water in overall water-resources management in the region include:

1. A reduction of nonbeneficial discharge of water.
2. A more uniform water supply.
3. A means of meeting growing in-region water needs while satisfying the growing demand outside the region for water from the Colorado River.
4. A means of developing water supplies without the adverse effects on the environment that are commonly associated with large surface-water facilities.
5. A means of alleviating the salinity problems in the water in the Colorado River system.
6. Full use of the total water resource of the region.

The ground-water resources of those areas cited as having a potential for including ground water in overall water-resources management have had only cursory or reconnaissance study. Considerable more study, including the use of models, is needed to fully evaluate this potential and to determine the most feasible development options.

There is a great potential for use of underground space, such as for fluid-waste disposal and development of underground storage reservoirs and chambers, in the region. Special detailed studies are needed to locate and evaluate the potential sites.

SELECTED REFERENCES

[Astericks identify references that contain most useful information to water planning in various parts of the region]

Baltz, E. H., 1967, Stratigraphy and regional tectonic implications of part of Upper Cretaceous and Tertiary rocks, east-central San Juan Basin, New Mexico: U.S. Geol. Survey Prof. Paper 552, 101 p.

*Baltz, E. H., and West, S. W., 1967, Ground-water resources of the southern part of Jicarilla Apache Indian Reservation and adjacent areas, New Mexico: U.S. Geol. Survey Water-Supply Paper 1576-H, 89 p.

Berry, D. W., 1960, Geology and ground-water resources of the Rawlins area, Carbon County, Wyoming: U.S. Geol. Survey Water-Supply Paper 1458, 74 p.

*Bjorklund, L. J., 1969, Reconnaissance of the ground-water resources of the upper Fremont River valley, Wayne County, Utah: Utah Dept. Nat. Resources Tech. Pub. 22, 54 p.

Boettcher, A. J., 1971, Evaluation of the water supply at six sites in the Curecanti Recreation Area, southwestern Colorado: U.S. Geol. Survey open-file report, 28 p.

*______1972, Ground-water occurrence in the northern and central parts of western Colorado: Colorado Water Conserv. Board Water-Resources Circ. 15, 50 p.

Carter, W. D., and Gualtieri, J. L., 1965, Geology and uranium-vanadium deposits of the La Sal quadrangle, San Juan County, Utah, and Montrose County, Colorado: U.S. Geol. Survey Prof. Paper 508, 82 p.

*Case, F. O. [chm.], and others, 1971, Comprehensive framework study, Upper Colorado Region, Appendix VI (land resources and use): Pacific Southwest Interagency Committee, Water Resources Council open-file report, 142 p.

*Coffin, D. L., Welder, F. A., and Glanzman, R. K., 1971, Geohydrology of Piceance Creek structural basin between the White and Colorado Rivers, northwestern Colorado: U.S. Geol. Survey Hydrol. Inv. Atlas HA-370.

Coffin, D. L., Welder, F. A., Glanzman, R. K., and Dutton, X. W., 1968, Geohydrologic data from the Piceance Creek basin between the White and Colorado Rivers, northwestern Colorado: Colorado Water Conserv. Board Ground-Water Circ. 12, 38 p.

Connor, J. G., Mitchell, C. G., and others, 1958, A compilation of chemical quality data for ground and surface waters in Utah: Utah State Engineer Tech. Pub. 10, 276 p.

Cooley, M. E., 1965, Stratigraphic sections and records of springs in the Glen Canyon region of Utah and Arizona: Musem of Northern Arizona Tech. Ser. 6, 140 p.

Cooley, M. E., Akers, J. P., and Stevens, P. R., 1964, Geohydrologic data in the Navajo and Hopi Indian Reservations, Arizona, New Mexico, and Utah, Part III, Selected lithologic logs, drillers' logs, and stratigraphic sections: Arizona State Land Dept. Water-Resources Rept. 12-C, 157 p.

*Cooley, M. E., Harshbarger, J. W., Akers, J. P., and Hardt, W. F., 1969, Regional hydrology of the Navajo and Hopi Indian Reservations, Arizona, New Mexico, and Utah, *with a section on* Vegetation, by O. N. Hicks: U.S. Geol. Survey Prof. Paper 521-A, 61 p.

Cooley, M. E., and others, 1966, Geohydrologic data in the Navajo and Hopi Indian Reservations, Arizona, New Mexico, and Utah, Part IV, Maps showing locations of wells, springs, and stratigraphic sections: Arizona State Land Dept. Water-Resources Rept. 12-D.

*Cordova, R. M., 1964, Hydrogeologic reconnaissance of part of the head waters area of the Price River, Utah: Utah Geol. and Mineralog. Survey Water-Resources Bull. 4, 26 p.

Davis, G. E., Hardt, W. F., Thompson, L. K., and Cooley, M. E., 1963, Geohydrologic data in the Navajo and Hopi Indian Reservations,

Arizona, New Mexico, and Utah, Part I, Records of ground-water supplies: Arizona State Land Dept. Water-Resources Rept. 12-A, 159 p.

Eisenlohr, W. S., and others, 1962, Explorations for water supplies on the public domain, 1960: U.S. Geol. Survey Circ. 461, 28 p.

*Feltis, R. D., 1966, Water from bedrock in the Colorado Plateau of Utah: Utah State Engineer Tech. Pub. 15, 82 p.

Feltis, R. D., and Goode, H. D., 1961, Production and use of fresh water from the Ashley Valley oil field, Uintah County, Utah, *in* Geological Survey Research 1961: U.S. Geol. Survey Prof. Paper 424-C, p. C90-C93.

Fenneman, N. M., 1931, Physiography of the Western United States: New York, McGraw-Hill Book Co., 534 p.

———1946, Physical divisions of the United States: U.S. Geol. Survey map, scale 1 : 7,000,000.

Feth, J. H., and others, 1965, Preliminary map oof the conterminous United States showing depth to and quality of shallowest ground water containing more than 1,000 parts per million dissolved solids: U.S. Geol. Survey Hydrol. Inv. Atlas HA-199.

Goode, H. D., 1958, The geology and distribution of aquifers in the southwestern part of San Juan County, Utah: U.S. Geol. Survey open-file report, 9 p.

*———1964, Reconnaissance of water resources of part of western Kane County, Utah: Utah Geol. and Mineralog. Survey Water-Resources Bull. 5, 63 p.

*———1966, Second reconnaissance of the water resources in western Kane County, Utah: Utah Geol. and Mineralog. Survey Water-Resources Bull. 8, 44 p.

*———1969, Reconnaissance appraisal of the water resources near Escalante, Garfield County, Utah: Utah Geol. and Mineralog. Survey Water-Resources Bull. 11, 38 p.

Goode, H. D., and Feltis, R. D., 1962, Water production from oil wells of the Uinta Basin, Uintah and Duchesne Counties, Utah: Utah Geol. and Mineralog. Survey Water-Resources Bull. 1, 30 p.

Gregg, D. O., Meyer, E. L., Targy, M. M., and Moulder, E. A., 1961, Public water supplies of Colorado, 1959-60: Colorado State Univ. Agr. Expt. Sta. Gen. Ser. Rept. 757, 128 p.

Gregory, H. E., 1916, The Navajo country — A geographic and hydrographic reconnaissance of parts of Arizona, New Mexico, and Utah: U.S. Geol. Survey Water-Supply Paper 380, 219 p.

———1951, The geology and geography of the Paunsaugunt region, Utah: U.S. Geol. Survey Prof. Paper 226, 116 p.

*Hagen, R. H. [chm.], and others, 1971, Comprehensive framework study, Upper Colorado Region, Appendix XV (water quality, pollution control, and health factors): Pacific Southwest Interagency Committee, Water Resources Council open-file report, 219 p.

Hale, W. E., Reiland, L. J., and Beverage, J. P., 1965, Characteristics of the water supply in New Mexico: New Mexico State Engineer Tech. Rept. 31, 131 p.

Hall, W. J., Jr., 1968, Geology of southwestern North Park and vicinity, Colorado: U.S. Geol. Survey Bull. 1257, 119 p.

Halpenny, L. C., 1954, Preliminary report on the ground-water resources of the Navajo and Hopi Indian Reservations, Arizona, New Mexico, and Utah: New Mexico Geol. Soc. Guidebook of the South and West Sides of the San Juan Basin, New Mexico and Arizona, p. 147-151.

Halpenny, L. C., and Harshbarger, J. W., 1952, Ground-water resources of the Navajo and Hopi Indian Reservations, Arizona, New Mexico, and Utah: Geol. Soc. America Bull., v. 63, p. 1330.

Handy, A. H., Mower, R. W., and Sandberg, G. W., 1969, Changes in chemical quality of ground water in three areas in the Great Basin, Utah, *in* Geological Survey Research 1969: U.S. Geol. Survey Prof. Paper 650-D, p. D228-D234.

Hanley, G. W. [chm.], and others, 1971, Comprehensive framework study, Upper Colorado Region, Appendix XIV (electric power): Pacific Southwest Interagency Committee, Water Resources Council open-file report, 92 p.

Harshbarger, J. W., and Repenning, C. A., 1954, Water resources of the Chuska Mountains area, Navajo Indian Reservation, Arizona and New Mexico, *with a section on* Quality of water, by J. L. Hatchett: U.S. Geol. Survey Circ. 308,, 16 p.

Haycock, E. B., and others, 1970, Interim report on the Utah water plan: Utah Div. Water Resources, Div. Planning Staff Rept. 6, 70 p.

*Hedlund, J. D. [chm.], and others, 1971, Comprehensive framework study, Upper Colorado Region, Appendix V (water resources): Pacific Southwest Interagency Committee, Water Resources Council open-file report, 66 p.

Hood, J. W., and Kister, L. R., 1962, Saline water resources of New Mexico: U.S. Geol. Survey Water-Supply Paper 1601, 70 p.

Hunt, C. B., Averitt, Paul, and Miller, R. L., 1953, Geology and geography of the Henry Mountains region, Utah: U.S. Geol. Survey Prof. Paper 228, 234 p.

Hurr, R. T., Wilson, W. W., Welder, F. A., and Emerson, R. L., 1969, Records of selected wells and springs in the Rulison Project area, Garfield and Mesa Counties, Colorado: U.S. Geol. Survey open-file report, Rulison-2, 17 p.

*Iorns, W. V., Hembree, C. H., and Oakland, G. L., 1965, Water resources of the Upper Colorado River Basin — Technical report: U.S. Geol. Survey Prof. Paper 441, 370 p.

Iorns, W. V., Hembree, C. H., Phoenix, D. A., and Oakland, G. L., 1964, Water resources of the Upper Colorado River Basin — Basic data: U.S. Geol. Survey Prof. Paper 442, 1036 p.

*Irwin, J. H., 1966, Geology and availability of ground water on the Ute Mountain Indian Reservation, Colorado and New Mexico: U.S. Geol. Survey Water-Supply Paper 1576-G, p. G1-G109.

Johnson, Michael, and Lundeen, E. W., 1967, Alamitos barrier project — Résumé of geohydrologic investigation and status of barrier construction: Eng. Geology, v. 4, no. 1, p. 37-49.

Kister, L. R., and Hatchett, J. L., 1963, Geohydrologic data in the Navajo and Hopi Indian Reservations, Arizona, New Mexico, and Utah, Part II, Selected chemical analyses of the ground water: Arizona State Land Dept. Water-Resources Rept. 12-B, 58 p.

Kohler, M. A., Nordenson, T. J., and Baker, D. R., 1959, Evaporation maps for the United States: U.S. Weather Bur. Tech. Paper, 37.

Kohout, F. A., 1970, Reorientation of our saline water resources thinking: Water Resources Research, v. 6, no. 5, p. 1442-1446.

Kume, Jack, Emmett, L. F., Bader, J. S., and Vorhis, R. C., 1974, Ground-water contamination, Central States — A preliminary inventory: U.S. Geol. Survey Circ. 00, 000 p. (in press).

Lofgren, B. E., 1954, Ground-water possibilities of bedrock aquifers in southwestern Utah, *in* Progress report on selected ground-water basins in Utah: Utah State Engineer Tech. Pub. 9, p. 105-118.

*Lohman, S. W., 1965, Geology and artesian-water supply, Grand Junction area, Colorado: U.S. Geol. Survey Prof. Paper 451, 149 p.

Madison, R. J., and Waddell, K. M., 1973, Chemical quality of surface water in the Flaming Gorge Reservoir area, Wyoming and Utah: U.S. Geol. Survey Water-Supply Paper 2009-C, 18 p.

Marine, I. W., 1962, Water-supply possibilities at Capitol Reef National Monument, Utah: U.S. Geol. Survey Water-Supply Paper 1475-G, p. 201-208.

———1963, Ground-water resources of the Bryce Canyon National Park area, Utah: U.S. Geol. Survey Water-Supply Paper 1475-M, p. 441-485.

*Maxwell, J. D., Bridges, B. L., Barker, D. A., and Moore, L. G., 1971, Hydrology of the eastern portion of the south slopes of the Uinta Mountains, Utah: Utah Dept. Nat. Resources Inf. Bull. 21, 54 p.

McConaghy, J. A., and Colburn, G. W., 1964, Records of wells in Colorado: Colorado Water Conserv. Board Basic-Data Release 17, 384 p.

McGavock, E. H., Edmonds, R. J., Gillespie, E. L., and Halpenny, P. C., 1966, Geohydrologic data in the Navajo and Hopi Indian Reservations, Arizona, New Mexico, and Utah, Part 1A, Supplemental records of ground-water supplies: Arizona State Land Dept. Water-Resources Rept. 12-E, 55 p.

McGuinness, C. L., 1963, The role of ground water in the national water situation: U.S. Geol. Survey Water-Supply Paper 1800, 1,121 p.

Miser, H. D., 1924, The San Juan Canyon, southwestern Utah — A geographic and hydrographic reconnaissance: U.S. Geol. Survey Water-Supply Paper 538, 80 p.

Moore, J. E., 1971, Integrating the use of ground water into water-resources planning: Minutes of Missouri Basin Interagency meeting, June 23, 1971.

Mower, R. W., and Nace, R. L., 1957, Water consumption by water-loving plants in Malad Valley, Oneida County, Idaho: U.S. Geol. Survey Water-Supply Paper 1412, 33 p.

Mundorff, J. C., 1970, Major thermal springs of Utah: Utah Geol. and Mineralog. Survey Water-Resources Bull. 13, 60 p.

______1971, Nonthermal springs of Utah: Utal Geol. and Mineralog. Survey Water-Resources Bull. 16, 70 p.

______1972, Reconnaissance of chemical quality of surface water and fluvial sediment in the Price River basin, Utah: Utah Dept. Nat. Resources Tech. Pub. 39, 55 p.

Murray, C. R., and Reeves, E. B., 1972, Estimated use of water in the United States in 1970: U.S. Geol. Survey Circ. 676, 37 p.

New Mexico State Engineer, 1967, Water resources of New Mexico, Occurrence, development, and use: New Mexico State Planning Office report, 321 p.

Oak Ridge National Laboratory and Office of Saline Water, 1970, Office of Saline Water special report of status of desalting, Comprehensive framework study: U.S. Dept. Interior open-file report, 111 p.

Oriel, S. S., 1963, Preliminary geologic map of the Fort Hill quadrangle, Lincoln County, Wyoming: U.S. Geol. Survey Oil and Gas Inv. Map OM-212.

Pearl, R. H., 1972, Geothermal resources of Colorado: Colorado Geol. Survey Spec. Pub. 2, 39 p.

Poland, J. F., and Davis, G. H., 1969, Land subsidence due to withdrawal of fluids: Geol. Soc. America Reviews in Engineering Geology, v. 2, p. 187-269.

Price, Don, Hart, D. H., and Foxworthy, B. L., 1965, Artificial recharge in Oregon and Washington, 1962: U.S. Geol. Survey Water-Supply Paper 1594-C, p. C1-C65.

Rapp, J. R., 1959, Reconnaissance of the geology and water resources of the Farmington area, San Juan County, New Mexico: U.S. Geol. Survey open-file report, 13 p.

*Robinove, C. J., and Cummings, T. R., 1963, Ground-water resources and geology of the Lyman-Mountain View area, Uinta County, Wyoming: U.S. Geol. Survey Water-Supply Paper 1669-E, p. E1-E43.

Robinson, T. W., 1958, Phreatophytes: U.S. Geol. Survey Water-Supply Paper 1423, 84 p.

______1968, Areal extent of phreatophytes and hydrophytes in the Western States (section on Arizona, Colorado, New Mexico, and Utah): U.S. Geol. Survey open-file report, 47 p.

*Sumsion, C. T., 1971a, Geology and water resources of the Spanish Valley area, Grand and San Juan Counties, Utah: Utah Dept. Nat. Resources Tech. Pub. 32, 45 p.

*______1971b, Water-resources investigations in Dinosaur National Monument, Utah-Colorado, fiscal year 1970: U.S. Geol. Survey open-file report, 52 p.

*Sumsion, C. T., and Bolke, E. L., 1972, Water resources of part of Canyonlands National Park, southeastern Utah: U.S. Geol. Survey open-file report, 75 p.

Theis, C. V., 1940, The source of water derived from wells: Civil Eng., v. 10, no. 5, p. 277-280.

Thomas, H. E., 1952, Hydrologic reconnaissance of the Green River in Utah and Colorado: U.S. Geol. Survey Circ. 129, 32 p.

______1962, Water and the Southwest — What is the future?: U.S. Geol. Survey Circ. 469, 15 p.

Thomas, H. E., and Wilson, M. T., 1952, Determination of total evapotranspiration in Ashley Valley, Utah, by the inflow-outflow method: U.S. Geol. Survey open-file report, 15 p.

U.S. Bureau of Reclamation and New Mexico Interstate Stream Commission, (no date), Preliminary situation assessment report, New Mexico State water plan: U.S. Bur. Reclamation open-file report, 166 p.

*Voegeli, P. T., Sr., 1965, Ground-water resources of North Park and Middle Park, Colorado — A reconnaissance: U.S. Geol. Survey Water-Supply Paper 1809-G, p. G1-G54.

Waring, G. A., 1935, Ground-water resources of northwestern New Mexico: U.S. Geol. Survey open-file report, 160 p.

Waring, G. A., and Knechtel, M. M., 1935, Ground water in part of southeastern Utah and southwestern Colorado: U.S. Geol. Survey open-file report, 119 p.

Waring, G. A. (revised by R. R. Blankenship and Ray Bentall), 1965, Thermal springs of the United States and other countries of the world — A summary: U.S. Geol. Survey Prof. Paper 492, 383 p.

Weir, J. E., Jr., 1970, Geohydrology of the area near WOSCO [Western Oil Shale Corp.] exploratory hole number 1, Uintah County, Utah: U.S. Geol. Survey open-file report, 27 p.

*Welder, G. E., 1968, Ground-water reconnaissance of the Green River basin, southwestern Wyoming: U.S. Geol. Survey Hydrol. Inv. Atlas HA-290.

Welder, F. A., 1971, Ground-water reconnaissance of selected sites in Rocky Mountain National Park and Shadow Mountain National Recreation Area: U.S. Geol. Survey open-file report, 47 p.

*Welder, G. E., and McGreevy, L. J., 1966, Ground-water reconnaissance of the Great Divide and Washakie Basins and some adjacent areas, southwestern Wyoming: U.S. Geol. Survey Hydrol. Inv. Atlas HA-219.

West, S. W., 1957, Possibility of developing a ground-water supply at the Chaco Canyon National Monument, San Juan County, New Mexico: U.S. Geol. Survey open-file report, 32 p.

Wilson, W. W., 1965, Pumping tests in Colorado: Colorado Water Conserv. Board Circ. 11, 361 p.

Wyoming State Engineer, 1970, Water and related land resources of the Green River basin, Wyoming: Wyoming Planning Program Rept. 3, 167 p.

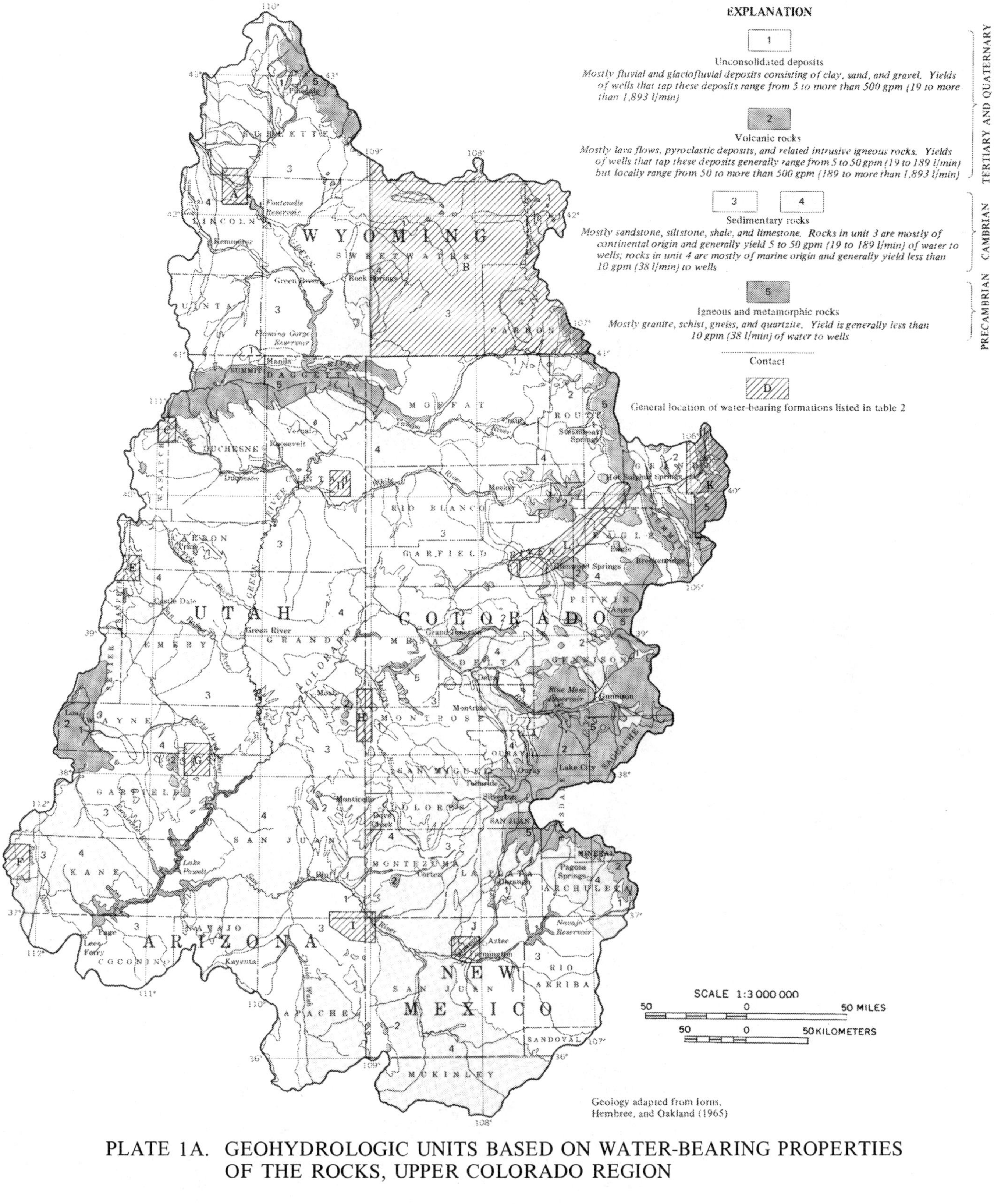

PLATE 1A. GEOHYDROLOGIC UNITS BASED ON WATER-BEARING PROPERTIES OF THE ROCKS, UPPER COLORADO REGION

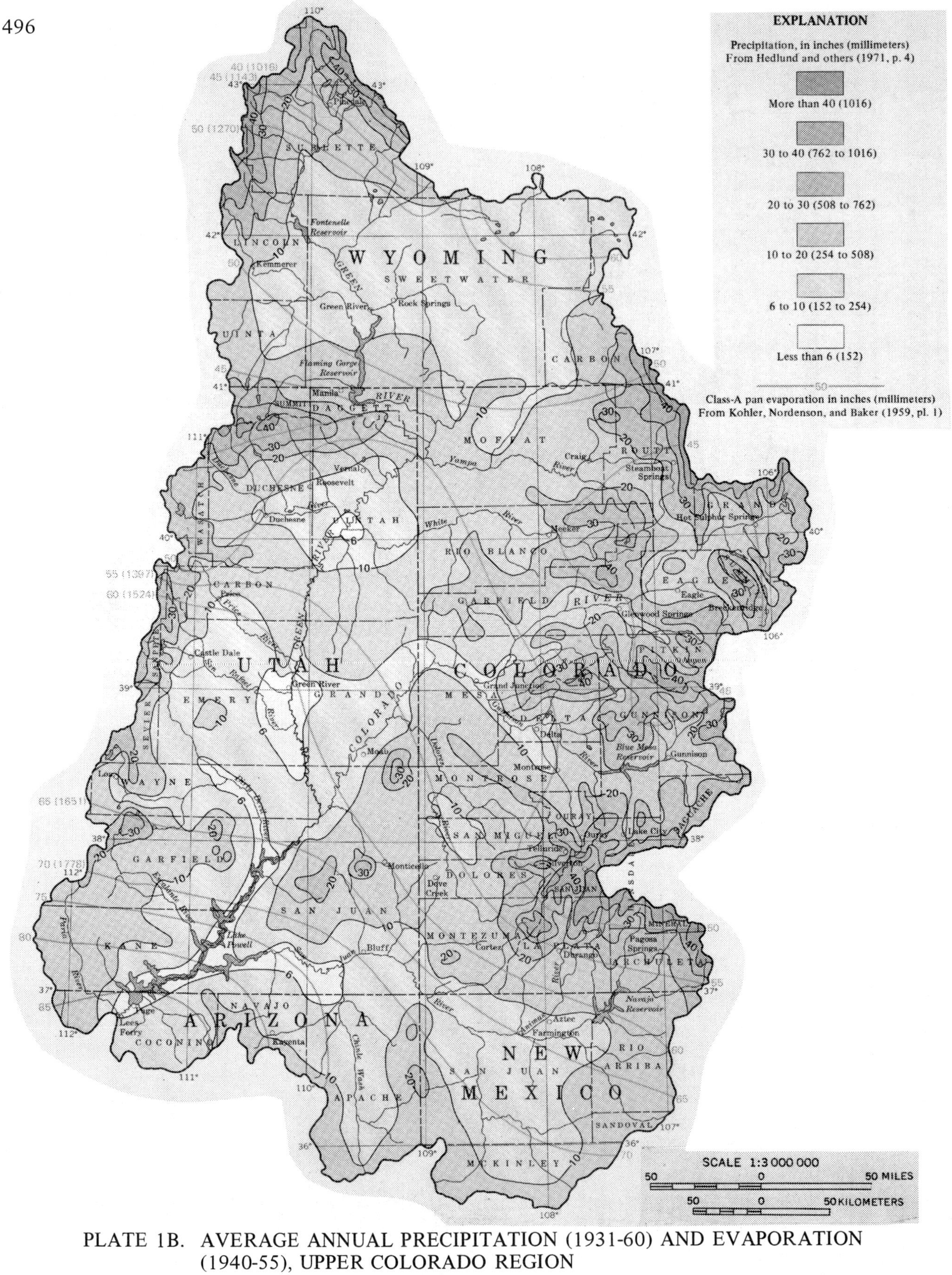

PLATE 1B. AVERAGE ANNUAL PRECIPITATION (1931-60) AND EVAPORATION (1940-55), UPPER COLORADO REGION

PLATE 1C. AVERAGE ANNUAL RUNOFF (1914-65) IN THE UPPER COLORADO REGION

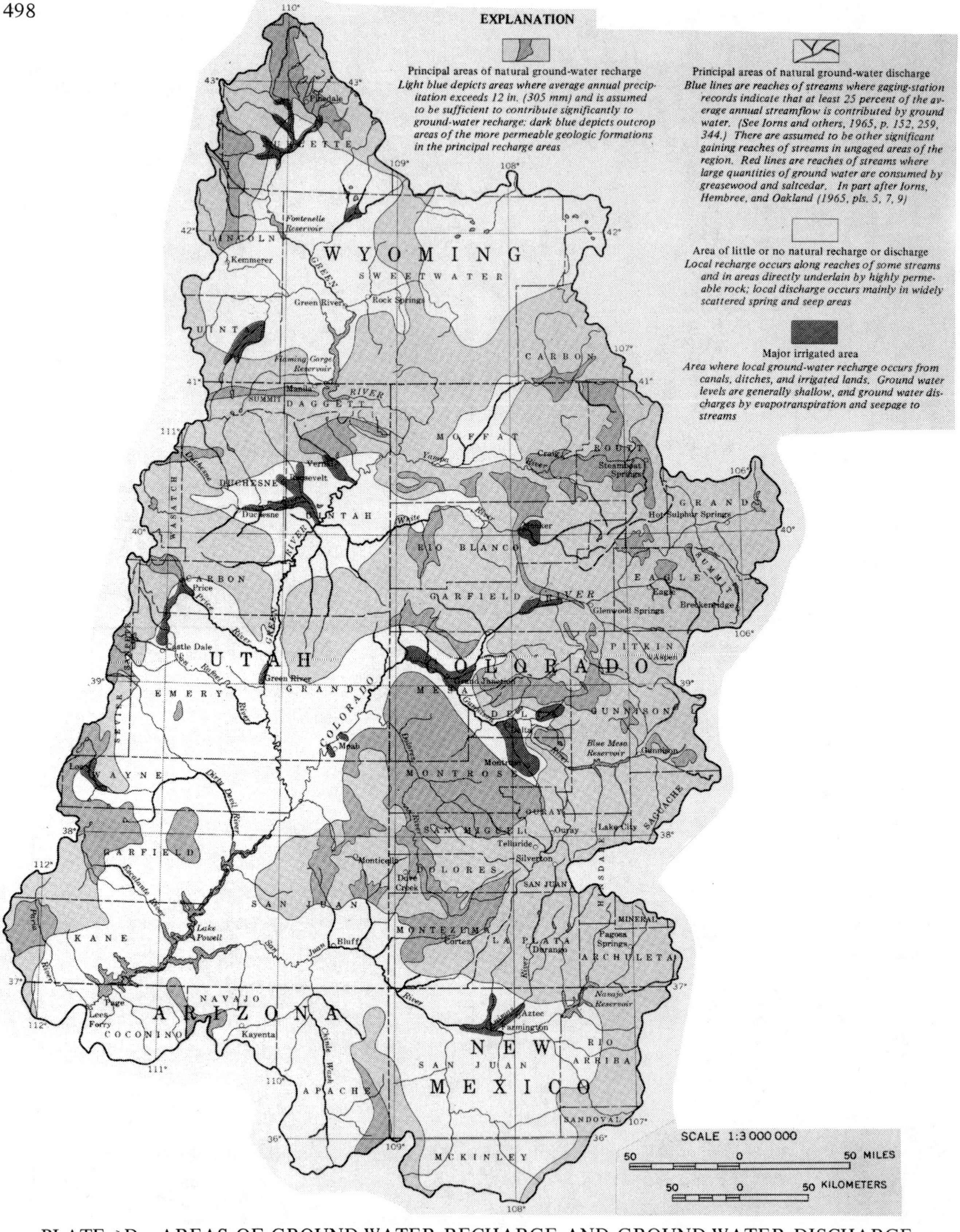

PLATE 1D. AREAS OF GROUND-WATER RECHARGE AND GROUND-WATER DISCHARGE, UPPER COLORADO REGION

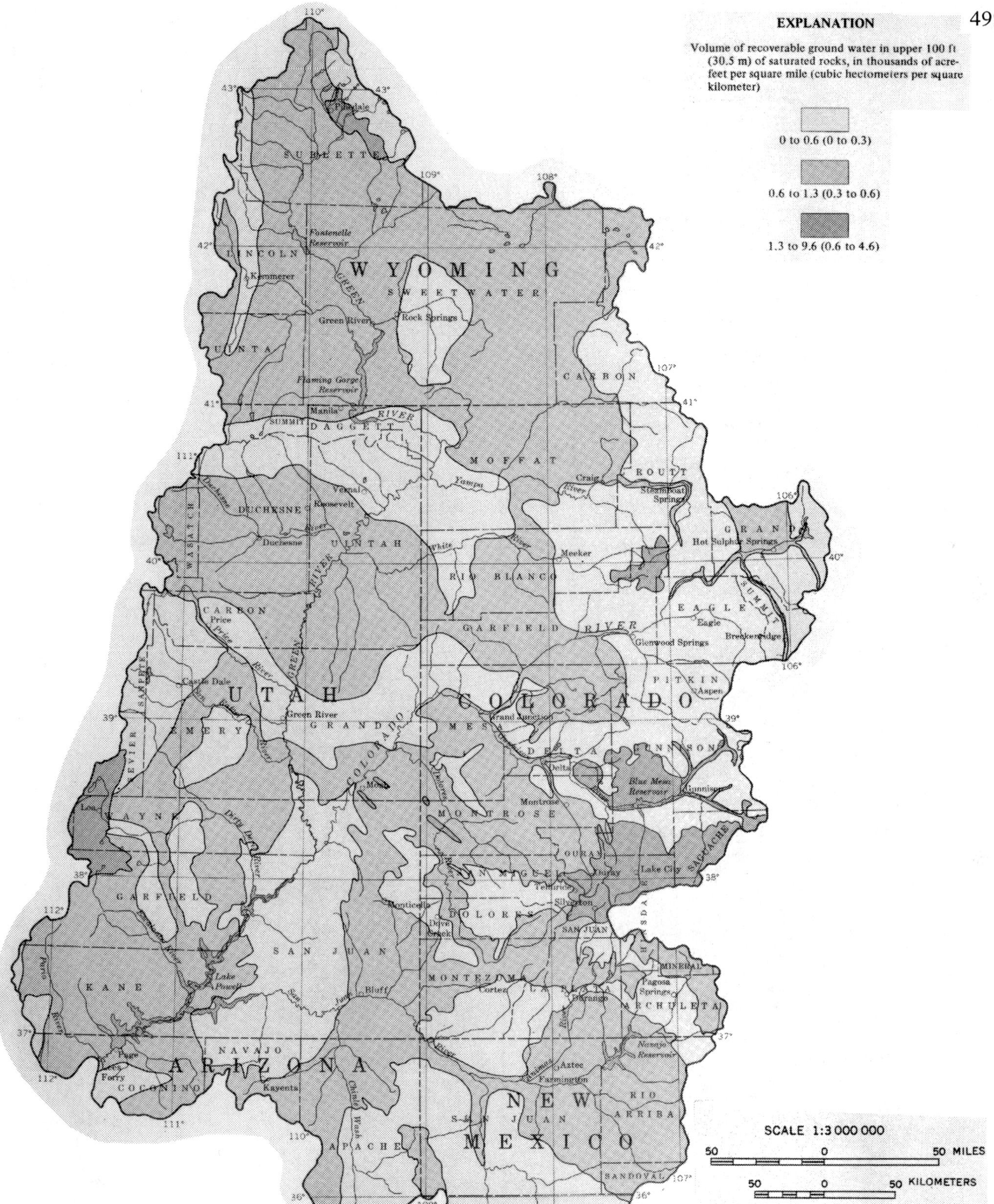

PLATE 1E. AREAS OF RECOVERABLE GROUND WATER IN THE UPPER COLORADO REGION

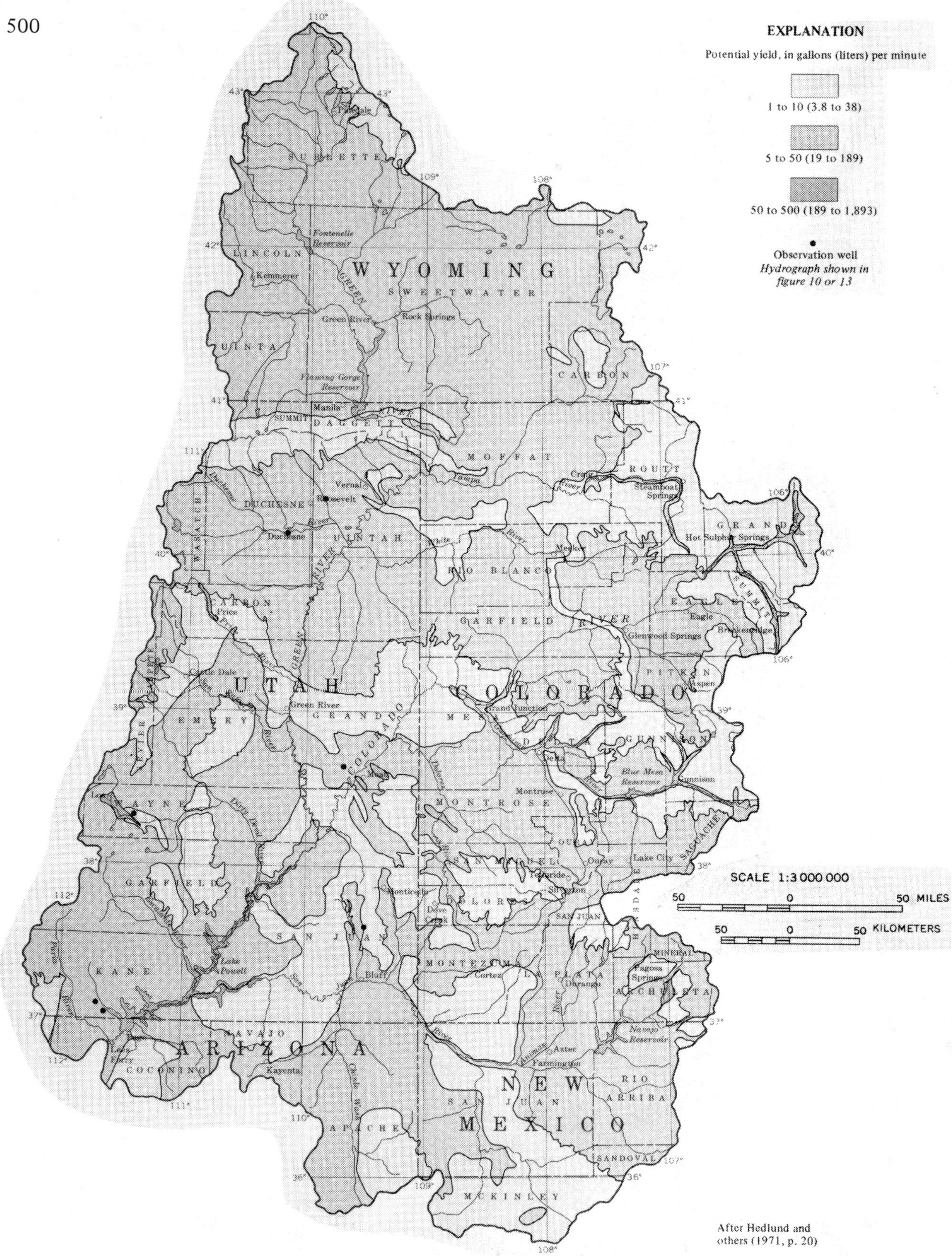

PLATE 1F. POTENTIAL YIELDS TO PROPERLY LOCATED, PROPERLY CONSTRUCTED WELLS IN THE UPPER COLORADO REGION

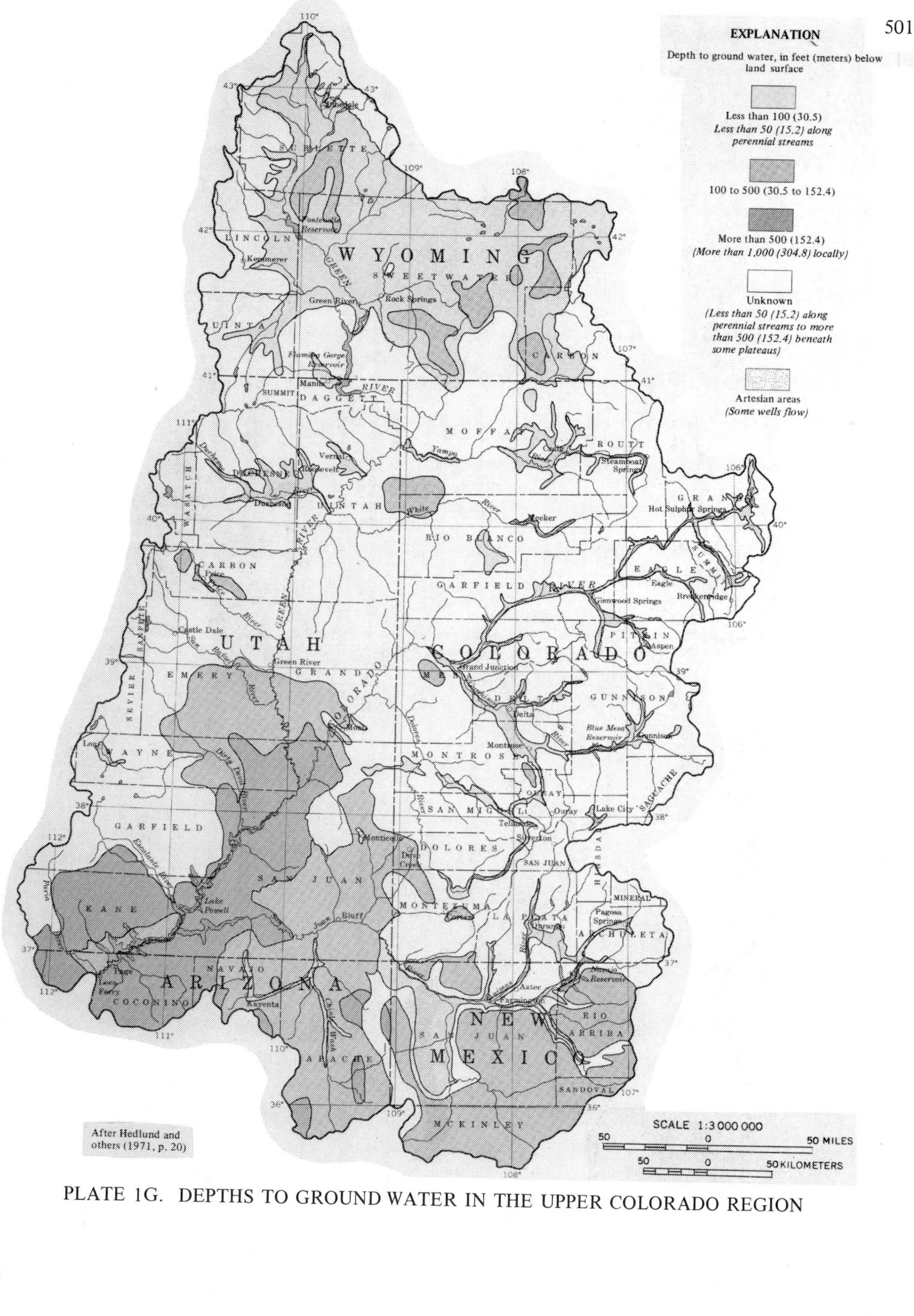

PLATE 1G. DEPTHS TO GROUND WATER IN THE UPPER COLORADO REGION

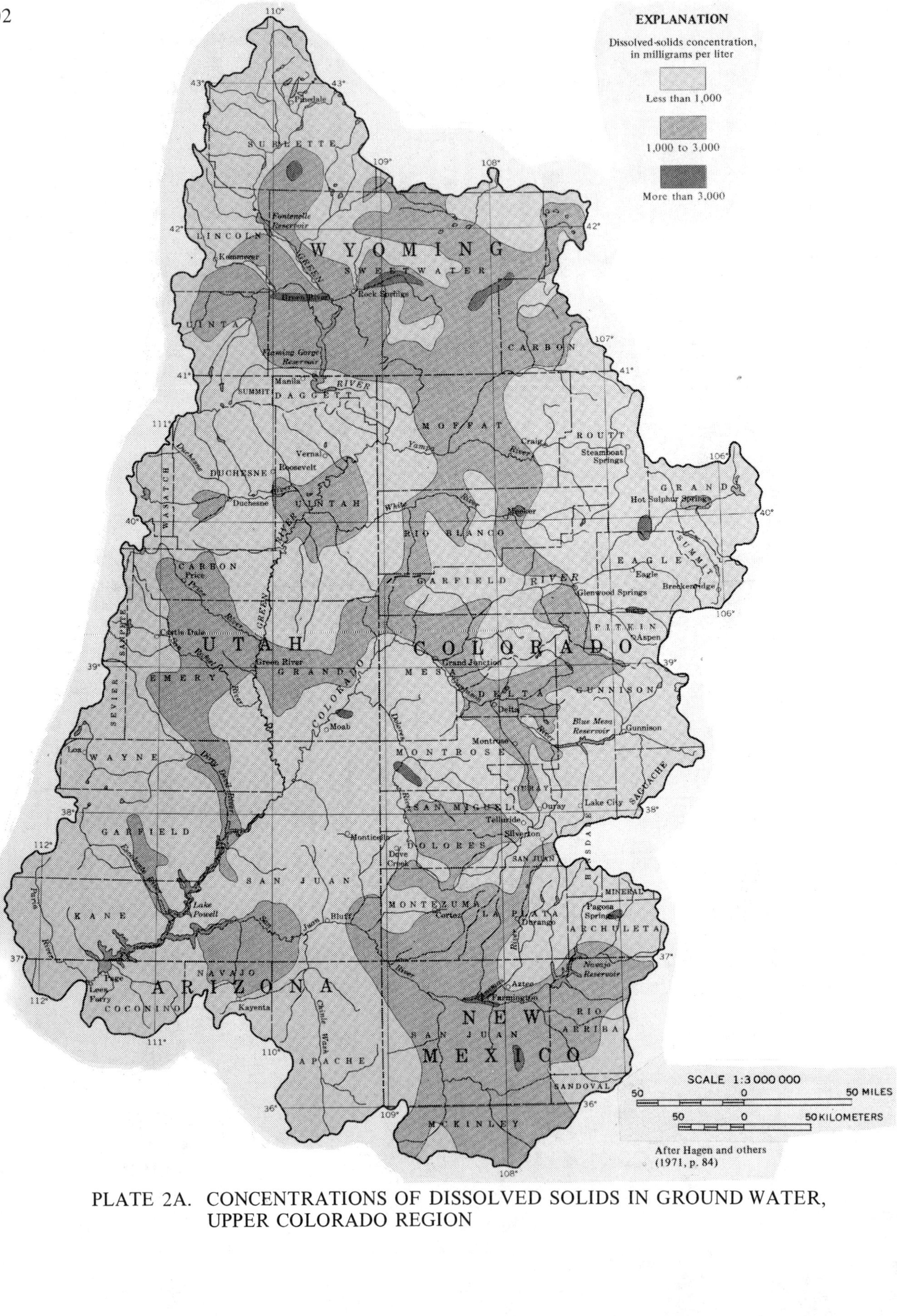

PLATE 2A. CONCENTRATIONS OF DISSOLVED SOLIDS IN GROUND WATER, UPPER COLORADO REGION

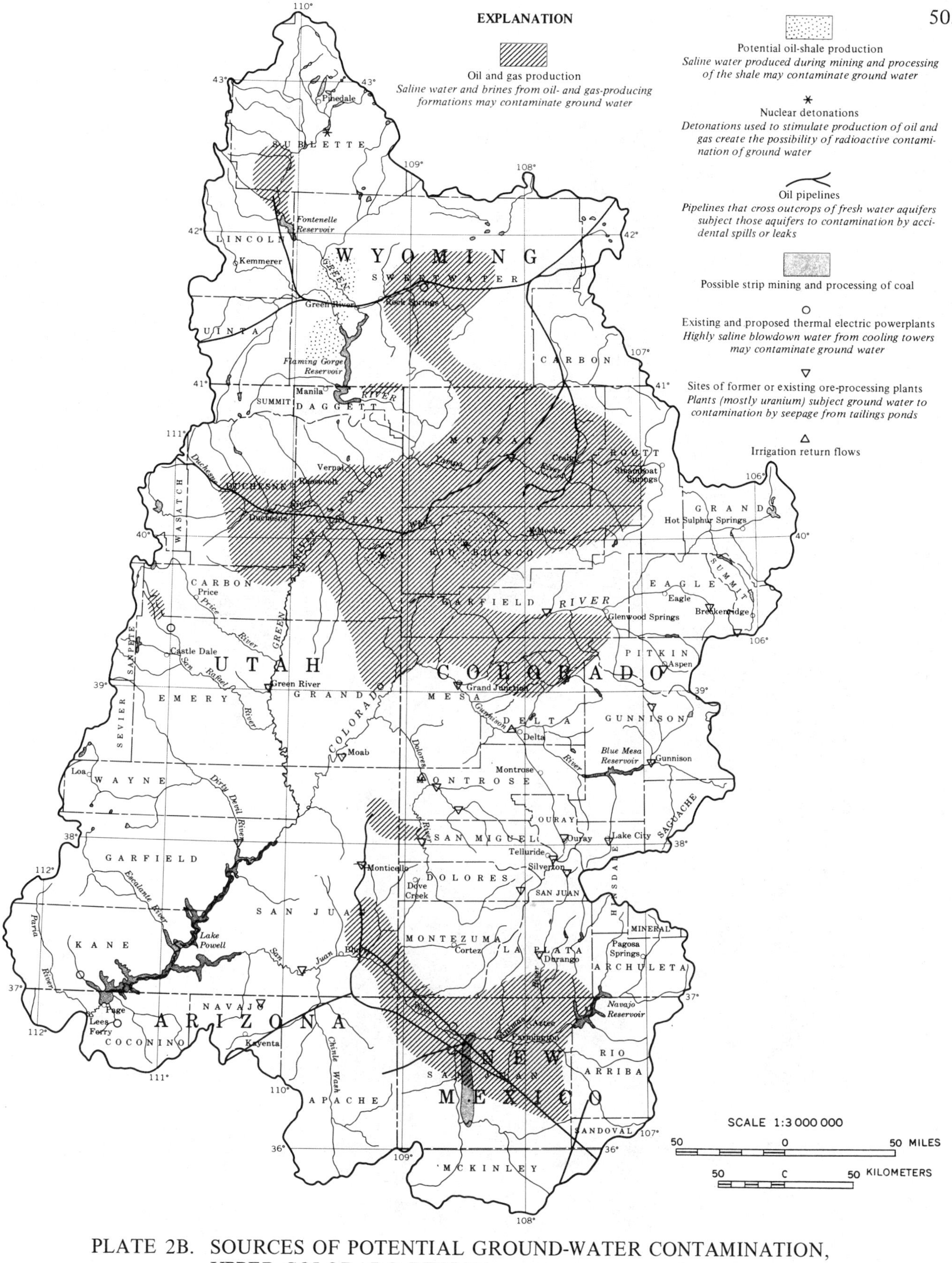

PLATE 2B. SOURCES OF POTENTIAL GROUND-WATER CONTAMINATION, UPPER COLORADO REGION

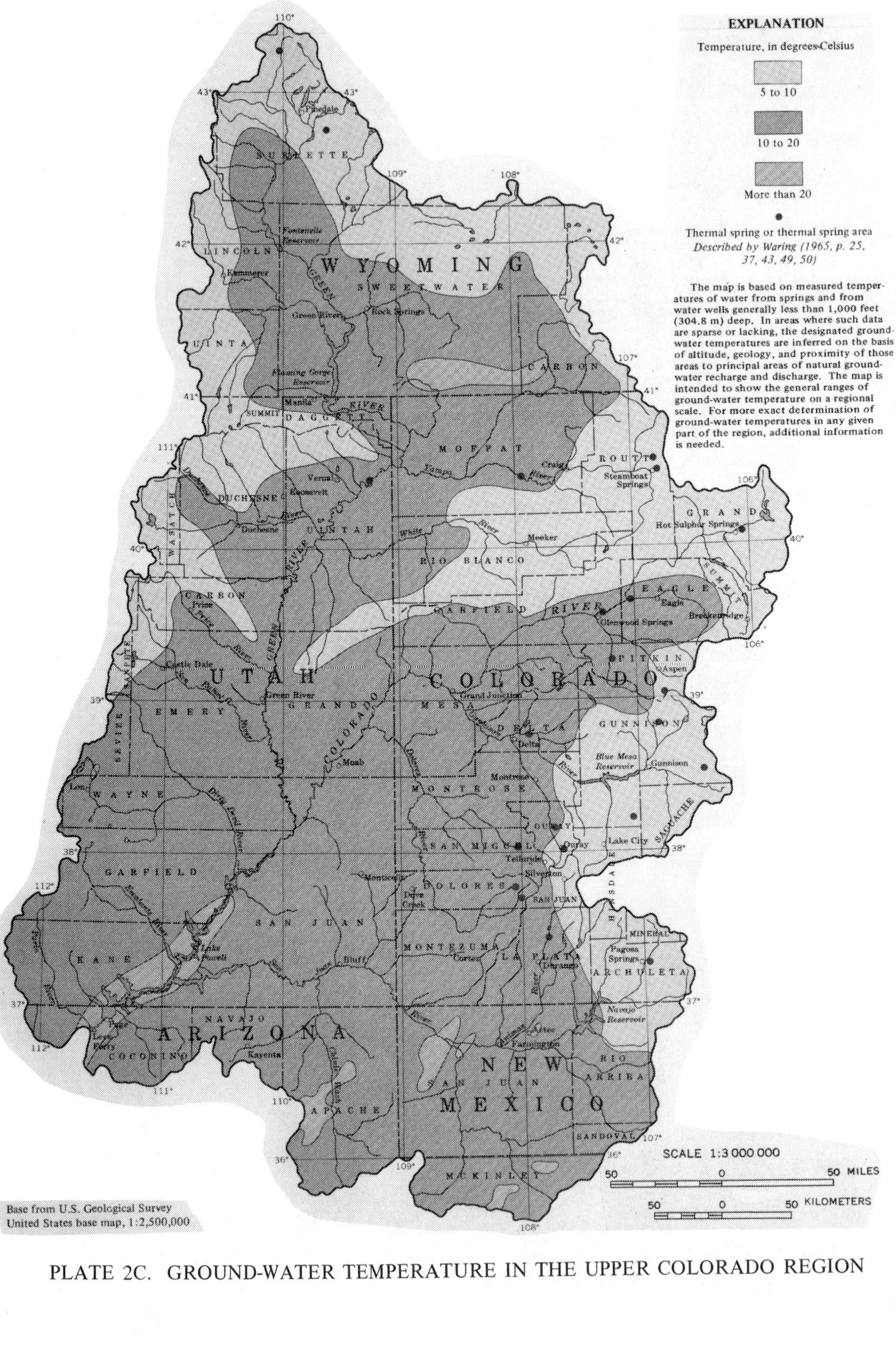

PLATE 2C. GROUND-WATER TEMPERATURE IN THE UPPER COLORADO REGION

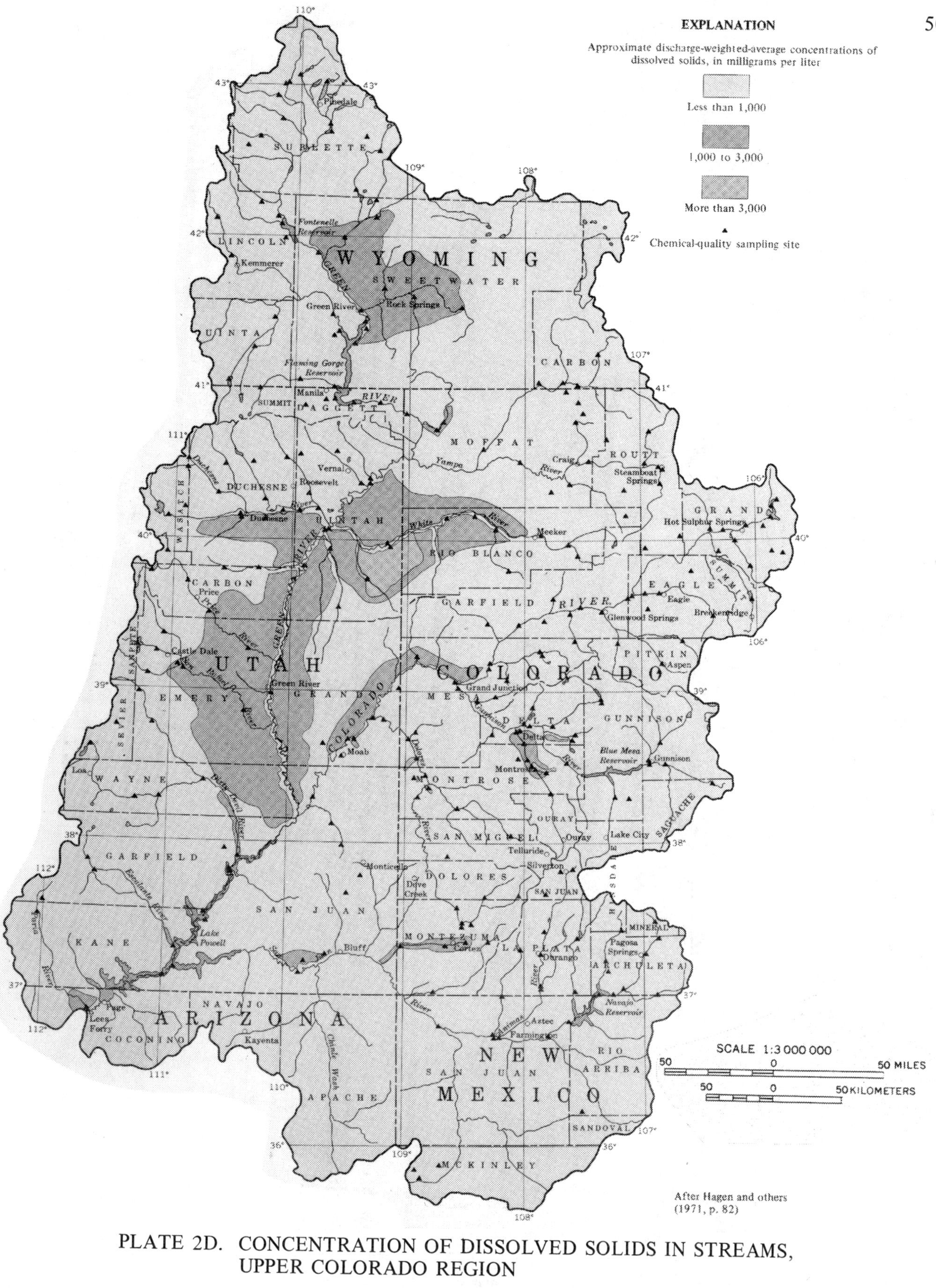

PLATE 2D. CONCENTRATION OF DISSOLVED SOLIDS IN STREAMS, UPPER COLORADO REGION

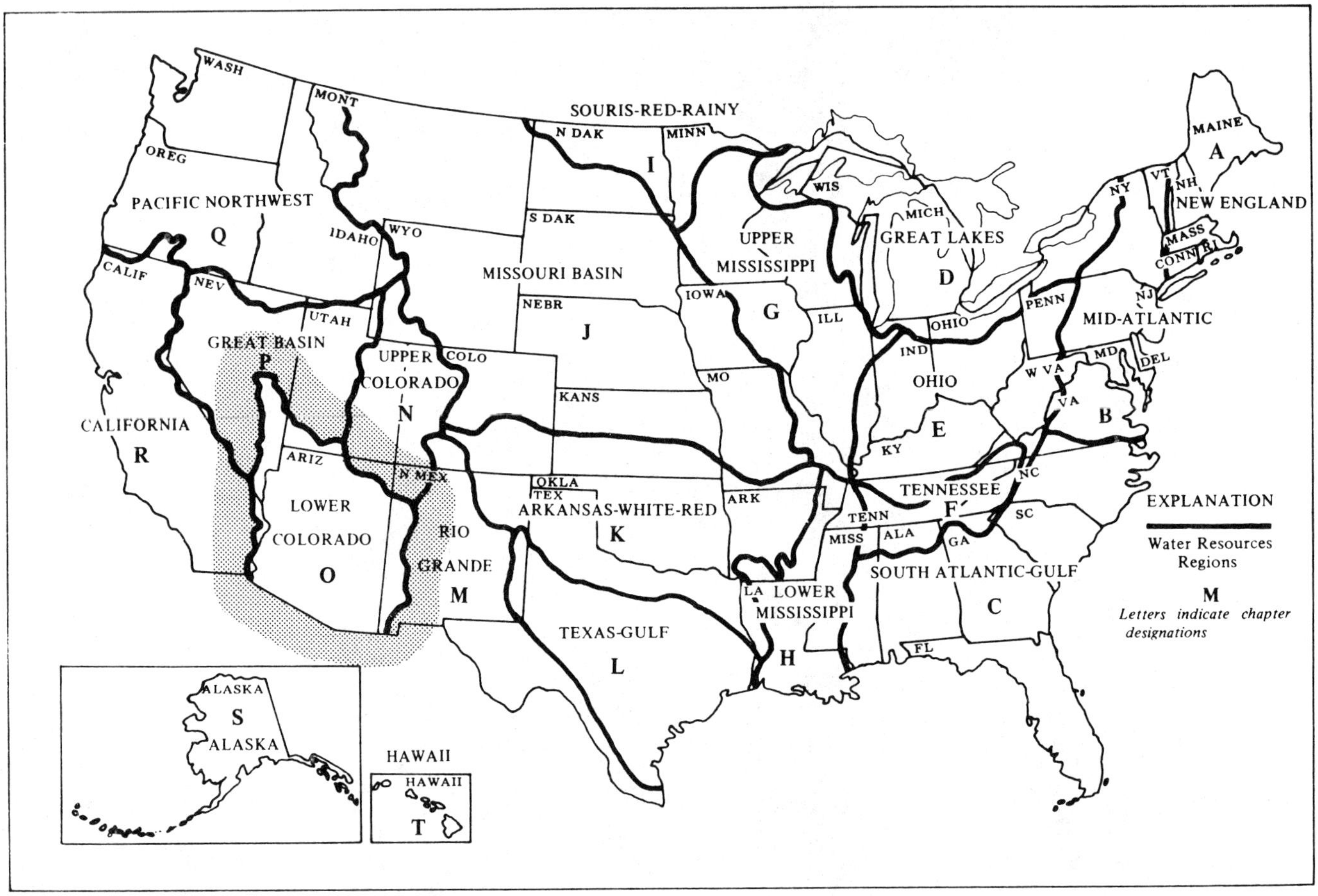

O – Location Map for the Lower Colorado Region

O

LOWER COLORADO REGION

By E. S. DAVIDSON

ABSTRACT

This report summarizes ground-water availability in the lower Colorado region and discusses the potential for greater ground-water development and increased efficiency of water use.

The climate in the most highly developed southwestern part of the region is warm and dry and that in the northeastern part is cool and moist to dry. Although the regional average annual precipitation is only about 14 inches and most streambeds are dry during most of the year, about 1.5 billion acre-feet of ground water of moderate to good chemical quality is stored in aquifers of the region. Much of the water use is founded on pumped withdrawal of ground water. However, in most of the southwestern part of the region pumpage and consumptive use are in excess of replenishment, resulting in declining water levels. In the southwestern part of the region, water levels generally are from 200–500 feet below land surface and in large areas are less than 200 feet. Large-diameter water wells in this part of the region commonly produce 500–1,500 gallons per minute of water. In the northeastern part of the region, water levels generally are more than 500 feet below land surface, but in some large areas water levels are from 200–500 feet, and locally are less than 200 feet below the surface. In the northeastern part of the region, water wells generally have lower yields than those to the southwest. The yields range from a few to 2,000 gallons per minute, but most wells yield from 10 to less than 500 gallons per minute. In the southwestern part of the region about 1 billion acre-feet of ground water is recoverable from storage from the water table to a depth of 700 feet below the land surface. In the northeast about 150 million acre-feet can be recovered by dewatering a 100-foot-thick section of a typical aquifer (16 million-acre area). The estimated current annual rate of ground-water depletion, occurring almost entirely in the southwestern part of the region, is 2.4–3.2 million acre-feet per year. Almost 6 million acre-feet is pumped each year—about 90 percent for agriculture, 6 percent for public supply, and 3 percent for industrial use. The percentage of use for public supply and industry is increasing each year.

The potential for greater development of ground water in the southwestern part of the region is constrained by land subsidence, earth cracks, increasing costs of pumping and transportation, and moderate to poor chemical quality of water. More ground water can be developed in the northeastern part of the region, where the major constraint is pumping cost owing to low to moderate well yields and depth to water. Some benefits can be realized everywhere in the region through changes in current use and greater efficiencies of use. Additional supplies may be made available by capture of natural evapotranspiration. Increasing the efficiency of use is possible hydrologically but, in the near term, is more expensive than increasing ground-water development. Decrease of irrigation, change to water-saving methods of irrigation, use of saline water, decrease of per capita public-supply use, and more reuse of water in almost every type of use could help extend the supply and thereby reduce the current rate of ground-water depletion. Financial problems have not yet caused an overall decrease in pumpage, but, locally, operating costs or partial dewatering of the aquifer has eliminated or decreased withdrawal. Current water laws in all States of the region, except Arizona, control or allocate the use of ground water.

INTRODUCTION

The lower Colorado region is in the semiarid to arid Southwestern United States and comprises most of Arizona and adjacent small parts of California, Nevada, New Mexico, and Utah (fig. 1). The region has been divided into three socioeconomic subregions for planning purposes, and the socioeconomic subregions are superimposed on the water provinces (fig. 1). In general, the region is the Colorado River drainage basin downstream from Lees Ferry, Ariz. The Colorado River and the tributary Little Colorado, Virgin, and Gila Rivers and their tributaries drain most of the region. A few basins are drained internally, and small areas in the south and southeast drain directly southward into Mexico and the Gulf of California. The total drainage area of the lower Colorado River upstream from the Mexican border near Yuma is about 135,000 mi², of which about 114,000 mi² is in Arizona.

The Southwestern part of the region generally is warm to hot (fig. 2), and rainfall and cloud cover are sparse. Owing partly to air-conditioning, population growth since 1950 makes it one of the fastest growing areas in the Nation. The central and northeastern parts are cool to warm and receive from 8 to about 40 in. per year of rainfall. The population of the region in 1975 was between about 2.5 and 3 million. The growing season in much of the southwestern part of the region is very long, and two or three crops can be harvested from one plot during the year. However, precipitation is insufficient to mature crops, necessitating irrigation with surface water and pumped ground water.

The economy of Arizona and the Las Vegas, Nev., area dominate the economy of the region. The major sources of income in Arizona are from agricultural production, 21 percent; manufacturing, mainly light indus-

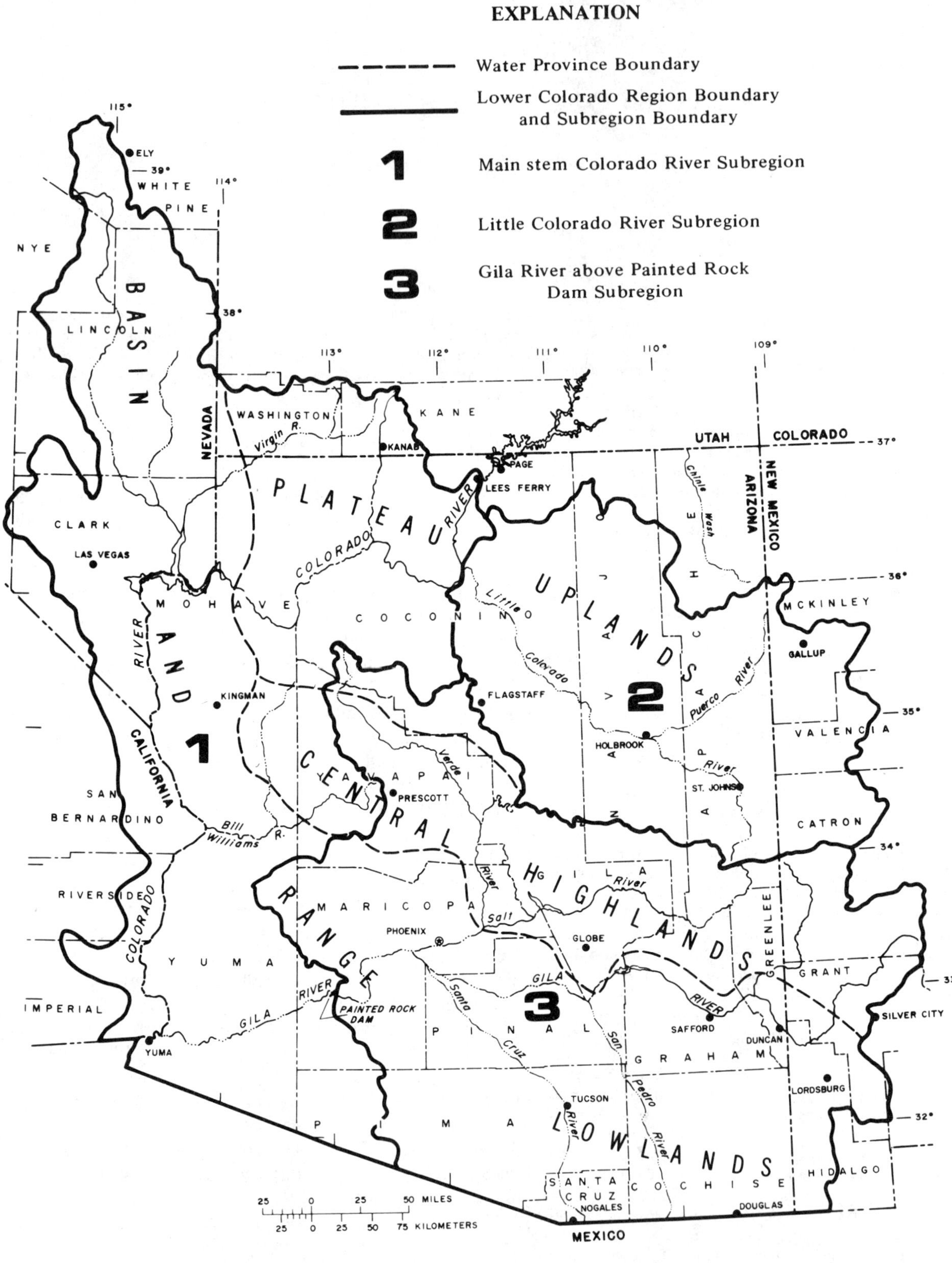

FIGURE 1.—Water provinces and subregions in the lower Colorado region.

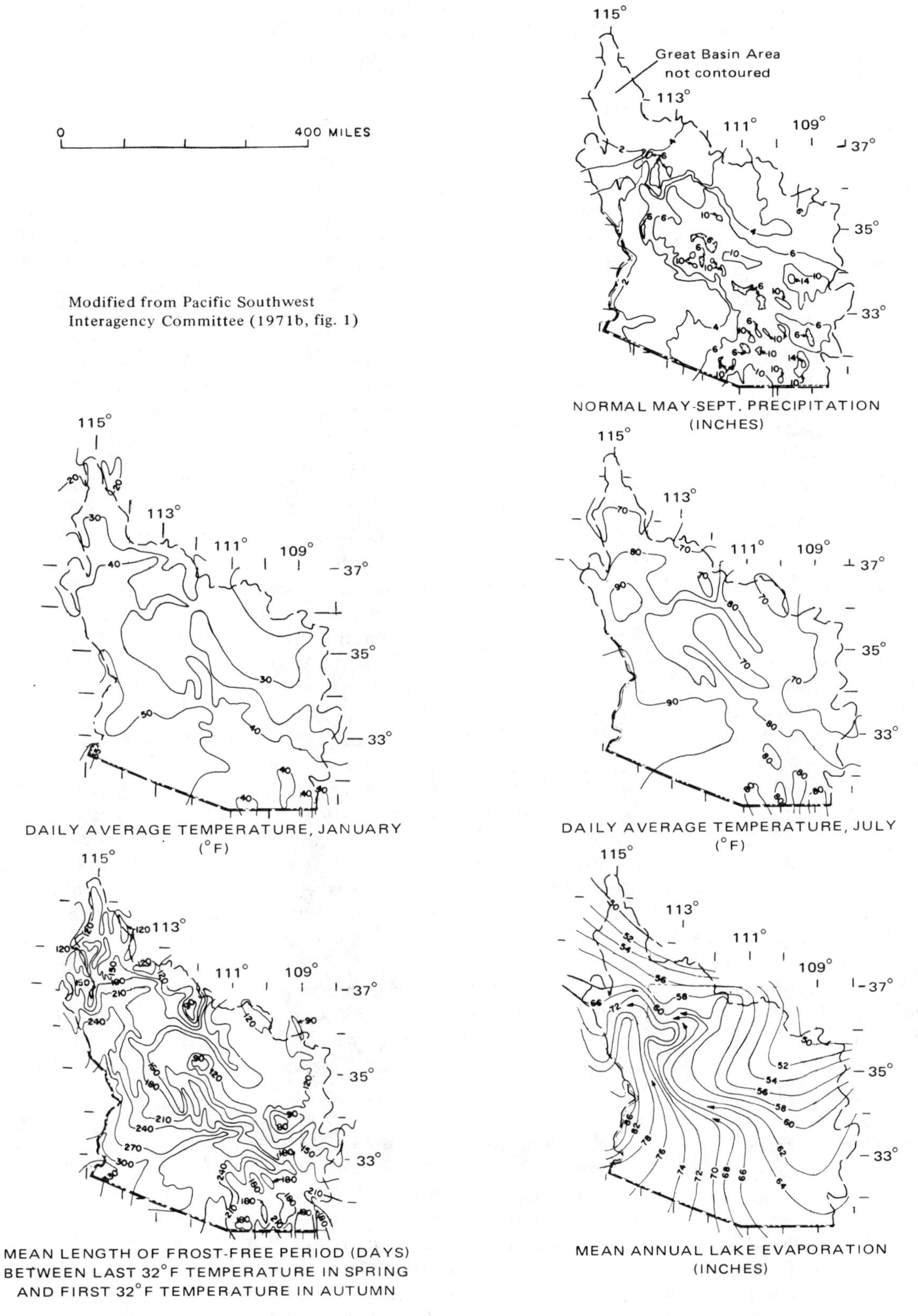

FIGURE 2.—Climatic data for the lower Colorado region.

try and high-technology industry, 41 percent; mining, 25 percent; and tourism, 13 percent (Valley National Bank of Arizona, 1975). The total income from these sources is about $5¼ billion per year.

Crops generate about 40 percent of the total agricultural income, and crop production is the largest consumer of water in the region. Consumption for public supply and mining is significant, and production of electrical energy is becoming a significant water-consuming industry.

PURPOSE

The purpose of this report is to describe the utility of ground water in the lower Colorado region in a format useful to administrators and policymakers who have the responsiblity of water-resource development or management. In much of the Nation, ground water is a little-utilized resource, and discussion of its utility is more pertinent nationally than in the semiarid Southwest, where ground water has long been utilized and is overdeveloped in many areas. However, alternative schemes for use and the potential deleterious effect of overuse need to be examined; discussion of these factors is the principal purpose of this report.

SCOPE

Ground water availability, use, depletion, potential for greater development and management options are summarized in this report. Basic ground water occurrence information, such as ground water chemical quality, recoverable quantities of water in storage and potential aquifer yields also are summarized. Included also is a discussion of the use of underground space for storage of miscellaneous fluids, such as natural gas and liquid toxic or radioactive wastes. The data presented are taken from many comprehensive reports, which are referenced at the end of this paper. Not all necessary ground-water information can be easily measured, and calculations, such as those for recharge and depletion, that seriously affect water administration and planning are briefly described.

Although the main emphasis in this report is on the potential usefulness of the ground-water supply, how the supply can be managed, and the negative factors that might result from increased use of ground water, effective and efficient use of the supply is complicated more often by legal problems and by public sentiment and opinion than by technical considerations. Therefore, schemes that are technically possible or even technically desirable frequently prove impossible to implement because of other considerations. Nevertheless, these other factors change more rapidly in time than do physical conditions and technical analyses. Accordingly, the report contains no more than passing reference to the sociologic or economic factors that may impact on the wise use of the resource.

ASSESSMENT OF GROUND WATER

The lower Colorado region comprises three water provinces, which have major differences in geology, physiography, altitude, climate, and availability and use of ground water. These water provinces are: The Basin and Range lowlands province in the southern and western parts of the region; the Plateau uplands province in the northeastern part of the region; and the Central highlands province, a transitional zone between the other two provinces (fig. 1). Ground-water use and storage calculations are presented for the socioeconomic subregions that overlap the water provinces.

The Basin and Range lowlands province contains most of the population, is most intensively developed, and is characterized by north- to northwestward-trending broad and gently sloping basins that surround and separate steep and rugged mountains. The basin surfaces are at altitudes of 100–4,500 ft above mean sea level, and the ranges crest from 1,000 to 4,000 ft higher. The average annual precipitation ranges from a low of 3 in. near Yuma to 35 in. in mountainous areas (fig. 3; Green and Sellers, 1964). Because of the latitude, altitude, and physiography, thc lowlands is the warmest province; it has the longest growing season, the most available water, and more land suitable for agriculture than the other two provinces.

The sparsely settled Plateau uplands province is characterized by plateaus, canyons, buttes, and mesas and contains a few dispersed volcanic mountains. Most of the province is 5,000–7,000 ft above mean sea level; the canyons are as low as 4,000 ft and the highest peaks are about 13,000 ft above sea level. Average annual precipitation ranges from 6 in. to 30 in. (fig. 3; Green and Sellers, 1964). The climate is cooler and the growing season shorter than in the Basin and Range lowlands (fig. 2); the land is not as suitable for farming, either because of topography or lack of sufficient quantity or usable quality of water. Ranching and recreation are the dominant land uses.

The Central highlands province is mountainous and very sparsely settled. The topography is precipitous, and the altitude ranges from 2,500 to 11,000 ft above mean sea level. Because of the mountainous terrain, the highlands are cool and receive the most precipitation. Precipitation ranges from 16 to 40 in. per year (fig. 3; Green and Sellers, 1964). Much of the streamflow used in the adjacent lowlands and uplands provinces originates in this province. Ranching, recreation, and mining are the dominant land uses.

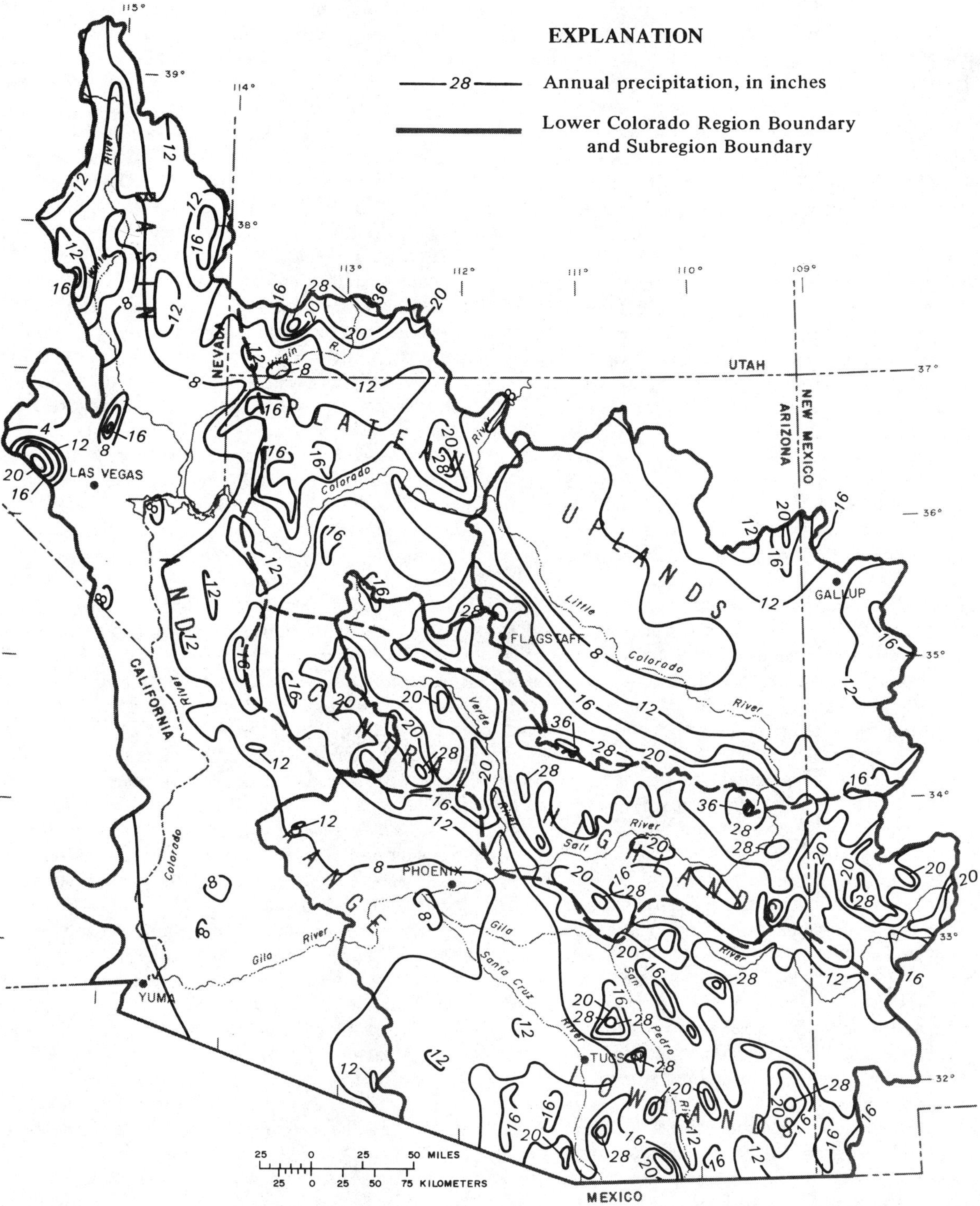

Precipitation data from Pacific Southwest Interagency Committee, 1971b

FIGURE 3.—Normal annual precipitation in the lower Colorado region.

SUMMARY OF GROUND-WATER AVAILABILITY AND CHEMICAL QUALITY

The Basin and Range lowlands province contains extensive highly developed aquifers (pl. 1), which store vast quantities of good-quality water that can be pumped at rates sufficient for economic irrigated agriculture. In addition, large quantities of streamflow are stored in reservoirs and used for agriculture. The total consumptive use of ground water is greater than the renewal of the supply, thus causing a steady decline of water levels in the aquifers.

The extensive thick sediment that fills the basins are the aquifers into which precipitation and streamflow infiltrate and are stored as ground water. The basins are filled to depths of 5,000 to at least 10,000 ft with discrete layers to poorly sorted deposits of clay, silt, sand and gravel. The mountains contain only small amounts of ground water; however, precipitation increases in direct proportion to altitude, and a substantial part of the water recharged to the aquifers originates as runoff of precipitation from mountainous terrain. The mountains in Arizona and New Mexico are composed of granitic, volcanic, and some sedimentary rocks. In Nevada, many of the ranges are limestone and associated sedimentary rocks.

In general, the coarser material in the basins is found near the mountains, and fine-grained material is deposited along along the axes and in the deeper parts of the basins. Volcanic rocks and evaporite deposits are interbedded with the clastic sediment in many basins. Most of these deposits are weakly to moderately consolidated, and the more deeply buried deposits are the more strongly cemented. Unconsolidated sand and gravel occur along streams and as a blanketing deposit; some are being deposited at the present time by active streams (Cooley and Davidson, 1963).

The more permeable sand and gravel beds in the uppermost parts of the basins are the most extensively developed sources of ground water; saturated and permeable sand and gravel beds along stream channels or in blanketing deposits yield the greatest amount of water to wells, but the vast quantity of underlying thick sediment stores the greatest volume of water. In most basins, the water-yielding beds are hydraulically connected and the contained water is unconfined at shallow depths; however, in some places, and generally as depth increases, water-bearing beds are more completely separated by the less permeable beds, and semiconfined artesian conditions prevail.

The chemical quality of most ground water is suitable for most purposes, either with no beneficiation or with minor treatment, such as softening. The dissolved-solids concentrations range from less than 100 to more than 100,000 mg/L and generally are less than 1,000 mg/L. Ground water that contains more than 1,000 mg/L dissolved solids occurs mainly along and near the Gila River from Safford to Yuma, Ariz.; along the southernmost reach of the Colorado River; in the southeastern part of Nevada; and near Wilcox, Casa Grande, and Tucson (pl. 2). The dissolved-solids concentrations in ground water differ areally and with depth owing to differences in the chemical character of the aquifers. For example, 1,000-ft-deep wells in the Casa Grande area penetrate mainly sand and gravel and minor thicknesses of gypsum and salty clay. Studies of water in these wells show that significant amounts of dissolved solids originate near the gypsum and salty beds. Because many wells in the region are open to the entire section of sedimentary deposits penetrated by the well, poor-quality water from such beds is mixed with better-quality water and the final well-discharge product may be so poor as to make the water nearly unusable. Additionally, because of such well construction, ground water having high concentrations of dissolved solids may migrate through the well and into other parts of the aquifer, thereby contaminating ground water formerly of good quality.

Generally because of increased residence time or because of greater opportunity for ground water to pass through and dissolve chemicals from the enclosing rocks, ground water at depth contains more dissolved solids than that nearer the surface. In some places, the concentration decreases with depth concurrent with a change in water type. In the Willcox and Tucson areas, in Pinal County, and in the Beardsley area of northwest Maricopa County, Ariz., the water grades with increasing depth from a calcium bicarbonate to a sodium bicarbonate type (Kister, 1973). Near Tucson, ground water shows a decrease in calcium relative to sodium with increasing depth, and water at a depth of 2,000 ft contains less dissolved solids than ground water nearer the surface (Kister, 1973; Laney, 1972).

The chemical quality of ground water contained in rocks of the mountain blocks generally is similar to that of precipitation and surface flow and contains a lower concentration of dissolved solids than ground water in the sedimentary rocks of the basins. The dissolved-solids concentrations generally are less than 1,000 mg/L. However, well yields are small compared with those in aquifers of the basin, and generally wells in mountainous areas are adequate only for individual domestic or stock supplies.

The Plateau uplands province is underlain chiefly by consolidated sedimentary rocks, which consist mainly of sandstone, siltstone, claystone, and limestone. The sandstone and limestone form the chief aquifers in the area (pl. 2); the siltstone and claystone are nearly impermeable and form confining beds throughout most of the area. Where water-bearing beds of sandstone and limestone alternate with the confining beds, the water in the aquifers is under artesian pressure. Sand and gravel deposits along major streams form isolated "shoestring" aquifers of limited areal extent. The water in the "shoestring" aquifers is unconfined. The quantity and chemical quality of water are dependent mainly on precipitation and streamflow, but in some places these aquifers are supplied by ground water discharging from the underlying consolidated sedimentary rocks. Ground water generally can be produced in the eastern part of the uplands, but adequate supplies are not easily available in the western part (pl. 2). Despite the large amount of ground water available in the eastern part, low well yields and poor to fair chemical quality of water restrict its use (McGavock and Edmonds, 1974).

The dissolved-solids concentrations in ground water in the rocks of the uplands range from 90 to more than 60,000 mg/L (Kister, 1973). Ground water in the N and C aquifers contains the least dissolved solids (pls. 1 and 2). The greatest concentrations are in ground water in the Black Mesa area and in aquifers along the Zuni River and the lower reach of the Little Colorado River (pl. 2). Although gradation between chemical types of ground water is common, most water with less than 500 mg/L of dissolved solids is a calcium or sodium bicarbonate type, and most with more than 500 mg/L is a sodium or calcium sulfate or sodium chloride type. Very highly mineralized water, however, commonly is a bicarbonate sulfate type (Kister, 1973).

The Central highlands province is similar geologically to the Basin and Range lowlands except that the mountains dominate the adjacent basins. Crystalline igneous and metamorphic rocks form much of the Central highlands; they do not store large amounts of ground water per unit area. Small amounts of ground water can be obtained from fractures in these rocks and from the thin sand and gravel deposits along streams that drain the mountainous terrain. Where geologic conditions are favorable in the highlands province, extensive bedded deposits of volcanic rocks and some of the few sediment-filled basins store and yield large amounts of water; the greatest water production is obtained from wells tapping these rocks.

Ground water draining from the rocks in and near the Central highlands is the source of perennial flow in the Gila, Salt, and Verde Rivers. The dissolved-solids concentration of the water generally is less than 1,000 mg/L; however, several springs yield saline water. Clifton Hot Springs yield sodium chloride type water containing more than 9,000 mg/L dissolved solids to the San Francisco River (Kister, 1973; pl. 2). Springs along the Salt River yield sodium-chloride type water that contains more than 37,000 mg/L dissolved solids, and springs on the tributary White River yield sodium chloride water that contains more than 8,000 mg/L dissolved solids.

DEPTH TO WATER AND CHANGE IN WATER LEVELS

The depth to water in a well directly affects the cost of producing ground water from that well; if the depth to water increases over time, the cost of production also increases, and eventually the increased cost may restrict the use. In most of the lower Colorado region, ground water is mined or removed from storage. In a few places, mostly along perennial streams, infiltration of streamflow is sufficient to replace the water withdrawn, and there is no long-term removal from storage or decline in water level. In these places water levels decline in response to withdrawal and rise during a wet season or in response to a decrease in pumped withdrawal.

In much of the region ground water is pumped from storage in excess of replacement, and the pumping depth to water increases in direct proportion to the volume of water pumped and the physical character of the aquifer. The most severe water-level declines are in the extensively developed Basin and Range lowlands province. Here, as water levels decline, the rate of decline commonly increases even though the pumping rate is held constant, because the aquifers tend to be more firmly cemented and less porous and permeable with depth. This combination of water-level lowering caused by removal of water from storage and the increased rate of lowering per unit of production tends to increase the cost of pumping at a geometric rather than arithmetic rate. In addition, the currently increasing cost of energy is accelerating the increase in cost of pumping. Incidence of land subsidence and earth cracks that are related to water-level decline also are a factor in the total cost of continued ground-water depletion. These hazards are relatively minor now, but increased ground-water decline probably will increase the cost to mitigate the damage attributable to subsidence.

To compare water-level changes from one year to another, the depth to water usually is measured just prior to the pumping season when water levels have recovered or nearly recovered to a uniform or virtually nonchanging level. This is called the "static" water level, even though the levels are known to be constantly changing with time. The water levels under pumping conditions generally are 100–150 ft lower than static levels in areas where large-capacity wells are producing in the range of 1,000–1,500 gal/min. The following discussion of depth to ground water and change in water level refers to the static water level.

In most of the Basin and Range lowlands province, the

depth to water in 1975 (pl. 3) was less than 500 ft below the land surface, and in a large part of the area the depth to water was less than 200 ft. Water levels are shallowest along the major stream channels in the lowest parts of the basins. In a few areas, the water level is at or very near the surface, and some wells that tap artesian aquifers flow. In heavily pumped areas, the maximum annual declines are as great as 10 ft and commonly range from 2 to 6 ft (Brown, 1976, sheet 2; Babcock, 1976, sheet 2). Water-level declines since pumping started in the early 1920's generally are greater than 150 ft in most moderately to heavily pumped areas and locally are as great as 400 ft. Land subsidence of inches to about 10 feet has accompanied withdrawal of ground water in much of central Arizona and in the Las Vegas area in Nevada. The volume of subsidence may be from 5 percent to slightly more than 30 percent of the volume of water withdrawn. In the basins, subsidence is accompanied by earth cracks or fissures, inches to feet wide and tens of feet to many miles long. The land on either side of some cracks is vertically displaced as much as 3 ft. The cracks present a hazard to highways, railroads, pipelines, buildings, etc. Some potential locales for earth cracks have been identified, but generally, potential fissure locales cannot yet be accurately predicted.

In the Plateau uplands province, the depth to water is greater than 500 ft in much of the area and is less than 200 ft mainly along the Little Colorado and tributary Puerco and Zuni Rivers (pl. 3). Near Flagstaff, Ariz., the depth to water in municipal wells is 1,100–1,300 ft below the land surface and to the west is as great as 2,840 ft (Babcock, 1976, sheet 2). Because withdrawal in the uplands does not generally exceed recharge, water levels generally show no decline. Declines of about 2 ft per year have occurred in small areas near Snowflake, Ariz., and along the Little Colorado River northeast of Snowflake.

In the mountainous areas of the Central highlands province, water levels in wells range from a few tens to about 300 ft below the land surface. In the parts of the area underlain by volcanic rocks, depths to water are from a few feet along major streams to more than 500 ft elsewhere. In the few sediment-filled basins, the water levels generally are less than 200 ft below land surface along streams and deeper away from the streams. Any change in water levels in this province generally is temporary because of ample recharge, except for a few areas where water levels decline about 1–2 ft per year (Brown, 1976, sheet 2; Babcock, 1976, sheet 2).

YIELD OF WATER TO WELLS

The greatest yield of water to wells in the lower Colorado region is in the Basin and Range lowlands province (Brown, 1976, sheet 3; Babcock, 1976, sheet 2). Large-diameter water wells tapping the permeable sand and gravel beds in the basins and along the Colorado River are capable of yielding more than 2,500 gal/min, although most wells in the basins are pumped at 500 to 1,500 gal/min. In the mountainous parts of the province, well yields generally are less than 100 gal/min and commonly are less than 10 gal/min.

The major aquifers of the Plateau uplands are capable of yielding at least 10 gal/min of water to wells, and locally irrigation wells produce as much as 2,000 gal/min. Wells in the sand and gravel beds along streams generally can produce 100 gal/min or more from shallow depths. In contrast, wells tapping the sandstone and limestone aquifers generally must be 500–1,000 ft deep to produce the same yield. In some areas it may be possible to produce, at high cost, good-quality water from depths of thousands of feet, but the deeper units have not yet been adequately prospected.

In the Central highlands province, well yields generally range from about 10 gal/min in the mountainous areas to 1,000 gal/min in places along streams or where wells are in thick saturated sediment or volcanic rock.

RECHARGE AND DEPLETION

Recharge is that part of precipitation that eventually reaches and is stored in the ground-water reservoir. Such factors as the permeability of near-surface materials, configuration of the land surface, intensity and distribution of precipitation, areal extent of the aquifer, volume of streamflow, amount and type of land cover, and potential for evaporation of water from the land surface control modify the amount of precipitation that ultimately may become ground-water recharge. Modification of the water regime in an area by ground-water withdrawal, removal of riparian growth, dam construction, or removal of gravel from the streambed may have a substantial effect on recharge. These factors may either increase or decrease recharge. If recharge is increased, the increase may be at the expense of the natural loss to evaporation and transpiration, making more of the original precipitation available for man's use.

The effect of these factors in the lower Colorado region is substantial, but it is so variable in space and time that the amount of recharge in most places can only be estimated crudely. In most studies of aquifers in the region, recharge is estimated indirectly by evaluating the ground-water budget in the study area. About 100 million acre-ft of precipitation (Pacific Southwest Interagency Committee, 1971b, p. 6) falls on the region annually (fig. 3), but only about 1 percent of this volume is estimated to recharge the ground-water aquifers (Arizona Westwide Study Team, 1973, p. 32).

In the lower Colorado region either recharge or depletion of storage must be estimated in order to plan effective long-term use of ground water. This is particularly true in planning use of ground water for a public supply, in which case the resource presumably should never be completely depleted unless replacement sources are available. The direct estimation of recharge is based on algebraic addition of a series of reasonable approximations, and nearly always the result must be compared with and adjusted to values derived from indirect methods. The direct method generally requires estimation of underflow, estimation of the infiltration of precipitation, measurement of infiltration of streamflow from the major streams, and estimation of infiltration from smaller streams. All these "direct" functions require knowledge of the precipitation, streamflow, and physical characteristics of the aquifer and the overlying earth. More commonly than not, such knowledge is incomplete, therefore, recharge estimates are only crudely accurate. If the ground-water discharge can be closely calculated, another method is commonly used; this method is based on the premise that over a long period of time prior to withdrawal, storage is stabilized and outflow from the aquifer is equivalent to inflow. If the surface and underground outflow of ground water can be closely calculated or measured and if the loss to evaporation and transpiration can be reasonably approximated, then the sum of the two outflow estimates may be a reasonably close estimation of recharge. Generally, this approach requires a complete knowledge of the streamflow budget in the area and is most accurate where streamflow infiltrates to the aquifer (losing streams). If the streams are fed both by ground water and by water from overland runoff (gaining streams), the estimate is more complicated and generally less accurate.

In many recent water-planning studies, recharge is ignored, and an estimate of ground-water depletion is used instead to relate both to ground-water storage and withdrawal. In this calculation, the pumpage less estimated depletion is equivalent to the sum of natural recharge and return to the aquifer of applied ground water. (Infiltration of applied surface water is here defined as recharge.) This method requires close measurement of all water used in an area, and it also requires accurate measurement and calculation of the decrease in water content of the dewatered part of the aquifer. Again, most of the volume calculations are estimates based largely on empirical solutions and the investigator's experience so the calculations are only as accurate as these two factors allow.

The basins in the Basin and Range province are recharged by infiltration of runoff in the main stream channels, by infiltration of runoff along small streams at the mountain fronts, by infiltration of excess applied irrigation water from surface-water sources, by infiltration of surface water transported in unlined irrigation canals, by underflow from upstream basins, and possibly by a small amount of direct penetration of precipitation. It is estimated that 3–6 percent of the precipitation that falls on the mountains infiltrates the sand and gravel deposits along small streams at the mountain fronts and probably recharges the ground-water reservoir (Brown, 1976). In some areas where "saturated" conditions exist because of long-term application of irrigation water, a large part of the surface water that flows through unlined canals and some that is applied to fields infiltrate to the ground-water reservoir. In many areas, however, the amount of recharge from this source probably is negligible. In some basins the ground-water reservoir is recharged by underflow from upstream basins. The interbasin movement of ground water is recharge to the lower basin but is discharge from the upper basin.

In the Plateau uplands province, the water-bearing rocks are recharged mainly in their areas of outcrop by infiltration of runoff from precipitation and snowmelt. A smaller part of the recharge is gained from infiltration of precipitation, snowmelt, and runoff on overlying rock units. The recharge areas—mostly in the southern and eastern parts of the uplands—are generally more than 6,000 ft above mean sea level, where the precipitation is more than 15 in. per year (fig. 3).

Most of the rocks in the Central highlands province are recharged from direct precipitation. Ground water in the few basins is recharged similarly to the basins in the Basin and Range province.

Aquifers in the lower Colorado region have not been investigated in sufficient detail to allow better than approximate estimates of the recharge. Therefore, recharge has been derived from calculations comparing pumpage, depletion, recharge, and return flow. Harshbarger and others (1966, p. 5, 6) estimated that in the late 1950's the annual statewide pumpage was about 5 million acre-ft, of which about 1 million acre-ft was from the renewable supply (recharge), and that the annual ground-water depletion was about 4 million acre-ft. The staff of the Arizona Water Commission (1975) estimated that for normalized 1970 values the statewide annual pumpage was about 5 million acre-ft, the annual recharge was about 0.2 million acre-ft (Arizona Water Commission, table 21, p. 113-114), and annual ground-water depletion was 2.2 million acre-ft (Arizona Water Commission, p. 88). The pumpage is about the same in both periods, and the discrepancy between the two depletion and recharge calculations lies partly in estimates and definition of the amount of applied water that returned to the aquifer.

A study team approached the same problem for the lower Colorado region in the late 1960's (Pacific Southwest Interagency Committee, 1971b). This team concluded that information was not adequate to estimate recharge but that estimation of the depletion quantity was adequate for planning purposes. The annual pumpage was about 5 million acre-ft and the annual depletion was estimated as 2.5 million acre-ft (Pacific Southwest Interagency Committee, 1971b, p. 30) and 2.6 million acre-ft (Arizona Westwide Study Team, 1973, p. 30).

In 1975 and 1976 members of the Arizona Water Resources Division staff (S. G. Brown, written and oral commun., 1976) calculated depletion and recharge using 1972 and 1973 annual pumpage figures of almost 5 million acre-ft. They found that depletion was about 2.4–3.2 million acre-ft per year and that recharge was about 1 million acre-ft per year. Again the discrepancy with other studies is due to a smaller allowance for infiltration of applied water to the underlying aquifer. In all the above calculations, pumpage was greater than depletion and these factors were the primary estimates. The difference between pumpage and depletion was then assigned to recharge and return from applied water. Natural losses are now a minor part of the equation in heavily pumped areas; hence, natural outflow and its originating recharge supply are not included in the above calculations.

STORAGE

The alluvial aquifers in the lower Colorado region contain vast amounts of physically recoverable ground water in storage (table 1). Along the main stem of the Colorado River and in the Gila River drainage basin, about 1 billion acre-ft of ground water is estimated to be recoverable from storage in the aquifers from the water table to a depth of about 700 ft below the land surface; in addition, about 150 million acre-ft of recoverable ground water is stored in a typical 100-ft-thick section of aquifer in the Little Colorado River subregion of the Plateau uplands province (pl. 1). These large quantities are in marked contrast to the usable capacity of the principal surface-water reservoirs; their capacity is 32 million acre-ft in the lower Colorado region (Pacific Southwest Interagency Committee, 1971b, p. 10). However, ground-water storage has accumulated over hundreds to thousands of years, but the surface-water storage is replenished yearly.

DEGRADATION OF GROUND WATER

In places where the ground-water aquifer has low capacity and is near the surface, degradation or pollution

TABLE 1.—*Estimated volume of recoverable ground water in storage in the main alluvial aquifers in the lower Colorado region*

[Modified from Brown (1976). Numbers rounded. See figure 1 for subregion boundaries]

Subregion and State	Volume of recoverable ground water, in millions of acre-feet				
	From land surface to 200 ft below land surface[1]	From water table to 100 ft below water table	From water table to 700 ft below land surface	From 700 ft to 1,200 ft below land surface	From water table to 1,200 ft below land surface
Main Stem Colorado River Subregion:					
Arizona	27	56	290	140	430
Nevada	12	53	140	45	180
Utah[2]	----	----	----	----	----
California[3]	----	----	----	----	----
Rounded total for subregion	40	110	430	190	610
Little Colorado River Subregion:	----	[4]150	----	----	----
Gila River Above Painted Rock Dam Subregion:					
Arizona	58	96	520	260	780
New Mexico	12	18	78	15	93
Rounded total for subregion	70	110	600	280	870
Rounded total for lower Colorado region (excludes Little Colorado River subregion)	110	220	1,030	470	1,480

[1]Depth to water is 200 feet or less below the land surface in 13,600 square miles—10,900 squares miles in Arizona, 1,500 square miles in Nevada, and 1,200 square miles in New Mexico. The total square miles constitutes about 10 percent of the region.
[2]The quantity of ground water stored in the alluvial aquifers in Utah is minor; the quantity in consolidated rocks is not calculated.
[3]The quantity of ground water in the California segment is poorly known; because it is considered to be minor, it was not calculated for this study.
[4]The Little Colorado River subregion is underlain by the D, N, and C aquifers, which are mainly consolidated rocks; ground-water storage in the alluvial aquifers is minor. About 16 million acres is underlain by the three aquifers, and about 150 million acre-feet of ground water is stored in a typical 100-foot-thick section of aquifer.

of the supply owing to recycling and infiltration of once-used water is potentially a serious deterrent to continued use of the supply. In such places, the degraded supply cannot easily be isolated by drilling deeper or elsewhere, or by selecting uncontaminated parts of the aquifer for a supply. The earliest use of ground water in Arizona caused serious problems in places, either because of waterlogging of the land or because of degradation of the chemical quality of the water by recycling. The degradation of chemical quality results primarily from evaporation of the applied water, subsequent concentration of salts in the residual, and then recharge of the salt-laden water to the aquifer (Harshbarger and others, 1966, p. 4–23). The problem of contamination of the salt-laden water typically has been avoided by drilling deeper wells and drawing better-quality water from deeper parts of the aquifer. Additionally, decline of water levels eliminated waterlogging and reduced the volume of salt that accumulated owing to evaporation from waterlogged land. Degradation of chemical quality of the water by use for irrigation now seems to be a minor problem in the lower Colorado region. In fact, in most of the area it is not possible to detect a major degradation of chemical quality that clearly results from return of water applied at the surface. However, in places, some of this return irrigation water is trapped temporarily in beds perched above the main aquifer, and the chemical quality of the perched return water is poorer than that of the original applied water.

Other potential water-pollution problems exist locally. In particular, sewage contains about 300 mg/L more dissolved solids and has more nitrogen and phosphorus than the original supply. The waste-disposal ponds utilized by various industries, but principally in copper-ore beneficiation, contain water much more highly mineralized than the original supply. Infiltration of this waste water to an aquifer may cause degradation of the supply, depending on the chemical quality of ground water in the area of percolation. Inadvertent dumping of liquid chemicals and unknown leakage below ground from liquid chemical and hydrocarbon storage also causes pollution of the ground-water supply. To date (1977) most of these problems have been minor and highly localized.

SUMMARY OF GROUND-WATER USE

Almost 6 million acre-ft of ground water was pumped in the lower Colorado region in 1975 (table 2). The increase in pumpage for the region over the years parallels the pumpage increase in Arizona (fig. 4). The major use of this water is for irrigated agriculture (fig. 5). Industrial uses, especially for mining and power generation, are increasing, as is municipal use because of the increase in population in some parts of the region. However, in 1975 industrial and municipal uses accounted for less than 10 percent of the total (fig. 5). Rural, domestic, and livestock supplies accounted for only a minor amount of the total ground water pumped in the region.

The largest use of ground water is in the Basin and Range lowlands, particularly in the Gila River above Painted Rock Dam subregion (fig. 1, table 2). In 1975 almost 85 percent of the total region pumpage was extracted and applied in the Gila River subregion and most of that was pumped in Arizona. Of that total subregion

TABLE 2.—*Estimated ground-water pumpage by subregion and use in the lower Colorado region, 1975*

[Data from U.S. Bureau of Reclamation (1975) and from the files of the U.S. Geological Survey offices in Arizona, Nevada, New Mexico, and Utah. Numbers rounded to nearest 100 acre-feet except as indicated. See figure 1 for subregion boundaries]

Subregion and State	Pumpage, in acre-feet				
	Agriculture	Drainage associated with agriculture	Industrial	Public supply, domestic, and other	Total (rounded to nearest 500 acre-feet)
Main stem Colorado River subregion:					
Arizona	312,400	368,000	7,800	4,200	692,500
California	19,500	----	200	7,800	27,500
Nevada	48,800	----	15,700	60,300	125,000
Utah	10,500	----	1,000	1,000	12,500
Total	391,200	368,000	24,700	73,300	857,500
Little Colorado River subregion:					
Arizona	30,800	----	18,300	13,900	63,000
New Mexico	900	----	8,200	3,400	12,500
Total	31,700	----	26,,500	17,300	75,,500
Gila River above Painted Rock Dam subregion:					
Arizona	4,358,100	16,000	124,500	285,300	4,784,000
New Mexico	83,000	----	20	2,100	85,000
Total	4,441,100	16,000	124,520	287,400	4,869,000
Grand total	4,864,000	384,000	175,700	378,000	5,802,000

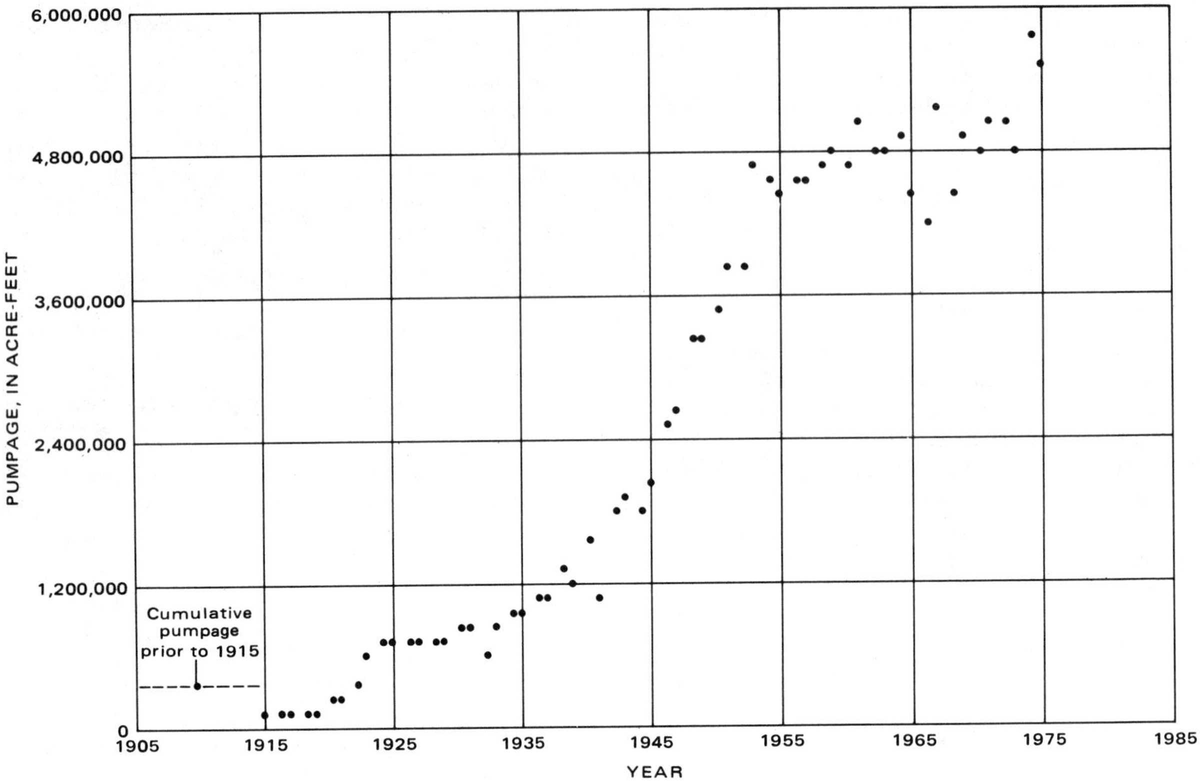

FIGURE 4.—Pumpage in Arizona from prior to 1915 to 1975.

pumpage of almost 4.9 million acre-ft, about 8 percent, 412,000 acre-ft, was used for industrial and municipal purposes. Some small part was pumped for drainage and for rural, domestic, livestock, and miscellaneous uses; 91 percent, 4.4 million acre-ft, was used for irrigation of crops. The next largest use is in the main stem Colorado River subregion, where almost 860,000 acre-ft was pumped in 1975. As in the Basin and Range lowlands, the major use (about 90 percent) was for irrigation, including a substantial amount of pumpage to drain waterlogged land to permit irrigated agriculture. The pumpage in the Little Colorado River subregion, which encompasses most of the Plateau uplands water province (fig. 1), was only about 75,000 acre-ft in 1975 (table 2) and agriculture accounted for about 41 percent—considerably less than in the other two subregions.

Slightly more than 175,000 acre-ft of ground water was used in 1975 for industrial use, and more than half this was used for mining. Most of the rest was used for energy production.

About 165 million tons of copper ore was mined and milled in Arizona in 1972 (U.S. Bureau of Mines, 1974, p. 75, 76). Slightly more than 153 million tons was concentrated and 12 million tons treated for extracting copper in heap-leach or vat-leach operations. Average use was 260 gal of water per ton in the concentration process, including transport of tailings, and at least 125 gal per ton in the heap- and vat-leach operations (L. C. Halpenny, written commun., 1975). This water, known as "makeup" water, is the amount of new water that must be supplied to the mining operation. Water used in the concentrating process is in a closed system, and consumptive losses are small, except for the water used to transport residual fine-grained waste rock to a "tailings pond" disposal site. According to Halpenny, about 100–260 gal of water per ton (of ore) is not recoverable from the tailings pond, owing to evaporation or retention of water in the finely ground waste rock. Water used in the leach operation is evaporated, and the loss is 125 gal per ton or greater. The consumptive use of ground water for mining is not well known, but in 1975 it may have been in the range of 75,000–130,000 acre-ft.

Water used in energy production is becoming a more significant consumptive use every year in the lower Colorado region; evaporation of water to condense exhaust steam accounts for the greatest consumption (Davis and Wood, 1974, p. 3). The average rate of water use per unit of energy capacity is about 15 acre-ft per year per megawatt (1,000 kilowatts) for fossil-fueled plants, 22 acre-ft per year per megawatt for nuclear-fuel plants, and 48

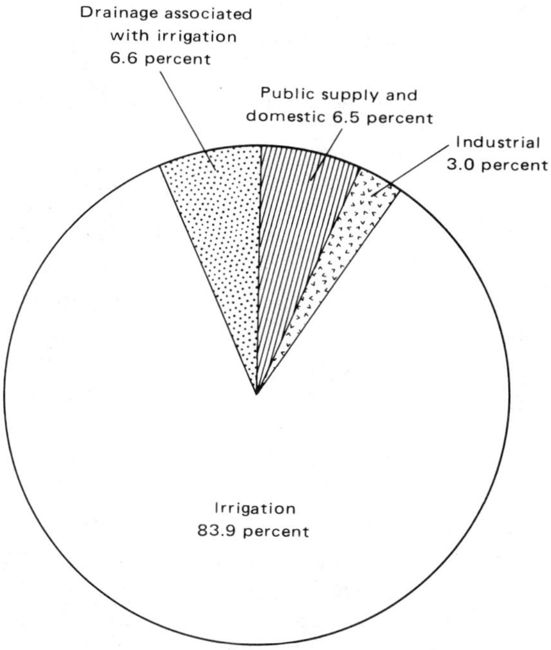

FIGURE 5.—Major uses of ground water pumped in the lower Colorado region in 1975.

acre-ft per year per megawatt for geothermal power-plants (Davis and Wood, 1974, p. 8). Water consumption in the region for electric power in 1965 was reported as 9,600 acre-ft (Pacific Southwest Interagency Committee, 1971c, p. 129), and by 1970 in Arizona alone the consumption had increased to 36,000 acre-ft (Murray and Reeves, 1972, p. 24). Ground water made up slightly more than 95 percent of the consumption. Since then, consumption of ground water has increased, and two new plants using Colorado River water have started in operation. The consumption in the region in 1975 was about 67,000 acre-ft, of which about 35,000 acre-ft, or 52 percent, was supplied from ground water.

Yearly increases in peak demands of 400 to almost 1,400 megawatts are estimated by the Electric Power Work Group of the framework study for the period 1970 to 2000 (Pacific Southwest Interagency Committee, 1971c, p. 49) compared to increases of 160–180 megawatts in the period 1955–65. Energy capacity estimates of the Economic Work Group that contributed to the same framework study are in the lowermost range (Pacific Southwest Interagency Committee, 1971a, p. 140) of peak demand predicted by the energy group. Some of the power to meet this demand will be generated outside the lower Colorado region.

The water demand anticipated by the Energy Work Group for 1980 is about 36,000 acre-ft and in the year 2000 is 106,000 acre-ft (Pacific Southwest Interagency Committee, 1971b, p. 36, 38; 1971c, p. 129). A water demand of 37,800 acre-ft in 1980—plus 7,400 acre-ft of initial supply for pumpback-power generation—was calculated by the Economics Work Group (Pacific Southwest Interagency Committee, 1971c, p. 144). However, in the year 2000, the calculation based on the economic analysis is a consumptive use of only 38,800 acre-ft—plus 9,300 acre-ft for initial supply in pumpback-power generation (Pacific Southwest Interagency Committee, 1971c, p. 144)—or less than 40 percent of the amount predicted by the Energy Work Group (Pacific Southwest Interagency Committee, 1971c). Both estimates may be low; on the basis of ongoing power-generation station construction and announced plans for new construction, a consumptive use of at least 100,000–150,000 acre-ft per year should be anticipated. Estimates prepared for the Energy Research and Development Administration predict a 10-fold growth between 1975 and 2000 in water consumption for energy development (MITRE Corp., 1977, p xiv).

POTENTIAL OF GROUND WATER TO SUPPORT MORE DEVELOPMENT

The history of water development in the lower Colorado region is one of surface-water supplies being "supplemented" by water from storage in the ground-water reservoirs. The State of Arizona projected alternative levels of water development that are thought to bracket the range of possible future consumptive use in the State (Arizona Water Commission, 1977). The report concludes that "water supplies of the State would either have to be augmented, or ground-water overdraft increased substantially," or that "uses be reduced to a level that can be sustained with existing dependable supplies without resorting to appreciable ground-water overdraft" (Arizona Water Commission, 1977, p. xiii). The projections (table 3) are for consumptive use of all water estimated for 1970 and projected to 1990 and 2020. Projections for the parts of Nevada, California, Utah, and New Mexico in the region might show the same ratio of increase for urban use, mineral production, and electric

TABLE 3.—*Projections of annual water depletion in Arizona*

[All values in thousands of acre-feet per year. Source: Arizona Water Commission (1977, p. 2–5, 67–72)]

Consumptive water use	Estimated for 1970	Projected for 1990	Projected for 2020
Urban	328	423–568	542–950
Steam-electric generation	20	109–178	248–787
Mineral production	131	236–337	265–841
Agricultural	4,300		
Highest		4,600	4,500
"Zero overdraft"		3,800	2,500

generation, probably resulting in a reduction in agricultural use. In 1975 the use of ground water was more than 1.5 times that of surface water, and the potential for further economic or large-scale ground-water development was constrained by location and availability, depth to water, poor chemical quality, and potential hazards.

Large-scale withdrawal of ground water presents some problems, particularly in the Basin and Range lowlands. Land subsidence has occurred in the Las Vegas area in Nevada and in Arizona. Earth cracks, many miles long and that range from a few inches to several feet wide, are widespread in central Arizona, mainly on the perimeter of the areas of greatest water-level decline and particularly where much of the withdrawn ground water was stored under partially confined (semiartesian) conditions (Schumann, 1974). Increased pumping of ground water in these areas will increase the problems of subsidence and earth cracks. Likewise, continued large-scale withdrawal of ground water from areas not yet affected probably will cause additional land subsidence and earth cracks. In some areas in this part of the lower Colorado region, a part of the recoverable ground water in storage contains dissolved solids substantially in excess of 1,000 mg/L (plate 1) and requires treatment to make it suitable for most uses. Some areas where large amounts of ground water are stored are remote from areas of potential use. Use of this water requires either transport of the water or very high cost of new developments in the immediate vicinity of the supply. Large amounts of water can be obtained from some of the deeper basins, but much of the untapped ground-water reserves are at depths of more than 500 ft. Use of this water is costly and would require detailed well design and spacing and the construction of much deeper wells than currently exist in most areas.

In the Plateau uplands the few alluvial aquifers do not store large amounts of ground water. However, the sandstone and limestone aquifers store very large quantities of recoverable ground water, but the potential for large-scale future use is limited by low to moderate well yields.

In the Central highlands the storage of ground water is not large per unit area except in a few basins. In these basins and in some parts of the Uplands province in eastern Arizona and western New Mexico, where volcanic rocks dominate, water supplies are large and dependable enough to support small communities and industries. The water supply in the Central highlands province probably is adequate for most foreseeable developments, largely because the terrain is not conducive to large-scale agriculture or heavy industry, except mining. The mining industry may be able to design for lower water consumption than is the current normal practice. Water use for tourism and recreational facilities is the largest use of ground and surface water in the area.

EXTENSION OF THE GROUND-WATER SUPPLY

Other than importation of surface or ground water to places of current use, more water may be made available by physically increasing the supply and by increasing efficiency of use. Schemes to increase the supply include efforts to increase precipitation and to increase recharge of surface water that might otherwise be lost directly to the atmosphere. Increases in efficiency imply conservation in use of water and reduction of natural uncontrolled evapotranspiration losses of both ground water and surface water.

INCREASING THE SUPPLY

Importation of water from regions beyond the Colorado River basin is being considered, and the Central Arizona Canal now is being constructed to carry Arizona's entitlement of Colorado River flow into the central part of the State. The U.S. Bureau of Reclamation is conducting weather modification experiments in an attempt to determine whether or not precipitation can be increased on a regional basis. Except for the mention of these major programs, a discussion of increasing the total water supply of the region by importation of surface water or by weather modification is beyond the scope of this report.

Techniques for increasing the supply of ground water involve transfer of water consumption from the present natural or virtually uncontrolled state to a use controlled by man. The total supply would not be increased, but the proportion available for man's use may be increased.

INCREASING NATURAL RECHARGE

In most of the region, ground-water withdrawal has lowered water levels. In places where ground water now evaporates at the surface, such as along the middle reaches in the Little Colorado River, ground-water lowering is a practical first step in reducing uncontrolled losses. Lowering of a water level that is near surface or that feeds a flowing stream will decrease not only the natural water loss to evapotranspiration but also the surface flow and underflow downstream. In the circumstance of original near-surface water levels, which was fairly common along streams in the region, much floodflow ran off downstream because of lack of storage space under the streambed; lowering of water levels creates storage space in the dewatered part of the aquifer, following which floodflow can recharge the aquifer. Assuming adequate space is available, the limit to the increase in recharge is relative to the total flow and to the relation between the rate of streamflow and the rate of infiltration to the aquifer (Burkham, 1970b). Where the original water level everywhere is deeper than about 25 ft, the probable evapotranspiration loss is small, and space is available to accept floodflow infiltration. Additional lowering of the water level will provide more storage space for recharge, but it will not alter the rate of infiltration. In an area where the water level intersects the streambed or is near surface and is tapped by plants, attempts to increase recharge without lowering water levels probably would only increase uncontrolled loss.

Schemes to increase natural recharge along streams generally require alteration of stream courses or diversion of streamflow to areas other than the channel (American Society of Civil Engineers, 1949, p. 61, 62; Laverty, 1945; Richter and Chun, 1959; Todd, 1959). Flood-control dams, coupled with controlled release, can successfully increase recharge rates (Babcock and Cushing, 1942), but many of these dams are economically justifiable only if they provide flood protection. Recharge also may be increased by diverting streamflow into secondary channels, into unlined canals, or into recharge pits. Simple disking or breaking of the crusted surface of major arroyos following floodflow would also increase the recharge from medium and low flows after the water has cleared.

Increasing streamflow infiltration in one area may only change the location of recharge and fail to increase the total amount of recharge. This condition usually prevails when the amount of streamflow is small. However, if the downstream site of recharge is an area where the water is evaporated or contaminated by mixing with poor-quality water, induced upstream recharge might be practical.

ARTIFICIAL RECHARGE

Artificial recharge is feasible mainly in areas where surface water is abundant and unused, and where there is adequate underground storage space to accommodate the recharged water. The lower Colorado region is deficient in surface flow, and nearly every drop of available surface water is utilized or appropriated. The legal status and ownership of surface water is clear, but the status of recharged ground water is not in many places. This problem may be a major impediment to artificial recharge in Arizona, where surface and ground water have differing legal status.

Artificial recharge is practiced on a small scale in Phoenix, Ariz., where water is pumped out for cooling purposes and returned through recharge wells. Another method of recharging water to underground storage is the diversion of temporarily surplus water in a reservoir to streambeds where the recharge mechanism operates under natural conditions. Water recharged in this manner may be recovered by wells. The Salt River dam and reservoir system in the Phoenix area so operates when release of water is necessary to provide additional storage for control of predicted flood runoff. Almost all the water released during one period was recharged to the aquifer in the Phoenix area (Briggs and Werho, 1966).

Although artificial recharge is a management tool for increasing the amount of water in storage and for decreasing the loss of water stored on the surface, currently very little water is surplus or available to be recharged. Artificial recharge may be used effectively to change the location of current natural recharge.

WATERSHED MANAGEMENT—INCREASING RUNOFF AND DECREASING EVAPOTRANSPIRATION

In the lower Colorado region, more than 95 percent of the precipitation is lost through evaporation and transpiration. If the amount of water lost to the atmosphere could be reduced even by a small percentage, the amount of usable surface water and ground water would be markedly increased. Since publication of the Barr reports in 1956 (Barr, 1956a, 1956b), a great deal of effort has been put into research and pilot programs in Arizona to assess the potential for increasing water yield by vegetation management on watersheds.

Small-scale experiments in mixed conifer and ponderosa pine forests, pinyon-juniper woodlands, and chaparral vegetation zones indicate that removal and (or) replacement of the original vegetation will increase the water yield. Operational and experimental programs in control or removal of riparian plants also result in an increase in water yield. However, results of the water-yield experiments cannot be extended to larger areas with certainty, and it is not known that the increased water yield will survive transmission losses prior to flowing into a surface reservoir or recharging a usable ground-water aquifer. The cost of management maintenance makes the additional runoff costly, and replacement of cleared woody vegetation by grass or forbs reduces the original increase in runoff.

Ffolliott and Thorud (1974, p. 34–36) concluded that water yield could be increased in mixed conifer, ponderosa pine forest, chaparral, and riparian vegetation zones, and they recommended increased data collection and research to evaluate the experimental results. Their analysis of existing data indicated that in Arizona about 0.5–1.2 million acre-ft of potential increased water yield could result from treatment of 6 million acres of mixed conifer, ponderosa pine forest, and chaparral vegetation. They estimated a potential yield of 2 acre-ft per acre for management of 300,000 acres of riparian vegetation for a total annual potential of another 600,000 acre-ft (Ffolliott and Thorud, 1974, p. 22).

The use of watershed management solely to increase ground water in most of the region does not seem promising, owing to land treatment and maintenance costs and the distance between the additional runoff and the ground-water storage area. Additional water might be derived through range and forest management practices designed to increase runoff, but the program costs seem to require surface storage or nearby use of this water rather than increased ground-water recharge.

INCREASING THE EFFICIENCY IN USE OF WATER

The greatest potential and the most economical method of extending the ground-water and total water supply probably is to increase the efficiency of water use. In general, efficiencies should reduce the inadvertent or uncontrolled loss of water to the atmosphere or to places where it cannot be recovered for years or even decades. Multiple reuse of water is practiced generally in the region. In many cases, the reuse is accidental or inadvertent, but the mining industry and utilities in particular make an effort to reuse water. At this time, most conscious reuse is dependent on economics rather than simple conservation; however, water becomes more expensive each year, and more and more large users are instituting water-recycling systems. The extent and management of recycling are carefully planned because water becomes more highly charged with dissolved solids following each cycle. Eventually, particularly in cooling uses, recycled water has so high a concentration of salts that it must be removed from the system.

Conservation can prevent or delay some of the problems associated with ground-water exploitation. Planned reuse of sewage and water used for cooling and slurry transport and innovative design that reduces or eliminates large-scale use of water are necessary to conserve the supply. Efficiency of irrigation can be improved markedly through canal lining and different methods of application, such as sprinkling and drip irrigation methods. In some areas in the region irrigated lands have been forced out of production by urbanization or by municipal purchase; their associated ground-water supply is used to meet less water-demanding urban needs.

The capital cost of instituting conservation practices must be met now largely by individual entities because of current legal impediments and lack of areawide management options. The threat to economic, industrial, and population growth caused by lack of conservation is only dimly perceived, and the threat is not deemed as immediate a problem as capital expenditures. An attempt in 1976 by the City Council of Tucson, Ariz., to institute conservation practices through a pricing mechanism resulted in a recall referendum and a replacement of the council majority.

CROP IRRIGATION

The greatest demand on water in the region is for irrigated agriculture; irrigation accounted for about 90 percent of ground water (fig. 5) and total use in 1975. In Arizona about 7 million acre-ft of water was applied on about 1.2 million acres of cropland in 1970 (Arizona Crop and Livestock Reporting Service, 1974, p. 1) for an average annual use of almost 6 acre-ft per acre of harvested crop. In 1973, 1.4 million acres of crops was grown using about 7 million acre-ft of water for an average of about 5½ acre-ft per acre of crop. The actual applications of water for irrigation probably ranged from about 2½ acre-ft per acre for some pasture and hay crops to about 40 acre-ft per acre for citrus crops grown on very permeable ground in the Yuma area. The consumptive use of water by most crops grown in the region ranges from about 1½ acre-ft to about 6 acre-ft of water per acre (Erie and others, 1965). The soil water required for growth of most forage, fiber, fruit, and oil crops in 40–50 in. per year or growing season and for most grain and vegetable crops is 16–25 in. per growing season (Erie and others, 1965). However, considerable additional irrigation water per acre is evaporated because of water applications necessary to encourage germination, control pests, protect from frost, and for cooling to improve quality or preserve crispness (lettuce). (See Erie and others, 1965, p. 6). In the area of greatest agricultural development, the annual potential evapotranspiration ranges from 35 to 50 in. per year (Buol, 1964).

Methods of accurately measuring the total water consumed in crop growth require a complete accounting of water applied; such an accounting is much too cumbersome, expensive, and time consuming to be applied generally. Therefore, total consumption has been estimated by use of equations derived from prior research where a budget-accounting process was used for control. Methods suggested by Thornthwaite (1948), Blaney and Morin (1942), and Blaney and Criddle (1950, 1962) use empirical formulas to calculate evapotranspiration. These formulas are in common use primarily because climatic data necessary in the calculations are available. A more accurate and physically valid method was presented by Penman (1948, 1956a, 1956b), but the necessary data for solution are not commonly available in existing meteorological records, not are they easily obtained. The value of the Thornthwaite and Blaney-Criddle methods in determining total consumption of water is dependent on prudent irrigation. If fields or parts of fields are "overwatered" in respect to plant growth needs, then the estimates of consumption are likely to be low.

Efficiency of water delivery and application is becoming a more significant part of agricultural expense, and more and more effort and capital investment is being put into improving the efficiency of applying water to crops. If water costs are not great, the irrigation systems are rather crude and are subject to large conveyance losses and inefficiencies in water application; the greatest losses are by seepage from ditches and wastage by evaporation of excess applied water and of tailwater that accumulates in the lower ends of the irrigated fields. In the last 20 years, owing largely to the cost of delivering water and secondarily to the shortage of water, more and more irrigators have installed sealed transmission ditches and systems for pumping back "tailwater" from the ends of the fields. However, change is slow, owing to the relatively low value of the product and the high initial cost of a new irrigation system to replace one already in existence.

The ideal arrangement for water conservation is application of just enough water to meet the growth needs of the plant and enough excess to flush downward the salts that accumulate in the soil because of transpiration of water by the plants and evaporation. The ideal system would eliminate unnecessary evaporation and seepage. Most crops are irrigated by flooding the entire field or by running water in furrows, but in parts of Arizona where water cost is a factor, sprinkler systems and drip irrigation are replacing the flood method. In the Yuma area, as reported by Lucas (1975), flood irrigation of citrus trees required 11 acre-ft per acre per year, but a change to drip irrigation dramatically lowered the demand to 6 acre-ft; saving of water also was achieved by changing from flood to sprinkler irrigation of grain crops (Lucas, 1975, p. 24–25). Further, according to results reported by Lucas (1975), the capital investment and maintenance cost of drip or sprinkler irrigation is less than that of flood irrigation.

The operational benefits in 1974–75 justified costs of shifting to a drip irrigation system in Yuma; there, installation of a flood irrigation system was about $520 per acre, whereas installation of more efficient drip irrigation in the same area was about $400 an acre, or $120 per acre less. Similar savings were reported using sprinkler irrigation instead of flood irrigation. In most places in the region where large quantities of water are used, the 1970–75 cost of water was in the range of $3 to $15 per acre-ft. However, the project to deliver 1.2 million acre-ft of Colorado River water to central Arizona will cost $2.1 billion for construction, and operation and maintenance costs are estimated at about $100 per acre-ft (Lucas, 1975; City of Tucson, 1974, p. 54–61, 83–97). Because of the low costs now current in most places, an economic incentive for improvement of irrigation efficiency by installing a new irrigation system on presently irrigated lands will have to rely on reduction of costs other than that of water.

An alternative crop growth and economic incentive to reduce water consumption was suggested by Kelso, Martin, and Mack (1973) reporting on an 8-year study financed by the Rockefeller Foundation and the University of Arizona. They (Kelso and others, 1973, p. 26, 27) indicated that

> scarce water supplies are, by legal devices and by reasons of location, locked into uses in which the marginal value of the product of the water is extremely low—approaching zero in some cases. * * * curtailing water uses producing lowest net income * * * expanding * * * higher * * * net income per unit of water, can release growth of the Arizona economy from all restraint by water, well into the twenty-first century.

A later report (Kelso and others, 1974) and several public addresses by the authors and their colleagues (oral commun., 1973–76) indicated that it is more economical to import many high-water-consumption crops, such as alfalfa and forage, than to produce them in Arizona. Some of the water used to grow these animal feed stocks could be converted to higher value uses, but, more importantly, total water use could be drastically reduced with no measurable negative impact on the economy of the lower Colorado region.

PUBLIC SUPPLIES

Consumptive use of water in public supply can be reduced substantially. Consumptive-use calculations trace most of the consumption to outside watering of lawns and plants rather than to inside-home use. Home, city, town, and subdivision designs that reduce the desire or need for abundant outside growth or that substitute low-water-need desert growth can halve the current 180–200 gal per day per capita use and 90–100 gal per day per capita consumption. However, the use of water for public supply is only 6.5 percent of the total ground-water use (fig. 5) and only about 5 percent of all the water used (1975) in the region.

SEWAGE EFFLUENT

Sewage effluent is discharged onto permeable areas where it infiltrates to ground water or is reused for irrigation or industrial purposes. In some places the effluent is discharged into a flowing stream or a body of water; for instance, some of the treated effluent from the Las Vegas area is discharged via Las Vegas Wash into Lake Mead. The dissolved-solids concentration of the effluent may be less than that of the native ground water in the area of discharge. This condition holds for the Phoenix, Tucson, and Las Vegas areas. In other places, particularly the smaller cities and towns, the reverse is true. The principal disadvantage of recharged sewage effluent is its concentrations of boron, nitrate, and phosphate. Sufficient concentrations of boron are injurious to growing plants, nitrate concentrations greater than 10 mg/L are a health threat to infants (National Academy of Sciences and National Academy of Engineering, 1973, p. 73), and nitrate and phosphate compounds encourage unwanted vegetal growth where effluent wets surface soil or enters surface-water bodies.

In most parts of the region sewage effluent is considered a cheaply available source of water, and considerable effort has gone into reusing the resource. The Tucson metropolitan area now uses effluent to water a park and hopes to use a major part for industrial uses and possibly for irrigation of crops. Effluent from the Phoenix metropolitan area is used for irrigation, and effluent from Las Vegas is used for irrigation of crops and a minor amount in power generation. Effluent from most other communities in the region is not consciously reused and is either discharged to flowing streams, recharged to ground water, or evaporated.

The trend in the region seems to be toward greater use of sewage effluent in industrial processes, such as copper-ore beneficiation, power generation, and cooling. Much effluent is used to irrigate forage crops, but owing to cost and conservation factors that might give higher priority to public supply and industrial use, irrigation usage seems likely to diminish.

INDUSTRIAL USES

Most industrial water-use systems are designed to make the maximum use of water prior to its loss by evaporation and, in general, the amount of evaporation is a necessary and integral part of the industrial process. The greatest uses are for mining and milling of copper ore and for generation of electric power.

As noted in the section on "Summary of Ground-Water Use," an average of 260 gal of water per ton of ore is used in the ore-concentration process, and half to all this water is not recovered in most operations. The dominant loss of this water is associated with hydraulic transport of finely ground waste rock to tailings ponds. Prior to 1975, many waste-rock tailings ponds were so constructed that seepage through the bottom of the pond was a substantial part of water loss from the system. Many new tailings ponds now are effectively sealed by first coating the raw earth surface with fine slimes produced in the milling operation. This procedure might reduce the rate of water consumption per ton to 100–125 gal or less.

Water used in most steam-electric generation stations in the lower Colorado region is recycled, and nearly all new plants utilize water to complete consumption. Water consumption in thermal and geothermal powerplants probably cannot be lowered owing to present high efficiencies or operating constraints. Consumption of water in nuclear power-generation plants is considerably greater than that in thermal plants, and the consumption can be reduced (Davis and Wood, 1974, p. 8); however, no nuclear plants were operating in the region as of 1978 and the first plant, near Phoenix, is scheduled to start in 1982.

POTENTIAL USE OF SALINE WATER

Saline water generally is not used in the lower Colorado region. Large quantities of saline ground water occur, and as much as 50,000–70,000 acre-ft per year may be lost to evaporation or to uncontrolled transpiration by plants. However, the cost of treatment to allow use of saline water is so high as to be prohibitive if other water sources are available.

In northern Arizona, considerable water is lost to evapotranspiration along the Little Colorado River (fig. 1). There, saline water underlies freshwater, and a mixture of both is near surface and adds to the river flow. Withdrawal of the fresh groundwater for projected expansion of a thermal powerplant in the area will reduce the head in the entire aquifer, and with time the flow of both freshwater and saline water to the surface will diminish. It is not now economic to use saline water in powerplants, but in this case, usage of freshwater will reduce loss of both freshwater and saline water.

In other parts of the region, large quantities of saline ground water are available, and as freshwater becomes more and more expensive to pump or transport, these saline-water bodies will be developed. Depending on the location and the quantity withdrawn, the quantities in storage (table 1) will assure a 30- to 100-year supply or perhaps longer. Large reserves of saline ground water are present along the Little Colorado, Gila, and the Salt Rivers in Arizona, and small to moderate amounts are present along other smaller rivers in the region (Feth and others, 1965). The opportunity to produce large permanent supplies is poor, but the prospect for some short-term developments, roughly 30-year, is good. Additionally, use of saline water will keep such water from migrating into and contaminating adjacent bodies of freshwater. This use is particularly indicated where the ground-water flow is from the area of saline water toward the area of freshwater, and the flow rate is increased with time because of removal of freshwater.

In some areas near cities in Arizona, ground water contains concentrations of nitrate, fluoride, and chromium in excess of the recommended or mandatory drinking-water standards of the National Academy of Sciences and National Academy of Engineering (1973). Some of this water is too highly mineralized to be used directly, but it could be mixed with better-quality water and used for public supply.

CONCEPTS OF GROUND-WATER USE

MINING

Ground water is used in the region mainly according to water need and the economic formula of return versus cost. In much of the region where water levels were near the land surface at the turn of the century or later, it has been profitable and necessary to pump the amount of water needed to sustain the surface operation. Because most ground-water reservoirs in the lower Colorado region store vast amounts of water and receive relatively small amounts of recharge, the region generally has pumped and consumed more ground water than is replenished. The result of this imbalance is steady depletion of stored ground water, as reflected by water levels that decline about 2–10 ft each year throughout much of the developed part of the region. The depletion is termed "mining" because of its near parallelism with mining and permanent depletion of ore bodies. Actually, unlike an ore body, a ground-water aquifer theoretically can be returned nearly to its original saturated condition through artificially increased recharge or through reduction of use; however, time, expense, and other problems may prove artificial or induced recharge impractical in many circumstances.

In recent years much thought has been given by legislators, scientists, and water managers in the region to reconciling economic gain, conservation interests, and guarantees of water supplies necessary for all major uses. Rationally, one might expect that public supplies would be guaranteed to last forever, either by restriction of uses other than public supply, by importation, or by restriction on public-supply use. For example, in places where public supplies of ground water are threatened, it might be financially and politically possible to limit ground-water consumption to the sum of recharge plus a small fraction of ground-water storage. However, in areas where public supplies are not a critical part of the usage, an alternative plan similar to New Mexico law (Chalmers, 1974, p. 73–85) could regulate usage so as to guarantee that the life of the water supply would support an economic return on the surface development. The above concepts imply that (1) "mining" or development of stored ground water is a practical process and the resource should be developed as a mineral property—mined and planned for depletion over a known period of years, and (2) public-supply consumption in towns and municipalities should be balanced to the renewable and assured water inflow. This last condition is not easily attained because most people equate growth with prosperity and believe that a chronic water shortage can be remedied by importation of water. Yet some towns, principally in Arizona, have been severely limited in growth owing to the lack of water and the high cost of water importation.

LEGAL CONSTRAINTS

The scope of this report allows only a discussion of the major legal constraints on ground-water use in the region. Arizona's ground-water law is close to English rule, or as stated by Chalmers (1974, p. 5):

> The English or common law rule is a doctrine based on the theory of absolute ownership, whereby water beneath the land is considered to be strictly a part of the land in which the water is located and therefore belonging to that particular land's owner. So long as no indication of malice exists, such a property owner can utilize any amount of water from his land without liability for harm to neighboring property owners. The English rule was based on conditions in England, and on deductions from absolute private rights in land, supported by the theory that a landowner owns everything above and below his land * * *.

Chalmers (1974, p. 51) further stated: "As of 1973, the Arizona Groundwater Code's structure encourages landowners in areas of serious depletion to pump at the maximum rate economically feasible so as to 'get theirs' before their neighbors do."

Nevada administers ground water conjunctively with its surface water under an appropriation scheme (Chalmers, 1974, p. 72), and "as a general policy, the Nevada Code has restricted ground-water withdrawals to the average annual replenishment or 'safe yield' " (Chalmers, 1974, p. 71).

New Mexico's ground-water code provides: "The waters of underground streams, channels, artesian basins, reservoirs or lakes, having reasonably ascertainable boundaries, are hereby declared to be public waters and to belong to the public and to be subject to appropriation for beneficial use" (Chalmers, 1974, p. 74). The State Engineer now has authority to regulate use to protect prior appropriators and yet make full beneficial and economic use of the resource (Chalmers, 1974, p. 73–85).

Utah manages ground water through a prior appropriation doctrine and is able to restrict use. According to Chalmers (1974, p. 104):

> The Utah groundwater law, * * * has followed a chronological path, from the English rule to the correlative rights doctrine to the present prior appropriation doctrine. The present doctrinal basis permitted the elimination of the historical distinction between groundwaters, with the result that waters of all sources are declared to be public property. Since all prospective appropriators must apply for a permit, the State Engineer in his process of review is in an excellent position to administer the use of groundwater resources. Similarly, conflicting claims may also lead to a determination of the adequacy of such supplies, and to a priority-based reapportionment, and to either a restricted or full closure status for such an area. Adequate statutory flexibility, providing for accumulation of needed data and possible appointment of commissioners, exists within the code, as well as enforcement provisions.
>
> The Utah approach appears to be a reasonably effective effort at managing that state's groundwater resources in an equitable manner that respects public and private interests. The fact that this approach

was implemented nearly four decades ago, and that there was not then in existence situations of extensive mining of aquifers, has undoubtedly made administration of the code less painful than would be implementation at this date.

In summary, of all the States in the region, Arizona is the only State that cannot easily control ground-water use—even ground-water depletion that may do economic damage to a prior appropriator—and cannot easily resolve problems where pumping may deplete surface flow. Kelso, Martin, and Mack (1973, p. 27) commented that

* * * the Arizona problem is more a problem of the lack of manmade institutions (policies) for developing and transferring water than a problem of physically short supplies. At least, the problem can be resolved more cheaply for many years to come if it is approached through institutional (policy) reform relating to water transfer rather than through development and/or importation of additional water supplies.

USE OF UNDERGROUND STORAGE SPACE

In recent years, underground space has been used for storage of resource materials or hazardous wastes. The most common use of underground space has been temporary storage of resource materials, such as natural gas and other hydrocarbons or freshwater; these liquids and gases are stored during periods of surplus and withdrawn during periods of deficiency. Underground space also is used for disposal of liquid waste, comprising caustic, toxic, or radioactive fluids. Generally, the quantities of wastes are small and do not require as much space as the resource materials.

The lower Colorado region has been explored along present routes of natural gas pipelines for geologic environments that might be favorable for storage of natural gas. The requirements for preventing uncontrolled escape of stored gas from the host rock are rigorous, and only a few areas in the Basin and Range province show promise. The El Paso Gas Co. is involved in a joint salt-removal and gas-storage scheme near Phoenix, Ariz., where salt is dissolved from a thick deposit by cycling water, using injection and recovery wells. Salt then is recovered on the surface in sealed evaporation ponds. The space created in the the deposit by salt solution and removal is used for "leak-proof" pressurized storage of natural gas. Another scheme to create natural gas storage space has been proposed for the Kingman area in Arizona. There, it is proposed to create a large cavity in large salt body by an atomic explosion. Other salt or gypsum-anhydrite deposits occur in the basins of the Basin and Range province and presumably also could be used for storage of certain fluids or gasses.

The Plateau uplands province of the region presents more geologic target areas for storage of natural gas and hydrocarbons, but the exploration and development of storage capability may be more costly than that developed to date in the Basin and Range province.

Excess water can be stored beneath the surface in much or nearly all of the lower Colorado region, but technically, the Basin and Range province contains the most practical sites. There, water can be recharged along stream courses; the recharge rate is high and efficient, and with proper planning most of the water can be recovered with very little deterioration of chemical quality. The same scheme can be used in the Colorado Plateaus area, but the location must be carefully chosen so as to yield maximum infiltration rates and yet allow complete and practical rates of recovery of the stored water. Additionally, more care is needed in this province to avoid contamination by native saline water.

The current dominant waste storage is near-surface landfill disposal of solid waste from urban areas. Storage of hazardous or environmentally objectionable materials—such as radioactive waste products, brines, and various industrial materials—has not yet occurred to any great extent in the region.

Disposal of solid urban-derived waste products should not be a major threat to public health or to ground water, but two factors in the region interfere with wise disposal. First, most of this waste is stored in existing large holes in the earth, and in this region these holes typically are "borrow pits" where gravel and sand were removed. Nearly all the gravel and sand pits are along stream courses, and, therefore, most solid waste disposal areas are subject to exhumation when the stream course changes slightly. Such changes are extremely common, particularly in the flood plains along which the landfills are located. Second, because most of the region is semiarid and humidity in the earth is low, the solid waste materials does not disintegrate or revert to soil. Newspapers buried in the last century have been recovered in good condition despite long interment. One solution is to locate solid waste sites in areas removed from major stream courses—in pits developed for the purpose of disposal or following extraction of brick clay. The danger of exhumation in such areas is minimal. Another solution is recycling of the waste into usable products or energy.

Disposal of radioactive wastes may be feasible in thick unsaturated materials between the land surface and water-saturated beds. The feasibility, criteria, and problems are discussed by Winograd (1974, p. 884–894), who suggested the remote Nevada Test Site of southern Nevada for radioactive-waste storage. Other less hazardous materials produced in amounts that are not so large as to be unmanageable might be stored in similar geologic sites in the region.

Other environmentally objectionable fluids, such as brines from desalinization operations or from electric powerplants, probably can be safely and effectively injected into aquifers that contain water too salty to be utilized. Some industrial wastes also could be stored in these aquifers, but the hazard to health from uncontrolled leakage of these materials would have to be carefully assessed. Currently, most of these fluids are evaporated in place, and the solid waste is allowed to accumulate in the evaporation ponds. In some places the less toxic effluents are discharged into municipal sewage-treatment plants.

SUMMARY

Ground water has been developed extensively in the lower Colorado region and large quantities still are available for use. At the present time, particularly in Arizona, most management and planning for ground-water use has not considered whether or not ground-water withdrawal should be limited to the average annual replenishment or whether it should be pumped out to total depletion and the consequences planned for and accepted. In Arizona, the largest part of the lower Colorado region, withdrawal generally occurs to meet a need, and only total depletion or excessive cost has a bearing on ground-water management. Regionally, mixed Federal, State, and private land ownership and legal conflicts or statutes also place constraints on ground-water management. These factors frequently have proved to be significant in ground-water development and management.

In the Arizona and Las Vegas, Nev., parts of the Basin and Range lowlands province, more ground water is pumped out annually than is replenished by recharge. This condition is causing ground-water-level declines, increased pumping costs, land subsidence, and earth cracks. However, even in this highly developed province, there are places where large volumes of ground water are stored. The water in these places generally is at greater depth and may be more saline than the water in the more highly pumped areas. This potential water supply may be developed eventually. Each State in the region is considering the development of ground water in its State water plan.

The northeastern part of the region, comprising the Central highlands and Plateau uplands provinces, also contains large quantities of ground water, and, on a regional basis, the annual ground-water use does not exceed replenishment. However, areas of intensive use, such as the Flagstaff and Williams areas in Arizona, do have inadequate supplies, particularly during the summer. More ground water can be developed in this part of the region, but it will be more costly per unit of water than most ground water now pumped in the southwestern part of the area.

States in the lower Colorado region are completing State water plans that assess the water supply and accompanying problems of development, that predict possible future needs, and that provide for several future water-management options. Adequate, continuing, and up-to-date technical information about ground-water resources in the region will be necessary to select from management options that increase in complexity with time. As the water supplies become more scarce and more valuable, good management can help prevent some of the conflict that will be inevitable over the use of the water.

SELECTED REFERENCES

American Society of Civil Engineers, 1949, Hydrology handbook: Am. Soc. Civil Engineers, Manual of Engineering Practice, no. 28, 184 p.

Arizona Crop and Livestock Reporting Service, 1974, Cropland atlas of Arizona: Phoenix, Arizona Crop and Livestock Reporting Service duplicated report, 68 p.

Arizona Water Commission, 1975, Inventory of resource and uses—phase 1, Arizona State water plan: Arizona Water Comm. report, 224 p.

——— 1977, Inventory of resource and uses—phase 2, Alternative futures: Arizona Water Comm. report, 145 p.

Arizona Westwide Study Team, 1973, Western United States water plan, State of Arizona: Boulder City, Nev., U.S. Bur. Reclamation, 218 p.

Babcock, H. M., 1976, Annual summary of ground-water conditions in Arizona, spring 1974 to spring 1975: U.S. Geol. Survey Water-Resources Inv. 76–59, 2 sheets.

Babcock, H. M., and Cushing, E. M., 1942, Recharge to ground water from floods in a typical desert wash, Pinal County, Arizona: Am. Geophys. Union Trans., 23d Ann. Mtg., pt. 1, p. 49–56.

Barr, G. W., ed., 1956a, Recovering rainfall, Part I: Arizona Univ., Dept. Agr. Economics report, 33 p.

———1956b, Recovering rainfall, Part II: Arizona Univ., Dept. Agr. Economics report, 218 p.

Blaney, H. F., and Criddle, W. D., 1950, Determining water requirements in irrigated areas from climatological and irrigation data: U.S. Soil Conserv. Service SCS–TP–96, 48 p.

———1962, Determining consumptive use and irrigation water requirements: U.S. Dept. Argriculture Tech. Bull. 1275, 59 p.

Blaney, H.F., and Morin, K. V., 1942, Evaporation and consumptive use of water: empirical formulas: Am. Geophy. Union Trans., v. 23, p. 76–82.

Briggs, P. C., and Werho, L. L., 1966, Infiltration and recharge from the flow of April 1965 in the Salt River near Phoenix, Arizona: Arizona State Land Dept. Water-Resources Rept. 29, 12 p.

Brown, S. G., 1976, Preliminary maps showing ground-water resources in the lower Colorado River region, Arizona, Nevada, New Mexico, and Utah: U.S. Geol. Survey Hydrol. Inv. Atlas HA–542, 3 sheets.

Bull, W. B., and Poland, J. F., 1975, Land subsidence due to ground-water withdrawal in the Los Banos-Kettleman City area, California, Part 3. Interrelations of water-level change, change in aquifer-system thickness, and subsidence: U.S. Geol. Survey Prof. Paper 437–G, 62 p.

Buol, S. W., 1964, Calculated actual and potential evapotranspiration in Arizona: Arizona Univ., Agr. Expt. Sta. Tech. Bull. 162, 48 p.

Burkham, D. E., 1970a, A method for relating infiltration rates to streamflow rates in perched streams, *in* Geological Survey research 1970: U.S. Geol. Survey Prof. Paper 700–D, p. D266–D271.

———1970b, Depletion of streamflow by infiltration in the main channels of the Tucson basin, southeastern Arizona: U.S. Geol. Survey Water-Supply Paper 1939–B, 36 p.

Chalmers, J. R., 1974, Southwestern groundwater law—A textual and bibliographic interpretation: Arizona Univ., Arid Lands Resource Inf. Paper 4, 229 p.

City of Tucson, 1974, The Central Arizona Project—A staff report to the Metropolitan Utilities Management Agency Board and the Mayor and Council of the city of Tucson: City of Tucson duplicated report, 133 p.

Cooley, M. E., and Davidson, E. S., 1963, The Mogollon Highlands—their influence on Mesozoic and Cenozoic erosion and sedimentation: Arizona Geol. Soc. Digest, v. 6, p. 7–35.

Davis, G. H., and Wood, L. A., 1974, Water demands for expanding energy development: U.S. Geol. Survey Circ. 703, 14 p.

Erie, L. J., French, O. F., and Harris, Karl, 1965, Consumptive use of water by crops in Arizona: Arizona Univ., Agr. Expt. Sta. Tech. Bull. 169, 41 p.

Feth, J. H., and others, 1965, Preliminary map of the conterminous United States showing depth to and quality of shallowest ground water containing more than 1,000 parts per million dissolved solids: U.S. Geol. Survey Hydrol. Inv. Atlas HA–199, 2 sheets, text.

Ffolliott, P. F., and Thorud, D. B., 1974, Vegetation management for increased water yield in Arizona: Arizona Univ., Agr. Expt. Sta. Tech. Bull 215, 38 p.

Green, C. R., and Sellers, W. D., eds., 1964, Arizona climate: Tucson, Arizona Univ. Press, 503 p.

Harshbarger, J. W., Lewis, D. D., Skibitzke, H. E., Heckler, W. L., and Kister, L. R., 1966, Arizona water: U.S. Geol. Survey Water-Supply paper 1648, 85 p.

Hillel, Daniel, 1971, Soil and Water—Physical principles and processes: New York, Academic Press, 288 p.

Kelso, M. M., Martin, W. E., and Mack, L. E., 1973, Water supplies and economic growth in an arid environment—An Arizona case study: Tucson, Arizona Univ. Press, 327 p.

Kelso, M. M., and others, 1974, Economic implication of public water policy in Arizona: U.S. National Tech. Inf. Service PB 241 513, 133 p.

Kister, L. R., 1973, Quality of ground water in the lower Colorado River region, Arizona, Nevada, New Mexico, and Utah: U.S. Geol. Survey Hydrol. Inv. Atlas HA–478, 2 sheets.

Laney, L. J., 1972, Chemical quality of the water in the Tucson basin, Arizona: U.S. Geol. Survey Water-supply Paper 1939–D, 46 p.

Laverty, F. B., 1945, Correlating flood control and water supply—Los Angeles coastal plain: Am. Soc. Civil Engineers Proc., v. 71, no. 6, p. 831–848.

Levings, G. W., and Farrar, C. D., 1977a, Maps showing ground-water conditions in the Monument Valley and northern part of the Black Mesa areas, Navajo, Apache, and Coconino Counties, Arizona—1976: U.S. Geol. Survey Water-Resources Inv. 77–44, maps.

———1977b, Maps showing ground-water conditions in the southern part of the Black Mesa area, Navajo, Apache, and Coconino Counties, Arizona—1976: U.S. Geol. Survey Water-Resources Inv. 77–41, maps.

———1977c, Maps showing ground-water conditions in the southern part of the Chinle area, Apache County, Arizona—1976: U.S. Geol. Survey Water-Resources Inv. 77–50, maps.

Lofgren, B. E., 1968, Analysis of stresses causing land subsidence, *in* Geological survey research 1968: U.S. Geol. Survey Prof. Paper 600–B, P. B219-B225.

Lucas, Ken, 1975, Thanks to sprinklers, Yuma desert alive with wheat: Arizona Farmer-Ranchman, v. 54, no. 5, p. 24–25.

McGavock, E. H., and Edmonds, R. J., 1974, Availability of ground water for irrigation, municipal, or industrial use in the Navajo and Hopi Indian Reservations, Arizona, New Mexico, and Utah: U.S. Geol. Survey Misc. Inv. Ser. Map I–878, 4 sheets.

Mann, L. J., 1976, Ground-water resources and water use in southern Navajo County, Arizona: Arizona Water Comm. Bull. 10, 106 p.

———1977, Maps showing ground-water conditions in the Puerco-Zuni area, Apache and Navajo Counties, Arizona—1975: U.S. Geol. Survey Water-Resources Inv. 77–5, maps.

Meyers, J. S., 1962, Evaporation from the 17 Western States: U.S. Geol. Survey Prof. Paper 272–D, 100 p.

MITRE Corporation, 1977, Annual environmental analysis report—An environmental analysis of energy technologies using the assumptions of the National Energy Plans—Volume 1, Technical summary: MITRE Corp. Rept., September.

Murray, C. R., and Reeves, E. B., 1972, Estimated use of water in the United States in 1970: U.S. Geol. Survey Circ. 676, 37 p.

National Academy of Sciences and National Academy of Engineering, 1973, Water quality criteria 1972: U.S. Environmental Protection Agency Rept. EPA–R3–73–033, 594 p.

Pacific Southwest Interagency Committee, 1971a, Lower Colorado Region Comprehensive Framework Study, Appendix IV, Economic base and projection for 1980, 2000, 2020: Pacific Southwest Interagency Comm. report, 181 p.

———1971b, Lower Colorado Region Comprehensive Framework Study, Appendix V, Water resources: Pacific Southwest Interagency Comm. report, 115 p.

———1971c, Lower Colorado Region Comprehensive Framework Study, Appendix XIV, Electric power: Pacific Southwest Interagency Comm. report, 144 p.

Penman, H. L., 1948, Natural evaporation from open water, bare soil, and grass: Royal Soc. London Proc., sec. A, v. 193, p. 120–145.

———1956a, Estimating evaporation: Am. Geophys. Union Trans., v. 37, no. 1, p. 43–46.

———1956b, Evaporation, an introductory survey: Netherland Jour. Agricultural Science, v. 4, p. 9–29.

Poland, J. F., and Davis, G. H., 1969, Land subsidence due to withdrawal of fluids, *in* Varnes, D. J., and Kiersch, George, eds., Reviews in engineering geology—Volume 2: Boulder, Colo., Geol. Soc. America, p. 187–269.

Richter, R.C., and Chun, Y. D., 1959, Artificial recharge of groundwater reservoirs in California: Am. Soc. Civil Engineers Proc., Irrigation and Drainage Div. Jour., v. 85, no. IR4, p. 1–7.

Schumann, H. H., 1974, Land subsidence and earth fissures in alluvial deposits in the Phoenix area, Arizona: U.S. Geol. Survey Misc. Inv. Ser. Map I–845–H, 1 sheet.

Schumann, H. H., and Poland, J. F., 1970, Land subsidence, earth fissures, and groundwater withdrawal in south-central Arizona, U.S.A., *in* Land subsidence: Tokyo, Internat. Assoc. Sci. Hydrology Pub. 88, v. 1, p. 295–302.

Thornthwaite, C. W., 1948, An approach toward a rational classification of climate: Geog. Rev., v. 38, no. 1, p. 55–94.

Todd, D. K., 1959, Annotated bibliography on artificial recharge of ground water through 1954: U.S. Geol. Survey Water-Supply Paper 1477, 115 p.

U.S. Bureau of Mines, 1974, Area reports—domestic, *in* Minerals yearbook, 1972, Volume II: U.S. Bur. Mines, 805 p.

U.S. Bureau of Reclamation, 1975, Compilation of records in accordance with Article V of the decree of the Supreme Court of the United States in Arizona v. California dated March 9, 1964: Boulder City, Nev., U.S. Bur. Reclamation, 40 p.

U.S. Geological Survey, 1971, Map showing the general depth to ground water in New Mexico: U.S. Geol. Survey open-file map.

Valley National Bank of Arizona, 1974, Arizona statistical review: Valley Natl. Bank of Arizona, 30th ann. ed., 70 p.

———1975, Arizona statistical review: Valley Natl. Bank of Arizona, 31st ann. ed., 72 p.

Winograd, I. J., 1974, Radioactive waste storage in the arid zone: Am. Geophys. Union Trans., v. 55, no. 10, p. 884–894.

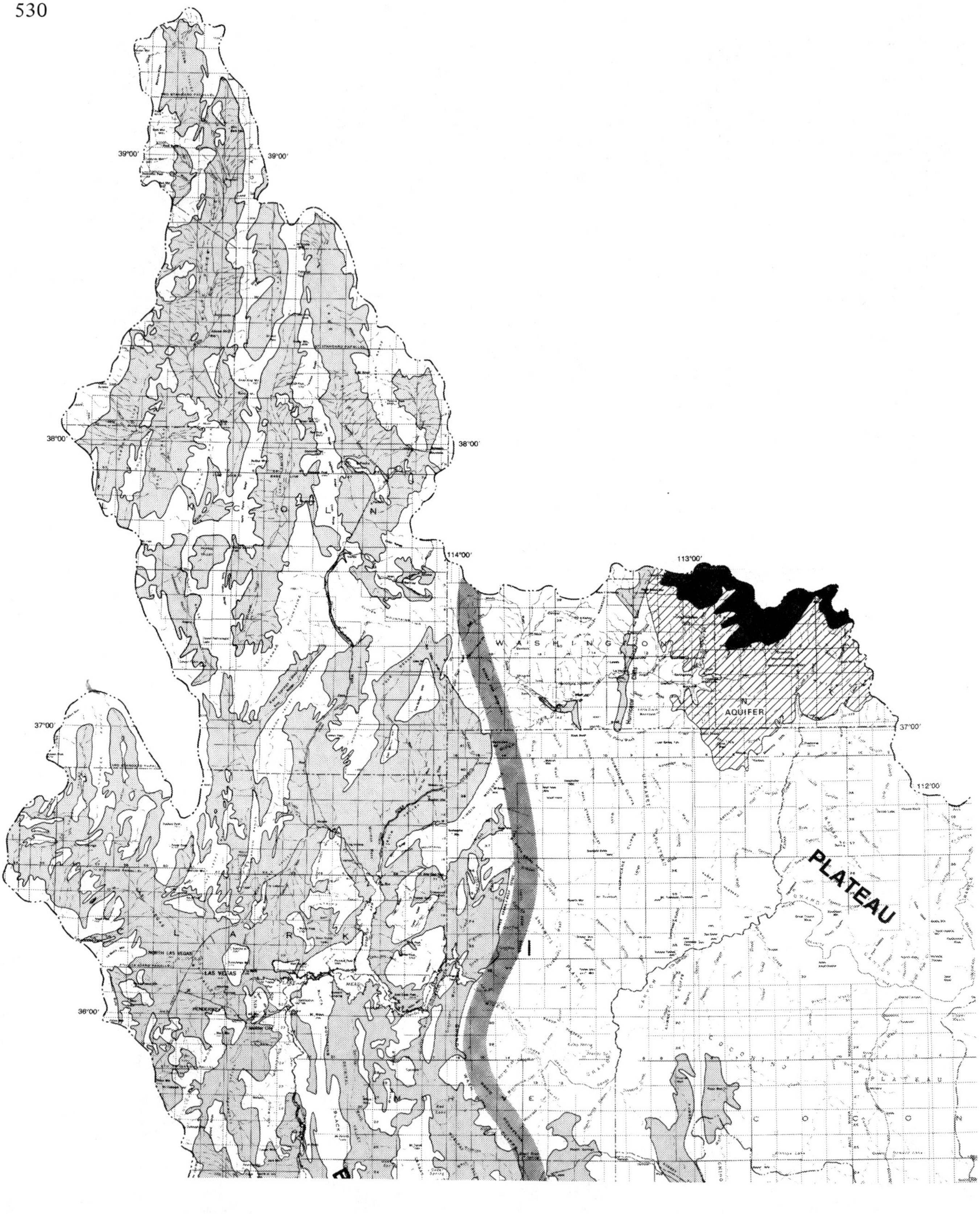

PLATE 1A. GENERALIZED GEOHYDROLOGIC FRAMEWORK IN THE LOWER COLORADO REGION (NORTHWEST AREA)

EXPLANATION

AQUIFERS OF THE BASIN AND RANGE LOWLANDS AND CENTRAL HIGHLANDS PROVINCES

ALLUVIUM – Includes mainly sedimentary deposits of uncemented to strongly cemented gravel, sand, and silt; in places the sedimentary deposits are interbedded with volcanic rocks. Yields moderate to large quantities of water to wells

HARD-ROCK AREA – Yields only small amounts of water to wells; where fractured, yields small to moderate quantities of water to springs

AQUIFERS OF THE PLATEAU UPLANDS PROVINCE

D AQUIFER – Yields small quantities of water to wells. Includes the Dakota Sandstone, Morrison Formation, and Cow Springs and Entrada Sandstones (Cretaceous and Jurassic age)

N AQUIFER – Yields small quantities of water to wells. Includes the Navajo Sandstone, Kayenta and Moenave Formations, and the Lukachukai Member of the Wingate Sandstone (Jurassic and Triassic age)

C AQUIFER – Yields small to moderate quantities of water to wells, where fractured, the aquifer system yields moderate to large quantities of water to wells. Includes the Shinarump Member of the Chinle Formation, Moenkopi Formation, Kaibab Limestone, Coconino and De Chelly Sandstones, and upper part of the Cutler Formation (Triassic and Permian age)

OTHER AQUIFERS – Include the Wahweap Sandstone, Straight Cliffs Sandstone, and Kaiparowits Formation of Cretaceous age and the Wasatch Formation of Tertiary age; yield water to wells in the part of the province that is drained by the Virgin River and in the part from the Kolob Terrace east to Pink Cliffs in Washington and Kane Counties, Utah. Yield small to moderate quantities of water to springs

D, N, AND C AQUIFER

N AND C AQUIFER

D AND C AQUIFER

Area where data are insufficient to delineate extent of aquifer systems or where formations are generally dry

BOUNDARY OF SUBREGION
1. Main stem Colorado River
2. Little Colorado River
3. Gila River above Painted Rock Dam

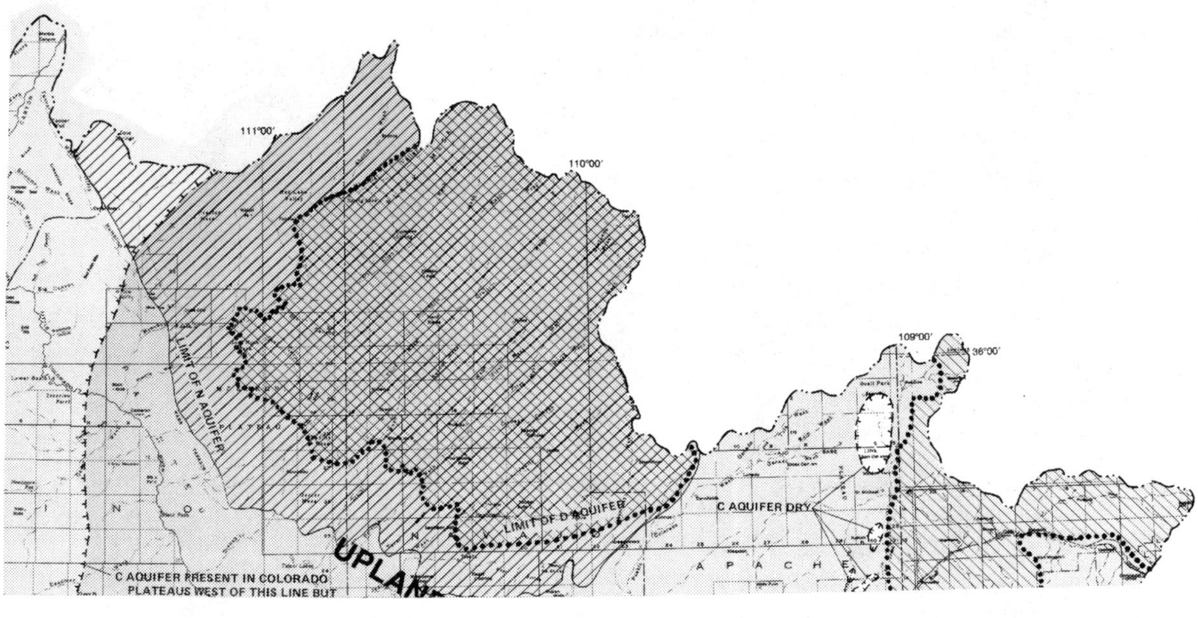

PLATE 1B. GENERALIZED GEOHYDROLOGIC FRAMEWORK IN THE LOWER COLORADO REGION (NORTHEAST AREA)

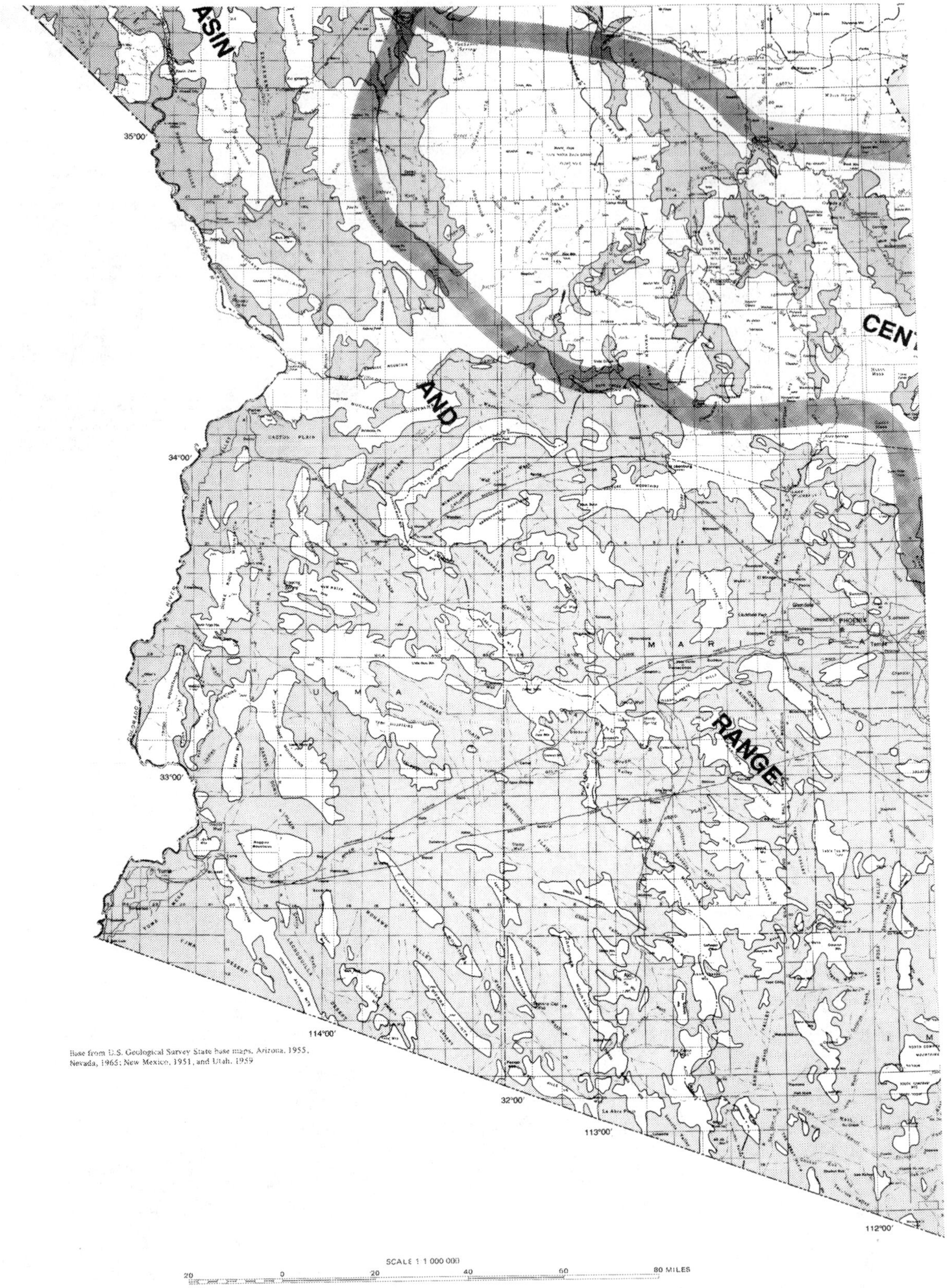

PLATE 1C. GENERALIZED GEOHYDROLOGIC FRAMEWORK IN THE LOWER COLORADO REGION (SOUTHWEST AREA)

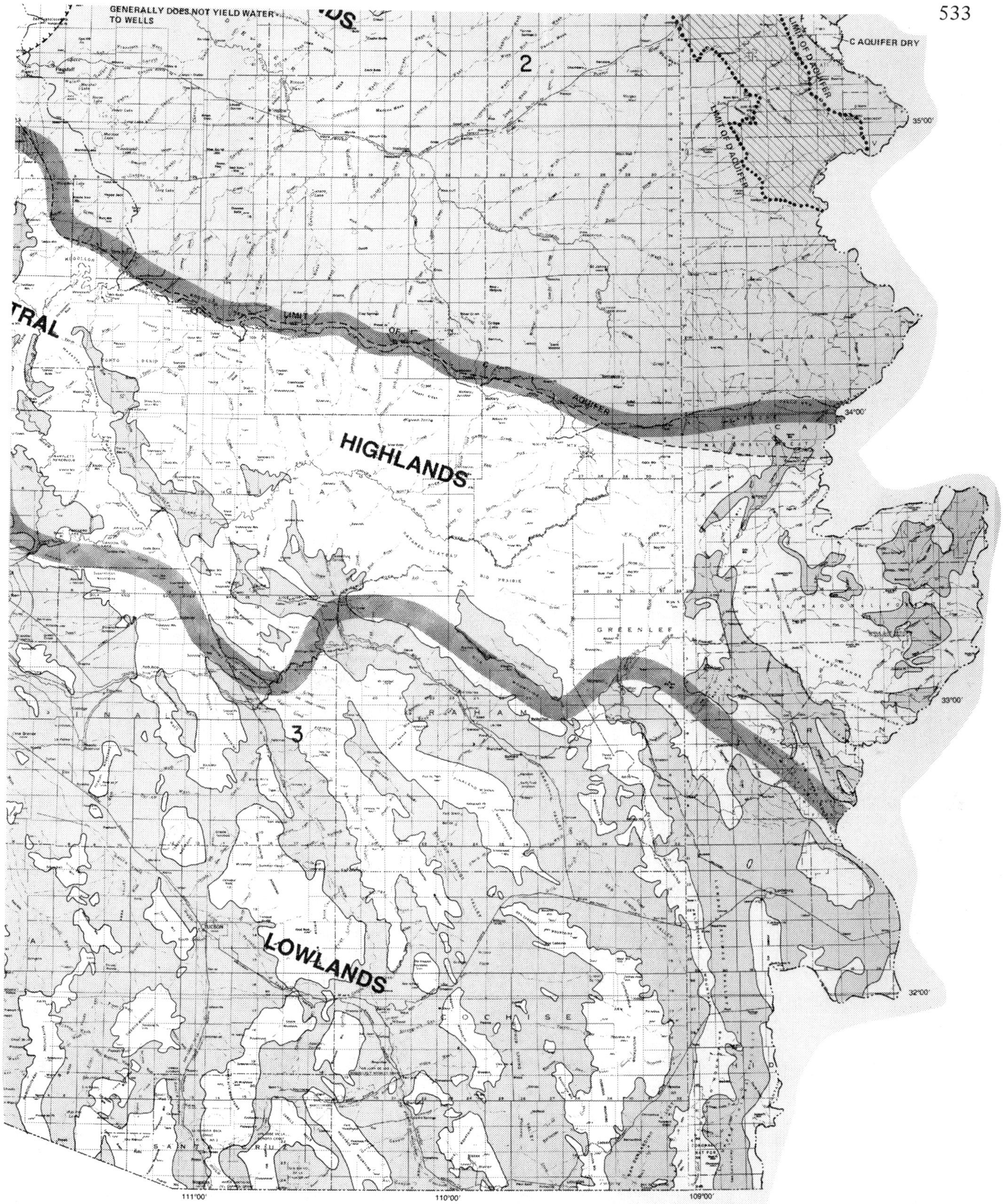

Modified from Brown (1976)

PLATE 1D. GENERALIZED GEOHYDROLOGIC FRAMEWORK IN THE LOWER COLORADO REGION (SOUTHEAST AREA)

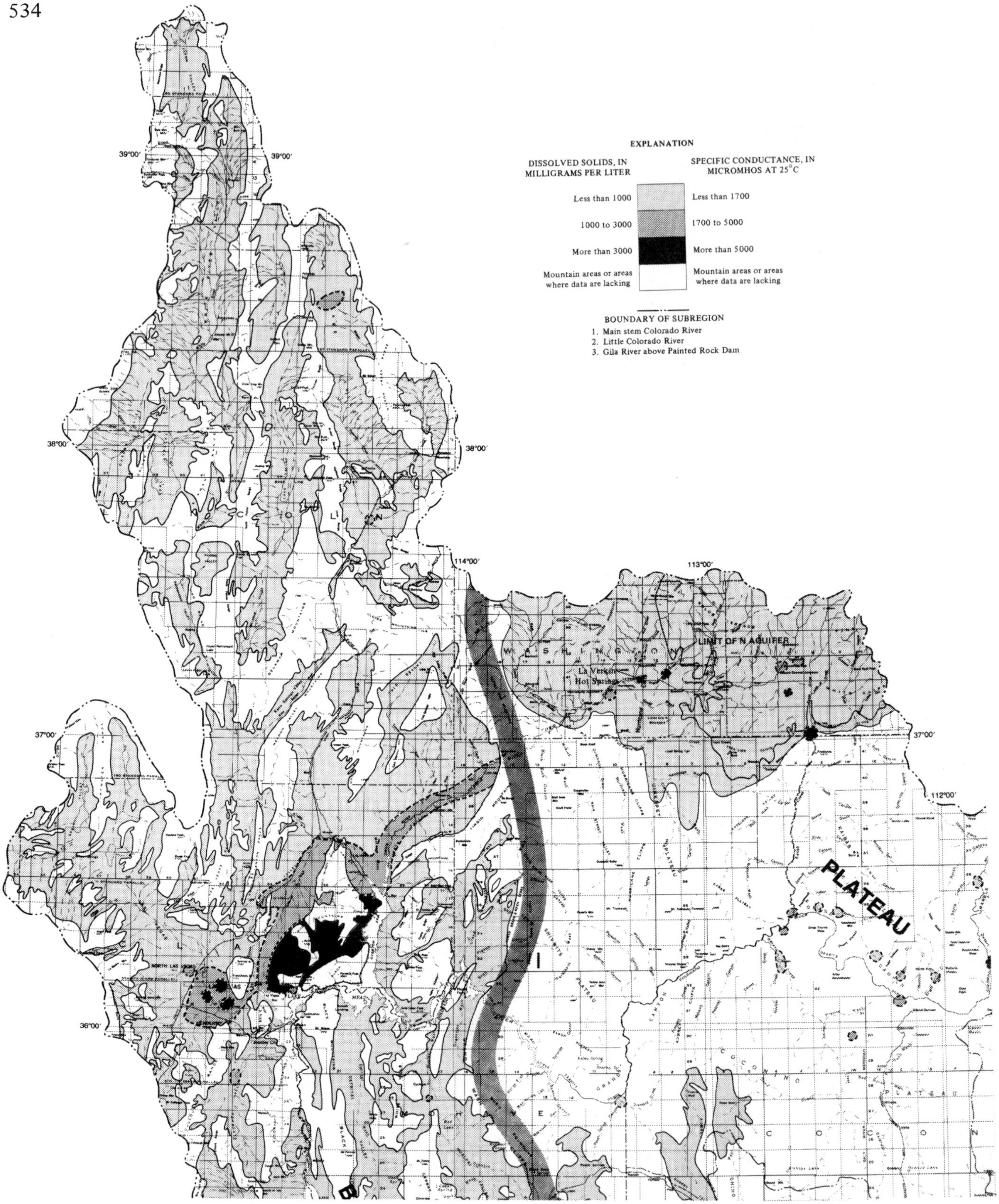

PLATE 2A. GENERALIZED DISTRIBUTION OF DISSOLVED SOLIDS IN GROUND WATER, LOWER COLORADO REGION (NORTHWEST AREA)

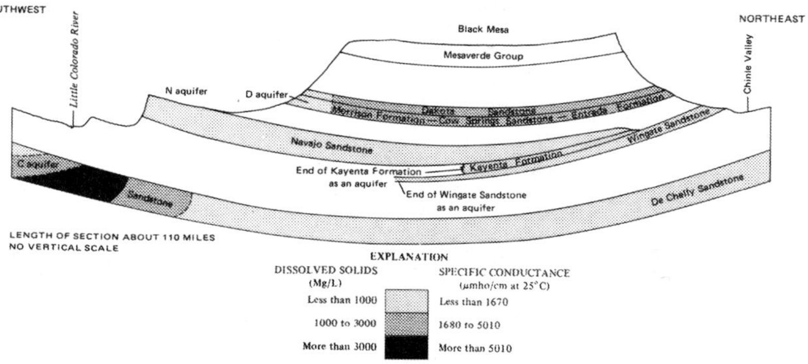

DIAGRAMMATIC SECTION SHOWING THE GENERAL QUALITY OF WATER FROM THE D, N, AND C AQUIFERS IN THE PLATEAU UPLANDS PROVINCE

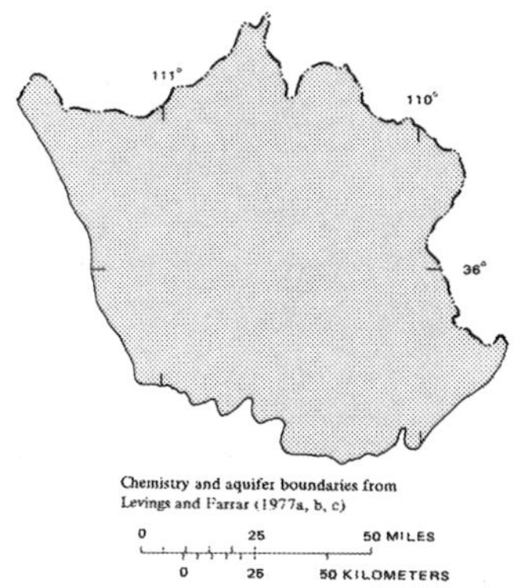

EXTENT AND DISSOLVED-SOLIDS CONTENT OF GROUND WATER IN THE N AQUIFER

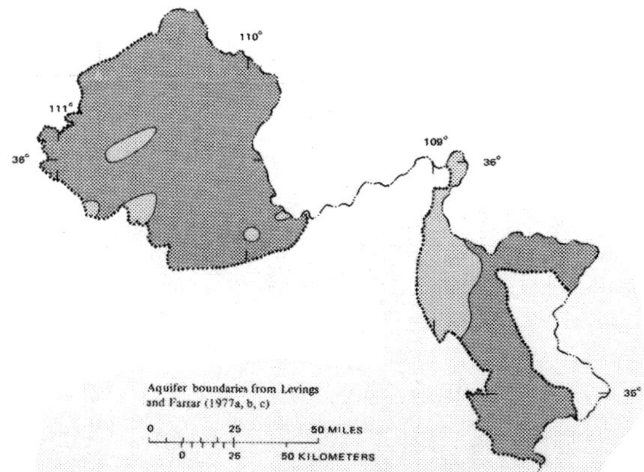

EXTENT AND DISSOLVED-SOLIDS CONTENT OF GROUND WATER IN THE D AQUIFER

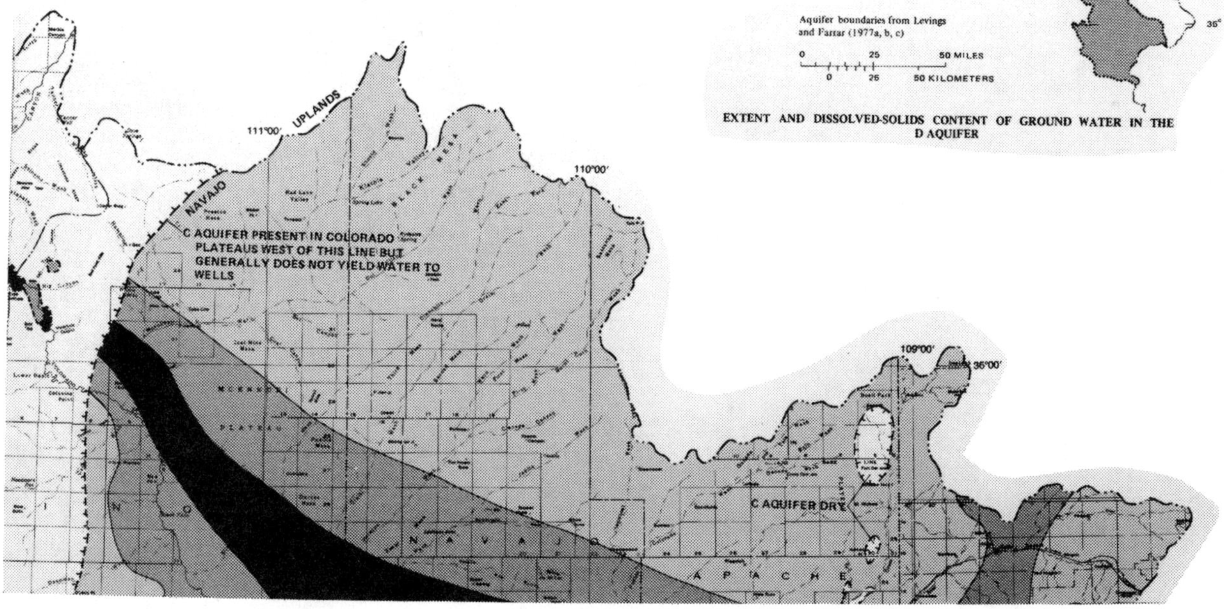

PLATE 2B. GENERALIZED DISTRIBUTION OF DISSOLVED SOLIDS IN GROUND WATER, LOWER COLORADO REGION (NORTHEAST AREA)

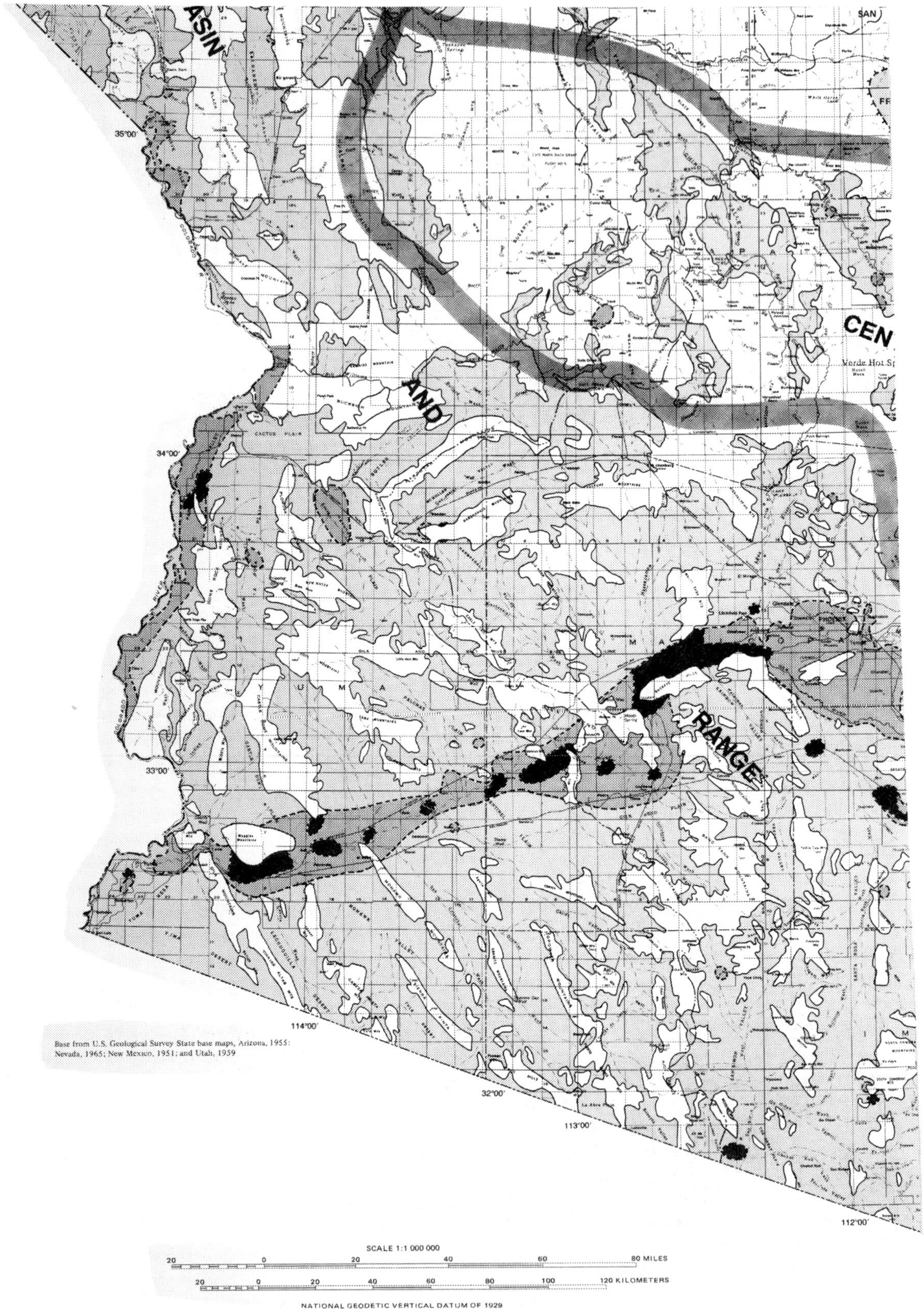

PLATE 2C. GENERALIZED DISTRIBUTION OF DISSOLVED SOLIDS IN GROUND WATER, LOWER COLORADO REGION (SOUTHWEST AREA)

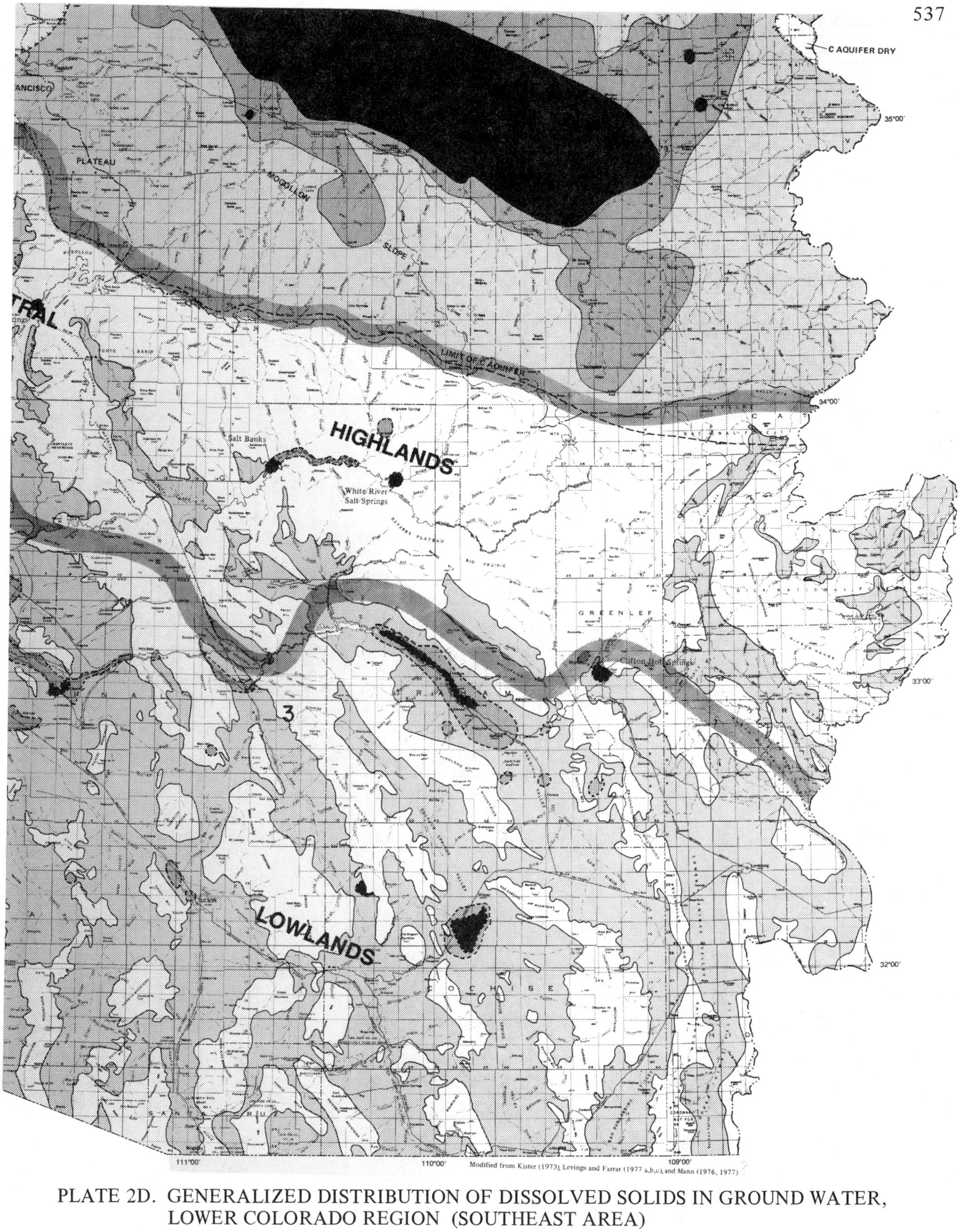

Modified from Kister (1973), Levings and Farrar (1977 a,b,c), and Mann (1976, 1977)

PLATE 2D. GENERALIZED DISTRIBUTION OF DISSOLVED SOLIDS IN GROUND WATER, LOWER COLORADO REGION (SOUTHEAST AREA)

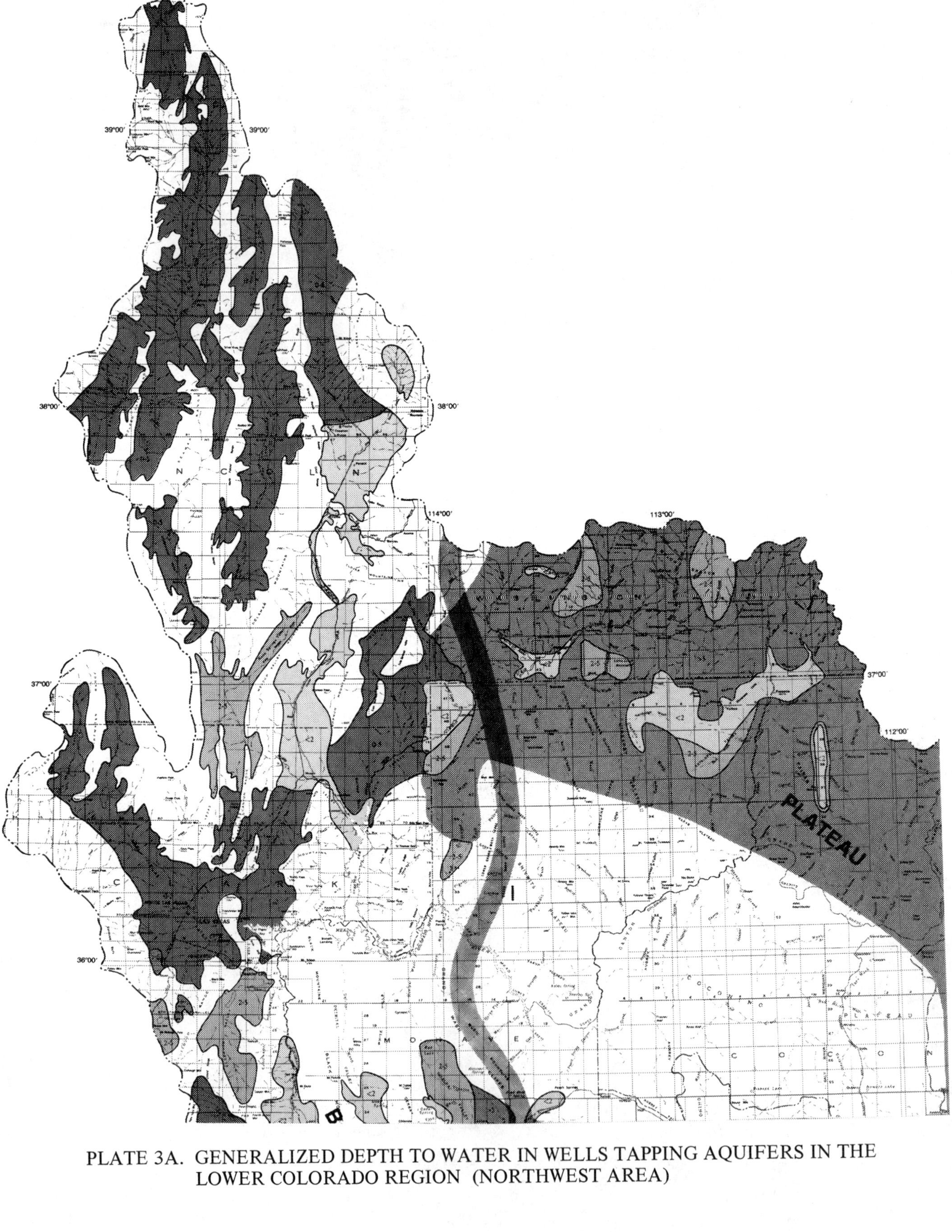

PLATE 3A. GENERALIZED DEPTH TO WATER IN WELLS TAPPING AQUIFERS IN THE LOWER COLORADO REGION (NORTHWEST AREA)

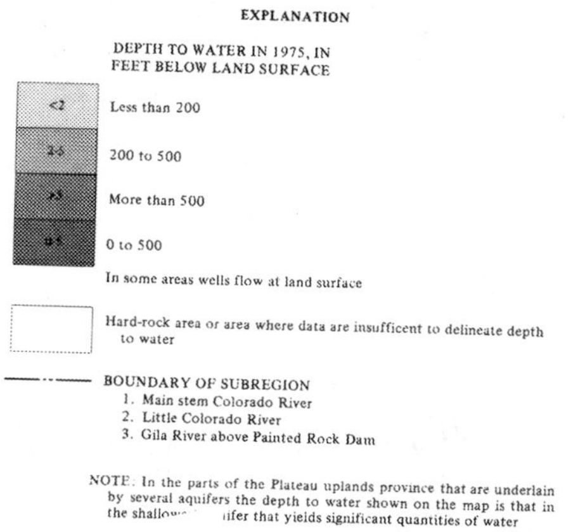

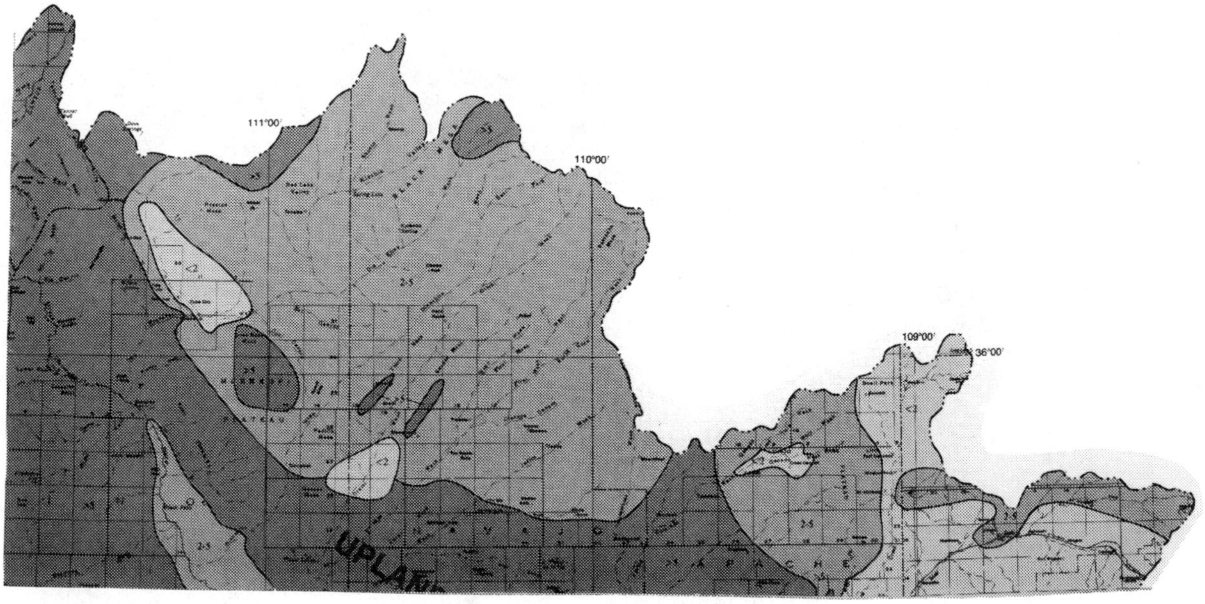

PLATE 3B. GENERALIZED DEPTH TO WATER IN WELLS TAPPING AQUIFERS IN THE LOWER COLORADO REGION (NORTHEAST AREA)

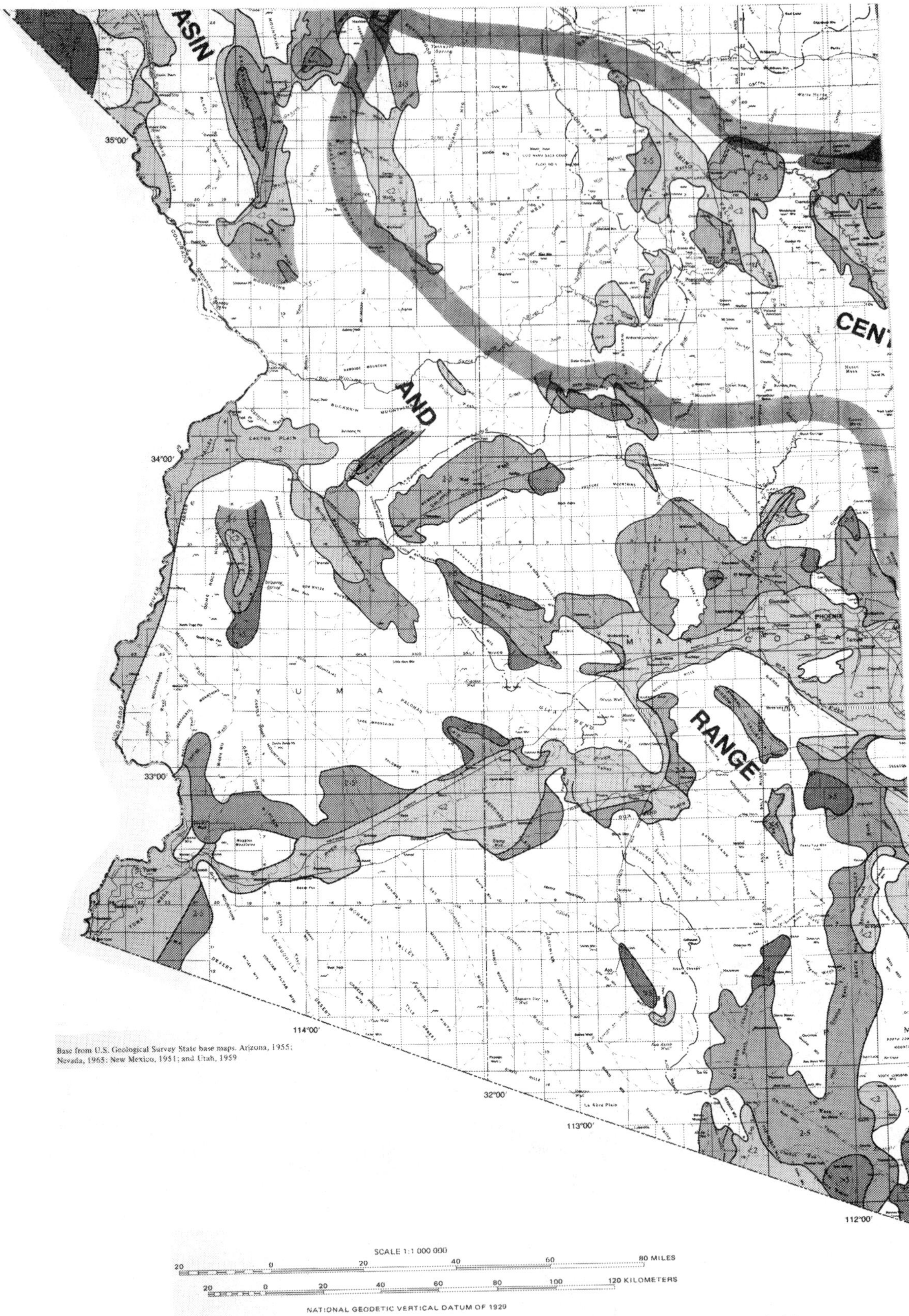

PLATE 3C. GENERALIZED DEPTH TO WATER IN WELLS TAPPING AQUIFERS IN THE LOWER COLORADO REGION (SOUTHWEST AREA)

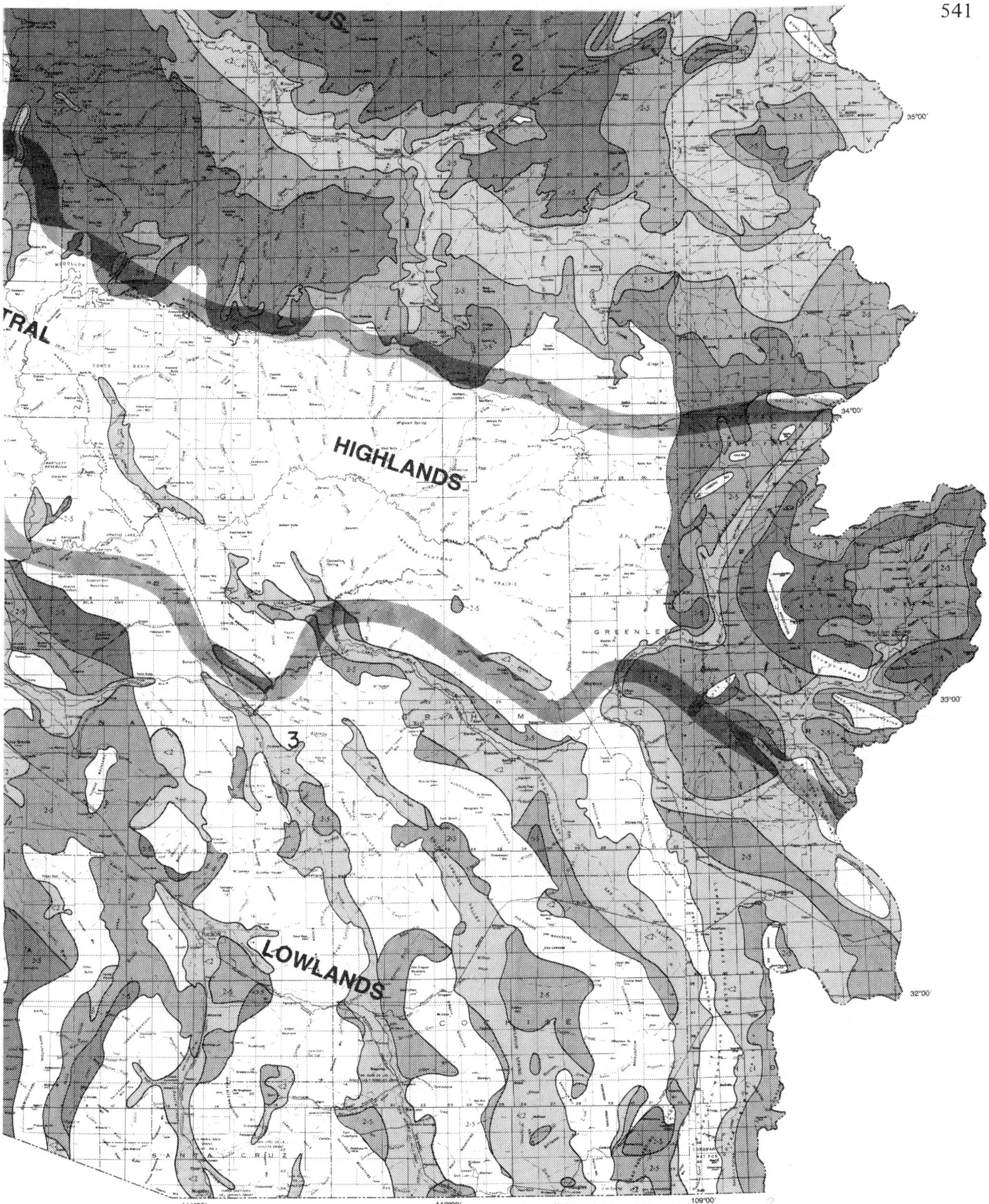

Hydrology from Brown (1976), modified by E.B. Hodges, 1977, from data from the files of the U.S. Geological Survey and U.S. Geological Survey (1971)

PLATE 3D. GENERALIZED DEPTH TO WATER IN WELLS TAPPING AQUIFERS IN THE LOWER COLORADO REGION (SOUTHEAST AREA)

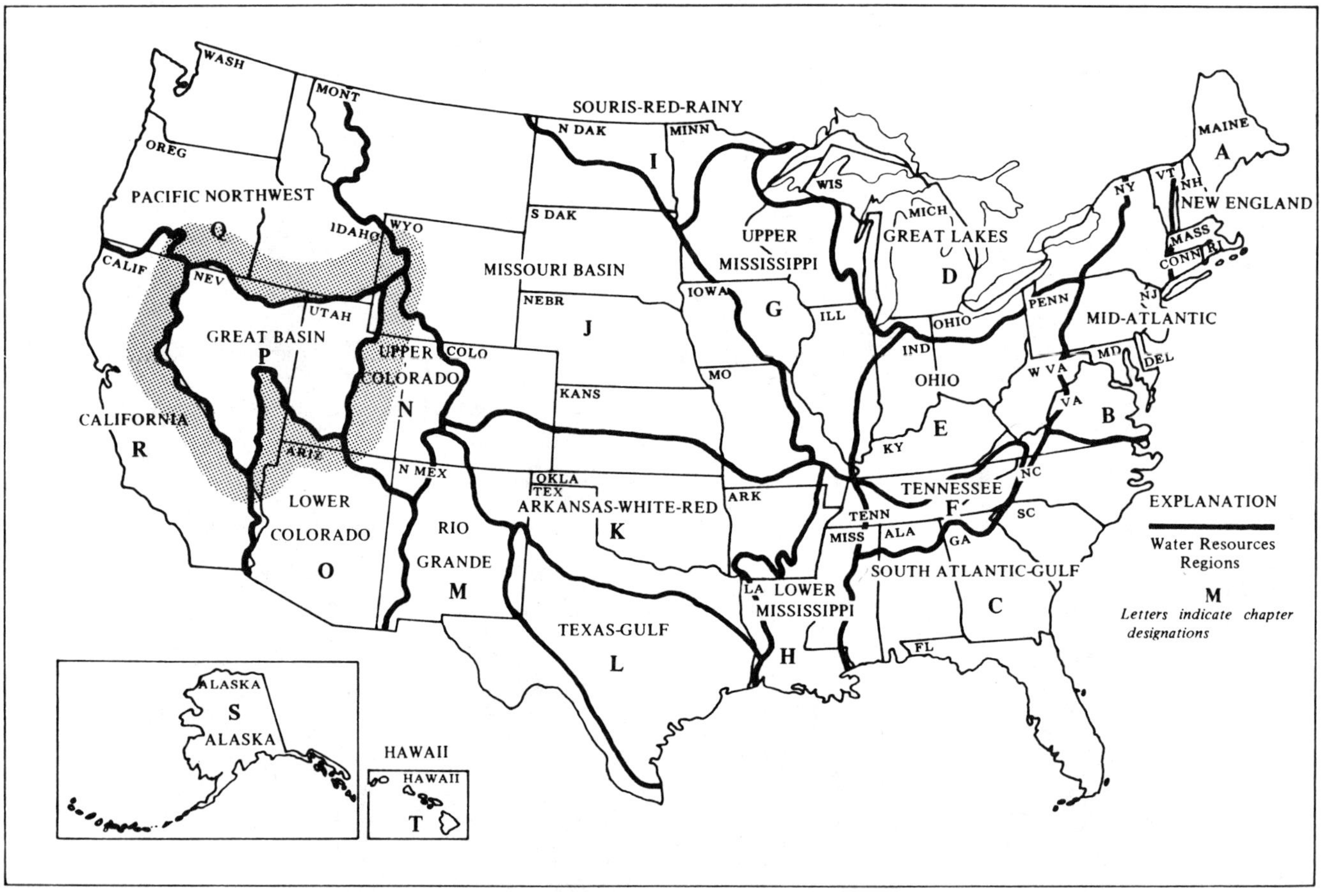

P — Location Map for the Great Basin Region

P ....

GREAT BASIN REGION

By THOMAS E. EAKIN, DON PRICE, and J. R. HARRILL

ABSTRACT

Ground-water withdrawals by wells in the Great Basin Region were about 1.1 million acre-feet (1,360 cubic hectometres) in 1970. Most of these withdrawals were from 87 of the 234 hydrographic areas in the region. Withdrawals ranged from about 1,000 acre-feet (1.2 cubic hectometres) to more than 100,000 acre-feet (123 cubic hectometres). Jordan Valley, which includes Salt Lake City, had the largest withdrawal, about 115,000 acre-feet (142 cubic hectometres).

An appraisal of the regional ground-water resource indicates the region could sustain an annual net pumpage of about 2.6 million acre-feet (3,200 cubic hectometres). Larger withdrawals could be sustained if only part of the pumped water was used consumptively, if conflicts with existing surface-water rights are resolved, and if extensive treatment, artificial recharge, and reuse of water prove feasible. Ground water stored in the upper 100 feet (30 metres) of saturated deposits of the valley ground-water reservoirs is estimated to be on the order of 300 million acre-feet (370,000 cubic hectometres). Total ground-water storage probably exceeds several billion acre-feet; however, much of this could not be developed within economic feasibility expected over the next several decades.

Only a few areas of the Great Basin Region have been studied in detail sufficient to enable adequate design of an areawide ground-water development. These areas already have been developed. As of 1973 data for broadly outlining the ground-water resources of the region had been obtained. However, if large-scale planned development is to become a reality, a program for obtaining adequate hydrologic and related data would be a prerequisite. Ideally, the data should be obtained in time to be available for the successively more intensive levels of planning required to implement developments.

INTRODUCTION

Development of ground-water resources has traditionally been a piecemeal process. In a given area, ground-water supplies are typically developed by individuals, industries, and municipalities to meet specific needs. The magnitude and distribution of pumping is dependent largely on the needs and locations of individual users. If combined net withdrawals of ground water do not exceed the average long-term replenishment (recharge) to the ground-water system, problems caused by development may be minimal. In some parts of the United States, net ground-water withdrawals greatly exceed the recharge, and ground-water resources are being mined—extracted at rates in excess of rates of replenishment. More often, however, problems occur because of local overdevelopment, the concentration of pumping in too small an area. This results in excessive pumping lifts and may lead to reduced well yields and deterioration in water quality. Detrimental land subsidence has occurred in some areas because of local overdevelopment. In most cases, the ground-water system was not well understood in advance.

The purpose of this report is to: (1) Outline the overall ground-water resources of the Great Basin Region, (2) describe the regional ground-water use as of 1970, (3) evaluate the potential for ground-water development in the Great Basin Region, (4) identify some information necessary to plan ground-water development, and (5) discuss some options for ground-water development.

The ground-water resources of the Great Basin Region consist of two components: (1) A large volume of water stored in alluvial and consolidated-rock reservoirs, and (2) an annual replenishment which is much smaller than the total volume of water in storage. That part of the total resource that may be successfully developed is limited by various constraints. Specific constraints vary with each area but always include: (1) Physical constraints imposed by the natural system, (2) economic constraints based on the cost of water in relation to benefits derived from use, and (3) legal and administrative constraints imposed by local, State, and Federal Governments. If the ground-water system has been adequately defined in advance, a supply system can be planned, constructed, and operated to provide optimum use of water within limitations posed by the preceding constraints.

All numerical values in this report are given in English units. For those who prefer to use metric units, the conversion factors are given on preliminary page IV.

Chemical concentration and water temperature are given only in metric units. Chemical concentration is given in milligrams per litre. For concentrations less than 7,000 mg/l (milligrams per litre), the numerical values are about the same as those for concentrations given in parts per million.

THE LAND

The Basin and Range physiographic province (Fenneman and Johnson, 1946) is an area of about 200,000 square miles (518,000 km^2) of internal drainages; it has no surface outlet to the ocean. The Great Basin Region, as described in this report, is that part of this physio-

TABLE 1.—*Size and principal features of subregions in the Great Basin Region*

Subregion	Area (thousands of mi²)	Principal drainages[1] Major	Principal drainages[1] Secondary	Principal sinks	Number of hydrographic areas in subregion
Bear River	7.5	Bear River (8)	Logan River (1)	None—drains to Great Salt Lake subregion.	9
Great Salt Lake	29.0	None	Ogden River (2), Weber River (3), Provo River (2), Jordan River (2).	Great Salt Lake	37
Sevier Lake	16.2	Sevier River (4)	San Pitch River (2), Beaver River (2).	Sevier Lake	14
Humboldt	30.0	Humboldt River (16)	Little Humboldt River (2) Quinn River (5)[2] Kings River (1).	Humboldt Sink (with overflow to Carson Sink).	59
Central Lahontan	11.5	Truckee River (7), Carson River (5), Walker River (7).		Pyramid Lake, Carson Sink, Walker Lake.	33
Tonopah	49.7	None	None	None—Many small sinks in individual valleys.	82
Region (rounded)	144	....	....	....	234

[1]Number in parentheses is the number of hydrographic areas traversed or bounded by the river.
[2]Perennial flow in only one area.

graphic province whose drainage is into Nevada and Utah.[1] It includes parts of California, Idaho, and Wyoming and covers about 70 percent of the physiographic province (about 144,000 mi², or 373,000 km²). Boundaries and general features of the region are shown on plate 1*A*.

The Great Basin Region is characterized by generally parallel, north- to northeast-trending mountain ranges that are separated by broad alluviated desert basins. The mountain ranges commonly are 40 to 80 miles (64 to 128 km) long and are rather regularly spaced 15 to 25 miles (24 to 40 km) apart. Their crests are commonly 3,000 to 5,000 feet (915 to 1,525 m) above the adjacent valley floors. The higher ranges, which are in central and eastern Nevada and along the east and west margins of the region, have crests more than 10,000 feet (3,050 m) above sea level; elsewhere, the altitudes of the mountain areas generally are less than 9,000 feet (2,740 m).

Altitudes of the valley floors range from about 2,100 feet (640 m) in the Amargosa Desert to about 7,000 feet (2,130 m) in central Nevada. However, the altitudes of the floors of most valleys are between 4,000 and 6,000 feet (1,220 and 1,830 m). Typically, there is an intermediate slope from the valley floors to the bordering mountain ranges. The slope, sometimes called the alluvial apron, is formed either by coalescing alluvial fans or by sedimentary materials thinly mantling eroded bedrock surfaces. Gradients of these slopes generally range from a few tens of feet to several hundreds of feet per mile.

There are 234 valleys in the Great Basin Region. Many are topographically closed; however, others are interconnected and drain directly or by way of eight major river systems (Bear, Weber, Jordan, Sevier, Humboldt, Carson, Truckee, and Walker) into five major terminal lakes or sinks. These sinks are the Great Salt Lake and Sevier Lake in Utah, and Humboldt-Carson Sink, and Pyramid and Walker Lakes in Nevada (pl. 1*A*). Black Rock Desert in Humboldt subregion, a sixth major terminal sink, is associated with the Quinn River, a secondary river system. Great Salt Lake, Pyramid Lake, and Walker Lake are remnants of two large lakes that occupied much of the region about 10,000 to 20,000 years ago—Lake Bonneville that covered about 20,000 square miles (52,000 km²) in Utah and Lake Lahontan that covered about 8,000 square miles (20,700 km²) in Nevada.

The region is divided into six subregions (pl. 1*A*). Each subregion is further divided into hydrographic areas which provide the basic units used to present hydrologic information. Table 1 lists the subregions and some of their principal features. Plate 1*B* shows the 234 hydrographic areas in the Great Basin Region and lists area names. Some hydrographic areas listed are parts of larger hydrologic systems, such as areas tributary to the principal drainages. Other areas have smaller, self-contained hydrographic systems that would be virtually unaffected by development in adjacent valleys. This variability between areas and the boundary effects of mountain ranges which

[1]The boundary of the region described in this report differs from the boundary of the recently completed Great Basin Region Comprehensive Framework Study (Pacific Southwest Inter-Agency Committee, 1971a, b, c), which follows the California and Oregon State lines and includes some closed basins whose subsurface flow drains to the Lower Colorado River Region.

commonly separate valleys complicate regional evaluation of the ground-water resources. An adequate analysis must take the above factors into account. In this report, quantitative information will be presented graphically on an area-by-area basis and then summarized in tabular form by subregions. Selected information for each hydrographic area is listed in table 8.

The variability in hydrographic areas is one of the characteristics of the region. Adequate description of the detailed differences of some 234 areas is not possible in this report. Instead, areas are classified on the basis of selected characteristics, in seven general groups. Areas within each group have enough similarity that knowledge gained in the more intensively developed areas in each group may have significant transfer value in planning development of remaining areas.

The classification used in this report is an extension of the scheme used by Snyder (1962) to propose a hydrologic classification of valleys in the Great Basin. Valleys are classified primarily on two parameters—the topographic nature of area boundaries and the degree of ground-water drainage.

Topographically, areas are classified as either open or closed. A topographically open area is one where there is surface inflow, or surface outflow, or both. In contrast, no surface flow crosses the boundary of a topographically closed basin.

The degree of ground-water drainage is described in terms of the ground-water regimen of the area. The following three categories are used:

1. Undrained: There is no discernible subsurface leakage to adjacent areas. All ground-water discharge is by evapotranspiration from areas of shallow ground water, areas of spring-supported vegetation, or bare ground.
2. Partly drained: There is discernible subsurface outflow from the area; however, magnitude of the outflow is insufficient to completely drain the area. Consequently, significant ground-water evapotranspiration occurs in areas of shallow ground water. This is a transitional category and includes valleys which vary from those where ground-water discharge is primarily by evapotranspiration to those where ground-water discharge is primarily by subsurface outflow.
3. Drained: Virtually all ground-water discharge is by subsurface outflow. This outflow has lowered the water table so that there are no appreciable areas of shallow ground water and, consequently, no significant ground-water evapotranspiration.

The category for partly drained topographically open areas is divided into three subcategories based on relation between the surface-water and ground-water regimens in an area.

Table 2 shows the classification scheme used in this report. The classification of hydrographic areas in the Great Basin Region is shown on plate 1*B*.

THE CLIMATE

The climate of the Great Basin Region is highly variable, owing to large variations in altitude, the wide range in latitude, and the presence of numerous mountain ranges; generally, it is arid to semiarid. Average annual precipitation ranges from less than 5 inches to about 16 inches (127 to 406 mm) on the valley floors and from about 16 inches (406 mm) to more than 60 inches (1,524 mm) in the mountains (pl. 1C). Estimated annual precipitation over the entire region averages about 11 inches (279) mm), or about 88 million acre-feet (108,500 hm^3; table 3). Most of the precipitation occurs during the winter and provides the mountain snowpack which melts and runs off in the spring and summer as a major component of the region's total water supply.

Average annual temperature ranges from about minus 1°C (30°F) in some high northern valleys to about 16°C (60°F) in the extreme southern valleys. One of the more characteristic features is the wide range between daily maximum and minimum temperatures. The daily range exceeds 17°C (30°F) in most valleys and 28°C (50°F) in some valleys of western Nevada. Large variations in temperature also occur within short distances, owing to the wide variations in altitude. The average length of the growing season in the principal agricultural areas of the region ranges from about 100 to 175 days; it exceeds 200 days in the extreme south but is less than 30 days in some higher, northern mountain valleys.

Average annual humidity ranges from about 30 to 40 percent over most of the region and is about 20 percent in the extreme south. The low humidity, coupled with abundant sunshine and light to moderate winds produce very rapid evaporation. Average annual lake evaporation ranges from about 45 in. (1,143 mm) in the north to more than 90 in (2,286 mm) in the extreme south (Kohler and others, 1959).

POPULATION AND INDUSTRY

The Great Basin Region is sparsely populated. Population densities range from about 0.4 person per square mile in the Tonopah subregion to about 30 persons per square mile in the Great Salt Lake subregion. The greatest concentrations of population are in the Provo–Salt Lake City–Ogden area of Utah and the Reno-Carson City area of Nevada (fig. 1). In general, population is concentrated in the river segment or perennial stream valleys shown on plate 1*B*. Total population for the region in 1970 was about 1.2 million people; the distribution of the population by hydrologic subregion, together with projected increases to the year 2020, are listed in figure 1.

TABLE 2.—*Classification of areas in the Great Basin*

Ground-water regimen	Topographic nature of boundary: Closed	Topographic nature of boundary: Open
Undrained	*Closed basins.*—Areas with no surface flow across boundary; all ground-water discharge ultimately by evapotranspiration. Typified by valleys with wet playas and phreatophyte stands in shallow ground-water areas. Ground-water gradient toward center, or low part, of valley.	*Sinks.*—Surface and (or) subsurface inflow across boundaries. All ground-water discharge ultimately by evapotranspiration. Most have terminal lakes or wet playas which are large in proportion to the size of the area. Valley may contain a large volume of saline ground water. (Undrained valley with only surface outflow not recognized at present in the Great Basin. Lake Tahoe and Bear Lake Valleys nearly fit this category.)
Partly drained	*Topographically closed areas.*—No surface inflow or outflow, but discernible subsurface outflow. Valley has moist playa and stand of phreatophytes. Area of ground-water discharge may be small in comparison to undrained basins of similar size. Ground-water gradients may indicate subsurface outflow, if wells are strategically located.	*Tributary areas.*—(divided into three subcategories to allow for various relations between the ground-water and surface-water regimen): *Arid area.*—Surface outflow and (or) inflow typically in ephemeral channels. Some subsurface outflow, but significant areas of ground-water evapotranspiration. Hydrologic regimen dominated by the ground-water flow system. *Perennial stream areas.*—Surface outflow and (or) inflow. Perennial streams prominent. Streams generally do not flow through area. Section or boundary of area may be traversed surface-water system dominates the hydrologic regimen of the area. Both are important. *River segment areas.*—Surface inflow and (or) outflow by way of perennial streams or rivers. Most subsurface inflow and (or) outflow typically underflow in stream-channel deposits. Areas of ground-water evapotranspiration generally along flood plain. Hydrologic regimen dominated by surface-water system. Most areas in this category are either headwater areas or river-reach segments of principal drainages.
Drained	*Drained area (closed).*[1]—No surface flow across boundary; virtually all ground-water discharge is by subsurface outflow. No significant areas of ground-water evapotranspiration. Shallowest depth to water is more than 50 feet. Playas dry and may exhibit large desiccation cracks.	*Drained area (open).*[1]—Surface inflow and (or) outflow, virtually all ground-water discharge is by subsurface outflow. No significant areas of ground-water evapotranspiration. Shallowest depth to water is more than 50 feet. Playas (if present) dry and may exhibit large desiccation cracks.

[1]For purposes of this report, topographically closed and open drained valleys are grouped together because most interarea surface flow is by way of ephemeral streams. Significant flow occurs infrequently.

TABLE 3.—*Estimated annual precipitation by subregion, Great Basin Region*

Subregion	Precipitation (millions of acre-ft)	Average precipitation (in.)
Bear River	7.7	19.2
Great Salt Lake	19.3	12.4
Sevier Lake	12.5	14.6
Humboldt	16.0	10.1
Central Lahontan	8.0	9.3
Tonopah	24.7	9.6
Region (rounded)	88	11

The economy of the Great Basin Region is based chiefly on mining (mostly metallic ores), farming, and livestock. Recently, light manufacturing and warehousing have also contributed to the economy of parts of the region. Most of the towns and cities grew from early farming, mining, and ore-processing communities, and from early railroad settlements. Since the early 1940's defense facilities, industries, tourism, and recreation have been major contributors to the economy of the region.

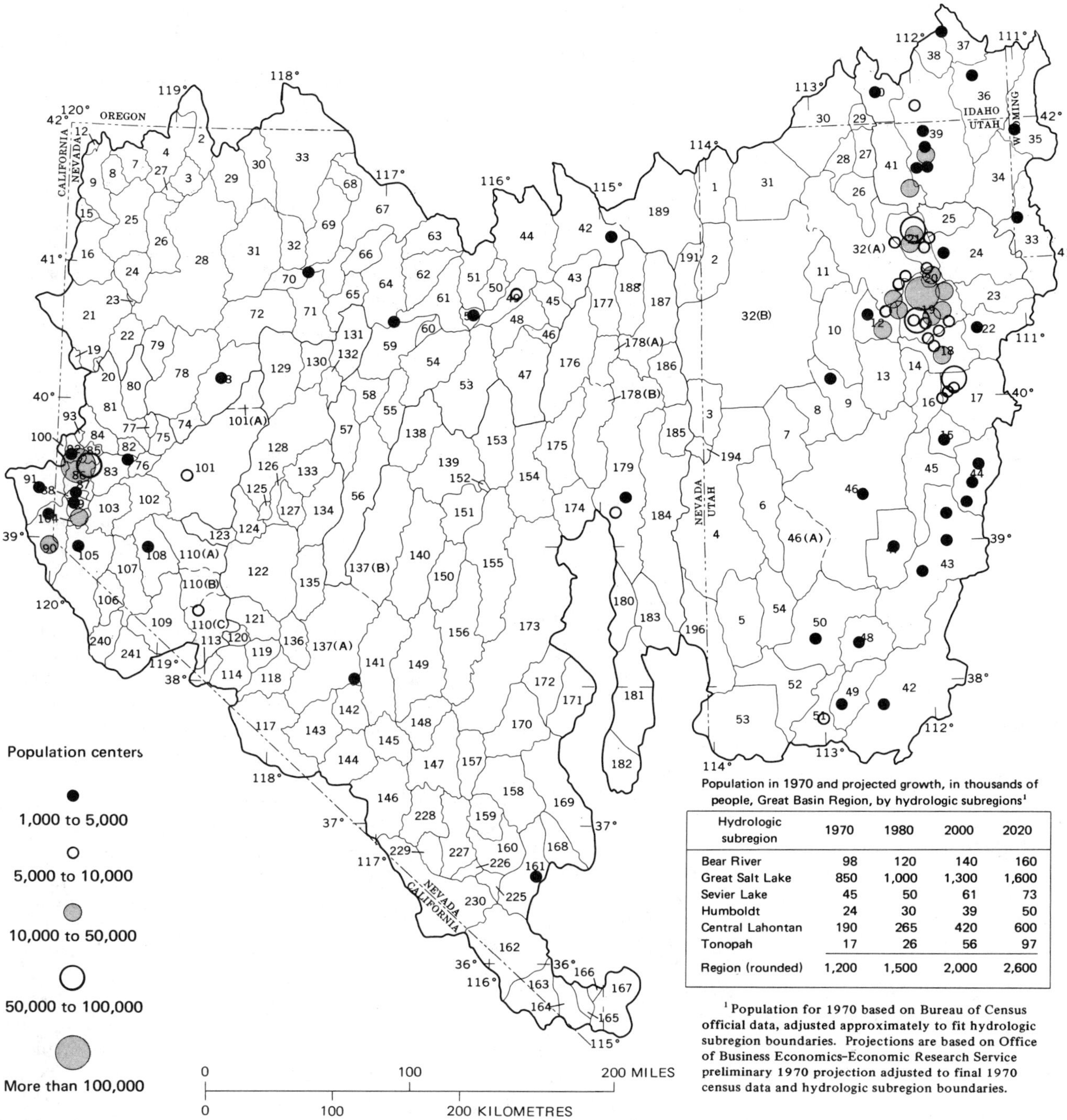

Population in 1970 and projected growth, in thousands of people, Great Basin Region, by hydrologic subregions[1]

Hydrologic subregion	1970	1980	2000	2020
Bear River	98	120	140	160
Great Salt Lake	850	1,000	1,300	1,600
Sevier Lake	45	50	61	73
Humboldt	24	30	39	50
Central Lahontan	190	265	420	600
Tonopah	17	26	56	97
Region (rounded)	1,200	1,500	2,000	2,600

[1] Population for 1970 based on Bureau of Census official data, adjusted approximately to fit hydrologic subregion boundaries. Projections are based on Office of Business Economics-Economic Research Service preliminary 1970 projection adjusted to final 1970 census data and hydrologic subregion boundaries.

FIGURE 1.—Distribution of population in the Great Basin Region, 1970.

HISTORY OF WATER-RESOURCE DEVELOPMENT

Most of the developed water supply in the Great Basin is used for irrigation, which began during the summer of 1847, when Mormon pioneers arrived in Jordan Valley, Utah. Water was diverted by structures consisting of simple rock and brush dams across small mountain streams from which ditches conveyed it to nearby fields. Subsequent immigrants to the region generally settled along mountain streams or close to springs, where water could be readily obtained.

As the demand for water increased, the water systems were improved. Small storage reservoirs were built, and supplies from larger streams were developed. With further demand, private developments were supplemented by construction of Federal projects after the passage of the Reclamation Act in 1902. These projects were constructed principally on the major stream systems and generally included large storage reservoirs to regulate flow to downstream areas of use.

The first substantial development of water supplies for nonirrigation purposes dates back to the early mining boom in Nevada, which began about 1849. Providing water for the mining industry presented a somewhat different problem than for irrigation. Generally, land suitable for irrigation was developed adjacent to, or relatively near, a source of supply, whereas mines commonly were distant from the nearest source of water. Mining towns and the requisite supporting supplies and services had to be near the mines. The situation commonly led to development of the nearest supply of water and transporting it to the mines and towns. On occasion, the water was transported many miles by tank wagons or pipelines. The development of some of these supplies involved complex and very imaginative engineering designs and were very costly (Shamberger, 1971, 1972).

Water supplies for public, municipal, and industrial use evolved partly by incorporation with large-scale irrigation systems, partly as an extension of the water supply for the mining industry, and partly from a need by the railroads and other industrial activities.

Initially, most ground-water supplies were obtained from dug wells by use of hand-drawn buckets. Windmills and small mechanical pumps came into use near the end of the 1800's. As pump designs improved, fuel and electrical power became cheaper and more plentiful, and more effective well-drilling and developing methods were used, large-yield wells capable of operating economically with high pumping lifts became a reality. These advances made ground water more attractive to develop for irrigation, public supply, and industrial needs, especially in areas where surface-water supplies were not readily available or where virtually all surface water was appropriated. Consequently, use of ground water has increased steadily, and as of 1970 more than 1 million acre-feet (1,233 hm^3) per year of ground water was withdrawn from wells in the Great Basin Region.

WATER SUPPLY

The water supply of the Great Basin Region is derived from precipitation, which falls within the region boundaries, and from some imported water. Much of the average precipitation of about 88 million acre-feet (108,500 hm^3) per year (table 3) is evaporated from the soil near the place where it falls. The remainder supplies streamflow and recharges ground water. The average annual replenishment to surface water and ground water makes up the renewable water supply of the region. It does not include stored ground water that is available for use on a one-time basis.

A rough estimate of the renewable water supply of the region is made as follows:

1. The contribution from the principal drainages is estimated by selecting a gaging station near the point of maximum annual flow on each major stream. Flow past the gage plus evapotranspiration from both streamflow and ground water above the gage compose the renewable supply generated above each station. Figure 2 shows locations of the gaging stations and upstream areas of streamflow depletion (evapotranspiration) above the stations.
2. The average flow of secondary and minor streams is estimated from available streamflow records and by indirect methods.
3. Natural ground-water discharge from areas not recharged by perennial streams was estimated. This required a subjective judgment as to which areas were supported by recharge from rivers and streams. Insufficient data or errors in judgment may result in some water included in streamflow estimates being counted again as natural ground-water discharge.
4. The sum of the first three items is a rough estimate of the renewable water supply of the Great Basin Region. Errors caused by counting some water twice probably are small in terms of the total.

Estimates of the above components are listed in table 4. The total of about 9.7 million acre-feet (11,960 hm^3) includes about 0.1 million acre-feet (123 hm^3) of water imported from the Colorado River basin. Undepleted water from the principal streams, from some secondary streams, and from some ground water flows to terminal lakes.

Of the several components of renewable water supply, ground water is the most widely distributed in the region. It is the only component stored in natural reservoirs and readily available during dry periods. Accordingly, ground water offers significant potential to meet part of the water needs of the region.

The degree to which the supply can be ultimately developed depends on what the user is willing and able to pay for water and to what degree the use will adversely affect the environment. Maximum utilization of the water supply in the region would require management of ground-water and surface-water resources in such a way that use of one would complement the other. Also maximum utilization would include use of return flows where possible.

GROUND-WATER RESOURCES

In the arid Great Basin Region, ground water is a vital constituent of the total water resource. The average ground-water recharge is only about 5 percent of the total precipitation and is only about half as large as the average annual runoff. However, both precipitation and stream-

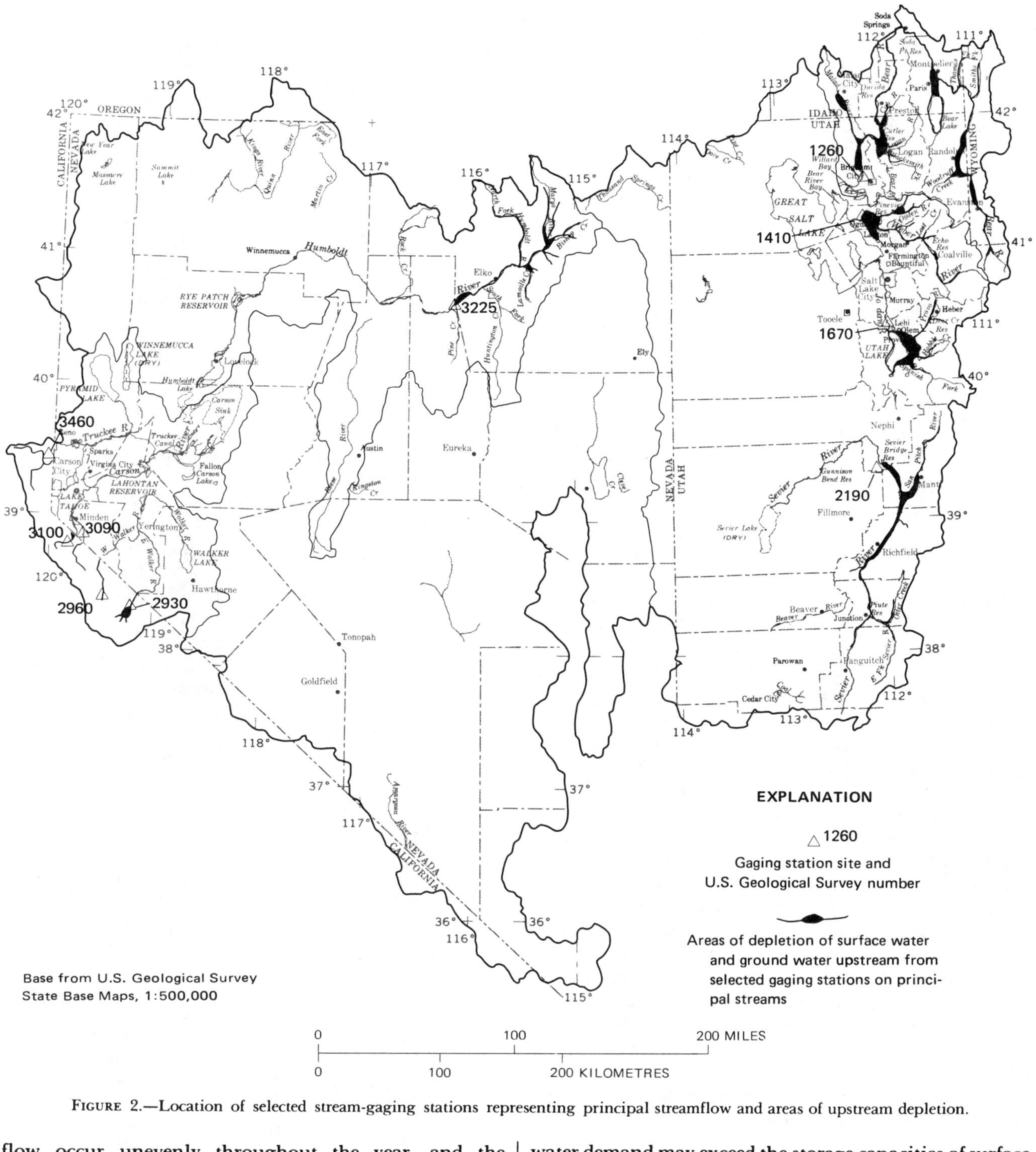

FIGURE 2.—Location of selected stream-gaging stations representing principal streamflow and areas of upstream depletion.

flow occur unevenly throughout the year, and the amounts vary from year to year. Most of the precipitation falls during the spring and winter months, and most of the streamflow occurs during the spring runoff. Consequently, storage and conveyance facilities must be provided to store water and distribute it during periods of heavy summer demand. During prolonged dry periods, water demand may exceed the storage capacities of surface reservoirs and water shortages may result.

In contrast, ground-water reservoirs typically have large volumes and store sufficient water to supply heavy demands during prolonged dry periods. Most valleys in the Great Basin Region have ground-water reservoirs, but comparatively few have abundant surface-water

TABLE 4.—*Summary of renewable water supply in the Great Basin Region*[1]

[Thousands of acre-feet per year]

Subregion	Principal streams		Secondary and minor streams	Natural ground-water discharge[4]	Combined supply
	Gaged streamflow[2]	Upstream depletion[3]			
Bear River	838	1,150	102	40	2,130
Great Salt Lake	517	1,009	388	630	2,544
Sevier Lake	139	539	234	328	1,240
Humboldt	254	300	300	480	1,334
Central Lahontan	1,120	[5]176	108	104	1,508
Tonopah	.....	.....	290	680	970
Region (rounded)	2,900	3,200	1,400	2,300	9,700

[1]Modified from Pacific Southwest Inter-Agency Committee (1971b, table 23).
[2]Annual average for 1931-60 reference period modified for the 1965 level of development. Other values are considered to represent long-term annual averages.
[3]Supplied from surface water and ground water undifferentiated.
[4]Generally excluding ground water supplied from principal streams.
[5]Included in this report due to adjustment in Regional boundary from that used in the Great Basin Region Comprehensive Framework Study.

supplies. In these arid areas, ground water is the only source of water that might be developed on a large scale.

CONSOLIDATED-ROCK RESERVOIRS

A large amount of ground water is stored in consolidated rocks that occur both in the mountains and beneath the valley alluvium. These ground-water reservoirs generally are not continuous and are extremely difficult to evaluate. Local barriers formed by faults or rock units of low permeability produce a complex pattern of perched ground-water bodies in the mountains. These perched ground-water bodies supply numerous springs and late-season flow of mountain streams.

Carbonate rocks and volcanic rocks compose the consolidated-rock reservoirs. (See pl. 1*D* for areas underlain by these rocks.) Carbonate rocks in most areas are highly permeable and transmit substantial quantities of water to large springs in eastern Nevada and western Utah. In southern Tonopah subregion, carbonate rocks transmit water in several multi-valley ground-water systems.

In some areas, volcanic rocks have fracture and interflow openings that store and transmit large quantities of ground water, which sustain the flow of many streams and of numerous springs near bedrock-alluvial contacts. Locally, volcanic rocks transmit water readily, such as the Quaternary basalt in Soda Creek and Gem Valleys in the Bear River subregion. Other areas with known volcanic-rock aquifers include Pavant Valley in the Sevier Lake subregion, Curlew Valley in the Great Salt Lake subregion, Winnemucca area in the Humboldt subregion, and Fallon area in the Central Lahontan subregion.

VALLEY GROUND-WATER RESERVOIRS

Valley ground-water reservoirs are composed of alluvial deposits which partly fill the structural depressions that form intermontane basins. These deposits generally contain sand and gravel aquifers which, in most places, provide the only supply of ground water available for large-scale development. Distribution of the principal valley ground-water reservoirs is shown on plate 1*D*.

Ground water occurs in the porous valley alluvium in a zone of continuous saturation. The water table is near the land surface in topographically low parts of most valleys. Depth to water increases toward the mountains and may be several hundred feet beneath the upper parts of some alluvial fans. The position of the water table is regulated by surface-water altitudes in valleys that contain lakes or large streams that flow through the valleys. Artesian conditions occur where layers of silt and clay are abundant enough to confine or partly confine water in underlying deposits. This condition is common in the lower parts of most valleys. The most extensive artesian conditions occur in the eastern part of the region. This results in part from the extensive silt and clay deposits that formed during high stages of Lake Bonneville.

Detailed information on thickness of valley ground-water reservoirs generally is not available. Maximum thickness probably varies greatly from area to area. Available information indicates maximum thicknesses of more than 1,000 feet (300 m) are common in the larger valleys, but the information available is insufficient to make any generalized statement about the smaller areas.

STORED GROUND WATER

A large volume of water is in transient storage in valley ground water reservoirs throughout the Great Basin Region (pl. 1*D*). The recoverable quantity stored to a selected depth below the water table may be estimated as the product of an area, a selected depth, and the specific yield of the deposits. The depth selected for this study is the uppermost 100 feet (30 m) of saturation. Volumes of water are estimated without regard to water quality.

For purposes of this report, average specific yield of the upper 100 feet (30 m) of saturated deposits was assumed to be 10 percent, except in areas of extensive fine-grained deposits, such as the Salt Lake Desert, Black Rock Desert, and Carson Sink. Specific yield is assumed to average 5 percent in parts of these areas. Specific yield along some river flood-plain areas, such as in parts of the Bear River basin, is assumed to average 15 percent. These averages probably give reasonably representative estimates for subregion areas. Specific yield locally ranges from less than 5 percent to more than 30 percent. Higher values are associated with the well-sorted sand and gravel in stream channels or with the beach deposits of ancient lakes. Lower values are associated with the silt and clay in lake and playa deposits. Thus, different values are used for specific areas where they are known as a result of detailed local studies.

In areas where artesian conditions are extensive, significant pumping may occur before heads in the artesian aquifers are drawn down sufficiently to induce any lowering of the water table. In time, however, if artesian heads are permanently lowered, deposits weil drain in response to pumping and a quantity of water equal to the specific yield will be recovered from storage.

Plate 1*G* shows the distribution of ground-water storage in the Great Basin Region. The tabulation on the figure summarizes estimates of recoverable ground-water storage in the upper 100 feet (30 m) of saturated deposits by subregions. Estimates for individual hydrographic areas are listed in table 8 at the end of the report. For planning and development purposes, these quantities can be expressed as an annual rate for a specified number of years. For example, the regional total is about 300 million acre-feet (370,000 hm^3). This quantity is equivalent to a withdrawal rate of 6 million acre-feet (7,400 hm^3) a year for a 50-year period, excluding that part of the annual supply made available by salvage of natural discharge.

The estimate of recoverable stored ground water is only a fraction of the total amount of stored ground water in the region. Saturated alluvium occurs to depths of several thousand feet beneath some valleys. Ground water also occurs in consolidated rocks in the mountains and beneath the valleys. Undoubtedly, the total stored ground water amounts to several billion acre-feet, but data are too few for a firm estimate to be made. As economic need arises, studies probably will be made to determine the amount of ground water stored to greater depths in the valley ground-water reservoirs and elsewhere.

GROUND-WATER FLOW SYSTEMS

Ground water percolates along the path of least resistance from areas of high head to areas of lower head. In arid basins water recharges the ground-water system in or near the mountains and moves downgradient to areas of spring flow and ground-water evapotranspiration in the low part of the valley. The various types of areas shown on plate 1*B* and variations within types precludes choice of any one ground-water flow system to describe a "typical valley" in the Great Basin Region. Instead, examples are presented for selected types of areas described in table 2.

Ground-water flow systems are evaluated on two levels. The first level considers the factors that control flow between basins and regional patterns of ground-water flow. The second level considers those factors that regulate subsurface flow within a single basin. This involves geologic and topographic factors and, where perennial streams are present, relations between the ground water and surface water.

Two conditions must be met before significant interbasin flow can occur. Consolidated rocks separating areas must be permeable enough to transmit appreciable amounts of water and a hydraulic gradient must exist between two areas. Hydraulic continuity and a gradient may extend across more than two valleys and result in a regional flow system where all or part of the ground-water recharge from several valleys drains to a common sink and ultimately is consumed by evapotranspiration. Figure 3 illustrates some ground-water flow systems present in the great Basin Region.

Factors that affect the flow system in any one valley are the relative locations of recharge and discharge areas and geologic factors that influence the capability of valley-fill deposits to transmit water. Recharge areas are typically in the mountains and on the upper parts of alluvial fans, whereas discharge areas commonly are in the low parts of the areas near the central parts of the valleys. The manner in which ground water moves between these areas in regulated by the geologic properties of the valley-fill deposits. One condition that strongly affects the flow regimen is the interlayering of deposits having different permeabilities. For example, sand and gravel aquifers of high permeability may be underlain and overlain by silt, clay, or mudflow deposits of low permeability. On a smaller scale, lake-bed deposits may contain thin layers of interbedded silt and clay. The result is that in most places it is much easier for water to move horizontally than vertically. Another factor that strongly affects the ground-water flow regimen is the variation of permeability within a valley. Generally, composition of the valley fill grades from highly permeable coarse-grained deposits along the margins of a valley to poorly permeable fine-grained deposits near the center of a valley. Exceptions to this generalized pattern may be mudflow deposits of low permeability near the mountains, highly permeable stream-channel deposits in the central part of a valley, or extensive lake-bed deposits of low permeability.

The ground-water system is also affected by perennial streams. A stream may lose or gain water by seepage or have no loss or gain, depending on the relative water-surface altitude in the stream, the altitude of the water table, and the permeability of the bed and banks. Altitudes of both the water table and the water surface of the stream vary with time, so during the course of a year all three conditions may occur in parts of some stream channels. The segments of the principal streams upstream from the gage where depletion was estimated in

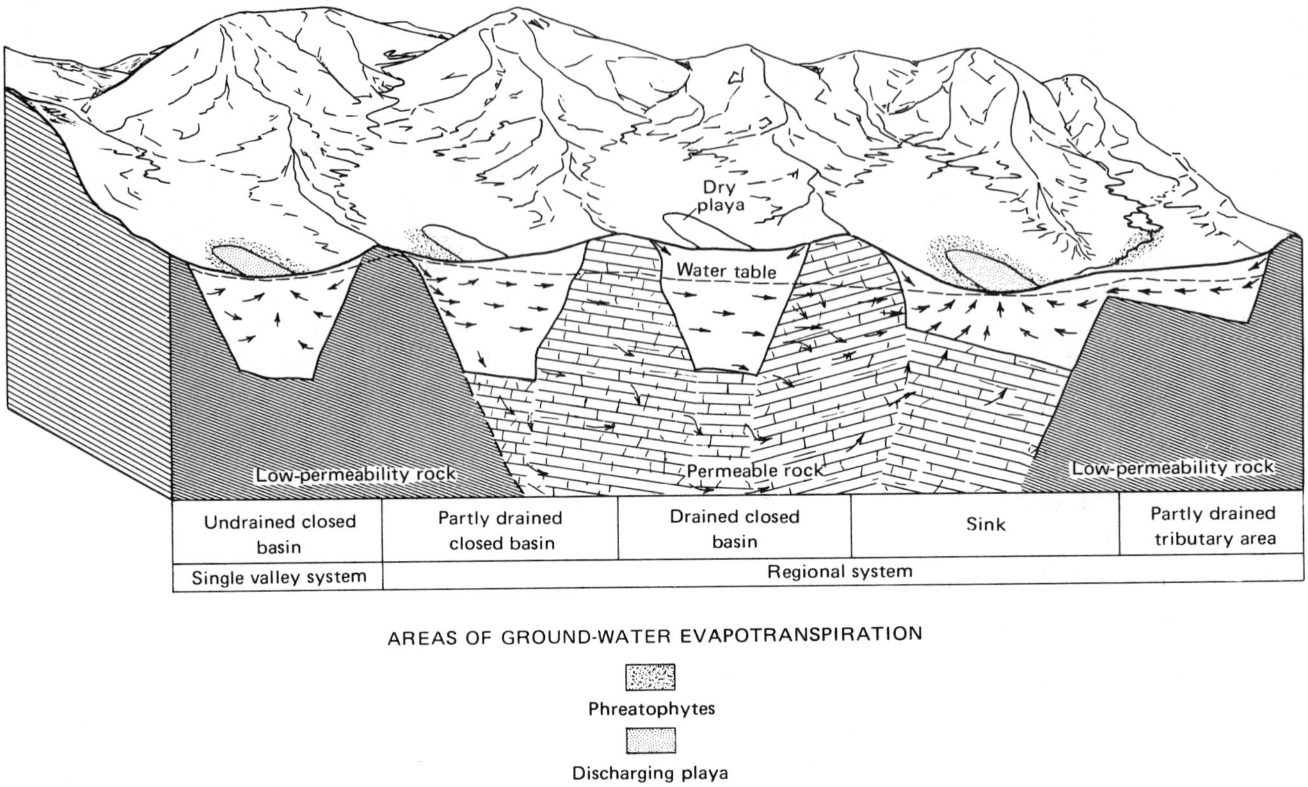

FIGURE 3.—Common ground-water flow systems.

figure 2 are generally gaining reaches. The lower reaches of these streams are generally losing streams; however, some segments may gain due to natural inflow or irrigation return flows.

Figure 4 shows diagrammatically some of the common relationships between ground water and surface water in the Great Basin.

The principal lakes receive inflow from ground water as well as precipitation and surface flow. The water levels of these lakes in turn control ground-water levels in adjacent saturated deposits. Lakes that discharge into rivers, such as Utah Lake and Bear Lake, maintain fairly uniform ground-water levels in adjacent areas. Variations in water supply caused by climatic or other factors commonly result in variations in outflow from the lake while the lake stage and adjacent ground-water levels remain fairly constant. Lakes that have no outlets, such as Great Salt Lake, Pyramid Lake, and Walker Lake, discharge water entirely by evaporation. Water levels in those lakes fluctuate in response to variations of evaporation and inflow. Sustained periods of water-level decline in those lakes result in consequent depletions of adjacent ground-water storage by either evaporation or drainage into the declining lake. For example, water levels in Walker Lake declined 105 feet (32 m) between 1908 and 1965, resulting in a depletion of ground-water storage in the adjacent valley deposits of about 150,000 acre-feet (185 hm^3) (Everett and Rush, 1967, p. 24).

INFLOW TO THE GROUND-WATER SYSTEM

Precipitation ultimately is the source of virtually all ground-water recharge in the Great Basin. Most natural recharge is generated from precipitation that falls in the mountains; however, recharge may occur either in the mountains or in the valleys, where streams lose water by infiltration into permeable valley-fill deposits.

Available recharge estimates for individual valleys are listed in table 8. Estimates are not available for all areas in the Great Basin Region. However, available information suggests that ground-water recharge generally is between about 3 and 7 percent of the total precipitation on an individual drainage basin. Using these percentages, average annual recharge to the entire region is estimated to be within the range of about 3 to 6 million acre-feet (3,700 to 7,400 hm^3) per year.

OUTFLOW FROM GROUND-WATER SYSTEM

Outflow from ground-water systems, or discharge, occurs in four principal ways: (1) By evapotranspiration in areas of phreatophytes, (2) by direct evaporation from bare soil where the capillary fringe is near the land surface, (3) by discharge from springs or directly to streams, and (4) by subsurface outflow to adjacent areas.

The principal areas of natural ground-water discharge in the Great Basin Region are shown on plate 1*E*. Table 5 summarizes estimates of natural ground-water dis-

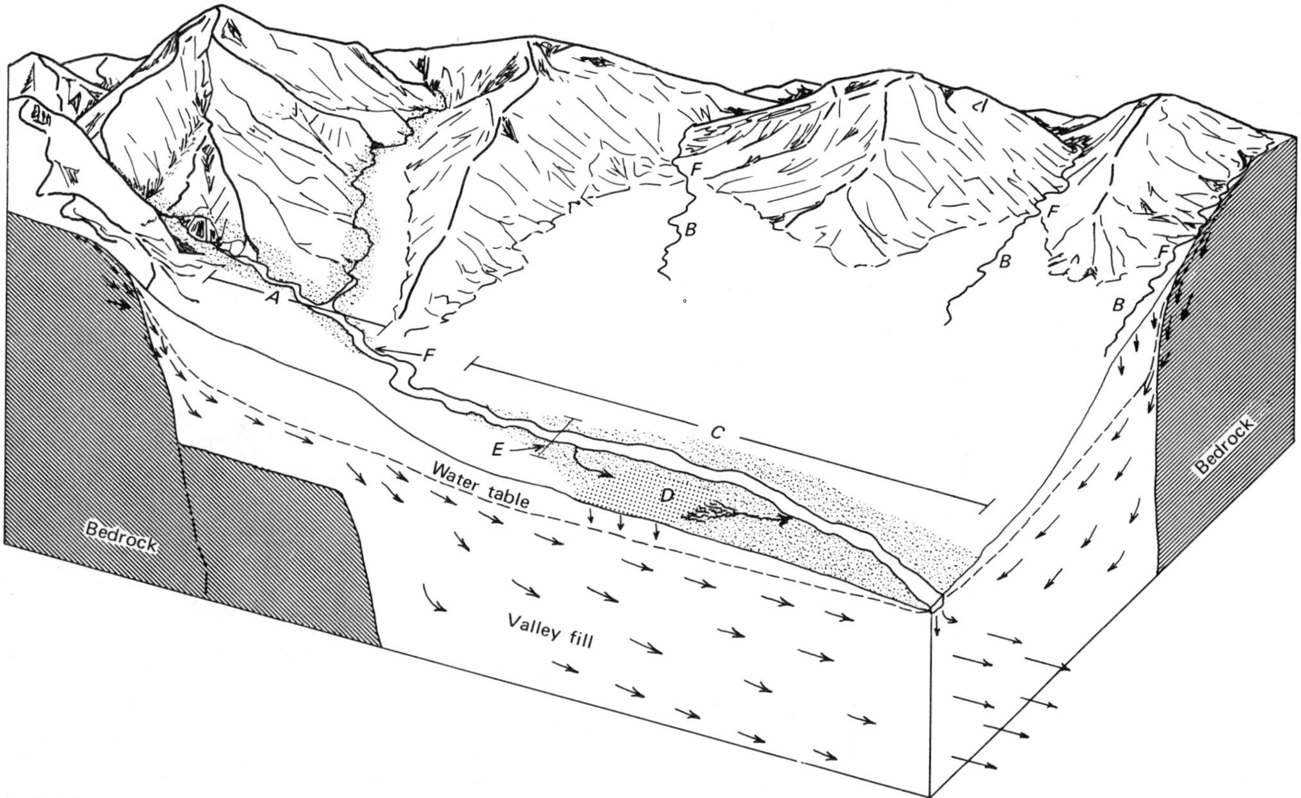

A, Gaining reach, net gain from ground-water inflow although in localized areas stream may recharge wet meadows along flood plain. Hydraulic continuity is maintained between stream and ground-water reservoir. Pumping can affect streamflow by inducing stream recharge or by diverting ground-water inflow which would have contributed to streamflow.

B, Minor tributary streams, may be perennial in the mountains but become losing ephemeral streams on the alluvial fans. Pumping will not affect the flow of these streams because hydraulic continuity is not maintained between streams and the principal ground-water reservoir. These streams are the only ones present in arid basins.

C, Losing reach, net loss in flow due to surface water diversions and seepage to ground water. Local sections may lose or gain depending on hydraulic gradient between stream and ground-water reservoir. Gradient may reverse during certain times of the year. Hydraulic continuity is maintained between stream and ground-water reservoir. Pumping can affect streamflow by inducing recharge or by diverting irrigation return flows.

D, Irrigated area, some return flow from irrigation water recharges ground water.

E, Flood plain, hydrologic regimen of this area dominated by the river. Water table fluctuates in response to changes in river stage and diversions. Area commonly covered by phreatophytes (shown by random dot pattern).

F, Approximate point of maximum stream flow.

FIGURE 4.—Common relationships between ground water and surface water in the Great Basin Region. Not to scale.

charge for valleys exclusive of the areas of upstream depletion (fig. 2). It includes published estimates for many valleys, and provisional estimates for areas where no previous estimates were available. Estimates for specific areas are listed in table 8. It was not possible to make estimates of ground-water discharge separate from surface-water discharge for that part of the principal stream systems identified as upstream depletion areas. Thus, total ground-water discharge is the 2.3 million acre-feet (2,840 hm^3) per year of ground-water discharge estimated in table 5 plus that part of the 3.2 million acre-feet (3,950 hm^3) per year of upstream depletion (table 4) that is supplied from ground water.

GROUND-WATER FLOW BETWEEN REGIONS

The quantity of interregion flow is small in relation to the total water supply; however, it may be a significant part of the hydrologic budget in some valleys along the regional border. Table 6 summarizes estimates of the principal occurrence of interregion ground-water flow.

TABLE 5.—*Estimated natural ground-water discharge exclusive of areas of upstream depletion*[1]

[Values significant to two figures]

Subregion	Number of valleys	Approximate area of natural ground-water discharge (thousands of acres)	Quantity of natural ground-water discharge (thousands of acre-ft/yr)
Bear River	1	70	40
Great Salt Lake	17	2,400	[2]630
Sevier Lake	11	920	330
Humboldt	32	1,700	480
Central Lahontan	19	320	[2]100
Tonopah	74	2,400	680
Region (rounded)	154	7,800	2,300

[1]Areas of upstream depletion are shown in figure 2. As used here, such areas refer to flood plains of principal streams and their upstream tributaries, other upstream irrigated areas supplied from that stream system, and any water exported from that area. In those areas, a combined estimate of evapotranspiration from both surface-water and ground-water sources was made, because ground-water and surface-water components could not be separated. The combined estimate was about 3.2 million acre-feet per year (pl. 1*G*).

[2]Excludes that supplied from streamflow, such as 170,000 acre-feet supplied by principal and secondary streams in Jordan Valley, Great Salt Lake subregion, and 101,000 acre-feet from principal streams in Central Lahontan subregion.

TABLE 6.—*Summary of estimates of ground-water flow between regions*

[Acre-feet per year]

Region	Inflow to the Great Basin (+)	Outflow from the Great Basin (-)	Net inter-region flow
Columbia-North Pacific	([1])	([1])	([1])
California	14,000	[2]24,000	-10,000
Lower Colorado	0	18,000	-18,000
Upper Colorado	10,000	7,000	+3,000
	24,000	49,000	-25,000

[1]Inflow to Soda Springs Valley, Bear River subregion, and outflow from Gem Valley to the Columbia-North Pacific Region may be on the order of several tens of thousands of acre-feet a year; a net gain to the Great Basin Region is believed to occur.

[2]Includes underflow from Nevada part of Amargosa Desert, which may be on the order of 3,000 acre-feet per year (Walker and Eakin, 1963), and from Nevada part of Pahrump Valley, about 12,000 acre-feet per year (Malmberg, 1967).

CHEMICAL QUALITY OF GROUND WATER

The chemical quality of ground water in the Great Basin Region ranges from fresh (less than 1,000 mg/l, milligrams per litre, dissolved solids) to brine (more than 35,000 mg/l dissolved solids). Waters between these extremes are classed as saline, as follows: slightly saline (1,000 to 3,000 mg/l dissolved solids), moderately saline (3,000 to 10,000 mg/l dissolved solids), and very saline (10,000 to 35,000 mg/l dissolved solids). Generally, in sheds and alluvial aprons at the margins of most valleys, the ground water is fresh. Saline water occurs locally near some thermal springs and in areas where the aquifer includes rocks containing large amounts of soluble salts, such as in parts of the middle Sevier River area. The general distribution of the dissolved-solids concentration of ground water is shown on plate 1*F*. Other areas can be delineated in greater detail as more data become available.

In sink areas, such as the Great Salt Lake, Sevier Lake, and Carson Sink, the dissolved-solids concentrations may exceed that of ocean water. The ground water beneath the playas of smaller closed valleys may be brackish but ordinarily does not reach the concentrations found in the major terminal sinks. Common assumption is that the poor quality water extends to substantial depths below the playas. However, this is not necessarily so. For example, water from a well 1,200 feet (366 m) deep on the northern margin of the playa in Railroad Valley, Tonopah subregion, has a dissolved-solids concentration of less than 350 mg/l. This quality apparently reflects the deep circulation in the valley fill. In valleys with subsurface discharge into a regional ground-water system, the water throughout the valley-fill reservoir generally is fresh.

Individual constituents may present a quality problem even though the dissolved-solids concentration may be low. In much of the Tonopah subregion, the fluoride concentration in ground water exceeds Public Health Service recommended standards for drinking water (U.S. Public Health Service, 1962). The distribution of high-fluoride ground water may be associated with volcanic tuff that is extensive in that subregion.

GEOTHERMAL ENERGY

Geothermal areas are those in which above-normal temperatures occur below land surface and in which there is a reasonable possibility of finding reservoir rocks that will yield steam or heated water to wells. The fluid involved is either fresh or saline ground water. The high-temperature water generally is localized and occupies only a part of the ground-water reservoir in which it occurs. Because the thermal fluid is ground water, the States consider development of geothermal resources to be subject to State ground-water laws.

Many, but not all, geothermal areas are indicated by hot springs. In some areas, the thermal water does not reach the land surface but instead circulates below ground in various kinds of conduit and aquifer systems. Some of the more favorable areas for potential development in the Western United States have been designated as known geothermal resources areas (KGRA's) under the Geothermal Steam Act of 1970 (Godwin and others, 1971, and the Federal Register). Of the 33 areas listed (Godwin and others, 1971, p. 2), 13 are in the Nevada part and 2 in the Utah part of the Great Basin Region (fig. 5).

Federal lands in KGRA's can be leased for geothermal prospecting only by competitive bidding. Interest in several geothermal areas, such as Steamboat Hot Springs (near Reno), Bradys Hot Springs, Beowawe Hot Springs, and Monte Neva areas in the western Great Basin Region has included test drilling within the past 10–15 years. Geothermal steam-generated electricity has not been produced in commercial quantity to date in the region. However, at various times, some of the springs have been used for limited space-heating, swimming pools, agriculture, and industrial processing.

Before substantial commercial development of the

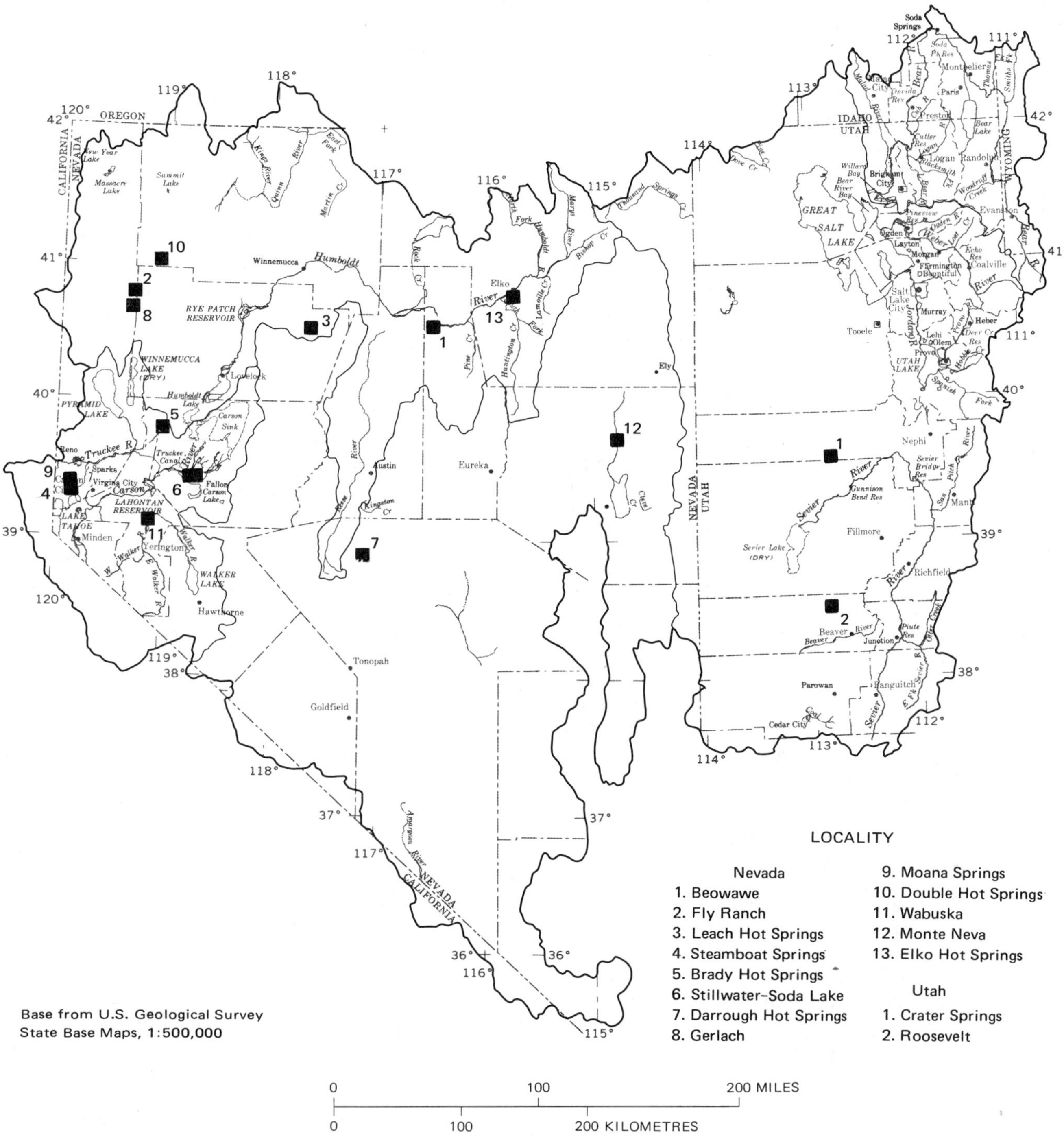

FIGURE 5.—Locations of known geothermal areas in the Great Basin Region.

energy of the geothermal areas is undertaken, a better evaluation of the reservoirs, their capacity to produce energy, either as steam or hot water, and improvement in the technology of using the geothermal resources are probably needed.

YIELD OF THE GROUND-WATER SYSTEM

Knowledge of how much water an area can produce without creating undesirable effects is necessary to plan the orderly development of ground-water supplies. However, determination of this quantity is difficult because of the requirement to define undesirable effects in advance and then determine the degree to which they may be tolerated in the future. Thus, yield of a ground-water system is determined in part by hydrologic factors that control cause-and-effect relationships and in part by nonhydrologic constraints based on the degree to which specified effects are to be tolerated.

Some of the more significant hydrologic factors are (1) the configuration of the ground-water reservoir, (2) the ability of deposits to yield water to wells, (3) the ability of deposits to store water, (4) the distribution of water quality, and (5) the location and quantity of ground-water recharge and discharge. In most alluvial basins these factors vary considerably from place to place. Consequently, the location of pumping has a great effect in determining the results of development in any given case.

The nonhydrologic constraints vary from area to area. Examples of two constraints present in some areas of the Great Basin Region are as follows:

1. Regulation of pumping to protect existing surface-water rights. Heavy pumping near streams which are in hydraulic continuity with the ground-water reservoir will eventually induce additional recharge from streamflow at the expense of water already appropriated for use. If one of the primary objectives in managing the area's water resources is to protect preexisting surface-water rights, and if no satisfactory arrangement for retribution can be worked out between users, then ground-water development near streams may necessarily be curtailed.
2. Regulating development when net pumpage reaches the approximate level of natural inflow to and outflow from the ground-water system. In some arid basins the chance of inducing additional recharge by pumping and the probability of obtaining imported water as a supplemental supply in the future are nil. Limiting net pumping to a rate about equal to the natural ground-water inflow and outflow is a possible alternative in regulating development so that sustained mining will not occur. Even so, if pumping is not strategically distributed with regard to the flow system, large quantities of water will eventually be removed from storage. However, pumping could be relocated within the area without reducing the level of development, and problems that may occur will be postponed for many years and will be of lesser magnitude than if sustained mining were allowed.

Application of the above constraints limits development below the level possible if water were allowed to be depleted from storage on a planned basis or if the surface water and ground water were managed conjunctively as a source of supply. The amount of ground-water storage available for consumption on a one-time basis greatly exceeds the annual rate of replenishment. The degree to which storage should be depleted and the rate at which depletion should be allowed are questions which arise in every area where large-scale ground-water development occurs. Possibilities range between two extremes. One is realization of maximum economic returns over a given period of time. This would require a large-scale planned depletion of storage. The other extreme is conservation of a near maximum amount of storage for future generations. This would restrict development to limited pumping near areas of natural discharge. One development alternative would be a compromise which allowed significant economic returns and still conserved adequate storage for possible future use. Such a development would also have to be feasible within constraints posed by State water laws. The above statements give some idea of the difficulties involved in estimating meaningful values of ground-water yield.

Pumped ground water must be derived from one of three possible sources—depletion of storage, additional recharge induced by pumping, or capture of existing natural discharge.

The distribution of ground-water storage shown on plate 1*G* gives an approximate indication of the magnitude of water in storage potentially available to be consumed on a one-time basis. The amount that can actually be used will be determined by economic and legal factors. It could be more or less than the quantities tabulated on plate 1*G*.

The various types of hydrographic areas in the Great Basin Region preclude the use of one concept of yield to approximate the annual rate of net pumpage that could be sustained for an indefinite period of time. This rate is much less than the volume of storage indicated on plate 1*G*, but it ultimately is the limiting factor to sustained ground-water development. The preceding statement assumes that no additional water is imported at some time in the future and that average climatic conditions over the next several centuries will be about the same as those indicated by available records.

In arid areas, such as the closed basins listed in table 2, no appreciable recharge can be induced by pumping. Consequently, after a period of storage depletion, ground-water development will be limited to the amount of natural discharge that can be captured by wells. This rate has been referred to as the perennial yield. The perennial yield of a ground-water reservoir is defined as the maximum amount of water of usable chemical quality that can be withdrawn each year for an indefinite period of time. It generally cannot exceed the natural discharge from the reservoir and in a practical sense is limited to that part of the natural discharge that economically can be captured for use.

In wetter areas, such as the perennial stream areas and river reach areas listed in table 2, ground water and surface water are interconnected. The one cannot be developed without some effect on the other. The yield of these areas has been considered in terms of a system yield (Worts and Malmberg, 1966, p. 39). The yield of a discrete hydrologic system is the maximum amount of surface and ground water of usable chemical quality that can be obtained and consumed each year from sources within the system for an indefinite period of time. The system yield cannot be more than the natural inflow to or the outflow from the system. Under practical conditions of development, system yield is limited to the maximum amount of surface water, ground water, and water-vapor outflow that can be captured economically each year for beneficial use. In this report, where a river system has been divided into several hydrographic areas, system yield has been apportioned among the various areas based on the distribution of flow as of 1970.

Estimates of perennial yield or system yield for individual areas have been grouped into five categories. Plate 1*H* shows the distribution of these categories in the Great Basin Region. A summary of yield estimates is also shown on plate 1*H*. Provisional estimates were used for 48 areas so the tabulation must be considered only as approximations to provide a relative concept of magnitude. The total yield of about 8.7 million acre-feet (10,700 hm^3) per year is smaller than the combined renewable supply of about 9.7 million acre-feet (11,960 hm^3) per year (table 4). It would not be feasible to capture the entire water supply for beneficial use. Total perennial yield of all arid areas is about 1.8 million acre-feet (2,220 hm^3) per year. This includes about 0.2 million acre-feet (250 hm^3) per year of water that may require treatment to be suitable for most uses. The amount of pumping desirable from system-yield areas depends largely upon the point at which pumping causes adverse effects on existing surface-water rights. The optimum rate is not known. However, if wells are strategically situated and seasonal pumping is properly regulated, probably 0.8 million acre-feet (990 hm^3) per year could be pumped without significant adverse effects. Thus, the ground-water system in the Great Basin Region can sustain a net pumpage of at least 2.6 million acre-feet (3,200 hm^3) per year. Higher sustained withdrawals are possible if conflicts with existing surface-water rights are resolved and if extensive treatment and reuse are feasible. More detailed information on the yield of specific areas is presented in the references listed in table 8.

GROUND-WATER DEVELOPMENT IN 1970

Ground-water development ordinarily refers to the use of wells to obtain water. Wells have been used extensively to produce water for domestic, stock, municipal, industrial, and irrigation purposes. Flowing wells have been obtained in many areas, such as Malad and Cache Valleys in the Bear River subregion, parts of the East Shore area, Utah Valley, and Jordan Valley in the Great Salt Lake subregion, Pavant, Parowan, and Sevier Desert in the Sevier Lake subregion, and elsewhere in the eastern subregions.

TABLE 7.—*Yield characteristics of large-capacity wells in the Great Basin Region*

Subregion	Well yields			
	Number of areas	Number of wells	Range in yield (gal/min)	Average yield (gal/min, rounded)
Bear River	5	169	80–4,650	1,060
Great Salt Lake	19	611	10–8,600	930
Sevier Lake	11	677	20–3,500	840
Humboldt	17	213	90–4,000	1,700
Central Lahontan	9	108	12–4,400	1,500
Tonopah	16	210	76–3,300	1,200
Region	......	1,988	10–8,600	1,050
	Specific capacity			
	Number of areas	Number of wells	Range in specific capacity [(gal/min)/ft]	Average specific capacity [(gal/min)/ft]
Bear River	5	130	1–1,500	116
Great Salt Lake	19	430	0.2–1,370	48
Sevier Lake	12	182	0.2–2,900	70
Humboldt	15	160	3–200	28
Central Lahontan	8	100	0.1–160	33
Tonopah	8	112	2–260	32
Region	......	1,114	0.1–2,900	43

Flowing wells occur in a few areas in the western subregions, such as Smith, Mason, and Carson Valleys, and Truckee Meadows in Central Lahontan subregion, parts of the flood plain of the Humboldt River in Humboldt subregion, and the lower parts of Diamond, Railroad, and Spring Valleys in Tonopah subregion.

Large-capacity pumped wells have accounted for most of the annual withdrawals of ground water. Individual yields of these wells are as much as 8,600 gal/min (540 l/s), and the specific capacities may be as much as 3,000 (gal/min)/ft [620(l/s)/m] of drawdown. The average pumping rate, however, is about 1,000 gal/min (63 l/s), according to an analysis of nearly 2,000 large-capacity wells in the region. Table 7 shows the yield characteristics of large-capacity wells by subregion.

In 1970 water was withdrawn by wells in about 230 areas. Withdrawals in more than 60 of these areas ranged from 1,000 to 10,000 acre-feet (1.2 to 12 hm^3). In 23 areas withdrawals exceeded 10,000 acre-feet (12 hm^3). The greatest single area of ground-water withdrawal was Jordan Valley, with 115,000 acre-feet (142 hm^3). Total withdrawal in the region in 1970 is estimated to be about 1.1 million acre-feet (1,360 hm^3). A summary of withdrawals and the distribution of the principal areas of withdrawal are shown in figure 6.

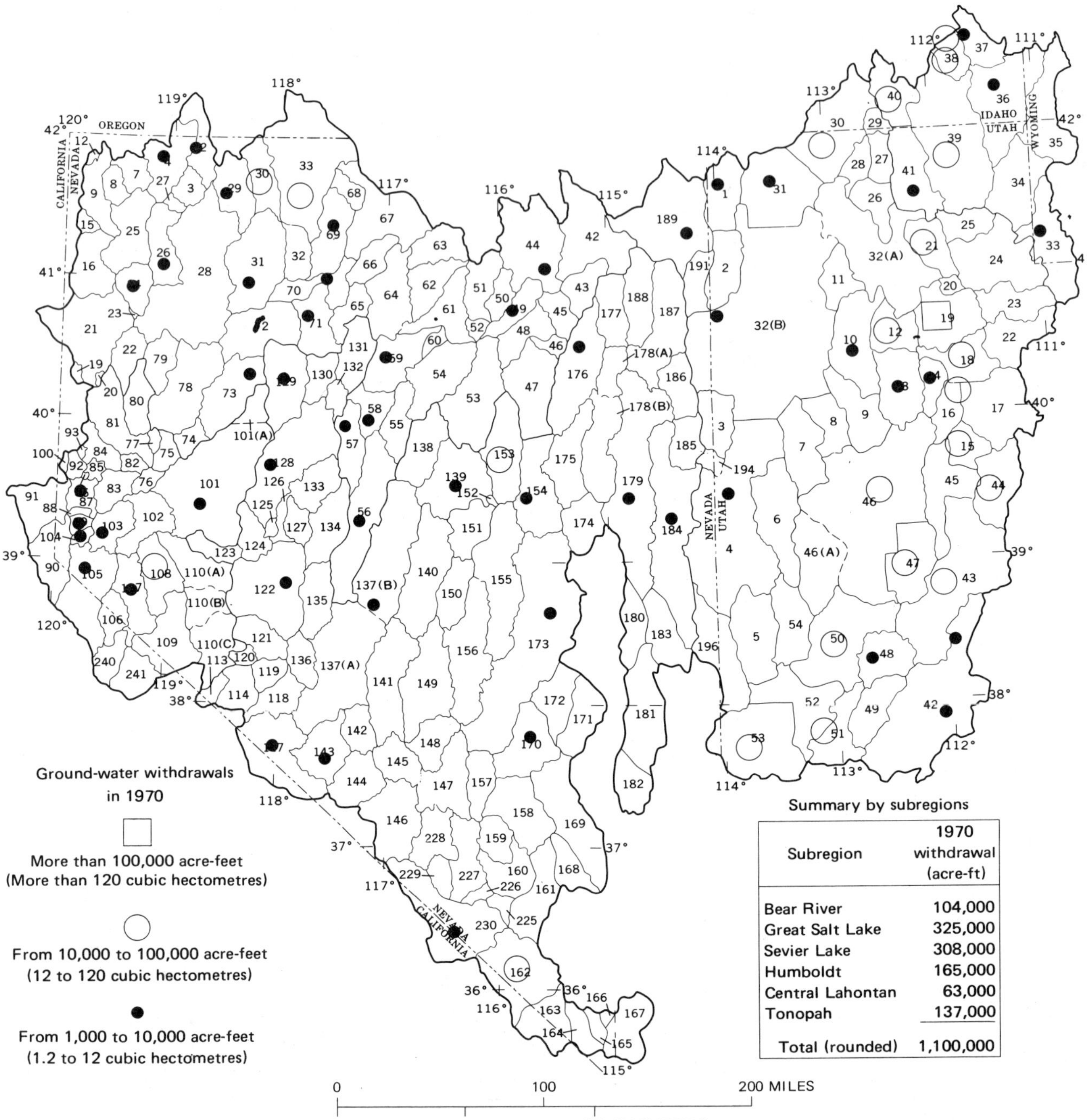

Summary by subregions

Subregion	1970 withdrawal (acre-ft)
Bear River	104,000
Great Salt Lake	325,000
Sevier Lake	308,000
Humboldt	165,000
Central Lahontan	63,000
Tonopah	137,000
Total (rounded)	1,100,000

FIGURE 6.—Distribution of ground-water pumpage in 1970.

LEGAL FACTORS AND LARGE-SCALE GROUND-WATER DEVELOPMENT

Water rights are granted by the several States in the Great Basin Region under the appropriation doctrine (Pacific Southwest Inter-Agency Committee, 1971a, p. 194–200). Initially, the doctrine applied only to surface water, but later it was extended to include ground water.

Large-scale surface-water development has been successfully accomplished within the framework of the appropriation doctrine. Planned large-scale ground-water development also should be possible. The development from either source, however, will require solutions to various legal problems.

Some conflicts have arisen in water-resources law that may have a bearing on any particular ground-water development; for example, the question of the relationship between the States' water laws and the Federal "reserved water doctrine." The States' water laws imply that all water in the State is subject to appropriation. The Federal "reserved water doctrine" is based on the principle that when land is reserved by the Federal Government, the water necessary to use the land for the intended purpose is also reserved.

The Federal Government controls more than three-quarters of the land of the Great Basin Region. Possibly 80 percent of the usable runoff of the region is generated on the mountainous parts of the Federal land, and a similar percentage of the ground-water recharge may be derived from those lands. It is evident, then, that any planned large-scale ground-water development for a particular area could be affected substantially if a legal decision involving both the State water laws and the Federal "reserved water doctrine" were required.

Areas may be limited in some degree to further ground-water development under existing State water laws. The limitations ordinarily apply to areas where the magnitude of existing development is significant, although they also apply to areas preserved for specific uses in the future. Areas having some limitation on additional development are shown in figure 7. Any proposal for substantial additional water-resources development in those areas needs to carefully consider the legal limitations on the development.

POTENTIAL FOR DEVELOPMENT

Ground water in the Great Basin Region was withdrawn primarily through large-capacity wells at an annual rate of about 1.1 million acre-feet (1,360 hm^3) per year in 1970. Plate 1*H* shows that the region probably can sustain an annual net pumpage of at least 2.6 million acre-feet (3,200 hm^3). Thus, an additional 1.5 million acre-feet (1,850 hm^3) per year of ground water is potentially available. Some of this water is now used for low-value purposes, such as subirrigated saline pastures. Development of the full additional 1.5 million acre-feet (1,850 hm^3) per year of ground water potentially available would require intensive management and water treatment in some areas, detailed evaluation of surface-water–ground-water relations and resolution of legal conflicts in other areas, and encouragement of small-scale developments in areas with low yields.

The amount of ground water that may be developed from one-time storage reserve is highly dependent upon the constraints imposed. The annual amount could be large, perhaps several million acre-feet, if legally allowed with no concern as to how long it might be economically possible to withdraw the ground water in the particular areas of development.

If an amortization period is utilized, for an interval of say 50 years, a substantial annual supply could be provided. Thus, 1.0 million acre-feet (1,233 hm^3) of ground water might be withdrawn annually for 50 years from a one-time storage reserve by planned development in 20 areas, supplying an average of 50,000 acre-feet (62 hm^3) per year each. Obviously, the finite time limit may present serious problems to a permanent community if an alternative supply were not available when needed. Even if the ground water systems continue to be legally restricted to the sustained-yield concept, considerable storage is used. Ground-water storage will always be reduced until the well-field withdrawals consist only of intercepted natural recharge or of captured natural discharge.

An evaluation of the relative potential for moderate to large-scale development of ground-water resources was made using an arbitrary scheme based on yield and ground-water storage of individual hydrographic areas. The following factors were considered:

1. Ultimately, yield is the limiting factor in determining the long-term viability of a large-scale ground-water development. Consequently, yield was given more weight than storage. Where a system yield was involved, no consideration was given to the relative proportions of surface water and ground water.
2. Stored ground water represents water available for use on a time-limited basis. A large volume of water per foot of dewatering also indicates a large area of valley fill. This allows for more options in the location and spacing of any ground-water development.

In conjunctive-use areas, ample ground-water storage would be valuable during droughts or short-term emergencies. In arid valleys storage might be depleted on a planned basis and then pumping decreased to a rate compatable with the yield. Because of these options, the availability of stored ground water was also incorporated into the rating scheme. If water quality was not suitable for most uses or if depth to water was prohibitive, storage was not considered.

The actual procedures were as follows: The distribution of yield and storage shown on plate 1*G* and 1*H* provided the basic information. The categories shown on each map were assigned index numbers from 1 to 5. If there was no restriction on quality or depth to water, the index of potential for each area was computed as the sum of the yield-index number and half the storage-index number. If there was a restriction due to water quality or depth to water, the index of potential was assumed equal to the yield-index number. The distribution of areas of poor water quality is shown on plate 1*F*. Valleys where excessive depth to water would hinder large-scale development are the drained areas on plate 1*B*. Each area was rated, and the ratings were grouped into four categories.

Plate 1*I* shows the distribution of relative potential for development in the Great Basin Region. Numerical magnitudes were not assigned to the categories because requirements of any proposed development need to be

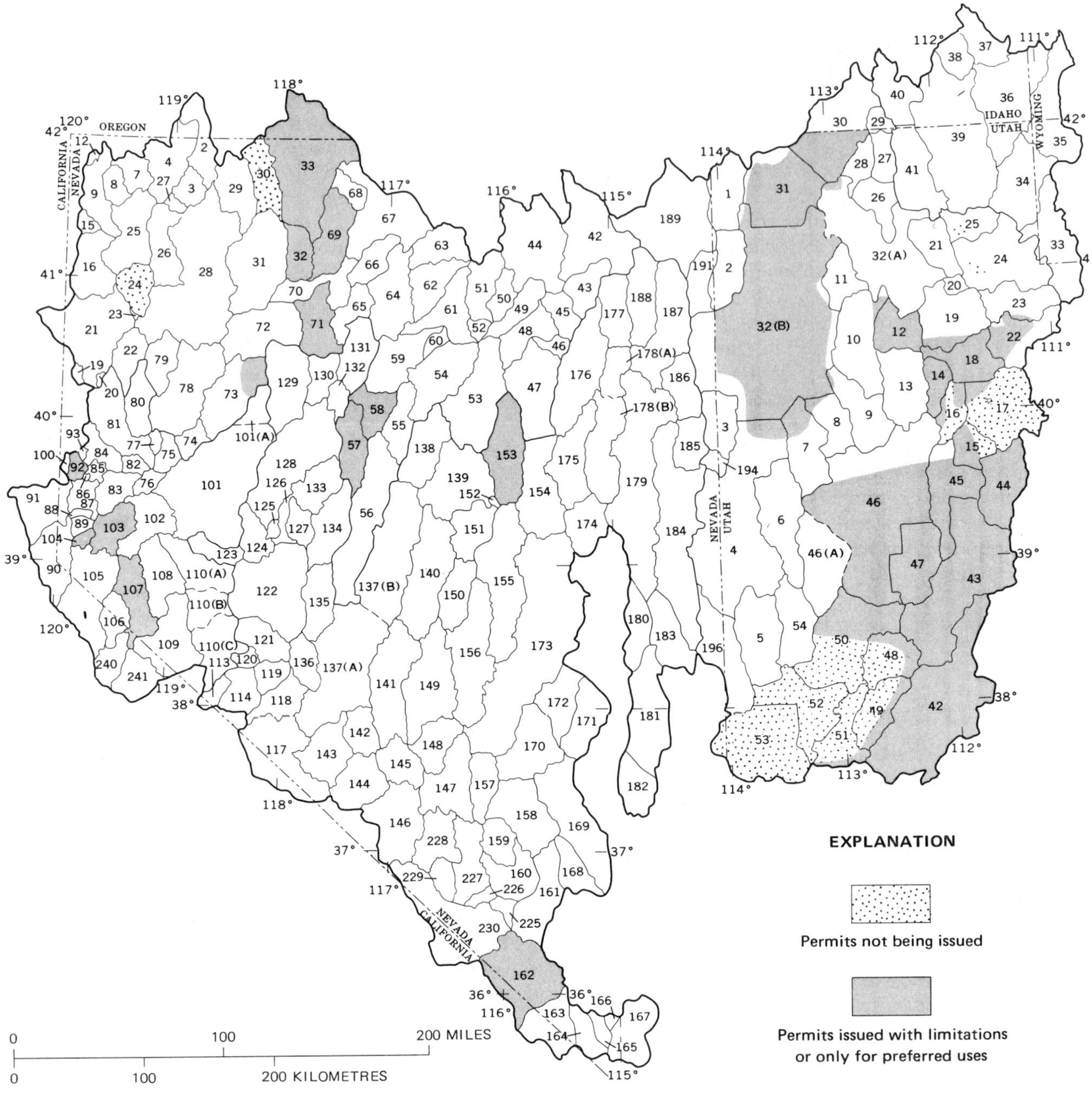

FIGURE 7.—Critical ground-water areas (as of 1973).

evaluated in terms of the more detailed information in the references for each area listed in table 8.

The above rating scheme was based only on the availability of water; however, many other factors affect the suitability of an area for the objectives of a specific development. Some restrictions on development are shown in figure on plate 1*I*. Many areas which have the highest potential for development correspond with the existing centers of population (fig. 1) and the distribution of pumpage (fig. 6). The several valleys which have a high development potential with regard to water and are still relatively undeveloped generally either have a short growing season, which would hinder agricultural development, or are remote.

INFORMATION TO PLAN GROUND-WATER DEVELOPMENT

Planning for additional water development and improved operation of present developments has been increasing at all levels of government. Few planners are completely familiar with all elements that make up a plan. Water supply is a significant element in nearly all planning of areal development. In general, many planners may only want to know that the necessary water supply can be provided in the quantity and quality desired at an acceptable cost. Others specialize in finding the most effective way to provide the water required. These specialists must have detailed technical information about the resource.

Where surface-water resources are to provide the supply, much information on water quantities, structural design, and concepts for managing large-scale water development is available. In contrast, few areas in this country have had the full benefit of advance planning and implementation of large-scale ground-water development. In some places, lack of experience may have resulted in not utilizing ground water to the best advantage.

The supply developed from the ground-water system may be the only water available; it may be the supply for a period of years until another supply can be developed and imported to the area of use; or, it may be used jointly with surface-water supplies. As a part of a joint-use development, the ground-water supply may provide a significant part of the normal supply or be used for special purposes, such as meeting peak-demand requirements, supplementing supplies during seasonal low-flow periods, and to supply demands through drought years.

Opportunities exist to make more effective use of ground water based on a sound knowledge of the total water system. A specific plan may call for full utilization of a ground-water system as a single major project, or it may call for development of ground water as a coordinated series of smaller projects designed to provide increased supply as the demand increases. The plan may be implemented under public or private auspices as deemed preferable. There needs to be a rough balance between the ability to predict future water demands, the detail of planning initiated, and the type of hydrologic information collected. Information needs vary, depending on the character of an area and its stage of development. Examples of data needs for several types of valleys are presented in the following paragraphs.

Sufficient information is needed for an undeveloped arid valley, where development is not anticipated in the near future, to provide provisional estimates of the area's ground-water resource. When grouped with estimates from adjacent areas, this information is useful in forming long-range plans and policies for the orderly development of water resources throughout large areas, such as a county, State, or river basin. At the local level there is a need to be able to judge whether an area has sufficient potential to be considered as a possible site for developments as demands for water arise. To the extent possible, conditions in the natural system should be documented to provide a basis for comparison in the future. Generally, these data needs can be satisfied by a reconnaissance-level study which roughly delineates boundaries of the principal aquifers or ground-water reservoirs; indicates areas of ground-water recharge and discharge; determines if there is inflow from or outflow to adjacent areas; estimates the approximate magnitude of ground-water recharge, discharge, and storage; and documents the depth to water in wells, chemical quality of ground water; magnitude of spring discharge, and types and distribution of phreatophytes under near natural conditions. These brief studies provide a great deal of information relative to their cost; however, when significant development occurs, additional information is needed.

When significant development is anticipated in an area, such as the arid valley discussed in the preceding paragraph, the data needs are more specific. Reconnaissance-level information generally provides estimates of the total resources of an area with little regard as to how they are distributed within the area. However, the manner in which the ground-water system will respond to a specific development is strongly influenced by its location with respect to various hydrologic features of the system. Consequently, information needs to be developed regarding the distribution of water-bearing deposits (primarily sand and gravel), extensive deposits of clay, transmissivity, specific yield or storage coefficient, areas of confined and unconfined water, recharge, discharge, chemical quality of ground water, subsurface inflow or outflow, if any, and the location of barriers to ground-water movement.

These data can be used to evaluate results expected to occur under various distributions and rates of pumping. Information developed would be useful in planning the orderly development of an area and in optimizing use of the ground-water resource.

Information needs for a river-segment area where significant development is anticipated include all items mentioned for an arid valley plus specific data on the relation between surface and ground water. Some additional items needed are surface-water inflow and outflow, flow durations, seepage gains or losses during low flow, location of areas where irrigation diversions cause significant secondary recharge, delineation of flood plains, transmissivity and specific yield of flood-plain deposits, and subsurface extent of the flood-plain deposits. Specific studies may be required to evaluate the degree to which various distributions of pumping will affect streamflow.

In all studies, establishment of a network of observation wells at the earliest possible opportunity provides information on water-level changes with time that is essential to evaluating cause-and-effect relations and refining reconnaissance-level estimates.

Figure 8 shows the level of available ground-water information in the Great Basin Region as of 1973. Most of the area has been covered by reconnaissance studies; how-

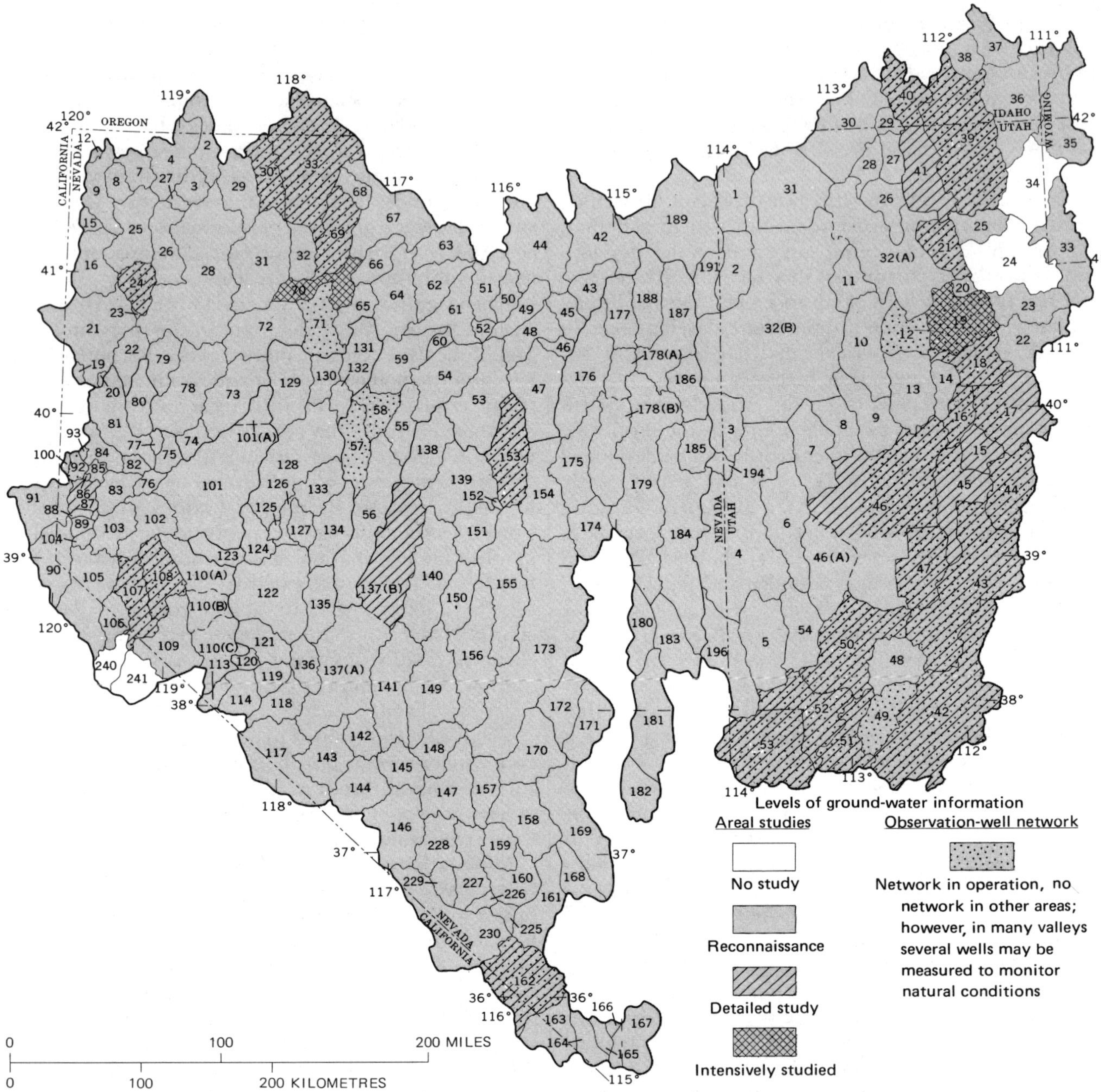

FIGURE 8.—Levels of ground-water information in the Great Basin Region, 1973.

ever, some earlier studies are mainly descriptive and contain few quantitative estimates.

The information available as of 1973 is sufficient to crudely identify the distribution, magnitude, and quality of ground-water resources throughout the region. Thus, the first phase of progressive ground-water data acquisition is nearing completion. The next phase of study needs to emphasize acquisition of time-series data to provide a base for defining changes in storage, discharge, and recharge to the ground-water system. This will allow for detailed studies to refine quantitative aspects of the ground-water system in areas where there is significant development. Where full development or overdevelopment of the water resource is foreseen, intensive studies may be needed to determine how the resource can be managed most efficiently to meet demands.

OPTIONS FOR GROUND-WATER DEVELOPMENT

The preceding sections of this report have made a broad appraisal of ground-water resources of the Great Basin Region, shown relative potentials for development of hydrographic areas, and briefly discussed the information necessary for planned developments. This section will illustrate some options available for the utilization of ground water in the region. A factor which might significantly limit the options available in a given area is the set of constraints imposed by existing development. If development has proceeded without much regard to the hydrologic regimen of an area, optimum utilization of water might not be obtained without some modification of the development.

The degree to which surface and ground water are each developed depends on conditions in individual areas. The wetter areas on the region are the river reach, prominent stream, and some sink areas shown on plate 1*B*. In these valleys, joint development of both surface and ground water may be the best way the system can be managed and optimum use of water obtained. Most closed basins, partly drained valleys other than river-reach and prominent stream areas, and some sinks have scant surface-water supplies and ground water is the only source suitable for large-scale development. The following sections illustrate some of the options available.

DEVELOPMENT OF A SUPPLEMENTAL GROUND-WATER SUPPLY

TO MEET SEASONAL DEMANDS

In areas where surface water is the principal source of supply for irrigation and other uses, the peak demand comes during periods of low flow and a seasonal water deficiency results. The usual procedure has been to construct dams which provide storage on local streams and alleviate these short-term deficiencies. In some areas, such facilities are not feasible, in other areas the supply is already completely adjudicated and highly regulated. An alternative to further regulation is to develop ground water to augment the surface supply during high-demand periods.

The Sevier Desert area, Utah, is used as an example. There is an average demand of about 160,000 acre-feet (197 hm^3) per year, primarily for agricultural purposes. System yield is about 186,000 acre-feet (229 hm^3) per year so adequate water is available to meet the annual demand. However, the August to September supply from surface-water sources is only on the order of 20,000 acre-feet (25 hm^3), but demand during this period is on the order of about 80,000 acre-feet (99 hm^3). Thus, during average and below average years a 2-month seasonal deficiency of at least 60,000 acre-feet (74 hm^3) occurs. Ground water could provide adequate water to cover this deficiency and assure a firm supply of 160,000 acre-feet (197 hm^3) per year. Any local overdraft caused by seasonal pumping could be made up by recharge during the nonpumping months or during subsequent wetter years. A properly designed well field would reduce the potential for local overdraft. Much of the ground-water withdrawal would be drawn primarily from nonbeneficial natural discharge.

Upstream areas of the Humboldt River basin, Nevada, are other places where ground water could supplement seasonal irrigation demands. Streamflow is unregulated; consequently, substantial areas of pasture or native hay receive irrigation water only in the spring and early summer. If wells were drilled to produce late-summer irrigation water, some parts of this area might sustain higher value crops, such as alfalfa. Wells could be located so that pumping would have minimal effects on downstream surface-water rights. Evaluation of the location and of the amount of seasonal pumping possible would require detailed study but might result in a significant increase in crop value.

TO OFFSET DEFICIENCIES DURING DROUGHTS OF 1 OR MORE YEARS

During droughts of 1 year to several years, the surface-water supply for an area may be inadequate, even where the supply is well regulated by storage and imports. In a densely populated area, such intermediate-term surface-water deficiencies may become critical. Larger storage facilities and imports are the usual means of minimizing such shortages until the cost of providing carryover storage for periods of several years becomes prohibitive. Planned development of supplemental ground water could be an alternative approach to the problem. The general prerequisite for this type of development is that sufficient stored ground water be available to supply the temporary depletion and that the stored water be of suitable quality for the intended use. Also, well yields would have to be adequate for the anticipated pumping, and wells would have to be distributed so pumping would have minimal effects on surface-water supplies.

Jordan Valley, Utah, is an example of an area where such a system might be beneficial. System yield plus water imported from outside the area exceeds 600,000 acre-feet (740 hm^3) per year. During 1964–68 the annual withdrawal of water for all uses averaged about 580,000 acre-feet (715 hm^3) per year. Of this amount, about 78 percent was from surface-water sources, and the remainder was from ground water.

The annual supply required to supplement surface water during a dry period of one to several years might be as much as 410,000 acre-feet (506 hm^3)—the difference between the minimum annual recorded surface flow of 170,000 acre-feet (210 hm^3) and the average annual demand of 580,000 acre-feet (715 hm^3). However, the net withdrawals might be somewhat less than this amount.

For example, if about 160,000 acre-feet (197 hm^3) of ground water were used for irrigation, roughly 25 percent, or 40,000 acre-feet (49 hm^3), would be returned to the ground-water system by deep percolation. The net ground-water withdrawal thus would be 370,000 acre-feet (456 hm^3) during years of minimum surface-water supplies. Some of the water withdrawn for industry, public supply, and other uses also probably would return to the ground-water body. Nevertheless, the net withdrawals still would be substantial and periodically would put a severe stress on the ground-water system. Ideally, the required well field or fields would be distributed and designed to minimize potential problems of local overdraft, depletion of flow in the lower reaches of streams and wetland areas, and local deterioration of water quality.

About 2 million acre-feet (2,470 hm^3) of water is stored in the upper 100 feet (30 m) of saturated valley fill, so ample stored water is available to supply a deficit of 370,000 acre-feet (456 hm^3) per year for several years. Withdrawals of this magnitude would last only during the dry periods and recovery probably could be expected during the intervening wet cycles. Recovery could be augmented by artificial recharge if necessary by utilizing excessive wet-period streamflow.

A system of this type also would be available to provide an emergency public supply during a disaster that might destroy or cause extensive damage to the surface-water-supply system.

DEVELOPMENT BY PUMPING GROUND WATER ONLY

WITHDRAWAL PLANNED TO MAINTAIN SOME NATURAL DISCHARGE

In many areas in the Great Basin Region, some natural ground-water discharge is beneficial. It may be desirable to maintain this discharge when the area is developed, particularly where extensive areas of wildlife habitat are sustained primarily by ground-water discharge. Ruby Valley, Nev., is an example of this type of area. The valley is a topographically closed basin about 60 miles (97 km) long and 8 to 12 miles (13 to 19 km) wide. It receives some subsurface inflow from adjacent Huntington Valley. Perennial yield is estimated to be about 58,000 acre-feet (72 hm^3) per year, and each foot of dewatering throughout the valley-fill reservoir would provide about 33,000 acre-feet (41 hm^3) of stored ground water. Because of these large quantities of water, the valley is classed among those with the highest potential for development (p. *1I*). However, a large marsh in the south half of the valley has been designated as the Ruby Lake National Wildlife Refuge. Almost half the ground-water resource is consumed by evapotranspiration in and near the Refuge. Any large-scale ground-water development would have to be strategically situated in the northern part of the valley, where pumping would have minimal effect on the marsh area. Consumptive use of the pumped water may eventually have to be limited to an amount equal to evapotranspiration in the northern part of the valley that could be captured by pumping. Eakin and others (1951, p. 90) estimated that about 20,000 acre-feet (25 hm^3) per year could be captured by pumping in the north end of the valley. If this water was used primarily for irrigation, perhaps one-third of the pumpage would not be consumed and would recirculate to the ground-water reservoir. Under these conditions, an annual pumpage of nearly 30,000 acre-feet (37 hm^3) could be sustained in the northern part of the valley, and the Wildlife Refuge at the south end could be maintained at near-natural conditions. Detailed cause and effect studies would be required after some significant development, but before the valley approached full development. Analytical models of the ground-water reservoir would be useful in predicting long-term effects of pumping, including salt buildup from recycling, and in refining the initial estimate of allowable pumpage.

DEVELOPMENT DESIGNED TO CAPTURE ALL NATURAL DISCHARGE

Where ground water is the principal source of supply and availability of water is the limiting factor to development, it is desirable to plan developments so that all natural discharge will be captured. Ideally, this would involve situating wells strategically in and near areas of ground-water discharge so that the natural discharge could be captured with only minimal water-level declines and storage depletion. In reality, most ground-water discharge areas are associated with concentrations of poor-quality water and saline soils. Consequently, pumping generally should be located as strategically as possible within the constraints posed by water and soil conditions. Consumption of pumped ground water may eventually have to be limited to the amount of discharge that economically can be captured (the perennial yield). Temporary higher withdrawals are possible hydrologically but may result in legal problems. Regulation of pumpage is at the discretion of the appropriate State Engineer.

Hualapai Flat, Nev., is an area where this option has been employed. The area is a small topographically closed basin on the northwest flank of the Black Rock Desert. Perennial yield is about 6,700 acre-feet (8.3 hm^3) per year, and each foot of dewatering of the valley-fill reservoir will provide about 3,500 acre-feet (4.3 hm^3) of stored water. Intensive agricultural development began in about 1960 and by 1967 the gross annual pumpage was about 11,000 acre-feet (13.6 hm^3) per year. However, only about two-thirds of the gross pumpage was consumed (7,400 acre-ft or 9.1 hm^3 per year) the remainder was recirculated to the ground-water reservoir (Harrill, 1969, p. 41). Allowing for some errors in estimates, the consumption of pumped water was about equal to the perennial yield. At this same time, a decision was made by the Nevada State Engineer to restrict development, and no permits to pump ground water for additional irrigated acreages have been issued for several years. Pumping was situated primarily with respect to land availability and good soil and is concentrated at one end of the valley. Ultimately, some pumping may have to be relocated in order to capture all available natural discharge. Also, problems in salt balance may develop in the pumping areas. However, regulation of pumpage early in the course of the area's development has probably minimized future problems and insured a moderate-sized, viable agricultural development for many years.

UTILIZATION OF GROUND-WATER STORAGE

Stored ground water is a valuable component of the region's water resource. There are about 300 million acre-feet (370,000 hm^3) of recoverable water stored in the upper 100 feet (30 m) of saturated valley deposits. This water may be withdrawn as part of a planned depletion, it may be utilized principally during periods of drought or peak demand, or it may be held in reserve for possible future use.

USE AS A TRANSITIONAL RESERVE

This method of utilizing stored ground water is based on the concept that prior to any development a ground-water reservoir is in a state of dynamic equilibrium, where average recharge to the system equals average discharge, and there is no long-term change in storage. Any pumping will create an imbalance where water is pumped from storage until water levels decline sufficiently to cause natural discharge to decrease or recharge to increase and return the system to a new equilibrium where recharge equals natural discharge (if any) plus pumpage. In most desert valleys, there are no appreciable areas where recharge can be increased by pumping. If wells are strategically located in and near areas of natural discharge, the water that must be removed from storage before a new equilibrium can be obtained will be minimal. The minimal quantity of water that must be pumped from storage before an arid ground-water basin can attain a new equilibrium when developed at a rate equal to the perennial yield has been called the transitional storage reserve (Worts, 1967, p. 50). This quantity is a property of the ground-water basin and may be useful in long-term planning.

Diamond Valley, Nev., is an area where some planned storage depletion may be an option for the future. The valley is about 50 miles (80 km) long and averages about 12 miles (19 km) wide. All natural discharge is in the north half of the valley. Perennial yield is about 30,000 acre-feet (37 hm^3) per year, and each foot of dewatering of the valley-fill reservoir will produce about 28,000 acre-feet (35 hm^3) of stored water. Transitional storage reserve is about 1,400,000 acre-feet (1,730 hm^3). The area has been developed for agricultural use. Permits to pump ground water were issued for large tracts of land at the south end of the valley. In 1969 about 23,000 acre-feet (28 hm^3) of water was pumped from this part of the valley. Future increases are anticipated because of increased demand for hay and grain and because of the recent availability of electric power in the area. Most of the pumping is at least 10 miles (16 km) south of the area of natural discharge and a local overdraft is probable before all natural discharge is captured by pumping from the present area of development.

One alternative in regulating storage depletion in the south end of the valley would be to allow a depletion equal to the transitional storage reserve to occur and then redistribute pumping with more regard to the natural system. If net pumpage (consumption) were held to 30,000 acre-feet (37 hm^3) per year, the time required to deplete the storage can be approximated by the following formula (Worts, 1967, p. 52):

$$\text{Net annual pumping rate} = \frac{\text{Transitional storage reserve}}{\text{Time, in years}} + \frac{\text{Perennial yield}}{2}$$

For the Diamond Valley area, time equals about 93 years. Thus, water could be depleted from the south end of the valley at a rate of 30,000 acre-feet (37 hm^3) per year for almost 100 years before the transitional storage reserve would be depleted. At that time it might be desirable to redistribute pumping as the average decline of non-pumping water levels in the south half of the valley would be about 100 feet (30 m); greater declines would be expected near centers of pumping. When a new equilibrium is finally attained, the total depletion of storage would only be slightly more than if the original pumping had been in or near the discharge areas. If desired, this same volume of storage could be depleted by pumping at a higher rate for a shorter period of time. After significant depletion has occurred, analytical models could be used to refine the initial estimates of time and to evaluate the optimum distribution of pumpage.

USE TO SUPPLY A TIME-LIMITED DEVELOPMENT

Ground water may be withdrawn at a rate greater than the perennial yield of the ground-water system for a limited period of time. This results in depletion of storage; however, it is not uncommon for an amount equal to several hundred times the perennial yield of the ground-water system to be stored at economically recoverable depths. Thus, in many areas, appreciable amounts of water might be removed from storage before seriously adverse economic effects occur. Ultimately, however, withdrawals would have to be limited to the approximate perennial yield of the ground-water system under the State laws or policies. Time-limited developments could be either (1) semipermanent developments, which could be economically terminated or reduced at the end of a given period of time, or (2) permanent developments, which must obtain additional water at the end of a given period of time.

The Beryl-Enterprise area, Utah, is used to illustrate this option. This valley in southwestern Utah has a warm climate and has been intensively developed for agriculture. Perennial yield is about 6,000 acre-feet (7.4 hm³) per year and pumpage in 1972 was about 77,000 acre-feet (95 hm³) per year. The valley contains about 5 million acre-feet (6,165 hm³) of recoverable stored water in the upper 200 feet (61 m) of the ground-water reservoir.

At the present time there is a large overdraft in the basin; eventually, all economically recoverable storage will be depleted. At that time either the pumping draft will have to be reduced to approximately the perennial yield or water will have to be imported. For purposes of illustration, assume the economically recoverable storage is about 5 million acre-feet (6,165 hm³) and that 50 years was determined to be the optimum amortization period in which to deplete this storage. The formula (Worts. 1967, p. 52) by which the average annual quantity of depletion can be estimated is:

$$\text{Net annual pumping rate } (Q) = \frac{\text{Depletable storage}}{\text{years}} + \frac{\text{Perennial yield}}{2}.$$

which for the area is

$$Q = 100{,}000 \;+\; 3{,}000,$$

or about 100,000 acre-feet (123 hm³) a year (rounded).

If the development was principally for agricultural purposes, the local return flow to the ground-water system probably would be about 25 percent of the pumpage. Because the computation of water available from reuse is a geometric progression, a gross pumpage of about 130,000 acre-feet (160 hm³) per year could be maintained throughout the 50-year amortization period. Degradation in water quality due to reuse would be partly offset by mixing with water withdrawn from storage for the first time.

This average annual rate could be maintained for the selected period if the distribution of the well field or well fields were such as to permit full depletion of the storage area used to define the quantity of depletable storage. Development of this magnitude will result in important changes in the preexisting conditions. The following changes are the most likely ones to occur:

1. Extensive changes in water quality over a period of time. This may be considered as a constraint on the development. However, treatment ranging from minimal to desalination could be incorporated in the development plan to maintain the quality of the supply.
2. Natural areas of ground-water discharge may be dried up. Thus, parts or all of natural wetlands and phreatophyte vegetation may be desiccated. This would adversely affect some wildlife habitat. Antecedent knowledge of this potential problem may permit devising a development plan whose effects would be within acceptable limits.
3. Substantial pumping from the ground-water reservoir may result in land subsidence that locally may have adverse effects. Land subsidence due to withdrawal of ground water is generally more severe in close proximity to well fields. The most serious problems would occur where the well fields are located in metropolitan areas—that is, in populous areas with their associated concentrations of buildings, water and sewer lines, and other structures potentially subject to damage by subsidence. However, a planned ground-water development would locate wells where subsidence would have minimal adverse effects.

FOR USE AS A FUNCTIONAL RESERVOIR

Throughout the Great Basin Region, most of the water supply is generated during the winter and spring; however, the highest demand occurs during the late summer. Consequently, dams have been built on most of the principal rivers so that regulating reservoirs can distribute the water more in accord with demand. Most of the principal rivers are highly regulated, and for many of them, construction of additional surface-water storage facilities may not be practical. Also, many areas do not contain adequate sites for the construction of surface-water storage facilities. In these areas, utilization of the valley ground-water reservoir as a managed storage reservoir may be a desirable alternative. The examples already given for the Sevier Desert and Jordan Valley illustrate two ways in which ground-water reservoirs may be used as functional storage reservoirs.

SECONDARY OPTIONS

The preceding sections have illustrated ways that ground-water resources may be utilized in selected areas with characteristics typical of many valleys in the Great Basin Region. The discussion has dealt principally with the primary option selected in each area. There are additional options which may be utilized to increase efficiency of water use and overcome problems posed by specific conditions in individual areas. Examples of four of these are reuse of water, treatment of water, redistribution of water, and augmenting ground-water storage. These are the options most commonly considered; however, there are others that may be more important in specific areas.

REUSE OF WATER

In arid areas with limited water supplies, reuse of water is one means of extending the available supply. Each use generally results in some degradation of quality, and treatment may be required. In a simple form, planned reuse may consist of no more than a series of uses where lower quality water is acceptable for each successive use. The most common example of this type is use for municipal purposes, secondary treatment, and then use of the effluent for irrigation. More sophisticated schemes may require elaborate tertiary treatment which would enable reuse for high-quality purposes. Unplanned reuse generally occurs in irrigated areas where some unconsumed irrigation water becomes secondary recharge. The maximum reuse possible varies with the quantity consumed in use. If it is assumed that water will be used and reused until it is consumed, the rate at which a given inflow of new water can be used may be estimated as the rate of new water input divided by the consumption (expressed as a decimal fraction). The following examples, from areas already discussed, illustrate how reuse may be applied to planned ground-water developments.

Example A.—Full use of perennial yield:

Area—Hualapai Flat;

Perennial yield—7,000 acre-feet (8.6 hm^3) per year;

Use—Irrigation, about 2/3 (0.67) of pumpage consumed in use;

$$\text{Maximum sustained pumping rate} = \frac{\text{Perennial yield}}{\text{Consumption}}$$

$$= \frac{7{,}000}{0.67} = 10{,}000 \text{ acre-feet } (12 \text{ hm}^3) \text{ per year (rounded).}$$

Example B.—Storage depletion:

Area—Beryl-Enterprise area;

Rate of depletion—100,000 acre-feet (123 hm^3) per year;

Use—Irrigation, about ¾ (0.75) of pumpage consumed in use;

$$\text{Pumping rate} = \frac{\text{Rate of depletion}}{\text{Consumption}} = \frac{100{,}000}{0.75}$$

$$= 130{,}000 \text{ acre-feet } (160 \text{ hm}^3) \text{ per year (rounded).}$$

TREATMENT OF WATER

Treatment is another option that may be used to extend usefulness of the available water. Water treatment in conjunction with reuse has already been mentioned. If significant development occurs in some sink areas where water quality is marginal for most uses, then some treatment may be required. Facilities available range from large plants used for public-supply systems to small units suitable for use by individuals. The type and degree of treatment required, if any, vary with each area. Generally, specialized technical assistance is available to assist in planning a development that would use a specific type of water treatment.

REDISTRIBUTION OF WATER

Commonly, developments are located on the basis of accessibility, land ownership, soil conditions, or climate. Where ground-water development is poorly situated with respect to the occurrence of water, and local overdraft is anticipated, relocation of pumping and transportation of water to the existing area of development may be more desirable than attempting to relocate the development.

Occasionally, it may even be necessary to import water from outside the area. If this type of development is anticipated, cause and effect relationships at the proposed pumping sites, along transmission lines, and at development where the water is to be used need to be evaluated.

AUGMENTING GROUND-WATER STORAGE

In most valley ground-water reservoirs in the Great Basin Region, the level of stored ground water responds to the variations of natural recharge and discharge. A succession of years of significantly higher than average recharge will raise the level of storage until the increased natural recharge is balanced by an increased natural discharge. Conversely, the level of storage will lower following years of reduced recharge until the natural discharge and recharge are in balance. However, in most ground-water reservoirs, natural variations in stored water are minor.

In the Great Basin Region, part of the annual pumpage is supplied from ground-water storage. Some of the depletion is temporary and is necessary for the natural system to adjust to well-field development. Part, however, is a local or basin overdraft and is a continuing depletion. Examples of areas of overdraft are in the Beryl-Enterprise district in Sevier Lake subregion, Quinn River Valley in Humboldt subregion, and Pahrump Valley in Tonopah subregion.

Areas of overdraft might benefit from artificial recharge. However, in each of the named areas, no local supply is available, nor is there a convenient nearby source for possible importation. Several conditions are required for artificial recharge to be successful: (1) Underground space must be available for storing the additional water; (2) the physical and chemical characteristics of the rock units in and above the ground-water reservoir must be suitable to accept water either from surface spreading grounds or through wells; (3) additional water must be available; (4) the water to be used must be chemically suitable; and (5) the cost should be within the range of "willingness and ability to pay."

Artificial recharge might be advantageous where ground-water development significantly depletes stored ground water. Properly incorporated into planning a large-scale ground-water development, it could permit significant augmentation of the natural yield of the system.

In drained valleys, the depth to water is as much as several hundred feet below land surface. In these areas, natural ground-water discharge is by underflow and not by evapotranspiration. Obviously, a large volume of underground space is available between the water table and land surface. However, artificial recharge would raise the level in these reservoirs and increase discharge by increasing underflow. Consequently, it may not be feasible to augment storage in these areas even though the increased discharge would be augmenting recharge to a downgradient valley. The time lag involved in these flow systems may be many years. Much more needs to be known of the details of regional ground-water systems before decisions can be reached as to whether plans for their augmentation are practical and economically sound.

Underground space might be used for purposes other than for artificial recharge. For example, in southern Tonopah subregion the deep water levels in the valleys have been a significant factor in using the area for underground nuclear-detonation experiments. In other areas, brine or other waste disposal might be proposed by underground emplacement. However, any proposed use of underground space ideally should require specific investigations be made to determine the suitability of the area for the specific use contemplated. Few, if any, areas in the region have been evaluated for such purposes to date.

CLOSING STATEMENT

Ground water has been developed in the past in response to individual needs. Supplies from individual wells or groups of wells have been planned to a degree, but fuller development of the ground-water basins may not have been considered. Many ground-water basins or areas are now being managed through administration of ground-water laws, or resulting legal actions, particularly in the Nevada and Utah parts of the Great Basin Region. However, full development of some areas may not be possible without drastic adjustments in the distribution of existing withdrawals in these areas.

The present trend of formulating State water plans is leading to more emphasis on achieving near optimal use of the limited water resources of the States in the region. The ground-water potential for additional development is receiving substantial attention in this planning. If an effective planning effort is to be made, more information is needed on the ground-water resources, the potential for additional development, the conditions to be met for successful development, and the planning, implementation, operation, and maintenance of ground-water supply systems.

The development of additional ground water could be of significant benefit to the Great Basin Region. Some idea of the approximate value of water may be obtained by the following illustrations:

1. At the current level of water use, irrigation accounts for about 80 percent of the withdrawals. A common yearly rate of water application for irrigation is about 4 acre-feet per acre. If it is assumed that ground-water withdrawals for irrigation were planned for an additional 1 million acre-feet (1,233 hm^3) annually, then about 250,000 acres (1,010 km^2 could be irrigated. Or,
2. Current levels of municipal and industrial use in the region range from about 200 to 400 gallons (760 to 1,510 litres) per day per capita. If it is assumed that an average use of 300 gallons (1,140 litres) per day per capita will be adequate, then additional ground-water withdrawals of 1 million acre-feet (1,233 hm^3) could support a population of more than 2 million people.

Development of ground water could be of substantial benefit in helping to meet future demands for water in the Great Basin Region, and sound planning for its most effective development is dependent upon the availability and adequacy of the information on this valuable resource at the time of planning.

REFERENCES

Eakin, T. E., and others, 1951, Contributions to the hydrology of eastern Nevada: Nevada State Engineer Water Resources Bull. 12, p. 79–91.

Everett, D. E., and Rush, F. E., 1967, A brief appraisal of the water resources of the Walker Lake area, Mineral, Lyon, and Churchill Counties, Nevada: Nevada Dept. Conserv. and Nat. Resources Water Resources–Reconn. Ser. Rept. 40, 44 p.

Fenneman, N. M., and Johnson, D. W., 1946, Physical division of the United States: U.S. Geol. Survey Map.

Godwin, L. H., and others, 1971, Classification of Federal lands valuable for geothermal steam and associated geothermal resources: U.S. Geol. Survey Circ. 647, 18 p.

Harrill, J. R., 1969, Hydrologic response to irrigation pumping in Hualapai Flat, Washoe, Pershing, and Humboldt Counties, Nevada, 1960–67: Nevada Dept. Conserv. and Nat. Resources Water Resources Bull. 37, 75 p.

Kohler, M. A., Nordenson, T. J., and Baker, D. R., 1959, Evaporation maps for the United States: U.S. Dept. Commerce, Weather Bur. Tech. Paper 37, 13 p.

Malmberg, G. T., 1967, Hydrology of the valley-fill and carbonate-rock reservoirs, Pahrump Valley, Nevada-California: U.S. Geol. Survey Water-Supply Paper 1832, 47 p.

Pacific Southwest Inter-Agency Committee, 1971a, Great Basin Region Comprehensive Framework Study, Appendix III, Legal and Institutional Environments: Salt Lake City, Water Resources Council, 227 p.

_____1971b, Great Basin Region Comprehensive Framework Study, Appendix V, Water Resources: Salt Lake City, Water Resources Council, 109 p.

_____1971c, Great Basin Region Comprehensive Framework Study, Appendix XV, Water Quality, Pollution Control and Health Factors: Salt Lake City, Water Resources Council, 164 p.

Shamberger, H. A., 1971, The story of Rawhide, Mineral County, Nevada: Nevada Dept. Conserv. and Nat. Resources, Historic mining camps of Nevada, Rept. 2, 50 p.

_____1972, The story of the water supply for the Comstock: U.S. Geol. Survey Prof. Paper 779, 53 p.

Snyder, C. T., 1962, A hydrologic classification of valleys in the Great Basin, Western United States: Internat. Assoc. Sci. Hydrology Bull., v. 7, no. 3, p. 53–59.

U.S. Public Health Service, 1962, Drinking-water standards, 1962: U.S. Public Health Service Pub. 956, 61 p.

Walker, G. E., and Eakin, T. E., 1963, Geology and ground water of Amargosa Desert, Nevada-California: Nevada Dept. Conserv. and Nat. Resources Ground-Water Resources–Reconn. Ser. Rept. 14, 58 p.

Worts, G. F., Jr., 1967, The available water supply, *in* Rush, F. E., and Glancy, P. A., Water-resources appraisal of the Warm Springs-Lemmon Valley area, Washoe County, Nevada: Nevada Dept. Conserv. and Nat. Resources, Water Resources–Reconn. Ser. Rept. 43, p. 48–53.

Worts, G. F., Jr., and Malmberg, G. T., 1966, Hydrologic appraisal of Eagle Valley, Ormsby County, Nevada: Nevada Dept. Conserv. and Nat. Resources, Water Resources–Reconn. Ser. Rept. 39, 55 p.

TABLE 8 AND TABLE 8 REFERENCES

A regional quantitative appraisal of the ground-water resources in the Great Basin Region is given in an earlier section of this report entitled "Yield of the Ground-Water System." Table 8 presents quantitative estimates of various parameters of the ground-water system in the hydrographic areas of the region. The general locations of these hydrographic areas are shown on plate 1*B*.

TABLE 8.—*Summary of selected ground-water resources data in the Great Basin Region*[1]

[L, large amount; Sm, small amount; S, some; M, minor amount; leaders (...) indicate no data]

Hydrographic area number and name[2]	Approximate area (mi²)	Approximate altitude of valley floor	Average growing season (days)	Average annual precipitation (acre-ft)	Ground-water recharge from precipitation (acre-ft/yr)[3]	Ground-water inflow (acre-ft/yr)	Ground-water outflow (acre-ft/yr)	Ground-water evapotranspiration (acre-ft/yr)	Ground water in storage (acre-ft/ft)[4]	Report reference (See table references)
WESTERN GREAT BASIN										
Humboldt subregion										
2 Continental Lake valley	214	4,200	......	100,000	4,000	......	0	10,500	3,800	R22.
3 Gridley Lake valley	195	4,500	......	98,000	[3]4,500	......	0	3,000	2,300	R22.
4 Virgin Valley	494	4,800	42	230,000	7,000	......	0	6,000	420	R22.
7 Swan Lake valley	226	5,700	42	130,000	[3]6,700	0	0	0	M	R15.
8 Massacre Lake valley	176	5,700	42	97,000	3,500	0	2,000	2,500	1,400	R15.
9 Long Valley	433	5,600	42	240,000	[3]10,000	5,000	0	11,000	10,000	R15.
12 Mosquito Valley	32	5,700	42	16,000	700	0	0	1,600	470	R15.
15 Boulder Valley	88	5,700	42	52,000	[3]2,700	0	2,000	< 2,700	600	R15.
16 Duck Lake valley	533	4,700	75	270,000	9,000	0	0	7,000	5,600	R17.
19 Dry Valley	39	4,200	......	14,000	200	0	180	20	1,000	R44.
20 Sano Valley	12	4,000	160	3,100	< 10	0	0	30	200	R44.
21 Smoke Creek Desert	980	3,900	160	440,000	13,000	380	0	19,000	20,000	R44.
22 San Emidio Desert	305	4,000	160	100,000	2,100	0	< 300	3,000	8,400	R44.
23 Granite basin	9	5,000	......	6,000	[3]400	0	M	0	50	R20.
24 Hualapai Flat	315	4,100	150	170,000	7,000	0	400	6,300	3,500	R11, B37.
25 High Rock Lake valley	665	5,000	......	435,000	13,000	0	9,000	750	610	R20.
26 Mud Meadow	495	4,000	......	220,000	8,000	9,000	1,500	11,000	8,500	R20.
27 Summit Lake valley	60	5,900	......	43,000	[3]4,200	0	S	M	630	R20, R22.
28 Black Rock Desert	2,179	4,000	179	840,000	20,000	4,700	M	35,000	56,000	R20.
29 Pine Forest Valley	528	4,000	77	260,000	10,000	250	2,700	11,000	18,000	R4.
30 Kings River valley	413	4,200	88	260,000	[3]15,000	} 300	} 100	16,000	20,000	B31.
31 Desert Valley	1,052	4,200	......	370,000	5,000		} 150	10,000	40,000	R7.
32 Silver State Valley	313	4,200	......	140,000	1,400	4,500	100	5,800	16,000	B34.
33 Quinn River valley	1,224	4,300	112	880,000	[3]62,000	M	4,700	51,000	42,000	B34.
42 Marys River area	1,073	5,600	......	700,000	[3]54,000	...... }			{ [5]8,000	B32.
43 Starr Valley area	332	6,000	......	230,000	[3]26,000	...... }	M	83,000	{ [5]11,000	B32.
44 North Fork area	1,110	5,400	......	750,000	[3]58,000	M }			{ [5]5,000	B32.
45 Lamoille Valley	257	5,400	140	180,000	[3]36,000	0	......	......	[5]8,000	B32.
46 South Fork area	99	5,600	100	98,000	[3]4,000	0	600	3,000	2,400	R35, B32.
47 Huntington Valley	787	5,500	90	550,000	[3]14,000	0	10,400	14,000	32,000	R35, B32.
48 Dixie Creek–Tenmile Creek area	392	5,400	100	240,000	[3]13,000	1,000	9,000	4,000	16,000	R35, B32.
49 Elko segment	314	5,100	103	170,000	[3]7,400	......	......	13,000	[5]6,000	B32.
50 Susie Creek area	223	5,000	......	130,000	[3]8,000	...... }	M	6,100	{ [5]300	B32.
51 Maggie Creek area	396	5,300	......	240,000	[3]16,000	...... }			{ [5]1,000	B32.
52 Marys Creek area	61	5,200	......	34,000	1,500	......	......	......	[5]300	B32.
53 Pine Valley	1,002	5,400	105	660,000	[3]50,000	0	9,300	15,000	20,000	R2, B32.
54 Crescent Valley	752	5,000	110	430,000	[3]13,000	> 300	M	12,000	15,000	B15, B32.
55 Carico Lake valley	376	5,100	120	160,000	[3]4,300	0	300	3,800	8,000	R37, B32.
56 Upper Reese River valley	1,138	5,800	117	700,000	[3]37,000	0	500	37,000	12,000	R31, B32.
57 Antelope Valley	452	5,000	120	260,000	[3]11,000	......	6,000	500	5,000	R19, B32.
58 Middle Reese River valley	319	4,900	120	170,000	7,000	6,500	9,000	3,000	5,000	R19, B32.
59 Lower Reese River valley	588	4,700	120	280,000	14,000	17,000 }	3,000	22,000	{ [5]17,000	B32.
60 Whirlwind Valley	94	4,800	......	45,000	1,700	...... }			{ [5]800	B32.
61 Boulder Flat	544	4,700	......	240,000	17,000	M	2,000	30,000	[5]13,000	B32.
62 Rock Creek valley	444	4,900	......	240,000	9,000	......	...... }	2,800	{ [5]100	B32.
63 Willow Creek valley	405	5,100	......	250,000	[3]15,000	......	...... }		{ [5]100	B32.

64 Clovers area	702	4,500	120	300,000	[3]9,000	......			[5]24,000	B32.
65 Pumpernickel Valley	299	4,500	......	130,000	3,400	......	1,000	72,000	[5]6,000	B32.
66 Kelly Creek area	301	4,400	......	130,000	4,000	......			[5]8,000	B32.
67 Little Humboldt Valley	975	4,600	110	500,000	[3]21,000	0	300	4,000	8,000	B32, B39.
68 Hardscrabble area	167	5,200	110	120,000	[3]9,000	0	M	M	< 100	B32, B39.
69 Paradise Valley	600	4,500	120	900,000	10,000	300	3,500	40,000	36,000	B32, B39.
70 Winnemucca segment	435	4,400	141	170,000	4,400	9,000	3,000	16,000	15,000	B19, B20, B22, B24, B27, B32.
71 Grass Valley	520	4,400	130	250,000	12,000	......	4,000	13,000	15,000	R29, B32.
72 Imlay area	771	4,200	128	300,000	7,000	3,000	1,000	7,400	8,000	R5, B32.
73 Lovelock Valley	635	4,000	128	260,000	3,200	1,000	S	3,100	20,000	R32, B32.
74 White Plains	164	3,900	......	51,000	M	S	S	......	4,200	R59.
78 Granite Springs valley	967	4,000	150–170	350,000	3,500	1,000	0	4,400	26,000	R55.
79 Kumiva Valley	333	4,500	150–160	120,000	1,000		1,000	0	10,000	R55.
Central Lahontan subregion										
75 Bradys Hot Springs area	178	4,200	150–170	59,000	160	1,200	0	3,000	3,500	R55.
76 Fernley area	120	4,200	......	43,000	600	0	5,800	......	4,200	B17, R57.
77 Fireball Valley	58	4,700	150–170	21,000	200	......	200	0	1,300	R55.
80 Winnemucca Lake valley	371	3,800	137	130,000	8,000	400	0	> 5,000	9,600	B15, R57.
81 Pyramid Lake valley	672	3,800	137	270,000	6,600	350	350	......	[6]19,000	R57.
82 Dodge Flat	92	4,200	......	43,000	1,400	2,800	150	......	2,600	R57.
83 Tracy segment	285	4,300	......	110,000	6,000	2,100	700	......	1,000	R57.
84 Warm Springs valley	247	4,300	140	130,000	6,000	0	≤ 200	1,500	4,200	R43, R57.
85 Spanish Springs valley	76	4,500	140	30,000	600	0	100	900	1,700	R43, R57
86 Sun Valley	10	4,700	140	4,000	50	0	25	2	200	R43, R57.
87 Truckee Meadows	203	4,500	155	160,000	[3]27,000	1,100	M	......	4,500	W1779, R57.
88 Pleasant Valley	39	4,500	......	46,000	[3]10,000	50	300	......	300	R57.
89 Washoe Valley	82	5,100	120	87,000	[3]15,000	0	50	8,500	2,700	R41, R57.
90 Lake Tahoe basin	139	6,200	......	180,000	[3]45,000	0	0	M	[6]300	W1972.
91 Truckee Canyon segment	84	4,900	......	110,000	[3]27,000	400	700	......	400	R57.
92 Lemmon Valley	93	5,000	130	48,000	2,100	......	......	1,200	2,300	R43.
93 Antelope Valley	18	5,200	130	9,000	300	0	S	0	470	R43.
100 Cold Spring valley	30	5,100	130	18,000	[3]900	0	M	130	450	R43.
101 Carson Desert	2,182	3,900	127	720,000	2,000	M?	M?	......	85,000	R59.
A. Packard Wash		......	......	......	......	......	......	......	......	R59.
102 Churchill Valley	480	4,200	......	170,000	1,300	M?	M?	......	7,400	R59.
103 Dayton Valley	369	4,400	......	180,000	7,900	M?	M?	......	4,400	R59.
104 Eagle Valley	69	4,700	119	58,000	[3]8,700	0	M	4,000	2,000	R39, R59.
105 Carson Valley	419	4,800	114	270,000	[3]25,000	3,000	M	......	7,100	P417–F, R59.
106 Antelope Valley	115	5,000	......	69,000	[3]5,000	S	200	5,700	2,000	R53.
107 Smith Valley	479	4,700	120	270,000	[3]21,000	200	500	......	9,800	W1228.
108 Mason Valley	516	4,500	118	160,000	2,000	500	1,500	57,000	29,000	B38.
109 East Walker area	586	6,800	......	250,000	12,000	S	150	6,500	8,000	R53.
110 Walker Lake valley	1,350	4,300	136	......	......	......	......	......	......	R40.
A. Schurz subarea	502	4,200	136	160,000	500	1,400	S	17,000	15,000	R40.
B. Lake subarea	307	4,000	......	......	600	S	0	800	[6]1,000	R40.
C. Whiskey Flat Hawthorne subarea	541	4,800	......	210,000	5,400	300	S	4,600	9,000	R40.
240 Upper West Walker River		......	......	......	......	......	......	......	[5]100	
241 Bridgeport Valley		......	......	......	......	......	......	......	[5]2,000	
113 Huntoon Valley	97	5,800	100–150	43,000	800	0	300	300	1,200	R52.
114 Teels Marsh valley	323	5,000	170–200	120,000	1,300	< 300	0	1,400	2,600	R52.
117 Fish Lake valley	706	4,800	96+	270,000	7,300	S	> 200	22,000	16,000	B11.
118 Columbus Salt Marsh valley	370	4,600	186	100,000	700	> 200	0	4,000	5,300	R52.

See footnotes at end of table, p. G35.

TABLE 8.—*Summary of selected ground-water resources data in the Great Basin Region*—Continued

Hydrographic area number and name[2]	Approximate area (mi²)	Approximate altitude of valley floor	Average growing season (days)	Average annual precipitation (acre-ft)	Ground-water recharge from precipitation (acre-ft/yr)[3]	Ground-water inflow (acre-ft/yr)	Ground-water outflow (acre-ft/yr)	Ground-water evapotranspiration (acre-ft/yr)	Ground water in storage (acre-ft/ft)[4]	Report reference (See table 8 references)
WESTERN GREAT BASIN—Continued										
Central Lahontan subregion—Continued										
119 Rhodes Salt Marsh valley	199	4,600	170–200	59,000	500	400	0	1,000	3,400	R52.
120 Garfield Flat	92	5,700	100–150	34,000	300	0	300	0	1,500	R52.
121 Soda Springs valley	376	4,600	170–200	110,000	700	500	900	330	7,100	R52.
122 Gabbs Valley	1,277	4,300	100–120	520,000	5,000	0	0	> 3,700	16,000	R9.
123 Rawhide Flats	227	4,000	......	75,000	150	350	0	800	600	R40.
124 Fairview Valley	285	4,200	......	100,000	500	0	500	0	7,800	R23.
125 Stingaree Valley	43	4,400	......	16,000	110	M	M		1,300	R23.
126 Cowkick Valley	110	4,700	......	44,000	800	M	M	400	1,700	R23.
127 Eastgate Valley area	216	4,800	......	100,000	[3]4,000	0	M		1,900	R23.
128 Dixie Valley	1,303	3,600	220	460,000	6,000	1,800	0	16,500	35,000	R23.
129 Buena Vista Valley	742	4,100	110	310,000	10,000	0	0	12,500	24,000	B13.
130 Pleasant Valley	285	4,400	......	110,000	3,000	0	800	2,200	6,200	R23.
131 Buffalo Valley	504	4,700	120	240,000	12,000	0	8,000	4,000	17,000	R32.
132 Jersey Valley	142	4,200	......	56,000	800	0	500	0	1,600	R23.
133 Edwards Creek valley	416	5,200	120	190,000	8,000	0	0	7,300	7,000	R26.
134 Smith Creek valley	582	6,100	......	280,000	12,000	0	0	6,600	15,000	R28.
135 Ione Valley	460	6,000	......	230,000	8,000	0	2,000	1,300	13,000	R28.
136 Monte Cristo Valley	284	5,400	100–150	94,000	500	0	0	400	7,200	R52.
137 Big Smoky Valley	2,926	......	......	......	......	......	......	......	......	B41.
A. Tonopah Flat	1,603	4,800	150	580,000	12,000	2,000	8,000	6,000	70,000	B41.
B. Northern part	1,323	5,500	130	740,000	[3]65,000	0	0	64,000	50,000	B41.
138 Grass Valley	595	5,700	120	290,000	13,000	0	0	12,000	16,000	R37.
139 Kobeh Valley	868	6,200	100	560,000	11,000	6,000	M	15,000	27,000	R30.
140 Monitor Valley	1,038	6,500	100	510,000	21,300	2,000	8,000	11,200	20,000	R30.
141 Ralston Valley	971	5,600	144	360,000	5,000	3,000	5,500	2,500	27,000	R12, R45.
142 Alkali Spring valley (Esmeralda)	313	5,000	140	100,000	100	5,500	5,000	400	13,000	R45.
143 Clayton Valley	555	4,400	150	180,000	1,500	13.000	0	24,000	13,000	R45.
144 Lida Valley	535	5,000	140	170,000	500	200	700	0	15.000	R45.
145 Stonewall Flat	381	4,800	140	110,000	100	S?	0	200	8,200	R45.
146 Sarcobatus Flat	812	4,100	150	190,000	1,200	1,300	3,000	500	24,000	R10, R54.
147 Gold Flat	684	5,200	......	250,000	3,800	0	0	3,800	16,000	R54.
148 Cactus Flat	403	5,400	......	130,000	600	0	0	600	14,000	R54.
149 Stone Cabin Valley	985	5,700	144	350,000	5,000	0	2,000	3,000	22,000	R12, R45.
150 Little Fish Lake valley	434	6,600	75–100	230,000	11,000	0	10,000	200	8,000	R38.
151 Antelope Valley (Eureka and Nye)	444	6,200	100	190,000	4,100	0	4,200	S	12,000	R30.
152 Stevens basin	17	7,200	100	8,500	200	0	0	200	500	R30.
153 Diamond Valley	752	5,900	100	400,000	[3]21,000	9,000	30,000	0	28,000	R6, B35.
154 Newark Valley	801	5,900	80–100	410,000	17,500	1,000	18,500	0	15,000	R1.
155 Little Smoky Valley	1,158	6,100	75–150	450,000	5,600	S	1,900	1,200	25,000	R38.
156 Hot Creek valley	1,036	5,300	150	390,000	7,000	> 200	4,600	S	23,000	R38.
157 Kawich Valley	350	5,500	......	150,000	3,500	1,000	0	4,500	9,600	B12, R54.
158 Emigrant Valley	104	4,600	......	284,000	3,200	0	0	3,200	16,000	R54.
159 Yucca Flat	305	4,000	......	100,000	700	0	0	700	5,200	R54.
160 Frenchman Flat	463	3,200	......	150,000	100	33,000	0	33,000	7,900	R54.
161 Indian Springs valley	655	3,200	......	270,000	10,000	22,000	32,000	M	18,000	R54.

162 Pahrump Valley	789	2,800	......	420,000	22,000	0	13,000	10,000	23,000	W1832.
163 Mesquite Valley	236	2,600	200–250	90,000	1,400	700	M	2,200	7,000	R46.
164 Ivanpah Valley	235	2,700	......	81,000	700	800	1,500	0	7,400	R46.
165 Jean Lake valley	96	2,800	......	32,000	100	1,500	> 100	0	3,200	R46.
166 Hidden Valley	34	3,100	......	11,000	M	M	M	0	800	R46.
167 Eldorado Valley	530	1,800	275	190,000	1,100	M	1,100	0	14,000	R36.
168 Three Lakes valley	298	3,600	......	110,000	2,000	6,000	8,000	0	8,300	R54.
169 Tikapoo valley	998	3,400	......	380,000	6,000	2,600	8,600	0	21,500	R54.
170 Penoyer Valley	700	5,000	......	270,000	4,300	0	0	6,400	22,000	B12.
171 Coal Valley	460	5,000	150	170,000	2,000	8,000	10,000	M	15,000	R18, B33.
172 Garden Valley	493	5,500	150	230,000	10,000	0	8,000	2,000	15,000	R18, B33.
173 Railroad Valley	2,752	4,900	......	1,240,000	52,000	S	1,000	50,000	81,000	B12.
174 Jakes Valley	422	6,400	......	240,000	17,000	8,000	25,000	0	9,800	B33.
175 Long Valley	651	6,100	100	250,000	10,000	0	8,000	2,200	16,000	R3, B33.
176 Ruby Valley	1,004	6,000	107	720,000	[3]68,000	10,800	0	53,000	33,000	B12.
177 Clover Valley	464	5,700	100	260,000	[3]21,000	0?	M	19,000	15,000	B12.
178 Butte Valley	1,010	......	100–130	......	......	.....	.....	.....	.....	R49.
A. Northern part	271	6,100	......	140,000	3,900	0	800	6,900	9,800	R49.
B. Southern part	739	6,300	......	420,000	[3]15,000	0	?	11,000	22,000	R49.
179 Steptoe Valley	1,942	5,900	119	1,200,000	[3]85,000	0	S	70,000	50,000	R42.
180 Cave Valley	362	6,100	......	220,000	[3]14,000	0	14,000	200	10,000	R13, B33.
181 Dry Lake valley	882	4,800	150	340,000	5,000	0	5,000	M	28,000	R16, B33.
182 Delamar Valley	383	4,600	150	140,000	1,000	5,000	6,000	M	12,000	R16, B33.
183 Lake Valley	557	6,000	100	290,000	13,000	0	3,000	8,500	18,000	R24.
184 Spring Valley	1,661	5,700	100	960,000	[3]75,000	2,000	4,000	70,000	42,000	R33.
185 Tippett Valley	345	5,700	110	160,000	6,900	0	7,000	0	11,000	R56.
186 Antelope Valley	395	5,600	110	170,000	4,700	3,300	7,900	100	9,900	B12, R56.
187 Goshute Valley	954	5,600	100	440,000	11,000	S	2,300	10,000	22,000	B12.
188 Independence Valley	562	5,600	100	250,000	9,300		0	9,500	18,000	B12.
225 Mercury Valley	110	3,200	......	38,000	250	16,000	17,000	0	M	R14, R54
226 Rock Valley	82	3,300	......	26,000	30	17,000	17,000	0	1,500	R14, R54.
227 Forty Mile Canyon	519	3,500	......	188,000	2,300	13,000	15,300	0	7,400	R14, R54
228 Oasis Valley	460	3,800	184	150,000	1,000	2,500	1,500	2,000	4,000	R10, R54.
229 Crater Flat	182	3,200	......	61,000	220	1,500	1,700	0	3,500	R14, R54.
230 Amargosa Desert	896	2,600	180–200	240,000	600	44,000	19,000	24,000	35,000	R14, R54.

EASTERN GREAT BASIN

Great Salt Lake subregion

189 Thousand Springs valley	1,446	5,600	100-110	587,000	12,300	> 18,000	>20,500	>5,700	38,000	R47.
191 Pilot Creek valley	326	4,600	110	130,000	2,400	1,000	300	4,600	11,000	R56.
194 Pleasant Valley	75	6,200	......	54,000	4,800	0	3,000	M	420	R34.
196 Hamlin Valley	413	5,800	......	260,000	10,000	4,000	10,000	400	12,000	R34.
1 Grouse Creek valley	430	4,800	132	276,000	14,000	......	2,000	11,000	1,600	T29, T42.
2 Pilot Valley	470	4,200	130	184,000	3,400	......	......	7,400	5,000	T41, T42.
3 Deep Creek valley	440	5,500	95	290,000	17,000	......	......	15,000	3,200	T24.
4 Snake Valley	2,700	4,900	152	1,420,000	44,000	30,000	10,000	69,000	107,000	T14.
5 Pine Valley		......	......	......	......	......	......	......	[5]12,000	W1475–N.
6 White Valley		......	......	......	......	......	......	......	[5]14,000	W277, W1475–N.
7 Fish Springs Flat	400	4,400	......	......	......	......	......	......	[5]12,000	W277, W1475–N.
8 Dugway Valley		......	......	......	......	......	......	......	[5]13,000	W1475–N.
9 Government Creek valley		......	......	......	......	......	......	......	[5]7,000	W1475–N.
10 Skull Valley	880	4,700	140	490,000	30,000–50,000	......	800–1,600	23,000–37,000	23,000	T18.
11 Sink Valley	150	4,400	205	54,000	1,000	......	1,000	0	3,700	T26, T42.
12 Tooele Valley	400	4,500	209	[7]200,000	100,000	Sm	......	40,000	[5]14,000	T12.

See footnotes at end of table, p. G35.

TABLE 8.—*Summary of selected ground-water resources data in the Great Basin Region*—Continued

Hydrographic area number and name[2]	Approximate area (mi²)	Approximate altitude of valley floor	Average growing season (days)	Average annual precipitation (acre-ft)	Ground-water recharge from precipitation (acre-ft/yr)[3]	Ground-water inflow (acre-ft/yr)	Ground-water out flow (acre-ft/yr)	Ground-water evapotranspiration (acre-ft/yr)	Ground water in storage (acre-ft/ft)[4]	Report reference (See table 8 references)
EASTERN GREAT BASIN—Continued										
Great Salt Lake subregion—Continued										
13 Rush Valley	730	5,200	195	550,000	34,000	......	5,000	27,000	16,000	T23.
14 Cedar Valley	300	5,000	120	[9]151,000	24,000	......	10,000	......	[5]8,000	T16.
15 Northern Juab Valley	120	5,000	148	......	......	Sm	Sm	18,000+	[5]5,000	T17.
16 Goshen Valley	} 400	} 4,800	......	......	[8]30,000	Sm	4,000(?)	27,000	[9]15,000	} T28.
17 Southern Utah Lake valley	}	} 4,800	......	......	[8]120,000	......	4,000	55,000	[6] [9]7,500	}
18 Northern Utah Lake valley	225	4,800	......	......	[8]150,000+	......	2,500 +	13,000+	[5] [6]8,000	T11.
19 Jordan Valley	500	4,700	......	......	[8]361,800	2,500	4,000	60,000	13,000	T31.
20 Bountiful district	} 450	4,600	}	......	......	......	......	......	3,000	} T5, T35, P518.
21 Weber Delta district	}		}	......	[8]70,000	......	......	......	26,000	}
22 Heber Valley	330	5,700	......	......	[8]86,000	30,000	17,000	11,000	2,800	T27.
23 Rhodes Valley		......	......	......	[8]22,000	......	....	10,000	3,100	T27.
24 Morgan Valley		......	......	......	......	......	......	......	[5]1,000±	
25 Ogden Valley		4,900	......	......	[8]23,000+	......	2,100	1,600	[5]1,000	W796–D.
26 Promontory Mountains area	360	4,300	170	240,000	12,000	15,000	9,000	14,000	7,600	T38, T42.
27 Blue Creek valley	220	4,800	122	184,000	14,000	......	5,500	200	2,000	T37, T42.
28 Hansel Valley and northern Rozel Flat	240	4,600	122	160,000	8,000	......	1,000	7,600	650	T33, T42.
29 Pocatello Valley		5,100	......	......	......	......	......	......	[5]2,000	W333.
30 Curlew Valley	1,200	4,500	122	868,000	75,600	......	Sm	[10]34,000	[10]10,000	T25, T42.
31 Park Valley area	1,050	4,900	149	520,000	24,000	......	8,000	16,000	5,000	T30, T42.
32 Great Salt Lake Desert		......	......	......	......	......	......	......	......	W1475–N, T42.
A. Lake subarea		......	......	......	......	......	......	......	([6])	
B. Western Desert		......	......	......	......	......	......	......	[5]80,000	
Bear River subregion										
33 Evanston area		....	....	....	....	......	......	......	[5]6,000	
34 Randolf area		....	....	....	....	......	......	......	[5]9,000	
35 Smith's Fork area		6,800	....	....	....	......	......	......	[5]9,000	W1539–U.
36 Bear Lake Valley		....	....	....	....	......	......	......	[5] [6]9,000	01969.
37 Soda Springs–China Hat area		6,000	....	....	....	......	......	......	[5]4,000	01969.
38 Gem Valley–Portneur area		5,500	....	....	....	......	......	......	[5]7,000	01969.
39 Cache Valley	1,840	4,700	150	......	[7]280,000	......	Sm	......	25,000	T36.
40 Malad Valley, Idaho	485	4,600	......	......	50,000(?)	......	3,400	28,500	3,000	W1412, 01969
41 Lower Malad Valley–Bear River Bay area		....	....	....	....	3,400	......	......	[5]19,000	

Sevier Lake subregion

42 Upper Sevier River basin	2,400	6,000–7,000	60–100	......	......	......	1,000–2,000	43,500	5,000	W1836.
43 Central Sevier River basin	2,800	5,100–6,000	120	......	......	1,000–2,000	Sm	100,000	6,200	W1787.
44 Sanpete Valley	850	5,500	128	......	......	L	2,200	113,000	18,000	W1896.
45 Yuba Dam–Leamington Canyon area	900	5,000	100–135	......	......	......	2,200	......	[5]10,000	W1848.
46 Sevier Desert	3,000	4,700	149	......	[5]169,800	16,200	< 5,000	135,000–175,000	70,000	W1854.
A. Dry Lake subarea		......	......	......	......	......	......	......	[5]12,000	
47 Pavant Valley	800	4,700	155	......	66,000	SM	14,000	24,000	[11]31,800	W1794.
48 Beaver Valley	200	5,800	......	......	......	......	100	......	[5]4,000	T13, W217.
49 Parowan Valley	150	5,800	150	......	......	......	3,000	......	[5]8,000	T13, W993.
50 Milford Valley		......	......	......	......	......	......	......	[5]29,000	T13, W659–A.
51 Cedar City Valley	300	5,500	150	......	[9]30,000	3,000	2,100	......	[5]15,000	T13, W993.
52 Lund district		......	......	......	......	......	......	......	[5]20,000	T13.
53 Beryl-Enterprise district	400	5,200	......	......	[9]5,000	......	Sm	5,000	[5]25,000	T13, W659–A.
54 Wah Wah Valley		......	......	......	......	......	......	......	[5]8,000	W277.

[1]Tabular material for Nevada from Office of the Nevada State Engineer, 1971, Water For Nevada, Nevada's Water Resources Water Planning Report 3.

[2]In Nevada, valleys are grouped by hydrographic regions as used by the Nevada State Engineer. These regions are shown in Nevada Planning Report 3, which also shows the hydrographic area numbers used in this report. Areas in Nevada outside the Great Basin Region also have assigned numbers; this is why some numbers are missing in this table. A separate numbering system is used in areas in Utah, Idaho, and Wyoming.

[3]Most estimates of recharge are based on an empirical relationship between precipitation and ground-water recharge. Estimates preceded by a footnote indicate areas where some potential recharge may be rejected and actual recharge may be less than estimated.

[4]Determined for upper 100 feet of saturated valley fill unless noted otherwise.

[5]Estimated by the authors for use in this report.

[6]Does not include storage beneath lake areas.

[7]Includes precipitation in recharge areas only.

[8]Includes recharge from surface water originating outside the drainage basin.

[9]Estimated from determinations of storage in upper 400 feet of valley fill.

[10]Utah part of valley only.

[11]Estimated from determinations of storage in upper 500 feet of saturated valley-fill materials.

TABLE 8 REFERENCES

PUBLICATIONS OF THE U.S. GEOLOGICAL SURVEY

PROFESSIONAL PAPERS

P417–F..............A water budget for Carson Valley, Nevada: A. M. Piper, 1969.

P518..................Lake Bonneville—Geology and hydrology of the Weber Delta district, including Ogden, Utah: J. H. Feth and others, 1966.

WATER-SUPPLY PAPERS

W217.................Water resources of Beaver Valley, Utah: W. T. Lee, 1908.

W277.................Ground water in Juab, Millard, and Iron Counties, Utah: O. E. Meinzer, 1911.

W333.................Ground water in Box Elder and Tooele Counties, Utah: Everett Carpenter, 1913.

W659–A............A method of estimating ground-water supplies based on discharge by plants and evaporation from soil—Results of investigations in Escalante Valley, Utah: W. N. White, 1932.

W796–D............Geology and ground-water resources of Ogden Valley, Utah: R. M. Leggette and G. H. Taylor, 1937.

W993.................Geology and ground-water resources of Cedar City and Parowan valleys, Iron County, Utah: H. E. Thomas and G. H. Taylor, 1946.

W1228...............Geology and water resources of Smith Valley, Lyon and Douglas Counties, Nevada: O. J. Loeltz and T. E. Eakin, 1953.

W1412...............Water consumption by water-loving plants in the Malad Valley, Oneida County, Idaho: R. W. Mower and R. L. Nace, 1957.

W1475–N..........Hydrology and stock-watering development on the public domain of western Utah: C. T. Snyder, 1963.

W1539–U..........Availability of ground water in the Bear River valley, Wyoming: C. J. Robinove and D. W. Berry, *with a section on chemical quality of the water,* by J. G. Connor, 1963.

W1779...............Evaluation of hydrogeology and hydrochemistry of Truckee Meadows area, Washoe County, Nevada: Philip Cohen and O. J. Loeltz, 1964.

W1787...............Ground-water conditions and storage in the central Sevier Valley, Utah: R. A. Young and C. H. Carpenter, 1965.

W1794...............Ground-water resources of Pavant Valley, Utah: R. W. Mower, 1965.

W1832...............Hydrology of the valley fill and carbonate-rock reservoirs, Pahrump Valley, Nevada–California: G. T. Malmberg, 1967.

W1836...............Ground-water conditions and geologic reconnaissance in the upper Sevier River basin, Utah: C. H. Carpenter, G. B. Robinson, Jr., and L. B. Bjorklund, 1967.

W1848...............Ground-water resources of the Sevier River basin between Yuba Dam and Leamington Canyon, Utah: L. J. Bjorklund and G. B. Robinson, Jr., 1968.

W1854...............Ground-water hydrology of the Sevier Desert, Utah: R. W. Mower and R. D. Feltis, 1968.

W1896...............Ground-water hydrology of the San Pitch River drainage basin, Sanpete County, Utah: G. B. Robinson, Jr., 1971.

W1972...............The Lake Tahoe Basin, California-Nevada: J. R. Crippen and B. R. Pavelka, 1970.

OPEN-FILE REPORTS

O1969 .... Availability of ground water for large-scale use in the Malad Valley-Bear River areas of southeastern Idaho—An initial assessment: W. L. Burnham, A. H. Harder, and N. P. Dion, 1969.

PUBLICATIONS OF THE NEVADA DEPARTMENT OF CONSERVATION AND NATURAL RESOURCES

WATER RESOURCES—RECONNAISSANCE SERIES

R1 .... Newark Valley: T. E. Eakin, 1960.
R2 .... Pine Valley: T. E. Eakin, 1961.
R3 .... Long Valley [White Pine County]: T. E. Eakin, 1961.
R4 .... Pine Forest Valley: W. C. Sinclair, 1962.
R5 .... Imlay area: T. E. Eakin, 1962.
R6 .... Diamond Valley: T. E. Eakin, 1962.
R7 .... Desert Valley: W. C. Sinclair, 1962.
R9 .... Gabbs Valley: T. E. Eakin, 1962.
R10 .... Sarcobatus Flat and Oasis Valley: G. T. Malmberg and T. E. Eakin, 1962.
R11 .... Hualapai Flat: W. C. Sinclair, 1962.
R12 .... Ralston and Stonecabin Valleys: T. E. Eakin, 1962.
R13 .... Cave Valley: T. E. Eakin, 1962.
R14 .... Amargosa Desert: G. E. Walker and T. E. Eakin, 1963.
R15 .... Long Valley—Massacre Lake region: W. C. Sinclair, 1963.
R16 .... Dry Lake and Delamar Valleys: T. E. Eakin, 1963.
R17 .... Duck Lake Valley: W. C. Sinclair, 1963.
R18 .... Garden and Coal Valleys: T..E. Eakin, 1963.
R19 .... Antelope and Middle Reese River Valleys: E. G. Crosthwaite, 1963.
R20 .... Black Rock Desert area: W. C. Sinclair, 1963.
R22 .... Pueblo Valley—Continental Lake region: W. C. Sinclair, 1963.
R23 .... Dixie—Fairview Valley area: Philip Cohen and D. E. Everett, 1963.
R24 .... Lake Valley: F. E. Rush and T. E. Eakin, 1963.
R26 .... Edwards Creek Valley: D. E. Everett, 1964.
R28 .... Smith Creek and Ione Valleys: D. E. Everett and F. E. Rush, 1964.
R29 .... Grass Valley [Pershing County]: Philip Cohen, 1964.
R30 .... Monitor, Antelope, and Kobeh Valleys: F. E. Rush and D. E. Everett, 1965.

The following reports are water-resources appraisals of the specified areas.

R31 .... Upper Reese Valley: T. E. Eakin, D. O. Moore, and D. E. Everett, 1965.
R32 .... Lovelock Valley: D. E. Everett and F. E. Rush, 1965.
R33 .... Spring Valley: F. E. Rush and S. A. T. Kazmi, 1965.
R34 .... Snake Valley area: J. W. Hood and F. E. Rush, 1965.
R35 .... Huntington Valley area: F. E. Rush and D. E. Everett, 1965.
R36 .... Eldorado–Piute Valley area: F. E. Rush and C. J. Huxel, Jr., 1966.
R37 .... Grass and Carico Lake Valleys: D. E. Everett and F. E. Rush, 1966.
R38 .... Little Fish Lake, Hot Creek, and Little Smoky Valleys: F. E. Rush and D. E. Everett, 1966.
R39 .... Eagle Valley: G. F. Worts, Jr. and G. T. Malmberg, 1966.
R40 .... Walker Lake area: D. E. Everett and F. E. Rush, 1967.
R41 .... Washoe Valley: F. E. Rush, 1967.
R42 .... Steptoe Valley: T. E. Eakin and J. L. Hughes, 1967.
R43 .... Warm Springs–Lemmon Valley area: F. E. Rush and P. A. Glancy, 1967.
R44 .... Smoke Creek and San Emidio Deserts: P. A. Glancy, 1967.
R45 .... Clayton Valley–Stonewall Flat area: F. E. Rush, 1968.
R46 .... Mesquite–Ivanpah Valley area: P. A. Glancy, 1968.
R47 .... Thousand Springs Valley: F. E. Rush, 1968.
R49 .... Butte Valley, P. A. Glancy: 1968.
R52 .... Columbus Salt Marsh–Soda Spring Valley area: A. S. Van Denburgh and P. A. Glancy, 1970.
R53 .... Antelope Valley and East Walker River area: P. A. Glancy, 1971.
R54 .... The Nevada test site area: F. E. Rush, 1971.
R55 .... Granite Springs Valley area: J. R. Harrill, 1971.
R56 .... Pilot Creek area: J. R. Harrill, 1971.
R57 .... Truckee River basin: A. S. Van Denburgh, R. D. Lamke, and J. L. Hughes, 1973.
R59 .... Carson River basin: P. A. Glancy and T. L. Katzer, 1976 (in press).

WATER-RESOURCES BULLETINS

B11 .... Preliminary report on ground water in the Fish Lake Valley, Nevada and California: T. E. Eakin, 1950.
B12 .... Contributions to the hydrology of eastern Nevada: T. E. Eakin, G. B. Maxey, T. W. Robinson, J. C. Fredericks, and O. J. Loeltz, 1951.
B13 .... Geology and ground-water resources of Buena Vista Valley, Pershing County, Nevada: O. J. Loeltz and D. A. Phoenix, 1955.
B15 .... Ground-water potentialities in the Crescent Valley, Eureka and Lander Counties, Nevada: C. P. Zones, 1961.
B17 .... Ground-water conditions in the Fernley-Wadsworth area, Churchill, Lyon, Storey, and Washoe Counties, Nevada: W. C. Sinclair and O. J. Loeltz, 1963.
B19 .... Preliminary results of hydrogeochemical studies in the Humboldt River Valley near Winnemucca, Nevada: Philip Cohen, 1962.
B20 .... A summary of specific-yield and particle-size relation of Quaternary alluvium, Humboldt River valley, Nevada: Philip Cohen, 1963.
B22 .... Preliminary results of hydrogeologic investigations in the valley of the Humboldt River near Winnemucca, Nevada: Philip Cohen, 1964.
B24 .... An evaluation of the water resources of the Humboldt River valley near Winnemucca, Nevada: Philip Cohen, 1963.
B27 .... Water in the Humboldt River valley near Winnemucca, Nevada: Philip Cohen, 1964.
B31 .... The effects of pumping on the hydrology of Kings River valley, Humboldt County, Nevada, 1957-64: G. T. Malmberg and G. F. Worts, Jr., 1966.
B32 .... Hydrologic reconnaissance of the Humboldt River basin, Nevada: T. E. Eakin and R. D. Lamke, 1966.
B33 .... A regional interbasin ground-water system in the White River area, southeastern Nevada: T. E. Eakin, 1966.
B34 .... Effects of irrigation development on the water supply of Quinn River Valley area, Nevada and Oregon, 1950–64: C. J. Huxel, Jr. 1966.
B35 .... Hydrologic response to irrigation pumping in Diamond Valley, Eureka and Elko Counties, Nevada, 1950–65: J. R. Harrill, 1968.
B37 .... Hydrologic response to irrigation pumping in Hualapai Flat, Washoe, Pershing, and Churchill Counties, Nevada, 1960–67: J. R. Harrill, 1969.

B38.....................Water resources and development in Mason Valley, Lyon County, Nevada, 1948-65: C. J. Huxel, 1969.

B39.....................Effects of pumping ground water in Paradise Valley on the hydrologic regimen of the Little Humboldt River basin, Humboldt and Elko Counties, Nevada: J. R. Harrill, 1970.

B41.....................Water-resources of Big Smokey Valley, Lander, Nye, and Esmeralda Counties, Nevada: F. E. Rush and C. V. Schroer, 1970.

PUBLICATIONS OF THE UTAH DEPARTMENT OF NATURAL RESOURCES (DIVISION OF WATER RIGHTS)

TECHNICAL PUBLICATIONS

T5.....................Ground water in the East Shore area, Utah, Part 1, Bountiful district, Davis County: H. E. Thomas and W. B. Nelson, 1948. (In Utah State Engineer 26th Bienn. Rept., p. 53-206).

T11.....................Ground water in northern Utah Valley, Utah—A progress report for the period 1948-63: R. M. Cordova and Seymour Subitzky, 1965.

T12.....................Reevaluation of the ground-water resources of Tooele Valley, Utah: J. S. Gates, 1965.

T13.....................Ground-water resources of selected basins in southwestern Utah: G. W. Sandberg, 1966.

T14.....................Water-resources appriasal of the Snake Valley area, Utah and Nevada: J. W. Hood and F. E. Rush, 1965.

T16.....................Ground-water conditions in Cedar Valley, Utah County, Utah: R. D. Feltis, 1967.

T17.....................Ground-water resources of northern Juab Valley, Utah: L. J. Bjorklund, 1967.

T18.....................Hydrologic reconnaissance of Skull Valley, Tooele County, Utah: J. W. Hood and K. M. Waddell, 1968.

T23.....................Hydrologic reconnaissance of Rush Valley, Tooele County, Utah: J. W. Hood, Don Price, and K. M. Waddell, 1969.

T24.....................Hydrologic reconnaissance of Deep Creek valley, Tooele and Juab Counties, Utah, and Elko and White Pine Counties, Nevada: J. W. Hood and K. M. Waddell, 1969.

T25.....................Hydrologic reconnaissance of Curlew Valley, Utah and Idaho: E. L. Bolke and Don Price, 1969.

T26.....................Hydrologic reconnaissance of Sink Valley, Tooele and Box Elder Counties, Utah: Don Price and E. L. Bolke, 1970.

T27.....................Water resources of the Heber-Kamas Park City area, north-central Utah: C. H. Baker, Jr., 1970.

T28.....................Ground-water conditions in southern Utah Valley and Goshen Valley, Utah: R. M. Cordova, 1970.

T29.....................Hydrologic reconnaissance of Grouse Creek valley, Box Elder County, Utah: J. W. Hood and Don Price, 1970.

T30.....................Hydrologic reconnaissance of the Park Valley area, Box Elder County, Utah: J. W. Hood, 1971.

T31.....................Water resources of Salt Lake County, Utah: A. G. Hely, R. W. Mower, and C. A. Harr, 1971.

T33.....................Hydrologic reconnaissance of Hansel Valley and northern Rozel Flat, Box Elder County, Utah: J. W. Hood, 1971.

T35.....................Ground-water conditions in the East Shore area, Box Elder, Davis, and Weber Counties, Utah: E. L. Bolke and K. M. Waddell, 1972.

T36.....................Ground-water resources of Cache Valley, Utah and Idaho: L. J. Bjorklund and L. J. McGreevy, 1971.

T37.....................Hydrologic reconnaissance of the Blue Creek Valley area, Box Elder County, Utah: E. L. Bolke and Don Price, 1972.

T38.....................Hydrologic reconnaissance of the Promontory Mountains area, Box Elder County, Utah: J. W. Hood, 1976 (in press).

T41.....................Hydrologic reconnaissance of Pilot Valley, Utah and Nevada: J. C. Stephens and J. W. Hood, 1973.

T42.....................Hydrologic reconnaissance of the northern Great Salt Lake Desert and summary hydrologic reconnaissance of northwestern Utah: Jerry C. Stephens, 1974.

CENTRAL LAHONTON

75. Bradys Hot Springs area
76. Fernley area
77. Fireball Valley
80. Winnemucca Lake valley
81. Pyramid Lake valley
82. Dodge Flat
83. Tracy segment
84. Warm Springs valley
85. Spanish Springs valley
86. Sun Valley
87. Truckee Meadows
88. Pleasant Valley
89. Washoe Valley
90. Lake Tahoe basin
91. Truckee Canyon segment
92. Lemmon Valley
93. Antelope Valley
100. Cold Spring valley
101. Carson Desert
  A. Packard Wash
102. Churchill Valley
103. Dayton Valley
104. Eagle Valley
105. Carson Valley
106. Antelope Valley
107. Smith Valley
108. Mason Valley
109. East Walker area
110. Walker Lake valley
  A. Schurz subarea
  B. Lake subarea
  C. Wiskey Flat Hawthorne subarea
240. Upper West Walker River
241. Bridgeport Valley

HUMBOLDT

2. Continental Lake valley
3. Gridley Lake valley
4. Virgin Valley
7. Swan Lake valley
8. Massacre Lake valley
9. Long Valley
12. Mosquito Valley
15. Boulder Valley
16. Duck Lake valley
19. Dry Valley
20. Sano Valley
21. Smoke Creek Desert
22. San Emidio Desert
23. Granite basin
24. Hualapai Flat
25. High Rock Lake valley
26. Mud Meadow
27. Summit Lake valley
28. Black Rock Desert
29. Pine Forest Valley
30. Kings River valley
31. Desert Valley
32. Silver State Valley
33. Quinn River valley
42. Marys River area
43. Starr Valley area
44. North Fork area
45. Lamoille Valley
46. South Fork area
47. Huntington Valley
48. Dixie Creek–Tenmile Creek area
49. Elko segment
50. Susie Creek area
51. Maggie Creek area
52. Marys Creek area
53. Pine Valley
54. Crescent Valley
55. Carico Lake valley
56. Upper Reese River valley
57. Antelope Valley
58. Middle Reese River valley
59. Lower Reese River valley
60. Whirlwind Valley
61. Boulder Flat
62. Rock Creek valley
63. Willow Creek valley
64. Clovers area
65. Pumpernickel Valley
66. Kelly Creek area
67. Little Humboldt Valley
68. Hardscrabble area
69. Paradise Valley
70. Winnemucca segment
71. Grass Valley
72. Imlay area
73. Lovelock Valley
74. White Plains
78. Granite Springs valley
79. Kumiva Valley

TONOPAH

113. Huntoon Valley
114. Teels Marsh valley
117. Fish Lake valley
118. Columbus Salt Marsh valley
119. Rhodes Salt Marsh valley
120. Garfield Flat
121. Soda Springs valley
122. Gabbs Valley
123. Rawhide Flats
124. Fairview Valley
125. Stingaree Valley
126. Cowkick Valley
127. Eastgate Valley area
128. Dixie Valley
129. Buena Vista Valley
130. Pleasant Valley
131. Buffalo Valley
132. Jersey Valley
133. Edwards Creek valley
134. Smith Creek valley
135. Ione Valley
136. Monte Cristo Valley
137. Big Smokey Valley
  A. Tonopah Flat
  B. Northern part
138. Grass Valley
139. Kobeh Valley
140. Monitor Valley
141. Ralston Valley
142. Alkali Spring valley (Esmeralda)
143. Clayton Valley
144. Lida Valley
145. Stonewall Flat
146. Sarcobatus Flat
147. Gold Flat
148. Cactus Flat
149. Stone Cabin Valley
150. Little Fish Lake valley
151. Antelope Valley (Eureka and Nye)
152. Stevens basin
153. Diamond Valley
154. Newark Valley
155. Little Smoky Valley
156. Hot Creek valley
157. Kawich Valley
158. Emigrant Valley
159. Yucca Flat
160. Frenchman Flat
161. Indian Springs valley
162. Pahrump Valley
163. Mesquite Valley
164. Ivanpah Valley
165. Jean Lake valley
166. Hidden Valley
167. Eldorado Valley
168. Three Lakes valley
169. Tikapoo Valley
170. Penoyer Valley
171. Coal Valley
172. Garden Valley
173. Railroad Valley
174. Jakes Valley
175. Long Valley
176. Ruby Valley
177. Clover Valley
178. Butte Valley
  A. Northern part
  B. Southern part
179. Steptoe Valley
180. Cave Valley
181. Dry Lake valley
182. Delamar Valley
183. Lake Valley
184. Spring Valley
185. Tippett Valley
186. Antelope Valley
187. Goshute Valley
188. Independence Valley
225. Mercury Valley
226. Rock Valley
227. Forty Mile Canyon
228. Oasis Valley
229. Crater Flat
230. Amargosa Desert

BEAR RIVER

33. Evanston area
34. Randolf area
35. Smith's Fork area
36. Bear Lake valley
37. Soda Springs–China Hat area
38. Gem Valley–Portneur area
39. Cache Valley
40. Malad Valley, Idaho
41. Lower Malad Valley–Bear River Bay area.

GREAT SALT LAKE

189. Thousand Springs valley
191. Pilot Creek valley
194. Pleasant Valley
196. Hamlin Valley
1. Grouse Creek valley
2. Pilot Valley
3. Deep Creek valley
4. Snake Valley
5. Pine Valley
6. White Valley
7. Fish Springs Flat
8. Dugway Valley
9. Government Creek valley
10. Skull Valley
11. Sink Valley
12. Tooele Valley
13. Rush Valley
14. Cedar Valley
15. Northern Juab Valley
16. Goshen Valley
17. Southern Utah Lake valley
18. Northern Utah Lake valley
19. Jordan Valley
20. Bountiful district
21. Weber Delta district
22. Heber Valley
23. Rhodes Valley
24. Morgan Valley
25. Ogden Valley
26. Promontory Mountains area
27. Blue Creek valley
28. Hansel Valley and northern Rozel Flat
29. Pocatello Valley
30. Curlew Valley
31. Park Valley area
32. Great Salt Lake Desert
  A. Lake subarea
  B. Western Desert

SEVIER LAKE

42. Upper Sevier River basin
43. Central Sevier River basin
44. Sanpete Valley
45. Yuba Dam–Leamington Canyon area
46. Sevier Desert
  A. Dry Lake subarea
47. Pavant Valley
48. Beaver Valley
49. Parowan Valley
50. Milford Valley
51. Cedar City Valley
52. Lund district
53. Beryl-Enterprise district
54. Wah Wah Valley

LEGEND FOR PLATES 1A, 1B, 1G, 1H, and 1I OF THE GREAT BASIN REGION

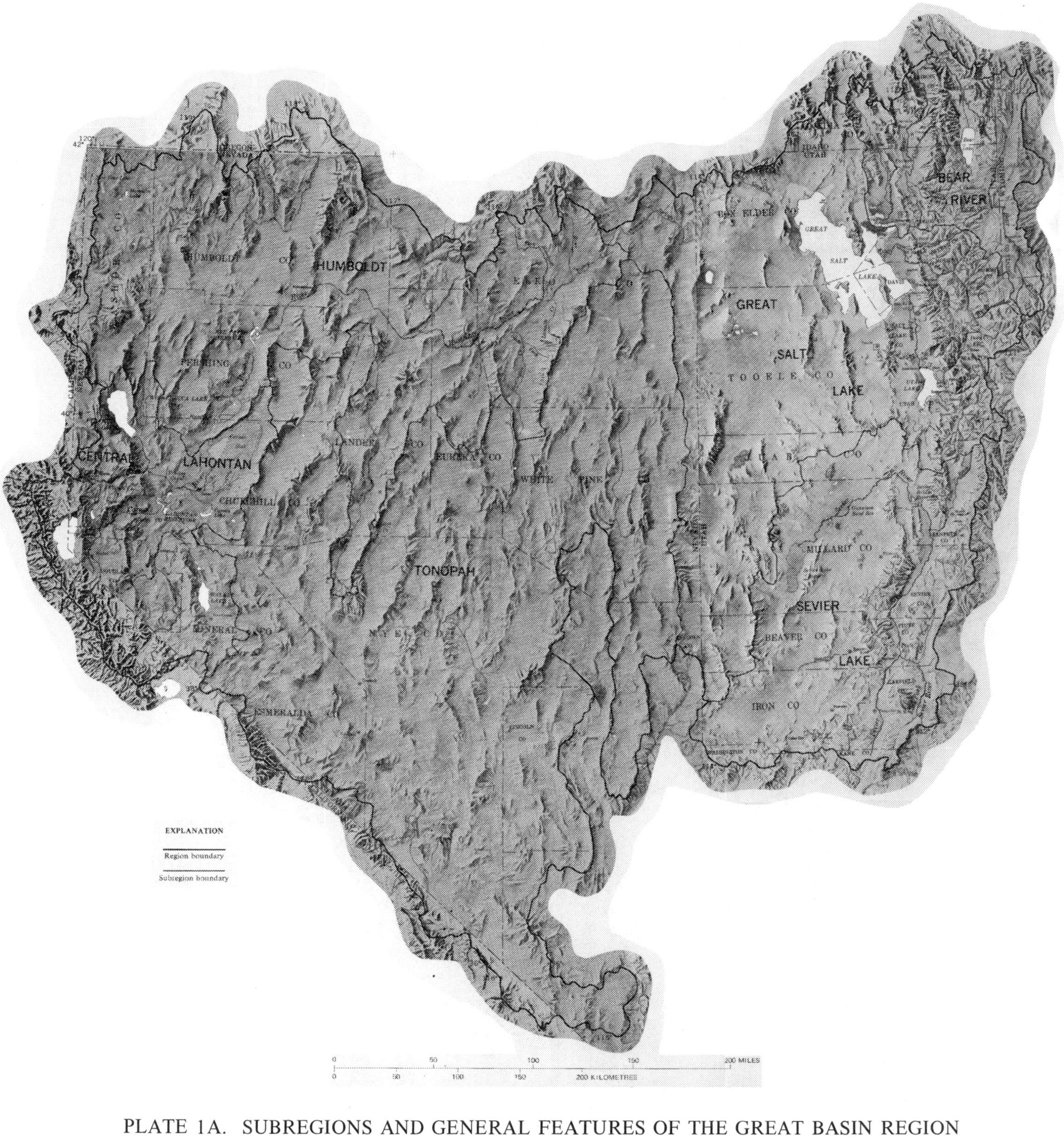

PLATE 1A. SUBREGIONS AND GENERAL FEATURES OF THE GREAT BASIN REGION

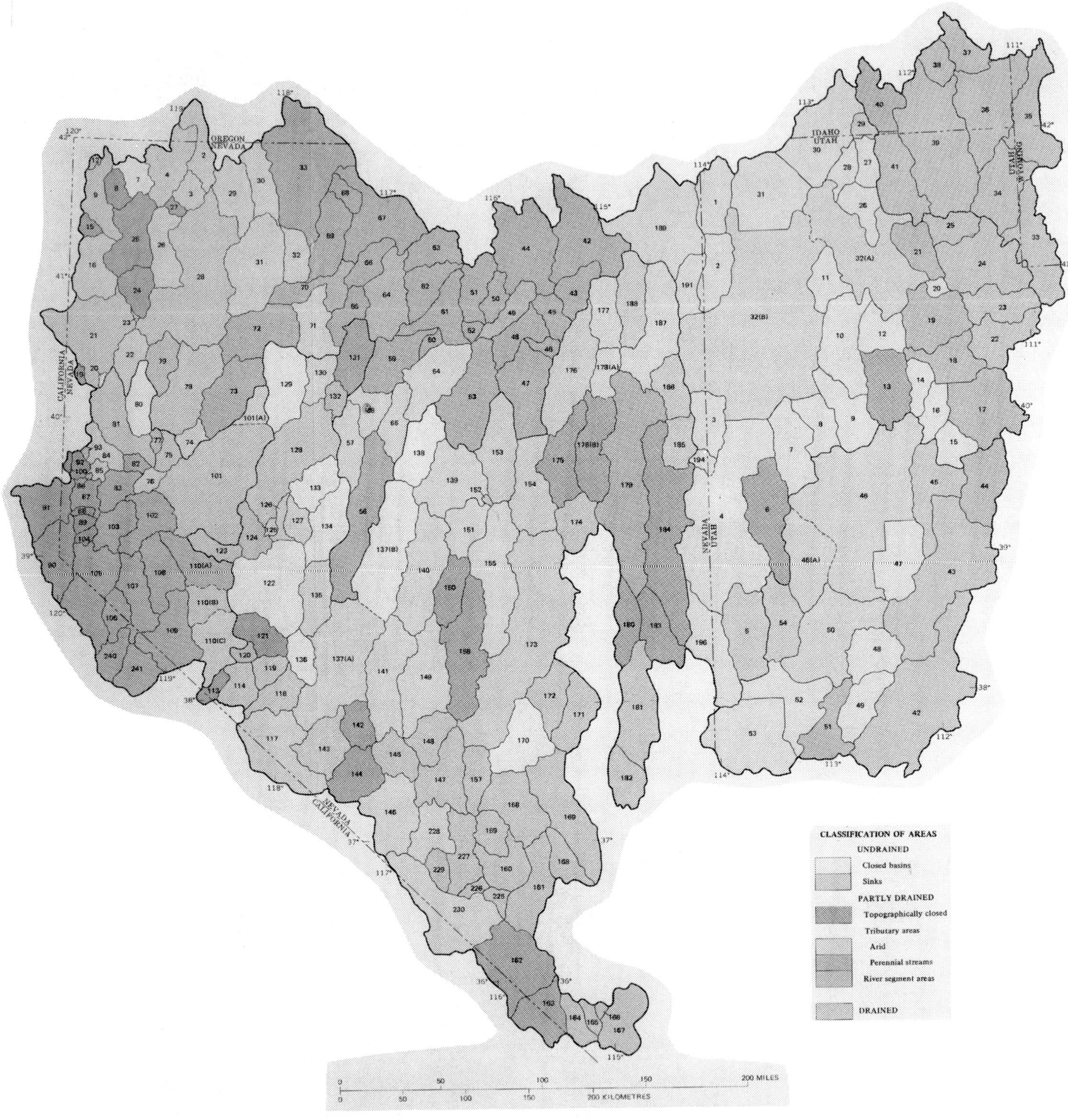

PLATE 1B. INDEX MAP AND CLASSIFICATION OF VALLEYS IN THE GREAT BASIN REGION

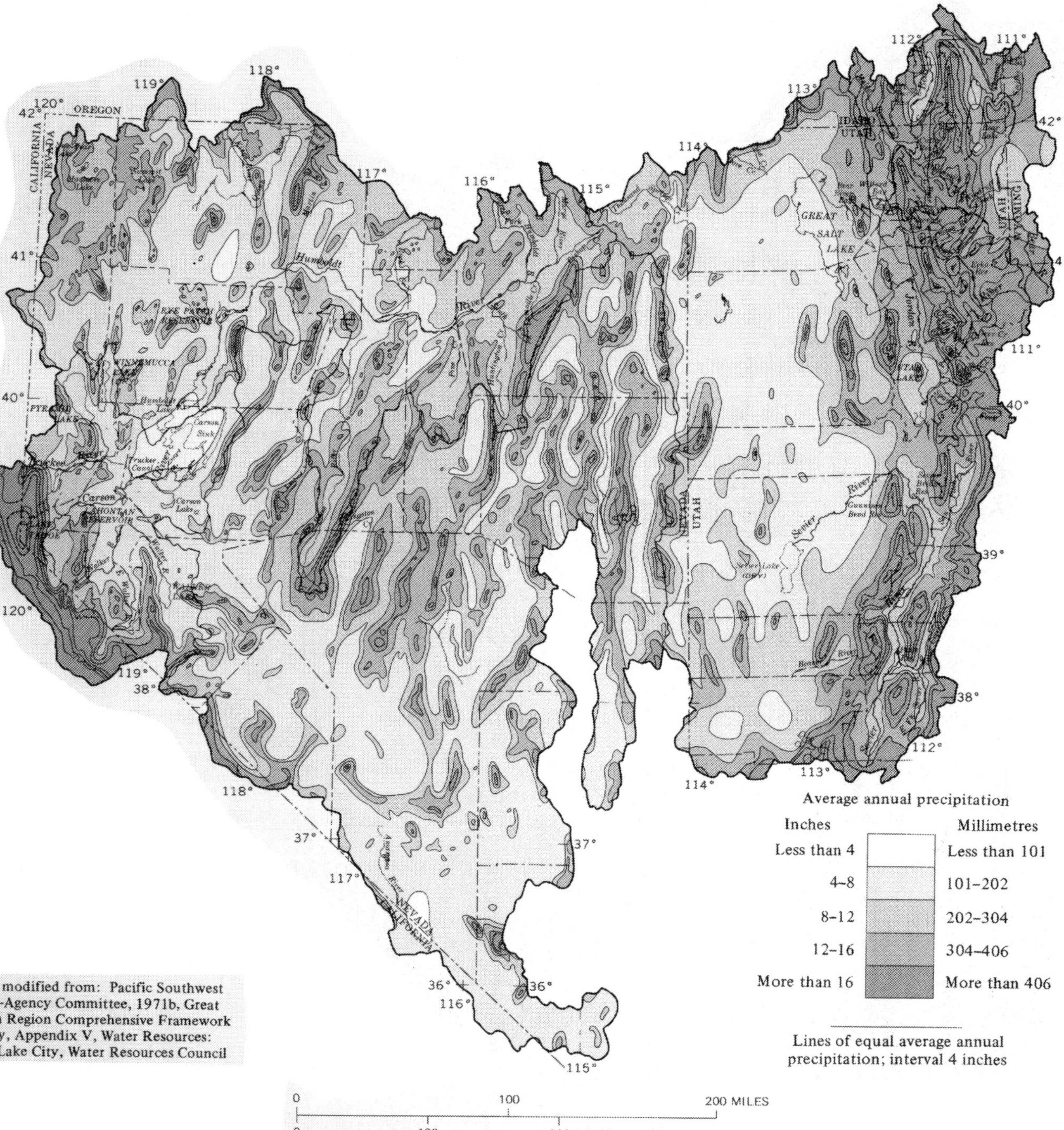

Data modified from: Pacific Southwest Inter-Agency Committee, 1971b, Great Basin Region Comprehensive Framework Study, Appendix V, Water Resources: Salt Lake City, Water Resources Council

PLATE 1C. AVERAGE ANNUAL PRECIPITATION IN THE GREAT BASIN REGION

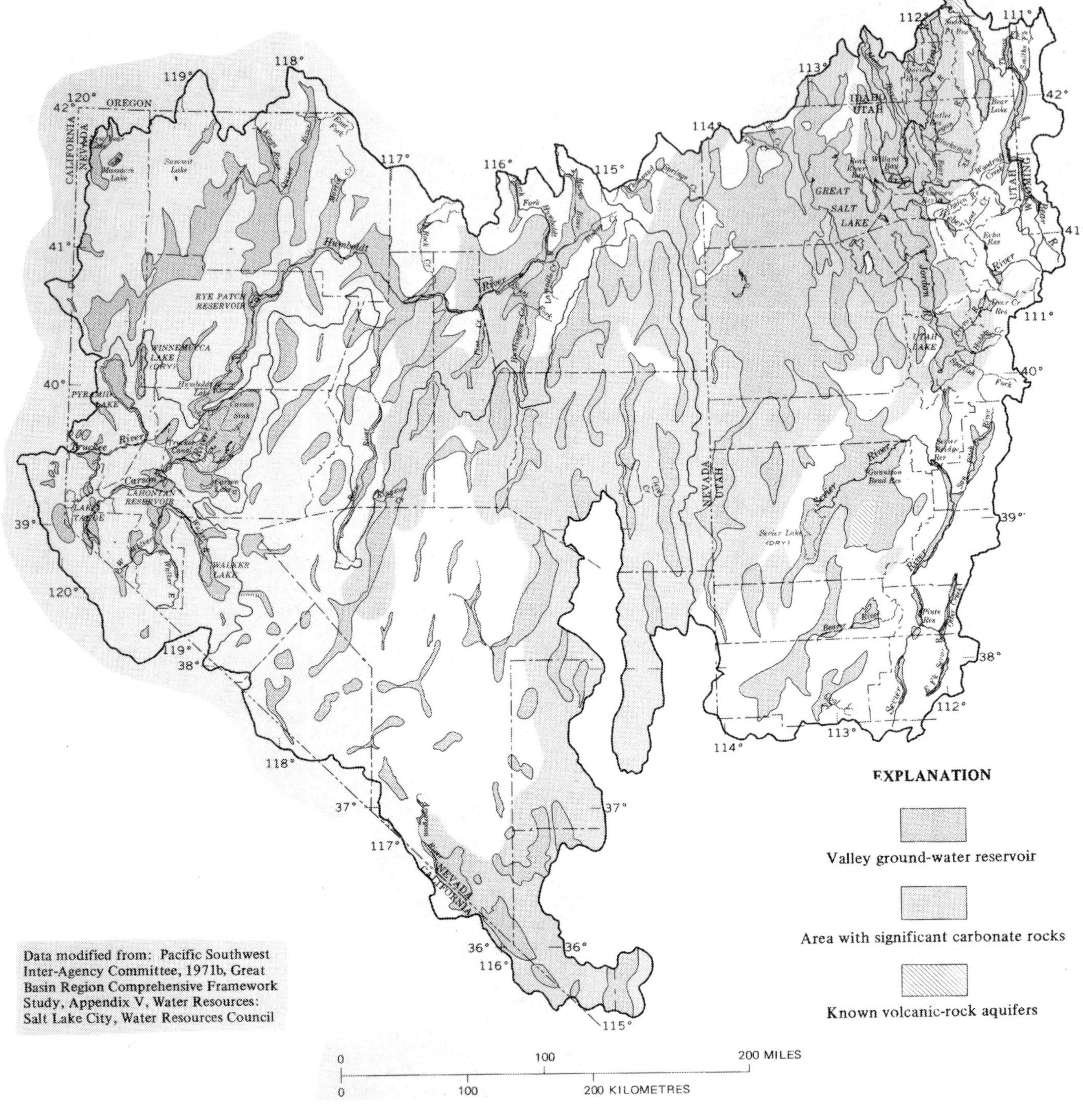

PLATE 1D. PRINCIPAL GROUND-WATER RESERVOIRS IN THE GREAT BASIN REGION

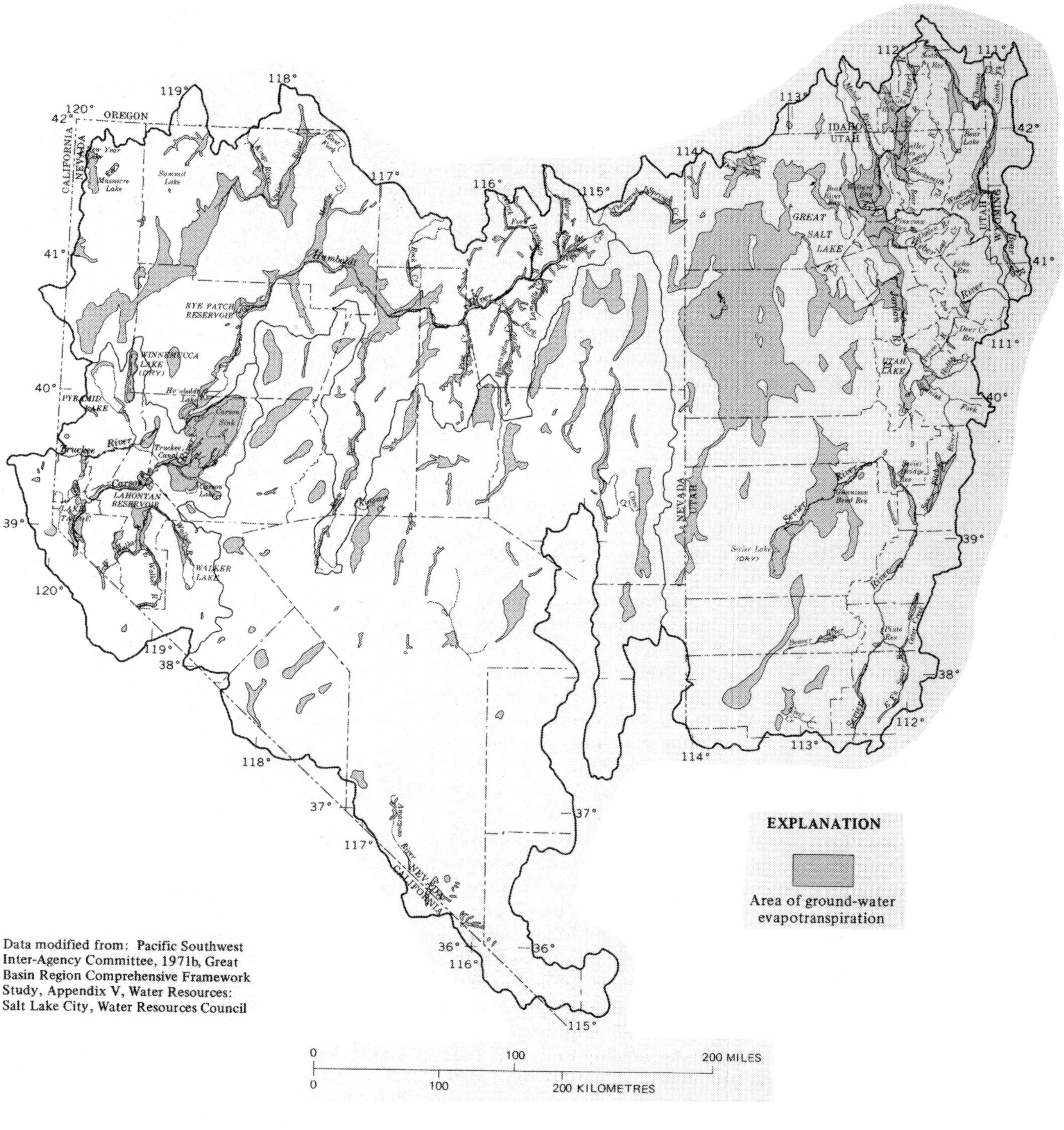

Data modified from: Pacific Southwest Inter-Agency Committee, 1971b, Great Basin Region Comprehensive Framework Study, Appendix V, Water Resources: Salt Lake City, Water Resources Council

PLATE 1E. AREAS OF NATURAL GROUND-WATER DISCHARGE IN THE GREAT BASIN REGION

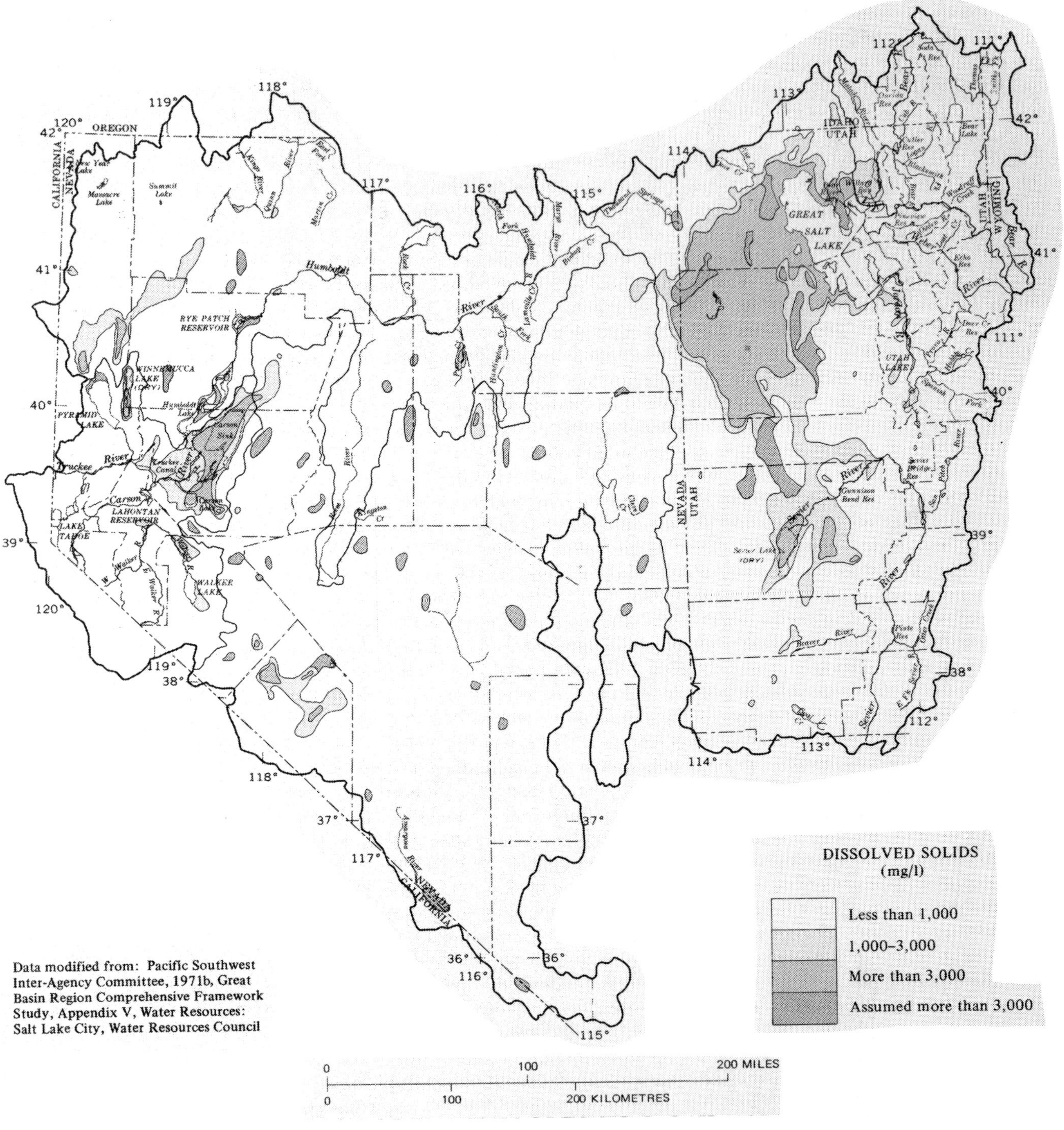

Data modified from: Pacific Southwest Inter-Agency Committee, 1971b, Great Basin Region Comprehensive Framework Study, Appendix V, Water Resources: Salt Lake City, Water Resources Council

PLATE 1F. DISSOLVED-SOLIDS CONCENTRATION OF GROUND-WATER IN THE GREAT BASIN REGION

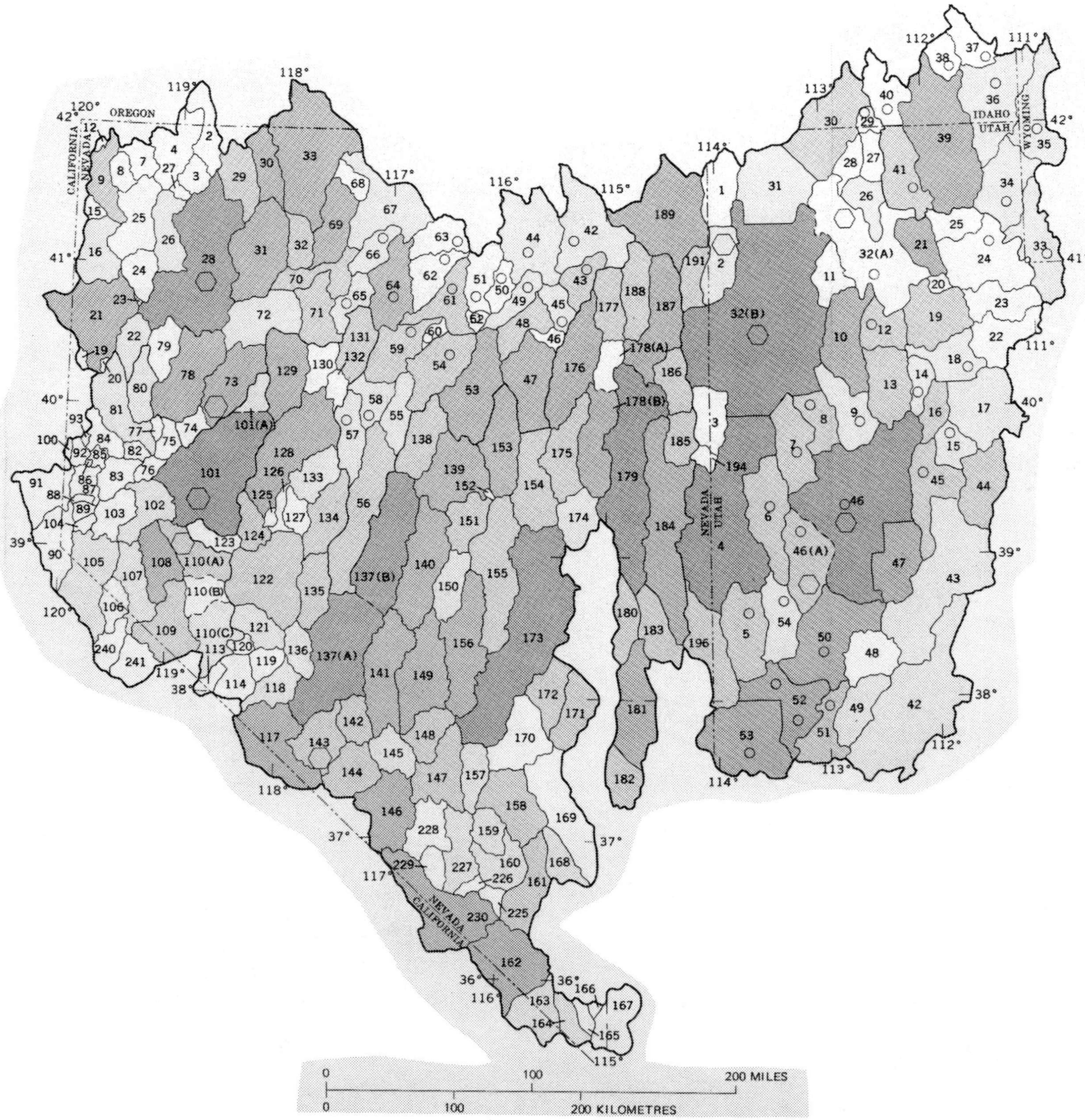

ESTIMATED RECOVERABLE GROUND WATER IN UPPER 100 FEET OF SATURATED VALLEY GROUND-WATER RESERVOIR DEPOSITS

Subregion	Area (millions of acres)	Storage (millions of acre-ft)
Bear River	1.2	9
Great Salt Lake	8.0	43
Sevier Lake	2.7	24
Humboldt	5.4	62
Central Lahontan	2.2	24
Tonopah	13.0	130
Region (rounded)		300

GROUND-WATER STORAGE

Thousands of acre-feet per foot dewatered based on estimates for upper 100 feet

- Less than 5
- 5–10
- 10–20
- 20–50
- More than 50

○ Provisional estimate made by authors for use in this report. Other estimates from reports listed in table 8.

⬡ Quality of stored water may be undesirable for many uses

PLATE 1G. GROUND-WATER STORAGE IN THE GREAT BASIN REGION

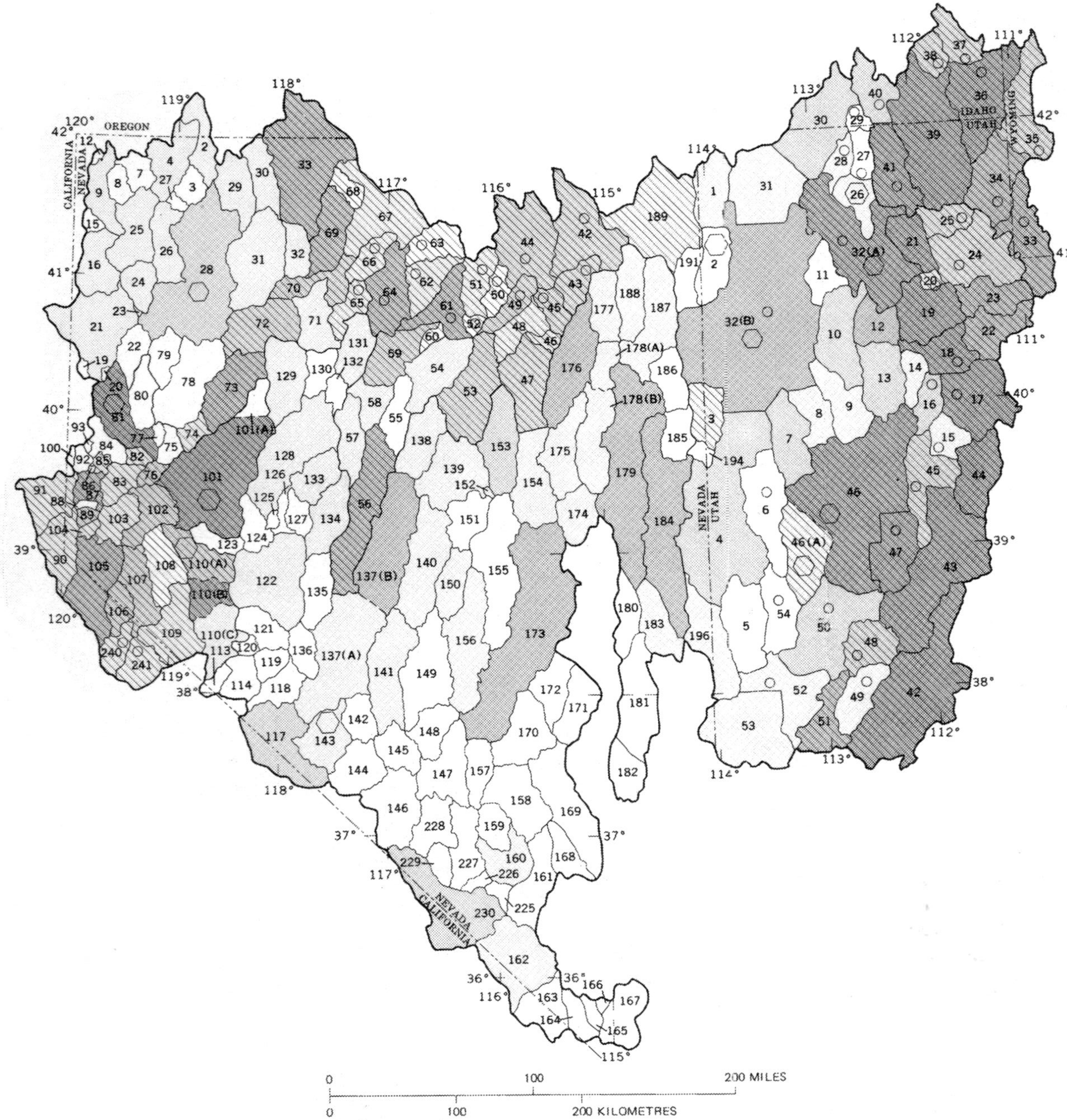

PLATE 1H. YIELD IN THE GREAT BASIN REGION

SUMMARY OF YIELD ESTIMATES

[Summary includes provisional estimates for 48 areas. Subregion totals may be revised by future studies]

Subregion	Million acre-feet per year		
	Perennial yield	System yield	Total
Bear River	--------	1.2	1.2
Great Salt Lake	0.6	[a]2.6	3.2
Sevier Lake	.1	1.0	1.1
Humboldt	.3	.9	1.2
Central Lahontan	Small	1.3	1.3
Tonopah	.8	--------	.8
Total	[b]1.8	[c]7.0	[d]8.8

[a]Includes inflow from Bear River subregion.

[b]Includes about 0.2 million acre-feet per year that may require treatment to be acceptable for most uses.

[c]Includes about 1.5 million acre-feet per year surface inflow to terminal lakes. Optimum rate of ground-water withdrawal not known; probably at least 0.8 million acre-feet per year could be pumped without significant adverse effects on existing surface-water rights.

[d]Total yield of the ground-water system is at least 2.6 million acre-feet per year. Larger withdrawals are possible if conflicts with surface-water rights are resolved and reuse and treatment are feasible.

YIELD

Thousands of acre-feet per year

Less than 5

5–25

25–50

50–100

More than 100

Hachures indicate system yield (the maximum amount of surface and ground water that can be obtained each year from sources within the system for an indefinite period of time); estimate assumes that outflow to downstream areas will be maintained at about the same level as in 1970. No hachures indicate perennial yield

⬡ Quality of water may be undesirable for many uses

○ Provisional estimate made by authors for use in this report. Other estimates from reports listed in table 8

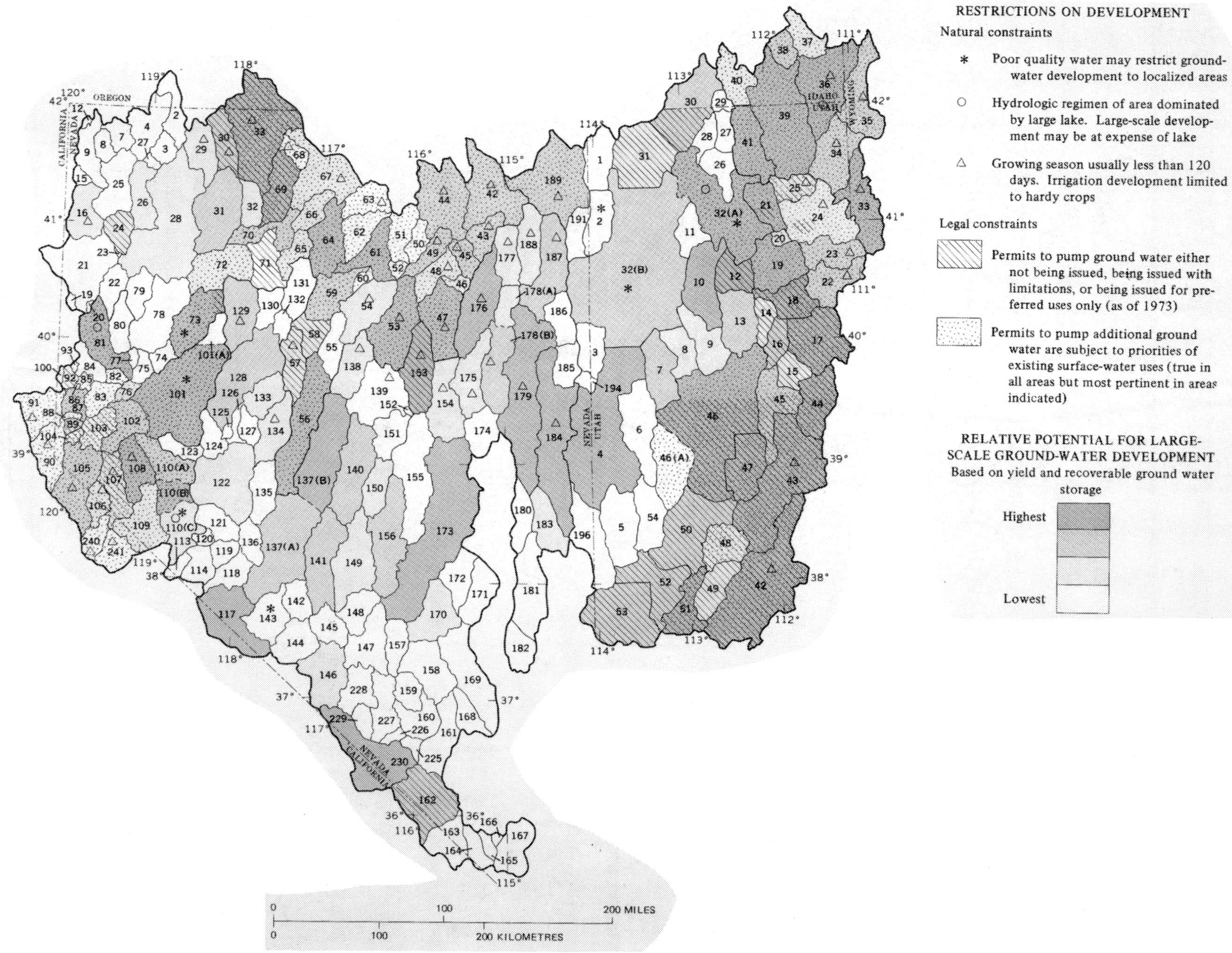

PLATE 11. POTENTIAL FOR GROUND-WATER DEVELOPMENT IN THE GREAT BASIN REGION

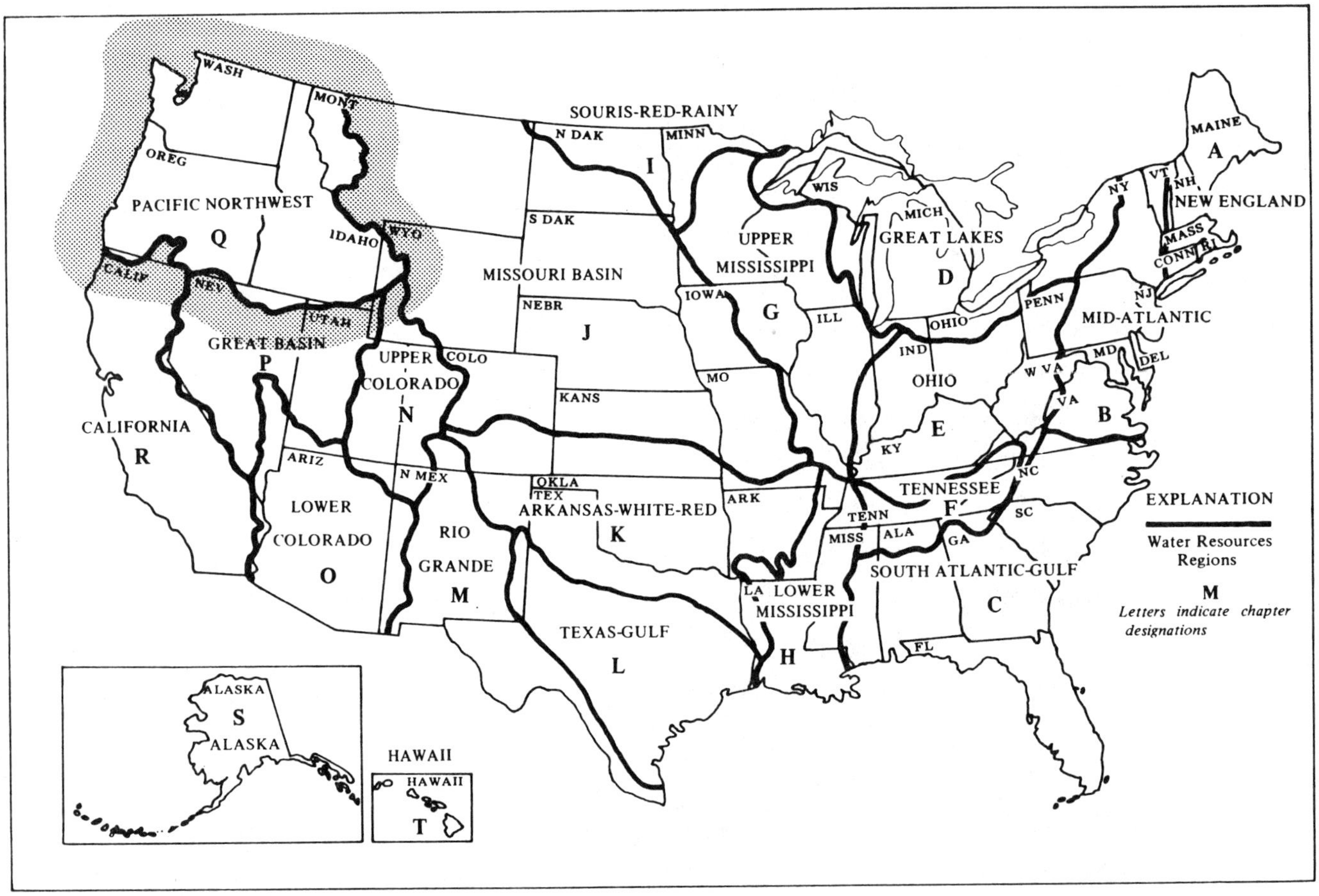

Q – Location Map for the Pacific Northwest Region

Q ....

PACIFIC NORTHWEST REGION

By BRUCE L. FOXWORTHY

ABSTRACT

The Pacific Northwest Region's ground-water reservoirs are capable of providing large additional fresh-water supplies; these reservoirs become more important as undeveloped surface-storage sites and unapportioned surface-water supplies dwindle. Withdrawals of fresh water from all surface and underground sources are increasing; they may rise from the rate of 30 billion gallons per day in 1970 to about 60 billion gallons per day in 2020. By 1975 the withdrawal of ground water had increased 70 percent over the 1970 rate and accounted for 22 percent of total fresh-water withdrawal. Substantial increases in ground-water withdrawal must continue if projected water demands are to be met in the future.

Large variations exist in the availability of and needs for ground water in the region, largely because of the great variety of landforms, climate, and earth materials. More than one-half the region is underlain by rock materials capable of yielding ground water to wells at moderate to large rates; six extensive areas are identified as major ground-water reservoirs. The most significant current problems pertaining to one or more of these ground-water reservoirs are: (1) Progressively declining water levels; (2) water-quality problems; (3) waterlogging; (4) inadequate information; and (5) competition for available supplies. In the future these same problems are expected to persist and generally worsen (especially water-quality deterioration) in most of the major reservoir areas.

Management opportunities in the region include: (1) Development of new supplies and additional uses of ground water; (2) protection and enhancement of water quality; (3) reduction of waterlogging; (4) energy development from some ground-water reservoirs; (5) improving access to the ground water; (6) increased use of underground space for storage and disposal; and (7) greater use of advanced management and conservation techniques. Conjunctive use of surface and ground water to provide greater available supplies probably is the most promising water-management opportunity. However, if the full potential of the ground-water resources is to be realized, important constraints, including present water-right structures and serious deficiencies in information, must be overcome.

INTRODUCTION

The Pacific Northwest, a region largely drained by The Columbia River, the second biggest river in the Nation, has the reputation for having abundant water. The long-term average discharge of stream water from the region has been estimated at 384,000 cubic feet per second—much greater than from most other U.S. regions of comparable area (Columbia-North Pacific Technical Staff, 1970b, p. 18). In spite of the apparent large total supply, however, water is not always available where and when it is needed. The distribution of runoff throughout the region is extremely uneven, and large variations in streamflow occur on a monthly, yearly, and long-term basis. Serious competition exists between irrigation diversions and important instream uses of surface water, and thousands of square miles of irrigated farmland are subject to water shortages. Although additional supplies of surface water can be developed at many places, problems associated with surface-water supplies are likely to worsen in the future as demands for water increase.

The Pacific Northwest Region's ground-water reservoirs, on the other hand, are capable of providing large additional fresh-water supplies, represent huge amounts of storage space for surplus runoff, and have the potential for greatly alleviating present and foreseeable water-supply problems. As undeveloped surface-storage sites and unapportioned surface-water supplies dwindle, the ground-water reservoirs of the region should become the focus of water-management schemes. Traditionally, however, the ground-water resources have been accorded a subordinate role in regional resource planning, owing partly to the major importance of surface water in the region and partly, no doubt, to a general lack of understanding of the potential value of the ground-water resource and of the underground space containing it.

The purpose of this report is to summarize the general status and significance of the region's ground-water resources, some of the likely demands upon them in the future, and major problems and constraints associated with their use, and to evaluate some of the ways in which the ground-water reservoirs can be further utilized. The intended audience includes, but is not restricted to, officials responsible for resource development and administration, and experienced in water-resource planning and management, but with somewhat limited familiarity with the ground-water resource. The report is intended to focus attention on the importance of, and opportunities for, ground-water management and conjunctive use of ground waters and surface waters to meet water needs under the changing conditions that seem likely within the next several decades.

The data on which this report is based were gleaned mainly from reports of many earlier studies. The U.S. Geological Survey has made water-resource studies in the region since about 1900, largely in cooperation with State, local, and other Federal agencies. Many of the reports of those studies are listed in the section on "Selected References," along with several reports of studies by other agencies or individuals on subjects that are especially pertinent to this report. Information derived from these earlier reports was augmented by the writer's experience on several hydrologic studies in this region and by contributions from ground-water hydrologists currently working in the individual States of the region. Contributors included E. G. Crosthwaite and A. M. La Sala, Jr., Boise, Idaho; A. R. Leonard and J. H. Robison, Portland, Oreg.; R. G. McMurtrey and J. A. Moreland, Montana; and K. L. Walters, Tacoma, Wash.

Especially helpful to the present study were two earlier reports containing water-resources data and interpretations of regional and subregional scope, compiled mostly by colleagues in the U.S. Geological Survey. Those reports are: "The role of water in shaping the economy of the Pacific Northwest" (Bodhaine and others, 1965); and the "Water Resources" appendix (app. 5) of the "Columbia-North Pacific Region comprehensive framework study of water and related lands" (Columbia-North Pacific Technical Staff, 1970b), compiled largely by G. L. Bodhaine and M. J. Mundorff. The reader is referred to the latter report, especially, for subregional summaries of ground-water resources and conditions, including the types and productivity of water-bearing rock materials in the different parts of the region.

The present report is designed to be a regional part of a broad-perspective analysis by the U.S. Geological Survey of the ground-water resources of the nation. This part, previously designated the Columbia-North Pacific Region and now designated the Pacific Northwest Region, is shown in figure 1. This region includes all the Columbia River drainage basin in the United States, the drainage areas of coastal streams in Oregon and Washington, and the closed basins in south-central Oregon. This region of about 274,400 square miles (271,430 mi^2 of land and 2,980 mi^2 of water) comprises all of Washington, most of Idaho and Oregon, parts of Montana, Nevada and Wyoming, and a very small part of Utah. It is bounded on the east by the Rocky Mountains, on the north by the Canadian border, on the west by the Pacific Ocean, and on the south mainly by the California and Nevada state lines and the Snake River drainage divide. (See also fig. 4.)

THE CHALLENGE OF MEETING FUTURE WATER REQUIREMENTS

The requirements for water in the Pacific Northwest Region are increasing, and will become a progressively greater challenge to water-management skills and regulatory innovation. For a variety of reasons, the ground-water resources seem destined to play an increasing role in meeting the future water requirements.

PROJECTIONS OF WATER NEEDS

Projections of future water needs in various parts of the region are available from the report of the "Columbia-North Pacific Region Comprehensive Framework Study of Water and Related Land," by the Columbia-North Pacific Technical Staff—hereafter shortened to CNPTS (1972). Those projections are summarized in figure 2, along with compilations of water use during 1970 and 1975 obtained by canvass of water managers and others (Murray and Reeves, 1972, 1977). As shown in the figure, total water diversions and withdrawals are strongly dominated by irrigation requirements, and these requirements are expected to dominate in the future, also. The water withdrawn for irrigation in 1970 was about eight times greater than the combined withdrawals for industrial, municipal, and rural-domestic uses. As projected, water for irrigation would still represent the dominant need by 2020, amounting to about 85 percent of the total diversions within the region (fig. 2). The projected irrigation requirements by 2020 would represent a nearly 90-percent increase over the 1970 diversions for that use and would be equivalent to about 20 percent of the total stream discharge from the region (fig. 3).

The projections of irrigation water needs in figure 2 include not only diversions to irrigate cropland and pasture, but also water for nonfarm irrigation (such as parks, golf courses, and cemeteries), and a predicted need for irrigation of forest crops. Expected irrigation of forest lands will constitute an essentially new type of irrigation need, based on predictions of an increased wood-fiber demand and downward trend in future amounts of natural forest land. Irrigation to assure tree survival during the early years and to establish forests on dry or marginal lands is seen as one way to maintain or even increase wood-fiber yields from the region (CNPTS, 1971c, p. 28). Estimates in the

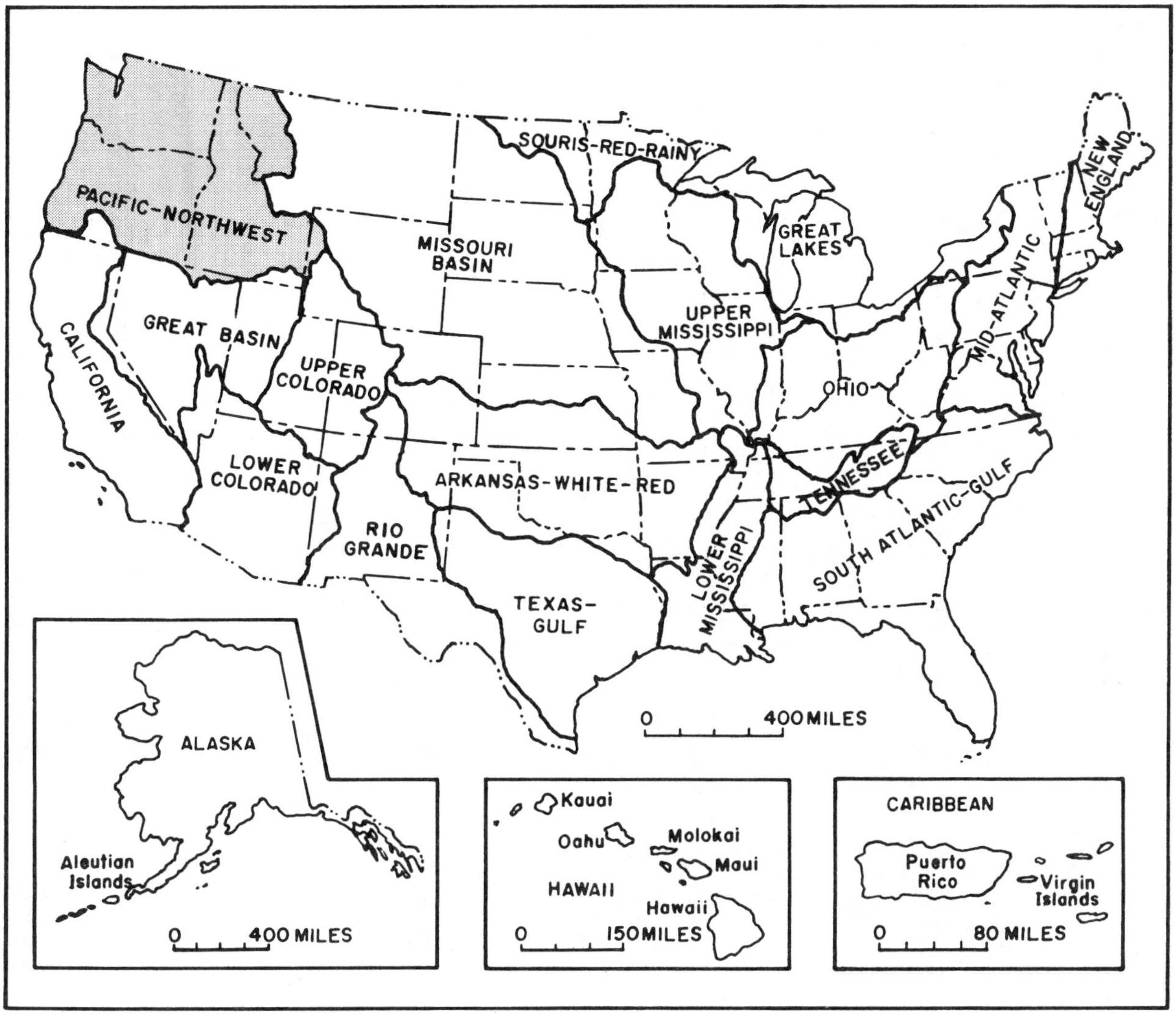

FIGURE 1.—Map showing the Pacific-Northwest Region and other regions of similar ground-water appraisals.

"framework" study indicate that roughly 5 million acres of forest land in the region might respond favorably if water were available.

Industrial and municipal water requirements, which constituted 6 and 4 percent respectively of the total withdrawals in 1970, were projected (fig. 2) to increase at rates much greater than the irrigation requirements. By the year 2020, needs for industry were expected to increase nearly 170 percent, and those for municipal supplies by more than 160 percent over the 1970 amounts. Even after those sizeable increases, however, the industrial and municipal water needs would amount to only 8 and 6 percent respectively of the projected total by 2020. Future needs for industrial and municipal water supplies are expected to be greatest in areas where these uses are already concentrated—namely, the Willamette-Puget Sound Trough and the vicinities of metropolitan centers throughout the rest of the region.

Rural-domestic water requirements were expected to grow generally apace with the projected total increases, remaining at less than 1 percent of the totals.

Sizeable demands for water that were not foreseen in the "framework" study may develop in the region, such as water for processing and transport of coal or other new uses associated with energy production. However, the needs for such purposes are expected to be restricted to only a few parts of the region, and the total of such unforeseen demands is expected to be relatively small.

The projections of water requirements (fig. 2), like all projections, are based on certain assumptions. Some of

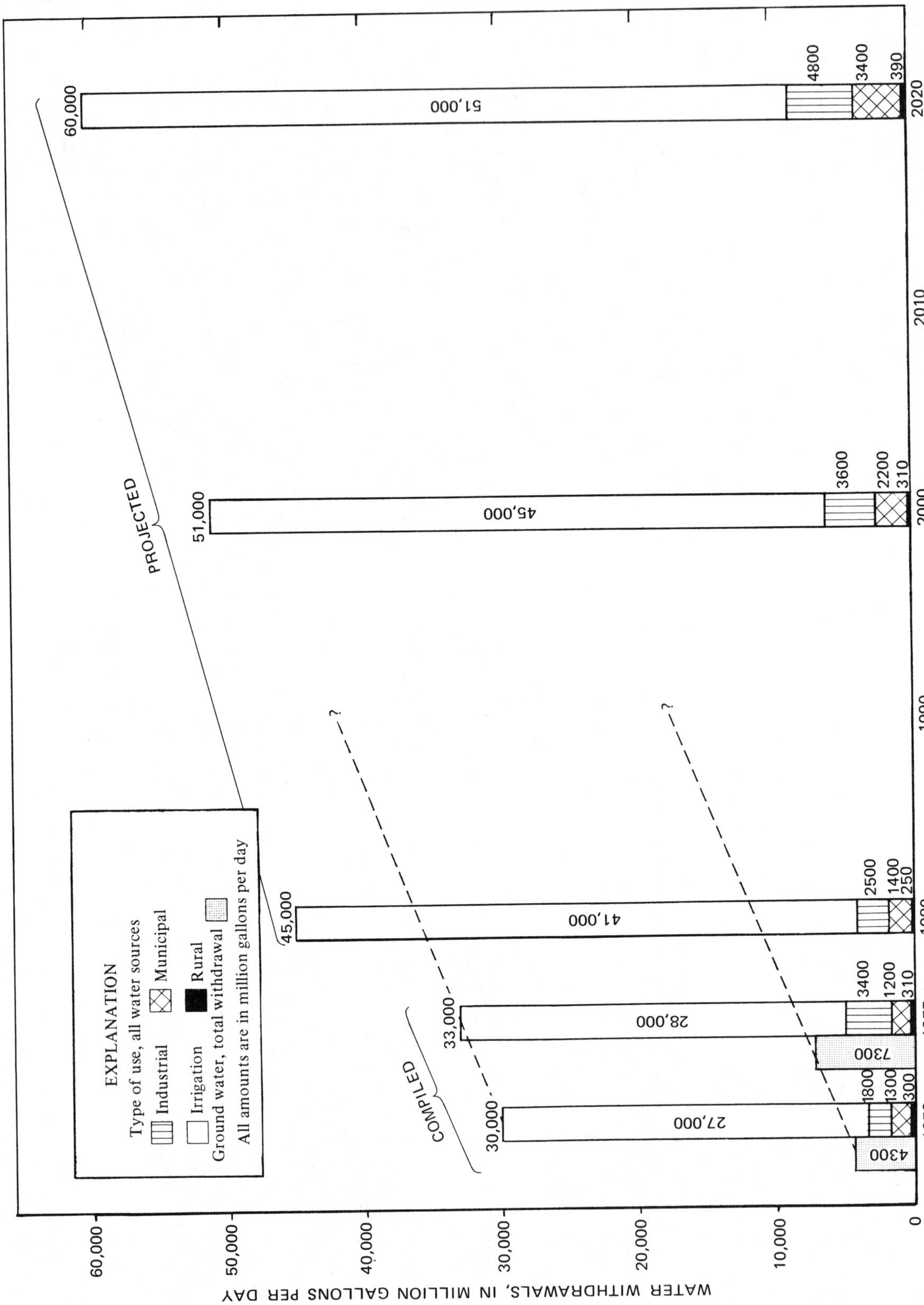

FIGURE 2.—Pacific Northwest water withdrawals compiled by U.S. Geological Survey for 1970 and 1975 and those projected by Columbia-North Pacific Technical Staff for 1980, 2000, and 2020. Sources: For 1970, Murray and Reeves (1972); for 1975, Murray and Reeves (1977); for 1980–2020, CNPTS (1972). Possible differences in reporting for the categories shown here as "municipal" and "rural" have no significant effect on the indicated relationships. Withdrawals for power generation are not included.

these are discussed in appendix 6 of the "framework" study report (CNPTS, 1971a, p. 5, 6). Most of the assumptions involved extension of socioeconomic conditions that prevailed prior to about 1965. The assumptions do not take into account certain shifts away from some earlier trends during the decade 1966–75, especially in the last half of the decade. Also, significant international events and shifts in national policies have occurred which, along with changes in socioeconomic trends, probably will affect future water requirements in the region.

Although a detailed analysis of the earlier assumptions and subsequent departures from them is beyond the scope of this report, certain near-term probabilities seem fairly obvious. These include a general slowdown in population growth, domestic commerce, and industry below the assumed trends for the region, with attendant slowing of related water needs.

Future energy sources and the fate of industries dependent upon them seem uncertain. The major industries are agriculture and food processing, lumbering and forest products, services (including tourism), minerals, metals, chemicals, and manufacturing. Some industries doubtless will be affected adversely, but deployment of industry (for environmental, energy, or other reasons) may create unexpected demands for water supplies at various places.

In judging the continued usefulness of the water-need projections from the "framework" study (fig. 2), requirements for irrigation are the key factor. In 1970 about 7.5 million acres were irrigated—roughly one-third of the total crop land at that time. The Snake River Plain accounted for more than half the total irrigated acreage, and the Columbia Plateau accounted for the second greatest fraction. Sizeable acreages are irrigated in all major parts of the region; even the farms in the more humid parts west of the Cascade Range can produce more and better crops with timely irrigation. About one-fourth of the land irrigated in 1970 reportedly had late-season water shortages. The dry lands that were classed as potentially irrigable totaled some 33 million acres, more than 40 percent of which were judged capable of producing high yields of climatically adapted crops (CNPTS, 1972, p. 85). Irrigated land was forecast to increase by 6 million acres, reaching a total of 13.5 million acres by 2020—still only one-third of the total irrigable land in the region.

The projected growth trend in irrigated farming also seems likely to be affected by counteracting influences. The depressing effects of economic slowdown, reduced population growth, and sharply higher farming expenses may be largely offset, or even dominated, by the spurring effect of strong export markets for farm products. Conversely, rising energy costs seem likely to constrain large-scale water withdrawals that involve high pumping lifts.

A comparison of water withdrawals in 1970 and 1975 (fig. 2) provides some insight into the durability of the projections for later years, although the extrapolations of a trend based on withdrawals for only those two years is tenuous, at best. A linear projection of the increase in withdrawals between 1970 and 1975 (upper dashed line in fig. 2) suggests that the total withdrawals by 1980 will fall far short of the projection for that year. This is not surprising, inasmuch as that forecasted growth in withdrawals would have required "a decade of unprecedented irrigation development during the 1970's" (CNPTS, 1971a, p. 45). Beyond 1980, however, the 1970–75 trend line is aimed generally toward the tops of the bars representing total withdrawals for 2000 and 2020; this trend line therefore suggests that the earlier forecasts of total withdrawals (largely irrigation withdrawals) for those years are still reasonable.

There are other reasons to expect that the projected needs for irrigation water by the years 2000 and 2020 may still be reasonable even though the needs for 1980 were overestimated in the earlier analysis:

1. The strong advantages that irrigation provides in the farming of the Pacific Northwest Region;
2. exports apparently will need to be maintained at high levels to offset U.S. purchases of foreign petroleum and other products;
3. worldwide needs for food and fiber resources continue to grow;
4. increased opportunities for international commerce with Communist-bloc nations including broadened markets for food and fiber products from the region.

Although irrigation use by 1980 apparently will fall far short of the projected amount, the withdrawals for nonirrigation uses in 1975 (fig. 2) had already exceeded the projections for 1980, owing largely to a 90-percent increase in withdrawal by industries.

In the aggregate, the water-need projections from the "framework" study seem to provide adequate guidance for the present study with regard to the general magnitude of future additional water requirements. Because of the aforementioned changing conditions that were not foreseen at the time the projections were made, the estimates of region-wide requirements by 2020 may be somewhat less reliable for the industrial and municipal categories than for the irrigation and rural-domestic categories.

SIGNIFICANCE OF GROUND WATER IN THE REGIONAL WATER SUPPLY

The Pacific Northwest Region's ground-water system may be thought of as a series of separate reservoirs, the upper parts of which are intersected by the surface-water system. The upper few hundred feet of the ground-water reservoirs serve the essential functions of moderating streamflow and sustaining the dry-weather flow of streams, feeding springs, and supplying water for most wells. Water in the deeper parts of the ground-water system is less available for use at present; however, the deeper zones may become very important in the future for subsurface disposal of wastes or as more water is required.

Ground water is by far the largest part of the region's freshwater resource, although most of it is hidden. Ground water fills the pore spaces and crevices in rock materials beneath all parts of the region down to depths of many thousands of feet. Very deep lying ground water, however, is commonly salty and is costly to extract because the rock materials at depth tend to have low permeability, or hydraulic conductivity. Furthermore, the renewability of the deep ground water is generally very limited; because of the low conductivity, water moves very slowly underground and commonly must move considerable distances from sources of replenishment at or near the land surface. Even discounting the very deepest water, however, the amount of fresh ground water in storage in the upper 1,000 feet of saturated rock materials is estimated to be about 160,000 billion ft^3 (cubic feet) or more than 13 times the average yearly outflow from the region through streams, which is about 12,100 billion ft^3. For a variety of reasons, not all this fresh ground water can be considered available for man's uses; the estimated amount of fresh ground water in "recoverable" storage at any time probably is on the order of 24,000 billion ft^3 (CNPTS, 1970b, p. 85), or about double the average yearly stream discharge from the region. The ground water receives an estimated 100 million acre-feet (CNPTS, 1970b, p. 89), or about 4,000 billion ft^3, of annual recharge, most of which ultimately discharges to streams and becomes part of the stream discharge. The estimated amounts of inflow to, and outflow from, the region are shown in relation to ground-water storage in figure 3.

The ground water and surface water are inseparable parts of the region's hydrologic system, with good hydraulic interconnection in many areas. This continuity between the ground water and surface water provides opportunities for greater beneficial use of the water resources but, at the same time, makes efficient management of the resources a difficult task under prevailing water laws, which tend to treat each of these two phases of water occurrence separately. Although greater utilization of the ground-water reservoirs cannot add to the total water supply (the inflow amounts in fig. 3), it can increase the dependability and the proportion of the total supply that is available for man's use. The improvement in dependability accrues through greater use of the reservoir function of the ground-water system. The increase in available supply results from the salvage of water that otherwise would evaporate or discharge naturally without being put to beneficial use. In contrast to these benefits, greater use of the ground water depletes the total water supply only to the extent that additional water from underground evaporates or is transported outside the region or area of management.

The available ground water is not evenly distributed throughout the region, as is shown schematically in figure 3, but varies greatly in availability from place to place. Ground water can be obtained nearly every place in the region, but at some places the quantities obtainable from a well are small. Also, in some areas a well must be more than 1,000 feet deep to obtain a dependable water supply. As stated by Mundorff (CNPTS, 1970b, p. 69),

> About 42 percent of the region is underlain by aquifer units with low porosity or permeability that generally will yield only small supplies of ground water. It is fortunate that these units are confined mainly to thinly populated (mostly mountainous) areas where water needs are comparatively small and surface-water supplies are plentiful.

The ground-water reservoirs that are most important in resource-management considerations are those capable of yielding water to wells at moderate or large rates and, therefore, which are large enough to be of public significance. Discussions in later sections of this report emphasize the major ground-water reservoirs that comprise the more extensive and productive aquifers in the region.

Up to the present time, man's use of the ground-water resources in the Pacific Northwest Region has been relatively small in comparison with the total amount available, and use of the ground-water reservoirs must increase greatly if the projected water needs (fig. 2) are to be met fully (or even largely) in the future. The average regionwide rate of ground-water withdrawal in 1975 was estimated to be about 7,300 Mgal/d of which about 4,500 Mgal/d were for irrigation and 2,800 Mgal/d for municipal, industrial, and rural water supplies. The ground-water withdrawal during 1975 constituted about 22 percent of the 33,000 Mgal/d diverted and withdrawn for all the uses listed in figure 2.

From 1970 to 1975, the withdrawal of ground water apparently increased by the largest amount in the history of the region. The estimated increase was 70 percent of the 1970 amount of 3,000 Mgal/d—the same as

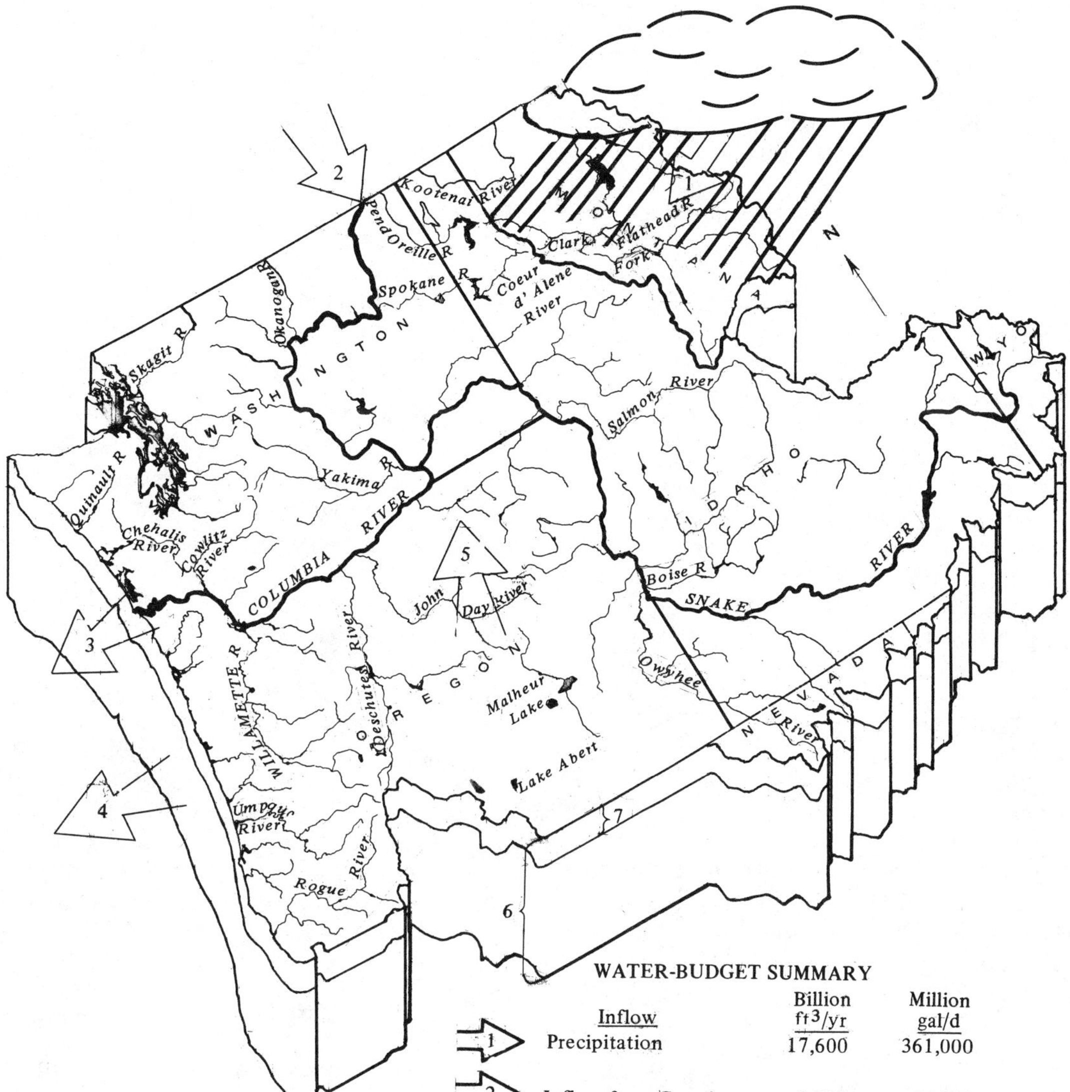

WATER-BUDGET SUMMARY

	Billion ft^3/yr	Million gal/d
Inflow		
1 Precipitation	17,600	361,000
2 Inflow from Canada	2,300	47,000
Totals (rounded)	20,000	410,000
Outflow		
3 Stream discharge	12,100	248,000
4 Ground-water discharge to saltwater	300	6,000
5 Evapotranspiration	7,600	156,000
Totals (rounded)	20,000	410,000

GROUND WATER IN STORAGE

	Billion ft^3
6 Total in upper 1,000 feet	160,000
7 Amount "recoverable" (fraction of total—not intended to show depth of wells)	24,000

FIGURE 3.—Generalized water budget and volumes of groundwater in storage for the region (data largely from Columbia-North Pacific Technical Staff, 1972, and appendices).

the increase in total withdrawal from all freshwater sources (fig. 2). The difference in indicated withdrawal rates apparently represents a legitimate, generally progressive increase during the 5-year period rather than an anomalous difference caused, for example, by drastically different amounts of rainfall during the irrigation seasons in the two years. Most of the increase was in Idaho, including shift of about 1,400 Mgal/d of irrigation withdrawals from surface-water to ground-water sources, and an increase of more than 1,500 Mgal/d in withdrawals by self-supplied industries.

Further increases in use of the ground-water reservoirs in the region seem inevitable for several reasons:

1. The dry-season flows of most streams east of the Cascade Range are fully appropriated (some are overappropriated).
2. Most cost-effective surface reservoir sites already have been used; therefore, further deliberate storage of water for later utilization must be largely in the ground-water reservoirs.
3. As ground-water information and water-management techniques develop further, management of the stream systems and ground-water reservoirs conjunctively will become more practical and effective.

Even in cases where choices exist, use of ground-water reservoirs has important advantages over use of surface-water reservoirs. For example, the initial costs connected with management of subsurface reservoirs commonly are much lower than for surface-water reservoirs. Also, ground-water bodies are less subject to evaporation, pollution, and temperature changes, and do not present hazards of downstream flooding.

GEOGRAPHIC FACTORS AFFECTING THE GROUND-WATER RESOURCES

Because the distribution, availability, and abundance of ground water in the Pacific Northwest Region are largely controlled by the landforms, climate, and earth materials, a general understanding of these features is necessary as a foundation for more specific discussions of the ground-water resources.

If one word could be used to describe the geography of the Pacific Northwest Region, that word probably would be "variety." The region is a composite of contrasts in physical and cultural geography. In population it ranges from dense urban centers, such as Seattle and Portland, to large tracts of land with no permanent human inhabitants. Its climate is dominated mostly by marine air masses but at times is influenced by continental air. It includes the wettest spot in the 48 conterminous states and some of the driest places in the Nation. The region contains examples of most of the landforms known to man, at altitudes ranging from sea level to more than 14,000 feet. Its vegetation assemblages range from rain forest to alpine to desert. It is not surprising, therefore, that great variety exists in the availability of and needs for the region's water resources, including its large and valuable ground-water resources.

LANDFORMS AND DRAINAGE

The landforms of the Pacific Northwest Region largely control the availability and use of ground water as well as the patterns of replenishment from precipitation and of drainage. The features of greatest significance to this ground-water appraisal include: The mountains, which receive much of the precipitation and divide the region into climatic zones; the flatter lands—valley floors, plains, and plateaus—which contain most of the population and account for most of the ground-water use; and the streams, which variously recharge or drain the ground-water reservoirs. The general physiographic divisions of the region are shown in figure 4. Detailed discussion of the physiographic features is beyond the scope of this report but can be found in CNPTS (1970b) from which this figure was taken.

One important feature of the landforms that is not shown in figure 4 is the altitude of the flatter lands. The altitude strongly influences the growing season which, along with other factors such as soil productivity and growing-season precipitation, largely controls the demands for irrigation water. In order of increasing altitude, the areas of extensive flatter land labeled in figure 4 are: Willamette-Puget Sound Trough (most populous part), Columbia Plateau, Snake River Plain, and Oregon Closed Basins. At the two extremes, the Willamette-Puget Sound Trough has much land below the 500-foot altitude and an average growing season commonly greater than 160 days per year; the Oregon Closed Basins area has much land above 4,000 feet where frost may occur during any clear night of the year. The Columbia Plateau and Snake River Plain are intermediate in growing season as they are in altitude.

PRECIPITATION AND REPLENISHMENT PATTERNS

Precipitation over the Pacific Northwest Region (fig. 5) is the ultimate source of all the water except inflow from Canada. The characteristics of the precipitation (mainly the rates, seasonal distribution, and amounts as rain and snow) control the water that is available to recharge (replenish) the ground-water reservoirs, either by direct infiltration at the land surface or by seepage from streams. The actual recharge from precipitation, of course, is also controlled by other factors, principal of which are the character of the soil and of the rock materials that constitute the ground-water reservoirs.

The most abundant annual precipitation in the conterminous 48 States occurs at high altitudes in the Olympic Mountains (northern part of Coast Range, fig. 4), where it exceeds 200 inches. Eastward from the crest of the Olympics and the coastal mountains to the south, precipitation decreases to about 35 inches along the Willamette-Puget Sound Trough, then rises to 120

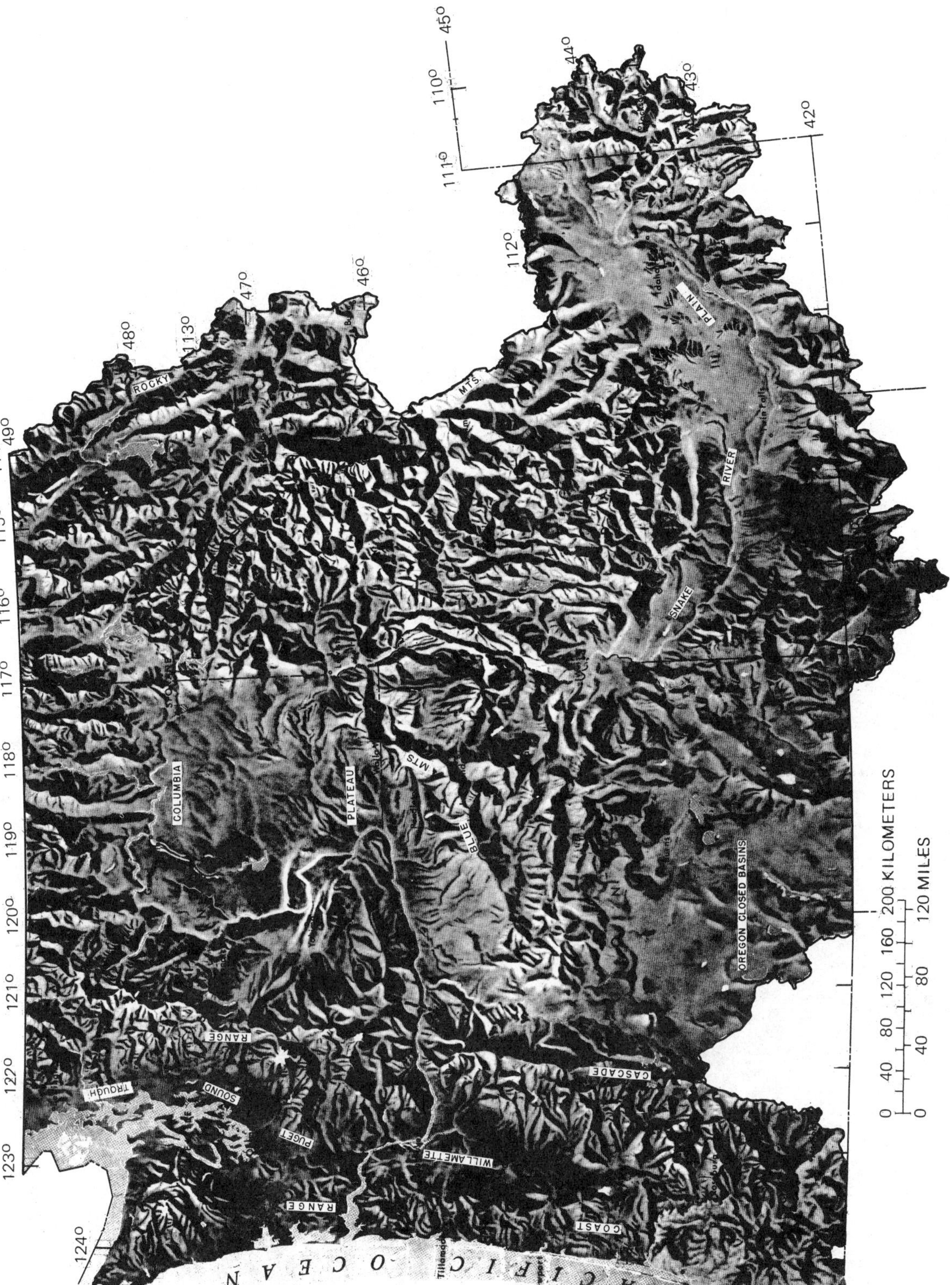

FIGURE 4.—Major land and water features of the region (Columbia-North Pacific Technical Staff, 1970b, fig. 2).

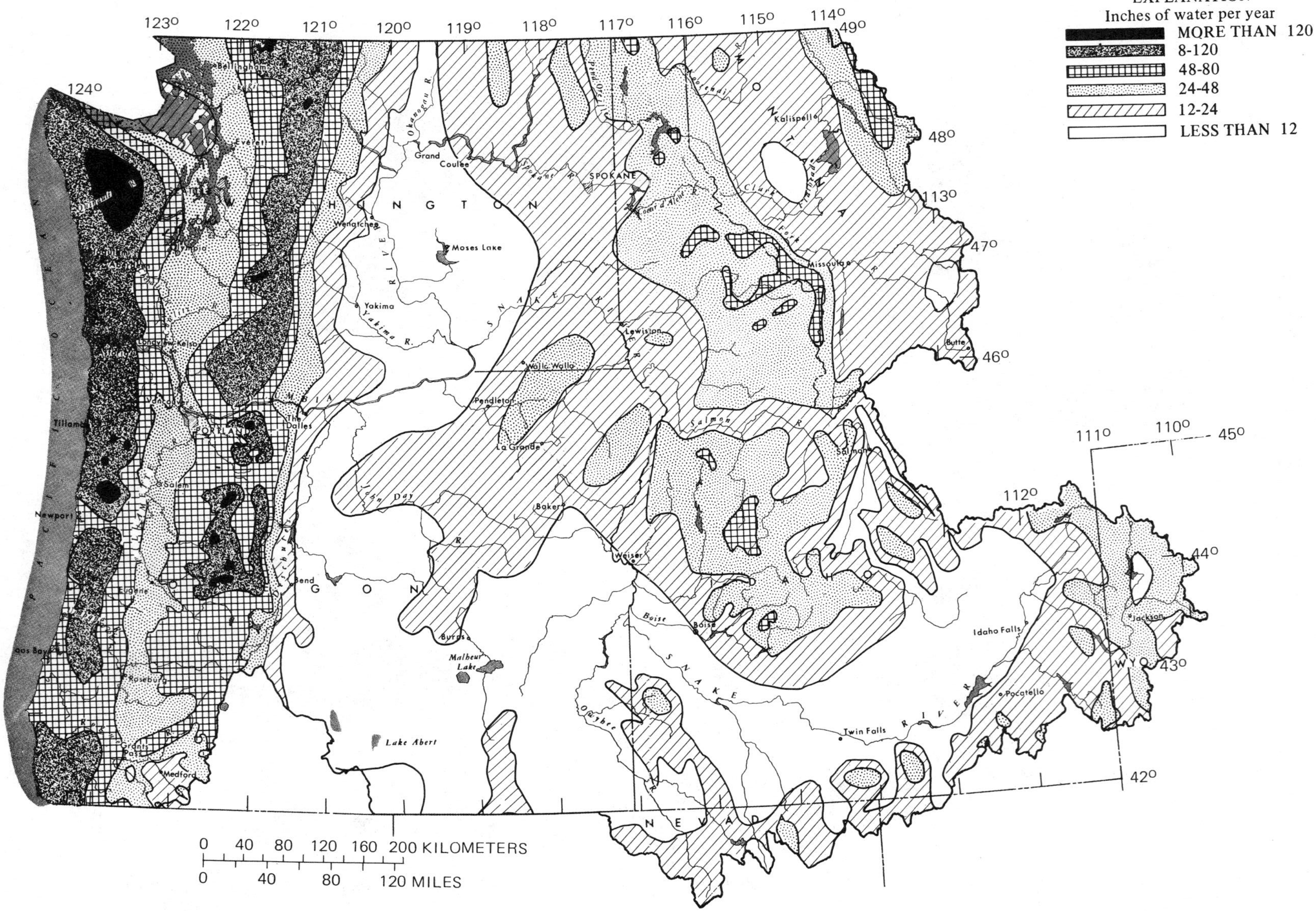

FIGURE 5.—Generalized distribution of annual precipitation in the region (adapted from Columbia-North Pacific Technical Staff, 1970b, fig. 3).

inches or more toward the crest of the Cascade Range.

East of the Cascade Range in Washington and Oregon the annual precipitation decreases sharply to about 10 inches or less in the valleys and on the plateaus, but generally increases again toward the east, where it reaches 40–50 inches in mountain areas.

The Snake River Plain receives from 6 to 15 inches of annual precipitation, and mountains surrounding the plain receive as much as 40 inches. In northern Idaho annual precipitation ranges from about 16 to 70 inches. Western Montana receives about 12 inches per year in some intermontane valleys but more than 150 inches at places in Glacier National Park (too small to show in fig. 5). Relatively dry belts exist in the rain shadows of the mountain ranges, and the greatest precipitation generally falls on the western, windward slopes.

Extreme variations occur in the distribution of annual runoff in the region. In the western part, large amounts of precipitation generate an average annual runoff that exceeds 100 inches in a substantial area of the Coast and Cascade Ranges. In contrast, annual runoff averages less than 1 inch in much of eastern Washington, eastern Oregon, and southern Idaho. Annual runoff greater than 40 inches is rare east of the Cascade Range. Patterns and amounts of runoff in various parts of the region are further described in CNPTS (1970b).

In areas receiving less than 12 inches, much of the precipitation water is likely to run off or evaporate soon after it falls or, if it infiltrates below land surface, to be held in the soil and be used up by vegetation. In these areas, the amount of precipitation that percolates down to the ground-water system probably is equivalent to less than one inch of rainfall except where the infiltration capacity of the soil or rocks at land surface is unusually favorable. Where infiltration is poor, and in some drier areas, precipitation might recharge the ground water only during unusually wet periods.

In areas shown as receiving less than 12 inches of precipitation per year (about 27 percent of the region), crops generally cannot be grown without irrigation. Exceptions occur in some areas that are subirrigated by shallow ground water, and in areas that are too small to be differentiated in figure 5 and which have slightly greater precipitation. Also, at places where soil conditions are especially favorable in depth and texture to store the moisture, dry-land farming practices have evolved which allow successful production of grain crops in alternate years on land that apparently receives only 10–12 inches of precipitation per year. In general, where soil and other conditions are favorable for farming, these areas potentially would benefit most of all areas in the region from the availability of water for irrigation.

The areas receiving 12–24 inches (about 34 percent of the region) are intermediate in regard to potential recharge for ground water and water needs for farming. The places that receive precipitation amounts in the upper part of this range (generally closer to the higher-precipitation areas in fig. 5) commonly have a few inches potentially available for recharging the ground water. Actual recharge is more likely to be limited by land slope and infiltration capacity of soils rather than by precipitation amounts. In these places, crops that can be grown by dry-land farming methods are somewhat more varied and considerably more dependable and productive than in drier areas.

Areas receiving more than 24 inches of precipitation per year may be generally considered to have abundant water available at the land surface for recharging the ground water. Some of these areas, including lowland parts of the Willamette-Puget Sound Trough and the floors of many smaller valleys, are underlain by permeable rock materials and receive large amounts of recharge. However, in some of the areas of abundant precipitation, actual recharge may be insignificant. Many of these areas, especially east of the Cascade Range, consist of mountains where the combination of steep slopes and soils of low permeability induces a large proportion of surface runoff from the precipitation. (Compare figs. 4 and 5.) Therefore, except in areas of permeable rocks in the high Cascades of Oregon and southern Washington (the High Cascades ground-water reservoir, table 1 and fig. 8), the local recharge in the mountainous areas of abundant precipitation is relatively unimportant to the manageable parts of the regional ground-water resources. Also, for several reasons including rugged terrain, generally poor or absent soils, and available surface water, the higher areas of abundant precipitation generally do not require additional ground water for irrigation or other purposes. The higher areas of abundant precipitation are significant to the management of ground-water mainly in that they generate stream runoff which flows to lower, generally drier parts of the region and contributes to the recharge of major ground-water reservoirs located there.

The seasonal distribution and type of precipitation also affect the availability of water and opportunity for recharge of the ground water. Although average yearly precipitation ranges widely, the seasonal distribution of precipitation is similar throughout most of the region. Except for parts of western Montana and north-central and southeastern Idaho, the region has distinctly dry summers and wet winters. About one-third, or less, of the annual precipitation generally occurs during the 6-month period April-September, and less than 15 percent during the latter 3 months of that period (fig. 6). About two-thirds occurs during October-March, when it generally falls as rain at the lower altitutdes and as snow at the higher altitutdes. Abundant snow on many of the high mountains has maintained glaciers and perennial snowfields that are important sources and regulators of streamflow. The heaviest annual snowfalls—sometimes totaling more than 60 feet—are in the higher parts of the Cascade Range. In most mountain areas at altitudes above 6,000 feet, snow cover commonly persists from early November until early June. Melting of the abundant snow maintains flow at relatively high levels in many of the streams throughout the spring; in some places the high flow does not end until early July.

At the stations shown in figure 6, precipitation is greatest during the winter but differs significantly in amounts. Astoria, Oregon, receives 13 times as much precipitation in December as does Yakima, Washington. In some eastern parts of the region, however, the seasonal distribution of precipitation is different, and the differences between amounts during the summer and winter are generally less distinct. At some lowland precipitation stations in western Montana and north-central and southeastern Idaho more than one-half the annual precipitation falls during the period April-September.

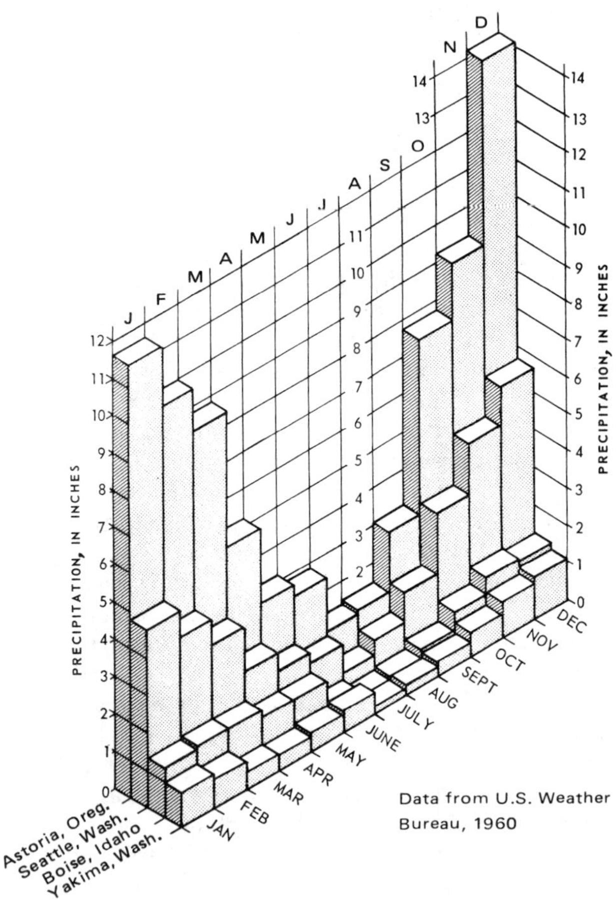

FIGURE 6.—Seasonal distribution of precipitation at selected weather stations in the region (Bodhaine and others, 1965, fig. 8).

The coincidence of the wettest period with the cooler, nongrowing season is highly favorable for recharge to the ground-water reservoirs. During the cooler months, the losses of precipitation to evaporation and vegetation are minimal, and the opportunity for available water to infiltrate the land surface is greatly enhanced. Conversely, the small percentage of precipitation that arrives during the growing season causes greater water needs for irrigation and summertime municipal uses than would occur in comparable situations in most other regions of the Nation. The distinctly dry summer period is the reason why even the crops grown west of the Cascades benefit greatly from irrigation.

Evaluation of the supplies of water that can be expected in the future, and of future water needs as well, must include the possibility of protracted periods of drought or above-normal precipitation. Previous analyses of long-term variations in precipitation, including determination of prehistoric climatic trends based on tree growth rings, have yielded the following conclusions (Bodhaine and others, 1965, p. 31):

1. Periods during which precipitation will be considerably above or below the average for several successive years can be expected occasionally in any part of the region.
2. The cumulative effects of several closely spaced periods of deficient or abundant precipitation will be greatest in the drier parts of the region.
3. During the period 1287–1937, there was no persistent trend toward a wetter or drier climate in eastern Oregon.
4. Although significant variations of the climate (including precipitation) have occurred throughout much of eastern Oregon (and probably beyond), no rhythmic cycle was evident or could be predicted on the basis of tree rings.
5. A general drought in eastern Oregon, which began in 1918 and lasted until about 1940, was the most severe that the area had experienced during the 650 years prior to 1937.

One of the unstated assumptions related to the projections of future water needs (fig. 2) is that the patterns of precipitation will continue in the future about the same as those in the past.

GEOLOGIC SOURCES AND DISTRIBUTION OF GROUND WATER

THE PRODUCTIVE AQUIFERS

The main focus of this report is on the more than half of the Pacific Northwest Region that is underlain by rock materials known, or believed from geologic evidence, to be capable of yielding ground water to wells at moderate to large rates. These are the areas where proper planning and effective water management, based on adequate data and sound hydrologic principles, along with the removal of nonessential constraints, can produce a significant increase in the amount of water available for man's use. These are also the areas where the ground-water reservoirs are most likely to be economic sources of supplies for irrigation, industrial, and municipal uses.

The major ground-water reservoirs can be classified mainly in three groups—those made up of (1) predominantly sedimentary rocks (including deposits from streams, glaciers, and the wind), (2) predominantly volcanic rocks, and (3) combinations of sedimentary and volcanic rocks. The generalized distribution of these types of aquifers where they are capable of yielding 100 gallons per minute or more of water to individual wells (but without regard to depth of well or amount of pumping lift) is shown in figure 7.

Differences in the occurrence and movement of ground water in the two main classes of rocks—sedimentary and volcanic—are significant to the functions and management of these rocks as ground-water reservoirs. The sedimentary rocks yield water to wells mainly from layers of sand and gravel, which are the kinds of sedimentary rocks in this region that are most likely to yield water at moderate to high rates. Less permeable sedimentary materials, such as clay and silt, are more common in the region and, although they yield water to wells at lower rates, they serve the extremely valuable function of long-term storage that can drain slowly to adjacent permeable aquifers. These fine-grained deposits have a relatively high porosity, but the pore spaces are so small that considerable force (in the form of hydraulic pressure or weight of overlying rocks) is required to dislodge much of the water they contain. The saturated space within the permeable sedimentary rocks consists mostly of the primary pore space left between the particles of rock materials as they were deposited. Those sedimentary rocks in the region that have been hardened by such natural processes as compaction or cementation may develop secondary joints and fractures which transmit water, but the benefits of such secondary openings commonly do not offset the loss of pore space and permeability that results from the rock-hardening processes.

The volcanic rocks yield water mainly from permeable zones that occur at or near the contacts between some flow layers; central and lower parts of the flow layers are nearly always denser and much less permeable. The abundance, causes, and nature of the permeable zones differs greatly. The zones tend to occur less commonly in rocks of the Columbia Plateau (fig. 4) than in rocks of the Snake River Plain and younger volcanic rocks. (See table 1, "Remarks".) They commonly result from the incomplete filling of openings at the top of one flow layer (such as rubble, cinders, or scoria) by the lava of a succeeding flow. Other permeable interflow zones have other origins, including deposition of sedimentary materials during the times between lava outpourings.

The volcanic ground-water reservoirs commonly yield water to wells at high rates but generally can yield less water from storage than the same volume of permeable sedimentary rocks. This is because the total volume of openings in the volcanic rocks is nearly always less than that in productive sedimentary rocks; moreover, many openings in the volcanic rock are likely to be poorly connected.

Both the sedimentary and the volcanic reservoir rocks are layered and, consequently, movement of water through the rocks along the beds (mainly horizontal) is less restricted than movement across the beds (mainly vertical). In the sedimentary reservoirs, vertical movement between the permeable beds (mostly sand and gravel) is commonly impeded by intervening layers of silt, clay, or naturally cemented materials. In the volcanic rocks, vertical movement is impeded by the dense and poorly permeable rock which generally constitutes the bulk of each flow layer. Even the dense parts of the flow layers, however, generally contain enough fracture and joint openings to allow some movement of water, although leakage from one permeable interflow zone to another may be greatly restricted. Cross-bed leakage between permeable zones in both the sedimentary and volcanic rocks is "short-circuited" by many wells that are either unlined (uncased) in the interval between permeable zones, or have casing that is open to more than one of the zones.

Some large bodies of the saturated permeable rocks shown in figure 7 are at such great depths, or in such high or rugged terrains, that imminent development is unlikely. However, at many places, as in the Cascade Range in Oregon and southern Washington, these remote aquifers are very valuable because of the large quantities of water they supply to streams. Also, some of these remote aquifers are likely to be future sources of water for increased forest irrigation, (See table 1, High Cascades ground-water reservoir.)

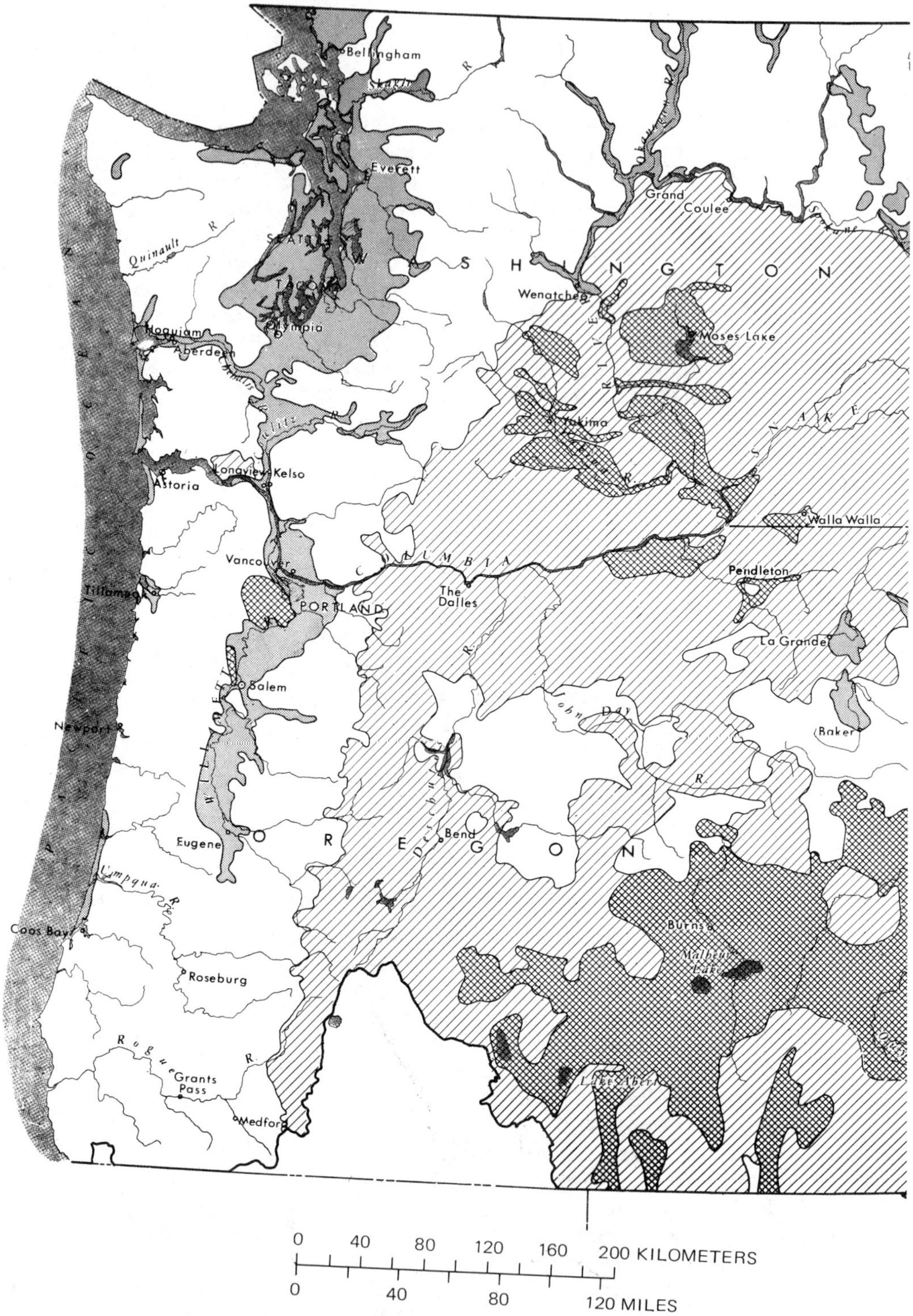

FIGURE 7.—Generalized distribution and types of major aquifers

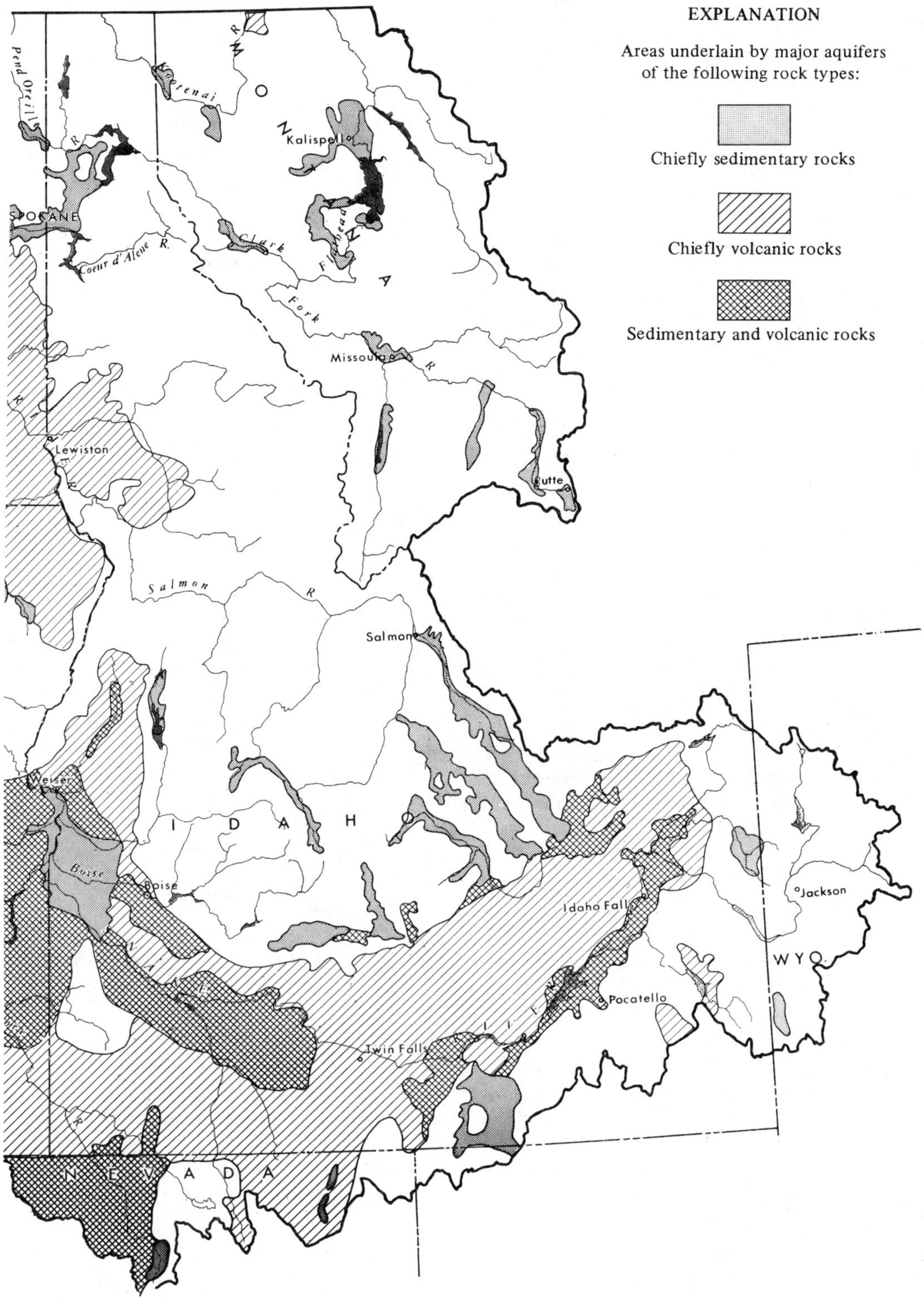

(Columbia-North Pacific Technical Staff, 1970b, fig. 20).

DESCRIPTION OF THE GROUND-WATER RESERVOIRS

Extensive areas in the region are similar enough in climate, topography, water uses, and character of the aquifers to be considered as a unit for discussion of ground-water problems and management opportunities. Six such areas, referred to in this report as major ground-water reservoirs, are shown in figure 8. The general characteristics of these major ground-water reservoirs are summarized in table 1, along with associated problems and water-management opportunities.

The grouping of areas into the major ground-water reservoirs is not intended to imply that all parts of the designated reservoirs are hydraulically interconnected or can be managed as a unit, or that water necessarily moves across the state boundaries that cross some of the designated reservoirs. Unfortunately, the natural extent of these major ground-water reservoirs does not fit drainage-basin or other subregional boundaries and, therefore, the reservoir areas are difficult to relate to the summary values found in earlier regional reports. However, a few general statements can be made. For example, the major ground-water reservoirs probably receive more than one-half the estimated 4,000 billion ft^3 of annual recharge (p. 6); they contain about two-thirds of the region's ground water in recoverable storage (fig. 3), or 16,000 billion ft^3; and the ground-water withdrawal from them in 1970 (about 3,600 Mgal/d) accounted for more than four-fifths of the total for the region that year (fig. 2). More detailed information about various parts of designated reservoir areas is available in reports listed in "Selected References," especially in the aforementioned Appendix 5 of the "framework" study (CPNTS, 1970b). The notes in table 1 help relate the major ground-water reservoirs to the subregions used in that earlier study.

Besides those areas shown in figure 8, there are many smaller ground-water reservoirs in the region that are capable of yielding water at rates suitable for industrial, irrigation, and public supplies. Some of these, including scattered areas along the coastlines of the Pacific Ocean and along major streams, are shown in figure 7, but others are too small to be shown at the scale of that map. The smaller size of these ground-water reservoirs does not reflect their relative importance in the region; their aggregate volume is several hundred cubic miles, and at many places they are the only usable sources of fresh ground water. Moreover, they include some of the areas most likely to benefit from water-management possibilities. Therefore, although they are not listed individually in this report, they are included as a category in table 1 and are mentioned in the following summary discussions of ground-water problems and management opportunities. Some of these smaller reservoir areas were described in greater detail by Mundorff (CPNTS, 1970b).

MAJOR GROUND-WATER PROBLEMS

The most significant problems pertaining to ground water, which are listed briefly for each of the ground-water reservoirs in table 1, fall into five general categories. These are: (1) Progressively declining water levels; (2) ground-water quality problems; (3) waterlogging; (4) inadequate information, and (5) competition for available supplies.

PROGRESSIVELY DECLINING WATER LEVELS

A decline in ground-water levels is an inevitable consequence of withdrawing water from wells—it does not necessarily represent a long-term depletion of the ground-water reservoir, otherwise called ground-water mining. The long-term significance of water-level declines depends on several interrelated conditions, among which are the year-to-year rate and persistence of decline, the size of the ground-water reservoir in relation to the area where the decline occurs, and the hydrologic and economic consequences of continuing decline. Short-term declines in levels can have beneficial water-conservation effects, such as the salvage of water otherwise lost as evapotranspiration from waterlogged land, or creation of storage space in the ground-water reservoir for subsequent filling by surplus stream water. Even if water-level decline can be identified as a long-term depletion of a ground-water reservoir, the results may not be all bad, especially if there is no compelling reason to preserve the ground water and if undesirable consequences of the depletion (such as reduction of streamflow or ultimate disruption of local economy based on irrigation farming) can be tolerated. Each declining water level must be judged on its own special conditions.

Persistent declines of ground-water levels have occurred in several areas of the region, and major depletion of the ground-water reservoir has been projected for a few areas. The declines are all in areas of sizable pumping for irrigation, although pumping for municipal and industrial uses has contributed to the declines in some areas. Most of the declines involve the volcanic aquifers, and most are in the arid and semiarid parts of the region. Significant, persistent declines have occurred in the Willamette Valley, the High Desert, the Columbia Plateau, and the Southern Idaho major ground-water reservoirs shown in figure 8. Locations of the areas of decline are shown in figure 9, and information about the declines is summarized in table 2.

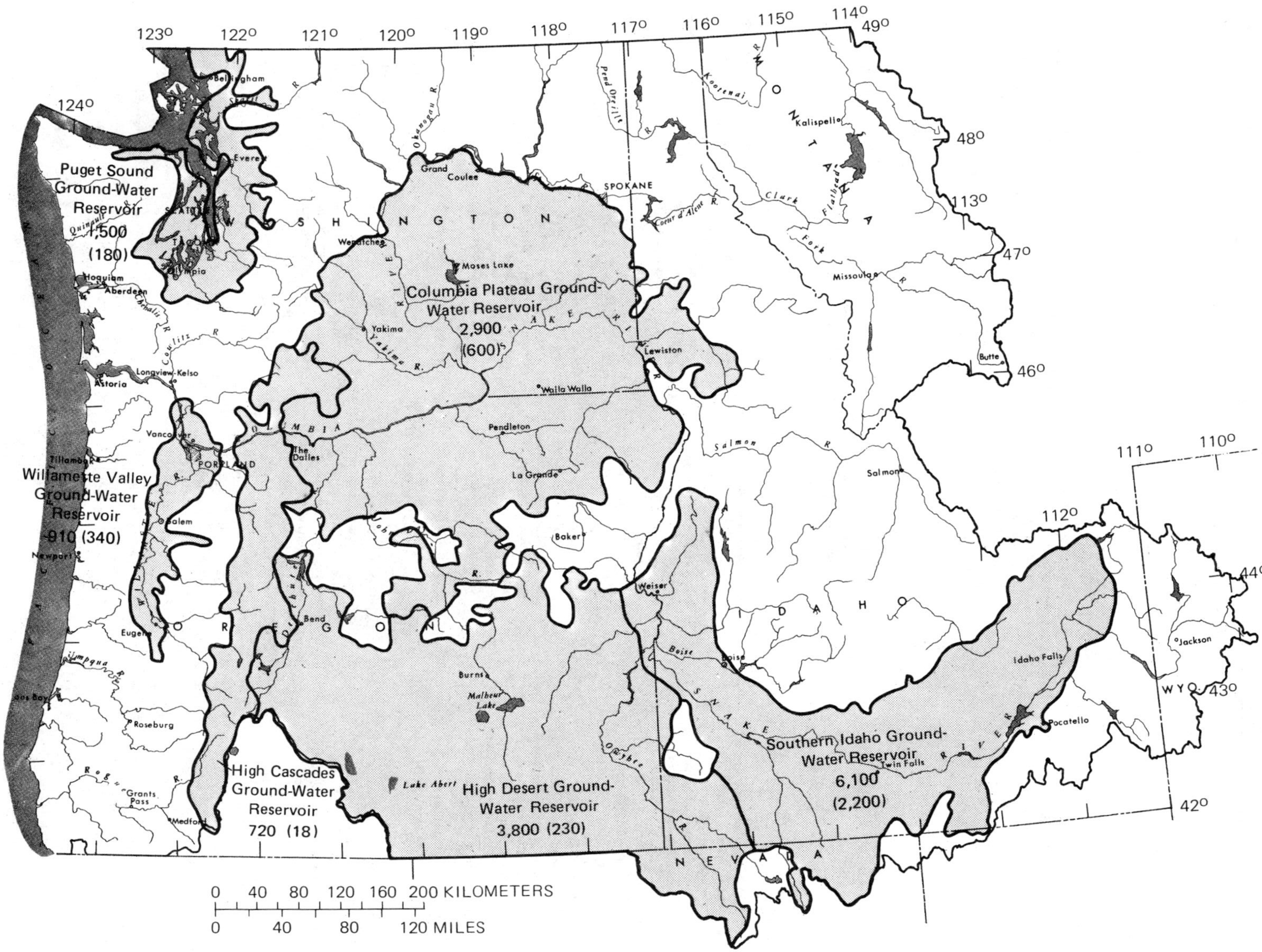

FIGURE 8.—Areas designated as major ground-water reservoirs, estimates of recoverable water in each (billion cubic feet), and 1970 pumpage (in parentheses—million gallons per day).

TABLE 1—*Summary of conditions in major ground-water*

Designation used in this report	Major aquifer[1] materials and yields to wells	Recharge to major aquifers	Discharge modes and withdrawal	Largest existing use of the ground water
Puget Sound ground-water reservoir[2]	Sedimentary rocks deposited by streams or glaciers; moderate to very large rates of yield.	Infiltration of precipitation and streamflow; generally ample in relation to pumpage.	Seepage to streams and marine waters; springflow along slopes. Pumpage is small in relation to recharge.	Public supplies, largely supplemental to stream supplies.
Willamette Valley ground-water reservoir[3]	Sedimentary rocks deposited by streams and glaciers and (in northern part) interflow zones in basalt; alluvial gravel along Columbia River has very large rates of yield.	Mainly by infiltration of precipitation water; flood waters from major streams recharge adjacent aquifers; irrigation augments natural recharge, Recharge generally ample, except for basalt aquifers.	Seepage to Columbia and Willamette Rivers and their major tributaries. Pumpage from sedimentary aquifers is small in relation to recharge.	Industrial supplies, mostly in metropolitan area near Columbia River,
High Cascades ground-water reservoir[4]	"Young" volcanic rocks; very large rates of yield.	Infiltration of precipitation water; recharge rates greatest of all the major ground-water reservoirs in the region.	Seepage to streams, which have very high dry-season flows, and springflow. Very little pumpage.	Rural domestic and stock supplies.
High Desert ground-water reservoir[5]	Sedimentary rocks deposited from streams and lakes; "young" volcanic rocks; older sedimentary and volcanic rocks. Rates of yield are moderate to very large.	Infiltration of precipitation and of runoff from mountains within and bordering the area. Away from the mountains, recharge may occur only occasionally.	Evapotranspiration, springflow, and (in northern parts of the area) seepage to streams. Pumpage very small in relation to ground water in storage and average recharge.	Irrigation.
Columbia Plateau ground-water reservoir[6]	Volcanic rocks of the Columbia River Group (and similar rocks in Oregon); alluvial deposits from modern streams and from Glacial-Age meltwater floods; older sedimentary basin-fill deposits. Rates of yield are moderate to very large.	Infiltration of rain and snowmelt water from land surface and stream channels; infiltration of irrigation water. Recharge in non-irrigated areas is limited by dry climate; recharge incidental to irrigation can be many times the natural recharge. At places, recharge probably occurs only during unusually wet periods.	Seepage to streams; springflow and seepage at land surface; evapotranspiration. Pumpage is generally small in relation to total recharge and amount of ground-water in storage, but tends to be concentrated.	Irrigation.
Southern Idaho ground-water reservoir[7]	Young basaltic volcanic rocks; young sedimentary deposits; older volcanic and sedimentary rocks. Rates of yield are moderate to very large.	Natural recharge is mainly by seepage from streams during their high-flow periods (especially streams that carry water into the area from adjacent uplands), and by direct infiltration of rain and snowmelt water. Water lost from the irrigation process is a large source of recharge, the largest source at places.	Springflow and seepage to streams; evapotranspiration; pumpage for irrigation is very large, but is causing water-level declines in only a few areas.	Irrigation.
Smaller ground-water reservoirs[8]	Mainly young sedimentary deposits (alluvial fan, floodplain, glacial outwash, and lake deposits; dune and beach sand); some volcanic rocks.	Infiltration of precipitation water; seepage from streams during their high-flow periods; seepage from irrigated soils and conveyance ditches.	Seepage to streams during their low-flow periods; evapotranspiration; springflow; seepage to the ocean. In relation to the amounts in the individual ground-water reservoirs, pumpage ranges from very small to large.	Generally irrigation.

NOTES: 1. Arbitrarily designated as capable of yielding water to a properly constructed well at a rate of 100 gpm or more, from a depth that is reasonable for common drilling methods and modern pumping equipment.
2. Lies entirely within Subregion 11 of the CNPTS study.
3. Comprises southern part of Subregion 8 and lowland parts of Subregion 9 of the CNPTS study.
4. Comprises mountainous parts of Subregions 3, 7, 8, 9, and 10 of the CNPTS study.

reservoirs (General location and extent shown in fig. 8)

Major problems pertaining to ground water			
Existing	Foreseeable	Management opportunities	Remarks
Troublesome amounts of dissolved iron in some aquifers; local deterioration of water quality resulting from waste-disposal practices; lack of adequate information in critical near-shore areas; seasonal waterlogging.	Continued local deterioration of water quality resulting from waste disposal; local salt-water intrusion resulting from increased near-shore pumping.	Reduction of pollution from surface or shallow waste disposal; greater use of ground water as supplies for near-shore development and for irrigation; use of deep, confined salty-water aquifers for waste disposal or storage of freshwater and natural gas.	Very large additional supplies of ground water are available; development has been retarded by general abundance of good-quality stream waters. Withdrawal for public supplies is about equal to combined withdrawals for industrial, irrigation, and rural-domestic uses. Ground-water resources in many island and peninsular areas are extremely valuable but inadequately appraised.
Limited distribution of large-yield aquifers; excessive dissolved minerals at places, including salty water; deterioration of water quality resulting from disposal of household wastes and water used for air conditioning; declining water levels in basalt aquifers, local waterlogging.	Continued local deterioration of water quality resulting from waste-disposal practices; continued depletion of water in basalt aquifers.	Reduction of pollution from waste disposal; promotion of well designs best suited for improved yields from marginal sedimentary aquifers; additional pumpage from alluvial aquifers; possible use of salty-water aquifers for subsurface storage or disposal of fluids; pumping from waterlogged areas as drainage measure.	Very large additional supplies are available, chiefly from alluvial aquifers, especially those along Columbia River. Industrial and irrigation uses constitute about 90 percent of total ground-water use. Development will be constrained by limited availability of large-yield aquifers, water-quality problems, and declining levels in basalt aquifers.
Lack of adequate information.	Some undesirable dissolved minerals may be present locally.	Forest irrigation in areas of favorable topography and soils; drainage, by artificial tunnels, for hydropower generation; controlled drainage for support of low-level flow or quality of streams.	Ground water is of good to excellent quality and recharge is adequate to support almost any conceivable withdrawals. Development will be severely restricted by remoteness, short growing season, and great depth to water table in many places. May have potential for geothermal development at places.
Water quality, generally not as good as in other major ground-water reservoirs; lack of adequate information; waterlogging in some basin-floor areas.	Major problems unlikely until development increases considerably. Intensive pumping for irrigation could easily deplete aquifers that receive little recharge. More waterlogging is likely if some basin-floor areas are irrigated.	Greater development of ground water in most areas; use of deep aquifers for storage or disposal of fluids. Possible development of geothermal ground water.	Young volcanic aquifers are capable of the largest yields to wells, but much of the available ground water is stored in sedimentary rocks. Availability of ground water is limited by great depth to water in many places and meager recharge. Reservoir system is complicated by abundant faulting. Development will be constrained by short growing season, remoteness, by locally poor water quality, and by risks in prospecting for large-yield aquifers owing to inadequate information.
Declining water levels and local depletion of the ground-water reservoir; competition for available supplies; waterlogging.	Continuing and growing areas of water-level decline; some actual depletion of water in aquifers of limited extent and recharge; increased waterlogging caused by increased irrigation.	Additional ground-water withdrawal, especially in deeper aquifers and away from areas of concentrated pumping; greater use of sedimentary aquifers; conjunctive use of surface water and ground water; deep subsurface waste disposal; lining of leaking well bores and improvement of new-well construction; artificial recharge using surplus stream water; salvage of some evapotranspiration discharge; reducing and forestalling waterlogging.	In much of the area the basalt aquifer is the only available source of rural-domestic, stock, and public water supplies. Where recharge is small, the bsalt aquifer is very susceptible to overpumping and depletion because of the large rate of yield and very small storage capacity. The ground-water is generally of good to excellent quality; hardness and dissolved-iron content are greater than desirable at places.
Conflicts between ground-water withdrawals and surface-water rights; waterlogging; lack of adequate information about ground-water availability at places.	More areas of water-level decline; continuing waterlogging; increasing conflicts with surface-water rights; increased occurrences of local pollution of ground water	Withdrawal of water from untapped parts of the ground-water reservoir; conjunctive use of surface water and ground water; artificial recharge with surplus stream water; salvage of some evapotranspiration discharge; reducing and forestalling waterlogging.	The young volcanic rock aquifer of this area is one of the most extensive and productive aquifers in the U.S.; it is considerably more permeable and porous than most basalt of the Columbia Plateau ground-water reservoir, and its storage capacity may be on the order of 20 times greater, volume for volume. Water from the younger volcanic and sedimentary rocks usually has satisfactory quality for most uses. Some of the hottest (90°C) and most mineralized ground water in the region is found in some of the older, less permeable rocks.
Inadequate total supply; conflicts between ground-water uses and surface-water rights; waterlogging; unsatisfactory water quality.	Increased competition for available water supplies; spread of ground-water pollution; salt-water intrusion near the Pacific Ocean.	More beneficial use of the total water supply through conjunctive use and management of both surface water and ground water; reducing and forestalling local pollution and waterlogging; salvage of some evapotranspiration discharge.	The aggregate volume of the reservoir rock materials in areas of this category is several hundred cubic miles. In many places these are the only usable sources of fresh ground water; examples are deposits along the coastlines of the Pacific Ocean and Puget Sound, and stream-valley deposits in widely scattered areas that otherwise are underlain by impermeable rocks. The locations of these smaller areas can be found by comparing figures 7 and 8. They are described in greater detail by Mundorff (CNPTS, 1970b).

5. Comprises areas in southwestern part of Subregion 5, southeastern part of Subregion 7, and all of Subregion 12, of the CNPTS study.
6. Comprises areas in southern part of Subregion 2, eastern part of Subregion 3, western part of Subregion 6, and northern part of Subregion 7, of the CNPTS study.
7. Comprises areas in central and southern parts of Subregion 4, eastern part of Subregion 5, and southern part of Subregion 6, of the CNPTS study.
8. Mainly in Subregions 1, 2, 4, 6, and 10, of the CNPTS study.

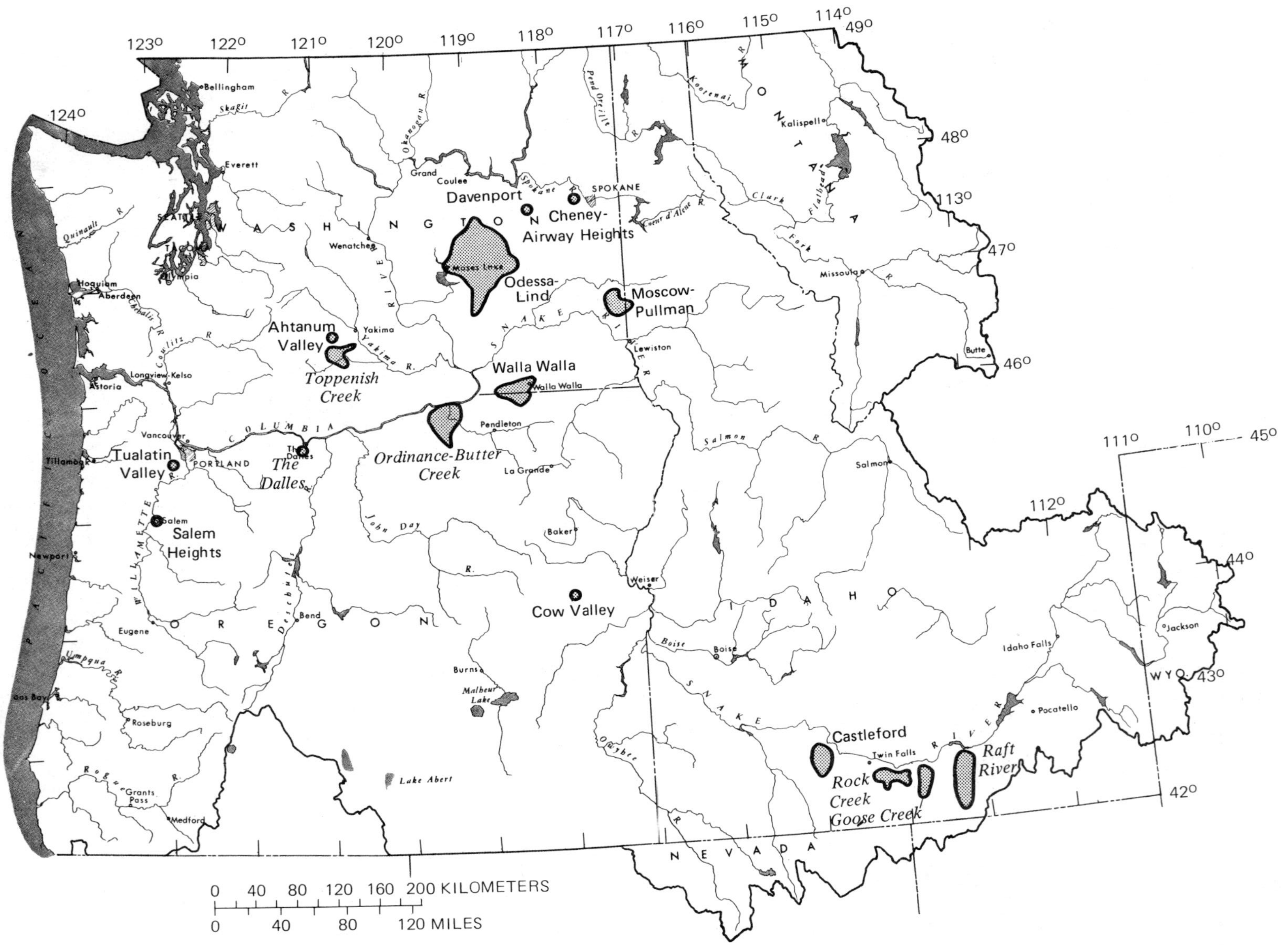

FIGURE 9.—Location of areas where significant declines in ground-water levels have occurred. Rates and duration of decline are listed in table 2. Circles represent areas where extent of decline is indefinite or too small to be shown precisely at this scale.

TABLE 2. *Summary of information about water-level declines*
[Areas shown in figure 9. Values represent net differences between yearly high water levels:]

Area and reference report	Decline	Aquifer(s)	Main use(s) of water	Remarks
Salem Heights (Foxworthy, 1970)	As much as 8 ft/yr, at least 38 ft total by 1961; thereafter, levels have risen gradually.	Basalt (small extent)	Public supply, industrial	Alternative public water supply was provided in 1961; aquifer later used mainly for artificial recharge and supplying peak demands.
Tualatin Valley (Bartholomew and others, 1973)	As much as 12 ft/yr; more than 40 ft total, early 1960's to 1974; thereafter, levels have generally risen.	Basalt	Irrigation, public supply	Declared to be a critical ground-water area by Oregon State Engineer in 1974; that same year, the Wolf Creek Irrigation Project began supplying water from Tualatin River and ground-water use declined.
Cow Valley (Brown and Newcomb, 1962)	As much as about 3 ft/yr by 1959, slower by late 1960's.	Basalt and overlying sand and gravel	Irrigation	Declared to be a critical ground-water area by Oregon State Engineer in 1959; pumpage restricted thereafter.
The Dalles (Bartholomew and others, 1973	As much as about 6 ft/yr by 1965; levels largely stabilized since 1970.	Basalt (faulted and folded)	Irrigation, industrial, public supply	Declared to be a critical ground-water area by Oregon State Engineer in 1959. Substitute water supply from Columbia River was provided by The Dalles Irrigation Project beginning in 1967.
Ordnance-Butter Creek (Bartholomew and others, 1973)	Ranged from less than 4 to about 40 ft/yr, 1961–73.	Basalt and overlying gravel	Irrigation, industrial, public supply	Declared to be a critical ground-water area by Oregon State Engineer in 1976.
Walla Walla (MacNish and others, 1973)	As much as 15 ft/yr; total as much as 150 ft since the late 1890's.	Basalt	Irrigation, industrial, public supply	Rate of decline is expected to slow greatly unless pumpage continues to increase substantially.
Ahtanum Valley (Foxworthy, 1962)	As much as 3½ ft/yr, 1953–57.	Basalt	Irrigation	Additional large-yield wells tapping the basalt have been restricted by Washington Department of Ecology.
Toppenish Creek (U.S. Geol. Survey, 1975)	As much as 7 ft/yr, 1960–70	Basalt	Irrigation	Yakima Tribal Council was considering (1977) prohibiting additional large-yield wells tapping the basalt, and further regulation of pumping.
Odessa-Lind (Luzier and Skrivan, 1975)	As much as 40 ft total, 1967–71.	Basalt	Irrigation	Washington Department of Ecology is trying to regulate pumping so that progressive decline does not exceed about 10 ft/yr.
Davenport (Luzier and Burt, 1974)	Began about 1966; amounted to about 17 ft (3.4 ft/yr) during the period 1972–77.	Basalt	Irrigation	Declines have been monitored, but no regulation of pumping had been imposed by 1977.
Cheney-Airway Heights (Luzier and Burt, 1974)	Generally less than 5 ft/yr.	Basalt (relatively thin)	Irrigation, public supply	The basalt (only aquifer available) is unusually susceptible to total depletion.
Moscow-Pullman (Barker, 1979)	Began in 1890's; total more than 100 ft at places, but averaged only about 1 ft/yr.	Basalt and interbedded sand	Public supply, irrigation	Declines are expected to slow unless future pumping continues to increase.
Raft River Valley (Sisco, 1974)	Began by at least the early 1950's; has been as much as about 5 ft/yr.	Basalt and sedimentary rocks	Irrigation	Idaho State Reclamation Engineer declared the valley to be a critical area in 1963, and restricted the drilling of additional wells; declines subsequently slowed or stopped.
Rock Creek and Goose Creek (Sisco, 1974)	As much as 22 ft/yr (total greater than 200 ft) 1961–72 in limestone aquifer; as much as 5 ft/yr in other aquifers.	Basalt and other volcanic, alluvial, and limestone rocks	Irrigation, industrial	Idaho State Reclamation Engineer stopped issuing well permits in 1962; thereafter, declines slowed or stopped.
Castleford (Sisco, 1974)	Mostly in the range of 2–4 ft/yr; as much as 60 ft total since the 1960's.	Volcanic rocks	Irrigation	Idaho Department of Water Resources closed the area in 1974 to further ground-water development.

WATER-QUALITY PROBLEMS

Water quality is presently not a significant limiting factor to the use of water from the major ground-water reservoirs. Although at places some dissolved constituents are found in concentrations greater than desirable for some purposes, most of the ground water is considered good to excellent for common use. Saline ground water is known to underlie several hundred square miles of the major ground-water reservoirs, but it is generally at considerable depth in areas where other sources of fresh ground water are available. Contamination resulting from man's activities is a growing problem, but only in a few areas has it reached proportions that cause serious limitations to continued use of the ground water.

The data that are presently available on ground-water quality pertain mostly to inorganic chemical constituents of the water that occur in varying amounts under natural conditions. In this respect, the data are mostly indicative of "baseline" conditions. Data on occurrences of organic compounds, which include most pesticides and many industrial products, are gradually becoming increasingly available, but are commonly related only to known or suspected cases of ground-water contamination.

PROBLEMS OF NATURAL ORIGIN

The most significant problems involving natural ground-water quality are:

DISSOLVED IRON

Probably the most widely bothersome natural constituent in the ground water of the region is dissolved iron, sometimes in combination with manganese. It occurs in undesirable concentrations at places in all the major ground-water reservoirs.

Dissolved iron does not affect the usefulness of ground water for irrigation, but causes problems in some industrial uses of the water and in household use where it may impart undesirable taste and color to drinking water and cause staining of plumbing fixtures and laundry. Some type of treatment (usually simple) to remove or reduce the effects of dissolved iron is a common adjunct to the development of ground-water supplies for public or individual domestic uses in the region.

SALINITY

Ground waters containing enough dissolved minerals to be considered "saline" waters probably underlie much of the area of the major groundwater reservoirs but are troublesome in only a few places. Well waters with dissolved-solids concentrations of 1,000 mg/L (milligrams per liter) or more have been found mostly in the Puget Sound, Willamette Valley, High Desert, and Southern Idaho ground-water reservoirs. Ground waters having dissolved-solids concentrations greater than 1,000 mg/L are mainly in older rocks that are not permeable enough to be important aquifers, and almost all these areas also have freshwater aquifers capable of larger yields. The saline ground waters are tapped by wells generally deeper than 100 feet, but in a few coastal areas of the Puget Sound ground-water reservoir, salty water is pumped from shallow wells. Also, saline ground water from older rocks locally migrates into shallow aquifers, and even discharges at the land surface as mineral springs in the High Desert and Southern Idaho ground-water reservoirs. In the latter reservoir area most of the springs that yield water with dissolved-solids concentrations greater than 1,000 mg/L are thermal springs.

Throughout the region, well waters having dissolved solids concentrations of 3,000 mg/L or more are rare, and generally represent seawater intrusion (Puget Sound ground-water reservoir) or thermal ground water (High Desert and Southern Idaho ground-water reservoirs).

OTHER DISSOLVED CONSTITUENTS

In addition to iron, a few other dissolved constituents occur naturally in waters of the major ground-water reservoirs in undesirably high concentrations. These occurrences are generally rather localized, in terms of aquifer zones as well as areal extent, and are mostly in the arid parts of the region. They tend to be near the margins of the major ground-water reservoirs, where the important aquifers are close to older rocks which generally are the sources of the undesirable constituents. These occurrences include concentrations of sodium and boron that are excessive for irrigation waters, mostly at places in the area of the High Desert ground-water reservoir, and concentrations of fluoride that exceed national drinking water standards, mainly at places in the High Desert and the Columbia Plateau ground-water reservoirs and a few places in thermal waters in the Southern Idaho ground-water reservoir (National Academy of Sciences and National Academy of Engineering, 1973).

The only known naturally occurring dissolved constituent in the region posing a health threat to humans is arsenic in ground water that was used for domestic supplies in the southern part of the Willamette Valley ground-water reservoir. However, that threat was recognized (Goldblatt and others, 1963) and alternative supplies developed, and the occurrence is now significant mainly as a historical reminder.

PROBLEMS RELATED TO MAN'S ACTIVITIES

Degradation of ground-water quality as a result of man's activities is a greater problem in this region than the naturally occurring water-quality characteristics, but it has not reached the major proportions found in some other regions. Degradation of the quality is, as would be expected, related largely to population distribution, being best known and most varied in cause where people and industries are concentrated.

CONTAMINATION FROM THE LAND SURFACE

Known and suspected occurrences of ground-water contamination from sources at the land surface are numerous, but so far no major ground-water supplies have been abandoned for this reason. The sources of such contamination differ widely, and include such things as solid-waste landfills, spilled chemicals and petroleum fuels, wastes from mining and milling operations, waste-treatment lagoons, feedlots, large wood-waste disposal sites, and fertilizer and pesticide applications. This type of contamination occurs at places in all the major ground-water reservoirs, except possibly in the High Cascades, and in many of the smaller ground-water reservoirs.

In the more populous areas west of the Cascade Range, solid-waste disposal sites are among the more obvious sources of contaminants. In these areas water from the abundant precipitation enters and percolates through the disposal sites, dissolving and carrying material from the wastes and becoming a generally noxious liquid called "leachate." As with other types of contamination from the land surface, migration of the leachate down to the ground water may take many years, and even covered disposal sites remain as pollution sources for years after their operation is stopped (Wilson, 1975).

Considering the incidences of known and suspected contamination, the relatively small number of apparent problems seems remarkable. The principal reasons seem to be:

1. Public-health officials and others have been alert in recognizing and reacting to some cases of contamination.
2. The soil and shallow rock materials adsorb or filter some contaminants.
3. Dissolved contaminants, moving through the unsaturated soil and the ground water, do not generally disperse widely but rather tend to move along narrow paths that make detection and sampling difficult; thus, some problems have not been recognized or have been underestimated.
4. The contaminants generally enter ground water of good quality which is capable of assimilating and diluting the contaminants to acceptable concentrations.
5. Most contaminants enter shallow aquifers and may discharge into streams without reaching the deeper aquifers that are favored as public water supplies.
6. Rate of movement is so slow that contaminants have not been detected, even at fairly short distances from the sources.

HOUSEHOLD SEPTIC SYSTEMS

Urban and suburban growth, which intensified during the 1940's, 50's, and 60's and which still continues, has involved a pattern of waste disposal that is having a significant impact on ground-water quality. In the developing areas, disposal of household wastes usually was initially through individual septic-tank systems or cesspools, the effluent from which percolated downward toward the ground water. Communal sewer systems and treatment plants normally were installed only where soil conditions prevented acceptable functioning of the individual disposal systems, or after housing densities and attendant waste discharges caused public-health hazards too great to ignore. In the meantime, contaminants from the household wastes have accumulated in the soils and in the upper part of the ground-water system.

Although a proper septic tank system is designed to destroy most or all bacteria in the household wastes, inorganic and some organic compounds pass through the system in the effluent and enter the soil. Where the homes and disposal systems are densely spaced, large concentrations of these compounds have been added to the soil column. Because of the usually slow rates of liquid movement underground, years or even decades may pass before the effluent from the septic tanks in a new housing development shows up as measurable contamination in a well or stream fed by the ground water. However, in this region those years and decades since the most intensive suburban growth are adding up, and increasing concentrations of certain dissolved solids from household sewage can be found in the shallow ground water beneath such areas. Several lakes in the Puget Sound ground-water reservoir area that are fed by shallow ground water have deteriorated for this reason. Also, contamination of the ground water beneath a suburban area east of Portland, Oregon has been reported and attributed to septic-tank effluent (Quan and others, 1974). Studies of ground-water quality have been or are being made in several other areas in anticipation of the problem, including areas southwest of Tacoma, Washington (Lakewood area), east of Spokane, Washington (Spokane Valley), near Boise,

Idaho (Boise Valley), and southeast of Kalispell, Montana (Swan-Avon Valley).

SALTWATER INTRUSION

In coastal areas of the region, including areas of the Puget Sound ground-water reservoir and smaller ground-water reservoirs along the coasts of Oregon and Washington, the known intrusions of seawater into freshwater aquifers are surprisingly few. This is largely due to the general abundance of precipitation to recharge the coastal aquifers. Also, the pumping from the smaller coastal ground-water reservoirs has not yet reached major proportions and, in Oregon, few productive aquifers are in contact with seawater.

At present, the aquifers contaminated by modern seawater are probably outnumbered by those contaminated by relict seawater contained in extensive deposits of older marine sedimentary rocks. In the Willamette Valley ground-water reservoir and some smaller ground-water reservoirs west of the Cascade Range, intensive pumping of wells that tap younger aquifers has locally caused migration of salty water from those adjacent older rocks.

PROBLEMS RELATED TO ARTIFICIAL RECHARGE AND DISPOSAL WELLS

In Portland, Oregon, and a few places in the Puget Sound area, ground water, returned underground following its use for heating or cooling, has resulted in local water-quality problems. Not only is the water returned to the aquifers at a drastically different temperature, but, in several places, aeration of the recharge water has promoted the growth of nuisance bacteria which reduce the permeability of the aquifers in the vicinity of the recharge well.

Related problems result from the use of wells for disposal of surplus irrigation water and waterborne wastes from households and other sources. Such disposal is workable only where the aquifer materials are extremely permeable, so that they can assimilate the waste products without rapidly clogging. Disposal wells have been used in the region for many years, mainly in the High Desert and especially in the Southern Idaho ground-water reservoirs. In the latter reservoir there are thousands of disposal wells. According to Whitehead (1974, p. 1) and Seitz, La Sala, and Moreland (1977, p. 2), most of the disposal wells in the eastern part of the Southern Idaho ground-water reservoir are for disposal of household wastes, but most of the water injected is from irrigation tail water and street runoff. The widespread use of disposal wells is considered to be clearly a potential threat to the quality of water in the major basalt aquifer which is the source of almost all the water used locally for household and community supplies. Experiments reported by Graham, Clapp, and Putkey (1977) suggest that some bacteria introduced through disposal wells may survive for a month or more after disposal stops. There is some local concern that the waste water may increase the nutrient content of the ground water that discharges into the Snake River and into lakes and ponds in canyon-bottom lands and contribute to occasional algal blooms in those waters.

WATERLOGGING

Waterlogging, the buildup of ground water to such a shallow depth that it hampers the use or productivity of agricultural land, occurs in all the major ground-water reservoir areas except the High Cascades, and in areas of many of the smaller ground-water reservoirs, as well. The region reportedly had about 2.5 million acres of "wet cropland" in 1966 (CNPTS, 1971b, p. 17).

In the more humid parts of the region, especially west of the Cascade Range, waterlogging is mainly a naturally occurring condition. It is most troublesome in low-lying, typically floodplain areas where the water table is never far below the land surface. Planting and crop production are delayed because the water table rises as a result of seasonally abundant precipitation, flooding of through-flowing streams, or runoff from adjacent lands, and the soil drains slowly.

In the drier parts of the region, the problems of waterlogging are characteristically found in flat or gently sloping areas that are irrigated with diverted stream water. In some places, the saturated zone that extends to the land surface and waterlogs the soil is temporarily perched above the general water table by a relatively impermeable material (for example clay, cemented gravel, or the top of the basalt). In many places, however, the application of irrigation water has raised the general water table to near the land surface, so that all the soil and underlying rock materials are saturated. The areas most susceptible to waterlogging by ground water are the localities of natural discharge of upwelling ground water—low-lying floodplains and the lowermost parts of closed basins or faultbounded ground-water "compartments." Some of these areas of ground-water discharge were waterlogged under natural, preirrigation conditions, but irrigation extended and worsened the waterlogging. Most, if not all, of the major irrigation projects have had associated waterlogging problems. In the Columbia Basin Irrigation Project of central Washington, for example, the seepage from canals distributing the Columbia River water and the application of the water for irrigation have caused the water table to rise more than 100 feet at places, waterlogging a few thousand acres of land.

INADEQUATE INFORMATION

Inadequate information about the ground-water reservoirs and their role in the total water resources of the region presently restricts the ability to deal effectively with the known and foreseeable problems, and hampers future management opportunities. There are two main levels of information needs: One is for the public officials, consultants, and others having major responsibilities in planning for, developing, regulating, or conserving the water resources; the second is for the direct user of the water resource—the irrigator, plant manager, water superintendent, and the general public. The two information levels largely overlap, but not entirely. The following discussion pertains mainly to the first level, and includes needs for new basic information about aquifers and their potential, needs for more quantitative information, and needs for water-quality information.

Probably no truly unknown additional major freshwater aquifers exist in the region, but the knowledge about every aquifer is incomplete, and for some aquifers is very sketchy, particularly in the more arid, sparsely inhabited parts of Idaho, Oregon, and Nevada. Geologic mapping in many of those areas has been of a reconnaissance type; consequently, although the general character of the aquifers is known, their actual potential can only be inferred. Because these areas are mostly in arid parts of the region, lack of firm water supplies is a severe and obvious deterrent to most types of development. Occasional private prospecting efforts to develop ground-water supplies in these areas during the 1960's and 1970's have been farily successful, although often unduly expensive. Much greater and more productive use of some of these areas could be expected if definitive information about ground-water potential were available. Information about potential aquifers in the deeper parts of some of the major groundwater reservoirs and on ground-water availability in some of the coastal areas of the region is also lacking. Until the ground-water conditions, including the geologic controls on the occurrence and movement of ground water, have been studied in greater detail, development of additional supplies of ground water will be largely a prospecting venture, with associated risks and expense.

For many of the ground-water reservoir areas, information is available about the general occurrence of the ground water and character of the aquifers, but water-management needs call for more quantitative data. Such data needs are likely to include volumes of a ground-water reservoir and of water it contains, amounts of pumpage and resultant water-level change, and numerical values for hydraulic characteristics of the aquifers. Quantitative data are considerably more difficult and expensive to obtain than more general information, commonly requiring the drilling of wells, the complex testing of aquifers, and the metering of discharges of large-yield wells. However, there is no substitute for the quantitative information in some types of water-management procedures, such as the use of mathematical models to predict future changes in the ground-water reservoirs (see section on "Management using Predictive Models").

A type of information that is increasingly needed, especially in the developing areas, is information on the quality of the ground water. More information is needed about the inorganic chemical constituents (the kinds of data that now are available for many ground waters), especially for previously unexplored aquifers and those susceptible to contamination. In addition, programs of reliable sampling and analysis for noxious organic compounds are needed at many places where such compounds present a threat to valuable ground-water sources—for example, in some areas of pesticide application, industrial waste disposal, and chemical spills.

COMPETITION FOR AVAILABLE SUPPLIES

Of the major ground-water reservoirs, the competition for available water supplies is a serious problem only with respect to the Columbia Plateau and the Southern Idaho ground-water reservoirs. It is also a major regional problem because: (1) It also occurs in areas of several smaller ground-water reservoirs; (2) it involves most, if not all, the States comprising the region, (3) the amount of water involved in the competition is large, and (4) the problem is critical to prospects for fullest and most effective use of the total water resource.

In the area of Columbia Plateau ground-water reservoir, the competition for available supplies has produced problems of interference between pumping wells, conflicts between ground-water and surface-water rights, and litigation between Indian and non-Indian users of ground water and surface water. The early, uncontrolled development of limited ground-water supplies caused rapidly declining ground-water levels that led to pumping restrictions in areas listed in table 2.

In the area of the Southern Idaho ground-water reservoir, surface and ground water are closely related at many places. Development of large supplies of ground water in the eastern part of the area affects to some degree return flows to the Snake River farther west;

this interrelationship is also true in some of the basins flanking the Snake River Plain. Therefore, large withdrawals of ground water may have an adverse affect on the surface-water supply, and are, or will be, in conflict with surface-water rights.

ANTICIPATED OR POTENTIAL PROBLEMS

Problems foreseeable in the future for the Puget Sound ground-water reservoir include local deterioration of water quality as a result of waste-disposal practices, and seawater intrusion associated with additional pumping in near-shore areas. The growth of these two problems seems to be an inevitable consequence of present waste-disposal practices and developmental trends in near-shore areas. Some of the better aquifers in this part of the region lie beneath areas that have fairly dense spacing of household septic-tank disposal systems. Doubtless the combination of abundant recharge to dilute the wastes and the slow movement of water underground has helped limit the effects of potential pollution from these waste-disposal systems. However, pollutants now in the soil and shallow rock materials will continue to migrate downward to the ground-water and increasing effects of these pollution sources can be expected in the future despite whatever near-term management steps are likely to be taken. Local problems of seawater intrusion are also a likely future consequence of indiscriminate development of ground water in near-shore areas. This problem may be troublesome at places but will be somewhat self-limiting; pumping that induces landward movement of the salty water generally decreases as the wells "go salty."

For the Willamette Valley ground-water reservoir, foreseeable problems include further depletion of water in the basalt aquifers and water-quality deterioration in areas of waste disposal by landfills and by densely spaced septic-tank systems.

No special problems are foreseen in greater development of the ground-water of the High Cascades ground-water reservoir. Meager water-quality information indicates that the ground water is of good to excellent quality, but experience in similar terranes suggests that undesirable dissolved minerals might be present locally. The available recharge is adequate to support almost any conceivable direct withdrawals of ground water with little impact on the low-level flows of the streams that drain these aquifers. The short growing season and the remoteness of the areas underlain by this reservoir suggest little possibility that much of this supply will be needed within the mountains in the near future, and most of the reservoir probably will remain an untapped reserve for many years.

Future problems in the area of the High Desert ground-water reservoir are likely to be largely the same as present problems—namely, limitations on ground-water supplies imposed by meager recharge, locally poor water quality, and waterlogging. In addition, areas of intensive pumping may develop or enlarge, and result in significant water-level declines, perhaps to the extent of long-term depletion of the ground water in localized areas (table 1). Future major ground-water development may tend to be areally concentrated for two reasons: (1) A relatively small amount of land in the area is privately owned, and (2) successful wells tend to foster prospecting for ground water in their immediate vicinity. Management of the ground-water reservoir, however, is not likely to be a major problem until development proceeds to a considerably greater extent than at present.

In the future, problems related to ground water in the Columbia Plateau ground-water area are likely to be of the same kinds as at present (table 1) but more extensive and in new places. Problems of declining water levels are expected to become more widespread as development of water from the basalt progresses. Because the basalt has relatively small storage capacity, any intensive withdrawals from wells tapping the basalt can be expected to cause sizable water-level declines. Examples of areas now undergoing such development, where the increased pumping from the basalt is likely to cause significant water-level declines within a few years, are the Moxee Valley and Horse Heaven Hills areas (not shown on included maps) both in south-central Washington, respectively east and southeast of Yakima (fig. 7). In areas of presently declining levels, the declines are likely to continue until limited by such factors as excessive pumping lifts, importation of surface-water supplies (as in The Dalles area, Oregon, table 2), legal restriction on pumping withdrawal, or actual dewatering of the available aquifers—a possibility in the Cheney-Airway Heights area, eastern Washington (table 2; also Luzier and Burt, 1974, p. 47). Where the depleted zones include shallow aquifers which supply most of the rural domestic and stock water and the dry-season flow of small streams, troublesome reductions may occur in the streamflow and in these rural well-water supplies. In areas having many wells that penetrate deeply into the aquifer system and are open to more than one aquifer zone, the continual leakage from one zone to another through the well bores may contribute to declines in some aquifer zones.

Future ground-water problems in the Southern Idaho ground-water reservoir area probably will be related mainly to further declining water levels (local overdrafts on the ground-water reservoir), increasing conflicts with surface-water rights, spreading of water-logging associated with the growth of irrigated farming, and increasing contamination. If withdrawals for irrigation follow the predicted increasing trend, declining ground-water levels can be expected to show up in new places and perhaps resume in some of the areas already noted (table 2). The decrease of ground water in storage, reflected by water-level declines, is likely to affect the low flow of streams and, therefore, the availability of surface water for use under existing rights. The greatest undesirable impacts on surface-water rights likely would result from development of additional large supplies of ground water in the eastern and northern parts of the reservoir system.

Soluble residues and waste products from intensive agriculture and processing of agricultural products are expected to be sources of increasing ground-water contamination. Most organic compounds tend to be filtered out or adsorbed in their passage through the first few feet of soil; however, dissolved salts, including the major constituents of fertilizers and some pesticides, infiltrate the soil and percolate downward readily with the rainwater and irrigation water. Some of the best evidence for this type of ground-water-quality deterioration is in the extensive Columbia Basin Irrigation Project (Columbia Plateau ground-water reservoir; Van Denburgh and Santos, 1965), but the effects are being felt and can be expected to increase also in the Southern Idaho ground-water reservoir and elsewhere in the more arid parts of the region.

The major water-related problems that are foreseen for the areas of the smaller ground-water reservoirs are increasing conflicts in the use of surface water and ground water and increasing water-quality problems.

Some of the smaller ground-water reservoirs, such as isolated coastal sand deposits or valley-filling alluvial deposits in areas of impermeable bedrock, are extremely valuable as sources of water but, because of their relatively small volume, have limited waste-assimilating capability. The preservation of good-quality ground water under conditions of expected growth in these reservoir areas will require some difficult (and probably expensive) choices regarding waste-disposal practices and land use. An example is in the Spokane Valley-Rathdrum Prairie area, Washington-Idaho, east and northeast of Spokane (fig. 7; Drost and Sites, 1978). Because of the importance of the ground-water reservoir in that area as a source of public water supplies and the lack of suitable alternative supplies, the U.S. Environmental Protection Agency, in 1978, designated the reservoir as a "sole-source aquifer" (U.S. Environmental Protection Agency, 1978). Such a designation restricts future waste-disposal practices and land uses that might degrade the ground water in that area.

By 1977, the intrusion of sea water in the western part of the region was not a widespread problem. However, despite strong trends to protect these shorelands through State and Federal legislation, the projected development in these areas doubtless will lead to intensified local ground-water withdrawal that will be accompanied by problems of potential seawater intrusion. Elsewhere in the region, the induced migration of saline water into the freshwater aquifers of the small ground-water reservoirs is not expected to be significant unless pumping is intensive in areas adjacent to older saline-water-bearing rocks. Saline-water problems also may accompany future development of geothermal energy in the smaller reservoir areas.

MANAGEMENT OPPORTUNITIES AND CONSTRAINTS

Management opportunities presented by ground-water reservoirs in the region are discussed below, along with related constraints and problems. As in foregoing discussions, the major ground-water reservoirs are emphasized, but most of these management opportunities and constraints apply also to the smaller ground-water reservoirs.

DEVELOPMENT OF ADDITIONAL SUPPLIES

Additional large supplies of water can be developed from each of the major ground-water reservoirs, but most of the reservoir areas have their own set of constraints on such further development. Additional major supplies, large enough for industrial, irrigation, and municipal uses, are available from presently utilized aquifers at places in all the major reservoir areas, but especially in parts of the Puget Sound and Willamette Valley ground-water reservoirs.

In the Puget Sound area, underutilized glacial and alluvial aquifers are capable of meeting much of the foreseeable increase in water demand, such as additional supplemental irrigation water and new industrial and community supplies. Some coastal aquifers contain a considerable resource of fresh ground water that is now flowing to the sea without being used beneficially; however, the optimum use of this resource will require improved information about the coastal aquifers and adroit management to minimize problems of seawater intrusion.

In the Willamette Valley ground-water reservoir, large additional supplies of ground water also are available, chiefly from the alluvial deposits. Not only is the present withdrawal only a small fraction of the natural recharge, but recharge might locally be increased by increasing withdrawals. That is, if more storage space were provided in the aquifer by intensified pumping during the summer months, that space would be largely refilled by abundant recharge water during winter and spring. Also, for many of the large-yield wells near the Columbia River and other major streams, much of the water they produce is induced infiltration from the rivers. That induced infiltration could be increased—at the expense of river flow, if course. Along favorable reaches of the Columbia River, for example, yields of about 50–100 million gallons per day might be obtained per mile of river length (Mundorff, 1964, p. 93).

At places near the margins of the Columbia Plateau and Southern Idaho ground-water reservoirs, large bodies of saturated older sedimentary rock materials occur beneath, overlying, or interfingered with basalt aquifers. These sedimentary rocks can be used more extensively as sources of ground water not only by tapping them directly with effective wells (see section on "Improved Design and Construction of Wells"), but also by inducing these reservoir rocks to drain into adjacent, more permeable basalt aquifers. Such drainage might be accomplished by purposely causing large water-level declines in the adjacent basalt aquifers to accelerate the migration of water from the sedimentary materials.

Where ground-water reservoirs are already heavily stressed by pumping, greatly increased pumping from the same aquifers will tend to increase or create problems of declining water levels and conflicts between water users. It can also accelerate the spread of any ground-water contaminants by increasing rates of ground-water flow in the heavily pumped parts of the ground-water reservoir.

In some areas, increased pumping from presently utilized aquifers probably will include pumping to reduce waterlogging (see section on "Reduction of Waterlogging") and to salvage ground water now used by vegetation. Common water-loving vegetation, phreatophytes and hydrophytes, do not grow extensively in the region, but they are abundant enough in the lowlands and along stream courses to consume a significant part of the summer water supply. In the ground-water reservoir areas east of the Cascade Range, where evapotranspiration rates are high and summertime water supplies are especially valuable, some of this water may be economically salvageable. Phreatophytes may be removed by uprooting, mowing, or spraying with growth-retardant chemicals. An alternative, and generally more effective, method of reducing evapotranspiration is to remove the water supply by draining marshes or pumping wells to lower the water table below the root zone.

Destroying the natural vegetation in order to increase the water yield, however, would eliminate possible benefits of the vegetation and might create other problems. For example, the plants may be a valuable habitat for wildlife and shelter for livestock. Moreover, they may inhibit evaporation losses from surface water by shading the water and thereby reducing its temperature. Also, removing all the natural vegetation might create problems of increased flood damage and erosion.

DEEPER AQUIFERS OF KNOWN POTENTIAL

In some of the major ground-water reservoirs, productive aquifers are known to occur at depths greater than the presently utilized aquifers, and thus are in the category of proven reserves. This is especially true for parts of the Puget Sound, Columbia Plateau, and Southern Idaho ground-water reservoirs. Constraints on development of these deeper aquifers include high costs of drilling the required deep wells, less desirable water quality (at places in the Puget Sound and Southern Idaho areas), high pumping lifts, and present availability of more accessible alternative water supplies. The temperature of the ground water increases with depth, and the higher temperature is generally considered to be a constraint to the use of the water for municipal supplies, some industrial purposes and, if sufficiently high, for irrigation.

Three points regarding the future development of deep aquifers in the region should be emphasized. First, in many of the reservoir areas, the total depth of productive aquifer materials, being controlled by the geology, is fairly well known. In these areas, hydrologists can say with some assurance that highly productive aquifers do not occur at greater depths. Second, the productivity of sedimentary aquifers especially, and volcanic aquifers to a lesser extent, locally tends to decrease with greater depth. Third, the withdrawal of water from deeper zones in a presently utilized ground-water reservoir does not necessarily constitute "new" water; although it may increase the short-term supply, in the long term it must be counted as part of the total draft on the reservoir.

NEW SOURCES OF SUPPLY

All the major and many of the smaller ground-water reservoirs offer opportunities for finding and developing water supplies from presently unknown or unproven aquifer zones. This is especially true of the sparsely inhabited areas, including much of the High Cascades and High Desert, and parts of the Columbia Plateau and Southern Idaho ground-water reservoirs. Much of this sparsely populated land has low foreseeable demands for ground water because of high altitude or rugged terrain, short growing season, scarcity of arable land or privately owned land, and adequacy of surface water (at places). A further constraint on ground water development in unproven areas will be the high cost and financial risk of prospecting for new sources of supply. Also, especially in the High Desert area and in deeper zones elsewhere, undesirable water quality may be a constraint. Demands for ground water doubtless will increase somewhat as additional information about the potential reservoirs becomes available, and as new uses develop such as forest irrigation, fish-hatchery supply, and energy production (see section on "Geothermal Energy"). Also, in some remote areas where streams are contiguous to permeable aquifers with a large storage capacity, ground-water supplies might be economically developed to supplement the low flow of the stream during relatively short but intense periods of drought.

CONTAMINATION ABATEMENT AND AVOIDANCE

Significant opportunities exist for reducing presently occurring contamination of ground water and for avoiding future contamination in critical areas of ground-water need. These opportunities exist mainly in the Puget Sound, Willamette Valley, Columbia Plateau, and Southern Idaho ground-water reservoirs, and several smaller reservoir areas.

Effective abatement of presently occurring contamination must start with identification of the areas undergoing the contamination and determination of the sources and degree of contamination before appropriate regulatory or management steps are formulated and undertaken. A number of areas of ground-water contamination in the region—mainly areas of solid-waste disposal, concentrated septic-tank systems, disposal wells, and chemical spills—have been or are being studied to some degree, and some monitoring of ground-water quality in impacted areas has been going on for decades. However, few of these studies have yielded information that is adequate for such purposes as deciding between alternative abatement schemes or estimating future die-away rates of contamination following abatement. For such purposes data are needed, for example, on: Rates and modes of contaminant movement in the ground-water system and the unsaturated zone above the ground water; adsorption of the contaminants on the soil and rock materials; and the ultimate discharge of the contaminants from the ground-water system. Much of this information must be obtained through test drilling and detailed subsurface sampling.

Actual abatement steps might include stopping a specific waste-disposal operation, changing the locality or method of waste disposal, clean-up of accidental spills, pumping or artificial recharge to change flow directions of the contaminated ground water, or even the direct pumping of the contaminated water for disposal elsewhere. In the humid parts of the region west of the Cascade Range, the only practical way to reduce contamination from landfill disposal may be to collect as much as possible of the leachate and other drainage from the disposal sites and treat these waste waters to make them acceptable for redisposal by dilution in large streams or marine waters.

Avoidance of future ground-water contamination should focus on the parts of the ground-water reservoirs that are most important to protect and retain in a high-quality state, such as aquifers that are sources of present or anticipated public supplies, aquifers that are the only feasible source of water supply, or those clearly advantageous economically or otherwise as present or future high-quality supplies. For example, ground-water resources are extremely important in marine island and peninsular areas where sites for surface reservoirs are limited, stream supplies are inadequate or subject to pollution, and where great costs would be incurred in piping water from dependable mainland sources.

Because of the importance of considering future needs in decisions about protection of the ground water, such decisions are best made within the framework of comprehensive land-use and water-use plans. After a decision that a particular ground-water body should be protected, the steps necessary to avoid ground-water contamination must be taken by the agency having jurisdiction over use of the overlying land or recharge areas. Depending on the ownership and jurisdiction of the land, the protective measures may include: Land zoning to control population density; prohibition of waste disposal; open-space designation of identified recharge areas, or "watershed" designation with accompanying restriction on entry into recharge areas. Also, because withdrawal of ground water can accelerate the spread of contaminants, the protective measures might include restrictions on pumping wells in areas subject to salt-water contamination.

In many areas where contaminants are being added to the subsurface, remedial action apparently will not be started before there is obvious evidence that significant contamination of valuable ground-water supplies has occurred. Future impacts of the contamination might be lessened in those areas if the onset of the contamination can be detected as soon as possible. This will require identification of such areas (if they have not been identified already), studies to determine the degree or threat of the contamination and the subsurface conditions in the threatened areas, and monitoring programs to detect the onset and trace the movement of contamination. In most cases, existing wells will not suffice for an effective monitoring program—special test-observation wells probably will be needed also to provide information not available from supply wells. The monitoring programs in such areas may be needed for decades into the future.

Careful monitoring of the underground disposal of potentially hazardous contaminants—for example, the radioactive wastes in the Hanford, Washington (Columbia Plateau ground-water reservoir) and the Arco, Idaho (Southern Idaho ground-water reservoir) areas—are required by existing environmental protection laws.

IMPROVED DESIGN AND CONSTRUCTION OF WELLS

Well-construction practices in this region can be substantially improved to the advantage of future development and management of the ground-water reservoirs. Many wells that presently yield less than enough for irrigation, industrial, or public supplies could be brought up to that range through better well construction. In general, the finer the saturated material, the more the yield of the well depends on good design and construction.

Techniques that have been used to increase well efficiency and productivity, and that hold promise for wider application in the region, include:

(1). Recently developed rock-drilling methods and equipment;
(2). use of well screens and graded gravel or sand "packs" for greatest yields from sedimentary aquifers;
(3). use of special drilling fluids that reduce clogging of aquifer pores;
(4). greater use of well-development techniques (surging, chemical treatment, water jets, and so forth) in completion of wells; use of these techniques for wells tapping volcanic, as well as sedimentary, aquifers.

Although some of these techniques are applicable mainly to new well construction, several can be used beneficially on wells that are being rehabilitated or deepened.

One conspicuous opportunity for improvement is the reduction of unwanted vertical ground-water leakage because of "short circuiting" through unlined or leaky well bores (see section on "The Productive Aquifers"). The best way to prevent this cross-bed leakage is to design and construct wells using well casing cemented to the rock materials so as to isolate the different aquifer zones; the casing should be open only to the zone or zones actually to be developed. Installation of casing in existing unlined wells is likely to be much more troublesome, but may be necessary at places in order to reduce ground-water contamination (Foxworthy, 1970, p. 14) or alleviate the draining of aquifer zones tapped for domestic supplies (Luzier and Burt, 1974, p. 44).

The recognition of deficiencies in many of the older wells has led the State water-management agencies in the region to formulate and recommend, and in some cases legally require, minimum standards for water-well design and construction. Moreover, in some areas of intensive ground-water management, such as the Odessa-Lind area of Washington, certain well designs (specified depths of well casing) are made a condition for approval of additional wells. As problems of water-level declines and ground-water contamination increase in the region, the imposition of well-design requirements by State and local water-management agencies is likely to increase.

REDUCTION OF WATERLOGGING

Waterlogging might be reduced by greater pumping from aquifers in some waterlogged areas, and gravity drains within the saturated zone also can be useful; at other places, a reduction in the application of irrigation water may be the only practical corrective measure. At some places a combination of these measures may be the most effective. Pumping from wells to reduce waterlogging is effective only if: (1) Aquifers of at least moderate permeability can be tapped at relatively shallow depths, and (2) the pumping can induce substantial downward migration of water from the waterlogged zone or can intercept significant amounts of naturally upwelling ground water. Moreover, for such remedial pumping to be economically feasible in most cases, there must be beneficial uses for the water pumped from the shallow aquifers. "Relief wells" and drainage tunnels to reduce waterlogging have been used in the area of the Southern Idaho ground-water reservoir for several decades, and relief wells, alone or combined with drains, have been suggested for a waterlogged area south of Yakima, Wash. (fig. 7; Mundorff and others, 1977, p. 81).

The widespread buildup of ground-water levels and occurrences of waterlogging in areas of major irrigation projects strongly suggest that future irrigation projects should be designed, if possible, for importation of only part of the water supply needed, with the balance of the supply to be obtained from aquifers recharged by the irrigation. This approach would not only reduce the amount of surface-water diversion required but would also tend to reduce waterlogging problems.

SUBSURFACE STORAGE

Underground space has been used in the region for some common and some uncommon types of waste disposal. It has also been used to store freshwater and natural gas. The following are brief discussions of the opportunities for and experiences in use of the underground space for these and other types of subsurface storage.

ARTIFICIAL RECHARGE

Artificial recharge—the addition of water to the ground-water reservoir as a result of man's activities—is important to the concept of total water-resources management. It offers possibilities of manipulating input to the ground-water reservoirs in addition to the manipulation of withdrawals that is afforded by regulation of pumping. In the broadest sense, artificial recharge includes not only planned replenishment but also additions to the ground water that are incidental to other activities; it includes not only good-quality water but also the waste water that is added.

In this sense, the most significant type of artificial recharge in the region is that occurring as a result of irrigation. Seepage from irrigated lands and conveyance ditches occurs in all areas where irrigation is extensive, but it is most significant in the more arid parts of the region. Irrigation is the major form of recharge in parts of the Columbia Plateau and Southern Idaho ground-water reservoirs and in some smaller ground-water reservoirs. Recharge from this source is so great at places that waterlogging has become a problem (see section on "Waterlogging").

A related type of artificial recharge is surface spreading, which consists of diverting water from a stream and allowing it to flow over or stand upon permeable material so that it may infiltrate to the ground-water body. Spreading can be accomplished by simply flooding flat areas in which the water can be retained for substantial periods of time, by holding the water in artificially constructed basins or in natural ponds, or by directing the recharge water into ditches or furrows. Surface spreading was used with reasonable success to augment natural recharge for an alluvial aquifer tapped by municipal wells at Richland, Wash. (Price and others, 1965, p. C5–C19), and occurs incidental to the operation of storm-water retention basins in the area of the Puget Sound ground-water reservoir and other urban areas in the region.

A method of increasing the recharge to aquifers, at the expense of flow in contiguous streams, is the deliberate drawdown of ground-water levels in shallow aquifers adjacent to streams so as to induce or increase infiltration of water from the streams into the aquifers (see section on "Development of Additional Supplies"). This practice, called induced infiltration, is fairly common, and is used largely to obtain water for municipal and industrial supplies that requires less treatment than if they were taken directly from the streams. The gain in well supplies by this method, of course, is offset by a corresponding reduction in the flow of the affected streams. However, at places where the wells can be located sufficiently far from the affected streams, the reduction in streamflow caused by the seasonal pumping from wells may be delayed until after periods of low streamflow. This method offers important opportunities for increasing the amount of water available during periods of low streamflow by utilizing even relatively small ground-water reservoirs. Moreover, it is a key to practical conjunctive use of surface and ground-water resources (see section on "Conjunctive Use of Ground Water and Surface Water).

In addition to the disposal of storm-runoff and waste waters through wells (see section on "Problems Related to Artificial Recharge and Disposal Wells"), which is most significant in the areas of highly permeable young basalt rocks (Southern Idaho and High Desert reservoirs), injection of good-quality water through wells for later recovery from the aquifers has been attempted elsewhere in the region. Experiments and trial injection through wells, mainly in the Willamette Valley and Columbia Plateau ground-water reservoir areas, have utilized aquifers of moderate to high permeability, though not as permeable nor nearly as porous as the young basalt. These experiments had limited success. They indicated that the artificial recharge through wells generally is technically feasible, but it requires sediment-free water of good chemical quality, careful injection techniques, and occasional cleaning of the injection well for practical long-term success (Foxworthy and Bryant, 1967; Price and others, 1965). Also, not all the water recharged artificially was recoverable for later use.

Artificial recharge, to be a feasible water-management aid, requires three basic conditions: (1) The aquifer must be at least moderately permeable and must have adequate storage capacity to receive the water; (2) the supply of recharge water must be of adequate quantity and suitable quality; and (3) the water must be economically recoverable from the ground-water reservoir and of a quality suitable for the intended use. Economic considerations may be as complex as the technical aspects in judging the feasibility of artificial recharge. They include consideration of the economic benefits of sustaining or improving the ground water, cost of the artificial recharge operations, value of the water added to the aquifers, and relative value of the recharge water if used for other purposes, such as hydropower generation.

Conditions favorable for artificial recharge probably can be found in several parts of the region, including some of the smaller ground-water reservoirs. In many valleys and basins, there are terraces underlain by young sedimentary rocks which could be potentially productive aquifers except that they are mostly above the normal water table. The water-yielding capability of these terrace deposits depends not only on the character of the rock materials, but also on their saturated thickness. At some places the terrace deposits were unsaturated under natural conditions but have become partly saturated from irrigation water. At other places, the unsaturated deposits await only the artificial addition of water to become valuable storage reservoirs.

Artificial recharge seems to be a strong possibility for increasing the usable supplies of ground water in the Southern Idaho ground-water reservoir. No planned artificial recharge is being done in the reservoir area; however, recharge incidental to irrigation is a major factor in the hydrologic regimen. Mundorff (1962), in his study of the feasibility of artificial recharge of aquifers underlying the Snake River Plain, concluded that surplus water for recharging was available and that artificial recharge was hydrologically feasible. There are several places where water could be diverted to flow by gravity to recharge sites on the Snake River Plain. Opportunities also exist for artificial recharge in basins flanking the Snake River Plain. In some areas, surplus river water is available locally to recharge the aquifers; in other areas no local surplus water is available and recharge water would need to be diverted from one of the major rivers.

Several parts of the Columbia Basin ground-water reservoir area are included in plans for additional irrigation with surface water. These include the East High division of the Columbia Basin Irrigation Project, east-central Washington, and lands adjacent to the Columbia River in south-central Washington and north-central Oregon. These or similar expansions of irrigation using surface water are likely to provide artificial recharge to aquifers underlying the irrigated areas, thereby improving the productivity and dependability of such aquifers. In addition, opportunities may exist for more deliberate artificial recharge using irrigation canals to convey surplus water to recharge sites during the nonirrigation seasons (Garrett and Londquist, 1972, p. 15, 16).

WASTES

Many types of wastes have been committed to subsurface disposal in the region. They range from the least troublesome types, such as burials of nondegrading construction materials, to wastes that are toxic or otherwise dangerous, including the radioactive wastes disposed of at sites near the Hanford nuclear facility, Washington and INEL (the Idaho National Engineering Laboratory), Idaho.

In general, the more hazardous the waste, the farther away it should be from the realm of living things (biosphere); this often requires disposal deep underground. Successful disposal of wastes in underground space with the least undesirable consequences requires thorough knowledge of the geologic framework and of the path that the ground water—and any contaminants derived from the wastes—takes through that framework. Obviously, the more hazardous the waste, the more important this knowledge. Exhaustive studies and elaborate test drilling have been conducted and are being continued in the vicinity of the Hanford nuclear facility, to determine the feasibility of storing high-level hazardous wastes deep in volcanic rocks. At INEL, also, many observation wells have been drilled to monitor the movement of radioactive wastes.

Some rocks, but not all, are worthy of consideration for subsurface disposal of hazardous wastes in areas of the major ground-water reservoirs. Included are deep-lying, older sedimentary rocks containing salty water in the Puget Sound and Willamette Valley ground-water reservoirs, and deep zones of the basalt or underlying rocks in the Columbia Plateau and Southern Idaho ground-water reservoirs. In addition, parts of the High Desert ground-water reservoir seem to offer possibilities. This ground-water reservoir includes a number of closed basins that have no surface outflow and probably insignificant ground-water outflow; there the discharge of water is nearly all by evapotranspiration or from wells. Owing to the remoteness and lack of substantial outflow from much of the area, opportunity may exist for storage of wastes in deep saline aquifers, but additional subsurface studies are needed for verification.

NATURAL GAS

Deep, confined salty-water aquifers in the Puget Sound and Willamette Valley ground-water reservoirs may offer unusual opportunities for underground storage of natural gas in proximity to major urban centers. Confined zones at depths of about 1,800 to 2,100 feet have been used for the storage of many billions of cubic feet of natural gas in a small area a few miles south of the Puget Sound ground-water reservoir. The further use of deep confined aquifers for storage of this type should be based on greatly increased knowledge about subsurface conditions in the areas involved, including deep circulation patterns of the ground water.

IMPROVED MANAGEMENT AND CONSERVATION

Many opportunities exist for improved management and conservation of proven, available water resources. These include greater conjunctive use of ground water and surface water, conservation of existing resources through changes in conveyance and use, and management of water resources with the aid of predictive models of ground-water flow systems. Of these, greater conjunctive use of ground water and surface water probably offer the largest benefits, but this management opportunity also has the greatest problems and constraints.

CONJUNCTIVE USE OF GROUND WATER AND SURFACE WATER

The integral relationship between ground water and surface water is most apparent in the smaller ground-water reservoirs, where the best aquifers commonly are well connected hydraulically to the local stream system. Because of this degree of interconnection, and because the ground-water reservoirs feed the streams through most of the year, any change that has a major effect on the ground water, such as intensive pumping or changes in recharge, generally has a rapid and obvious effect on the streams. The same interrelationship also exists for the larger ground-water reservoirs, creating conflicts between use of wells and stream water (see section on "Deeper Aquifers of Known Potential"), and dooming to failure any attempt to manage effectively either the ground water or surface water separately. Water flowing through an area commonly passes back and forth between the aquifers and the stream channels. For example, in the area of the Southern Idaho ground-water reservoir, all outflow ultimately is through the Snake River; however, throughout the reservoir area the river alternately gains and loses in several reaches before finally collecting all surface and ground-water discharge near the west end of the reservoir area. Some water that is diverted for irrigation cycles back into the aquifer system at least three times as it moves from the upper end to the lower end of the Southern Idaho ground-water reservoir.

In this region, largely because of the highly seasonal nature of the precipitation and snowmelt (p. 11), some runoff commonly escapes from most areas without apparent beneficial use. Management of the streamflow and aquifers conjunctively might provide the flexibility to salvage and store for later use some of this runoff. Such management might include intensive pumping for the deliberate drawdown of ground-water levels during periods of low stream flow to make space for additional recharge water available during periods of high runoff, provided this can be accomplished without serious reduction of streamflow during critical periods.

Conjunctive use of ground water and surface water probably represents the most widely available opportunity in the region for effectively managing the available water supplies. The benefits possible from this management approach are effectively blocked, however, by existing water-rights laws. In many areas efficient management will require flexibility to allow the source of water supplies to be shifted from stream to underground sources and back again in accordance with water availability. It also may require the readjudication of withdrawal periods and rates of withdrawal.

EFFICIENCIES IN CONVEYANCE AND USE

For some parts of the region, no matter how much emphasis is given to water supplies, the supply-demand equation cannot be balanced without sensible and effective conservation of the water. Fortunately, opportunities are available for conserving the ground water; some of these—reduction of water-loving plants, artificial recharge, and repair of leaky wells—were discussed previously.

A conservation opportunity of considerable promise is a relatively new method of irrigation—drip irrigation. Great savings of water are possible because much soil that is wetted during the conveyance and application of water in other irrigation methods remains dry during drip irrigation. Savings are also possible in fertilizer and some pesticides, which can be applied in the irrigation water, and in weed control. Other advantages include the virtual elimination of irrigation return flow with its degrading effect on water quality, avoidance of waterlogging resulting from excessive water application or irrigation return flow and, for some crops, a reduced need for level land. Constraints on widespread use of drip irrigation include capital investment required for pipelines and drip nozzles, major investments that already have been made in other irrigation systems, possibility of salt build-up in the soil, and some problems of clogging of the drip nozzles with particles such as weed seeds and with chemical precipitates.

Ground water not only is suitable for drip irrigation, it is a preferred supply. One major reason is that ground water usually can be delivered free of weed seeds, spores, algae, or other clogging agents. Sand or silt in water from some wells however, may be a problem, as in many surface-water supplies. Moreover, the smaller demand rate for irrigation water for this method favors the decentralizing of supply wells—that is, the use of several widely separated, smaller-yield wells with short, efficient supply lines instead of one large-yield well pumping into long supply lines or inefficient distribution ditches. The decentralizing of supply wells may have economic benefits even in areas where large-yield aquifers are available. In addition, the smaller water demand for the drip method makes irrigation with ground water much more attractive for areas where aquifers do not have the proven capability for the large yields normally required for irrigation by other methods.

Any other means of increasing efficiency in the use of the irrigation water or decreasing need for water application can also help stretch supplies of surface and ground water. Such improvements might include greater use of efficient sprinkler systems, or techniques of fertilizer application and soil treatment to reduce the need for excess water to flush the soil. Because of the huge amount of the total water withdrawal that is dedicated to irrigation (fig. 2), even small individual savings of irrigation water could be valuable. For example, a saving of only 18 percent in the average irrigation-water diversion or withdrawal in 1975 would have conserved water in an amount equivalent to all other withdrawals combined for that year (5 billion gallons per day).

MANAGEMENT USING PREDICTIVE MODELS

Research in the field of ground-water hydrology has, within the last two decades, developed a set of powerful tools for assisting water management—models of ground-water flow systems that allow prediction of conditions in the ground-water reservoir under different schemes of development. These models were made practical by recent advances in electronics and high-speed computers. Predictive models have been developed and used for several ground-water reservoirs in the region and are possible for other areas as well.

Ground-water models have been used to predict water-level changes likely to result from different rates and patterns of pumping withdrawal in the Columbia Plateau (about 10 models, some overlapping); the Southern Idaho (9 models); the Willamette Valley (1 model); and a few smaller ground-water reservoirs. Some models have been deliberately developed on the basis of preliminary or sketchy information in order to define sharply the data needed for a more reliable model to be used for management guidance. In addition, some models are being developed principally to predict the movement of contaminants in the ground water. A few of the ground-water models reflect the interrelationships of the streams and aquifer systems and, therefore, could guide management programs for conjunctive use of ground water and surface water (see section on "Conjunctive Use of Ground Water and Surface Water").

Constraints on the greater use of ground-water models include: The large investments in facilities, data handling, and time usually required for development of a reliable, effective model; lack of adequate data for modeling many ground-water reservoirs, a situation that often can be overcome only by collecting field data, which is expensive and time-consuming; a shortage of ground-water hydrologists with the skills, support, and facilities necessary for successful modeling of complex hydrologic systems; and the lack of assurance that, even with the best expectations, a reliable model can be developed that is suitable for the intended use. Despite these constraints, however, predictive ground-water models remain one of the more useful water-management tools available for problem areas that warrant, and are amenable to, modeling. Once calibrated and verified against actual conditions, such models can be used indefinitely, with appropriate modifications, for study of a large variety of water-management problems.

ENERGY-RELATED OPPORTUNITIES AND CONSTRAINTS

Events of the past few years have brought into sharp focus the dependency of the Pacific Northwest Region and the Nation on abundant energy. Accordingly, no contemporary analysis of resource-management opportunities and constraints would be complete without considering those related to energy.

Opportunities for energy-related uses of ground-water in the region appear at present to be few but may be substantial in the future. They include geothermal energy development; use of ground water in mining, processing and (or) movement of coal; small amounts of additional hydropower generation using ground water; and use of the ground water for disposal and (or) storage of heat. Ground-water problems associated with these possible uses include depletion of supplies and degradation of water quality.

GEOTHERMAL ENERGY

Naturally occurring hot ground water has been used in this region for several decades for irrigation and space heating, and is now being explored for wider applications, including generation of electrical power. Warm ground water is used for irrigation in many areas east of the Cascade Range, and thermal ground water also is used to heat buildings in the Boise and Sun Valley areas, Idaho.[1]

The term "geothermal resources" is commonly used to mean ground water or rocks hot enough to be of

[1]In the Klamath Falls area, Oregon, which is adjacent to but outside this region, the use of hot ground water for heating home, commercial, and institutional buildings is said to be the largest use of thermal ground water for space heating in the United States.

economic value—for example, to produce electrical power by releasing steam through drill holes for operation of turbine generators. About 120,000 acres of land in the region has been classed as being within KGRA's ("known geothermal resource areas"), that is, areas worthy of careful evaluation for geothermal potential (Godwin and others, 1971). These KGRA's are mainly in the vicinity of the volcanoes of the Cascade Range and in the areas of the High Cascades, High Desert, and Southern Idaho ground-water reservoirs. To date (1977), by far the most intensive drilling and testing has been done in the Raft River basin, Idaho (fig. 9). There, several dozen holes have been drilled for subsurface samples, heat-flow measurement, and production of hot water or steam. The deepest known geothermal test well in the region (6,543 feet) is in this basin.

None of the geothermal zones tapped has yielded water hotter than about 150°C, a temperature generally considered too low for efficient steam-power generation. However, experiments are planned at the Raft River basin site to investigate the practical applications of this low-temperature geothermal water for both electrical power generation and nonelectrical heat uses in industries related to agriculture and food processing. The experimental project is based on an integrated "energy park" concept, in which heat remaining after one use of the hot water would be used in other applications requiring successively lower temperatures. If this experimental project is successful, it may lead to more "energy parks" across southern Idaho and southeastern Oregon.

The hot water from the Raft River basin is similar to most other thermal waters of the region, in that it is saline enough to be a potential contaminant to other water resources. The dissolved-solids concentration in water from one core hole in that area was 6,650 mg/L in September 1974 (U.S. Geological Survey data).

HYDROPOWER GENERATION WITH GROUND WATER

One unusual, though small-scale, energy-related use of ground water in this region is for the generation of electrical power. In the Thousand Springs area of the Snake River canyon near Twin Falls, Idaho, the discharge from springs high on the canyon wall is so large that the water is collected in flumes and used to drive generators located on the canyon floor. The combination of copious spring flow and discharge of the water at a relatively high elevation is needed for this use of the ground water; the combination exists in few other areas of the region. However, opportunities probably exist in parts of the High Cascades ground-water reservoir, where very permeable young volcanic rocks underlie steep land and receive abundant recharge from precipitation. The abundant yield of the young volcanic rocks was confirmed in Linn County, Oregon, where a tunnel was driven to drain these rocks and produce extra water for hydropower generation (R. B. Sanderson, U.S. Geol. Survey, written communication).

GROUND WATER AND COAL

Coal is the only fossil fuel known to occur in commercial quantities in the Pacific Northwest Region. Most of the coal is in and near the area of the Puget Sound ground-water reservoir, and in the eastern foothills of the Cascade Range. The estimated reserves in Washington are sizable—more than 6 billion (short) tons, but the reserves consist largely of subbituminous coal and lignite (about 4.3 billion tons; Livingston, 1974, p. 41). In addition to the relatively low ranking and heat content of most of the coal, the coal beds have been contorted by folding and faulting so that they commonly are steeply dipping, discontinuous, or sheared. In only a few areas are dips of the beds gentle enough to permit strip mining. Despite these problems, Washington has had some coal production almost continuously since the late 1800's, the largest being a strip-mining operation near Centralia, Wash. (south of the Puget Sound ground-water reservoir); this operation supplies as much as 2.6 million tons a year to a nearby steam-electric generating plant (Livingston, 1974, p. 41).

Washington's coal has two important attributes that partly offset the aforementioned problems. One is that it is "low-sulfur" coal and, therefore, it offers some economic advantages in meeting environmental-protection requirements during mining and use. The other is that most of the coal occurs in reasonably close proximity to abundant water resources. This affords the possibility of using water for some forms of remote mining of the deep coal beds (hydraulic fracturing or flushing the broken coal to the surface), for conveying the coal in slurry pipelines, or for in-place gasification of the coal.

At a few places near the coal beds, ground water could be produced at rates that would be useful in such processes, but the main impact of future coal development on ground water is likely to be a degradation of water quality. Water-quality impacts of coal-mine drainage and possible increased production of coal in Washington are being assessed by the U.S. Geological Survey (Packard and Haushild, 1977).

SUBSURFACE STORAGE OF HEAT

Water has a very high specific heat, meaning that a comparatively large amount of heat can be added to, or taken from, water for a given change in the water's temperature. Also, slowly moving ground water and the rocks that enclose it conduct heat very slowly. These conditions make ground water advantageous for the storage of heat for later recovery. Similarly, the ground-water can be a sink for waste heat, although the cooling or assimilation rates using ground water normally cannot equal the rates possible with disposal of heat to streams or the atmosphere. Although there has been some use in the region of ground water for heat storage and disposal, much greater opportunities exist for utilizing the subsurface for the deliberate storage of heat.

Ground water has been used for disposal and storage of heat mainly in the Puget Sound and Willamette Valley ground-water reservoir areas, mostly in the Portland, Oreg. area (Price and others, 1965, p. 20–33). Although problems of clogging of the aquifers have been a constraint to the continued or more widespread use of this type of subsurface storage and disposal, these problems can be largely overcome by improved methods such as closed-loop transfer of the water, selection of permeable aquifers, and use of the best well-construction methods for injection wells. As energy costs rise, reverse-cycle heat-pump systems, using ground water as the heat reservoir, are likely to become more popular for homes and other buildings.

In addition to the saturated rocks available to serve as a heat reservoir, large bodies of permeable but unsaturated rocks are available, especially in arid parts of the region, that might serve for storage of local sources of heat—for example, heat from solar-collection systems or waste heat from power-generation facilities.

SUMMARY OF NEEDS FOR ADDITIONAL INFORMATION

The need for additional information is a recurring theme in this appraisal. State officials who were contacted, contributing hydrologists, technical reviewers, as well as this writer, generally agree that improved management of the ground-water resources in the Pacific Northwest Region is seriously hampered by information inadequacies. The scarcity of unappropriated surface water and accelerating uses of ground water are increasing the water-related problems and intensifying the need for ground-water management. Some difficult and far-reaching decisions are being made or must be made about the region's ground-water resources within the next decade. Without sound technical information, those decisions must be based largely on supposition and conjecture.

Opinions differ as to which kinds of information are most needed and in what degree of detail. Judgments of this nature will be important because the total amount and variety of ground-water-related information that could be beneficially used within the next decade seem far beyond the present capabilities of hydrologists and others to supply. Efficiencies in the use of the limited technical manpower can be fostered by anticipating problems and maintaining an orderly program of investigations and research to provide information before water problems reach crisis proportions.

The water managers of the region need ground-water information for the following purposes:

1. To assess the resource—a recurring need that, for some parts of the region, has not been met even the first time;
2. to detect changes with time;
3. to make sound management responses to current problems;
4. to identify developing or future problems, and to formulate strategies for forestalling, circumventing, or mitigating the problems;
5. to gain public understanding of and support for the needs to manage the ground water;
6. to combine scientific knowledge with economic, social, and political considerations as a basis for formulating water-resources policies.

The kinds of ground-water data, investigations, and research that are needed by the water managers and others in the coming decade range widely, in accordance with the varied settings and problems of the region. Discussion of the full range of information needs is beyond the scope of this summary section, but the following general statements seem pertinent:

1. Ground-water-quality data, evaluations, and monitoring will be needed increasingly.
2. More complete and reliable information on ground-water withdrawals and resultant water-level declines is needed for all the major and many smaller ground-water reservoirs.
3. Although reconnaissance-type data will continue to be adequate for some purposes, more quantitative, detailed information will be required, especially in problem areas, areas destined for conjunctive use of surface and ground water, and areas considered for disposal of wastes.
4. More perspective will be needed of the total ground-water reservoir or basin—the geologic framework and controls on the ground water; places and amounts of recharge and discharge; ground-water flow patterns (including deep parts of the reservoirs); hydraulic characteristics of the rock mate-

rials; and quality of the ground water throughout.

5. The more detailed, extensive, and quantitative information will be more difficult and expensive to obtain, in general. Some information about deeper parts of the ground-water reservoirs can be obtained only by the drilling and testing of expensive, deep wells. Though not substitutes for the deep-well information, geologic mapping and geophysical surveys likely will be used increasingly to extend the information returns from deep test-observation wells.
6. As the ground-water management problems increase in complexity, they will require reliable interpretations of greater sophistication and incisiveness. Mathematical models of ground-water flow systems will remain as powerful interpretive tools and will maintain a high level of demand for quantitative basic data.
7. Presentation and dissemination of the ground-water information in a clear and understandable form will continue to be nearly as important, for some purposes, as the information itself.

SELECTED REFERENCES

Barker, R. A., 1979, Digital simulation and geohydrology of a basalt aquifer system in the Pullman-Moscow basin, Washington and Idaho: Washington Dept. Ecology Water-Supply Bull. 48.

Barker, R. A., and MacNish, R. D., 1976, Digital model of the gravel aquifer, Walla Walla River basin, Washington and Oregon: Washington Dept. Ecology Water-Supply Bull. 45, 49 p.

Bartholomew, W. S., and DeBow, Robert, 1967, Ground-water levels, 1966: Oregon Ground Water Rept. 12.

———1970, Ground-water levels, 1967–68: Oregon Ground Water Rept. 15.

Bartholomew, W. S., Graham, Monte E., and Feusner, John, 1973, Ground-water levels, 1968–72: Oregon Ground Water Rept. 18, 134 p.

Bodhaine, G. L., Foxworthy, B. L., Santos, J. F., and Cummans, J. E., 1965, The role of water in shaping the economy of the Pacific Northwest: U.S. Bonneville Power Adm. Pacific Northwest Econ. Base Study for Power Market, v. 2, pt. 10, 218 p.

Boettcher, A. J., and Gosling, A. W., 1977, Water resources of the Clark Fork basin above St. Regis, Montana: Montana Bur. Mines and Geol. Bull. 104, 28 p.

Brown, S. G., 1963, Problems of utilizing ground water in the westside business district of Portland, Oregon: U.S. Geol. Survey Water-Supply Paper 1619–O, 42 p.

Brown, S. G., and Newcomb, R. C., 1962, Ground-water resources of Cow Valley, Malheur County, Oregon: U.S. Geol. Survey Water-Supply Paper 1619–M, 38 p.

Burnham, W. L., Crosthwaite, E. G., Walker, E. H., Norvitch, R. F., and Waite, H. A., 1966, Ground-water conditions in Idaho: Idaho Dept. of Reclamation Water Information Bull. No. 1, 60 p.

Butcher, W. R., Crosby, J. W., III, Whittlesey, N. K., Bassett, D. L., Weakly, E. C., Muller, A. C., and Kloster, L. D., 1971, Long-run costs and policy implications of adjusting to a declining water supply in eastern Washington: Washington Water Research Center Rept. 9, 106 p.

Coffin, D. L., Breitkrietz, Alex, and McMurtrey, R. G., 1971, Surficial geology and water resources of the Tobacco and upper Stillwater River valleys, northwestern Montana: Montana Bur. Mines and Geology Bull. 81, 48 p.

Columbia-North Pacific Technical Staff, 1970a, Land and mineral resources *in* Columbia-North Pacific comprehensive framework study of water and related lands: Vancouver, Wash., Pacific Northwest River Basins Comm., app. 4, 383 p.

———1970b, Water resources *in* Columbia-North Pacific comprehensive framework study of water and related lands: Vancouver, Wash., Pacific Northwest River Basins Comm., app. 5, 1022 p.

———1971a, Economic base & projections *in* Columbia-North Pacific comprehensive framework study of water and related lands: Vancouver, Wash., Pacific Northwest River Basins Comm., app. 6, 187 p.

———1971b, Land measures & watershed protection *in* Columbia-North Pacific comprehensive framework study of water and related lands: Vancouver, Wash., Pacific Northwest River Basins Comm., app. 8, 735 p.

———1971c, Irrigation *in* Columbia-North Pacific comprehensive framework study of water and related lands: Vancouver, Wash., Pacific Northwest River Basins Comm., app. 9, 343 p.

———1972, Columbia-North Pacific comprehensive framework study of water and related lands: Vancouver, Wash., Pacific Northwest River Basins Comm., 373 p.

Crosthwaite, E. G., 1957, Ground-water possibilities south of the Snake River between Twin Falls and Pocatello, Idaho: U.S. Geol. Survey Water-Supply Paper 1460–C, p. 99–145.

———1968, Water resources of the Goose Creek-Rock Creek area, Idaho, Utah, and Nevada: Idaho Dept. Water Resources Water Inf. Bull. 8, 73 p.

de Sonneville, J. L. J., 1974, Development of a digital ground-water model with application to aquifers in Idaho: Moscow, Idaho, Univ. of Idaho, unpublished Ph D. thesis, 228 p.

Dion, N. P., 1972, Some effects of land use changes on the shallow ground water system in the Boise-Nampa area, Idaho: Idaho Dept. Water Resources Water Inf. Bull. 26, 47 p.

Drost, B. W., and Sites, H. R., 1978, Spokane Valley-Rathdrum Prairie aquifer, Washington and Idaho: U.S. Geol. Survey Open-File Report 77–829, 79 p.

Dyer, K. L., and Young, H. W., 1971, A reconnaissance of the quality of water from irrigation wells and springs in the Snake Plain aquifer, southeastern Idaho: U.S. Geol. Survey Open-File Report, 29 p.

Feth, J. H., and others, 1965, Preliminary map of the conterminous United States, showing depth to and quality of shallowest ground water containing more than 1,000 parts per million dissolved solids: U.S. Geol. Survey Hydrol. Inv. Atlas HA–199, scale 1:3,168,000.

Foxworthy, Bruce L., 1962, Geology and ground-water resources of the Ahtanum Valley, Yakima County, Washington: U.S. Geol. Survey Water-Supply Paper 1598, 99 p.

———1970, Hydrologic conditions and artificial recharge through a well in the Salem Heights area of Salem, Oregon: U.S. Geol. Survey Water-Supply Paper 1594–F, 56 p.

Foxworthy, B. L., and Bryant, C. T., 1967, Artificial recharge through a well tapping basalt aquifers at The Dalles, Oregon: U.S. Geol. Survey Water-Supply Paper 1594–E, 55 p.

Garrett, A. A., and Londquist, C. J., 1972, Feasibility of artificially recharging basalt aquifers in eastern Washington: U.S. Geol.

Survey Open-File Report, 42 p.

Godwin, L. H., Haigler, L. B., Rioux, R. L., White, D. E., Muffler, L. J. P., and Wayland, R. G., 1971, Classification of public lands valuable for geothermal steam and associated geothermal resources: U.S. Geol. Survey Circ. 647, 18 p.

Goldblatt, E. L., Van Denburgh, A. S., and Marsland, R. A., 1963, The unusual and widespread occurrence of arsenic in well waters of Lane County, Oregon: Lane County Dept. Health, 24 p.

Graham, William G., Clapp, Darrel W., and Putkey, Thomas A., 1977, Irrigation wastewater disposal well studies, Snake Plain aquifer: U.S. Environmental Protection Agency Ecological Research Series, EPA-600/3-77-071, 51 p.

Gregg, D. O., and Laird, L. B., 1975, General outline of the water resources of the Toppenish Creek basin, Yakima Indian Reservation, Washington: U.S. Geol. Survey Open-File Report 75-19, 37 p.

Hart, D. H., 1954, List of ground-water sources in Oregon known to yield mineralized water (over 1,000 ppm dissolved solids or 60 percent sodium): U.S. Geol. Survey Open-File Report, 14 p.

Hart, D. H., and Newcomb, R. C., 1965, Geology and ground water of the Tualatin Valley, Oregon: U.S. Geol. Survey Water-Supply Paper 1697, 172 p.

Kimmel, G. E., 1963, Contamination by sea-water intrusion along the Puget Sound, Washington, an area having abundant precipitation: U.S. Geol. Survey Prof. Paper 475-B, p. B182-B185.

Konizeski, R. L., Brietkrietz, Alex, and McMurtrey, R. G., 1968, Geology and ground-water resources of the Kalispell Valley, northwestern Montana: Montana Bur. Mines and Geology Bull. 68, 42 p.

Laird, L. B., and Walters, K. L., 1967, Municipal, industrial, and irrigation water use in Washington, 1965: U.S. Geol. Survey Open-File Report, 13 p.

La Sala, A. M., Jr., and Doty, G. C., 1971, Preliminary evaluation of hydrologic factors related to radioactive waste storage in basaltic rocks at the Hanford Reservation, Washington: U.S. Geol. Survey Open-File Report, 68 p.

Livingston, Vaughn E., Jr., 1974, Energy resources of Washington: Washington Div. Geol. and Earth Resources, Information Circ. No. 50, 158 p.

Luzier, J. E., and Burt, R. J., 1974, Hydrology of basalt aquifers and depletion of ground water in east-central Washington: Washington Dept. Ecology Water-Supply Bull. 33, 53 p.

Luzier, J. E., and Skrivan, J. A., 1975, Digital-simulation and projection of water-level declines in basalt aquifers of the Odessa-Lind area, east-central Washington: U.S. Geol. Survey Water-Supply Paper 2036, 48 p.

McGuinness, C. L., 1963, The role of ground water in the national water situation: U.S. Geol. Survey Water-Supply Paper 1800, 1121 p.

MacKichan, K. A., and Kammerer, J. C., 1961, Estimated use of water in the United States, 1960: U.S. Geol. Survey Circ. 456, 26 p.

McMurtrey, R. G., Konizeski, R. L., and Brietkrietz, Alex, 1965, Geology and ground-water resources of the Missoula Basin, Montana: Montana Bur. Mines and Geol. Bull. 47, 35 p.

MacNish, R. D., and Barker, R. A., 1976, Digital model of a basalt aquifer system, Walla Walla River basin, Washington and Oregon: Washington Dept. Ecology Water-Supply Bull. 44, 73 p.

MacNish, R. D., Myers, D. A., and Barker, R. A., 1973, Appraisal of ground-water availability and management projections, Walla Walla River basin, Washington and Oregon: Washington Dept. Ecology Water-Supply Bull. 37, 25 p.

Madison, R. J., 1966, Water-quality data in the Willamette Basin, Oregon, 1910-64: U.S. Geol. Survey Open-File Report., 40 p.

Moreland, Joe A., 1976, Digital-model analysis of the effects of water-use alternatives on spring discharges, Gooding and Jerome Counties, Idaho: Idaho Dept. of Water Resources Water Information Bull. 42, 46 p.

Mundorff, M. J., 1962, The feasibility of artificial recharge in the Snake River Basin, Idaho: U.S. Geol. Survey Open-File Report, 49 p.

——— 1964, Geology and ground-water conditions of Clark County, Washington, with a description of a major alluvial aquifer along the Columbia River: U.S. Geol. Survey Water-Supply Paper 1600, 268 p.

Mundorff, M. J., Crosthwaite, E. G., and Kilburn, Chabot, 1964, Ground water for irrigation in the Snake River Basin in Idaho: U.S. Geol. Survey Water-Supply Paper 1654, 224 p.

Mundorff, M. J., MacNish, R. D., and Cline, D. R., 1977, Water resources of the Satus Creek basin, Yakima Indian Reservation, Washington: U.S. Geol. Survey Open-File Report 76-685, 102 p.

Mundorff, M. J., and Sisco, H. G., 1963, Ground water in the Raft River Basin, Idaho, with special reference to irrigation use: U.S. Geol. Survey Water-Supply Paper 1619-CC, 23 p.

Murray, C. R., 1968, Estimated use of water in the United States: U.S. Geol. Survey Circ. 556, 53 p.

Murray, C. R., and Reeves, E. B., 1972, Estimated use of water in the United States in 1970: U.S. Geol. Survey Circ. 676, 37 p.

——— 1977, Estimated use of water in the United States in 1975: U.S. Geol. Survey Circ. 765, 39 p.

Nace, R. L., Deutsch, Morris, and Voegeli, P. T., 1956, Geography, geology, and water resources of the National Reactor Testing Station, Idaho: U.S. Geol. Survey Open-File Report, Pt. 2, geography and geology, 225 p.

Nace, R. L., Stewart, J. W., Walton, W. C., and others, 1959, Geography, geology, and water resources of the National Reactor Testing Station, Idaho, Pt. 3, hydrology and water resources: U.S. Geol. Survey Open-File Report, 253 p.

National Academy of Sciences and National Academy of Engineering, 1973 (1974), Water quality criteria, 1972: U.S. Government Printing Office, 594 p.

Newcomb, R. C., 1961a, Ground water in the western part of the Cow Creek and Soldier Creek Grazing Units, Malheur County, Oregon: U.S. Geol. Survey Water-Supply Paper 1475-E, p. 159-172.

——— 1961b, Storage of ground water behind subsurface dams in the Columbia River Basalt, Washington, Oregon, and Idaho: U.S. Geol. Survey Prof. Paper 383-A, 15 p.

——— 1969, Effect of tectonic structure on the occurrence of ground water in the basalt of the Columbia River Group of The Dalles area, Oregon and Washington: U.S. Geol. Survey Prof. Paper 383-C, 33 p.

———1972, Quality of the ground water in basalt of the Columbia River Group, Washington, Oregon, and Idaho: U.S. Geol. Survey Water-Supply Paper 1999-N, 71 p.

Newcomb, R. C., and Strand, J. R., 1972, Geology and ground-water characteristics of the Hanford Reservation of the Atomic Energy Commission, Washington: U.S. Geol. Survey Prof. Paper 717, 78 p.

Packard, F. A., and Haushild, W. L., 1977, Reconnaissance evaluation of water resources for hydraulic coal mining, Wilkeson-Carbonado mining district, Washington: U.S. Geol. Survey Open-File Report, 19 p.

Parker, G. G., Jr., 1971, Municipal, industrial, and irrigation water use in Washington, 1970: U.S. Geol. Survey Open-File Report, 21 p.

Piper, A. M., 1932, Geology and ground-water resources of The Dalles region, Oregon: U.S. Geol. Survey Water-Supply Paper 659-B, p. 107-189.

——— 1969, Disposal of liquid wastes by injection underground—Neither myth nor millennium: U.S. Geol. Survey Circ. 631, 15 p.

Price, C. E., 1961, Artificial recharge through a well tapping basalt aquifers, Walla Walla area, Washington: U.S. Geol. Survey Water-Supply Paper 1594–A, 33 p.

Price, Don, Hart, D. H., and Foxworthy, B. L., 1965, Artificial recharge in Oregon and Washington: U.S. Geol. Survey Water-Supply Paper 1594–C, 65 p.

Puget Sound Task Force, 1970, Hydrology and natural environment, App. 3 *in* Comprehensive study of water and related land resources, Puget Sound and adjacent waters, State of Washington: Vancouver, Wash., Pacific Northwest River Basins Comm., 205 p.

Quan, E. L., Sweet, H. R., and Illian, Joseph R., 1974, Subsurface sewage disposal and contamination of ground water in East Portland, Oregon: Ground Water, v. 12, no. 6, p. 356–368.

Robertson, J. B., and Barraclough, J. T., 1973, Radioactive and chemical waste transport in ground water at the National Reactor Testing Station, Idaho, *in* Am. Assoc. Petrol. Geologists, Internat. Symp. Underground Waste Management and Artificial Recharge, 2d., New Orleans, La., 1973, Preprints, v. 1, p. 291–322.

Robertson, J. B., Schoen, Robert, and Barraclough, J. T., 1973, Influence of liquid waste disposal on geochemistry of water at the National Reactor Testing Station, Idaho, 1952–70: U.S. Geol. Survey Open-File Report, 31 p.

Robison, J. H., 1968, Estimated existing and potential ground-water storage in major drainage basins in Oregon: U.S. Geol. Survey Open-File Report, 13 p.

——— 1971, Hydrology of basalt aquifers in the Hermiston-Ordnance area, Umatilla and Morrow Counties, Oregon: U.S. Geol. Survey Hydrol. Inv. Atlas HA–387.

Rorabaugh, M. I., and Simons, W. D., 1966, Exploration of methods of relating ground water to surface water, Columbia River basin, second phase: U.S. Geol. Survey Open-File Report, 99 p.

Rorabaugh, M. I., Simons, W. D., Garrett, A. A., and McMurtrey, R. G., 1966, Exploration of methods of relating ground water to surface water, Columbia River basin, first phase, *with a section on* Direct computation of ground-water outflow by electric analog, by B. J. Bermes: U.S. Geol. Survey Open-File Report, 124 p.

Sceva, J. E., 1966, A brief description of the ground-water conditions in the Ordnance area, Morrow and Umatilla Counties, Oregon: Oregon Ground-Water Rept. 11, 43 p.

——— 1968, Liquid waste disposal in the lava terrane of central Oregon: Federal Water Pollution Control Authority Rept. FR–4, 66 p. (with 96 p. app.).

Sceva, J. E., and DeBow, Robert, 1965, Ground-water levels, 1964: Oregon Ground Water Report 5, 109 p.

——— 1966, Ground-water levels, 1965: Oregon Ground Water Rept. 9, 111 p.

Schneider, W. J., 1970, Hydrologic implications of solid-waste disposal: U.S. Geol. Survey Circ. 601–F, 10 p.

Seitz, H. R., LaSala, A. M., Jr., and Moreland, J. A., 1977, Effects of drain wells on the ground-water quality of the western Snake Plain aquifer, Idaho: U.S. Geol. Survey Open-File Report 76–673, 34 p.

Sisco, H. G., 1974, Ground-water levels and well records for current observation wells in Idaho, 1922–73, Parts A, B, and C: U.S. Geol. Survey basic-data release, 309 p. (A), 495 p. (B), 430 p. (C).

Sisco, H. G., and Whitehead, R. L., 1969, Ground-water levels in Idaho: Idaho Dept. Water Resources Water Inf. Bull. 11, 75 p.

——— 1970, Ground-water levels in Idaho, 1970: Idaho Dept. Water Resources Water Inf. Bull. 18, 75 p.

Stearns, H. T., Crandall, Lynn, and Steward, W. G., 1938, Geology and ground-water resources of the Snake River Plain in southeastern Idaho: U.S. Geol. Survey Water-Supply Paper 774, 268 p.

Tanaka, H. H., Hansen, A. J., Jr., and Skrivan, J. A., 1974, Digital-model study of ground-water hydrology, Columbia Basin Irrigation Project area, Washington: Washington Dept. Ecology Water-Supply Bull. 40, 60 p.

Tracy, J. V., and Dion, N. P., 1976, Evaluation of ground-water contamination from cleaning explosive-projectile casings at the Bangor Annex, Kitsap County, Washington, Phase 2: U.S. Geol. Survey Water Resources Inv. 62–75, 44 p.

U.S. Environmental Protection Agency, 1978: Federal Register, v. 43, no. 28, Thursday, February 9, 1978, p. 5566–5567.

U.S. Geological Survey, 1971, Ground-water levels in Idaho, 1971: Idaho Dept. Water Resources Water Inf. Bull. 24, 77 p.

——— 1972, Ground-water levels in Idaho, 1972: Idaho Dept. Water Resources Water Inf. Bull. 27, 51 p.

——— 1975, Water resources of the Toppenish Creek basin, Yakima Indian Reservation, Washington: U.S. Geol. Survey Water Resources Inv. 42–74, 144 p.

Van Denburgh, A. S., and Santos, J. F., 1965, Ground water in Washington: Its chemical and physical quality: Washington Div. Water Resources Water-Supply Bull. 24, 93 p.

Walker, E. H., 1960, The quality of ground water in the upper Snake River basin: Washington State Univ. Inst. of Technol., Proc. Water Quality Conf., p. 79–93.

Walker, E. H., Dutcher, L. C., Decker, S. O., and Dyer, K. L., 1970, The Raft River basin, Idaho-Utah, as of 1966–A reappraisal of ground-water development: Idaho Dept. Water Resources Water Inf. Bull. 19, 95 p.

Walters, K. L., 1971, Reconnaissance of sea-water intrusion along coastal Washington, 1966–68: Washington Dept. Water Resources Water-Supply Bull. 32, 208 p.

Waring, G. A., 1965, Thermal springs of the United States and other countries of the world—a summary: U.S. Geol. Survey Prof. Paper 492, 383 p.

Whitehead, R. L., 1974, Chemical and physical data for disposal wells, eastern Snake River Plain, Idaho: Idaho Dept. Water Resources Water Inf. Bull. 39, 31 p.

Whitehead, R. L., and Sisco, H. G., 1968, Ground-water levels in Idaho, 1968: Idaho Dept. Water Resources Water Inf. Bull. 5, 67 p.

Wilson, R. T., 1975, Solid-waste disposal sites in relation to water resources in the Seattle-Tacoma urban complex and vicinity, Washington: U.S. Geol. Survey Open-File Report 75–344.

Young, H. W., and Harenberg, W. A., 1971, Ground-water pumpage from the Snake Plain aquifer, southeastern Idaho: Idaho Dept. Water Resouces Water Inf. Bull. 23, 28 p.

Young, H. W., and Mitchell, J. C., 1973, Geothermal investigations in Idaho, Pt. 1, Geochemistry and geologic setting of selected thermal water: Idaho Dept. Water Resources Water Inf. Bull. 30, 43 p.

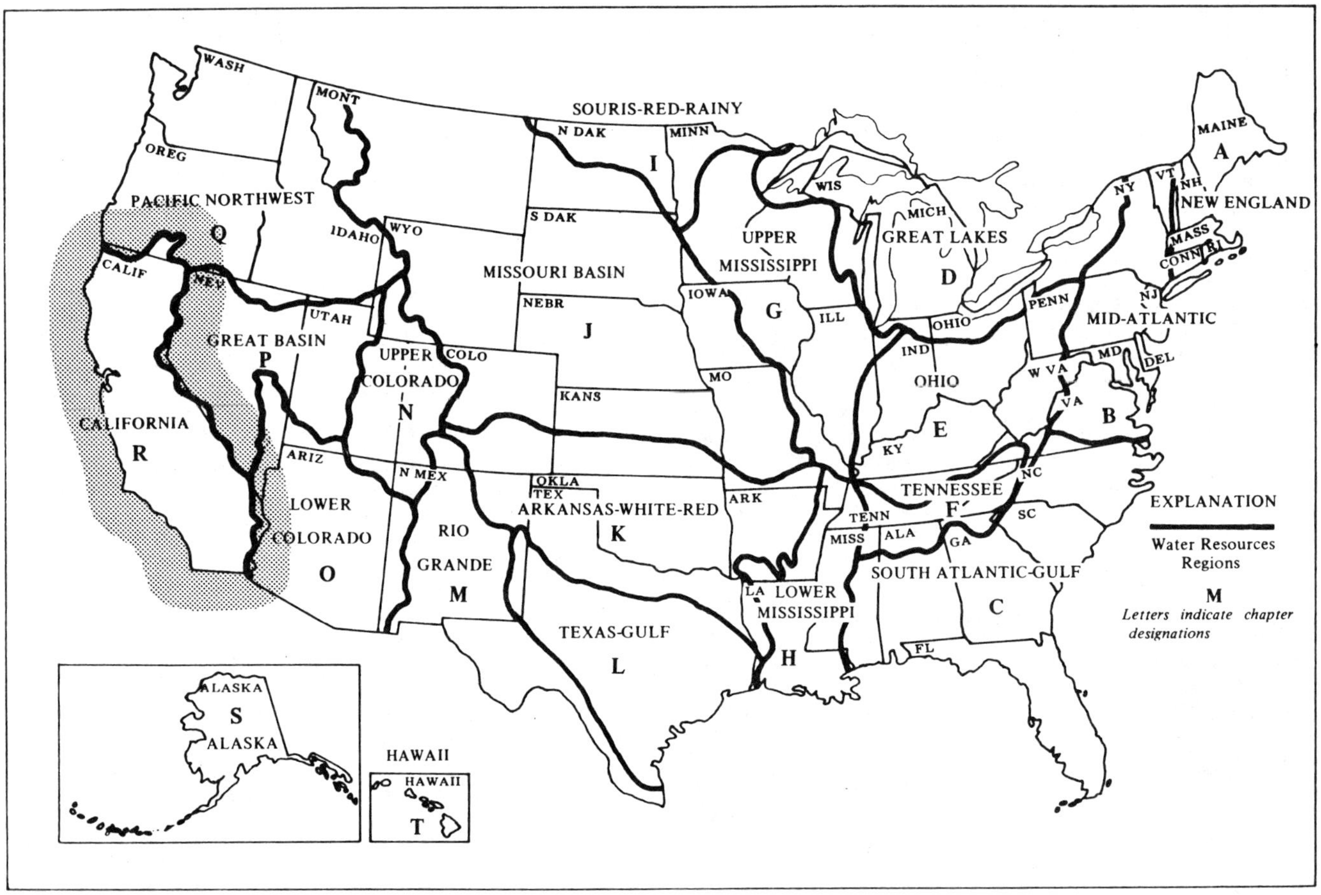

R – Location Map for the California Region

R

CALIFORNIA REGION

By H. E. THOMAS and D. A. PHOENIX

ABSTRACT

Most people in the California Region live in a semiarid or arid climate, with precipitation less than the potential evapotranspiration—environments of perennial water deficiency. The deficiency becomes most onerous during the characteristically rainless summers and during recurrent droughts that may continue for 10–20 years. However, water from winter rain and snow can be stored for use during the dry summer months, and water stored during a wet climatic period can be used in a succeeding dry period; moreover, perennial deficiency can be overcome by bringing water from areas of perennial surplus. Ground-water reservoirs have especial significance in arid and semiarid regions as repositories where water is stored or can be stored with minimum loss by evaporation.

Nearly all the ground-water reservoirs of the California Region are in alluvial sediments of valleys and plains that flank the mountain ranges. The largest, underlying the vast Central Valley, occupies 10 percent of the area of the region, has an estimated usable capacity exceeding 100 million acre-feet (125 cubic kilometres), and has an annual pumpage from wells of about 13 million acre-feet (16 cubic kilometres). Another 10 percent of the region is occupied by 55 developed ground-water reservoirs that are widely distributed; aggregate annual pumpage from them is about 3½ million acre-feet (4 cubic kilometres). In the southeastern desert about 60 ground-water reservoirs occupy still another 10 percent of the region; these have been explored only enough to show that most have some usable water, but current use is negligible. In northeastern California and adjacent Oregon and Nevada, ground-water reservoirs are identified only in valleys and lowlands where wells are feasible, but basaltic rocks of the Cascade Range and Modoc Plateau are excellent aquifers distributed over an area constituting about 15 percent of the region. In sum, slightly less than half the California Region is underlain by ground-water reservoirs, either in valley fill or in volcanic rocks, which can yield significant quantities of water to wells.

The rest of the California Region includes the mountains, canyons, slopes, and foothills of the Sierra Nevada, Coast Ranges, and Basin Ranges, whose consolidated rocks and products of their weathering may be permeable locally but are not generally so. Here, the prevailing method of ground-water development is still mostly trial and error, and while in many places a well can yield enough water for a family, some families might have to do without amenities such as flush toilets and automatic washers.

For more than half a century the California Region has led all others in North America in pumping of ground water as well as in the area, variety, yield, and export of crops irrigated by water from wells. It has led in the development and use of deep-well turbine pumps for large yield and in the drilling of water wells to great depths. At the same time, such developments have resulted in the elimination of artesian pressures that produced thousands of flowing wells in the 19th century and led to the wide distribution of "falling water tables." Also, California was first to induce encroachment of seawater into wells (in 1906); first to recognize subsidence of land caused by pumping from wells (in 1933), generating news about land sinking in San Jose, Long Beach, and along the Delta-Mendota and Friant-Kern Canals; and first to experience pollution of ground-water reservoirs by brines, chemicals, industrial wastes, and petroleum byproducts including gasoline. The region has led in research in several fields leading to solution of many of these problems.

Ground-water problems developed rapidly after World War II with booming population, agriculture, industry, and water demand during several years of regionwide drought. Water levels in wells trended downward almost everywhere as a natural effect of the drought and at accelerated rates in areas of pumping for new enterprises or to supplement subnormal surface-water supplies. The declines in many pumping areas exceeded 100 feet (30 metres), and in some confined aquifers the potentiometric surface was drawn down more than 330 feet (100 metres). The depletion of ground-water storage has had "permanent" side effects, including subsidence of the land exceeding 10 feet (3 metres) in extensive areas, and seawater intrusion that ended the useful lives of many wells along the coast and as much as 6 miles (10 kilometres) inland. Some problems have been solved, but these solutions have at times created other problems. Many ground-water reservoirs have gone through one or more stages—exploration for productive aquifers, exploitation and development for use of the water, restriction to the perennial supply or "safe" yield, importation of surface water, artificial recharge of ground water, conjunctive use of surface and ground water, protection of water quality, and integrated management of use and disposal of water. This evolutionary sequence is unique for each reservoir, and so generalizations become difficult in a regional appraisal; also, the changes with time are significant and varied, and knowledge of prior events is a prerequisite in an appraisal of the resource in a specific year.

As of 1970, water levels in many wells had risen significantly from the minimum levels of record reached during the 1960's or earlier; only in areas of new development and in desert areas of long-continued "mining" of nonreplenished water was ground-water storage still being depleted. Land subsidence has continued at diminishing rates and practically has come to a halt in some areas; invading seawater has been stopped or nudged back in most places where problems were significant. The current, favorable situation has been helped by climatic variations, from drought in 1945-52 and exceedingly dry years in 1959 and 1961 to above-normal precipitation in 1969 and 1970; but most of the serious problems have been solved by human efforts, including especially the implementation of the California Water Plan, transporting water from areas of perennial surplus to areas where it is used in lieu of ground water or for ground-water replenishment. All major urban areas now import water to supplement or replace the water pumped from wells. Extensive agricultural areas that formerly were irrigated solely by ground water now obtain some of their water from surface reservoirs and canals, especially in the Central Valley. With surface water available as an alternative supply, well owners can view their ground water with complacency. But complacency can lead to neglect and carelessness and consequent deterioration of the ground-water resource by pollution.

Claiming heritage from the English Common Law, the existing California law grants to the landowner (riparian) and private enterprise (appropriator) rights to the water stored in ground-water reservoirs or discharged from them, including the base flow of streams. Ground-water development has been by private and local enterprise, and the California legislature has protected and encouraged local responsibility, control, and management of ground water. As to surface water, a constitutional amendment in 1928 limited riparian rights to the quantities of water that were "reasonably required for the beneficial use to be served." The surpluses have become public waters which are collected, stored, transported, and delivered under various contracts by Federal, State, and other agencies. The agencies have not stored water underground because of uncertainty as to their rights, but some local agencies have been encouraged with favorable pricing schedules to undertake the artificial recharge and management of ground-water reservoirs. Thus, conjunctive use of surface and ground water has become a matter of interagency negotiation.

Of all the constraints on effective use of ground-water reservoirs, the most formidable may be the attitudes of people. Assurance of water supply is vital in areas of water deficiency, and Government has assumed increasing responsibility for the welfare of people in these areas. Unfortunately, when Government provides this assurance, most beneficiaries demand continued subsidy to the exclusion of perhaps cheaper private development. Indeed, as the water resources are presently segregated—with private rights predominant in ground water and public interest dominant in surface water—ground-water development has suffered for lack of public concern. The region has the scientific and technologic capability for effective use of ground-water reservoirs, as shown by the achievements and programs of several districts, but many districts are not organized or staffed for such comprehensive management and will need assistance and scientific expertise available from State and Federal agencies. Those agencies, in turn, may not have the scientific data that are essential to prevent haphazard activities and to enable programs to be organized for the most effective and attractive utilization of the water resources. In these days of increasing concern over pollution, existing data are generally inadequate to assess the natural deterioration of ground waters as a basis for defining pollution.

REGIONAL ENVIRONMENT

The California Region (fig. 1) comprises coastal valleys and the Coast Ranges, interior valleys and the Sierra Nevada, volcanic terranes and desert areas, fertile lands and wastelands, and the Pacific Ocean, which Tannehill (1947) called "the monster in the backyard" because of its great involvement in the climate and therefore the water resources of the region. By comparison with other water-resource regions of the country, the California Region is outstanding in the variety of its environments, ranging from very wet to very dry, very low to very high, surplus of sunshine to surplus of fog, prevailing high winds to prevailing calm. Despite this diversity, the entire region shares the characteristic that summers are dry. About 85 percent of the annual precipitation can be expected in the 6 months November to April. In most of the region, the rainfall in the 4 hottest months (June-September) is less than 5 percent of the annual total.

To meet the continuing requirements of life, some water from the rainy season must be available throughout the dry summer (fig. 2). Native vegetation is able to subsist on water collected by deep perennial root systems; annuals must ripen their seeds after making their vegetative growth in spring before the soil water available to their roots is exhausted. Animals have the advantage of mobility to search out the places where suitable water is available in the dry season. For mankind, seasonal aridity has provided various perils of occupancy, epitomized by the name Death Valley—the region's lowest, hottest, and driest valley.

PERILS OF OCCUPANCY

In the northern part of the California Region and in the high Sierra Nevada, the mean annual precipitation of more than 40 inches (1,000 mm) exceeds the annual evapotranspiration, and the perennial water surplus eventually becomes runoff in streams. But perennial water deficiency characterizes all of southern California and practically all the lowlands farther north—lands where the majority of the people in the region live. This water deficiency is a natural condition of the climate that is related to but distinct from water demands or water requirements of man and his crops and other activities. As pointed out by Thornthwaite (1948, p. 56); "The vegetation of the desert is sparse and uses little water because water is deficient***. When water supply increases, as in a desert irrigation project, evapotranspiration rises to maximum that depends only on the climate. This we may call 'potential evapotranspiration,' as distinct from actual evapotranspiration."

By Thornthwaite's empirical formula, potential evapotranspiration can be estimated for any locality if the mean monthly temperature and the latitude are known. In the San Joaquin Valley and the desert east of the southern Sierra Nevada and southern Coast Ranges, the deficiency ranges from 20 to 40 inches (500 to 1,000 mm) annually, equivalent to 1–2 Mgal/d/mi^2 (million U.S. gallons per day per square mile) (1,460 to 2,920 m^3 per day per km^2) (Piper, 1965); the deficiency is greater along the southeast border of the region (fig. 3). Studies (Cruff and Thompson, 1967) indicate that the Thornthwaite formula gives consistently low estimates of potential evapotranspiration in dry regions; thus actual water deficiencies may be greater than those indicated in figure 3.

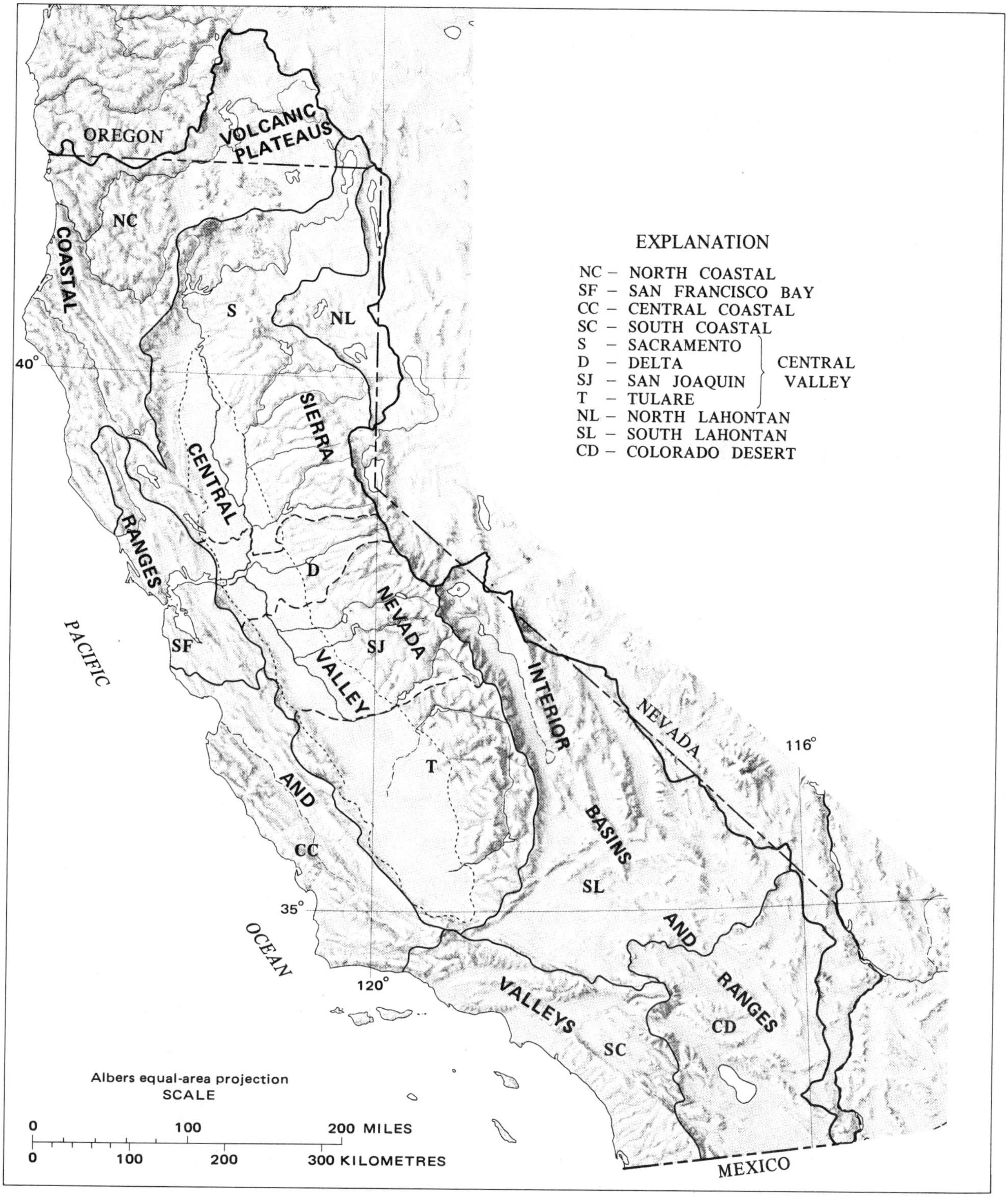

FIGURE 1.—Subregions and landforms of California Region

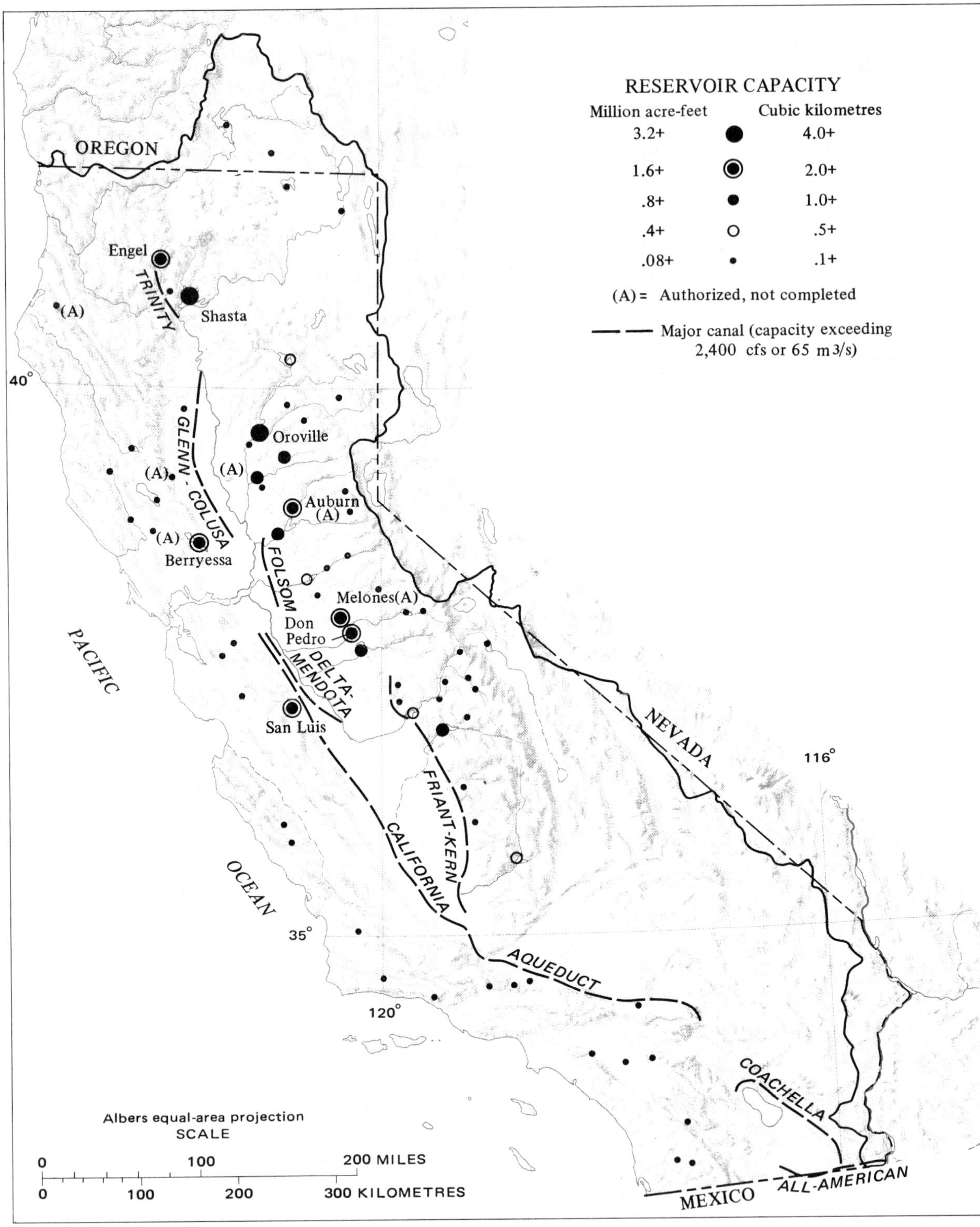

FIGURE 2.—Major surface reservoirs and canals in the California Region.

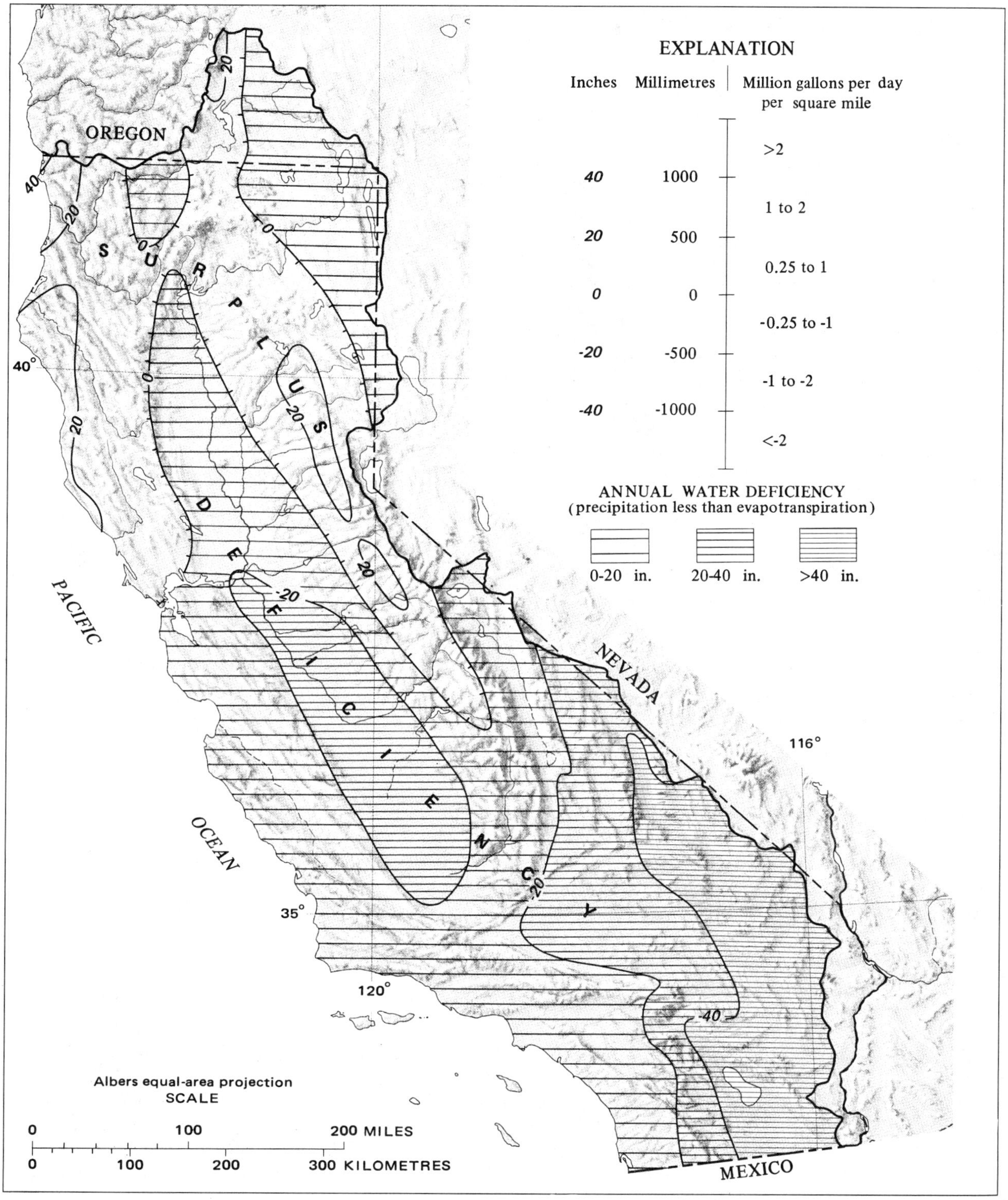

FIGURE 3.—Water surplus and water deficiency in the California Region

According to climatologic data the greatest water deficiency in the region is in Death Valley, where the average annual rainfall is 1.6 inches (40 mm) but evaporation from a standard land pan is more than 140 inches (3,500 mm). Annual evaporation from lakes and reservoirs in the California Region (Kohler and others, 1959) ranges from less than 28 inches (730 mm) along the north coast to more than 86 inches (2,200 mm) in the southeastern deserts including Death Valley. In the valleys of southern California where mean annual precipitation is less than 16 inches (400 mm) the June-September rainfall is negligible, and summers are dry and hot. In the northern part of the region, the summer rainfall is greater but still is less than 5 percent of the annual total; the summers are relatively cool, and many forms of life subsist on water from the preceding wet months, supplemented by occasional rain and fog.

The average annual precipitation and evapotranspiration and the mean seasonal characteristics of precipitation are based on records for the 30-year "normal" period 1931-60. But in every locality the precipitation varies markedly from year to year. At San Diego the normal precipitation is 10.4 inches (265 mm), but during a century of record its precipitation has ranged from 27.6 inches (700 mm) in 1884 to 3.4 inches (85 mm) in 1953. In the desert, the maximum annual rainfall at Indio (in 1939) was 35 times the minimum in 1923. It is the deviations from average precipitation, particularly the variations that cause outstanding floods and droughts, that are perilous and sometimes disastrous in the region.

The "normal" period (1931-60) includes 5 years of the 1924–35 drought, which was more severe in the northern part of the region than in the southern part. During the next 10 years, however, precipitation was above normal in all parts of the region, and the records included 2 years of highs unsurpassed in the 20th century to date, 1938, the wettest year, and 1941, the year of greatest runoff, including disastrous floods in the normally deficient Los Angeles area. In the years 1945-49 (fig. 4), precipitation was only about 75 percent of normal, and all parts of the region had a deficiency equivalent to one year's precipitation. Since 1950, above-average precipitation in the north has made up for that "lost" year, but continued less-than-normal precipitation in the San Joaquin Valley produced an accumulated deficiency of about 2 years of average rainfall by 1968. In southern California, precipitation was also prevailingly less than average, and the accumulated deficiency as of 1970 was equivalent to nearly 4 years of average rainfall. Official records indicate that comparable dry periods have occurred in Los Angeles and San Diego in 1917–34, 1894–1904, and 1870–83 and, according to diaries of missionaries, 1843–59, 1822–32, and 1793–1809, with intervening wetter periods (Thomas, 1962). Although the entire region is subject to relatively long periods during which precipitation and runoff have been materially above or below the long-term mean, southern California is subject to more severe and longer variations than is the northern part of the region.

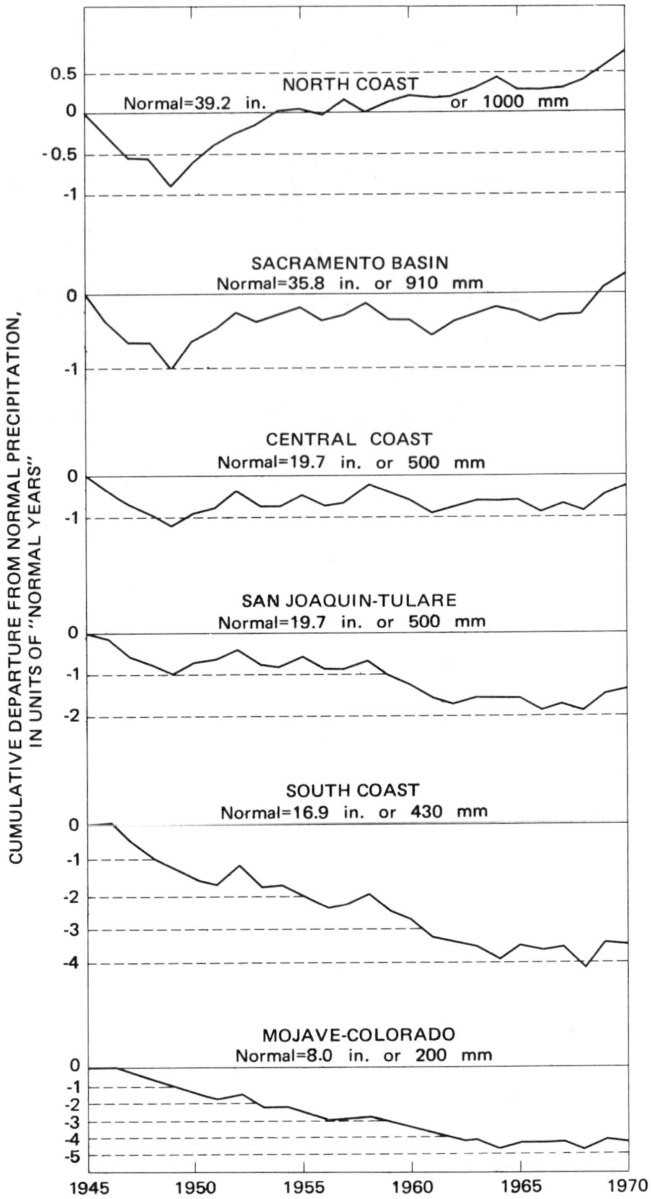

FIGURE 4.—Precipitation trends by subregions, 1945–70.

AIDS TO SURVIVAL

The mountain-and-valley topography greatly influences the distribution of precipitation over the region and has also influenced the distribution of human population, which is concentrated in precipitation-deficient valleys. These flat lowlands and valleys so highly prized for agriculture, commerce, industry, and urban dwelling have always relied on the neighboring mountains for runoff as surface water or ground water which can supplement the scant rainfall of the valleys.

In the Coast Ranges and low desert mountains, streamflow is predominantly storm runoff produced by and almost concurrent with intense or lengthy rainstorms; in the desert most stream channels carry water only after exceptional rainstorms. The streamflow in the Coast Ranges is closely related to rainfall and more than 95 percent of the annual total occurs during the 8 rainy months. Runoff during the rainy season can be stored for use during the dry season, and such storage occurs naturally in areas such as snowfields and lakes.

Snow may accumulate in areas of the Coast Ranges and Sierra Nevada above an altitude of 4,000 feet (1,200 m); above 6,000 feet (1,800 m) the seasonal snowfall commonly exceeds 78 inches (2 m). The snow in the Sierra generally accumulates until April, and its melting produces an annual freshet that reaches a peak in May; 20–40 percent of the annual runoff of streams draining the High Sierra occurs in the 4 summer months. The snow disappears during the summer, and by September and October these streams reach their minimum flow of the year.

The numerous natural lakes of the California Region are mostly in high mountains that have a perennial water surplus. They store water during the period of greatest inflow and release it gradually to sustain the runoff, perhaps throughout the year. Many of these lakes are now dammed to increase the storage, and many manmade lakes have been created by dams in areas of perennial surplus to store water during the periods of greatest inflow and release it as desired, particularly to meet demands during the dry season.

Many natural and manmade lakes are, however, in areas of perennial water deficiency, an environment in which evaporation may be equal to the inflow, and so there is no outflow; or outflow may exceed the inflow until the lake dries up. For example, Goose Lake in the northeastern part of the region had an area of about 185 square miles (480 km^2) and overflowed to the Pit River in 1869 and again in 1881, but the lake has had no outflow for many years, and during a drought beginning in 1924, the lake dwindled until it dried up in 1930. In the southeast desert most of the closed basins have a "dry lake" in their lowest part, where water accumulates by inflow from exceptional storms and then is dissipated by evaporation. Other lakes have sufficient inflow to be perennial, but as water is evaporated the dissolved salts remain behind and the water in the lake becomes increasingly saline.

Many large artificial lakes and reservoirs have been created in areas of water deficiency to store water in times of surplus for use in places of need. This storage of water for man's use is usually expensive, and as with natural lakes, evaporation from the lake surface reduces both the quantity and the quality of the water stored. Although the cost of surface storage has long been accepted by the State of California, it was cited by Conkling (1946, p. 279) as a basis for recommending alternative use of subsurface storage wherever feasible:

> A study of evaporation losses from hypothetical surface reservoirs on all streams in California indicates that, no matter how large the reservoir capacity, streams of erratic annual and cyclic flow will yield for useful purposes no more than 50 percent or 60 percent of the annual average discharge because the remainder will be lost, over the years, by evaporation from the excessive water surface of the reservoir necessary to impound the water of the infrequent years of large discharge.

SIGNIFICANCE OF GROUND-WATER RESERVOIRS

Ground-water reservoirs have always been significant because of their capability of storing usable water, perennially secure from loss, but especially so during the rainless summers and in the perennially water-deficient areas of the California Region. Because surface water in streams, lakes, marshes, and snowfields and soil moisture are vulnerable to evapotranspiration and progressive depletion between storms, ground water is the most reliable source of potable water in many parts of the region and in fact is the only source in some areas.

Natural sources of ground-water discharge are springs, seepage areas, lakes, and base flow of perennial streams, used by Indians, missions, and early settlers. As population increased during the 19th century, more ground water was located by digging wells, tunneling in sloping lands, and boring artesian wells, chiefly for domestic use. Large-scale development and use of ground water became possible in the 20th century, with advancing technology in well drilling and motor-driven pumps. In 1950, irrigation wells pumped 10 million acre-feet (12 km^3) equivalent to the volume of surface water used within the region and almost as much ground water as was used for irrigation in all the rest of the United States.

A century of development proved that ground water is distributed widely, though not uniformly, in the region, especially under the valley floors where it is withdrawn by thousands of large and productive wells and also in many foothill and mountain areas where wells and springs yield water sufficient for domestic use. The storage of potable water in ground-water reservoirs is many times as great as the capacity of all natural lakes plus artificial reservoirs, existing and contemplated. From early days ground-water development has also created problems in many areas—"falling water tables," overdraft and progressive depletion of storage, seawater encroachment and deterioration of quality in freshwater aquifers, and compaction of water-bearing sediments and subsidence of the overlying land.

As development and use of ground water increased, people in many water-deficient areas became concerned about inadequacy or impermanence of their ground-water supply and about obtaining water from other sources, and thus a State Water Plan was proposed in a report to the 1931 legislature (California Department of Water Resources, 1930). The later and present California Water Plan was far more comprehensive and pointed out (California Department of Water Resources, 1957a, p. 242) the necessity for "the development and use of vast regulatory and carry over storage capacity, both surface and underground, in order to attain the degree of conservation required to meet water needs under ultimate conditions of development; and the construction and operation of a major system of works to convey the regulated excess waters from areas of inherent surplus to areas of inherent deficiency."

The implementation of the California Water Plan to date has been concerned chiefly with facilities for storage and transport of surface water from places of surplus in order to satisfy water demands in various areas in the immediate future. The region in 1970 had 11 surface reservoirs of capacity greater than 800,000 acre-feet (1 km^3); their combined usable capacity was about 23 million acre-feet (28 km^3) slightly less than that of Lake Mead on the Colorado River. There are 60 smaller reservoirs with capacity greater than 80,000 acre-feet (100 million m^3), and these increase the surface-water storage in the region to more than 40 million acre-feet (50 km^3) (fig. 2), equivalent to 55 percent of the average annual runoff. Many of the reservoirs in the California Region were created before environmental awareness became fashionable; thanks to good planning, good fortune, and public acceptance, these reservoirs have caused far less environmental disruption than has been reported as a result of superdams on other continents, and many are significant enhancements. Figure 2 shows the major components of the Central Valley Project and State Water Project, including several that were authorized but not yet under construction in 1970.

There is increasing recognition that ground-water reservoirs are significant not only for the water they contain but also for the space in which water may be stored for years with minimum loss. This is expressed in a recent summary of planning within the California Water Plan (California Department of Water Resources, 1970c, p. 72):

> Ground water in storage and ground water storage capacity constitute an extremely valuable resource at present and will continue to be in the future. The value of ground-water resources lies in the use of ground water in storage and under-ground storage capacity (1) to provide regulation of natural replenishment, and (2) operated coordinatedly with both local and imported surface supplies, to effect the most economical use of total available storage, both surface and underground, as an integrated system.

GROUND-WATER RESERVOIRS

The term "reservoir," from the French, has been used especially for artificial lakes and some natural lakes. Implicit in this usage is the delaying and storing—reserving—of surface water in its flow through the hydrologic cycle. If one of these lakes were filled with gravel or sand, its water storage would be reduced by two-thirds, and the rate of movement of water from points of inflow to outflow would be reduced; it would then have the characteristics of a ground-water reservoir. Ground-water reservoirs exist in nature, but their extent, depth, capacity, and the variations of the quality of the water in them are known only as knowledge is acquired in the course of prospecting for, developing, and withdrawing water from them. Most ground-water reservoirs contain far more water in storage than the volume that flows through them annually; however, only this flow-through volume is the renewable resource, the sub-surface phase of the hydrologic cycle.

As used in this appraisal, a ground-water reservoir may contain several water-bearing zones, or aquifers, at various depths and having varied thickness and areal extent. Aquifer is more specifically a geologic term, referring to a rock formation or group, or part of a formation, that is water bearing. Basin, a term in common use for surface-water drainage, may also be applied to a geographic area containing a well-defined ground-water reservoir or part thereof.

NATURAL HISTORY

Most of the ground-water reservoirs of the California region are in the valleys and plains that receive runoff and debris from the mountains. The longest and highest mountain range is the Sierra Nevada, a broad tilted block of relatively impermeable igneous and metamorphic rocks extending from the south end of the Cascade Range at Mt. Lassen southward 400 miles (640 km). To the west, the great Central Valley of similar length and breadth is composed partly of stream-borne sediments that now contain freshwater to depths of 400 to more than 4,000 feet (120 to 1,200 m) below sea level.

The Central Valley ground-water reservoir includes

numerous gravel-and-sand aquifers, formed by stream-carried sediments from the mountains, separated by interstream sediments of chiefly silt and clay; it also includes deep confined aquifers which are separated from shallow aquifers by extensive beds of clay. Also, the pattern of precipitation and surface inflow—generally greater in the north, less to the south—is responsible for other variations in ground-water occurrence. This ground-water reservoir is thus a complex and heterogeneous mass, too large to consider conveniently as a unit and yet with sufficient unity that division on the basis of ground-water characteristics is difficult. For overall water-resource studies and planning in the California Region, the Central Valley drainage basin is divisable into at least two subregions—the Sacramento basin to the north and the San Joaquin Basin southeast of the valley's outlet to San Francisco Bay; it has been divided additionally to form the Delta, where the valley floor is approximately at sea level, and the Tulare Basin of interior drainage at the south end of the Central Valley (fig. 1).

The Coast Ranges comprise folded and faulted sedimentary and metamorphic rocks generally parallel to the Pacific coastline. Most of the ground-water reservoirs are in the intervening structural valleys and coastal plains; some are along the streams that drain, traverse, or bypass various ranges as they flow toward the ocean. The coastal valleys and mountains are included in four subregions—North Coastal, San Francisco Bay, Central Coastal, and South Coastal—of the California Region, each comprising drainage basins of several streams debouching into the Pacific Ocean. The North Coastal subregion has the greatest precipitation and runoff; its ground-water reservoirs are recharged each rainy season and maintain perennial flow of streams in the rainless season. Water deficiency is prevalent in the South Coastal subregion, where ground-water reservoirs are recharged in wet seasons but where the water may remain underground as it moves toward the ocean, appearing at the surface only where it encounters faults or other barriers.

East of the Sierra Nevada and the Transverse Ranges farther south, the California Region includes the Mojave Desert, the Colorado Desert, and the valleys and ranges of the Great Basin in California, designated the North and South Lahontan subregions. Separated from the Pacific Ocean by formidable mountain barriers, these areas are the most arid lands of the region, all in basins of interior drainage except for a narrow zone along the through-flowing Colorado River. The Mojave, Owens, and Susan Rivers are the largest of the few streams with headwaters in high mountains that yield perennial inflow to these arid basins. Other stream channels are dry except when rare torrential storms cause floods of short duration or during brief snowmelt seasons. Ground-water reservoirs are found under the valleys and plains and may be recharged chiefly by intense storms or flood runoff. Discharge from these ground-water reservoirs may be by springs, by evapotranspiration where water is at shallow depth, or by subsurface movement toward a lower valley. For example, some water in the Mojave River may recharge ground water and continue downgradient as subsurface flow through half a dozen valleys, perhaps eventually contributing to the "Badwater" of Death Valley.

In the northeastern part of the California Region, the Modoc Plateau consists of a thick accumulation of lava flows and tuffs and small volcanic cones. Many of these volcanic rocks are excellent aquifers, readily recharged by precipitation and permeable enough to store and subsequently discharge water at numerous large springs, although most of the plateau is in the rain shadow of the Sierra-Cascades and is semiarid. The total area of these volcanic rocks exceeds 10,000 square miles (26,000 km^2), which is chiefly highlands and slopes undesirable for drilling wells or using the water from them. The ground-water reservoirs outlined in this volcanic area (pl. 1) include only the plains or valleys where ground-water development is active or likely. Some of these valleys are in the Klamath River basin (North Coastal subregion), and others are in the Sacramento River basin and subregion or in the North Lahontan subregion, part of the Great Basin.

HISTORY OF DEVELOPMENT AND USE

The original and aboriginal inhabitants of the California Region adapted well to their environment, hunting, fishing, and taking food and water where they found them. In rainless summers and in deserts, the Indians necessarily depended upon ground water that came to the surface as the base flow of streams or that was found in springs, waterholes, or plants.

The arrival of the Spaniards in 1769 saw the introduction of traditions and practices that had been developed in arid and semiarid regions where the value of land depended upon the availability of ground or surface water. The first dams, reservoirs, aqueducts, and wells in the region were constructed at the California missions, and gardens, vineyards, orchards, and field crops were planted and irrigated. Development and use of ground water at the missions was minor: Cienegas discharged a copious supply for the San Gabriel and San Fernando Missions, wells were dug at several other missions when stream supplies became inadequate, and at the San Antonio Mission a noria (undershot water wheel) was constructed to lift water from a dug well for irrigation of a garden and vineyard.

The independence of Mexico in 1824 and the secularization of the missions in the following decade marked the beginning of the era of the Californios, Spanish families who had obtained about 30 grants of land from the Spanish government prior to 1822 and nearly 800 grants from the Mexicans by 1846 (Pitt, 1966). The maximum grant to an individual was 11 square leagues (about 99 square miles or 250 km^2), including 1 of irrigable land, 4 for dry farming, and 6 for grazing. By 1849 an estimated 200 families of Californios had been granted 14 million acres (56,700 km^2) of prime land, especially in the South Coastal, Central Coastal, and San Francisco Bay subregions.

The Mexican War, which ended with the cession of California to the United States in February 1848, brought new settlers to the California Region with a contrasting culture of Anglo-Saxon traditions and customs. Their esteem for the yeoman who could wrest a decent livelihood by rainfall agriculture on 160 acres and their equal distaste for those who left resources idle or failed to make the most of them prompted many of these newcomers to move onto the rancho lands and assert squatter's rights on the basis of their occupancy. Many obtained title to land through legislation, litigation, purchase, or financial or legal manipulation; however, most of the grants to Californios were eventually accepted. Titles were confirmed by Federal courts for about 600 of the grants with a total area about 9 million acres (36,400 km^2), and a few have remained virtually intact to the present.

The influx of settlers was accelerated tremendously after the discovery of gold in the Sierra Nevada in January 1848. In the gold country men staked their claims, built their camps, and appropriated water and wood and brought them to their workings for their operations, all as trespassers on the public land. The miners acted like the squatters on agricultural lands, but subsequent generations recognize that they applied the doctrine of prior application—first in time is first in right—to the water and to the mining claims. This law of water subsequently became the water law in most other States of the arid West.

In 1850, before California became a State, the Legislature enacted a statute declaring the common law of England "the rule of decision in all the courts of the State." This was a triumph of English over Spanish culture, and it also imposed upon an arid and semiarid region the traditions of a well-settled and orderly humid region. The common law had little to say about squatter's rights or appropriative rights but had strong support for landownership. It included the doctrine that each owner of land bordering a stream had a right to the natural flow of the stream, undiminished in quantity and unimpaired in quality, and that each riparian owner, but no nonriparian owner, had the right to use the water and this right was not lost by nonuse. Conflicts soon arose between riparians and appropriators of water. The riparian rights to natural flow could restrict or prevent the storage of winter runoff from rain or of spring runoff from melting snow to be used for irrigation when natural streamflow was negligible. Consequently, construction of large dams on California streams came slowly, years later than the first large dams in any other western State. It took a constitutional amendment (article XIV, sec. 3) in 1928 to limit riparian rights to the quantities of water that were "reasonably required for the beneficial use to be served."

The common law also included the English rule that a landowner owns the percolating water beneath his land and can do as he pleases with it.[1] Thus to nonriparian landowners, and to riparians who could not store water to use when they needed it most, ground water would be a lifesaver if it could be obtained in quantities sufficient for irrigation, the predominant use.

During the 19th century the most productive wells were flowing artesian wells: only a few in the 1860's but increasing to thousands in the next decade so that the Legislature in 1878 enacted a statute to prevent waste from flowing wells. In 1890 about 3,200 flowing wells irrigated 4 percent of the 1 million irrigated acres (4,000 km^2) in the region. In 1894 a well was pumped by electric motor, and by 1906 some 600 pumped wells in San Joaquin Valley had a combined yield of 7,660 ft^3/sec (cubic feet per second) (220m^3/s), more than five times the yield of the 520 flowing irrigation wells.

In the course of this water development, wells started interfering, artesian pressures declined, and litigation ensued, during which it became evident that the common-law doctrine of absolute ownership of ground water was unsuited to California conditions.[2] A leading decision by the California Supreme Court (Katz v. Walkinshaw, 1903) resulted in the California doctrine that the owners of land overlying a ground-water reservoir have correlative rights in a common supply and that each owner is limited to reasonable beneficial use of the water. If a ground-water reservoir cannot yield perennially enough water to irrigate the overlying

[1]In contrast to this doctrine that percolating water belongs to the overlying landowner, the States of Nevada and Oregon have declared that all waters within their boundaries belong to the public and are subject to appropriation for beneficial use as provided by statute. Because of the contrasting doctrines of ground-water rights in the several States, the Water Resources Council has drawn the boundaries of the "California Region" to include all closed basins of the Great Basin that lie entirely within the State of California and also the interstate closed basins that drain California. Thus the boundary between the California Region and the Great Basin Region (pl. 1) is a hydrologic boundary that approximates the State boundary without dissecting ground-water reservoirs or stream drainage basins.

[2]This was a doctrine in the State of California, not the portions of the California Region that lie within Nevada and Oregon, which long ago repudiated the common-law doctrines of water rights.

land, this doctrine permits all owners to share in the deficiency. Furthermore, if an appropriator is pumping water from the reservoir for export, the landowner's right extends only to the water necessary for use on his land, and the appropriator may take any surplus.

In 1939 more than 50,000 wells were used to irrigate 1½ million acres (6,000 km^2) and to sustain another 1 million acres through the summer when stream supplies were minimal. By 1950 the total irrigated area in the region exceeded 6½ million acres (26,000 km^2); 45 percent was irrigated solely and 18 percent partly by water from 80,000 wells. This increasing use of water was accompanied by "falling water tables" in practically all areas of development and complicating problems in many places. Many of these problems have been solved by obtaining water from alternative sources, generally surface water, and reducing withdrawals of ground water. But restriction or regulation of pumping raised questions as to the respective water rights of appropriators and of overlying landowners including both users and nonusers of the water.

A California Supreme Court decision, Pasadena v. Alhambra(1949) added a new principle to the correlative doctrine. Pasadena and Alhambra are cities that pump water from the Raymond Basin, a 40-square-mile (100-km^2) area in the northwestern part of the San Gabriel Valley ground-water reservoir; Pasadena uses its supply within the basin, and Alhambra exports its supply. Both are classed as appropriators, for even though a city overlies a ground-water basin it does not own the land occupied by the people to whom it supplies water. In California, water may be appropriated for public use or for export so long as there is a surplus not used by overlying landowners, but evidence showed that beginning in 1913 there was no surplus; nevertheless, appropriators and overlying landowners continued to pump, the annual total averaging about 133 percent of the calculated "safe" yield of the basin. The decision was that pumpers who caused the overdraft acquired prescriptive rights by infringing upon superior overlying and appropriative rights for more than 5 years; however, the earlier users, continuing to pump to meet their needs, gained mutual prescriptive rights by infringing on the infringers. The "safe" yield was achieved by reducing proportionately the withdrawals of all users, and overlying owners who had taken no water lost their rights to water.

If the city fathers of Pasadena and Alhambra had known what was going on hydrologically, they might have called a meeting of all landowners of the Raymond Basin back in 1913 to announce that the surplus was just about gone, express appreciation for use of those surpluses all these years, and start looking for a new supply. Then it would have been up to the landowners with their correlative rights to work out ways to supply water to increasing numbers of people owning smaller and smaller plots of land, with all sharing the water deficiency; instead, the appropriators continued pumping for decades before going to court. The subsequent decision has become the prevailing law: An appropriative taking of water which is not surplus is wrongful but may ripen into a prescriptive right where the use is open and notorious, hostile and adverse to the original owner, continuous and uninterrupted for the statutory period of 5 years, and under claim of right. This principle has been the basis for adjudication of rights or for negotiation among right holders in several ground-water reservoirs in southern California and serves to specify the rights of each in the local resource.

Under the basic premise that ground water is part and parcel of the land, the development and use of ground water have been almost exclusively by private enterprise and local or community initiative. Particularly in early stages and in many places right down to the present, this development has been in response to the settlement, cultivation, and urbanization of the land, and thus haphazard and unplanned.

CHARACTERISTICS OF INDIVIDUAL RESERVOIRS

The characteristics of individual reservoirs are known chiefly from the wells and other excavations that have been dug, drilled, or tunneled into the earth. Some desert areas are devoid of inhabitants and wells, and geologic reconnaissances have provided the only information as to rocks and rock materials that may be water bearing; other areas presently uninhabited have been the sites of mining, agricultural, or other activities long ago, with or without record as to the sources of water supply. At the other extreme are ground-water reservoirs in which thousands of wells provide quantitative data as to aquifers and their boundaries, storage, and transmissivity. These best-known reservoirs have a history that may span several decades and perhaps a century, in which development has proceeded from exploration and exploitation to regulation and management. Moreover, each ground-water reservoir that has been identified and named in the California Region (California Department of Water Resources, 1952) can be classified on the basis of its current stage of development.

The uniqueness of individual ground-water reservoirs is the basis of recommendations that data-collection programs be organized on a basin-by-basin basis (Dutcher, 1972), and the California Legislature approves the established practice of keeping the development and management of ground-water reservoirs under local control. At times, however, the local agencies may need project reconnaissance studies

(Peters 1972b) to show the engineering and economic feasibility of a specific management program; these studies require very specific geologic and hydrologic information for the free ground-water zone and for each major confined zone to provide a hydrologic balance or model validation. For such tasks some agencies call upon expert consultants; others utilize the services of specialists in the California Department of Water Resources or the U.S. Geological Survey, under a cooperative or cost-sharing agreement.

EXPLORATION

If the total annual withdrawal is less than 8,100 acre-feet (10 million m^3 or 0.01 km^3), the ground water reservoir is here categorized as in the exploratory stage. Most reservoirs are included in the tabulation of undeveloped ground-water reservoirs in table 1, and all are shown on the regional map (pl.1). Some of these reservoirs are in uninhabited areas; most are in sparsely inhabited areas in which several wells have been dug or drilled chiefly for domestic or stock supplies. Some are tapped by a few large-capacity wells used for irrigation, military, industrial, or other purpose.

The 79 undeveloped ground-water reservoirs in the California Region range in size from 40 to 1,870 square miles (100 to 4,900 km^2), with an aggregate area of about 19,000 square miles (50,000 km^2). Data concerning these undeveloped reservoirs come chiefly from wells and may include well logs, measurements of water level or well discharge, and chemical analyses of sampled water. These data, plus any information from geologic maps and reports, are generally sufficient only for a precursory or cursory hydrologic study (Peters, 1972b). They are too meager to describe the areal extent, depth, and boundaries of aquifers or the storage, flow, or quality of the water in them in the detail necessary for sound development. In these reservoirs the natural equilibrium, inflow and outflow, is presumed to be unaffected by the negligible withdrawals from wells; the principal means of natural outflow from these reservoirs, known or presumed, is indicated in table 1.

EXPLOITATION

Exploitation of reservoirs is marked by successful development of large production wells, increasing numbers of wells, and increasing rates of pumping. The developed ground-water reservoirs of the California Region, including chiefly those having annual pumpage greater than 8,100 acre-feet (0.01 km^3), are given in table 2, with summary data on pumpage, estimated usable storage capacity, and water quality. Altogether 55 reservoirs are tabulated, underlying a total area of 30,000 square miles (77,000 km^2) of which half is in the Central Valley (table 3).

REGULATION

The fact or fear of overdevelopment, or the use of water at rates greater than the natural replenishment, generally results in restriction of pumping or additional development, in some instances voluntarily by agreement among water users or landowners, in other instances involuntarily by law. Under the doctrine of appropriation in most Western States, water belongs to the public and may be appropriated for beneficial use so long as there is surplus or "unappropriated" water; thereafter, the State may restrict or prevent additional development so that total withdrawals do not exceed the natural recharge.

In the State of California, regulation is unique both in timing and method. Ground-water development, by private enterprise, has not been subject to regulation until the need is established by clear evidence of overdraft, as presented during litigation calling for adjudication of ground-water rights. Because such adjudication requires a great volume and variety of quantitative data and study, the court may refer the case to the State Water Resources Control Board as referee (the court reference procedure) to be responsible for these studies. Effective regulation is likely to require reduction in pumping and denial of water to established users.

At the time of the leading decision placing restrictions on pumpage (Pasadena v. Alhambra, 1949), water imported from the Colorado River was available to those whose ground-water supplies must be reduced, and it is practically axiomatic that alternative sources of water shall be available as a supplement where it is necessary to reduce withdrawals from a ground-water reservoir to reach a "safe" yield. After rights are adjudicated, the extractions from the reservoir are monitored by a court-appointed watermaster who makes annual reports on water conditions, ground-water extractions and recharge, and imported water. Currently, watermaster service is provided for regulation of pumping in the West Coast Basin and Central Basin of the Los Angeles coastal plain (State Basin No. 4–11 on pl. 1), the upper Los Angeles River area of San Fernando Valley (No. 4–12), and the Raymond Basin of San Gabriel Valley (No. 4–13) (California Department of Water Resources, 1970d, e; 1971b, c).

MANAGEMENT

To achieve a "safe" yield from a ground-water reservoir or to make the best of limited natural resources, management may be imposed by public agencies or districts as authorized and encouraged by the State. These local water agencies are generally "users cooperatives" (Bain and others, 1966), most of which are

TABLE 1.—*Undeveloped ground-water reservoirs (withdrawals less than 8,100 acre-feet (10 million m^3) per year)*

[From Bader (1969) and California Region Framework Study Committee (1971), revised and updated]

State No.: Ground-water reservoir is numbered after California Division of Water Resources (1952) for California and by the Geological Survey for Oregon. Number before dash indicates California Regional Water Quality Control Board as follows: 1, North Coastal; 2, San Francisco; 3, Central Coastal; 4, Los Angeles; 5, Central Valley; 6, Lahontan; 7, Colorado River; 8, Santa Ana; 9, San Diego.
County: Where state name has asterisk (*), the ground-water reservoir is interstate, and the data given apply only to that part of the reservoirs within that state.
Natural outflow: ET, evapotranspiration; GW, ground-water outflow.

State No.	County	Ground-water reservoir	Estimated area (km^2)	Natural outflow	Exploration data			
					Year	Number of wells	Depth explored (m)	Range in dissolved solids (mg/l)
		North Coastal subregion						
		Klamath River Basin:						
0–1	Klamath, Oreg.*	Klamath Marsh	600	Williamson River	1970	12	205	50–110
0–5	do	Langell Valley	180	Lost River	1970	10	150	130–170
0–6	do	Upper Klamath Valley	750	Klamath River	1954	13	145	80–100
0–7	do	Poe Valley	80	Lost River	1970	9	135	140–270
1–4	Siskiyou, Calif.*	Shasta Valley	650	Shasta River	1953	50	215	160–980
1–5	do	Scott River Valley	210	Scott River	1953	5	65	30–420
		San Francisco Bay subregion						
		Russian River basin:						
1–17	Sonoma, Calif.	Alexander Valley	100	Russian River	1954	40	135	220–1,300
		Central Coastal subregion						
		Closed basin:						
3–19	San Luis Obispo, Calif.	Carrizo Plain	700	ET-Soda Lake	1954	8	305	340–4,300
		Sacramento Basin subregion						
5–1	Lake, Oreg.*	Goose Lake Valley	410	ET-Goose Lake	1948	10	915	120–1,460
	Modoc, Calif.*	do	490	do	1964	15	230	100–450
		Pit River basin:						
5–2	Modoc, Calif.	Alturas basin	230	Pit River	1964	25	305	150–500
5–4	Modoc and Shasta, Calif.	Big Valley	260	do	1964	25	365	150–1,380
5–5	Shasta, Calif.	Fall River Valley	260	Fall River	1964	50	215	100–550
		Feather River basin:						
5–12	Plumas and Sierra, Calif	Sierra Valley	360	Feather River	1964	20	425	120–1,400
		San Joaquin Basin subregion						
5–23	San Benito, Calif.	Panoche Valley	130	Panoche Creek		--	120	--
5–25	Kern, Calif.	Kern River Valley	70	Kern River		(Partly overlain by Isabella Reservoir)		
		North Lahontan subregion						
		Closed basins:						
6–1	Modoc, Calif.*	Surprise Valley	910	ET-Alkali Lakes	1954	60	245	165–2,000
	Washoe, Nev.*	do	30					
6–2	Modoc, Calif.	Madeline Plains	700	ET-Tule Lake	1964	10	260	100–245
6–4	Lassen, Calif.*	Honey Lake Valley	1,270	ET-Honey Lake	1964	100	425	175–1,350
	Washoe, Nev.*	do	490	do	1967	10	120	170–5,000
		South Lahontan subregion						
		Closed basins:						
6–9	Mono, Calif.*	Mono Valley	520	ET-Mono Lake	1960	2	290	2,000–brine
	Mineral, Nev.*	do	80					
6–10	Mono, Calif.	Adobe Lake Valley	160	ET-Adobe Lake	1962	4	10	130–280
6–11	do	Long Valley	260	Lake Crowley	1954	8	25	90–1,500
6–13	Inyo, Calif.	Centennial (Black Springs) Valley.	130	GW to 6–12	1954	--	--	360
6–15	do	Deep Springs Valley	100	ET-Deep Springs Lake	1955	4	235	--
6–16	do	Eureka Valley	410	ET-playa	1955	1	115	550
6–17	do	Saline Valley	540	ET-Salt Lake	1955	1	15	3,700–brine
6–18	Inyo and San Bernardino, Calif.	Death Valley	3,420	ET-Badwater	1961	7	305	550–brine
6–19	San Bernardino, Calif.	Wingate Valley	180		1953	--	--	660
6–20	Inyo, Calif.*	Middle Amargosa basin	1,350	ET-playa	1962	20	145	300–2,900
6–21	San Bernardino, Calif.	Lower Kingston (Valjean) Valley.	750	GW to 6–18	1954	1	No water to 130 ft	5,300–8,600
6–22	do	Upper Kingston (Shadow) Valley.	700	GW to 6–21	1961	10	120	340–1,100
6–23	do	Riggs Valley	260	GW to 6–21	1954	1	45	1,740
6–24	do	Red Pass Valley	390	ET-Red Pass Lake	1944	1	115	--
6–25	do	Bicycle Valley	310	ET-Bicycle Lake	1955	1	135	610–brine
6–26	do	Avawatz Valley	180	ET-Drinkwater Lake		--	--	--
6–27	do	Leach Valley	180	ET-Leach Lake	1917	--	--	300–700
6–29	Inyo, San Bernardino, Calif.	Mesquite Valley	310	ET-Mesquite Lake	1959	22	335	300–6,300
6–30	San Bernardino, Calif.*	Ivanpah Valley	780	ET-Ivanpah Lake	1960	19	260	290–2,200
6–31	San Bernardino, Calif.	Kelso Valley	960	GW to 6–33	1961	2	195	250–750
6–32	do	Broadwell Valley	310	ET-Broadwell Lake	1883	1	330	470–1,260
6–33	do	Soda Lake Valley	1,530	ET-Soda Lake GW to 6–34.	1961	20	150	240–3,400
6–34	do	Silver Lake Valley	100	ET-Silver Lake GW to 6–23.	1954	3	55	1,100–1,740
6–35	do	Cronise Valley	390	ET-Cronise Lake	1961	13	230	450–3,100
6–36	do	Langford Valley	130	ET-Langford Lake	1958	7	160	470–640

TABLE 1.—*Undeveloped ground-water reservoirs (withdrawals less than 8,100 acre-feet (10 million m^3) per year)*—Continued

State No.	County	Ground-water reservoir	Estimated area (km^2)	Natural outflow	Exploration data			
					Year	Number of wells	Depth explored (m)	Range in dissolved solids (mg/l)
			South Lahontan subregion—Continued					
6–37	do	Coyote Lake Valley	390	ET-Coyote Lake	1961	5	175	300–2,500
3–38	do	Caves Canyon Valley	260	GW to 6–33	1961	5	65	200–1,300
6–39	do	Troy Valley	340	ET–Troy Lake	1961	20	120	280–6,500
6–43	do	El Mirage Valley	310	GW to 6–42	1961	75	295	320–2,600
6–49	do	Superior Valley	440	ET-Superior Lake	1956	25	115	360–2,300
6–50	do	Cuddeback Valley	340	ET-Cuddeback Lake	1956	30	90	370–4,700
6–51	do	Pilot Knob Valley	520	GW to 6–58	1918	1	--	400
6–55	Inyo, Calif.	Coso Valley	130	GW to 6–54	1946	1	35	----------
6–56	do	Rose Valley	160	ET-playa		6	55	150–1,300
6–57	do	Darwin Valley	180		1954	3	75	350–750
6–58	do	Panamint Valley, Brown Mt. Valley (6–76).	930	ET-Panamint Lake	1955	3	300	780–brine
6–71	San Bernardino, Calif.	Lost Lake Valley	100	ET-Lost Lake	1953	0	--	360
6–79	Inyo and San Bernardino, Calif.	California Valley	210	ET-playa	1953	2	15	350–500
			Colorado Desert subregion					
		Closed Basins:						
7–1	San Bernardino, Calif.	Lanfair Valley	730	GW to 7–2	1952	18	270	230–2,000
7–2	do	Fenner Valley	1,860	GW to 7–7	1952	5	285	280–870
7–3	do	Ward Valley	1,990	ET–Danby Lake	1952	2	315	330–brine
7–5	Riverside, Calif.	Chuckwalla Valley	2,250	ET-Palm Lake	1952	10	185	270–brine
7–6	do	Pinto Basin	800	GW to 7–5	1937	3	170	120–830
7–7	San Bernardino, Calif.	Cadiz Valley	1,110	ET-Cadiz Lake	1910	1	105	610–brine
7–8	do	Bristol Valley	1,840	ET-Bristol Lake	1910	8	695	290–brine
7–9	do	Dale Valley	670	ET-Dale Lake	1952	10	--	1,060–brine
7–10	do	Twentynine Palms Valley, Copper Mt. Valley (7–11), Warren Valley (7–12).	960	GW to 7–9	1952	50	150	100–1,180
7–13	do	Deadman Valley, Ames Valley (7–16).	750	GW to 7–10	1952	35	--	180–600
7–14	do	Lavic Valley	100	ET-Lavic Lake	1917	1	40	1,680
7–15	do	Bessemer Valley	180	ET-Galway Lake		--	--	--------
7–17	do	Means Valley	100	ET-Means Lake		1	--	--------
7–18	do	Johnson Valley	360	GW to 7–17	1952	1	45	340–610
7–22	Imperial, Calif.	West Salton Sea basin	750	Salton Sea	1950	3		2,260–brine
7–23	do	Clark Valley	100	ET-Clark Lake		--	--	--------
7–25	San Diego, Calif.	Ocotillo Valley	210	GW to 7–30	1952	6	65	700
7–27	do	San Felipe (Earthquake) Valley.	160	San Felipe Creek	1952	3	--	1,060
7–28	do	Vallecito and Carrizo Valleys.	310	GW to 7–30		--	--	----------
7–29	Imperial, Calif.	Coyote Wells Valley	260	GW to 7–30	1948	6	50	440–8,700
7–30	do	Imperial Valley	4,840	Salton Sea		80	335	690–7,500
7–31	Riverside, Calif.	Orocopia Valley	360	GW to 7–21	1952	--	--	350–1,500
7–32	do	Chocolate Valley	310	GW to 7–33		--	--	350–brine
7–33	Riverside and Imperial, Calif.	East Salton Sea basin	1,170	Salton Sea	1952	4	100	350–3,850

concerned chiefly or at least partly with collection, distribution and use, or contracting for surface water. Some water agencies are made up entirely of ground-water users, but several of these were organized to import supplemental supplies. The districts may include all or parts of ground-water reservoirs suitable for an effective program either of ground-water management or of conjunctive use of surface and ground water.

Ground-water management may involve the programming of withdrawals in space and time, the use and maintenance of barriers, and (or) artificial recharge of the reservoir. It is likely to be one aspect of overall water-management that involves conjunctive use of ground water and surface water, amelioration and maintenance of water quality, and disposal of undesirable and waste waters and soluble materials. In an area of natural water deficiency, effective water management is likely to require importation from areas of surplus; the major metropolitan areas have long been importers of water, and the principal purveyors of water from areas of surplus to areas of deficiency are State and Federal agencies.

PROBLEMS ASSOCIATED WITH USE OF GROUND WATER

A simple but very general problem of discharging wells is commonly called "falling water tables." All water discharged from a well must be balanced by a loss of water somewhere, and this loss is always to some extent, and may be largely, from storage in the aquifer (Theis, 1940). Thus some depletion of ground-water storage is an inevitable result of exploitation. Eventually the water withdrawn from the aquifer may be replaced by an equivalent amount of water of suitable quality, or the withdrawals from the well may cause increased recharge to the aquifer or decreased natural discharge from it so that the discharging well becomes a part of a new steady-state equilibrium. On the other hand, if anything else moves into the space dewatered

TABLE 2.—*Developed ground-water reservoirs*

[From Bader (1969) and California Region Framework Study Committee (1971), revised and updated]

State No.	County	Ground-water reservoir	Area (km²)	Aquifer: Depth zone (m)	Aquifer: Usable capacity (km³)	Withdrawals from wells: Year	Withdrawals from wells: Pumpage (10⁶m³)	Withdrawals from wells: Range in dissolved solids (mg/l)[1]
		North Coastal subregion						
		Smith River Basin:						
1–1	Del Norte, Calif.	Smith River plain	180	3–11	0.09	1968	7	30–200
		Klamath River Basin:						
0–2	Klamath, Oreg.	Sprague River valley	440	----	----	1970	31	80–230
0–3	do	Swan Lake Valley	120	----	----	1970	31	80–270
0–4	do	Yonna Valley	120	----	----	1970	16	110–270
0–8	do	Lower Klamath River valley	490	----	----	1970	100	130–880
1–2	Siskiyou, Calif.	do	410	----	----	1954	12	80–830
		Closed basin:						
1–3	do	Butte Valley	470	----	----	1953	26	110–1,900
		Eel River and Mad River basins:						
1–9	Humboldt, Calif.	Eureka plain, Mad River valley (1–8), Eel River valley (1–10).	600	3–12	.15	1962	18	50–2,000
		San Francisco Bay subregion						
		Russian River basin:						
1–15	Mendocino, Calif.	Ukiah Valley	180	3–15	0.04	1954	12	110–1,120
1–18	Sonoma, Calif.	Santa Rosa Valley and Healdsburg area.	470	3–60	1.2	1954	22	90–800
		San Francisco Bay:						
2–	do	Petaluma Valley	340	3–60	.25	1958	2	110–4,800
2–2.01	Napa, Calif.	Napa Valley	210	3–60	.30	1950	7	100–5,000
2–2.02	Sonoma, Calif.	Sonoma Valley	100	5–60	.05	1950	2	130–2,800
2–3	Solano, Calif.	Suisan-Fairfield Valley	670	3–60	.05	1949	10	300–1,350
2–4	Contra Costa, Calif.	Pittsburg plain	100	30–60	----	1931	10	480–2,060
2–5	do	Clayton Valley, Ygnacio Valley (2–6).	160	6–60	.15	1930	10	210–2,170
		Santa Clara Valley:						
2–9	Santa Clara, Calif.	South Bay	830	8–60	.95	1969	220	240–960
	Alameda, Calif.	East Bay	470	----	----	1970	50	300–7,000
2–10	do	Livermore Valley, Sunol Valley (2–11), San Ramon Valley (2–7).	570	8–60	.25	1970	25	290–2,800
		Central Coastal subregion						
		Soquel Creek basin:						
3–1	Santa Cruz, Calif.	Soquel-Aptos area	260	----	----	----	----	180–700
		Pajaro River basin:						
3–2	do	Pajaro Valley	360	6–90	0.03	1969	65	170–1,500
3–3	Santa Clara, Calif.	Llagas Valley	210	----	----	1969	60	250–550
	San Benito, Calif.	Hollister Valley	670	6–60	1.0	1960	135	280–2,550
		Salinas River basin:						
3–4.01 to 3–4.05	Monterey, Calif.	Salinas Valley	1,810	6–60	1.6	1960	370	240–3,000
3–4.06	San Luis Obispo, Calif	Paso Robles	2,330	15–75	2.1	1967	55	--------
		Santa Maria River basin:						
3–11	do	Arroyo Grande Valley	100	30–240	.15	1967	20	200–2,900
3–12	Santa Barbara, Calif.	Santa Maria Valley	520	6–60	1.2	1967	140	230–3,200
3–13	do	Cuyama Valley	600	----	.5	1967	80	400–5,000
		San Antonio Creek basin:						
3–14	do	San Antonio Creek valley	230	----	.35	1967	14	300–3,000
		Santa Ynez River basin:						
3–15	Santa Barbara, Calif	Santa Ynez River valley	670	6–75	1.2	1967	50	400–2,000
		Santa Barbara Coastal basins:						
3–17	do	Santa Barbara basin, Goleta Basin (3–16), Carpinteria Basin (3–18).	100	15–75	.2	1967	9	340–1,400
		South Coastal subregion						
4–4	Ventura, Calif.	Santa Clara River Valley	1,190	----	0.75	1951	----	270–4,700
		Los Angeles River and Santa Ana River basins:						
4–11	Los Angeles, Calif	Coastal plain	1,300	WT–370	4.9	1970	350	140–1,340
		Central Basin	(600)	----	----	1970	(275)	
		West Basin	(410)	----	----	1970	(75)	--------
4–12	do	San Fernando Valley	520	----	----	1970	135	220–2,130
4–13	do	San Gabriel Valley	520	0–490	11.	1965	250	110–1,000
		Raymond Basin	(100)	6–460	1.2	1961–69	35	150–700
4–14	San Bernardino, Calif.	Upper Santa Ana Valley	1,680	----	----	1965	630	100–1,000
		Bunker Hill–San Timoteo.	540	----	2.1			
		Chino-Riverside	1,110	6–210	6.8	----	----	--------
8–1	Orange, Calif.	Coastal Plain	930	----	----	1970	240	200–2,000
		San Jacinto River basin:						
8–5	Riverside, Calif.	San Jacinto basin	650	----	----	----	----	280–3,900
		Santa Margarita and adjacent basins:						
9–4	San Diego, Calif.	Lower Valley, San Mateo (9–2), San Onofre (9–3).	60	2–msl	.07	1966	11	180–1,600
9–5	do	Temecula Valley, Warner Valley (9–8).	100	----	.65	1961	12	250–5,000
		San Luis Rey basin:						
9–7	do	San Luis Rey Valley	100	6–35	.06	----	----	300–9,000

TABLE 2.—*Developed ground-water reservoirs*—Continued

State No.	County	Ground-water reservoir	Area (km²)	Aquifer: Depth zone (m)	Aquifer: Usable capacity (km³)	Withdrawals from wells: Year	Withdrawals from wells: Pumpage (10⁶m³)	Withdrawals from wells: Range in dissolved solids (mg/l)[1]
		South Coastal subregion—Continued						
9–12	do	San Dieguito River basin: San Dieguito area, Escondido area (9–9), San Pasqual Valley (9–10).	210	----	----	----	----	250–5,000
9–15	do	San Diego River basin: San Diego area, El Cajon (6–16).	130	3–35	.12	----	----	160–4,500
9–8	do	Warner Valley	100	6–65	.07	----	----	150–420
		Central Valley subregion						
5–15	Lake, Calif.	Sacramento River basin: Kelseyville Valley, Upper Lake (5–13), Scott Valley (5–14), Burns Valley (5–17).	130	3–30	0.09	1951	17	80–660
5–6	Shasta, Calif.	Redding basin	1,300	6–60	.15	1955	----	120–1,700
5–21	Several (Calif.)	Sacramento Valley	11,000	6–60	35	1964	3,080	110–2,800
5–22	San Joaquin, Calif.	Mokelumne area: Delta	3,110	----	----	1966	1,230	300–3,500
5–22	Several (Calif.)	San Joaquin River basin: San Joaquin Valley	13,000	6–60	69	1966	8,010	90–5,000
5–22	Kern, Kings, and Tulare, Calif.	Tulare closed basin: Tulare Basin	11,700	6–60	46	1966	3,700	120–2,400
		South Lahontan subregion						
6–12	Inyo and Mono, Calif.	Closed basins: Owens Valley	2,230	----	----	1970	40	100–brine
6–40	San Bernardino, Calif.	Lower Mojave River valley	780	0–90	5.8	1963	85	190–2,340
6–41	do	Middle Mojave River valley	1,090	0–90	11	1963	75	140–3,900
6–42	do	Upper Mojave River valley	1,550	0–90	9.9	1963	55	80–2,760
6–44	Kern and Los Angeles, Calif.	Antelope Valley	4,140	6–60	6.7	1960	380	120–7,700
5–27 5–28 6–45	Kern, Calif.	Divided basin: Cummings Valley: Tehachapi Valley West, Tehachapi Valley East.	130	----	----	1961	20	350–570
6–46	do	Closed basins: Fremont Valley	850	----	----	1958	40	350–brine
6–47	San Bernardino, Calif.	Harper Valley	1,320	6–65	8.6	1963	15	320–10,700
6–52	Inyo and San Bernardino.	Searles Valley	650	----	----	1962	6	8,000
			100	----	----	1962	12	350,000
6–54	Inyo, Kern, and San Bernardino.	Indian Wells Valley	1,350	6–65	.89	1968	15	140–brine
		Colorado Desert subregion						
7–19	San Bernardino	Closed basins: Lucerne Valley	830	----	----	1952	20	340–5,000
7–21	Riverside	Coachella Valley: Upper valley	620	WT–20	4.4	1958	55	149–1,000
		Artesian basin	520	----	----	----	130	750–3,200
7–24	San Diego	Borrego Valley	260	3–60	----	----	12	290–1,480

[1]Includes the range in dissolved-solids concentration in observation wells of the California statewide water quality monitoring program.

by pumping from wells, the problem may become more complicated and aggravating. Removal of water from unconsolidated sediments may induce compaction of the sediments, resulting in subsidence of the land surface. In coastal areas pumping may create gradients favorable to encroachment of seawater into freshwater aquifers, and in areas where freshwater aquifers are in proximity to saline or polluted water—in lakes, playas or deep underground—withdrawal of freshwater may induce flow of saline or polluted water into the aquifer.

Numerous problems are caused by the variety of uses of water by mankind. Water used for washing, processing, cooking, or drinking may subsequently enter a ground-water reservoir with any soluble material it has acquired during the nonconsumptive use, including solid wastes, fertilizers, pesticides, herbicides, or other chemicals that could create a polluting effect. Problems may also develop with the disposal of brines and other byproducts in the production of petroleum or the disposal of soluble salts, minerals, or fuels.

DEPLETION OF STORAGE

Although depletion of storage is an inevitable result of taking water from a well, replenishment may come from precipitation or streamflow or both, or in fact not at all, as shown for example by the long-term graphs of figure 5. The variations from year to year in rainfall and runoff have induced varying responses in the water levels of four wells in southern California.

In the Williams well in Upper Santa Ana Valley

TABLE 3.—*Summary of the ground-water reservoirs listed in tables 1 and 2*

Subregion	Undeveloped ground-water reservoirs (table 1)		Developed ground-water reservoirs (table 2)			
	Number	Estimated area (km²)	Number	Area (km²)	Usable capacity[1] (km³)	Annual pumpage (km³)
North Coastal	6	2,300	7	2,800	--	0.2
San Francisco Bay	1	100	9	4,000	3	.4
Central Coastal	1	700	10	7,900	9	1.0
South Coastal	0	----	13	7,800	26	1.7
Coastal basins	8	3,100	39	22,500	38	3.3
Tributary valleys	7	2,300	2	1,400	--	--
Sacramento Basin	--	----		10,800	--	3.1
Delta area	--	--	1	3,100	--	1.2
San Joaquin Basin	--	----		13,000	--	8.0
Tulare Basin	--	----		11,300	--	3.7
Central Valley	7	2,300	3	39,600	125	16
North Lahontan	3	3,400	0	----	--	--
South Lahontan	38	18,800	10	14,100	43	0.7
Colorado Desert	23	22,200	3	2,200	4	.2
Interior basins	64	44,400	13	16,300	47	0.9
Total	79	50,000	55	78,000	210	20

[1]Estimates of usable capacity have been made for only 37 of the 55 ground-water reservoirs listed, and these represent the investigators' judgment of what is technically and economically practical at the time.

(State Basin No. 4–14 on pl. 1), the water level declined 40 feet (12 m) or more during each of the drought periods 1895–1904 and 1924–36 but recovered during the subsequent wetter periods 1905–23 and 1937–45. Since 1945 the water level has declined nearly 165 feet (50 m), interrupted by temporary rises caused by exceptional recharge in that part of the reservoir during the wet years 1952, 1958, and 1969. In this valley artesian supplies were once so abundant that San Bernardino County was exempted from the State law enacted in 1878 prohibiting waste from flowing wells.

The well on the Oxnard Plain (Basin No. 4–4) indicates reduction in storage during the dry years that have been prevalent since 1945, and full replenishment during years of abundant runoff such as 1952, 1958, 1967, and 1969.

Another hydrograph in figure 5 is a composite record of two wells in Antelope Valley (Basin No. 6–44), which also had an extensive area of artesian flow and more than 200 flowing wells prior to 1908. By 1920 there were 250 pumped wells, and records available since 1922 indicate increasing pumping and a continuous decline of water levels in wells, not only throughout the drought of 1924–36 but during the subsequent wet years 1937–45 and continuing today. By 1946 the decline throughout the pumping district was at an average rate of 3 feet (1 m) a year, prompting the State Legislature to call for a survey. The report of that survey (Gleason, 1947) showed that pumpage exceeded recharge even in the wet years 1944 and 1945 and forecast it would exceed the recharge by more than 50,000 acre-feet (60 million m³) in 1947, and by greater amounts in subsequent years with expansion of irrigated acreage. For a quarter of a century, water levels have declined as predicted, at rates of 3 feet (1 m) or more per year in the pumping districts (Snyder, 1955; Powers, 1970).

In the Central Valley, the annual pumpage from wells has exceeded 12 million acre-feet (15 km³) for several years. In many parts of the valley, the water withdrawn has been replenished within months by infiltration of precipitation and surface water; in other areas replenishment has been ample in years of abundant precipitation and streamflow but not in years of drought; in some areas of the Central Valley pumping has caused progressive decline of water levels in wells and depletion of ground-water storage.

The principal areas of storage depletion in the Central Valley are outlined in figure 6. Those in the Sacramento Valley appear as composite cones of depression where pumping has lowered water levels by 30–80 feet (10–25 m), and the depressions extend below sea level; these areas of depletion are in the interstream areas between major tributaries flowing from the Sierra Nevada. Similar closed depressions are found between the Sierra streams entering the southern part of the Central Valley; they generally do not extend down to sea level, but the largest—between the Kings River and Kern River—is more than 65 feet (20 m) deep and has a length of about 125 miles (200 km). The most formidable decline in water levels, however, has been caused by pumping from the confined aquifers underlying the western part of San Joaquin Valley. Here, water levels are below sea level throughout an area extending from Tulare Lake northwestward 90 miles (140 km) toward Los Banos; in numerous wells they are more than 200 feet (60 m) below sea level. (See section "Land Subsidence.")

The depressed areas in the Central Valley, as in Antelope Valley, are noteworthy because water levels have declined more than 100 feet (30 m) under extensive areas, indicating considerable depletion. Water levels reached record lows in many wells during the 1960's, especially in 1961 and 1966, which were the driest of the decade, and were rising during the wetter years 1969 and 1970, although water levels continued to decline in areas near Sacramento and Stockton and in the southern part of the Tulare Basin (California Department of Water Resources, 1971a). In several irrigation districts in the Tulare Basin, however, water levels in wells have risen more than 65 feet (20 m) since 1951, following the arrival and availability for irrigation of water in the Friant-Kern Canal.

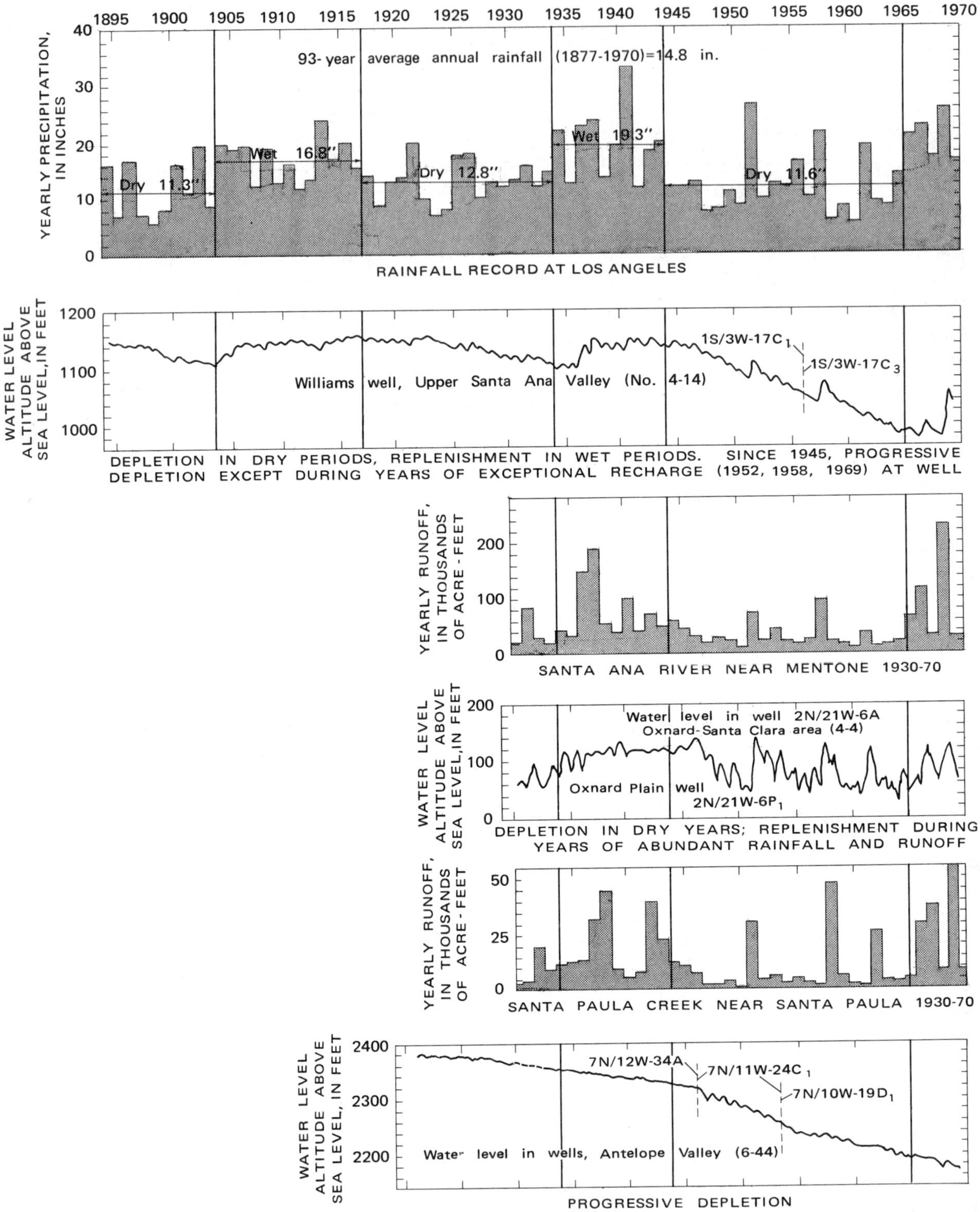

FIGURE. 5.—Effects of pumping and recharge on water levels in wells in southern California.

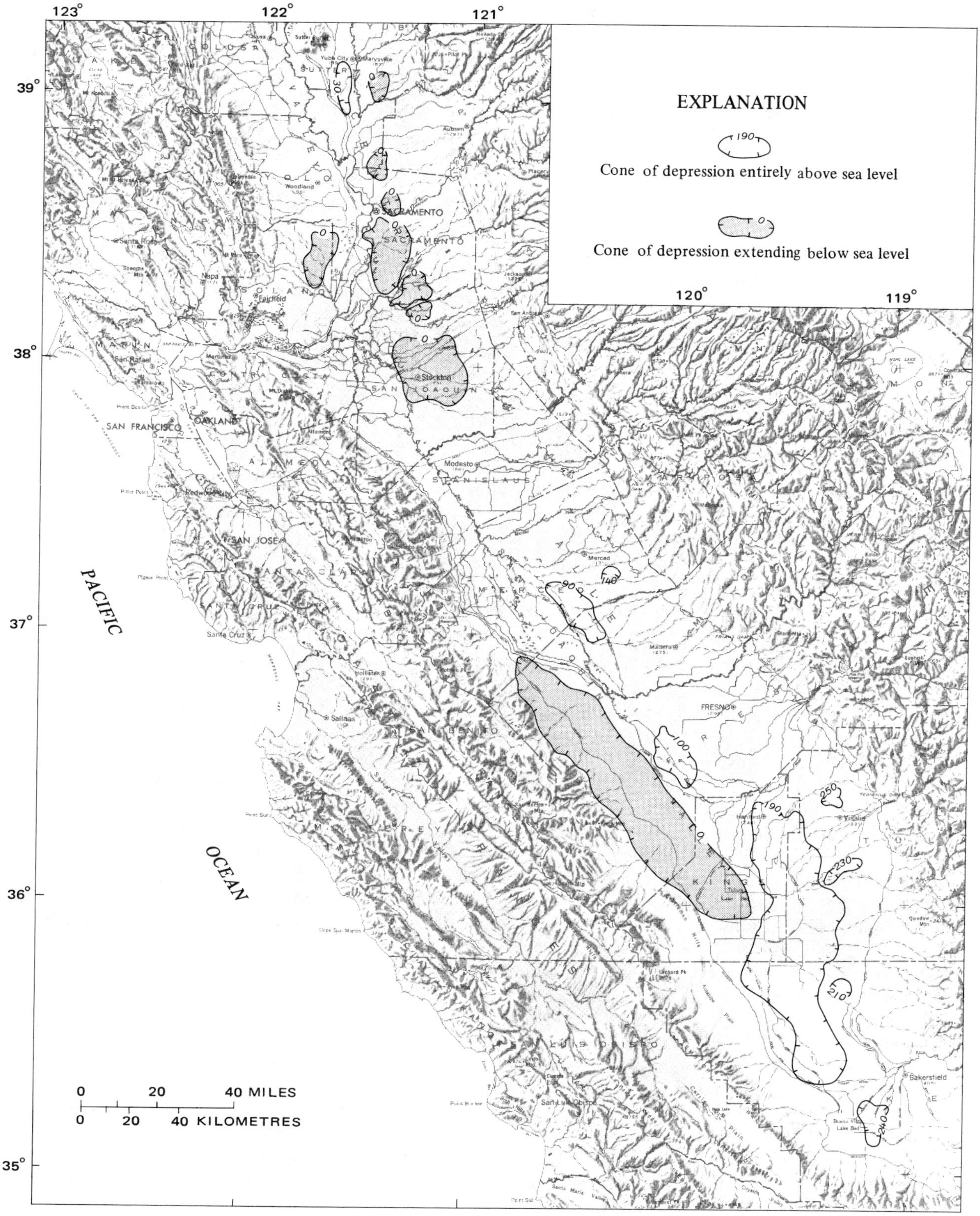

FIGURE 6.—Cones of depression caused by pumping from wells in the Central Valley.

In some ground-water reservoirs, depletion of storage has created other problems, as exemplified by the coastal plain (Basin No. 8–1) in Orange County (fig. 7). In 1944, after several wet years, the potentiometric surface had a natural gradient toward the ocean, with two small depressions caused by pumping near the coast. Ten years later the water levels in wells had declined moderately 10–30 feet (3–9 m), and then were below sea level under practically all the coastal plain. Thus the reservoirs were opened to seawater intrusion. (See section "Seawater Intrusion.")

LAND SUBSIDENCE

Significant land subsidence due to withdrawal of ground water has occurred in areas totaling about half of San Joaquin Valley (State Basin No. 5–22 on pl. 1) and half of Santa Clara Valley (No. 2–9) at the south end of San Francisco Bay. In Antelope Valley (No. 6–44), releveling indicates subsidence as great as 3 feet (90 cm). These areas of subsidence overlie confined aquifer systems in which the artesian head has been drawn down more than 100 feet (30 m) and in places more than 325 feet (100 m). The aquifer systems contain permeable sand and gravel interbedded with and overlain by silty and clayey materials which become compacted as water is pumped out, and the land surface sinks. The water squeezed out by compaction is permanently withdrawn, but because compaction of the permeable beds is slight, the effect upon the usable capacity of the aquifers is negligible. Subsidence may be occurring in other areas of pumping from confined aquifers in the California Region, but it is not sufficient to be observed without releveling.

In Santa Clara Valley, releveling in 1933 to bench marks in San Jose produced evidence—the first in the United States—of subsidence caused by ground-water withdrawal. As summarized by Poland (1969), land subsidence occurred where pumping drew down the artesian head in the confined aquifer system, but not where the water was unconfined. The subsidence ceased during the wet period 1936–43, when water levels rose as much as 80 feet (25 m), and resumed in 1948 after the artesian head declined below the 1935 minimum. The volume of land subsidence from 1934 to 1967 is computed to have been half a million acre-feet (0.6 km^3), equivalent to about 10 percent of the total pumpage in the 33-year period. Water imported from the Central Valley through the State's South Bay aqueduct became available in 1965, and annual imports increased to more than 120,000 acre-feet (150 million m^3) by 1970. As a result of the increasing imports and decreasing pumping, plus increased natural and artificial recharge, the artesian head in more than a hundred wells rose an average of about 56 feet (17 m) from 1967 to 1970, and the land subsidence practically ceased; nevertheless, no rebound, or aquifer expansion, took place. Records of subsidence in relation to water-level trends, pumpage, and imports in Santa Clara Valley are given in figure 8.

San Joaquin Valley has three principal areas of land subsidence due to ground-water withdrawals (Poland and others, 1973), the largest encompassing about 1,500 square miles (4,000 km^2) along the west slope of the valley between Los Banos and Kettleman City (fig. 9). Thousands of irrigation wells, many yielding more than 1,500 acre-feet (1.8 million m^3) a year, pump water from confined aquifers 300–3,000 feet (90–900 m) deep and have caused water levels to decline as much as 450 feet (140 m). The maximum subsidence exceeded 28 feet (8.5 m) by 1970; because the volume of subsidence was more than 8.1 million acre-feet (10 km^3) and the estimated gross pumpage about three times as great, approximately one-third of the water withdrawn has been derived from compaction of sediments.

In the Tulare-Wasco area of about 800 square miles (2,100 km^2), pumping from 1930 to 1951 caused as much as 10 feet (3 m) of subsidence, while water levels declined 230 feet (70 m). After importation of water through the Friant-Kern Canal began in 1951, pumping decreased in parts of the area, and by 1970 land subsidence had virtually ceased in about one-third of the area. In the Arvin-Maricopa area at the south end of San Joaquin Valley, at least 400 square miles (1,000 km^2) of irrigated land has subsided at rates of as much as 0.5 foot (15 cm) a year, and subsidence is continuing. This subsidence also is caused by pumping from wells in confined and semi-confined aquifers, chiefly since 1940, which has caused water levels to fall 3–13 feet (1–4 m) a year and as much as 400 feet (120 m) since 1929.

For all of San Joaquin Valley, the volume of subsidence, and therefore the reduction in pore space caused by ground-water depletion, exceeded 11 million acre-feet (13½ km^3) by 1966.

Not all the land subsidence in the Central Valley is related to ground-water withdrawal. Subsidence in some areas has occurred where the soil had never been thoroughly dry and in other areas where the soil had never been thoroughly wet until man came along with his reclamation of lands for agricultural use (Poland and Evenson, 1966, p. 244). The wetlands were in the delta at the confluence of the Sacramento and San Joaquin Rivers, where islands have been drained for cultivation and protected by levees; with the lowering of the water table, the organic soils have subsided 6–15 feet (2–5 m) because of drying, oxidation, shrinking, and wind erosion. At the other extreme, the drylands are the clayey alluvial-fan soils along the west side of the San Joaquin Valley, which have been moisture deficient ever since their deposition and which have subsided

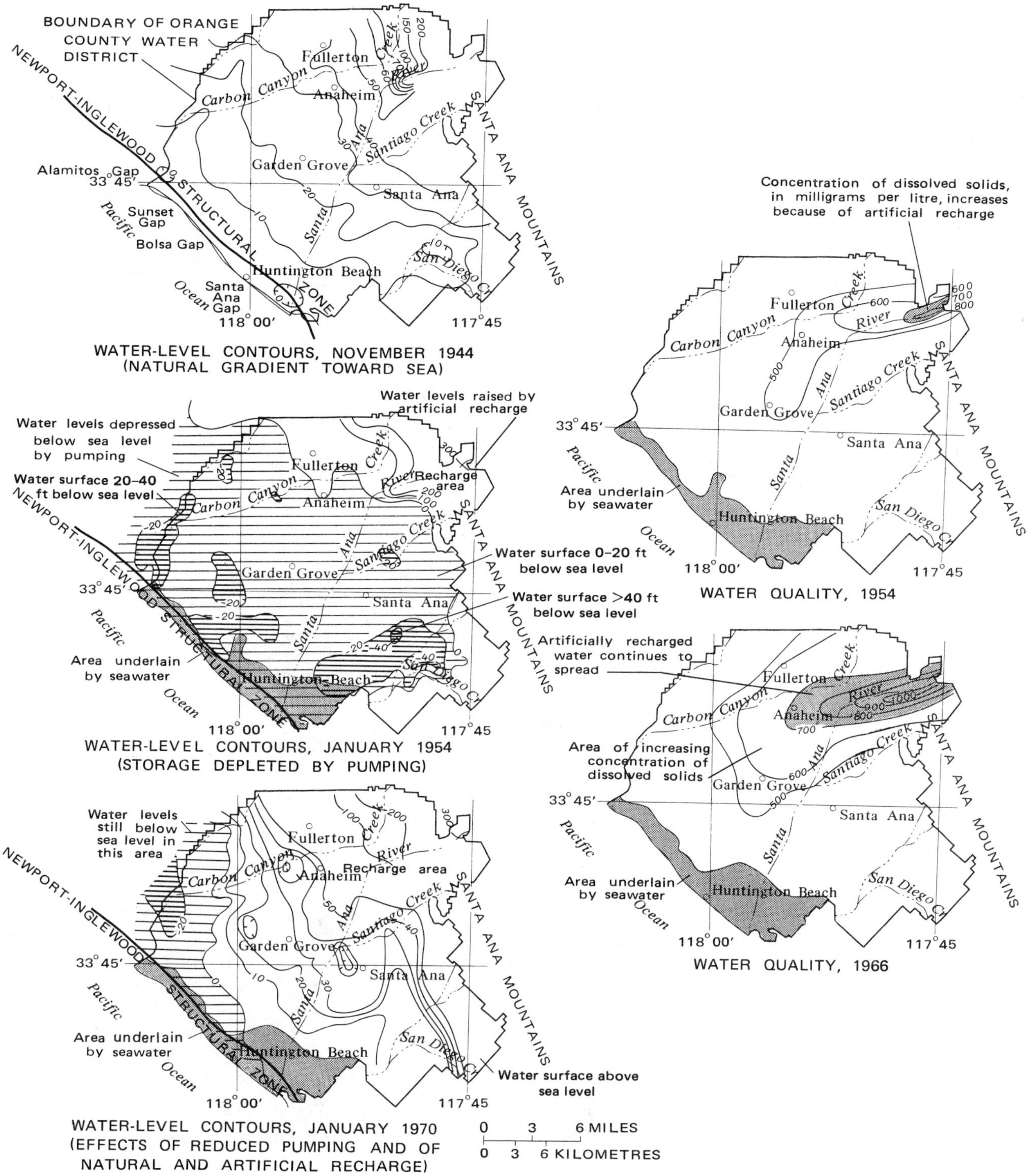

FIGURE 7.—Depletion and recovery of ground-water storage, Orange County coastal plain.

3–15 feet (1–5 m) in response to the first irrigation of the land.

Subsidence has caused failure of many wells owing to rupture of casings by the compaction. Also, subsidence

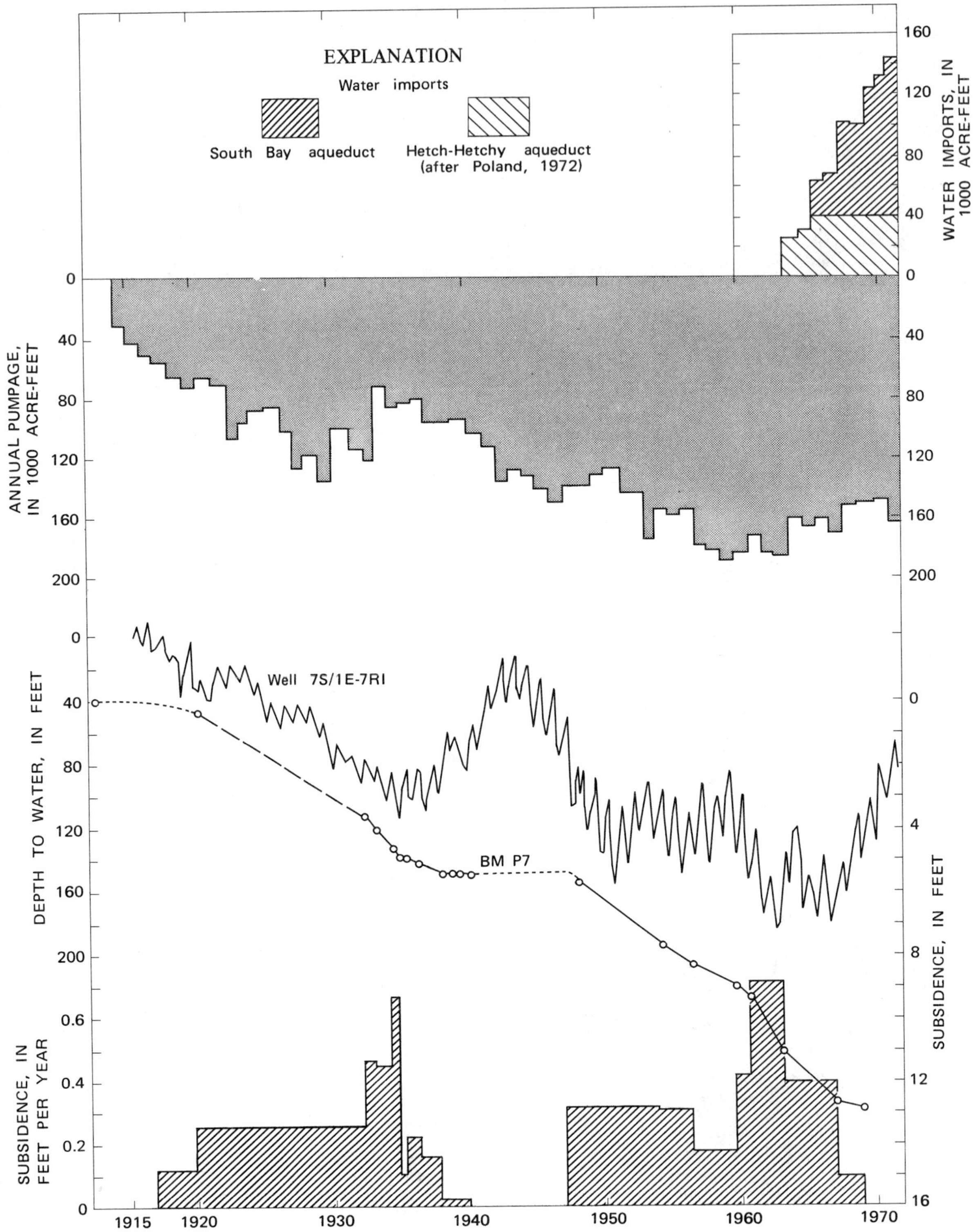

FIGURE 8—Pumpage and imports of water in Santa Clara Valley and correlative water-level changes and land subsidence, 1915–70.

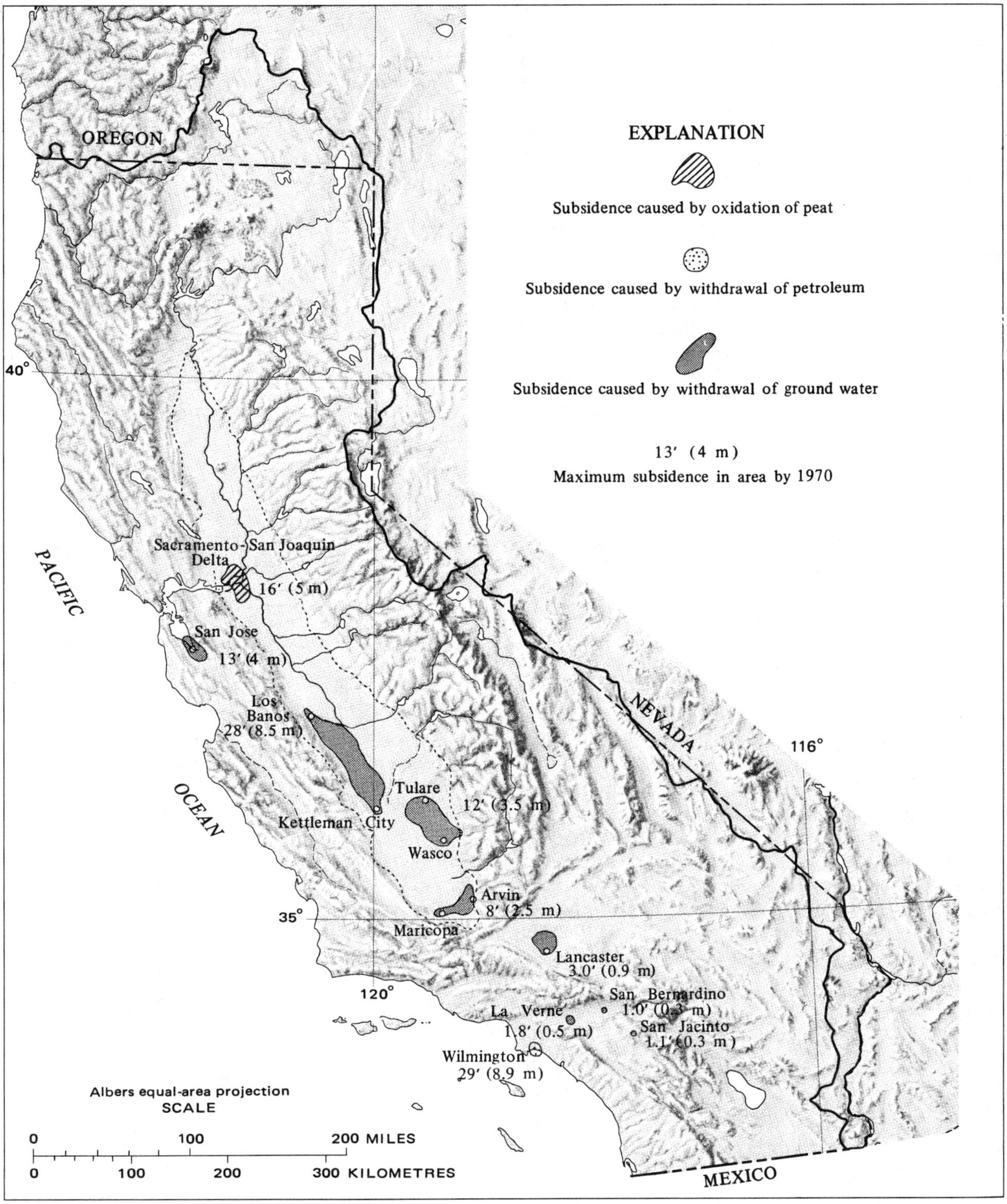

FIGURE 9—Areas of land subsidence in the California Region.

may increase the difficulty or expense of using the land, particularly along the seacoast if the land subsides below sea level.[3] However, the subsidence of Santa Clara Valley, although exceeding 10 feet (3 m) at some places, was less than 3 feet (1 m) along the south shore of San Francisco Bay. This made wetlands wetter, deepened the water in sloughs and channels entering the bay, and reduced the gradient and gravity flow of drains and pipes discharging into the bay.

The subsidence caused by ground-water depletion is generally slight in comparison to the breadth of area involved, and residents may be unaware of any change except as shown by releveling. Subsidence has probably occurred in several valleys in California where water levels have been lowered tens of feet, but has been noticed only where it disturbed canals or pipelines. For example, the Metropolitan aqueduct has dropped as much as 3 feet (1 m) near La Verne in San Gabriel Valley Basin (No. 4–13), and the San Diego aqueduct has dropped 1 foot (30 cm) near San Jacinto in San Jacinto Valley Basin (No. 8–5) (Miller and Singer, 1971).

In San Joaquin Valley the area of greatest subsidence is not in the trough of the valley, where it might affect the drainways of Fresno Slough and San Joaquin River, but along the west slope where streamflow is ephemeral. Here, both the Delta-Mendota Canal and the Friant-Kern Canal have required remedial work because of subsidence. Although the California Aqueduct passes through the centers of most rapid westside subsidence, future subsidence was estimated prior to construction so that protection from anticipated subsidence could be built into the aqueduct freeboard. Pumping from wells has approached 2 million acre-feet (2½ km^3) annually along the west slope of San Joaquin Valley, and the subsidence will continue while pumping continues. Water deliveries from the aqueduct to the Federal San Luis project (in the subsiding area) have increased from 200,000 acre-feet (0.25 km^3) in 1968 to 650,000 acre-feet (0.8 km^3) in 1971, and because pumping of ground water has decreased, the artesian head in the main pumped zone had risen an average of 56 feet (17 m) by December 1970 (Poland, 1972). Upon completion of the distribution systems, a maximum of 1.2 million acre-feet (1.5 km^3) can be imported into the San Luis project area.

[3]The most spectacular land subsidence in the California Region occurred in the Long Beach harbor area (State Basin No. 4–11 on pl. 1), not because of ground-water withdrawal but because of withdrawal of oil and gas in the Wilmington oil field (Poland and Davis, 1969). First noticed in 1940, the subsidence increased by 1962 to a maximum of 27 feet (8.2 m) at its center on Terminal Island and included an area of 25 square miles (65 km^2) that had subsided 2 feet (0.6 m) or more. The cost of remedial measures such as levees and fill to keep the sea from invading the subsiding lands and the repair of several hundred oil wells had exceeded $100 million by 1962, but the value of petroleum produced was far greater. Beginning in 1958, the oil zones were repressured by injecting saline water, which stopped the subsidence in much of the field by the end of 1962, and appreciable rebound was noted at several bench marks.

In summary, research and experience in areas of significant land subsidence provide a fair understanding of the causes and processes of that subsidence. Where subsidence is caused by ground-water withdrawals, reduction or cessation of withdrawals can slow down or halt the subsidence. Both in southern San Joaquin Valley and in Santa Clara Valley, pumping from wells has been reduced as substitute supplies became available from canals, and the problem of subsidence has been alleviated.

SEAWATER INTRUSION

Seawater intrusion occurs only where conditions are favorable to landward or upward movement of seawater, conditions that develop where ground-water levels are lowered below sea level by pumping from wells. Along the California coast, 263 ground-water basins or areas, large and small, (California Department of Water Resources, 1958a) are contiguous to the sea or to saline inland bays; as of 1955, evidence of seawater intrusion was found in 11 of these. In 70 other areas the ground water near the coast was more saline than water farther inland; eight of these areas had extensive ground-water development, and pumping from wells had lowered water levels below sea level in some places so that intrusion may have been suspected but was not proved. In 48 areas pumping from wells was not known to have reversed the natural seaward ground-water gradient, and no evidence of seawater intrusion was found. In 134 minuscule ground-water basins there was nearly no ground-water development and no likelihood that the natural seaward gradient was modified.

The 11 areas where seawater intrusion was confirmed in 1955 are given in table 4, which summarizes the location, extent, source, and human response to the saline intrusion. The northermost of these sites is around San Francisco Bay, where about half the runoff from the California Region encounters seawater. Every day during high tides, ocean water enters the bay through the Golden Gate, and the bay is characteristically saline as far east as the Carquinez Straits, but during the greatest of historic floods (1862), flow was continuous out of the bay into the ocean, and San Francisco Bay had freshwater fish for several months (Harding, 1960). In Suisun Bay, east of the straits, the water flowing from the Central Valley during the 19th century was naturally fresh enough to drink. With increasing diversions for irrigation, the freshwater flow diminished, especially in dry years such as 1924 and 1931, and incursions of saline water into the channels and sloughs of the delta occurred as far upstream as Stockton and Walnut Grove. During the 1924–34 drought, wells on the Pittsburg Plain (Basin No. 2–4 on pl. 1) pumped water whose fluctuations in salinity

TABLE 4.—*Seawater intrusion in 1955*

State No.	Ground-water reservoir	Aquifer intruded: Name and depth (m)	Aquifer intruded: Area (km²)	Aquifer intruded: Distance from sea (km)	Source of seawater	Solution
	San Francisco Bay subregion					
2–1	Petaluma Valley	Shallow	4	16	Tidal channels	Affected wells abandoned.
2–2	Napa Valley, Sonoma Valley	Alluvium <30	8	19	do	Do.
2–3	Suisun-Fairfield Plain	Alluvium	81	8	do	Do.
2–4	Pittsburg Plain	Alluvium >30	32	1½	Suisun Bay	Do.
2–9	Santa Clara Valley:					
	South Bay	Shallow	243	6	San Francisco Bay	Do.
	East Bay	Newark 20–45	162	10	do	210 wells abandoned.
	East Bay	Centerville 60	12	--	Shallow aquifer	50 wells abandoned, 30 defective wells repaired.
	Central Coastal subregion					
3–2	Pajaro Valley	75	16	1½	Pacific Ocean	50 wells abandoned.
3–4	Salinas Valley	55	40	4	do	100 wells abandoned, 20 wells deepened.
3–4	do	120	16	3	do	
	South Coastal subregion					
4–4	Oxnard Plain	Oxnard 45	32	3	Pacific Ocean	45 wells abandoned; injection wells; extraction wells.
4–11	Los Angeles Coastal Plain	Gaspur 60; Silverado 120.	65	3 5	do do	Artificial recharge; freshwater ridge; reduced withdrawal.
8–1	Orange Coastal Plain	Talbert 25	40	5	do	Artificial recharge; reduced withdrawal; combination barrier.
9–7	San Luis Rey Valley	<30	.8	--	Tidal channel	Artificial recharge.

correlated with those in a nearby slough of the Sacramento River (Tolman and others, 1931). The Contra Costa Canal now supplies water for industrial use in this area, and most of the wells are no longer in service. Farther west, along the north shore of the bay in Napa, Sonoma, and Petaluma Valleys (Basins Nos. 2–1, 2–2), seawater has intruded into pumped aquifers by infiltration of surface water in tidal channels, rather than by subsurface inflow from the bay.

In Santa Clara Valley along the south arm of San Francisco Bay, intrusion of salt water has been sufficiently widespread to cause abandonment of most wells in the shallow aquifer, but numerous shallow wells still yield usable water, indicating that the aquifer is heterogeneous in water quality as well as permeability characteristics. East of the bay in the vicinity of Niles and Hayward, saline water has evidently advanced 2–5 miles (3–8 km) into the shallow (Newark) aquifer on a broad front. The deterioration of quality in the deeper (Centerville) aquifer is spotty and has been attributed to faulty or abandoned wells that bring water from the Newark aquifer or to leakage through the separating aquicludes (California Department of Water Resources, 1960a, 1968d). At several of the monitored wells in this East Bay area, the salinity of water increased progressively during the 1960's, and the area of degraded water in the Centerville aquifer expanded slightly during 1970 (Alameda County Water District, 1971). The total pumpage in 1970, however, was about 15 percent less than the average prior to 1965, and recharge has been augmented artificially since 1963. (See section "Recharge in Urban Areas.") The 1970 water levels in several observation wells were 15–65 feet (5–20 m) higher than the minimums registered in 1961 or 1962 but were still below sea level (California Department of Water Resources, 1971a).

In Pajaro Valley along the Central Coast, the variation in quality of water from well to well in the producing aquifer indicates that some saline water may enter cones of depression from tidal channels or sloughs or ponds. In the Salinas Valley the confined aquifer near the seacoast appears to include segregated lenses of fresher and saltier water, responsible for variations in quality of water from wells (California Department of Water Resources, 1949, 1970f).

In the Oxnard Plain in the South Coastal subregion,

seawater intrusion continued at apparent rates of as much as 1,000 feet (300 m) a year in 1961 and 1962, especially at Port Hueneme and Point Mugu (California Department of Water Resources, 1965a). Many wells tapping the shallow (Oxnard) aquifer were abandoned in favor of wells in deeper aquifers not yet reached by seawater. The Ocean City Municipal Water District has injected water into the Oxnard aquifer through a line of wells in the "pumping trough" inland from the invading seawater (Price and Baker, 1963), using water from wells about 8 miles (13 km) farther inland; this program achieved an effect similar to rearrangement of pumping pattern, with no reduction in total withdrawal. An experimental extraction-type barrier was constructed near Port Hueneme in 1966 and operated for 2 years (California Department of Water Resources, 1970b), during which five wells extracted 9,000 acre-feet (11 million m^3) of brackish water and reduced the area underlain by degraded waters.

Along the San Luis Rey River in San Diego County where seawater intrusion was first noted in 1938 (California Department of Water Resources, 1950), the importation of water from the Colorado River has enabled the City of Oceanside and suburban agriculturists to restrict their pumping from wells, and reclaimed sewage is used to recharge the ground water; by achieving an approximate balance between recharge and discharge, the ground-water reservoir is apparently holding its own against seawater invasion, although pumping has drawn water levels below sea level in a trough 2–6 miles (3–10 km) from the ocean. Along the San Diego River salt-water encroachment from Mission Bay, noted in 1906, caused abandonment of wells in the Old Town pumping field. In the valleys of the San Dieguito and Tia Juana Rivers farther south, water levels near the coast are drawn seasonally below sea level, causing some seawater intrusion.

The coastal plain in Los Angeles and Orange Counties has been the scene of the most serious seawater intrusion and most comprehensive counter measures in the California Region. For more than half a century, the ground-water reservoir was pumped for the requirements of a progressively increasing population (Poland, 1959). By 1953, water levels in wells were below sea level in extensive areas (fig. 7), numerous wells near the coast had been abandoned because of increased salinity, and brackish water had been reported in some wells as much as 8 miles (13 km) inland (California Department of Water Resources, 1957b, 1958a). Not all the salt came from seawater, for beginning in the 1920's oil-field brines in several localities were dumped in ponds and surface channels and in some areas wells have discovered connate water. Most of the seawater intrusion has occurred west of the Newport-Inglewood zone of faulting and uplift, which generally impedes ground-water movement. This zone is less than 2 miles (3 km) from the coast in Orange County but as much as 7 miles (11 km) inland in Los Angeles County. The battle against the invading seawater is one phase of water-management activities by local agencies of countywide jurisdiction. (See section "Recharge in Urban Areas.") These agencies coordinate their anti-invasion efforts at Alamitos Gap along the county line but operate separately elsewhere.

In the coastal plain in Los Angeles County, pumpage has been reduced from 354,000 acre-feet (436.5 million m^3) in 1961 to court-limited extractions of 281,835 acre-feet (347.5 million m^3) since 1963 (Bookman and Edmonston, 1971). In 1970, however, water levels in wells reaching the main (Silverado) aquifer were still as much as 100 feet (30 m) below sea level in extensive areas, as low as they had been 10 years earlier. From 1953 to 1970, water spread for artificial recharge aggregating 1.6 million acre-feet (2 km^3) raised levels in extensive areas, but not in the West Basin, which is west of the Newport-Inglewood zone and where all the seawater intrusion has occurred.

Although the water levels in most of the West Basin are still below sea level, the sea has been repelled and cannot enter the main aquifer because of a continuous barrier ridge of freshwater created by importation of water from the Colorado River and injection in a line of 93 wells extending 11 miles (18 km) south from the Los Angeles International Airport to Palos Verdes Hills. By 1961, when 28,000 acre-feet (34.5 million m^3) had been injected into 12 wells, a wedge of invading seawater had been cut off and diluted, and in 1970, after injections had aggregated about 300,000 acre-feet (370 million m^3), water in the aquifer east of the barrier was generally of usable quality. Meanwhile, barriers are being similarly designed and developed to halt seawater invasion in Dominguez Gap east of Palos Verdes Hills and in Alamitos Gap at the Orange County line.

In Orange County most ground water seaward of the Newport-Inglewood barrier zone is saline in shallow aquifers and brackish in the main (Silverado) aquifer 200 feet (60 m) or more below sea level. Freshwater naturally moved across this barrier through Alamitos, Sunset, Bolsa, and Santa Ana Gaps, forming stream channels, alluvial aquifers, peat bogs, swamps, and marshes as it moved (California Department of Water Resources, 1968a). The direction of flow was reversed during the 1950's, when the water levels were below sea level in wells at the coast and for more than 6 miles (10 km) inland (fig. 7). As corrective measures, annual pumpage was substantially reduced, and Colorado River water was imported and spread for artificial recharge (Crooke and Toups, 1962). As of November

1970, water levels in wells were above sea level in all parts of the county except in the vicinity of Alamitos Gap, where a barrier ridge was maintained to repel further intrusion of seawater (Cofer, 1971). At the Santa Ana Gap, the county's combination barrier project (California Department of Water Resources, 1966b) includes a line of wells in the Newport-Inglewood zone about 2 miles (3 km) from the sea which extract brackish water and return it to sea and also a line of wells 2 miles (3 km) farther inland where freshwater is injected to create a barrier ridge.

In the Central Valley, a large body of saline water of marine origin underlies the freshwater aquifers of San Joaquin Valley at depths generally greater than 1,000 feet (300 m) but is as little as 400 feet (120 m) deep in the Modesto-Turlock-Waterford area, where gas wells yield water with 4,000 mg/l (milligrams per litre) dissolved solids including 2,400 mg/l chloride (California Department of Water Resources, 1960b). South of the Delta-Mendota canal near Firebaugh and Mendota, water pumped from wells has 1,700 mg/l dissolved solids, 350 mg/l chloride, and 750 mg/l sulfate. In the islands of the Delta, fresh ground water may be in lenses surrounded by brackish water at depths less than 100 feet (30 m) (California Department of Water Resources, 1965c). These may be connate waters but modified in composition during subsequent millennia so that they are different from modern seawater. In the vicinity of Stockton, many wells obtain freshwater from shallow aquifers, but studies in the 1950's (California Department of Water Resources, 1955b) indicated little or no vertical mixing of freshwater with underlying saline water. Nor was there migration of the saline water into the cone of depression formed by pumping for irrigation farther east, perhaps because of the extractions of saline water for industrial use in the western part of Stockton.

In the developed ground-water reservoirs where seawater intrusion was suspected in 1955 (fig. 10), no notable crises have developed in subsequent years. Along the North Coast, Crescent City and Eureka have freshwater in aquifers above and below sea level and salty water in sediments adjacent to bays and tidal sloughs. Wells can yield freshwater perennially from aquifers that are replenished by major rivers as they flow to the sea, although their yield will become salty if pumping induces inflow of ocean water, but this has not happened to a significant degree. Water-well standards have been recommended (California Department of Water Resources, 1966a) to minimize the vulnerability of freshwater aquifers to deterioration by water of inferior quality. Along the Central Coast some wells at Santa Cruz are close enough to the sea to be similarly vulnerable. Farther south, pumping in Arroyo Grande Valley can induce seawater intrusion, but it has not happened yet (California Department of Water Resources, 1970g).

Along the South Coast, three small ground-water reservoirs underlie narrow coastal plains in the vicinity of Santa Barbara. During the drought of 1945–55, ground-water levels were lowered below sea level in these basins; however, the reservoirs are separated from the Pacific Ocean by impermeable rocks along the coast, and thus no breakthrough of seawater occurred. With completion of Cachuma Dam on the Santa Ynez River in 1955 and deliveries of water through the Tecolote Tunnel, pumping from these ground-water reservoirs has been reduced, and they constitute reserves available whenever surface supplies are diminished unduly by drought.

The ground-water reservoir under the Lompoc Plain near the mouth of the Santa Ynez River is not known to be separated from the Pacific Ocean by any barrier. Pumping for irrigation has been greatest in years of drought and minimum streamflow, notably 1946–51 and 1959–61, but even in these years the water level in wells nearest the coast was at least 3 feet (1 m) above sea level. Water has been diverted from the Santa Ynez River basin for municipal supply in the Santa Barbara area; the ground-water reservoir gains when the Cachuma reservoir stores flood water in wet years (1958, 1962, 1969), and it is released at rates suitable for ground-water replenishment. The Santa Maria Valley farther north is similar in many respects, for it has no natural barrier to seawater intrusion and substantial pumping for irrigation has not depressed water levels below sea level. The Twichell Reservoir in the headwaters provides controlled releases of water that increase the ground-water replenishment. Thus, these ground-water reservoirs of Santa Barbara County have been rendered less vulnerable to seawater intrusion by integrated management and conjunctive use of surface-water and ground-water resources.

RESERVOIR POLLUTION

Broadly speaking, pollution and contamination have about the same meaning. The American Heritage Dictionary (1969) states: "Pollution: 1. The contamination of soil, water, or the atmosphere by the discharge of noxious substances." The laws to control pollution or protect quality of water, however, include definitions that may differentiate between pollution and contamination in order to identify human perils or human responsibilities more specifically.[4] The Federal Water

[4]The California Water Code (sec. 13050) defines pollution as "an alteration of the quality of the waters of the state by waste to a degree which unreasonably affects (1) such waters for beneficial uses, or (2) facilities which serve such beneficial uses." Contamination is defined as "impairment of the quality of waters of the state by waste to a degree which creates a hazard to the public health through poisoning or through the spread of disease."

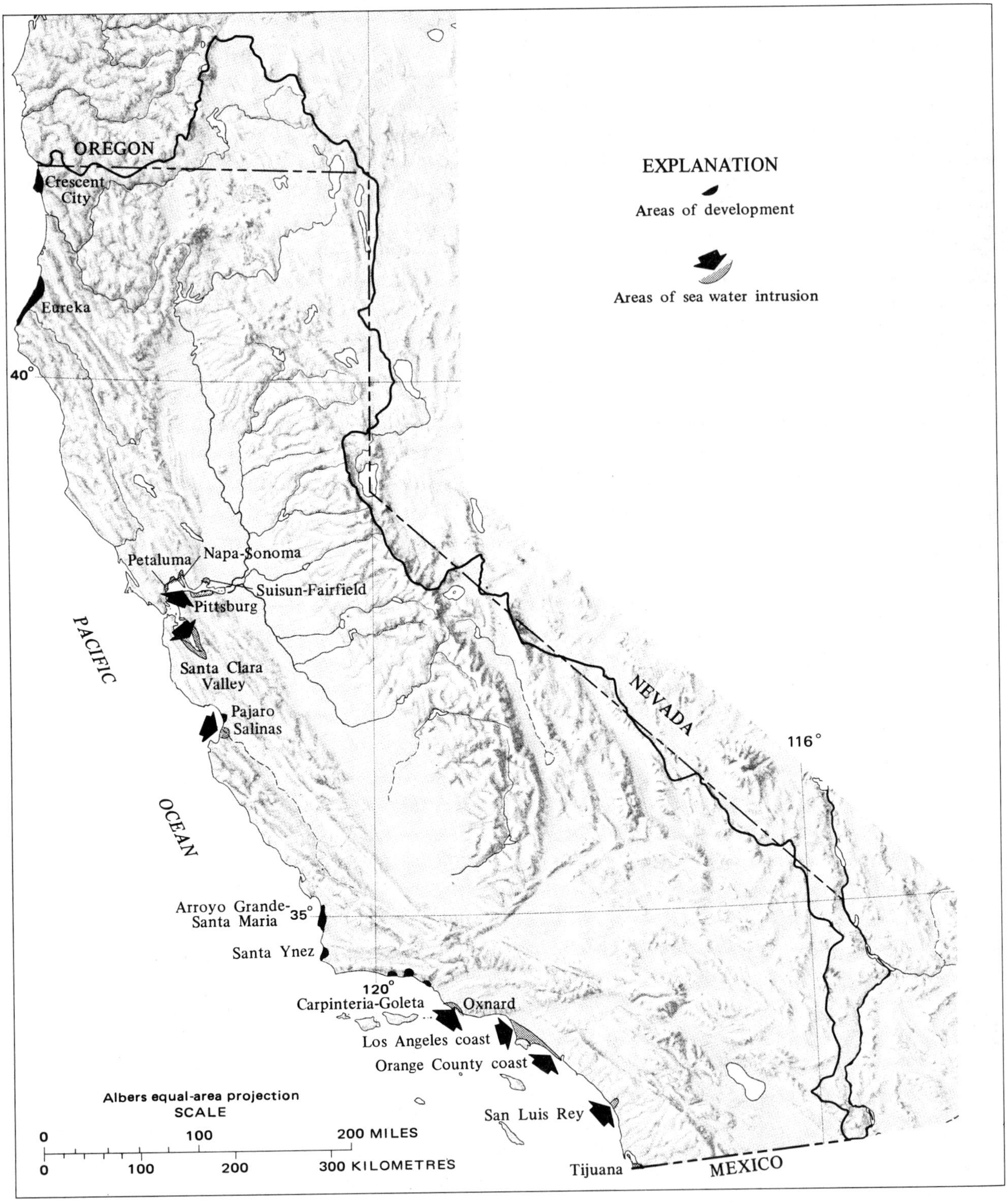

FIGURE 10.—Seawater intrusion and coastal ground-water reservoirs.

Pollution Control Act Amendment of 1972 (PL 92–500) states in its Title V, section 502:

(19) The term "pollution" means the manmade or man-induced alteration of the chemical, physical, biological, and radiological integrity of water.

(6) The term "pollutant" means dredged spoil, solid waste, incinerator residue, sewage, garbage, sewage sludge, munitions, chemical wastes, biological material, radioactive materials, rock, sand, cellar dirt, and industrial, municipal and agricultural waste discharged into water.

The Safe Drinking Water Act of 1974 (PL 93–523), included as Title XIV of the Public Health Service Act, is concerned with protection of public health from harmful impurities in water, whether manmade or man induced or natural, and it states in its section 1401: "(9) The term 'contaminant' means any physical, chemical, biological, or radiological substance or matter in water."

In this appraisal, in accordance with these definitions, "contaminants" include all impurities in water, and "contamination" is the process of "impurification," whether by act of man, act of nature, or act of God; "pollutants" are limited to the impurities for which man is responsible, and "pollution" is the process by which man contributes to deterioration of water quality.

Ground water has been polluted in many places in the California Region by improper construction, use, or abandonment of wells (United States Public Health Service, 1961), and to alleviate this problem minimum well construction and sealing standards have been published (California Department of Water Resources, 1968b). Pollution of ground-water reservoirs generally is difficult to detect or trace because of the deterioration of water quality is likely to be gradual in time and space. Also, many of the human activities and processes that cause pollution have their counterparts in natural processes that cause deterioration of water quality. Thus the identification, quantification, and description of pollution by mankind requires background knowledge of the natural deterioration of the water resources.

NATURAL DETERIORATION

Degradation is a geologic process, a general lowering of the earth's land surfaces by erosion chiefly by water of the lithosphere where it is exposed to the atmosphere. Within the range of temperatures at the earth's surface, water may exist as solid, liquid, or gas; more than 97 percent is habitually in liquid form, in the oceans, forming the hydrosphere that covers more than 70 percent of the lithosphere, and another 2 percent is solid, in polar icecaps. The process of degradation is accomplished by solar energy that evaporates water from the oceans and carries it in air currents over the continents and islands, where it falls as rain or snow upon the land surface. Then the forces of gravity may move it to run off overland or to infiltrate into the soil and percolate downward to become ground water, or to flow in streams or underground and eventually reach an ocean. In the course of this hydrologic cycle, water may break solid rock by freezing and thawing, dissolve rock materials and carry them in solution, dislodge small particles and carry them in suspension, undercut larger fragments and lubricate their downhill passage, move larger rocks along the beds of streams, and thus move the lithosphere, piece by piece and ion by ion, to lower levels and ultimately to the ocean unless epeirogenic or orogenic forces intervene.

The degradation of the continents and islands is a gradual, progressive, and inexorable process carried on by water in the hydrologic cycle. As to the water, in its plan of salvation, one part in 10,000 evaporates each year and returns to earth purified. This is the preferred water among people, because they are among the terrestrial life that has become adapted to freshwater. But in the process of degradation of the land, there is a natural deterioration of the water—by accumulation of impurities—for which man is not responsible and which is therefore not properly termed pollution.

The atmosphere habitually contains only about seven of each million molecules of the water on earth, but every year about 380 molecules per million are evaporated into the atmosphere and then dropped again as precipitation. The water of precipitation is characteristically the purest water in the hydrologic cycle, but even so it may collect from less than 1 to several hundred milligrams of dissolved material per litre of water during its fall through the atmosphere. At the earth's surface some of this water may infiltrate into the ground, and because of water's capacity for rock weathering and organic reactions in the presence of oxygen and carbon dioxide, as soil moisture it is characteristically more mineralized than the water of precipitation.

In the degradation process, overland runoff and flow in streams are the dominant factors in erosion and in transportation of sediment, floating materials, dissolved gases, and organisms. Ground water, on the other hand, generally is the prime gatherer of soluble minerals, and the solids it dissolves depend mostly upon the chemistry of the minerals contacted and the already present dissolved load. In both surface water and ground water the solutes of the degradation process correlate with amount of precipitation; that is, for the United States as a whole, Rainwater (1962, pls. 1, 2) showed that streams draining areas of perennial water surplus generally contain calcium magnesium bicarbonate waters with less than 340 mg/l dissolved solids. This "normal" condition occurs during periods of low flow, when much of the water in streams is base flow from ground-water reservoirs; therefore, an abundance of water is able to dilute the products of natural degradation, and people have long counted on—or at least hoped for—such dilution as a solution to their pollution of water.

Like the rest of the country, the California region has water of high quality in streams whose headwaters are in areas of perennial water surplus. Ground water is also of high quality and has low mineralization in those places that receive runoff from the areas of surplus. Tables 1 and 2 include columns giving the range of dissolved solids in the well water analyzed in each reservoir. In all but 14 of the 134 reservoirs, at least some wells obtain water that meets the U.S. Public Health Service recommended standard for drinking water of not exceeding 500 mg/l dissolved solids. These water resources of low mineral content are highly prized and have been developed for public supplies by the largest cities in the region (table 5). Throughout the area of perennial water surplus and in many of the valleys that receive those surpluses, ground water has generally less than 300 mg/l dissolved solids, chiefly calcium, magnesium, and bicarbonate (fig. 11).

Areas of perennial water deficiency characteristically have insufficient outflow to the ocean to carry all the solids that have been dissolved in the water, and extensive areas of this kind are in closed basins devoid of outflow. Increasing aridity is reflected in increasing mineralization of the ground water, and the closed basins in the desert may contain some highly saline ground water. In these basins the waters may be changed significantly by meteorological extremes: An exceptional storm may produce flood waters to be stored on the surface or underground, and this water may dissolve salts that had been precipitated or may dilute waters already accumulated, as shown by periodic analyses of waters from some wells in the Mojave River valley (State No. 6–40) (Miller, 1969).

Unfortunately, any water near the land surface may be lost by evapotranspiration, leaving dissolved minerals more concentrated in the water remaining or residual mineral evaporites. For example, in an area of perhaps 15,000 square miles (40,000 km^2) in southern California and Nevada, the water that moves toward the lowest part of Death Valley generally increases in salinity during this movement and exceeds 10 times seawater salinity near Badwater, the lowest point of a saltpan occupying 200 square miles (500 km^2); moreover, the changes in composition of the brines as they move into the saltpan are generally those expected from the solubilities of each salt in pure water (Hunt and Robinson, 1966). Some variations in composition are traceable to various source rocks, as for example the cottonball borates that were transported by 20-mule teams; these have been derived from such streams as the Owens River and the Mojave River, which carry small quantities of dissolved boron from mountain headwaters into the desert.

The South Lahontan and Colorado Desert subregions of eastern California have many closed basins with saltpans (pl. 1) smaller than Death Valley but similar to it in that they are underlain by salt and briny ground water; many are in the midst of extensive areas where the ground water occurs in similar saline and brackish zones (California Department of Water Resources, 1964c). Some zonation is also evident in the ground

TABLE 5.—*Megalopolitan water supplies*

[Adapted from Durfor and Becker (1964)]

Source	City	Population served (millions)	Chemical constituents, in milligrams per litre							
			SiO_2	HCO_3+CO_3	Ca+Mg	SO_4	Na+K	Cl	Hardness as $CaCO_3$	Dissolved solids
1	San Francisco	1.6	2	4	1	1	1	1	3	10
2	Oakland	1.0	10	22	9	0	3	5	23	40
3	Sacramento	.2	21	69	17	7	11	8	52	110
4	Los Angeles	1.4	31	123	29	23	35	16	81	212
5	Fresno	.15	68	118	26	6	21	6	82	221
6	Long Beach	.2	20	187	14	13	72	20	37	229
7	Orange County	.5	--	155–270	60–165	35–255	30–100	15–105	170–470	285–875
8	San Jose	.4	31	237	69	23	30	32	203	335
9	San Diego	.3	12	74	95	290	104	92	282	666

Sources and date of sample:
1. Hetch Hetchy Reservoir on Tuolumne River, 7/8/61.
2. Pardee Reservoir on Mokelumne River, 1/30/62.
3. Filter plant on Sacramento River, 7/26/61.
4. San Fernando Reservoir, Owens Valley aqueduct, 7/25/61.
5. Composite of 58 wells, San Joaquin Valley, 7/19/61.
6. Composite of 35 wells, Los Angeles coastal plain, 7/60–7/61 average.
7. Range in 130 wells, Orange coastal plain (California Department of Public Health, 1962), 1961–62.
8. Composite of 119 wells, Santa Clara Valley, 7/20/61.
9. Alvarado Treatment Plant, Inc., San Diego River and Colorado River, 1/24/62.

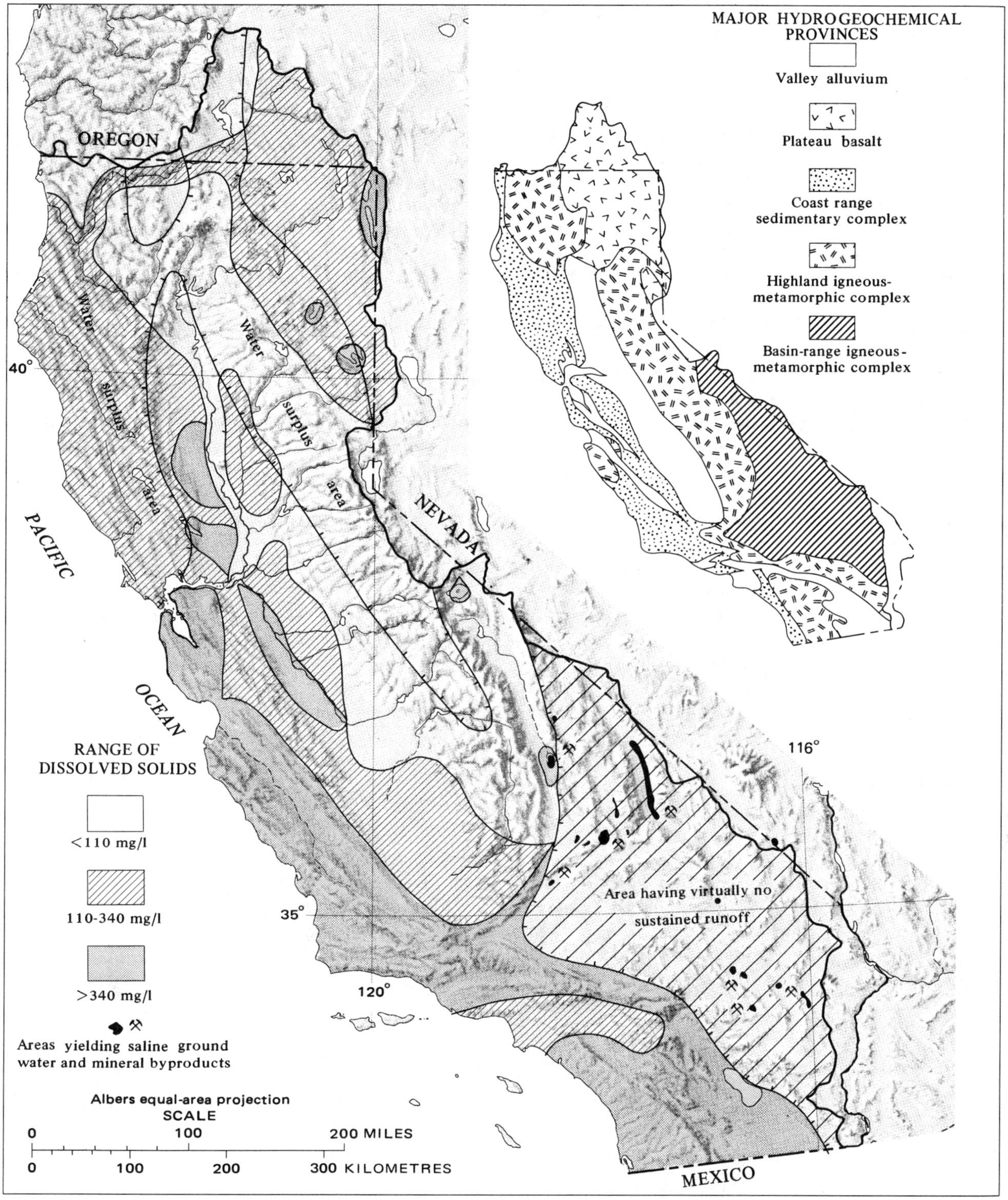

FIGURE 11.—Chemical characteristics of the base flow of streams

water in the closed Tulare Basin at the south end of the Central Valley, where water from the Sierra Nevada is a calcium bicarbonate type with less than 100 mg/l dissolved solids; sodium, sulfate, and chloride increase to the west and are dominant in the mineralized waters along the west slope. Although saltpans might be expected in the lowest areas, the Tulare and Buena Vista lake beds, outflow has evidently been sufficient to carry the more soluble salts over into the San Joaquin River, even during the past century; however, even though the river is throughflowing, sodium, chloride, and sulfate have accumulated in ground water in the western and lowest parts of San Joaquin Valley. A significant increase of chlorides also is found in the ground water along the lower western part of the Delta. A reconnaissance study of minor tributaries to the San Joaquin and Sacramento Rivers in the lower part of the valley (Richardson and Rantz, 1961) indicates that the alluvial materials are sufficiently permeable to permit significant interchange of surface water and ground water; consequently ground-water recharge may be at rates as great as 3 feet (1m) per day, and some seasons the streams gain comparable amounts from ground-water discharge. In this area of abundant evaporites and saline residues, this interchange of waters contributes to the deterioration of both.

In the four subregions along the Pacific Coast, outflow to the ocean from each drainage basin is likely to occur at some time during the winter rainy season; however, in summer most streams cease flowing for several weeks or months, and the larger streams generally contain only the base flow discharged from ground-water reservoirs, except where there are releases for fish or navigation, where irrigation water is stream transported, or where return flows augment stream flow. Significant concentrations of sulfate, sodium, and (or) chloride are characteristic of base flows of streams draining the areas of perennial water deficiency south of San Francisco Bay.

Several coastal streams have two or more ground-water reservoirs within their drainage basins, separated by impermeable rocks and canyons but connected by streams so that the discharge from one ground-water reservoir may become the recharge for another. Nowhere are chemical analyses found to document adequately either the natural hydrologic cycle or the pollution achieved by man, but the analyses may help in determining the relative contributions of streamflow, local precipitation, and artificial recharge to ground water in storage. Thus along the Napa River (table 6) the midvalley and downvalley ground waters, perhaps because of some recharge from precipitation, are more dilute than those near Calistoga, which are in marine sediments. In Salinas Valley also the dissolved solids in ground water decrease downvalley, where the precipitation is greater. But in several basins along the South Coast, where water deficiency is more pronounced and perennial, the ground water downstream has greater dissolved solids than in midstream well fields, which in turn are more mineralized than ground waters farther upstream.

In the Santa Ana River drainage basin, salinity of the confined and semiconfined ground water increases downstream to the Santa Ana Canyon. At the entrance to the canyon, ground water is forced to the surface by bedrock and becomes available for infiltration downstream. In 1954, the ground water underlying most of the inland sector of the coastal plain of Orange County, including the recharge area below Santa Ana Canyon, ranged from 400 to 600 mg/l dissolved solids. But in 1966, the dissolved-solids concentration in much of the basin was greater than 600 mg/l and exceeded 1,000 mg/l in the area of artificial recharge below Santa Ana Canyon (Moreland and Singer, 1969). The quality of ground water deteriorated because of artificial recharge by water imported from the Colorado River and by the base flow of the Santa Ana River.

AGRICULTURAL POLLUTION

As a natural effect of water in contact with rocks and soil, dissolved solids are commonplace in natural waters. However, when arid lands are irrigated with water from surface reservoirs created in the desert or with water imported from other river basins, these dissolved solids remain in the water, not consumed by

TABLE 6.—*Dissolved solids in ground water along coastal streams*

[Values in milligrams per litre; data from California Department of Water Resources, 1949, 1958b, 1964a, 1967b, 1970a]

Subregion	Upstream well field	Midstream well field	Downstream well field
San Francisco Bay:			
Napa River	Calistoga, 550–600	Rutherford, 200–250	Napa, 300–500.
Central Coastal:			
Salinas River	King City, 1,000–1,500	Gonzales, 1,000–1,300	Nashau, 500–700.
Santa Ynez River	Cachuma, 550–575	Buellton, 800	Lompoc, 1,500–2,000.
South Coastal:			
Santa Clara River	Piru, 700–900	Santa Paula, 800–1,000	Oxnard, 800–900.
Upper Santa Ana River	San Bernardino, 200–300	Riverside, 500–700	Corona, 850–1,000.
Lower Santa Ana River	Santa Ana Canyon, 1,000	Garden Grove, 650–700	Bolsa, 300–500.
San Luis Rey River	Pauma Valley, 300–350	Rancho SLR, 550–1,500	San Luis Rey, 1,500–3,000.

evaporation and transpiration, and thus increase the concentration of salts in the soil, in water percolating to a ground-water reservoir, and in surface-return flows. In this way both the quantity and quality of water resources are modified. Irrigation, even though it may simulate natural processes, tends to deteriorate the water, and thereby man causes pollution.

In irrigation—the dominant use of ground water in the California Region—the water used nonconsumptively may percolate downward to the water table or it may run off in ditches and stream channels. It may carry minerals dissolved from the soil zone as well as identical minerals dissolved by infiltrating rainfall or overland flow during the rainy season. In the Lompoc area of Santa Ynez Valley, Evenson (1965) estimated that since 1960 the withdrawal for irrigation and return by percolation may have doubled the concentration of chloride and sulfate ions in the shallow ground water. In this and other basins having seasonal outflow of water, artificial storage and the use of that water for irrigation can accelerate the natural degradation process, increasing the salinity of the shallow ground water or of the surface water in periods of low flow. Although human activities have doubtless contributed to the deterioration of shallow ground water in many places, how much manmade pollution has been added to natural contamination cannot generally be determined with certainty.

About one-twelfth of the natural runoff of the California Region comes from the San Joaquin River drainage basin, but the outflow is only half as great because of evapotranspiration. Degradation of ground water and surface water as they move through the valley is indicated by data summarized in table 7. As shown, the flow of the San Joaquin River below Friant Dam decreases by natural losses and diversions for irrigation and then is increased somewhat by return flows, both quantity of water and content of dissolved solids. The dissolved load continues to increase downstream, but the concentration is less because of dilution by tributary inflows. Although some of this degradation is natural, some results from the use of the water for irrigation, and the increases in recent years both in chlorides and in dissolved solids indicate increasing agricultural pollution by man. The ground water is of highest quality along the east slope and has the greatest content of dissolved solids in the lower part of the valley.

Beginning in June 1951, the Central Valley project has imported water from the Delta into lower San Joaquin Valley; in the first 7 full years of operation, the Delta-Mendota Canal carried 6.8 million acre-feet (8.3 km^3). In several places this imported water has diluted the surface water or the ground water in the lower San Joaquin Valley, but the ultimate effect of the increased

TABLE 7.—*Degradation and pollution of water in San Joaquin Valley*
[Based on averages compiled from California Department of Water Resources (1960b, p. 116, 134–141); quantities in milligrams per litre]

A. Surface Water (San Joaquin River)

Locality	Chloride		Dissolved solids	
	1938–51	Increase by 1955–59	1938–51	Increase by 1955–59
Friant Dam	3		35	
Mendota (average flow decreases 75 percent; imported water enters river)	3	+60	45	+245
Fremont Ford (average flow increases 50 percent	120	+110	430	+260
Newman (average flow doubled by inflow of Merced River)	85	+85	355	+270
Grayson (average flow doubled by inflow of Tuolomne and Stanislaus Rivers)	100	+40	400	+120

B. Ground Water

Locality	Chloride prior to 1951	Dissolved solids	
		Prior to 1951	Change by 1955–59
Madera	40	280	None
Firebaugh	90	370	
Dos Palos	350	800	+200
Los Banos	100	500	
Gustine	55	530	+100
Los Banos Creek	425	1,270	
Turlock	40	300	None
Hills Ferry	380	1,000	−130
Patterson (shallow)	400	1,900	None
Patterson (deep)	90	1,000	None

supply for irrigation of these arid lands has been greater evapotranspiration and greater degradation of the ground water in the valley and of the surface-water outflow (fig. 12). The measured increases (table 7) of dissolved solids and chloride since 1951 may be evidence of pollution traceable to man's development and use of the water resources. Nevertheless, too little is known of the patterns and rates of flow of the water in the area to be certain.

Imports from the Colorado River to the California Region generally carry more than 800 mg/l dissolved solids, perhaps twice as much as the native water in most areas where the imported waters are used. Imperial Valley, the chief importer for irrigation, receives about a ton of salts in each acre-foot of water. To maintain a salt balance suitable for crops, it is necessary to irrigate with enough water to flush the salts from the soil through an intricate system of drains to eventual disposal in the Salton Sea. In Coachella Valley an underlying ground-water reservoir was the sole source of water for irrigation prior to 1948 and yielded about 100,000 acre-feet (125 million m^3) in 1937, chiefly from artesian wells southeast of the town of Indio where the valley is below sea level. Imports from the Colorado River began in 1948 and increased to more than 300,000 acre-feet (370 million m^3) after 1952, of which more than 20 percent soon reached the Salton Sea via surface drains (California Department of Water

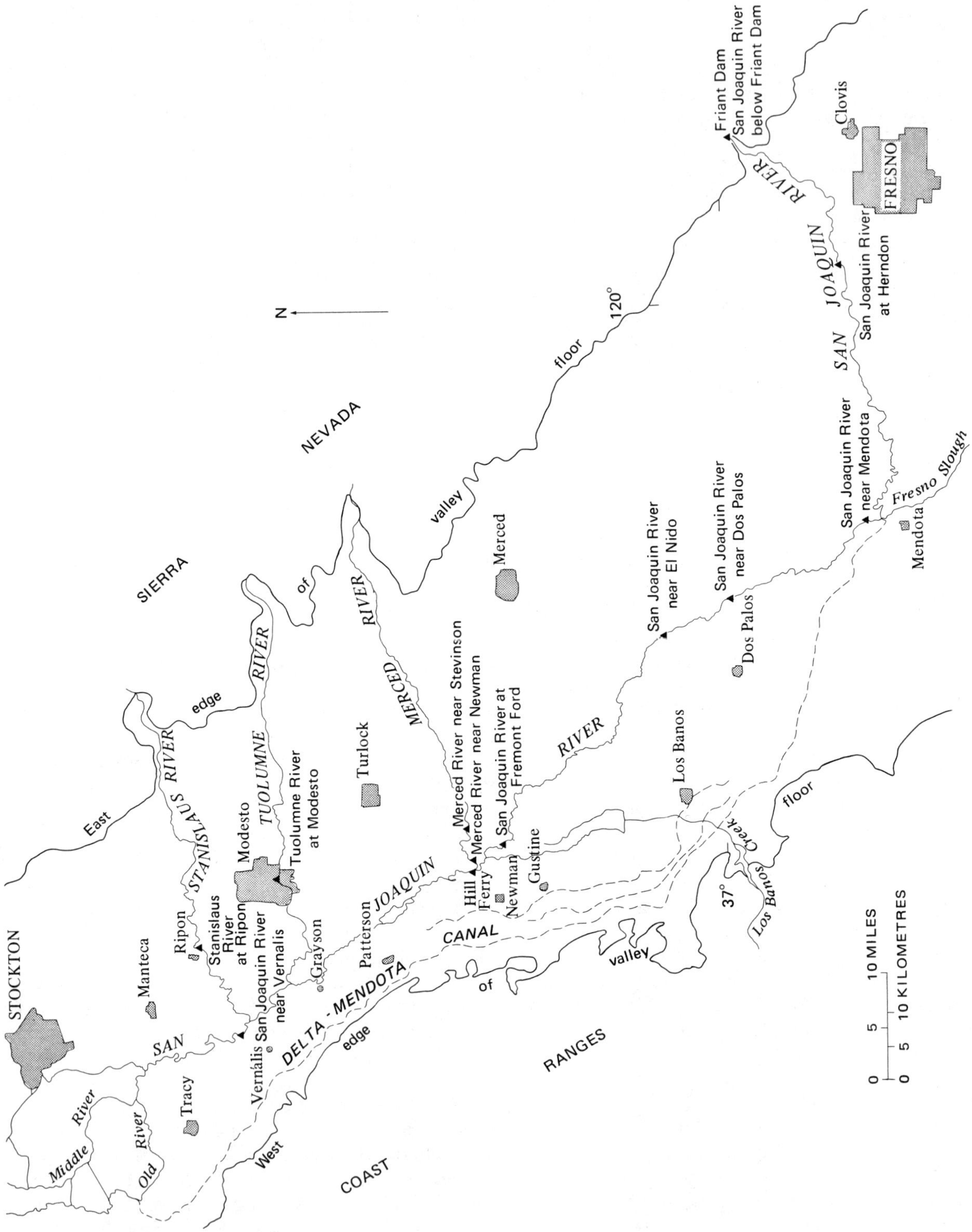

FIGURE 12.—Cities, towns, rivers, and gaging stations along the San Joaquin River.

Resources, 1964d). The progressive decline of water levels in artesian wells was slowed or halted as withdrawals from wells were reduced. Ground-water storage was increased by downward percolation of the imported water into a semiperched zone above the confined aquifer; however, with the mineralization of the Colorado River and the addition of the residual salts from irrigation and fertilizers, the shallow ground water now contains dissolved solids exceeding 3,000 mg/l and is unsuitable for domestic or agricultural use.

Agriculture has many byproducts and end products—at food packing and processing plants, feed lots, barnyards, and septic tanks—which may pollute water. Some are noxious, others foul smelling, but all are generally biodegradable, eventually being used in other living processes or reduced to their basic components of oxygen, nitrogen, water, and carbon dioxide. Notable exceptions are the persistent chemicals used as fertilizers, pesticides or herbicides, or cleaning agents, including some that are very soluble in water and others only slightly so. Presumably some of these chemicals applied over the land surface are dissolved in water that percolates downward and into ground-water reservoirs, and those used in households may similarly percolate from septic tanks. But quantitative data are meager and are limited to studies of small areas where people have been concerned about possible health hazards.

An example of an investigation of agricultural pollution is that concerning nitrates in ground water in the Tulare Basin of the Central Valley. Nitrate concentration exceeds U.S. Public Health Service drinking water standards in water from numerous wells in the irrigated areas of the Tulare Basin, especially from the Kaweah River southward to Bakersfield and beyond (California Department of Water Resources, 1970h). The town of Delano is in the midst of this area and is exceptional in the investigation and monitoring that has been undertaken to discriminate the agricultural pollution from the natural degradation of the water. As reported by the California Department of Water Resources (1968f), water from many wells near Delano have nitrates in excess of 45 mg/l. For half a century wells were the sole source of water for irrigation, and water levels were lowered as much as 250 feet (75 m) by 1950; however, since 1951, the Friant-Kern Canal has provided some of the water for irrigation, the rate of pumping has been reduced, and water levels have recovered as much as 165 feet (50 m). Fertilizers applied in the Delano area at annual rates as great as 1,000 pounds per acre (or kilograms per hectare) are suspect as a likely source for the nitrates, which are most concentrated in areas (1) that have been irrigated for the longest period of time and (2) under which ground water has risen markedly in recent years as surface water has been applied for irrigation. On the other hand, the nitrates may be of natural origin, at least in part, for analyses have indicated excessive concentrations of nitrate in several wells in 1945 and earlier years. Elsewhere in the Central Valley excessive nitrate concentrations are far less common, although the geologic and hydrologic conditions are similar. However, excessive nitrate has been reported in several ground-water reservoirs along the central coast (California Department of Water Resources, 1971d).

Public concern over the effects of pesticide use led to a study of the chlorinated hydrocarbon DDT, applied to a test plot of alkaline clay loam near Dos Palos in western San Joaquin Valley (California Department of Water Resources, 1968h). Although the land had never been cultivated prior to the test, the soil contained fairly high concentrations of DDT to depths of more than 6 feet (2 m), probably from nearby windblown aerial spray. The test plot was flooded three times for periods of 75–125 days each and at intervals of 9 months. Before the first two floodings only, DDT was applied at rates of 2–4 pounds per acre (or kilograms per hectare), slightly higher than normal for the area; with the third flooding rice was grown and harvested. The effluent from the first two floodings, both as surface runoff and in tile drains, contained considerably less than 1 part per billion of DDT, indicating its slight solubility and strong affinity for the soil, which retained most of it. The effluent after the third flooding, however, contained about 10 times as much DDT as that after the second flooding, an ominous indication of accumulation, stability, and long-term response of the toxic substance. The study was unfortunately not continued long enough for a full evaluation of the relative significance of leaching and of degradation by biological and chemical reactions. No comparable studies have been made in recharge areas of ground-water reservoirs to show the extent of entrainment of chlorinated hydrocarbons by downward percolating water.

URBAN POLLUTION

The problems of pollution of water in urban areas are not unlike some of the pollution problems in rural areas. Although a concentration of people produces a large volume of biodegradable wastes, they cannot be broken down within the space and time available on a city lot; also, the importation and use of mineral and chemical substances are likely to increase the volume and variety of dissolved materials in the water in urban areas, as do the chemicals spread over the land or utilized in households in rural areas.

In many rural areas of sufficient permeability, dug wells and septic tanks have been commonplace for domestic water supply and disposal, the septic tank holding the waste water until depleted of its organic materials by chemical reduction and then releasing it to the soil zone and perhaps to ground water. Although this is a usable recycling system, it requires cleaning and rehabilitation, becomes less satisfactory in suburban crowding, and thus is generally replaced by sewerage systems in urban regions. If the sewerage system is restricted to collecting pipes for raw sewage and does not include plants for primary and secondary treatment of the sewage, it is inferior to the septic tanks it replaces, for even though it may carry away the wastes of individual landowners, pollution abatement is nil. The organic wastes then may pollute streams and canals, soils, lakes and ponds, bays and estuaries, and the ocean but probably not ground water, because most living organisms cannot survive for long the anaerobic conditions as in a ground-water reservoir.

Standard sewage treatment is intended to remove most of the biodegradable components but not the dissolved inorganic solids. As a general average or rule of thumb (McGauhey, 1971), domestic use of water is likely to add about 300 mg/l to the original dissolved solids, and if the water is recycled, another 300 mg/l is added. This manmade increase in dissolved solids constitutes pollution of the resource that receives the sewage effluent, whether stream or lake or ground water. Studies in several areas (table 8) show sewage effluent, for the period of record, to have increased the maximum observed dissolved-solids load in ground-water reservoirs by 1,200 mg/l. In the study of San Bernardino (California Water Resources Control Board, 1965) undertaken when housewives everywhere were using the persistent foaming ABS (alkylbenzenesulfonate) detergent, water-quality factors were excellent tracers of ground-water movement.

In some urban areas where the shallow ground water is more mineralized than that at greater depth, the salinity distribution may have resulted from natural processes of degradation, expecially in areas of perennial deficiency where salt residues from evaporation are redissolved by storm waters as they infiltrate into the ground or in areas where irrigation could have caused leaching and downward migration of dissolved minerals. In Ventura County (Basin No. 4–4 on pl. 1) and coastal Los Angeles County (No. 4–11), the quality of water in shallow aquifers is generally inferior to that at greater depth (California Department of Water Resources 1965b, 1968c; California Department of Public Health, 1968). The relative importance of pollution and natural degradation was not determined, but strict water-well standards were recommended to prevent influx of inferior water to the deeper aquifers.

Garbage, rubbish, and other solid wastes (known collectively as refuse) are potential pollutants of the water if they are soluble. Most refuse disposal in the California Region is in sanitary landfills operated by cities, towns, and counties. Many of these are in areas overlying ground-water reservoirs, and the pollution of ground water by decomposition products is a hazard. The effects of solid-waste disposal upon the quality of ground water in several localities in southern California are summarized in table 8 from studies by the

TABLE 8.—*Ground-water reservoir pollution by municipal wastes*

[Data for sewage treatment plants from California Department of Water Resources, 1965c, 1968g; California Water Resources Control Board, 1961, 1965; Miller, 1969. Data for sanitary landfills from California Department of Water Resources, 1969b]

State No. (pl. 1)	Valley	Year	City	Pollutants	Maximum observed increase (mg/l)	Maximum observed extent	
						mi	km
			Sewage-treatment plants				
4–11	Coastal plain	1945	Alhambra	Phenols	+0.1	6	10
6–40	Lower Mojave	1960	Barstow	Phenols	+1.1	4	6
4–14	Upper Santa Ana	1963	San Bernardino	Alkylbenzenesulfonate	+2.5	10	16
5–22	San Joaquin	1963	Fresno	Dissolved solids	+400	2	3
				NO_3	+30		
				Alkylbenzenesulfonate	+.3	.1	.2
3–15	Santa Ynez	1966	Lompoc	Dissolved solids	+200	2	2
				Cl	+50		
			Sanitary landfills				
4–14	Upper Santa Ana	1952–54	Riverside	Dissolved solids	+500	0.6	1
				Cl	+190		
4–13	San Gabriel	1958–61	Monrovia	CO_2	+1,200	.3	.5
4–13	San Gabriel	1962–65	Azusa	Dissolved solids	+160	.3	.5
				CO_2	+120		
				NO_3	+15		
4–12	San Fernando	1962–65	Glendale (Scholl Canyon)	Dissolved solids	+1,200	0	0
				Fe	+250		
				Mn	+1,200		
				Zn	+1,000		

California Department of Water Resources (1969b).

INDUSTRIAL POLLUTION

Most industries depend upon municipal systems for their water supply and disposal, and about half the volume of sewage treated in municipal facilities comes from industrial sources. The pollutants from numerous industries, for example, food processing, are mostly biodegradable, but other pollutants, such as those included in table 8 are troublesome to incorrigible in these treatment processes.

Many large industrial plants have developed their own water-supply and disposal facilities, which may include treatment before use and before return of the used water to public resources. The principal industrial use of water in the California Region is for cooling, especially for thermal-electric power generation which uses more than 12 million acre-feet (15 km^3) annually, 90 percent of which is brackish or saline (Murray, 1968, p. 30). In self-supplied manufacturing, industries reportedly have used about 917,000 acre-feet (1.1 km^3) of freshwater annually (California Department of Water Resources, 1964e), but in 1965 the estimated total (Murray, 1968) was only 530 mgd (0.75 km^3). These figures on industrial use are very tentative, but even less is known about industrial water pollution, for the subject has received little attention from industry and is of apparently little interest to the general public.

The California Department of Water Resources survey of industrial uses in 1957–59 showed that the principal uses of freshwater in manufacturing were approximately 25 percent for food and drink processing, 22 percent for lumber and wood products, 16 percent for petroleum refining, 10 percent for chemical products, 8 percent for metals and machinery, and about 8 percent for paper products. One-third of the manufacturing use of water was in southern California, and almost as much in the San Francisco Bay area.

Examples of industrial waste pollution of ground water are common but generally lack the quantitative data essential to evaluate the pollution. Three examples of pollution by disposal of water-softener regeneration brines have been reported (California Water Resources Control Board, 1961) and are summarized in table 9.

Solid wastes in the California Region amount to about 4½ pounds (2 kilograms) of refuse daily per capita, of which 15 percent is garbage, 30 percent paper and cartons, and 10 percent glass and metal containers. This personal refuse, however, is only about 10 percent of the total solid wastes of an industrial civilization, and most of these solid wastes are disposed of on lands owned or leased by the individual companies. Such landfill may make land of natural lakes, bays, and streambeds; plains or smooth slopes in rough country; or relief on level lands. But by modifying the land surface, these landfills also modify the hydrologic cycle in the locality, and soluble material may be dissolved and pollute ground water or surface water.

MINERAL-RESOURCE DEVELOPMENT

The development and use of the land is primarily concerned with resources other than water, including the soils for agriculture, fossil fuels, metallic and other minerals for manufacturing, rocks and sediments for construction, and the land surface for occupancy. In such development water becomes a byproduct that may be of some benefit or may be detrimental (Thomas, 1969). Water may constitute a nuisance if it harms the development and use of the earth's resources and invades recognized property rights. Conversely, the development and use of the land resources may have

TABLE 9.—*Ground-water reservoir pollution by industrial byproducts*

[Data from California Department of Water Resources, 1955a, 1968g; California Water Resources Control Board, 1961; Piper and Garrett (1955); Williams and Wilder (1971)]

State No.	Valley	Locality	Year	Pollutants	Maximum observed increase (mg/l)		Maximum observed extent		Remarks
							mi	km	
4–4	Oxnard–Santa Clara.	Saugus	1942–49	Water softener regeneration brine.	Dissolved solids	+5,000	0.4	0.6	Quality "normal" by 1953.
5–22	San Joaquin	Fresno	1953	do	do	+1,400	.1	.2	
					Cl	+750			
					Na	+400			
3–15	Santa Ynez	Lompoc	1952	do	Dissolved solids	+400	1	1.6	
4–11	Coastal plain	Dominguez Gap.	1928–48	Oil-field brine	do	+3,200	1	1.6	
					Cl	+1,700			
					Na	+600			
4–11	Coastal plain	Norwalk	1953	do	Dissolved solids		1	1.6	
					Cl				
					Na				
5–22	San Joaquin	Raisin City.	1948–53	do	Dissolved solids	+40	1	1.6	
					Cl	+15			
			1954–55	do	Cl	+2,600	.1	.2	Increase in same well in 7 months.
4–12	San Fernando	Glendale (Forest Lawn).	1968–71	Gasoline (broken pipe).			.3	.5	190 m^3 free gasoline removed; remanent taste and odor.

effects on water that interfere with the rights of others or that result in a nuisance.[5]

Ground water occurs with several minerals and fuels of economic value and may be useful in the development or processing of the resources, although it must eventually be disposed of somewhere. Some of these waters are of a quality suitable for other water users, but many are so mineralized that they would pollute the water in streams or ground-water reservoirs.

PETROLEUM

In California during 1959 (Musser, 1959, tables 1, 5, 6, 7), about 39,000 wells yielded 310 million barrels (50 million m^3) of crude petroleum, along with 800 million barrels of water generally saline enough to be classed as brine. About one-third of this brine was returned underground through 465 wells, either for water-flooding in the producing zone or for disposal. Sixty-nine thousand acre-feet (85 million m^3) of brine was left to be dumped into seepage ponds, stream channels or dry valleys, or other convenient location. In some places these least-cost methods of disposal had continued for decades, but in others they soon became unacceptable to water users and were discontinued. In many localities the ground-water reservoirs have been polluted by oil-field brines, and quantitative data are available for some, as summarized in table 9. The brines for surface disposal in San Joaquin Valley alone had aggregated 55,000 acre-feet (69 million m^3) in 1959; in several oilfields, the native ground water ranges from mediocre to unfit for use (California Department of Water Resources, 1956), but other oilfields, of which the Raisin City field is an example (California Department of Water Resources, 1955a), have been developed by going through the ground-water reservoir.

By 1972 most of California's oil-field brines were being recycled, increasing the yield of petroleum and reducing the pollution of ground-water reservoirs (California Division of Oil and Gas, 1972, tables 1, 5, 6, 7). In that year 39,600 wells produced 325 million barrels (52 million m^3) of oil and 1,700 million barrels (280 million m^3) of brine, but 95 percent of that brine was returned underground, 1,300 million barrels (210 million m^3) was injected through 11,800 wells for water flooding or as steam and hot water for secondary oil recovery, and an additional 340 million barrels (54 million m^3) was returned in 270 disposal wells. That left only 60 million barrels or 8,000 acre-feet (10 million m^3) of brine as a potential pollutant to freshwater resources.

EVAPORITES AND BRINES

Several desert basins east of the Sierra Nevada and the high mountains farther south (fig. 11) contain saline deposits of economic value, including common salt (sodium chloride), trona (hydrous sodium carbonate), salt cake or Glaubers salt (sodium sulfate), potash (potassium carbonate), borax (sodium borate), and bromide and lithium salts. These highly soluble salts were dissolved perhaps in a dominantly volcanic terrane by water at times and places of surplus and brought by water into an area of perennial water deficiency, where by natural processes of degradation the water became increasingly unfit for use and eventually disappeared. Some of the oldest deposits are the borax beds at Boron, accumulated in Miocene lakes 20 million years ago, dried and kept dry in subsequent time. In the present drainage systems, the degradation process is continuing today. The most favorable time for accumulation in these desert basins, however, was during the pluvial-glacial stages of the Pleistocene Epoch, when water was sufficiently abundant to form numerous large and deep lakes in which long-continued evaporation left saline deposits that extended to considerable depth in some basins.

Every basin of interior drainage is likely to have a playa in its lowest part where floodwater can accumulate with its clastics and then evaporate or where ground water is shallow enough to evaporate and leave soluble salts. Economic value of these salts has depended partly upon the degree to which they have been separated from clastics and from each other. Four playas within a hundred kilometres of the Colorado River have extensive crystal bodies that are practically free of clastics (Ver Planck, 1958). At Danby Lake in Ward Valley (State No. 7–3), nearly pure halite was quarried from 1890 intermittently until 1942. Rock salt was produced at Bristol Lake in Bristol Valley (No. 7–8) beginning in 1909; however, the brine there also has calcium chloride which is readily separated from the common salt by solar evaporation, and this has been the sole product since 1931. Cadiz Lake (No. 7–7) 22 miles (35 km) to the southeast has a similar sodium chloride–calcium chloride brine, but none has been produced. On the other hand, Dale Lake (No. 7–9) 25 miles (40 km) south of Bristol Lake has abundant sodium sulfate with the chloride and is readily separated at low temperature while the common salt

[5]As broadly defined by the California Water Code (sec. 13050), a nuisance is "anything which (1) is injurious to health or is indecent or offensive to the senses, or an obstruction to the free use of property, so as to interfere with the comfortable enjoyment of life or property, and (2) affects at the same time an entire community or neighborhood or any considerable number of persons, although the extent of the annoyance or damage inflicted upon individuals may be unequal, and (3) occurs during or as a result of the treatment or disposal of wastes." The nuisance factor of water is not within the scope of this appraisal of water resources.

remains in solution; both commodities were produced in the 10 years ending in 1948.

Several other playas have histories of commercial production of common salt. In Saline Valley (No. 6–17) freshwater was used to dissolve the rock salt, which then was recrystallized in place, but production ceased after 1930, victim of transportation costs and the Depression. Generally the desert producers of salt have lost out to the producers of salt by evaporation of seawater close to cities. One survivor in the desert has been at Koehn Lake in Fremont Valley (No. 6–46), nearest to the Los Angeles area, where salt is recovered from inflowing floodwaters that become brine and then salt residue by solar evaporation; also, brine comes from wells as much as 100 feet (30 m) deep. Within 1 mile (1½ km) southwest of the playa, however, the ground water from flowing wells is fresh enough to be used for irrigation (California Department of Water Resources, 1969a).

In Searles Valley (No. 6–52) the degradation of ground water has proceeded to the ultimate in concentration, variety, and economic value of dissolved solids (U.S. Geological Survey, 1966, p. 104, 111, 385–391). This playa too has a crystal body free of clastics extending over 7,500 acres (30 km^2), with nearly pure rock salt to a depth of 10–15 feet (3–5 m). Beneath the halite is a zone of crystalline salts averaging 60 feet (20 m) thick, impermeable clay less than 15 feet (5 m) thick, and a deeper saline zone 30 feet (10 m) thick, all extending over an area exceeding 25,000 acres (100 km^2). Brines saturate both saline zones and account for about 40 percent of their total volume, and the brines in both zones are pumped selectively from wells for commercial production. The brines are distinctive in that they are complex alkaline waters with 350 parts dissolved solids per thousand of water, of which about 20 percent is Na_2SO_4, 14 percent Na_2CO_3, 14 percent KCl, 5 percent borax, some commercial bromide and lithium salts, and more than 45 percent common salt that is unwanted and sluiced back to the playa. These percentages vary from well to well, from deep well to shallow well, and from time to time in the same well, and so production can be modified somewhat to fit market demand. Before processing, the brine may be enriched by solar evaporation, which preferentially crystallizes NaCl and adds to the halite crust.

The Searles Lake playa is the nearly dessicated remnant of a Pleistocene lake which was part of the Owens River drainage system and which accumulated the dissolved salts from a large area of the Sierra Nevada. Owens Lake has not overflowed for thousands of years and would need to rise 250 feet (75 m) to overflow, but ancient shorelines show that it has done so; the escaping water formed a lake in Searles Valley for extended periods, and that lake occasionally rose high enough to overflow into Panamint Valley (State No. 6–58). The dissolved solids that accumulated in Searles Valley as the water evaporated are a nonrenewable resource, at least until the next pluvial-glacial age.

The ground water in the crystal body, however, appears to be replenishable and sufficient to have sustained the industrial development of the past 45 years, even though the normal annual rainfall at Trona is only 4 inches (100 mm) ranging from 0.4 inch (10 mm) in 1953 to 9 inches (230 mm) in 1941. The brine that is pumped from wells goes through various evaporators and heaters and coolers as it is divested of its desirable salines; some water returns to the playa, but some is lost in the processing. Beyond the crystal body in the valley alluvium, brackish water is pumped at Valley Wells for washing and processing, and this pumpage has created a cone of depression that draws more mineralized water toward those wells as groundwater storage is depleted. No water suitable for domestic use has been found in the valley, and the residents depend upon water imported via pipeline from Indian Wells Valley (No. 6–54).

In Owens Valley (No. 6–12) human activities have increased the degradation of the water and accumulation of salines. Today the brines of the Owens Lake playa contain 31 percent dissolved solids (10 times seawater salinity), with a higher concentration of Na_2CO_3 (the only substance in commercial production) and a lower concentration of KCl and borax than the Searles brine. A century ago, Owens Lake had an area of 72,000 acres (300 km^2) and a volume of 2.4 million acre-feet (3 km^3). By 1913 it had dwindled to one-fourth of that volume because of increasing diversions from Owens River for irrigation. In subsequent years, when practically the entire riverflow was diverted for the Los Angeles municipal supply, the lake became an artificial playa.

SAND AND GRAVEL

In contrast to the water-deficient environment that is essential for the preservation of evaporites, sand and gravel are deposited in environments of water surplus, notably along the channels of major streams and especially those issuing from mountains (U.S. Geological Survey, 1966, p. 361–68; California Division of Mines and Geology, 1961, 1964, 1968). Many water-laid deposits of sand and gravel, now in arid or semarid areas, accumulated under conditions of greater volumes of water than now present, whether in ancient rivers or melting glaciers, in deltas, lakes, or bays; or in littoral currents. Many others are along the channels and in the valleys of the present stream systems and are closely related to them (fig. 13).

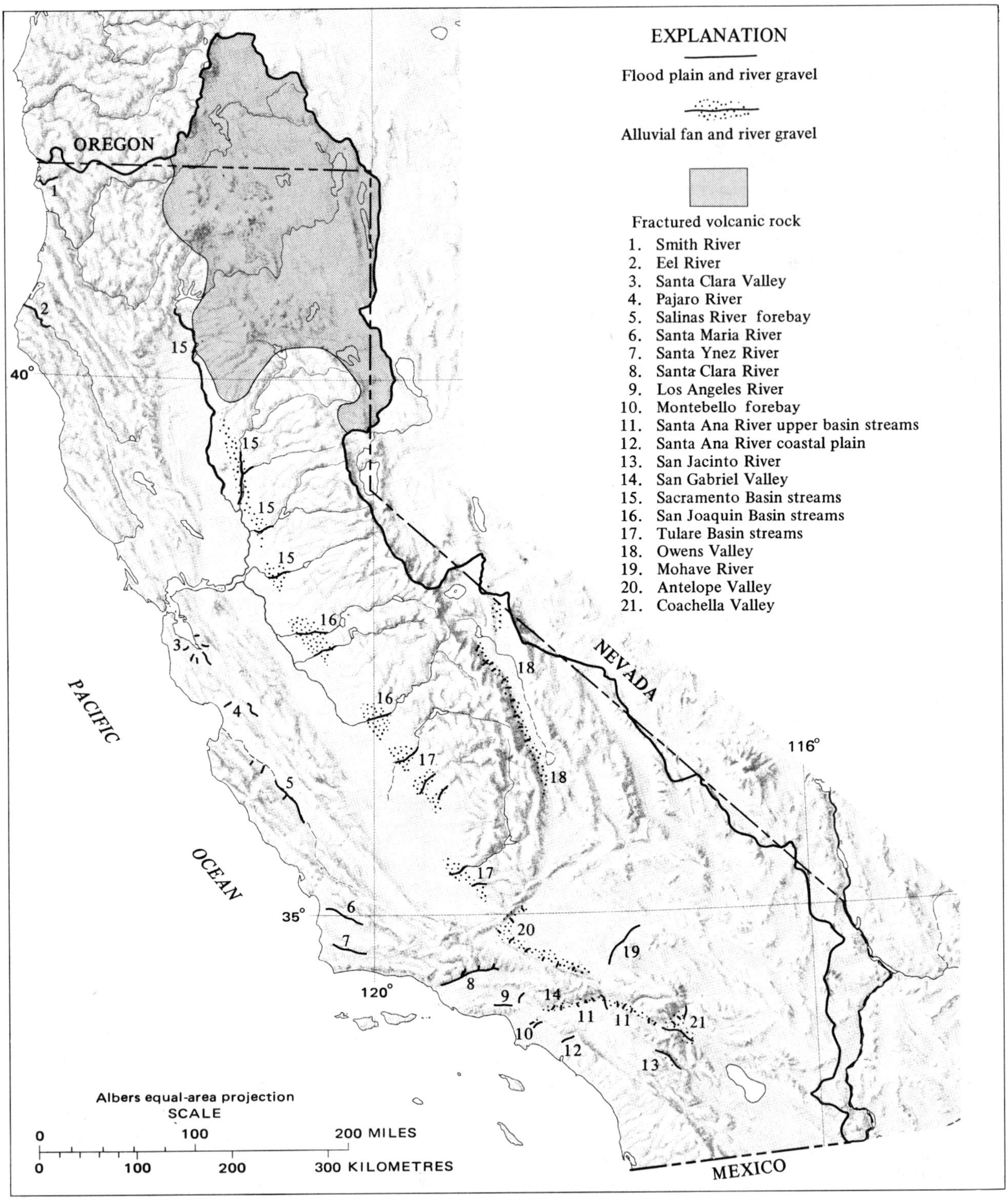

FIGURE 13—Outcrops of sand and gravel and other permeable rocks

Much of the sand and gravel at the land surface and therefore most accessible for use is also in areas where underlying ground water is naturally or can be artifically replenished, and in several urban areas gravel pits have become excellent sites for artificial recharge. In the construction of off-channel spreading basins for artificial recharge (see section "Recharge in Urban Areas"), the Orange County Water District obtains revenue from sand and gravel excavators, which materially reduces the cost of the basin construction. In the Central Valley many of the principal sand-and-gravel producing areas are in the zones of ground-water recharge along such streams as the Sacramento, San Joaquin, Kings, and Kern Rivers. (See section "Recharge in Central Valley.")

GEOTHERMAL ENERGY

In several parts of the California Region, the circulation of ground water and steam has produced reservoirs of stored heat at depths shallow enough for economic recovery by means of wells. Currently the only production of geothermal energy is from "dry" steam at The Geysers (Koenig, 1969), in the northern Coast Ranges about 12 miles (20 km) southwest of Clear Lake (fig. 14), where in 40 wells 1,600–8,100 feet (500–2,500 m) deep the cool shallow ground water is sealed off. Condensing turbines are used for generation of electric power from the steam, and estimates of the field potential are as much as 1,000 megawatts, of which 180 is presently developed. The condensate water carries ammonia and boron in solution above permissible levels and is therefore discharged into disposal wells.

In other parts of the California Region, the geothermal resource consists of hot water underground. In the South Lahontan subregion in the headwaters of Owens River, Casa Diablo is an extensive area where temperatures as high as 180°C have been measured in wells and where hot springs discharging water averaging 11 mg/l boron have been an important source of the saline deposits of Searles Lake. In the Colorado Desert subregion an even larger geothermal resource underlies the Salton Sea and the Imperial and Mexicali Valleys to the south (Rex, 1971). Because some of the brine in the Colorado Desert area is very concentrated, surface disposal would cause unacceptable pollution of the Salton Sea; as suggested by Rex, some of the heat might be used for desalination of some of the brine, although numerous obstacles to such development are recognized.

OPPORTUNITIES FOR UTILIZING GROUND-WATER RESERVOIRS IN OVERALL WATER MANAGEMENT

As stated by the Committee on Water (1966, p. 48–49),

> A review of current efforts to manage water to serve the needs and desires of man reveals that all aspects of water management would be improved by planning that would maintain flexibility for the future, foreclose as few choices as practicable, and put fresh demands on science to predict consequences and to provide alternatives to meet changing needs***. The essential facts now are becoming well known, and the basic hydrologic problems have been defined. Still lacking is a broad recognition by scientists and engineers as well as policy makers that advances in the knowledge of water and its possible uses not only have changed the character of water problems, but have made it possible to deal with these problems in a greater variety of ways and more effectively than in the past.

Ground water has long been the alternative to surface water as a natural source of supply, especially valuable to the nonriparian landowner who is denied the use of surface water by the riparian doctrine. Large-scale development of ground water in the California Region began nearly a century ago, and wells now supply about 40 percent of the total need for freshwater. Many ground-water reservoirs, however, have a history of exploitation, overdevelopment, and depletion; others have not yielded supplies of satisfactory quality or quantity, or they do not look promising and have not been explored. It is, therefore, to be expected that some people would mark ground water as the alternative least likely to succeed and dismiss it by stating that "in the California region ground water has generally been developed to its maximum" (California Region Framework Study Committee, 1971, p. 48).

In comprehensive planning that considers a range of alternatives to meet changing needs, ground-water reservoirs may have an accessory but nevertheless essential role. Desalination of seawater will require surface or subsurface water-storage facilities whenever and wherever the demand is less than the production of desalted water; for inland areas where the cost of pipelines to the ocean would be prohibitive, desalination requires saline or brackish water as raw material and feasible disposal of residual brines, and these requirements demand saline ground-water reservoirs for both supply and disposal. Reclamation of nonconsumptively used "waste" water will also generally require storage facilities for the water so reclaimed until the time comes to use it again. Weather modification may lead to significant increases in precipitation but not from the cloudless skies that prevail during droughts when water is needed most; thus it will not be an alternative to the use of surface or underground reservoirs but must rely on them as tools for management of the augmented water resources. Watershed management, whether it uses impermeable surfaces to increase runoff or other techniques to increase infiltration, will doubtless depend on surface or underground storage of the water so yielded. Furthermore, the evapotranspiration that is labeled "nonbeneficial consumptive use" may be reduced by various means, one of which is management of storage in underlying ground-water reservoirs.

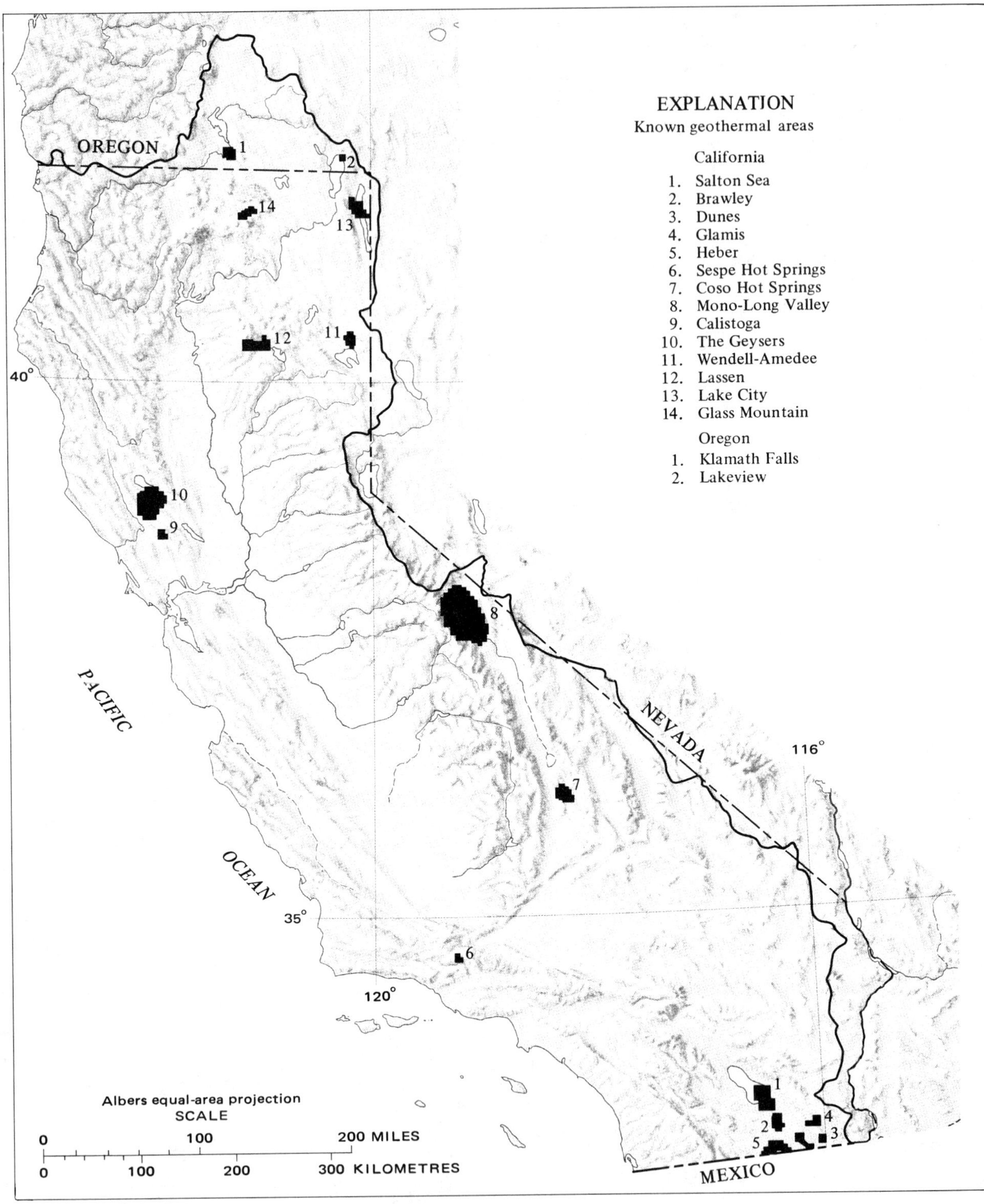

FIGURE 14.—Known geothermal areas in the California Region.

Each of these alternatives has the prime purpose of increasing the supply of usable freshwater; each requires expenditures in energy, manhours, equipment, and therefore money. If the water produced were stored in surface reservoirs for future use, substantial volumes must be lost by evaporation all year round in water-deficient areas (fig. 3) and throughout the California Region during the largely rainless summers. Ground-water reservoirs may become the preferred alternative, because the losses from them by evapotranspiration are characteristically far less. Also, ground-water reservoirs can be maintained and used in urban areas without denying or degrading the surface for other uses and without danger of disruption, spilling, or flooding disaster during earthquakes.

Artificial recharge is a prerequisite for effective management of underground space. Without it any replenishment of ground-water supplies is dependent upon the vagaries of nature, except for inadvertent human contributions from overirrigation or disposal of nonconsumptively used and polluted waters. Unless we have the capability of storing surpluses of water underground when they are available for use when needed, the use of ground-water reservoirs cannot exceed the limits set by nature (Committee on Ground Water, 1961, p. 92).

Where ground-water reservoirs can be operated as warehouses for accepting, holding, and releasing water as required, ground-water management becomes one aspect of integrated water management, involving conjunctive use of all water resources. Aside from warehousing of water in subsurface space, efficient water management requires the maintenance of the water with minimum deterioration. This depends upon the successful solution of problems in a variety of categories under the subject of "reservoir pollution." (See sections under "Reservoir Pollution.") Here the duties of management are likely to include monitoring, licensing, prohibiting, and punishing offenders.

ARTIFICIAL RECHARGE

Artificial recharge is the process of intentionally replenishing ground water through works provided for that purpose (Committee on Ground Water, 1961, p. 72). Ground water is replenished in many places by irrigation in exceedingly permeable soils, in natural recharge areas where the irrigator knows he will add to ground-water storage, and in some places by flooding to leach soluble salts from saline or alkaline soils; but these do not qualify as artificial recharge, nor do the various means of disposal of nonconsumptively used water as in septic tanks, pits and sumps, and seepage ponds.

As early as 1895 floodwaters were spread over the alluvial fan at the mouth of San Antonio Canyon to sustain the many flowing wells in the Upper Santa Ana Valley (No. 4–14) in southern California. The practice spread to other parts of the valley and was so successful that downstream users obtained an injunction to prohibit artificial recharge from local waters until adjudication of the waters of the river basin had been completed (Orange County Water District v. Chino Basin Municipal Water District). The Upper Santa Ana Valley provides excellent opportunities for artificial recharge (Moreland, 1972)— 7,500 acres (3,000 hectares) of unlined channels plus 60 off-channel basins aggregating 2,600 acres (1,050 hectares) with a recharge capability exceeding 3,200 cubic feet per second (90 m^3/s). The beds of major streams could absorb the water of a 5-year flood (peak 5–20 times the mean flow), and most streambeds could absorb waters of a 10-year flood (peak 10–100 times the mean flow).

As of 1958 (Richter and Chun, 1961), 54 agencies were practicing artificial recharge in California, chiefly in the South Coastal, San Francisco Bay, and Tulare Basin subregions. About 65 percent of the projects were basins or pits, and these received about 60 percent of the artificially recharged water; 23 percent used natural channels or artificial ditches or furrows, which accounted for nearly 40 percent of the total artificial recharge. Twelve projects used recharge wells, accounting for only 1 percent of the water applied. Artificial recharge totaled about 630,000 acre-feet (780 million m^3) in 1958, of which more than one-fourth came from the Colorado River. In 1962 and again in 1963, after a year when rainfall at Los Angeles was less than 5 inches (125 mm) and the least in a century, about 360,000 acre-feet (440 million m^3) of Colorado River water was used for artificial recharge. The Los Angeles County Flood Control District for nearly 40 years has used various types of water to replenish ground-water reservoirs in the South Coastal subregion and has probably more experience and more complete records on artificial recharge than any other single agency (Bianchi and Muckel, 1970).

RECHARGE IN URBAN AREAS

The artificial recharge reported for 1970 (California Department of Water Resources, 1971a) and summarized in table 10 was chiefly in the South Coastal and San Francisco Bay subregions, in four of the six most populous counties in California. In the San Francisco Bay area, most of the artificial recharge was by water of local origin—storm runoff delayed along natural channels or diverted into recharge basins—but 40 percent came from the Central Valley via the South Bay aqueduct. In the Los Angeles metropolitan area, most of the artificial recharge was by water from the Colorado River, but this amounted to only about 15 percent of the total Colorado import in 1970.

All the metropolitan areas of the California Region

TABLE 10.—*Artificial recharge reported for 1970*

Subregion	Source	Quantity (million m³)
San Francisco Bay	Local	107
	Imported	57
		164
Central Coastal	Local	33
South Coastal	Local	217
	Imported	[1]229
	Reclaimed	21
		[1]467
Tulare Basin	Channels	258
	Spreading basins	132
	Replenishing irrigation	43
		433
San Joaquin Basin	Channels	411
	Spreading basins	55
	Replenishing irrigation	471
		937
Total		2,030

[1]Includes 35 thousand acre-feet (43 million m³) of water from Colorado River injected into 120 recharge wells along saltwater barriers.

are in valleys and lowlands containing permeable alluvial fans, channels, and soils that are excellent for ground-water recharge. Storage of water in a metropolitan area has always been essential for distribution to people when they need it, but artificial surface-water storage in the metropolitan areas is costly because of exorbitant land values. Ground-water reservoirs have helped solve this problem. The urban areas thus have both the capabilities and incentives for artificial recharge of underlying ground-water reservoirs. They may, however, reduce or prevent natural recharge of underlying ground-water reservoirs because of impermeable roofs, streets, and pavements and the gutters, drains, and storm sewers that take rainwater away.

In southern California divergent achievements are shown on the coastal plain in Orange and Los Angeles counties. The principal area of artificial recharge by the Orange County Water District is along the Santa Ana River in the forebay area[6] below Santa Ana Canyon (fig. 7). Similarly the Montebello forebay, along San Gabriel River and Rio Hondo below the Whittier Narrows, is a major area of artificial recharge by the Los Angeles County Flood Control District. Farther west, the Los Angeles River also has a forebay which once served as a natural recharge area for ground water of the coastal plain; however, the urban sprawl of downtown Los Angeles into this area has made the dispatch of floodwaters the first priority. The channel has been paved with concrete, and the recharge to the Los Angeles forebay has been drastically reduced.

RECHARGE IN CENTRAL VALLEY

The annual pumpage from wells in the urbanized South Coastal and San Francisco Bay subregions is only about 10 percent of the total ground-water withdrawal in the California Region. By contrast, the Central Valley accounts for about 80 percent of the regional pumpage, predominantly in the San Joaquin Valley and the closed Tulare Basin, and in addition is estimated (California Division of Water Resources, 1970c, p. 69) to have received more natural recharge than the total recharge to all other ground-water reservoirs in the California Region. The form of the water table and other potentiometric surfaces of the ground-water reservoir indicates that recharge occurs chiefly along the east slope of the valley and especially from streams draining the Sierra Nevada. Several of these streams flow in channels and over alluvial fans with excellent recharge areas of permeable sand and gravel. (See fig. 13).

The Tulare Basin is especially well suited for artificial recharge by spreading, for the Kern, Tule, and Kaweah Rivers have shallow natural channels readily available for disposal of surplus water. Since 1938, the North Kern Water Storage District has spread surplus water when available for artificial recharge; it reported spreading 250,000 acre-feet (0.3 km^3) in the 3 years 1956–58, when a million acre-feet (1.3 km^3) was released to Kern River from Isabella Reservoir. The Kaweah Delta Water Conservation District and the Lower Tule River Irrigation District have also used surplus waters for artificial recharge in natural channels.

The streams that enter San Joaquin Valley north of the Kaweah River flow in trenches incised in the alluvial fans and plains of the valley. Some channels are in materials so impermeable that surface reservoirs are maintained on them with negligible loss by seepage; some flow over thin beds of sand and gravel which absorb so little water as to be negligible in recharge. To deliver water from these streams to the valley lands for irrigation has required major detention and conveyance works, but many of the irrigated lands and the stream channels and canals in the vicinity are excellent for artificial recharge of the underlying ground-water reservoir. Several agencies have used the channels of Ash and Berenda Sloughs and Chowchilla and Fresno Rivers for artificial recharge, and several agencies have purchased from the Bureau of Reclamation large amounts of Class II water (the nonfirm supply) from San Joaquin River when available through the Friant-Kern

[6]In his Salinas Basin investigation (California Department of Water Resources, 1946), Russel Simpson divided the valley floor into five areas for analytical purposes, of which one was designated the "forebay" area—an area of coarse-textured alluvial soils and underlying materials in which ground water is unconfined; its ground water comes chiefly from the Salinas River and from ground water in the upper valley area, and it moves northwestward into confined aquifers in the pressure area, the principal area of pumping for irrigation. The term "forebay" has been accepted in California as appropriate for similar areas along several other streams where ground water is in permeable alluvium, unconfined and readily recharged from the stream, but then moves into confining conditions.

Canal or Madera Canal for water spreading or for "replenishment irrigation." By using stream supplies for irrigation and for replenishment of ground water in years of plenty and pumping from wells for irrigation in years of scarcity as well as purchasing canal supplies from the Bureau of Reclamation, several agencies in the San Joaquin Valley are well set for efficient water management.

The permeable materials in the San Joaquin Valley in stream channels, canals, alluvial fans and plains, and soils, which are the most suitable for ground-water recharge, are similar to the permeable materials also abundant and widely distributed in the Sacramento Valley where less is recorded, however, about the achievements and potentialities in ground-water recharge, for water supply is less critical. For example, levees are generally constructed to protect cultivated or urban land from flooding, rather than to hold water on the land for recharge.

Most of the Central Valley has fine-textured soils, with suitable retention of water for crops but not permeable enough for ground-water recharge. In extensive areas where ground water has been pumped for irrigation, the replenishment has been less than the withdrawal over extended periods, and the result has been progressive depletion of storage. (See fig. 6.) The need to know more about recharge has been recognized for many years, and the Central Valley has a long history of research and experiments in artificial recharge. Experiments beginning in 1944 showed (Muckel, 1959) that the rate of infiltration could be increased and sustained over longer periods by various treatments, notably by Bermuda grass or other vegetative cover, cotton gin trash, and other organic residues. Test plots also were used for studies of soil structure and microbiology and the effects of soluble materials and of sediment. It was concluded (Schiff and Johnson, 1966) that successful artificial recharge depends upon the characteristics of the soil and substrata and that the depth, shape, and extent of subsurface control layers determine the usable storage capacity, buildup of the water table, and the likelihood of drainage problems. These conclusions are in accord with findings in many parts of the world that the best recharge areas are generally not ideal for agriculture because they are too permeable to hold soil water for crops. Nevertheless the Central Valley includes extensive areas where both artificial recharge and crops can be sustained by replenishment irrigation, and since 1958 the research and experiments have been increasingly concerned with selection and development of recharge areas and with replenishment irrigation.

Beds of clay or other rock material impermeable enough to create perched ground-water bodies above them or confined (artesian) water beneath them will also inhibit artificial recharge. In the San Joaquin Valley, the Corcoran Clay Member of the Tulare Formation is sufficiently widespread to impede recharge to the confined aquifers that supply large quantities of water for irrigation. The confined aquifers are not replenished as an incidental benefit of irrigation by surface water, even if the water is used lavishly. Artificial recharge of confined aquifers has, however, been achieved directly through recharge wells. The Lindsay-Strathmore Irrigation District in the Tulare Basin was successful in recharging through wells during the first 4 months of 1932, when 1,900 acre-feet (2.3 million m^3) of the winter runoff of the Kaweah River was filtered, chlorinated, and injected into 75 irrigation wells. This water was subsequently pumped out and could be identified by its lower temperature. The experiment lost no wells, pumps, or water and was a success, but it was not repeated because of legal difficulties. Currently 15 wells are being used for artificial recharge by the Delano-Earlimart District in the Tulare Basin.

CONJUNCTIVE USE OF GROUND WATER AND SURFACE WATER

Conjunctive use of ground water and surface water has long been recognized as essential in the California Water Plan. Studies made during the formulation of that plan (California Department of Water Resources, 1957a) indicated that its objectives could not be achieved without full and careful use of the ground-water resources. See also Berry (1962, p. 3.): "The answer for the future thus lies in the full development and use of our ground-water basins, both for conservation of local supplies and for seasonal and long-term cyclic regulation of imported water supplies. This will involve the planned use of ground-water storage in conjunction with surface storage facilities." Several years earlier, the U.S. Bureau of Reclamation (1949, p. 214) had indicated the importance of ground-water development in relation to its Central Valley Project: "In planning and constructing the necessary works [Friant-Kern and Madera Canals] special attention must be given to the problem of using ground-water reservoirs to best advantage. Only by the full use of these underground basins can the irrigable areas of east side upper San Joaquin Valley be developed completely." In California, however, private property rights have been asserted and protected, particularly as to ground water, stemming from the common-law maxim "Cujus est solum ejus est usque ad coelum et ad infernos"—roughly, the landlord owns everything above and beneath his land from heaven to hell. Federal or State agencies thus can lose control of and title to the water they put into ground-water reservoirs: "Leakage from the canal would be quite effective, but how would we collect for it?" (Bain and others, 1966, p. 414). The Bureau of Reclamation, supplying water under contract to several local agencies in San Joaquin Valley, necessarily lines its canals with concrete where they traverse the natural recharge areas of ground-water reservoirs, to prevent "loss" by seepage.

Nor do State agencies have managerial authority over ground-water reservoirs. At a panel discussion of practical considerations in implementing public policy (McGauhey, 1967, p. 78), moderator Harvey Banks asked John Teerink, Deputy Director of the California Department of Water Resources: "How can we bring about the necessity of coordinated operation of long aqueducts and ground water basins to even out aqueduct flows without undue interference with local control of ground water basins?" Mr. Teerink replied: "In determining the need for regulatory storage along the California aqueduct, we looked for surface storage reservoir sites. We did consider that ground storage was a real possibility. But there did not exist, and there does not exist today, any means by which the State can involve itself in ground water basin management, so we had to go to surface storage."

Krieger and Banks (1962) noted that Californians have made relatively few attempts to stretch the available water supplies and have tended to squander them; they pointed out that one means of checking this waste would be a basin management program of ample scope to maximize the use of the State's ground-water basins. This "demands the immediate attention of our courts, lawmakers, local governing bodies, and water distributing entities, the skill and resources of lawyers, engineers, geologists, economists, financiers, and political scientists***." Eight years later the California Water Plan still faced the same impediment (California Department of Water Resources, 1970c, p. 72): "Full realization of such integrated surface water-ground water system operations in areas where the ground water resource is available will require legal and legislative action and social and political acceptance."

This action may be delayed yet awhile. Fortunately, the California legislature has generally supported local initiative in ground-water basin management and also the conjunctive use of surface and ground water. It has passed special legislation when required for the effective operations of public agencies in populous areas, notably the Orange County Water District (Basin No. 8–1 on pl. 1), as well as in rural areas such as the small mountain-rimmed basins included in the Tehachapi–Cummings County Water District (Nos. 5–28, 6–45) (Porter, 1967). Conjunctive use of surface and ground water currently depends heavily upon the conjunctive operations of local agencies, whose dominant concern is ground water, and Federal and State agencies, whose dominant concern is surface water.

With these social and legal handicaps, conjunctive use of surface water and ground water has, nevertheless, been achieved in many places. It has long been recognized and practiced by local agencies whose user-members have rights in modest though significant local resources of both surface water and ground water, for example in Los Angeles, Santa Clara, and Ventura Counties. Smith (1962) summarized the history of the organization of public districts in Santa Clara County (Basin Nos. 2–9, 3–3) and concluded optimistically on future organizations. In other places varying degrees of conjunctive use have been achieved by the actions of (1) carriers and wholesalers of surface water (Federal and State agencies, large municipalities, and regional agencies such as the Metropolitan Water District of Southern California) and (2) local agencies (county water districts, irrigation districts, water conservation districts, water storage districts, and water replenishment districts) who are purchasers, distributors, and retailers of water and who generally act as users' cooperatives in developing water for a specific group of users. These wholesalers and user cooperatives together form a complex and intricate water industry which has been analyzed by Bain, Caves, and Margolis (1966).

Conjunctive use is possible in many areas because the costs of surface water can be set nearly equal to the costs of pumping ground water and can be manipulated to encourage use of surface water in periods of abundance. Thus the Bureau of Reclamation has set the price of Class II (nonfirm) water supply at less than half the price for Class I water, which encourages replenishment irrigation, and Class II water may be available in the preirrigation season for artificial recharge; contracts requiring payment whether the surface water is used or not will encourage use of surface water from canals rather than ground water from wells. Since 1960, the Metropolitan Water District of Southern California has offered Colorado River water for underground replenishment or for agricultural use at prices substantially lower (in 1970 and 1971 about 50 percent) than domestic rates for untreated water.

REDUCTION OF LOSSES BY EVAPORATION

Throughout the California Region, surface reservoirs lose water by evaporation during the dry summer season. In areas of perennial water surplus, this loss is more than offset by precipitation upon the reservoir area during the wet season, resulting in a net gain in an average year. At numerous reservoirs along north coastal streams or in the Sierra Nevada, therefore, the annual evaporation is less than or only slightly greater than average annual precipitation; Lake Shasta on the Sacramento River and Clair Engel Lake on the Trinity River are the largest of these.

In the drainage basin tributary to the Central Valley, 11 of the largest surface reservoirs are close to the valley and at altitudes less than 1,100 feet (330 m) (table 11). These reservoirs are in localities where annual rainfall generally is less than 20 inches (500 mm); evaporation from a standard land pan is more than 60 inches (1,500 mm) and exceeds 108 inches (2,700 mm) a year at San Luis Reservoir, the highest rate measured in the region except in Death Valley. No published calculations or estimates of the volume of evaporation are available from any of these reservoirs, but one reservoir for which such data exist in the southwestern United States is Lake Mead, with a record of reservoir evaporation beginning in 1952. In the first 11 years after records began, the average storage at Lake Mead was about 18 million acre-feet (22 km^3), average reservoir area 118 thousand acres (29 km^2), and annual evaporation about 850,000 acre-feet (1 km^3). After storage began in upstream Lake Powell in 1963, the storage in Lake Mead was reduced to an average of about 15 million acre-feet (18½ km^3), and the annual evaporation decreased 20 percent to about 660,000 acre-feet (0.7 km^3) in 1964–70, inclusive.

Evaporation per unit of water surface at Lake Mead is probably greater than that at the large reservoirs in the Central Valley, for pan evaporation at Boulder City and at Pierce's Ferry, Nev., is greater than that at San Luis Reservoir. Also, it is to be expected that the California reservoirs will ordinarily contain far less than their capacity during the summer months when the rate of evaporation is greatest. Nevertheless, storage in these large lowland reservoirs requires significant sacrifices of water by evaporation. For efficient water management in arid regions, the surface storage of water should be held to minimums of water area and length of exposure, particularly during the hot dry summers.

RECLAMATION OF NONCONSUMPTIVELY USED WATER

Water reclamation may be achieved by processes that are also used in sewage treatment, waste-water treatment, water renovation, or pollution control, although the quality of water reached by reclamation is normally considerably higher than that resulting from treatment intended merely to satisfy water-pollution control standards. Reclamation may include removal of nutrients from urban waste water or agricultural return flows, or other partial demineralization. But the chief objective of water reclamation is, as pointed out by McGauhey (1971, p. 172), the idea that the user of water should return it to the resource essentially unchanged.

TABLE 11.—*Large surface reservoirs in Central Valley*

Basin	Reservoir	Water surface at usable capacity		Water storage (km^3)
		Altitude (m)	Area (km^2)	
Sacramento	Oroville	274	61	4.29
	Folsom	142	45	1.25
	Marysville[1]	104	45	1.23
	Berryessa	134	85	1.97
Delta	New Hogan	217	16	.41
	Comanche	72	32	.53
San Joaquin	New Melones[2]	332	49	2.96
	New Exchequer	264	28	1.27
	New Don Pedro	253	53	2.50
	Millerton (Friant)	176	20	.64
	San Luis and O'Neill	166	61	2.59
Tulare	Pine Flat	293	24	1.29
Total			519	20.93

[1]Authorized for future construction.
[2]Under construction in 1975.

Most of the water now reclaimed in the California Region is returned to ground-water reservoirs by artificial recharge, but substantial quantities are reused directly for agriculture, industry, and recreation (table 12). The Los Angeles metropolitan area has the greatest concentration of population in the California Region and also leads in sewage production, averaging 100 gallons (380 l) per day per capita. Los Angeles County produces about 700 Mgal/d (2.6 million m^3 per day), and Orange County one-fifth as much, for an annual total approaching a million acre-feet (1.2 km^3). Sewage-treatment plants on the coastal plain and near the coast discharge their effluent into the ocean, but many of the inland treatment plants from the time of their construction have produced waters which percolated through the soil and might enter ground-water reservoirs. The City of Pomona has long maintained a water renovation plant which currently produces about 16 thousand acre-feet (20 million m^3) annually of water with dissolved solids less than 700 mg/l, which is used for irrigation and experimentation.

As population and water demand increased after

TABLE 12.—*Reclaimed water*

Number of plants	Principle use of reclaimed water	Quantity (million m^3)
	South Coastal subregion, 1966 (McGauhey, 1971, p. 167)	
9	Ground-water recharge by spreading	293
9	Ground-water recharge by injection (barrier to seawater intrusion)	153
4	Industrial use	86
13	Recreational use (parks and golf courses)	18
4	Recreational use (ponds for fishing, boating)[1]	14
6	Agricultural use (irrigation	6
		570
	San Francisco Bay subregion, 1970 (California Division of Water Resources, 1971a)	
7	Recreational use (parks and golf courses)	2.0
6	Agricultural use (irrigation)	2.6
		4.6

[1]Also, swimming and water-contact sports at Santee, San Diego County.

World War II, ground-water reservoirs were depleted by pumping and intruded by seawater along the coast, and increasing amounts of water were imported from the Colorado River. Studies during the 1950's led to the conclusion that 40 percent of the sewage then produced could be economically reclaimed, but the mineral quality of the rest was too poor for salvage (California Department of Water Resources, 1961). Thus, the Whittier Narrows Water Reclamation Plant—with Los Angeles County as banker, the Sanitation District of Los Angeles County as designer and operator, Central and West Basin Water Replenishment District as sole customer, and Los Angeles County Flood Control District as water spreader—began operating in 1962 and currently produces for ground-water recharge about 12 Mgal/d (530 l/s) of water comparable with or superior to the water imported from the Colorado River as to dissolved solids, hardness, and price (Parkhurst, 1969). The Los Angeles County Sanitation Districts have now constructed two additional reclamation plants that increase the production to 60 Mgal/d (85 million m^3 annually) and have plans for reclamation exceeding 400 Mgal/d (1.5 million m^3 per day).

The City of Los Angeles has constructed a pilot plant for tertiary treatment of 5 Mgal/d (7 million m^3 annually) of the waste waters from its Hyperion Plant. The reclaimed water, with not more than 700 mg/l dissolved solids and 200 mg/l hardness, will be used for injection into wells at the north end of the city's coastal barrier (Bookman and Edmonston, 1971). (See section "Seawater Intrusion.") Orange County too has been planning for water reclamation and has reached the stage of implementation of its Water Factory 21 in Fountain Valley (Cofer, 1972). This factory includes a plant to reclaim 15 Mgal/d (21 million m^3 annually) of effluent from a secondary treatment plant, reducing dissolved solids from 1,300 to 1,100 mg/l and hardness from 400 to 200 mg/l by tertiary treatment. This reclaimed water will then be blended with an equivalent amount of seawater to be desalted at an adjacent plant, and the product will be put into the ground-water reservoir through a series of injection wells. The 15-Mgal/d reclamation plant and desalination plant are envisioned as prototypes of 200 Mgal/d (276 million m^3 annually) factories that may ultimately be required for southern California.

Water reclamation is further from reality in the metropolitan area around San Francisco Bay, which still receives a considerable volume of untreated wastes; some studies have proposed a water reclamation plant near the head of the bay to collect waste water from several counties (California Region Framework Study Committee, 1971, p. 51).

MAINTENANCE OF WATER QUALITY

Nearly half the surface water and ground water withdrawn for use in the California Region is used nonconsumptively and then becomes wastewater which generally is polluted by dissolved solids, organic residues, sediment, floating debris, or heat. With increasing population and urbanization, ground-water reservoirs are more and more vulnerable to recharge by waste waters from individuals, communities, farms, and industries; the California legislature has seen this potential pollution both as an aspect of the general problem of water pollution and as a specific problem of protecting the ground-water resources. Rather consistently, its policy has been that water-quality control can be most effectively administered regionally and that the basic responsibility for ground-water development and management should rest with local people. Consequently, the Dickey Water Pollution Act of 1949 and later the Porter-Cologne Water Quality Control Act of 1969 set up nine regional water-quality control boards and a coordinating State board.

The Porter-Dolwig Ground Water Basin Protection Act of 1961 provided for studies and plans for "correction and prevention of irreparable damage to, or impaired use of, the ground water basins of this State." This provision was amended in 1967 to require, for local projects, a cooperative agreement between the State and the local agency and substantial participation or cost sharing, or both, by the local agency; thus the California Department of Water Resources can provide technical assistance in assessing the present status of and in considering plans for management of quality and quantity of ground water. The Porter-Cologne Water Quality Control Act of 1969 gave increased powers to the California regional water quality control boards, which may now require adjudication of ground-water rights in order to protect water quality. Also after having the essential data and findings, a regional board may file actions to restrict pumping and impose a physical solution to prevent destruction or irreparable injury to the quality of ground water in a reservoir. Another act of the legislature requires that water well standards be established on the basis of the recommendations of the California Department of Water Resources in areas where necessary to protect water quality; however, the standards are to be established and enforced by the counties and cities in the area.

Clearly, local people can have a dominant voice in the maintenance of water quality in ground-water reservoirs in counties and public districts as owners and users of the land overlying the ground-water reservoir. At the local level, there may be conflict of interest between ground-water users and those who obtain their supplies from other sources but have wastes to dispose of; however, by monitoring the quality in various localities and by providing scientific data and expertise to local agencies, State and Federal agencies should eventually overcome the advantage that the waste-makers have achieved by tradition and the common law.

SELECTED REFERENCES

Alameda County Water District, 1971, Summary report on ground water conditions: 16 p. 7 pl.

Bader, J. S., 1969, Ground-water data as of 1967: U.S. Geol. Survey open-file reports—North Coastal subregion, 11 p; San Francisco Bay subregion, 12 p; Central Coastal subregion, 16 p; South Coastal subregion, 21 p; Sacramento Basin subregion, 16 p; San Joaquin Basin subregion, 16 p; North Lahontan subregion, 8 p; South Lahontan subregion, 25 p; Colorado Desert subregion, 19 p; Summary, California region, 32 p.

Bain, J. S., Caves, R. E., and Margolis, Julius, 1966, Northern California's water industry: Baltimore, Johns-Hopkins Press, 729 p.

Berry, W. L., 1962, Ground water in California's future: Conf. on Ground Water Recharge, 3d, Berkeley 1961, p. 1–14.

Bianchi, W. C., and Muckel, D. C., 1970, Ground-water recharge hydrology: Agr. Research Service Pub. 41–161, 62 p.

Bookman, Max, and Edmonston, R. M., 1971, Annual survey report on ground water replenishment, Central and West Basin Water Replenishment District: Downey, Calif., 80 p.

California Department of Public Health, 1962, California domestic water supplies: Bur. Sanitary Eng., 76 p.

———1968, Investigation of ground water quality problems in Central Basin of Los Angeles Co.: Bur. Sanitary Eng.

California Department (or Division) of Water Resources, 1930, Report to the Legislature on State Water Plan: California Dept. Water Resources Bull. 25, 204 p.

———1946, Salinas Basin investigation: California Dept. Water Resources Bull. 52, 246 p.

———1949, Salinas Basin investigation, basic data: California Dept. Water Resources Bull. 52a, 432 p., 10 pl.

———1950, Sea-water intrusion into ground water basins bordering the California Coast and inland bays: Water Pollution Inv. Rept. 1, 23 p., 2 tables.

———1952, Ground water basins in California: Water-Quality Inv. Rept. 3, 44 p.

———1955a, Oil field waste water disposal, Raisin City Oil Field, Fresno County: Water-Quality Inv. rept., 26 p.

———1955b, Quality of ground water in the Stockton area, San Joaquin County: Water-Quality Inv. Rept. 7, 51 p., 21 pl.

———1956, Geologic and waste disposal investigation, Lost Hills Oil Field, Kern County: Interdepartment commun. 11 p.

———1957a, The California Water Plan: California Dept. Water Resources Bull. 3, 272 p.

———1957b, Sea water intrusion, Report of Los Angeles Flood Control District: California Dept. Water Resources Bull. 63, app. A, 142 p.

———1958a, Sea water intrusion in California: California Dept. Water Resources Bull. 63, 92 p.

———1958b, Quality of ground waters in California 1955–1956: California Dept. Water Resources Bull. 66, 228 p.

———1958c, Reclamation of water from sewage and industrial wastes: California Dept. Water Resources Bull. 68, 44 p.

———1960a, Intrusion of salt water, southern Alameda County: California Dept. Water Resources Bull. 81, 74 p.

———1960b, Lower San Joaquin Valley water quality investigation: California Dept. Water Resources Bull. 89, 189 p., 3 app., 12 pl.

———1961, Feasibility of reclamation of water from wastes in the Los Angeles metropolitan area: California Dept. Water Resources Bull. 80, 155 p., 22 pl.

———1964a, Quality of ground waters in California 1961–1962: California Dept. Water Resources Bull. 66–62, 429 p.

———1964b, Water well standards, Alameda County: California Dept. Water Resources Bull. 74–2, 94 p.

———1964c, Ground water occurrence and quality, Lahontan region: California Dept. Water Resources Bull. 106–1, 486 p.

———1964d, Coachella Valley investigation: California Dept. Water Resources Bull. 108, 180 p.

———1964e, Water use by manufacturing industries in California: California Dept. Water Resources Bull. 124, 154 p.

———1965a, Sea water intrusion, Oxnard Plain: California Dept. Water Resources Bull. 63–1, 100 p.

———1965b, Water well standards, Los Angeles County: California Dept. Water Resources Bull. 74–7, 106 p.

———1965c, Water well standards, San Joaquin County: California Dept. Water Resources Bull. 74–5, 114 p.

———1965d, California State Water Project water supply contracts: California Dept. Water Resources Bull. 141, v. 1, 702 p., v. 2, 598 p.

———1966a, Water well standards, Del Norte County: California Dept. Water Resources Bull. 74–3, 132 p.

———1966b, Santa Ana Gap salinity barrier: California Dept. Water Resources Bull. 147–1, 194 p.

———1966c, Implementation of the California Water Plan: California Dept. Water Resources Bull. 160–66, 143 p.

———1967a, Mojave River ground water investigation: California Dept. Water Resources Bull. 84, 166 p.

———1967b, Ground water occurrence and quality, San Diego region: California Dept. Water Resources Bull. 106–2, 250 p.

———1968a, Sea water intrusion, Bolsa- Sunset area: California Dept. Water Resources Bull. 63–2, 186 p.

———1968b, Water well standards, State of California: California Dept. Water Resources Bull. 74, 224 p.

———1968c, Water well standards, Fresno County: California Dept. Water Resources Bull. 74–6, 16 p.

———1968d, Water well standards, Ventura County: California Dept. Water Resources Bull. 74–9, 84 p.

———1968e, South Bay, Fremont study area, ground water resources: California Dept. Water Resources Bull. 118–1, 136 p.

———1968f, Delano nitrate investigation: California Dept. Water Resources Bull. 146, 196 p.

———1968g, Investigation of waste discharges in the Lompoc Basin: Memo. rept., 85 p.

———1968h, The fate of pesticides applied to irrigated agricultural land: California Dept. Water Resources Bull. 174–1, 30 p.

———1969a, Water wells and springs in the Fremont Valley area: California Dept. Water Resources Bull. 91–16, 157 p.

———1969b, Sanitary land fill studies: California Dept. Water Resources Bull. 147–5, 122 p.

———1970a, Hydrologic data, 1969: California Dept. Water Resources Bull. 130–69, v. 2, Northeastern California, 531 p; v. 5, Southern California, 555 p.

———1970b, Oxnard Basin extraction-type barrier: California Dept. Water Resources Bull. 147–6, 157 p.

———1970c, The California Water Plan—outlook in 1970: California Dept. Water Resources Bull. 160–70, 170 p.

———1970d, Watermaster service in the Raymond Basin: California Dept. Water Resources Bull. 178–70, 58 p.

———1970e, Watermaster service in the West Coast Basin: California Dept. Water Resources Bull. 179-70, 121 p.

———1970f, Sea water intrusion, Lower Salinas Valley: Progress rept., 28 p.

———1970g, Sea water intrusion, Pismo-Guadalupe area: California Dept. Water Resources Bull. 63–3, 82 p.

———1970h, Nitrates in ground waters of San Joaquin Valley: Memorandum rept., 44 p., 4 maps.

———1971a, Hydrologic data, 1970: California Dept. Water Resources Bull. 130–70, v. I, North Coastal area, 55 p.; v. II, Northeastern California; v. III, Central Coastal area, 137 p.; v. IV, San Joaquin Valley, 223 p.; v. V, Southern California.

———1971b, Watermaster service in the Central Basin: California Dept. Water Resources Bull. 180–70, 151 p.

———1971c, Watermaster service in the upper Los Angeles River area: 89 p.

———1971d, Nitrates in ground waters of the Central Coast area: Memo. rept., 15 p.

California Division of Mines and Geology, 1961, Sand and gravel in California, Part A—Northern California: California Div. Mines and Geology Bull. 180A, 38 p.

———1964, Sand and gravel in California, Part B—Central California: California Div. Mines and Geology Bull. 180B, 58 p.

———1968, Sand and gravel in California, Part C—Southern California: California Div. Mines and Geology Bull. 180C, 56 p.

California Division of Oil and Gas, 1972, Oil and gas statistics—statistical tables: v. 58, no. 2, p. 64–151.

California Region Framework Study Committee, 1971, Comprehensive framework study; App. 5, water resources: Pacific Southwest Inter-Agency Comm. 339 p., 41 maps.

California State Oil and Gas Supervisor, 1959 and 1972, Summary of operations, California oil fields: California Div. Oil and Gas, 45th and 58th ann. rep. pt. 2, tables 1, 5, 6, and 7.

California Water Resources Control Board, 1961, Effects of refuse dumps on ground water quality, California Water Resources Control Board Pub. 24, 101 p.

———1965, Dispersion and persistence of synthetic detergents in ground water, San Bernardino and Riverside Counties: California Water Resources Control Board Pub. 30, 97 p.

Cofer, J. R., 1971, Ground water conditions, water supply, and basin utilization: Orange County Water Dist. Engineer's Rept., 53 p., 11 pl.

———1972, Orange County Water District's Water Factory 21: American Soc. Civil Engineers Nat. Water Resources Eng. Mtg., Atlanta, Preprint 1636, 28 p.

Committee on Ground Water, 1961, Ground water basin management: American Soc. Civil Engineers Manual Eng. Practice No. 40, 160 p.

Committee on Water, 1966, Alternatives of water management: Natl. Acad. Sci., Natl. Resource Council Pub. 1408, 51 p.

Conkling, Harold, 1946, Utilization of ground-water storage in stream-system development: Am. Soc. Civil Engineers Trans., v. 111, p. 275–354.

Crooke, H. W., and Toups, J. M., 1962, Ground water basin management and artificial recharge in Orange County: Conf. on Ground Water Recharge, 3d, Berkeley 1961, Proc., 13 p.

Cruff, R. W., and Thompson, T. H., 1967, A comparison of methods of estimating potential evapotranspiration from climatological data in arid and subhumid regions: U.S. Geol. Survey Water-Supply Paper 1839–M, 28p.

Davis, G. H., Green, J. H., Olmsted, F. H., and Brown, D. W., 1959, Ground-water conditions and storage capacity in the San Joaquin Valley, California: U.S. Geol. Survey Water-Supply Paper 1469, 277 p.

Durfor, C. N., and Becker, Edith, 1964, Public water supplies of the 100 largest cities in the United States, 1962: U.S. Geol. Survey Water-Supply Paper 1812, p. 99-118.

Dutcher, L. C., 1972, Proposed criteria for design of a data collection system for ground-water hydrology in California: Water Resources Research, v. 8, p. 188–193.

Evenson, R. E., 1965, Suitability of irrigation water and changes in ground-water quality in the Lompoc subarea of the Santa Ynez River basin: U.S. Geol. Survey Water-Supply Paper 1960–S, 20 p.

Federal Water Quality Administration, 1970, The economics of clean water: U.S. Dept. Interior, v. 1, p. 18.

Gleason, G. B., 1947, Report to Assembly of the State Legislature on water supply of Antelope Valley*** pursuant to House Resolution 101, 20 p.

Harding, S. T., 1960, Water in California: Palo Alto, N-P Publications, 231 p.

Hunt, C. B., and Robinson, T. W., 1966, Hydrologic basin, Death Valley, California: U.S. Geol. Survey Prof. Paper 494–B, 103 p.

Katz v. Walkinshaw, 141 California 116, 135, 74 Pac. 766 (1903).

Koenig, J. B., 1969, The Geysers geothermal field 1969: California Div. Mines and Geology, Mineral Inf. Service, v. 22, p. 123–128.

Kohler, M. A., Nordenson, T. J., and Baker, D. R., 1959, Evaporation maps for the United States: U.S. Weather Bur. Tech. Paper 37, 13 p.

Krieger, J. A., and Banks, H. O., 1962, Ground water basin management: 50 California Law Rev., p. 56–77.

Lofgren, B. E., 1969, Field measurement of aquifer-system compaction, San Joaquin Valley, California: Internat. Assoc. Sci. Hydrology, Colloque de Tokyo, p. 273–284.

McGauhey, P. H., coordinator, 1967, Practical considerations in implementing public policy, panel discussion: Conf. on Ground Water Recharge, 6th, Berkeley 1967, Proc., p. 34–83.

———1971, Waste water reclamation–urban and agricultural, *in* Seckler, David, ed., California Water: Berkeley, Univ. California Press, p. 161–173.

Miller, G. A., 1969, Water resources of the Marine Corps Supply area, Barstow, Calif. : U.S. Geol. Survey open-file report, p. 33–40.

Miller, R. E., and Singer, J. A., 1971, Subsidence in the Bunker Hill–San Timoteo area, Southern California: U.S. Geol. Survey open-file report, 27 p.

Moreland, J. A., 1972, Artificial recharge in the Upper Santa Ana Valley, Calif.: U.S. Geol. Survey open-file report, 82 p.

Moreland, J. A., and Singer, J. A., 1969, Evaluation of water-quality monitoring in the Orange County Water District, California: U.S. Geol. Survey open-file report, 27 p.

Muckel, D. C., 1959, Replenishment of ground water supplies by artificial means: U.S. Dept. Agriculture Tech. Bull. 1195, 51 p.

Murray, C. R., 1968, Estimated water use in the United States, 1965: U.S. Geol. Survey Circ. 556, 53 p.

Musser, E. H., 1959, Resume of oil field operations in 1959: California Div. Oil and Gas—Summ. Operations, v. 45, no. 2, p. 51–82.

Olmsted, F. H., and Davis, G. H., 1961, Geologic features and ground-water storage capacity of the Sacramento Valley, California: U.S. Geol. Survey Water-Supply Paper 1497, 236 p.

Parkhurst, J. D., 1969, Progress report, Los Angeles County Sanitation District's water reclamation program: Conf. on Ground Water Recharge, 7th, Los Angeles 1969, Proc., 10 p.

Pasadena v. Alhambra, 33 California (2d) 908, 207 (2d) 17 (1949).

Peters, H. J., 1972a, Ground water management: Water Resources Bull. v. 8, p. 188–197.

———1972b, Criteria for ground water level data networks for hydrologic and modeling purposes: Water Resources Research, v. 8, p. 194–200.

Piper, A. M., 1965, Has the United States enough water?: U.S. Geol. Survey Water-Supply Paper 1797, 27 p.

Piper, A. M., Garrett, A. A., and others, 1955, Native and contaminated waters in the Long Beach–Santa Ana area: U.S. Geol. Survey Water-Supply Paper 1136, 315 p.

Pitt, Leonard, 1966, The decline of the Californios: Berkeley, Univ. California Press, p. 1–25, 83–120.

Poland, J. F., 1959, Hydrology of the Long Beach–Santa Ana area, California: U.S. Geol. Survey Water-Supply Paper 1471, 252 p.

———1969, Land subsidence and aquifer-system compaction, Santa Clara Valley, California: Internat. Assoc. Sci. Hydrology, Colloque de Tokyo, p. 12–21.

———1972, Subsidence and its control, *in* Underground waste management and environmental implications—a symposium: Am. Assoc. Petroleum Geologists Mem. 18, p. 50–71.

Poland, J. F., and Davis, G. H., 1969, Land subsidence due to withdrawal of fluids: Geol. Soc. America, Eng. Geol. Rev., v. 2, p. 187–269.

Poland, J. F., Lofgren, B. E., Ireland, R. L., and Pugh, R. G., 1975, Land subsidence in the San Joaquin Valley, California, as of 1972, U.S. Geological Survey Prof. Paper 437–H, 78 p.

Poland, J. F., and Evenson, R. E., 1966, Hydrogeology and land subsidence, Great Central Valley, California; *in* Bailey, E. H., Geology of northern California: California Div. Mines Bull. 190, p. 239–247.

Porter, C. V., 1967, Implementing policies through legislation: Conf. on Ground Water Recharge, 6th, Berkeley 1967, Proc., p. 26–33.

Powell, J. W., 1962, Report on the lands of the arid region of the United States: Cambridge, Mass., Harvard Univ. Press, 196 p.

Powers, W. R. III, 1970, Water-resources inventory Antelope Valley–East Kern Water Agency area: U.S. Geol. Survey open-file report, 15 p.

Price, W. P., and Baker, C. K., 1963, Ground water recharge activities in United Water Conservation District of Ventura Co.: Conf. on Ground Water Recharge, 4th, Berkeley 1963, Proc. 6 p.

Rainwater, F. H., 1962, Stream composition of the conterminous United States: U.S. Geol. Survey Hydrol. Inv. Atlas HA–61, 3 pl.

Rex, R. W., 1971, Geothermal resources in the Imperial Valley, *in* Seckler, David, ed., California Water: Berkeley, Los Angeles, Univ. California Press, p. 190–205.

Richardson, Donald, and Rantz, S. E., 1961, Interchange of surface and ground water along tributary streams in the Central Valley, California: U.S. Geol. Survey open-file report, 253 p.

Richter, R. C., and Chun, R. Y. D., 1961, Artificial recharge in California: Am. Soc. Civil Engineers Trans., v. 126, pt. 3, p. 742–762.

Schiff, Leonard, and Johnson, K. L., 1966, Some physical and chemical considerations in artificial ground-water recharge: Conf. on Ground Water Recharge 5th, Los Angeles 1965, Proc., p. H–60–70.

Smith, S. C., 1962, The public district in integrating ground and surface water management: California Agricultural Experimental Sta., Giannini Found. Research rept. 252, 135 p.

Snyder, J. H., 1955, Ground water in California—the experience of Antelope Valley: California Agricultural Experimental Sta., Giannini Found. Ground Water Studies No. 2, 171 p.

Tannehill, I. R., 1947, Drought, its causes and effects: Princeton, N.J., Princeton Univ. Press, 254 p.

Theis, C. V., 1940, The source of water derived from wells— essential factors controlling the response of an aquifer to development: Civil Eng., v. 10, p. 277–280.

Thomas, H. E., 1962, The meteorologic phenomenon of drought in the Southwest: U.S. Geol. Survey Prof. Paper 372–A, 43 p.

———1969, Water as a resource and as a nuisance: Highway Research Rec. 271, p. 17–28.

Thornthwaite, C. W., 1948, An approach toward a rational classification of climate: Geog. Rev., v. 38, p. 55–94.

Tolman, C. F., Hyde, B. C., and Killingsworth, C., 1931, Ground water supply and saline contamination of Pittsburg and adjacent industrial area: Stanford Univ., 66 p. 8 figs.

U.S. Bureau of Reclamation, 1949, Central Valley Basin: U.S. 81st Cong., Senate Doc. 113, 430 p.

United States Geological Survey, 1966, Mineral and water resources of California, Part I: U.S. 89th Cong., Senate Comm. on Interior and Insular Affairs, Comm. Print, p. 1–450.

U.S. Public Health Service, 1961, Ground water contamination: U.S. Public Health Service Tech. Rept. W61–5.

U.S. Weather Bureau, 1963, Decennial census of United States, monthly average for state climatic divisions, 1931–60, California: U.S. Weather Bur. Climatography of the United States, no. 84–4, 3 p.

Ver Planck, W. E., 1958, Salt in California: California Div. Mines Bull. 175, 164 p.

Waananen, A. O., and Bean, R. T., 1966, Mineral and Water Resources of California, Part 2: U.S. 89th Cong., Senate Comm. on Interior and Insular Affairs, Comm. Print, p. 451–647.

Williams, D. E., and Wilder, D. G., 1971, Gasoline pollution of a ground–water reservoir—a case history: Ground Water, v. 9, no. 6, p. 50–56.

5-21

Developed ground-water reservoirs

State No.	Name
0-2	Sprague River valley
0-3	Swan Lake Valley
0-4	Yonna Valley
0-8	Lower Klamath Valley
1-1	Smith River Plain
1-2	Lower Klamath Valley
1-3	Butte Valley
1-9	Eureka area
1-15	Ukiah Valley
1-18	Santa Rosa-Healdsburg area
2-1	Petuluma Valley
2-2	Napa Valley
	Sonoma Valley
2-3	Suisun-Fairfield area
2-4	Pittsburg Plain
2-5	Clayton and Ygnacio Valleys
2-9	Santa Clara Valley
	South Bay
	East Bay
2-10	Livermore Valley
3-1	Soquel-Aptos area
3-2	Pajaro Valley
3-3	Llagas Valley
	Hollister area
3-4	Salinas Valley
	Paso Robles area
3-11	Arroyo Grande Valley
3-12	Santa Maria Valley
3-13	Cuyama Valley
3-14	San Antonio Valley
3-15	Santa Ynez Valley
3-17	Carpinteria-Goleta area
4-4	Oxnard-Santa Clara area
4-11	Coastal Plain
	Central Basin
	West Basin
4-12	San Fernando Valley
4-13	San Gabriel Valley
	Raymond Basin
4-14	Upper Santa Ana Valley
	Bunker Hill-San Timoteo
	Chino-Riverside
5-15	Clear Lake area
5-6	Redding
5-21	Sacramento Valley
5-22	Delta
5-22	San Joaquin Valley
5-22	Tulare Basin
5-25	Kern River valley
5-28	Tehachapi-Cummings area
6-12	Owens Valley
6-40	Lower Mojave Valley
6-41	Middle Mojave Valley
6-42	Upper Mojave Valley
6-44	Antelope Valley
6-45	Tehachapi-Cummings area
6-46	Fremont Valley
6-47	Harper Valley
6-52	Searles Valley
6-54	Indian Wells Valley
7-19	Lucerne Valley
7-21	Coachella Valley
	Upper Valley
	Artesian Basin
7-24	Borrego Valley
7-26	Coahuila-Terwilliger Valley
8-1	Coastal Plain
8-5	San Jacinto Valley
9-4	Lower valley(s)
9-5	Temecula Valley
9-6	Coahuila-Terwilliger Valley
9-7	San Luis Rey Valley
9-8	Warner Valley
9-12	San Dieguito-Escondido area
9-15	San Diego-El Cajon area

6-18

Undeveloped ground-water reservoirs

State No.	Name
0-1	Klamath Marsh
0-5	Langell Valley
0-6	Upper Klamath Valley
0-7	Poe Valley
1-4	Shasta Valley
1-5	Scott Valley
1-17	Alexander Valley
3-19	Carrizo Plain
5-1	Goose Lake Valley
5-2	Alturas Basin
5-4	Big Valley
5-5	Fall River valley
5-12	Sierra Valley
5-23	Panoche Valley
6-1	Surprise Valley
6-2	Madeline Plains
6-4	Honey Lake Valley
6-9	Mono Valley
6-10	Adobe Valley
6-11	Long Valley
6-13	Centennial (Black Springs) Valley
6-15	Deep Springs Valley
6-16	Eureka Valley
6-17	Saline Valley
6-18	Death Valley
6-19	Wingate Valley
6-20	Amargosa Valley
6-21	Lower Kingston (Valjean) Valley
6-22	Upper Kingston (Shadow) Valley
6-23	Riggs Valley
6-24	Red Pass Valley
6-25	Bicycle Valley
6-26	Avawatz Valley
6-27	Leach Valley
6-29	Mesquite Valley
6-30	Ivanpah Valley
6-31	Kelson Valley
6-32	Broadwell Valley
6-33	Soda Lake Valley
6-34	Silver Lake Valley
6-35	Cronise Valley
6-36	Langford Valley
6-37	Coyote Lake Valley
6-38	Caves Canyon Valley
6-39	Troy Valley
6-43	El Mirage Valley
6-49	Superior Valley
6-50	Cuddeback Valley
6-51	Pilot Knob Valley
6-55	Coso Valley
6-56	Rose Valley
6-57	Darwin Valley
6-58	Panamint Valley
6-71	Lost Lake Valley
6-79	California Valley
7-1	Lanfair Valley
7-2	Fenner Valley
7-3	Ward Valley
7-5	Chuckwalla Valley
7-6	Pinto Basin
7-7	Cadiz Valley
7-8	Bristol Valley
7-9	Dale Valley
7-10	Twentynine Palms Valley
7-13	Deadman (Ames) Valley
7-14	Lavic Valley
7-15	Bessemer Valley
7-17	Means Valley
7-18	Johnson Valley
7-22	West Salton Basin
7-23	Clark Valley
7-25	Ocotillo Valley
7-27	San Felipe (Earthquake) Valley
7-28	Vallecito-Carrizo Valley
7-29	Coyote Wells Valley
7-30	Imperial Valley
7-31	Orocopia Valley
7-32	Chocolate Valley
7-33	East Salton Basin

HYDROLOGIC SUBREGIONS

NC	North Coastal
SF	San Francisco Bay
CC	Central Coastal
SC	South Coastal
S	Sacramento
SJ	San Joaquin
NL	North Lahontan
SL	South Lahontan
CD	Colorado Desert

GROUND-WATER RESERVOIRS IN THE CALIFORNIA REGION SHOWN ON PLATES 1A, 1B, 1C and 1D

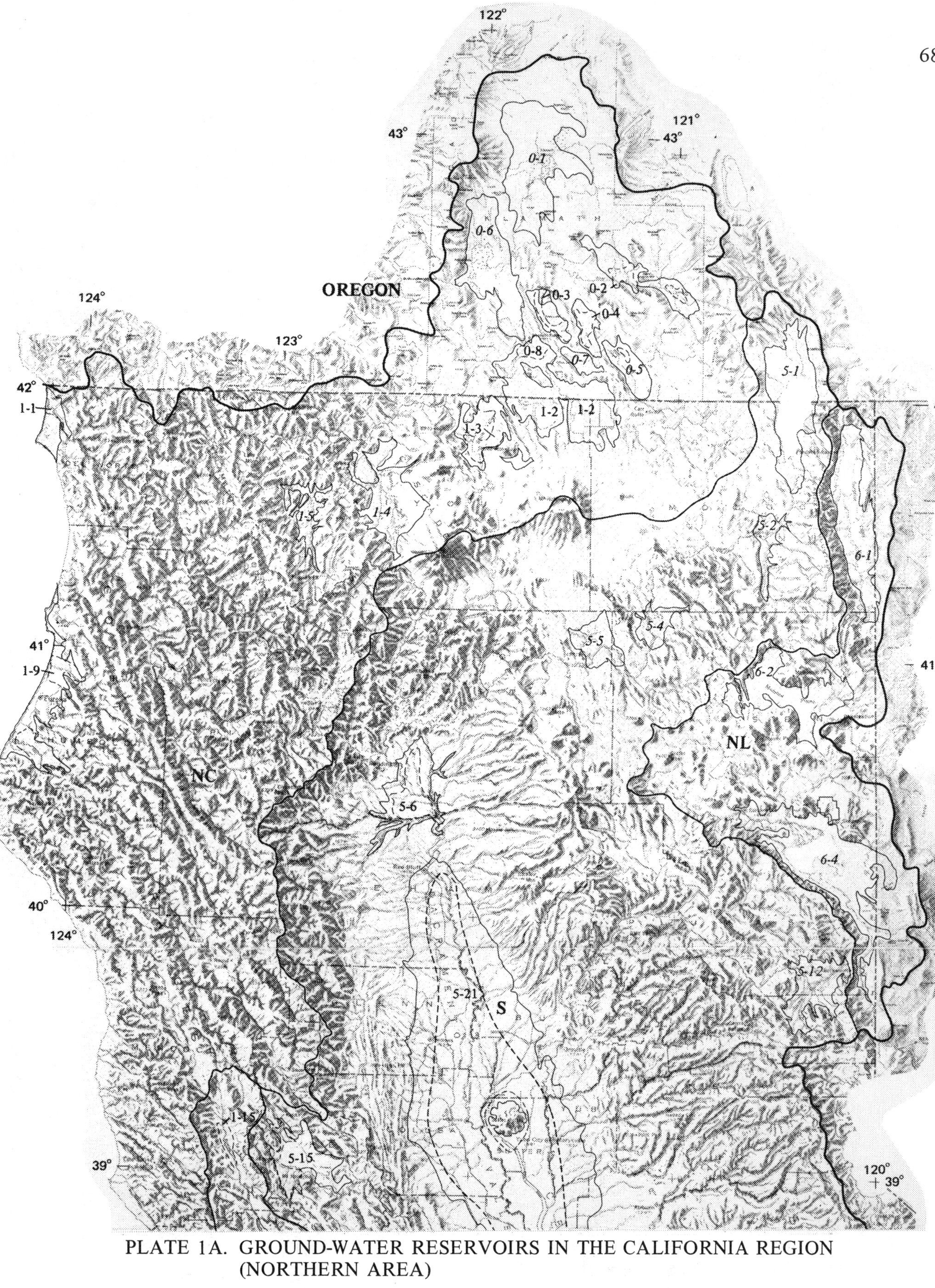

PLATE 1A. GROUND-WATER RESERVOIRS IN THE CALIFORNIA REGION (NORTHERN AREA)

PLATE 1B. GROUND-WATER RESERVOIRS IN THE CALIFORNIA REGION
(WEST CENTRAL AREA)

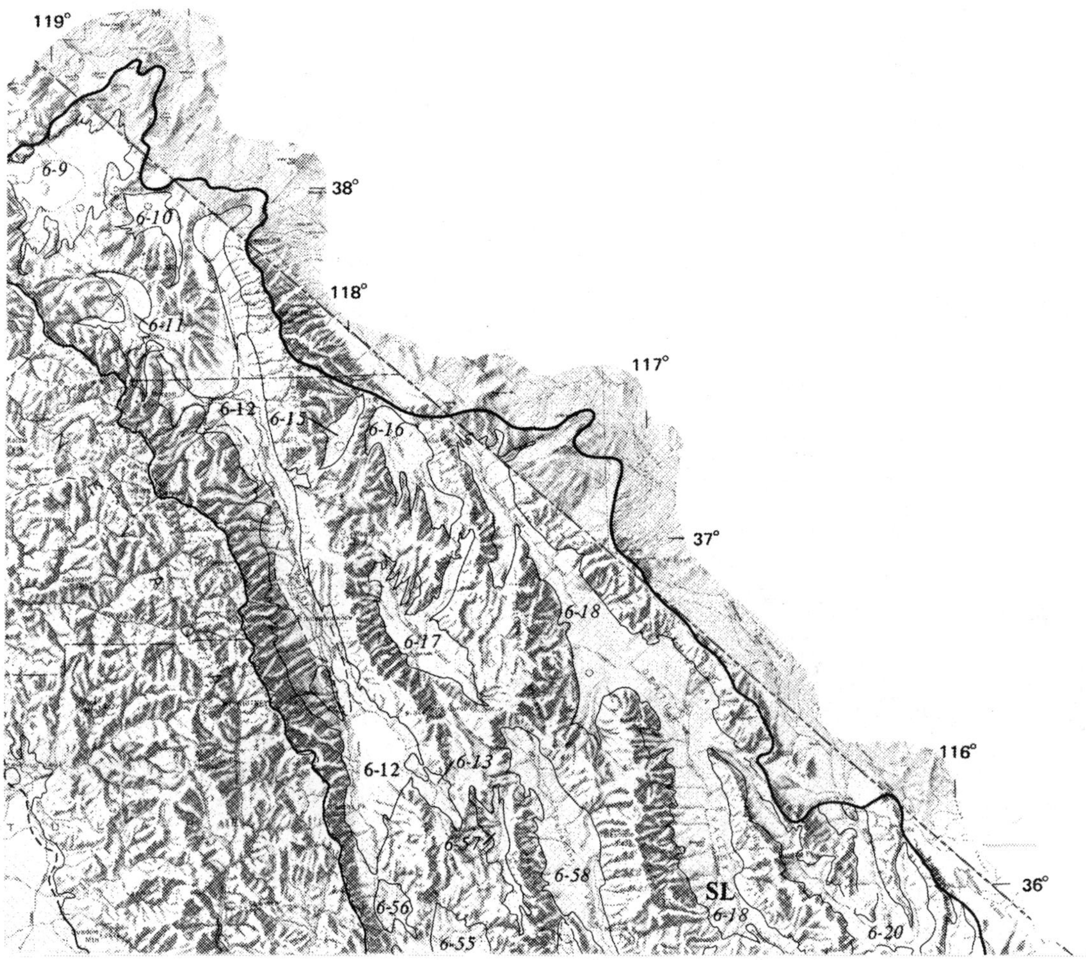

SCALE 1:2 000 000

50 0 50 100 MILES

50 0 50 100 KILOMETRES

DATUM IS MEAN SEA LEVEL

PLATE 1C. GROUND-WATER RESERVOIRS IN THE CALIFORNIA REGION (EAST CENTRAL AREA)

PLATE 1D. GROUND-WATER RESERVOIRS IN THE CALIFORNIA REGION (SOUTHERN AREA)

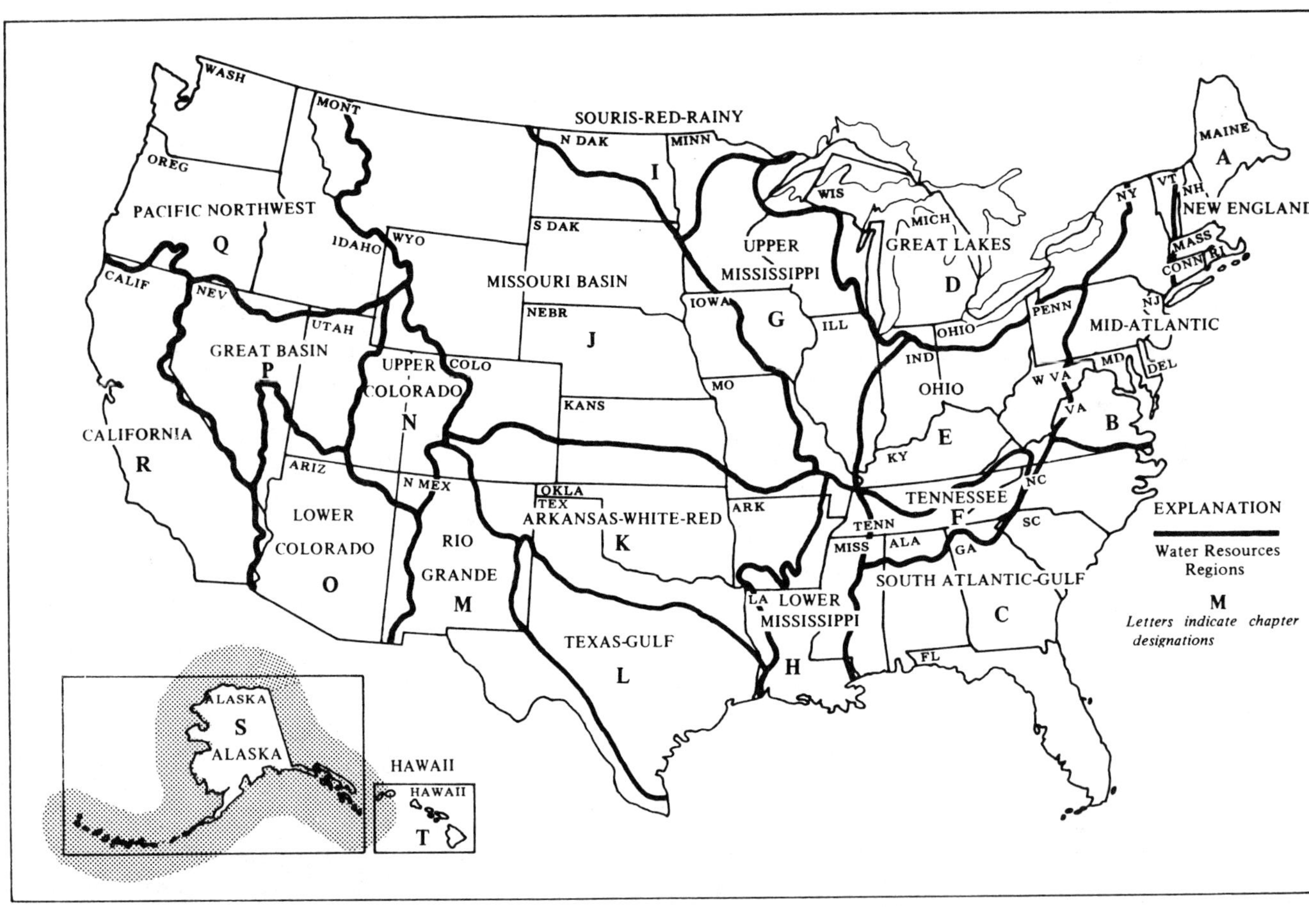

S – Location Map for the Alaska Region

S ....

ALASKA REGION

By CHESTER ZENONE and GARY S. ANDERSON

ABSTRACT

Alaska has enormous surface-water resources, but many of the streams are frozen for most of the year and most contain glacial silt that makes them unacceptable for human use. These factors lend special significance to ground water as a water-supply source, even though perennially frozen ground (permafrost) profoundly modifies ground-water flow systems in much of Alaska north of the maritime southern coast and southeastern panhandle areas. Frozen ground is a virtually impermeable layer that restricts recharge, discharge, and movement of ground water, acts as a confining layer and limits the volume of unconsolidated deposits and bedrock in which water may be stored.

Ground water is an untested resource in most of Alaska, but in many areas potential development of ground water far exceeds current use. Alluvium of major river valleys, such as the Yukon, Tanana, Kuskokwim and Susitna Rivers, probably contains the most extensive aquifers in the State. Large amounts of ground water are also stored in glacial outwash aquifers that underlie coastal basins and valleys, such as those at Kenai and Anchorage in the Cook Inlet lowland. Individual wells yielding more than 1,000 gallons per minute have been developed in the Tanana River valley, Cook Inlet lowland, and the coastal valleys at Seward and Juneau. Comparable yields should be possible in other areas that have similar geohydrologic environments. No major aquifers have been identified in glacial and glaciolacustrine deposits of interior valleys or in deltaic deposits. Major bedrock aquifers have been identified only in carbonate rocks of the Brooks Range and on the north side of the Alaska Range. Springs issuing from the carbonate rocks of the Brooks Range have discharges as great as 16,000 gallons per minute.

Most ground-water recharge occurs beneath reaches of stream channels that are losing flow to the ground-water system. Most ground-water discharge also takes place along reaches of stream channels. This discharge augments streamflows during summer and maintains low flows during winter when there is no surface-water runoff. On the basis of a streamflow hydrograph separation technique and using the 60 percent flow-duration value as an indicator of ground-water discharge, it is estimated that 25 percent of the total volume of streamflow in Alaska (exclusive of coastal, maritime environments) is contributed by ground-water discharge.

The thawing of frozen ground in the permafrost regions of Alaska causes construction and engineering problems. Disturbance of the ground surface disrupts the natural thermal equilibrium and tends to thaw part of the permafrost. Thawing can cause loss of strength, a decrease in volume, and an increase in erosion potential, particularly if the frozen ground is fine grained and poorly drained.

Present deficiencies in the ground-water information base are obvious limiting factors to ground-water development in Alaska. There is a need to extend the ground-water data-collection network and to pursue special research into the quantitative aspects of ground-water hydrology in cold regions, particularly the continuous permafrost zone.

INTRODUCTION

An oft-repeated, probably overworked, but still true statement about Alaska is that it contains America's last remaining true wilderness areas—areas about which there are many unknowns. There are,however, enough "knowns" in Alaska to suggest the presence of vast reserves of mineral resources. The inevitable future development of Alaska's mineral wealth, and of its forestry, fishery, and agricultural potentials will create a great demand for water, itself a critical resource.

The early, sparse developments and settlements in Alaska created few water-resource problems. Demands for water and water-resource information began with hydraulic mining for placer gold in interior and western Alaska in the early 1900's and increased thereafter with the need for water power in the south-central region and water supply for the pulp and paper industry in southeastern Alaska (Waananen and Giles, 1964). The steady growth of population and development since World War II has created more critical water-supply needs and problems, particularly at Anchorage and Kenai. Detailed hydrologic studies have been made in and around these major population centers and at several of the larger settlements along the State's limited highway system, but there are still large areas where data are insufficient to allow more than a general evaluation of water-resource potentials.

Surface water is abundant in Alaska, but the presence of glacial silt in many streams and drastic reduction or even cessation of stream-flow caused by freezing of surface water during winter may make it an unacceptable or unreliable supply source. These factors lend special significance to ground water. At the present time, however, because of the sparse hydrologic data base, evaluations of local ground-water resources must be founded largely on basic principles of ground-water occurrence and a general knowledge of regional climatic effects on the hydrologic environment.

The purpose of this report is to present a regional assessment of Alaska's ground water, an important resource that should be considered in any water development plans. The assessment includes discussions of factors that influence the occurrence, movement, and availability of ground water, and of the problems that may accompany its development and use.

No new data were acquired specifically for this report, and the present authors have drawn extensively from summaries and compilations by earlier investigators of Alaska's water resources, most notably those by Cederstrom (1952), Hopkins and others (1955), Waananen and Giles (1964), Williams (1970a), Feulner, Childers, and Norman (1971) and Balding (1976). Figure 1 is an index map of the geographic locations mentioned in the text.

CLIMATE

Alaska has a great geographic extent, spanning lat 54° to 71° N. and long 130° to 172° W.; consequently it has a large range in climate. The State is divided into four climatic zones—arctic, continental, transition, and maritime—on the basis of temperature and precipitation (Watson, 1959, and fig. 2). Mean annual temperatures range from 10°F in the arctic zone to 45°F in the maritime zone; extremes range from a high of 100°F to a low of −80°F, both of which occur in the continental zone. Recorded annual precipitation ranges from about 5 in. (inches) in the arctic zone to 300 in. along the southeast coast within the maritime zone. Heavy precipitation and relatively low temperatures in the coastal mountains of the southeastern and south-central parts of the State favor the existence of glaciers and icefields. Nearly 30 ft (feet) of snow, the deepest winter snowpack ever reported for an established long-term measurement site, has been measured at the 4,430-foot level of a small glacier near the west side of Prince William Sound (L. R. Mayo and D. C. Trabant, U.S. Geol. Survey, written commun., 1977). Perennial ice now covers 28,100 square miles (mi^2) of Alaska and is an integral part of the hydrologic regimen where it occurs.

PERMAFROST

Permafrost, or perennially frozen ground, is present throughout the State except for a narrow strip along the southern and southeastern coasts. Permafrost is the result of both the present climate and colder climates that occurred intermittently in the past (Williams, 1970a). Permafrost is discontinuous or occurs as isolated masses in the southern and interior parts of the State and is continuous further north on the northern Seward Peninsula, in the Brooks Range, the Arctic Foothills, and the Arctic Coastal Plain (fig. 3). However, even in the continuous permafrost areas, unfrozen zones generally are present under deep lakes and in the alluvium adjacent to major rivers.

THE HYDROLOGIC REGIMEN

The principles of movement of water through the hydrologic cycle, via precipitation, surface runoff, infiltration, ground-water flow, and evapotranspiration, are as valid in Alaska as in more temperate regions, but the rate of occurrence of transient hydrologic events and their distribution in time may differ greatly. Therefore, although the progression of events is not unique to subarctic and arctic regions, a flurry of hydrologic activity is concentrated in a short summer period after many months of hydrologic dormancy (Brandon, 1964).

Except in the maritime zones of south-central and southeastern Alaska, subfreezing air temperatures cause precipitation to fall as snow for 6–9 months of the year and even year-round in the high mountain regions. The water content of that snow is stored on the surface for several months until Alaska's spring melt or "breakup" period begins, rather than being immediately available as runoff to streams and as recharge to the ground-water body. This seasonal storage of snow gives to hydrographs of streamflow in interior and northern areas of Alaska a characteristic prolonged winter recession followed by a rapid rise when snowmelt begins in the spring. In those areas where only a thin snow cover accumulates, the ground may freeze to a depth of many feet during winter. This frozen layer acts as a relatively impermeable barrier that temporarily restricts ground-water recharge and promotes rapid runoff of snowmelt. As soon as melt runoff begins, rapid downward thawing occurs in the ground beneath meltwater channels, and ground-water recharge can then take place (Cederstrom, 1963, p. 25).

Water is stored for long periods of time in glaciers and icefields. The melting of snow and ice in glaciated areas, therefore, provides a water source not directly related in time to local precipitation. For example, the net long-term melting of glacier ice may contribute 5 percent of the total water yield in the Tanana River basin in central Alaska (Anderson, 1970). The meltwater has a regulatory or moderating effect on streamflow variability and, in turn, on ground-water recharge and discharge along alluviated glacial valleys. Given similar basin characteristics such as size, configuration, topography, and geology, streamflow variability is generally greater for nonglacial streams than for glacier-fed streams (fig. 4).

The presence of permafrost on Alaska's Arctic Coastal Plain results in a hydrologic paradox in that the area is semiarid, having less than 10 in. annual precipitation, but it contains a myriad of lakes. A study of the hydrology of a drainage basin near Barrow, Alaska, showed that summer precipitation was ap-

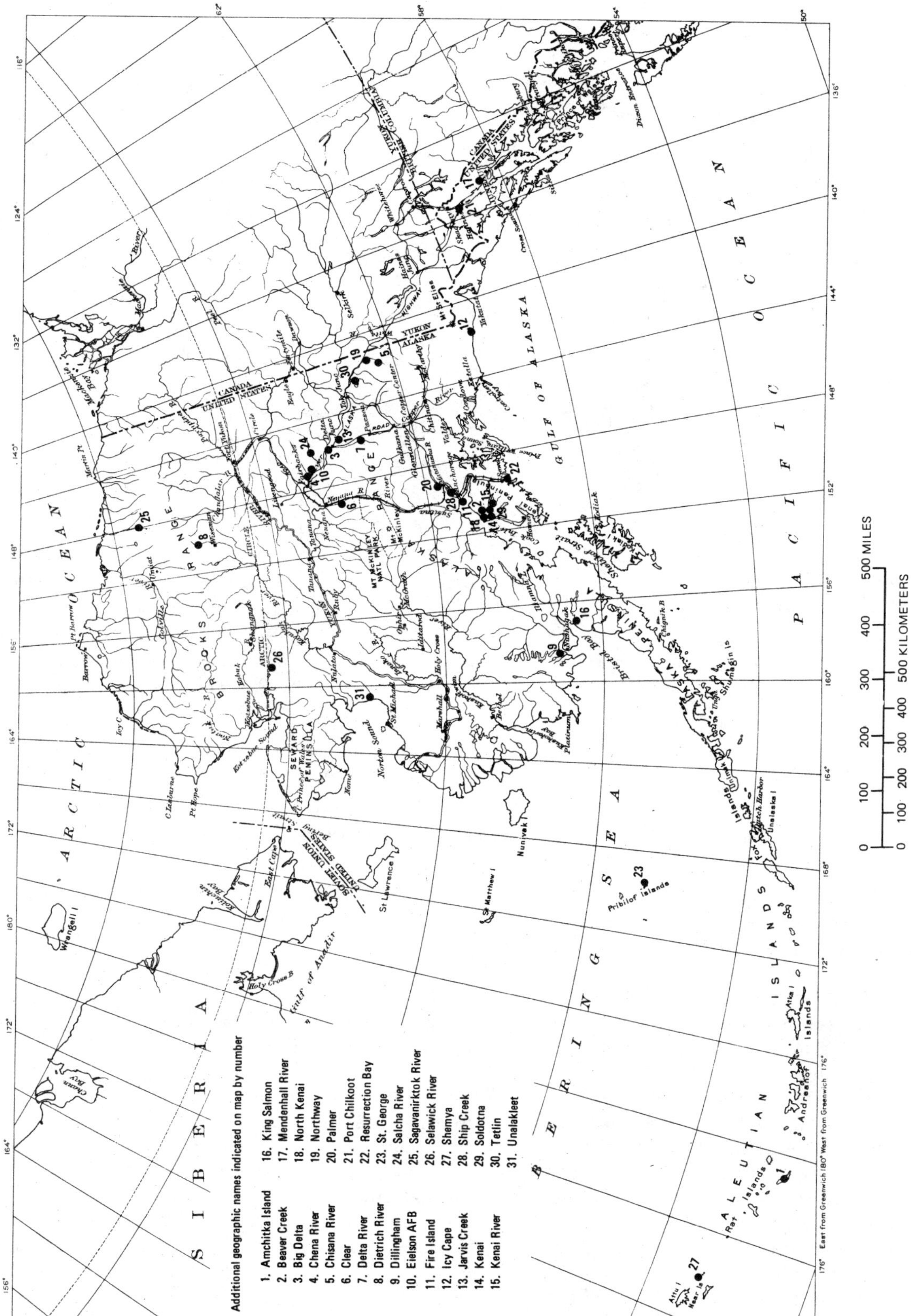

FIGURE 1.—Geographic locations mentioned in text.

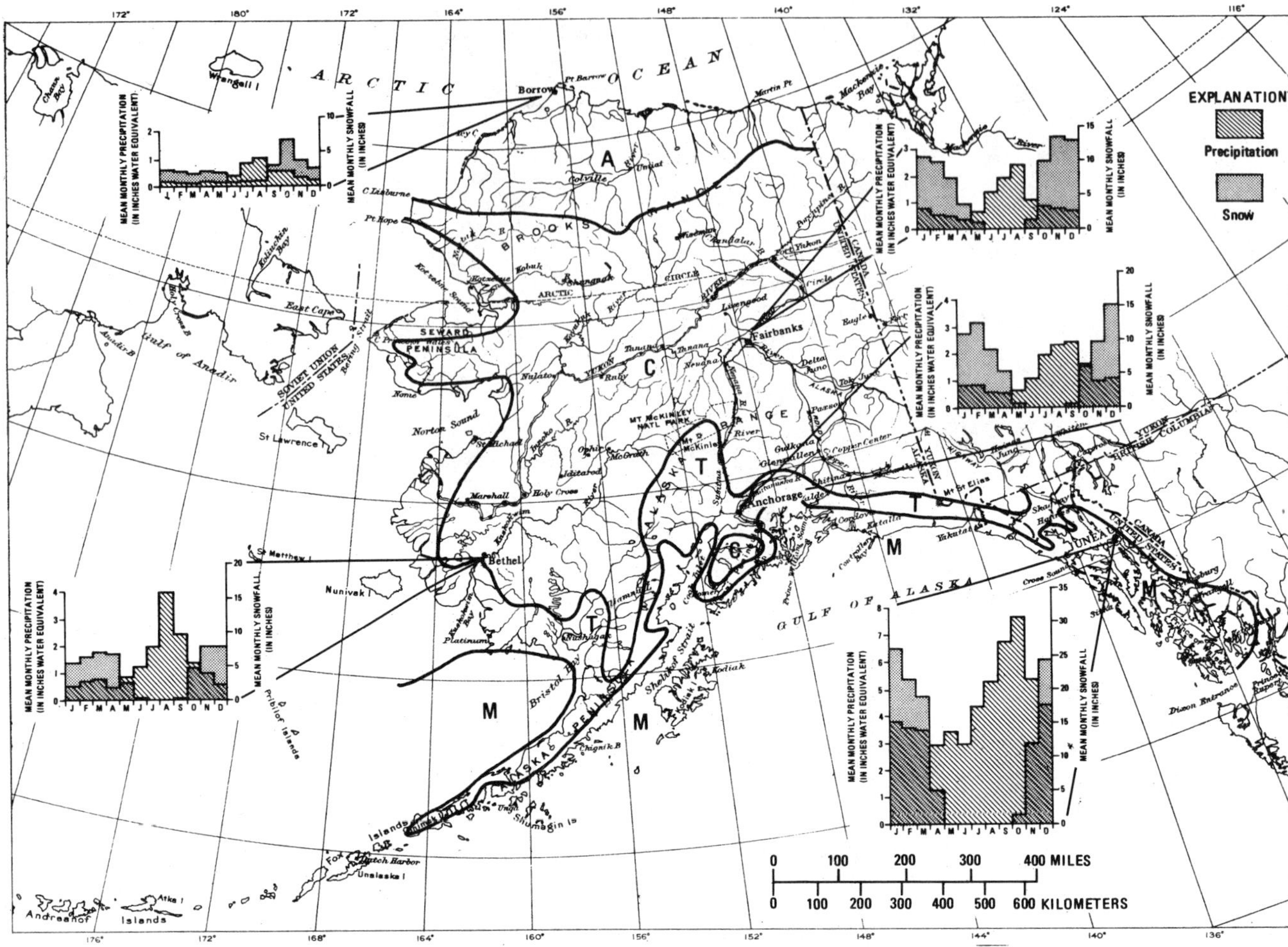

FIGURE 2.—Climatic zones of Alaska. A represents Arctic; C, continental; T, transition, and M, maritime.

proximately balanced by evapotranspiration, and that evaporation from lake surfaces was two to three times precipitation during the short (60–90 days) summer thaw season (Brown and others, 1968). However, runoff from melting ice and snow stored for the remainder of the year compensates for this summer deficit. Seasonal thawing of saturated frozen ground, the active layer, may be an important factor in maintaining summer lake levels (Marsh and Woo, 1977); presumably this release of water is also a source of surface runoff. The low permeability of frozen ground beneath the active layer restricts recharge and effectively isolates the near-surface ground-water system from the regional subpermafrost layer.

In summary, the special characteristics of the hydrologic environment in Alaska, as well as in other arctic and subarctic areas, result from the influence of climatic conditions on the physical state (solid or liquid) of water and its rate of movement through the hydrologic system.

ASSESSMENT OF THE GROUND-WATER RESOURCE

REGIONAL OCCURRENCE

The occurrence of ground water in Alaska is allied to the geologic and physiographic framework and is influenced by the distribution of permafrost in all but the south coastal regions. The effects of permafrost on ground-water systems decrease progressively from the continuous permafrost zone of northern Alaska to the southern limit of permafrost. In the continuous permafrost zone, the sediments are almost entirely frozen to depths as great as 2,000 ft (Howitt and Clegg, 1970). Ground water is usually present below the base of the permafrost, or locally, above permafrost at places where convective heatflow from surface-water bodies lowers the upper surface of permafrost (the permafrost table) below the depth of seasonal freezing (fig.5). In the discontinuous permafrost zone, the fro-

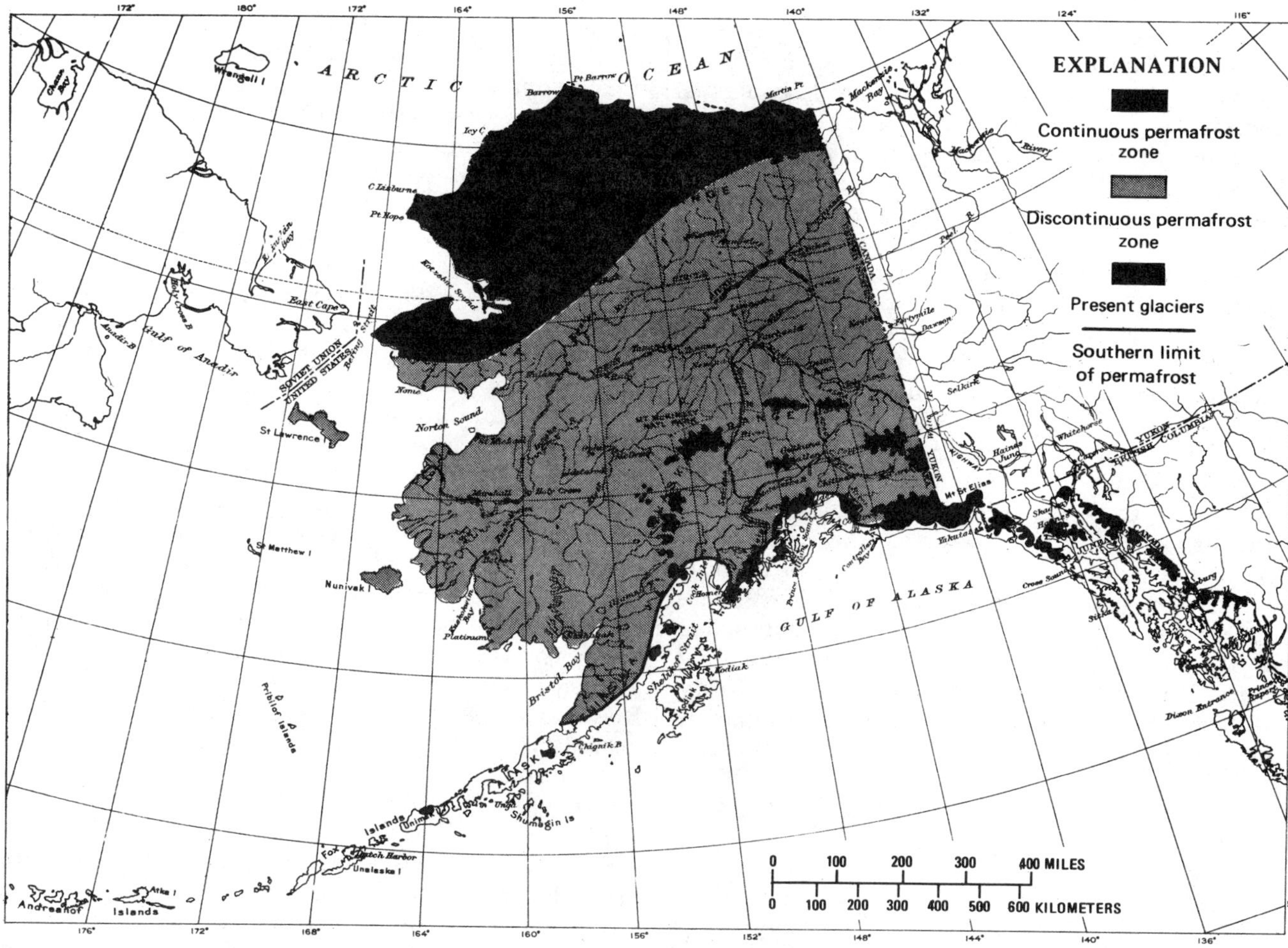

FIGURE 3.—Permafrost zones (Williams, 1970a) and location of present glaciers.

zen ground is warmer than in the continuous permafrost zone, and larger, interconnected or areally extensive unfrozen areas are present. However, even within this zone of "warm" permafrost, its local distribution is determined largely by subsurface drainage and surface insulation (Hopkins and others, 1955, p. 117), which are in turn controlled by such factors as slope, geologic materials, and vegetation cover. For example, in the Fairbanks area (fig. 6*A*) deposits of silt are extensively frozen and effectively confine ground water in the principal aquifers. However, in the Big Delta area (fig. 6*B*), which has a climate similar to that of Fairbanks, the deposits generally consist of coarse sand and gravel, and permafrost is present as isolated bodies which have little effect on the occurrence of ground water.

Alaska's geologic framework has been compared to that of western Canada and the western United States (fig. 7, insert). Major landforms include three great mountain ranges: the Brooks, Alaskan, and Coastal; a broad interior lowland drained by large rivers and containing scattered highlands and plateaus; and large coastal plains, valleys, and river deltas (fig. 7). Covering the rocks of the mountain ranges and those of smaller mountains and uplands is the drift of glaciers that occupied these highlands during the Pleistocene Epoch. Outwash deposits derived from and extending beyond the limits of existing and former glaciers and materials originally of alluvial rather than glacial origin occupy the interior valleys and mantle the older rocks in large areas of the interior lowland.

Four generalized geohydrologic environments are recognized in Alaska (Williams, 1970a, p. 5): (1) alluvium of river valleys, including the flood plains, terraces, and alluvial fans of both major valleys and upland and mountain valleys, (2) glacial and glaciolacustrine deposits of the interior valleys, (3) coastal-lowland deposits, and (4) bedrock of the uplands and mountains (fig. 8).

ALLUVIUM OF RIVER VALLEYS

Of the major geohydrologic environments in Alaska, alluvium of river valleys, including the flood plains, terraces, and alluvial fans of both major valleys and smaller mountain and upland valleys, contains the greatest volume of ground water in storage. Alluvial deposits also have the greatest recharge potential because they are hydraulically connected to the extensive surface-water drainage system. The Yukon, Tanana, Kuskokwim, Kobuk, and Susitna River basins are the major ground- and surface-water systems. In the lower Tanana basin, for example, the maximum known thickness of alluvium is 2,000 ft (Anderson, 1970). In the lower coastal reaches of large river valleys such as the Susitna, the alluvial deposits, perhaps only a few tens of feet thick, are interbedded with thick lacustrine, glacial, and glaciomarine deposits.

In the continuous permafrost region, all the deposits may be frozen except those adjacent to river channels and large lakes. Williams (1970b) refers to subsurface data from two wells near the Colville River at Umiat and cites Black (1955, p. 119) to suggest that unfrozen zones (perforating taliks) may extend completely through the permafrost along the major rivers north of the Brooks Range. Williams (1970b, and written commun., 1977) also suggests that a more or less continuous unfrozen zone is present beneath the Colville River. Data from recent Geological Survey test borings do not refute the concept of perforating taliks, but they indicate that such unfrozen zones are isolated under deeper parts of the stream channel and do not form a continuous unfrozen conduit for ground water beneath the river (C.E. Sloan, oral commun., 1977). Even if such streambed aquifers are present in the continuous permafrost zone, frozen ground severely reduces the storage capacity of the aquifers in the zone as a whole.

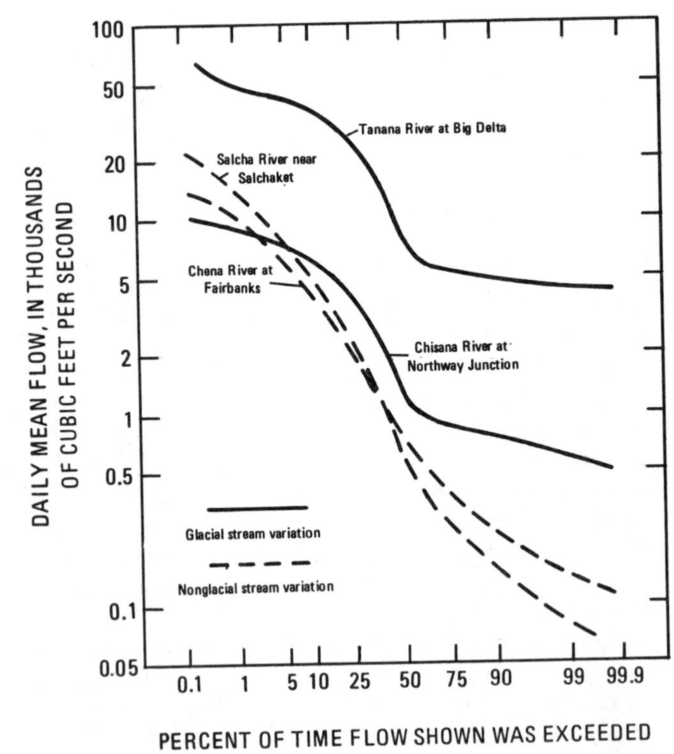

FIGURE 4.—Variability of flow in glacial versus nonglacial streams (modified from Anderson, 1970). 1. Tanana River at Big Delta, 2. Salcha River near Salchaket, 3. Chena River at Fairbanks, and 4. Chisana River at Northway Junction.

In the discontinuous permafrost zone, the volume of frozen ground does not significantly reduce the storage capacity of alluvial aquifers along the major rivers—the Yukon, Tanana, and Kuskokwim—because the thickness of permafrost is generally small relative to the total thickness of unconsolidated deposits. Alluvium along smaller tributary valleys may be mostly frozen, however, and there are areas within the major valleys, such as the Yukon Flats and the vicinity of Northway-Tetlin in the Upper Tanana River valley where frozen ground does greatly reduce ground-water storage capacity (Williams, 1970a, and written commun., 1977).

The major ground-water bodies lie beneath the base of permafrost in the discontinuous permafrost zone. At Fairbanks, productive water-bearing deposits are present beneath permafrost that varies widely and irregularly in thickness and has a maximum reported thickness of 265 ft (Anderson, 1970). Recharge and discharge of these alluvial aquifers takes place chiefly along stream channels and through lake beds and other thawed zones that perforate the permafrost.

GLACIAL AND GLACIOLACUSTRINE DEPOSITS OF INTERIOR VALLEYS

Glacial deposits consist largely of till laid down during the Pleistocene epoch as the glaciers advanced from major mountain ranges into the interior plains and lowlands (fig. 7). During the retreat of glaciers, the till was covered by permeable water-worked deposits, some of which were buried by till of later glacier advances. Where the glaciers advanced into proglacial lakes, beds and lenses of clay, silt, sand, gravel, and stony clay or silt are interbedded with the till. The sediments are predominantly fine grained and are frozen in northern Alaska and at the higher altitudes of the south-central part of the State. The largest deposits of glacial and glaciolacustrine sediments are in the Copper River lowland. Glacial deposits of smaller extent occur in valleys of the southern Brooks Range and the Alaska Range. Although these deposits are

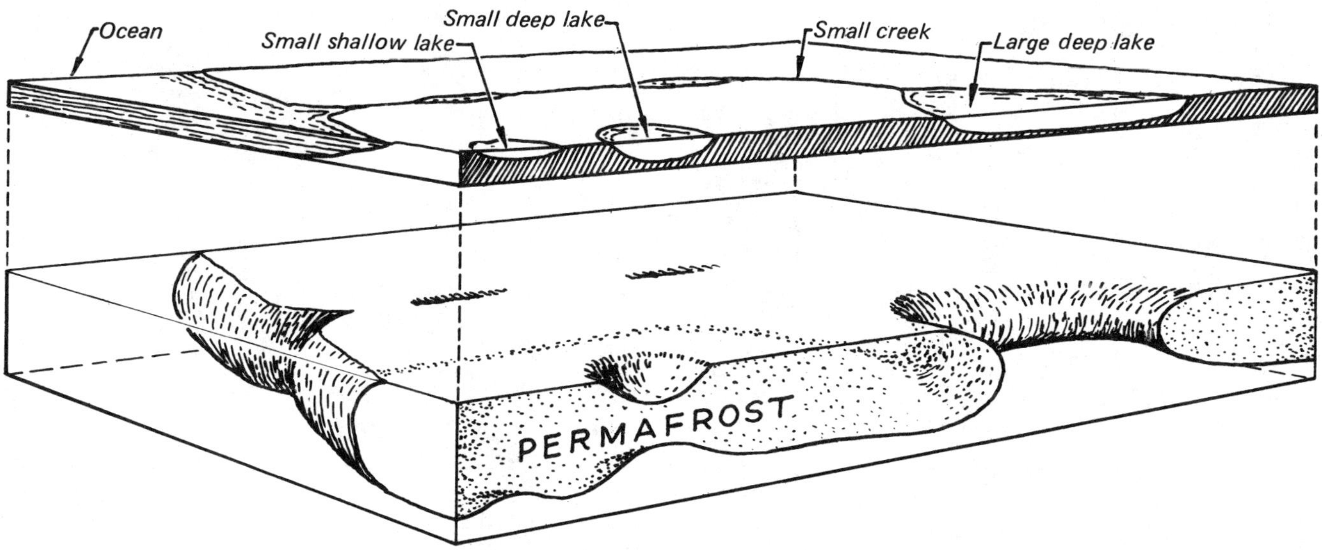

FIGURE 5.—Schematic representation of the effect of bodies of water on the configuration of permafrost in the continuous permafrost zone (Williams, 1970a, modified from Lachenbruch and others, 1962).

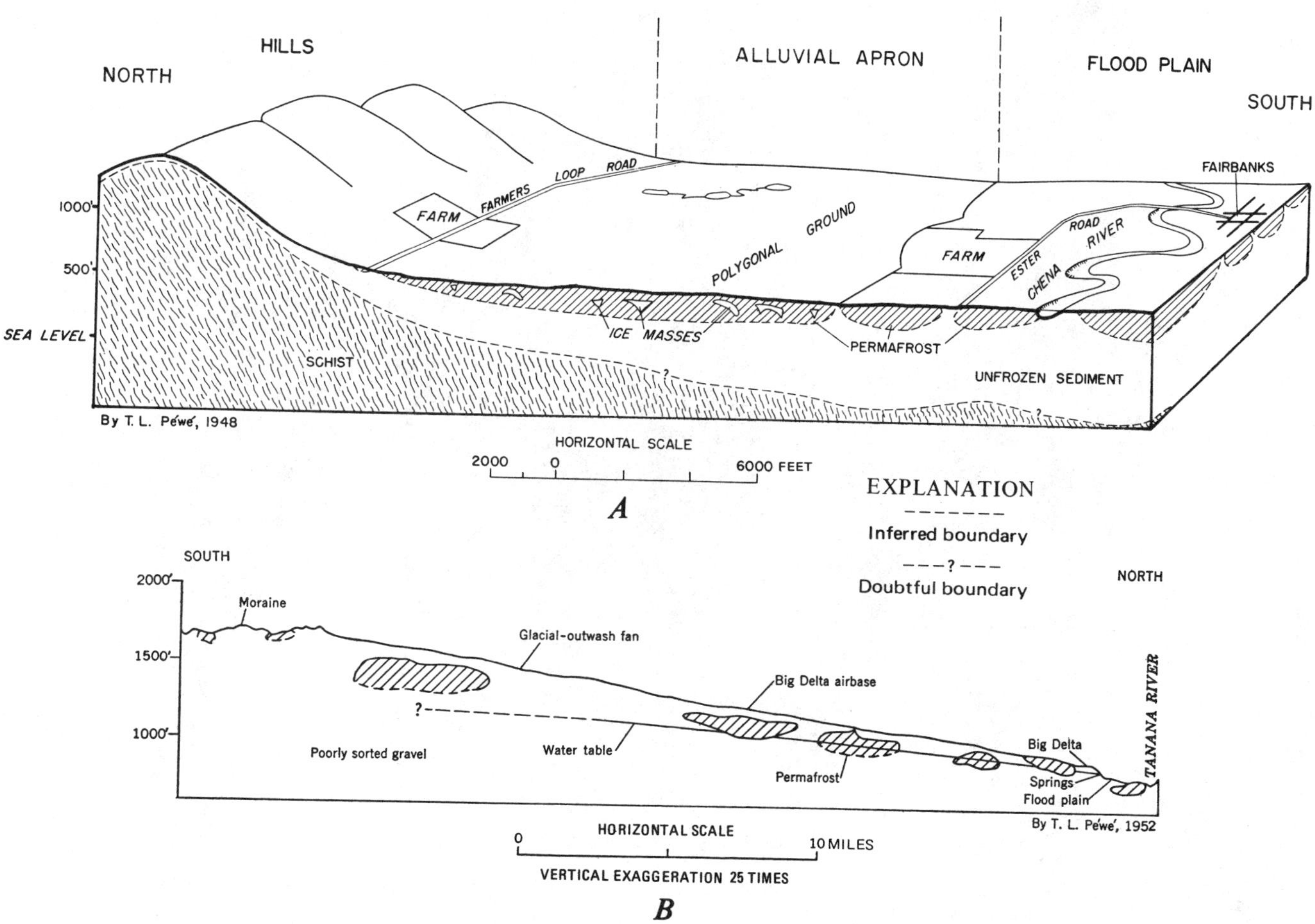

FIGURE 6.—Diagrammatic sketches showing: *A*, character and distribution of permafrost in the Fairbanks area; *B*, cross section of permafrost and ground water in the Big Delta area (from Hopkins and others, 1955).

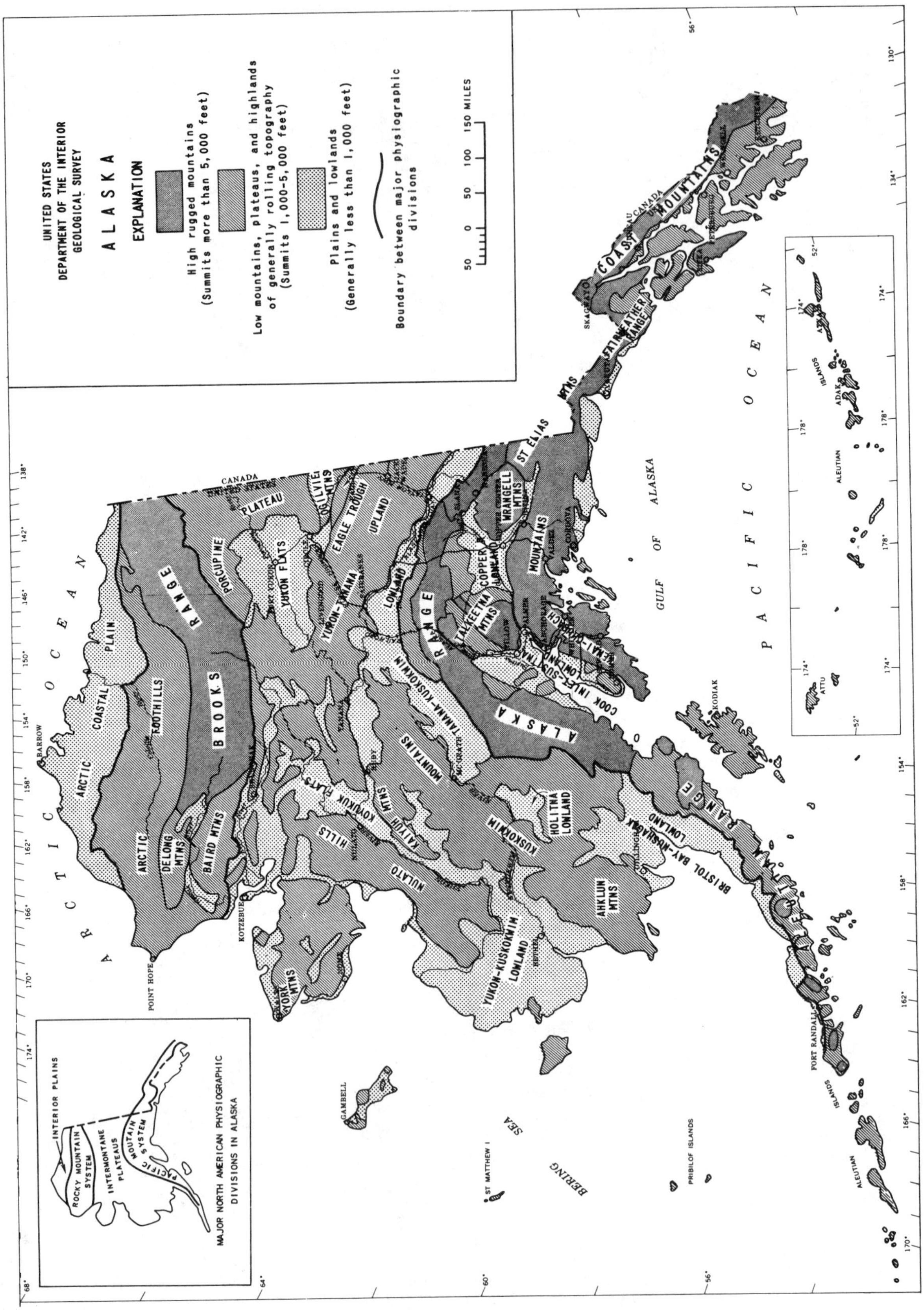

FIGURE 7.—Physiographic provinces of Alaska (Wahrhaftig, 1965).

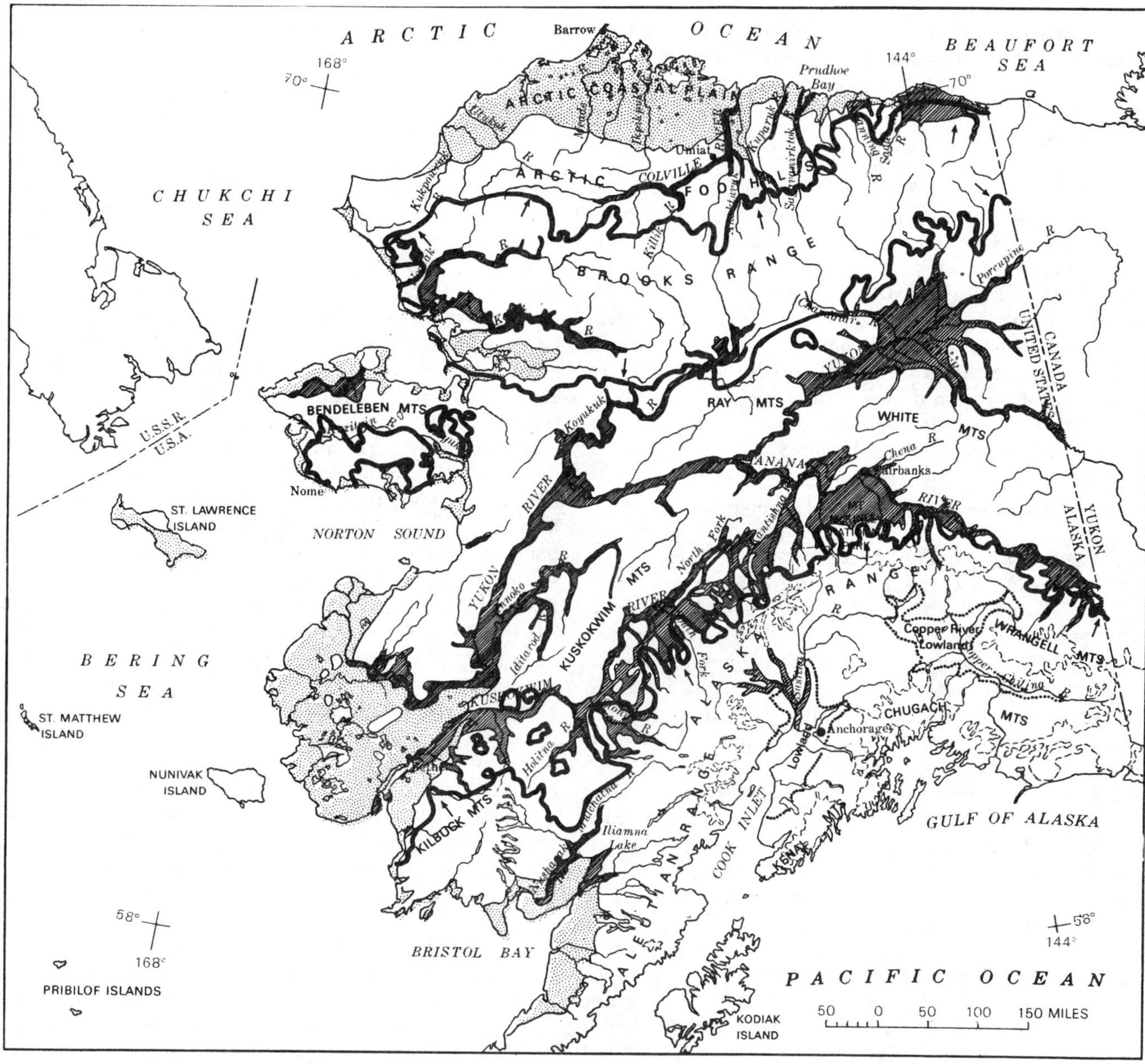

EXPLANATION

Alluvium of major valleys
Sand, gravel, and silt of flood plains, low terraces, and alluvial fans

Coastal-lowland deposits
Chiefly silt and sand, and subordinate amounts of gravel; includes sand and gravel of bars, beaches and spits, and sand, silt, and gravel of deltas. Within limits of Pleistocene glaciation, include till, glaciomarine, and glacioestuarine deposits

Bedrock of mountains and uplands
Chiefly bedrock mantled locally by weathered bedrock rubble reworked by frost action, alluvium, and eolian deposits, and, within limit of Pleistocene glaciation, by glacial deposits

Limit of Pleistocene glaciation
Unconsolidated deposits within this limit may include till, sand and gravel, and silt and clay of glacial, glaciolacustrine, glaciomarine, or glacioestuarine origin. (Coulter and others, 1965)

Cook Inlet and Copper River Lowlands
Approximate upper limit of lacustrine silt, clay, sand, gravel, and stony silt and clay of extensive glacial lakes in Copper River Lowland and of glaciolacustrine, glacioestuarine, and glaciomarine deposits in the Cook Inlet lowland

Major glaciers and ice fields

Contact

FIGURE 8.—Geology and major geohydrologic environments of Alaska (Williams, 1970a; modified from Coulter and others, 1965; and Karlstrom and others, 1964). Note: Southeastern Alaska, the Alaska Peninsula and the Aleutian Islands fall virtually entirely within the category "Bedrock of mountains and uplands."

widely distributed, they have not been extensively explored for ground water and probably do not contain major aquifers except in southern Alaska, where permafrost either is absent or occurs only as small isolated masses.

COASTAL LOWLAND DEPOSITS

The coastal lowland deposits of Alaska reflect various combinations of several depositional environments. They can be generalized to include: (1) constructional shoreline features formed on land or in shallow water, including coastal plains, deltas, bars, spits, and beaches, and (2) coastal basins and valleys whose basic configuration was produced chiefly by structural or erosional forces and which subsequently were filled by sediments from both marine and nonmarine (glacial and alluvial) sources.

CONSTRUCTIONAL SHORELINE DEPOSITS

Coastal plains.—The Arctic Coastal Plain is the largest coastal lowland deposit in Alaska; it is underlain by as much as 200 ft of unconsolidated silt, sand, and gravel (Black, 1964). Permafrost generally extends from the land surface through the entire thickness of the unconsolidated sediments and into the underlying bedrock. In the Prudhoe Bay area, the permafrost is approximately 2,000 ft thick (Howitt and Clegg, 1970). Permafrost restricts ground water to local, discontinuous thawed zones under deep lakes and river and to places where springs issuing from bedrock maintain a thawed conduit. The water from bedrock beneath permafrost is generally brackish to saline. To the southwest, the coastal-plain deposits at Nome have been extensively explored during placer gold-mining operations. Some ground water is present there, but its movement and storage are restricted by permafrost. A narrow, discontinuous coastal plain borders the Gulf of Alaska from the Copper River southeast to Icy Cape. Exploratory drilling for oil and gas in this area revealed that alluvial, glacial, marine, and beach sediments overlie great thicknesses of sedimentary rocks (Plafker, 1967; Miller, 1971), but the hydrologic characteristics of the deposits were not reported. Farther southeast at the city of Yakutat, which is on a glacial moraine, one water well was drilled through 335 ft of unconsolidated, predominantly fine-grained glacial sediments that included 60 ft of coarse-grained sand and gravel. Petroleum-exploration wells on this coastal plain are reported to have reached bedrock at depths of less than 500 ft. Because of the high precipitation and lack of permafrost in this area, recharge rates are assumed to be high. However, freshwater aquifers are probably quite thin because of the low relief and proximity to the Gulf of Alaska.

Deltas.—Deltas of the Colville, Noatak, Kobuk, and Selawik Rivers and the compound delta of the Yukon and Kuskokwim Rivers form a large part of the coastal lowlands of northern and western Alaska. Smaller deltas are formed by several other rivers in south-central and south-eastern Alaska; the largest of these is the Copper River delta. Subsurface data are available only from the Yukon-Kuskokwim and Copper River deltas. Sediments of the deltas consist of sand, silt, and thin beds of gravel. Yukon-Kuskokwim deltaic sediments are nearly 1,000 ft thick at the site of an exploratory oil well near Bethel (Williams, 1970a). The base of deltaic sediments near the mouth of the Copper River was not reached in a 200-ft-deep well east of Cordova. Permafrost is widespread in the upper part of the deposits of the Yukon-Kuskokwim delta. It extends from near the surface to a maximum known depth of 603 ft near Bethel. However, a comparison of river stage and ground-water levels at Bethel (Waller, 1957) suggests an hydraulic connection between the river and ground-water system, and thus a lack of permafrost beneath the river. If the sediments of major deltas are in hydraulic connection with their associated rivers, the potential for induced ground-water recharge from streamflow is enormous. Ground-water availability in the deltas is limited, then, by predominantly fine-grained sediments and to a greater extent by permafrost in northern and northwestern Alaska. Additionally, the thickness of the freshwater aquifer is determined by the position of the freshwater-saltwater interface, which is probably very shallow near the distal edge of the deltas.

Bars, spits, and beaches.—These depositional forms border the entire coastline of Alaska. They may contain only a small amount of ground water, but its presence is significant at many Native villages and military sites which require an economical but not necessarily large source. The distribution and availability of fresh ground water in these deposits depends on several factors. Along the Arctic Ocean coastline, ground water is perched on permafrost and the thickness of the water body depends on the rate of recharge from rainfall and snowmelt, the depth to the permafrost, and the time since storm waves last inundated the area. In areas without permafrost, the presence of fresh ground water depends on the rate of recharge and discharge and consequently the depth to the freshwater-saltwater interface. Along the Gulf of Alaska coastline, abundant recharge from precipitation, freshwater lagoons, and streamflow create conditions favorable for the occurrence of shallow ground water in permeable coastal sediments.

DEPOSITS OF THE COASTAL BASINS AND VALLEYS

Deposits of the coastal basins and valleys consist of till interbedded with sand, silt, clay, and gravel of fluvial and marine origin. The largest and best-known ground-water system of this type is that of the Cook Inlet lowland, particularly in the Kenai and Anchorage areas. Petroleum-exploration records indicate that unconsolidated sediments reach a maximum thickness of more than 4,000 ft in the Cook Inlet lowland, but in most wells the thickness is less than 1,000 ft. The sediments are predominantly interbedded glacial till, fine-grained glaciolacustrine deposits and more permeable water-worked deposits of sand and gravel. Throughout the Cook Inlet lowland, small quantities of ground water occur under unconfined conditions near the surface or are confined in thin water-bearing lenses within the till or glaciolacustrine deposits. Greater volumes of recoverable ground water are found in coarse-grained alluvial deposits, ranging in thickness from 10 to 100 ft, such as those forming the confined glacial-outwash aquifers at Anchorage. These deposits may be of limited areal extent, however, and their location is unpredictable without exhaustive subsurface exploration. Ground-water conditions in the Bristol Bay lowland are inferred to be similar to those in the Cook Inlet lowland, except for the effects of more extensive and thicker permafrost. Permafrost is present in the Bristol Bay lowland in approximately one-fourth of 600 wells and borings for which records are available; the maximum reported thickness of frozen ground is 175 ft (Williams, 1970a). The potential for recharge of the ground-water system by induced infiltration of streamflow through unfrozen, permeable channel and floodplain sediments could be realized by ground-water development efforts near the active stream channel. Away from the central part of the river valleys of the Cook Inlet and Bristol Bay lowlands, recharge takes place principally by infiltration of rainfall and snowmelt, and recharge rates are highly variable because of the complex stratigraphy and wide range of permeability of glacial sediments.

Ground-water supplies can be developed from alluvial fills at the heads of fiords along the Gulf of Alaska. In the Seward area at the head of Resurrection Bay, unconsolidated sediments have been penetrated by drill holes to a depth of 480 ft and are estimated from geophysical surveys to be more than 1,000 ft thick. The sediments consist of glacial till, glaciofluvial sand and gravel, marine deposits of fine sand and silt, and stream-deposited sand and gravel. Recharge rates to these deposits are assumed to be high because streams crossing the alluvial fans that border the basin lose a large part of their flow to the ground-water system. The volume of fresh water within the deposits is not known, but fresh water was present several hundred feet below sea level in wells drilled near the coastline. The freshwater aquifers are confined beneath marine silt and clay. At Valdez, where ground-water conditions are assumed to be similar to those at Seward, unconsolidated sediments have been penetrated to a depth of 250 ft and may be as much as 600 ft thick (Coulter and Migliaccio, 1966).

Alluvium-filled coastal valleys in mountainous southeastern Alaska are generally smaller than those of the Gulf of Alaska coast. However, Barnwell and Boning (1968) reported the presence of a fresh ground-water reservoir as thick as 700 ft in the Mendenhall River valley near Juneau.

BEDROCK

In approximately 75 percent of Alaska, glacial and alluvial deposits are thin, poorly permeable, or absent. In such areas, appreciable amounts of ground water occur only in aquifers consisting of consolidated rocks. The occurrence of ground water in bedrock is described here under four general rock types: carbonate rocks, sandstone, volcanic rocks, and metamorphic and intrusive igneous rocks.

CARBONATE ROCKS

Although carbonate rocks are quite extensive in southeastern Alaska, they have not been explored and utilized there as a water source. The Brooks Range area, which lies entirely in the continuous permafrost zone, also contains extensive carbonate-rock deposits. Ground-water movement in these rocks apparently has developed solution cavities and conduits through which the water flows to the surface. Conclusions concerning the occurrence and movement of ground water in carbonate rocks of the Brooks Range are inferred from measurements and observations of spring flow in both the eastern Brooks Range (Childers and others, 1973, 1977) and the Baird and DeLong Mountains area in the western Brooks Range (C. E. Sloan, U.S. Geol. Survey, written commun., 1976). The hydrology of springs in the latter area was described by Waller (1966) and by Piper (1966). Craig and McCart (1974) studied both the northern and southern flanks of the Brooks Range and reported the existence of approximately 25 springs which emerge from carbonate rocks. Some springs in the Brooks Range have discharges as great as 16,000 gal/min.

SANDSTONE

Sandstone and alternating strata of indurated sand, silt, and clay are widespread throughout the State.

Although these deposits have been extensively explored for oil, gas, and coal, few hydrologic data have been reported. Such deposits have been explored for ground water in the western Kenai Peninsula, where they are poor aquifers because of low permeability.

VOLCANIC ROCKS

Volcanic rocks are widespread in Alaska, but they are especially prevalent in some areas in the western part of the State, on the Aleutian Chain, Alaska Peninsula, the Seward Peninsula, and in the Wrangell Mountains. The occurrence of ground water in volcanic rocks is highly variable, and only a few wells are known to produce water from them. On Shemya Island near the west end of the Aleutian Chain, the volcanic rocks are dense and unfractured; consequently they are poor aquifers (Feulner and others, 1976). By comparison, the volcanic rocks on St. George Island have much greater permeability because of layering of flows, interbeds of gravel or scoria, and cooling fractures within the rocks (Anderson, 1976). Yet the occurrence of fresh ground water on St. George Island is limited, not because the rocks do not have the capacity to receive and store water, but rather because they are so permeable that the ground water drains rapidly to the sea. The presence of many springs suggests that basaltic lava flows in the mountains and uplands of the northern Seward Peninsula are promising sources of ground water (Hopkins and others, 1955, p. 122).

METAMORPHIC AND IGNEOUS ROCKS

These rocks make up the cores of the principal ranges of Alaska. Ground water occurs chiefly in faults and fractures within these rocks. Some areas of fractured slate on Kodiak Island yield a modest amount [50–100 gal/min (gallons per minute)] of water (S. H. Jones, Chester Zenone, and R. J. Madison, U.S. Geol. Survey, written commun., 1977). Elsewhere, wells in such rocks generally produce less than 10 gal/min. Probably the most intensive development of bedrock aquifers is in the uplands near Fairbanks (fractured schist), Anchorage (slate and metagreywacke), and a few places on the Kenai Peninsula and southeastern Alaska. Ground water is generally present only in weathered zones at the bedrock surface and in fractures. The probability of penetrating zones of secondary porosity below a depth of 300–500 ft diminishes rapidly, and many deep wells are reported to be dry.

GROUND-WATER STORAGE AND AVAILABILITY

It is estimated that 1.5×10^{14} cubic feet of ground water, with a dissolved-solids concentration of less than 3,000 mg/L (milligrams per liter), is stored in aquifers in Alaska (A. J. Feulner, written commun., 1975). Estimates of the amount of ground water in storage are academic, however, if they cannot be related to the quantity that can be transmitted by the aquifer and economically developed. Aquifer transmissivity, or water-transmission property, which varies greatly in Alaska and is strongly influenced by permafrost, exerts the major control on the availability of ground water that is present. For example, the porosity of predominantly fine-grained glacial and glaciolacustrine sediments is large, and a great volume of water is stored in the sediments. However, the amount of water that can be efficiently recovered is limited by low permeability of the fine-grained sediments themselves and by the limited extent of coarser, more permeable horizons within those sediments. In permafrost zones, even the coarse-grained potential aquifers may be frozen.

There have been attempts at ground-water development in most areas of the State. However, only at Anchorage and perhaps at Fairbanks, Kenai, and Juneau is the basic framework of the ground-water system reasonably well known. Elsewhere, data are too scarce to describe the areal extent, thickness, and hydraulic boundaries of the aquifers. In the areas for which data are scarce, ground-water availability is deduced from reports of the few wells that have been drilled and from what is known of the lithology, permafrost distribution, potential recharge, and base flow of streams, which reflects ground-water discharge (see later section). This paucity of data, combined with the known complexity and wide range of geologic and hydrologic conditions in Alaska, permits preparation of only a very generalized map of ground-water availability on a regional basis. Potential yields of properly constructed (screened and adequately developed) wells are indicated in figure 9. Exceptions to the indicated yields can and do occur; the map is intended to represent average conditions only.

Individual yields greater than 1,000 gal/min have been obtained from wells in alluvial deposits in the Tanana valley (at Fairbanks, Eielson AFB, Big Delta, and Clear), Cook Inlet lowland (Anchorage and Kenai), and in the glacier-carved coastal valleys at Juneau and Seward. Development of comparable well yields should be possible in other areas that have similar geohydrologic environments and comparable permafrost conditions, that is, the major alluvial valleys, large alluvial fans, coastal valleys and basins, and perhaps in some volcanic rocks of the Seward and Alaska Peninsulas. Attempts to develop ground water in the carbonate rocks of the Brooks Range have been largely unsuccessful because of thick (600–800 ft)

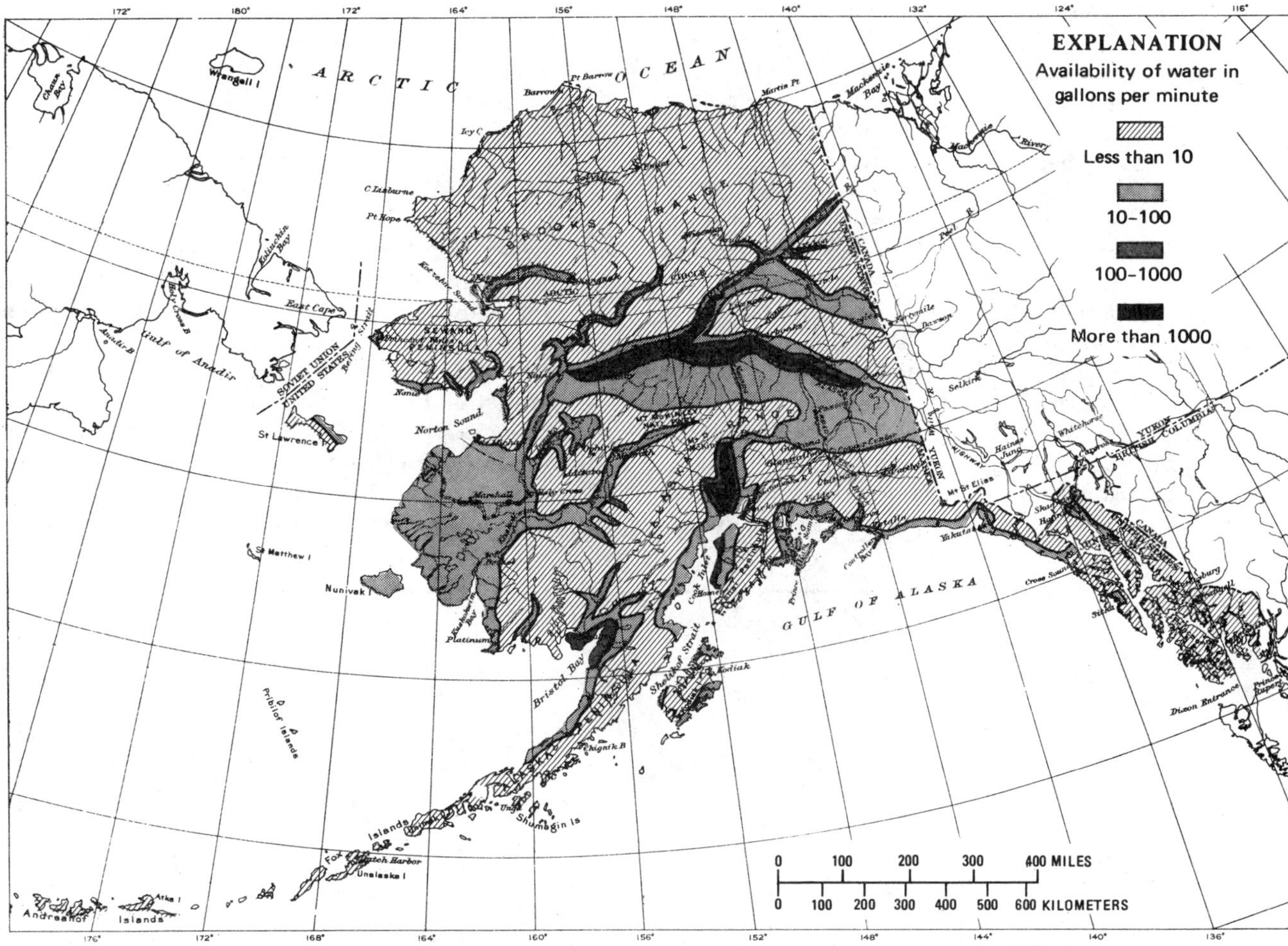

FIGURE 9.—Generalized availability of ground water (Feulner and others, 1971).

permafrost (J. R. Williams, U.S. Geol. Survey, written commun., 1977), but flows ranging from 500 to as much as 16,000 gal/min have been measured at springs issuing from those rocks. Yields of 100–150 gal/min have been reported from deltaic sediments of the Yukon-Kuskokwim Rivers. Larger yields should be recoverable from permeable sand and gravel zones within the several hundred feet of unfrozen material beneath the base of permafrost. Well yields of less than 10 to as much as 20 gal/min are common for most areas of igneous or metamorphic bedrock, glacial and glaciolacustrine deposits (Copper River lowland), and in the permafrost regions of nothern and northwestern Alaska.

GROUND-WATER RECHARGE

Ground water is considered to be a renewable resource because practically all usable supplies are ultimately derived from precipitation. Total volumes of water available for recharge on an annual basis depend in part on, and thus vary with, annual amounts of precipitation. Ground-water reservoirs are replenished either by direct infiltration of precipitation or by percolation from surface-water bodies. Recharge rates also vary with surficial geologic conditions. A perennial supply of ground water is generally assumed to be available if the amount discharged from the aquifers does not exceed the rate of replenishment.

The highest ground-water recharge rates in Alaska occur beneath alluvial stream channels that are losing flow to the ground-water system. Infiltration through the permeable sediments in these channels and adjacent flood plains probably accounts for most of the total volume of recharge. Direct recharge from precipitation is not significant in the uplands of Alaska. Precipitation is not abundant except in the coastal areas and in the higher mountains of the interior (fig. 2). The winter precipitation in the mountains is temporarily stored as snow or ice and there is only minor surface-water loss to evapotranspiration or ground-water infiltration. Snowmelt and icemelt from the mountains, as well as most of the surface-water runoff from bedrock and permafrost areas, is eventually concentrated in the major alluvial valleys.

The relationship between ground water and surface water in Ship Creek near Anchorage was discussed by Weeks (1970) (fig. 10). Stream-channel losses for Ship Creek have been calculated, on the basis of seepage investigations, to be approximately 4 ft^3/s (cubic feet per second) per mile of channel; this is equivalent to a vertical infiltration rate of 1.0 to 1.5 ft/d (feet per day) within the wetted perimeter of the channel. At Jarvis Creek near Big Delta, Anderson (1970) reported streamflow losses of 10 ft^3/s per mile of channel, for a vertical infiltration rate of 1.5–4 ft/d. These infiltration rates are comparable to those measured in artificial recharge experiments on the Ship Creek alluvial fan near Anchorage (Anderson, 1977) where a rate of 1.4 ft/d was measured in one basin and a rate as great as 6 ft/d was estimated for a second basin.

Ground-water recharge from the infiltration of precipitation and snowmelt in the Anchorage area is estimated, from analog- and digital-model results, to be 5–9 in/yr (inches per year) or about 30–50 percent of the mean annual precipitation (R. S. George, U.S. Geol. Survey, oral commun., 1976). Such infiltration along the Chugach Mountains front is considered to be a significant source of recharge for the confined ground-water system at Anchorage. In the Beaver Creek valley near Kenai, where the drainage basin has low relief and the surficial deposits consist of generally fine-grained glacial outwash, the estimated mean annual ground-water recharge from precipitation is 4 in/yr or about 25 percent of the mean annual precipitation. This estimate for Beaver Creek basin

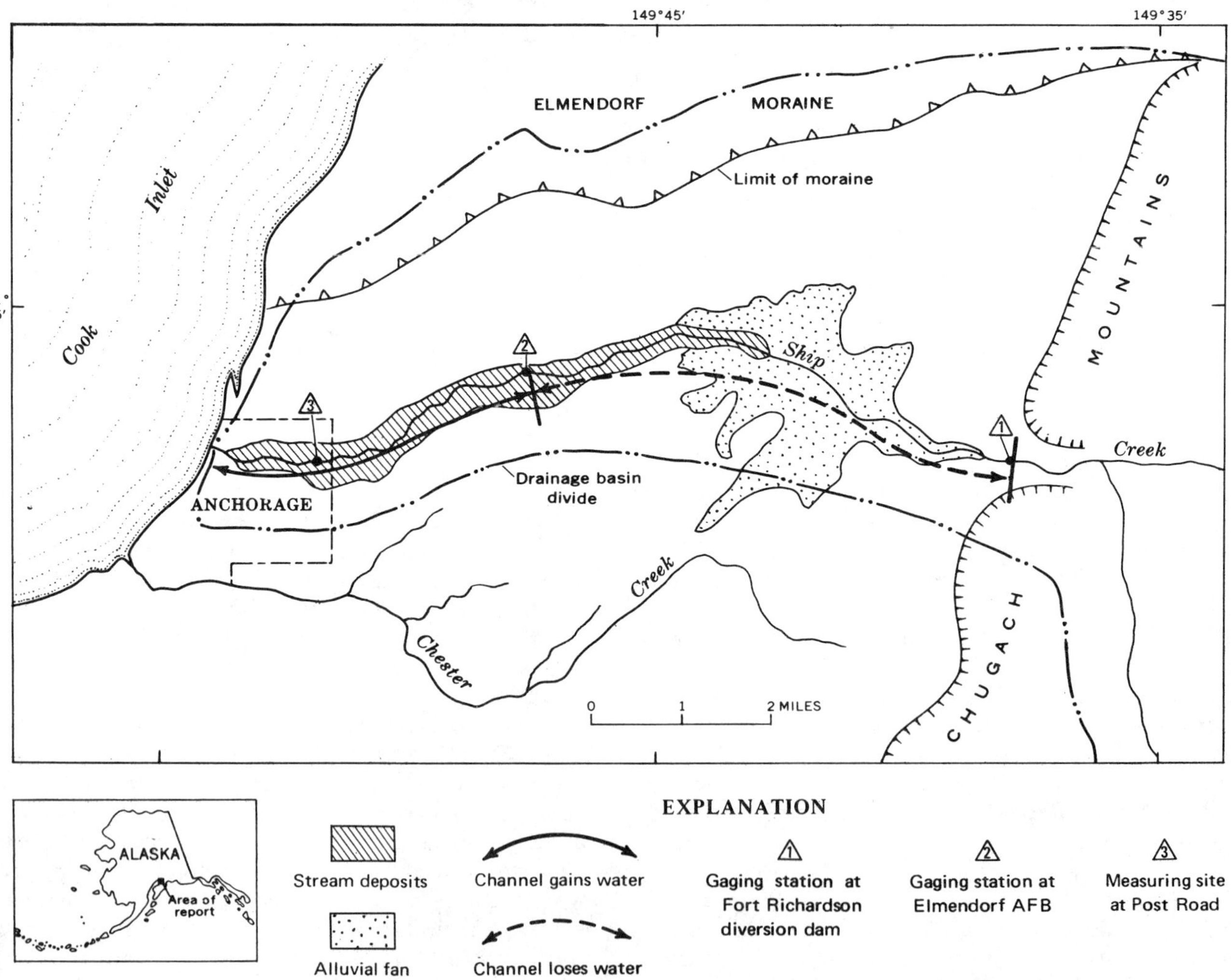

FIGURE 10.—Lower part of Ship Creek basin showing gaining and losing reaches of the stream channel (Weeks, 1970).

was derived from a streamflow hydrograph separation (Anderson and Jones, 1972). In digital-model simulations of the North Kenai area (Dames and Moore, 1976), it was estimated that about 40 percent of annual precipitation, or 7–9 in/yr, ultimately becomes part of the ground-water body. In North Kenai, the surficial deposits are coarser and better drained than those of Beaver Creek valley and there is virtually no surface-water runoff.

Over most of the land area of Alaska, either permafrost or poorly permeable bedrock is present at or near the surface, and there is very little infiltration from precipitation.

GROUND-WATER DISCHARGE

Ground water is discharged both naturally and artificially. It is discharged naturally from water-table aquifers as seeps or springs, as ground-water runoff to streams whose channels have cut below the water table, and by evaporation and transpiration. Natural discharge from artesian aquifers occurs by vertical leakage through confining beds or at springs where the aquifers crop out. Both types of aquifers also discharge freshwater along coastlines. The winter flow of many Alaskan streams is maintained by discharge of ground water. In interior and northern regions, this discharge is most conspicuously displayed by the annual formation of extensive icings[1] along stream channels and flood plains (Sloan and others, 1976). Artificial discharge of aquifers is accomplished principally by pumping from wells. Shallow, unconfined ground water is sometimes discharged along ditches excavated to drain water-logged areas.

A great quantity of ground-water discharge in Alaska occurs naturally as inflow to streams. Because little information is available on aquifer lithology and ground-water hydraulics, probably the best quantitative estimate of regional ground-water budgets in Alaska can be made from analyses of streamflow records that are available for most areas of the State. Separation of the ground-water component (base flow) of a streamflow hydrograph is a technique commonly used to estimate ground-water discharge (Linsley and others, 1975, p. 230 and fig. 8–7). Under natural conditions, a long-term equilibrium exists between ground water entering and leaving an aquifer system. Ground-water discharge estimated from streamflow data should thus approach the long-term average ground-water recharge in the area contributing to streamflow.

Base flows of Alaskan streams, estimated from simple hydrograph-separation techniques, correspond to points on the flow-duration (cumulative flow-frequency) curve within a range of streamflows that are exceeded 50–75 percent of the time. For this report, the streamflow equalled or exceeded 60 percent of the time is taken as an indicator or index of natural ground-water discharge. Stated another way, 40 percent of the time streamflow is supported entirely by ground-water discharge. Actual ground-water discharge to many Alaskan streams may be greater than that indicated by the 60 percent flow parameter, however, because there is virtually no direct surface-water runoff during about half the year in most of the State.

The regional distribution of ground-water unit runoff and the total ground-water discharge for hydrologic subregions of the State, estimated on the basis of the 60-percent duration of flow at gaging stations, are shown in figure 11 and table 1, respectively. These estimates (exclusive of Bristol Bay drainage and the maritime areas) were made by comparing relationships between mean annual and 60-percent duration flows within hydrologically similar regions and then estimating the 60-percent flows for the corresponding mean annual basin runoffs reported by Balding (1976). The estimate of ground-water discharge for the Kotzebue Sound subregion seems anomalously high for an area underlain by permafrost. This value was based on analysis of streamflow data for the Kobuk River, which may not be representative of the entire subregion. However, the headwaters of the Kobuk River basin contain many springs that discharge ground water from limestone rocks of the Brooks Range.

In the Bristol Bay subregion, the 60-percent flow duration estimating techniques gave a ground-water discharge rate of 47,000 ft^3/sec, or unit base flow of 15.9 in. This estimate was not considered to be representative of ground-water discharge because all gaged basins in the subregion contain considerable lake storage that augments winter streamflow. For this subregion, then, a ground-water discharge value equal to 25 percent of streamflow, which is the average value for the State if Bristol Bay and the maritime areas are excluded, was selected as a first approximation.

Reliable estimates of ground-water discharge based on streamflow data cannot be made for southeastern Alaska and the Gulf of Alaska coast, where large amounts of precipitation fall (see fig. 2) and most streams receive snowmelt throughout the year. These

[1]An icing is a buildup of surface ice formed by freezing of water that discharges to the surface from seeps and springs, or from streams whose channel flow is constricted by freezing and forced to flow above the original "normal" winter ice cover. The summer flow of several large rivers in northeastern Alaska is augmented by the melting of very large icings.

areas are characterized by steep, mountainous and (or) glaciated terrain underlain by igneous and metamorphic rocks that have low permeability. In most places only a thin mantle of glacial and alluvial deposits is present. Ground-water discharge from these areas probably would not increase the total estimated ground-water discharge for the remainder of Alaska by more than 20 percent. Excluding estimates for the Bristol Bay, southern coastal and southeastern Alaska areas for the reasons noted above, total ground-water discharge (equal to recharge) for the remainder of the State is estimated to be 159,200 ft^3/s, or 100,000 Mgal/day.

WATER QUALITY

A data base adequate to describe areal variations in the chemical quality of ground water is available for only a few places, such as Juneau, Kenai, Anchorage, and Fairbanks, and data is virtually nonexistent for large areas of the State. Known dissolved-solids concentrations range from less than 25 mg/L in shallow ground-water in stream channel alluvium, to as much as 64,000 mg/L in saline waters in shallow coastal wells, but most of the sampled ground water contains less than 400 mg/L dissolved solids (table 2) and is acceptable for most uses. In inland areas, calcium bicarbonate- or calcium magnesium bicarbonate-type water is most common. Water of sodium bicarbonate or sodium chloride type is present in coastal areas.

The quality of ground water causes problems, however, such as those related to saline waters in coastal areas and in springs and deep artesian wells in the Copper River lowland, relatively high mineralization of subpermafrost water in some areas, and magnesium sulfate bicarbonate-type waters from bedrock in the Tanana River valley. Iron is present in objectionable concentrations in a large proportion of shallow wells in most areas of the State. Undesirable concentrations of other constituents also occur in ground water in Alaska. Chemical analyses (U.S. Geol. Sur-

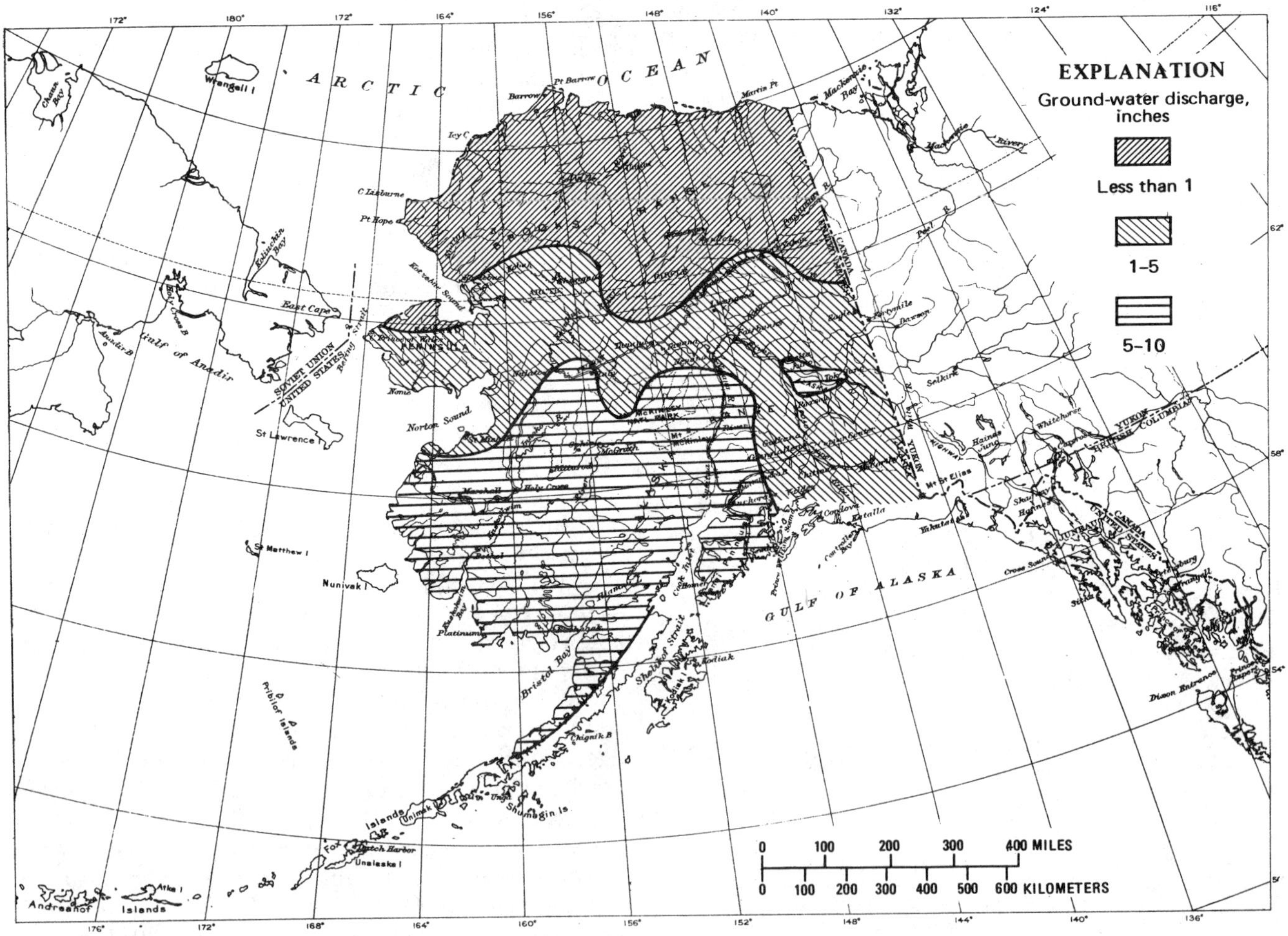

FIGURE 11.—Mean annual ground-water component of streamflow.

TABLE 1.—*Summary of surface-water and ground-water discharge by hydrologic subregions and subareas of Alaska*

[Mean annual surface-water runoff estimated following method by Balding (1976). Estimated ground-water discharge determined using flow exceeded 60 percent of the time (from flow duration curves) as an index, except for values in parentheses, which are approximations discussed in text]

Subregions and subareas	Area (mi^2)	Mean annual surface-water runoff: From within unit	Mean annual surface-water runoff: Including upstream inflow	Estimated ground-water discharge: Total ft^3/s	Estimated ground-water discharge: Inches of water per unit area
1. Arctic:					
1:1 West Arctic	31,000	14,800	—	200	0.1
1:2 Colville	24,000	14,000	—	200	.1
1:3 East Arctic	26,000	15,200	—	200	.1
2. Northwest:					
2:1 Kotzebue Sound	41,000	59,100	—	10,600	3.5
2:2 Norton Sound	26,000	18,500	—	1,900	1.0
3. Yukon:					
3:1 Lower Yukon	38,000	43,000	262,000	18,200	6.5
3:2 Central Yukon	19,000	28,300	219,000	11,000	7.9
3:3 Koyukuk	33,000	23,100	—	2,000	.8
3:4 Upper Yukon	60,000	34,700	128,000	7,000	1.6
3:5 Tanana	45,000	39,200	39,600	12,600	3.8
3:6 Upper Yukon—Canada	9,000	6,140	—	1,700	2.6
4. Southwest:					
4:1 Kuskokwin Bay	58,000	71,200	—	31,200	7.3
4:2 Bristol Bay	40,000	120,000	—	(30,000)	(10.0)
4:3 Aleutian	11,000	22,400	—	no estimate	—
5. Southcentral:					
5:1 Kodiak-Shelikof	11,000	54,200	—	no estimate	—
5:2 Cook Inlet	38,000	91,300	—	19,600	7.0
5:3 Gulf of Alaska	34,000	61,800	—	12,800	5.1
6. Southeast	42,000	388,500	458,000	no estimate	—
Totals		1,105,000		[1]159,200	

[1]Equivalent to 10 × 10⁴ million gallons per day.

TABLE 2.—*Summary of DS (dissolved-solids) concentrations data in ground waters of Alaska*

[Normal range is defined as ± one standard deviation of the mean; approximately 68 percent of all values would fall within this range]

Area	Number of DS values	Average DS (mg/L)	Normal range of DS (mg/L)
Juneau (McConaghy, 1969)	141	199	26– 372
Haines-Port Chilkoot (McConaghy, 1970)	14	209	66– 352
Copper River lowland (U.S. Geol. Survey, unpub. data)			
Wells less than 100 ft deep	11	535	55–1015
Wells greater than 100 ft deep	13	1936	382–3490
King Salmon (Feulner, 1963)	16	158	94– 222
Bethel (Waller, 1957, and U.S. Geol. Survey, unpub. data)	36	157	124– 190
Homer (Waller, 1963)	27	314	139– 489
Kenai-Soldotna (Anderson and Jones, 1971)	140	209	50– 368
Anchorage (Donaldson, and others, 1975)	567	159	93– 225
Fairbanks (U.S. Geol. Survey, unpub. data)	502	227	112– 342
Nome (Waller and Mathur, 1962)	8	354	188– 520

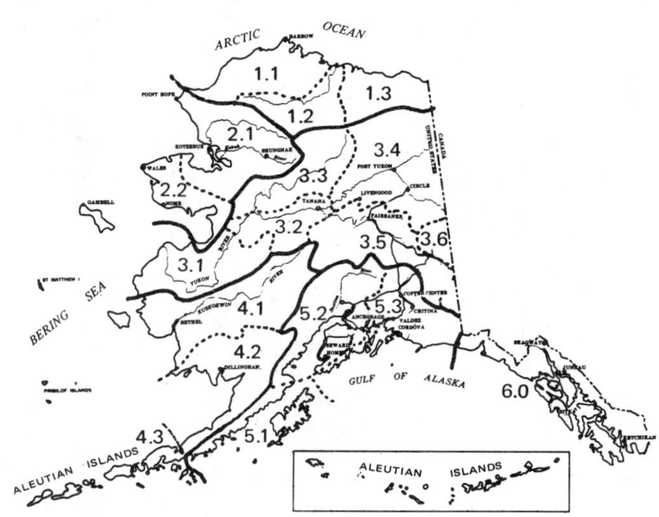

vey, unpub. data) show undesirably high concentrations of mercury in the ground water in the Yukon-Kuskokwim delta area, and high concentrations of nitrate are reported in occur locally in the Palmer and Fairbanks areas. Other examples are: organic materials in ground water of alluvial stream channels and flood plains, and arsenic in ground water in the bedrock hills north of Fairbanks (Hawkins, 1976) and in glaciolacustrine clay at Anchorage and Kenai.

The role of permafrost in directly imparting a particular type of mineralization to ground water is probably minor. Reduction of the rate of ground-water movement by permafrost provides a longer residence time for reactions between the water and the enclosing rocks than in regions without permafrost (Williams and van Everdingen, 1973). Chemical reaction rates and saturation concentrations of some constituents are also affected by low water temperatures in permafrost areas. Ground water beneath permafrost is usually of nearly constant quality at a particular site, although it may differ in composition from one area to another. Ground water above permafrost is subject to variation in quality due to seasonal reduction in recharge and possibly also because of fractionation due to freezing of part of the aquifer. Feulner and Schupp (1963) found such a seasonal variation in ground water developed by infiltration gallery at Cape Lisburne in northwestern Alaska.

Streams that receive ground-water inflow show a seasonal variation in concentration of dissolved minerals because of the changing proportions of ground water and surface water that contribute to streamflow (Brandon, 1964). Thus surface-water quality data, particularly during low-flow (or base-flow) periods, can be used as an indication of regional ground-water quality. Because ground water is generally more highly mineralized than surface water, the highest concentrations of dissolved solids occur during low-flow periods. Dissolved-solids concentration of a stream can be still further increased during low-flow periods in winter as freezing causes rejection of impurities from the ice phase into the remaining water (Osterkamp and others, 1975). This phenomenon can cause the winter flow of streams to have an even higher concentration of dissolved minerals than the inflowing ground water.

Dissolved-solids concentration is considered to be a useful index of overall water quality, but available data do not provide representative coverage for winter streamflow in Alaska. However, specific conductance, which is dependent on dissolved-solids content, is a characteristic commonly measured in field water-quality determinations. Ranges in specific conductance values measured during base-flow periods, November through April, for streams in the various hydrologic subregions of Alaska are given in table 3.

CURRENT DEVELOPMENT AND USE

Water use in Alaska in 1975, exclusive of large nonconsumptive uses by fish hatcheries, pulp mills, and by hydroelectric power-generating facilities, was estimated to be about 100 Mgal/d (million gallons per day) (A. J. Feulner, written commun., 1976, and table 4). Nearly half of that total was ground water. Ground water supplies an even higher proportion of total demand in some public-supply systems (50–60 percent at Anchorage) and is the only source in others (Fairbanks and Juneau). Most private domestic supply in the Anchorage and Fairbanks areas is from ground water.

TABLE 3.—*Summary of specific conductance values for base flow of streams in Alaska*

[Base-flow period November 1 to April 30 when little or no surface runoff is occurring]

Hydrologic subregion and subarea	Range of specific conductance (μmhos/cm²) of base flow
Arctic	225–350
Northwest	150–250
Yukon:	
Lower and Central Yukon	250–300
Koyukuk	350–400
Upper Yukon and Tanana	250–400
Southwest:	
Kuskokwim Bay	100–240
Bristol Bay	[1]50–100
South-central:	
Cook Inlet	100–300
Gulf of Alaska	100–250
Copper River Lowland	300–500
Southeast	(Base flow not defined)

[1]Influenced by significant volume of lake storage.

TABLE 4.—*Water use in Alaska in 1975*

[Does not include 80 Mgal/d nonconsumptive pulpmill use or 907 Mgal/d nonwithdrawal hydroelectric power use. Table based on revision of data compiled by A. J. Feulner, U.S. Geological Survey. Values in Mgal/d (millions of gallons per day)]

Use category	Ground water	Surface water	Total
Public supply	35	21	[1]56
Rural use:			
Domestic	6.2	3.0	9.2
Livestock	.03	.08	.11
Irrigation	—	.02	.02
Electric powerplant cooling	2.20	19	21
Industrial	5.0	10	15
Total	48	53	101

[1]Includes unspecified industrial uses in Fairbanks, Anchorage, Juneau, and other smaller communities.

Surface water provides most of the supply to the small population of the Arctic Slope and Northwest regions of the State (see index map on table 1). Wells supply distribution systems, chiefly for domestic use, at some small villages and at the larger communities of Nome and Unalakleet. A military installation at Cape Lisburne formerly used about 5,000 gal/d (gallons per day); this water was provided mainly by infiltration galleries installed in the alluvium of stream valleys or in fractured bedrock. Galleries have also been used to develop small ground-water supplies from streambed and flood-plain alluvium at several sites along the trans-Alaska oil pipeline route.

Ground water is the principal source of supply in the Yukon subregion. Virtually all the population, and thus most of the water use, is within the Tanana and Delta River basins; Fairbanks and nearby military installations use approximately 11–12 Mgal/d. Springs, especially thermal springs, have been used locally for many years for small farming operations and as summer resorts.

In Alaska's Southwest region, ground water provides part of the community water supply at Dillingham, and wells have been drilled at many small villages. Springs, wells, and galleries are used for water supply at military installations on Shemya and Amchitka Islands in the Aleutians.

The largest concentration of population in Alaska, and consequently the greatest development and use of water, is in the South-central region. Ground water provides about 150,000 gal/d at Palmer and about one-half of the 40 Mgal/d of water used in the Anchorage area for domestic, industrial, and commercial purposes. At Anchorage, the largest municipal wells yield up to 2,000 gal/min. There has been considerable development of ground water in the Kenai-North Kenai industrial area. Public supply systems and private, domestic water demands at Kenai and nearby Soldotna are also exclusively served by ground water. Total ground-water withdrawal in the Kenai and Soldotna areas is about 4 Mgal/d.

Ground water is the principal source of domestic, commercial, and industrial supply at Juneau, Sitka, Skagway, and at several other major communities in southeastern Alaska. The large water needs of the wood pulp processing industry in that region, however, are met almost exclusively by surface water.

PROBLEMS IN THE DEVELOPMENT AND USE OF GROUND WATER

Man impinges on the domain of ground water when he drills and pumps wells, uses aquifers as waste-disposal sites whether intentionally or accidentally, and in Arctic and subArctic areas, when he constructs buildings, roads, and pipelines on and across zones of ice-rich permafrost. The development of mineral and fuel deposits and of geothermal resources usually involves ground water. These activities, along with the development and use of ground water itself, may lead to the types of problems described below.

DECLINE OF WATER LEVELS

Pumping from a well or closely-spaced group of wells creates a cone of depression in the water level of an unconfined aquifer or in the potentiometric surface of a confined aquifer. The decline in water level or potentiometric surface may be small and temporary. However, if ground-water withdrawal exceeds recharge, the decline will become progressively greater and will result in increasingly greater pumping lifts, higher pumping costs, and ultimately, depletion of the resource.

In Anchorage, a significant cone of depression has formed in the potentiometric surface of the artesian aquifer near the Municipal well field, where withdrawal in 1975 averaged 4.2 Mgal/d. The potentiometric surface has not been permanently lowered, but fluctuates in response to pumping rates (fig. 12). Water-level data collected during a short, nonpumping period in 1969 indicate that if all ground-water withdrawal from the confined aquifer were to cease, the potentiometric surface would recover to within a few feet of the 1955 pre-pumping levels within a few months. Production of additional ground water from confined sand and gravel aquifers in the Anchorage area is currently limited by drawdown of water levels near existing wells (Barnwell and others, 1972) rather than by available resources of the aquifers.

Other areas in Alaska where artesian pressure reductions are caused by the pumping of high-yield municipal or industrial wells include North Kenai, Kanai, Soldotna, and Cordova. In the North Kenai area, a decline of lake levels is also thought to be caused at least partly by the lowering of the potentiometric surface in the confined aquifer, with consequent leakage of unconfined water downward to the confined system. Anderson and Jones (1972) reported maximum declines of 6 ft at three lakes near North Kenai industrial plants that pump large volumes (3 Mgal/d) of ground water.

Declines in ground-water levels have been observed in the upland area north of Fairbanks where low-yielding wells produce water from fractures and small perched zones in bedrock (G. L. Nelson, U.S. Geol. Survey, written commun., 1976). In this area the aquifers are unconfined and the water-level declines represent a reduction in storage. This loss of storage may be due partly to a decrease in natural recharge caused by below-average precipitation in Fairbanks over the past 4 years, but pumping from an increasing number of privately owned wells apparently is having some effect. Locally, well owners have reported water-level declines of 20–35 ft since 1973; a decline of 7 ft in a recent 17-month period was recorded at a U.S. Geological Survey observation wells in the area.

DEPLETION OF STREAMFLOW

In areas where streams are hydraulically connected to the ground-water reservoir, pumping water from nearby wells may cause interception of potential streamflow or diversion of water from the stream. In Alaska this interconnection of surface and ground waters may have significant effects on streamflow, particularly in winter when no surface runoff is occurring.

Some communities and individuals utilize water from shallow wells drilled in stream-valley alluvium, but there are few quantitative data relating to streamflow depletion caused by these ground-water withdrawals. Infiltration galleries near streams provide water supplies at some military installations in Alaska (Feulner, 1964; Feulner and Williams, 1967). Galleries or shallow (20 ft or less) wells near active stream channels were used for water supply at former construction camps along the Dietrich and Sagavanirktok River sections of the trans-Alaska oil pipeline route (Sherman, 1973; J. R. Williams and C. E. Sloan, oral commun., 1976). These ground-water

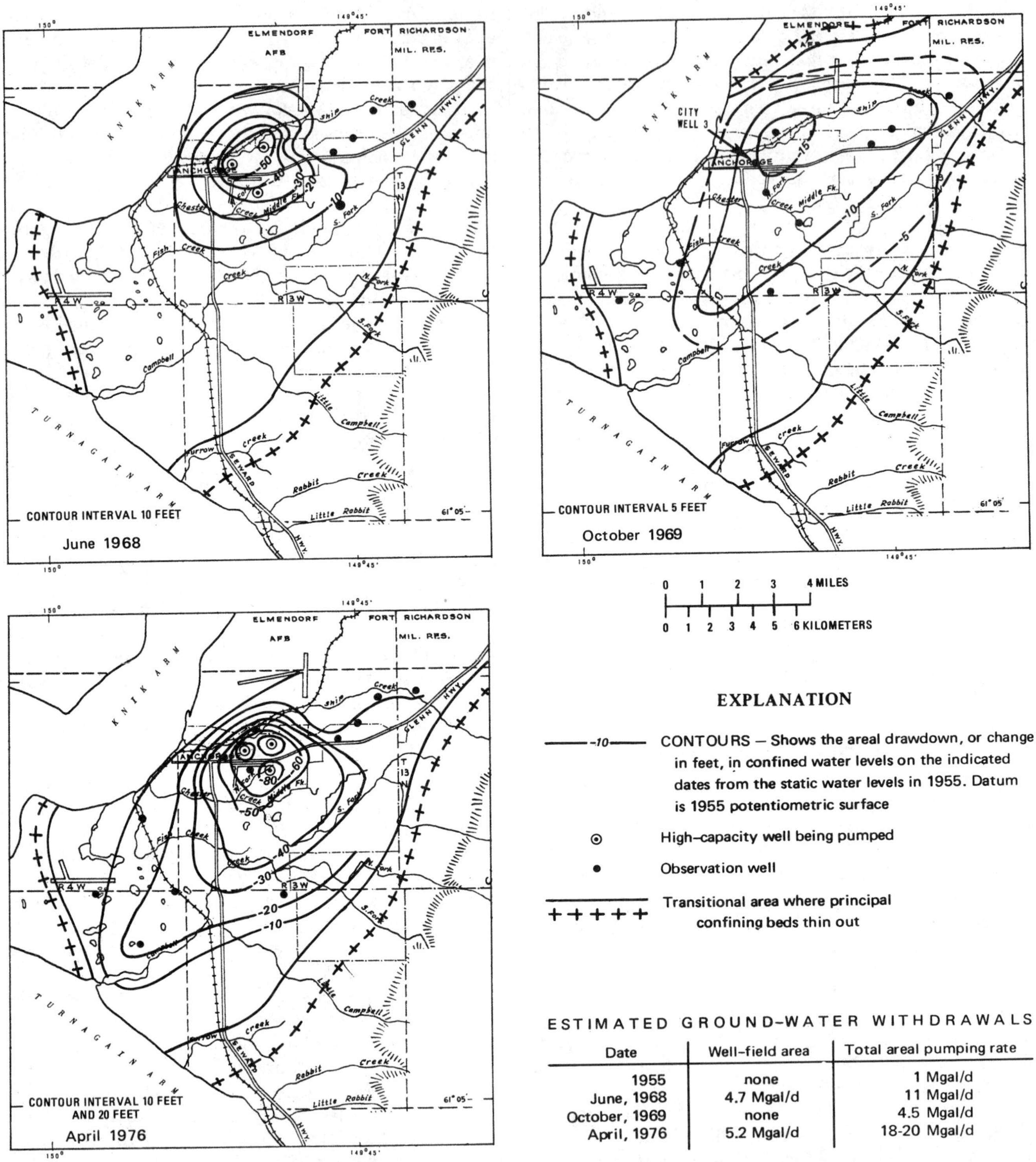

ESTIMATED GROUND-WATER WITHDRAWALS

Date	Well-field area	Total areal pumping rate
1955	none	1 Mgal/d
June, 1968	4.7 Mgal/d	11 Mgal/d
October, 1969	none	4.5 Mgal/d
April, 1976	5.2 Mgal/d	18-20 Mgal/d

FIGURE 12.—Changes in water levels caused by pumping from artesian aquifers at the Anchorage Municipal well field (modified from Barnwell and others, 1972).

developments rely amost exclusively on recharge from the nearby surface-water source; consequently, some streamflow depletion inevitably occurs. This depletion can be significant in winter when no surface runoff is occurring and all water is pumped from storage in deep pools or from alluvium under the pools.

The relationship of surface water and ground water along the channel of Ship Creek near Anchorage was discussed earlier (p. 14). Water required for the proposed expansion of fish-rearing facilities along lower Ship Creek would rely chiefly on ground water recovered from shallow wells and infiltration galleries in streamside alluvium (Freethey, 1976). Hypothetical streamflow depletion curves, derived by L. D. Patrick and others (U.S. Geol. Survey, unpub. data 1976–77), show that the proposed development would ultimately reduce streamflow by an amount nearly equal to the ground-water withdrawal. The streamflow could be totally depleted during winter low-flow periods.

POLLUTION OF THE GROUND-WATER RESERVOIR

Deterioration of water quality caused by man's own activities is commonly termed pollution or contamination. The cone of depression and resulting change in gradient of the water table or potentiometric surface caused by pumping from wells can induce the movement of pollutants already in the aquifer toward the pumping center. If this occurs, wells may have to be abandoned and new water sources sought. The slow movement of ground water and the slow process of natural dilution make the effects of water pollution long-lasting, even though pumping or the discharge of pollutants to the ground water stops.

Pollution may be caused by actions or land-use practices not directly involving ground-water withdrawal. The concentration of people and activities that accompanies urbanization promotes the generation of large amounts of waste materials. In some cases, the wastes are disposed of or stored at sites where there is a potential for ground-water pollution. Ground water has been polluted locally in the Anchorage area by solid-waste disposal directly into lakes which are hydraulically connected to the ground-water body or by surface disposal at places where the water table is at very shallow depth (Zenone and others, 1975). Solid-waste disposal in frozen ground in the Fairbanks area is in effect a storage of material from which a contaminating leachate could be produced if the frozen mass of waste and adjacent permafrost were to be naturally or artificially thawed (Straughn, 1972).

Local pollution of ground water has been caused by septic tank effluent in unsewered areas of Anchorage (Dearborn and Barnwell, 1975) and at Fairbanks (G. L. Nelson, oral commun., 1976). The present problem area at Anchorage is the mountain slope environment east of the city where bedrock is mantled by only a thin, poorly permeable layer of unconsolidated glacial materials. The quality of water of some private water supplies has also been impaired by drainage of surface water into wells that were either improperly constructed or had damaged casing at the ground surface. Other potential sources of ground-water pollution in heavily urbanized areas are leakage of sanitary sewers and infiltration of runoff from streets (deicing chemicals) and lawns (fertilizers). The disposal of primary-treated sewage effluent, either through injection wells or by surface spreading and infiltration, is being considered as an alternative to the present ocean outfall at Anchorage's sewage-treatment plant. Such practices create yet another potential for ground-water pollution, even though they have been tested elsewhere and shown to be at least partly successful methods of purifying and recycling wastewater (Bouwer, Lance and Riggs, 1974; Bouwer, Rice and Escarcega, 1974; Satterwhite and others, 1976).

The practice of using lagoons for wastewater disposal and treatment at remote construction camps and villages throughout Alaska is a potential source of pollution for ground-water supplies. Some of these lagoons are too close to shallow water-supply wells and galleries. Successful operation of the wells and galleries may depend on the permeability and porosity of the same small body of unfrozen material as that being recharged by the lagoons. Accidental oil or fuel spills that inevitably occur during major construction projects, and disposal of petrochemical wastes, oil-field brines, and drilling muds are other potential sources of pollution of ground water.

SALINE WATER INTRUSION

In coastal areas, aquifers that contain fresh water may be hydraulically connected to the ocean or other saline water body. Under natural conditions the hydraulic gradient or direction of ground-water flow is generally toward the coast (fig. 13a). If ground-water withdrawal lowers the water table or potentiometric surface so that the natural gradient is reversed, saline water will begin to invade the freshwater aquifer (fig. 13b). The potential for saltwater intrusion at Anchorage has been discussed by Barnwell and others (1972, p. 45), but the probability of its occurrence due to

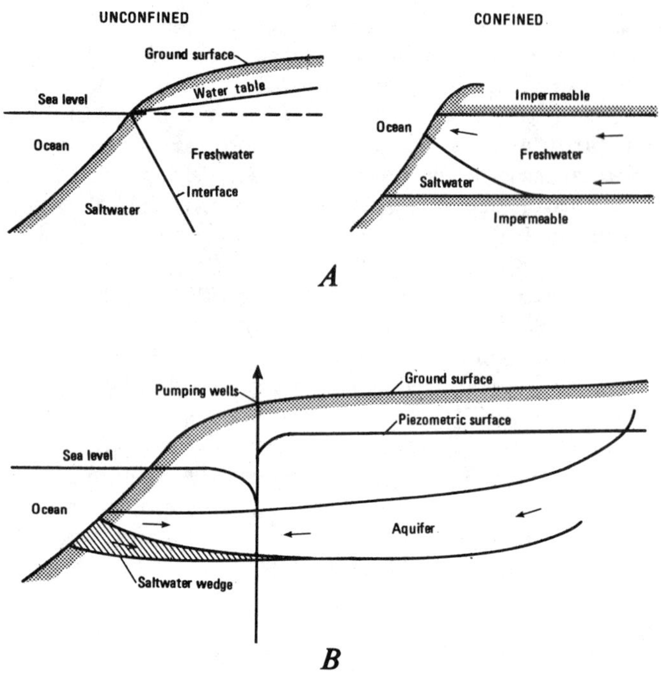

FIGURE 13.—Aquifers that contain fresh water may be hydraulically connected to a saline water body and may be intruded by saltwater. *A*. Idealized sketch of freshwater and saltwater distributions in coastal aquifers illustrates Ghyben-Herzberg relation. *B*. Effect of pumping on saltwater wedge movement into a coastal confined aquifer.

ground-water withdrawals at present (1978) pumping sites and rates is considered remote. Existing wells are already being pumped at the maximum rate allowed by well construction and drawdown limitations. The center of the cone of depression at the Municipal well field is several miles inland and artesian water levels near the coast were 20–30 ft above sea level in 1976 (fig. 12). The two major public water suppliers in the Anchorage area are aware that the potential for saltwater intrusion must be considered in selecting new well locations.

In island settings and on offshore bars and spits, fresh ground water generally occurs as an unconfined, lens-shaped body floating on sea water. The Ghyben-Herzberg principle (see Todd, 1959, p. 278) states that there is a 40-ft-thick zone of freshwater below sea level per foot of water-table altitude above sea level. This situation exists at St. George Island, in the Pribilof Islands in the Bering Sea (Anderson, 1976) and at Fire Island near Anchorage (A. J. Feulner, U.S. Geol. Survey, written commun., 1964). Other areas where ground water probably occurs only as thin lenses are at locations such as Unalakleet, Kotzebue, and the lowermost fringes of the Yukon and Kuskokwim River deltas in the coastal areas of western Alaska. Under natural or equilibrium conditions, a transition zone of brackish water separates the freshwater from the saltwater below.

The withdrawal of ground water from the fresh lens and the resulting lowering of the water table may disturb the natural balance and induce movement of saltwater into the freshwater-bearing formations. For example, saltwater intrusion has apparently occurred because of pumping from bedrock wells near the coast on Kodiak Island (F. and S. Ogden, drilling contractors at Kodiak, written commun., 1976). It follows from the Ghyben-Herzberg principle that each foot of lowering of the water table in the freshwater lens will induce a rise of 40 ft in the top of the underlying saltwater body. Successful development of ground water under such conditions depends largely upon control of pumping to avoid creating excessive drawdown.

All major population centers except Fairbanks are in coastal areas. Imminent development of Alaska's offshore petroleum and other resources will insure that future growth, including an increasing demand for water, will be concentrated in coastal areas, in part because of the need for transportation bases from which those resources will be shipped outside the State. Any development of ground-water supplies in those areas will require consideration of the potential for saltwater intrusion.

In the Copper River basin area, saline water is pumped from wells and issues from several springs and mud cones. The connate origin of the saline water (Grantz and others, 1962) has been confirmed by exploratory drill holes for oil (Williams, 1970a). Barnes (1970) believes the water to be a product of low-grade metamorphism of marine sediments. Relatively impermeable glaciolacustrine sediments confine the saline aquifers and separate them from overlying potable-water aquifers (Williams, 1970a). Dissolved-solids concentrations of the waters range from 760 to 35,000 mg/L. Dissolved-solids concentration generally increases with the depth of the well. The potential exists for degradation of the shallow, fresh (or at least less saline) water should excessive pumping cause up-coning of the deeper, highly saline water or leakage through the confining beds.

CONSTRUCTION AND ENGINEERING PROBLEMS RELATED TO THE PRESENCE OF GROUND WATER

Construction or installation of roadways, foundations, and pipelines is generally more difficult, more costly, and more time consuming if saturated soil must be dealt with. The soil may have to be dewatered temporarily during construction, though permanent drainage may be required for some structures. Rapid urban development during dry years in areas that normally have a high water table (such as parts of Anchorage) could result in flooding of basements and failure of foundations when water levels return to normal. Similar problems could result from recovery of a water table that was temporarily lowered because of ground-water withdrawal or high infiltration rates to sewer and storm-drainage lines.

The strength of unconsolidated materials such as sand and silt is reduced when they are saturated. Damage to and even collapse of foundations at Fairbanks during the disastrous 1967 flood was caused by a decrease in bearing capacity of the soil as the water table rose (Childers and others, 1972). A rising ground-water table that accompanies a rise in river stage can damage foundations on flood-plain deposits even though the river is below flood stage.

Landslides may be triggered by imposition of loads of buildings on slopes or bluff areas, or by seismic shaking of liquifiable deposits in earthquake-prone areas such as south-central Alaska. Liquifaction of saturated sand and silt lenses in the Bootlegger Cove clay was considered to be a major factor in causing landslides at Anchorage during the 1964 earthquake (Shannon and Wilson, 1964). Severe damage by submarine landsliding and ground displacement at Valdez was also attributed to spontaneous liquifaction of saturated sand (Coulter and Migliaccio, 1966).

Normal construction and maintenance problems in saturated ground are magnified if the ground is frozen, as it is in the permafrost zones of Alaska.[2] Permafrost is generally in a natural state of thermal equilibrium with the present climate, even though it may be a relic of former, colder periods. Disruption of this equilibrium caused by disturbance of the ground surface or vegetation cover by vehicular movement, construction and maintenance activities, and by heat produced within structures or by operation of facilities tends either to raise the mean annual ground-surface temperature or to increase the amplitude of seasonal variations in ground temperature. Either result will degrade or thaw part of the permafrost. A discussion of the specific engineering problems caused by the melting of permafrost is beyond the scope of this paper. Those problems are discussed in detail by Ferrians, Kachadoorian and Greene (1969), Lachenbruch (1970), and by Linell (1973). The development and application of unique engineering practice in permafrost is described by Ferrians and others (1969); Linell and Johnston (1973); and by Sanger (1969).

[2]This discussion is included under the assumption that ice-rich permafrost can be considered to be "frozen ground water" when in fact some subsurface ice may have formed as hoarfrost or buried icings that were never actually ground water. Meinzer (1923, p. 22) specifically excluded subsurface ice as a type of ground water.

CONSIDERATIONS IN RESOURCE DEVELOPMENT

The environmental effects of ground-water development are not always known or anticipated, even in areas where ground-water use is most intensive. In the Anchorage area, artesian flow to the surface at many private wells was reduced or eliminated coincident with initiation of large-scale Municipal pumping in the late 1950's. Effects such as these are obvious and measurable; the more subtle or long-term effects are not completely understood. These might be, for example, the potential effect of land-drainage activities or ground-water pumpage on lake levels, wetlands, and the base flow of streams that are hydraulically connected to the producing aquifer. Development of ground water invariably produces some changes in other elements of the hydrologic system. However, the amount of change that will be tolerated must be defined if conflicts of use are to be held to acceptable limits. Paraphrasing Anderson and Jones (1972, p. 74, 75), who drew principally from a discussion by Lohman (1972, p. 62): To determine whether moderate quantities of water can be withdrawn from a given ground-water reservoir is generally not difficult. But to determine the maximum amount of water that can be withdrawn without producing an undesirable result requires detailed knowledge of the geologic framework of the hydrologic system and planning decisions as to the extent that undesirable results can be tolerated. This is difficult to do because normally the potential of a hydrologic basin cannot be determined in advance of development. However, the hydrologic information that is available concerning the geohydrologic system and the cause-effect relationships of ground-water withdrawals can be used to construct models that will predict aquifer response. Even after a model of the aquifer has been constructed and shown to be reliable for predicting hydrologic responses to pumping, the optimum yield cannot be estimated until the water planners or users define precisely the undesirable results that are to be avoided.

Some questions that are relevant to water-resources development (specifically ground-water withdrawal) in Alaska include:

1. What is a tolerable water-level drawdown, and over what areal extent can it occur?
2. What is the relative value of swamp lands? Should they be preserved as natural habitat areas, or should the lowering of water levels be allowed in order to make the land more suitable for construction or agriculture?
3. Can changes in lake levels be tolerated; if so, to what extent?
4. Can the reduction of natural ground-water discharge be tolerated, and what will be the ef-

fect on streamflow maintained by the reduced discharge? Will such a reduction affect instream flow requirements of salmon migration, degrade spawning habitat, or cause movement of the freshwater-saltwater interface in coastal areas?

5. Should temporary overdraft (pumpage in excess of recharge) of the hydrologic basin be permitted, provided the effects are not catastrophic, if use of ground water contributes to development of an economic base that would later be able to support a more expensive, integrated surface-water/ground-water system?

As a basis for management decision, it may be assumed that optimum yield of ground water will be achieved when the average recharge is balanced by the average discharge and when the discharge for nonbeneficial purposes has been reduced to a minimum. In most years, however, such a balance will not be achieved. During dry years, ground water is drawn from storage, water levels decline, and recharge from lakes and streams may increase. During wet years, lakes are filled, streams flow above normal discharge, and the ground-water reservoir is replenished. Maximum use of ground-water reservoirs cannot be achieved by holding the storage constant; it is achieved by drawing on storage during dry years, but only to the extent that the storage can be replenished during wet years.

Superimposed on the natural hydrologic factors which must guide the decisions of the water-resource manager are the legal aspects of water appropriation and use. All the waters of Alaska, both surface and subsurface, are considered to be the property of the public and thus subject to appropriation. The doctrine of prior appropriation was applied to early territorial decisions in order to settle water-use conflicts between competing users of the available water supply. The principles of the appropriation system were carried forward in the State Constitution and the water rights code which was adopted in 1966 (Dewsnup and others, 1973). The code is now (1977) being reviewed and will probably be revised and updated to meet the changing needs of a rapidly developing area. Although the State presently claims public ownership of all the waters of Alaska, there remain conflicts of ownership and rights to water use that will have to be considered as part of the final implementation of the Alaska Native Settlement Act of 1971. Disputes have already arisen over established water uses on lands now being claimed by the various Native corporations (Robert Bursiel, Alaska Division of Lands, oral commun., 1976). The adjudication of water rights and management of the resource will also have to be accomplished within a checkerboard pattern of projected land uses and political subdivisions that does not correspond with either surface-water or ground-water drainage divides.

SIGNIFICANCE OF THE GROUND-WATER RESOURCE

Advantages of ground water over surface water are common to both arctic and temperate regions. Ground water is widely if not evenly distributed, has less annual variation in temperature and chemical quality, and is less subject to pollution than is surface water. In addition, the subsurface environment provides a vast storage reservoir which is not as subject as surface water to changes in the volume of storage caused by short-term extremes in climate. It is this storage factor that enhances the importance of ground water in Alaska, where seasonal freezing reduces and in some cases eliminates the availability of surface water.

Alaska's ground-water resource has been virtually untouched over vast areas. Permafrost severely restricts the availability of ground water in northern Alaska, but for the remainder of the State potential development far exceeds current use. Probably the greatest potential for development of ground water is in the Tanana River valley and to a lesser extent in the Yukon and Kuskokwim valleys and their major tributaries.

CONJUNCTIVE USE OF GROUND WATER AND SURFACE WATER

Conjunctive use of ground water and surface water should be considered in water-supply development in Alaska. In some areas use of these dual sources has been a necessity rather than a consciously applied principle of management. Surface-water flows in the Arctic are drastically reduced in early winter and may even cease during mid- to late winter. If a stream does maintain year-round flow, domestic, commercial, or industrial water needs that outgrow the supply will require a search for supplemental sources. Ground water may be able to provide the storage necessary to carry the community over the low-flow period of its surface-water source.

An excellent example of conjunctive use of ground and surface sources is the Municipal water supply at Anchorage. Direct withdrawal from Ship Creek, from nearby springs, and from an infiltration gallery in the gravels beneath the channel of lower Ship Creek provided Anchorage's first water supply. About 1946 a reservoir was built on the upper reaches of Ship

Creek, and for a time the city was supplied entirely by surface water. By 1956 the capacity of the reservoir was taxed by the demand from growing civilian and military populations, and development of a well field near the city center was begun to increase the water supply. Since that time all additional development has been from ground water, which now provides about one-half of the total Municipal water utility's demand of 16 Mgal/d. An added benefit from the use of dual sources at Anchorage has been a significant reduction in the incidence of freezing water mains, a common winter occurrence when surface water alone supplied the city (Nyman, 1958).

Locations favorable for large-yield (1–2 Mgal/d) wells are nearly exhausted in the Anchorage area, and various plans have been proposed recently for more complete utilization of excess summer flows of Ship Creek. Among these plans are the use of off-stream storage basins, diversion and treatment of water near the mouth of the creek (U.S. Army Corps of Engineers, 1976), and artificial recharge of the Ship Creek alluvial fan deposits (Barnwell and others, 1972, p. 46; Anderson, 1977).

NEED FOR ADDITIONAL INFORMATION ON GROUND-WATER RESOURCES

The deficiency of the ground-water data base has been referred to throughout this report. That deficiency is an obvious limiting factor for ground-water development in Alaska. The potential of the resource and the advantages of its utilization can best be realized if investigations are made in anticipation of future water demands. Important ground-water information needs in Alaska may be placed in two broad categories.

BASIC DATA

Several government agencies, private organizations, and individuals collect and compile ground-water data in Alaska. Such studies usually are made within a limited area in order to analyze or resolve a specific problem of water availability or water quality for a community or industrial venture; consequently, there remain large gaps in the ground-water data network for Alaska. Collection of hydrologic data in remote areas is time consuming and expensive and there may be neither present nor anticipated demand for the data. Collection of some data has been possible through Federal and State government funding made available to analyze environmental impacts of proposed development. Additional subsurface hydrologic information is available from activities such as pipeline trenching, and exploratory, foundation, and seismic shot-hole drilling. These records are not generally recognized as being valuable to the hydrologist and are not collected systematically, nor are they generally available to the scientific community.

There is a critical need, on a Statewide basis to (a) identify the sources and types of ground-water data being collected, (b) design a system and schedule to collect and process the data and make them accessible to interested parties, and (c) evaluate the adequacy of the data base on a continuing basis in order to recommend the collection of new information where deficiencies exist. Efforts to meet this ground-water data need were initiated in 1977 as part of a cooperative water-resources study between the Alaska Department of Natural Resources and the U.S. Geological Survey.

HYDROLOGIC STUDY AND RESEARCH

Alaska's generally unexplored and undeveloped ground-water resource has been the subject of several descriptive reviews, including this one. More important than simple description and data collection, however, is the need to understand the physical role of ground water in the hydrologic cycle of arctic and subarctic areas. Quantitative investigations of the hydrology of cold regions, particularly in the continuous permafrost zones, will help to explain the important "hydrological linkage" between surface and ground waters (Guymon, 1973). Suggested study and research areas include the following:

1. Field studies of evapotranspiration and soil moisture relations. Few such studies have been conducted in Alaska, and most empirical formulas, developed for temperate climates, do not apply in high-latitude locations.
2. Studies of the hydrology of springs and icings, and the identification of areas for aquifer recharge and discharge through unfrozen zones in permafrost. These studies could provide important information and methods for predicting locations of water-supply sources, the potential occurrence of icings, and the effect of permafrost on base flow of streams (Williams, 1970a). Resistivity and seismic-refraction methods could be used to supplement existing data in determining the configuration of permafrost, particularly adjacent to surface-water bodies (Hoekstra and McNeill, 1973).
3. Studies of the hydraulics of flow through porous media at subfreezing temperatures. Results of aquifer tests should be considered in terms of boundaries imposed by permafrost

as well as those related to lithology and rock structure. The validity of the common assumptions that permafrost contains water only in the solid state and that it is impermeable and thus prevents ground-water movement has been questioned (Williams and van Everdingen, 1973). Quantitative studies could resolve many uncertainties and permit better prediction of aquifer response in permafrost regions.

4. Continuation of geothermal studies of the type pioneered by Lachenbruch, Brewer, Greene, and Marshall (1962) in order to refine understanding of the thermal regimen and its effect on the formation, stability, and degradation of permafrost. Results of such studies could be applied in creating aquifers by thawing frozen ground as was attempted in the alluvium of Selin Creek at Cape Lisburne (Feulner and Williams, 1967).
5. Investigations of the chemistry of ground ice and unfrozen brines, and the seasonal variation and selective concentration of chemical constituents in ground water associated with permafrost. Information gained could be helpful in the solution or avoidance of water-quality or pollution problems. Reports of fresh potable water encountered during drilling for petroleum at offshore sites in Cook Inlet emphasize the need to evaluate the effects of brine injection for secondary recovery on potential water-supply aquifers.

SUMMARY AND CONCLUSIONS

Alaska's ground water is a large resource, but one that is virtually unexplored and undeveloped. Estimated average annual ground-water discharge, which is assumed to approximate recharge on a long-term basis, is 100,000 Mgal/d. In 1975, ground-water use in Alaska amounted to only 50 Mgal/d.

Probably only a small fraction of the total ground-water discharge can ever be captured for man's use, but this fraction represents a renewable resource which can play a significant role in the future development of the State. Alaska's population, currently less than one-half million, will inevitably increase as the demand for and development of the State's vast mineral and petroleum reserves increase. Much of that development, particularly for petroleum and natural gas, will take place in the Arctic Foothills and Arctic Coastal Plain. The occurrence of ground water in those areas is severely limited by continuous permafrost. However, the major population centers and greatest demand for water will remain in the southern half of Alaska where the influence of frozen ground on the occurrence and availability of ground water is comparatively minor and where potential ground-water development is many times greater than current use.

In the southern half of Alaska, conditions favorable for ground-water development occur over large areas. Wells in the major river valleys, alluvial fans, and some coastal valleys and basins commonly yield as much as 1,000 gal/min, and similar yields are possible from wells in volcanic rocks at some locations. At many places in these areas, annual ground-water recharge is sufficient to sustain high rates of continuous withdrawal, and, at these places, the ground-water resource is adequate for present and future needs. At other places, available ground water may be insufficient to meet needs but, in conjunction with surface water, may permit optimal development of the total water resource.

The development of additional supplies of ground water will circumvent many of the problems currently associated with the use of surface water, such as extreme seasonal variability of supply, the need for impoundment reservoirs, and the presence of objectionable amounts of glacial silt in many streams. The use of ground water, however, also entails some problems. Local water-level declines, depletion of streamflow, saltwater intrusion into aquifers, and natural or man-caused pollution of ground-water reservoirs are all significant problems at some places in Alaska and are potential problems at other places.

The ground-water resource of Alaska thus presents both challenge and opportunity. Through the acquisition of new information from field studies and the application of new and existing data in analytical models and interpretive reports, the information needed for effective management can be made available, and ground water can take on a vital role in the overall water-resource planning and management for Alaska.

REFERENCES CITED

Anderson, G. S., 1970, Hydrologic reconnaissance of the Tanana Basin, central Alaska: U.S. Geol. Survey Hydrol. Inv. Atlas HA–319.

———1976, Water-resources reconnaissance of St. George Island, Pribilof Islands, Alaska: U.S. Geol. Survey Water-Resources Inv. 6–76, 15 p.

———1977, Artificial recharge experiments on the Ship Creek alluvial fan, Anchorage, Alaska: U.S. Geol. Survey Water-Resources Inv. 77–38, 45 p.

Anderson, G. S., and Jones, S. H., 1971, Hydrologic data of the Kenai-Soldotna area, Alaska: U.S. Geol. Survey open-file report, 58 p.

———1972, Water resources of the Kenai-Soldotna area, Alaska, U.S. Geol. Survey open-file report, 81 p.

Balding, G. O., 1976, Water availability, quality and use in Alaska (A supplementary study to Phase II, Activity 2, Specific Problem Analysis portion of the Water Resources Council's 1975 National Assessment of Water and Related Land Resources): U.S. Geol. Survey open-file report 76–513, 236 p.

Barnes, Ivan, 1970, Metamorphic waters from the Pacific Tectonic Belt of the west coast of the United States: Science, May 22, 1970, vol. 168, p. 973–975.

Barnwell, W. W., and Boning, C. W., 1968, Water resources and surficial geology of the Mendenhall Valley, Alaska: U.S. Geol. Survey Hydrol Inv. Atlas HA–259.

Barnwell, W. W., George, R. S., Dearborn, L. L., Weeks, J. B., and Zenone, Chester, 1972, Water for Anchorage—An atlas of the water resources of the Anchorage area, Alaska: Anchorage, Alaska, pub. by city of Anchorage and Greater Anchorage Area Borough, 77 p.

Black, R. F., 1955, Arctic Slope, *in* Hopkins, D. M., Karlstrom, T. N. V., and others, Permafrost and ground water in Alaska: U.S. Geol. Survey Prof. Paper 264–F, p. F118, F119.

———1964, Gubik formation of Quaternary age in northern Alaska: U.S. Geol. Survey Prof. Paper 302–C, p. C59–C91.

Bouwer, Herman, Lance, J. C., and Riggs, M. S., 1974, High-rate land treatment II: Water quality and economic aspects of the Flushing Meadows project: Jour. Water Pollution Control Federation, v. 46, no. 5, p. 844–859.

Bouwer, Herman, Rice, R. C., and Escarcega, E. D., 1974, High-rate land treatment I: Infiltration and hydraulic aspects of the Flushing Meadows project: Jour. Water Pollution Control Federation, v. 46, no. 5, p. 835–843.

Brandon, L. V., 1964, Ground-water hydrology and water supply in the District of MacKenzie, Yukon Territory, and adjoining parts of British Columbia: Canada Geol. Survey, Dept. of Mines and Technical Surveys, Paper 64–39, 102 p.

Brown, Jerry, Dingman, S. L., and Lewellen, R. I., 1968, Hydrology of a drainage basin on the Alaskan Coastal Plain: U.S. Army Material Command Cold Regions Research and Eng. Lab., Research Rept. 240, 18 p.

Cederstrom, D. J., 1952, Summary of ground-water development in Alaska, 1950: U.S. Geol. Survey Circ. 169, 37 p.

———1963, Ground-water resources of the Fairbanks area, Alaska: U.S. Geol. Survey Water-Supply Paper 1590, 84 p.

Childers, J. M., Meckel, J. P., and Anderson, G. S., 1972, Floods of August 1967 in east-central Alaska: U.S. Geol. Survey Water-Supply Paper 1880–A, 77 p.

Childers, J. M., Sloan, C. E., and Meckel, J. P., 1973, Hydrologic reconnaissance of streams and springs in eastern Brooks Range, Alaska—July 1972: U.S. Geol. Survey open-file report, 25 p.

———1977, Hydrologic reconnaissance of the eastern North Slope, Alaska, 1975: U.S. Geol. Survey open-file report 77–492, 65 p.

Coulter, H. W., and Migliaccio, R. R., 1966, Effects of the earthquake of March 27, 1964, at Valdez, Alaska: U.S. Geol. Survey Prof. Paper 542–C, p. C1–C36.

Coulter, H. W., and others, 1965, Map showing extent of glaciations in Alaska: U.S. Geol. Survey Misc. Geol. Inv. Map I-415, scale 1:2,500,000, 1 sheet.

Craig, P., and McCart, P. J., 1974, Fall spawning and overwintering areas of fish populations along routes of the proposed pipeline between Prudhoe Bay and the MacKenzie Delta, 1972–73, Chap. 3 *in* McCart, P. J., ed., Fisheries research associated with proposed gas pipeline route in Alaska, Yukon and Northwest Territories: Arctic Gas Biological Rept. Ser., v. 15, Canadian Arctic Gas Study, Ltd. and Alaska Arctic Gas Study Co., 37 p.

Dames and Moore, 1976, Ground-water investigation—Interrelationship between aquifers and surface-water regimes, North Kenai area, Alaska: Dames and Moore Consulting Engineers, for Collier Carbon and Chemical Corp., 92 p.

Dearborn, L. L., and Barnwell, W. W., 1975, Hydrology for land-use planning—The Hillside area, Anchorage, Alaska: U.S. Geol. Survey open-file report 75–105, 46 p.

Dewsnup, R. L., Jensen, D. W., eds., and Swenson, R. W., assoc. ed., 1973, A Summary-Digest of State Water Laws: Arlington, Va., Natl. Water Comm., 826 p.

Donaldson, D. E., Still, P. J., and Zenone, Chester, 1975, Water quality data for Anchorage and vicinity, Alaska, 1948–1973: U.S. Geol. Survey open-file report, 59 p.

Ferrians, O. J., Jr., Kachadoorian, Reuben, and Greene, G. W., 1969, Permafrost and related engineering problems in Alaska: U.S. Geol. Survey Prof. Paper 678, 37 p.

Feulner, A. J., 1963, Data on wells in the King Salmon area, Alaska: U.S. Geol. Survey Hydrol. Data Rept. 24, 18 p.

———1964, Galleries and their use for development of shallow ground-water supplies, with special reference to Alaska: U.S. Geol. Survey Water-Supply Paper 1809–E, 16 p.

Feulner, A. J., Childers, J. M., and Norman, V. W., 1971, Water resources of Alaska: U.S. Geol. Survey open-file rept., 60 p.

Feulner, A. J., and Schupp, R. G., 1963, Seasonal changes in the chemical quality of shallow ground water in northwestern Alaska, *in* Geological Survey Research 1963: U.S. Geol. Survey Prof. Paper 475–B, p. B189–B191.

Feulner, A. J., and Williams, J. R., 1967, Development of a ground-water supply at Cape Lisburne, Alaska, by modification of the thermal regime of permafrost, *in* Geological Survey Research 1966: U.S. Geol. Survey Prof. Paper 575–B, p. B199–B202.

Feulner, A. J., Zenone, Chester, and Reed, K. M., 1976, Geohydrology and water supply, Shemya Island, Alaska: U.S. Geol. Survey open-file report 76–82, map.

Freethey, G. W., 1976, Preliminary report on water availability in the lower Ship Creek basin, Anchorage, Alaska—with special reference to the fish hatchery on Fort Richardson and a proposed fish hatchery site near the Elmendorf Air Force Base powerplant: U.S. Geol. Survey Water-Resources Inv. 48–75, 21 p.

Grantz, Arthur, White, D. E., Whitehead, H. C., and Tagg, A. R., 1962, Saline springs, Copper River lowland, Alaska: Am. Assoc. Petroleum Geologists Bull., v. 46, no. 11, p. 1990–2002.

Guymon, G. L., 1973, Why study subsurface water in Alaska? [abs.]: EOS, vol. 54, no. 11, p. 1089; also *in* Guymon, G. L., 1974, Suprapermafrost water: Univ. of Alaska, Inst. of Water Resources, Report No. IWR–53, 11 p.

Hawkins, D. B., 1976, Environmental path of arsenic in ground water: Univ. of Alaska Inst. of Water Resources Report No. IWR–77, 7 p.

Hoekstra, Peter, and McNeill, Duncan, 1973, Electromagnetic probing of permafrost, *in* Permafrost, the North American contribution to the second international conference of the National Academy of Sciences, Yakutsk, USSR, July 1973, Proc.: Washington, D. C., Natl. Acad. of Sci., p. 517–526.

Hopkins, D. M., Karlstrom, T. N. V., and others, 1955, Permafrost and ground water in Alaska: U.S. Geol. Survey Prof. Paper 264–F, p. F113–F146.

Howitt, Frank, and Clegg, M. W., 1970, Permafrost in Prudhoe Bay field—Geology and physical characteristics [abs.]: Am. Assoc. Petroleum Geologist Bull., v. 54, no. 12, p. 2487.

Karlstrom, T. N. V., and others, 1964, Surficial geology of Alaska: U.S. Geol. Survey Misc. Geol. Inv. Map I–357.

Lachenbruch, A. H., 1970, Some estimates of the thermal effects of a heated pipeline in permafrost: U.S. Geol. Survey Circ. 632, 23 p.

Lachenbruch, A. H., Brewer, M. C., Greene, G. W., and Marshall, B. V., 1962, Temperatures in permafrost, *in* Temperature—its measurement and control in science and industry: New York, Reinhold Pub. Corp., v. 3, pt. 1, p. 791–803.

Linell, K. A., 1973, Risk of uncontrolled flow from wells through permafrost, *in* Permafrost, the North American contribution to the second international conference of the National Academy of Sciences, Yakutsk, USSR, July 1973, Proc.: Washington, D.C., Natl. Acad. of Sci., p. 462–468.

Linell, K. A., and Johnston, G. H., 1973, Engineering design and construction in permafrost regions, A review, *in* Permafrost, the North American contribution to the second international conference of the National Academy of Sciences, Yakutsk, USSR, July 1973, Proc.: Washington, D.C., Natl. Acad. of Sci., p. 553–575.

Linsley, R. K., Kohler, M. A., and Paulhus, J. L. H., 1975, Hydrology for Engineers, 2d ed.: McGraw-Hill, 482 p.

Lohman, S. W., 1972, Ground-water hydraulics: U.S. Geol. Survey Prof. Paper 708, 70 p.

McConaghy, J. A., 1969, Hydrologic data of the Juneau Borough, Alaska: U.S. Geol. Survey open-file report, 77 p.

———1970, Reconnaissance of water resources in the Haines-Port Chilkoot area, Alaska: U.S. Geol. Survey Circ. 626, 16 p.

Marsh, P., and Woo, M. K., 1977, The water balance of a small pond in the high Arctic: Arctic, vol. 30, no. 2, p. 109–117.

Meinzer, O. E., 1923, Outline of ground-water hydrology: U.S. Geol. Survey Water-Supply Paper 494, 71 p.

Miller, D. J., 1971, Geologic map of the Yakataga District, Gulf of Alaska Tertiary province, Alaska: U.S. Geol. Survey Misc. Geol. Inv., map I–610, 1:125,000.

Nyman, Frank, 1958, Well water protects Anchorage water system: Johnson National Drillers Journal, v. 30, no. 1, p. 8–11.

Osterkamp, T. E., Gilfilian, R. E., and Benson, C. S., 1975, Observations of stage, discharge, pH, and electrical conductivity during periods of ice formation in a small subarctic stream: Water Resources Research, v. 11, no. 2, p. 268–272.

Piper, A. M., 1966, Potential effects of Project Chariot on local water supplies, northwestern Alaska: U.S. Geol. Survey Prof. Paper 539, 45 p.

Plafker, George, 1967, Geologic map of the Gulf of Alaska Tertiary province, Alaska: U.S. Geol. Survey Misc. Geol. Inv. Map I–484, scale 1:500,000.

Sanger, F. J., 1969, Foundation of structures in cold regions: U.S. Army Cold Regions Research and Eng. Lab., Monograph III–C4, 91 p.

Satterwhite, M. B., Stewart, G. L., Condike, B. J., and Vlach, E., 1976, Rapid infiltration of primary sewage effluent at Fort Devens, Massachusetts: U.S. Army, Cold Regions Research Engineering Laboratory, Report 76–48, 34 p.

Shannon and Wilson, Inc., 1964, Report on Anchorage area soil studies, Alaska, to U.S. Army Engineer District, Anchorage, Alaska: Shannon and Wilson, Inc., Seattle, Washington, 109 p.

Sherman, R. G., 1973, A ground-water supply for an oil camp near Prudhoe Bay, *in* Permafrost, the North American contribution to the second international conference of the National Academy of Sciences, Yakutsk, USSR, July 1973, Proc.: Washington, D.C., Natl. Acad. Sci, p. 469–472.

Sloan, C. E., Zenone, Chester, and Mayo, L. R., 1976, Icings along the trans-Alaska pipeline route: U.S. Geol. Survey Prof. Paper 979, 31 p.

Straughn, R. O., 1972, The sanitary landfill in the sub-Arctic: Arctic, v. 25, no. 1, p. 40–48.

Todd, D. K., 1959, Ground-Water Hydrology: New York, John Wiley and Sons, 336 p.

U.S. Army Corps of Engineers, 1976, Metropolitan Anchorage urban study - Plan of study: U.S. Army, Corps of Engineers, Alaska District, 106 p.

Waananen, A. O., and Giles, G. C., 1964, Water resources, *in* Mineral and water resources of Alaska: U.S. Cong. 88th, 2nd sess., [Senate Comm. Interior and Insular Affairs] Comm. Print. 31–068, p. 149–179.

Wahrhaftig, Clyde, 1965, Physiographic divisions of Alaska: U.S. Geol. Survey Prof. Paper 482, 52 p.

Waller, R. M., 1957, Ground water and permafrost at Bethel, Alaska: U.S. Geol. Survey Hydrol. Data Rept. 2, 11 p.

———1963, Data on wells in the Homer area, Alaska: U.S. Geol. Survey Hydrol. Data Rept. 23, 24 p.

———1966, Ground-water hydrology, Chap. 7 *in* Wilimovsky, N. J., and Wolfe, J. N., eds., Environment of the Cape Thompson region, Alaska: U.S. Atomic Energy Comm., Div. Tech. Inf., Rept. PNE–481, p. 115–124.

Waller, R. M., and Mathur, S. P., 1962, Data on water supplies at Nome, Alaska: U.S. Geol. Survey Hydrol. Data Rept. 17, 13 p.

Watson, C. E., 1959, Climates of the States—Alaska: U.S. Dept. of Commerce, Weather Bureau (now National Weather Service), *in* Climatography of the United States, no. 60–49.

Weeks, J. B., 1970, The relationship between surface and ground water in Ship Creek near Anchorage, Alaska, *in* Geol. Survey Research 1970: U.S. Geol. Survey Prof. Paper 700–B, p. B224–B226.

Williams, J. R., 1970a, Ground water in permafrost regions of Alaska: U.S. Geol. Survey Prof. Paper 696, 83 p. [Contains comprehensive bibliography.]

———1970b, A review of water resources of the Umiat area, northern Alaska: U.S. Geol. Survey Circular 636, 8 p.

Williams, J. R., and van Everdingen, R. O., 1973, Ground-water investigations in permafrost regions of North America, A review, *in* Permafrost, the North American contribution to the second international conference of the National Academy of Sciences, Yakutsk, USSR, July 1973, Proc.: Washington, D.C., Natl. Acad. Sci., p. 435–446.

Zenone, Chester, Donaldson, D. E., and Grunwaldt, J. J., 1975, Ground-water quality beneath solid-waste disposal sites at Anchorage, Alaska: Ground Water, v. 13, no. 2, p. 182–190.

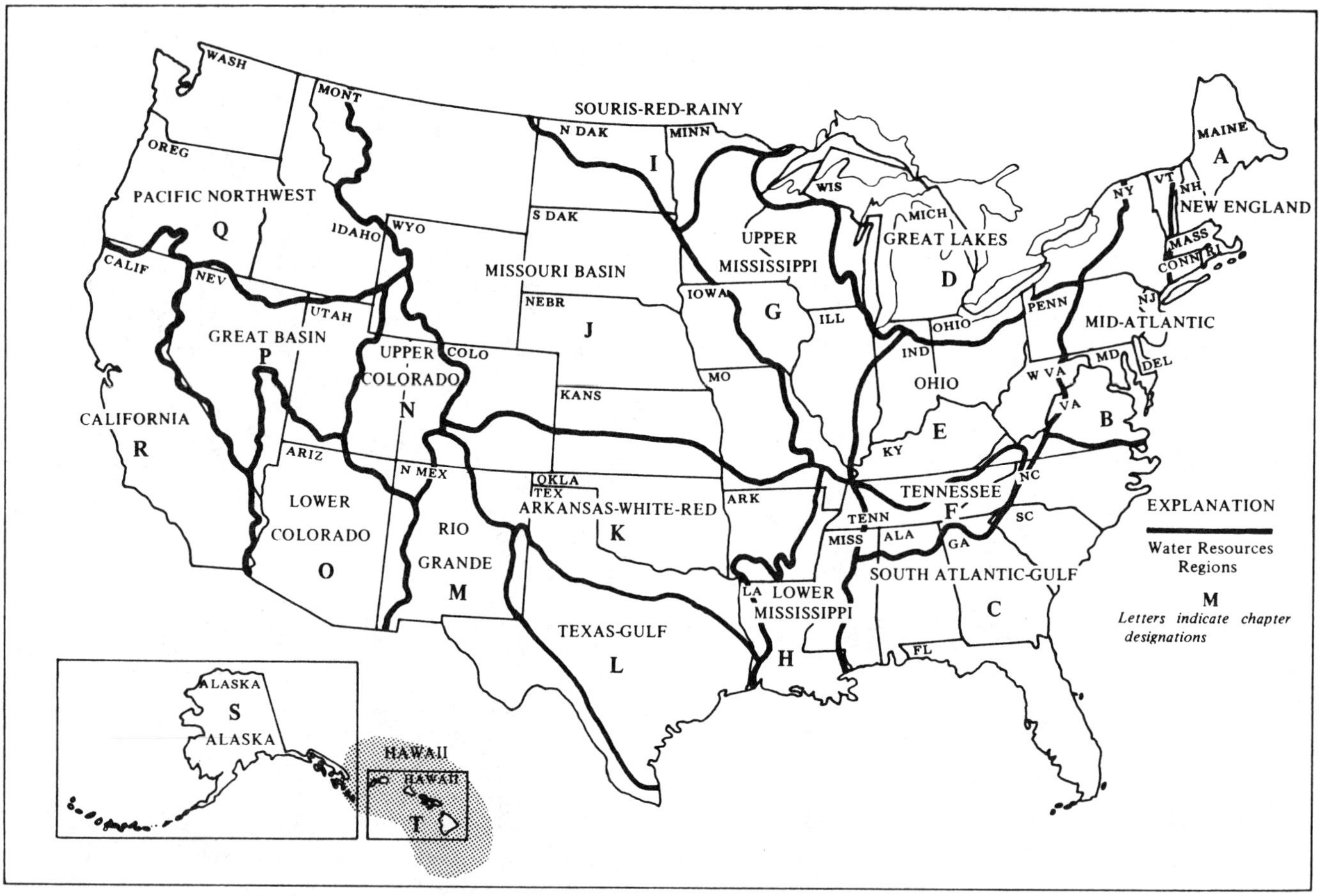

T – Location Map for the Hawaii Region

T ....

HAWAII REGION

By K. J. TAKASAKI

ABSTRACT

The water resources of the Hawaii Region, taken as a whole, are far greater than foreseeable future demands on them, but this is not so for the individual islands. Each and every island is independent with respect to water supply, and the occurrence and availability of water vary widely from island to island.

The ground-water resources offer better prospects for supplying additional water needs in the future than the surface-water resources. Most of the surface supplies that are easy to develop have been fully utilized where needed, and conduits and reservoirs necessary to develop new or additional supplies would generally require large and perhaps prohibitive outlays of capital. In 1975, ground water supplied 46 percent, and surface water 54 percent of the water needs but, in the years ahead, these percentages will likely be reversed as more ground-water development takes place. Total water use, in 1975, averaged about 1,775 million gallons per day, of which about 810 million gallons per day was ground water. The total water use is divided into public supply, 11 percent; self-supplied industrial use, 23 percent; and agricultural, 66 percent.

Rainfall is the principal source of ground-water recharge. Local mean annual rainfall ranges from less than 20 inches to more than 300 inches, with the annual average rainfall on the large islands exposed to the trade winds being slightly more than 73 inches and that on the small islands situated in the rain shadow of the larger islands being less than 26 inches. Ground-water recharge has been estimated at about 2,400 billion gallons per year (6.5 billion gallons per day) or roughly 30 percent of the rainfall.

Most fresh ground water in the region is stored below sea level in porous lava flows, much of it as basal-water lenses floating on saline ground water, as distinguished from dike-impounded water in the interior of the islands. The basal-water lens is maintained by recharge, which, if reduced, leads to thinning of the lens and subsequent encroachment of seawater. Seawater is the biggest pollutant of freshwater, and many of the ground-water problems are, in some way, associated with the encroachment of saline water induced by development.

The major problem areas include the entire island of Oahu, south Kohala-Kona coast on the island of Hawaii, Lahaina District in Maui, and the Koloa and Kekaha-Mana areas in Kauai.

INTRODUCTION

The purpose of this report is to provide pertinent ground-water information and to relate this information to the potential role of ground water in meeting future water needs in the Hawaii Region. The report also focuses on some problems related to the hydrology of ground water in overall water supply and management, and on some of the deficiencies in information and understanding that hinder the optimum development of the ground-water resources.

Ground water in the Hawaii Region is a large and valuable resource. Despite the large overall supply, ground water is not available in some areas, and much of it, especially that which is of suitable quality for domestic use, has to be imported by pipeline from distant areas. In 1975, only about 20 percent of the approximately 810 million gallons per day (Mgal/d) of ground water developed was used for public supplies. As a result of development, the quality of the ground water has deteriorated at some places, but water of less than potable quality can be tolerated in uses such as irrigation of sugarcane and cooling.

Most recharge to ground water occurs in the wet interior mountains, generally upgradient from lower-lying developed areas where wastes and subsequent waste disposals are more likely to occur. This favorable situation is gradually changing for the worse as land developments encroach into the principal recharge areas. Some deterioration of the ground water is to be expected, the degree of which will depend greatly on how much deterioration is allowed by policies and decisions related to land use, land development, and waste-disposal practices. In most cases, the prevention of pollution by eliminating the pollutant at the source would be a more effective course to follow than attempting detection and surveillance of the pollutant after it reaches the ground-water reservoir.

NOTEWORTHY GROUND-WATER DEVELOPMENT EVENTS

In the early years of development in the Hawaiian Islands, water supply was generally not a major problem, except in the Honolulu area. Land and surface water were abundant at places within each of the larger islands, and development was concentrated in these areas. The early Hawaiians settled near perennial streams and springs and diverted water for their highly developed agriculture. Later, sugarcane was grown in areas where rainfall was sufficient or where water could be easily diverted from perennial streams. In the water-scarce areas, the land remained largely undeveloped and unsettled.

Domestic water needed for these early developments was not, as a rule, developed separately but was taken from the available irrigation supply. The water, thus obtained, was generally of good quality or was considered so because most of it came from sources in the wet interior areas and was usually diverted from ditches and streams upgradient from the developed areas.

Some ground water, other than from springs, was developed by the early Hawaiians from shallow wells dug at or near the beach. Most of the water thus developed was brackish.

By the middle 1850's, the population of Honolulu had grown to nearly 12,000. Whaling in the Pacific had reached its peak in activity, and Honolulu had become a principal port of supply. Thus, the city of Honolulu, unlike the rest of the Region, was even then experiencing a water-supply problem. Water was piped from distant surface-water sources but the supply could barely meet the demands of the population and of the ships. The availability of freshwater had, in fact, become a severely limiting factor to the growth of Honolulu (Honolulu Board of Water Supply, 1963).

In 1879, a well tapping a basal ground-water aquifer under sufficient artesian head to flow freely was drilled in the arid land west of Pearl Harbor in Oahu. Many wells followed, first in Oahu and a little later in the other islands. The presence of a cheap and dependable ground-water supply in many of the dry leeward areas had immediate impact on the Region's development. By 1890, water from artesian wells provided a large share of the water needs for Honolulu, thereby removing the city's principal constraint to growth. By 1920, the cultivation of sugarcane had shifted from the generally wet windward areas to the drier leeward areas, and lands heretofore deemed useless became very much in demand. Sugarcane cultivation had become the principal economic base for the Region. By 1975, more than 1,000 wells had been drilled or dug, most of them to develop water for the irrigation of sugarcane.

Rapid and indiscriminate drilling of wells followed the first successful well and continued virtually unabated for nearly 50 years, especially in southern Oahu. During this time, water from wells near the coast in southern Oahu, once of domestic quality, became brackish. Artesian head, once as much as 43 feet above sea level, steadily declined to about 25 feet above sea level. Concern over this decline led to legislation and the establishment of an authority to curb waste and to manage the seemingly deteriorating artesian water supply in Honolulu.

Competition for the ground-water supply in Oahu did not become a serious problem until the Second World War, when the first significant transfer of sugarcane lands to other uses occurred. During the Second World War, large areas in the western part of Honolulu, abutting the Pearl Harbor area, were turned over to the military for their use. The military use necessitated the drilling of new wells to provide the domestic-quality water required. After the war urbanization increasingly replaced sugarcane lands, and the ensuing competition for water by the various users led to unplanned development in many areas. Problems related to water quality, rights, and management became apparent at this time and still persist.

The advent of statehood for Hawaii, in 1959, led to increased tourism and hastened the growth of the population of southern Oahu. More agricultural land was urbanized. Meanwhile, considerations of possible impact on the environment of further water and land development added to the complexity of problems.

Except for the intensive growth in sugarcane cultivation, the development and population growth in the other islands did not parallel that in Oahu in the years following the end of the Second World War. Sugarcane cultivation increased somewhat during this period, but owing to many technological advances, fewer workers were needed and the population on all the other islands declined. It was not until sometime after the coming of statehood and the growth of tourism in Oahu that resort areas and resort-residential complexes began to expand in the other islands. Tourism now is the fastest growing segment of the economy in all the islands. The ever-increasing demand for water, especially for domestic-quality ground water, in the heretofore undeveloped dry leeward areas that are favored for resort development, is a problem that needs to be solved if the tourist industry is to continue to grow.

REGIONAL SETTING

GEOLOGIC FRAMEWORK

The Hawaiian Archipelago, which makes up the Hawaii Region, is a group of shoals, reefs, and islands trending northwest to southeast more than 1,500 miles across the Pacific Ocean (fig. 1). The Hawaiian Islands, which lie at the southeastern end of the archipelago, constitute more than 99 percent of the region's total land area. Only these islands are considered in the appraisal of the region's ground-water resources.

The Hawaiian Islands are the tops of shield volcanoes rising from the ocean floor, the oldest is Kauai in the northwest part, and the youngest, the island of Hawaii in the southeast part. Each of the islands consists of one to five volcanic domes, the bulk of which is composed of thousands of basaltic lava flows. The lavas issued in repeated outpourings from narrow zones of

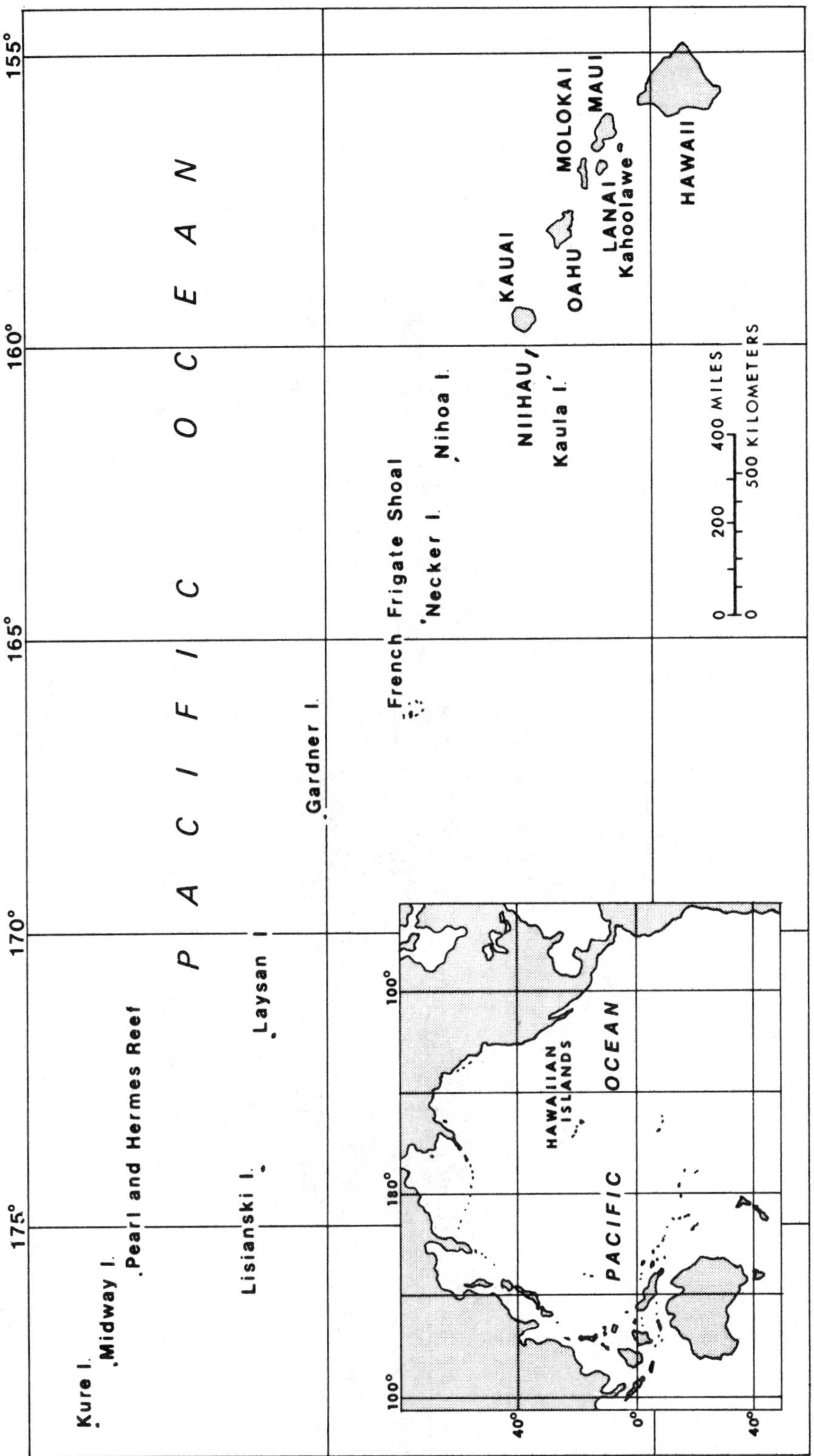

FIGURE 1.—The Hawaiian Archipelago. The small inset shows the location of the islands in the Pacific Ocean.

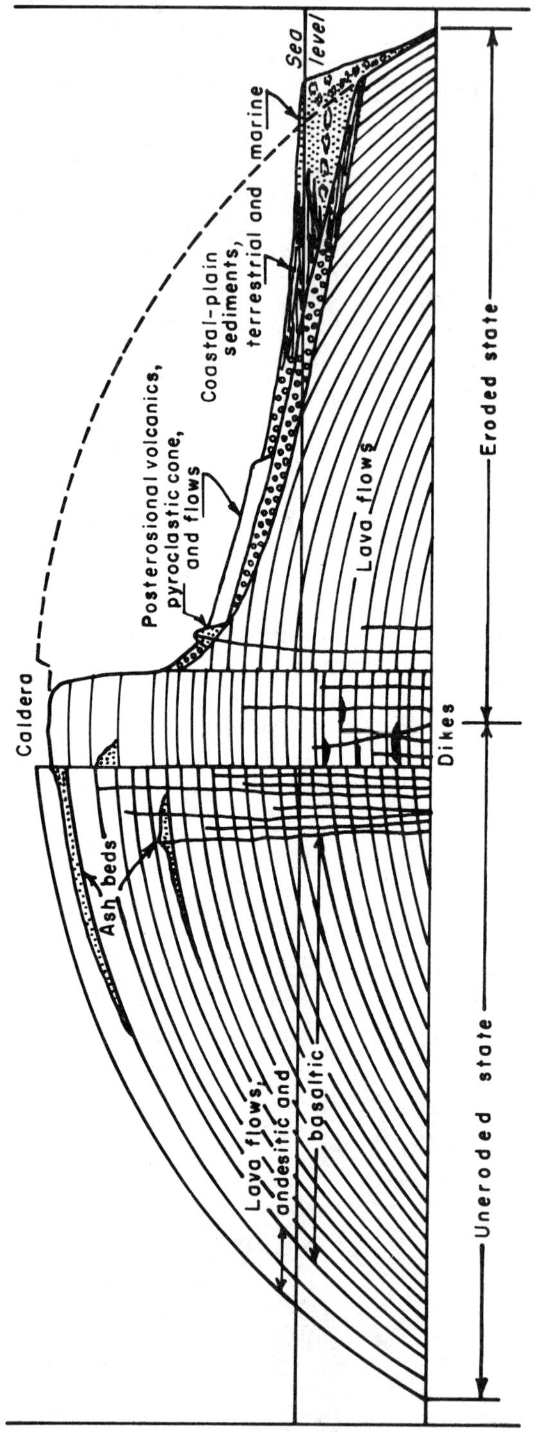

FIGURE 2.—Geologic structure of an idealized Hawaiian volcanic dome (from Cox, 1954).

fissures associated with each volcano, first below sea level, then above it, to form huge mountain masses. The basaltic lavas that were extruded above sea level are generally thin-bedded, highly clinkery, and highly permeable. In contrast, the lavas extruded in water were quickly chilled, are less clinkery, more massive, and are likely to be less permeable. All of the islands have sunk, to some extent, to adjust isostatically for their great weight on the earth's crust. Consequently, the highly permeable lava flows, which were originally extruded above sea level, now extend some distance below it. This rock assemblage of highly permeable basaltic lava flows, intruded in part by dikes in the zones of fissure and free of dikes outside of them, makes up the principal reservoirs for ground water in the Hawaiian Islands.

A diagrammatic cross section showing the geologic structure of a typical Hawaiian volcanic island in the form of an idealized volcanic dome is shown in figure 2. Rocks other than the basaltic flank flows, although widespread in places, generally do not comprise important reservoirs of ground water.

HYDROGRAPHIC AREAS

In order to facilitate a study of the water resources, the six larger islands were subdivided into hydrographic areas by the Hawaii Water Authority in 1959. The boundaries of the areas were based on topography and generally outline the major surface drainage basins (fig. 3). The boundaries of ground-water reservoirs could be better delineated and described if they were based on geology. In order to take advantage of the results of the previous studies, however, the hydrographic areas established in the 1959 study will be used in this appraisal.

RAINFALL

Rainfall is the principal source of recharge and its quantity and variability are significant in determining the extent and quality of ground water. Open-ocean rainfall near the Hawaiian Islands is about 25 inches per year compared with an average of about 70 inches on land. The islands, by their presence, thus increase the open-ocean rainfall by 45 inches per year or by 5,000 billion gallons (14 billion gallons per day). However, because most of the rainfall is orographic, that is, closely related to topography and exposure, the spatial variability of annual rainfall is extremely high.

Annual average rainfall for the islands ranges from less than 35 inches to more than 100 inches per year. Local mean annual rainfall within the larger islands, however, ranges from less than 20 inches to more than 300 inches per year. Variability of annual rainfall generally is greater where rainfall is low because rain types other than those due to trade wind, although small in total quantity, generally fall heavily during short and infrequent storms. These non-trade-wind rains make up most of the rainfall in the open-ocean and low-altitude areas, and their contribution to the recharge of ground water is generally small. The distribution of rainfall in the islands is shown in figure 4.

Owing to the large spatial variation in the trade-wind rainfall, there is a marked inverse relation between the areas of water supply and the areas of water demand. The orographic effects on rainfall are shown by cross-sectional sketches in figure 5.

The high volcanic domes referred to in figure 5 are Mauna Kea and Mauna Loa on the island of Hawaii and Haleakala on Maui. Examples of medium high domes are the Koolau dome on Oahu, the West Maui dome, and the island of Kauai. The low domes include the islands of Kahoolawe and Niihau.

EVAPOTRANSPIRATION

The amount of evapotranspiration approaches that of evaporation from a free-water surface wherever water is constantly available to plants or crops, owing either to the proximity of ground water or to high frequency of rainfall or irrigation. Research in Hawaii, mainly by the sugar and pineapple industries, shows that, although evapotranspiration differs widely from place to place, annual variations at any one place are small. The average pan evaporation in the larger islands ranges from less than 20 inches to more than 80 inches per year, and in the windy dry areas, to more than 90 inches per year. In Hawaii, where rainfall or an estimate of it is the only climatic datum generally available, the relation of rainfall to pan evaporation and to evapotranspiration recognized by Cox (Campbell and others, 1959) and Mink (1962b) is often the only tool available in studies involving the water budget of large drainage basins.

SURFACE RUNOFF AND INFILTRATION

Hawaiian streams, in general, are short and steep, and runoff depends largely on the intensity and duration of rainfall. Exceptions are streams in which runoff is made up largely of ground-water or swamp discharge. In areas where the infiltration capacity of the surface rocks is especially high, most rainfall is quickly absorbed, and there may be no runoff, except during intense storms. In some of these areas of high absorption rates overland runoff is so low that stream patterns have not yet been developed.

Where rainfall is plentiful and well distributed during the year, as in the wet interior mountains, streams are generally perennial and abundant. Where rainfall is light, as in the dry leeward areas, streams are ephemeral and flow only at infrequent intervals during heavy storms.

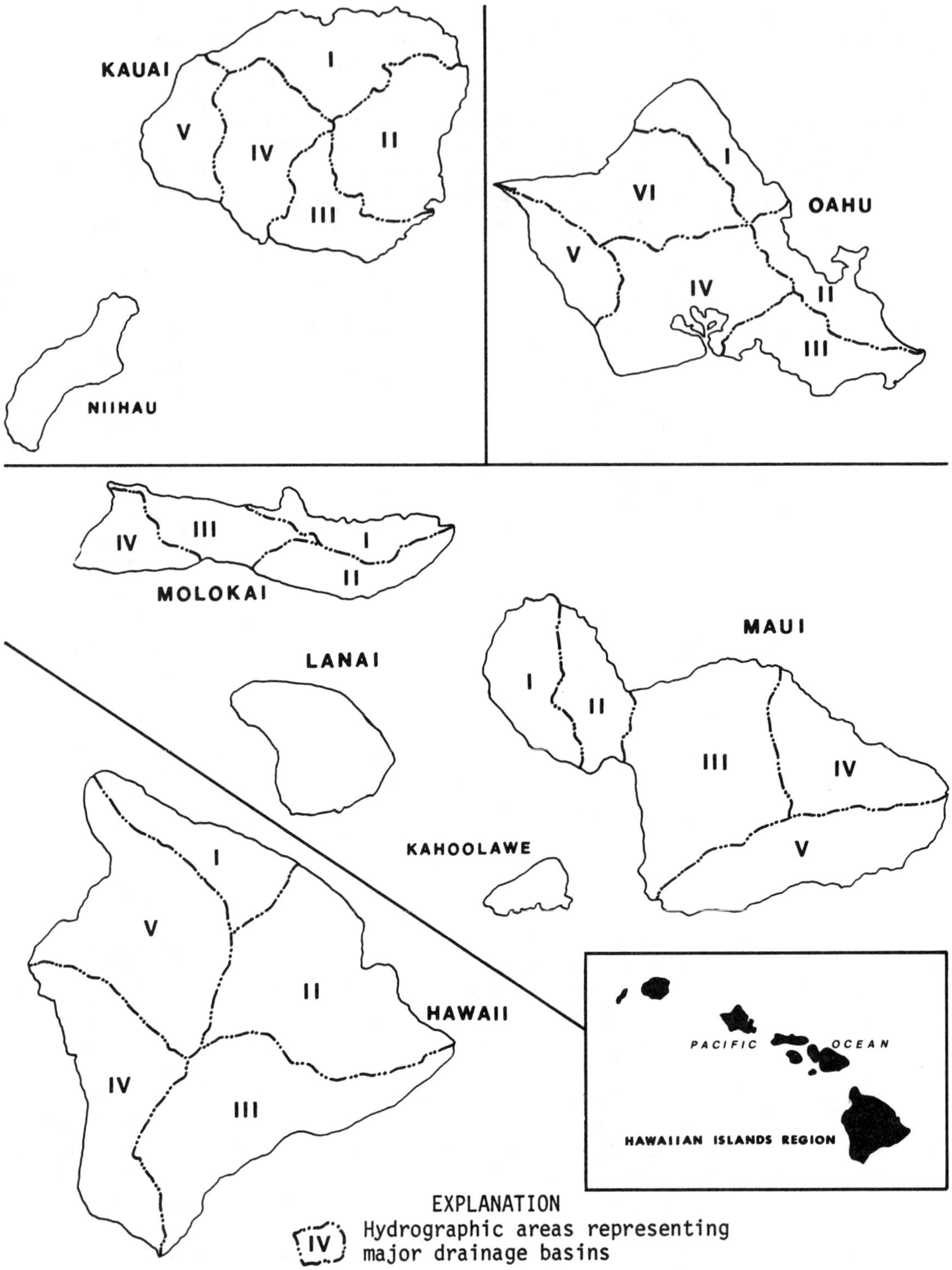

FIGURE 3.—Hydrographic areas of the Hawaiian Islands representing major drainage basins (from Hawaii Water Authority, 1959).

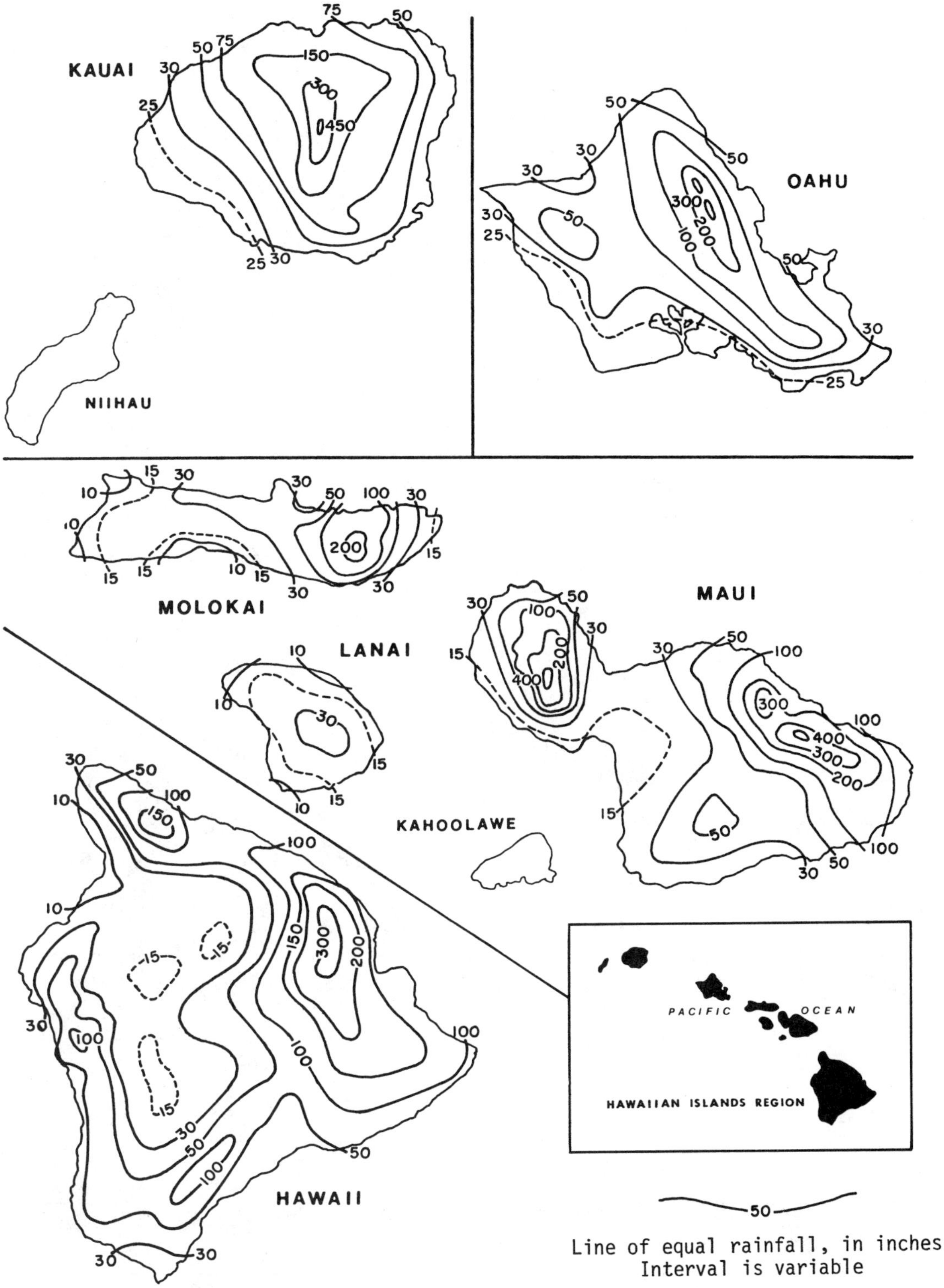

Data from Blumenstock, 1961

FIGURE 4.—Map of the Hawaiian Islands showing lines of equal annual average rainfall in inches.

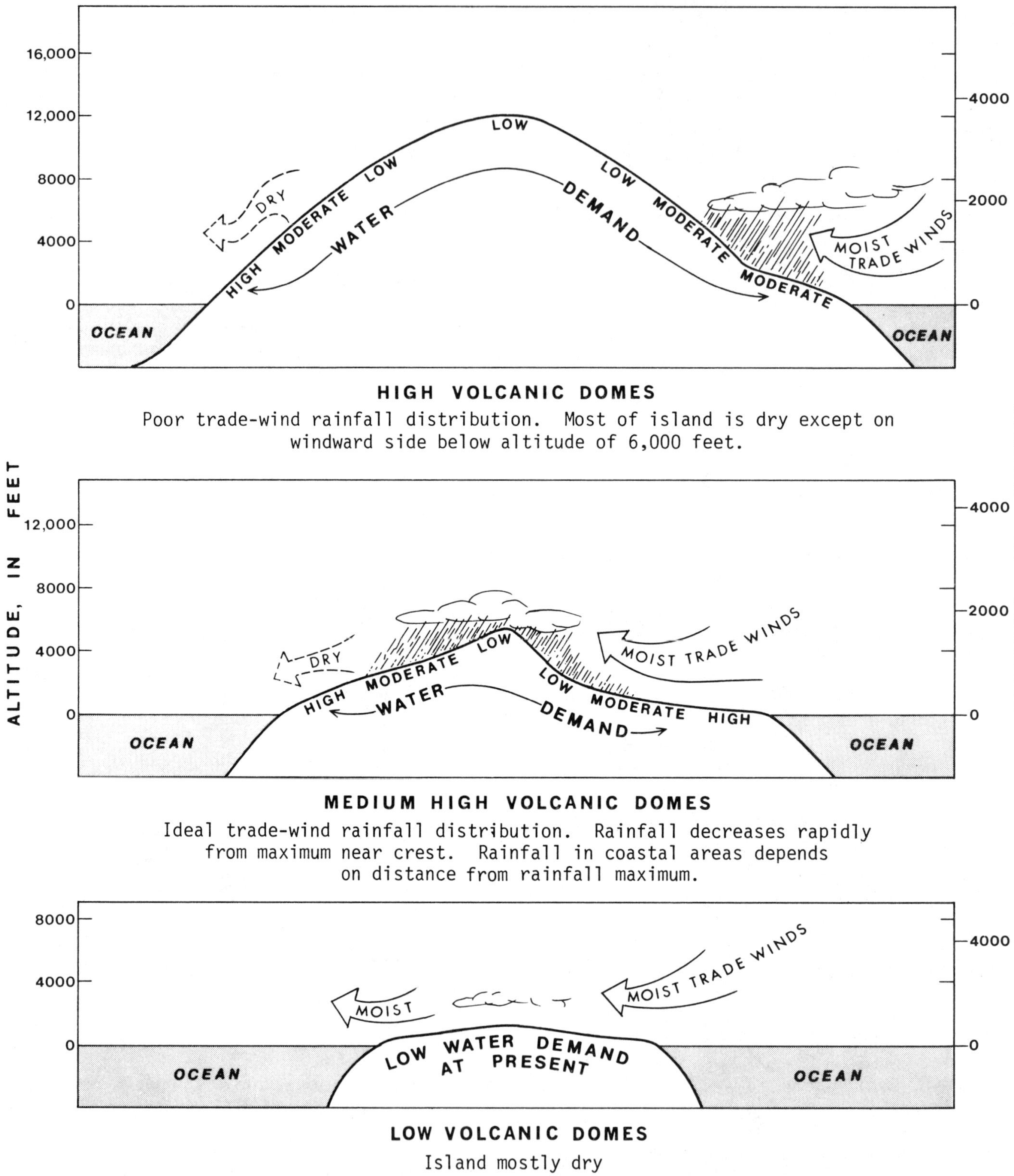

FIGURE 5.—Orographic effects on rainfall.

DISTRIBUTION OF RAINFALL

Rainfall is the principal source of freshwater. Some of it evaporates or is transpired by plants, some replenishes soil moisture, some runs off, and the remainder percolates to ground-water reservoirs. The distribution of the rainfall, however, is not the same everywhere and depends largely on the quantity and variability, and the absorption ability of the land surface. The distribution of rainfall by islands is shown in figure 6.

An example of the change in the distribution of rainfall with change in rainfall is shown in figure 7. The area considered in figure 7 consists roughly of the eastern two-thirds of hydrographic area IV on Oahu. The surface rocks are mostly highly permeable flank flows of the Koolau Volcanic Series. The data used to construct the figure were taken from Hirashima (1971) and Dale and Takasaki (1976).

Table 1 gives a rough estimation of the distribution of rainfall in the hydrographic areas of the region. This table was adapted from a study element report on the surface-water and ground-water resources prepared for the Hawaii Water Resources Regional Study (1975). At best, the figures on distribution of rainfall listed in table 1 are gross estimates and should be regarded as such. They do, however, reflect production possibilities of surface water and ground water in the hydrographic areas, based on known or inferred conditions of rainfall, surface geology, and the geologic framework of the subsurface.

Ground-water pumpage and the quantity of water transported by pipelines and ditches from and into the various hydrographic areas are also listed in table 1.

WATER USE

WATER USE IN 1975

Water use in the Hawaii Region in 1975 averaged about 1,775 Mgal/d, of which about 810 Mgal/d, or 46 percent, was derived from ground-water sources. Table 2 shows the breakdown in use and sources by islands.

Public-supply water use includes general domestic, commercial and limited industrial use.

Self-supplied industrial water use is limited to water systems that are privately owned and operated by industry. The bulk of this use consists of process water used by the sugar industry and of cooling water for the electrical power industry. Industrial water supplied by municipal systems is not included. Brackish ground water used for cooling purposes is included.

The largest use of water in the Region is for irrigation (1,175 Mgal/d) and, of this amount, as much as 90 percent is used for irrigation of sugarcane in dry sunny areas. Sugar yields are highest where sugarcane is irrigated and water use may be as much as 12 acre feet per acre per year. In many places, the irrigation water is imported from distant wet areas. Upwards of 500 Mgal/d of irrigation water is diverted from wet to dry areas, 360 Mgal/d of this quantity from one hydrographic area to another.

The impact on the ground-water resources in the wet areas, resulting from the exportation of water, is usually not pronounced because much of the water exported represents natural ground-water discharge. The impact, however, on ground water in the dry areas to which the water is delivered for irrigation, is pronounced. When used for irrigation between furrows, as much as 60 percent of the irrigated water applied filters into the ground and recharges the underlying ground-water reservoir. In many dry areas, this return of irrigation water represents the principal source of recharge to ground water, the effect of which is to maintain a usable ground-water resource where none may have existed previously.

FUTURE WATER-USE PATTERNS

Most of the current demands for water are in the dry sunny areas where sugarcane yields are highest and where most of the population resides. No change is foreseen in this pattern for the future, except that the demand will increase significantly, owing to the development of larger tourist-oriented and residential complexes in presently little-developed coastal areas. In some areas, the demand is already larger than the supply and must be met by importing water. This is likely to be the situation for other areas of high growth in the future. Some of these problems could be avoided by locating new industries in the areas of plentiful water supply. Predictions of future water-use patterns and demands based on projections of past trends may be unreliable, however, owing to rapid changes in the principal economic base, shifts in population, and changes in life styles.

MUNICIPAL

Per-capita water-use figures based on current population samplings are generally not reliable for future projections. The higher proportion of use by the tourist and related industries, changes in life styles, and more widespread use of luxury home appliances are the reasons for possible discrepancies in projecting water use.

AGRICULTURAL

Many factors other than crop type affect the use pattern of irrigation water. Water use for sugarcane, for instance, is dependent on the type, slope, and profile

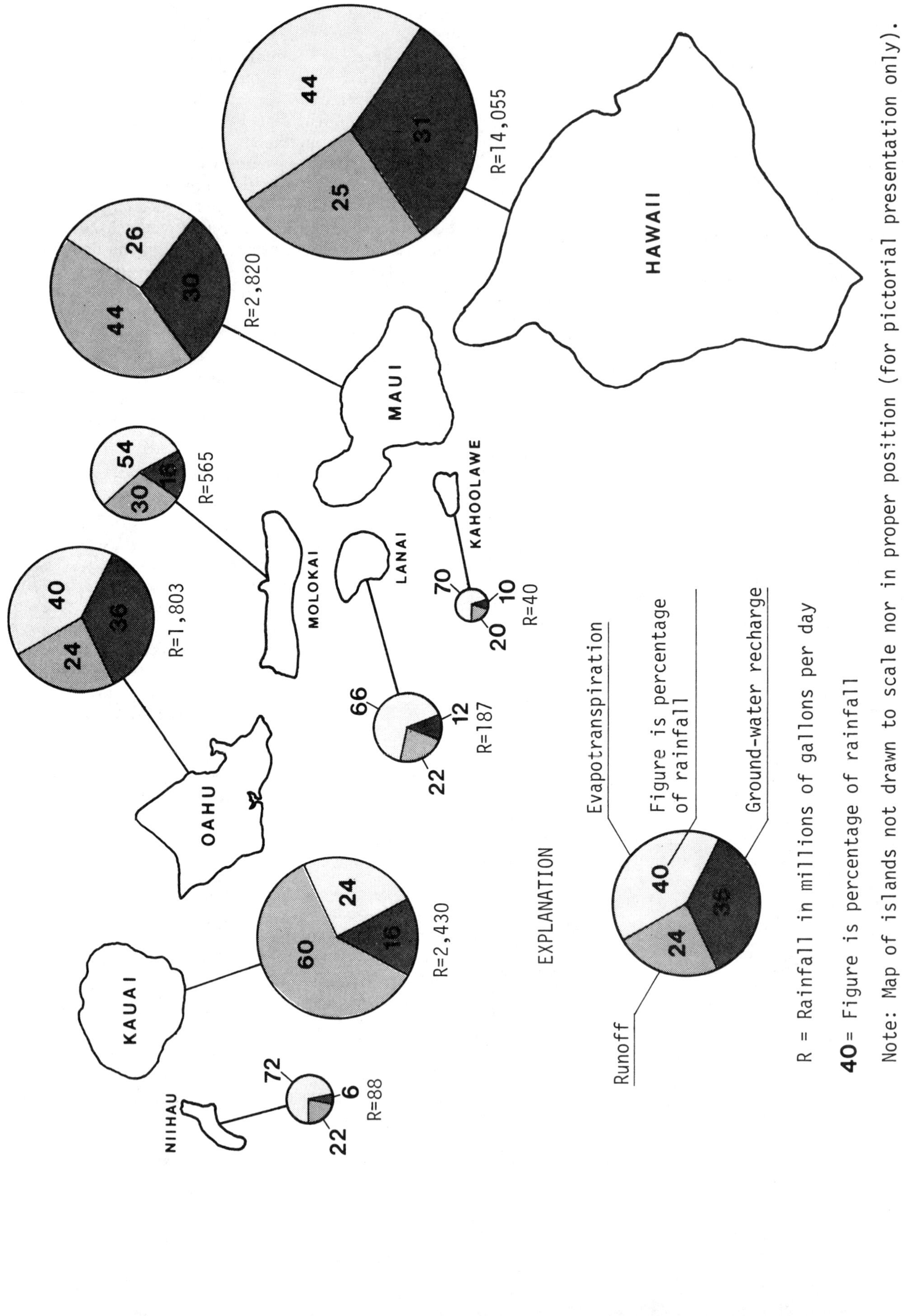

FIGURE 6.—Distribution of rainfall by islands.

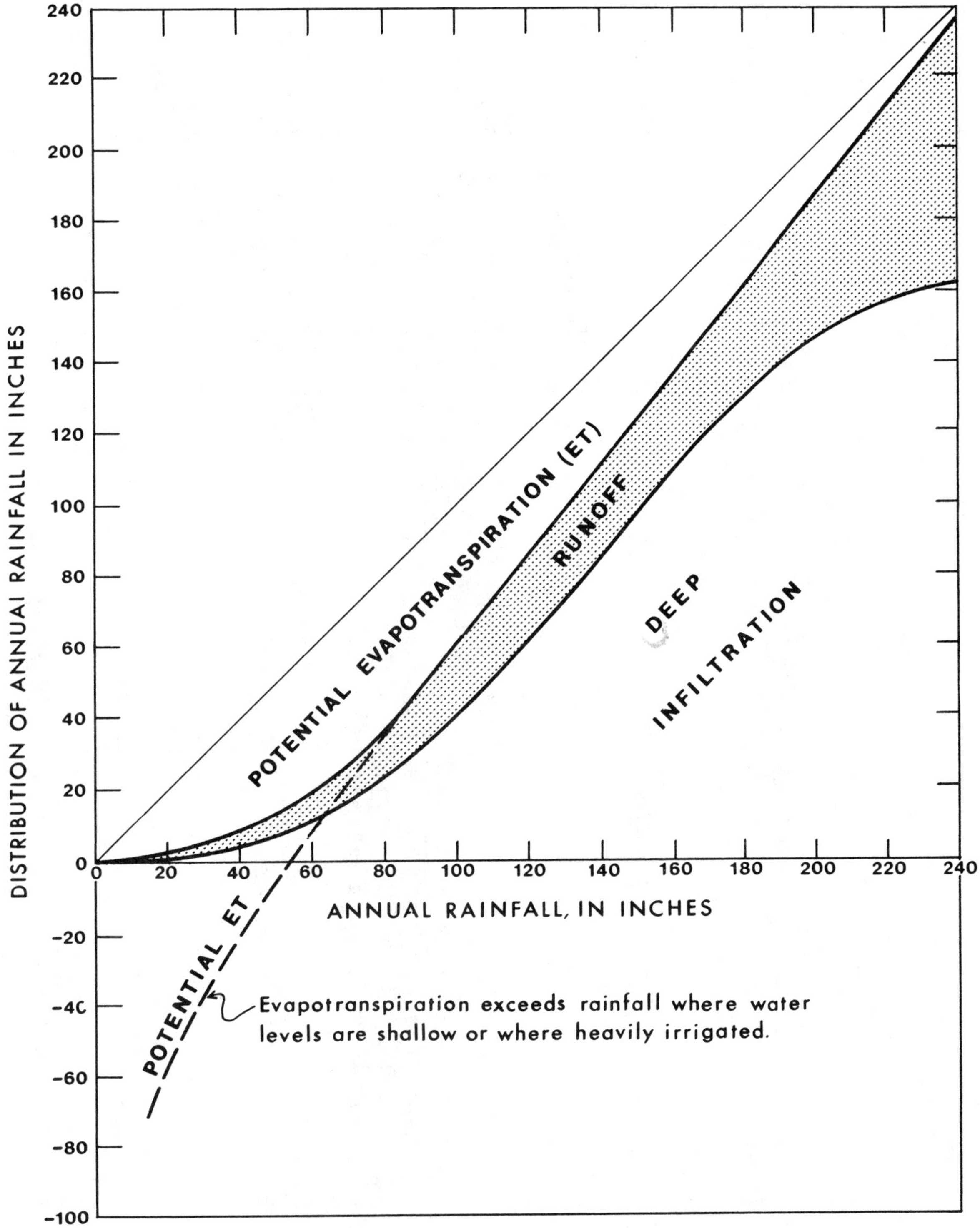

FIGURE 7.—Distribution of rainfall as related to rainfall range in leeward Koolau Range.

TABLE 1.—*Distribution of rainfall, in million gallons per day, to evapotranspiration, runoff, and ground-water recharge*

Island	Hydrographic area representing major drainage basin	Distribution of rainfall							1975 ground-water quantities (approximate)		
		Total rainfall	Evapotranspiration	Percentage of rainfall	Runoff	Percentage of rainfall	Ground-water recharge	Percentage of rainfall	Exported	Imported	Withdrawn from wells
Large islands exposed to trade winds											
Hawaii	I	1,430	695	49	430	30	305	21	5	1	5
	II	7,335	1,730	24	2,510	34	3,095	42	0	0	18
	III	2,340	[1]1,705	73	235	10	400	17	0	0	8
	IV	1,790	[1]1,265	71	180	10	345	19	0	0	3
	V	1,160	[1]745	64	180	16	235	20	1	5	1
	Totals (rounded)	14,100	6,200	44	3,500	25	4,300	31	6	6	35
Maui	I	340	125	37	145	41	70	21	0	0	60
	II	370	130	35	175	47	65	18	20	0	26
	III	685	215	31	325	47	145	21	0	180	170
	IV	925	145	16	310	34	470	51	160	0	7
	V	500	145	29	270	54	85	17	0	0	1
	Totals	2,820	760	27	1,225	43	835	30	180	180	264
Molokai	I	230	30	13	150	65	50	22	5	0	<1
	II	175	125	71	15	9	35	20	1	1	<1
	[2]III, IV	160	150	94	5	3	5	3	0	5	<1
	Totals	565	305	54	170	30	90	16	6	6	1
Oahu	I	270	90	33	85	31	95	35	3	1	8
	II	255	115	45	100	39	40	16	25	3	42
	III	235	120	51	30	13	85	36	0	22	56
	IV	425	105	25	70	16	250	59	24	25	260
	V	98	77	79	15	15	6	6	0	2	6
	VI	520	210	40	130	25	180	35	1	0	48
	Totals (rounded)	1,800	715	40	430	24	655	36	53	53	420
Kauai	I	910	160	18	[3]705	77	[4]45	5	18	0	2
	II	710	219	31	[3]455	64	[4]36	5	11	18	2
	III	280	75	27	[3]195	70	[4]10	4	25	11	12
	IV	414	89	21	[3]300	72	[4]25	6	55	25	2
	V	116	50	43	58	50	8	7	0	55	32
	Totals (rounded)	2,430	595	24	[3]1,715	70	120	5	109	109	50
Kauai	Adjusted totals	2,430	595	24	[5]1,455	60	380	16	109	109	
Large islands	Totals (rounded)	21,700	8,520	39	6,820	31	6,340	30	354	354	770
Small islands in rain-shadow of large islands											
Kahoolawe	Total	40	28	70	8	20	4	10	0	0	0
Lanai	Total	187	124	66	40	22	23	12	0	0	2
Niihau	Total	88	63	72	20	23	5	5	0	0	0
Small islands	Totals (rounded)	315	215	68	70	22	30	10	0	0	2
Region	Grand totals (rounded)	22,000	8,730	40	6,800	31	6,460	29	357	357	772

[1]Probably too high owing to infrequency of storms which provide much of rainfall total.
[2]Hydrographic areas combined owing to low rainfall density in each area.
[3]Includes large quantity of ground-water inflow, see footnote 5.
[4]Too low; ground water included with runoff.
[5]Reduced by 15 percent and added to ground water.

TABLE 2.—*Water use in Hawaii Region, in million gallons per day*
[From water-supply study-element report, Hawaii Water Resources Regional Study (1975)]

Island	Water use				Source	
	Public supplies	Self-supplied industrial	Agri-cultural	Total	Surface water	Ground water
Hawaii	15	117	31	163	134	29
Maui	13	85	467	565	305	260
Molokai	0.4	0	2	2.4	2	1
Oahu	161	130	321	612	152	460
Kauai	7	72	352	431	371	60
Lanai	0.3	0	2.2	2.5	0	2
Kahoolawe	0	0	0	0	0	0
Niihau	0	0	0	0	0	0
	197	404	1,175	1,776	964	812

of the soil, climatic parameters (rainfall, etc.), the length of fallow period, and the availability and quality of water. Available information is generally not adequate for planning future irrigation needs owing to large-scale shifts from furrow to drip irrigation.

INDUSTRIAL

At present most self-supplied industrial water is used as cooling water for air-conditioning and for electric power generation. Potential industries that may require large quantities of water include the mining of manganese or other ores from the sea floor, the extraction of minerals from seawater, or the utilization of geothermal energy resources.

SURFACE WATER VERSUS GROUND WATER

The major source of usable surface water is the fair-weather or base flow of streams. Much of this water is transported from wet to dry areas in extensive ditch systems which were constructed early in this century. More such water could be developed but only at high cost because new ditches or enlargements of the present ditches would be needed. Other constraints that could restrict future development and transfer of water include existing water rights, environmental considerations, and State and Federal statutory regulations.

Generally, there are few constraints other than cost in the onsite development of ground water, except where land development is extensive such as in Oahu and possibly in the Lahaina area in Maui. Areas of high water demand in coastal areas generally coincide with areas underlain by poor-quality basal ground water. In many areas, availability of good-quality ground water would necessitate development 3 to 4 miles farther inland to an altitude of 1,000 feet or higher. A well at these altitudes would have to be drilled to below sea level and the water pumped up from near sea level to the land surface.

In areas where ground-water and surface-water sources are closely integrated and where the withdrawal of ground water is quickly reflected in reduced surface-water flow, the need to maintain streamflow may restrict ground-water development. Knowledge of the hydraulic characteristics of the aquifers and the nature of the stream-aquifer interconnection are needed in order to estimate safe limits for development in these areas.

Compliance with the Federal Safe Drinking Water Act (PL 93-529, 1974) will have pronounced effect on managerial and planning decisions, especially in areas where both ground-water and surface-water sources are being considered as municipal supplies. The high costs of treating surface-water supplies must be weighed against the cost of developing new ground-water supplies.

GROUND-WATER RESERVOIRS

Most fresh ground water is stored near and below sea level to depths ranging to 1,000 feet or more below sea level. The principal fresh ground-water reservoirs consist of thin-bedded basaltic lava flows. These reservoirs contain interconnected water bodies that are impounded by dikes in the interior of the islands or are in dynamic equilibrium with the underlying saline ground water in the outer rims of the islands. Ground water in these settings is referred to, respectively, as dike-impounded water and basal water. Other water bodies, small in comparison, are perched above and isolated from these interconnected water bodies and are called perched water. The term high-level water is often used to describe both dike-impounded and perched water, or a combination of them when they occur at high altitudes.

A diagrammatic cross section showing occurrence and development of ground water in an idealized Hawaiian volcanic dome is shown in figure 8. The principal ground-water reservoirs in each of the Hawaiian Islands are outlined in figures 9 to 15. Estimates of ground-water recharge and ground-water draft in each of the hydrographic areas are also shown in figures 9 to 15 and table 1. A brief description of the principal reservoirs and their potential for additional development is given by hydrographic areas in table 3.

Ground-water development is generally most favorable in areas directly downslope from mountain areas of high rainfall and becomes less favorable with increasing distance away from these downslope areas.

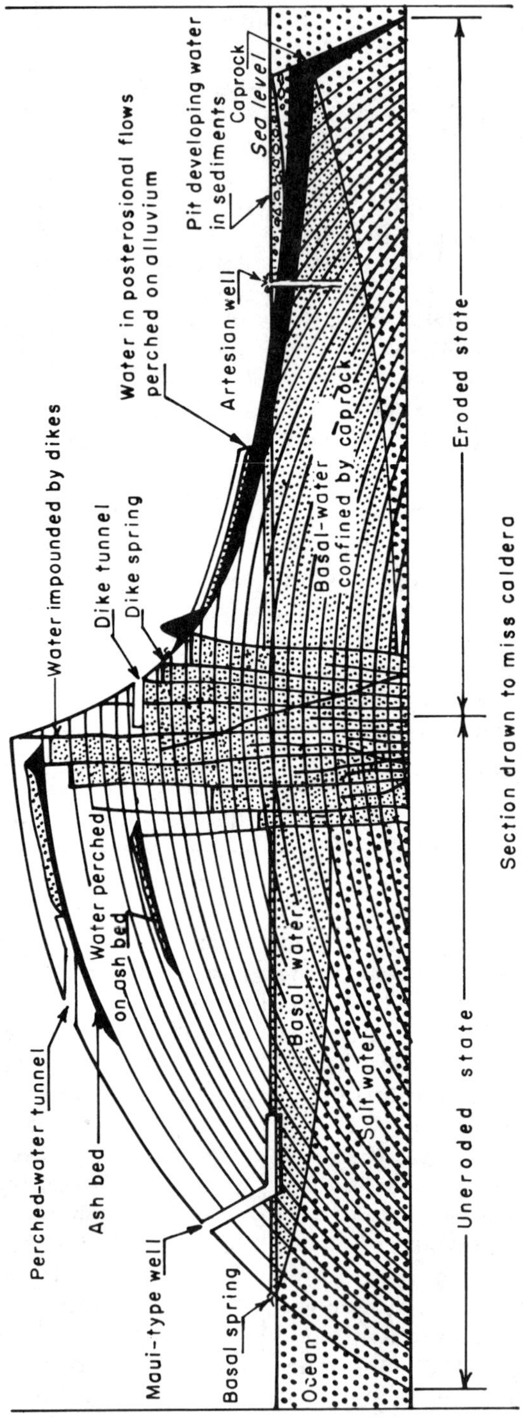

FIGURE 8.—Occurrence and development of ground water in an idealized Hawaiian volcanic dome (from Cox, 1954).

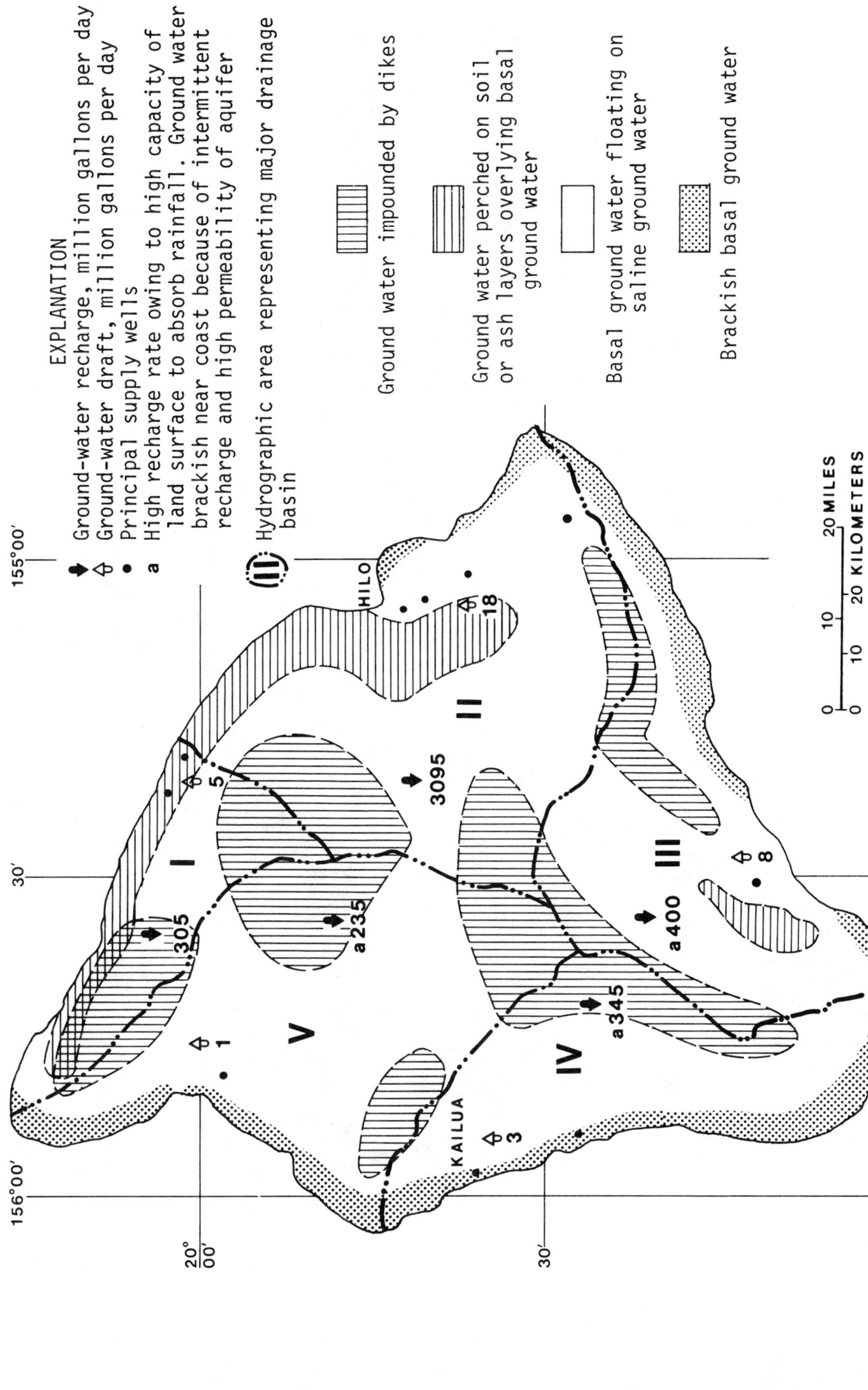

FIGURE 9.—Map of island of Hawaii showing approximate outline of ground-water reservoirs, recharge, 1975 draft, and principal supply wells by hydrographic areas representing major drainage basins. (Modified from Davis and Yamanaga, 1973.)

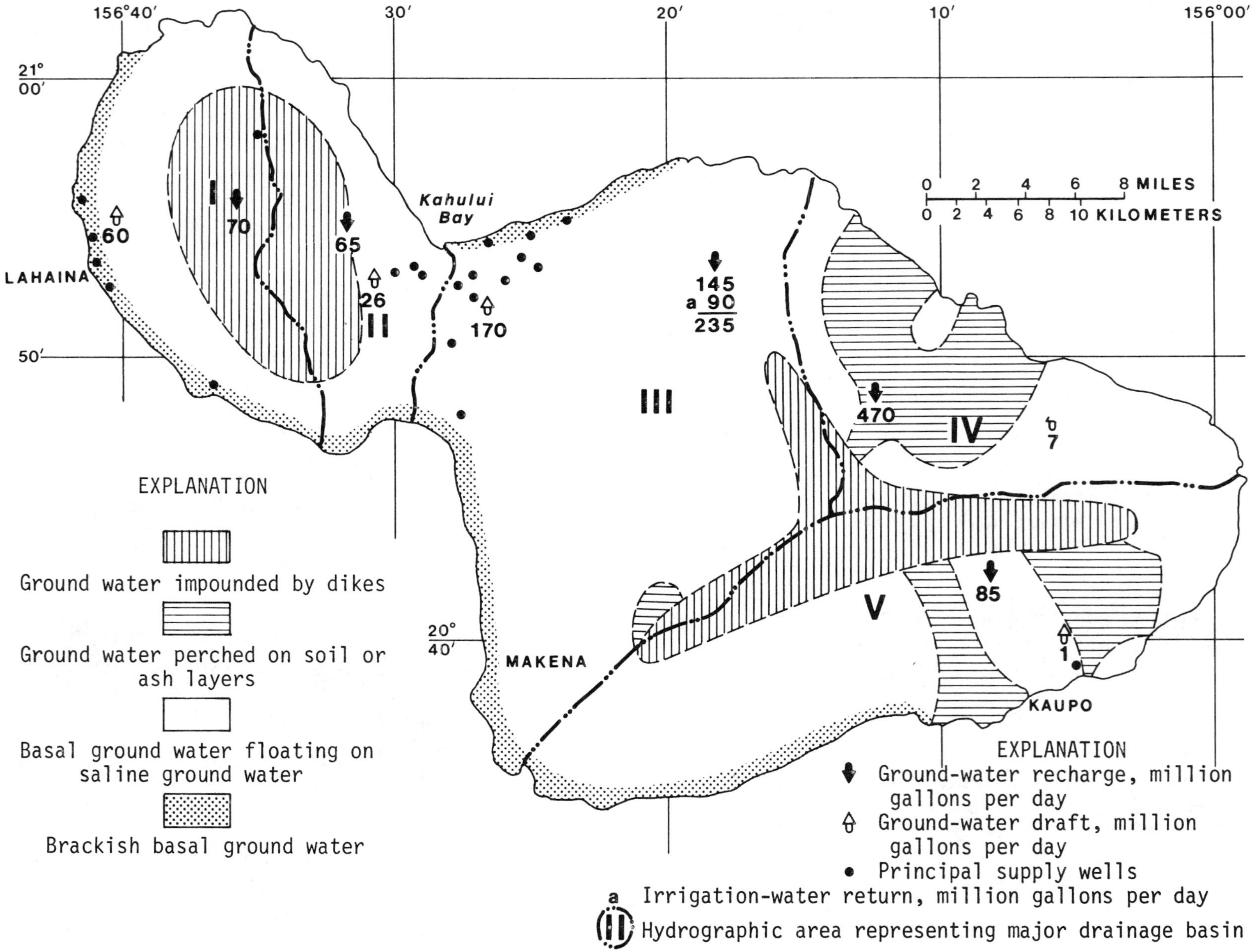

FIGURE 10.—Map of island of Maui showing approximate outline of ground-water reservoirs, recharge, 1975 draft, and principal supply wells by hydrographic areas representing major drainage basins. (Modified from Stearns and Macdonald, 1942.)

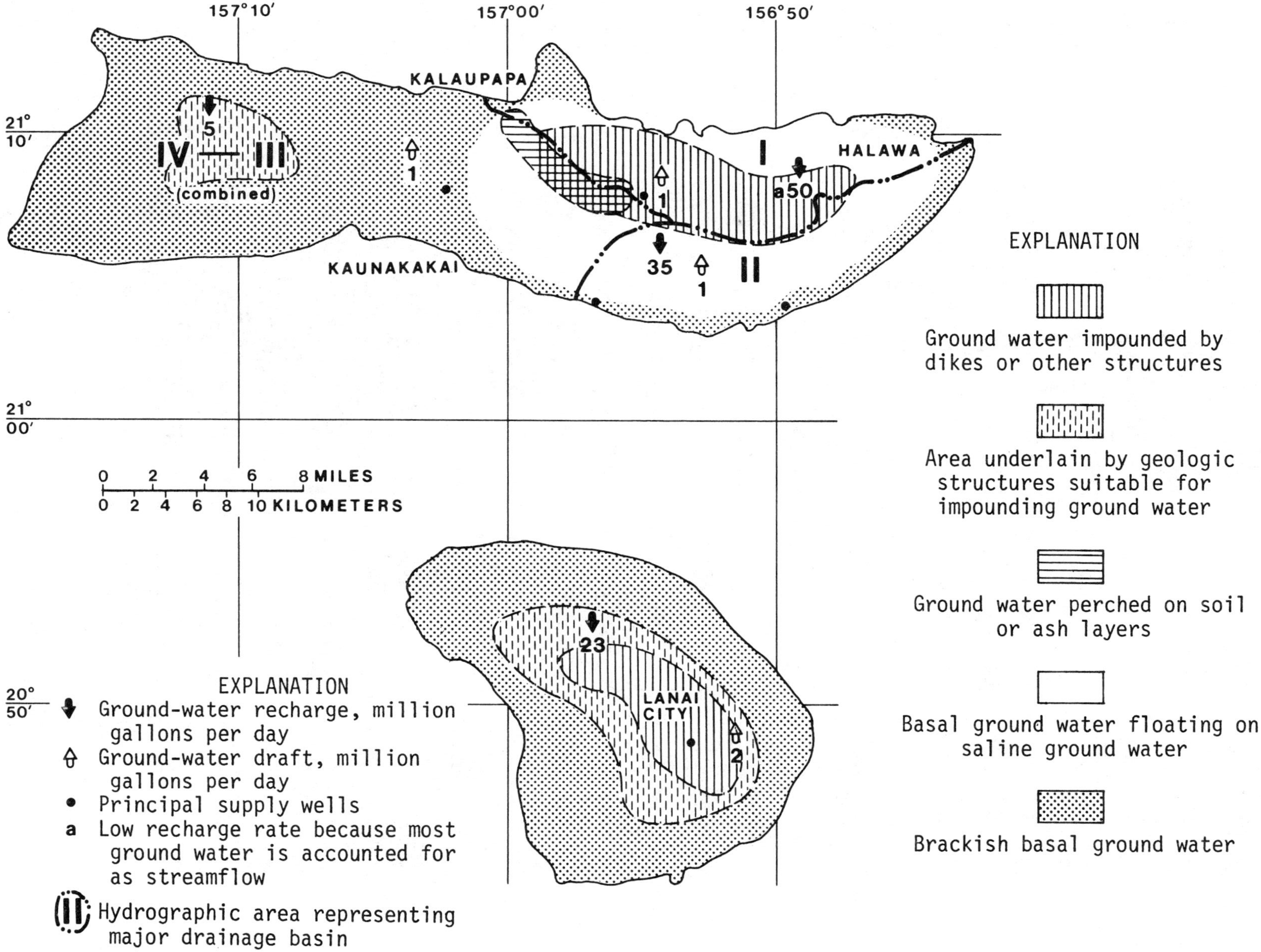

FIGURE 11.—Map of islands of Molokai and Lanai showing ground-water reservoirs, recharge, 1975 draft, and principal supply wells by hydrographic areas representing major drainage basins. (Molokai: modified from Stearns and Macdonald, 1947; Lanai: from S.P. Bowles, unpublished data, in 1973.)

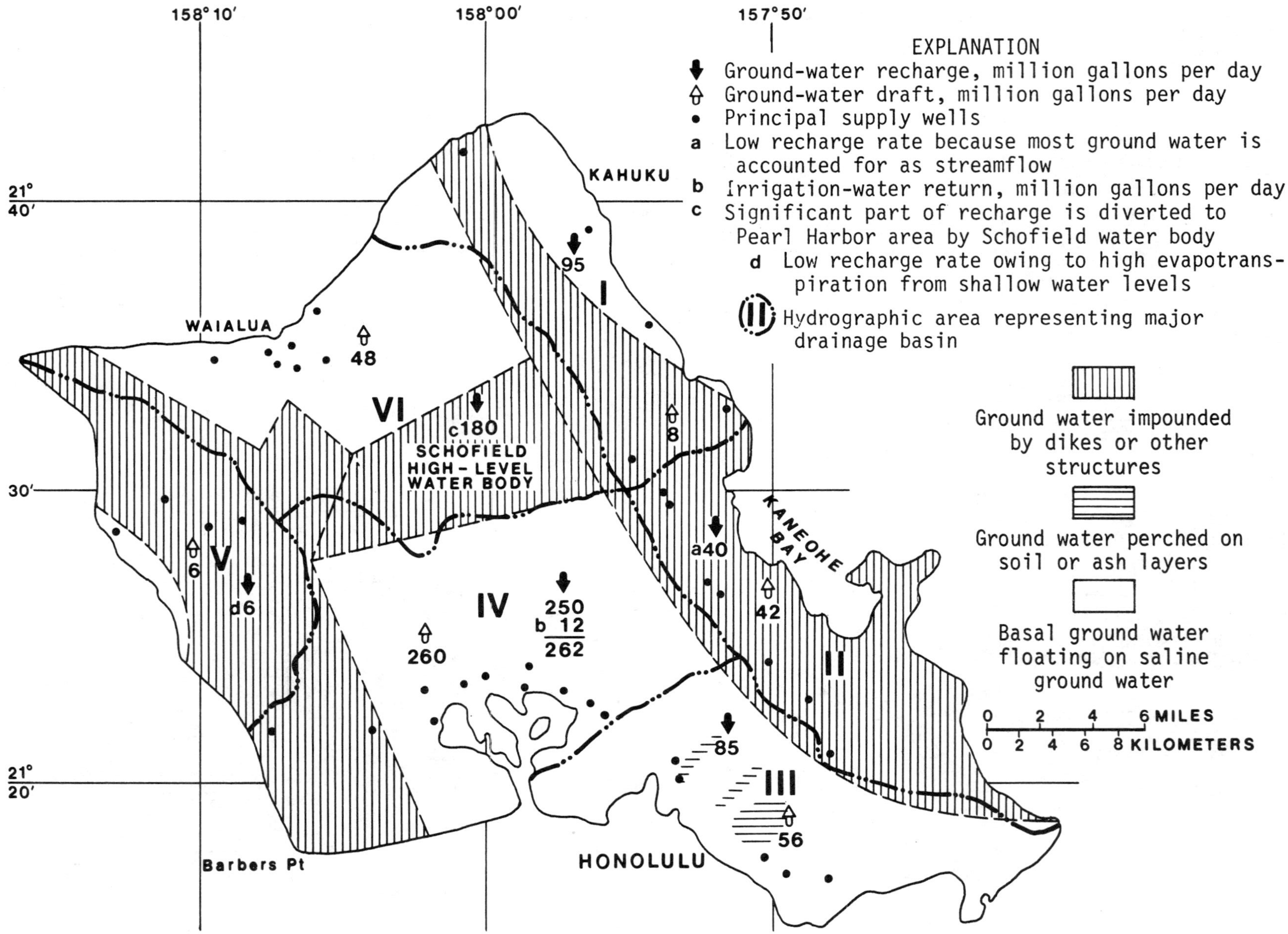

FIGURE 12.—Map of island of Oahu showing approximate outline of ground-water reservoirs, recharge, 1975 draft, and principal supply wells by hydrographic areas representing major drainage basins.

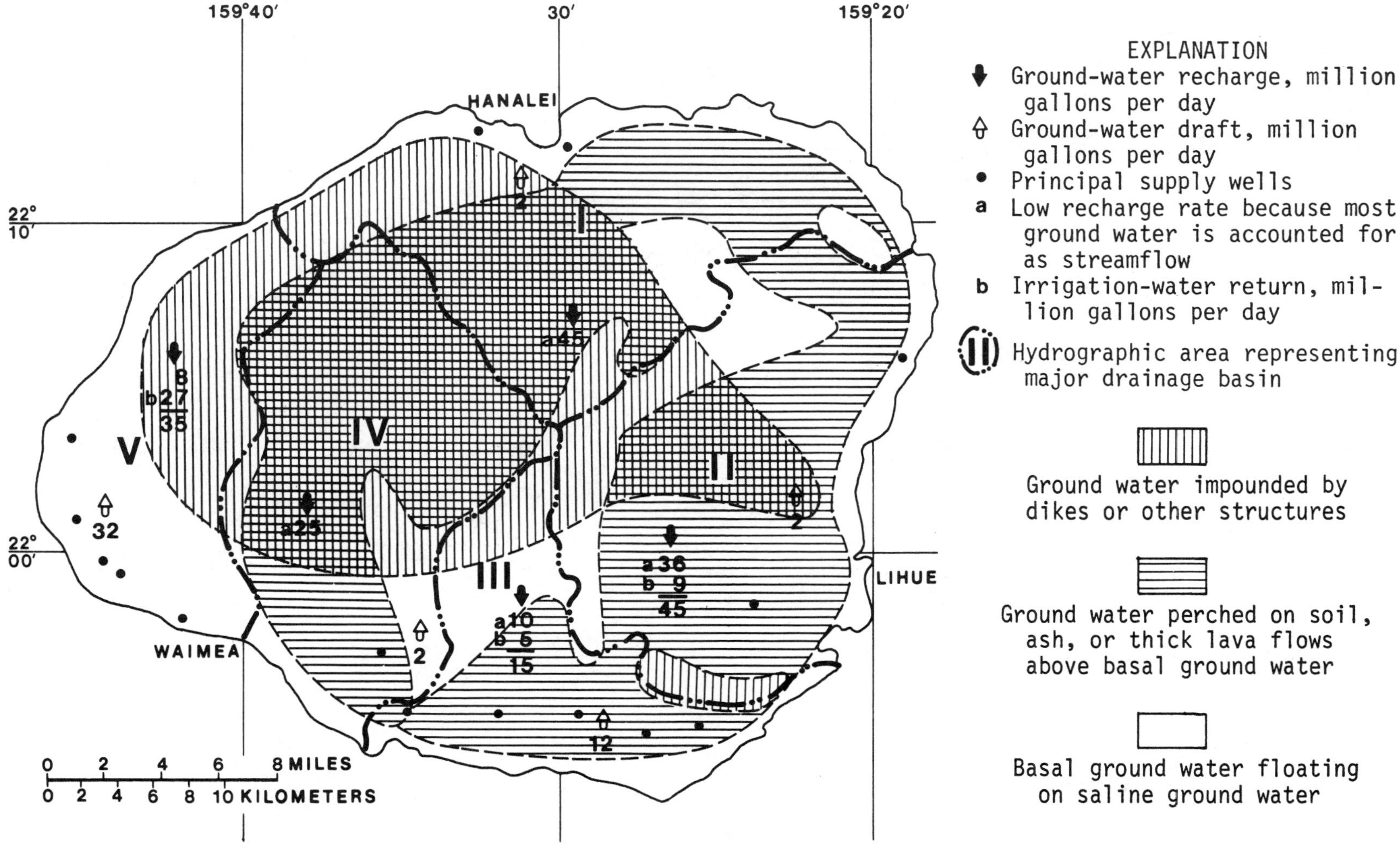

FIGURE 13.—Map of island of Kauai showing approximate outline of ground-water reservoirs, recharge, 1975 draft, and principal supply wells by hydrographic areas representing major drainage basins.

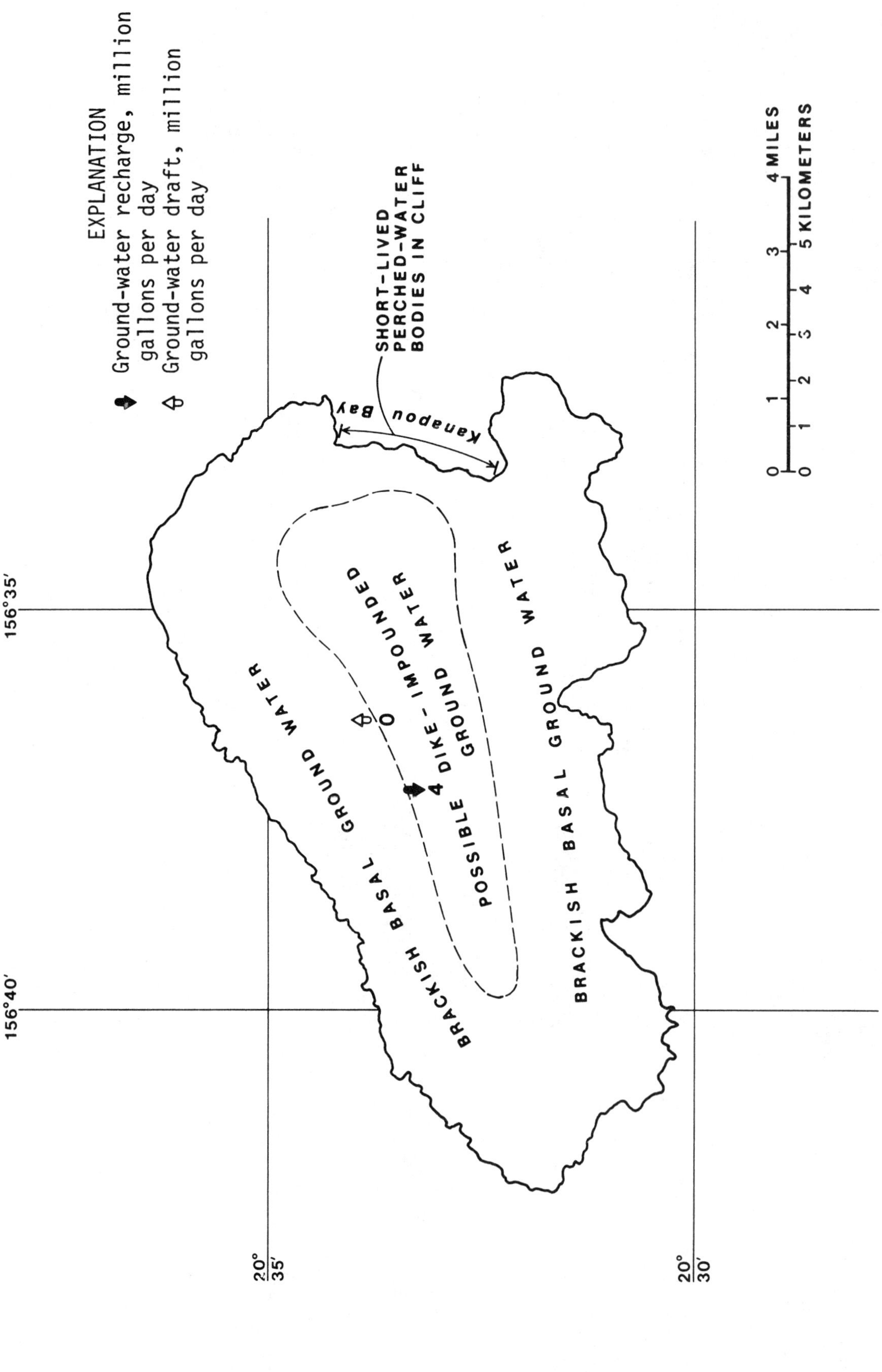

FIGURE 14.—Map of island of Kahoolawe showing approximate outline of ground-water reservoirs, recharge, and 1975 draft.

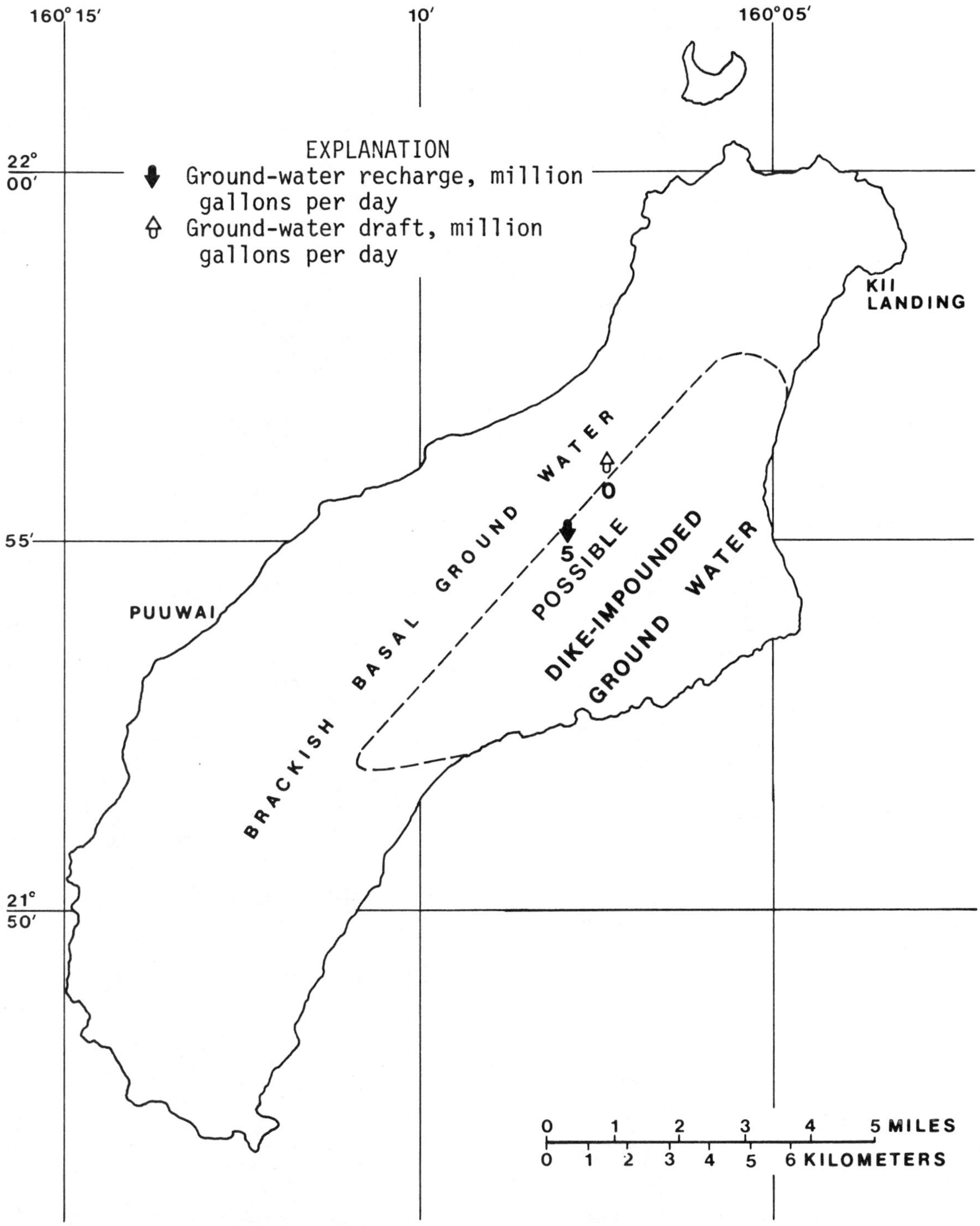

FIGURE 15.—Map of island of Niihau showing approximate outline of ground-water reservoirs, recharge, and 1975 draft.

This downslope direction generally coincides with the downslope flow direction of the lavas and the favorable direction of ground-water flow from the mountains and the sea. Owing to this generally favorable direction of ground-water flow, ground-water development is seriously limited in leeward areas where rainfall is heavy only on the windward side. The south Kohala-Kona coast on the island of Hawaii and the Kihei-Makena

TABLE 3.—*Principal ground-water reservoirs*

Hydrographic area representing major drainage basin	Principal ground-water reservoirs		Remarks	Potential for additional development
	Occurrence	Aquifer rock		
Hawaii I	Basal	Kohala Mt. lavas	Large, untapped source.	Good
	Dike impounded	do	Excellent domestic quality.	Good
	Basal	Mauna Kea lavas	Flow increases toward south.	Fair to good
Hawaii II	do	do	Large, little-tapped source.	Good
	do	Mauna Loa lavas	Very large, little-tapped source.	Good
Hawaii III	High level	do	Large impounded supply.	Good
	Basal	do	Large supply in eastern part.	Poor to good
Hawaii IV	do	do	Brackish near coast.	Poor to fair
	do	Hualalai Volcanic Series	do	Poor to fair
Hawaii V	do	do	do	Poor
	do	Mauna Loa lavas	do	Poor
	do	Mauna Kea lavas	do	Poor to fair
Maui I	Dike impounded	West Maui lavas	Excellent domestic quality.	Fair
	Basal	do	Poor quality near coast.	Poor
Maui II	Dike impounded	do	Excellent domestic quality.	Fair
	Basal	do	Excellent quality water.	Fair to good
Maui III	do	Haleakala lavas	Large recharge from return irrigation water, quality improves to the east.	Fair
Maui IV	Perched	Posterosional Haleakala lavas	Excellent quality water.	Good
	Basal	do	Fair to good quality in thin basal-water bodies.	Fair to good
Maui V	do	Haleakala lavas	Good quality in east part and poor in west part.	Poor to fair
Molokai I	Dike impounded	East Molokai Volcanic Series	Excellent quality water.	Good
Molokai II	Basal	do	Quality and flow improve toward east.	Fair
Molokai III, IV	do	East and West Molokai Volcanic Series	do	Poor to fair
Oahu I	Dike impounded	Koolau Volcanic Series	Development will reduce streamflow.	Fair
	Basal	do	Separated into two parts by dike zone.	Fair to good.
Oahu II	Dike impounded	do	Development will reduce streamflow.	Fair
Oahu III	Basal	do	Heavily tapped in coastal plain areas.	Poor
Oahu IV	High level	do	Schofield water body; development at expense of reduced flow to Pearl Harbor and Waialua areas.	Good
	Basal	do	Thick lens, heavily tapped.	Poor
	do	Waianae Volcanic Series	Heavily pumped for irrigation, poor general quality.	Poor
	do	Coralline	Heavily pumped for irrigation and recharged by return irrigation water.	Poor to fair
Oahu V	Dike impounded	Waianae Volcanic Series	Excellent quality in limited supply.	Poor to fair
	Basal	do	Brackish near coast.	Poor
Oahu VI	High level	Koolau Volcanic Series	Schofield water body, development at expense of reduced flow to Pearl Harbor and Waialua areas.	Fair to good
	Basal	do	Deep valleys restrict lateral flow.	Fair
	do	Waianae Volcanic Series	Thick lens.	Fair
Kauai I	do	Napali Formation	Large, little-tapped source, excellent quality.	Good
	Perched	Koloa Volcanic Series	Large, untapped source, yield poorly to wells.	Poor to fair
Kauai II	Basal	Napali Formation	Thick basal reservoirs, excellent quality.	Fair
	do	Koloa Volcanic Series	Large, little-tapped source, yield generally low to wells.	Poor to fair
	Perched	do	Mostly in small reservoirs, excellent quality.	Fair
Kauai III	Basal	Napali Formation	High water levels owing to impoundment by Koloa Volcanic Series.	Fair to good
	do	Koloa Volcanic Series	Brackish near coast.	Poor to fair
	Perched	do	Little-tapped source in eastern part.	Fair
Kauai IV	Basal	Makaweli Formation	High water levels.	Fair
Kauai V	do	Napali Formation	Significant recharge by return irrigation water.	Poor
Kahoolawe	do		Mostly brackish.	Poor
Lanai	Dike or fault impounded	Lanai Volcanic Series	Large storage, can be utilized for peaking.	Fair
Niihau	Basal		Mostly brackish.	Poor

area in Maui are such areas. In the high volcanic dome of Haleakala on Maui, for instance, more than two-thirds of the rain falls on the windward side. About 160 Mgal/d of water developed in this area is conveyed by ditches to irrigate sugarcane in the lower eastern slopes, but much more goes unused to the sea owing to the lack of adequate conduits.

Water for domestic use in east Maui is currently

transported from limited ground-water sources in west Maui and plans are to increase the amount of water transported for this purpose. From the standpoint of long-term optimum use of the ground-water resource in west Maui, development of the large but unused water resources available in the windward slopes of east Maui should be considered for use in east Maui.

OPTIMIZING GROUND-WATER DEVELOPMENT

The ground-water resources of the region, taken as a whole, are far greater than foreseeable future demands of the region, but this is not true for single basins and single islands. Each island is a separate entity and stands alone with respect to utilization of its ground-water resources. Within each island, ground water may be obtained from the three types of reservoirs described in the preceding section. Optimal development will occur, however, only by taking into account the relationships of discharge and recharge that may exist locally among the various ground-water bodies.

BASAL GROUND WATER

It is commonly not recognized that basal ground water is a highly fugitive, as well as renewable, resource, and that the ocean, to which it escapes, represents a sink of no recovery and no return. Many existing well fields utilize ground water inefficiently by concentrating the wells in small areas, thus causing large local drawdowns while permitting large amounts of ground water to escape unused to the sea in the intervening areas. The wells are concentrated spatially, partly for efficiency of distribution, but in part this has been done because of mistaken belief that they tap an underground river. Pumping may also be unnecessarily intermittent rather than steady in the belief that the longer the hiatus in pumping, the greater the buildup in storage.

Ideally, optimum development of basal ground water would occur by constantly pumping a line of wells normal to the flow paths, which are generally perpendicular to the coastline. Assuming that reliable estimates of ground-water recharge and hydraulic conductivity were obtained, the well spacing and pumping rates could be optimized. An optimum development plan is generally a compromise based on a theoretical scheme for maximum development, aquifer-test results, and economics. Land ownership and availability, however, may be the most important considerations in areas of high land cost and development.

Problems in implementing an optimum plan are few if the existing ground-water draft is small. The problems, however, increase manyfold where the draft is large and the existing development is inefficient as the result of poorly constructed wells, improperly spaced wells, and sustained localized overdraft. The remedy for existing problem areas may involve high costs in the abandonment of old wells and the drilling of new wells, in land acquisition for new well sites, and in the curtailment of ground-water withdrawals at some places. These costs must be weighed against the long-term benefits of optimum development.

At places where basal water is confined by sedimentary caprock and is under artesian pressure, optimum development may require reduction of the artesian heads and pumping at constant rates in order to minimize leakage through the caprock. This operating plan will result in permanent reduction of storage in the basal freshwater lens and possible intrusion of saline water into some existing wells. The costs of these consequences must be carefully considered.

In order that reliable estimates of the consequences can be made and the alternatives properly considered, there is a need for development of hydrologic models of the aquifers. These models can be used to predict results of implementation or nonimplementation of certain operation plans. The degree of success derived from the use of such models would depend entirely, however, on the amount and reliability of the hydrologic data used in constructing the models.

DIKE-IMPOUNDED WATER

A dike-impounded reservoir accumulates rainfall, stores it temporarily, and steadily leaks it to abutting basal reservoirs or to streams cutting into the reservoir. Development of dike-impounded reservoirs is attractive because hydraulic heads are high and the reservoirs are isolated from saline water. One major consequence, not fully understood or acknowledged, is that withdrawal of water from such a reservoir reduces its leakage to abutting basal reservoirs by the quantity developed. The most perplexing consequence, however, is the loss of hydraulic head in the reservoir and the tremendous waste of water from storage during construction of water-development tunnels below the top of the reservoir. During the tunneling period, which lasts for many months, all of the storage above the invert of the tunnel is depleted and flow from the tunnel is reduced to some steady rate. The reduction of storage caused by construction of eight tunnels in Oahu has been estimated at 25,800 Mgal (Takasaki and others, 1969), equivalent to the total ground-water withdrawal in Oahu for about 60 days. The combined steady-state flow of these eight tunnels is about 28 Mgal/d, which implies that it would require at least 920 days for natural recharge to restore the depleted storage.

The Honolulu Board of Water Supply recently drilled an inclined well to tap dike-impounded water in Oahu.

This technique permitted the development of dike-impounded water without the large initial and uncontrollable waste of stored water common to development by tunneling.

A significant part of the reduction in storage by tunneling could be restored by constructing bulkheads at the controlling dikes. Bulkheads have been installed in several tunnels, but only the one in a tunnel in Waihee Valley is effective in restoring water to its pretunnel level. Bulkheads in other tunnels in Oahu were not constructed at dikes that originally stored the most water; hence, they are only partly effective in the restoration of storage. The locations of dikes that control the most water can best be determined at the time of tunneling. If tunneling information is not available and storage is depleted, gain in flow between dikes, determined by measurements, should indicate the best sites for bulkheads. Bulkheads are most effective where single dikes control large quantities of water and where the contrast in permeability between lava flows and dikes is great. In a dike complex, the permeability contrast is too small or the dikes are too numerous for bulkheads to be very effective.

The storage above the tunnel in Waihee Valley has been estimated at 2,200 Mgal and capability of manipulating this amount of storage is of tremendous value in water-supply management. The remaining tunnels are operating at steady-state flow and storage is virtually nil. The restoration of significant amounts of storage above these tunnels would be invaluable. Optimum use of these tunnels would be to store water during the wet winter months for use during the dry summer months at rates much larger than the perennial steady-state flow of 28 Mgal/d. This scheme would eliminate excessive pumping of basal-water wells during the high-demand summer months, thus allowing pumping to be held at some constant rate throughout the year. The reverse is the current practice, involving near-constant withdrawal from dike-impounded reservoirs and manipulated light-winter to heavy-summer withdrawal from basal reservoirs.

SURFACE ACTIVITIES TO MAXIMIZE RECHARGE OF GROUND WATER

Owing to the general lack of surface-reservoir sites, surface activities designed to induce the recharge of ground water must be such that large surface storage is not essential. Possibilities are:

1. Construction of multipurpose flood control structures designed to maximize or to induce recharge.
2. Lowering near-surface ground-water levels by development to reduce evapotranspiration and induce recharge.
3. Agricultural and forest-management practices conducive to recharge and the enhancement of water quality.
4. Fog-drip inducement by artificial means for the purposes of recharge.
5. Artificial recharge of excess surface and irrigation water or sewage effluent by ponding and deep injection.
6. Land development conducive to preservation of water quality and increased recharge.
7. Maximum use and storage of water at high altitudes instead of dropping the water quickly to lower altitudes as is commonly practiced where abutting property owners discharge water individually.
8. Use of inexpensive or excess power generated by burning bagasse at sugar mills for pumping unused irrigation water for storage in reservoirs or for subsurface injection at high levels.

GROUND-WATER PROBLEMS

Water-supply and other water-assessment problems are rarely regional problems, but are, for the most part, single-island problems. Problems related to technology, water rights, use, pollution, and management of ground water, however, are generally regional in scope.

The ground-water problems common to the Region can be considered in three interrelated groups: (1) problems of inadequate information, (2) technological problems, and (3) institutional problems.

PROBLEMS ARISING FROM INADEQUATE INFORMATION

Inadequate information of the geologic framework, meterology, and streamflow often presents serious problems in interpretation of ground-water regimes. Inadequate geologic information leads to difficulty in describing the ground-water reservoirs, the flow within them, and the outflow from them. Inadequate meteorological and streamflow information leads to problems in determining quantitatively the recharge to and the discharge from the ground-water reservoirs.

Major ground-water problems arise from lack of information on the following aspects of the geologic framework:

1. The water-bearing properties of lava flows below sea level. Most of the Region's basal ground water, and the saline ground water that underlies it, occur in these rocks. Lava flows extruded below sea level are generally of low permeability; however, the aquifers that now occur below sea level in the Region are often highly permeable subaerial flows which were subsequently submerged. The extent of the submergence is not known nor is the thickness of the permeable lava flows below sea

level. This leads to difficulty in evaluating amounts of storage and the deep circulation pattern of the fresh and saline ground water.

2. The extent of water-bearing properties of near-shore sedimentary and volcanic rocks. The ensuing problems are those related to the discharge of polluted ground water at or near the shore resulting from the injection of wastes into these rocks.
3. The structure, extent, and density of occurrence of dikes in the dike-intruded zones. The occurrence and movement of dike-impounded waters could be better described if this information were available, and related problems pertaining to water rights and priorities and to environmental impacts that would result from development could be handled more intelligently.
4. The geology and hydrology of rift-zone areas of active or recently active volcanoes. Rift zones occur on the flanks of volcanoes in places where magma is extruded from long fissures. The rift zones become areas of dike intrusion as the magmas cool in the fissures. Hot waters of possible interest for geothermal development occur in these areas. A major concern is the determination of the sustained water yield and the potential energy yield as a basis for evaluating the economic potential of the geothermal waters.

Problems in evaluating ground-water recharge and discharge arise from inadequacies in the following meteorological and streamflow information:

1. The rainfall distribution in the wet interior areas where most recharge occurs.
2. The relation between actual and potential evapotranspiration in the moderately dry to moderately wet areas, especially where the potential evapotranspiration has been estimated as being equal to pan evaporation.
3. The contribution of fog drip to ground-water recharge. Preliminary studies in the region suggest that the contribution of fog drip to recharge may be significant either as direct infiltration or as a replacement for water lost by evaporation and transpiration.
4. The relation between basin rainfall and basin runoff in the wet interior areas where drainage basins are small and where the ground-water divide often differs significantly from the topographic divide.

TECHNOLOGICAL PROBLEMS

The major technical problems related to ground-water hydrology in the Region are those pertaining to basal ground water. Some of the more important are as follows:

1. The relation of basal ground-water levels and their natural and induced fluctuations to the storage, recharge, discharge, and development of the basal lens. More importantly, how to evaluate anticipated changes in rates of natural recharge and discharge at places where water levels are declining.
2. The general inadequacy of equations available for analyzing pumping-test data from partially penetrating wells, and also the application of the concept of transmissivity, which includes aquifer thickness, for determining hydraulic conductivity. The inadequacy is compounded for aquifers consisting of thin lenses, where pumping, even at moderate rates, causes rapid upconing of the saltwater interface.
3. The determination of the relation between horizontal and vertical hydraulic conductivities in basaltic lava-flow aquifers. This relation often determines the yield and, consequently, the optimum mode of development, especially from thin basal lenses where upconing of the underlying saline ground water is always a threat.
4. The evaluation, in space and time, of recharge to and discharge from storage in the part of the basal lens below sea level with change in water level above sea level.
5. The evaluation of tidal responses in basal lenses, especially in thin lenses, and the application of this information to the understanding of the mechanics of flow and the mixing process in the lenses, and in determining hydraulic conductivity. This may be the only applicable hydrologic analytical method in many areas in the Region where the major stress on the basal aquifer is not caused by pumping but by tidal fluctuations.
6. The analysis, in water-budget studies, of the large intermittent amounts of recharge to thin basal lenses as potentially recoverable increments of stored ground water. Intermittent recharge due to intense storm precipitation usually disperses quickly in thin basal lenses, particularly where these lenses occur in highly permeable recent lava flows.
7. Many of the problems mentioned earlier are related to the potential for degradation of water quality which, although not restricted to basal ground-water bodies, is most evident in these water bodies because of their widespread occurrence. Water quality is a critical factor which often limits the use and development of ground water. Overdevelopment results in saltwater upconing or encroachment, and man's activities may produce other contaminants which may degrade present

supplies. There is a need to identify areas where overdevelopment has occurred, where the activities of man may endanger present and future supplies, and where geologic and hydrologic conditions might permit the surface or subsurface disposal of wastes without undesirable environmental consequences.

INSTITUTIONAL PROBLEMS

Owing to recent interest in the quality of the environment and the need to comply with government regulations on water quality and waste disposal, the roles of government, business, and the public in ground-water management have been significantly altered. Government's managerial role includes research, control, conservation, development, monitoring, and planning. The relations of these public functions at various levels of the Federal, State, and County governments warrant review because of these recent changes. The jurisdictional lines between different levels of government and between agencies at the same level remain extremely complex.

Problems in water management related to priority and exchange or transfer of water may require government intervention and assistance for resolution so that water development can be better integrated and coordinated with land-use plans and comprehensive planning in general.

The allocation and management of water has been left largely to private agreements through user associations, which generally include land estates, agricultural interests, municipal suppliers, and the military. Problems are beginning to arise as water supplies locally become more scarce and as land developments affect recharge rates and quality of the water. To cope with these problems, the State, through the Ground Water Use Statutes (Chapter 177, Hawaii Revised Statutes, as amended, was reenacted by Act 122, of the 1961 Session Laws of Hawaii), is authorized to designate ground-water areas for regulation, protection, and control if conditions are found to exist that will endanger the quality or quantity of ground water in such areas. Technical problems will likely arise, however, in trying to establish that such conditions do exist.

Management problems are likely to arise because of changes in the hydrology caused by:

1. Intensive development of ground water, which would significantly reduce perennial flow in nearby streams.
2. Massive shifts to drip irrigation from furrow irrigation.
3. Extensive changes in land use and irrigation practice that would occur in the "Hanapepe" State Supreme Court decision (1973, McBryde Sugar Co., Ltd. v. Almer F. Robinson, et al.) is upheld and the transfer of surface water from one watershed to another is curtailed. The Hanapepe decision establishes that all surface-water rights belong to the State; appurtenant rights, as use rights, remain in existence; all other rights to the use of water are based on the riparian doctrine; and water rights are not transferable from one watershed to another. Clarification is still needed as to whether or not the state can transfer water from one watershed to another.
4. The 1973 State Supreme Court decision which declared the State owner of surplus water in Hawaii's rivers and streams (Hanapepe decision) was overturned as unconstitutional on October 26, 1977 by the Federal District Court in Honolulu (1977, Selwyn A. Robinson, et al. v. George K. Ariyoshi, et al.). The State may appeal the recent Federal Court ruling to the U.S. Supreme Court. The State did not seek title to the water rights in the initial suit.

The State government has developed a carrying-capacity concept for use as a growth management tool and framework for decisionmaking (1976). The carrying capacity of a region can be defined as the capacity of the region's environmental and resource systems to support a given or planned level of economic activity. The methodology has been formalized and integrated for a prototype study of the water supply of Oahu. Safe yields representing reasonable developable supplies, assuming present technology and economics, need to be translated into carrying capacities which can be extrapolated into the future. A major problem in the effective implementation of such a concept lies in the general lack of information needed to adequately evaluate the water resources.

CONCLUSIONS

The ground-water resources of the Hawaii Region offer the best prospects for meeting future water needs. In comparison to the surface-water resources, they appear to be less costly to develop and more dependable when developed. Development of ground water may also be more compatible with environmental quality considerations, existing water rights, and statutory regulations.

About 810 Mgal/d is currently withdrawn from the ground-water resources. This quantity represents about 4 percent of the estimated rainfall. In spite of this small percentage, many problems in the quantity and quality of the ground-water supply exist. Under natural conditions, these problems exist owing to the extremely uneven distribution of rainfall, size and shape of the islands, and the varying ability of the rocks to absorb and transport the water. The heavy local development needed to meet increasing demands and the deterioration of the ground-water supply owing to withdrawal, land development, and waste disposal contribute most to the existing problems.

Owing to the large spatial variation in the predominant trade-wind rainfall and the subsequent marked inverse relation in water availability between the areas of supply and areas of demand, most water-supply problems have been solved by the transport of water. With a significant increase in the number of conduits for water transport, many of the future supply problems can probably be best resolved in this way. The import of water from the wet to the dry areas, especially for irrigation, is a most desirable solution because it has the effect of widening the rainfall recharge area.

In many water-deficient areas, where water is unavailable for import or is too costly, supply problems would have to be resolved by measures that would reduce demand or enable the use of treated sewage effluent or brackish water for some purposes. For example, in coastal areas where additional development of good-quality ground water is not feasible, brackish ground water might be substituted in irrigation and industrial uses for the domestic-quality water now being used for these purposes. Water-supply problems that now exist in areas where sugarcane is heavily irrigated, such as in the Lahaina District in Maui and the Kekaha-Mana and Koloa areas in Kauai, could be relieved somewhat by similar types of water-exchange practices.

For the island of Oahu, where the demand for domestic-quality water has a high priority and where most easily obtainable water has already been developed, actions needed to resolve supply problems are more restrictive and specific.

There has been a tendency for large users of water to ignore the development of small sources of water, the primary reason being the difficulty or infeasibility of assimilating small scattered supplies in the existing distribution systems. However, in view of the pressing need, many excellent small sources available in windward Oahu, the Waianae area, and the southeastern end of the island should be seriously considered for development.

There is a general lack of manmade storage to meet summer peaking needs, and the prospects of significantly increasing the storage are nil. Lacking adequate storage, summer peaking needs are now met by increased pumping. This method is undesirable in that extreme pumping stresses are imposed in the basal-water bodies which are the principal sources of supply. Some of the summer peaking needs could be met by restoring dike-impounded storage in the Koolau Mountains which was lost by development. This additional storage and the use of water currently stored at high levels in the Schofield water body could be utilized principally for high-demand summer periods. This scheme would make it possible to reduce some of the extremely heavy summer pumping from the basal aquifers.

Some increase in the domestic water supply could be accomplished on Oahu by the exchange of poor quality water, such as brackish water or treated sewage effluent for domestic-quality water now used for irrigation. Large amounts of nonpotable water are available from treated sewage or from water in coastal sediments.

SELECTED REFERENCES

Adams, W. M., Peterson, F. L., Surendra, P. M., Lepley, L. K., Wareen, Clifton, and Huber, R. D., 1969, A hydrogeophysical survey from Kawaihae to Kailua-Kona, Hawaii: Water Resources Research Center, Univ. Hawaii, Honolulu, Tech. Rept. No. 32, 156 p.

Austin, H. A. R., and Stearns, H. T., 1954, Methods for development and delivery of water for irrigation of Hawaiian Homes Commission Lands at Hoolehua, Island of Molokai: Hawaii Water Authority, Hawaii Div. Water and Land Devel. Rept. R1, 57 p.

Belt, Collins, and Associates, 1969, A water source development plan for the Lahaina District, Island of Maui: Hawaii Div. Water and Land Devel. Rept. R33.

Blumenstock, D. I., 1961, Climates of the states—Hawaii: U.S. Weather Bureau Climatography of the United States, no. 60–51, 20 p.

Broadbent, E. W., 1969, An estimate of present and future sources and uses of water in the Lahaina-Kahana sector of West Maui: A report for AMFAC Properties, a division of AMFAC, Inc. December 1969.

Campbell, R. B., Chang, J., and Cox, D. C., 1959, Evapotranspiration of sugarcane in Hawaii as measured by in-field lysimeters in relation to climate: Internat. Soc. Sugarcane Technologists Cong., 10th, Mauritius Island, 1959, Proc., p. 637–649.

Caskey, M. C., 1968, The recharge of the Waikapu aquifer, Maui: Master's thesis, Univ. Hawaii, Honolulu, 75 p.

Chinn, S. S. W., 1965, Water-supply potential from an asphalt-lined catchment near Holualoa, Kona, Hawaii: U. S. Geol. Survey Water-Supply Paper 1809–P, 25 p.

Cox, D. C., 1954, Water development for Hawaiian sugarcane irrigation: The Hawaiian Planters' Rec., v. 54, p. 175–197.

Cox, D. C., Peterson, F. L., Adams, W. M., and Lao, Chester, 1969, Coastal evidences of ground-water conditions in the vicinity of Anaehoomalu and Lalamilo, South Kohala, Hawaii: Water Resources Research Center, Univ. Hawaii, Honolulu, Tech. Rept. No. 24, 53 p.

Dale, R. H., 1967, Land use and its effect on the basal water supply, Pearl Harbor area, Oahu, Hawaii, 1931–65: U.S. Geol. Survey Hydrol. Inv. Atlas HA–267.

Dale, R. H., and Takasaki, K. J., 1976, Probable effects of increasing pumpage from Schofield ground-water body, island of Oahu, Hawaii: U.S. Geol. Survey Water-Resources Inv. 76–47, 45 p.

Davis, D. A., and Yamanaga, George, 1973, Water resources summary, island of Hawaii: Hawaii Div. Water and Land Devel. Rept. R47, 42 p.

Davis, R. T., 1960, Major streams of Kauai and their utilization: Master's thesis, Univ. Hawaii, Honolulu.

Druecker, Michael, and Fan, Pow-foong, 1976, Hydrology and chemistry of ground water in Puna, Hawaii: Hawaii Inst. Geophysics, Univ. Hawaii, Honolulu, Contr. No. 732.

Engineering Science, Inc., Sunn, Low, Tom, and Hara, Inc., and Dillingham Environmental Co., 1972, Water quality program for Oahu with special emphasis on waste disposal: A report for the City and County of Honolulu, State of Hawaii, chap. I–XIV (a consortium).

Fischer, W. A., Davis, D. A., and Sousa, T. M., 1966, Fresh-water springs of Hawaii from infrared images: U.S. Geol. Survey Hydrol. Inv. Atlas HA–218.

Hargis, D. R., and Peterson, F. L., 1970, Artificial recharge practices in Hawaii: Water Resources Research Center, Univ. Hawaii, Honolulu, Tech. Rept. No. 43, 36 p.

Hawaii County Department of Water Supply, 1971, Water master plan, island of Hawaii.

Hawaii Division of Water and Land Development, 1961, An inventory of basic water resources data, Molokai: Rept. R15, 107 p.

———1962, Report on the Lihue water system, island of Kauai: Rept. R20, 37 p.

———1963, Improvements to County water system, Lahaina district, Maui: Rept. R21, 38 p.

———1965, A water development plan for South Kohala-Hamakua, island of Hawaii: Rept. R25, 104 p.

———1966, Water-supply investigations for Laupahoehoe water system, South Hamakua, island of Hawaii: Rept. R29, 39 p.

———1966, Water resources development, Molokai: Bull. B16, 69 p.

———1970, An inventory of basic water resources data, island of Hawaii: Rept. R34, 188 p.

———1970, Water for Kihei-Makena, island of Maui: Rept. R38, 89 p.

———1971, Legislation on ground water and wells, generally.

———1972, A general plan for domestic water, island of Kauai: Rept. R40, 121 p.

———1973, Pan evaporation in Hawaii, 1894–1970: Rept. R51, 82 p.

Hawaii State Department of Health, 1973, Water pollution control, Public Health Regulations: Honolulu, Hawaii, Chap. 37.

———1974, Water quality standards, Public Health Regulations: Honolulu, Hawaii, Chap. 37–A.

———1974, Solid waste management control, Public Health Regulations: Honolulu, Hawaii, Chap. 46.

Hawaii State Office of Environmental Quality Control, 1976, Carrying capacity prototype investigations in the State of Hawaii: Governor's Steering Committee for Carrying Capacity Studies, Honolulu, Hawaii, 42 p., App. A–C.

Hawaii Water Authority, 1959, Water resources in Hawaii: Hawaii Div. Water and Land Devel. Bull B14, 148 p.

Hawaii Water Resources Regional Study, 1975, Laws and institutions: Unpublished study element report, Honolulu, Hawaii, 82 p.

———1975, Surface and ground water resources: Unpublished study element report, Honolulu, Hawaii, 431 p.

———1975, Water supply: Unpublished study element report, Honolulu, Hawaii, 236 p.

———1976, Level B water supply planning, Case study, Island of Maui: Supplement to water supply study element report, 275 p.

Hirashima, G. T., 1962, Effect of the Haiku Tunnel on Kahaluu Stream, Oahu, Hawaii; *in* Geological Survey research 1962: U.S. Geol. Survey Prof. Paper 450–C., p. 118–120.

———1963, Aspects of ground-water storage and depletion along the Molokai Irrigation Tunnel, Molokai, Hawaii: Hawaii Div. Water and Land Devel. Circ. C20, 21 p.

———1963, Influence of water-development tunnels on streamflow-groundwater relations in Haiku-Kahaluu area, Oahu, Hawaii: Hawaii Div. Water and Land Devel. Circ. C21, 11 p.

———1965, Flow characteristics of selected streams in Hawaii: Hawaii Div. Water and Land Devel. Rept. R27, 114 p.

———1965, Effects of water withdrawals by tunnels, Waihee Valley, Oahu, Hawaii: Hawaii Div. Water and Land Devel. Rept. R28, 38 p.

———1970, Availability of streamflow for recharge of the basal aquifer in the Pearl Harbor area, Hawaii: U.S. Geol. Survey Water-Supply Paper 1999–B, 26 p.

———1971, Tunnels and dikes of the Koolau Range, Oahu, Hawaii, and their effect on storage depletion and movement of ground water: U.S. Geol. Survey Water-Supply Paper 1999–M, 21 p.

Honolulu Board of Water Supply, 1963, Oahu water plan, 68 p.

———1971, 2020 plan

———1972, Rules and regulations for the protection, development and conservation of water resources.

———1974, Proceedings of the conference on the status of ground-water resources in southern Oahu.

Hufen, T. H., 1975, A geohydrologic investigation of Honolulu's basal water based on isotopic and chemical analyses of water samples: Ph.D. dissert., Univ. Hawaii, Honolulu, 160 p.

Hutchins, W. A., 1946, The Hawaiian system of water rights: A report to the U.S. Dept. of Agriculture and Honolulu Board of Water Supply, 278 p.

Lao, Chester, and Adams, W. M., 1968, Potential water development of the Kahua area: Water Resources Research Center, Univ. Hawaii, Honolulu, Tech. Rept. 19, 16 p.

Lau, L. S., 1976, Water problems and research needs for Hawaii, 1975: Water Resources Research Center, Univ. Hawaii, Honolulu, Tech. Rept. 95, 77 p.

Lum, Daniel, 1969, Preliminary report on geohydrologic exploration for deep well disposal of effluent, Waimanalo sewage treatment plant, Oahu, Hawaii: Hawaii Div. Water and Land Devel. Circ. C54, 14 p.

Macdonald, G. A., 1947, Bibliography of the geology and ground-water resources of the island of Hawaii: Hawaii Div. Hydrography Bull. 10, 191 p.

Macdonald, G. A., Davis, D. A., and Cox, D. C., 1960, Geology and ground-water resources of the island of Kauai, Hawaii: Hawaii Div. Hydrography Bull. 13, 212 p.

Martin, W. F., and Pierce, C. H., 1913, Water resources of Hawaii, 1909–1911: U.S. Geol. Survey Water-Supply Paper 318, 552 p.

Meinzer, O. E., 1930, Ground water in the Hawaiian Islands, *in* Stearns, H. T., and Clark, W. O., Geology and water resources of the Kau District, Hawaii (including parts of Kilauea and Mauna Loa Volcanoes): U.S. Geol. Survey Water-Supply Paper 616, p. 1–28.

Mink, J. F., 1960, Distribution pattern of rainfall in the leeward Koolau Mountains, Oahu, Hawaii: Jour. Geophys. Research, v. 65, no. 9, p. 2869–2876.

———1962a, Excessive irrigation and the soils and ground water of Oahu, Hawaii: Science, v. 135, no. 3504, p. 672–673.

———1962b, Rainfall and runoff in the leeward Koolau Mountains, Oahu, Hawaii: Pacific Sci., v. 16, no. 2, p. 147–159.

———1964, Ground water temperatures in a tropical island environment: Jour. Geophys. Research, v. 69, no. 24, p. 5225–5230.

Palmer, H. S., 1927, The geology of the Honolulu artesian system: Honolulu Sewer and Water Comm. Rept., Supp., 68 p.

Pfund, R. T., and Steller, D. L., 1971, Bibliography of water resources of the Hawaiian Islands: Water Resources Research Center, Univ. Hawaii, Honolulu, Tech. Rept. No. 50, 142 p.

Pfund, R. T., and Wickes, J. W., 1975, Bibliography of water resources of the Hawaiian Islands: Water Resources Research Center, Univ. Hawaii, Honolulu, Tech. Rept. No. 88, 111 p.

Rosenau, J. C., Lubke, E. R., and Nakahara, R. H., 1971, Water resources of north-central Oahu, Hawaii: U.S. Geol. Survey Water-Supply Paper 1899–D, 40 p.

Saito Engineering Consultants, Inc., CH2M Hill, Inc., and Stearns, H. T., 1974, Central Maui water study for the development of sources, transmission lines, and storage reservoirs: A report for Maui Board of Water Supply, March 1974, 24 p.

Soroos, R. L., 1973, Determination of hydraulic conductivity of some Oahu aquifers with step-drawdown test data: Master's thesis, Univ. Hawaii, Honolulu, 57 p.

Stearns, H. T., 1939, Geologic map and guide of Oahu, Hawaii: Hawaii Div. Hydrography Bull. 2, 75 p., 6 pl.

———1940, Supplement to the geology and ground-water resources of the island of Oahu, Hawaii with chapters by J. H. Swartz and G. A. Macdonald: Hawaii Div. Hydrography Bull. 5, 164 p.

———1940, Geology and ground-water resources of Lanai and Kahoolawe, Hawaii: Hawaii Div. Hydrography Bull. 6, 176 p.

———1946, Geology of the Hawaiian Islands: Hawaii Div. Hydrography Bull. 8, 73 p.

———1947, Geology and ground-water resources of the island of Niihau, Hawaii, with a chapter on petrography by G. A. Macdonald: Hawaii Div. Hydrography Bull. 12, 53 p.

Stearns, H. T., and Clark, W. O., 1930, Geology and water resources of the Kau District, Hawaii, *with a chapter on* Ground water in the Hawaiian Islands by O. E. Meinzer: U.S. Geol. Survey Water-Supply Paper 616, 194 p., 33 pls, 9 figs., 3 maps.

Stearns, H. T., and Macdonald, G. A., 1942, Geology and ground-water resources of the island of Maui, Hawaii: Hawaii Div, Hydrography Bull. 7, 344 p.

———1946, Geology and ground-water resources of the island of Hawaii: Hawaii Div. Hydrography Bull. 9, 363 p.

———1947, Geology and ground-water resources of the island of Molokai, Hawaii: Hawaii Div. Hydrography Bull. 11, 113 p.

Stearns, H. T., and Vaksvik, K. N., 1935, Geology and ground-water resources of the island of Oahu, Hawaii: Hawaii Div. Hydrography Bull. 1, 479 p.

———1938, Records of the drilled wells on Oahu, Hawaii: Hawaii Div. Hydrography Bull. 4, 213 p.

Stearns, N. D., 1935, Annotated bibliography and index of the geology and water supply of the island of Oahu, Hawaii: Hawaii Div. Hydrography Bull. 3, 75 p.

Sunn, Low, Tom, and Hara, Inc., 1963, General plans on water, sewage, and drainage for the County of Kauai, State of Hawaii: 350 p.

———1973, Water quality management plan as related to waste treatment for the County of Kauai: Honolulu, Hawaii, p. 95–97.

Swain, L. A., 1973, Chemical quality of ground water in Hawaii: Hawaii Div. Water and Land Devel. Rept. R48, 54 p.

Swartz, J. H., 1939, Geophysical investigations in (the) Hawaiian Islands: Am. Geophys. Union Trans., 20th Ann. Mtg. pt. 3, p. 292–298.

Takasaki, K. J., 1971, Ground water in the Waianae District, Oahu, Hawaii: U.S. Geol. Survey Hydrol. Inv. Atlas HA–358.

———1971, Preliminary report on the water resources of southeast Maui: Hawaii Div. Water and Land Devel. Circ. C58, 37 p.

———1972, Preliminary report on the water resources of central Maui: Hawaii Div. Water and Land Devel. Circ. C62, 59 p.

———1974, Hydrologic conditions related to subsurface and surface disposal of wastes in Hawaii: U.S. Geol. Survey open-file report, WRI 1–74, 5 sheets.

———1976, Elements needed in design of a ground-water-quality monitoring network in the Hawaiian Islands: U.S. Geol. Survey Water-Supply Paper 2041, 23 p.

Takasaki, K. J., Hirashima, G. T., and Lubke, E. R., 1969, Water resources of windward Oahu, Hawaii: U.S. Geol. Survey Water-Supply Paper 1894, 119 p.

Takasaki, K. J., and Valenciano, Santos, 1969, Water in the Kahuku area, Oahu, Hawaii: U.S. Geol. Survey Water-Supply Paper 1874, 59 p.

Takasaki, K. J., and Yamanaga, George, 1970, Preliminary report on the water resources of northeast Maui: Hawaii Div. Water and Land Devel. Circ. C60, 41 p.

Taliaferro, W. J., 1959, Rainfall of the Hawaiian Islands: Hawaii Water Authority, Hawaii Div. Water and Land Devel. Rept. R12, 394 p.

Tenorio, P. A., Young, R. H. F., Burbank, N. C., and Lau, L. S., 1970, Identification of irrigation return water in the sub-surface, Phase III, Kahuku, Oahu, and Kahului and Lahaina, Maui: Water Resources Research Center, Univ. Hawaii, Honolulu, Tech. Rept. No. 44, 55 p.

R. M. Towill Corp., 1971, Sewerage master plan for the County of Maui, State of Hawaii: 254 p.

University of Hawaii, 1974, Water for Hawaii, Summary proceedings—Environmental conferences on the public understanding of science for Hawaii: Honolulu, Hawaii, 110 p.

Visher, F. N., and Mink, J. F., 1964, Ground-water resources in southern Oahu, Hawaii: U.S. Geol. Survey Water-Supply Paper 1778, 133 p.

Watson, L. J., 1963, The valuation of water in Hawaii: Appraisal and valuation manual of the American Society of Appraisers, 1962–1963.

———1964, Development of ground water in Hawaii: Am. Soc. Civil Engineers Proc., Jour. Hydraulics Div., Paper 4144, 17 p.

Wentworth, C. K., 1951, Geology and ground-water resources of the Honolulu-Pearl Harbor area, Oahu, Hawaii: Board of Water Supply, City and County of Honolulu, 111 p.

Williams, J. A., and Soroos, R. L., 1973, Evaluation of methods of pumping test analyses for application to Hawaiian aquifers: Water Resources Research Center, Univ. Hawaii, Honolulu, Tech. Rept. No. 70, 159 p.

Yamanaga, George, and Huxel, C. J., 1969, Preliminary report on the water resources of the Lahaina District, Maui: Hawaii Div. Water and Land Devel. Circ. C51, 47 p.

———1970, Preliminary report on the water resources of the Wailuku area, Maui: Hawaii Div. Water and Land Devel. Circ. C61, 43 p.

APPENDIX

Metric-English Equivalents

Metric unit		English equivalent	
Length			
millimetre (mm)	=	0.03937	inch (in)
metre (m)	=	3.28	feet (ft)
kilometre (km)	=	.62	mile (mi)
Area			
square metre (m^2)	=	10.76	square feet (ft^2)
square kilometre (km^2)	=	.386	square mile (mi^2)
hectare (ha)	=	2.47	acres
Volume			
cubic centimetre (cm^3)	=	0.061	cubic inch (in^3)
litre (l)	=	61.03	cubic inches
cubic metre (m^3)	=	35.31	cubic feet (ft^3)
cubic metre	=	.00081	acre-foot (acre-ft)
cubic hectometre (hm^3)	=	810.7	acre-feet
litre	=	2.113	pints (pt)
litre	=	1.06	quarts (qt)
litre	=	.26	gallon (gal)
cubic metre	=	.00026	million gallons (Mgal or 10^6 gal)
cubic metre	=	6.290	barrels (bbl) (1 bbl = 42 gal)
Weight			
gram (g)	=	0.035	ounce, avoirdupois (oz avdp)
gram	=	.0022	pound, avoirdupois (lb avdp)
tonne (t)	=	1.1	tons, short (2,000 lb)
tonne	=	.98	ton, long (2,240 lb)
Specific combinations			
kilogram per square centimetre (kg/cm^2)	=	0.96	atmosphere (atm)
kilogram per square centimetre	=	.98	bar (0.9869 atm)
cubic metre per second (m^3/s)	=	35.3	cubic feet per second (ft^3/s)
Specific combinations—Continued			
litre per second (l/s)	=	.0353	cubic foot per second
cubic metre per second per square kilometre [$(m^3/s)/km^2$]	=	91.47	cubic feet per second per square mile [$(ft^3/s)/mi^2$]
metre per day (m/d)	=	3.28	feet per day (hydraulic conductivity) (ft/d)
metre per kilometre (m/km)	=	5.28	feet per mile (ft/mi)
kilometre per hour (km/h)	=	.9113	foot per second (ft/s)
metre per second (m/s)	=	3.28	feet per second
metre squared per day (m^2/d)	=	10.764	feet squared per day (ft^2/d) (transmissivity)
cubic metre per second (m^3/s)	=	22.826	million gallons per day (Mgal/d)
cubic metre per minute (m^3/min)	=	264.2	gallons per minute (gal/min)
litre per second (l/s)	=	15.85	gallons per minute
litre per second per metre [(l/s)/m]	=	4.83	gallons per minute per foot [(gal/min)/ft]
kilometre per hour (km/h)	=	.62	mile per hour (mi/h)
metre per second (m/s)	=	2.237	miles per hour
gram per cubic centimetre (g/cm^3)	=	62.43	pounds per cubic foot (lb/ft^3)
gram per square centimetre (g/cm^2)	=	2.048	pounds per square foot (lb/ft^2)
gram per square centimetre	=	.0142	pound per square inch (lb/in^2)
Temperature			
degree Celsius (°C)	=	1.8	degrees Fahrenheit (°F)
degrees Celsius (temperature)	=	[(1.8 × °C) + 32]	degrees Fahrenheit